Edition

D1261362

Boundary Value Problems and Singular Pseudo-differential Operators

Boundary Value Problems and Singular Pseudo-differential Operators

Bert-Wolfgang Schulze
Universität Potsdam, Germany

JOHN WILEY & SONS

Chichester · New York · Weinheim · Brisbane · Singapore · Toronto

Copyright © 1998 by   John Wiley & Sons Ltd,
Baffins Lane, Chichester,
West Sussex PO19 1UD, England

National   01243 779777
International   (+44) 1243 779777

e-mail (for orders and customer service enquiries): cs-books@wiley.co.uk

Visit our Home Page on http://www.wiley.co.uk or http://www.wiley.com

Other Wiley Editorial Offices

John Wiley & Sons Inc, 605 Third Avenue,
New York, NY 10158-0012, USA

Wiley-VCH Verlag GmbH, Pappelallee 3,
D-69469 Weinheim, Germany

Jacaranda Wiley Ltd, 33 Park Road, Milton,
Queensland 4064, Australia

John Wiley & Sons (Asia) Pte Ltd, 2 Clementi Loop #02-01,
Jin Xing Distripark, Singapore 129809

John Wiley & Sons (Canada) Ltd, 22 Worcester Road,
Rexdale, Ontario M9W 1L1, Canada

Library of Congress Cataloging-in-Publication Data

Schulze, Bert-Wolfgang.
  Boundary value problems and singular pseudo-differential operators
  B-W Schulze
    p. cm.
  Includes bibliographical references and index.
  ISBN 0 471 97557 5
  1. Pseudodifferential operators. 2. Boundary value problems.
I. Title
QA329.7.S37 1998   97-31699
515'.7242–dc21   CIP

British Library Cataloguing in Publication Data

A Catalog record for this book is available from the British Library

ISBN 0 471 97557 5

Produced from camera-ready copy supplied by the author
Printed and bound in Great Britain by Biddles Ltd, Guildford and Kings Lynn
This book is printed on acid-free paper responsibly manufactured from sustainable forestry,
in which at least two trees are planted for each one used for paper production.

Contents

Preface

Pseudo–differential operators generalise the family of differential operators in a natural way. While the inverse of a differential operator is not itself a differential operator, it can be understood as a pseudo–differential one; indeed, the "near–inverses", or parametrices, of elliptic differential operators provide basic examples of pseudo–differential operators. In this context, the construction of a parametrix on a C^∞ manifold is particularly simple and has the virtue — once the mapping properties of pseudo–differential operators between appropriate Sobolev spaces are known — of offering a concise proof of the regularity of solutions to elliptic equations.

The ellipticity of a pseudo–differential operator on a closed C^∞ manifold is equivalent to its having the Fredholm property in the scale of Sobolev spaces. The resulting index of a classical operator turns out to depend on the stable homotopy class of its homogeneous principal symbol. This is the background of the famous Atiyah–Singer theorem, which evaluates the analytic index in topological terms.

Pseudo–differential operators and the field of micro–local analysis built upon them have enormously enriched the general theory of partial differential equations. A classical period, since the end of the Fifties, featured the creation of the basic structures and the discovery of many interesting relations to other areas of mathematics, in particular, to geometry, topology, mathematical physics, numerical analysis, and various branches of applied mathematics. Such interactions always led to the observation of new phenomena which, in turn, stimulated the further development of micro–local analysis. Over the years the classical elements of the calculus have been refined into a transparent and accessible form, and there are many monographs devoted to these topics, cf. Hörmander [54], Treves [147], Kumano–go [61].

The present book studies pseudo–differential operators on singular spaces and develops the concept of ellipticity in operator algebras on manifolds with boundary, with conical or edge singularities, or with non–compact "exits to infinity". We will often speak about manifolds with singularities. These "geometric singularities" give rise to certain singular symbols of the operators, i.e., to symbols marked by a special "degenerate" behaviour near the singular points on the manifold.

The problem of studying partial differential operators on manifolds with singularities is also interesting for parabolic, hyperbolic or more general types of operators, including non–linear ones, as these arise in concrete models and applications. The corresponding operator theories are much less complete and, in some cases, practically unknown. Nevertheless many ideas from the elliptic theory apply in a modified form to parabolic problems. Moreover, experience has shown that precise insight into the linear elliptic theory is a rich source of ideas for more complicated equations, and that, conversely, the various applications often require new investigations in the elliptic theory.

In particular, for manifolds with singularities this is much more than a philosophical observation. In this case the theory even has to formulate the correct notion of ellipticity, a notion which is by no means a priori evident. The corners and edges of polyhedra, for example, give rise to ever more symbols for additional trace and potential data along the lower–dimensional skeletons, where the ellipticities are higher analogues of the Shapiro–Lopatinskij condition in boundary value problems. Moreover, surprising new structures appear, even for the simplest singularities of conical or edge type. They integrate large areas of classical analysis into a hierarchy of symbolic and operator levels. In particular, cone theory employs the calculus of standard pseudo–differential operators on a closed compact C^∞ manifold in a parameter–dependent form as an operator–valued symbolic structure of global character along the cone base. The theory on a manifold with edges builds upon this by using a parameter–dependent calculus of cone pseudo–differential operators (over an infinite cone).

A further novelty in the concept of ellipticity near edges (that is, singular sets of dimension at least one) is the generally rather difficult structure of the asymptotics of solutions with respect to the edge variables. These asymptotics are generated by the poles of meromorphic operator–valued functions on the edges, so that the multiplicities may be variable, i.e., the poles may be branching. The Sobolev–space smoothness of the coefficients of the asymptotics depends on the real part of the poles, hence their smoothness may also be branching. The kernels of elliptic operators that are C^∞ in the interior inherit a rich variety of asymptotic information, dependent on the given operator. This finally leads to the problem of inventing adequate classes of smoothing operators within the algebras that include asymptotics. In this context it is remarkable that there appear systems of ideals of smoothing operators that are governed by operator–valued symbols. Note that for the corresponding objects in classical (pseudo–differential) boundary value problems with the transmission property these phenomena are not so relevant, since the asymptotics in such boundary value problems mean smoothness of solutions up to the boundary. Nevertheless the Green operators in the boundary value theory form a particularly interesting operator ideal of the above kind.

In order to illustrate the strategy to obtain the operator theories on manifolds with higher singularities we want to sketch the geometric idea of generating the singularities by an iterative construction. Starting from a given space X, say with conical singularities, we obtain a corner of next higher order by forming a cone with base X. In a further step a wedge of higher order emerges as a Cartesian product between the cone and an open set of $\mathbb{R}^q$. By gluing together such pieces of spaces with singularities we can construct more general configurations with singularities and then start the procedure of forming cones and wedges all over again. In this way there appear not only (warped) polyhedra but also networks or cell complexes with components of different dimensions, for instance hedgehog–shaped spaces or foliants of manifolds which are identified along lower–dimensional subspaces, and, in particular, all types of spaces which occur as lower–dimensional skeletons of such piece–wise smooth structures.

Other examples are Cartesian products of spaces so constructed. Moreover, for every X with such a piece–wise smooth geometry that is embedded into a Riemannian manifold M of any codimension, $M \setminus X$ also has piece–wise smooth geometry. Such configurations are of interest, for instance, in crack problems. Furthermore, if $\chi : M \to N$ is a diffeomorphism, then the image $\chi X \subset N$ likewise has a piece–wise C^∞ geometry.

The singular manifolds in the present geometric sense are non–compact after removing certain thin singular subsets. The typical operators on them are linked to the particular choice of a compact–ification, together with certain metric information. Non–compactness in the methods used to treat these operators also plays a role in connection with specific exits to infinity, where the manifold may look like bagpipes, for example with infinite cones as pipes.

The most elementary singular configuration is the half axis $\overline{\mathbb{R}}_+$; it can be regarded as having both a conical singularity and an exit to infinity. The associated wedge with edge $\mathbb{R}^q$ is then the half space $\overline{\mathbb{R}}_+ \times \mathbb{R}^q$, which we view as the local model of a manifold with boundary. In other words, in this case the inner normal (with respect to a Riemannian metric) plays the role of the model cone and the boundary of the edge. It is thus natural to expect, and it is indeed the case, that the pseudo–differential analysis of boundary value problems is subsumed under the analysis on manifolds with edges.

Now the idea to establish the corresponding operator classes including the associated natural scales of spaces is iteratively to formulate a hierarchy of operator–valued symbolic structures, based on parameter–dependent variants of theories already achieved in the spaces with lower orders of singularity. In other words, every operator theory in this hierarchy, first being of independent interest to solving the concrete problems on a given singular space, is at the same time an ingredient of the symbolic structure for the calculus on a higher level. Some such theories (e.g., of singular integral operators on a half line or of boundary value problems) are already the subject of separate monographs, and the theory on a manifold with edges is much more general than the pseudo–differential analysis of boundary value problems. This shows that the project of establishing a hierarchy of operator structures on manifolds with higher singularities leads, after only a few steps to what might be termed monograph–valued symbols! This is certainly an exciting program but one hard to fulfill. A realistic approach to the analysis on a manifold with arbitrary orders of singularity requires that the theory be organised in an axiomatic form, where the iteration first starts with conical and edge singularities and in which the supporting principles from there form the guideline for the following steps. It turns out that such a program, realised in this work, also contributes new elements to the more classical parts, such as boundary value problems.

The axiomatic ideas contain formal procedures to obtain from a given operator algebra on some space with singularities a new one over the cone generated with that base by means of a machinery that might be called "conification". Every conification contributes to the calculus a new global symbolic level along the given base and a new weight in the scales of Sobolev spaces. A suitably conified algebra on a corresponding infinite model cone then serves as the starting object for another procedure, called "edgification". This leads to an algebra over a wedge, which is locally the Cartesian product between the given model cone and an open subset in $\mathbb{R}^q$, the edge. Every edgification step contributes new "edge–trace" and "edge–potential" conditions that are corresponding entries in operator block matrices, as well as pseudo–differential operators along the edge itself. Invariance discussions for the conified and edgified algebras and appropriate globalisations then allow one to start the process again.

The concept of the conification of an operator algebra was introduced in [119], building upon ideas in [118]. Roughly speaking the conification is a pseudo–differential calculus along the cone axis $\mathbb{R}_+$, based on the Mellin transform, with symbols taking values in the given operator algebra on the base X. This general method, also termed the "Mellin–order–reduction approach", contains further specific ingredients for the axiomatic descriptions. The essential key words in this context are Fuchs–type (or higher corner–degenerate) symbols, Mellin operator conventions (Mellin quantisations), holomorphic and meromorphic operator functions, kernel cut–off schemes, weighted cone Sobolev spaces, analytic functionals, discrete and continuous asymptotics, conormal symbols, and Green operators. The conification of an operator theory on a space X gives rise to a cone algebra on the open stretched cone $\mathbb{R}_+ \times X =: X^\wedge$ with that base and on any space with conical singularities, where (for the stretched object) $X^\wedge$ is the local model.

The edgification of an operator algebra was introduced in [119], after some previous work in [117], [118], and then continued in [121]; cf. also [24]. It may be understood as a kind of twisted product between the cone algebra on $X^\wedge$ and the standard pseudo–differential algebra on an open set $\Omega \subseteq \mathbb{R}^q$. Standard means that the symbols are taken from the classes $S^\mu(\Omega \times \mathbb{R}^q)$ and that the operator convention relies on the Fourier transform in $\mathbb{R}^q$. The (open, stretched) wedge $X^\wedge \times \Omega$, the local model of a manifold with edges of dimension q, is considered in a corresponding splitting of coordinates $(t, x, y) \in \mathbb{R}_+ \times X \times \Omega$. The wedge Sobolev spaces are formulated in anisotropic terms that preserve in the interior the standard local Sobolev spaces. The anisotropic formulations are possible by means of a strongly continuous group of isomorphisms acting on the weighted cone Sobolev spaces on $X^\wedge$. The edgification of the cone algebra on $X^\wedge$ is a pseudo–differential calculus along the edge Ω, based on the Fourier transform, with symbols taking values as operators in the cone algebra. The formulations of the operator–valued symbols also contain the above group of isomorphisms. Essential key words in this "Fourier–edge approach" are edge–degenerate symbols, parameter–dependent Mellin operator conventions, parameter–dependent kernel cut–off schemes, smoothing Mellin and Green symbols, weighted wedge Sobolev spaces, discrete, continuous and variable discrete (branching) asymptotics, trace and potential operators along the edge, and operator–valued edge symbols.

The edgification of a cone algebra on $X^\wedge$ gives rise to a wedge algebra on the open stretched wedge $X^\wedge \times \Omega$ with the given model cone.

Let us comment in this connection on why the Mellin transform is the relevant integral transform here, whenever a new cone axis $\mathbb{R}_+$ appears. Of course, we could always invoke the simple diffeomorphism $\log : \mathbb{R}_+ \to \mathbb{R}$ and pass to the Fourier transform in the new coordinates. However, in the original axial variable, the calculus needs the Mellin and the Fourier transform at the same time. The link between

the Fourier and the Mellin description of operators by means of Mellin
operator conventions belongs to the essential technical points of the
calculus. A logarithmic coordinate change would destroy this relation,
and it seems in fact much more natural to employ the Mellin transform, though the elements of the Mellin–based version of the pseudo–
differential calculus have to be established and accepted as a tool. An
additional advantage of the Mellin calculus is that we control smoothness in the axial variable t up to $t = 0$ under all operations. By substituting, for instance, polar coordinates into a differential operator with
C^∞ coefficients in $\mathbb{R}^m \ni \tilde{x}$, $\tilde{x} \to (t, x)$ for $t = |\tilde{x}|$, $x = \tilde{x}/|\tilde{x}|$, $\tilde{x} \neq 0$,
we obtain a t–dependence of coefficients in the transformed operator
with smoothness up to $t = 0$, except for a weight factor in front of the
operator. In this sense, every point in the C^∞ part of the configuration
may be interpreted as a fictitious conical singularity. Since the (cone-)
calculus in polar coordinates is compatible with the usual one from the
C^∞ calculus, it would not be practical to keep in mind a transformation to an infinite cylinder for practically every point. Moreover, the
higher singularities have many $\overline{\mathbb{R}}_+$–directions. So the multiple use of
the logarithm would dissolve the configuration into a system of infinite
shells with a system of rules to identify different sorts of points at infinity; the intuitive geometry would be completely destroyed as a result.
For this and many other reasons, especially the natural homogeneity
properties of the Mellin transform under rescaling, we do not dismiss
the Mellin picture.

Ellipticity and the Fredholm property, including parametrix constructions within pseudo–differential algebras on manifolds with singularities form the foundation on which to attempt a general index
theory on manifolds with singularities as a continuation of the classical index theory in the spirit of the Atiyah–Singer theorem. Among
the interesting problems in this area are the K–theoretic interpretation of the stable homotopy classes of elliptic symbolic tuples and the
connection of this method to other known approaches to index theory,
in particular, to analytical index formulas in the sense of Fedosov's
work [29]. One may expect the pseudo–differential analysis on manifolds with singularities to lead to further relations between analysis,
geometry and topology, and in fact there is now a growing stream of
investigations in this direction, cf. the references below.

On the other hand, the operators on singular or non–compact manifolds are degenerate in a typical way, and their theory has uncovered an
enormous variety of new analytical effects arising in concrete models in
engineering, applied sciences and numerical analysis. Piece–wise C^∞
configurations, discontinuous symbols, and boundary or transmission
conditions are typical for many applications in mechanics, elasticity

theory, crack theory, heat conduction, scattering theory and other areas of mathematical physics. From the point of view of realistic models the singular structures are no less classical than the smooth ones and there is a vast literature of such concrete investigations, motivated by applications, which has developed over the last century, cf. Grisvard [46], Dauge [19] and the references there. The variety of applications suggests yet another reason to seek insight into unifying structural theories.

The present book may be understood as a further contribution to this program. The pseudo–differential analysis is arranged in a way that earlier essential achievements of the literature fit into the concept or can be interpreted within our methods. This concerns, in particular, the theory of singular integral operators, cf. Gochberg, Krupnik [41]; Sobolev type problems, cf. Sternin [141]; Višik and Eskin's work on pseudo–differential boundary value problems, cf. Višik, Eskin [150], [151], Eskin [25], Rempel, Schulze [96]; further, Boutet de Monvel's calculus of boundary value problems for operators with the transmission property, cf. Boutet de Monvel [10], [9], Rempel, Schulze [91] and Grubb [48]; and finally, Kondrat'ev's approach to operators on manifolds with conical singularities, cf. Kondrat'ev [58]. Let us mention also the methods and ideas of Agranovič, Višik [1], Grushin [49], Parenti [82], Feygin [34], Cordes [17], Lockhart, McOwen [69] Rabinovich [89], Melrose [77], Melrose and Mendoza [78], Müller [80]. In recent years this field has seen further progress, particularly for the case when the base of the model cone has a boundary, cf. Schrohe, Schulze [106], [107], [109], [108], [111], [110], Rabinovich, Schulze, Tarkhanov [90]; moreover, for an axiomatic description of norm closures of the algebras, cf. Mantlik [72],[71]; for the program to derive index formulas, cf. Gilkey [40] and the bibliography there, Plamenevskij, Rozenblum [87], Brüning, Seeley [11], Melrose [77], Piazza [83], Rozenblum [97], Lesch [63], Fedosov, Schulze, Tarkhanov [30], [31], [32], [33], Melrose, Nistor [79], Schrohe, Seiler [112], Schulze, Sternin, Shatalov [130]; or for the integration of Maslov's techniques in the study of asymptotics of solutions, cf. Sternin, Shatalov [144], [143], Schulze, Sternin, Shatalov [128], [143].

In the literature there exist numerous alternative approaches for studying various aspects of pseudo–differential analysis on singular and non–compact manifolds; some accounts bearing many relations to our theory can be found in Cordes [17], [18], Plamenevskij [85], [86], Unterberger, Upmeier [149].

In recent years the interest in the analysis on singular spaces has increased enormously, and it was not possible here to cite all papers and books in this direction. In our choice of references we have emphasised, in particular, the pseudo–differential approach. More references on the traditional analysis for differential operators near singularities may be found in Kondrat'ev, Oleynik [59], Grisvard [46], Maz'ja, Kozlov, Rossmann [74].

The reader who compares the intersection of the present exposition with [119], [122], [24] will find here many improvements and new details which make the theory more accessible. The process of reducing the tower of structures to simpler axiomatic ideas and also of deepening the basic tools is by no means finished and there are, in fact, many unsolved and interesting problems for the future.

Acknowledgement: The author thanks O. Liess (University of Bologna), F. Mantlik (University of Dortmund) and Ch. Dorschfeldt, M. Gerisch, J. Gil, U. Grieme, M. Korey, E. Schrohe, J. Seiler (University of Potsdam) for valuable remarks on the manuscript.

Potsdam, Germany   Bert–Wolfgang Schulze
November 1997

Chapter 1

Pseudo–differential operators

In this book pseudo–differential operators in the context of Section 1.1 are referred to as the standard calculus. There follows in Section 1.2 a parameter–dependent version, in Section 1.3 a generalisation to operator–valued symbols and in Section 1.4 a calculus with control up to conical exits to infinity.

1.1 Elements of the classical calculus

1.1.1 Tools in functional analysis

This section will introduce basic definitions and auxiliary material. $\mathbb{Z}$ will denote the set of integers, $\mathbb{N}$ the subset of non–negative integers (including zero), $\mathbb{R}$ the real axis, and $\mathbb{C}$ the complex plane. If $\alpha = (\alpha_1, \ldots, \alpha_n) \in \mathbb{N}^n$ is a multi–index we set $|\alpha| = \alpha_1 + \cdots + \alpha_n$, $\alpha! = \alpha_1! \ldots \alpha_n!$, furthermore

$$x^\alpha = x_1^{\alpha_1} \ldots x_n^{\alpha_n} \quad \text{for} \quad x = (x_1, \ldots, x_n) \in \mathbb{R}^n,$$

$$\partial_x^\alpha = \left(\frac{\partial}{\partial x_1}\right)^{\alpha_1} \ldots \left(\frac{\partial}{\partial x_n}\right)^{\alpha_n}, \quad D_x^\alpha = \left(\frac{1}{i}\frac{\partial}{\partial x_1}\right)^{\alpha_1} \ldots \left(\frac{1}{i}\frac{\partial}{\partial x_n}\right)^{\alpha_n}$$

with $i = \sqrt{-1}$. We will also write $\partial_{x_k} = \frac{\partial}{\partial x_k}$, $D_{x_k} = -i\frac{\partial}{\partial x_k}$. Moreover, $\alpha \leq \beta$ for $\alpha, \beta \in \mathbb{N}^n$ means $\alpha_j \leq \beta_j$ for $j = 1, \ldots, n$. If $dx = dx_1 \ldots dx_n$ is the Lebesgue measure in $\mathbb{R}^n$ we set

$$(u, v) = \int_{\mathbb{R}^n} u(x)\overline{v(x)}\,dx, \qquad \langle u, v \rangle = \int_{\mathbb{R}^n} u(x)v(x)\,dx \qquad (1.1.1)$$

for functions u, v such that the corresponding integrals are finite. $L^2(\mathbb{R}^n)$ will denote the space of all (complex–valued) square integrable functions in $\mathbb{R}^n$ with the scalar product $(u,v)_{L^2(\mathbb{R}^n)} = (u,v)$. More generally $L^p(\mathbb{R}^n)$, $1 \le p < \infty$, is defined as the space of all u with

$$\|u\|_{L^p(\mathbb{R}^n)} = \left\{ \int_{\mathbb{R}^n} |u(x)|^p dx \right\}^{\frac{1}{p}} < \infty.$$

Analogously we have the spaces $L^p(\Omega)$ for any open $\Omega \subseteq \mathbb{R}^n$ defined as the set of all (classes of) u with

$$\|u\|_{L^p(\Omega)} = \left\{ \int_{\Omega} |u(x)|^p dx \right\}^{\frac{1}{p}} < \infty.$$

The support of u, written supp u, is the complement of the largest open set where u vanishes. $L^p_{comp}(\Omega)$ is then the space of all elements in $L^p(\Omega)$ with compact support and $L^p_{loc}(\Omega)$ the space of all (classes of) measurable functions in Ω that restrict to elements in $L^p(\Omega_0)$ for every open Ω_0 with compact $\overline{\Omega}_0 \subset \Omega$. If $\Omega \subseteq \mathbb{R}^n$ is an open set then $C(\Omega)$ is defined as the space of all continuous functions in Ω. In order to indicate that a set $K \subset \Omega$ is compact we will often write $K \subset\subset \Omega$. The space $C(\Omega)$ is Fréchet with the system of semi–norms $C(\Omega) \ni u \to \sup_{x \in K} |u(x)|$ for arbitrary $K \subset\subset \Omega$. Furthermore $C^k(\Omega)$ for $k \in \mathbb{N}$ will denote the subspace of all $u \in C(\Omega)$ with $D_x^\alpha u \in C(\Omega)$ for all $\alpha \in \mathbb{N}^n$ with $|\alpha| \le k$. This is a Fréchet space with the semi–norm system

$$C^k(\Omega) \ni u \to \sup_{\substack{x \in K \\ |\alpha| \le k}} |D_x^\alpha u(x)| \quad \text{for arbitrary} \quad K \subset\subset \Omega.$$

Finally $C^\infty(\Omega) = \bigcap_{k \in \mathbb{N}} C^k(\Omega)$ is also Fréchet with the system of semi–norms from $C^k(\Omega)$ for all $k \in \mathbb{N}$. Thus $C^\infty(\Omega) = \varprojlim_{k \in \mathbb{N}} C^k(\Omega)$, with $\varprojlim$ as the projective limit. Here and in future we shall freely use the standard notions from topological vector spaces, cf. Schäfer [98]. For every $K \subset\subset \mathbb{R}^n$ we denote by $C_0^\infty(K)$ the space of all $u \in C^\infty(\mathbb{R}^n)$ with supp $u \subseteq K$. For $K \subset\subset \Omega$ we have a natural embedding $C_0^\infty(K) \subset C^\infty(\Omega)$. We then define

$$C_0^\infty(\Omega) = \varinjlim_{K \subset\subset \Omega} C_0^\infty(K)$$

with $\varinjlim$ as the inductive limit.

Let $\mathbb{R}^n_+ = \{x = (x_1, \ldots, x_n) \in \mathbb{R}^n : x_n > 0\}$. Then we set

$$C^\infty(\overline{\mathbb{R}^n_+}) = \Big\{ u \in C^\infty(\mathbb{R}^n_+) : \text{ there exists an } \tilde{u} \in C^\infty(\mathbb{R}^n)$$

$$\text{with } u = \tilde{u}|_{\mathbb{R}^n_+} \Big\},$$

and analogously

$$C_0^\infty(\overline{\mathbb{R}}_+^n) = \left\{ u \in C^\infty(\mathbb{R}_+^n) : \text{ there exists an } \tilde{u} \in C_0^\infty(\mathbb{R}^n) \right.$$

$$\left. \text{with } u = \tilde{u}|_{\mathbb{R}_+^n} \right\}.$$

We will often use excision and cut–off functions. A $\chi \in C^\infty(\mathbb{R}^n)$ is called an excision function (with respect to $x = 0$) if $\chi(x) = 0$ for $|x| < c_0$, $\chi(x) = 1$ for $|x| > c_1$ for constants $0 < c_0 < c_1 < \infty$. A cut–off function ω is an element in $C_0^\infty(\overline{\mathbb{R}}_+^n)$ such that $\omega(x) = 1$ for $x_n < \varepsilon_0$, $\omega(x_n) = 0$ for $x_n > \varepsilon_1$ for certain $0 < \varepsilon_0 < \varepsilon_1 < \infty$. We will usually assume that excision and cut–off functions are real–valued. Let $\mathcal{S}(\mathbb{R}^n) \subset C^\infty(\mathbb{R}^n)$ be the subspace of all u such that

$$\sup_{x \in \mathbb{R}^n} |x^\beta D_x^\alpha u(x)| \qquad \text{for all } \alpha, \beta \in \mathbb{N}^n \qquad (1.1.2)$$

are finite. $\mathcal{S}(\mathbb{R}^n)$ is called the Schwartz space of rapidly decreasing functions. It is a Fréchet space with the semi–norm system (1.1.2). An equivalent semi–norm system for the Fréchet topology in $\mathcal{S}(\mathbb{R}^n)$ is

$$\|x^\beta D_x^\alpha u(x)\|_{L^2(\mathbb{R}^n)} \qquad \text{for all } \alpha, \beta \in \mathbb{N}^n. \qquad (1.1.3)$$

We will set

$$\mathcal{S}(\overline{\mathbb{R}}_+^n) = \left\{ u \in C^\infty(\mathbb{R}_+^n) : \text{ there is a } \tilde{u} \in \mathcal{S}(\mathbb{R}^n) \text{ such that } u = \tilde{u}|_{\mathbb{R}_+^n} \right\}.$$

This is also a Fréchet space in a natural way.

$\mathcal{S}'(\mathbb{R}^n)$, the dual of $\mathcal{S}(\mathbb{R}^n)$, is called the space of temperate distributions in $\mathbb{R}^n$. If $\Omega \subseteq \mathbb{R}^n$ is open, the dual $(C_0^\infty(\Omega))' =: \mathcal{D}'(\Omega)$ is the space of all distributions in Ω and $(C^\infty(\Omega))' =: \mathcal{E}'(\Omega)$ the subspace of all $u \in \mathcal{D}'(\Omega)$ for which the support supp u is compact. The bilinear pairing between elements $u \in \mathcal{D}'(\Omega)$, $\varphi \in C_0^\infty(\Omega)$ will be denoted by $\langle u, \varphi \rangle$ or likewise $\langle \varphi, u \rangle$. An example is the Dirac distribution δ_y concentrated at $y \in \Omega$, given by $\langle \delta_y, \varphi \rangle = \varphi(y)$. The complement of the largest open set outside of which a distribution u is a C^∞ function is called singular support of u, written sing supp u. References for the standard properties of distributions are Gelfand, Šilov [35], Treves [146], Hörmander [54], Vol. 1. We will not recall all generalities here. Let us only mention that every $u \in L^1_{\text{loc}}(\Omega)$ represents an element in $\mathcal{D}'(\Omega)$ by

$$C_0^\infty(\Omega) \ni \varphi \to \langle u, \varphi \rangle := \int_\Omega u(x)\varphi(x)\, dx. \qquad (1.1.4)$$

Then, the operator $\mathcal{M}_\psi$ of multiplication by a function $\psi \in C^\infty(\Omega)$ has an extension to $\mathcal{D}'(\Omega)$ by $\langle \mathcal{M}_\psi u, \varphi \rangle = \langle u, \mathcal{M}_\psi \varphi \rangle$, $\mathcal{M}_\psi \varphi = \psi \varphi$. The differentiations D_x^α extend from $C^\infty(\Omega)$ to $\mathcal{D}'(\Omega)$ by $\langle D_x^\alpha u, \varphi \rangle := \langle u, (-D_x)^\alpha \varphi \rangle$ which is motivated by integrating by parts when $u \in C^\infty(\Omega)$.

Next recall the Fourier transform

$$(Fu)(\xi) = \int_{\mathbb{R}^n} e^{-ix\xi} u(x) \, dx, \qquad (1.1.5)$$

with $x\xi = \sum_{j=1}^n x_j \xi_j$.

Theorem 1.1.1 *The Fourier transform induces an isomorphism*

$$F : \mathcal{S}(\mathbb{R}^n) \to \mathcal{S}(\mathbb{R}^n)$$

with the inverse

$$\left(F^{-1} g \right)(x) = (2\pi)^{-n} \int_{\mathbb{R}^n} e^{ix\xi} g(\xi) \, d\xi. \qquad (1.1.6)$$

We have

$$\langle Fu, f \rangle = \langle u, Ff \rangle \qquad (1.1.7)$$

and

$$(Fu, f)_{L^2(\mathbb{R}^n_\xi)} = (2\pi)^n \left(u, F^{-1} f \right)_{L^2(\mathbb{R}^n_x)} \qquad (1.1.8)$$

for all $u, f \in \mathcal{S}(\mathbb{R}^n)$.

Corollary 1.1.2 *The formula* (1.1.7) *admits an extension of F to an isomorphism*

$$F : \mathcal{S}'(\mathbb{R}^n) \to \mathcal{S}'(\mathbb{R}^n). \qquad (1.1.9)$$

Moreover, (1.1.8) *for $f = Fv$ shows that F extends by continuity to an isomorphism*

$$F : L^2(\mathbb{R}^n) \to L^2(\mathbb{R}^n)$$

where $\|Fu\|_{L^2(\mathbb{R}^n)} = (2\pi)^{\frac{n}{2}} \|u\|_{L^2(\mathbb{R}^n)}$.

We will also write

$$\hat{u}(\xi) = (Fu)(\xi), \qquad d\xi = (2\pi)^{-n} d\xi.$$

ξ is often called the covariable to x. Set

$$\langle \xi \rangle = \left(1 + |\xi|^2 \right)^{\frac{1}{2}}.$$

Moreover, $[\xi]$ will denote a real–valued C^∞ function in $\mathbb{R}^n$ with $[\xi] > 0$ for all $\xi \in \mathbb{R}^n$ and

$$[\xi] = |\xi| \qquad \text{for all } |\xi| > c \qquad\qquad (1.1.10)$$

for some $c > 0$.

If E, F are locally convex vector spaces, $\mathcal{L}(E, F)$ will denote the space of all linear continuous operators $A : E \to F$. If $E = F$ we will also write $\mathcal{L}(E)$ instead of $\mathcal{L}(E, F)$.

Theorem 1.1.3 *Let $A \in \mathcal{L}(L^2(\mathbb{R}^n))$, then the following conditions are equivalent:*

(i) *A is an integral operator with kernel $k_A(x, x')$ in $\mathcal{S}(\mathbb{R}^n \times \mathbb{R}^n)$, i.e.*

$$Au(x) = \int k_A(x, x') u(x') \, dx'$$

for all $u \in L^2(\mathbb{R}^n)$.

(ii) *A and its adjoint A^* induce continuous operators*

$$A : L^2(\mathbb{R}^n) \to \mathcal{S}(\mathbb{R}^n), \qquad A^* : L^2(\mathbb{R}^n) \to \mathcal{S}(\mathbb{R}^n).$$

Let $U \subseteq \mathbb{C}$ be an open set and $\mathcal{A}(U)$ the space of all holomorphic functions $h(z)$ in U, endowed with the Fréchet topology defined by the system of semi–norms

$$\sup_{z \in K} |h(z)| \qquad \text{for all } K \Subset U.$$

The dual $\mathcal{A}'(U)$ is called the space of analytic functionals in U. Let $K \Subset \mathbb{C}$ be fixed and $f(z) \in \mathcal{A}(\mathbb{C} \setminus K)$. Choose an arbitrary (say C^∞) curve L surrounding K clockwise and form

$$\langle \zeta_f, h \rangle := \int_L h(z) f(z) \, dz \qquad \text{for } h \in \mathcal{A}(\mathbb{C}). \qquad (1.1.11)$$

Then $\zeta_f \in \mathcal{A}'(\mathbb{C})$, and ζ_f is carried by K in the sense of the definition of a carrier, cf. Hörmander [54]. We will denote by $\mathcal{A}'(K)$ the space of all analytic functionals in $\mathbb{C}$ carried by K. Every $\zeta \in \mathcal{A}'(K)$ can be written as $\zeta = \zeta_f$ for an $f(z) \in \mathcal{A}(\mathbb{C} \setminus K)$. Below in Section 2.3.2 we will return to a more thorough exposition on analytic functionals.

Definition 1.1.4 *Let E, F be Fréchet spaces, embedded as vector sub-spaces in a topological Hausdorff vector space H. Then the non–direct Fréchet sum $E + F$ is defined as*

$$E + F = \{e + f : \ e \in E, f \in F\}$$

endowed with the Fréchet topology from

$$E + F \cong E \oplus F/\Delta, \ where \ \Delta = \{(e, -e) : \ e \in E \cap F\}.$$

If A is an algebra, E a Fréchet space which is a left module over A, i.e., the elements $a \in A$ induce by multiplication $e \to ae$ linear operators $a : E \to E$, with the usual algebraic rules, then we set

$$[a]E = \{\text{closure of } \{ae : \ e \in E\} \text{ in } E\} \tag{1.1.12}$$

for every $a \in A$. In an analogous manner we employ notations like $E[b]$, when E is a right B–module for $b \in B$ or $[a]E[b]$ when E is a left A–module and a right B–module.

Proposition 1.1.5 *Let E, F and X be Fréchet spaces, and assume that E, F are vector subspaces of a topological Hausdorff space H. Suppose $A : E + F \to X$ is a linear mapping which restricts to continuous operators $T_E \in \mathcal{L}(E, X)$ and $T_F \in \mathcal{L}(F, X)$. Then $T \in \mathcal{L}(E + F, X)$.*

Let E, F be vector spaces (over $\mathbb{C}$). Then we define the algebraic tensor product $E \otimes_a F$ between E, F as the vector space consisting of all equivalence classes of finite sums of arbitrary length $\sum_{m=0}^{N} e_m \otimes f_m$ for $e_m \in E$, $f_m \in F$ where the equivalence relation identifies sums which can be reduced to each other by rearranging factors on the left and the right of $\otimes$ by means of the algebraic operations in E and F, respectively. For instance, $e_1 \otimes f + e_2 \otimes f \sim e \otimes f$ when $e = e_1 + e_2$, or $e \otimes f_1 + e \otimes f_2 \sim e \otimes f$ when $f = f_1 + f_2$.

Let E and F be Fréchet spaces with semi–norm systems $\{p_j\}_{j \in \mathbb{N}}$ and $\{q_k\}_{k \in \mathbb{N}}$, respectively, which define the corresponding Fréchet topologies. Define for every fixed j, k

$$(p_j \otimes_\pi q_k)(h) = \inf \sum p_j(e_m) q_k(f_m)$$

for $h \in E \otimes_a F$ with the infimum being taken over all representations of h as a sum $\sum e_m \otimes f_m$ of finite length with $e_m \in E$, $f_m \in F$. Then $\{p_j \otimes_\pi q_k\}_{j,k \in \mathbb{N}}$ is a semi–norm system on $E \otimes_a F$. Let us denote by $E \otimes_\pi F$ the completion of $E \otimes_a F$ with respect to the locally convex topology that is defined by $\{p_j \otimes_\pi q_k\}_{j,k \in \mathbb{N}}$. Then $E \otimes_\pi F$ is a Fréchet space, called the projective tensor product of E and F.

If E, F, G are Fréchet spaces we can form $(E \otimes_\pi F) \otimes_\pi G$ which is the same as $E \otimes_\pi (F \otimes_\pi G)$, written $E \otimes_\pi F \otimes_\pi G$.

Theorem 1.1.6 *Let E, F be Fréchet spaces. Then each element $h \in$
$E \otimes_\pi F$ has a representation as a convergent sum*

$$h = \sum_{m=0}^{\infty} \lambda_m e_m \otimes f_m$$

*with sequences $\{\lambda_m\}_{m\in\mathbb{N}} \subset \mathbb{C}$, $\sum_{m=0}^{\infty} |\lambda_m| < \infty$, and $\{e_m\}_{m\in\mathbb{N}} \subset E$,
$\{f_m\}_{m\in\mathbb{N}} \subset F$ with $e_m \to 0$ in E, $f_m \to 0$ in F for $m \to \infty$.*

Remark 1.1.7 *There is also a variant of Theorem 1.1.6 for projective
tensor products between finitely many Fréchet spaces. For instance,
every $h \in E \otimes_\pi F \otimes_\pi G$ for Fréchet spaces E, F, G can be written as a
convergent sum*

$$\sum_{m=0}^{\infty} \lambda_m e_m \otimes f_m \otimes g_m$$

*with sequences $\{\lambda_m\}_{m\in\mathbb{N}} \subset \mathbb{C}$, $\sum |\lambda_m| < \infty$, and $\{e_m\}_{m\in\mathbb{N}} \subset E$,
$\{f_m\}_{m\in\mathbb{N}} \subset F$, $\{g_m\}_{m\in\mathbb{N}} \subset G$ tending to zero in the corresponding
spaces for $m \to \infty$.*

Proposition 1.1.8 *Let E, F, G be Fréchet spaces, where the assumptions to define non–direct sums are satisfied in the corresponding cases.
Then there are canonical isomorphisms*

$$(E + F) \otimes_\pi G = E \otimes_\pi G + F \otimes_\pi G,$$
$$G \otimes_\pi (E + F) = G \otimes_\pi E + G \otimes_\pi F.$$

1.1.2 Oscillatory integrals and pseudo–differential operators

Let

$$A = \sum_{|\alpha| \le \mu} a_\alpha(x) D_x^\alpha$$

be a differential operator in a domain $\Omega \subseteq \mathbb{R}^n$ with coefficients $a_\alpha(x) \in$
$C^\infty(\Omega)$, regarded as an operator $A : C_0^\infty(\Omega) \to C_0^\infty(\Omega)$. Then A can
be expressed by the Fourier transform F as

$$A = F^{-1} a(x, \xi) F \quad \text{with} \quad a(x, \xi) = \sum_{|\alpha| \le \mu} a_\alpha(x) \xi^\alpha,$$

using the elementary identity $D_x^\alpha = F^{-1} \xi^\alpha F$. Thus

$$Au(x) = \int e^{ix\xi} a(x, \xi) \left\{ \int e^{-ix'\xi} u(x') \, dx' \right\} d\xi. \qquad (1.1.13)$$

This gives us a relation between A and its so–called complete symbol $a(x,\xi)$. The operator A is determined by $a(x,\xi)$. Conversely, by knowing $A : C^\infty(\Omega) \to C^\infty(\Omega)$, the complete symbol $a(x,\xi)$ can be recovered as

$$a(x,\xi) = e^{-ix\xi} A e^{ix'\xi},$$

where A acts with respect to x'.

Now the idea behind pseudo–differential operators is just to allow more general symbols than polynomials in ξ in the defining relation (1.1.13). For the calculus it will be convenient to write (1.1.13) formally as a double integral, interpreted below as a so–called oscillatory integral. It will also be natural to permit so–called amplitude functions $a(x,x',\xi)$ that are C^∞–dependent on $(x,x') \in \Omega \times \Omega$. In other words pseudo–differential operators will be of the form

$$Au(x) = \iint e^{i(x-x')\xi} a(x,x',\xi) u(x')dx'd\xi. \tag{1.1.14}$$

We begin by studying the appropriate classes of amplitude functions; for simplicity we also refer to these as symbols. Initially we make no restriction on the dimensions of the variable and covariable spaces: We assume that x varies over an open set $\Omega \subseteq \mathbb{R}^m$ and ξ over all of $\mathbb{R}^n$, for arbitrary m,n. Later we shall often suppose that $m = n$ or $m = 2n$.

Definition 1.1.9 *Let $\Omega \subseteq \mathbb{R}^m$ be open and $\mu \in \mathbb{R}$. Then $S^\mu(\Omega \times \mathbb{R}^n)$ denotes the space of all $a(x,\xi) \in C^\infty(\Omega \times \mathbb{R}^n)$ such that*

$$|D_x^\alpha D_\xi^\beta a(x,\xi)| \leq c(1 + |\xi|)^{\mu-|\beta|} \tag{1.1.15}$$

for all $\alpha \in \mathbb{N}^m$, $\beta \in \mathbb{N}^n$, $x \in K$, with arbitrary $K \subset\subset \Omega$, $\xi \in \mathbb{R}^n$, with constants $c = c(\alpha,\beta,K) > 0$. μ is called the order of the symbol a, written $\mu = \operatorname{ord} a$.

The best possible constants c in (1.1.15) form a semi–norm system on the space $S^\mu(\Omega \times \mathbb{R}^n)$:

$$a \to \sup_{\substack{x \in K \\ \xi \in \mathbb{R}^n}} |D_x^\alpha D_\xi^\beta a(x,\xi)|(1 + |\xi|)^{-\mu+|\beta|}$$

It is clear that $a(x,\xi)$ belongs to $S^\mu(\Omega \times \mathbb{R}^n)$ just in case the estimates (1.1.15) are satisfied for a countable system of compact sets $K \subset \Omega$ and all α,β. It suffices to take all closed balls $\subset \Omega$ of rational radius and centres with rational coordinates. It is then easy to verify that $S^\mu(\Omega \times \mathbb{R}^n)$ is a Fréchet space with this semi–norm system. Furthermore the subspace $S^\mu(\mathbb{R}^n)$, defined as the set of all x–independent elements of $S^\mu(\Omega \times \mathbb{R}^n)$, is closed in the induced topology.

There are natural continuous embeddings

$$S^\mu(\Omega \times \mathbb{R}^n) \hookrightarrow S^\nu(\Omega \times \mathbb{R}^n) \qquad \text{for } \mu \leq \nu.$$

We set $S^{-\infty}(\Omega \times \mathbb{R}^n) = \bigcap_{\mu \in \mathbb{R}} S^\mu(\Omega \times \mathbb{R}^n)$. Analogously we get the space $S^{-\infty}(\mathbb{R}^n)$. Observe that we have

$$S^{-\infty}(\mathbb{R}^n) = \mathcal{S}(\mathbb{R}^n),$$

and $S^\mu(\mathbb{R}^n)$ is contained in $\mathcal{S}'(\mathbb{R}^n)$ for every μ.

Remark 1.1.10 *We have*

$$S^\mu(\Omega \times \mathbb{R}^n) = C^\infty(\Omega, S^\mu(\mathbb{R}^n)) = C^\infty(\Omega) \otimes_\pi S^\mu(\mathbb{R}^n)$$

with $\otimes_\pi$ as the completed projective tensor product between the corresponding spaces. In particular,

$$S^{-\infty}(\Omega \times \mathbb{R}^n) = C^\infty(\Omega, \mathcal{S}(\mathbb{R}^n)) = C^\infty(\Omega) \otimes_\pi \mathcal{S}(\mathbb{R}^n).$$

Recall that a $\chi(\xi) \in C^\infty(\mathbb{R}^n)$ is called an excision function if $\chi(\xi) = 0$ for $|\xi| < c_0$, $\chi(\xi) = 1$ for $|\xi| > c_1$ with constants $0 < c_0 < c_1 < \infty$.

A function $f(\xi) \in C^\infty(\mathbb{R}^n \setminus \{0\})$ is called homogeneous of order $\mu \in \mathbb{R}$ if $f(\lambda\xi) = \lambda^\mu f(\xi)$ for all $\lambda > 0$, $\xi \in \mathbb{R}^n \setminus \{0\}$. Let

$$S^{(\mu)}(\Omega \times (\mathbb{R}^n \setminus \{0\})) \qquad (1.1.16)$$

be the space of all $f(x, \xi) \in C^\infty(\Omega \times (\mathbb{R}^n \setminus \{0\}))$ that are homogeneous in ξ of order μ for all $x \in \Omega$. Then

$$D_x^\alpha D_\xi^\beta S^{(\mu)}(\Omega \times (\mathbb{R}^n \setminus \{0\})) \subseteq S^{\mu - |\beta|}(\Omega \times (\mathbb{R}^n \setminus \{0\}))$$

for all $\alpha \in \mathbb{N}^m$, $\beta \in \mathbb{N}^n$. Note that there is a canonical isomorphism

$$S^{(\mu)}(\Omega \times (\mathbb{R}^n \setminus \{0\})) \cong C^\infty(\Omega \times S^{n-1}),$$

$S^{n-1} = \{\xi \in \mathbb{R}^n : |\xi| = 1\}$. It is given by restricting $f(x, \xi)$ to $|\xi| = 1$. The inverse mapping is $f(x, \xi/|\xi|) \to |\xi|^\mu f(x, \xi/|\xi|)$, the extension of an element in $C^\infty(\Omega \times S^{n-1})$ by homogeneity μ to arbitrary $\xi \neq 0$.

Remark 1.1.11 *We have (with obvious notation)*

$$\chi S^{(\mu)}(\Omega \times (\mathbb{R}^n \setminus \{0\})) \subset S^\mu(\Omega \times \mathbb{R}^n)$$

for every excision function $\chi(\xi)$.

Definition 1.1.12 *A symbol* $a(x,\xi) \in S^\mu(\Omega \times \mathbb{R}^n)$ *is called classical if there is a sequence* $a_{(\mu-j)}(x,\xi) \in S^{(\mu-j)}(\Omega \times (\mathbb{R}^n \setminus \{0\}))$, $j \in \mathbb{N}$, *such that for any excision function* $\chi(\xi)$

$$a(x,\xi) - \sum_{j=0}^{N} \chi(\xi) a_{(\mu-j)}(x,\xi) \in S^{\mu-(N+1)}(\Omega \times \mathbb{R}^n) \qquad (1.1.17)$$

for all $N \in \mathbb{N}$. *We denote by* $S_{\mathrm{cl}}^\mu(\Omega \times \mathbb{R}^n)$ *the space of all classical symbols of order* μ.

In view of $a_{(\mu)}(x,\xi) = \lim_{\lambda \to \infty} \lambda^{-\mu} a(x, \lambda\xi)$ we can recover $a_{(\mu)}$ in a unique way. Applying the same procedure to $a_1(x,\xi) = a(x,\xi) - \chi(\xi) a_{(\mu)}(x,\xi) \in S_{\mathrm{cl}}^{\mu-1}(\Omega \times \mathbb{R}^n)$ for the order $\mu - 1$ we obtain $a_{(\mu-1)}(x,\xi)$. Thus we obtain a sequence of homogeneous components $a_{(\mu-j)}(x,\xi)$, $j \in \mathbb{N}$. This gives us linear mappings

$$\eta_j : S_{\mathrm{cl}}^\mu(\Omega \times \mathbb{R}^n) \to S^{(\mu-j)}(\Omega \times (\mathbb{R}^n \setminus \{0\})), \qquad j \in \mathbb{N}. \qquad (1.1.18)$$

The relation (1.1.17) defines linear mappings

$$\delta_N : S_{\mathrm{cl}}^\mu(\Omega \times \mathbb{R}^n) \to S^{\mu-(N+1)}(\Omega \times \mathbb{R}^n), \qquad N \in \mathbb{N}. \qquad (1.1.19)$$

We endow $S_{\mathrm{cl}}^\mu(\Omega \times \mathbb{R}^n)$ with the topology of the projective limit with respect to (1.1.18), (1.1.19) for all $j, N \in \mathbb{N}$. Then $S_{\mathrm{cl}}^\mu(\Omega \times \mathbb{R}^n)$ is a nuclear Fréchet space. The topology is independent of the particular choice of the excision function χ involved in (1.1.19).

Remark 1.1.13 *The subspace* $S_{\mathrm{cl}}^\mu(\mathbb{R}^n)$ *of* x–*independent elements of* $S_{\mathrm{cl}}^\mu(\Omega \times \mathbb{R}^n)$ *is closed in the induced topology, and we have analogously to Remark* 1.1.10 *the relations*

$$S_{\mathrm{cl}}^\mu(\Omega \times \mathbb{R}^n) = C^\infty(\Omega, S_{\mathrm{cl}}^\mu(\mathbb{R}^n)) = C^\infty(\Omega) \otimes_\pi S_{\mathrm{cl}}^\mu(\mathbb{R}^n).$$

Remark 1.1.14 *From the definitions we obtain*

$$D_x^\alpha D_\xi^\beta S^\mu(\Omega \times \mathbb{R}^n) \subset S^{\mu-|\beta|}(\Omega \times \mathbb{R}^n)$$

for all $\alpha \in \mathbb{N}^m$, $\beta \in \mathbb{N}^n$, *and (with obvious notation)*

$$S^\mu(\Omega \times \mathbb{R}^n) S^\nu(\Omega \times \mathbb{R}^n) \subset S^{\mu+\nu}(\Omega \times \mathbb{R}^n)$$

for all $\mu, \nu \in \mathbb{R}$. *Analogous relations hold for the spaces of classical symbols.*

As noted above $S^\mu(\mathbb{R}^n)$ can be interpreted as a subspace of $\mathcal{S}'(\mathbb{R}^n)$ via $\langle a, \varphi \rangle = \int a(\xi)\varphi(\xi)d\xi$ for $\varphi \in \mathcal{S}(\mathbb{R}^n)$, $a \in S^\mu(\mathbb{R}^n)$. Thus the Fourier transform (and its inverse) can be applied to symbols. We obtain, in particular,

$$F^{-1}_{\xi \to \zeta} : S^\mu(\mathbb{R}^n_\xi) \to \mathcal{S}'(\mathbb{R}^n_\zeta). \tag{1.1.20}$$

Set $k(a)(\zeta) = \left(F^{-1}_{\xi \to \zeta} a \right)(\zeta)$. Similarly we can form

$$k(a)(x, \zeta) = \left(F^{-1}_{\xi \to \zeta} a \right)(x, \zeta) \tag{1.1.21}$$

for $a(x, \xi) \in S^\mu(\Omega \times \mathbb{R}^n)$. It can easily be verified that for every $r \in \mathbb{N}$ we have

$$k(a)(x, \zeta) \in C^\infty(\Omega, C^r(\mathbb{R}^n)) \quad \text{for} \quad \mu < -n - r. \tag{1.1.22}$$

Proposition 1.1.15 *We have*

$$\text{sing supp } k(a) \subseteq \{0\} \tag{1.1.23}$$

for every $a(\xi) \in S^\mu(\mathbb{R}^n)$, $\mu \in \mathbb{R}$. Moreover, if $\chi(\zeta)$ is an arbitrary excision function (i.e., $\chi(\zeta) \in C^\infty(\mathbb{R}^n)$, $\chi(\zeta) = 0$ near $\zeta = 0$, $\chi(\zeta) = 1$ outside another neighbourhood of $\zeta = 0$) then

$$\chi(\zeta)k(a)(\zeta) \in \mathcal{S}(\mathbb{R}^n). \tag{1.1.24}$$

Proof. The Fourier transform satisfies the identities

$$F^{-1}_{\xi \to \zeta}(\xi^\gamma D^\delta_\xi a)(\zeta) = D^\gamma_\zeta (-\zeta)^\delta \left(F^{-1}_{\xi \to \zeta} a \right)(\zeta)$$

for arbitrary $a \in \mathcal{S}'(\mathbb{R}^n)$, $\gamma, \delta \in \mathbb{N}^n$. In particular,

$$k(\xi^\gamma(-\Delta_\xi)^N a)(\zeta) = |\zeta|^{2N} D^\gamma_\zeta k(a)(\zeta) \tag{1.1.25}$$

for every N. From $(-\Delta_\xi)^N a \in S^{\mu-2N}(\mathbb{R}^n)$ and (1.1.22) we obtain $k((-\Delta_\xi)^N a)(\zeta) = |\zeta|^{2N} k(a)(\zeta) \in C^r(\mathbb{R}^n)$ for $\mu - 2N < -n - r$. Since N is arbitrary, it follows that $k(a)(\zeta) \in C^\infty(\mathbb{R}^n \setminus \{0\})$, i.e., (1.1.23). Finally we have $\xi^\gamma(-\Delta_\xi)^N a(\xi) \in S^{\mu-2N+|\gamma|}(\mathbb{R}^n)$ for every $\gamma \in \mathbb{N}^n$, $N \in \mathbb{N}$. For N satisfying $\mu - 2N + |\gamma| < -n$ we get $|k(\xi^\gamma(-\Delta_\xi)^N a)(\zeta)| < c$ with a constant $c = c(\gamma, N)$, for all $\zeta \in \mathbb{R}^n$. In other words (1.1.25) gives us

$$\sup_{\zeta \in \mathbb{R}^n} \left| \chi(\zeta)|\zeta|^{2N} D^\gamma_\zeta k(a)(\zeta) \right| < \infty$$

for all γ and sufficiently large $N = N(\gamma)$. This implies the same estimates for all N. From that we easily obtain $\chi(\zeta)k(a)(\zeta) \in \mathcal{S}(\mathbb{R}^n)$. $\square$

Theorem 1.1.16 *Let $a_j(x,\xi) \in S^{\mu_j}(\Omega \times \mathbb{R}^n)$, $j \in \mathbb{N}$, be an arbitrary sequence, where $\mu_j \to -\infty$ as $j \to \infty$. Then there is a symbol $a(x,\xi) \in S^\mu(\Omega \times \mathbb{R}^n)$ with $\mu = \max_{j\in\mathbb{N}}\{\mu_j\}$ such that for every M there is an $N(M)$ such that for all $N \geq N(M)$*

$$a(x,\xi) - \sum_{j=0}^{N} a_j(x,\xi) \in S^{\mu-M}(\Omega \times \mathbb{R}^n).$$

The element $a(x,\xi) \in S^\mu(\Omega \times \mathbb{R}^n)$ is uniquely determined by this property $\mathrm{mod}\, S^{-\infty}(\Omega \times \mathbb{R}^n)$.

We call any such $a(x,\xi)$ asymptotic sum of the $a_j(x,\xi)$, $j \in \mathbb{N}$, and write

$$a(x,\xi) \sim \sum_{j=0}^{\infty} a_j(x,\xi).$$

Asymptotic expansions will occur in this book in various modifications. The following proposition gives a scheme to carry out asymptotic sums.

Proposition 1.1.17 *Let E^j, $j \in \mathbb{N}$, be a sequence of Fréchet spaces with continuous embeddings $E^{j+1} \hookrightarrow E^j$ for all j. Set $E^\infty = \varprojlim_{j\in\mathbb{N}} E^j$. Assume that there exists a c–dependent family of linear operators*

$$\chi^j(c) : E^j \to E^j \qquad \text{for all } j \in \mathbb{N}, \tag{1.1.26}$$

$c \in \mathbb{R}_+$, with the following properties:

(i) *We have for every $j \in \mathbb{N}$*

$$e - \chi^j(c)e \in E^\infty \qquad \text{for all } c \in \mathbb{R}_+, \ e \in E^j. \tag{1.1.27}$$

(ii) *The diagrams*

$$\begin{array}{ccc} E^{j+1} & \longleftarrow & E^j \\ \chi^{j+1}(c)\downarrow & & \downarrow\chi^j(c) \\ E^{j+1} & \longleftarrow & E^j \end{array} \tag{1.1.28}$$

commute for all $j \in \mathbb{N}$, $c \in \mathbb{R}_+$.

(iii) *If $\{r_k^j\}_{k\in\mathbb{N}}$ is a semi–norm system that defines the Fréchet topology in E^j then for arbitrary fixed $j,k \in \mathbb{N}$ there exists an $l(j,k) \geq j$ such that $f \in E^m$ for every $m \geq l(j,k)$ implies*

$$r_k^j(\chi^m(c)f) \to 0 \quad \text{as} \quad c \to \infty. \tag{1.1.29}$$

Then, for every sequence $e_j \in E^j$, $j \in \mathbb{N}$, there exists a sequence of constants $c_j \in \mathbb{R}_+$, such that

$$\sum_{j=k}^{\infty} \chi^j(c_j)e_j \ \text{converges in } E^k \qquad (1.1.30)$$

for every $k \in \mathbb{N}$. In other words $e := \sum_{j=0}^{\infty} \chi^j(c_j)e_j$ converges in E^0 and has the property

$$e - \sum_{j=0}^{N} e_j \in E^{N+1} \qquad \text{for all } N \in \mathbb{N}. \qquad (1.1.31)$$

e is unique $\bmod E^\infty$, and we write $e \sim \sum_{j=0}^{\infty} e_j$.

Proof. Without loss of generality we may assume that the semi–norms r_k^j satisfy the inequalities $r_{k+1}^j(e) \geq r_k^j(e)$ for all $e \in E^j$ and all k, j. From (iii) we know $r_k^0(\chi^m(c)e_m) \to 0$ for $c \to \infty$ when $m \geq l(0, k)$. Hence we find constants $c_{0,k,m} > 0$ such that $r_k^0(\chi^m(c)e_m) < 2^{-m}$ for $m \geq l(0, k)$, $c \geq c_{0,k,m}$. For $m < l(0, k)$ we may fix $c_{0,k,m}$ in an arbitrary way. Then $\sum_{m=0}^{\infty} \chi^m(c_{0,k,m})e_m$ converges with respect to the semi–norm r_k^0. If we set $c_{0,m} = \max\{c_{0,k,m} : 0 \leq k \leq m\}$ then $c_{0,m} \geq c_{0,k,m}$ for all $m \geq k$. Thus, for every fixed k, we have

$$\sum_{m=0}^{\infty} r_k^0(\chi^m(c_{0,m})e_m) = \sum_{m=0}^{k-1} r_k^0(\chi^m(c_{0,m})e_m) + \sum_{m=k}^{\infty} r_k^0(\chi^m(c_{0,m})e_m)$$

$$\leq \sum_{m=0}^{k-1} r_k^0(\chi^m(c_{0,m})e_m) + \sum_{m=k}^{\infty} 2^{-m} < \infty,$$

i.e. $\sum_{m=0}^{\infty} \chi^m(c_{0,m})e_m$ converges in E^0. We now proceed inductively and assume that the constants $c_{j-1,m}$ are defined for $j \geq 1$ such that $\sum_{m=n}^{\infty} \chi^m(c_{j-1,m})e_m$ converges in E^n for all $n = 0, \ldots, j-1$. Using the assumption (iii) we obtain $r_k^j(\chi^m(c)e_m) \to 0$ for $c \to \infty$ and $m \geq l(j, k)$. Hence we find $c_{j,k,m} > 0$ with $r_k^j(\chi^m(c)e_m) < 2^{-m}$ for $m \geq l(j, k)$, $c \geq c_{j,k,m}$. Again we fix $c_{j,k,m}$ for $m < l(j, k)$ in an arbitrary way. We now set $c_{j,m} = \max\{c_{n,k,m} : 0 \leq n \leq j, 0 \leq k \leq m\}$ which implies $c_{j,m} \geq c_{j,k,m}$ for all $m \geq k$, and $c_{j,m} \geq c_{n,m}$ for $n \leq j$, $m \geq k$. Then, for every k,

$$\sum_{m=j}^{\infty} r_k^j(\chi^m(c_{j,m})e_m) = \sum_{m=j}^{k-1} r_k^j(\chi^m(c_{j,m})e_m) + \sum_{m=k}^{\infty} r_k^j(\chi^m(c_{j,m})e_m)$$

$$\leq \sum_{m=j}^{k-1} r_k^j(\chi^m(c_{j,m})e_m) + \sum_{m=k}^{\infty} 2^{-m} < \infty,$$

i.e. $\sum_{m=n}^{\infty} \chi^m(c_{j,m})e_m$ converges in E^n for all $0 \le n \le j$. The final step is that we set $c_m = \max\{c_{j,m} : 0 \le j \le m\}$. We claim that $\sum_{m=j}^{\infty} r_k^j(\chi^m(c_m)e_m) < \infty$ for all j,k. In fact, we can write

$$\sum_{m=j}^{\infty} r_k^j(\chi^m(c_m)e_m) = \sum_{m=j}^{j+k-1} r_k^j(\chi^m(c_m)e_m) + \sum_{m=j+k}^{\infty} r_k^j(\chi^m(c_m)e_m)$$

$$= \sum_{m=j}^{j+k-1} r_k^j(\chi^m(c_m)e_m) + \sum_{m=j+k}^{\infty} 2^{-m} < \infty,$$

since $c_m \ge c_{j,k} \ge c_{j,k,m}$ for $m \ge j+k$. Thus we have obtained (1.1.30). Using (1.1.27) we get (1.1.31). $\qquad\square$

Proof of Theorem 1.1.16. Let us set $E^j = S^{\mu_j}(\Omega \times \mathbb{R}^n)$, $j \in \mathbb{N}$. Fix a countable system of compact sets $\{K_l\}_{l\in\mathbb{N}}$ with $K_l \subseteq K_{l+1} \subset \Omega$ for all l and $\bigcup_l K_l = \Omega$. Then

$$r_k^j(a) = \sup_{\substack{x\in K_k,\ \xi\in\mathbb{R}^n \\ |\alpha|+|\beta|\le k}} (1+|\xi|)^{-\mu_j+|\beta|}|D_x^\alpha D_\xi^\beta a(x,\xi)|, \qquad k \in \mathbb{N}$$

is a semi-norm system for the Fréchet topology of E^j. Let $\chi(\xi)$ be an excision function and define $\chi^j(c) : E^j \to E^j$ as the operator of multiplication by $\chi(c^{-1}\xi)$. Then (1.1.27) is satisfied and (1.1.28) commutes. In order to apply Proposition 1.1.17 it suffices to check that for every j and every $K\subset\subset\Omega$, $\alpha \in \mathbb{N}^m$, $\beta \in \mathbb{N}^n$ there is an l such that $b(x,\xi) \in S^\nu(\Omega \times \mathbb{R}^n)$ for $\nu \le \mu_l$ implies

$$\sup_{\substack{x\in K \\ \xi\in\mathbb{R}^n}} (1+|\xi|)^{-\mu_j+|\beta|}|D_x^\alpha D_\xi^\beta\{\chi(c^{-1}\xi)b(x,\xi)\}| \to 0 \qquad (1.1.32)$$

for $c \to \infty$. Once we have this we can represent the asymptotic sum as a convergent series

$$a(x,\xi) = \sum_{j=0}^{\infty} \chi(c_j^{-1}\xi)a_j(x,\xi) \qquad (1.1.33)$$

with constants c_j tending to infinity sufficiently fast as $j \to \infty$. Let us show that (1.1.32) holds for arbitrary $\nu < \mu_j$. We have

$$\sup(1+|\xi|)^{-\mu_j+|\beta|}|D_x^\alpha D_\xi^\beta \chi(c^{-1}\xi)b(x,\xi)|$$
$$= \sup(1+|\xi|)^{\nu-\mu_j}(1+|\xi|)^{-\nu+|\beta|}|D_x^\alpha D_\xi^\beta \chi(c^{-1}\xi)b(x,\xi)| \qquad (1.1.34)$$

with sup being taken over $x \in K$, $\xi \in \mathbb{R}^n$. Since $\chi(\xi)$ is an excision function, we have $\chi(\xi) = 0$ for $|\xi| < R$ with some $R > 0$. Then $\chi(c^{-1}\xi)$ vanishes for $|c^{-1}\xi| < R$. Thus it is permitted to take sup over $|\xi| \geq Rc$, and hence

$$\sup(1 + |\xi|)^{-\mu_j + |\beta|}|D_x^\alpha D_\xi^\beta \chi(c^{-1}\xi)b(x,\xi)| \leq D(c)(1 + Rc)^{\nu - \mu_j}$$

with $D(c) = \sup(1 + |\xi|)^{-\nu + |\beta|}|D_x^\alpha D_\xi^\beta \chi(c^{-1}\xi)b(x,\xi)|$. However it is rather trivial that $D(c)$ is uniformly bounded for $c \geq \varepsilon$ for every $\varepsilon > 0$. This gives us (1.1.32). □

Definition 1.1.18 *A function $\varphi(x,\xi) \in C^\infty(\Omega \times (\mathbb{R}^n \setminus \{0\}))$ is called a phase function if*

(i) $\varphi(x,\xi)$ *is real–valued,*

(ii) $\varphi(x, \lambda\xi) = \lambda\varphi(x,\xi)$ *for all $\lambda \in \mathbb{R}_+$, $(x,\xi) \in \Omega \times (\mathbb{R}^n \setminus \{0\})$,*

(iii) $d_{x,\xi}\varphi(x,\xi) \neq 0$ *for all $(x,\xi) \in \Omega \times (\mathbb{R}^n \setminus \{0\})$ (here $d_{x,\xi}$ means the gradient with respect to the corresponding variables).*

We now consider so–called oscillatory integrals

$$I_\varphi(au) = \int_{\mathbb{R}^n} \int_\Omega e^{i\varphi(x,\xi)}a(x,\xi)u(x)\,dxd\xi \qquad (1.1.35)$$

with open $\Omega \subseteq \mathbb{R}^m \ni x$, a phase function $\varphi(x,\xi)$, an $a(x,\xi) \in S^\mu(\Omega \times \mathbb{R}^n)$, called amplitude function, and $u \in C^\infty(\Omega)$. The applications below will concern the case $m = 2n$.

Note that (1.1.35) converges as a double integral for $\mu < -n$.

Lemma 1.1.19 *If $\varphi(x,\xi)$ is a phase function on $\Omega \times (\mathbb{R}^n \setminus \{0\})$ there exists an operator*

$$L = \sum_{j=1}^m a_j(x,\xi)\frac{\partial}{\partial x_j} + \sum_{k=1}^n b_k(x,\xi)\frac{\partial}{\partial \xi_k} + c(x,\xi)$$

with $a_j \in S^{-1}(\Omega \times \mathbb{R}^n)$, $b_k(x,\xi) \in S^0(\Omega \times \mathbb{R}^n)$, $c(x,\xi) \in S^{-1}(\Omega \times \mathbb{R}^n)$, such that the formal transposed ${}^tL$, acting on $v(x,\xi)$ as

$${}^tLv = -\sum_{j=1}^m \frac{\partial}{\partial x_j}(a_jv) - \sum_{k=1}^n \frac{\partial}{\partial \xi_k}(b_kv) + cv$$

satisfies

$${}^tLe^{i\varphi} = e^{i\varphi}. \qquad (1.1.36)$$

Proof. Using (iii) of Definition 1.1.18 we can form the function

$$\psi(x,\xi) = \left\{ \sum_{j=1}^{m} |\frac{\partial\varphi}{\partial x_j}|^2 + \sum_{k=1}^{n} |\xi|^2 |\frac{\partial\varphi}{\partial \xi_k}|^2 \right\}^{-1}$$

that belongs to $C^\infty(\Omega \times (\mathbb{R}^n \setminus \{0\}))$. In view of (ii) it follows that $\psi(x,\lambda\xi) = \lambda^{-2}\psi(x,\xi)$ for all $\lambda \in \mathbb{R}_+$, $(x,\xi) \in \Omega \times (\mathbb{R}^n \setminus \{0\})$. We now have the obvious identity

$$-i\psi \left\{ \sum_{j=1}^{m} \frac{\partial\varphi}{\partial x_j}\frac{\partial}{\partial x_j} + \sum_{k=1}^{n} |\xi|^2 \frac{\partial\varphi}{\partial \xi_k}\frac{\partial}{\partial \xi_k} \right\} e^{i\varphi} = e^{i\varphi} \qquad (1.1.37)$$

for $\xi \neq 0$. In order to remove the singularity at $\xi = 0$ we choose an $\omega(\xi) \in C_0^\infty(\mathbb{R}^n)$ with $\omega(\xi) = 1$ in an open neighbourhood of $\xi = 0$. Then $e^{i\varphi} = (1-\omega)e^{i\varphi}+\omega e^{i\varphi}$ gives us together with (1.1.37) a differential operator

$$-i\psi(1 - \omega) \left\{ \sum_{j=1}^{m} \frac{\partial\varphi}{\partial x_j}\frac{\partial}{\partial x_j} + \sum_{k=1}^{n} |\xi|^2 \frac{\partial\varphi}{\partial \xi_k}\frac{\partial}{\partial \xi_k} \right\} + \omega =: M$$

with coefficients in $C^\infty(\Omega \times \mathbb{R}^n)$ satisfying $Me^{i\varphi} = e^{i\varphi}$. Because of the homogeneity of φ the coefficients at $\frac{\partial}{\partial \xi_k}$ are symbols in S^0, cf. Remark 1.1.14. Further ω is of order -1 (here even $-\infty$). It suffices now to set $L = {}^t M$ and to observe that the above orders of the coefficients remain preserved under forming the transposed operator. □

It follows that $L^N S^\mu(\Omega \times \mathbb{R}^n) \subset S^{\mu-N}(\Omega \times \mathbb{R}^n)$ for every N. Integrating by parts in (1.1.35) yields

$$I_\varphi(au) = \iint e^{i\varphi(x,\xi)} L^N(a(x,\xi)u(x)) \, dx d\xi \qquad (1.1.38)$$

for $\mu < -n$. Now (1.1.38) is a possible definition of $I_\varphi(au)$ for arbitrary $\mu \in \mathbb{R}$ when we choose N so large that $\mu - N < -n$.

Proposition 1.1.20 *Let $a(x,\xi) \in S^\mu(\Omega \times \mathbb{R}^n)$, $\mu \in \mathbb{R}$, and $a_\varepsilon(x,\xi) = \omega(\varepsilon\xi)a(x,\xi)$, $\varepsilon > 0$, with some $\omega(\xi) \in C_0^\infty(\mathbb{R}^n)$, $\omega(\xi) = 1$ in a neighbourhood of $\xi = 0$. Then*

$$\iint e^{i\varphi(x,\xi)} L^N(a(x,\xi)u(x)) \, dx d\xi = \lim_{\varepsilon \to 0} I_\varphi(a_\varepsilon u)$$

for every N with $\mu - N < -n$.

Proof. We have $a_\varepsilon(x, \xi) \in S^{-\infty}(\Omega \times \mathbb{R}^n)$. Thus $I_\varphi(a_\varepsilon u)$ exists as an ordinary integral. At the same time (1.1.38) can be applied to a_ε, i.e.

$$I_\varphi(a_\varepsilon u) = \iint e^{i\varphi(x,\xi)} L^N(\omega(\varepsilon\xi)a(x,\xi)u(x))dx d\xi. \qquad (1.1.39)$$

In view of $|D_\xi^\alpha \omega(\varepsilon\xi)| \leq c_\alpha(1 + |\xi|)^{-|\alpha|}$ for all α, with constants c_α independent of ε for $0 < \varepsilon \leq 1$, we may (by the Lebesgue theorem on dominated convergence) pass to the limit for $\varepsilon \to 0$ under the integral on the right of (1.1.39). This gives us the assertion. $\qquad \square$

Thus, for arbitrary $\mu \in \mathbb{R}$, we can define $I_\varphi(au) = \lim_{\varepsilon \to 0} I_\varphi(a_\varepsilon u)$ which is equivalent to (1.1.38). This regularisation of the (in general) divergent integrals is called the method of oscillatory integrals. It can easily be adapted to extend the standard rules such as integrations by parts. This will tacitly be used from now on.

Let us set $\langle w(a), u \rangle = I_\varphi(au)$ for $u \in C_0^\infty(\Omega)$ and fixed $a(x, \xi) \in S^\mu(\Omega \times \mathbb{R}^n)$. Then $w(a) \in \mathcal{D}'(\Omega)$. Define

$$C_\varphi = \{(x, \xi) \in \Omega \times (\mathbb{R}^n \setminus \{0\}) : d_\xi\varphi(x, \xi) = 0\} \qquad (1.1.40)$$

(recall that $d_\xi\varphi$ is the vector of derivatives $(\frac{\partial \varphi}{\partial \xi_1}, \ldots, \frac{\partial \varphi}{\partial \xi_n})$). If $\pi : \Omega \times (\mathbb{R}^n \setminus \{0\}) \to \Omega$ denotes the canonical projection $\pi(x, \xi) = x$, then

$$\text{sing supp} \, w(a) \subseteq \pi C_\varphi. \qquad (1.1.41)$$

In fact, the latter relation is equivalent to $w(a) \in C^\infty(\Omega \setminus \pi C_\varphi)$, and hence it suffices to show $\langle w(a), u \rangle = \int w(a; x)u(x) \, dx$ for every $u \in C_0^\infty(\Omega \setminus \pi C_\phi)$ with some function $w(a; x) \in C^\infty(\Omega \setminus \pi C_\phi)$. We may set

$$w(a; x) = \int e^{i\varphi(x,\xi)} a(x, \xi) \, d\xi \qquad \text{for } x \in \Omega \setminus \pi C_\varphi.$$

In view of the definition of C_φ this can be interpreted as an oscillatory integral with respect to ξ, dependent on the parameter x, analogously to the above (1.1.38) with a convergent integral for large N. But the differentiation with respect to x preserves convergence of integrals.

Next we turn to the case $\Omega = U \times V$ with open $U \subseteq \mathbb{R}^p$, $V \subseteq \mathbb{R}^{p'}$ and variables $(x, x') \in U \times V$.

Definition 1.1.21 *Let $\varphi(x, x', \xi)$ be a phase function on $U \times V \times (\mathbb{R}^n \setminus \{0\})$. Then φ is called an operator phase function if $d_{x,\xi}\varphi(x, x', \xi) \neq 0$ and $d_{x',\xi}\varphi(x, x', \xi) \neq 0$ for all $(x, x', \xi) \in U \times V \times (\mathbb{R}^n \setminus \{0\})$.*

18 PSEUDO–DIFFERENTIAL OPERATORS

Let $\varphi(x, x', \xi)$ be a fixed phase function on $U \times V \times (\mathbb{R}^n \setminus \{0\})$, and $a(x, x', \xi) \in S^\mu(U \times V \times \mathbb{R}^n)$. Then $w(a) \in \mathcal{D}'(U \times V)$ can be regarded as the distributional kernel of an operator $A : C_0^\infty(V) \to \mathcal{D}'(U)$, i.e. $\langle Ag, f \rangle = \langle w(a), fg \rangle$ for $f \in C_0^\infty(U)$, $g \in C_0^\infty(V)$.

The operator A is called properly supported if for arbitrary $K \subset\subset U$, $L \subset\subset V$ the sets $(K \times V) \cap \operatorname{supp} w(a)$ and $(U \times L) \cap \operatorname{supp} w(a)$ are compact. If a distribution in $U \times V$ has such a support we will also say that the support is proper.

Proposition 1.1.22 *Let $\varphi(x, x', \xi)$ be an operator phase function on $U \times V \times (\mathbb{R}^n \setminus \{0\})$, and $a(x, x', \xi) \in S^\mu(U \times V \times \mathbb{R}^n)$. Then A induces a continuous operator*

$$A : C_0^\infty(V) \to C^\infty(U) \tag{1.1.42}$$

which extends to

$$A : \mathcal{E}'(V) \to \mathcal{D}'(U), \tag{1.1.43}$$

continuous in the w^*–topology of corresponding dual spaces. If A is properly supported, then we obtain $A : C_0^\infty(V) \to C_0^\infty(U)$ with extensions as $A : C^\infty(V) \to C^\infty(U)$, $A : \mathcal{E}'(V) \to \mathcal{E}'(U)$ and $A : \mathcal{D}'(V) \to \mathcal{D}'(U)$.

Proof. The continuity of (1.1.42) follows by interpreting

$$Ag(x) = \iint e^{i\varphi(x,x',\xi)} a(x, x', \xi) g(x') \, dx' d\xi$$

as an oscillatory integral with respect to (x', ξ), with the parameter x. This can obviously be differentiated with respect to x. The transposed of A

$$^tAf(x') = \iint e^{i\varphi(x,x',\xi)} a(x, x', \xi) f(x) \, dx d\xi$$

induces analogously a continuous operator $^tA : C_0^\infty(V) \to C^\infty(U)$. Thus we can define the extension (1.1.43) by $\langle Ag, f \rangle = \langle g, {}^tAf \rangle$ for $g \in \mathcal{E}'(V)$, $f \in C^\infty(U)$. The other assertions in the properly supported case are evident. $\qquad\Box$

Theorem 1.1.23 *Under the conditions of Proposition 1.1.22 we have*

$$\operatorname{sing\,supp} Au \subseteq (\pi C_\varphi) \circ \operatorname{sing\,supp} u.$$

Here $W \circ M$ for subsets $W \subseteq U \times V$, $M \subseteq V$ is defined as the set of all $x \in U$ for which $(x, x') \in W$ for some $x' \in M$.

Proof. Write $u = v + g$ with $g \in C_0^\infty(V)$ and v being concentrated in a small neighbourhood of sing supp u. Then, because of $Ag \in C^\infty(U)$, it suffices to show sing supp $Av \subseteq \pi C_\varphi \circ M$ for $M = \text{supp } v$. Let $K \subset\subset U$ be arbitrary with $K \times M \subset (U \times V) \setminus \pi C_\varphi$. Since $(U \times V) \setminus \pi C_\varphi$ is open there are open neighbourhoods U_0 of K, V_0 of M such that $U_0 \times V_0 \subseteq (U \times V) \setminus \pi C_\varphi$. It suffices now to obtain $Av \in C^\infty(V_0)$. However this is clear, since the distributional kernel of A is C^∞ over $(U \times V) \setminus \pi C_\varphi$, cf. (1.1.41). □

Example 1.1.24 *Let $\Omega \subseteq \mathbb{R}^n$ be an open set. Then $\varphi(x, x', \xi) = (x - x')\xi$ is an operator phase function on $\Omega \times \Omega \times (\mathbb{R}^n \setminus \{0\})$. In this case $C_\varphi = (\text{diag } \Omega \times \Omega) \times (\mathbb{R}^n \setminus \{0\})$ with $\text{diag } \Omega \times \Omega = \{(x, x) : x \in \Omega\}$.*

Definition 1.1.25 *Let $\Omega \subseteq \mathbb{R}^n$ be open and $\mu \in \mathbb{R}$. An operator of the form*

$$\text{Op}(a)u(x) = \iint e^{i(x-x')\xi} a(x, x', \xi) u(x') \, dx' d\xi \qquad (1.1.44)$$

with a symbol (or amplitude function) $a(x, x', \xi) \in S^\mu(\Omega \times \Omega \times \mathbb{R}^n)$ is called a pseudo–differential operator on Ω of order μ. We set

$$L^\mu(\Omega) = \{\text{Op}(a) : a(x, x', \xi) \in S^\mu(\Omega \times \Omega \times \mathbb{R}^n)\}, \qquad (1.1.45)$$
$$L_{\text{cl}}^\mu(\Omega) = \{\text{Op}(a) : a(x, x', \xi) \in S_{\text{cl}}^\mu(\Omega \times \Omega \times \mathbb{R}^n)\}. \qquad (1.1.46)$$

The elements of (1.1.46) are called classical pseudo–differential operators.

According to the above discussion every $A \in L^\mu(\Omega)$ has a distributional kernel $K_A \in \mathcal{D}'(\Omega \times \Omega)$. From (1.1.41) we know that

$$\text{sing supp } K_A \subseteq \text{diag } \Omega \times \Omega. \qquad (1.1.47)$$

On the other hand we can form

$$k(a)(x, x', \beta) = \int e^{i\beta\xi} a(x, x', \xi) \, d\xi, \qquad (1.1.48)$$

cf. (1.1.21) and Proposition 1.1.15. Then

$$K_A(x, x') = k(a)(x, x', x - x') \quad \text{for} \quad A = \text{Op}(a).$$

Proposition 1.1.26 *The space $L^{-\infty}(\Omega) := \bigcap_{\mu \in \mathbb{R}} L^\mu(\Omega)$ coincides with the space of all integral operators*

$$Cu(x) = \int_\Omega c(x, x') u(x') \, dx' \qquad (1.1.49)$$

with kernels $c(x, x') \in C^\infty(\Omega \times \Omega)$.

Proof. If $A \in L^{-\infty}(\Omega)$ is given then to every $r \in \mathbb{N}$ there is an $a_r(x, x', \xi) \in S^{\mu(r)}(\Omega \times \Omega \times \mathbb{R}^n)$ for $\mu(r) < -n-r$ such that $A = \mathrm{Op}(a_r)$. Then $K_A(x, x') = k(a_r)(x, x', x - x') \in C^r(\Omega \times \Omega)$, cf. (1.1.22). Hence $K_A(x, x') \in C^\infty(\Omega \times \Omega)$. Conversely to every $c(x, x') \in C^\infty(\Omega \times \Omega)$ we can form an $a(x, x', \xi) \in S^{-\infty}(\Omega \times \Omega \times \mathbb{R}^n)$ such that (1.1.49) equals $\mathrm{Op}(a)$. In fact, it suffices to set

$$a(x, x', \xi) = e^{i(x-x')\xi} c(x, x') \psi(\xi)$$

for an arbitrary $\psi \in C_0^\infty(\mathbb{R}^n)$ satisfying $\int \psi(\xi) \, d\xi = 1$. $\qquad\square$

Proposition 1.1.27 *Every $A \in L^\mu(\Omega)$ can be written in the form $A = A_0 + C$, where $A_0 \in L^\mu(\Omega)$ is properly supported and $C \in L^{-\infty}(\Omega)$.*

Proof. Let us choose an arbitrary function $\omega(x, x') \in C^\infty(\Omega \times \Omega)$ that equals 1 in an open neighbourhood of diag $\Omega \times \Omega$ and such that both $(\Omega \times M') \cap \mathrm{supp}\,\omega$ and $(M \times \Omega) \cap \mathrm{supp}\,\omega$ are compact for arbitrary $M' \subset\subset \Omega$, $M \subset\subset \Omega$. Then, for $A = \mathrm{Op}(a)$ with $a(x, x', \xi) \in S^\mu(\Omega \times \Omega \times \mathbb{R}^n)$ we can set $A_0 = \mathrm{Op}(\omega a)$ which is properly supported, and $C = \mathrm{Op}((1-\omega)a)$. The kernel of C is $K_C = (1 - \omega(x, x'))K_A(x, x') \in C^\infty(\Omega \times \Omega)$, cf. (1.1.47). $\qquad\square$

Remark 1.1.28 *From Proposition 1.1.22 we know that every $A \in L^\mu(\Omega)$ gives rise to a continuous operator*

$$A : C_0^\infty(\Omega) \to C^\infty(\Omega)$$

which extends to an $A : \mathcal{E}'(\Omega) \to \mathcal{D}'(\Omega)$. If A is properly supported then A induces continuous operators

$$A : C_0^\infty(\Omega) \to C_0^\infty(\Omega), \quad A : C^\infty(\Omega) \to C^\infty(\Omega).$$

Furthermore A extends to

$$A : \mathcal{E}'(\Omega) \to \mathcal{E}'(\Omega), \quad A : \mathcal{D}'(\Omega) \to \mathcal{D}'(\Omega).$$

1.1.3   Pseudo–differential calculus

An element $a(x, \xi) \in S^\mu(\Omega \times \mathbb{R}^n)$ is called a complete symbol of $A \in L^\mu(\Omega)$ if $A - \mathrm{Op}(a) \in L^{-\infty}(\Omega)$. To prove the existence of complete symbols we will use the following result:

Lemma 1.1.29 *Let $a(x, x', \xi) \in S^\mu(\Omega \times \Omega \times \mathbb{R}^n)$ be a symbol with the property $|x - x'|^{-2N} a(x, x', \xi) \in S^\mu(\Omega \times \Omega \times \mathbb{R}^n)$ for some $N \in \mathbb{N}$. Then there exists an $a_N(x, x', \xi) \in S^{\mu-2N}(\Omega \times \Omega \times \mathbb{R}^n)$ such that $\mathrm{Op}(a) = \mathrm{Op}(a_N)$.*

Proof. Using $e^{i(x-x')\xi} = |x-x'|^{-2N}(-\Delta_\xi)^N e^{i(x-x')\xi}$ and integration by parts it follows that $\mathrm{Op}(a) = \mathrm{Op}(a_N)$ with $a_N(x,x',\xi) = (-\Delta_\xi)^N |x - x'|^{-2N} a(x,x',\xi) \in S^{\mu-2N}(\Omega \times \Omega \times \mathbb{R}^n)$. $\square$

Theorem 1.1.30 *To every* $a(x,x',\xi) \in S^\mu(\Omega \times \Omega \times \mathbb{R}^n)$ *there is an* $\boldsymbol{a}(x,\xi) \in S^\mu(\Omega \times \mathbb{R}^n)$ *with* $\mathrm{Op}(a) = \mathrm{Op}(\boldsymbol{a}) \bmod L^{-\infty}(\Omega)$ *and* $\boldsymbol{a}(x,\xi)$ *admits the asymptotic expansion*

$$\boldsymbol{a}(x,\xi) \sim \sum_\alpha \frac{1}{\alpha!} D_\xi^\alpha \partial_{x'}^\alpha a(x,x',\xi)|_{x'=x}. \tag{1.1.50}$$

In other words every $A \in L^\mu(\Omega)$ *has a complete symbol. Moreover, there is a* $\boldsymbol{b}(x',\xi) \in S^\mu(\Omega \times \mathbb{R}^n)$ *with* $\mathrm{Op}(a) = \mathrm{Op}(\boldsymbol{b}) \bmod L^{-\infty}(\Omega)$ *that has the asymptotic expansion*

$$\boldsymbol{b}(x',\xi) \sim \sum_\alpha \frac{1}{\alpha!} (-D_\xi)^\alpha \partial_x^\alpha a(x,x',\xi)|_{x=x'}. \tag{1.1.51}$$

Proof. Applying the Taylor expansion of $a(x,x',\xi)$ on diag $\Omega \times \Omega$ we obtain

$$a(x,x',\xi) = \sum_{|\alpha| \le M} \frac{1}{\alpha!}(x'-x)^\alpha \partial_{x'}^\alpha a(x,x',\xi)|_{x'=x} + r_M(x,x',\xi) \tag{1.1.52}$$

with a remainder term $r_M(x,x',\xi) \in S^\mu(\Omega \times \Omega \times \mathbb{R}^n)$. For every N we can choose M so large that $|x-x'|^{-2N} r_M(x,x',\xi) \in S^\mu(\Omega \times \Omega \times \mathbb{R}^n)$. Applying Op on both sides of (1.1.52) we obtain from Lemma 1.1.29, together with the identity $\mathrm{Op}((x'-x)^\alpha b) = \mathrm{Op}(D_\xi^\alpha b)$ for any symbol b,

$$\mathrm{Op}(a) = \sum_{|\alpha| \le M} \frac{1}{\alpha!} \mathrm{Op}(D_\xi^\alpha \partial_{x'}^\alpha a|_{x'=x}) + \mathrm{Op}(a_N)$$

with some $a_N \in S^{\mu-2N}(\Omega \times \Omega \times \mathbb{R}^n)$. If we form (1.1.50) by carrying out the asymptotic sum we obtain immediately $\mathrm{Op}(a) - \mathrm{Op}(\boldsymbol{a}) = \mathrm{Op}(\tilde{a}_N)$ for another $\tilde{a}_N \in S^{\mu-2N}(\Omega \times \Omega \times \mathbb{R}^n)$. This is true for every N and hence $\mathrm{Op}(a) - \mathrm{Op}(\boldsymbol{a}) \in L^{-\infty}(\Omega)$. The second statement can be proved in an analogous manner by interchanging the role of x and x'. $\square$

Theorem 1.1.31 *Let* $A \in L^\mu(\Omega)$ *and* A^* *its formal adjoint, defined by*

$$(Au,v) = (u,A^*v) \quad \text{for all } u,v \in C_0^\infty(\Omega) \tag{1.1.53}$$

with the L^2 *scalar product* $(\cdot,\cdot)$. *Then* $A^* \in L^\mu(\Omega)$. *If* $A = \mathrm{Op}(a) \bmod L^{-\infty}(\Omega)$, *for some* $a(x,\xi) \in S^\mu(\Omega \times \mathbb{R}^n)$, *we have* $A^* = \mathrm{Op}(a^*) \bmod L^{-\infty}(\Omega)$ *with an* $a^*(x,\xi) \in S^\mu(\Omega \times \mathbb{R}^n)$,

$$a^*(x,\xi) \sim \sum_\alpha \frac{1}{\alpha!} D_\xi^\alpha \partial_x^\alpha \overline{a(x,\xi)}. \tag{1.1.54}$$

Analogously, if $^t\!A$ is the formal transposed of A with respect to $\langle\cdot,\cdot\rangle$
we have $^t\!A \in L^\mu(\Omega)$, and $^t\!A = \mathrm{Op}(c) \bmod L^{-\infty}(\Omega)$, with $c(x,\xi) \in$
$S^\mu(\Omega \times \mathbb{R}^n)$,

$$c(x,\xi) \sim \sum_\alpha \frac{1}{\alpha!} D_\xi^\alpha \partial_x^\alpha a(x,-\xi). \tag{1.1.55}$$

Proof. The behaviour of operators in $L^{-\infty}(\Omega)$ under formal adjoints
is obvious. Thus we may assume $A = \mathrm{Op}(a)$ with a complete symbol
$a(x,\xi)$. Then

$$A^* v(x) = \iint e^{i(x-x')\xi} \overline{a(x',\xi)} v(x')\, dx'd\xi$$

shows $A^* \in L^\mu(\Omega)$ including (1.1.54) as a consequence of Theorem
1.1.30. The arguments for the transposed operator are analogous. $\quad\Box$

Theorem 1.1.32 *Let $A \in L^\mu(\Omega)$, $B \in L^\nu(\Omega)$ and A or B be prop-*
erly supported. Then $AB \in L^{\mu+\nu}(\Omega)$. If $A = \mathrm{Op}(a) \bmod L^{-\infty}(\Omega)$,
$a(x,\xi) \in S^\mu(\Omega \times \mathbb{R}^n)$, $B = \mathrm{Op}(b) \bmod L^{-\infty}(\Omega)$, $b(x,\xi) \in S^\nu(\Omega \times \mathbb{R}^n)$,
then $AB = \mathrm{Op}(c) \bmod L^{-\infty}(\Omega)$ for a symbol $c(x,\xi) \in S^{\mu+\nu}(\Omega \times \mathbb{R}^n)$,

$$c(x,\xi) \sim \sum_\alpha \frac{1}{\alpha!} (D_\xi^\alpha a(x,\xi)) \partial_x^\alpha b(x,\xi). \tag{1.1.56}$$

Proof. Let us assume, for instance, that A is properly supported.
Write B in the form $B = \mathrm{Op}(b') \bmod L^{-\infty}(\Omega)$ for a symbol $b'(x',\xi) \in$
$S^\nu(\Omega \times \mathbb{R}^n)$. Then, according to (1.1.51),

$$b'(x',\xi) \sim \sum_\alpha \frac{1}{\alpha!} (-D_\xi)^\alpha \partial_x^\alpha b(x',\xi). \tag{1.1.57}$$

The relations $\mathrm{Op}(b')u(y) = F_{\xi\to y}^{-1} F_{x'\to\xi}\{b'(x',\xi)u(x')\}$ and $\mathrm{Op}(a)v(x) =$
$F_{\xi\to x}^{-1} a(x,\xi) \{F_{y\to\xi} v(y)\}$ yield

$$\mathrm{Op}(a)\,\mathrm{Op}(b')u(x) = \iint e^{i(x-x')\xi} a(x,\xi) b'(x',\xi) u(x')\, dx'd\xi.$$

In order to obtain an asymptotic formula for $c(x,\xi)$ it suffices to apply
(1.1.50) to $a(x,\xi)b'(x',\xi)$, which gives us

$$c(x,\xi) \sim \sum_\alpha \frac{1}{\alpha!} \partial_\xi^\alpha \{a(x,\xi) D_x^\alpha b'(x,\xi)\}$$

$$\sim \sum_{\alpha,\beta} \frac{1}{\alpha!\beta!} \partial_\xi^\alpha \{a(x,\xi)(-\partial_\xi)^\beta D_x^{\alpha+\beta} b(x,\xi)\}.$$

It was used in the second relation (1.1.57). Now we employ the Leibniz rule

$$\partial_\xi^\alpha(f(\xi)g(\xi)) = \sum_{\gamma+\delta=\alpha} \frac{\alpha!}{\gamma!\delta!}\{\partial_\xi^\gamma f(\xi)\}\{\partial_\xi^\delta g(\xi)\}$$

and the binomial formula

$$(x+y)^\varrho = \sum_{\beta+\delta=\varrho} \frac{\varrho!}{\beta!\delta!} x^\delta y^\beta$$

for arbitrary $x, y \in \mathbb{R}^n$, $\varrho \in \mathbb{N}^n$. Then, in particular, for $x = -y = e = (1, \ldots, 1)$ we get

$$(e-e)^\varrho = \sum_{\beta+\delta=\varrho} \frac{\varrho!}{\beta!\delta!} e^\delta(-e)^\beta = \varrho! \sum_{\beta+\delta=\varrho} \frac{(-1)^{|\beta|}}{\beta!\delta!} = \delta_{0,\varrho}$$

for $\delta_{0,\varrho} = 1$ for $\varrho = (0, \ldots, 0)$, $\delta_{0,\varrho} = 0$ for $\varrho \neq 0$. Thus

$$c(x,\xi) \sim \sum_{\substack{\alpha,\beta,\gamma,\delta \\ \gamma+\delta=\alpha}} \frac{1}{\beta!\gamma!\delta!}\{\partial_\xi^\gamma a(x,\xi)\}\{(-\partial_\xi)^\beta \partial_\xi^\delta D_x^{\alpha+\beta} b(x,\xi)\}$$

$$= \sum_{\beta,\gamma,\delta} \frac{(-1)^{|\beta|}}{\beta!\gamma!\delta!}\{\partial_\xi^\gamma a(x,\xi)\}\{\partial_\xi^{\beta+\delta} D_x^{\beta+\gamma+\delta} b(x,\xi)\}$$

$$= \sum_\gamma \frac{1}{\gamma!} \sum_\varrho \Big\{ \sum_{\beta+\delta=\varrho} \frac{(-1)^{|\beta|}}{\beta!\delta!} \Big\} [\partial_\xi^\gamma a(x,\xi)][\partial_\xi^\varrho D_x^{\varrho+\gamma} b(x,\xi)]$$

just simplifies to (1.1.56).   □

The asymptotic sum in (1.1.56) will also be called the Leibniz product between a and b, written

$$c(x,\xi) = a(x,\xi)\#b(x,\xi). \tag{1.1.58}$$

$c(x,\xi)$ is unique $\mathrm{mod} S^{-\infty}(\Omega \times \mathbb{R}^n)$. Let us set $e_\xi(x) = e^{ix\xi}$ and consider a properly supported operator $A \in L^\mu(\Omega)$. Then

$$a(x,\xi) := e_{-\xi}(x)Ae_\xi \tag{1.1.59}$$

is a C^∞ function in (x,ξ), cf. Proposition 1.1.22.

Theorem 1.1.33 *Let $A \in L^\mu(\Omega)$ be properly supported. Then (1.1.59) is a complete symbol of A satisfying $A = \mathrm{Op}(a)$. If the operator is given as $A = \mathrm{Op}(a)$ for an $a(x,x',\xi) \in S^\mu(\Omega \times \Omega \times \mathbb{R}^n)$ then (1.1.59) has the asymptotic expansion (1.1.50).*

Proof. Let us apply A to $u \in C_0^\infty(\Omega)$ and write $u(x) = \int e_\xi(x)\hat{u}(\xi)\,d\xi$ (Fourier inversion formula). Then $Au(x) = \int (Ae_\xi)(x)\hat{u}(\xi)\,d\xi$, because the integrals converge in $C^\infty(\Omega)$, cf. Proposition 1.1.22. Thus

$$
\begin{aligned}
Au(x) &= \int e^{ix\xi}\{e_{-\xi}(x)(Ae_\xi)(x)\}\hat{u}(\xi)\,d\xi \\
&= \int e^{ix\xi}\boldsymbol{a}(x,\xi)\Big\{\int e^{-ix'\xi}u(x')dx'\Big\}\,d\xi.
\end{aligned}
$$

We will show $\boldsymbol{a}(x,\xi) \in S^\mu(\Omega \times \mathbb{R}^n)$. Then $\boldsymbol{a}(x,\xi)$ is a complete symbol of A. Since A is properly supported, we may start with $A = \mathrm{Op}(a)$ for an $a(x,x',\xi) \in S^\mu(\Omega \times \Omega \times \mathbb{R}^n)$ which has proper support in (x,x'). We have

$$
\boldsymbol{a}(x,\xi) = \iint e^{-ix\xi}e^{i(x-x')\eta}a(x,x',\eta)e^{ix'\xi}\,dx'd\eta \tag{1.1.60}
$$

as an iterated integral, where the integration in x' is taken over a compact set when $x \in M$ for any $M \subset\subset \Omega$. The expression (1.1.60) can be regarded as an oscillatory integral dependent on the parameter $x \in M$. Substituting $z = x' - x$, $\zeta = \eta - \xi$ we obtain

$$
\boldsymbol{a}(x,\xi) = \iint a(x,x+z,\xi+\zeta)e^{-iz\zeta}\,dzd\zeta.
$$

Taylor expansion of $a(x,x+z,\xi+\zeta)$ in ζ at $\zeta = 0$ gives us

$$
a(x,x+z,\xi+\zeta) = \sum_{|\alpha|\leq N-1}\frac{1}{\alpha!}(\partial_\xi^\alpha a)(x,x+z,\xi)\zeta^\alpha + r_N(x,x+z,\xi,\zeta)
$$

with

$$
r_N(x,x+z,\xi,\zeta) = \sum_{|\alpha|=N}\frac{N\zeta^\alpha}{\alpha!}\int_0^1 (1-t)^{N-1}(\partial_\xi^\alpha a)(x,x+z,\xi+t\zeta)\,dt.
$$

From the Fourier inversion formula it follows that

$$
\iint \partial_\xi^\alpha a(x,x+z,\xi)\zeta^\alpha e^{-iz\zeta}\,dzd\zeta = \partial_\xi^\alpha D_z^\alpha a(x,x+z,\xi)|_{z=0}.
$$

Thus

$$
\boldsymbol{a}(x,\xi) = b_N(x,\xi) + \iint r_N(x,x+z,\xi,\zeta)e^{-iz\zeta}\,dzd\zeta \tag{1.1.61}
$$

for $b_N(x,\xi) = \sum_{|\alpha|\leq N-1}\frac{1}{\alpha!}\partial_\xi^\alpha D_{x'}^\alpha a(x,x',\xi)|_{x'=x} \in S^\mu(\Omega \times \mathbb{R}^n)$. If we show that the second summand on the right of (1.1.61) is a symbol

$c_N(x, \xi)$ of order μ_N tending to $-\infty$ as $N \to \infty$, then we obtain $a(x, \xi) \in S^\mu(\Omega \times \mathbb{R}^n)$ and at the same time the asserted asymptotic expansion. We have

$$c_N(x, \xi) =$$

$$\sum_{|\alpha|=N} \frac{N}{\alpha!} \int_0^1 (1-t)^{N-1} \left\{ \iint e^{-iz\zeta} \zeta^\alpha \partial_\xi^\alpha a(x, x+z, \xi+t\zeta)\, dz d\zeta \right\} dt.$$

Let us consider

$$r_{\alpha,t}(x, \xi) = \iint e^{-iz\zeta} \zeta^\alpha \partial_\xi^\alpha a(x, x+z, \xi+t\zeta)\, dz d\zeta.$$

Integration by parts yields

$$r_{\alpha,t}(x, \xi) = \iint e^{-iz\zeta} \partial_\xi^\alpha D_z^\alpha a(x, x+z, \xi+t\zeta)\, dz d\zeta.$$

Choose an $m \in \mathbb{N}$ and apply again integration by parts, using

$$\langle \zeta \rangle^{-2m} \langle D_z \rangle^{2m} e^{-iz\zeta} = e^{-iz\zeta}.$$

Then $r_{\alpha,t}(x, \xi)$ equals a finite sum of expressions of the form

$$r_{\alpha,\beta,t}(x, \xi) = \iint e^{-iz\zeta} \langle \zeta \rangle^{-2m} \partial_\xi^\alpha D_z^{\alpha+\beta} a(x, x+z, \xi+t\zeta)\, dz d\zeta,$$

where $|\beta| \leq 2m$. Let us write $r_{\alpha,\beta,t} = p_{\alpha,\beta,t} + q_{\alpha,\beta,t}$, where $p_{\alpha,\beta,t}$ is defined as the integral over $\{(z, \zeta) : |\zeta| \leq |\xi|/2\}$ and $q_{\alpha,\beta,t}$ as the integral over the complement. If $|\zeta| \leq |\xi|/2$ then $|\xi|/2 \leq |\xi+t\zeta| \leq \frac{3}{2}|\xi|$. Since the integration in $p_{\alpha,\beta,t}$ with respect to ζ is taken over a domain of measure $\leq c|\xi|^n$ for a constant c, we get

$$|p_{\alpha,\beta,t}(x, \xi)| \leq c\langle \xi \rangle^{\mu-N+n} \qquad (1.1.62)$$

for another constant c, not dependent on ξ and t. For $|\zeta| > |\xi|/2$ we have

$$|\partial_\xi^\alpha D_z^{\alpha+\beta} a(x, x+z, \xi+t\zeta)| \leq c\langle \zeta \rangle^{\mu-N}$$

for $\mu - N \geq 0$ and $\leq \text{const}$ for $\mu - N < 0$. Thus, if $h := \max\{\mu - N, 0\}$, we get

$$|q_{\alpha,\beta,t}(x, \xi)| \leq c \int_{|\zeta|>|\xi|/2} \langle \zeta \rangle^{-2m} \langle \zeta \rangle^h d\zeta.$$

For m so large that $h - 2m + n + 1 < 0$ it follows that

$$|q_{\alpha,\beta,t}(x, \xi)| \leq c\langle \xi \rangle^{h-2m+n+1} \int \langle \zeta \rangle^{-n-1} d\zeta \leq c\langle \xi \rangle^{h-2m+n+1} \qquad (1.1.63)$$

with different constants c that are independent of ξ and t, $0 \leq t \leq 1$. Since m is arbitrary, (1.1.62), (1.1.63) imply the required estimate for the remainder, with $\mu_N = \mu - N + n \to -\infty$ as $N \to \infty$. $\square$

Theorem 1.1.34 For open $\Omega \subseteq \mathbb{R}^n$ and $\mu \in \mathbb{R}$ the map

$$\mathrm{Op} : S^\mu(\Omega \times \mathbb{R}^n) \to L^\mu(\Omega)$$

induces an (algebraic) isomorphism

$$S^\mu(\Omega \times \mathbb{R}^n)/S^{-\infty}(\Omega \times \mathbb{R}^n) \cong L^\mu(\Omega)/L^{-\infty}(\Omega).$$

Proof. From Theorem 1.1.30 we know that every $A \in L^\mu(\Omega)$ has the form $A = \mathrm{Op}(a) + C$ with $a(x,\xi) \in S^\mu(\Omega \times \mathbb{R}^n)$, $C \in L^{-\infty}(\Omega)$. Further $\mathrm{Op} : S^{-\infty} \to L^{-\infty}$. Hence Op induces a surjective mapping $S^\mu/S^{-\infty} \to L^\mu/L^{-\infty}$. On the other hand to every $a_1(x,\xi) \in S^\mu(\Omega \times \mathbb{R}^n)$ we can form $A_0 = \mathrm{Op}(\omega a_1)$ with a cut–off function $\omega(x,x')$ as in the proof of Proposition 1.1.27. Applying Theorem 1.1.33 to the properly supported A_0 we get $a(x,\xi) = e_{-\xi} A_0 e_\xi \in S^\mu(\Omega \times \mathbb{R}^n)$, $A_0 = \mathrm{Op}(a)$, and for $a(x,x',\xi) := \omega(x,x')a_1(x,\xi)$ the asymptotic expansion (1.1.50). Since all items with $\alpha \neq 0$ disappear, it follows that $a(x,\xi) - a_1(x,\xi) \in S^{-\infty}(\Omega \times \mathbb{R}^n)$ and hence Op also induces an injective mapping $S^\mu/S^{-\infty} \to L^\mu/L^{-\infty}$. $\square$

Corollary 1.1.35 To every $A \in L^\mu_{\mathrm{cl}}(\Omega)$ we can recover an $a(x,\xi) \in S^\mu_{\mathrm{cl}}(\Omega \times \mathbb{R}^n)$ with $\mathrm{Op}(a) - A \in L^{-\infty}(\Omega)$, and the homogeneous components $a_{(\mu-j)}(x,\xi)$, $j \in \mathbb{N}$, of $a(x,\xi)$ are uniquely determined by A. Then $\sigma^\mu_\psi(A) := a_{(\mu)}(x,\xi)$ gives us a linear mapping

$$\sigma^\mu_\psi : L^\mu_{\mathrm{cl}}(\Omega) \to S^{(\mu)}(\Omega \times (\mathbb{R}^n \setminus \{0\}))$$

that is surjective, and $\ker \sigma_\psi = L^{\mu-1}_{\mathrm{cl}}(\Omega)$.

Remark 1.1.36 Let $A \in L^\mu_{\mathrm{cl}}(\Omega)$, $B \in L^\nu_{\mathrm{cl}}(\Omega)$. Then $A^* \in L^\mu_{\mathrm{cl}}(\Omega)$, $AB \in L^{\mu+\nu}_{\mathrm{cl}}(\Omega)$ (if one factor is properly supported) and

$$\sigma^\mu_\psi(A^*) = \overline{\sigma^\mu_\psi(A)}, \qquad \sigma^{\mu+\nu}_\psi(AB) = \sigma^\mu_\psi(A)\sigma^\nu_\psi(B).$$

Further, as a consequence of (1.1.56), we have

$$\sigma^{\mu+\nu-1}_\psi(AB - BA) = \frac{1}{i}\{\sigma^\mu_\psi(A), \sigma^\nu_\psi(B)\}$$

with the Poisson bracket

$$\{f,g\} = \sum_{j=1}^n \left(\frac{\partial f}{\partial \xi_j}\frac{\partial g}{\partial x_j} - \frac{\partial f}{\partial x_j}\frac{\partial g}{\partial \xi_j} \right)$$

between the functions $f(x,\xi)$, $g(x,\xi)$.

Let us fix a relatively closed set $\Delta \subset \Omega \times \Omega$ that contains an open neighbourhood of diag $\Omega \times \Omega$ and which is proper. Denote by $L^\mu(\Omega)_\Delta$ the subspace of all $A \in L^\mu(\Omega)$ with supp $K_A \subseteq \Delta$. Write simply $\sigma^\mu(A) = e_{-\xi}(x)(Ae_\xi)(x)$ for any properly supported A and let $S^\mu(\Omega \times \mathbb{R}^n)_\Delta = \{\sigma^\mu(A)(x,\xi) : A \in L^\mu(\Omega)_\Delta\}$. Then

$$\sigma^\mu : L^\mu(\Omega)_\Delta \to S^\mu(\Omega \times \mathbb{R}^n)_\Delta \qquad (1.1.64)$$

is an isomorphism. In fact, the way to obtain $\sigma^\mu(A)$ by (1.1.59) is unique, and conversely we have $A = \mathrm{Op}(\sigma^\mu(A))$. The space $S^\mu(\Omega \times \mathbb{R}^n)_\Delta$ is obviously closed in $S^\mu(\Omega \times \mathbb{R}^n)$, i.e. it is Fréchet in a canonical way. We now endow the space $L^\mu(\Omega)_\Delta$ with the Fréchet topology by (1.1.64).

From Proposition 1.1.27 we get the decomposition

$$L^\mu(\Omega) = L^\mu(\Omega)_\Delta + L^{-\infty}(\Omega), \qquad (1.1.65)$$

which is a non–direct sum of vector spaces. The space $L^{-\infty}(\Omega)$ of all smoothing operators is isomorphic to $C^\infty(\Omega \times \Omega)$, cf. Proposition 1.1.26. Hence it has a canonical Fréchet topology. In the space (1.1.65) we then introduce the Fréchet topology from the non–direct sum, cf. Section 1.1.1. It can easily be proved that it is independent of the particular choice of Δ.

In an analogous manner we proceed for $L^\mu_{\mathrm{cl}}(\Omega)_\Delta := L^\mu_{\mathrm{cl}}(\Omega) \cap L^\mu(\Omega)_\Delta$, using the restriction of (1.1.64) to an isomorphism from $L^\mu_{\mathrm{cl}}(\Omega)_\Delta$ to a closed subspace of $S^\mu_{\mathrm{cl}}(\Omega)_\Delta$ that induces a Fréchet topology in $L^\mu_{\mathrm{cl}}(\Omega)_\Delta$. Then $L^\mu_{\mathrm{cl}}(\Omega) = L^\mu_{\mathrm{cl}}(\Omega)_\Delta + L^{-\infty}(\Omega)$ is Fréchet in the topology of the non–direct sum, and the topology is independent of the particular Δ, again.

Theorem 1.1.37 *Let $\varphi(x, x', \xi)$ be a phase function on $\Omega \times \Omega \times (\mathbb{R}^n \setminus \{0\})$ that is linear in ξ and satisfies $\pi C_\varphi = \mathrm{diag}\,\Omega \times \Omega$, cf. (1.1.40). Then*

$$C_0^\infty(\Omega) \ni u \to \iint e^{i\varphi(x,x',\xi)} a(x, x', \xi) u(x')\, dx' d\xi$$

for $a(x, x', \xi) \in S^\mu(\Omega \times \Omega \times \mathbb{R}^n)$ represents an operator in $L^\mu(\Omega)$.

Proof. We will show first the following result. There exists a neighbourhood V of diag $\Omega \times \Omega$ in $\Omega \times \Omega$ and a C^∞ map $g : V \to GL(n, \mathbb{R})$ such that

$$\varphi(x, x', g(x, x')\xi) = (x - x')\xi \quad \text{for all} \quad (x, x') \in V. \qquad (1.1.66)$$

Here

$$g(x, x) = {}^t(d_{x,\xi}\varphi(x, x', \xi)|_{x'=x})^{-1}, \qquad (1.1.67)$$

i.e.,

$$\det g(x,x)\det d_{x,\xi}\varphi(x,x',\xi)|_{x'=x} = 1 \qquad (1.1.68)$$

with the matrix $d_{x,\xi}\varphi$ of derivatives $\partial_{x_j}\partial_{\xi_k}\varphi$. Since φ is linear in ξ, we have $\varphi(x,x',\xi) = \sum_{j=1}^{n}\varphi_j(x,x')\xi_j$ with $\varphi_j(x,x) = 0$, $j = 1,\ldots,n$. Further $\varphi_j(x,x') = 0$ for all $j = 1,\ldots,n$ implies $x = x'$. Moreover,

$$\det\left(\frac{\partial\varphi_j(x,x')}{\partial x_k}\right)|_{x'=x} \neq 0. \qquad (1.1.69)$$

In order to verify (1.1.69) we observe that the differentiation of the relation $\varphi(x,x,\xi) = 0$ with respect to x yields $d_x\varphi|_{x'=x} = -d_{x'}\varphi|_{x'=x}$. By assumption on φ we have $d_{x,x',\xi}\varphi \neq 0$ for $\xi \neq 0$. Then $d_\xi\varphi(x,x,\xi) = 0$ implies $d_x\varphi(x,x,\xi) \neq 0$. Thus for every $\xi \neq 0$ there is a k such that $\sum_{j=1}^{n}\frac{\partial\varphi_j}{\partial x_k}\xi_j|_{x'=x} \neq 0$ which implies (1.1.69). From Hadamard's Lemma we get for $\Phi(x,x') := \{\varphi_1(x,x'),\ldots,\varphi_n(x,x')\}$ on some neighbourhood $\tilde{V}$ of diag $\Omega \times \Omega$

$$\Phi(x,x') = G(x,x')(x-x')$$

with a C^∞ matrix function $G(x,x') = (G_{jk}(x,x'))_{j,k=1,\ldots,n}$ on $\tilde{V}$. Here

$$G_{jk}(x,x) = \frac{\partial\varphi_j(x,x')}{\partial x_k}|_{x'=x}. \qquad (1.1.70)$$

The relations (1.1.69) and (1.1.70) show that there is a neighbourhood V of diag $\Omega \times \Omega$ such that $G(x,x')$ is non–singular for all $(x,x') \in V$. If we set $g(x,x') = {}^tG^{-1}(x,x')$ then (1.1.66) follows from

$$\begin{aligned}
\varphi(x,x',g(x,x')\xi) &= \langle \Phi(x,x'), {}^tG^{-1}(x,x')\xi\rangle \\
&= \langle G(x,x')(x-x'), {}^tG^{-1}(x,x')\xi\rangle \\
&= (x-x')\xi.
\end{aligned}$$

The formula (1.1.68) is a consequence of (1.1.70).

Now let us look at the operator

$$Au(x) = \iint e^{i\varphi(x,x',\xi)}a(x,x',\xi)u(x')\,dx'd\xi.$$

The distributional kernel $K_A(x,x')$ of A satisfies

$$\text{sing supp}\, K_A \subseteq \text{diag}\,\Omega \times \Omega,$$

cf. analogously (1.1.47) and (1.1.41). Thus, if $\omega(x,x')$ is a function in $C^\infty(\Omega \times \Omega)$ with proper support in V and identically equal to 1 close to diag $\Omega \times \Omega$, the operator

$$Cu(x) = \iint e^{i\varphi(x,x',\xi)}(1 - \omega(x,x'))a(x,x',\xi)u(x')\,dx'd\xi$$

belongs to $L^{-\infty}(\Omega)$, cf. Proposition 1.1.26. Hence it remains to characterise

$$A_0 u(x) = \iint e^{i\varphi(x,x',\xi)}\omega(x,x')a(x,x',\xi)u(x')\,dx'd\xi.$$

By inserting (1.1.66) and substituting $\eta = g(x,x')\xi$ we obtain

$$A_0 u(x) = \iint e^{i(x-x')\eta}\omega(x,x')a(x,x',g(x,x')\eta)|\det g(x,x')|u(x')\,dx'd\eta.$$

A_0 belongs to $L^\mu(\Omega)$ because of $a(x,x',g(x,x')\eta) \in S^\mu(\Omega\times\Omega\times\mathbb{R}^n_\eta)$. $\quad\square$

Our next objective is the invariance of pseudo–differential operators under coordinate diffeomorphisms. Let $U \subseteq \mathbb{R}^n_x$, $V \subseteq \mathbb{R}^n_y$ be open sets and $\kappa : U \to V$ be a diffeomorphism. Then we have the function pull–backs $\kappa^* : C_0^\infty(V) \to C_0^\infty(U)$, $\kappa^* : C^\infty(V) \to C^\infty(U)$. Thus, every $A \in L^\mu(U)$, when regarded as an operator $A : C_0^\infty(U) \to C^\infty(U)$, induces

$$\kappa_* A := (\kappa^*)^{-1}A\kappa^* : C_0^\infty(V) \to C^\infty(V),$$

called the operator push–forward of A with respect to κ.

Let $d\kappa(x)$ be the Jacobi matrix of κ at $x \in U$ and set $\delta(x,z) = \kappa(z) - \kappa(x) - d\kappa(x)(z-x)$. Then the expression

$$\Phi_\alpha(x,\eta) = D_z^\alpha e^{i\delta(x,z)\eta}|_{z=x}, \qquad \alpha \in \mathbb{N}^n \qquad (1.1.71)$$

is a polynomial in η of degree $\leq |\alpha|/2$. Hence $a(x,\xi) \in S^\mu(\Omega \times \mathbb{R}^n_\xi)$ implies $(\partial_\xi^\alpha a)(x,{}^t d\kappa(x)\eta)\Phi_\alpha(x,\eta) \in S^{\mu-|\alpha|/2}(\Omega \times \mathbb{R}^n_\eta)$.

Theorem 1.1.38 *The operator push–forward under a diffeomorphism $\kappa : U \to V$ induces isomorphisms*

$$\kappa_* : L^\mu(U) \to L^\mu(V), \quad A \to (\kappa^*)^{-1}A\kappa^* = \kappa_* A$$

for all $\mu \in \mathbb{R}$. If $A \in L^\mu(U)$ is written as $A = \mathrm{Op}(a) \bmod L^{-\infty}(U)$ with a symbol $a(x,\xi) \in S^\mu(U \times \mathbb{R}^n)$, then $B = \kappa_* A$ has the form $B = \mathrm{Op}(b) \bmod L^{-\infty}(V)$ for a symbol $b(y,\eta) \in S^\mu(V \times \mathbb{R}^n)$ that has the asymptotic expansion

$$b(y,\eta)|_{y=\kappa(x)} \sim \sum_\alpha \frac{1}{\alpha!}(\partial_\xi^\alpha a)(x,{}^t d\kappa(x)\eta)\Phi_\alpha(x,\eta). \qquad (1.1.72)$$

Proof. First we have $\kappa_* : L^{-\infty}(U) \to L^{-\infty}(V)$. Thus we may assume

$$Au(x) = \iint e^{i(x-x')\xi}a(x,\xi)u(x')\,dx'd\xi.$$

Set $u = \kappa^* v$, $x = \kappa^{-1}(y)$, $x' = \kappa^{-1}(y')$. Then

$$Bv(y) = \iint e^{i\varphi(y,y',\xi)} a(\kappa^{-1}(y),\xi)|\det d\kappa^{-1}(y')|v(y')\,dy'd\xi$$

with the phase function $\varphi(y,y',\xi) = (\kappa^{-1}(y) - \kappa^{-1}(y'))\xi$ that satisfies the assumptions of Theorem 1.1.37. Thus $B \in L^\mu(V)$. If $g(y,y')$ is the matrix function from the proof of Theorem 1.1.37 we can write $\varphi(y,y',\xi) = (y - y')g^{-1}(y,y')\xi$. It follows that

$$Bv(y) = \iint e^{i(y-y')\eta} a(\kappa^{-1}(y), g(y,y')\eta)h(y,y')v(y')\,dy'd\eta$$

for $\eta = g^{-1}(y,y')\xi$, $h(y,y') = |\det g(y,y')||\det d\kappa^{-1}(y')|$. Using Theorem 1.1.30 we obtain the asymptotic expansion

$$b(y,\eta) \sim \sum_\alpha \frac{1}{\alpha!} D_\eta^\alpha \partial_{y'}^\alpha a(\kappa^{-1}(y), g(y,y')\eta)h(y,y')|_{y'=y} \qquad (1.1.73)$$

for a complete symbol of B. The summands with the multi-index α consist of terms of the form

$$c(y,y')\eta^\gamma (\partial_\xi^\beta a)(\kappa^{-1}(y), g(y,y')\eta)$$

with $c(y,y')$ only dependent on κ, not on a, and the multi-indices γ, β satisfy

$$|\beta| \leq 2|\alpha|, \qquad |\gamma| + |\alpha| \leq |\beta|.$$

The first estimate is obvious, while the second one follows from the fact that $\partial_{y'}$ in the αth item of (1.1.73) does not change $|\beta| - |\gamma|$ whereas the derivatives ∂_η raise $|\beta| - |\gamma|$ by 1. The formula (1.1.68) implies in our case $g(y,y) = ({}^t d\kappa^{-1}(y))^{-1}$. We now rearrange (1.1.73) by taking together all items with the same β. This yields a sum of the form

$$b(y,\eta) \sim \sum_\beta \frac{1}{\beta!}(\partial_\xi^\beta a)(\kappa^{-1}(y), ({}^t d\kappa^{-1}(y))^{-1}\eta)\Psi_\beta(y,\eta),$$

where $\Psi_\beta(y,\eta)$ are polynomials in η of degree $\leq |\beta|/2$, not dependent on the symbol of A, where $\Psi_0 = 1$. Inserting $y = \kappa(x)$ we obtain

$$b(y,\eta)|_{y=\kappa(x)} = \sum_\beta \frac{1}{\beta!}(\partial_\xi^\beta a)(x, {}^t d\kappa(x)\eta)\Phi_\beta(x,\eta),$$

where $\Phi_\beta(x,\eta)$ are polynomials in η of degree $\leq |\beta|/2$, with coefficients in $C^\infty(U)$. Since these polynomials are independent of the given special operator, they can be calculated by inserting differential operators A. In that case we have

$$b(y,\eta)|_{y=\kappa(x)} = e^{-iy\eta}Be^{iy\eta}|_{y=\kappa(x)} = e^{-i\kappa(x)\eta}a(z,D_z)e^{i\kappa(z)\eta}|_{z=x}. \quad (1.1.74)$$

We now write $\kappa(z) = \kappa(x) + d\kappa(x)(z-x) + \delta(x,z)$ and insert $\kappa(z)\eta = \kappa(x)\eta + z\,{}^td\kappa(x)\eta + \delta(x,z)\eta - x\,{}^td\kappa(x)\eta$ in (1.1.73). Then

$$b(y,\eta)|_{y=\kappa(x)} = e^{-ix^t d\kappa(x)\eta}a(z,D_z)\{e^{iz^t d\kappa(x)\eta}e^{i\delta(x,z)\eta}\}|_{z=x}.$$

Next we apply the Leibniz rule for a differential operator $a(x,D_x) = \sum_{|\alpha|\leq\mu} a_\alpha(x)D_x^\alpha$ with $a(x,\xi) = \sum_{|\alpha|\leq\mu} a_\alpha(x)\xi^\alpha$, namely

$$a(x,D_x)(uv)(x) = \sum_\alpha \{(\partial_\xi^\alpha a)(x,D_x)u(x)\}D_x^\alpha v(x).$$

Then, using $a(x,D_x)e^{ix\xi} = e^{ix\xi}a(x,\xi)$, we just obtain the required formula (1.1.71) for $\Phi_\alpha(x,\eta)$. $\qquad\Box$

Remark 1.1.39 *Let* $a(x,\xi) \in S^\mu(U\times\mathbb{R}^n)$, *and denote by* $(\kappa_*a)(y,\eta)$ *any choice in* $S^\mu(V\times\mathbb{R}^n)$ *with*

$$(\kappa_*a)(y,\eta)|_{y=\kappa(x)} \sim \sum_\alpha \frac{1}{\alpha!}(\partial_\xi^\alpha a)(x,{}^td\kappa(x)\eta)\Phi_\alpha(x,\eta).$$

Then, for $a(x,\xi)\in S^\mu(U\times\mathbb{R}^n)$, $b(x,\xi)\in S^\nu(U\times\mathbb{R}^n)$ *it follows that*

$$\kappa_*(a\#b)(y,\eta) - (\kappa_*a)(y,\eta)\#(\kappa_*b)(y,\eta) \in S^{-\infty}(V\times\mathbb{R}^n).$$

This is an immediate consequence of

$$\kappa_*(\mathrm{Op}(a)\,\mathrm{Op}(b)) = \kappa_*\,\mathrm{Op}(a)\kappa_*\,\mathrm{Op}(b)$$

and of Theorem 1.1.34.

Remark 1.1.40 $A \in L^\mu_{\mathrm{cl}}(U)$ *implies* $\kappa_*A \in L^\mu_{\mathrm{cl}}(V)$ *and*

$$\sigma_\psi^\mu(\kappa_*A)(y,\eta) = \sigma_\psi^\mu(A)(x,\xi) \qquad \text{for } (y,\eta) = (\kappa(x),{}^td\kappa^{-1}(x)\xi).$$

Lemma 1.1.41 *(Peetre's inequality). We have*

$$(1+|\xi-\eta|)^s \leq (1+|\eta|)^s(1+|\xi|)^{|s|} \qquad (1.1.75)$$

for all $\xi,\eta \in \mathbb{R}^n$ *and every* $s\in\mathbb{R}$.

Proof. The inequality (1.1.75) for $s \geq 0$ is an obvious consequence of

$$1 + |\xi + \eta| \leq 1 + |\xi| + |\eta| \leq (1 + |\eta|)(1 + |\xi|). \tag{1.1.76}$$

(1.1.76) implies for $\eta = -(\zeta + \xi)$

$$1 + |\zeta| \leq (1 + |\zeta + \xi|)(1 + |\xi|). \tag{1.1.77}$$

This yields for $s \leq 0$ the estimates $(1+|\zeta|)^{-s} \leq (1+|\zeta+\xi|)^{-s}(1+|\xi|)^{-s}$ and hence for $\eta = -\zeta$ the assertion. $\qquad\qquad\square$

Using (1.1.75) in the form

$$(1 + |\xi + \eta|)^s \leq (1 + |\eta|)^s (1 + |\xi|)^{|s|}$$

for all $\xi, \eta \in \mathbb{R}^n$ we get for $\xi - \eta$ instead of ξ the inequalities

$$\left(\frac{1 + |\xi|}{1 + |\eta|}\right)^s \leq (1 + |\xi - \eta|)^{|s|} \tag{1.1.78}$$

as well as

$$\left(\frac{1 + |\eta|}{1 + |\xi|}\right)^s \leq (1 + |\xi - \eta|)^{|s|} \tag{1.1.79}$$

for all $\xi, \eta \in \mathbb{R}^n$ and every $s \in \mathbb{R}$.

Let us now return to the consideration of

$$k(a)(x, x', \zeta) = (F_{\xi \to \zeta}^{-1} a)(x, x', \zeta) \tag{1.1.80}$$

for $a(x, x', \xi) \in S^\mu(\Omega \times \Omega \times \mathbb{R}^n)$, cf. (1.1.21). Recall that $k(a)(x, x', x - x')$ is the distributional kernel of $\text{Op}(a)$. We want to deepen the relation between a and $k(a)$ in terms of a so–called kernel cut–off construction. The dependence on $(x, x') \in \Omega \times \Omega$ will not cause specific effects; so we first consider symbols with constant coefficients.

The inverse Fourier transform $F_{\xi \to \zeta}^{-1} : \mathcal{S}'(\mathbb{R}_\xi^n) \to \mathcal{S}'(\mathbb{R}_\zeta^n)$ restricts to an injective operator

$$F_{\xi \to \zeta}^{-1} : S^\mu(\mathbb{R}_\xi^n) \to \mathcal{S}'(\mathbb{R}_\zeta^n) \tag{1.1.81}$$

and to an isomorphism

$$F_{\xi \to \zeta}^{-1} : S^{-\infty}(\mathbb{R}_\xi^n) \underset{\cong}{\to} \mathcal{S}(\mathbb{R}_\zeta^n), \tag{1.1.82}$$

using $S^{-\infty}(\mathbb{R}^n) = \mathcal{S}(\mathbb{R}^n)$.

Let $\varphi \in \mathcal{S}(\mathbb{R}_\zeta^n)$ and set

$$H(\varphi) = F_{\zeta \to \xi} \varphi(\zeta) F_{\xi \to \zeta}^{-1}.$$

Theorem 1.1.42 *Let the function* $\varphi(\zeta) \in \mathcal{S}(\mathbb{R}^n)$ *be fixed. Then the mapping* $a(\xi) \to (H(\varphi)a)(\xi)$ *induces a continuous operator*

$$H(\varphi) : S^\mu(\mathbb{R}^n) \to S^\mu(\mathbb{R}^n), \qquad (1.1.83)$$

for every $\mu \in \mathbb{R}$. *Furthermore,* $\varphi(\zeta) \to (H(\varphi)a)(\xi)$ *is a continuous operator*

$$\mathcal{S}(\mathbb{R}^n_\zeta) \to S^\mu(\mathbb{R}^n_\xi) \qquad (1.1.84)$$

for every $a(\xi) \in S^\mu(\mathbb{R}^n)$.

Proof. Set $b = H(\varphi)a$, i.e. $b(\xi) = \int e^{-i\xi\zeta}\varphi(\zeta)k(a)(\zeta)\,d\zeta$. The first task is to show $|D_\xi^\alpha b(\xi)| \le c(1 + |\xi|)^{\mu - |\alpha|}$ for every $\alpha \in \mathbb{N}^n$ with some $c = c(\alpha) > 0$, for all $\xi \in \mathbb{R}^n$. We have

$$D_\xi^\alpha b(\xi) = \int e^{-i\xi\zeta}\varphi(\zeta)(-\zeta)^\alpha k(a)(\zeta)\,d\zeta$$
$$= \int e^{-i\xi\zeta}\varphi(\zeta)k(D_\xi^\alpha a)(\zeta)\,d\zeta.$$

Using $F(uv) = (Fu)*(Fv)$, it follows for $f(\xi) = (F_{\zeta\to\xi}\varphi)(\xi)$, $a^{(\alpha)}(\xi) := D_\xi^\alpha a(\xi)$ that

$$D_\xi^\alpha b(\xi) = \int f(\xi - \eta)a^{(\alpha)}(\eta)\,d\eta. \qquad (1.1.85)$$

Since $f(\xi)$ is a Schwartz function, we obtain

$$|D_\xi^\alpha b(\xi)| \le \int |f(\xi - \eta)||a^{(\alpha)}(\eta)|\,d\eta$$
$$\le c_N \int (1 + |\xi - \eta|)^{-N}(1 + |\eta|)^{\mu - |\alpha|}\,d\eta$$

for every $N \in \mathbb{N}$, with a suitable constant $c_N = c_N(f, \alpha)$. Applying Peetre's inequality we get

$$(1 + |\xi - \eta|)^{-\mu + |\alpha|} \le (1 + |\eta|)^{-\mu + |\alpha|}(1 + |\xi|)^{\mu - |\alpha|} \qquad \text{for } \mu - |\alpha| \ge 0,$$
$$(1 + |\xi - \eta|)^{\mu - |\alpha|} \le (1 + |\eta|)^{-\mu + |\alpha|}(1 + |\xi|)^{\mu - |\alpha|} \qquad \text{for } \mu - |\alpha| \le 0.$$

This gives us

$$|D_\xi^\alpha b(\xi)| \le c_N(1 + |\xi|)^{\mu - |\alpha|} \int (1 + |\xi - \eta|)^{-N + \mu - |\alpha|}\,d\eta$$
$$\text{for } \mu - |\alpha| \ge 0,$$

$$|D_\xi^\alpha b(\xi)| \le c_N (1 + |\xi|)^{\mu - |\alpha|} \int (1 + |\xi - \eta|)^{-N - \mu + |\alpha|} \, d\eta$$

$$\text{for } \mu - |\alpha| \le 0.$$

Choosing N so large that $\max\{-N - \mu + |\alpha|, -N + \mu - |\alpha|\} < -n$ we get the asserted estimates, namely $|D_\xi^\alpha b(\xi)| \le d_N(\varphi, \alpha)(1 + |\xi|)^{\mu - |\alpha|}$, with constants $d_N(\varphi, \alpha)$ that can be estimated by a finite number of semi–norms with respect to φ in the Schwartz space. This gives us the continuity of the operator $H(\varphi)$ in the sense (1.1.83). At the same time we get the continuity of (1.1.84). □

Definition 1.1.43 *Define $S^\mu(\mathbb{C}^n)_{\mathrm{hol}}$, $\mu \in \mathbb{R}$, as the set of all $a(z) \in \mathcal{A}(\mathbb{C}^n)$ with*

$$a(\xi + i\eta) \in S^\mu(\mathbb{R}_\xi^n) \tag{1.1.86}$$

for all $\eta \in \mathbb{R}^n$, uniformly in $|\eta| \le c$ for every $c > 0$. More generally if $\mathbb{R}^n = \mathbb{R}^p \times \mathbb{R}^q \ni \xi = (\xi', \xi'')$, $1 \le q \le n$, then $S^\mu(\mathbb{R}^p \times \mathbb{C}^q)_{\mathrm{hol}}$ is the subspace of all $a(\xi', z) \in C^\infty(\mathbb{R}_{\xi'}^p, \mathcal{A}(\mathbb{C}_z^q))$ with

$$a_{\eta''}(\xi) := a(\xi', \xi'' + i\eta'') \in S^\mu(\mathbb{R}_\xi^n) \tag{1.1.87}$$

for all $\eta'' \in \mathbb{R}^q$, uniformly in $|\eta''| < c$ for every $c > 0$. In an analogous manner we define the spaces $S_{\mathrm{cl}}^\mu(\mathbb{R}^p \times \mathbb{C}^n)_{\mathrm{hol}}$ and $S_{\mathrm{cl}}^\mu(\mathbb{C}^q)_{\mathrm{hol}}$, respectively.

Remark 1.1.44 *Let $\{r_k\}_{k \in \mathbb{N}}$ be a semi–norm system for the Fréchet topology of $S^\mu(\mathbb{R}_\xi^n)$. Then*

$$a(\xi', \xi'' + i\eta'') \to \sup_{|\eta''| \le l} r_k(a_{\eta''}), \qquad k, l \in \mathbb{N}$$

is a semi–norm system on $S^\mu(\mathbb{R}^p \times \mathbb{C}^q)_{\mathrm{hol}}$ in which this space is Fréchet. Thus we get

$$S^\mu(\Omega \times \mathbb{R}^p \times \mathbb{C}^q)_{\mathrm{hol}} := C^\infty(\Omega, S^\mu(\mathbb{R}^p \times \mathbb{C}^q)_{\mathrm{hol}})$$

for any open set $\Omega \subseteq \mathbb{R}^m$ and analogously $S_{\mathrm{cl}}^\mu(\Omega \times \mathbb{R}^p \times \mathbb{C}^q)_{\mathrm{hol}}$.

Theorem 1.1.45 *Let $\psi(\zeta) \in C_0^\infty(\mathbb{R}^n)$ with $\psi(\zeta) = 1$ in a neighbourhood of $\zeta = 0$. Then $H(\psi)$ (acting with respect to the covariables) induces a continuous operator*

$$H(\psi) : S^\mu(\Omega \times \mathbb{R}^n) \to S^\mu(\Omega \times \mathbb{C}^n)_{\mathrm{hol}},$$

and we have $a(x, \xi) - H(\psi)a(x, \xi) \in S^{-\infty}(\Omega \times \mathbb{R}^n)$ for every $a(x, \xi) \in S^\mu(\Omega \times \mathbb{R}^n)$.

Proof. Let us restrict ourselves to x–independent symbols; the general case is completely analogous. First $H(\psi)a(\xi)$ extends to a function in $\mathcal{A}(\mathbb{C}^n)$, since it is the inverse Fourier transform of a distribution with compact support in ζ. From (1.1.24) we get $(1 - \psi(\zeta))k(a)(\zeta) \in \mathcal{S}(\mathbb{R}^n)$ and hence

$$a(\xi) - H(\psi)a(\xi) = F_{\zeta \to \xi}((1 - \psi(\zeta))k(a)(\zeta)) \in \mathcal{S}(\mathbb{R}^n) = S^{-\infty}(\mathbb{R}^n).$$

This yields $H(\psi)a \in S^\mu(\mathbb{R}^n)$, cf. Theorem 1.1.42. Now

$$H(\psi)a(\xi + i\eta) = \int e^{-i\zeta(\xi + i\eta)}\psi(\zeta)k(a)(\zeta)\, d\zeta = H(\psi_\eta)a(\xi)$$

for $\psi_\eta(\zeta) := e^{\zeta \eta}\psi(\zeta) \in C_0^\infty(\mathbb{R}^n)$. Then the continuity of (1.1.84) yields the assertion. $\square$

Theorem 1.1.42 has the following useful generalisation.

Theorem 1.1.46 *Let $\varphi''(\zeta'') \in \mathcal{S}(\mathbb{R}^q)$ for $1 \leq q \leq n$ be fixed. Then*

$$H(\varphi'')a(\xi) = F_{\zeta'' \to \xi''}\varphi''(\zeta'')F^{-1}_{\xi'' \to \zeta''}a(\xi', \xi'')$$

induces a continuous operator

$$H(\varphi'') : S^\mu(\mathbb{R}^n) \to S^\mu(\mathbb{R}^n)$$

for every $\mu \in \mathbb{R}$. Furthermore, $\varphi''(\zeta'') \to (H(\varphi'')a)(\xi)$ is a continuous operator

$$\mathcal{S}(\mathbb{R}^q_{\zeta''}) \to S^\mu(\mathbb{R}^n_\xi)$$

for every $a(\xi) \in S^\mu(\mathbb{R}^n)$.

Proof. Setting $b = H(\varphi'')a$ and $f''(\xi'') = (F_{\zeta'' \to \xi''}\varphi)(\xi'')$ we get similarly to the proof of Theorem 1.1.42

$$|D_\xi^\alpha b(\xi)| = \left| \int f''(\xi'' - \eta'')(D_\xi^\alpha a)(\xi', \eta'')\, d\eta'' \right|$$

$$\leq c_N \int (1 + |\xi'' - \eta''|)^{-N}(1 + |\xi', \eta''|)^{\mu - |\alpha|}\, d\eta''.$$

Using

$$(1 + |\xi'' - \eta''|)^{-\mu + |\alpha|} = (1 + |(\xi', \xi'') - (\xi', \eta'')|)^{-\mu + |\alpha|}$$

$$\leq (1 + |\xi', \eta''|)^{-\mu + |\alpha|}(1 + |\xi', \xi''|)^{\mu - |\alpha|}$$

$$\text{for } \mu - |\alpha| \geq 0$$

and analogously

$$(1 + |\xi'' - \eta''|)^{\mu - |\alpha|} \leq (1 + |\xi', \eta''|)^{-\mu + |\alpha|}(1 + |\xi', \xi''|)^{\mu - |\alpha|}$$
$$\text{for } \mu - |\alpha| \leq 0$$

it follows that

$$|D_\xi^\alpha b(\xi)| \leq c_N (1 + |\xi|)^{\mu - |\alpha|} \int (1 + |\xi'' - \eta''|)^{-N + \mu - |\alpha|} \, d\eta''$$
$$\text{for } \mu - |\alpha| \geq 0,$$

$$|D_\xi^\alpha b(\xi)| \leq c_N (1 + |\xi|)^{\mu - |\alpha|} \int (1 + |\xi'' - \eta''|)^{-N - \mu + |\alpha|} \, d\eta''$$
$$\text{for } \mu - |\alpha| \leq 0.$$

Now we can proceed as in the proof of Theorem 1.1.42, which shows the assertion.   □

Theorem 1.1.47 *Let* $\zeta = (\zeta', \zeta'') \in \mathbb{R}^{p+q}$, $p + q = n$, $1 \leq q \leq n$, *and choose a* $\psi''(\zeta'') \in C_0^\infty(\mathbb{R}^q)$ *that equals* 1 *in a neighbourhood of* $\zeta'' = 0$. *Then the operator*

$$H(\psi'') : S^\mu(\Omega \times \mathbb{R}^n) \to S^\mu(\Omega \times \mathbb{R}^p \times \mathbb{C}^q)_{\text{hol}}$$

is continuous, and $a(x, \xi) \in S^\mu(\Omega \times \mathbb{R}^n)$.

Note that $H(\psi'')$ in Theorem 1.1.47 has the same structure as the corresponding operator from Theorem 1.1.45, however, here with respect to ξ'', while the variables ξ' remain untouched because $F_{\zeta' \to \xi'}^{-1} \circ F_{\xi' \to \zeta'} = \text{id}$. The proof of Theorem 1.1.47 is analogous to that of Theorem 1.1.45.

Remark 1.1.48 *It can easily be proved that* $(\psi'', a) \to H(\psi'')a$ *induces a separately continuous operator*

$$C_0^\infty(\mathbb{R}^q) \times S^\mu(\Omega \times \mathbb{R}^n) \to S^\mu(\Omega \times \mathbb{R}^p \times \mathbb{C}^q)_{\text{hol}},$$

and analogously for classical symbols.

Remark 1.1.49 *Let* $\varphi \in \mathcal{S}(\mathbb{R}^n)$ *and consider a symbol of the form* $b(x, x', \xi) := \varphi(x - x')a(x, x', \xi)$ *with* $a(x, x', \xi) \in S^\mu(\Omega \times \Omega \times \mathbb{R}^n)$. *Then, if we define*

$$a_\varphi(x, x', \xi) = F_{\zeta \to \xi}\varphi(\zeta)k(a)(x, x', \zeta)$$

which belongs to $S^\mu(\Omega \times \Omega \times \mathbb{R}^n)$, *it follows that*

$$K(a_\varphi)(x, x', \zeta) = \varphi(\zeta)k(a)(x, x', \zeta)$$

and

$$\text{Op}(b) = \text{Op}(a_\varphi).$$

Remark 1.1.50 *Theorem 1.1.42 holds in analogous form also for classical symbols. Also the spaces $S_{\mathrm{cl}}^{\mu}(\Omega \times \mathbb{R}^p \times \mathbb{C}^q)_{\mathrm{hol}}$ can be defined in a straightforward way. Then, in particular, we get corresponding analogues of the Theorems 1.1.45, 1.1.47.*

Remark 1.1.51 *Using the notation of Theorem 1.1.47 we have the following approximation property of the kernel cut–off construction. Set $\psi_R''(\xi'') = \psi''(\xi''/R)$ for $R > 0$. Then for every $a(x,\xi)$ and $a_R(x,\xi) := H(\psi_R'')a(x,\xi)$ (with the right hand side being interpreted as the restriction to $\mathbb{R}_{\xi}^{\eta} = \mathbb{R}^q \times \mathbb{R}^q$), $a_R(x,\xi) \in S^{\mu}(\Omega \times \mathbb{R}^n)$, we have*

$$a_R(x,\xi) \to a(x,\xi) \qquad \text{for } R \to \infty$$

in the space $S^{\mu}(\Omega \times \mathbb{R}^n)$. An analogous result holds for classical symbols.

1.1.4 Continuity in Sobolev spaces

As noted above in Remark 1.1.28 every $A \in L^{\mu}(\Omega)$ for open $\Omega \subseteq \mathbb{R}^n$ can be regarded as a continuous operator

$$A : C_0^{\infty}(\Omega) \to C^{\infty}(\Omega).$$

The present section studies the question of extending A to continuous operators between Sobolev spaces. We shall not recall here the standard material on Sobolev spaces in a self–contained form but only recall some basic facts. The Sobolev space $H^s(\mathbb{R}^n)$ of smoothness $s \in \mathbb{R}$ is defined as the closure of $C_0^{\infty}(\mathbb{R}^n)$ with respect to the norm

$$\|u\|_s := \left\{ \int_{\mathbb{R}^n} (1 + |\xi|^2)^s |(Fu)(\xi)|^2 d\xi \right\}^{\frac{1}{2}},$$

with the Fourier transform $F = F_{x \to \xi}$ in $\mathbb{R}^n$. Alternatively we could say that $H^s(\mathbb{R}^n)$ is the subspace of all $u \in \mathcal{S}'(\mathbb{R}^n)$ for which $\|u\|_s$ is finite, cf. (1.1.9). In particular, we have

$$H^0(\mathbb{R}^n) = L^2(\mathbb{R}^n).$$

The operator norm in $\mathcal{L}(H^s(\mathbb{R}^n), H^r(\mathbb{R}^n))$ will be denoted by $\| \cdot \|_{s,r}$.

Proposition 1.1.52 *Let $a(\xi) \in S^{\mu}(\mathbb{R}^n)$, $\mu \in \mathbb{R}$, and $A = \mathrm{Op}(a)$. Then A extends to a continuous operator $A : H^s(\mathbb{R}^n) \to H^{s-\mu}(\mathbb{R}^n)$ and $a \to A$ induces a continuous operator $S^{\mu}(\mathbb{R}^n) \to \mathcal{L}(H^s(\mathbb{R}^n), H^{s-\mu}(\mathbb{R}^n))$ for every $s \in \mathbb{R}$.*

Proof. It suffices to show $\|Au\|_{s-\mu} \leq c\|u\|_s$ for all $u \in C_0^\infty(\mathbb{R}^n)$ with $c = \sup(1+|\xi|^2)^{-\frac{\mu}{2}}|a(\xi)|$. We have $A = F^{-1}aF$ and hence, for $r^\nu(\xi) := (1+|\xi|^2)^{\frac{\nu}{2}}$, $\nu \in \mathbb{R}$,

$$\|Au\|_{s-\mu}^2 = \int r^{2(s-\mu)}(\xi)|F(F^{-1}a(\xi)Fu)(\xi)|^2 \, d\xi$$

$$= \int r^{2(s-\mu)}(\xi)|a(\xi)|^2|Fu(\xi)|^2 \, d\xi$$

$$\leq c^2\|u\|_s^2.$$

$\square$

It is obvious that for every $a(\xi) \in C^\infty(\mathbb{R}^n)$ with $c_1 r^\mu(\xi) \leq |a(\xi)| \leq c_2 r^\mu(\xi)$ for all $\xi \in \mathbb{R}^n$ with constants $c_1, c_2 > 0$ the associated operator $\mathrm{Op}(a) = F_{\xi \to x}^{-1}a(\xi)F_{x' \to \xi}$ induces isomorphisms

$$\mathrm{Op}(a) : H^s(\mathbb{R}^n) \to H^{s-\mu}(\mathbb{R}^n)$$

for all $s \in \mathbb{R}$, where $\mathrm{Op}(a)^{-1} = \mathrm{Op}(a^{-1})$. In particular,

$$\|u\|_\mu \sim \|\mathrm{Op}(a)u\|_{L^2(\mathbb{R}^n)}, \tag{1.1.88}$$

where $\sim$ indicates equivalence of norms.

Lemma 1.1.53 *For every* $\mu \in \mathbb{R}$ *we have*

$$\|u\|_\mu \sim \sup_{\substack{v \in C_0^\infty(\mathbb{R}^n) \\ v \neq 0}} \frac{|\langle u, v \rangle|}{\|v\|_{-\mu}}. \tag{1.1.89}$$

Proof. First we have

$$\|g\|_{L^2(\mathbb{R}^n)} \sim \sup_{\substack{f \in V \\ f \neq 0}} \frac{|\langle g, f \rangle|}{\|f\|_{L^2(\mathbb{R}^n)}} \tag{1.1.90}$$

for every dense vector subspace V of $L^2(\mathbb{R}^n)$. In particular, we can take $V = \mathrm{Op}(r^{-\mu})C_0^\infty(\mathbb{R}^n)$. To prove (1.1.89) it suffices to show the equivalence of norms for $u \in C_0^\infty(\mathbb{R}^n)$. Set $g = \mathrm{Op}(r^\mu)u$, $f = \mathrm{Op}(r^{-\mu})v$ for $u, v \in C_0^\infty(\mathbb{R}^n)$. Then

$$\langle g, f \rangle = \langle \mathrm{Op}(r^\mu)u, \mathrm{Op}(r^{-\mu})v \rangle = \langle u, v \rangle.$$

This yields together with (1.1.90)

$$\|\mathrm{Op}(r^\mu)u\|_{L^2(\mathbb{R}^n)} \sim \sup_{\substack{v \in C_0^\infty(\mathbb{R}^n) \\ v \neq 0}} \frac{|\langle u, v \rangle|}{\|\mathrm{Op}(r^{-\mu})v\|_{L^2(\mathbb{R}^n)}}$$

which is our assertion because of (1.1.88). $\square$

The next result is the continuity of $A = \mathrm{Op}(a)$ for a symbol $a(x,\xi) \in C^\infty(\mathbb{R}^n \times \mathbb{R}^n)$ satisfying

$$\pi(a) := \sup_{\substack{(x,\xi)\in\mathbb{R}^{2n} \\ \alpha,\beta\leq(1,\dots,1)}} |D_x^\alpha D_\xi^\beta a(x,\xi)| < \infty. \tag{1.1.91}$$

Here, as usual,

$$\mathrm{Op}(a)u(x) = \int e^{ix\xi} a(x,\xi)\hat{u}(\xi)\,d\xi, \tag{1.1.92}$$

though $a(x,\xi)$ does not belong in general to the symbol class $S^0(\mathbb{R}^n \times \mathbb{R}^n)$; $\hat{u}(\xi) = (F_{x\to\xi}u)(\xi)$. Let us consider

$$(Au,v) = \iint \overline{v(x)} e^{ix\xi} a(x,\xi)\hat{u}(\xi)\,d\xi dx$$

for $u,v \in C_0^\infty(\mathbb{R}^n)$. Choose a function $\omega(x,\xi) \in C_0^\infty(\mathbb{R}^n \times \mathbb{R}^n)$ with $\omega(x,\xi) = 1$ in an open neighbourhood of $(x,\xi) = 0$, and set

$$a_\varepsilon(x,\xi) = \omega(\varepsilon x, \varepsilon\xi)a(x,\xi) \qquad \text{for } \varepsilon > 0. \tag{1.1.93}$$

Then $a_\varepsilon(x,\xi) \in C_0^\infty(\mathbb{R}^n \times \mathbb{R}^n)$, and $a_\varepsilon(x,\xi)$ is uniformly bounded in $(x,\xi) \in \mathbb{R}^{2n}$, $\varepsilon > 0$. Moreover, $a_\varepsilon(x,\xi) \to a(x,\xi)$ for $\varepsilon \to 0$ for every (x,ξ). From the Lebesgue theorem on dominated convergence it follows that

$$(\mathrm{Op}(a)u,v) = \lim_{\varepsilon\to 0}(\mathrm{Op}(a_\varepsilon)u,v). \tag{1.1.94}$$

Observe also that

$$\sup_{(x,\xi)\in\mathbb{R}^{2n}} |D_x^\alpha D_\xi^\beta a_\varepsilon(x,\xi)| \leq c_{\alpha\beta} \sup_{\substack{(x,\xi)\in\mathbb{R}^{2n} \\ \lambda\leq\alpha,\kappa\leq\beta}} |D_x^\lambda D_\xi^\kappa a(x,\xi)| \tag{1.1.95}$$

holds with constants $c_{\alpha\beta}$ independent of $\varepsilon \in (0,1)$. In particular,

$$\pi(a_\varepsilon) \leq c\pi(a) \qquad \text{for } \varepsilon \in (0,1) \tag{1.1.96}$$

with an ε–independent constant $c > 0$.

Theorem 1.1.54 *Every $A = \mathrm{Op}(a)$ with $a(x,\xi) \in C^\infty(\mathbb{R}^n \times \mathbb{R}^n)$ satisfying the estimate* (1.1.91) *induces a continuous operator*

$$A : L^2(\mathbb{R}^n) \to L^2(\mathbb{R}^n)$$

with

$$\|A\|_{\mathcal{L}(L^2(\mathbb{R}^n))} \leq c\pi(a) \tag{1.1.97}$$

for some constant $c > 0$.

Proof. We first assume $a(x,\xi) \in C_0^\infty(\mathbb{R}^{2n})$. Let $u, v \in C_0^\infty(\mathbb{R}^n)$ and consider

$$(Au, v) = \int \overline{v(x)} Au(x)\, dx$$

$$= \iiint e^{i(x-y)\xi} \overline{v(x)} a(x,\xi) u(y)\, dy\, d\xi\, dx$$

$$= \iiiint e^{i(x-y)\xi - ix\eta} \overline{\hat{v}(\eta)} a(x,\xi) u(y)\, dy\, d\xi\, dx\, d\eta. \qquad (1.1.98)$$

For $\delta \in \mathbb{N}^n$ we form $(i+\xi)^\delta = (i+\xi_1)^{\delta_1} \cdot \ldots \cdot (i+\xi_n)^{\delta_n}$, $i = \sqrt{-1}$, and analogously $(i+D_\xi)^\delta$. Then

$$(i + D_\xi)^\delta = (i + x - y)^\delta e^{i(x-y)\xi}.$$

It follows that

$$(Au, v)$$
$$= \iiiint \frac{(i+D_\xi)^\delta}{(i+x-y)^\delta} \frac{(i+D_x)^\delta}{(i+\xi-\eta)^\delta} e^{i(x-y)\xi - ix\eta} \overline{\hat{v}(\eta)} a(x,\xi) u(y)$$
$$dy\, d\xi\, dx\, d\eta$$
$$= \iiiint (i+D_x)^\delta e^{i(x-y)\xi - ix\eta} \frac{\overline{\hat{v}(-\eta)}}{(i+\xi-\eta)^\delta} \frac{u(y)}{(i+x-y)^\delta} (i - D_\xi)^\delta$$
$$a(x,\xi)\, dy\, d\xi\, dx\, d\eta.$$

Here we have applied integration by parts in ξ and inserted $\overline{\hat{v}(\eta)} = \overline{\hat{v}(-\eta)}$. Integration by parts in x yields

$$(Au, v) \qquad\qquad\qquad\qquad\qquad\qquad\qquad\qquad (1.1.99)$$
$$= \iiiint e^{i(x-y)\xi - ix\eta} \frac{\overline{\hat{v}(-\eta)}}{(i+\xi-\eta)^\delta} (i - D_\xi)^\delta$$
$$\left\{ \frac{u(y)}{(i+x-y)^\delta} (i-D_\xi)^\delta a(x,\xi) \right\} dy\, d\xi\, dx\, d\eta$$
$$= \sum_{\beta+\gamma=\delta} c(\beta,\gamma,\delta) \iiiint e^{i(x-y)\xi - ix\eta} \frac{\overline{\hat{v}(-\eta)}}{(i+\xi-\eta)^\delta}$$
$$\left\{ (i-D_x)^\beta \frac{u(y)}{(i+x-y)^\delta} \right\} \left\{ (i-D_x)^\gamma (i-D_\xi)^\delta a(x,\xi) \right\} dy\, d\xi\, dx\, d\eta$$
$$= \sum_{\beta+\gamma=\delta} c(\beta,\gamma,\delta) \iint g_\delta(x,\xi) f_{\delta\beta}(x,\xi) e^{ix\xi} (i-D_x)^\gamma (i-D_\xi)^\delta$$
$$a(x,\xi)\, dx\, d\xi \qquad\qquad\qquad\qquad\qquad\qquad (1.1.100)$$

for suitable constants $c(\beta, \gamma, \delta)$, where

$$g_\delta(x, \xi) = \int e^{-ix\eta} \frac{\hat{\bar{v}}(-\eta)}{(i + \xi - \eta)^\delta} \, d\eta,$$

$$f_{\delta\beta}(x, \xi) = \int e^{-iy\xi} (i - D_x)^\beta \frac{u(y)}{(i + x - y)^\delta} \, dy.$$

We will show below the estimates

$$\|g_\delta\|_{L^2(\mathbb{R}^{2n})} \leq c_\delta \|v\|_{L^2(\mathbb{R}^n)}, \tag{1.1.101}$$

$$\|f_{\delta\beta}\|_{L^2(\mathbb{R}^{2n})} \leq c_{\delta\beta} \|u\|_{L^2(\mathbb{R}^n)} \tag{1.1.102}$$

for all $\delta \geq (1, \ldots, 1)$, with constants c_δ, $c_{\delta\beta}$ only dependent on δ, β. Using (1.1.101), (1.1.102) we get from (1.1.100)

$$|(v, Au)| \tag{1.1.103}$$

$$\leq \sum_{\beta+\gamma=\delta} c(\beta, \gamma, \delta) \iint |g_\delta(x, \xi) f_{\delta\beta}(x, \xi)(i - D_x)^\gamma (i - D_\xi)^\delta a(x, \xi)| dx d\xi$$

$$\leq \tilde{c}_\delta \sup_{\substack{(x, \xi) \in \mathbb{R}^{2n} \\ \alpha, \beta \leq \delta}} |D_x^\alpha D_\xi^\beta a(x, \xi)| \|g_\delta\|_{L^2(\mathbb{R}^{2n})} \|f_{\delta\beta}\|_{L^2(\mathbb{R}^{2n})}$$

$$\leq c_\delta \sup_{\substack{(x, \xi) \in \mathbb{R}^{2n} \\ \alpha, \beta \leq \delta}} |D_x^\alpha D_\xi^\beta a(x, \xi)| \|v\|_{L^2(\mathbb{R}^n)} \|u\|_{L^2(\mathbb{R}^n)} \tag{1.1.104}$$

for suitable constants $\tilde{c}_\delta$, c_δ. Now let $a(x, \xi)$ be the given symbol satisfying the estimate (1.1.91) and apply the above arguments to $a_\varepsilon(x, \xi)$. Then (1.1.94) and (1.100) for $\delta = (1, \ldots, 1)$ yield

$$|(Au, v)| = \lim_{\varepsilon \to 0} |(\mathrm{Op}(a_\varepsilon)u, v)|$$

$$\leq \tilde{c} \lim_{\varepsilon \to 0} \pi(a_\varepsilon) \|v\|_{L^2(\mathbb{R}^n)} \|u\|_{L^2(\mathbb{R}^n)}$$

$$\leq c\pi(a) \|v\|_{L^2(\mathbb{R}^n)} \|u\|_{L^2(\mathbb{R}^n)}$$

with constants $\tilde{c}$, c. Since $C_0^\infty(\mathbb{R}^n)$ is dense in $L^2(\mathbb{R}^n)$ this yields (1.1.96). It remains to show (1.1.101) and (1.1.102). Let us consider (1.1.101). We have

$$\|g_\delta(x, \xi)\|_{L^2(\mathbb{R}^{2n})}^2 = \int \|g_\delta(x, \xi)\|_{L^2(\mathbb{R}_x^n)}^2 \, d\xi$$

$$= (2\pi)^n \int \|(F_{x \to \eta}^{-1} g_\delta)(\eta, \xi)\|_{L^2(\mathbb{R}_\eta^n)}^2 \, d\xi$$

$$= \int \|(i + \xi - \eta)^{-\delta} \hat{\bar{v}}(-\eta)\|_{L^2(\mathbb{R}_\eta^n)}^2 \, d\xi.$$

Here we have used Plancherel's formula. It follows that

$$\|g_\delta(x,\xi)\|^2_{L^2(\mathbb{R}^{2n})} = \iint |(i + \xi - \eta)^{-\delta}|^2 |\hat{\bar{v}}(-\eta)|^2 \, d\eta d\xi$$

$$= (2\pi)^n \int |(i + \zeta)^{-\delta}|^2 \, d\zeta \|v\|^2_{L^2(\mathbb{R}^n)}$$

$$= c_\delta^2 \|v\|^2_{L^2(\mathbb{R}^n)}$$

with $c_\delta^2 = \int |(i + \zeta)^{-\delta}|^2 \, d\zeta < \infty$ for $\delta \geq (1, \ldots, 1)$. The proof of (1.1.102) is very similar and will be omitted. $\qquad\square$

Theorem 1.1.55 *Let $a(x,\xi) \in S^\mu(\mathbb{R}^n \times \mathbb{R}^n)$ be a symbol with $a(x,\xi) = 0$ for $x \notin K$ for some compact set $K \subset \mathbb{R}^n$. Then $A = \mathrm{Op}(a)$ has a continuous extension*

$$A : H^s(\mathbb{R}^n) \to H^{s-\mu}(\mathbb{R}^n)$$

for every $s \in \mathbb{R}$ and we have $\|A\|_{s,s-\mu} \leq \tilde{c} c_s$ for a constant $\tilde{c} > 0$ and

$$c_s = \sup_{\xi \in \mathbb{R}^n} (1 + |\xi|)^{-\mu} \int |(1 - \Delta_x)^M a(x,\xi)| \, dx$$

for any $M \in \mathbb{N}$ with $M > \dfrac{(n + |s - \mu|)}{2}$.

Proof. The function $b(\zeta,\xi) = \int e^{-ix\zeta} a(x,\xi) \, dx$ satisfies

$$|b(\zeta,\xi)| \leq c_N (1 + |\zeta|)^{-N} (1 + |\xi|)^\mu \qquad (1.1.105)$$

for every $N \in \mathbb{N}$ with some constant $c_N > 0$. In fact, we have $\zeta^\alpha b(\zeta,\xi) = \int e^{-ix\zeta} D_x^\alpha a(x,\xi) \, dx$ for every $\alpha \in \mathbb{N}^n$ which gives us for arbitrary M

$$|(1 + |\zeta|^2)^M b(\zeta,\xi)| = |\int e^{-ix\zeta}(1 - \Delta_x)^M a(x,\xi) \, dx| \leq c(1 + |\xi|)^\mu \qquad (1.1.106)$$

with an M–dependent constant $c > 0$. Since $(1 + |\zeta|)^N (1 + |\zeta|^2)^{-M} <$ const for $M \geq \frac{N}{2}$, we get immediately (1.1.105).

Next we set $K(\xi,\eta) = b(\eta - \xi, \xi)(1 + |\xi|)^{-s}(1 + |\eta|)^{s-\mu}$ and observe that

$$\int |K(\xi,\eta)| \, d\xi \leq c, \qquad \int |K(\xi,\eta)| \, d\eta \leq c \qquad (1.1.107)$$

for all η and ξ, respectively, for some constant $c > 0$. In fact, (1.1.105) gives us together with (1.1.79) the estimate

$$|K(\xi,\eta)| \leq c_N (1 + |\xi - \eta|)^{-N} \left(\frac{1 + |\eta|}{1 + |\xi|}\right)^{s-\mu} \leq c_N (1 + |\xi - \eta|)^{-N+|s-\mu|}$$

which yields (1.1.107) when we choose $N > n + |s - \mu|$. Applying now the Fourier transform $F_{x \to \eta}$ to $Au(x)$, $u \in C_0^\infty(\mathbb{R}^n)$, we obtain

$$(\widehat{Au})(\eta) = \iint e^{ix(\xi - \eta)} a(x, \xi) \hat{u}(\xi) \, d\xi dx = \int b(\eta - \xi, \xi) \hat{u}(\xi) \, d\xi.$$

This yields for any $v \in H^{\mu - s}(\mathbb{R}^n)$

$$\int (\widehat{Au})(\eta) \hat{v}(\eta) \, d\eta = \iint b(\eta - \xi, \xi) \hat{v}(\eta) \hat{u}(\xi) \, d\xi d\eta$$

$$= \iint K(\xi, \eta) \hat{v}(\eta)(1 + |\eta|)^{\mu - s} \hat{u}(\xi)(1 + |\xi|)^s \, d\xi d\eta.$$

We then obtain

$$|\langle Au, v \rangle| = |\langle \widehat{Au}, \hat{v} \rangle|$$

$$\leq (2\pi)^{-n} \left\{ \iint |K(\xi, \eta)|(1 + |\eta|)^{2(\mu - s)} |\hat{v}(\eta)|^2 \, d\xi d\eta \right\}^{\frac{1}{2}}$$

$$\cdot \left\{ \iint |K(\xi, \eta)|(1 + |\xi|)^{2s} |\hat{u}(\xi)|^2 \, d\xi d\eta \right\}^{\frac{1}{2}}$$

$$\leq c \|v\|_{\mu - s} \|u\|_s.$$

In the latter estimate we have employed (1.1.107). Hence it follows that $\sup \frac{|\langle Au, v \rangle|}{\|v\|_{\mu - s}} \leq c \|u\|_s$, where sup is taken over all $v \in H^{\mu - s}(\mathbb{R}^n)$, $v \neq 0$. This yields $\|Au\|_{s - \mu} \leq c \|u\|_s$, cf. (1.1.88), for some $c > 0$. The latter constant is of the form $c = \tilde{c} c_N$ with $\tilde{c} > 0$ and c_N from (1.1.105). For c_N we have from (1.1.106)

$$c_N = \sup_{\xi \in \mathbb{R}^n} (1 + |\xi|)^{-\mu} \int |(1 - \Delta_x)^M a(x, \xi)| dx$$

for any $M \geq \frac{N}{2} > \frac{(n + |s - \mu|)}{2}$. This completes the proof of Theorem 1.1.55. □

Corollary 1.1.56 *The operator $\mathcal{M}_\varphi$ of multiplication by $\varphi \in C_0^\infty(\mathbb{R}^n)$ induces continuous operators*

$$\mathcal{M}_\varphi : H^s(\mathbb{R}^n) \to H^s(\mathbb{R}^n) \qquad (1.1.108)$$

for all $s \in \mathbb{R}$. Moreover, $\varphi \to \mathcal{M}_\varphi$ represents a continuous operator $C_0^\infty(\mathbb{R}^n) \to \mathcal{L}(H^s(\mathbb{R}^n))$ for every $s \in \mathbb{R}$.

In fact, $\varphi(x) \in C_0^\infty(\mathbb{R}^n)$ can be regarded as an element in $S^0(\mathbb{R}_x^n \times \mathbb{R}_\xi^n)$ with compact support with respect to x. Then $\mathcal{M}_\varphi$ corresponds to $\text{Op}(\varphi)$, and we can apply Theorem 1.1.55 which shows at the same time the continuity of $C_0^\infty(\mathbb{R}^n) \to \mathcal{L}(H^s(\mathbb{R}^n))$.

Remark 1.1.57 *Theorem* 1.1.55 *has an obvious generalisation to symbols* $a(x, \xi) \in \mathcal{S}(\mathbb{R}^n, S^\mu(\mathbb{R}^n))$. *Then, in particular,* $\mathcal{M}_\varphi$ *for* $\varphi \in \mathcal{S}(\mathbb{R}^n)$ *induces continuous operators* (1.1.108) *and the corresponding mapping* $\mathcal{S}(\mathbb{R}^n) \to \mathcal{L}(H^s(\mathbb{R}^n))$ *is continuous for every* $s \in \mathbb{R}$.

If $\Omega \subseteq \mathbb{R}^n$ is open we denote by $H^s_{\mathrm{loc}}(\Omega)$ the subspace of all $u \in \mathcal{D}'(\Omega)$ such that $\varphi u \in H^s(\mathbb{R}^n)$ for every $\varphi \in C_0^\infty(\Omega)$ (here φu is interpreted as the corresponding extension by 0 from Ω to $\mathbb{R}^n$). Moreover, $H^s_{\mathrm{comp}}(\Omega)$ denotes the subspace of all $u \in H^s(\mathbb{R}^n)$ with compact supp $u \subset \Omega$ (the elements are then regarded as distributions on Ω). $u \to \|\varphi u\|_s$, $\varphi \in C_0^\infty(\Omega)$, represents a semi–norm system on $H^s_{\mathrm{loc}}(\Omega)$. This defines a Fréchet topology in the space $H^s_{\mathrm{loc}}(\Omega)$. Moreover, $H^s_{\mathrm{comp}}(\Omega)$ may be regarded as a (countable) inductive limit of the spaces $H^s(K) := \{u \in H^s_{\mathrm{comp}}(\Omega) : \text{supp}\, u \subseteq K\}$ for compact $K \subset\subset \Omega$. Every $H^s(K)$ can be interpreted as a closed subspace of $H^s(\mathbb{R}^n)$. The following result is then an easy consequence of the above Theorem 1.1.55.

Theorem 1.1.58 *Each* $A \in L^\mu(\Omega)$ *induces continuous operators*

$$A : H^s_{\mathrm{comp}}(\Omega) \to H^{s-\mu}_{\mathrm{loc}}(\Omega)$$

for all $s \in \mathbb{R}$. *If* A *is properly supported the corresponding operators*

$$A : H^s_{\mathrm{comp}}(\Omega) \to H^{s-\mu}_{\mathrm{comp}}(\Omega), \qquad A : H^s_{\mathrm{loc}}(\Omega) \to H^{s-\mu}_{\mathrm{loc}}(\Omega)$$

are continuous for all $s \in \mathbb{R}$.

Theorem 1.1.59 *Let* $\kappa : \Omega \to \tilde{\Omega}$ *be a diffeomorphism. Then the pullback* $\kappa^* : C^\infty(\tilde{\Omega}) \to C^\infty(\Omega)$ *extends to an isomorphism*

$$\kappa_s^* : H^s_{\mathrm{loc}}(\tilde{\Omega}) \to H^s_{\mathrm{loc}}(\Omega)$$

for every $s \in \mathbb{R}$.

Proof. First it is clear that the result is true for $s = 0$, since in this case $H^0_{\mathrm{loc}}(\Omega) = L^2_{\mathrm{loc}}(\Omega)$ which is the space of all $u \in \mathcal{D}'(\Omega)$ with $\varphi u \in L^2(\mathbb{R}^n)$ for every $\varphi \in C_0^\infty(\Omega)$. Moreover, it obviously suffices to show that κ^* induces isomorphisms

$$\kappa_s^* : H^s(\tilde{K}) \to H^s(K)$$

for any subdomain $U \subset \Omega$ with C^∞ boundary and compact $K = \overline{U}$, $\tilde{K} = \kappa(K)$, $H^s(K)$ being the closure of $C_0^\infty(K)$ in $H^s(\mathbb{R}^n)$. Choose a properly supported pseudo–differential operator $R \in L^s(\mathbb{R}^n)$ with the symbol $(1 + |\xi|^2)^{\frac{s}{2}}$ and a properly supported pseudo–differential operator $R^{-1} \in L^{-s}(\mathbb{R}^n)$ with the symbol $(1 + |\xi|^2)^{-\frac{s}{2}}$. Then we have

$RR^{-1} - 1$, $R^{-1}R - 1 \in L^{-\infty}(\mathbb{R}^n)$. Moreover, we set $\tilde{R} = \kappa_* R$, $\tilde{R}^{-1} = \kappa_* R^{-1}$, where κ_* is the operator push–forward in the sense of Theorem 1.1.38. Then $\tilde{R}$, $\tilde{R}^{-1}$ are also properly supported, and $\kappa_*(R^{-1}R - 1) = \kappa_* R^{-1} \kappa_* R - 1$ shows that we also have $\tilde{R}\tilde{R}^{-1} - 1$, $\tilde{R}^{-1}\tilde{R} - 1 \in L^{-\infty}(\mathbb{R}^n)$. If κ^* denotes the function pull–back, we have by definition

$$\tilde{R} = (\kappa^*)^{-1} R \kappa^*, \qquad \tilde{R}^{-1} = (\kappa^*)^{-1} R^{-1} \kappa^*.$$

Let

$$\kappa_0^* : L^2_{\text{loc}}(\mathbb{R}^n) \to L^2_{\text{loc}}(\mathbb{R}^n)$$

be the pull–back on corresponding L^2_{loc} spaces. Then, in view of Theorem 1.1.58,

$$\kappa_{(s)}^* := R^{-1} \kappa_0^* \tilde{R} : H^s_{\text{comp}}(\mathbb{R}^n) \to H^s_{\text{comp}}(\mathbb{R}^n)$$

is a continuous operator. It can be regarded as extension by continuity of the operator $R^{-1}\kappa^* \tilde{R} : C_0^\infty(\mathbb{R}^n) \to C_0^\infty(\mathbb{R}^n)$. In view of $\tilde{R} = (\kappa^*)^{-1} R \kappa^*$ the latter operator equals $(1 + C)\kappa^*$ with a certain (properly supported) operator $C \in L^{-\infty}(\mathbb{R}^n)$. In other words the operator $(1 + C)\kappa^*$ with the function pull–back κ^* extends by continuity to $\kappa_{(s)}^*$. In view of $\kappa^* H^s_{\text{loc}}(\mathbb{R}^n) \subset \mathcal{E}'(\mathbb{R}^n)$ and since C is smoothing, the operator $C\kappa^*$ has an extension by continuity

$$C\kappa^* : H^s_{\text{comp}}(\mathbb{R}^n) \to H^\infty_{\text{comp}}(\mathbb{R}^n).$$

Thus $\kappa^* = (1 + C)\kappa^* - C\kappa^*$, first being given on $C_0^\infty(\mathbb{R}^n)$, extends to a continuous operator $\kappa_s^* : H^s_{\text{comp}}(\mathbb{R}^n) \to H^s_{\text{comp}}(\mathbb{R}^n)$. In the converse direction we can argue in an analogous manner, such that κ_s^* is actually an isomorphism. □

1.1.5 Pseudo–differential operators on C^∞ manifolds

We assume in this book that the C^∞ manifolds in consideration are of finite dimension, paracompact and oriented. Every C^∞ manifold X is endowed with a Riemannian metric that is kept fixed; dx denotes the associated positive measure. On X we have the space $C^\infty(X)$ of infinitely differentiable functions in its natural Fréchet topology. $C_0^\infty(K) := \{u \in C^\infty(X) : \text{supp}\, u \subseteq K\}$ for compact $K \subseteq X$ is a Fréchet subspace, and we set $C_0^\infty(X) = \varinjlim C_0^\infty(K)$ with the inductive limit over all compact $K \subseteq X$. Moreover, $\mathcal{D}'(X)$ $(= C_0^\infty(X)')$ denotes the space of distributions on X and $\mathcal{E}'(X)$ $(= C^\infty(X)')$ the subspace of elements with compact support.

TX denotes the tangent bundle , T^*X the cotangent bundle on X, and SX, S^*X the associated sphere bundles induced by the Riemannian metric.

Vector bundles E on X are assumed to be C^∞, i.e., the transition mappings between trivialisations over intersections of coordinate neighbourhoods are C^∞. The space of C^∞ sections in E is denoted by $C^\infty(X,E)$, and $C_0^\infty(X,E)$ is the subspace of elements with compact support. We mainly talk about complex vector bundles E, and we fix a Hermitean metric in every E. Let $\mathcal{U} = \{U_j\}_{j\in\mathbb{Z}}$ be a locally finite open covering of X and fix a subordinate partition of unity $\{\varphi_j\}_{j\in\mathbb{Z}}$ on X. Moreover choose a system of functions $\{\psi_j\}_{j\in\mathbb{Z}}$, $\psi_j \in C_0^\infty(U_j)$, with $\varphi_j\psi_j = \varphi_j$ for all j. Let

$$\kappa_j : U_j \to \Omega_j, \qquad j \in \mathbb{Z}$$

be a system of charts with open $\Omega_j \subseteq \mathbb{R}^n$, where we assume that every Ω_j is homeomorphic to a ball. If E is a vector bundle on X we consider the system of restrictions $E|_{U_j}$ together with fixed trivialisations

$$\tau_j : E|_{U_j} \to \Omega_j \times \mathbb{C}^k, \qquad j \in \mathbb{Z} \qquad (1.1.109)$$

that induce on the zero sections just the mappings κ_j. For every $u \in C^\infty(X,E)$ we can form $\tau_{j*}(u|_{U_j}) \in C^\infty(\Omega_j,\mathbb{C}^k)$ which is the push–forward of $u|_{U_j}$ under (1.1.109). Let us fix a compact subset $K \subset\subset X$. Then $H^s(K,E)$, $s \in \mathbb{R}$, is defined as the closure of $C_0^\infty(K,E) = \{u \in C_0^\infty(X,E) : \operatorname{supp} u \subseteq K\}$ with respect to the norm

$$\left\{ \sum_{\substack{j\in\mathbb{Z} \\ U_j\cap K\neq\emptyset}} \|\tau_{j*}(\varphi_j u)\|^2_{H^s(\mathbb{R}^n,\mathbb{C}^k)} \right\}^{\frac{1}{2}}.$$

Here $H^s(\mathbb{R}^n,\mathbb{C}^k) = H^s(\mathbb{R}^n) \otimes \mathbb{C}^k$. Set

$$H^s_{\mathrm{comp}}(X,E) = \varinjlim H^s(K,E),$$

where the inductive limit is taken over all compact $K \subseteq X$. The closure of $C^\infty(X,E)$ with respect to the system of semi–norms $u \to \|\tau_{j*}(\varphi_j u)\|_{H^s(\mathbb{R}^n,\mathbb{C}^k)}$, $j \in \mathbb{Z}$, is denoted by $H^s_{\mathrm{loc}}(X,E)$. This is a Fréchet space, independent of the particular choice of $\mathcal{U}$, $\{\varphi_j\}$, $\{\tau_j\}$. Also $H^s_{\mathrm{comp}}(X,E) \subset H^s_{\mathrm{loc}}(X,E)$ is correctly defined in the sense of independence of these data. Note that there is involved also the invariance of Sobolev distributions under diffeomorphisms, cf. Theorem 1.1.59. For the trivial bundle $E = X \times \mathbb{C}$ we denote the corresponding spaces by $H^s_{\mathrm{loc}}(X)$ and $H^s_{\mathrm{comp}}(X)$, respectively. If X is compact the difference between loc– and comp–spaces disappears, and in this case we will denote the corresponding spaces by $H^s(X, E)$. For $E = X \times \mathbb{C}$ we will write $H^s(X)$. The space $H^s(X, E)$ can be endowed with a scalar product $(\cdot, \cdot)_s$ such that $\|u\|_s = (u, u)_s^{\frac{1}{2}}$ (subscript s here denotes the norm in $H^s(X, E)$). Let us fix a particular choice in the case $s = 0$. The Hermitean structure of E yields an invariant sesquilinear pairing $\langle u(x), \overline{v}(x) \rangle_{(E,E^*)}$ for every $x \in X$. Here E^* is the dual bundle to E and we first assume $v \in C^\infty(X, E)$. Then we set

$$(u, v)_0 = \int_X \langle u(x), \overline{v}(x) \rangle_{(E,E^*)} \, dx. \qquad (1.1.110)$$

Since $C^\infty(X, E)$ is dense in $H^0(X, E)$, (1.1.110) extends to a scalar product in $H^0(X, E)$, where the norm induced by (1.1.110) is equivalent to $\| \cdot \|_{H^0(X,E)}$. The space $H^0(X, E)$ with this fixed scalar product is also denoted by $L^2(X, E)$. In particular, for $E = X \times \mathbb{C}$ we get the space $L^2(X)$. We can define a sesquilinear pairing

$$(\cdot, \cdot)_0 : C_0^\infty(X, E) \times C_0^\infty(X, E) \to \mathbb{C} \qquad (1.1.111)$$

by formula (1.1.110); here X is not necessarily compact.

Let us now turn to the definition of the spaces

$$L^\mu(X; E, F), \quad L_{\mathrm{cl}}^\mu(X; E, F) \qquad (1.1.112)$$

of pseudo–differential operators on X of order $\mu \in \mathbb{R}$, where E, F are vector bundles on X. As usual, subscript "cl" indicates classical pseudo–differential operators. First we set

$$L^{-\infty}(X; E, F) = \bigcap_{s \in \mathbb{R}} \mathcal{L}(H_{\mathrm{comp}}^s(X, E), H_{\mathrm{loc}}^\infty(X, F)).$$

Let $U \in \mathcal{U}$ and $\tau_U : E|_U \to \Omega \times \mathbb{C}^k$, $\varrho_U : F|_U \to \Omega \times \mathbb{C}^l$ be corresponding trivialisations, $\Omega \subseteq \mathbb{R}^n$ open. Recall that the trivialisations induce isomorphisms between the spaces of C^∞ sections on U and Ω, respectively. In particular, every $A_\Omega \in L^\mu(\Omega) \otimes \mathbb{C}^l \otimes \mathbb{C}^k$ gives rise to an operator

$$A_U = \varrho_U^{-1} A_\Omega \tau_{U^*} : C_0^\infty(U, E|_U) \to C^\infty(U, F|_U).$$

Now $L^\mu(X; E, F)$ is defined as the space of all operators $A : C_0^\infty(X, E) \to C^\infty(X, F)$ which have the form

$$A = \sum_{j \in \mathbb{Z}} \varphi_j \{ \varrho_{U_j}^{-1} A_j \tau_{U_j^*} \} \psi_j + C \qquad (1.1.113)$$

for arbitrary $C \in L^{-\infty}(X; E, F)$ and arbitrary $A_j \in L^\mu(\Omega_j) \otimes \mathbb{C}^l \otimes \mathbb{C}^k$. Here $\mathcal{U} = \{U_j\}_{j \in \mathbb{Z}}$, $\Omega_j = \kappa_j U_j$ and $\{\varphi_j\}_{j \in \mathbb{Z}}$, $\{\psi_j\}_{j \in \mathbb{Z}}$ are the function systems introduced above. The subspace $L^\mu_{\mathrm{cl}}(X; E, F)$ of classical pseudo–differential operators is defined by requiring $A_j \in L^\mu_{\mathrm{cl}}(\Omega_j) \otimes \mathbb{C}^l \otimes \mathbb{C}^k$, $j \in \mathbb{Z}$. Note that the local matrices of homogeneous principal symbols of A_j of order μ have a global interpretation as a bundle morphism

$$\sigma^\mu_\psi(A) : \pi^* E \to \pi^* F. \tag{1.1.114}$$

Here $\pi : T^* X \backslash 0 \to X$ is the canonical projection ($\backslash 0$ indicates the bundle minus the zero section), and π^* is the bundle pull–back. Points in $T^* X$ are often denoted by (x, ξ) though this refers to local coordinates. The symbol (1.1.114) corresponds to a family of linear mappings

$$\sigma^\mu_\psi(A)(x, \xi) : E_x \to F_x$$

between the fibres of E and F over $x \in X$ (indicated by subscripts x), and we have

$$\sigma^\mu_\psi(A)(x, \lambda \xi) = \lambda^\mu \sigma^\mu_\psi(A)(x, \xi)$$

for all $\lambda > 0$. Let

$$S^{(\mu)}(T^* X \setminus 0; E, F) \tag{1.1.115}$$

denote the space of all bundle morphisms $p : \pi^* E \to \pi^* F$ satisfying $p(x, \lambda \xi) = \lambda^\mu p(x, \xi)$ for all $\lambda > 0$. For $E = X \times \mathbb{C}$, $F = X \times \mathbb{C}$ we simply write

$$S^{(\mu)}(T^* X \setminus 0) \tag{1.1.116}$$

instead of (1.1.115). This is the space of all elements in $C^\infty(T^* X \setminus 0)$ with the corresponding homogeneity μ in ξ.

Remark 1.1.60 *Similarly to the above local constructions the operator spaces in* (1.1.112) *can be endowed with natural Fréchet space topologies. Let us briefly describe this for compact X. We have non–direct sums of Fréchet spaces*

$$L^\mu(X; E, F) = \sum_{j=1}^N [\varphi_j] L^\mu(X; E|_{U_j}, F|_{U_j})[\psi_j] + L^{-\infty}(X; E, F)$$

for $\mathcal{U} = \{U_1, \ldots, U_N\}$, the summands are Fréchet, and then also the spaces themselves. The same can be done for the subspaces of classical operators.

Let us give a global formulation of the distributional kernel of a pseudo–differential operator on X. First the volume bundle V of X is the real line bundle associated to the tangent bundle TX by the representation $a \to |\det a|$ of $GL(n, \mathbb{R})$. The smooth sections of V are the smooth measures on X. Every $v \in C_0^\infty(X, V)$ gives rise to the integral $\int v$ in an invariant manner. For any bundle E we set $E' = E^* \otimes V$ with the dual bundle E^*. There is then a natural pairing $(\cdot, \cdot) : E_x \times E'_x \to V_x$. To every $e \in C^\infty(X, E)$, $e' \in C^\infty(X, E')$ we get an $(e(x), e'(x)) \in C^\infty(X, V)$ that can be integrated over X when one factor has compact support. This gives rise to a bilinear map

$$C_0^\infty(X, E) \times C^\infty(X, E') \to \mathbb{C}$$

and hence to a linear map $C^\infty(X, E') \to C_0^\infty(X, E)'$, where the prime on the right denotes the dual of the topological vector space. This procedure can be applied to E' instead of E. If we then identify E'' with E we thus obtain a map $C^\infty(X, E) \to C_0^\infty(X, E')'$ which is injective. $C_0^\infty(X, E')'$ is called the space of distributional sections of E. Analogously a distributional section of E with compact support is an element of $C^\infty(X, E')'$. Similarly to earlier notation we set

$$\mathcal{D}'(X, E) = C_0^\infty(X, E')', \qquad \mathcal{E}'(X, E) = C^\infty(X, E')'.$$

If E, F are vector bundles on X we can form the external tensor product $E \boxtimes F$ on $X \times X$ which is defined as $p_1^* E \otimes p_2^* F$ with $p_i : X \times X \to X$ being the canonical projection to the ith factor, $i = 1, 2$. Now the Schwartz kernel K_A of a continuous linear operator $A : C_0^\infty(X, E) \to C^\infty(X, F)$ is an element of $\mathcal{D}'(X \times X, F \boxtimes E')$. It is uniquely determined by

$$\langle K_A, \varphi \otimes \psi \rangle = \langle \varphi, A\psi \rangle \quad \text{for } \varphi \in C_0^\infty(X, F'), \qquad \psi \in C_0^\infty(X, E).$$

We have

$$K_A|_{X \times X \setminus \text{diag} X \times X} \in C^\infty(X \times X \setminus \text{diag} X \times X, F \boxtimes E')$$

for every $A \in L^\mu(X; E, F)$. An operator $A \in L^\mu(X; E, F)$ is said to be properly supported if the support of K_A is proper, i.e. supp $K_A \cap p_i^{-1}(M)$ is compact for every compact set $M \subseteq X$, $i = 1, 2$. Every $A \in L^\mu(X; E, F)$ can be decomposed as $A = A_0 + C$ with properly supported A_0 and $C \in L^{-\infty}(X; E, F)$.

The following results are immediate analogues of the corresponding local ones.

Theorem 1.1.61 *The sequence*

$$0 \to L_{\mathrm{cl}}^{\mu-1}(X; E, F) \xrightarrow{\iota} L_{\mathrm{cl}}^{\mu}(X; E, F) \xrightarrow{\sigma_\psi^\mu} S^{(\mu)}(T^*X \setminus 0; E, F) \to 0$$

is exact; here ι is the canonical embedding.

Theorem 1.1.62 *Let $A_j \in L^{\mu_j}(X; E, F)$, $j \in \mathbb{N}$, be an arbitrary sequence with $\mu_j \to -\infty$ as $j \to \infty$. Then there is an $A \in L^\mu(X; E, F)$ for $\mu = \max\{\mu_j\}$ such that for every $M \in \mathbb{N}$ there is an $N(M)$ such that for all $N \geq N(M)$*

$$A - \sum_{j=0}^{N} A_j \in L^{\mu-M}(X; E, F)$$

holds. The operator A is unique $\mathrm{mod}\, L^{-\infty}(X; E, F)$.

Write $A \sim \sum_{j=0}^{\infty} A_j$ and call A (an) asymptotic sum of the A_j.

Proposition 1.1.63 *Let $C \in L^{-1}(X; E, E)$ be arbitrary. Then there is a properly supported operator $D \in L^{-1}(X; E, E)$ such that*

$$(1 + C)(1 + D) - 1, (1 + D)(1 + C) - 1 \in L^{-\infty}(X; E, E).$$

Proof. Writing $C = C_0 + K$ with properly supported $C_0 \in L^{-1}$ and $K \in L^{-\infty}$ we obtain $C_0^j \in L^{-j}$. Hence there exists the formal Neumann series

$$1 + \tilde{D} := \sum_{j=0}^{\infty} (-1)^j C_0^j$$

in the sense of an asymptotic sum with $\tilde{D} \in L^{-1}$, and every properly supported representative D of $\tilde{D}$ satisfies the assertion. $\square$

Theorem 1.1.64 *Every $A \in L^\mu(X; E, F)$ extends to a continuous operator*

$$A : H_{\mathrm{comp}}^s(X, E) \to H_{\mathrm{loc}}^{s-\mu}(X, F)$$

for every $s \in \mathbb{R}$. If A is properly supported then A induces continuous operators

$$A : H_{\mathrm{comp}}^s(X, E) \to H_{\mathrm{comp}}^{s-\mu}(X, F), \qquad A : H_{\mathrm{loc}}^s(X, E) \to H_{\mathrm{loc}}^{s-\mu}(X, F)$$

for all $s \in \mathbb{R}$. If X is compact we get continuous operators

$$A : H^s(X, E) \to H^{s-\mu}(X, F)$$

for all $s \in \mathbb{R}$.

Theorem 1.1.65 *Let $A \in L^\mu(X; \tilde{E}, F)$, $B \in L^\nu(X; E, \tilde{E})$ and A or B properly supported. Then $AB \in L^{\mu+\nu}(X; E, F)$. If A and B are classical then AB is also classical and we have*

$$\sigma_\psi^{\mu+\nu}(AB) = \sigma_\psi^\mu(A)\sigma_\psi^\nu(B).$$

To every $A \in L^\mu(X; E, F)$ we can define a formal adjoint operator A^* by

$$\int \langle Au, \overline{v} \rangle_{(F,F^*)}\, dx = \int \langle u, \overline{A^*v} \rangle_{(E,E^*)}\, dx$$

for all $u \in C_0^\infty(X, E)$, $v \in C_0^\infty(X, F)$ (cf. the pairing (1.1.111)).

Theorem 1.1.66 $A \in L^\mu(X; E, F)$ *implies* $A^* \in L^\mu(X; F, E)$. *If A is classical then so is A^* and we have*

$$\sigma_\psi^\mu(A^*) = \sigma_\psi^\mu(A)^*$$

where $*$ on the right is interpreted in the sense

$$\langle \sigma_\psi^\mu(A)(x,\xi)e, \overline{f} \rangle_{(F_x, F_x^*)} = \langle e, \overline{\sigma_\psi^\mu(A)^*(x,\xi)f} \rangle_{(E_x, E_x^*)}$$

for all $e \in E_x$, $f \in F_x$ and $(x,\xi) \in T^*X \setminus 0$.

1.1.6 Ellipticity, parametrices, Fredholm property

Definition 1.1.67 *A pseudo–differential operator $A \in L^\mu(X)$ on a C^∞ manifold X is called elliptic of order μ if for every chart $\chi : U \to \Omega \subseteq \mathbb{R}_x^n$ any complete symbol $a(x,\xi)$ of the operator $\chi_*(A|_U) \in L^\mu(\Omega)$ satisfies the estimates*

$$|a(x,\xi)| \geq c(1 + |\xi|)^\mu \qquad (1.1.117)$$

for all $x \in K$ with arbitrary $K \subset\subset \Omega$ and all $\xi \in \mathbb{R}^n$ with $|\xi| \geq R$ for some $R = R(K)$, for certain constants $c = c(K) > 0$.

In general, if E, F are vector bundles on X, an operator $A \in L^\mu(X; E, F)$ is called elliptic if for every chart χ the determinant $\det a(x,\xi)$ of any complete symbol $a(x,\xi)$ of $\chi_*(A|_U)$ satisfies the corresponding estimates $|\det a(x,\xi)| \geq c(1 + |\xi|)^\mu$. Here $a(x,\xi)$ is regarded as a $k \times k$–matrix of elements in $S^\mu(\Omega \times \mathbb{R}^n)$ with k being the fibre dimensions of E and F.

Note that the conditions are independent of the specific choice of the complete symbols $a(x,\xi) \in S^\mu(\Omega \times \mathbb{R}^n)$ of $\chi_*(A|_U)$. The estimates (1.1.117) imply the existence of a $b(x,\xi) \in S^{-\mu}(\Omega \times \mathbb{R}^n)$ satisfying the analogous estimates for the order $-\mu$ and

$$a(x,\xi)b(x,\xi) - 1 = 0 \bmod S^{-1}(\Omega \times \mathbb{R}^n). \qquad (1.1.118)$$

For K–independent R it suffices to set $b(x,\xi) = a^{-1}(x,\xi)\chi(\xi)$ for a suitable excision function χ. The simple modification to construct $b(x,\xi)$ in the general case is left to the reader. Since Ω plays the role of a coordinate patch and since the open covering of X can be replaced by another one with smaller neighbourhoods, we may assume that the constants R in the definition of ellipticity are independent of K.

Remark 1.1.68 *An operator* $A \in L^\mu_{cl}(X)$ *is elliptic if and only if the homogeneous principal symbol* $\sigma^\mu_\psi(A)(x,\xi)$ *never vanishes on* $T^*X \setminus 0$. *More generally,* $A \in L^\mu_{cl}(X; E, F)$ *for vector bundles* E, F *on* X *is elliptic if*

$$\sigma^\mu_\psi(A) : \pi^* E \to \pi^* F$$

is an isomorphism. Here $\pi : T^*X \setminus 0 \to X$ *is the canonical projection.*

Definition 1.1.69 *Let* $A \in L^\mu(X; E, F)$ *be given; then an operator* $B \in L^{-\mu}(X; F, E)$ *is called a parametrix of* A *if*

$$BA - 1 \in L^{-\infty}(X; E, E), \qquad AB - 1 \in L^{-\infty}(X; F, F). \quad (1.1.119)$$

(Here and in the sequel 1 will often denote the identity operator.)

In the compositions we tacitly assume that one factor is properly supported. In view of the observations in the preceding section every pseudo–differential operator is properly supported modulo smoothing ones. So we will assume, for instance, B to be properly supported.

Note that for any other parametrix $\tilde{B}$ of A the relation $B - \tilde{B} \in L^{-\infty}(X; F, E)$ is satisfied.

If only the first (or the second) condition of (1.1.119) holds then B is called a left (or a right) parametrix of A. If A has a left and a right parametrix B_l and B_r, then $B_l - B_r \in L^{-\infty}(X; F, E)$.

Theorem 1.1.70 *Each elliptic operator* $A \in L^\mu(X; E, F)$ *has a parametrix* $B \in L^{-\mu}(X; F, E)$. *Moreover,* $A \in L^\mu_{cl}(X; E, F)$ *implies* $B \in L^{-\mu}_{cl}(X; F, E)$, *and the leading order symbols of* A *and* B *are reciprocal to one another:*

$$\sigma^{-\mu}_\psi(B) = \sigma^\mu_\psi(A)^{-1}. \quad (1.1.120)$$

Proof. Let $\chi : U \to \Omega$ be a chart on X and choose a complete symbol $a(x,\xi) \in S^\mu$ of $A_\Omega := \chi_*(A|_U)$. Then we can form a symbol $b(x,\xi) \in S^{-\mu}$ satisfying the relation (1.1.118). Let $B_\Omega \in L^{-\mu}$ be a properly supported pseudo–differential operator with $b(x,\xi)$ as complete symbol. Then $B_\Omega A_\Omega - 1 \in L^{-1}$, $A_\Omega B_\Omega - 1 \in L^{-1}$ (for brevity we have omitted indicating obvious data such as the corresponding bundles). Using the notation from (1.1.113) we get an operator

$$B_0 = \sum_{j \in \mathbb{Z}} \varphi_j \{ \tau_{U_j^*}^{-1} B_{\Omega_j} \varrho_{U_j^*} \} \psi_j \in L^{-\mu}(X; F, E).$$

It satisfies $B_0 A - 1 \in L^{-1}(X; E, E)$, $A B_0 - 1 \in L^{-1}(X; F, F)$. According to Proposition 1.1.63 we find an operator $D \in L^{-1}(X; E, E)$ such that $(1+D) B_0 A = 1 \mod L^{-\infty}(X; E, E)$ holds. Hence $B_l = (1+D) B_0$ is a left parametrix of A. In an analogous manner we can construct a right parametrix. Thus $B := B_l$ is a parametrix. Since classical operators remain preserved in all constructions, B is classical when A is classical, and the symbol rule (1.1.120) is an immediate consequence of Theorem 1.1.65 and Theorem 1.1.61. □

Theorem 1.1.71 *Let $A \in L^\mu(X; E, F)$ be elliptic and*

$$Au = f \in H_{\text{loc}}^r(X, F)$$

for some $u \in H_{\text{comp}}^{-\infty}(X, E)$ with any $r \in \mathbb{R}$. Then $u \in H_{\text{comp}}^{r+\mu}(X, E)$.

Proof. From Theorem 1.1.70 we have a properly supported parametrix B of A. Then $C = BA - 1 \in L^{-\infty}(X; E, E)$. From $C : H_{\text{comp}}^{-\infty}(X, E) \to H_{\text{loc}}^\infty(X, E)$, $B : H_{\text{loc}}^r(X, F) \to H_{\text{loc}}^{r+\mu}(X, E)$ it follows that $BAu = u + Cu = Bf \in H_{\text{loc}}^{r+\mu}(X, E)$ and hence $u \in H_{\text{loc}}^{r+\mu}(X, E)$ because $H_{\text{loc}}^\infty(X, E) \subset H_{\text{loc}}^t(X, E)$ for every $t \in \mathbb{R}$. Since u has by assumption compact support, we get $u \in H_{\text{comp}}^{r+\mu}(X, E)$. □

Next we pass to the Fredholm property of elliptic pseudo–differential operators on a closed compact C^∞ manifold. Let us formulate some standard material from the functional analytic background.

If $A : H_1 \to H_2$ is a linear map between vector spaces then kernel and cokernel of A are defined as $\ker A = \{ u \in H_1 : Au = 0 \}$ and $\operatorname{coker} A = H_2 / \operatorname{im} A$, respectively, $\operatorname{im} A = \{ v \in H_2 : v = Au$ for some $u \in H_1 \}$.

In the following H_1, H_2 will be Banach spaces in which the norms can equivalently be defined in terms of Hilbert space scalar products. We will briefly talk about Hilbert spaces though the particular choice of scalar products is often inessential. Moreover, we will assume that all Hilbert spaces in question are separable.

Definition 1.1.72 *Let H_1, H_2 be Hilbert spaces. Then, an $A \in \mathcal{L}(H_1, H_2)$ is called a Fredholm operator if*

$$\dim \ker A < \infty, \qquad \dim \operatorname{coker} A < \infty.$$

The number

$$\operatorname{ind} A = \dim \ker A - \dim \operatorname{coker} A \qquad (1.1.121)$$

is called the index of A.

We will denote by $\mathcal{F}(H_1, H_2)$ the subset of all Fredholm operators of $\mathcal{L}(H_1, H_2)$. Moreover let

$$\mathcal{F}_-(H_1, H_2) = \{A \in \mathcal{L}(H_1, H_2) : \dim \ker A = 0\},$$
$$\mathcal{F}_+(H_1, H_2) = \{A \in \mathcal{L}(H_1, H_2) : \dim \operatorname{coker} A = 0\}.$$

In the case $H = H_1 = H_2$ we will also write $\mathcal{F}(H)$, $\mathcal{F}_-(H)$ and $\mathcal{F}_+(H)$, respectively.

Theorem 1.1.73 *The sets* $\mathcal{F}(H_1, H_2)$, $\mathcal{F}_-(H_1, H_2)$ *and* $\mathcal{F}_+(H_1, H_2)$ *are open in* $\mathcal{L}(H_1, H_2)$ *with respect to the norm topology.*

Theorem 1.1.74 *An operator* $A \in \mathcal{L}(H_1, H_2)$ *belongs to* $\mathcal{F}(H_1, H_2)$ *if and only if there exists an operator* $P \in \mathcal{L}(H_2, H_1)$ *such that* $PA - 1$ *and* $AP - 1$ *are compact operators in* H_1 *and* H_2, *respectively.*

The operator P is also called a parametrix of A. Analogously, if $PA - 1$ ($AP - 1$) is compact we talk about a left (right) parametrix.

Note that $P + C$ for every compact $C : H_2 \to H_1$ is a parametrix once P is a parametrix of A. Moreover, every two parametrices of A are equal modulo compact operators. If P_l is a left and P_r a right parametrix of A the $P_l - P_r$ is a compact operator, and P_l, P_r are parametrices of A.

Theorem 1.1.75 $A \in \mathcal{F}(H_1, H_2)$ *and* $C : H_1 \to H_2$ *compact implies* $A + C \in \mathcal{F}(H_1, H_2)$ *and* $\operatorname{ind}(A + C) = \operatorname{ind} A$.

Theorem 1.1.76 $A \in \mathcal{F}(H_0, H_2)$, $B \in \mathcal{F}(H_1, H_0)$ *implies* $AB \in \mathcal{F}(H_1, H_2)$ *and*

$$\operatorname{ind} AB = \operatorname{ind} A + \operatorname{ind} B.$$

In particular, if A is a Fredholm operator and P a parametrix of A, then

$$\operatorname{ind} P = -\operatorname{ind} A.$$

The adjoint A^* of an operator A between Hilbert spaces refers to a fixed choice of the scalar products. Since $\operatorname{ind} \ker A = \dim \operatorname{coker} A^*$, $\operatorname{ind} \operatorname{coker} A = \dim \ker A^*$, it follows that

Proposition 1.1.77 $A \in \mathcal{F}(H_1, H_2)$ *implies* $A^* \in \mathcal{F}(H_2, H_1)$ *and* $\operatorname{ind} A = -\operatorname{ind} A^*$.

Let us now return to pseudo–differential operators on X. From now on in this section we assume that X is a closed compact C^∞ manifold.

Theorem 1.1.78 *If an operator $A \in L^\mu(X; E, F)$, $\mu \in \mathbb{R}$, is given, the following conditions are equivalent:*

(i) *A is elliptic.*

(ii) *The operator*

$$A : H^s(X, E) \to H^{s-\mu}(X, F) \qquad (1.1.122)$$

is Fredholm for some $s \in \mathbb{R}$.

If A is elliptic, then (1.1.122) *is a Fredholm operator for every $s \in \mathbb{R}$. Moreover, A has a parametrix $B \in L^{-\mu}(X; F, E)$ in the sense of Definition 1.1.69.*

Proof. Let A be elliptic. Then, in view of Theorem 1.1.70, there is a parametrix $B \in L^{-\mu}(X; F, E)$. Since X is compact, the operators $BA - 1 \in L^{-\infty}(X; E, E)$ and $AB - 1 \in L^{-\infty}(X; F, F)$ are compact in $H^s(X, E)$ and $H^{s-\mu}(X, F)$, respectively. From Theorem 1.1.74 it follows then that (1.1.122) is a Fredholm operator. This holds for every $s \in \mathbb{R}$. Thus we have proved (i) $\Rightarrow$ (ii) together with the last statement of the theorem. The direction (ii) $\Rightarrow$ (i) for classical operators will be postponed to the end of this section. The non–classical case will be omitted here, though incidentally below we shall use it tacitly. Technical details for a more general situation can be found, e.g., in Grieme [45]. □

Remark 1.1.79 *Let $A \in L^\mu(X; E, F)$ be elliptic and form the operator* (1.1.122). *Then there are finite–dimensional subspaces $N_+ \subset C^\infty(X, E)$, $N_- \subset C^\infty(X, F)$ such that*

$$N_+ = \ker A, N_- + \operatorname{im} A = H^{s-\mu}(X, F), \quad N_- \cap \operatorname{im} A = \{0\}$$

for every $s \in \mathbb{R}$. In particular, $\operatorname{ind} A = \dim N_+ - \dim N_-$ is independent of s.

Remark 1.1.80 *Let the operator $A \in L^\mu(X; E, F)$ be elliptic and $A^* \in L^\mu(X; F, E)$ the formal adjoint of A defined by*

$$(Au, v)_{L^2(X,F)} = (u, A^*v)_{L^2(X,E)}$$

for all $u \in C^\infty(X, E)$, $v \in C^\infty(X, F)$. Then A^* is also elliptic and we have $\operatorname{ind} A^* = \dim \operatorname{coker} A - \dim \ker A$. For the space N_- of Remark 1.1.79 we can set $N_- = \ker A^*$.

Proposition 1.1.81 *Let* $A \in L^{\mu}(X; E, F)$ *be elliptic and* $\operatorname{ind} A = 0$. *Then there exists a finite–dimensional operator* $G \in L^{-\infty}(X; E, F)$ *such that*

$$A + G : H^s(X, E) \to H^{s-\mu}(X, F)$$

is an isomorphism for every $s \in \mathbb{R}$.

Proof. Let $N = \dim \ker A^*$ and choose an arbitrary isomorphism $K : \mathbb{C}^N \to \ker A^*$. Then the row matrix of operators

$$\begin{pmatrix} A & K \end{pmatrix} : \begin{matrix} H^s(X, E) \\ \oplus \\ \mathbb{C}^N \end{matrix} \to H^{s-\mu}(X, F)$$

is surjective. Furthermore, $\ker \begin{pmatrix} A & K \end{pmatrix} = N$, and there exists another row matrix of operators

$$\begin{pmatrix} T & Q \end{pmatrix} : \begin{matrix} H^s(X, E) \\ \oplus \\ \mathbb{C}^N \end{matrix} \to \mathbb{C}^N$$

which restricts to an isomorphism $\begin{pmatrix} T & Q \end{pmatrix} : \ker \begin{pmatrix} A & K \end{pmatrix} \to \mathbb{C}^N$. Here $\ker \begin{pmatrix} A & K \end{pmatrix}$ is an N–dimensional subspace of $C^{\infty}(X, E) \oplus \mathbb{C}^N$. This follows from the relation $Au + Kv = 0$ for $(u, v) \in \ker \begin{pmatrix} A & K \end{pmatrix}$ and from the elliptic regularity. Let $(u_k, v_k)_{k=1,\dots,N}$ be an orthonormal base of $\ker \begin{pmatrix} A & K \end{pmatrix}$ with respect to $L^2(X, E) \oplus \mathbb{C}^N$. Then we can choose $\begin{pmatrix} T & Q \end{pmatrix}$ in the form

$$\begin{pmatrix} T & Q \end{pmatrix} \begin{pmatrix} f \\ g \end{pmatrix} = \sum_{k=1}^N u_k(f, u_k)_{L^2(X,E)} + \sum_{k=1}^N v_k(g, v_k)_{\mathbb{C}^N}.$$

The operator

$$\mathcal{A} = \begin{pmatrix} A & K \\ T & Q \end{pmatrix} : \begin{matrix} H^s(X, E) \\ \oplus \\ \mathbb{C}^N \end{matrix} \to \begin{matrix} H^{s-\mu}(X, F) \\ \oplus \\ \mathbb{C}^N \end{matrix}$$

is now an isomorphism. Without loss of generality we can suppose Q to be an isomorphism $\mathbb{C}^N \to \mathbb{C}^N$. Otherwise we could achieve this by a small perturbation of Q in the class of $N \times N$–matrices without violating the isomorphy property of $\mathcal{A}$. Set

$$\mathcal{C} = \begin{pmatrix} 1 & -KQ^{-1} \\ 0 & 1 \end{pmatrix}, \qquad \mathcal{D} = \begin{pmatrix} 1 & 0 \\ -Q^{-1}T & 1 \end{pmatrix}.$$

Then $\mathcal{C}\mathcal{A}\mathcal{D}$ is a diagonal matrix with the diagonal elements $A - KQ^{-1}T$ and Q. Since $\mathcal{A}$ was composed by isomorphisms we get, in particular, that

$$A - KQ^{-1}T : H^s(X, E) \to H^{s-\mu}(X, F)$$

is an isomorphism. However $G := -KQ^{-1}T \in L^{-\infty}(X; E, F)$ is of finite dimension. □

Theorem 1.1.82 *Let $A \in L^\mu(X; E, F)$ and assume that*

$$A : H^s(X, E) \to H^{s-\mu}(X, F) \qquad (1.1.123)$$

is an isomorphism for some $s \in \mathbb{R}$. Then, (1.1.123) is an isomorphism for all $s \in \mathbb{R}$, and we have $A^{-1} \in L^{-\mu}(X; F, E)$.

Proof. In view of the assumption, A is a Fredholm operator. Hence A is also elliptic because of Theorem 1.1.78. Since $\ker A$ and $\operatorname{coker} A$ are independent of s, the operator (1.1.123) is an isomorphism for all $s \in \mathbb{R}$. According to Theorem 1.1.78 there is a parametrix $B \in L^{-\mu}(X; F, E)$ of A. In view of $BA = 1 + C$ with $C \in L^{-\infty}(X; E, E)$ and because of the compactness of C in $H^s(X, E)$ we have $\operatorname{ind}(1+C) = 0$. Thus $0 = \operatorname{ind} BA = \operatorname{ind} B + \operatorname{ind} A$. It follows that $\operatorname{ind} B = 0$, since A is an isomorphism. Applying Proposition 1.1.81 we find a $G \in L^{-\infty}(X; F, E)$ such that $P := B + G : H^{s-\mu}(X, F) \to H^s(X, E)$ is an isomorphism. Hence $PA = 1 + D : H^s(X, E) \to H^s(X, E)$ is also an isomorphism. Here $D \in L^{-\infty}(X; E, E)$. It is now clear that $(1 + D)^{-1} = 1 + D_1$ with a certain $D_1 \in L^{-\infty}(X; E, E)$. Thus $A^{-1} = (1 + D_1)P \in L^{-\mu}(X; F, E)$. □

Theorem 1.1.83 *For every $\mu \in \mathbb{R}$ and every vector bundle E on X there exists an $R^\mu \in L^\mu_{\mathrm{cl}}(X; E, E)$ such that*

$$R^\mu : H^s(X, E) \to H^{s-\mu}(X, E)$$

is an isomorphism for all $s \in \mathbb{R}$.

Proof. Let us choose an arbitrary element

$$b^{\frac{\mu}{2}}(x, \xi) \in \mathcal{S}^{(\frac{\mu}{2})}(T^*X \setminus 0; E, E)$$

that induces an isomorphism $\pi^*E \to \pi^*E$. In order to find such a $b^{\frac{\mu}{2}}$ it suffices to choose an isomorphism $b_1^{\frac{\mu}{2}} : \pi_1^*E \to \pi_1^*E$ with $\pi_1 : S^*X \to X$, S^*X being the unit sphere bundle induced by T^*X and then to define $b^{\frac{\mu}{2}}(x, \xi) = |\xi|^{\frac{\mu}{2}}b(x, \xi/|\xi|)$ for arbitrary $(x, \xi) \in T^*X \setminus 0$. According to Theorem 1.1.61 there exists an operator $B^{\frac{\mu}{2}} \in L^{\frac{\mu}{2}}(X; E, E)$ with $\sigma_\psi^{\frac{\mu}{2}}(B^{\frac{\mu}{2}}) = b^{\frac{\mu}{2}}$. Let $(B^{\frac{\mu}{2}})^*$ be the formal adjoint of $B^{\frac{\mu}{2}}$. Then $\tilde{R}^\mu := B^{\frac{\mu}{2}}(B^{\frac{\mu}{2}})^* \in L^\mu(X; E, E)$ satisfies $(\tilde{R}^\mu)^* = \tilde{R}^\mu$ and hence, because of Remark 1.1.80, $\operatorname{ind} \tilde{R}^\mu = 0$. From Proposition 1.1.81 we get a $G \in L^{-\infty}(X; E, E)$ such that $R^\mu := \tilde{R}^\mu + G$ is as required. □

Note that $u \to \|R^s u\|_{L^2(X,E)}$ is an equivalent norm in $H^s(X, E)$ and that $(R^s u, R^s v)_{L^2(X,E)}$ is a scalar product in which $H^s(X, E)$ is a Hilbert space. We have

$$L^\mu(X; E, F) = L^0(X; E, F)R^\mu$$

for any fixed $R^\mu \in L^\mu_{\mathrm{cl}}(X; E, E)$ in the sense of Theorem 1.1.83 and analogously $L^\mu_{\mathrm{cl}}(X; E, F) = L^0_{\mathrm{cl}}(X; E, F)R^\mu$.

We now pass to the proof of (ii) $\Rightarrow$ (i) of Theorem 1.1.78. For convenience we first assume $A \in L^\mu(X)$. By assumption

$$A : H^s(X) \to H^{s-\mu}(X)$$

is then a Fredholm operator for a given $s \in \mathbb{R}$. Let B^ν, $\nu \in \mathbb{R}$, be the operator from the proof of Theorem 1.1.83 in the case $E = X \times \mathbb{C}$. Then $B^{-s} : H^0(X) \to H^s(X)$ and $B^{s-\mu} : H^{s-\mu}(X) \to H^0(X)$ are both elliptic and Fredholm operators. Hence $A_0 := B^{\mu-s}AB^s : H^0(X) \to H^0(X)$ is a Fredholm operator, and A is elliptic exactly when A_0 is elliptic. In other words, it suffices to consider the case $A \in L^0(X)$ and to show that the Fredholm property of

$$A : H^0(X) \to H^0(X) \qquad\qquad (1.1.124)$$

implies the ellipticity of A.

After a localisation argument we have to discuss operators in $L^2(\mathbb{R}^n) = H^0(\mathbb{R}^n)$, $n = \dim X$. Let us form the family of operators

$$(R_\lambda(x_0, \xi_0)u)(x) = \lambda^{\frac{n}{4}} e^{i\lambda x \xi_0} u(\lambda^{\frac{1}{2}}(x - x_0))$$

for $\lambda \in \mathbb{R}_+$, $(x_0, \xi_0) \in T^*\mathbb{R}^n \setminus 0$.

Lemma 1.1.84   (i) $R_\lambda(x_0, \xi_0)$ is a unitary operator in $L^2(\mathbb{R}^n)$.

(ii) For every $u \in L^2(\mathbb{R}^n)$ we have $R_\lambda(x_0, \xi_0)u \to 0$ weakly in $L^2(\mathbb{R}^n)$ when $\lambda \to \infty$.

Proof. Let us simply write R_λ instead of $R_\lambda(x_0, \xi_0)$.

(i) We have

$$\|R_\lambda u\|^2_{L^2(\mathbb{R}^n)} = \int \lambda^{\frac{n}{2}} |u(\lambda^{\frac{1}{2}}(x - x_0))|^2 \, dx$$

$$= \int |u(\tilde{x})|^2 \, d\tilde{x} = \|u\|^2_{L^2(\mathbb{R}^n)},$$

which shows that R_λ is an isometry. Moreover,

$$(R_\lambda^{-1} v)(x) = \lambda^{-\frac{n}{4}} e^{-i\lambda(x_0 + \lambda^{-\frac{1}{2}} x)\xi_0} v(x_0 + \lambda^{-\frac{1}{2}} x).$$

It follows now that

$$(R_\lambda u, v) = \int \lambda^{\frac{n}{4}} e^{i\lambda x \xi_0} u(\lambda^{\frac{1}{2}}(x - x_0)) \overline{v(x)} \, dx$$

$$= \int u(\tilde{x}) \lambda^{-\frac{n}{4}} e^{i\lambda(x_0 + \lambda^{-\frac{1}{2}}\tilde{x})} \overline{v(x_0 + \lambda^{-\frac{1}{2}}\tilde{x})} \, d\tilde{x},$$

where we have substituted $\tilde{x} = \lambda^{\frac{1}{2}}(x - x_0)$. Thus $R_\lambda^{-1} = R_\lambda^*$.

(ii) Let first $u, v \in C_0^\infty(\mathbb{R}^n)$ be arbitrary and fixed. Then

$$|(R_\lambda u, v)| \leq \lambda^{-\frac{n}{2}} \int |u(\tilde{x}) v(x_0 + \lambda^{-\frac{1}{2}}\tilde{x})| \, d\tilde{x}$$

gives us $|(R_\lambda u, v)| \to 0$ as $\lambda \to \infty$. For $v \in C_0^\infty(\mathbb{R}^n)$ and $u \in L^2(\mathbb{R}^n)$ we can choose a $u_\varepsilon \in C_0^\infty(\mathbb{R}^n)$ with $\|u - u_\varepsilon\|_{L^2(\mathbb{R}^n)} < \varepsilon$ which yields then

$$|(R_\lambda u, v)| \leq |(R_\lambda u_\varepsilon, v)| + |(R_\lambda(u - u_\varepsilon), v)|$$

$$\leq \varepsilon \|v\|_{L^2(\mathbb{R}^n)} + |(R_\lambda u_\varepsilon, v)|.$$

For any given $k \in \mathbb{N}$ we can assume $\varepsilon = \varepsilon(k)$ so small that $\varepsilon \|v\|_{L^2(\mathbb{R}^n)} < 2^{-k}$ holds. Moreover, we have from the first step of the proof $|(R_\lambda u_\varepsilon, v)| < 2^{-k}$ for all $\lambda > c(k)$ for a suitable constant $c(k) > 0$. This gives us also in this case $|(R_\lambda u, v)| \to 0$ as $\lambda \to \infty$. For arbitrary $u, v \in L^2(\mathbb{R}^n)$ we can choose a $v_\varepsilon \in C_0^\infty(\mathbb{R}^n)$ with $\|v - v_\varepsilon\|_{L^2(\mathbb{R}^n)} < \varepsilon$. We obtain

$$|(R_\lambda u, v)| \leq |(R_\lambda u, v - v_\varepsilon)| + |(R_\lambda u, v_\varepsilon)|$$

$$\leq \|u\|_{L^2(\mathbb{R}^n)} \|v - v_\varepsilon\|_{L^2(\mathbb{R}^n)} + |(R_\lambda u, v_\varepsilon)|$$

$$\leq \varepsilon \|u\|_{L^2(\mathbb{R}^n)} + |(R_\lambda u, v_\varepsilon)|.$$

For every $k \in \mathbb{N}$ we can choose $\varepsilon = \varepsilon(k)$ so small that $\varepsilon \|u\|_{L^2(\mathbb{R}^n)} < 2^{-k}$. Moreover, we find a constant $c(k)$ such that $|(R_\lambda u, v_\varepsilon)| < 2^{-k}$ for all $\lambda > c(k)$ which implies the assertion in general.

$\square$

Theorem 1.1.85 *Let $A \in L_{cl}^0(\mathbb{R}^n)$ be of compact support (i.e., $A|_{\mathbb{R}^n \setminus K} = 0$ for some compact set $K \subset \mathbb{R}^n$). Then, for every $u \in C_0^\infty(\mathbb{R}^n)$ we have*

$$\|\{R_\lambda^{-1}(x_0, \xi_0) A R_\lambda(x_0, \xi_0) - \sigma_\psi^0(A)(x_0, \xi_0)\} u\|_{L^2(\mathbb{R}^n)} \to 0 \qquad (1.1.125)$$

for $\lambda \to \infty$, for arbitrary fixed $(x_0, \xi_0) \in T^*\mathbb{R}^n \setminus 0$.

Proof. Write $a_0(x,\xi) = \sigma_\psi^0(A)(x,\xi)$ for short and define $a(x,\xi) = \chi(\xi)a_0(x,\xi)$ with some fixed excision function $\chi(\xi)$. Then $a(x,\xi)$ and $a_0(x,\xi)$ are of compact support with respect to $x \in \mathbb{R}^n$. Moreover, we have $A = \mathrm{Op}(a) + C$ with an operator $C \in L_{\mathrm{cl}}^{-1}(\mathbb{R}^n)$ which also has compact support. Then C is a compact operator in $L^2(\mathbb{R}^n)$. From Lemma 1.1.84 it follows that $\|R_\lambda^{-1} C R_\lambda u\|_{L^2(\mathbb{R}^n)} \to 0$ for $\lambda \to \infty$. Thus, C may be ignored and it suffices to assume $A = \mathrm{Op}(a)$. From

$$(F_{x'\to\xi}R_\lambda u)(\xi) = \lambda^{-\frac{n}{4}} e^{-ix_0(\xi-\lambda\xi_0)} \hat{u}(\lambda^{-\frac{1}{2}}(\xi-\lambda\xi_0))$$

we get

$$R_\lambda^{-1} \mathrm{Op}(a) R_\lambda u(x)$$

$$= R_\lambda^{-1} \int e^{ix\xi} a(x,\xi) \lambda^{-\frac{n}{4}} e^{-ix_0(\xi-\lambda\xi_0)} \hat{u}(\lambda^{-\frac{1}{2}}(\xi-\lambda\xi_0)) \, d\xi$$

$$= \lambda^{-\frac{n}{2}} \int e^{i(x_0+\lambda^{-\frac{1}{2}}x)\xi} e^{-i\lambda(x_0+\lambda^{-\frac{1}{2}}x)\xi_0} a(x_0+\lambda^{-\frac{1}{2}}x,\xi)$$

$$e^{-ix_0(\xi-\lambda\xi_0)} \hat{u}(\lambda^{-\frac{1}{2}}(\xi-\lambda\xi_0)) \, d\xi$$

$$= \lambda^{-\frac{n}{2}} \int e^{ix\lambda^{-\frac{1}{2}}(\xi-\lambda\xi_0)} a(x_0+\lambda^{-\frac{1}{2}}x,\xi) \hat{u}(\lambda^{-\frac{1}{2}}(\xi-\lambda\xi_0)) \, d\xi$$

$$= \int e^{ix\xi} a(x_0+\lambda^{-\frac{1}{2}}x, \lambda\xi_0+\lambda^{\frac{1}{2}}\xi) \hat{u}(\xi) \, d\xi$$

$$= \int e^{ix\xi} \chi(\lambda\xi_0+\lambda^{\frac{1}{2}}\xi) a_0(x_0+\lambda^{-\frac{1}{2}}x, \lambda\xi_0+\lambda^{\frac{1}{2}}\xi) \hat{u}(\xi) \, d\xi.$$

In the latter equation the homogeneity of $a_0 = \sigma_\psi^0(A)$ in ξ of order 0 was used. Let now (x,ξ) be fixed. Then

$$\chi(\lambda\xi_0+\lambda^{\frac{1}{2}}\xi) a_0(x_0+\lambda^{-\frac{1}{2}}x, \xi_0+\lambda^{-\frac{1}{2}}\xi) \hat{u}(\xi) \to a_0(x_0,\xi_0)\hat{u}(\xi)$$

for $\lambda \to \infty$, and we have

$$|\chi(\lambda\xi_0+\lambda^{\frac{1}{2}}\xi) a_0(x_0+\lambda^{-\frac{1}{2}}x, \xi_0+\lambda^{-\frac{1}{2}}\xi) - a_0(x_0,\xi_0)||\hat{u}(\xi)|$$
$$\leq 2 \sup_{x\in\mathbb{R}^n} |a_0(x,\xi)||\hat{u}(\xi)|.$$

Thus, for every fixed x, it follows that

$$|R_\lambda^{-1} \mathrm{Op}(a) R_\lambda u(x) - a_0(x_0,\xi_0)u(x)|$$

$$= |R_\lambda^{-1} \mathrm{Op}(a) R_\lambda u(x) - a_0(x_0,\xi_0) \int e^{ix\xi} \hat{u}(\xi) \, d\xi|$$

$$= \left| \int e^{ix\xi} \{\chi(\lambda\xi_0+\lambda^{\frac{1}{2}}\xi) a_0(x_0+\lambda^{-\frac{1}{2}}x, \xi_0+\lambda^{-\frac{1}{2}}\xi) - a_0(x_0,\xi_0)\} \hat{u}(\xi) \, d\xi \right|$$

$$\leq \int |\chi(\lambda\xi_0+\lambda^{\frac{1}{2}}\xi) a_0(x_0+\lambda^{-\frac{1}{2}}x, \xi_0+\lambda^{-\frac{1}{2}}\xi) - a_0(x_0,\xi_0)||\hat{u}(\xi)| \, d\xi$$

$$\to 0$$

for $\lambda \to \infty$. In order to prove (1.1.125) we apply integration by parts in $x^\alpha R_\lambda^{-1} \operatorname{Op}(a) R_\lambda u(x)$, $\alpha \in \mathbb{N}^n$, and obtain

$$x^\alpha R_\lambda^{-1} \operatorname{Op}(a) R_\lambda u(x) = \int e^{ix\xi} D_\xi^\alpha \{b_\lambda(x,\xi) \hat{u}(\xi)\} \, d\xi \qquad (1.1.126)$$

for $b_\lambda(x,\xi) := a(x_0 + \lambda^{-\frac{1}{2}} x, \lambda\xi_0 + \lambda^{\frac{1}{2}}\xi)$. Now $a(x,\xi) \in S^0$ implies $|D_\xi^\beta a(x,\xi)| \le c \langle \xi \rangle^{-|\beta|}$ for arbitrary $\beta \in \mathbb{N}^n$ and all x, ξ (recall that the support of $a(x,\xi)$ with respect to x is compact). Thus,

$$|D_\xi^\beta b_\lambda(x,\xi)| \le c\lambda^{\frac{|\beta|}{2}} \langle \lambda\xi_0 + \lambda^{\frac{1}{2}}\xi \rangle^{-|\beta|}$$

with $c = \sup_{x,\xi} \langle \xi \rangle^{|\beta|} |D_\xi^\beta a(x,\xi)|$. Using the inequality $(1 + |a - b|^2)^{-1} \le c(1 + |a|^2)(1 + |b|^2)^{-1}$ for all $a, b \in \mathbb{R}^n$ with a suitable constant $c > 0$ (which is a consequence of 1.1.4 Lemma 1.1.41 and setting $a = \lambda^{\frac{1}{2}}\xi$, $b = -\lambda\xi_0$, we obtain

$$\lambda^{\frac{|\beta|}{2}} (1 + |\lambda\xi_0 + \lambda^{\frac{1}{2}}\xi|^2)^{-\frac{|\beta|}{2}} \le c\lambda^{\frac{|\beta|}{2}} \left(\frac{1 + \lambda|\xi|^2}{1 + \lambda^2} \right)^{\frac{|\beta|}{2}} \le c(1 + |\xi|^2)^{\frac{|\beta|}{2}}.$$

For every N there is a constant c_N such that $|\hat{u}(\xi)| \le c_N \langle \xi \rangle^{-N}$. Using the Leibniz rule (1.1.126) we get for $N > |\alpha| + n$ the estimate

$$|x^\alpha R_\lambda^{-1} \operatorname{Op}(a) R_\lambda u(x)| \le c \int \langle \xi \rangle^{|\alpha| - N} d\xi < \infty$$

with a certain constant c which is independent of λ. Note that c is determined by finitely many semi–norms of the symbol $a(x,\xi)$ in S^0. Thus, for any $M > 2n$ we get

$$|R_\lambda^{-1} \operatorname{Op}(a) R_\lambda u(x)| \le c(1 + |x|^2)^{-M}$$

with a suitable λ–independent c. This shows that

$$\|\{R_\lambda^{-1}(x_0,\xi_0) \operatorname{Op}(a) R_\lambda(x_0,\xi_0) - a_0(x_0,\xi_0)\} u\|_{L^2(\mathbb{R}^n)} < \text{const}$$

with a λ–independent constant. Using now the pointwise convergence of $R_\lambda^{-1}(x_0,\xi_0) \operatorname{Op}(a) R_\lambda(x_0,\xi_0) u(x)$ to $a_0(x_0,\xi_0) u(x)$ for $\lambda \to \infty$ it follows from Lebesgue's theorem on bounded convergence the asserted convergence in $L^2(\mathbb{R}^n)$. $\qquad \square$

Proof of (ii) $\Rightarrow$ (i) of Theorem 1.1.78 (the classical case). We start with the Fredholm operator (1.1.124). Then, there is a continuous operator $B \in \mathcal{L}(H^0(X))$ such that $K_0 := BA - 1$ is compact. It follows that

$$\|v\|_{H^0(X)} = \|BAv - K_0v\|_{H^0(X)} \le c\{\|Av\|_{H^0(X)} + \|K_0v\|_{H^0(X)}\}$$

for all $v \in H^0(X)$ with a constant $c > 0$. Let $\chi : U \to \mathbb{R}^n$ be a chart on X and let V be another coordinate neighbourhood with $\overline{V} \subset U$. Assume $v \in C^\infty(X)$, $\operatorname{supp} v \subseteq \overline{V}$, and choose functions $\varphi, \psi \in C_0^\infty(U)$ with $\varphi\psi = \varphi$ and $\varphi v = v$. Then, the operator $K_1 := (1 - \psi)A\varphi$ is compact in $H^0(X)$, and we obtain

$$\|v\|_{H^0(X)} \le c\{\|\psi Av\|_{H^0(X)} + \|K_0 v\|_{H^0(X)} + \|K_1 v\|_{H^0(X)}\}. \quad (1.1.127)$$

Let us now fix an $\tilde{x}_0 \in V$, assume $\psi(\tilde{x}_0) = 1$, set $x_0 = \chi(\tilde{x}_0)$, and let $\xi_0 \in \mathbb{R}^n \setminus \{0\}$. Choose an arbitrary $u \in C_0^\infty(\mathbb{R}^n)$ such that $v_\lambda := \chi^* R_\lambda(x_0, \xi_0)u$ has its support in $\overline{V}$ for all $\lambda \ge 1$. Then, as a consequence of Lemma 1.1.84, we have $v_\lambda \to 0$ weakly in $H^0(X)$ for $\lambda \to \infty$. Thus,

$$C_\lambda := \sum_{i=0,1} \|K_i v_\lambda\|_{H^0(X)} \to 0 \qquad \text{for } \lambda \to \infty. \quad (1.1.128)$$

It follows from the estimate (1.1.127) that

$$\|(\chi^*)^{-1} v_\lambda\|_{L^2(\mathbb{R}^n)} \le c\{\|(\chi^*)^{-1}\psi Av_\lambda\|_{L^2(\mathbb{R}^n)} + C_\lambda\}$$

or

$$\|R_\lambda(x_0, \xi_0)u\|_{L^2(\mathbb{R}^n)} \le c\{\|A_0 R_\lambda(x_0, \xi_0)u\|_{L^2(\mathbb{R}^n)} + C_\lambda\}$$

for the push–forward $A_0 = (\chi^*)^{-1}\psi A\chi^*$ of ψA under χ and different constants c, independent of λ. The operator $A_0 \in L_{\mathrm{cl}}^0(\mathbb{R}^n)$ has compact support. Since $R_\lambda(x_0, \xi_0)$ is unitary, we obtain

$$\|u\|_{L^2(\mathbb{R}^n)} \le c\{\|R_\lambda^{-1}(x_0, \xi_0)A_0 R_\lambda(x_0, \xi_0)u\|_{L^2(\mathbb{R}^n)} + C_\lambda\}.$$

Theorem 1.1.85 yields for $\lambda \to \infty$

$$\|u\|_{L^2(\mathbb{R}^n)} \le c\|a_0(x_0, \xi_0)u\|_{L^2(\mathbb{R}^n)} = c|a_0(x_0, \xi_0)|\|u\|_{L^2(\mathbb{R}^n)}$$

for all $u \in C_0^\infty(\mathbb{R}^n)$ with the above properties. Thus $a_0(x_0, \xi_0) \ne 0$, which is the assertion in the scalar case. The consideration so far treated for convenience the case of trivial bundles of fibre dimension 1. The general case is completely analogous and left to the reader.

Now let $A \in L_{\mathrm{cl}}^\mu(X; E, F)$ be arbitrary, and let (1.1.122) be a Fredholm operator. The reduction to the order zero by using Theorem 1.1.83 allows us to restrict the consideration to $A \in L_{\mathrm{cl}}^0(X; E, F)$ and to $s = 0$. Now the arguments from the scalar case to prove the ellipticity can analogously be employed, by inserting locally vector–valued $u(x) \in C_0^\infty(\mathbb{R}^n) \otimes \mathbb{C}^k$, where k is the fibre dimension of E and F. This gives us finally

$$\|u\|_{L^2(\mathbb{R}^n, \mathbb{C}^k)} \le c\|a_0(x_0, \xi_0)u\|_{L^2(\mathbb{R}^n, \mathbb{C}^k)}$$

for all u under consideration. Hence we get the injectivity of $a_0(x_0, \xi_0)$ as a $k \times k$–matrix. Analogous arguments for the adjoint operator then yield the surjectivity of $a_0(x_0, \xi_0)$. $\qquad \square$

1.2 Parameter–dependent pseudo– differential operators

1.2.1 Symbols with parameters

Parameter–dependent pseudo–differential operators in the sense of the discussion here are generated by symbols of the class

$$S^\mu(\Omega \times \mathbb{R}^n_\xi \times \mathbb{R}^l_\lambda), \tag{1.2.1}$$

cf. Definition 1.1.9, for an open set $\Omega \subseteq \mathbb{R}^m$ (for $m = n$ or $m = 2n$), and $\lambda \in \mathbb{R}^l$ being interpreted as a parameter. In other words, (1.2.1) is the space of all $a(x, \xi, \lambda) \in C^\infty(\Omega \times \mathbb{R}^{n+l})$ such that

$$|D^\alpha_x D^\beta_{\xi,\lambda} a(x, \xi, \lambda)| \le c(1 + |\xi, \lambda|)^{\mu-|\beta|} \tag{1.2.2}$$

for all $\alpha \in \mathbb{N}^m$, $\beta \in \mathbb{N}^{n+l}$, $x \in K$, with arbitrary $K \subset\subset \Omega$, $(\xi, \lambda) \in \mathbb{R}^{n+l}$, with constants $c = c(\alpha, \beta, K) > 0$. Recall that we also have the subspace

$$S^\mu_{cl}(\Omega \times \mathbb{R}^n_\xi \times \mathbb{R}^l_\lambda) \tag{1.2.3}$$

of classical symbols in (ξ, λ). Instead of (1.2.1), (1.2.3) we also write

$$S^\mu(\Omega \times \mathbb{R}^n; \Lambda) \quad \text{and} \quad S^\mu_{cl}(\Omega \times \mathbb{R}^n; \Lambda),$$

respectively, where $\Lambda = \mathbb{R}^l$. Denote the subspace of x–independent symbols by

$$S^\mu(\mathbb{R}^n; \Lambda) \quad \text{and} \quad S^\mu_{cl}(\mathbb{R}^n; \Lambda),$$

respectively. Note that $a(x, \xi, \lambda) \in S^\mu(\Omega \times \mathbb{R}^n_\xi \times \mathbb{R}^l)$ implies $a(x, \xi, \lambda_0) \in S^\mu(\Omega \times \mathbb{R}^n_\xi)$ for every fixed $\lambda_0 \in \mathbb{R}^l$, and analogously for classical symbols. Every $a(x, \xi, \lambda) \in S^\mu_{cl}(\Omega \times \mathbb{R}^n \times \mathbb{R}^l)$ has a parameter–dependent homogeneous principal symbol $\sigma^\mu_\psi(a)(x, \xi, \lambda)$ of order μ, defined for $(\xi, \lambda) \neq 0$. Incidentally we also denote this by $\sigma^\mu_{\psi,\lambda}(a)$.

We might easily generalise the parameter set $\Lambda \subseteq \mathbb{R}^l$ to the case where there is a $\Lambda_1 \subset \mathbb{R}^l \setminus \{0\}$ with $\lambda \in \Lambda_1 \Rightarrow \delta\lambda \in \Lambda_1$ for all $\delta \in \mathbb{R}_+$ such that $\Lambda \setminus \Lambda_1$ is a bounded set. This will occasionally be used with the corresponding obvious modifications, though the main application here will be the case $\Lambda = \mathbb{R}^l$.

Remark 1.2.1 *The kernel cut–off constructions of Theorem 1.1.47 with respect to a group of covariables can be applied, in particular, to $\Lambda = \mathbb{R}^l$. Then the kernel cut–off operator*

$$H(\varphi) = F_{\varrho \to \lambda}\{\varphi(\varrho)(F^{-1}_{\lambda \to \varrho} a(x, \xi, \lambda))\}$$

for every $\varphi \in C_0^\infty(\mathbb{R}^l)$ *induces continuous operators*

$$H(\varphi) : S^\mu(\Omega \times \mathbb{R}^n \times \mathbb{R}^l) \to S^\mu(\Omega \times \mathbb{R}^n \times \mathbb{C}^l)_{\mathrm{hol}},$$

$$H(\varphi) : S^\mu_{\mathrm{cl}}(\Omega \times \mathbb{R}^n \times \mathbb{R}^l) \to S^\mu_{\mathrm{cl}}(\Omega \times \mathbb{R}^n \times \mathbb{C}^l)_{\mathrm{hol}}.$$

In particular, if $\varphi \equiv 1$ *in a neighbourhood of* $\varrho = 0$, *then*

$$(H(\varphi)a)(x,\xi,\lambda) - a(x,\xi,\lambda) \in S^{-\infty}(\Omega \times \mathbb{R}^n \times \mathbb{R}^l).$$

1.2.2 Pseudo–differential operators with parameters

Set

$$\mathrm{Op}(a)(\lambda)u(x) = \iint e^{i(x-x')\xi} a(x,x',\xi,\lambda)u(x')\,dx'd\xi$$

for $a(x,x',\xi,\lambda) \in S^\mu(\Omega \times \Omega \times \mathbb{R}^n; \Lambda)$, $\Omega \subseteq \mathbb{R}^n$ open, $u \in C_0^\infty(\Omega)$, $\lambda \in \Lambda$.

Definition 1.2.2 $L^\mu(\Omega; \Lambda)$ *for an open set* $\Omega \subseteq \mathbb{R}^n$, $\Lambda = \mathbb{R}^l$, $\mu \in \mathbb{R}$, *is the space of all* $\mathrm{Op}(a)(\lambda)$ *with* $a(x,x',\xi,\lambda) \in S^\mu(\Omega \times \Omega \times \mathbb{R}^n; \Lambda)$. *The elements of* $L^\mu(\Omega; \Lambda)$ *are called parameter–dependent pseudo–differential operators on* Ω *of order* μ. *The subspace* $L^\mu_{\mathrm{cl}}(\Omega; \Lambda)$ *of classical parameter–dependent pseudo–differential operators is defined by* $a(x,x',\xi,\lambda) \in S^\mu_{\mathrm{cl}}(\Omega \times \Omega \times \mathbb{R}^n; \Lambda)$.

Remark 1.2.3 $A(\lambda) \in L^\mu(\Omega; \Lambda)$ *(*$\in L^\mu_{\mathrm{cl}}(\Omega; \Lambda)$*) implies* $A(\lambda_0) \in L^\mu(\Omega)$ *(*$\in L^\mu_{\mathrm{cl}}(\Omega)$*) for every* $\lambda_0 \in \Lambda$.

The basic elements of the pseudo–differential calculus with parameters are very similar to those without parameters. Therefore, proofs will be omitted here, except when some aspect is specific for the parameter–dependent case.

First observe that

$$L^{-\infty}(\Omega; \Lambda) = \mathcal{S}(\Lambda, L^{-\infty}(\Omega)).$$

Here $\mathcal{S}(\Lambda, L^{-\infty}(\Omega))$ is the Schwartz space on Λ with values in $L^{-\infty}(\Omega)$, with $L^{-\infty}(\Omega)$ being endowed with the usual Fréchet topology. For every $\varphi, \psi \in C_0^\infty(\Omega)$ with $\mathrm{supp}\,\varphi \cap \mathrm{supp}\,\psi = \emptyset$ we have

$$\varphi L^\mu(\Omega; \Lambda)\psi \subseteq L^{-\infty}(\Omega; \Lambda).$$

Let $K \subset \Omega \times \Omega$ be an arbitrary proper subset, i.e., K is relatively closed and

$$K \cap \{(x,x') : x \in K_1, x' \in \Omega\}, \qquad K \cap \{(x,x') : x \in \Omega, x' \in K_2\}$$

are compact sets for arbitrary $K_1, K_2 \subset\subset \Omega$. Moreover, let diag $\Omega \times \Omega \subset$ int K, and denote by $L^\mu(\Omega; \Lambda)_K$ the subspace of all $A_0(\lambda) \in L^\mu(\Omega; \Lambda)$ such that the distributional kernel of $A_0(\lambda)$ is supported by K for every $\lambda \in \Lambda$. Analogously we form $L^\mu_{\mathrm{cl}}(\Omega; \Lambda)_K$.

Proposition 1.2.4 *For every proper* $K \subset \Omega \times \Omega$ *with* diag $\Omega \times \Omega \subset$ int K *we have*

$$L^\mu(\Omega; \Lambda) = L^\mu(\Omega; \Lambda)_K + L^{-\infty}(\Omega; \Lambda),$$
$$L^\mu_{\mathrm{cl}}(\Omega; \Lambda) = L^\mu_{\mathrm{cl}}(\Omega; \Lambda)_K + L^{-\infty}(\Omega; \Lambda).$$

Proposition 1.2.5 *For every proper* $K \subset \Omega \times \Omega$ *with* diag $\Omega \times \Omega \subset$ int K *there are closed subspaces* $S^\mu(\Omega \times \mathbb{R}^n; \Lambda)_K$ *and* $S^\mu_{\mathrm{cl}}(\Omega \times \mathbb{R}^n; \Lambda)_K$ *of* $S^\mu(\Omega \times \mathbb{R}^n; \Lambda)$ *and* $S^\mu_{\mathrm{cl}}(\Omega \times \mathbb{R}^n; \Lambda)$, *respectively, such that* $A(\lambda) \to e_{-\xi}(x)A(\lambda)e_\xi$ *for* $e_\xi(x) = e^{ix\xi}$ *induces isomorphisms*

$$L^\mu(\Omega; \Lambda)_K \to S^\mu(\Omega \times \mathbb{R}^n; \Lambda)_K,$$
$$L^\mu_{\mathrm{cl}}(\Omega; \Lambda)_K \to S^\mu_{\mathrm{cl}}(\Omega \times \mathbb{R}^n; \Lambda)_K.$$

The isomorphisms of Proposition 1.2.5 give rise to Fréchet topologies in $L^\mu(\Omega; \Lambda)_K$ and $L^\mu_{\mathrm{cl}}(\Omega; \Lambda)_K$. Then, Proposition 1.2.4 allows us to introduce in the spaces $L^\mu(\Omega; \Lambda)$, $L^\mu_{\mathrm{cl}}(\Omega; \Lambda)$ the corresponding Fréchet topologies of the non–direct sums. They are independent of the particular choice of K.

Proposition 1.2.6 *There are canonical isomorphisms*

$$L^\mu(\Omega; \Lambda)/L^{-\infty}(\Omega; \Lambda) \cong S^\mu(\Omega \times \mathbb{R}^n; \Lambda)/S^{-\infty}(\Omega \times \mathbb{R}^n; \Lambda),$$
$$L^\mu_{\mathrm{cl}}(\Omega; \Lambda)/L^{-\infty}(\Omega; \Lambda) \cong S^\mu_{\mathrm{cl}}(\Omega \times \mathbb{R}^n; \Lambda)/S^{-\infty}(\Omega \times \mathbb{R}^n; \Lambda).$$

If $A(\lambda) \in L^\mu(\Omega; \Lambda)$ is given, then every $\boldsymbol{a}(x, \xi, \lambda) \in S^\mu(\Omega \times \mathbb{R}^n; \Lambda)$ with $A(\lambda) - \mathrm{Op}(\boldsymbol{a})(\lambda) \in L^{-\infty}(\Omega, \Lambda)$ is called a complete symbol of $A(\lambda)$.

Remark 1.2.7 *Every complete symbol* $\boldsymbol{a}(x, \xi)$ *of* $\mathrm{Op}(a)(\lambda)$ *for*

$$a(x, x', \xi, \lambda) \in S^\mu(\Omega \times \Omega \times \mathbb{R}^n; \Lambda)$$

has the asymptotic expansion

$$\boldsymbol{a}(x, \xi, \lambda) \sim \sum_\alpha \frac{1}{\alpha!} D^\alpha_{x'} \partial^\alpha_\xi a(x, x', \xi, \lambda)|_{x'=x}.$$

Every $A(\lambda) \in L^\mu_{\mathrm{cl}}(\Omega; \Lambda)$ has a parameter–dependent homogeneous principal symbol of order μ

$$\sigma^\mu_{\psi, \lambda}(A) \in S^{(\mu)}(T^*\Omega \times \Lambda \setminus 0). \tag{1.2.4}$$

Here $T^*\Omega \times \Lambda \setminus 0$ consists of all $(x, \xi, \lambda) \in T^*\Omega \times \Lambda$ with $(\xi, \lambda) \neq 0$ and $S^{(\mu)}(T^*\Omega \times \Lambda \setminus 0)$ is the subspace of all $p(x, \xi, \lambda) \in C^\infty(T^*\Omega \times \Lambda \setminus 0)$ with $p(x, \delta\xi, \delta\lambda) = \delta^\mu p(x, \xi, \lambda)$ for all $\delta \in \mathbb{R}_+$, $(x, \xi, \lambda) \in T^*\Omega \times \Lambda \setminus 0$.

Remark 1.2.8 *The sequence*

$$0 \to L_{\mathrm{cl}}^{\mu-1}(\Omega;\Lambda) \xrightarrow{\iota} L_{\mathrm{cl}}^{\mu}(\Omega;\Lambda) \xrightarrow{\sigma} S^{(\mu)}(T^*\Omega \times \Lambda \setminus 0) \to 0$$

is exact, where ι is the canonical embedding and $\sigma = \sigma_{\psi,\lambda}^{\mu}$.

Proposition 1.2.9 *Let $A_j(\lambda) \in L^{\mu_j}(\Omega;\Lambda)$, $j \in \mathbb{N}$ be an arbitrary sequence with $\mu_j \to -\infty$ as $j \to \infty$. Then, there is an $A(\lambda) \in L^{\mu}(\Omega;\Lambda)$ for $\mu = \max\{\mu_j\}$, such that for every $M \in \mathbb{N}$ there is an $N(M) \in \mathbb{N}$ such that for every $N \geq N(M)$ we have*

$$A(\lambda) - \sum_{j=0}^{N} A_j(\lambda) \in L^{\mu-M}(\Omega;\Lambda).$$

$A(\lambda)$ is unique $\bmod L^{-\infty}(\Omega;\Lambda)$.

$A(\lambda)$ is called the asymptotic sum of the $A_j(\lambda)$, written $A(\lambda) \sim \sum_{j=0}^{\infty} A_j(\lambda)$.

Theorem 1.2.10 *The pointwise formal adjoint $A^*(\lambda)$ of an $A(\lambda) \in L^{\mu}(\Omega;\Lambda)$ $(\in L_{\mathrm{cl}}^{\mu}(\Omega;\Lambda))$ belongs to $L^{\mu}(\Omega;\Lambda)$ $(L_{\mathrm{cl}}^{\mu}(\Omega;\Lambda))$.*

Theorem 1.2.11 *Let $A(\lambda) \in L^{\mu}(\Omega;\Lambda)$, $B(\lambda) \in L^{\nu}(\Omega;\Lambda)$ and $A(\lambda)$ or $B(\lambda)$ be properly supported. Then $A(\lambda)B(\lambda) \in L^{\mu+\nu}(\Omega;\Lambda)$. The same is true of classical operators.*

Let $\Omega \subseteq \mathbb{R}_x^n$, $\tilde{\Omega} \subseteq \mathbb{R}_y^n$ be open sets and $\kappa : \Omega \to \tilde{\Omega}$ be a diffeomorphism. Then, for every fixed $\lambda \in \Lambda$ we have the push–forward of pseudo–differential operators

$$(\kappa_* A)(\lambda) = (\kappa^*)^{-1} A(\lambda)\kappa^* : C_0^{\infty}(\tilde{\Omega}) \to C^{\infty}(\Omega),$$

cf. Theorem 1.1.38.

Theorem 1.2.12 *The operator push–forward under $\kappa : \Omega \to \tilde{\Omega}$ induces an isomorphism*

$$\kappa^* : L^{\mu}(\Omega;\Lambda) \to L^{\mu}(\tilde{\Omega};\Lambda)$$

for every $\mu \in \mathbb{R}$. If $A(\lambda) \in L^{\mu}(\Omega;\Lambda)$ is written as $A(\lambda) = \mathrm{Op}(a)(\lambda)$ for an $a(x,\xi,\lambda) \in S^{\mu}(\Omega \times \mathbb{R}^n;\Lambda)$, then, $B(\lambda) = (\kappa_* A)(\lambda)$ has the form $B(\lambda) = \mathrm{Op}(b)(\lambda) \bmod L^{-\infty}(\tilde{\Omega};\Lambda)$ with a symbol $b(y,\eta,\lambda) \in S^{\mu}(\tilde{\Omega} \times \mathbb{R}^n;\Lambda)$ that has the asymptotic expansion

$$b(y,\eta,\lambda)|_{y=\kappa(x)} \sim \sum_{\alpha} \frac{1}{\alpha!}(\partial_{\xi}^{\alpha} a)(x,{}^t d\kappa(x)\eta,\lambda)\Phi_{\alpha}(x,\eta), \qquad (1.2.5)$$

with $\Phi_{\alpha}(x,\eta)$ being of the form (1.1.71).

Note that we also get isomorphisms

$$\kappa_* : L_{cl}^{\mu}(\Omega; \Lambda) \to L_{cl}^{\mu}(\tilde{\Omega}; \Lambda)$$

for all $\mu \in \mathbb{R}$.

Let X be a paracompact C^{∞} manifold, and $\kappa : U \to \Omega$ be a chart on X, with open $\Omega \subseteq \mathbb{R}^n$. Set

$$L^{\mu}(U; \Lambda) = (\kappa_*)^{-1}L^{\mu}(\Omega; \Lambda), \qquad (1.2.6)$$

where $((\kappa_*)^{-1}A)(\lambda) = \kappa^*A(\lambda)(\kappa^*)^{-1}$ is the inverse of the operator push–forward under κ. In view of Theorem 1.2.12 the space $L^{\mu}(U; \Lambda)$ is independent of the specific κ. Let us set

$$L^{-\infty}(X; \Lambda) = \mathcal{S}(\Lambda, L^{-\infty}(X)).$$

Moreover, fix a locally finite open covering $\mathcal{U} = \{U_j\}_{j \in \mathbb{Z}}$ on X, a subordinate partition of unity $\{\varphi_j\}_{j \in \mathbb{Z}}$ and a system $\{\psi_j\}_{j \in \mathbb{Z}}$ of functions $\psi_j \in C_0^{\infty}(U_j)$ such that $\varphi_j\psi_j = \varphi_j$ for all j. Let $\kappa_j : U_j \to \Omega_j$ be a system of charts, with $\Omega_j \subseteq \mathbb{R}^n$ open, $j \in \mathbb{Z}$.

Definition 1.2.13 $L^{\mu}(X; \Lambda)$, $\mu \in \mathbb{R}$, *is the space of all operator families*

$$A(\lambda) = \sum_{j \in \mathbb{Z}} \varphi_j A_j(\lambda)\psi_j + C(\lambda)$$

for arbitrary $A_j(\lambda) \in L^{\mu}(U_j; \Lambda)$, $j \in \mathbb{Z}$, *and* $C(\lambda) \in L^{-\infty}(X; \Lambda)$. *In an analogous manner we define* $L_{cl}^{\mu}(X; \Lambda)$.

This definition can easily be generalised to parameter–dependent spaces of pseudo–differential operators, acting on distributional sections of vector bundles E, F on X. This yields analogously to Section 1.1.5 the classes

$$L^{\mu}(X; E, F; \Lambda) \quad \text{and} \quad L_{cl}^{\mu}(X; E, F; \Lambda), \qquad (1.2.7)$$

respectively. This generalisation is straightforward and will not be commented on here in detail, though it will be used below and then tacitly employed.

Remark 1.2.14 *From* (1.2.6) *we get a Fréchet topology in the space* $L^{\mu}(U; \Lambda)$. *Also* $L^{-\infty}(X; \Lambda)$ *is a Fréchet space in a natural way. Then, using the notation from Section 1.1.1, we can form*

$$L^{\mu}(X; \Lambda) = \sum_{j=1}^{N}[\varphi_j]L^{\mu}(U_j; \Lambda)[\psi_j] + L^{-\infty}(X; \Lambda) \qquad (1.2.8)$$

for the case that X is compact and $\{U_1, \ldots, U_N\}$ a finite open covering. (1.2.8) allows us to define in $L^\mu(X, \Lambda)$ the Fréchet topology of the non-direct sum. If X is paracompact, we can define the mappings

$$L^\mu(X; \Lambda) \xrightarrow{r_j} L^\mu(U_j; \Lambda) \xrightarrow{\kappa_{j*}} L^\mu(\Omega_j; \Lambda) \qquad (1.2.9)$$

for all j, where r_j is the restriction operator to U_j, and endow $L^\mu(X; \Lambda)$ with the topology of the projective limit under all mappings (1.2.9). This yields a Fréchet topology in $L^\mu(X; \Lambda)$ that is independent of the specific choice of the charts and which is equivalent to that of (1.2.8) in the compact case. In an analogous manner we can proceed for $L^\mu_{cl}(X; \Lambda)$ to obtain a natural Fréchet topology in this space.

Remark 1.2.15 *Let $K \subset X \times X$ be a proper set, and let $L^\mu(X; \Lambda)_K$ be the subspace of all $A_0(\lambda) \in L^\mu(X; \Lambda)$ such that the distributional kernel of $A_0(\lambda)$ is supported by K for every $\lambda \in \Lambda$. Then*

$$L^\mu(X; \Lambda) = L^\mu(X; \Lambda)_K + L^{-\infty}(X; \Lambda).$$

An analogous relation holds for $L^\mu_{cl}(X; \Lambda)$.

Proposition 1.2.16 *If $A_j(\lambda) \in L^{\mu_j}(X; \Lambda)$, $j \in \mathbb{N}$, is an arbitrary sequence with $\mu_j \to -\infty$ as $j \to \infty$, then there is an $A(\lambda) \in L^\mu(X; \Lambda)$ for $\mu = \max\{\mu_j\}$, such that for every $M \in \mathbb{N}$ there is an $N(M) \in \mathbb{N}$ with*

$$A(\lambda) - \sum_{j=0}^{N} A_j(\lambda) \in L^{\mu-M}(X; \Lambda),$$

for all $N \geq N(M)$ and $A(\lambda)$ is unique $\mathrm{mod}\, L^{-\infty}(X; \Lambda)$. As usual we write $A(\lambda) \sim \sum_{j=0}^{\infty} A_j(\lambda)$ for the asymptotic sum $A(\lambda)$.

Theorem 1.2.17 *The sequence*

$$0 \to L^{\mu-1}_{cl}(X; \Lambda) \xrightarrow{\iota} L^\mu_{cl}(X; \Lambda) \xrightarrow{\sigma} S^{(\mu)}(T^*X \times \Lambda \setminus 0) \to 0$$

is exact, where ι is the canonical embedding and $\sigma = \sigma^\mu_{\psi,\lambda}$.

Theorem 1.2.18 *Let $A(\lambda) \in L^\mu(X; \Lambda)$, $B(\lambda) \in L^\nu(X; \Lambda)$ and $A(\lambda)$ or $B(\lambda)$ be properly supported. Then $A(\lambda)B(\lambda) \in L^{\mu+\nu}(X; \Lambda)$. For $A(\lambda) \in L^\mu_{cl}(X; \Lambda)$, $B(\lambda) \in L^\nu_{cl}(X; \Lambda)$ we obtain $A(\lambda)B(\lambda) \in L^{\mu+\nu}_{cl}(X; \Lambda)$ and*

$$\sigma^{\mu+\nu}_{\psi,\lambda}(A(\lambda)B(\lambda)) = \sigma^\mu_{\psi,\lambda}(A(\lambda))\sigma^\nu_{\psi,\lambda}(B(\lambda)).$$

Let X be compact, $A(\lambda) \in L^\mu(X; \Lambda)$, and consider the λ–dependent family of continuous operators

$$A(\lambda) : H^s(X) \to H^{s-\mu}(X),$$

$s \in \mathbb{R}$. Composing this with the continuous embedding $H^{s-\mu}(X) \to H^{s-\nu}(X)$, $\nu \geq \mu$, we get continuous operators

$$A(\lambda) : H^s(X) \to H^{s-\nu}(X),$$

$s \in \mathbb{R}$. It is interesting to estimate $\|A(\lambda)\|_{s,s-\nu}$ for growing $|\lambda|$; as usual $\| \cdot \|_{s,r}$ denotes the operator norm in $\mathcal{L}(H^s(X), H^r(X))$.

Theorem 1.2.19 *Let X be compact and $A(\lambda) \in L^\mu(X; \Lambda)$, $\mu \in \mathbb{R}$. Then, for every $\nu \geq \mu$, we have*

$$\|A(\lambda)\|_{s,s-\nu} \leq \begin{cases} c(1 + |\lambda|)^\mu & \text{for } \nu \geq 0, \\ c(1 + |\lambda|)^{\mu-\nu} & \text{for } \nu \leq 0, \end{cases}$$

for a constant $c = c(s, \nu)$.

Proof. Let us set

$$p(\lambda, \mu, \nu) = \begin{cases} (1 + |\lambda|)^\mu & \text{for } \nu \geq 0, \\ (1 + |\lambda|)^{\mu-\nu} & \text{for } \nu \leq 0. \end{cases}$$

We first show an analogous assertion for $A(\lambda) = \mathrm{Op}(a)(\lambda) \in L^\mu(\mathbb{R}^n; \Lambda)$ for $a(\xi, \lambda) \in S^\mu(\mathbb{R}^{n+l}_{\xi,\lambda})$. In this case we have

$$\|A(\lambda)u\|^2_{s-\nu} = \int (1 + |\xi|^2)^{s-\nu} |a(\xi, \lambda)\hat{u}(\xi)|^2 \, d\xi$$

$$\leq \{\sup_{\xi \in \mathbb{R}^n} (1 + |\xi|^2)^{-\nu} |a(\xi, \lambda)|^2\} \int (1 + |\xi|^2)^s |\hat{u}(\xi)|^2 \, d\xi$$

$$\leq c\{\sup_{\xi \in \mathbb{R}^n} (1 + |\xi|^2)^{-\nu} (1 + |\xi|^2 + |\lambda|^2)^\mu\} \|u\|^2_s,$$

where c is a constant dependent on a. In view of

$$\{\sup_{\xi \in \mathbb{R}^n} (1 + |\xi|^2)^{-\nu} (1 + |\xi|^2 + |\lambda|^2)^\mu\}^{\frac{1}{2}} \leq c_1 p(\lambda, \mu, \nu)$$

for $\nu \geq \mu$ with some $c_1 > 0$, we obtain

$$\|A(\lambda)\|_{s,s-\nu} \leq c(a)p(\lambda, \mu, \nu)$$

for a constant $c(a) > 0$. Here $c(a) \to 0$ as $a \to 0$ in the space $S^\mu(\mathbb{R}^{n+l}_{\xi,\lambda})$. Every $A(\lambda) \in L^\mu(X;\Lambda)$ can be written as

$$A(\lambda) = \sum_{j=1}^{N} \varphi_j A_j(\lambda) \psi_j + C(\lambda) \tag{1.2.10}$$

for certain $A_j(\lambda) \in L^\mu(U_j;\Lambda)$, $C(\lambda) \in L^{-\infty}(X;\Lambda)$. Here U_j runs over a finite open covering by coordinate neighbourhoods, $\{\varphi_1, \ldots, \varphi_N\}$ is a subordinate partition of unity, and $\{\psi_1, \ldots, \psi_N\}$ a system of functions $\psi_j \in C_0^\infty(U_j)$ with $\varphi_j \psi_j = \varphi_j$ for all j. Concerning $C(\lambda)$ it is easy to show that

$$C(\lambda) \in \bigcap_{s \in \mathbb{R}} \mathcal{S}(\Lambda; \mathcal{L}(H^s(X), H^r(X))) \tag{1.2.11}$$

for arbitrary $r \in \mathbb{R}$. The details will be omitted here. Then, it suffices to show the asserted estimate for $\varphi_j A_j(\lambda) \psi_j$ for every j. Set $U = U_j$ for fixed j and let $\kappa : U \to \mathbb{R}^n$ be a chart. Then the push–forward of the operator $\varphi_j A_j(\lambda) \psi_j$ under κ is of the form

$$\mathrm{Op}(a)(\lambda) \mathcal{M}_\psi + \tilde{C}(\lambda),$$

$\psi \in C_0^\infty(\mathbb{R}^n)$, where $a(x, \xi, \lambda) \in S^\mu(\mathbb{R}^n_x \times \mathbb{R}^n_\xi; \Lambda)$ has compact support with respect to x and $\tilde{C}(\lambda) \in L^{-\infty}(\mathbb{R}^n; \Lambda)$ vanishes outside a compact subset of $\mathbb{R}^n$. For $\tilde{C}(\lambda)$ we have an analogue of (1.2.11) in $\mathbb{R}^n$ which yields then a corresponding property of $(\kappa_*)^{-1}\tilde{C}(\lambda)$ on X. If we now show the estimate

$$\| \mathrm{Op}(a)(\lambda)\mathcal{M}_\psi \|_{\mathcal{L}(H^s(\mathbb{R}^n), H^{s-\nu}(\mathbb{R}^n))} \leq cp(\lambda, \mu, \nu), \tag{1.2.12}$$

then we obtain the corresponding estimate for $(\kappa_*)^{-1} \mathrm{Op}(a)(\lambda)\mathcal{M}_\psi$ on X. In other words we prove (1.2.12). There is a compact set $K \subset \mathbb{R}^n$ such that $a(x, \xi, \lambda) = 0$ for $x \notin K$. Thus

$$a(x, \xi, \lambda) \in C_0^\infty(K, S^\mu(\mathbb{R}^{n+l}_{\xi,\lambda})) = C_0^\infty(K) \otimes_\pi S^\mu(\mathbb{R}^{n+l}),$$

cf. Remark 1.1.10. In view of Theorem 1.1.6 we have a representation as a convergent sum

$$a(x, \xi, \lambda) = \sum_{k=0}^{\infty} \delta_k \beta_k(x) a_k(\xi, \lambda)$$

with $\delta_k \in \mathbb{C}$, $\sum |\delta_k| < \infty$, $\beta_k \in C_0^\infty(K)$, $a_k(\xi, \lambda) \in S^\mu(\mathbb{R}^{n+l}_{\xi,\lambda})$, $\beta_k \to 0$, $a_k \to 0$ in the corresponding spaces for $k \to \infty$. Then, for $\| \cdot \|_{s,r} =$

$\| \cdot \|_{\mathcal{L}(H^s(\mathbb{R}^n), H^r(\mathbb{R}^n))}$ we have

$$\| \operatorname{Op}(a)(\lambda) \mathcal{M}_\psi \|_{s,s-\nu} = \| \sum_{k=0}^{\infty} \delta_k \mathcal{M}_{\beta_k} \operatorname{Op}(a_k)(\lambda) \mathcal{M}_\psi \|_{s,s-\nu}$$

$$\leq \sum |\delta_k| \| \mathcal{M}_{\beta_k} \|_{s-\nu,s-\nu} \| \operatorname{Op}(a_k) \|_{s,s-\nu} \| \mathcal{M}_\psi \|_{s,s}.$$

From the first part of the proof we have $\| \operatorname{Op}(a_k) \|_{s,s-\nu} \leq c_k p(\lambda, \mu, \nu)$ with constants $c_k \to 0$ as $k \to \infty$. Moreover, using Corollary 1.1.56, we get $\| \mathcal{M}_\psi \|_{s,s} < \infty$ and $\| \mathcal{M}_{\beta_k} \|_{s-\nu,s-\nu} \to 0$ as $k \to \infty$. Thus it follows that

$$\| \operatorname{Op}(a)(\lambda) \mathcal{M}_\psi \|_{s,s-\nu} \leq \tilde{c} \Big\{ \sum_{k=0}^{\infty} |\delta_k| c_k \Big\} p(\lambda, \mu, \nu) \leq c p(\lambda, \mu, \nu)$$

with different constants $\tilde{c}, c > 0$.   □

Corollary 1.2.20 $A(\lambda) \in L^\mu(X; \Lambda)$, $\mu \leq 0$, *implies*

$$\| A(\lambda) \|_{0,0} \leq c(1 + |\lambda|)^\mu$$

for all $\lambda \in \Lambda$ for a constant $c > 0$.

1.2.3   Parameter–dependent ellipticity

Definition 1.2.21 *Let X be a paracompact C^∞ manifold. Then, $A \in L^\mu(X; \Lambda)$ is called parameter–dependent elliptic of order μ if for every chart $\chi_* : U \to \Omega \subseteq \mathbb{R}^n_x$ any complete parameter–dependent symbol $a(x, \xi, \lambda)$ of the operator $\chi_*(A(\lambda)|_U) \in L^\mu(\Omega; \Lambda)$ satisfies the estimates*

$$|a(x, \xi, \lambda)| \geq c(1 + |\xi, \lambda|)^\mu \qquad (1.2.13)$$

for all $x \in K$ and arbitrary $K \subset\subset \Omega$ and all $(\xi, \lambda) \in \mathbb{R}^{n+l}$ for $|\xi, \lambda| \geq R$ for some $R = R(K)$, and certain constants $c = c(K) > 0$.

In general, if E, F are vector bundles on X, an operator $A \in L^\mu(X; E, F; \Lambda)$ is called parameter–dependent elliptic of order μ if for every chart χ the determinant $\det a(x, \xi, \lambda)$ of any complete symbol of $\chi_*(A(\lambda)|_U)$ satisfies the corresponding estimates $|\det a(x, \xi, \lambda)| \geq c(1 + |\xi, \lambda|)^\mu$ for $|\xi, \lambda| > R$. Here $a(x, \xi, \lambda)$ is regarded as a $k \times k$-matrix of elements in $S^\mu(\Omega \times \mathbb{R}^n; \Lambda)$, where k is the fibre dimension of E and F.

To simplify notation we shall consider here the case of trivial bundles $E = X \times \mathbb{C}$, $F = X \times \mathbb{C}$. The general case is completely analogous and left to the reader.

The conditions in Definition 1.2.21 are independent of the specific choice of the complete symbols $a(x, \xi, \lambda) \in S^\mu(\Omega \times \mathbb{R}^n; \Lambda)$ of $\chi_*(A(\lambda)|_U)$. The estimates (1.2.13) imply the existence of a $b(x, \xi, \lambda) \in S^{-\mu}(\Omega \times \mathbb{R}^n; \Lambda)$ satisfying the analogous estimates for the order $-\mu$, such that

$$a(x, \xi, \lambda)b(x, \xi, \lambda) - 1 = 0 \mod S^{-1}(\Omega \times \mathbb{R}^n; \Lambda). \qquad (1.2.14)$$

Remark 1.2.22 *An operator* $A(\lambda) \in L^\mu_{\mathrm{cl}}(X; \Lambda)$ *is parameter–dependent elliptic if and only if* $\sigma^\mu_{\psi, \lambda}(A)$ *does not vanish on* $T^*X \times \Lambda \setminus 0 = \{(x, \xi, \lambda) \in T^*X \times \Lambda : (\xi, \lambda) \neq 0\}$.

Example 1.2.23 *Let us fix*

$$\mathcal{U} = \{U_j\}_{j \in \mathbb{Z}}, \qquad \{\varphi_j\}_{j \in \mathbb{Z}}, \qquad \{\psi_j\}_{j \in \mathbb{Z}}, \qquad \{\kappa_j\}_{j \in \mathbb{Z}}, \qquad (1.2.15)$$

where $\mathcal{U}$ *is a locally finite open covering of* X, $\{\varphi_j\}$ *a subordinate partition of unity,* $\psi_j \in C^\infty_0(U_j)$, $\varphi_j \psi_j = \varphi_j$, *and* $\kappa_j : U_j \to \mathbb{R}^n$ *are charts,* $j \in \mathbb{Z}$. *Set* $\Lambda = \mathbb{R}^l \ni \lambda$, $\Delta = \mathbb{R}^m \ni \delta$, *and define*

$$R^\mu_j(\lambda, \delta) = (\kappa_j^{-1})_* \operatorname{Op}((1 + |\xi|^2 + |\lambda|^2 + |\delta|^2)^{\frac{\mu}{2}}).$$

Then,

$$R^\mu(\lambda, \delta) := \sum_{j \in \mathbb{Z}} \varphi_j R^\mu_j(\lambda, \delta)\psi_j \in L^\mu_{\mathrm{cl}}(X; \Lambda \times \Delta)$$

is parameter–dependent elliptic of order μ, *with the parameters* (λ, δ). *For every fixed* $\delta_0 \in \Delta$ *we get*

$$R^\mu(\lambda) := R^\mu(\lambda, \delta_0) \in L^\mu_{\mathrm{cl}}(X; \Lambda),$$

which is parameter–dependent elliptic of order μ, *with the parameters* λ, *and we have*

$$\sigma^\mu_{\psi, \lambda}(R^\mu(\lambda)) = (|\xi|^2 + |\lambda|^2)^{\frac{\mu}{2}}.$$

Definition 1.2.24 *Let* $A(\lambda) \in L^\mu(X; \Lambda)$ *be given, then an operator* $B(\lambda) \in L^{-\mu}(X; \Lambda)$ *is called a parameter–dependent parametrix of* $A(\lambda)$ *if*

$$B(\lambda)A(\lambda) - 1, \qquad A(\lambda)B(\lambda) - 1 \in L^{-\infty}(X; \Lambda).$$

Theorem 1.2.25 *A parameter–dependent elliptic operator* $A(\lambda) \in L^\mu(X; \Lambda)$ *has a parameter–dependent parametrix* $B(\lambda) \in L^{-\mu}(X; \Lambda)$. *Moreover,* $A(\lambda) \in L^\mu_{\mathrm{cl}}(X; \Lambda)$ *implies* $B(\lambda) \in L^{-\mu}_{\mathrm{cl}}(X; \Lambda)$, *and*

$$\sigma^{-\mu}_{\psi, \lambda}(B) = \sigma^\mu_{\psi, \lambda}(A)^{-1}. \qquad (1.2.16)$$

Proof. Let $\kappa_j : U_j \to \mathbb{R}^n$ be a chart on X, and choose a $b_j(x, \xi, \lambda) \in S^{-\mu}(\mathbb{R}^n \times \mathbb{R}^n; \Lambda)$ with $a_j b_j - 1 \in S^{-1}(\mathbb{R}^n \times \mathbb{R}^n; \Lambda)$, cf. (1.2.14), where $a_j \in S^{\mu}(\mathbb{R}^n \times \mathbb{R}^n; \Lambda)$ is a complete symbol of $(\kappa_j)_* A(\lambda)|_{U_j}$. Denote by $B_0(\lambda)$ a properly supported representative of

$$\sum_{j \in \mathbb{Z}} \varphi_j \{(\kappa_j^{-1})_* \operatorname{Op}(b_j)(\lambda)\} \psi_j,$$

$B_0(\lambda) \in L^{\mu}(X; \Lambda)$, cf. Remark 1.2.15. Then, choose a properly supported representative $C_0(\lambda) \in L^{-1}(X; \Lambda)$ of $B_0(\lambda)A(\lambda) - 1$, and form the asymptotic sum $D_0(\lambda) \sim \sum_{j=1}^{\infty} (-1)^j C_0^j(\lambda)$, $D_0(\lambda) \in L^{-1}(X; \Lambda)$. Then $(1 + D_0(\lambda))B_0(\lambda) =: B(\lambda) \in L^{-\mu}(X; \Lambda)$ satisfies $B(\lambda)A(\lambda) - 1 \in L^{-\infty}(X; \Lambda)$. In an analogous manner we can proceed for $A(\lambda)B_0(\lambda) - 1$ and obtain in this way a $\tilde{B}(\lambda) \in L^{-\mu}(X; \Lambda)$ with $A(\lambda)\tilde{B}(\lambda) - 1 \in L^{-\infty}(X; \Lambda)$. It follows then easily that $B(\lambda) - \tilde{B}(\lambda) \in L^{-\infty}(X; \Lambda)$ and hence $B(\lambda)$ is a parameter–dependent parametrix of $A(\lambda)$. The relation (1.2.16) is an easy consequence of Remark 1.1.14 and Remark 1.2.8. □

Let X be a compact C^{∞} manifold, and let $A(\lambda) \in L^{\mu}(X; \Lambda)$ be parameter–dependent elliptic of order μ. Then, since $A(\lambda_0)$ is elliptic in the sense of Definition 1.1.67 for every fixed $\lambda_0 \in \Lambda$, we get a family

$$A(\lambda) : H^s(X) \to H^{s-\mu}(X) \tag{1.2.17}$$

of Fredholm operators, for every $s \in \mathbb{R}$, cf. Theorem 1.1.78.

Theorem 1.2.26 *Let X be compact, and let $A(\lambda) \in L^{\mu}(X; \Lambda)$ be parameter–dependent elliptic of order μ. Then, (1.2.17) is a Fredholm operator with index zero for every $s \in \mathbb{R}$, $\lambda \in \Lambda$, and there is a constant $c > 0$ such that (1.2.17) is an isomorphism for every $\lambda \in \Lambda$ with $|\lambda| \geq c$ and every $s \in \mathbb{R}$.*

Proof. According to Theorem 1.2.25 there is a parameter–dependent parametrix $B(\lambda) \in L^{-\mu}(X; \Lambda)$ of $A(\lambda)$, in other words

$$B(\lambda)A(\lambda) = 1 + C(\lambda)$$

for some $C(\lambda) \in \mathcal{S}(\Lambda, L^{-\infty}(X))$. It is now evident that there is an element $D(\lambda) \in \mathcal{S}(\Lambda, L^{-\infty}(X))$ such that

$$(1 + D(\lambda))(1 + C(\lambda)) - 1 \in C_0^{\infty}(\Lambda, L^{-\infty}(X)).$$

Thus we get for some $c > 0$ that $(1 + C(\lambda))^{-1} = 1 + D(\lambda)$ for all $|\lambda| \geq c$. Hence $(1 + D(\lambda))B(\lambda) \in L^{-\mu}(X; \Lambda)$ equals $A^{-1}(\lambda)$ for all $|\lambda| \geq c$. □

Remark 1.2.27 *Theorem 1.2.17 provides another method for constructing reductions of orders in $L^\mu_{\mathrm{cl}}(X)$ than that described above by Theorem 1.1.83. It suffices to choose a parameter–dependent elliptic operator $R^\mu(\lambda)$ of order μ and to fix $\lambda = \lambda_1$ for $|\lambda_1|$ sufficiently large. Then*

$$R^\mu(\lambda) : H^s(X) \to H^{s-\mu}(X) \qquad (1.2.18)$$

will be an isomorphism for $\lambda = \lambda_1$ for all $s \in \mathbb{R}$. Choosing $R^\mu(\lambda)$ as above with another parameter δ then it suffices to insert $|\delta|$ sufficiently large to obtain isomorphisms (1.2.18) for all $\lambda \in \Lambda$ and all $s \in \mathbb{R}$.

The following result can be proved by a modification of the technique used for the proof of Theorem 1.1.78, (ii) $\Rightarrow$ (i).

Theorem 1.2.28 *Let a family $A(\lambda) \in L^\mu_{\mathrm{cl}}(X; \Lambda)$ be given such that (1.2.17) is a Fredholm operator for one $s \in \mathbb{R}$ and all $\lambda \in \Lambda$. Assume that there are bounded families*

$$B_l(\lambda) : H^s(X) \to H^{s+\mu}(X),$$
$$B_r(\lambda) : H^{s-\mu}(X) \to H^s(X)$$

such that

$$B_l(\lambda)A(\lambda) - 1 \in C\left(\Lambda^+, \mathcal{K}(H^s(X))\right),$$
$$A(\lambda)B_r(\lambda) - 1 \in C\left(\Lambda^+, \mathcal{K}(H^{s-\mu}(X))\right)$$

holds, where $\mathcal{K}(\cdot)$ is the space of compact operators in the corresponding space and Λ^+ the one-point compactification of Λ. Then $A(\lambda)$ is parameter-dependent elliptic.

Theorem 1.2.29 *Let $A(\lambda) \in L^\mu_{\mathrm{cl}}(X; \Lambda)$ be parameter-dependent elliptic of order μ, and assume that (1.2.17) is an isomorphism for all $\lambda \in \Lambda$ and any fixed $s = s_0 \in \mathbb{R}$. Then (1.2.17) is an isomorphism for all $\lambda \in \Lambda$ and $s \in \mathbb{R}$, and we have $A^{-1}(\lambda) \in L^{-\mu}_{\mathrm{cl}}(X, \Lambda)$.*

The proof is an obvious analogue of that used for the above Theorem 1.1.82.

1.2.4 Meromorphic families of Fredholm operators

The parameter–dependent pseudo–differential operators will play the role of operator–valued symbols, where the parameters are the covariables, cf. Chapter 2 below. It will also be of interest to consider

extensions into the complexification of the parameter space as holomorphic operator families. This requires some auxiliary material on
operator families in general.

Let H_1, H_2 be (infinite-dimensional) Hilbert spaces, X a compact
topological space, and $a \in C(X, \mathcal{L}(H_1, H_2))$.

Proposition 1.2.30 *Let $a \in C(X, \mathcal{L}(H_1, H_2))$ be a family of Fredholm
operators $a(x) : H_1 \to H_2$. Then there exists an $N_- \in \mathbb{N}$ and a linear
operator $k : \mathbb{C}^{N_-} \to H_2$, such that*

$$(a(x), k) : \begin{matrix} H_1 \\ \oplus \\ \mathbb{C}^{N_-} \end{matrix} \to H_2 \qquad (1.2.19)$$

is surjective for every $x \in X$.

Proof. Without loss of generality we can set $H = H_1 = H_2$. To
every fixed $\tilde{x} \in X$ there exists an $N(\tilde{x}) \in \mathbb{N}$ and a linear operator
$k(\tilde{x}) : \mathbb{C}^{N(\tilde{x})} \to H$ such that $(a(x), k(\tilde{x})) : H \oplus \mathbb{C}^{N(\tilde{x})} \to H$ is surjective for $x = \tilde{x}$. It suffices to set $N(\tilde{x}) = \dim \operatorname{coker} a(\tilde{x})$ and to choose
an arbitrary map $k(\tilde{x})$ from $\mathbb{C}^{N(\tilde{x})}$ to a complement of $\operatorname{im} a(\tilde{x})$. Since
the surjective operators form an open set in the corresponding space
of linear continuous operators, there exists an open neighbourhood
$U(\tilde{x})$ of $\tilde{x}$ such that $(a(x), k(\tilde{x})) : H \oplus \mathbb{C}^{N(\tilde{x})} \to H$ is surjective for all
$a \in U(\tilde{x})$. Taking $\tilde{x}$ arbitrary we obtain in this way an open covering of
X by the neighbourhoods $U(\tilde{x})$, $\tilde{x} \in X$. Since X is compact, there exists a finite subcovering $\{U(x_1), \ldots, U(x_M)\}$ with corresponding points
$x_1, \ldots, x_M$. Setting $N_- = \sum_{j=1}^M N(x_j)$ and defining $k : \mathbb{C}^{N_-} \to H$ as
$k(c) = \sum_{j=1}^M k(x_j) c_j$ for $c = (c_1, \ldots, c_M)$, $c_j \in \mathbb{C}^{N(x_j)}$, we immediately
obtain the assertion. $\qquad \square$

The operator family (1.2.19) is Fredholm between $H_1 \oplus \mathbb{C}^{N_-}$ and H_2,
since k is finite-dimensional. For convenience we will assume in the
following discussion $H = H_1 = H_2$ and that X is arc-wise connected
(i.e., to every two points $x_0, x_1 \in X$ there exists a continuous map
$f : [0,1] \to X$ with $f(0) = x_0$, $f(1) = x_1$). The index

$$\operatorname{ind}(a(x), k) = \dim \ker(a(x), k) =: N_+$$

is then independent of x. The family of vector spaces

$$\tilde{J}_x = \{(u, x) \in H \oplus \mathbb{C}^{N_-} : (u, v) \in \ker(a(x), k)\}$$

is an N_+-dimensional vector subbundle of the trivial Hilbert bundle
$X \times (H \oplus \mathbb{C}^{N_-})$. We want to construct a system of trivialisations. Denote by $\pi(x)$ the orthogonal projection of $H \oplus \mathbb{C}^{N_-}$ to $\tilde{J}_x$. Moreover, let

$e : H \to H \oplus \mathbb{C}^{N-}$, $e' : \mathbb{C}^{N-} \to H \oplus \mathbb{C}^{N-}$ be the canonical embeddings, such that $H \oplus \mathbb{C}^{N-} = eH \oplus e'\mathbb{C}^{N-}$. Choose an arbitrary isomorphism $j(x) : \tilde{J}_x \to \mathbb{C}^{N+}$ and set

$$b(x) = j(x)\pi(x)e, \qquad d(x) = j(x)\pi(x)e'. \qquad (1.2.20)$$

Then

$$\begin{pmatrix} a(x) & k \\ b(x) & d(x) \end{pmatrix} : \begin{matrix} H \\ \oplus \\ \mathbb{C}^{N-} \end{matrix} \to \begin{matrix} H \\ \oplus \\ \mathbb{C}^{N+} \end{matrix}$$

is an isomorphism for every fixed $x \in X$. Since the isomorphisms form an open set in the norm topology,

$$\begin{pmatrix} a(x) & k \\ b(\tilde{x}) & d(\tilde{x}) \end{pmatrix} : \begin{matrix} H \\ \oplus \\ \mathbb{C}^{N-} \end{matrix} \to \begin{matrix} H \\ \oplus \\ \mathbb{C}^{N+} \end{matrix} \qquad (1.2.21)$$

is an isomorphism for all x in a certain neighbourhood $U(\tilde{x})$ of $\tilde{x}$. Now (1.2.21) is an isomorphism if both

$$(a(x), k) : \begin{matrix} H \\ \oplus \\ \mathbb{C}^{N-} \end{matrix} \to H \text{ and } (b(\tilde{x}), d(\tilde{x})) : \ker(a(x), k) \to \mathbb{C}^{N+}$$

are isomorphisms. In other words $(b(\tilde{x}), d(\tilde{x})) : \tilde{J}_x \to \mathbb{C}^{N+}$ is an isomorphism for all $x \in U(\tilde{x})$. This gives us a trivialisation $\tilde{J}|_{U(\tilde{x})} \to U(\tilde{x}) \times \mathbb{C}^{N+}$. Because $\tilde{x} \in X$ is arbitrary we get such trivialisations in a neighbourhood of every point.

Let $\mathrm{Vect}(X)$ denote the set of all finite–dimensional complex C–vector bundles over X. Recall that when X is a C^∞ manifold every isomorphy class can be represented by a C^∞ vector bundle. Denote by $K(X)$ the K–group of X, i.e., the set of all equivalence classes of pairs $(G, J) \in \mathrm{Vect}(X) \times \mathrm{Vect}(X)$, where (G, J) is equivalent to $(\tilde{G}, \tilde{J})$ if there are elements F, $\tilde{F} \in \mathrm{Vect}(X)$ such that $G \oplus F \cong \tilde{G} \oplus \tilde{F}$, $J \oplus F \cong \tilde{J} \oplus \tilde{F}$. Let $[G] - [J]$ be the equivalence class represented by (G, J).

To every family of Fredholm operators $a(x) : H_1 \to H_2$ on X we obtain by the above construction a pair of bundles $(\tilde{J}, X \times \mathbb{C}^{N-})$ on X. Then

$$\mathrm{ind}_X a := [\tilde{J}] - [X \times \mathbb{C}^{N-}] \in K(X)$$

is called the index element of a. It can easily be proved that it is independent of the particular choice of the involved data N_-, k. Operator block matrices

$$A = \begin{pmatrix} a & c \\ b & d \end{pmatrix} : \begin{matrix} H_1 \\ \oplus \\ L_1 \end{matrix} \to \begin{matrix} H_2 \\ \oplus \\ L_2 \end{matrix} \qquad (1.2.22)$$

will often occur in our exposition. For purposes below we want to formulate some simple observations. For convenience we consider Hilbert
spaces H_1, H_2, L_1, L_2, though obvious extensions of the assertions will
occasionally be used also for more general spaces. The identity operators in the various spaces are simply denoted by 1. Operators of the
form

$$K = \begin{pmatrix} 1 & k \\ 0 & 1 \end{pmatrix} : \begin{matrix} H \\ \oplus \\ L \end{matrix} \to \begin{matrix} H \\ \oplus \\ L \end{matrix}, \qquad T = \begin{pmatrix} 1 & 0 \\ t & 1 \end{pmatrix} : \begin{matrix} H \\ \oplus \\ L \end{matrix} \to \begin{matrix} H \\ \oplus \\ L \end{matrix}$$

are isomorphisms for arbitrary linear mappings $k : L \to H$, $t : H \to L$,
and the inverses are

$$K^{-1} = \begin{pmatrix} 1 & -k \\ 0 & 1 \end{pmatrix}, \qquad T^{-1} = \begin{pmatrix} 1 & 0 \\ -t & 1 \end{pmatrix}.$$

Note that

$$1 - kt : H \to H$$

is an isomorphism if and only if

$$1 - tk : L \to L$$

is an isomorphism. In fact, the mapping

$$C := \begin{pmatrix} 1 & k \\ t & 1 \end{pmatrix} : \begin{matrix} H \\ \oplus \\ L \end{matrix} \to \begin{matrix} H \\ \oplus \\ L \end{matrix}$$

is an isomorphism iff

$$C K^{-1} = \begin{pmatrix} 1 & 0 \\ t & 1 - tk \end{pmatrix} \quad \text{or} \quad C T^{-1} = \begin{pmatrix} 1 - kt & k \\ 0 & 1 \end{pmatrix}$$

are isomorphisms.

Consider now the operator A and assume that $a : H_1 \to H_2$ is an
isomorphism. Then we can form

$$\begin{pmatrix} 1 & 0 \\ -ba^{-1} & 1 \end{pmatrix} A \begin{pmatrix} 1 & -a^{-1}c \\ 0 & 1 \end{pmatrix} = \begin{pmatrix} a & 0 \\ 0 & r \end{pmatrix} : \begin{matrix} H_1 \\ \oplus \\ L_1 \end{matrix} \to \begin{matrix} H_2 \\ \oplus \\ L_2 \end{matrix} \qquad (1.2.23)$$

with

$$r = d - ba^{-1}c : L_1 \to L_2. \qquad (1.2.24)$$

The following result is purely algebraic.

Proposition 1.2.31 *Let* $a : H_1 \to H_2$ *be an isomorphism. Then* A *is an isomorphism if and only if* $r : L_1 \to L_2$ *is an isomorphism, and then*

$$A^{-1} = \begin{pmatrix} a^{-1} + a^{-1}cr^{-1}ba^{-1} & -a^{-1}cr^{-1} \\ -r^{-1}ba^{-1} & r^{-1} \end{pmatrix}. \qquad (1.2.25)$$

Proof. In view of (1.2.23) the operator A is an isomorphism if and only if $\operatorname{diag}(a, r)$ is an isomorphism which means that r is an isomorphism. The expression for A^{-1} is then an obvious consequence of the relation (1.2.23). □

Proposition 1.2.32 *Let* $a : H_1 \to H_2$ *be a Fredholm operator. Then* A *is a Fredholm operator if and only if* $r : L_1 \to L_2$ *is a Fredholm operator, and then*

$$\operatorname{ind} A = \operatorname{ind} a + \operatorname{ind} r.$$

The formula (1.2.25) *gives a parametrix of* A *if we insert for* a^{-1} *and* r^{-1} *parametrices of* a *and* r, *respectively.*

Proof. It is obvious that (1.2.23) holds modulo compact operators if we use a^{-1} in the sense of a parametrix. This implies the first assertion, including the index expression. The formula for a parametrix can be checked by calculating the compositions AA^{-1} and $A^{-1}A$, using the fact that aa^{-1}, $a^{-1}a$, rr^{-1}, $r^{-1}r$ are the identities modulo compact operators and that the compact operators form an ideal. □

Theorem 1.2.33 *Let* $a(z) \in \mathcal{A}(U, \mathcal{L}(H_1, H_2))$, $U \subseteq \mathbb{C}$ *open, be an operator function between Hilbert spaces* H_1, H_2. *Let* $a(z) : H_1 \to H_2$ *be a Fredholm operator for each* $z \in U$ *and let* $a(z)$ *be an isomorphism for at least one* $z = \tilde{z}$ *in every connected component of* U. *Then there is a countable set* $D = \bigcup_{j \in \mathbb{Z}} \{d_j\} \subset U$ *that intersects every compact subset* $K \subset U$ *in a finite set such that* $a(z)$ *is an isomorphism for all* $z \in U \setminus D$. *The inverse* $a^{-1}(z) \in \mathcal{A}(U \setminus D, \mathcal{L}(H_2, H_1))$ *extends to a meromorphic operator function on* U *with poles at* $d_j \in D$ *of certain multiplicities* m_j, $j \in \mathbb{Z}$, *and the Laurent coefficients at* $(z - d_j)^{-k}$, $1 \leq k \leq m_j$, *are finite–dimensional operators.*

Proof. Without loss of generality it suffices to assume $H = H_1 = H_2$ and that U is a connected open set with compact closure. The operator family $a(z)$ can be regarded as a homomorphism of Hilbert bundles $a : X \times H \to X \times H$ for $X = \overline{U}$ (if necessary U can be shrunk a little such that $a(z)$ is defined on the closure of U). From the assumptions it follows that $\operatorname{ind} a(z) = 0$ for all $z \in X$. We shall say that a function $h(z)$ is holomorphic on a compact subset K of U if it extends to a holomorphic function in an open neighbourhood of K. Applying the constructions in the proof of Proposition 1.2.30 we find for every compact subset K of U a finite–dimensional subspace $V \subset H$ (namely $\operatorname{im} k$) such that $(1-p)a(z) : K \times V^\perp$ is surjective for all $z \in K$. Here $p : H \to V$ is the orthogonal projection and $V^\perp = \ker p$. Let $W_z = \ker(1-p)a(z)$ and $q(z) : H \to W_z$ be the orthogonal projection, $W_z^\perp = \ker q(z)$. Then the families of subspaces W_z and $W_z^\perp$ constitute holomorphic vector bundles W and $W^\perp$, respectively, over K, with $X \times H = W^\perp \oplus W$. The operator function $a(z)$ can be identified with a morphism

$$
A = \begin{pmatrix} E & C \\ B & G \end{pmatrix} : \begin{array}{c} W^\perp \\ \oplus \\ W \end{array} \to \begin{array}{c} K \times V^\perp \\ \oplus \\ K \times V \end{array}
$$

The fibre dimension of W coincides with $M = \dim V$. By construction $E : W^\perp \to K \times V^\perp$ is an isomorphism and E^{-1} is holomorphic. Now

$$
T = \begin{pmatrix} 1 & 0 \\ -BE^{-1} & 1 \end{pmatrix} : \begin{array}{c} K \times V^\perp \\ \oplus \\ K \times V \end{array} \to \begin{array}{c} K \times V^\perp \\ \oplus \\ K \times V \end{array} ,
$$

$$
K = \begin{pmatrix} 1 & -E^{-1}C \\ 0 & 1 \end{pmatrix} : \begin{array}{c} W^\perp \\ \oplus \\ W \end{array} \to \begin{array}{c} W^\perp \\ \oplus \\ W \end{array}
$$

are holomorphic isomorphisms, and we have

$$
F := TAK = \begin{pmatrix} E & 0 \\ 0 & G - BE^{-1}C \end{pmatrix} .
$$

Thus, A is an isomorphism at precisely those $z \in K$, where the finite–dimensional morphism $G - BE^{-1}C : W \to K \times V$ induces an isomorphism $W_z \to V$. Our consideration is local in nature; so we may take a small disk for K. The bundle W is then trivial over K. Choose a holomorphic trivialisation $\tau : W \to K \times \mathbb{C}^M$ and a holomorphic isomorphism $\sigma : K \times V \to K \times \mathbb{C}^M$. Then $f(z) := \sigma(G - BE^{-1}C)\tau^{-1} : K \times \mathbb{C}^M \to K \times \mathbb{C}^M$ can be regarded as a holomorphic $M \times M$–matrix function. It follows then that $a(z)$ is an isomorphism exactly when $\det f(z) \neq 0$. By assumption there is a $\tilde{z} \in K$ where $\det f(\tilde{z}) \neq 0$. Since $\det f(z)$ is holomorphic, the compact set K contains at most finitely many points, where the determinant vanishes. Thus the first part of Theorem 1.2.33 is proved. For A^{-1} we obtain

$$
A^{-1} = K \begin{pmatrix} 1 & 0 \\ 0 & \tau^{-1} \end{pmatrix} \begin{pmatrix} E^{-1} & 0 \\ 0 & f^{-1} \end{pmatrix} \begin{pmatrix} 1 & 0 \\ 0 & \sigma \end{pmatrix} T.
$$

$f^{-1}(z)$ can be understood as a meromorphic $M \times M$–matrix function on U. Let $d \in D$ be a fixed pole of multiplicity $m + 1$. Then there is an open neighbourhood $\tilde{U}$ of d with $f^{-1}(z) = h(z) + l(z)$, $h(z) \in \mathcal{A}(\tilde{U}) \otimes \mathbb{C}^M \otimes \mathbb{C}^M$,

$$l(z) = \sum_{k=1}^{m} \gamma_k (z - d)^{-k}, \qquad \gamma_k \in \mathbb{C}^M \otimes \mathbb{C}^M.$$

Thus $a^{-1}(z)$ extends to a meromorphic operator function, and

$$K \begin{pmatrix} 1 & 0 \\ 0 & \tau^{-1} \end{pmatrix} \begin{pmatrix} 0 & 0 \\ 0 & \gamma_k \end{pmatrix} \begin{pmatrix} 1 & 0 \\ 0 & \sigma \end{pmatrix} T$$

is the coefficient at $(z - d)^{-k}$ of the Laurent expansion of a^{-1} at d. It is of finite dimension, since γ_k is an $M \times M$–matrix. □

Definition 1.2.34 $\mathcal{F}(U; H_1, H_2)$ *for an open set* $U \subseteq \mathbb{C}$ *and Hilbert spaces* H_1, H_2 *denotes the space of all meromorphic Fredholm families* $a(z) : H_1 \rightarrow H_2$ *on* U *with the following properties:*

(i) *There is a countable subset* $D \subset U$, *dependent on* a, *such that the intersection of* D *with every compact subset of* U *is finite and every* $d \in D$ *has an open neighbourhood* V *such that*

$$a(z) = \sum_{k=1}^{m} c_k (z - d)^{-k} + a_0(z), \qquad (1.2.26)$$

$z \in V$, *with some* $m = m(d) \in \mathbb{N} \setminus \{0\}$, *finite–dimensional operators* $c_k : H_1 \rightarrow H_2$, $1 \leq k \leq m$, *and a holomorphic family of Fredholm operators* $a_0(z) : H_1 \rightarrow H_2$, $z \in V$.

(ii) *There exists a* $z_1 \in U \setminus D$ *such that* $a(z_1) : H_1 \rightarrow H_2$ *is an isomorphism.*

We also write $\mathcal{F}(U; H)$ for $H_1 = H_2 = H$.

Proposition 1.2.35 *Let* $a \in \mathcal{F}(U; H_0, H_2)$, $b \in \mathcal{F}(U; H_1, H_0)$ *for Hilbert spaces* H_0, H_1, H_2. *Then* $ab \in \mathcal{F}(U; H_1, H_2)$.

Proof. The composition of meromorphic operator functions is meromorphic. Without loss of generality we suppose that the sets of poles of $a(z)$ and $b(z)$ coincide, since it was allowed that the above finite-dimensional Laurent coefficients of one factor vanish at a pole d of the other factor. Write in a neighbourhood V of a pole d

$$a(z) = \sum_{k=1}^{m} c_k (z - d)^{-k} + a_0(z), \qquad b(z) = \sum_{l=1}^{n} e_l (z - d)^{-l} + b_0(z)$$

with certain holomorphic Fredholm families $a_0(z)$, $b_0(z)$ in V and finite–dimensional c_k, e_l. Then we have

$$a(z)b(z) = \sum_{k=1}^{m} \sum_{l=1}^{n} c_k e_l (z-d)^{-(k+l)} + f(z) + \tilde{f}(z) + a_0(z)b_0(z)$$

$$(1.2.27)$$

for $f(z) = \sum_{k=1}^{m} c_k b_0(z)(z-d)^{-k}$, $\tilde{f}(z) = \sum_{l=1}^{n} a_0(z) e_l (z-d)^{-l}$. Let us form the Taylor expansion

$$b_0(z) = \sum_{p=0}^{m-1} \beta_p (z-d)^p + b_1(z)$$

with the corresponding $\beta_p \in \mathcal{L}(H_1, H_0)$, $b_1(z) \in \mathcal{A}(V, \mathcal{L}(H_1, H_0))$. Then

$$f(z) = \sum_{k=1}^{m} \sum_{p=0}^{m-1} c_k \beta_p (z-d)^{-k+p} + \sum_{k=1}^{m} c_k b_1(z).$$

The coefficients of all negative powers of $z - d$ are finite–dimensional operators, while the holomorphic part of $f(z)$ is a finite–dimensional operator function. For $\tilde{f}(z)$ we can argue in an analogous manner, using the Taylor expansion of $a_0(z)$ at $z = d$. Thus, since $a_0(z)b_0(z)$ plus a finite–dimensional holomorphic operator function is again Fredholm operator–valued, we see from (1.2.27) that $a(z)b(z)$ is of the asserted structure. □

Lemma 1.2.36 *Let $U \subseteq \mathbb{C}$ be an open neighbourhood of the origin, E a Banach space, and $c_1, \ldots, c_m \in \mathcal{L}(E)$ operators of finite rank. Let $l(z) \in \mathcal{A}(U, \mathcal{L}(E))$ and $M \subseteq E$ be a subspace of finite codimension such that $l(z)e = 0$ for all $e \in M$. Then there is an open neighbourhood V of 0 such that the meromorphic $\mathcal{L}(E)$–valued function $f(z) = 1 + l(z) + \sum_{k=1}^{m} c_k z^{-k}$ is invertible for all $z \in V \setminus \{0\}$. Then $f^{-1}(z) = 1 + l_1(z) + \sum_{k=1}^{\tilde{m}} \tilde{c}_k z^{-k}$ for $z \in V$ with $l_1(z) \in \mathcal{A}(V, \mathcal{L}(E))$ and finite–dimensional $\tilde{c}_k$.*

Proof. Let N be the intersection of the kernels of c_k. Then N is closed in E and has finite codimension. The same is true of $L = M \cap N$. Thus there is a finite–dimensional subspace C of E such that $E = L \oplus C$. Then $f(z)$ can be written as a block matrix

$$f(z) = \begin{pmatrix} 1 & p(z) \\ 0 & r(z) \end{pmatrix} : \begin{matrix} L \\ \oplus \\ C \end{matrix} \to \begin{matrix} L \\ \oplus \\ C \end{matrix}$$

since $e \in L$ implies $f(z)e = e$. Now $f(z)$ is invertible just in the case that $r(z) : C \to C$ is invertible. This is a meromorphic operator function which can be interpreted as an $n \times n$–matrix function for $n = \dim C$. This implies immediately the assertion, since

$$f^{-1}(z) = \begin{pmatrix} 1 & -p(z)r^{-1}(z) \\ 0 & r^{-1}(z) \end{pmatrix}$$

and $r^{-1}(z)$ is again meromorphic, where a possible pole at $z = 0$ implies the existence of a neighbourhood V without further poles. At the same time we see $f^{-1}(z) = 1 + l_1(z) + g_1(z)$ for some $l_1(z) \in \mathcal{A}(V, \mathcal{L}(E))$ and $g_1(z) = \sum_{k=1}^{\tilde{m}} \tilde{c}_k z^{-k}$ with finite–dimensional $\tilde{c}_k$.   □

Theorem 1.2.37 *To every $a(z) \in \mathcal{F}(U; H_1, H_2)$ there exists a $b(z) \in \mathcal{F}(U; H_2, H_1)$ such that $a(z)b(z) = 1$.*

Proof. Without loss of generality we can assume $H = H_1 = H_2$. Let $D \subset U$ be the set of poles of $a(z)$ and set $U_1 = U \setminus D$, $a_1 = a|_{U_1}$. Then a_1 satisfies on U_1 the conditions of Theorem 1.2.33 and hence $a_1^{-1} \in \mathcal{F}(U_1; H)$. It remains to prove that a_1^{-1} has no accumulation points near D. Let us fix an arbitrary $d \in D$. There is then an open neighbourhood U_0 of d with $U_0 \cap D = \{d\}$. For notational convenience we assume $d = 0$ (this can be achieved by considering $a(z-d)$ instead of $a(z)$). Then $a(z) = b(z) + g(z)$ with some holomorphic $b(z) \in \mathcal{F}(U_0; H)$ and $g(z) = \sum_{k=1}^{m} c_k z^{-k}$, with finite–dimensional coefficients c_k. Let us write $h(z) = b(z) - 1$. Then

$$a(z) = 1 + h(z) + g(z) = \{1 + h(z)\}\{1 + (1 + h(z))^{-1}g(z)\}.$$

From Theorem 1.2.33 we get $(1 + h(z))^{-1} \in \mathcal{F}(U_0; H)$. In view of Proposition 1.2.35 it suffices to characterise $1 + (1 + h)^{-1}g$. We have $(1 + h(z))^{-1}g(z) = l(z) + \tilde{g}(z)$ with $l(z) \in \mathcal{A}(U_0, \mathcal{L}(H))$, $\tilde{g}(z) = \sum_{k=1}^{n} \tilde{c}_k z^{-k}$ with finite–dimensional $\tilde{c}_k$. The spaces

$$N = \bigcap_{k=1}^{m} \ker c_k, \qquad \tilde{N} = \bigcap_{k=1}^{n} \ker \tilde{c}_k$$

are of finite codimension in H as well as $M = N \cap \tilde{N}$. We have $l(z)e = 0$ for all $e \in M$. In view of Lemma 1.2.36 there is an open neighbourhood V of 0 such that $1 + (1 + h)^{-1}g = 1 + l + \tilde{g}$ is invertible in $V \setminus \{0\}$. At the same time we see that the inverse extends to $z = 0$ of the desired structure.   □

We now turn to concrete examples of elements in $\mathcal{F}(U; H_1, H_2)$ for $U = \mathbb{C}$, $H_1 = H^s(X)$, $H_2 = H^{s-\mu}(X)$, $s, \mu \in \mathbb{R}$, for a closed compact C^{∞} manifold X. Set

$$\Gamma_\beta = \{z \in \mathbb{C} : \operatorname{Re} z = \beta\}$$

for $\beta \in \mathbb{R}$, and denote by $L^\mu(X; \Gamma_\beta)$ $(L^\mu_{cl}(X; \Gamma_\beta))$ the corresponding spaces of parameter–dependent pseudo–differential operators on X, with the parameter space $\Gamma_\beta \ni z$, identified with $\mathbb{R} \ni \operatorname{Im} z$.

Definition 1.2.38 $M^\mu_O(X)$, $\mu \in \mathbb{R}$, *denotes the subspace of all* $h(z) \in \mathcal{A}(\mathbb{C}, L^\mu_{cl}(X))$ *with the property* $h(z)|_{\Gamma_\beta} \in L^\mu_{cl}(X; \Gamma_\beta)$ *for all* $\beta \in \mathbb{R}$, *uniformly in* $c \leq \beta \leq c'$ *for arbitrary* $c \leq c'$. *In an analogous manner we define* $N^\mu_O(X)$ *in terms of the non–classical pseudo–differential operators* $L^\mu(\dots)$.

Both $M^\mu_O(X)$ and $N^\mu_O(X)$ are Fréchet spaces in natural way. Consider an operator $A \in \operatorname{Diff}^\mu(\mathbb{R}_+ \times X)$, $\mu \in \mathbb{N}$, of the form

$$A = \sum_{j=0}^{\mu} a_j(t) \left(-t\frac{\partial}{\partial t}\right)^j$$

with coefficients $a_j(t) \in C^\infty(\overline{\mathbb{R}}_+, \operatorname{Diff}^{\mu-j}(X))$, $j = 0, \dots, \mu$. Set

$$h(t, z) = \sum_{j=0}^{\mu} a_j(t) z^j.$$

Then $h(t_0, z) \in M^\mu_O(X)$ for every fixed $t_0 \in \overline{\mathbb{R}}_+$, and we have $h(t, z) \in C^\infty(\overline{\mathbb{R}}_+, M^\mu_O(X))$. Let $\sigma^\mu_\psi(A)(t, x, \tau, \xi)$ be the homogeneous principal symbol of A of order μ and assume $\sigma^\mu_\psi(A)(0, x, t^{-1}\tau, \xi) \neq 0$ for all $t \in \overline{\mathbb{R}}_+$, all x and $(\tau, \xi) \neq 0$. Then $h(0, z)|_{\Gamma_\beta}$ is parameter–dependent elliptic on X, with the parameter $\operatorname{Im} z$, uniformly in $c \leq \beta \leq c'$ for every $c \leq c'$. According to Theorem 1.2.26 for every $c \leq c'$ there exists an N such that

$$h(0, z) : H^s(X) \to H^{s-\mu}(X) \tag{1.2.28}$$

is an isomorphism for every z with $c \leq \operatorname{Re} z \leq c'$ and $|\operatorname{Im} z| \geq N$. Since (1.2.28) is a holomorphic family of Fredholm operators, we obtain $h(0, z) \in \mathcal{F}(\mathbb{C}; H^s(X), H^{s-\mu}(X))$. This holds for every $s \in \mathbb{R}$.

Remark 1.2.39 *Applying Theorem 1.2.37 to* $h(0, z)$ *we get* $h^{-1}(0, z) \in \mathcal{F}(\mathbb{C}; H^{s-\mu}(X), H^s(X))$ *for all* $s \in \mathbb{R}$. *Each strip* $\{c \leq \operatorname{Re} z \leq c'\}$ *contains only finitely many poles of* $h^{-1}(0, z)$. *If* p *is such a pole, say of multiplicity* $m + 1$, *then we have*

$$h^{-1}(0, z) = l(z) + \sum_{k=0}^{m} c_k (z - p)^{-(k+1)}$$

for some $l(z) \in M^\mu_O(X) \cap \mathcal{F}(\mathbb{C}; H^{s-\mu}(X), H^s(X))$, $s \in \mathbb{R}$, *and finite-dimensional operators* $c_k \in L^{-\infty}(X)$, $k = 0, \dots, m$.

Theorem 1.2.40 *To each $\mu \in \mathbb{R}$ and every $c, c' \in \mathbb{R}$, $c \leq c'$, there exists an element $b^\mu(z) \in M_O^\mu(X)$ such that*

$$b^\mu(z) : H^s(X) \to H^{s-\mu}(X) \qquad (1.2.29)$$

is a Fredholm operator for each $z \in \mathbb{C}$, $s \in \mathbb{R}$, and that (1.2.29) is an isomorphism for all z with $c \leq \operatorname{Re} z \leq c'$, $s \in \mathbb{R}$.

Theorem 1.2.40 is a consequence of the material from Section 1.2.4 below, in particular, of the so-called kernel cut-off constructions, cf. Section 1.2.5 and Schulze [119]. They also yield the following result.

Theorem 1.2.41 *Let $a(z) \in L_{\mathrm{cl}}^\mu(X; \Gamma_\beta)$ be an arbitrary parameter-dependent elliptic operator family, of order $\mu \in \mathbb{R}$, for any fixed $\beta \in \mathbb{R}$. Then there exists an $h(z) \in M_O^\mu(X)$ such that*

$$a(z) - h(z)|_{\Gamma_\beta} \in L^{-\infty}(X; \Gamma_\beta).$$

An analogous result holds for the non–classical operator families.

Remark 1.2.42 *The elements $h(z)$, obtained from Theorem 1.2.41, belong to $\mathcal{F}(\mathbb{C}; H^s(X), H^{s-\mu}(X))$ for all $s \in \mathbb{R}$.*

Let

$$P = \{(p_j, m_j, L_j)\}_{j \in \mathbb{Z}}$$

be an arbitrary sequence with $p_j \in \mathbb{C}$, $m_j \in \mathbb{N}$, $|\operatorname{Re} p_j| \to \infty$ as $|j| \to \infty$, and L_j a finite–dimensional subspace of finite–dimensional operators in $L^{-\infty}(X)$. Denote by $\mathbf{As}^\bullet(X)$ the set of all these sequences. Set $\pi_\mathbb{C} P = \{p_j\}_{j \in \mathbb{Z}}$. A $\pi_\mathbb{C} P$–excision function is any $\chi(z) \in C^\infty(\mathbb{C})$ with $\chi(z) = 0$ for $\operatorname{dist}(z, \pi_\mathbb{C} P) < \varepsilon_0$, $\chi(z) = 1$ for $\operatorname{dist}(z, \pi_\mathbb{C} P) > \varepsilon_1$ for certain $0 < \varepsilon_0 < \varepsilon_1 < \infty$.

Definition 1.2.43 $M_P^{-\infty}(X)$ *for $P \in \mathbf{As}^\bullet(X)$, denotes the subspace of all $h(z) \in \mathcal{A}(\mathbb{C} \setminus \pi_\mathbb{C} P, L^{-\infty}(X))$ which are meromorphic with poles at p_j of multiplicities $m + 1$ and Laurent expansions*

$$h(z) = \sum_{k=0}^{m_j} c_{jk} (z - p_j)^{-(k+1)} + l(z)$$

in a neighbourhood V_j of p_j with $l(z) \in \mathcal{A}(V_j, L^{-\infty}(X))$, $c_{jk} \in L_j$, $0 \leq k \leq m_j$, $j \in \mathbb{Z}$, and such that for any $\pi_\mathbb{C} P$–excision function $\chi(z)$

$$\chi(z) h(z)|_{\Gamma_\beta} \in L^{-\infty}(X; \Gamma_\beta)$$

for all $\beta \in \mathbb{R}$ and uniformly in $c \leq \beta \leq c'$ for every $c \leq c'$. The space $M_P^{-\infty}(X)$ is Fréchet in a canonical way. We then define

$$M_P^\mu(X) = M_O^\mu(X) + M_P^{-\infty}(X), \qquad N_P^\mu(X) = N_O^\mu(X) + M_P^{-\infty}(X)$$

endowed with the corresponding Fréchet topologies of the non–direct sums.

Proposition 1.2.44 $a \in M_P^\mu(X)$, $b \in M_Q^\nu(X)$ *for* $\mu, \nu \in \mathbb{R}$ *and* $P, Q \in \mathbf{As}^\bullet(X)$ *implies* $ab \in M_R^{\mu+\nu}(X)$ *for a certain* $R \in \mathbf{As}^\bullet(X)$. *An analogous result holds in the non–classical case.*

Proof. The arguments are analogous to those of the proof of Proposition 1.2.35. In addition we obtain here that the indicated subclasses are preserved under compositions, since the pseudo–differential operators as well as the parameter–dependent variants form algebras. In particular, it is easy to see from the proof of Proposition 1.2.35 how the spaces of finite–dimensional operators in the resulting R are generated. □

Theorem 1.2.45 *Let* $a(z) \in M_P^\mu(X)$ *for* $\mu \in \mathbb{R}$, $P \in \mathbf{As}^\bullet(X)$, *and suppose that for some* $\beta \in \mathbb{R}$ *with* $\Gamma_\beta \cap \pi_{\mathbb{C}} P = \emptyset$ *the operator family* $a(z)|_{\Gamma_\beta}$ *is parameter–dependent elliptic (with the parameter* $\mathrm{Im}\, z$*). Then there exists a* $b(z) \in M_Q^{-\mu}(X)$ *for some* $Q \in \mathbf{As}^\bullet(X)$ *such that* $a(z)b(z) = 1$ *for all* $z \in \mathbb{C}$*. An analogous result holds in the non–classical case.*

1.2.5   Kernel cut–off operators

We want to use Theorem 1.1.42 and Remark 1.1.50 to formulate a global analogue of the kernel cut–off constructions.

Definition 1.2.46 *Let* X *be a closed compact* C^∞ *manifold, and* $\mu \in \mathbb{R}$*. Then* $L^\mu(X; \mathbb{C}^l)_{\mathrm{hol}}$ *denotes the space of all operator functions* $a(\zeta) \in \mathcal{A}(\mathbb{C}_\zeta^l, L^\mu(X))$ *with the property*

$$a(\lambda + i\tau) \in L^\mu(X; \mathbb{R}_\lambda^l)$$

for all $\tau \in \mathbb{R}^l$ *and uniformly in* $\tau \in K$ *for every compact set* $K \subset \mathbb{R}^l$. *In analogous manner we define* $L_{\mathrm{cl}}^\mu(X; \mathbb{C}^l)_{\mathrm{hol}}$.

The spaces $L^\mu(X; \mathbb{C}^l)_{\mathrm{hol}}$ and $L_{\mathrm{cl}}^\mu(X; \mathbb{C}^l)_{\mathrm{hol}}$ have canonical Fréchet topologies; they follow immediately from the definition.

Theorem 1.2.47 *Let* $a(\lambda) \in L^\mu(X; \Lambda)$, $\Lambda \cong \mathbb{R}^l \ni \lambda$, *and suppose* $\varphi(y) \in C_0^\infty(\mathbb{R}_y^l)$*. Then the operator* $H(\varphi)$*, defined as*

$$(H(\varphi)a)(\lambda) := F_{y \to \lambda} \varphi(y)(F_{\lambda \to y}^{-1} a)(y)$$

induces a separately continuous operator

$$C_0^\infty(\mathbb{R}^l) \times L^\mu(X; \Lambda) \to L^\mu(X; \mathbb{C}^l)_{\mathrm{hol}},$$

and analogously for classical parameter–dependent operators.

Proof. If $a(\lambda)$ is supported in a coordinate neighbourhood U on X for all $\lambda \in \Lambda$ then the assertion can be reduced to the corresponding result on a symbolic level in local coordinates, where ξ'' from Remark 1.1.48 plays here the role of λ. Since an arbitrary $a(\lambda)$ is a finite sum of such operator families, modulo one in $L^{-\infty}(X;\Lambda)$, it suffices to show the assertion in this case. However this is very simple, because $L^{-\infty}(X;\Lambda) = \mathcal{S}(\Lambda) \otimes_{\pi} L^{-\infty}(X)$. Then it suffices to look at the calculation in the scalar analogue, namely $a(\lambda) \in \mathcal{S}(\Lambda)$. Define the operator $h(\varphi)$ by

$$h(\varphi)(a)(\lambda) := F_{y \to \lambda}\varphi(y)(F^{-1}_{\tilde{\lambda} \to y}a(\tilde{\lambda}))(y) \in \mathcal{A}(\mathbb{C}^l).$$

We then easily obtain $h(\lambda + i\tau) \in \mathcal{S}(\mathbb{R}^l_\lambda)$ uniformly for τ in compact sets $K \subset \mathbb{R}^l$. Then we can pass to the tensor product $h(\varphi) \otimes_\pi 1$ with the identity on $L^{-\infty}(X)$ which gives us the operator $H(\varphi)$ on $L^{-\infty}(X;\Lambda)$. $\square$

1.3   Pseudo–differential operators with operator–valued symbols

The ideas of the standard calculus of pseudo–differential operators from Section 1.1 can be viewed as a starting point of various generalisations. The applications to pseudo–differential operators on manifolds with singularities require a number of non–trivial modifications. These different variants will have to be localised, combined, globalised and iterated in many ways. So it will be convenient to fix notation for the most important generalisations that are to be used here. First there will occur the

Fourier–edge approach   (1.3.1)

based on operator–valued symbols between (say Banach) spaces E and $\tilde{E}$ equipped with strongly continuous groups of isomorphisms $\{\kappa_\lambda\}_{\lambda \in \mathbb{R}_+}$ and $\{\tilde{\kappa}_\lambda\}_{\lambda \in \mathbb{R}_+}$, respectively. This will be explained in detail below in the Sections 1.3.1–1.3.3. The Fourier–edge approach applies to pseudo–differential boundary value problems, cf. Chapter 4 and more generally to edge problems, cf. Chapter 3 . Furthermore we have the

Mellin–edge approach.   (1.3.2)

This refers to an $\mathbb{R}_+$–axis (a cone axis with the variable r) on which the Fourier transform from (1.3.1) is replaced by the Mellin transform on $\mathbb{R}_+$. The essential point is a control of operators and distributions for $r \to 0$. The Mellin–edge approach will be necessary for corner pseudo–differential operators.

A second group of generalisations will start from a global operator algebra on a base X of a cone, which yields the

$$\text{Fourier–order–reduction approach.} \qquad (1.3.3)$$

The definition of the symbol classes here relies on parameter–dependent order reducing operators from the theory on X with the parameters as covariables for the operators in (1.3.3). Moreover, there is the

$$\text{Mellin–order–reduction approach} \qquad (1.3.4)$$

which refers to an $\mathbb{R}_+$–coordinate with the Mellin transform instead of the Fourier transform of (1.3.3). Also here it is an aspect to control the operators and distributions for $t \to 0$ in a precise manner. Elements for (1.3.4) may be found in Schulze [119], [122].

1.3.1 Operator–valued symbols

If E is a Banach space then $\mathcal{L}_\sigma(E)$ will denote the space of all linear continuous operators $a : E \to E$, endowed with the strong operator topology. Recall that $a \to \|ae\|_E$, $e \in E$, is the semi–norm system in $\mathcal{L}(E) \ni a$ for the strong topology.

Let us consider a group

$$\{\kappa_\lambda\}_{\lambda \in \mathbb{R}_+} \in C(\mathbb{R}_+, \mathcal{L}_\sigma(E))$$

of isomorphisms $\kappa_\lambda : E \to E$, $\kappa_\lambda \kappa_\nu = \kappa_{\lambda\nu}$ for all $\lambda, \nu \in \mathbb{R}_+$. Set

$$K(\lambda) = \begin{cases} \lambda & \text{for } \lambda \geq 1, \\ \lambda^{-1} & \text{for } 0 < \lambda \leq 1. \end{cases} \qquad (1.3.5)$$

Proposition 1.3.1 *There are constants $c, M > 0$ with*

$$\|\kappa_\lambda\|_{\mathcal{L}(E)} \leq cK^M(\lambda) \text{ for all } \lambda \in \mathbb{R}_+.$$

Proof. By assumption for every $u \in E$ we have $\|\kappa_\lambda u\|_E \leq b$ for all $\lambda \in [e^{-1}, e]$, for a constant $b = b(u) > 0$. Then, applying the Banach–Steinhaus Theorem, we obtain $\|\kappa_\lambda\|_{\mathcal{L}(E)} \leq c$ for all $\lambda \in [e^{-1}, e]$ for a constant $c > 0$. This yields $\|\kappa_{\lambda^2}\|_{\mathcal{L}(E)} \leq \|\kappa_\lambda\|_{\mathcal{L}(E)}^2 \leq c^2$ for $\lambda \in [e^{-1}, e]$, i.e. $\|\kappa_\lambda\|_{\mathcal{L}(E)} \leq c^2$ for $\lambda \in [e^{-2}, e^2]$. By iterating the argument we get $\|\kappa_\lambda\|_{\mathcal{L}(E)} \leq c^k$ for $\lambda \in [e^{-k}, e^k]$ for every $k \in \mathbb{N}$. Thus

$$\|\kappa_\lambda\|_{\mathcal{L}(E)} \leq c^{1+\log K(\lambda)} \leq cK^M(\lambda)$$

for $M = \log c$, for every $\lambda \in \mathbb{R}_+$. $\qquad\qquad\square$

Example 1.3.2 *Let $E = H^s(\mathbb{R}^m)$, $s \in \mathbb{R}$, and*

$$(\kappa_\lambda u)(x) = \lambda^{\frac{m}{2}} u(\lambda x), \qquad \lambda \in \mathbb{R}_+, \qquad (1.3.6)$$

$u \in H^s(\mathbb{R}^m)$. Then $\{\kappa_\lambda\}_{\lambda \in \mathbb{R}_+}$ is a strongly continuous group of isomorphisms in $H^s(\mathbb{R}^m)$ for every $s \in \mathbb{R}$. Note that κ_λ is unitary in $L^2(\mathbb{R}^m)$ (with respect to the standard scalar product).

Note that (1.3.6) satisfies the relation

$$F(\kappa_\lambda u)(\xi) = \kappa_\lambda^{-1} F u(\xi) \qquad (1.3.7)$$

for the Fourier transform $F = F_{x \to \xi}$ in $\mathbb{R}^m$, where κ_λ^{-1} on the right hand side is taken with respect to ξ.

Let $\eta \to [\eta]$ be a fixed strictly positive function in $C^\infty(\mathbb{R}^q)$ with $[\eta] = |\eta|$ for $|\eta| > c$ for a constant $c > 0$. Set

$$\kappa(\eta) = \kappa_{[\eta]}$$

for a strongly continuous group of isomorphisms $\{\kappa_\lambda\}_{\lambda \in \mathbb{R}_+}$ in a Banach space E. Let $\tilde{E}$ be a second Banach space with a fixed strongly continuous group $\{\tilde{\kappa}_\lambda\}_{\lambda \in \mathbb{R}_+}$ of isomorphisms.

Definition 1.3.3 *Let $U \subseteq \mathbb{R}^p$ be an open set, $\mu \in \mathbb{R}$, and fix*

$$\{E, \{\kappa_\lambda\}_{\lambda \in \mathbb{R}_+}\}, \qquad \{\tilde{E}, \{\tilde{\kappa}_\lambda\}_{\lambda \in \mathbb{R}_+}\},$$

where E, $\tilde{E}$ are Banach spaces. Then

$$S^\mu(U \times \mathbb{R}^q; E, \tilde{E})$$

denotes the space of all $a(y, \eta) \in C^\infty(U \times \mathbb{R}^q, \mathcal{L}(E, \tilde{E}))$ satisfying

$$\|\tilde{\kappa}^{-1}(\eta)\{D_y^\alpha D_\eta^\beta a(y, \eta)\}\kappa(\eta)\|_{\mathcal{L}(E, \tilde{E})} \le c[\eta]^{\mu - |\beta|}$$

for all multi-indices $\alpha \in \mathbb{N}^p$, $\beta \in \mathbb{N}^q$, and all $y \in K$ for arbitrary $K \subset\subset U$, $\eta \in \mathbb{R}^q$, with constants $c = c(\alpha, \beta, K) > 0$. The elements of $S^\mu(U \times \mathbb{R}^q; E, \tilde{E})$ are called operator–valued symbols of order μ.

The space $\mathcal{L}(E, \tilde{E})$ is endowed with the norm topology. In our applications we mainly have $U = \Omega$ or $U = \Omega \times \Omega$ for an open set $\Omega \subseteq \mathbb{R}^q$. For $U = \Omega \times \Omega$ we also write (y, y') instead of y.

It is easy to verify that $S^\mu(U \times \mathbb{R}^q; E, \tilde{E})$ is independent of the specific choice of the function $[\eta]$. We could take as well

$$\langle \eta \rangle = (1 + |\eta|^2)^{\frac{1}{2}}.$$

The reason for $[\eta]$ is that the homogeneity in η for large $|\eta|$ is convenient in the operator–valued set–up below.

$S^\mu(U \times \mathbb{R}^q; E, \tilde{E})$ is a Fréchet space with the semi–norm system

$$a \to \sup_{\substack{y \in K \\ \eta \in \mathbb{R}^q}} [\eta]^{-\mu+|\beta|} \|\tilde{\kappa}^{-1}(\eta)\{D_y^\alpha D_\eta^\beta a(y, \eta)\}\kappa(\eta)\|_{\mathcal{L}(E,\tilde{E})},$$

$\alpha \in \mathbb{N}^p$, $\beta \in \mathbb{N}^q$, $K \subset\subset U$. The subspace

$$S^\mu(\mathbb{R}^q; E, \tilde{E})$$

of y–independent elements is then closed in the induced topology. Then we have

$$S^\mu(U \times \mathbb{R}^q; E, \tilde{E}) = C^\infty(U, S^\mu(\mathbb{R}^q; E, \tilde{E})), \qquad (1.3.8)$$

which is an immediate consequence of the definitions.

Proposition 1.3.4 *We have*

$$S^\mu(U \times \mathbb{R}^q; E, \tilde{E}) = C^\infty(U) \otimes_\pi S^\mu(\mathbb{R}^q; E, \tilde{E})$$

($\otimes_\pi$ is the completed projective tensor product).

Proof. The assertion is a consequence of (1.3.4) and of the nuclearity of the space $C^\infty(U)$. □

Note that the spaces

$$S^{-\infty}(U \times \mathbb{R}^q; E, \tilde{E}) = C^\infty(U, \mathcal{S}(\mathbb{R}^q, \mathcal{L}(E, \tilde{E}))), \qquad (1.3.9)$$
$$S^{-\infty}(\mathbb{R}^q; E, \tilde{E}) = \mathcal{S}(\mathbb{R}^q, \mathcal{L}(E, \tilde{E})) \qquad (1.3.10)$$

are independent of the particular choice of $\{\kappa_\lambda\}_{\lambda \in \mathbb{R}_+}$, $\{\tilde{\kappa}_\lambda\}_{\lambda \in \mathbb{R}_+}$. Here, like in the scalar theory, $S^{-\infty} = \bigcap_{\mu \in \mathbb{R}} S^\mu$.

Remark 1.3.5 *We have continuous embeddings*

$$S^\nu(U \times \mathbb{R}^q; E, \tilde{E}) \hookrightarrow S^\mu(U \times \mathbb{R}^q; E, \tilde{E})$$

for every $\mu, \nu \in \mathbb{R}$ with $\mu \geq \nu$. Moreover,

$$S^\mu(U \times \mathbb{R}^q; E_0, \tilde{E}) \cdot S^\varrho(U \times \mathbb{R}^q; E, E_0) \subseteq S^{\mu+\varrho}(U \times \mathbb{R}^q; E, \tilde{E})$$
$$(1.3.11)$$

with obvious meaning of notation, $\mu, \varrho \in \mathbb{R}$.

Remark 1.3.6 *If E is finite–dimensional we always set $\kappa_\lambda = \mathrm{id}_E$ for all $\lambda \in \mathbb{R}_+$. In particular, for $E = \tilde{E} = \mathbb{C}$ we have*

$$S^\mu(U \times \mathbb{R}^q; \mathbb{C}, \mathbb{C}) = S^\mu(U \times \mathbb{R}^q),$$

where the symbol space on the right equals that of Section 1.1.2. At the same time we can define an embedding

$$S^\mu(U \times \mathbb{R}^q) \hookrightarrow S^\mu(U \times \mathbb{R}^q; E, E)$$

for an arbitrary Banach space E with strongly continuous group $\{\kappa_\lambda\}_{\lambda \in \mathbb{R}}$ of isomorphisms by $p(y, \eta) \to p(y, \eta) \cdot \mathrm{id}_E$. This gives us as special cases of (1.3.11) *the relations*

$$S^\mu(U \times \mathbb{R}^q) \cdot S^\varrho(U \times \mathbb{R}^q; E, \tilde{E}) \subseteq S^{\mu+\varrho}(U \times \mathbb{R}^q; E, \tilde{E}),$$
$$S^\mu(U \times \mathbb{R}^q; E, \tilde{E}) \cdot S^\varrho(U \times \mathbb{R}^q) \subseteq S^{\mu+\varrho}(U \times \mathbb{R}^q; E, \tilde{E}).$$

for all $\mu, \varrho \in \mathbb{R}$ and arbitrary $E, \tilde{E}$ with the corresponding group actions.

Let

$$S^{(\mu)}(U \times (\mathbb{R}^q \setminus \{0\}); E, \tilde{E}) \qquad (1.3.12)$$

be the space of all $a(y, \eta) \in C^\infty(U \times (\mathbb{R}^q \setminus \{0\}), \mathcal{L}(E, \tilde{E}))$ satisfying

$$a(y, \lambda\eta) = \lambda^\mu \tilde{\kappa}_\lambda a(y, \eta) \kappa_\lambda^{-1} \qquad (1.3.13)$$

for all $\lambda \in \mathbb{R}$, $(y, \eta) \in U \times (\mathbb{R}^q \setminus \{0\})$. The elements in (1.3.12) are called (positively) homogeneous in η of order μ. Then, if $\chi(\eta)$ is an excision function in $\mathbb{R}^q$ (i.e., $\chi \in C^\infty(\mathbb{R}^q)$, $\chi(\eta) = 0$ for $|\eta| \leq c_0$, $\chi(\eta) = 1$ for $|\eta| \geq c_1$ with constants $0 < c_0 < c_1 < \infty$) we have

$$\chi(\eta) S^{(\mu)}(U \times (\mathbb{R}^q \setminus \{0\}); E, \tilde{E}) \subset S^\mu(U \times \mathbb{R}^q; E, \tilde{E})$$

(with the obvious meaning of notation on the left). $S^{(\mu)}(U \times (\mathbb{R}^q \setminus \{0\}); E, \tilde{E})$ is a Fréchet space in the topology induced by $C^\infty(U \times (\mathbb{R}^q \setminus \{0\}), \mathcal{L}(E, \tilde{E}))$.

Definition 1.3.7 *An element $a(y, \eta) \in S^\mu(U \times \mathbb{R}^q; E, \tilde{E})$ is called classical if there exists a sequence $a_{(\mu-j)}(y, \eta) \in S^{(\mu-j)}(U \times (\mathbb{R}^q \setminus \{0\}); E, \tilde{E})$, $j \in \mathbb{N}$, with*

$$a(y, \eta) - \chi(\eta) \sum_{j=0}^{N} a_{(\mu-j)}(y, \eta) \in S^{\mu-(N+1)}(U \times \mathbb{R}^q; E, \tilde{E})$$

for all $N \in \mathbb{N}$. Here $\chi(\eta)$ is any excision function. The subspace of all classical symbols of order μ is denoted by

$$S^\mu_{\mathrm{cl}}(U \times \mathbb{R}^q; E, \tilde{E}).$$

$a_{(\mu-j)}(y,\eta)$ is called the homogeneous component of $a(y,\eta)$ of order $\mu - j$. In particular, we set

$$\sigma_\wedge^\mu(a)(y,\eta) := a_{(\mu)}(y,\eta),$$

also called the homogeneous principal symbol of $a(y,\eta)$ of order μ. It can easily be proved that $a_{(\mu-j)}(y,\eta)$ is uniquely determined by $a(y,\eta)$ for all $j \in \mathbb{N}$. This gives rise to a sequence of maps

$$\sigma_j : S_{\mathrm{cl}}^\mu(U \times \mathbb{R}^q; E, \tilde{E}) \to S^{(\mu-j)}(U \times (\mathbb{R}^q \setminus \{0\}); E, \tilde{E}),$$

$j \in \mathbb{N}$. Setting $\varrho_k(a) = a(y,\eta) - \chi(\eta)\sum_{j=0}^k \sigma_j(a)(y,\eta)$, we get from Definition 1.3.7 a further sequence of maps

$$\varrho_k : S_{\mathrm{cl}}^\mu(U \times \mathbb{R}^q; E, \tilde{E}) \to S^{\mu-(k+1)}(U \times \mathbb{R}^q; E, \tilde{E}),$$

$k \in \mathbb{N}$. Endowing $S_{\mathrm{cl}}^\mu(U \times \mathbb{R}^q; E, \tilde{E})$ with the projective limit topology with respect to σ_j, ϱ_k for all j,k, we get a Fréchet topology in this space. Then $S_{\mathrm{cl}}^\mu(\mathbb{R}^q; E, \tilde{E})$, the subspace of all y-independent elements of $S_{\mathrm{cl}}^\mu(U \times \mathbb{R}^q; E, \tilde{E})$, is closed in the induced topology. Analogously to the above Proposition 1.3.4 we have

$$S_{\mathrm{cl}}^\mu(U \times \mathbb{R}^q; E, \tilde{E}) = C^\infty(U, S_{\mathrm{cl}}^\mu(\mathbb{R}^q; E, \tilde{E})) = C^\infty(U) \otimes_\pi S_{\mathrm{cl}}^\mu(\mathbb{R}^q; E, \tilde{E}).$$

Theorem 1.3.8 *Let $a_j(y,\eta) \in S^{\mu_j}(U \times \mathbb{R}^q; E, \tilde{E})$, $j \in \mathbb{N}$, be an arbitrary sequence with $\mu_j \to -\infty$ as $j \to \infty$. Then there exists an $a(y,\eta) \in S^\mu(U \times \mathbb{R}^q; E, \tilde{E})$ for $\mu = \max\{\mu_j\}$ such that for every $M \in \mathbb{N}$ there is an $N(M) \in \mathbb{N}$ such that for every $N \geq N(M)$*

$$a(y,\eta) - \sum_{j=0}^N a_j(y,\eta) \in S^{\mu-M}(U \times \mathbb{R}^q; E, \tilde{E}).$$

Each such $a(y,\eta)$ is unique mod $S^{-\infty}(U \times \mathbb{R}^q; E, \tilde{E})$. *Moreover, $a_j(y,\eta) \in S_{\mathrm{cl}}^{\mu-j}(U \times \mathbb{R}^q; E, \tilde{E})$, $j \in \mathbb{N}$, implies $a(y,\eta) \in S_{\mathrm{cl}}^\mu(U \times \mathbb{R}^q; E, \tilde{E})$.*

We set $a(y,\eta) \sim \sum_{j=0}^\infty a_j(y,\eta)$, called an asymptotic sum of the a_j, $j \in \mathbb{N}$.

We omit the proof of Theorem 1.3.8 which is completely analogous to that in the scalar case of Section 1.1.2. In particular, if $\chi(\eta)$ is an arbitrary excision function, there are constants $c_j > 0$, $j \in \mathbb{N}$, increasing sufficiently fast for $j \to \infty$, such that

$$a(y,\eta) = \sum_{j=0}^\infty \chi(\frac{\eta}{c_j})a_j(y,\eta) \tag{1.3.14}$$

converges in the space $S^\mu(U \times \mathbb{R}^q; E, \tilde{E})$ as well as $\sum_{j=N+1}^{\infty} \chi(\frac{\eta}{c_j}) a_j(y, \eta)$ in $S^{\mu-M}(U \times \mathbb{R}^q; E, \tilde{E})$ for every M (in the sense of the notation of Theorem 1.3.8), such that $a \sim \sum a_j$.

Remark 1.3.9 *In our concrete applications the spaces $E, \tilde{E}$ run over scales $\{E^s\}_{s \in \mathbb{R}}$, $\{\tilde{E}^r\}_{r \in \mathbb{R}}$ of Banach spaces with continuous embeddings $E^{s'} \hookrightarrow E^s$ for $s' \geq s$, where $\{\kappa_\lambda\}$ on $E^{s'}$ is the restriction of the corresponding group action on E^s for every $s' \geq s$, and similarly for $\{\tilde{E}^r\}_{r \in \mathbb{R}}$. For the spaces of symbols $a(y, \eta)$ of order μ we have*

$$a(y, \eta) \in S^\mu(U \times \mathbb{R}^q; E^s, \tilde{E}^{s-\mu}) \qquad \text{for all } s.$$

In addition the symbols have their values in certain algebras of pseudo-differential operators on M. It is then natural in such a situation that asymptotic sums should be carried out for all s at the same time and uniquely $\mod \bigcap_s S^{-\infty}(U \times \mathbb{R}^q; E^s, \tilde{E}^\infty) = \bigcap_{s,r} S^{-\infty}(U \times \mathbb{R}^q; E^s, \tilde{E}^r)$. However for $a_j(y, \eta) \in \bigcap_s S^{\mu_j}(U \times \mathbb{R}^q; E^s, \tilde{E}^{s-\mu_j})$ we have only $(1 - \chi)(\frac{\eta}{c_j}) a_j(y, \eta) \in \bigcap_s S^{-\infty}(U \times \mathbb{R}^q; E^s, \tilde{E}^{s-\mu_j})$ and we cannot expect that the latter symbol class equals $\bigcap_s S^{-\infty}(U \times \mathbb{R}^q; E^s, \tilde{E}^\infty)$. In concrete cases we can apply more subtle procedures than (1.3.14) to carry out asymptotic sums. Nevertheless for many purposes here in the abstract set–up the rule (1.3.14) is completely sufficient.

Example 1.3.10 *Let $\Omega \subseteq \mathbb{R}^q$ be an open set and $\varphi(x, y) \in C^\infty(\mathbb{R}^m \times \Omega)$ be a function with $\varphi(x, y) = 0$ for $|x| > c$ and all $y \in \Omega$, for a constant $c > 0$. Set $m_\varphi(y)u(x) = \varphi(x, y)u(x)$, $u \in H^s(\mathbb{R}^m)$. Then we have*

$$m_\varphi(y) \in S^0(\Omega \times \mathbb{R}^q; H^s(\mathbb{R}^m), H^s(\mathbb{R}^m)) \qquad (1.3.15)$$

for every $s \in \mathbb{R}$, with respect to the group (1.3.6).

The reader should not be confused by the fact that symbols may be independent of the covariable η though the symbol estimates contain some η–dependence. In fact, we have

$$\kappa^{-1}(\eta) m_\varphi(y) \kappa(\eta) u(x) = \varphi([\eta]^{-1} x, y) u(x),$$

such that (1.3.15) means that $\varphi([\eta]^{-1} x, y)$, interpreted as an operator of multiplication in $H^s(\mathbb{R}^m)$, satisfies $\|D_y^\alpha \varphi([\eta]^{-1} x, y)\|_{\mathcal{L}(H^s(\mathbb{R}^m))} \leq c$ for all $y \in K$ for arbitrary $K \subset\subset \Omega$ and $\alpha \in \mathbb{N}^q$, $c = c(\alpha, K)$.

Example 1.3.11 *Let $\omega(t)$ be a cut–off function and set $m(t, \tau)u(t) = \omega(t[\tau])u(t)$ for $u \in L^2(\mathbb{R}_+)$. Then*

$$m(t, \tau) \in S^0_{cl}(\overline{\mathbb{R}}_+ \times \mathbb{R}; L^2(\mathbb{R}_+), L^2(\mathbb{R}_+)).$$

Note that $\omega(t[\tau])$ can also be regarded as a scalar symbol in $S^0(\overline{\mathbb{R}}_+ \times \mathbb{R})$ with $\omega(t[\tau])|_{\mathbb{R}_+ \times \mathbb{R}} \in S^{-\infty}(\mathbb{R}_+ \times \mathbb{R})$.

Example 1.3.12 *Let*

$$A(x, y, D_x, D_y) = \sum_{|\alpha|+|\beta| \leq \mu} a_{\alpha\beta}(x, y) D_x^\alpha D_y^\beta$$

be a differential operator with coefficients $a_{\alpha\beta} \in C^\infty(\mathbb{R}^m \times \Omega)$, $(x, y) \in \mathbb{R}^m \times \Omega$, for an open set $\Omega \subseteq \mathbb{R}^q$. Assume, in particular, that $a_{\alpha\beta}$ is independent of x for $|x| > c$ for some $c > 0$. Then the operator family

$$a(y, \eta) := \sum_{|\alpha|+|\beta| \leq \mu} a_{\alpha\beta}(x, y) D_x^\alpha \eta^\beta : H^s(\mathbb{R}^m) \to H^{s-\mu}(\mathbb{R}^m)$$

represents an element in $S^\mu(\Omega \times \mathbb{R}^q; E, \tilde{E})$ for $E = H^s(\mathbb{R}^m)$, $\tilde{E} = H^{s-\mu}(\mathbb{R}^m)$, for arbitrary $s \in \mathbb{R}$, where $(\kappa_\lambda u)(x) = \lambda^{\frac{m}{2}} u(\lambda x)$, $\lambda \in \mathbb{R}_+$, both for u in E or $\tilde{E}$. If the coefficients $a_{\alpha\beta}$ are independent of x, then we have $a(y, \eta) \in S^\mu_{cl}(\Omega \times \mathbb{R}^q; E, \tilde{E})$ and

$$a_{(\mu-j)}(y, \eta) = \sum_{|\alpha|+|\beta|=\mu-j} a_{\alpha\beta}(y) D_x^\alpha \eta^\beta, \qquad j = 0, \ldots, \mu.$$

Remark 1.3.13 *To construct operator–valued symbols to a pair*

$$\{E, \{\kappa_\lambda\}_{\lambda \in \mathbb{R}_+}\}, \qquad \{\tilde{E}, \{\tilde{\kappa}_\lambda\}_{\lambda \in \mathbb{R}_+}\}$$

as in Definition 1.3.3 it is useful to have elements $a_0 \in \mathcal{L}(E, \tilde{E})$ with

$$\tilde{\kappa}_\lambda a_0 \kappa_\lambda^{-1} \in C^\infty(\mathbb{R}_+, \mathcal{L}(E, \tilde{E})). \tag{1.3.16}$$

Then $a(\eta) := [\eta]^\mu \tilde{\kappa}_{[\eta]} a_0 \kappa_{[\eta]}^{-1} \in S^\mu_{cl}(\mathbb{R}^q; E, \tilde{E})$. In applications it will also be of interest to have an isomorphism $a_0 : E \to \tilde{E}$ with the property (1.3.16). Then $a(\eta)$ has also values in the isomorphisms $E \to \tilde{E}$ for all η, and it follows that $a^{-1}(\eta) \in S^{-\mu}_{cl}(\mathbb{R}^q; \tilde{E}, E)$.

Concrete cases for which (1.3.16) holds can be constructed, for instance, in the set–up of standard pseudo–differential operators.

Proposition 1.3.14 *Let $p(\xi) \in S^\mu(\mathbb{R}^m)$ be a scalar symbol of order $\mu \in \mathbb{R}$ and*

$$\mathrm{Op}(p) : H^s(\mathbb{R}^m) \to H^{s-\mu}(\mathbb{R}^m)$$

be the associated pseudo–differential operator for any fixed $s \in \mathbb{R}$. Then, for $\{\kappa_\lambda\}_{\lambda \in \mathbb{R}_+}$, given by (1.3.6) in $E = H^s(\mathbb{R}^m)$ as well as in $\tilde{E} = H^{s-\mu}(\mathbb{R}^m)$, we have

$$\kappa_\lambda \mathrm{Op}(p) \kappa_\lambda^{-1} = \mathrm{Op}(p_\lambda)$$

for $p_\lambda(\xi) = p(\frac{\xi}{\lambda})$.

Proof. We have

$$
\kappa_\lambda \operatorname{Op}(p)\kappa_\lambda^{-1} u(x) = \iint e^{i(\lambda x - x')\xi} p(\xi) u(\lambda^{-1} x')\, dx' d\xi
$$

$$
= \iint e^{i(\lambda x - x'')\xi} p(\xi) u(x'') \lambda^{-m}\, dx'' d\xi
$$

$$
= \iint e^{i(x - x')\xi} p\left(\frac{\xi}{\lambda}\right) u(x')\, dx' d\xi.
$$

$\square$

Remark 1.3.15 *If we set* $p(\xi) = \langle\xi\rangle^\mu$ *then the operator* $\operatorname{Op}(p)$:
$H^s(\mathbb{R}^m) \to H^{s-\mu}(\mathbb{R}^m)$ *is an isomorphism for every* $s \in \mathbb{R}$. *Thus,
in this case,* $\operatorname{Op}(p_{[\eta]}) =: a(\eta)$ *is an example in the sense of the last part
of Remark* 1.3.13.

Example 1.3.16 *Let* $p(\xi,\eta) \in C^\infty((\mathbb{R}^m \setminus \{0\}) \times (\mathbb{R}^q \setminus \{0\}))$ *be a function satisfying*

$$
p(\lambda\xi, \lambda\eta) = \lambda^\mu p(\xi, \eta) \tag{1.3.17}
$$

for all $\lambda \in \mathbb{R}_+$, $(\xi,\eta) \in (\mathbb{R}^m \setminus \{0\}) \times (\mathbb{R}^q \setminus \{0\})$, *where* $\mu \geq 0$. *Set*

$$
a(\eta) u(x) = \iint e^{i(x-x')\xi} p(\xi, \eta) u(x')\, dx' d\xi.
$$

Then $\operatorname{Op}(p)(\eta) : H^s(\mathbb{R}^m) \to H^{s-\mu}(\mathbb{R}^m)$ *is continuous for every* $s \in \mathbb{R}$,
and we have

$$
a(\lambda\eta) = \lambda^\mu \kappa_\lambda a(\eta) \kappa_\lambda^{-1} \qquad \text{for all } \lambda \in \mathbb{R}_+, \tag{1.3.18}
$$

$\eta \in \mathbb{R}^q \setminus \{0\}$. *In fact, Proposition* 1.3.14 *shows that* $\kappa_\lambda a(\eta)\kappa_\lambda^{-1} u(x) =$
$\iint e^{i(x-x')\xi} p(\frac{\xi}{\lambda}, \eta) u(x')\, dx' d\xi$. *Using* (1.3.17) *this gives us the relation*
(1.3.18). *If we assume* $p(\xi, x) \in C^\infty((\mathbb{R}^m \times \mathbb{R}^q) \setminus \{0\})$ *with* (1.3.17) *for*
$\mu \in \mathbb{R}$, *we also obtain* (1.3.18).

Examples for the situation of Remark 1.3.13 can also be constructed
in terms of pseudo–differential operators on $\mathbb{R}_+$, based on the Mellin
transform

$$
Mu(z) = \int\limits_0^\infty t^{z-1} u(t)\, dt.
$$

If we insert first $u \in C_0^\infty(\mathbb{R}_+)$ then $Mu(z) \in \mathcal{A}(\mathbb{C})$, and we have
$Mu(z)|_{\Gamma_\beta} \in \mathcal{S}(\Gamma_\beta)$ for every $\beta \in \mathbb{R}$, for $\Gamma_\beta = \{z \in \mathbb{C} : \operatorname{Re} z = \beta\}$
and the Schwartz space $\mathcal{S}(\Gamma_\beta)$ on Γ_β. Analogously we have the space
$L^2(\Gamma_\beta)$ of square integrable functions on Γ_β. It is well–known that

$M : C_0^\infty(\mathbb{R}_+) \to \mathcal{A}(\mathbb{C})|_{\Gamma_{\frac{1}{2}}}$ extends to an isomorphism $M : L^2(\mathbb{R}_+) \to$ $L^2(\Gamma_{\frac{1}{2}})$. The inverse is given by the formula

$$(M^{-1}g)(t) = \frac{1}{2\pi i} \int\limits_{\Gamma_{\frac{1}{2}}} t^{-z} g(z) \, dz.$$

For every $p(z) \in S^0(\Gamma_{\frac{1}{2}})$, where $S^0(\Gamma_{\frac{1}{2}})$ is the space of symbols on $\Gamma_{\frac{1}{2}}$ of order 0, we can form the operator

$$\text{op}_M(p) := M_{z \to t}^{-1} p(z) M_{t' \to z} : L^2(\mathbb{R}_+) \to L^2(\mathbb{R}_+).$$

Setting $E = \tilde{E} = L^2(\mathbb{R}_+)$ and $(\kappa_\lambda u)(t) = \lambda^{\frac{1}{2}} u(\lambda t)$, $\lambda \in \mathbb{R}_+$, we have

$$\kappa_\lambda \, \text{op}_M(p)\kappa_\lambda^{-1} = \text{Op}_M(p) \qquad \text{for all } \lambda \in \mathbb{R}_+.$$

Let us now generalise the symbol spaces $S^\mu(U \times \mathbb{R}^q; E, \tilde{E})$ for Fréchet spaces E or $\tilde{E}$. First assume that E is a Banach space and that $\tilde{E}$ is the projective limit of the sequence of Banach spaces $\{\tilde{E}^k\}_{k \in \mathbb{N}}$ with continuous embeddings $\tilde{E}^{k+1} \hookrightarrow \tilde{E}^k$. Suppose further that $\tilde{E}^0$ has a strongly continuous group of isomorphisms $\{\tilde{\kappa}_\lambda^0\}_{k \in \mathbb{R}_+}$ which restricts to strongly continuous groups $\{\tilde{\kappa}_\lambda^k\}_{k \in \mathbb{R}_+}$ of isomorphisms on every $\tilde{E}^k$, $k \in \mathbb{N}$. Then the symbol spaces $S^\mu(U \times \mathbb{R}^q; E, \tilde{E}^k)$ are defined for each k, and we set

$$S^\mu(U \times \mathbb{R}^q; E, \tilde{E}) = \varprojlim_{k \in \mathbb{N}} S^\mu(U \times \mathbb{R}^q; E, \tilde{E}^k).$$

In an analogous manner we define

$$S_{\text{cl}}^\mu(U \times \mathbb{R}^q; E, \tilde{E}) = \varprojlim_{k \in \mathbb{N}} S_{\text{cl}}^\mu(U \times \mathbb{R}^q; E, \tilde{E}^k).$$

Next let also E be a Fréchet space, given as a projective limit of Banach spaces $\{E^k\}_{k \in \mathbb{N}}$ with continuous embeddings $E^{k+1} \hookrightarrow E^k$, $k \in \mathbb{N}$, and a group $\{\kappa_\lambda^k\}_{\lambda \in \mathbb{R}_+}$ on E^k for every k, where we assume analogous properties as for $\tilde{E}$.

$\mathcal{L}(E, \tilde{E})$ (the space of linear continuous operators $E \to \tilde{E}$) is the set of all linear mappings $b : E \to \tilde{E}$ such that for every $k \in \mathbb{N}$ there is an $l \in \mathbb{N}$ with

$$\|bu\|_{\tilde{E}^k} \leq c_{kl}\|u\|_{E^l} \qquad (1.3.19)$$

for all $u \in E$ for certain constants c_{kl}. Let us fix any map $r : \mathbb{N} \to \mathbb{N}$ and denote by $\mathcal{L}_r(E, \tilde{E})$ the subspace of all $b \in \mathcal{L}(E, \tilde{E})$ satisfying (1.3.19) for every pair $(k, l) = (k, r(k))$, $k \in \mathbb{N}$. Then we define

$$S_r^\mu(U \times \mathbb{R}^q; E, \tilde{E}) = \varprojlim_{k \in \mathbb{N}} S^\mu(U \times \mathbb{R}^q; E^{r(k)}, \tilde{E}^k),$$

and set
$$S^\mu(U \times \mathbb{R}^q; E, \tilde{E}) = \bigcup_r S_r^\mu(U \times \mathbb{R}^q; E, \tilde{E}),$$

where the union is taken over all maps $r : \mathbb{N} \to \mathbb{N}$. In an analogous manner we define the spaces $S_{cl}^\mu(U \times \mathbb{R}^q; E, \tilde{E})$ of classical symbols.

In order to illustrate the nature of homogeneity of operator functions we want to consider a particular class of examples. This will occur in more concrete form in Chapter 3 below. Let B and $\tilde{B}$ be Banach spaces and let

$$p_0(\tilde{\tau}, \tilde{\eta}, \tilde{\eta}') \in C^\infty(\mathbb{R}_{\tilde{\tau}} \times \mathbb{R}_{\eta, \tilde{\eta}}^{2q}, \mathcal{L}(B, \tilde{B}))$$

be an operator function satisfying

$$\sup_{|\tilde{\eta}| + |\tilde{\eta}'| \le R} \| D_{\tilde{\tau}}^k D_{\tilde{\eta}, \tilde{\eta}'}^\alpha p_0(\tilde{\tau}, \tilde{\eta}, \tilde{\eta}') \|_{\mathcal{L}(B, \tilde{B})} \le c(1 + |\tilde{\tau}|)^{\nu - k}$$

for all $k \in \mathbb{N}$, $\alpha \in \mathbb{N}^{2q}$ and all $\tilde{\tau} \in \mathbb{R}$, $R > 0$, with constants $c = c(k, \alpha, R) > 0$. Set $p(t, \tilde{\tau}, \tilde{\eta}, \tilde{\eta}') := t^{-\mu} p_0(\tilde{\tau}, \tilde{\eta}, \tilde{\eta}')$ and form

$$a(\eta)u(t) := \iint e^{i(t-t')\tau} p(t, t\tau, t\eta, t'\eta) u(t') \, dt' d\tau,$$

regarded as an operator family $a(\eta) : C_0^\infty(\mathbb{R}_+, B) \to C^\infty(\mathbb{R}_+, B)$, parametrised by $\eta \in \mathbb{R}^q$. Setting $(\kappa_\lambda u)(t) = \lambda^m u(\lambda t)$ for $\lambda \in \mathbb{R}_+$, with any fixed $m \in \mathbb{R}$, it follows that

$$a(\lambda\eta) = \lambda^\mu \kappa_\lambda a(\eta) \kappa_\lambda^{-1}$$

for all $\lambda \in \mathbb{R}_+$, $\eta \in \mathbb{R}^q$. Writing

$$f(r, r', \tau, \eta) := p(r^{-1}, r^{-1}\tau, r^{-1}\eta, (r')^{-1}\eta)$$

we have

$$f(\lambda r, \lambda r', \lambda\tau, \lambda\eta) = \lambda^\mu f(r, r', \tau, \eta)$$

for all $\lambda \in \mathbb{R}_+$, $r, r' \in \mathbb{R}_+$, $(\tau, \eta) \in \mathbb{R}^{1+q}$. Thus we can apply Euler's homogeneity relation

$$r \frac{\partial f}{\partial r} + r' \frac{\partial f}{\partial r'} + \tau \frac{\partial f}{\partial \tau} + \sum_{j=1}^q \eta_j \frac{\partial f}{\partial \eta_j} = \mu f. \qquad (1.3.20)$$

We want to derive an analogue of Euler's homogeneity relation for the operator family $a(\eta)$.

Proposition 1.3.17 *The operator family $a(\eta)$ satisfies the homogeneity relation*

$$\sum_{j=1}^{q} \eta_j \frac{\partial}{\partial \eta_j} a(\eta) = \mu a(\eta) + Aa(\eta) - a(\eta)A \qquad (1.3.21)$$

for the constant operator $A = t\frac{\partial}{\partial t}$.

Proof. Let us set $g(t, t', \tau, \eta) := f(t^{-1}, (t')^{-1}, \tau, \eta)$. Then (1.3.20) implies

$$\left\{ -t\frac{\partial}{\partial t} - t'\frac{\partial}{\partial t'} + \tau\frac{\partial}{\partial \tau} + \sum_{j=1}^{q} \eta_j \frac{\partial}{\partial \eta_j} \right\} g = \mu g. \qquad (1.3.22)$$

Thus

$$\left\{ \mu a(\eta) - \sum_{j=1}^{q} \eta_j \frac{\partial}{\partial \eta_j} a(\eta) \right\} u(t)$$

$$= \iint e^{i(t-t')\tau} \left(-t\frac{\partial}{\partial t} - t'\frac{\partial}{\partial t'} + \tau\frac{\partial}{\partial \tau} \right) g(t, t', \tau, \eta) u(t')\, dt'd\tau. \quad (1.3.23)$$

We have

$$\frac{\partial}{\partial t}(e^{it\tau}g) - i\tau e^{it\tau}g = e^{it\tau}\frac{\partial}{\partial t}g,$$

$$\frac{\partial}{\partial t'}(e^{-it'\tau}g) + i\tau e^{it'\tau}g = e^{-it'\tau}\frac{\partial}{\partial t'}g.$$

Thus the right hand side of (1.3.23) equals $-t\frac{\partial}{\partial t}a(\eta)u(t) + I_1 + I_2 + I_3$ for

$$I_1 = \iint e^{i(t-t')\tau} i(t - t')\tau g u(t')\, dt'd\tau$$

$$I_2 = \iint e^{i(t-t')\tau} g \left\{ \frac{\partial}{\partial t'} t' u(t') \right\}\, dt'd\tau,$$

$$I_3 = \iint e^{i(t-t')\tau} \left\{ \tau\frac{\partial}{\partial \tau} g \right\} u(t')\, dt'd\tau.$$

Using $\tau\frac{\partial}{\partial \tau}(e^{i(t-t')\tau}g) = i(t - t')\tau e^{i(t-t')\tau}g + e^{i(t-t')\tau}\tau\frac{\partial}{\partial \tau}g$ it follows that $I_3 = I_0 - I_1$, where

$$I_0 = \iint \tau\frac{\partial}{\partial \tau}(e^{i(t-t')\tau}g)u(t')\, dt'd\tau.$$

For I_2 we obtain

$$I_2 = a(\eta)u(t) + a(\eta)t\frac{\partial}{\partial t}u(t).$$

In I_0 we may replace $\tau\frac{\partial}{\partial\tau}$ by the operator $\frac{\partial}{\partial\tau} - \mathrm{id}$. Thus

$$I_0 = -a(\eta)u(t) + \iint \frac{\partial}{\partial\tau}\{\tau e^{i(t-t')\tau}g\}u(t')\,dt'd\tau.$$

The second term on the right of the latter equation vanishes. So $I_3 = -a(\eta)u(t) - I_1$ and hence

$$I_1 + I_2 + I_3 = I_1 + a(\eta)u(t) + a(\eta)Au(t) - a(\eta)u(t) - I_1 = a(\eta)Au(t).$$

Thus we have obtained

$$\mu a(\eta) - \sum_{j=1}^{q} \eta_j\frac{\partial}{\partial\eta_j}a(\eta) = -Aa(\eta) + Aa(\eta)$$

which is the assertion.   □

Homogeneity relations like (1.3.21) in an anisotropic set–up also play a role in the theory of parabolic pseudo–differential operators with operator–valued symbols or on manifolds with singularities, cf. Buchholz, Schulze [12], [13].

The remaining part of this Section 1.3.1 is due to T. Krainer, University of Potsdam, who contributed the following general characterisation of Euler's homogeneity relation for arbitrary Banach spaces E, $\tilde{E}$, with strongly continuous groups of isomorphisms $\{\kappa_\lambda\}_{\lambda\in\mathbb{R}_+}$, $\{\tilde{\kappa}_\lambda\}_{\lambda\in\mathbb{R}_+}$.
Set

$$Q(t) = \kappa_{\exp(t)}, \qquad \tilde{Q}(t) = \tilde{\kappa}_{\exp(t)}.$$

These are C_0–groups $Q : \mathbb{R} \to \mathcal{L}(E)$, $\tilde{Q} : \mathbb{R} \to \mathcal{L}(\tilde{E})$. Denote by A and $\tilde{A}$ the infinitesimal generators of Q and $\tilde{Q}$, respectively. Let $D(A)$, $D(\tilde{A})$ be the domains of A, $\tilde{A}$.

Theorem 1.3.18 *Let $S \subseteq D(A)$ be a subspace which is invariant under $\{\kappa_\lambda\}_{\lambda\in\mathbb{R}_+}$, i.e., $\kappa_\lambda S \subseteq S$ for all $\lambda \in \mathbb{R}_+$. Moreover, let S be dense in E. Suppose $a(\eta) \in C^\infty(\mathbb{R}^q \setminus \{0\}, \mathcal{L}(E, \tilde{E}))$ with $a(\eta)S \subseteq D(\tilde{A})$ for all $\eta \in \mathbb{R}^q \setminus \{0\}$. Then the following conditions are equivalent:*

(i) *The operator function $a(\eta)$ is homogeneous in the sense*

$$a(\lambda\eta) = \lambda^\mu \tilde{\kappa}_\lambda a(\eta)\kappa_\lambda^{-1}$$

for all $\lambda \in \mathbb{R}_+$ and all $\eta \in \mathbb{R}^q \setminus \{0\}$.

(ii) $a(\eta)$ *satisfies the relation*

$$\sum_{j=1}^{q} \eta_j \frac{\partial}{\partial \eta_j} a(\eta)e = \{\mu a(\eta) + \tilde{A}a(\eta) - a(\eta)A\}e$$

for all $e \in S$.

The proof will be omitted here; it is based on the following observation : Let E be a Banach space and $g(\varrho) \in C^1(\mathbb{R}_+, E)$ be a function satisfying

$$\left(\varrho \frac{\partial}{\partial \varrho}\right) g(\varrho) = \mu g(\varrho) \qquad \text{for all } \varrho \in \mathbb{R}_+. \tag{1.3.24}$$

Then $g(\varrho) = \varrho^\mu g(1)$. Conversely, for every $e \in E$ the function $g(\varrho) = \varrho^\mu e$, $g : \mathbb{R}_+ \to E$ has the property (1.3.24).

While the second statement is trivial, the first one follows from $\frac{\partial}{\partial \varrho}(\varrho^{-\mu} g(\varrho)) = 0$ which is a consequence of (1.3.24), i.e., $\varrho^{-\mu} g(\varrho) = $ const $=: e$ and so $g(\varrho) = \varrho^\mu e$, $e = g(1)$.

Another element of the proof is that when $Q : \mathbb{R} \to \mathcal{L}(E)$ is a C_0-group and A the infinitesimal generator, we have

$$\frac{d}{dt} Q(t)e = AQ(t)e = Q(t)Ae$$

for all $e \in S$.

1.3.2 Abstract wedge Sobolev spaces

As in the previous section we fix a Banach space E as a strongly continuous group $\{\kappa_\lambda\}_{\lambda \in \mathbb{R}_+}$ of isomorphisms in E.

Definition 1.3.19 *The abstract wedge Sobolev space* $\mathcal{W}^s(\mathbb{R}^q, E)$ *of smoothness* $s \in \mathbb{R}$ *(based on the fixed choice of* $\{\kappa_\lambda\}_{\lambda \in \mathbb{R}_+}$*) is the completion of* $S(\mathbb{R}^q, E)$ *with respect to the norm*

$$\|u\|_{\mathcal{W}^s(\mathbb{R}^q,E)} = \left\{ \int [\eta]^{2s} \|\kappa^{-1}(\eta)(F_{y \to \eta} u)(\eta)\|_E^2 \, d\eta \right\}^{\frac{1}{2}}, \tag{1.3.25}$$

where $F_{y \to \eta}$ *is the Fourier transform in* $\mathbb{R}^q$.

The abstract wedge Sobolev spaces were first introduced in [117] in connection with the pseudo-differential calculus on manifolds with edges, cf. also Chapter 3 below. For suitably specified parameter spaces E they are useful for many other problems, in particular, for

a concise description of edge asymptotics, cf. [119], [121], for boundary value problems, cf. [123], [125] and Schrohe, Schulze [106], [107],
transmission and crack problems [120], [81], or for corner–degenerate
operators, where the Fourier transform is replaced by the Mellin transform, cf. [121] and Dorschfeldt, Schulze [22]. Moreover, Hirschmann
[51] has investigated these spaces and obtained, in particular, interpolation results, cf. Theorem 1.3.37 below. Anisotropic variants of
wedge Sobolev spaces were studied in Buchholz, Schulze [12], [13] to
treat parabolic pseudo–differential operators.

In the special case $\kappa_\lambda = \mathrm{id}_E$ for all $\lambda \in \mathbb{R}_+$ we employ the more
common notation

$$H^s(\mathbb{R}^q, E)$$

instead of $\mathcal{W}^s(\mathbb{R}^q, E)$. The group $\{\kappa_\lambda\}_{\lambda \in \mathbb{R}_+}$ is given in each concrete
case; so ambiguities will not arise by our notation.

Observe that the operator $T = F_{\eta \to y}^{-1} \kappa^{-1}(\eta) F_{y' \to \eta}$ induces an isomorphism

$$T : \mathcal{W}^s(\mathbb{R}^q, E) \to H^s(\mathbb{R}^q, E)$$

for every $s \in \mathbb{R}$, where $\|u\|_{\mathcal{W}^s(\mathbb{R}^q, E)} = \|Tu\|_{H^s(\mathbb{R}^q, E)}$.

The operator T may be considered first on the subspace $\mathcal{F}(\mathbb{R}^q, E)$
of all $u \in \mathcal{S}'(\mathbb{R}^q, E)$ such that for every $N \in \mathbb{N}$ there is a constant c_N
with

$$\sup_{\eta \in \mathbb{R}^q} \|(Fu)(\eta)\|_E \leq c_N [\eta]^{-N}. \qquad (1.3.26)$$

$\mathcal{F}(\mathbb{R}^q, E)$ is a Fréchet space with the best constants c_N in (1.3.26) as
semi–norms, $N \in \mathbb{N}$. We have $\mathcal{F}(\mathbb{R}^q, E) \subset H^s(\mathbb{R}^q, E)$, $\mathcal{F}(\mathbb{R}^q, E) \subset$
$\mathcal{W}^s(\mathbb{R}^q, E)$, and $T : \mathcal{F}(\mathbb{R}^q, E) \to \mathcal{F}(\mathbb{R}^q, E)$ is an isomorphism. For
$E = \mathbb{C}$ we get, in particular, the space $\mathcal{F}(\mathbb{R}^q)$.

Remark 1.3.20 *If $\| \cdot \|_E$ and $\| \cdot \|'_E$ are equivalent norms in E then the
associated norms (1.3.25) are equivalent in $\mathcal{W}^s(\mathbb{R}^q, E)$. Also the specific choice of the function $\eta \to [\eta]$ is unimportant modulo equivalence
of norms. So we may (and will) assume without loss of generality that
$[\eta] \geq 1$ for all $\eta \in \mathbb{R}^q$.*

Remark 1.3.21 *We have continuous embeddings*

$$\mathcal{W}^s(\mathbb{R}^q, E) \hookrightarrow H^{s-M}(\mathbb{R}^q, E), \qquad H^s(\mathbb{R}^q, E) \hookrightarrow \mathcal{W}^{s-M}(\mathbb{R}^q, E)$$

*for all $s \in \mathbb{R}$, for a suitable constant M. This is a consequence of
Proposition 1.3.1. In particular, it follows that*

$$\mathcal{W}^\infty(\mathbb{R}^q, E) := \bigcap_{s \in \mathbb{R}} \mathcal{W}^s(\mathbb{R}^q, E) = H^\infty(\mathbb{R}^q, E).$$

Proposition 1.3.22 *Let* $\{E, \{\kappa_\lambda\}_{\lambda \in \mathbb{R}_+}\}$, $\{E', \{\kappa'_\lambda\}_{\lambda \in \mathbb{R}_+}\}$ *be Banach spaces with given group actions, and assume that* $b : E' \to E$ *is a continuous operator with*

$$\|\kappa_\lambda^{-1} b \kappa'_\lambda\|_{\mathcal{L}(E',E)} \leq c\lambda^\nu \qquad \text{for all } \lambda \geq 1$$

for some $\nu \in \mathbb{R}$ *and a constant* $c > 0$. *Then the induced operator* $B : \mathcal{S}(\mathbb{R}^q, E') \to \mathcal{S}(\mathbb{R}^q, E)$ *(defined by* $Bu = b \circ u$ *for* $u \in \mathcal{S}(\mathbb{R}^q, E')$, *regarded as a map* $\mathbb{R}^q \to E'$*) extends to a continuous operator*

$$B : \mathcal{W}^s(\mathbb{R}^q, E') \to \mathcal{W}^{s-\nu}(\mathbb{R}^q, E)$$

for every $s \in \mathbb{R}$.

Proof. We have for $F = F_{y \to \eta}$

$$\|Bu\|_{\mathcal{W}^{s-\nu}(\mathbb{R}^q, E)}^2$$

$$= \int [\eta]^{2(s-\nu)} \|\kappa^{-1}(\eta) F(b \circ u)(\eta)\|_E^2 \, d\eta$$

$$= \int [\eta]^{2(s-\nu)} \|\kappa^{-1}(\eta) b \circ (Fu)(\eta)\|_E^2 \, d\eta$$

$$= \int [\eta]^{2(s-\nu)} \|\kappa^{-1}(\eta) b \kappa'(\eta) \kappa'(\eta)^{-1} (Fu)(\eta)\|_E^2 \, d\eta$$

$$\leq \sup_\eta [\eta]^{-2\nu} \|\kappa^{-1}(\eta) b \kappa'(\eta)\|_{\mathcal{L}(E',E)}^2 \int [\eta]^{2s} \|\kappa'(\eta)^{-1} (Fu)(\eta)\|_E^2 \, d\eta$$

$$\leq c \|u\|_{\mathcal{W}^s(\mathbb{R}^q, E')}^2.$$

$\square$

Example 1.3.23 *Let us set* $E = H^s(\mathbb{R}^m)$ *for any fixed* $s \in \mathbb{R}$ *and*

$$(\kappa_\lambda u)(x) = \lambda^{\frac{m}{2}} u(\lambda x) \qquad \text{for all } \lambda \in \mathbb{R}_+, \tag{1.3.27}$$

$u \in H^s(\mathbb{R}^m)$. *Then we have* .

$$\mathcal{W}^s(\mathbb{R}^q, H^s(\mathbb{R}^m)) = H^s(\mathbb{R}^{m+q}). \tag{1.3.28}$$

In fact, writing

$$\|u\|_{H^s(\mathbb{R}^m)} = \left\{ \int (1 + |\xi|^2)^s |(F_{x \to \xi} u)(\xi)|^2 \, d\xi \right\}^{\frac{1}{2}}$$

and

$$\|f\|_{H^s(\mathbb{R}^{m+q})} = \left\{ \iint (|\xi|^2 + [\eta]^2)^s |(Ff)(\xi, \eta)|^2 \, d\xi d\eta \right\}^{\frac{1}{2}}$$

for $F = F_{(x,y)\to(\xi,\eta)}$ we obtain

$$\|f\|^2_{H^s(\mathbb{R}^{m+q})} = \iint [\eta]^{2s}(1 + (\frac{|\xi|}{[\eta]})^2)^s |(Ff)(\xi,\eta)|^2 \, d\xi d\eta$$

$$= \iint [\eta]^{2s}(1 + |\tilde{\xi}|^2)^s |(Ff)([\eta]\tilde{\xi},\eta)|^2 [\eta]^m d\tilde{\xi} \, d\eta$$

$$= \int [\eta]^{2s} \|\kappa^{-1}(\eta)(F_{y\to\eta}f)(\cdot,\eta)\|^2_{H^s(\mathbb{R}^m)} \, d\eta.$$

In the last equation we used the relation (1.3.7). Clearly, in these calculations we can first assume $f \in \mathcal{S}(\mathbb{R}^{m+q})$ and show equivalence of norms on the corresponding dense subspaces.

Note, in particular, that

$$\mathcal{W}^s(\mathbb{R}^q, \mathbb{C}) = H^s(\mathbb{R}^q), \tag{1.3.29}$$

where the group action on $E = \mathbb{C}$ is the identity for all $\lambda \in \mathbb{R}_+$.

Let $\Omega \subseteq \mathbb{R}^q$ be open, $a(y,y',\eta) \in S^\mu(\Omega \times \Omega \times \mathbb{R}^q; E, \tilde{E})$, $\mu \in \mathbb{R}$, with given Banach spaces E, $\tilde{E}$ and $\{\kappa_\lambda\}_{\lambda \in \mathbb{R}_+}$, $\{\tilde{\kappa}_\lambda\}_{\lambda \in \mathbb{R}_+}$. Set

$$\mathrm{Op}(a)u(y) = \iint e^{i(y-y')\eta}a(y,y',\eta)u(y') \, dy'd\eta, \tag{1.3.30}$$

$u \in C_0^\infty(\Omega, E)$. This is understood as an oscillatory integral, analogously to the scalar theory, cf. Section 1.1.2, and $\mathrm{Op}(a)$ induces a continuous operator

$$\mathrm{Op}(a) : C_0^\infty(\Omega, E) \to C^\infty(\Omega, \tilde{E}).$$

The following section will discuss the pseudo–differential calculus with operator–valued symbols in detail. In order to investigate the spaces $\mathcal{W}^s(\mathbb{R}^q, E)$, we only need some simple properties. First observe that when $a(\eta) \in S^\mu(\mathbb{R}^q; E, \tilde{E})$ is a symbol with constant coefficients,

$$\mathrm{Op}(a) : \mathcal{S}(\mathbb{R}^q, E) \to \mathcal{S}(\mathbb{R}^q, \tilde{E})$$

is a continuous operator.

Proposition 1.3.24 *The operator* $\mathrm{Op}(a)$ *for* $a(\eta) \in S^\mu(\mathbb{R}^q; E, \tilde{E})$, $\mu \in \mathbb{R}$, *extends to a continuous map*

$$\mathrm{Op}(a) : \mathcal{W}^s(\mathbb{R}^q, E) \to \mathcal{W}^{s-\mu}(\mathbb{R}^q, \tilde{E}),$$

and $a \to \mathrm{Op}(a)$ *represents a continuous operator*

$$S^\mu(\mathbb{R}^q; E, \tilde{E}) \to \mathcal{L}(\mathcal{W}^s(\mathbb{R}^q, E), \mathcal{W}^{s-\mu}(\mathbb{R}^q, \tilde{E}))$$

for each $s \in \mathbb{R}$.

Proof. We have for $\hat{u}(\eta) = (Fu)(\eta)$

$$\| \operatorname{Op}(a)u \|^2_{\mathcal{W}^{s-\mu}(\mathbb{R}^q, \tilde{E})}$$

$$= \int [\eta]^{2(s-\mu)} \| \tilde{\kappa}^{-1}(\eta) a(\eta) \hat{u}(\eta) \|^2_{\tilde{E}} \, d\eta$$

$$= \int [\eta]^{2(s-\mu)} \| \{ \tilde{\kappa}^{-1}(\eta) a(\eta) \kappa(\eta) \} \kappa^{-1}(\eta) \hat{u}(\eta) \|^2_{\tilde{E}} \, d\eta$$

$$\leq \int [\eta]^{2(s-\mu)} \| \tilde{\kappa}^{-1}(\eta) a(\eta) \kappa(\eta) \|^2_{\mathcal{L}(E, \tilde{E})} \| \kappa^{-1}(\eta) \hat{u}(\eta) \|^2_E \, d\eta$$

$$\leq \sup_{\eta \in \mathbb{R}^q} [\eta]^{-2\mu} \| \tilde{\kappa}^{-1}(\eta) a(\eta) \kappa(\eta) \|^2_{\mathcal{L}(E, \tilde{E})} \| u \|^2_{\mathcal{W}^s(\mathbb{R}^q, E)}.$$

From this we obtain the assertion, in particular,

$$\| \operatorname{Op}(a) \|_{\mathcal{L}(\mathcal{W}^s(\mathbb{R}^q, E), \mathcal{W}^{s-\mu}(\mathbb{R}^q, \tilde{E}))} \leq \sup_{\eta \in \mathbb{R}^q} [\eta]^{-\mu} \| \tilde{\kappa}^{-1}(\eta) a(\eta) \kappa(\eta) \|_{\mathcal{L}(E, \tilde{E})},$$

cf. Definition 1.3.3. □

Corollary 1.3.25 *Let* $p(\eta) \in S^\mu(\mathbb{R}^q)$ *be a scalar symbol of order* $\mu \in \mathbb{R}$, *with constant coefficients. Then*

$$\operatorname{Op}(p) : \mathcal{S}(\mathbb{R}^q, E) \to \mathcal{S}(\mathbb{R}^q, E)$$

extends to a continuous operator

$$\operatorname{Op}(p) : \mathcal{W}^s(\mathbb{R}^q, E) \to \mathcal{W}^{s-\mu}(\mathbb{R}^q, E),$$

and $p \to \operatorname{Op}(p)$ *represents a continuous operator*

$$S^\mu(\mathbb{R}^q) \to \mathcal{L}(\mathcal{W}^s(\mathbb{R}^q, E), \mathcal{W}^{s-\mu}(\mathbb{R}^q, E))$$

for each $s \in \mathbb{R}$.

Corollary 1.3.25 is a special case of Proposition 1.3.24, since $p(\eta)$ can be identified with $a(\eta) = p(\eta) \cdot \operatorname{id}_E$.

Corollary 1.3.26 *Let* $p(\eta) \in S^\mu(\mathbb{R}^q)$ *be a symbol with* $c_1[\eta]^\mu \leq |p(\eta)| \leq c_2[\eta]^\mu$ *for all* $\eta \in \mathbb{R}^q$, *for constants* $c_i > 0$, $i = 1, 2$. *Then*

$$\operatorname{Op}(p) : \mathcal{W}^s(\mathbb{R}^q, E) \to \mathcal{W}^{s-\mu}(\mathbb{R}^q, E)$$

is an isomorphism for all $s \in \mathbb{R}$.

From the point of view of abstract wedge Sobolev spaces the above statements are rather elementary. Let us derive some properties of the standard Sobolev spaces in the Euclidean space.

First observe that the restriction operator

$$r_0 : \mathcal{S}(\mathbb{R}^m) \to \mathbb{C}, \qquad r_0 u := u(0)$$

extends to a continuous operator

$$r_0 : H^s(\mathbb{R}^m) \to \mathbb{C} \qquad \text{for every } s > \frac{m}{2}. \qquad (1.3.31)$$

In fact, if we set $\hat{u}(\xi) = (F_{x \to \xi} u)(\xi)$, i.e., $u(x) = \int e^{ix\xi} \hat{u}(\xi) \, d\xi$, we have $u(0) = \int \hat{u}(\xi) \, d\xi$, and hence

$$|u(0)| \leq \int |\hat{u}(\xi)| \, d\xi = \int \langle \xi \rangle^{-s} \langle \xi \rangle^s |\hat{u}(\xi)| \, d\xi$$

$$\leq \left\{ \int \langle \xi \rangle^{2s} |\hat{u}(\xi)|^2 d\xi \int \langle \xi \rangle^{-2s} \, d\xi \right\}^{\frac{1}{2}} \leq c \|u\|_{H^s(\mathbb{R}^m)}$$

for $c = \{ \int \langle \xi \rangle^{-2s} \, d\xi \}^{\frac{1}{2}}$ which is finite for $s > \frac{m}{2}$. The map (1.3.31) can be interpreted as an operator–valued symbol between the spaces $E = H^s(\mathbb{R}^m)$ with (1.3.27) and $\tilde{E} = \mathbb{C}$ with $\tilde{\kappa}_\lambda = \mathrm{id}$ for all $\lambda \in \mathbb{R}_+$. In view of

$$\tilde{\kappa}_\lambda^{-1} r_0 \kappa_\lambda u = r_0 \lambda^{\frac{m}{2}} u(\lambda x) = \lambda^{\frac{m}{2}} r_0 u$$

for all $\lambda \in \mathbb{R}_+$ we get

$$r_0 \in S_{\mathrm{cl}}^{\frac{m}{2}}(\mathbb{R}^q; H^s(\mathbb{R}^m), \mathbb{C}) \qquad \text{for } s > \frac{m}{2}. \qquad (1.3.32)$$

Proposition 1.3.27 *The restriction operator*

$$R_0 : \mathcal{S}(\mathbb{R}^{m+q}) \to \mathcal{S}(\mathbb{R}^q), \qquad (R_0 f)(y) = f(0, y)$$

extends to a continuous operator

$$R_0 : H^s(\mathbb{R}^{m+q}) \to H^{s-\frac{m}{2}}(\mathbb{R}^q)$$

for every $s > \frac{m}{2}$ and $q \geq 1$.

Proof. We can write $(R_0 f)(y) = F_{\eta \to y}^{-1} r_0 F_{y' \to \eta} f = \mathrm{Op}(r_0) f$. Then (1.3.32) and Proposition 1.3.24 imply that $R_0 : \mathcal{W}^s(\mathbb{R}^q, E) \to \mathcal{W}^{s-\frac{m}{2}}(\mathbb{R}^q, \mathbb{C})$ is continuous for $E = H^s(\mathbb{R}^m)$ with the group action (1.3.27). Using (1.3.28), (1.3.29) we thus obtain the assertion. $\square$

More generally, if we set

$$t_\alpha u = r_0 D_x^\alpha u \qquad \text{for } \alpha \in \mathbb{N}^m, \; u \in H^s(\mathbb{R}^m),$$

we obtain easily $t_\alpha \in S_{\mathrm{cl}}^{\frac{m}{2}+|\alpha|}(\mathbb{R}^q; H^s(\mathbb{R}^m), \mathbb{C})$. Then

$$b_\alpha(\eta) := [\eta]^{-\frac{m}{2}-|\alpha|} t_\alpha \in S_{\mathrm{cl}}^0(\mathbb{R}^q; H^s(\mathbb{R}^m), \mathbb{C}).$$

Let us set $\tilde{E} = \times_{|\alpha|\leq k}\mathbb{C}$ for a fixed $k \in \mathbb{N}$. Then

$$b(\eta) := \{b_\alpha(\eta)\}_{|\alpha|\leq k} \in S^0_{cl}(\mathbb{R}^q; H^s(\mathbb{R}^m), \tilde{E}).$$

Moreover, choose a cut-off function $\omega(x)$ in $\mathbb{R}^m$, i.e., $\omega \in C_0^\infty(\mathbb{R}^m)$, $\omega(x) = 1$ for $|x| \leq \varepsilon$ for an $\varepsilon > 0$. Then

$$p_\alpha(\eta)c := [\eta]^{\frac{m}{2}}\frac{1}{\alpha!}([\eta]x)^\alpha\omega([\eta]x)c, \qquad c \in \mathbb{C}, \qquad (1.3.33)$$

interpreted as a family of maps $p_\alpha : \mathbb{C} \to H^s(\mathbb{R}^m)$, satisfies $p_\alpha(\lambda\eta) = \kappa_\lambda p_\alpha(\eta)$ for all $\lambda \geq 1$, $|\eta| \geq$ const. It follows that

$$p_\alpha(\eta) \in S^0_{cl}(\mathbb{R}^q; \mathbb{C}, H^s(\mathbb{R}^m))$$

for every $s \in \mathbb{R}$. Thus

$$p(\eta) := \{p_\alpha(\eta)\}_{|\alpha|\leq k} \in S^0_{cl}(\mathbb{R}^q; \tilde{E}, H^s(\mathbb{R}^m)),$$

and we have $b(\eta)p(\eta) = \mathrm{id}_{\tilde{E}}$ for all $\eta \in \mathbb{R}^q$. For the associated pseudo–differential operators

$$\mathrm{Op}(b) : \mathcal{W}^s(\mathbb{R}^q, H^s(\mathbb{R}^m)) \to \mathcal{W}^s(\mathbb{R}^q, \tilde{E}) = H^s(\mathbb{R}^q, \tilde{E}), \qquad (1.3.34)$$

$$\mathrm{Op}(p) : \mathcal{W}^s(\mathbb{R}^q, \tilde{E}) \to \mathcal{W}^s(\mathbb{R}^q, H^s(\mathbb{R}^m))$$

we obtain that

$$\mathrm{Op}(b)\,\mathrm{Op}(p) = \mathrm{Op}(bp) : H^s(\mathbb{R}^q, \tilde{E}) \to H^s(\mathbb{R}^q, \tilde{E})$$

is the identity operator for all $s > \frac{m}{2} + k$. In other words, $\mathrm{Op}(b)$ is surjective, and $\mathrm{Op}(p)$ is a right inverse of $\mathrm{Op}(b)$.

Theorem 1.3.28 *The vector of operators* $T = \{T_\alpha\}_{|\alpha|\leq k}$,

$$T_\alpha : \mathcal{S}(\mathbb{R}^{m+q}) \to \mathcal{S}(\mathbb{R}^q),$$

for $(T_\alpha f)(y) = (D_x^\alpha f)(0, y)$, $|\alpha| \leq k$, *extends to a continuous and surjective operator*

$$T : H^s(\mathbb{R}^{m+q}) \to \times_{|\alpha|\leq k} H^{s-\frac{m}{2}-|\alpha|}(\mathbb{R}^q)$$

for every $s > \frac{m}{2} + k$. *We have a direct decomposition*

$$H^s(\mathbb{R}^{m+q}) = \ker T \oplus V^s_{(k)}(\mathbb{R}^{m+q})$$

for

$$V_{(k)}^s(\mathbb{R}^{m+q})$$

$$= F_{\eta \to y}^{-1} \Big\{ \sum_{|\alpha| \le k} [\eta]^{\frac{m}{2}} ([\eta]x)^\alpha \omega([\eta]x) \hat{v}_\alpha(\eta) : v_\alpha(y) \in H^s(\mathbb{R}^q), |\alpha| \le k \Big\},$$

$\hat{v}_\alpha(\eta) = (F_{y \to \eta} v_\alpha)(\eta)$, *where $\omega(x)$ is an arbitrary fixed cut–off function in $\mathbb{R}^m$.*

Proof. Let us set $d(\eta) = \mathrm{diag}\{[\eta]^{\frac{m}{2}+|\alpha|}\}_{|\alpha| \le k}$ which means the diagonal matrix with the corresponding diagonal elements. Then

$$\mathrm{Op}(d) : H^s(\mathbb{R}^q, \tilde{E}) \to \underset{|\alpha| \le k}{\times} H^{s-\frac{m}{2}-|\alpha|}(\mathbb{R}^q)$$

is an isomorphism. Using(1.3.34) we get a surjective continuous operator

$$\mathrm{Op}(d)\,\mathrm{Op}(b) = \mathrm{Op}(db) = T : H^s(\mathbb{R}^{m+q}) \to \underset{|\alpha| \le k}{\times} H^{s-\frac{m}{2}-|\alpha|}(\mathbb{R}^q),$$

cf. (1.3.28). Setting $h(\eta) = p(\eta)d^{-1}(\eta)$ the operator

$$\mathrm{Op}(h) := \underset{|\alpha| \le k}{\times} H^{s-\frac{m}{2}-|\alpha|}(\mathbb{R}^q) \to H^s(\mathbb{R}^{m+q})$$

is a right inverse of T and hence $H^s(\mathbb{R}^{m+q}) = \ker T \oplus \mathrm{im}\,\mathrm{Op}(h)$. Thus it remains to show that $\mathrm{im}\,\mathrm{Op}(h) = V_{(k)}^s(\mathbb{R}^{m+q})$. The symbol of $h(\eta)$ is a row matrix of entries

$$h_\alpha(\eta) = [\eta]^{-|\alpha|} \frac{1}{\alpha!} ([\eta]x)^\alpha \omega([\eta]x) \in S_{\mathrm{cl}}^{-\frac{m}{2}-|\alpha|}(\mathbb{R}^q; \mathbb{C}, H^r(\mathbb{R}^m)),$$

$|\alpha| \le k$, for arbitrary $r \in \mathbb{R}$. Let us set, in particular, $r = -\frac{m}{2} - |\alpha|$. Then

$$\mathrm{im}\,\mathrm{Op}(h_\alpha) = F_{\eta \to y}^{-1}\{[\eta]^{-\alpha}([\eta]x)^\alpha \omega([\eta]x) \hat{w}_\alpha(\eta) : w_\alpha \in H^{s-\frac{m}{2}-|\alpha|}(\mathbb{R}^q)\}.$$

For $\hat{v}_\alpha(\eta) = [\eta]^{-\frac{m}{2}-|\alpha|} \hat{w}_\alpha(\eta)$, i.e., $v_\alpha = \mathrm{Op}([\eta]^{-\frac{m}{2}-|\alpha|}) w_\alpha \in H^s(\mathbb{R}^q)$ we get the assertion. $\square$

The above relations on classical Sobolev spaces permit modifications also for the case $\mathbb{R}_+$ instead of $\mathbb{R}^m$. Set

$$H^s(\mathbb{R}_+) = \{u|_{\mathbb{R}_+} : u \in H^s(\mathbb{R})\},$$

endowed with the topology of the quotient space

$$H^s(\mathbb{R}_+) \cong H^s(\mathbb{R})/H_0^s(\overline{\mathbb{R}}_-).$$

Here $H_0^s(\overline{\mathbb{R}}_\pm)$ is defined as the closure of $C_0^\infty(\mathbb{R}_\pm)$ in $H^s(\mathbb{R})$. We then have analogously

$$H^s(\mathbb{R}_-) = \{u|_{\mathbb{R}_-} : u \in H^s(\mathbb{R})\} \cong H^s(\mathbb{R})/H_0^s(\overline{\mathbb{R}}_+).$$

More generally, if $H_0^s(\overline{\mathbb{R}}_\pm^{1+q})$ is the closure of $C_0^\infty(\mathbb{R}_\pm^{1+q})$ in $H^s(\mathbb{R}^{1+q})$, where $\mathbb{R}_\pm^{1+q} = \{(t,y) \in \mathbb{R}^{1+q} : t \gtrless 0\}$, then we obtain

$$H^s(\mathbb{R}_\pm^{1+q}) := H^s(\mathbb{R}^{1+q})|_{\mathbb{R}_\pm^{1+q}} = H^s(\mathbb{R}^{1+q})/H_0^s(\overline{\mathbb{R}}_\mp^{1+q}).$$

Remark 1.3.29 $\kappa_\lambda : u(t) \to \lambda^{\frac{1}{2}}u(\lambda t)$, $\lambda \in \mathbb{R}_+$, *defines strongly continuous groups of isomorphisms both on $H^s(\mathbb{R}_+)$ and $H_0^s(\overline{\mathbb{R}}_+)$, and we have*

$$\mathcal{W}^s(\mathbb{R}^q, H^s(\mathbb{R}_+)) = H^s(\mathbb{R}_+^{1+q}),$$

$$\mathcal{W}^s(\mathbb{R}^q, H_0^s(\overline{\mathbb{R}}_+)) = H_0^s(\overline{\mathbb{R}}_+^{1+q}).$$

Analogous relations hold for $\mathbb{R}_-$.

Theorem 1.3.30 *The vector of operators $T = \{T_j\}_{j=0,\dots,k}$,*

$$T_j : \mathcal{S}(\overline{\mathbb{R}}_+^{1+q}) \to \mathcal{S}(\mathbb{R}^q),$$

$(T_j f)(y) = (D_t^j f)(0,y)$, $j = 0,\dots,k$, *extends to a continuous and surjective operator*

$$T : H^s(\mathbb{R}_+^{1+q}) \to \underset{j=0}{\overset{k}{\times}} H^{s-\frac{1}{2}-j}(\mathbb{R}^q)$$

for every $s > \frac{1}{2} + k$, and we have a direct decomposition

$$H^s(\mathbb{R}_+^{1+q}) = \ker T \oplus V_{(k)}^s(\mathbb{R}_+^{1+q})$$

for

$$V_{(k)}^s(\mathbb{R}_+^{1+q})$$

$$= F_{\eta \to y}^{-1}\Big\{\sum_{j=0}^{k}[\eta]^{\frac{1}{2}}([\eta]t)^j\omega([\eta]t)\hat{v}_j(\eta) : v_j \in H^s(\mathbb{R}^q), j = 0,\dots,k\Big\}.$$

Proof. This follows in an analogous manner to the above Theorem 1.3.28. □

It is well known that $\ker T = H_0^s(\overline{\mathbb{R}}_+^{1+q})$ for $\frac{1}{2} + k < s < \frac{3}{2} + k$.
We now return to the investigation of the spaces $\mathcal{W}^s(\mathbb{R}^q, E)$ in general. Let us set

$$\mathcal{S}'(\mathbb{R}^q, E) = \mathcal{L}(\mathcal{S}(\mathbb{R}^q), E),$$

regarded as the space of E-valued temperate distributions in $\mathbb{R}^q$. For every $u \in \mathcal{S}(\mathbb{R}^q, E)$, $\varphi \in \mathcal{S}(\mathbb{R}^q)$ we can define the pairing

$$\langle \varphi, u \rangle = \int\limits_{\mathbb{R}^q} \varphi(y) u(y) \, dy$$

which induces an embedding $\mathcal{S}(\mathbb{R}^q, E) \hookrightarrow \mathcal{S}'(\mathbb{R}^q, E)$.

Proposition 1.3.31 *We have* $\mathcal{W}^s(\mathbb{R}^q, E) \hookrightarrow \mathcal{S}'(\mathbb{R}^q, E)$ *for every* $s \in \mathbb{R}$.

Proof. Let $\varphi \in \mathcal{S}(\mathbb{R}^q)$, $u \in \mathcal{W}^s(\mathbb{R}^q, E)$ be given. Then, with various constants $c > 0$, we have for sufficiently large N

$$\|\langle \varphi, u \rangle\|_E = \| \int\limits_{\mathbb{R}^q} \varphi(y) u(y) dy \|_E = \| \int\limits_{\mathbb{R}^q} \hat{\varphi}(\eta) \hat{u}(\eta) \, d\eta \|_E$$

$$= \| \int [\eta]^N \hat{\varphi}(\eta)[\eta]^{-N} \hat{u}(\eta) \, d\eta \|_E$$

$$\leq \sup_{\eta} [\eta]^N |\hat{\varphi}(\eta)| \int [\eta]^{-N} \|\hat{u}(\eta)\|_E \, d\eta$$

$$\leq c \int [\eta]^{-N} \|\hat{u}(\eta)\|_E d\eta$$

$$\leq c \left\{ \int [\eta]^{-N} \, d\eta \right\}^{\frac{1}{2}} \left\{ \int [\eta]^{-N} \|\hat{u}(\eta)\|_E^2 \, d\eta \right\}^{\frac{1}{2}}$$

$$\leq c \left\{ \int [\eta]^{-N} \|\kappa(\eta) \kappa^{-1}(\eta) \hat{u}(\eta)\|_E^2 \, d\eta \right\}^{\frac{1}{2}}$$

$$\leq c \left\{ \int [\eta]^{-N+2M} \|\kappa^{-1}(\eta) \hat{u}(\eta)\|_E^2 \, d\eta \right\}^{\frac{1}{2}}.$$

Here we used Proposition 1.3.1 with the estimate $\|\kappa(\eta)\|_E \leq c[\eta]^M$ for a certain M. Choosing N so large that $-N + 2M \leq 2s$ we obtain

$$\|\langle \varphi, u \rangle\|_E \leq c \|u\|_{\mathcal{W}^s(\mathbb{R}^q, E)}.$$

Thus the pairing with u represents a continuous operator $\mathcal{S}(\mathbb{R}^q) \to E$. □

Remark 1.3.32 *The space* $\mathcal{W}^s(\mathbb{R}^q, E)$ *can equivalently be described as the subspace of all* $u \in \mathcal{S}'(\mathbb{R}^q, E)$ *for which* $\|u\|_{\mathcal{W}^s(\mathbb{R}^q, E)} < \infty$.

Proposition 1.3.33 *Let E be a Banach space with given $\{\kappa_\lambda\}_{\lambda \in \mathbb{R}_+}$ and E' a subspace, continuously embedded in E, with $\kappa'_\lambda := \kappa_\lambda|_{E'}$, $\lambda \in \mathbb{R}_+$, acting as a strongly continuous group of isomorphisms on E'. Then there are canonical continuous embeddings*

$$\mathcal{W}^{s'}(\mathbb{R}^q, E') \hookrightarrow \mathcal{W}^s(\mathbb{R}^q, E)$$

for all $s' \geq s$.

Proof. Proposition 1.3.22 can be applied for $\nu = 0$. This yields a continuous embedding $\mathcal{W}^{s'}(\mathbb{R}^q, E') \hookrightarrow \mathcal{W}^{s'}(\mathbb{R}^q, E)$. Moreover, from the above Definition 1.3.19 we get a continuous embedding $\mathcal{W}^{s'}(\mathbb{R}^q, E) \hookrightarrow \mathcal{W}^s(\mathbb{R}^q, E)$ for $s' \geq s$. $\square$

Theorem 1.3.34 *Let E be a Banach space and $\{\kappa_\lambda\}_{\lambda \in \mathbb{R}_+}$ be a strongly continuous group of isomorphisms on E. Then the operator $\mathcal{M}_\varphi$ of multiplication by $\varphi \in \mathcal{S}(\mathbb{R}^q)$ induces a continuous operator*

$$\mathcal{M}_\varphi : \mathcal{W}^s(\mathbb{R}^q, E) \to \mathcal{W}^s(\mathbb{R}^q, E)$$

for every $s \in \mathbb{R}$. Moreover, the mapping

$$\mathcal{S}(\mathbb{R}^q) \times \mathcal{W}^s(\mathbb{R}^q, E) \to \mathcal{W}^s(\mathbb{R}^q, E)$$

obtained by $(\varphi, u) \to \mathcal{M}_\varphi u$ is continuous for every s.

Proof. In order to prove the continuity of $\mathcal{M}_\varphi$ in $\mathcal{W}^s(\mathbb{R}^q, E)$ it suffices to show the corresponding norm estimate on $\mathcal{S}(\mathbb{R}^q, E)$ which is a dense subspace of $\mathcal{W}^s(\mathbb{R}^q, E)$. The Fourier transform $F = F_{y \to \eta}$ induces an isomorphism $F : \mathcal{S}(\mathbb{R}^q, E) \to \mathcal{S}(\mathbb{R}^q, E)$. Moreover, $\varphi \in \mathcal{S}(\mathbb{R}^q)$, $u \in \mathcal{S}(\mathbb{R}^q, E)$ implies $\varphi u \in \mathcal{S}(\mathbb{R}^q, E)$, and we have $F(\varphi u) = (F\varphi) * (Fu)$. Let us set

$$m(\eta) = \|[\eta]^s \kappa^{-1}(\eta) \int F\varphi(\eta - \xi) Fu(\xi)\, d\xi\|_E.$$

Then

$$\|\varphi u\|^2_{\mathcal{W}^s(\mathbb{R}^q, E)} = \int [\eta]^{2s} \|\kappa^{-1}(\eta) F(\varphi u)(\eta)\|^2_E\, d\eta = \|m\|^2_{L^2(\mathbb{R}^q)}.$$

For every $\xi, \eta \in \mathbb{R}^q$ and $s \in \mathbb{R}$ we have $[\eta]^s \leq c^{|s|}[\eta - \xi]^{|s|}[\xi]^s$ for a constant $c > 0$. Moreover, the function $K(\lambda)$ from (1.3.5) satisfies

$K([\eta]/[\eta]) \leq c[\eta - \xi]$. Applying Proposition 1.3.1 we obtain

$$
\begin{aligned}
m(\eta) &\leq c_s \|\kappa^{-1}(\eta) \int [\eta - \xi]^{|s|} \hat{\varphi}(\eta - \xi)[\xi]^s \hat{u}(\xi) \, d\xi \|_E \\
&= c_s \| \int \kappa^{-1}_{([\eta]/[\xi])}[\eta - \xi]^{|s|-k}[\eta - \xi]^k \hat{\varphi}(\eta - \xi)[\xi]^s \kappa^{-1}(\xi) \hat{u}(\xi) \, d\xi \|_E \\
&\leq \tilde{c}_s \int [\eta - \xi]^{|s|+M-k}[\eta - \xi]^k |\hat{\varphi}(\eta - \xi)|[\xi]^s \|\kappa^{-1}(\xi) \hat{u}(\xi)\|_E \, d\xi \\
&\leq \tilde{c}_s C_k(\varphi) \int [\eta - \xi]^{|s|+M-k}[\xi]^s \|\kappa^{-1}(\xi) \hat{u}(\xi)\|_E \, d\xi
\end{aligned}
\tag{1.3.35}
$$

for

$$
C_k(\varphi) = \sup_\xi [\xi]^k |\hat{\varphi}(\xi)|.
\tag{1.3.36}
$$

Let us set $g(\eta) = [\eta]^{|s|+M-k}$, $h(\eta) = [\eta]^s \|\kappa^{-1}(\xi) \hat{u}(\xi)\|_E$ and choose $k \in \mathbb{N}$ so large that $|s| + M - k < -q$. Then $g \in L^1(\mathbb{R}^q)$, $h \in L^2(\mathbb{R}^q)$, and we have $\|h\|_{L^2(\mathbb{R}^q)} = \|u\|_{\mathcal{W}^s(\mathbb{R}^q, E)}$. Writing the inequality (1.3.35) in the form

$$
m(\eta) \leq \tilde{c}_s C_k(\varphi)(g * h)(\eta) \qquad \text{for } \eta \in \mathbb{R}^q
$$

we can apply Young's inequality to see that

$$
\begin{aligned}
\|\varphi u\|_{\mathcal{W}^s(\mathbb{R}^q, E)} = \|m\|_{L^2(\mathbb{R}^q)} &\leq \tilde{c}_s C_k(\varphi) \|g * h\|_{L^2(\mathbb{R}^q)} \\
&\leq \tilde{c}_s C_k(\varphi) \|g\|_{L^1(\mathbb{R}^q)} \|h\|_{L^2(\mathbb{R}^q)} \\
&= C_s C_k(\varphi) \|u\|_{\mathcal{W}^s(\mathbb{R}^q, E)}.
\end{aligned}
$$

Thus $\mathcal{M}_\varphi$ is continuous. At the same time the latter estimate shows the continuity of $(\varphi, u) \to \varphi u$ which completes the proof. □

If we fix $s \in \mathbb{R}$ in Theorem 1.3.34 we can admit more general functions φ for the continuity of the operator $\mathcal{M}_\varphi : \mathcal{W}^s(\mathbb{R}^q, E) \to \mathcal{W}^s(\mathbb{R}^q, E)$. For the constant $C_k(\varphi)$ in the final estimate we have chosen $k > |s| + M + q$. Now let us write

$$
F\varphi(\eta) = \int e^{-iy\eta} \varphi(y) \, dy = \int e^{-iy\eta}(1 + |\eta|^2)^{-m}(1 - \Delta_y)^m \varphi(y) \, dy
$$

for any $m \in \mathbb{N}$. Then, using $\varphi \in \mathcal{S}(\mathbb{R}^q)$, we obtain

$$C_k(\varphi) \leq \sup_{\eta}[\eta]^k |F\varphi(\eta)|$$

$$= \sup_{\eta}[\eta]^k |\int e^{-iy\eta}(1+|\eta|^2)^{-m}(1+|y|^2)^{-n}(1+|y|^2)^n$$

$$(1-\Delta_y)^m\varphi(y)\,dy|$$

$$\leq \sup_{\eta}[\eta]^k(1+|\eta|^2)^{-m} \int (1+|y|^2)^{-n}\,dy$$

$$\times \sup_{y\in\mathbb{R}^q} |(1+|y|^2)^n(1-\Delta_y)^m\varphi(y)|$$

$$\leq c \sup_{y\in\mathbb{R}^q} |(1+|y|^2)^n(1-\Delta_y)^m\varphi(y)|$$

for a constant $c > 0$, where we assume $n > \frac{q}{2}$ and $k \leq 2m$. This yields the following:

Remark 1.3.35 *For every $s \in \mathbb{R}$ we have*

$$\|\mathcal{M}_\varphi\|_{\mathcal{L}(\mathcal{W}^s(\mathbb{R}^q,E))} \leq c \sup_{y\in\mathbb{R}^q} |(1+|y|^2)^n(1-\Delta_y)^m\varphi(y)|$$

for every $m, n \in \mathbb{N}$ with $|s| + M + q \leq 2m$, $n > \frac{q}{2}$, for a constant $c > 0$. Thus, if $F^{m,n}(\mathbb{R}^q)$ denotes the closure of $\mathcal{S}(\mathbb{R}^q)$ in the norm

$$\|\varphi\|_{F^{m,n}(\mathbb{R}^q)} = \sup_{y\in\mathbb{R}^q} |(1+|y|^2)^n(1-\Delta_y)^m\varphi(y)|$$

the multiplication operator $\mathcal{M}_\varphi$ is continuous in $\mathcal{W}^s(\mathbb{R}^q, E)$ for all $\varphi \in F^{m,n}(\mathbb{R}^q)$.

Definition 1.3.36 *If $\Omega \subseteq \mathbb{R}^q$ is an open set we define*

$$\mathcal{W}^s_{\text{comp}}(\Omega, E) = \{u \in \mathcal{W}^s(\mathbb{R}^q, E) : \text{ supp } u \subset \Omega \text{ compact}\},$$

with $\mathcal{W}^s_{\text{comp}}(\Omega, E)$ being identified with a subspace of $\mathcal{E}'(\Omega, E)$, and

$$\mathcal{W}^s_{\text{loc}}(\Omega, E)$$

$$= \{u \in \mathcal{D}'(\Omega, E) : \varphi u \in \mathcal{W}^s_{\text{comp}}(\Omega, E) \text{ for each } \varphi \in C_0^\infty(\Omega)\}.$$

In view of Proposition 1.3.31 and Theorem 1.3.34 the Definition 1.3.36 makes sense. $\mathcal{W}^s_{\text{loc}}(\Omega, E)$ is a Fréchet space with the semi–norm system

$$u \to \|\varphi u\|_{\mathcal{W}^s(\mathbb{R}^q,E)}, \qquad \varphi \in C_0^\infty(\Omega),$$

where a countable system of function φ suffices to decribe the Fréchet topology.

Let us set for short $\mathcal{W}^s(\mathbb{R}^q, E)_K = \{u \in \mathcal{W}^s(\mathbb{R}^q, E) : \operatorname{supp} u \subseteq K\}$ for a compact set $K \subset \mathbb{R}^q$. Then $\mathcal{W}^s(\mathbb{R}^q, E)_K$ is a closed subspace of $\mathcal{W}^s(\mathbb{R}^q, E)$. Moreover,

$$\mathcal{W}^s_{\mathrm{comp}}(\Omega, E) = \varinjlim{}_K \mathcal{W}^s(\mathbb{R}^q, E)_K,$$

where the inductive limit is taken over all $K \Subset \Omega$.

Let us now formulate without proof an interpolation property of the $\mathcal{W}^s(\mathbb{R}^q, E)$ spaces. A proof of this result may be found in Hirschmann [51].

Let E_0, E_1 be Banach spaces that are continuously embedded in a Hausdorff topological vector space H. According to the complex interpolation method, cf. [148], we then have an interpolation space $[E_0, E_1]_\theta$ for every $\theta \in [0, 1]$.

Assume that on the space $E_0 + E_1$ there is given a strongly continuous group $\{\kappa_\lambda\}_{\lambda \in \mathbb{R}_+}$ of isomorphisms which restricts to strongly continuous groups of isomorphisms $\{\kappa^i_\lambda\}_{\lambda \in \mathbb{R}_+}$ on E_i for $i = 0, 1$. We then have the space $\mathcal{W}^s(\mathbb{R}^q, E_i)$, $i = 0, 1$, for each $s \in \mathbb{R}$, and $[\mathcal{W}^{s_0}(\mathbb{R}^q, E_0), \mathcal{W}^{s_1}(\mathbb{R}^q, E_1)]_\theta$ is well–defined for every $s_0, s_1 \in \mathbb{R}$ and $\theta \in [0, 1]$. Moreover, $\{\kappa_\lambda\}_{\lambda \in \mathbb{R}_+}$ induces strongly continuous groups of isomorphisms $\{\kappa^\theta_\lambda\}_{\lambda \in \mathbb{R}_+}$ on $[E_0, E_1]_\theta$ for all $\theta \in [0, 1]$.

Theorem 1.3.37 *For every s_0, $s_1 \in R$ and $\theta \in [0, 1]$ we have*

$$[\mathcal{W}^{s_0}(\mathbb{R}^q, E_0), \mathcal{W}^{s_1}(\mathbb{R}^q, E_1)]_\theta = \mathcal{W}^s(\mathbb{R}^q, [E_0, E_1]_\theta)$$

for $s = (1 - \theta)s_0 + \theta s_1$.

Proposition 1.3.38 *Let E be a Banach space written as a non–direct sum $E = E_1 + E_2$ of Banach subspaces E_1, E_2, and let $\{\kappa_\lambda\}_{\lambda \in \mathbb{R}_+}$ be a strongly continuous group of isomorphisms on E_i for $i = 1, 2$. Then we have*

$$\mathcal{W}^s(\mathbb{R}^q, E) = \mathcal{W}^s(\mathbb{R}^q, E_1) + \mathcal{W}^s(\mathbb{R}^q, E_2), \quad s \in \mathbb{R}, \qquad (1.3.37)$$

in the sense of non–direct sums of Banach spaces.

Proof. Note that $\{\kappa_\lambda\}_{\lambda \in \mathbb{R}_+}$ induces strongly continuous groups of isomorphisms on $E_1 \oplus E_2$ and $E_1 \cap E_2$, where the latter space is endowed with the topology of the intersection. Then we can form the spaces

$$\mathcal{W}^s(\mathbb{R}^q, E_1 \oplus E_2) = \mathcal{W}^s(\mathbb{R}^q, E_1) \oplus \mathcal{W}^s(\mathbb{R}^q, E_2), \qquad (1.3.38)$$

$$\mathcal{W}^s(\mathbb{R}^q, E_1 \cap E_2) = \mathcal{W}^s(\mathbb{R}^q, E_1) \cap \mathcal{W}^s(\mathbb{R}^q, E_2). \qquad (1.3.39)$$

OPERATORS WITH OPERATOR-VALUED SYMBOLS 113

We have $\Delta := \{(e, -e) : e \in E_1 \cap E_2\} \simeq E_1 \cap E_2$, and $\Xi := \{(u, -u) : u \in \mathcal{W}^s(\mathbb{R}^q, E_1) \cap \mathcal{W}^s(\mathbb{R}^q, E_2)\} \simeq \mathcal{W}^s(\mathbb{R}^q, E_1 \cap E_2)$. Now

$$\mathcal{W}^s(\mathbb{R}^q, E_1) + \mathcal{W}^s(\mathbb{R}^q, E_2) = \mathcal{W}^s(\mathbb{R}^q, E_1 \oplus E_2)/\Xi$$

$$= \mathcal{W}^s(\mathbb{R}^q, E_1 \oplus E_2/\Delta)$$

$$= \mathcal{W}^s(\mathbb{R}^q, E_1 + E_2).$$

$\square$

Proposition 1.3.39 *Let E^0 be a Hilbert space and $\{\kappa_\lambda\}_{\lambda \in \mathbb{R}_+}$ be a strongly continuous group of unitary operators on E^0. Then, if*

$$\chi : \mathbb{R}^q \to \mathbb{R}^q$$

is a diffeomorphism satisfying $0 < c^{-1} \leq |\det(d\chi^{-1})(x)| \leq c$ for all $x \in \mathbb{R}^q$ for some constant $c \geq 1$, the pull-back $\chi^* : C_0^\infty(\mathbb{R}^q, E^0) \to C_0^\infty(\mathbb{R}^q, E^0)$ extends to an isomorphism

$$\chi^* : \mathcal{W}^0(\mathbb{R}^q, E^0) \to \mathcal{W}^0(\mathbb{R}^q, E^0).$$

Proof. Write $x = \chi(y)$. If $\{e_k\}_{k \in \mathbb{N}}$ is an orthonormal base of E^0 every $u(y) \in C_0^\infty(\mathbb{R}^q, E^0)$ can be written $u(y) = \sum_{k=0}^\infty u_k(y)e_k$ for $u_k \in C_0^\infty(\mathbb{R}^q)$. In view of the assumption that $\{\kappa_\lambda\}_{\lambda \in \mathbb{R}_+}$ is a unitary group we have

$$\|\chi^* u\|_{\mathcal{W}^0(\mathbb{R}^q, E^0)}^2 = \int \|F_{x \to \xi}(u \circ \chi)(\xi)\|_{E^0}^2 \, d\xi.$$

By the Plancherel Theorem we get

$$\|\chi^* u\|_{\mathcal{W}^0(\mathbb{R}^q, E^0)}^2 = \int_{\mathbb{R}^q} \|\sum_{k=0}^\infty e_k(F_{x \to \xi}(u \circ \chi))_k(\xi)\|_{E^0}^2 \, d\xi$$

$$= \sum_{k=0}^\infty \int_{\mathbb{R}^q} |(F_{x \to \xi}(u \circ \chi))_k(\xi)|^2 \, d\xi$$

$$= (2\pi)^q \sum_{k=0}^\infty \int_{\mathbb{R}^q} |(u \circ \chi)_k(x)|^2 \, dx$$

$$= (2\pi)^q \int_{\mathbb{R}^q} \|\chi^* u(x)\|_{E^0}^2 \, dx$$

$$= (2\pi)^q \int_{\mathbb{R}^q} \|u(y)\|_{E^0}^2 |\det(d\chi^{-1})(\chi(y))| \, dy.$$

Thus

$$\|\chi^* u\|^2_{\mathcal{W}^0(\mathbb{R}^q, E^0)} \leq c(2\pi)^q \int\limits_{\mathbb{R}^q} \|u(y)\|^2_{E^0}\, dy$$

$$= c(2\pi)^q \sum_{k=0}^{\infty} \int\limits_{\mathbb{R}^q} |u_k(y)|^2\, dy$$

$$= c \sum_{k=0}^{\infty} \int\limits_{\mathbb{R}^q} |(F_{y\to\eta} u_k)(\eta)|^2\, d\eta$$

$$= \int\limits_{\mathbb{R}^q} \|F_{y\to\eta} u(\eta)\|^2_{E^0}\, d\eta$$

$$= c\|u\|^2_{\mathcal{W}^0(\mathbb{R}^q, E^0)}.$$

Since $C_0^\infty(\mathbb{R}^q, E^0)$ is dense in $\mathcal{W}^0(\mathbb{R}^q, E^0)$, the assertion is proved. $\square$

Remark 1.3.40 *Let E^0 be a Hilbert space and $\{\kappa_\lambda\}_{\lambda\in\mathbb{R}_+}$ be a strongly continuous group of unitary operators on E^0. Then, if $\chi : \Omega \to \tilde{\Omega}$ is a diffeomorphism between open sets $\Omega,\ \tilde{\Omega} \subseteq \mathbb{R}^q$, it follows that $\chi^* : C^\infty(\tilde{\Omega}, E^0) \to C^\infty(\Omega, E^0)$ induces isomorphisms*

$$\chi^* : \mathcal{W}^0_{\mathrm{comp}}(\tilde{\Omega}, E^0) \to \mathcal{W}^0_{\mathrm{comp}}(\Omega, E^0),$$

$$\chi^* : \mathcal{W}^0_{\mathrm{loc}}(\tilde{\Omega}, E^0) \to \mathcal{W}^0_{\mathrm{loc}}(\Omega, E^0).$$

The proof of the comp case is an easy modification of the arguments for Proposition 1.3.39. The loc case is simple as well, since it suffices to argue in terms of φu for arbitrary $\varphi \in C_0^\infty(\tilde{\Omega})$, $u \in \mathcal{W}^0_{\mathrm{loc}}(\tilde{\Omega}, E^0)$, where $\varphi u \in \mathcal{W}^0_{\mathrm{comp}}(\tilde{\Omega}, E^0)$.

Definition 1.3.41 *Let E, E^0, E' be (separable) Hilbert spaces, $E \hookrightarrow E^0 \hookrightarrow E'$ continuous dense embeddings, and let $\{\kappa_\lambda\}_{\lambda\in\mathbb{R}_+}$, $\{\kappa^0_\lambda\}_{\lambda\in\mathbb{R}_+}$ and $\{\kappa'_\lambda\}_{\lambda\in\mathbb{R}_+}$ be strongly continuous groups of isomorphisms in E, E^0 and E', respectively, where $\{\kappa^0_\lambda\}_{\lambda\in\mathbb{R}_+}$ is unitary, and $\kappa_\lambda = \kappa^0_\lambda|_E$, $\kappa^0_\lambda = \kappa'_\lambda|_{E^0}$ for all $\lambda \in \mathbb{R}_+$. Let us set for short $\kappa_\lambda = \kappa^0_\lambda = \kappa'_\lambda$ if it is obvious where the actions are defined. We call*

$$\left\{ E, E^0, E'; \{\kappa_\lambda\}_{\lambda\in\mathbb{R}_+} \right\}$$

a Hilbert space triple with unitary actions.

Of course, $\{\kappa_\lambda\}_{\lambda\in\mathbb{R}_+}$ is not unitary on E and E' in general. An example is

$$\left\{ H^s(\mathbb{R}^n), H^0(\mathbb{R}^n), H^{-s}(\mathbb{R}^n); \{\kappa_\lambda\}_{\lambda\in\mathbb{R}_+} \right\}$$

for $s \geq 0$, with $(\kappa_\lambda u)(x) = \lambda^{\frac{n}{2}} u(\lambda x)$, $\lambda \in \mathbb{R}_+$.

Remark 1.3.42 *If* E, E_0 *are Hilbert spaces* $E \hookrightarrow E^0$ *continuous and dense, where a strongly continuous unitary group* $\{\kappa_\lambda^0\}_{\lambda \in \mathbb{R}_+}$ *on* E^0 *is given such that* $\kappa_\lambda := \kappa_\lambda^0|_E$ *is a strongly continuous group of isomorphisms on* E, *then we can generate a Hilbert space* E' *as the dual of* E *with respect to the non–degenerate sesquilinear pairing* $(\cdot,\cdot) : E \times E' \to \mathbb{C}$ *induced by extending* $(\cdot,\cdot)_{E^0} : E \times E \to \mathbb{C}$ *with the scalar product* $(\cdot,\cdot)_{E^0}$ *in* E^0. *It can be easily proved that then* $E^0 \hookrightarrow E'$ *is continuous and dense and that* $(\kappa_\lambda u, v) = (u, \kappa_\lambda' v)$ *for all* $u \in E$, $v \in E'$ *gives rise to a strongly continuous group of isomorphisms in* E' *with* $\kappa_\lambda^0 = \kappa_\lambda'|_{E^0}$. *Thus we get a Hilbert space triple with unitary actions.*

Remark 1.3.43 *Let* $\{E, E^0, E'; \{\kappa_\lambda\}_{\lambda \in \mathbb{R}_+}\}$ *be a Hilbert space triple with unitary actions. Let us form* $(\chi_\lambda u)(y) = \kappa_\lambda \lambda^{q/2} u(\lambda y)$ *for* $u \in \mathcal{S}(\mathbb{R}^q, E^0)$. *Then* $\{\chi_\lambda\}_{\lambda \in \mathbb{R}_+}$ *extends to a unitary group on* $\mathcal{W}^0(\mathbb{R}^q, E^0)$. *It restricts to a strongly continuous group of isomorphisms on the space* $\mathcal{W}^s(\mathbb{R}^q, E)$ *and extends to a strongly continuous group of isomorphisms on* $\mathcal{W}^{-s}(\mathbb{R}^q, E')$ *for every* $s \geq 0$. *Then*

$$\{\mathcal{W}^s(\mathbb{R}^q, E), \mathcal{W}^0(\mathbb{R}^q, E^0), \mathcal{W}^{-s}(\mathbb{R}^q, E'); \{\chi_\lambda\}_{\lambda \in \mathbb{R}_+}\}$$

is a Hilbert space triple with unitary actions.

Proposition 1.3.44 *Let* E *be endowed with* $\{\kappa_\lambda\}_{\lambda \in \mathbb{R}_+}$ *and* $\mathcal{W}^s(\mathbb{R}^q, E)$ *with* $\{\chi_\lambda\}_{\lambda \in \mathbb{R}_+}$ *as in Remark 1.3.43. Then we have*

$$\mathcal{W}^s(\mathbb{R}^p, \mathcal{W}^s(\mathbb{R}^q, E)) = \mathcal{W}^s(\mathbb{R}^{p+q}, E), \qquad s \in \mathbb{R}, \tag{1.3.40}$$

where the space on the left follows from Definition 1.3.19, with the dimension p *and* $\{\mathcal{W}^s(\mathbb{R}^q, E), \{\chi_\lambda\}_{\lambda \in \mathbb{R}_+}\}$ *instead of* q *and* $\{E, \{\kappa_\lambda\}_{\lambda \in \mathbb{R}_+}\}$, *respectively.*

Note that 1.3.40 is a generalisation of the above Example 1.3.23.

Proof. In this proof we indicate equivalence of norms by $\sim$, in particular,

$$\|f\|_{\mathcal{W}^s(\mathbb{R}^{p+q}, E)}$$

$$\sim \left\{ \iint ([\xi]^2 + |\eta|^2)^s \|\kappa^{-1}(\xi, \eta)(F_{(x,y) \to (\xi,\eta)} f)(\xi, \eta)\|_E^2 \, d\xi d\eta \right\}^{\frac{1}{2}},$$

$$\|u\|_{\mathcal{W}^s(\mathbb{R}^q, E)} \sim \left\{ \int (1 + |\eta|^2)^s \|\kappa^{-1}(\eta)(F_{y \to \eta} u)(\eta)\|_E^2 \, d\eta \right\}^{\frac{1}{2}}.$$

We now reformulate the norm for $f(x,y)$ as follows:

$$\|f\|^2_{\mathcal{W}^s(\mathbb{R}^{p+q},E)} = \iint [\xi]^{2s}\left(1 + \left(\frac{|\eta|}{[\xi]}\right)^2\right)^s \|\kappa^{-1}(\xi,\eta)\hat{f}(\xi,\eta)\|^2_E \, d\eta d\xi$$

$$= \int [\xi]^{2s} \int (1 + |\tilde{\eta}|^2)^s \|\kappa^{-1}(\xi,[\xi]\tilde{\eta})\hat{f}(\xi,[\xi]\tilde{\eta})\|^2_E [\xi]^q \, d\tilde{\eta}d\xi,$$

$$(1.3.41)$$

where $\hat{f} = Ff$ for $F = F_{(x,y)\to(\xi,\tau)}$. On the other hand we have

$$\|f\|^2_{\mathcal{W}^s(\mathbb{R}^p,\mathcal{W}^s(\mathbb{R}^q,E))} \qquad\qquad\qquad\qquad (1.3.42)$$

$$= \int [\xi]^{2s}\|\chi^{-1}(\xi)(F_{x\to\xi}f)(\xi,y)\|^2_{\mathcal{W}^s(\mathbb{R}^q,E)} \, d\xi$$

$$= \int [\xi]^{2s}\|\kappa^{-1}(\xi)[\xi]^{-\frac{q}{2}}(F_{x\to\xi}f)(\xi,[\xi]^{-1}y)\|^2_{\mathcal{W}^s(\mathbb{R}^q,E)} \, d\xi \qquad (1.3.43)$$

$$\sim \int [\xi]^{2s} \int (1 + |\eta|^2)^s \|\kappa^{-1}(\xi)\kappa^{-1}(\eta)[\xi]^{\frac{q}{2}}(Ff)(\xi,[\xi]\eta)\|^2_E \, d\eta d\xi.$$

$$(1.3.44)$$

In the latter equivalence we used the relation (1.3.7) and the above expression for the norm in $\mathcal{W}^s(\mathbb{R}^q, E)$.

In order to compare (1.3.41), (1.3.44) we observe

$$[\xi,[\xi]\eta] \sim (\langle\xi\rangle^2 + \langle\xi^2\rangle\langle\eta\rangle^2)^{\frac{1}{2}} \sim \langle\xi\rangle\langle\eta\rangle \sim [\xi][\eta],$$

i.e.,

$$c_1[\xi,[\xi]\eta] \le [\xi][\eta] \le c_2[\xi,[\xi]\eta]$$

for all $(\xi,\eta) \in \mathbb{R}^{p+q}$ for suitable constants c_1, c_2. Thus, in the expression (1.3.41) we may insert $\kappa^{-1}_{[\xi][\eta]}$ instead of $\kappa^{-1}(\xi,[\xi]\eta)$, up to equivalence of norms. Applying $\kappa^{-1}_{[\xi][\eta]} = \kappa^{-1}(\xi)\kappa^{-1}(\eta)$ we see that (1.3.41) is equivalent to (1.3.44).   □

The above constructions for the scalar Sobolev spaces can be generalised to the abstract wedge Sobolev spaces. Let us fix E and the group $\{\kappa_\lambda\}_{\lambda\in\mathbb{R}_+}$, and let M be the exponent from Proposition 1.3.1.

Proposition 1.3.45 *The operator of restriction $r_0 : \mathcal{S}(\mathbb{R}^m, E) \to E$, $r_0u = u(0)$, extends to a continuous operator*

$$r_0 : \mathcal{W}^s(\mathbb{R}^m, E) \to E \qquad\qquad (1.3.45)$$

for all $s > \frac{m}{2} + M$, and we have

$$r_0 = \lambda^{\frac{m}{2}} \kappa_\lambda r_0 \chi_\lambda^{-1} \quad \text{for all} \quad \lambda \in \mathbb{R}_+, \tag{1.3.46}$$

i.e., $r_0 \in S_{\mathrm{cl}}^{\frac{m}{2}}(\mathbb{R}^q; \mathcal{W}^s(\mathbb{R}^m, E), E)$ for arbitrary $q \in \mathbb{N}$.

Proof. From $u(x) = \int e^{ix\xi} \hat{u}(\xi) \, d\xi$ it follows that

$$\|u(0)\|_E \leq \int \|\hat{u}(\xi)\|_E \, d\xi = \int \langle\xi\rangle^{-s} \langle\xi\rangle^s \|\kappa(\xi)\kappa^{-1}(\xi)\hat{u}(\xi)\|_E \, d\xi$$

$$\leq \int \langle\xi\rangle^{-s+M} \langle\xi\rangle^s \|\kappa^{-1}(\xi)\hat{u}(\xi)\|_E \, d\xi$$

$$\leq \left(\int \langle\xi\rangle^{2(M-s)} \, d\xi \right)^{\frac{1}{2}} \left(\int \langle\xi\rangle^{2s} \|\kappa^{-1}(\xi)\hat{u}(\xi)\|_E^2 \, d\xi \right)^{\frac{1}{2}}$$

$$\leq c\|u\|_{\mathcal{W}^s(\mathbb{R}^m, E)}$$

for a constant $c > 0$, when $s > \frac{m}{2} + M$. This yields the extension (1.3.45). The homogeneity (1.3.46) is obvious, and hence r_0 is a classical operator–valued symbol of order $\frac{m}{2}$. □

Corollary 1.3.46 *The restriction operator*

$$R_0 : \mathcal{S}(\mathbb{R}^{m+q}, E) \to \mathcal{S}(\mathbb{R}^q, E), \qquad (R_0 f)(y) = f(0, y)$$

extends to a continuous operator

$$R_0 : \mathcal{W}^s(\mathbb{R}^{m+q}, E) \to \mathcal{W}^{s-\frac{m}{2}}(\mathbb{R}^q, E)$$

for every $s > \frac{m}{2} + M$ and $q \geq 1$.

The proof is similar to that of Proposition 1.3.27.

Analogously to (1.3.33) we can form a family of maps $p_\alpha(\eta) : E \to \mathcal{W}^s(\mathbb{R}^m, E)$, $\alpha \in \mathbb{N}^m$, by

$$p_\alpha(\eta)e = [\eta]^{\frac{m}{2}} \frac{1}{\alpha!}([\eta]x)^\alpha \omega([\eta]x)e.$$

Then the identity

$$p_\alpha(\lambda\eta)e = \lambda^{\frac{m}{2}} [\eta]^{\frac{m}{2}} \frac{1}{\alpha!}([\lambda\eta]x)^\alpha \omega([\lambda\eta]x)e$$

$$= [\eta]^{\frac{m}{2}} \frac{1}{\alpha!} \chi_\lambda([\eta]x)^\alpha \omega([\eta]x) \kappa_\lambda^{-1} e$$

$$= \chi_\lambda p_\alpha(\eta) \kappa_\lambda^{-1} e$$

gives us $p_\alpha(\eta) \in S_{\mathrm{cl}}^0(\mathbb{R}^q; E, \mathcal{W}^s(\mathbb{R}^m, E))$ for arbitrary $s \in \mathbb{R}$. We then obtain an analogue of Theorem 1.3.28. The obvious proof will be omitted.

Theorem 1.3.47 *The vector of operators*

$$T = \{T_\alpha\}_{|\alpha|\leq k} : \mathcal{S}(\mathbb{R}^{m+q}, E) \to \mathcal{S}(\mathbb{R}^q, E),$$

$(T_\alpha f)(y) = (D_x^\alpha f)(0, y)$, $|\alpha| \leq k$, *extends to a continuous surjective operator*

$$T : \mathcal{W}^s(\mathbb{R}^{m+q}, E) \to \underset{|\alpha|\leq k}{\times} \mathcal{W}^{s-\frac{m}{2}-|\alpha|}(\mathbb{R}^q, E)$$

for every $s > \frac{m}{2} + M + k$. *We have a direct decomposition*

$$\mathcal{W}^s(\mathbb{R}^{m+q}) = \ker T \oplus V_{(k)}^s(\mathbb{R}^{m+q}, E)$$

for

$$V_{(k)}^s(\mathbb{R}^{m+q}, E) = F_{\eta\to y}^{-1}\Big\{ \sum_{|\alpha|\leq k} [\eta]^{\frac{m}{2}}([\eta]x)^\alpha \omega([\eta]x)\hat{v}_\alpha(\eta) :$$

$$v_\alpha(y) \in \mathcal{W}^s(\mathbb{R}^q, E), |\alpha| \leq k \Big\}.$$

Example 1.3.48 *Let us give another explicit example of non–trivial groups* $\{\kappa_\lambda\}_{\lambda\in\mathbb{R}_+}$. *Let* X *be a closed compact* C^∞ *manifold and* v *be an arbitrary vector field on* X *(i.e., a* C^∞ *section in* TX*). Denote by* $\{\varphi_t\}_{t\in\mathbb{R}}$ *the associated group of diffeomorphisms of* X. *Then by* $u(x) \to u(\varphi_t^{-1}x)$ *we get a group of isomorphisms* $\Phi_t : H^s(X) \to H^s(X)$, $t \in \mathbb{R}$, *of the Sobolev spaces, for every* $s \in \mathbb{R}$. *If we set* $\lambda = e^t$ *and* $\kappa_\lambda \Phi_{\log \lambda}$, *we obtain a strongly continuous (multiplicative) group of isomorphisms* $\{\kappa_\lambda\}_{\lambda\in\mathbb{R}_+}$ *on* $H^s(X)$ *for every* s. *This allows us to form the associated spaces* $\mathcal{W}^s(\mathbb{R}^q, H^s(X))$, *which depend on the choice of the vector field* v.

1.3.3 Pseudo–differential calculus

Given an operator–valued symbol $a(y, y', \eta) \in S^\mu(\Omega \times \Omega \times \mathbb{R}^q; E, \tilde{E})$ for an open set $\Omega \subseteq \mathbb{R}^q$ and $\mu \in \mathbb{R}$, cf. Definition 1.3.3, and a function $u \in C_0^\infty(\Omega, E)$ we can form

$$\mathrm{Op}(a)u(y) = \iint e^{i(y-y')\eta} a(y, y', \eta)u(y')\, dy'd\eta. \tag{1.3.47}$$

The double integral (1.3.47) is interpreted in the sense of oscillatory integrals for the corresponding vector- or operator–valued functions. In contrast to the scalar case we have here the groups of isomorphisms

$\{\kappa_\lambda\}_{\lambda \in \mathbb{R}_+}$ and $\{\tilde{\kappa}_\lambda\}_{\lambda \in \mathbb{R}_+}$ in the spaces E and $\tilde{E}$, respectively. However the estimates

$$\|\kappa(\eta)\|_{\mathcal{L}(E)} \le c[\eta]^M, \qquad \|\tilde{\kappa}(\eta)\|_{\mathcal{L}(\tilde{E})} \le \tilde{c}[\eta]^{\tilde{M}}, \qquad (1.3.48)$$

$\eta \in \mathbb{R}^q$, for constants $c, \tilde{c}, M, \tilde{M} > 0$, cf. Proposition 1.3.1, allow us to carry out all essential constructions for oscillatory integrals in the operator–valued set–up. We will not elaborate here the analogues of the corresponding proofs but concentrate on the new aspects.

Definition 1.3.49 *Let* $\Omega \subseteq \mathbb{R}^q$ *be open,* $\mu \in \mathbb{R}$, *and set*

$$L^\mu(\Omega; E, \tilde{E}) = \left\{ \mathrm{Op}(a) : \; a(y, y', \eta) \in S^\mu(\Omega \times \Omega \times \mathbb{R}^q; E, \tilde{E}) \right\},$$

$$(1.3.49)$$

$$L^\mu_{\mathrm{cl}}(\Omega; E, \tilde{E}) = \left\{ \mathrm{Op}(a) : \; a(y, y', \eta) \in S^\mu_{\mathrm{cl}}(\Omega \times \Omega \times \mathbb{R}^q; E, \tilde{E}) \right\}.$$

$$(1.3.50)$$

Here E and $\tilde{E}$ are Banach spaces with fixed strongly continuous groups of isomorphisms $\{\kappa_\lambda\}_{\lambda \in \mathbb{R}_+}$ and $\{\tilde{\kappa}_\lambda\}_{\lambda \in \mathbb{R}_+}$, respectively. The elements in (1.3.49) are called pseudo–differential operators of order μ with operator–valued symbols (in the Fourier-edge approach), those in the subspace (1.3.50) are classical pseudo–differential operators.

We set

$$L^{-\infty}(\Omega; E, \tilde{E}) = \bigcap_{\mu \in \mathbb{R}} L^\mu(\Omega; E, \tilde{E}).$$

Note that for every $k \in \mathbb{N}$ there is a $\mu \in \mathbb{R}$ with

$$c(y, y') := \int e^{i(y-y')\eta} a(y, y', \eta) \, d\eta \in C^k(\Omega \times \Omega, \mathcal{L}(E, \tilde{E}))$$

for every $a(y, y', \eta) \in S^\mu(\Omega \times \Omega \times \mathbb{R}^q; E, \tilde{E})$. It suffices to choose $\mu < -q - k - M - \tilde{M}$ for the constants $M, \tilde{M}$ in (1.3.48). For every such $a(y, y', \eta)$ the associated $A = \mathrm{Op}(a)$ is an integral operator with kernel $c(y, y')$. This kernel is uniquely determined by A. Thus, if $b(y, y', \eta) \in S^\nu(\Omega \times \Omega \times \mathbb{R}^q; E, \tilde{E})$ is another symbol with $A = \mathrm{Op}(b)$ and $\nu \le \mu$, then the kernel of A is the same. Therefore $A \in L^{-\infty}(\Omega; E, \tilde{E})$ implies $c(y, y') \in C^\infty(\Omega \times \Omega, \mathcal{L}(E, \tilde{E}))$, conversely, if $c(y, y')$ is such a kernel, then the associated integral operator C can be written $C = \mathrm{Op}(a)$ for an $a(y, y', \eta) \in S^{-\infty}(\Omega \times \Omega \times \mathbb{R}^q; E, \tilde{E})$. It suffices to set

$$a(y, y', \eta) = e^{-i(y-y')\eta} c(y, y') \psi(\eta)$$

for an arbitrary $\psi(\eta) \in C_0^\infty(\mathbb{R}^q)$ satisfying $\int \psi(\eta) \, d\eta = 1$. Thus we obtained the following result:

Proposition 1.3.50 $L^{-\infty}(\Omega; E, \tilde{E})$ *coincides with the space of all integral operators of the form*

$$(Cu)(y) = \int c(y, y') u(y') \, dy'$$

for arbitrary $c(y, y') \in C^\infty(\Omega \times \Omega, \mathcal{L}(E, \tilde{E}))$.

We endow $L^{-\infty}(\Omega; E, \tilde{E})$ with the Fréchet topology of the space $C^\infty(\Omega \times \Omega, \mathcal{L}(E, \tilde{E}))$.

We say that an $a(y, \eta) \in S^\mu(\Omega \times \mathbb{R}^q; E, \tilde{E})$ is a complete symbol of $A \in L^\mu(\Omega; E, \tilde{E})$ if

$$A = \mathrm{Op}(a) \bmod L^{-\infty}(\Omega; E, \tilde{E}).$$

If $U \subseteq \mathbb{R}^n$ is open, we use the notation $\mathcal{D}'(U, E) = \mathcal{L}(C_0^\infty(U), E)$, $\mathcal{E}'(U, E) = \{ u \in \mathcal{D}'(U, E) : \mathrm{supp}\, u \text{ compact} \}$.

Remark 1.3.51 *Each* $A = \mathrm{Op}(a) \in L^\mu(\Omega; E, \tilde{E})$ *induces a continuous operator*

$$A : C_0^\infty(\Omega, E) \to C^\infty(\Omega, \tilde{E}).$$

Alternatively, by inserting $u(y) \in C_0^\infty(\Omega)$ *into* $\mathrm{Op}(a)$ *we obtain a continuous operator*

$$A_1 : C_0^\infty(\Omega) \to C^\infty(\Omega, \mathcal{L}(E, \tilde{E})). \tag{1.3.51}$$

Set $\langle K_A, u \otimes v \rangle = \int (A_1 u)(y) v(y) \, dy$ for $u, v \in C_0^\infty(\Omega)$. Then K_A extends to an element $K_A \in \mathcal{D}'(\Omega \times \Omega, \mathcal{L}(E, \tilde{E}))$. Thus every $A \in L^\mu(\Omega; E, \tilde{E})$ has a distributional kernel $K_A \in \mathcal{D}'((\Omega \times \Omega, \mathcal{L}(E, \tilde{E}))$. It has the property

$$\mathrm{sing\, supp}\, K_A \subseteq \mathrm{diag}\, \Omega \times \Omega. \tag{1.3.52}$$

An operator $A \in L^\mu(\Omega; E, \tilde{E})$ is called properly supported if for arbitrary compact sets $K_1, K_2 \subset\subset \Omega$ both $(K_1 \times \Omega) \cap \mathrm{supp}\, K_A$ and $(\Omega \times K_2) \cap \mathrm{supp}\, K_A$ are compact sets in $\Omega \times \Omega$. Let $\omega(y, y') \in C^\infty(\Omega \times \Omega)$ be any function which equals 1 in a neighbourhood of $\mathrm{diag}\, \Omega \times \Omega$ and for which $\mathrm{supp}\, \omega$ is proper in the above sense. Then (1.3.52) implies

$$c(y, y') := (1 - \omega(y, y')) K_A(y, y') \in C^\infty(\Omega \times \Omega, \mathcal{L}(E, \tilde{E})).$$

We have

$$K_A(y, y') = \int e^{i(y - y')\eta} a(y, y', \eta) \, d\eta$$

for $A = \mathrm{Op}(a)$, and $a(y, y', \eta) \in S^\mu(\Omega \times \Omega \times \mathbb{R}^q; E, \tilde{E})$ implies

$$\omega(y, y') a(y, y', \eta) \in S^\mu(\Omega \times \Omega \times \mathbb{R}^q; E, \tilde{E}).$$

Thus, analogously to Proposition 1.1.27 we obtain the following result:

Proposition 1.3.52 *Each* $A \in L^\mu(\Omega; E, \tilde{E})$ *can be written* $A = A_0 + C$, *where* $A_0 \in L^\mu(\Omega; E, \tilde{E})$ *is properly supported and* $C \in L^{-\infty}(\Omega; E, \tilde{E})$.

Proposition 1.3.53 *Each* $A \in L^\mu(\Omega; E, \tilde{E})$ *extends to an operator* $A : \mathcal{E}'(\Omega, E) \to \mathcal{D}'(\Omega, \tilde{E})$. *If* A *is properly supported then*

$$A : C_0^\infty(\Omega, E) \to C_0^\infty(\Omega, \tilde{E}), \quad A : C^\infty(\Omega, E) \to C^\infty(\Omega, \tilde{E})$$

are continuous. Furthermore, A *extends to*

$$A : \mathcal{E}'(\Omega, E) \to \mathcal{E}'(\Omega, \tilde{E}), \quad A : \mathcal{D}'(\Omega, E) \to \mathcal{D}'(\Omega, \tilde{E}).$$

In the interpretation (1.3.51) every properly supported operator $A \in L^\mu(\Omega; E, \tilde{E})$ induces continuous operators

$$C_0^\infty(\Omega) \to C_0^\infty(\Omega, \mathcal{L}(E, \tilde{E})), \quad C^\infty(\Omega) \to C^\infty(\Omega, \mathcal{L}(E, \tilde{E}))$$

as well as corresponding extensions to $\mathcal{E}'(\Omega)$ and $\mathcal{D}'(\Omega)$. For simplicity we use here and from now on the notation A instead of A_1.

For $e_\eta(y) := e^{iy\eta}$ we obtain, in particular, $\boldsymbol{a}(y,\eta) := e_{-\eta}(y)Ae_\eta \in C^\infty(\Omega \times \mathbb{R}^q; \mathcal{L}(E, \tilde{E}))$.

Theorem 1.3.54 *Each operator* $A \in L^\mu(\Omega; E, \tilde{E})$ *has a complete symbol* $\boldsymbol{a}(y,\eta) \in S^\mu(\Omega \times \mathbb{R}^q; E, \tilde{E})$. *If* A *is given in the form* $\mathrm{Op}(a)$ *for a symbol* $a(y,y',\eta) \in S^\mu(\Omega \times \Omega \times \mathbb{R}^q; E, \tilde{E})$ *then every complete symbol of* A *has the asymptotic expansion*

$$\boldsymbol{a}(y,\eta) \sim \sum_\alpha \frac{1}{\alpha!} D_\eta^\alpha \partial_{y'}^\alpha a(y,y',\eta)|_{y'=y}.$$

For every properly supported representative A_0 *of* A *mod* $L^{-\infty}(\Omega; E, \tilde{E})$ *we obtain by* $\boldsymbol{a}_0(y,\eta) = e_{-\eta}(y)A_0e_\eta$ *a complete symbol of* A, *and we have* $A_0 = \mathrm{Op}(\boldsymbol{a}_0)$.

From this we obtain, in particular, that every complete symbol of an operator in $L_{\mathrm{cl}}^\mu(\Omega; E, \tilde{E})$ belongs to $S_{\mathrm{cl}}^\mu(\Omega \times \mathbb{R}^q; E, \tilde{E})$.

Theorem 1.3.55 *The map* $A \to e_{-\eta}(y)A_0e_\eta$ *for any properly supported* A_0 *with* $A - A_0 \in L^{-\infty}(\Omega; E, \tilde{E})$ *induces isomorphisms*

$$L^\mu(\Omega; E, \tilde{E})/L^{-\infty}(\Omega; E, \tilde{E}) \simeq S^\mu(\Omega \times \mathbb{R}^q; E, \tilde{E})/S^{-\infty}(\Omega \times \mathbb{R}^q; E, \tilde{E}),$$

$$L_{\mathrm{cl}}^\mu(\Omega; E, \tilde{E})/L^{-\infty}(\Omega; E, \tilde{E}) \simeq S_{\mathrm{cl}}^\mu(\Omega \times \mathbb{R}^q; E, \tilde{E})/S^{-\infty}(\Omega \times \mathbb{R}^q; E, \tilde{E}).$$

Let us fix a relatively closed set $\Delta \subset \Omega \times \Omega$ which contains an open neighbourhood of $\operatorname{diag} \Omega \times \Omega$ and which is proper in the sense of the above notation. Let us set $L^\mu(\Omega; E, \tilde{E})_\Delta = \{A \in L^\mu(\Omega; E, \tilde{E}) : \operatorname{supp} K_A \subseteq \Delta\}$ and define $S^\mu(\Omega \times \mathbb{R}^q; E, \tilde{E}) = \{e_{-\eta}(y)Ae_\eta : A \in L^\mu(\Omega; E, \tilde{E})_\Delta\}$. According to Proposition 1.3.52 we have

$$L^\mu(\Omega; E, \tilde{E}) = L^\mu(\Omega; E, \tilde{E})_\Delta + L^{-\infty}(\Omega; E, \tilde{E}). \tag{1.3.53}$$

Moreover, Theorem 1.3.54 yields an isomorphism

$$\sigma^\mu : L^\mu(\Omega; E, \tilde{E})_\Delta \to S^\mu(\Omega \times \mathbb{R}^q; E, \tilde{E})_\Delta.$$

The space $S^\mu(\Omega \times \mathbb{R}^q; E, \tilde{E})_\Delta$ is closed in the topology induced by $S^\mu(\Omega \times \mathbb{R}^q; E, \tilde{E})$. Thus it is Fréchet, and hence $L^\mu(\Omega; E, \tilde{E})_\Delta$ can be endowed with the Fréchet topology induced by σ^μ. Then we introduce in $L^\mu(\Omega; E, \tilde{E})$ a Fréchet topology by the non-direct sum (1.3.53). This is independent of the specific choice of Δ.

In an analogous manner we can proceed for $L^\mu_{\mathrm{cl}}(\Omega; E, \tilde{E})$ which becomes a Fréchet space in a canonical way.

Note that when $a(y, \eta) \in S^\mu_{\mathrm{cl}}(\Omega \times \mathbb{R}^q; E, \tilde{E})$ is a complete symbol of an $A \in L^\mu_{\mathrm{cl}}(\Omega; E, \tilde{E})$, the sequence of homogeneous components $a_{(\mu-j)}(y, \eta) \in S^{(\mu-j)}(\Omega \times (\mathbb{R}^q \setminus \{0\}); E, \tilde{E})$, $j \in \mathbb{N}$, only depends on A. In other words we have a sequence of well-defined continuous operators

$$\sigma_\wedge^{\mu-j} : L^\mu_{\mathrm{cl}}(\Omega; E, \tilde{E}) \to S^{(\mu-j)}(\Omega \times (\mathbb{R}^q \setminus \{0\}); E, \tilde{E}),$$

$j \in \mathbb{N}$, by $a_{(\mu-j)}(y, \eta) =: \sigma_\wedge^{\mu-j}(A)(y, \eta)$. Similarly to the scalar theory we call $\sigma_\wedge^{\mu-j}(A)(y, \eta)$ the homogeneous principal symbol of A of order μ.

Theorem 1.3.56 *Let $A \in L^\mu(\Omega; E_0, \tilde{E})$, $B \in L^\nu(\Omega; E, E_0)$ and assume that A or B is properly supported. Then $AB \in L^{\mu+\nu}(\Omega; E, \tilde{E})$. If $A = \operatorname{Op}(a) \bmod L^{-\infty}(\Omega; E_0, \tilde{E})$, $B = \operatorname{Op}(b) \bmod L^{-\infty}(\Omega; E, E_0)$ for certain $a(y, \eta) \in S^\mu(\Omega \times \mathbb{R}^q; E_0, \tilde{E})$, $b(y, \eta) \in S^\nu(\Omega \times \mathbb{R}^q; E, E_0)$, then $AB = \operatorname{Op}(c) \bmod L^{-\infty}(\Omega; E, \tilde{E})$ for a $c(y, \eta) \in S^{\mu+\nu}(\Omega; E, \tilde{E})$ with asymptotic expansion*

$$c(y, \eta) \sim \sum_\alpha \frac{1}{\alpha!} \left(D_\eta^\alpha a(y, \eta) \right) \partial_y^\alpha b(y, \eta). \tag{1.3.54}$$

In particular, $A \in L^\mu_{\mathrm{cl}}(\Omega; E_0, \tilde{E})$, $B \in L^\nu_{\mathrm{cl}}(\Omega; E, E_0)$ implies $AB \in L^{\mu+\nu}_{\mathrm{cl}}(\Omega; E, \tilde{E})$, and we have

$$\sigma_\wedge^{\mu+\nu}(AB)(y, \eta) = \sigma_\wedge^\mu(A)(y, \eta)\sigma_\wedge^\nu(B)(y, \eta).$$

Analogously to the scalar theory we write for (1.3.54)

$$c(y, \eta) = a(y, \eta) \# b(y, \eta)$$

and call $c(y, \eta)$ the Leibniz product of $a(y, \eta)$ and $b(y, \eta)$ (which is unique mod $S^{-\infty}$).

Now let $\Omega \subseteq \mathbb{R}_y^q$, $\tilde{\Omega} \subseteq \mathbb{R}_{\tilde{y}}^q$ be open sets and $\chi : \Omega \to \tilde{\Omega}$ be a diffeomorphism. Then for every $A \in L^\mu(\Omega; E, \tilde{E})$ we can form the push-forward under χ

$$\chi_* A = (\chi^*)^{-1} A \chi^* : C_0^\infty(\tilde{\Omega}, E) \to C^\infty(\tilde{\Omega}, \tilde{E})$$

with the function pull-backs $\chi^* : C_0^\infty(\tilde{\Omega}, E) \to C_0^\infty(\Omega, E)$ and $\chi^* : C^\infty(\tilde{\Omega}, \tilde{E}) \to C^\infty(\Omega, \tilde{E})$, respectively.

We set

$$\Phi_\alpha(y, \tilde{\eta}) = D_z^\alpha e^{i\delta(y,z)\tilde{\eta}}\big|_{z=y}$$

for $\alpha \in \mathbb{N}^q$ and $\delta(y, z) = \chi(z) - \chi(y) - d\chi(y)(z - y)$, cf. (1.1.70).

Theorem 1.3.57 *The operator push-forward under a diffeomorphism $\chi : \Omega \to \tilde{\Omega}$ induces an isomorphism*

$$\chi_* : L^\mu(\Omega; E, \tilde{E}) \to L^\mu(\tilde{\Omega}; E, \tilde{E})$$

for every $\mu \in \mathbb{R}$. If $A \in L^\mu(\Omega; E, \tilde{E})$ is written in the form $A = \mathrm{Op}(a) \bmod L^{-\infty}(\Omega; E, \tilde{E})$ for an $a(y, \eta) \in S^\mu(\Omega \times \mathbb{R}^q; E, \tilde{E})$ then $\tilde{A} = \chi_* A$ is of the form $\tilde{A} = \mathrm{Op}(\tilde{a}) \bmod L^{-\infty}(\tilde{\Omega}; E, \tilde{E})$ for an $\tilde{a}(\tilde{y}, \tilde{\eta}) \in S^\mu(\tilde{\Omega} \times \mathbb{R}^q; E, \tilde{E})$ with assymptotic expansion

$$\tilde{a}(\tilde{y}, \tilde{\eta})\big|_{\tilde{y}=\chi(y)} \sim \sum_\alpha \frac{1}{\alpha!} (\partial_\eta^\alpha a)(y, {}^t d\chi(y)\tilde{\eta}) \Phi_\alpha(y, \tilde{\eta}).$$

In particular, $A \in L_{\mathrm{cl}}^\mu(\Omega; E, \tilde{E})$ implies $\chi_* A \in L_{\mathrm{cl}}^\mu(\tilde{\Omega}; E, \tilde{E})$, and we have

$$\sigma_\wedge^\mu(\chi_* A)(\tilde{y}, \tilde{\eta}) = \sigma_\wedge^\mu(A)(y, \eta)$$

for $(\tilde{y}, \tilde{\eta}) = (\chi(y), {}^t d\chi^{-1}(y)\eta)$.

Theorem 1.3.58 *Each $A \in L^\mu(\Omega; E, \tilde{E})$ induces a continuous operator*

$$A : \mathcal{W}_{\mathrm{comp}}^s(\Omega, E) \to \mathcal{W}_{\mathrm{loc}}^{s-\mu}(\Omega, \tilde{E})$$

for every $s \in \Omega$. If A is properly supported, then A induces continuous operators

$$A : \mathcal{W}_{\mathrm{comp}}^s(\Omega, E) \to \mathcal{W}_{\mathrm{comp}}^{s-\mu}(\Omega, \tilde{E}), \quad A : \mathcal{W}_{\mathrm{loc}}^s(\Omega, E) \to \mathcal{W}_{\mathrm{loc}}^{s-\mu}(\Omega, \tilde{E})$$

for all $s \in \mathbb{R}$.

Proof. By definition A can be written $A = \mathrm{Op}(a)$ for an $a(y, y', \eta) \in$ $S^\mu(\Omega \times \Omega \times \mathbb{R}^q; E, \tilde{E})$. In view of the definition of the spaces with subscripts comp, loc it suffices to show the continuity of the operator $\mathcal{M}_\varphi \mathrm{Op}(a) \mathcal{M}_\psi : \mathcal{W}^s(\mathbb{R}^q, E) \to \mathcal{W}^{s-\mu}(\mathbb{R}^q, \tilde{E})$ for arbitrary $\varphi(y), \psi(y) \in$ $C_0^\infty(\Omega)$. The symbol of the latter operator is $\varphi(y) a(y, y', \eta) \psi(y')$. In other words, we may assume that the given symbol has compact support in y and y'. If $C_0^\infty(K)$ denotes the Fréchet space of all $\varphi \in C^\infty(\mathbb{R}^q)$ with $\mathrm{supp}\, \varphi \subseteq K$ we thus have to show the assertion for $a(y, y', \eta) \in C_0^\infty(K) \otimes_\pi S^\mu(\mathbb{R}^q; E, \tilde{E}) \otimes_\pi C^\infty(K')$ for certain compact sets $K, K' \subset \mathbb{R}^q$. Applying Remark 1.1.7 we can write

$$a(y, y', \eta) = \sum_{m=0}^\infty \lambda_m \varphi_m(y) a_m(\eta) \psi_m(y'),$$

with $\sum |\lambda_m| < \infty$ and $\varphi_m \in C_0^\infty(K)$, $\psi_m \in C_0^\infty(K')$, $a_m(\eta) \in$ $S^\mu(\mathbb{R}^q; E, \tilde{E})$ tending to zero for $m \to \infty$ in the corresponding spaces. Then, applying Proposition 1.3.24 and Theorem 1.3.34,

$$\mathrm{Op}(a) = \sum_{m=0}^\infty \lambda_m \mathcal{M}_{\varphi_m} \mathrm{Op}(a_m) \mathcal{M}_{\psi_m}$$

converges in $\mathcal{L}(\mathcal{W}^s(\mathbb{R}^q, E), \mathcal{W}^{s-\mu}(\mathbb{R}^q, \tilde{E}))$. The asserted mapping properties for properly supported operators are evident. $\qquad \square$

We will need a slight modification of Theorem 1.3.58 for the case $A = \mathrm{Op}(a)$ for symbols $a(y, \eta)$ or $a(y', \eta)$. As usual we set

$$\mathrm{Op}(a)u(y) = \iint e^{i(y-y')\eta} a(y, y', \eta) u(y') \, dy' d\eta$$

for an amplitude function $a(y, y', \eta)$.

Theorem 1.3.59 *Let* $a(y, \eta) \in S^\mu(\mathbb{R}^q \times \mathbb{R}^q; E, \tilde{E})$, *and assume*

$$\|\tilde{\kappa}^{-1}(\eta)\{D_y^\alpha a(y, \eta)\}\kappa(\eta)\|_{\mathcal{L}(E, \tilde{E})} \leq c(1 + |y|^2)^{-n}[\eta]^\mu \qquad (1.3.55)$$

for any fixed $n \in \mathbb{R}$, $n > \frac{q}{2}$, *for all* $\alpha \in \mathbb{N}^q$ *with* $|\alpha| \leq 2l$, *where* $2l > \tilde{M} + |s - \mu| + q$ *for a given fixed* $s \in \mathbb{R}$, *for a constant* $c > 0$ *that is independent of* $(y, \eta) \in \mathbb{R}^q \times \mathbb{R}^q$. *Then* $\mathrm{Op}(a)$ *induces a continuous operator*

$$\mathrm{Op}(a) : \mathcal{W}^s(\mathbb{R}^q, E) \to \mathcal{W}^{s-\mu}(\mathbb{R}^q, \tilde{E}), \qquad (1.3.56)$$

and we have

$$\| \mathrm{Op}(a) \|_{\mathcal{L}(\mathcal{W}^s(\mathbb{R}^q, E), \mathcal{W}^{s-\mu}(\mathbb{R}^q, \tilde{E}))} \leq c_s c_l(a) \qquad (1.3.57)$$

for a constant c_s only dependent on s, and

$$c_l(a) = \sup_{y,\eta \in \mathbb{R}^q} \|\tilde{\kappa}^{-1}(\eta)\{(1+|y|^2)^n (1-\Delta_y)^l a(y,\eta)\}\kappa(\eta)\|_{\mathcal{L}(E,\tilde{E})}[\eta]^{-\mu}.$$

$$(1.3.58)$$

Analogously, let $a(y',\eta) \in S^\mu(\mathbb{R}^q \times \mathbb{R}^q; E, \tilde{E})$ satisfy the conditions (1.3.55) with respect to the (y',η)-variables for $n > \frac{q}{2}$, all $|\alpha| \leq 2l$ where $2l > M + |s| + q$ for a given fixed $s \in \mathbb{R}$, for a constant $c > 0$ independent of $(y',\eta) \in \mathbb{R}^q \times \mathbb{R}^q$. Then $\mathrm{Op}(a)$ induces a continuous operator (1.3.56), and we have (1.3.57) for c_s only dependent on s and $c_l(a)$ of the form (1.3.58), now with respect to (y',η).

Proof. First we have

$$\|\mathrm{Op}(a)u\|^2_{W^{s-\mu}(\mathbb{R}^q,\tilde{E})} = \int [\eta]^{2(s-\mu)}\|\tilde{\kappa}^{-1}(\eta)F\,\mathrm{Op}(a)u(\eta)\|^2_{\tilde{E}}\,d\eta.$$

Writing $\psi(\zeta,\eta) = F_{y\to\zeta}a(y,\eta)$ and $\hat{u}(\eta) = (F_{y\to\eta}u)(\eta)$ we have

$$(F_{y\to\eta}\,\mathrm{Op}(a)u)\,(\eta) = \int e^{-iy\eta}\left\{\int e^{iy\xi}a(y,\xi)\hat{u}(\xi)\,d\xi\right\}dy$$

$$= \iint e^{-iy(\eta-\xi)}a(y,\xi)\,dy\,\hat{u}(\xi)\,d\xi$$

$$= \int \psi(\eta-\xi,\xi)\hat{u}(\xi)\,d\xi.$$

We obtain

$$\|\mathrm{Op}(a)u\|^2_{W^{s-\mu}(\mathbb{R}^q,\tilde{E})} = \int [\eta]^{2(s-\mu)}\|\tilde{\kappa}^{-1}(\eta)\int \psi(\eta-\xi,\xi)\hat{u}(\xi)\,d\xi\|^2_{\tilde{E}}\,d\eta$$

$$= \|m\|^2_{L^2(\mathbb{R}^q_\eta)} \qquad (1.3.59)$$

for

$$m(\eta) = \|[\eta]^{s-\mu}\tilde{\kappa}^{-1}(\eta)\int \psi(\eta-\xi,\xi)\hat{u}(\xi)\,d\xi\|_{\tilde{E}}.$$

Using $[\eta]^{s-\mu} \leq c^{|s-\mu|}[\eta-\xi]^{|s-\mu|}[\xi]^{s-\mu}$ for a constant $c > 0$ we obtain

$$m(\eta) \leq c\|\tilde{\kappa}^{-1}(\eta)\int [\eta-\xi]^{|s-\mu|}\psi(\eta-\xi,\xi)[\xi]^{s-\mu}\hat{u}(\xi)\,d\xi\|_{\tilde{E}}.$$

In the sequel c will denote different suitable constants. Applying

$\|\tilde{\kappa}^{-1}_{[\eta]/[\xi]}\|_{\mathcal{L}(\tilde{E})} \le c[\eta - \xi]^{\tilde{M}}$ for constants c and $\tilde{M}$ it follows that

$$m(\eta) \tag{1.3.60}$$

$$\le c\| \int \tilde{\kappa}^{-1}_{[\eta]/[\xi]} \tilde{\kappa}^{-1}(\xi)[\eta - \xi]^{|s-\mu|} \psi(\eta - \xi, \xi) \kappa(\xi) \kappa^{-1}(\xi)[\xi]^{s-\mu} \hat{u}(\xi) \, d\xi \|_{\tilde{E}}$$

$$\le c \int \|\tilde{\kappa}^{-1}_{[\eta]/[\xi]}\|_{\mathcal{L}(\tilde{E})} [\eta - \xi]^{|s-\mu|} \|\tilde{\kappa}^{-1}(\xi)\psi(\eta - \xi, \xi)\kappa(\xi)\|_{\mathcal{L}(E,\tilde{E})} [\xi]^{s-\mu}$$

$$\times \|\kappa^{-1}(\xi)\hat{u}(\xi)\|_E \, d\xi$$

$$\le c \int [\eta - \xi]^{\tilde{M}+|s-\mu|} \|\tilde{\kappa}^{-1}(\xi)\psi(\eta - \xi, \xi)\kappa(\xi)\|_{\mathcal{L}(E,\tilde{E})}$$

$$\times [\xi]^{s-\mu} \|\kappa^{-1}(\xi)\hat{u}(\xi)\|_E \, d\xi. \tag{1.3.61}$$

For every $l \in \mathbb{N}$ we can write:

$$\psi(\zeta, \xi) = (1 + |\zeta|^2)^{-l} \int e^{-iy\zeta}(1 + |y|^2)^{-n}(1 + |y|^2)^n(1 - \Delta_y)^l a(y, \xi) \, dy,$$

cf. also the proof of Theorem 1.3.60. This yields for $n > \frac{q}{2}$

$$\|\tilde{\kappa}^{-1}(\xi)\psi(\zeta, \xi)\kappa(\xi)\|_{\mathcal{L}(E,\tilde{E})}$$

$$\le c(1 + |\zeta|^2)^{-l} \sup_{y\in\mathbb{R}^q} \|\tilde{\kappa}^{-1}(\xi)(1 + |y|^2)^n(1 - \Delta_y)^l a(y, \xi)\kappa(\xi)\|_{\mathcal{L}(E,\tilde{E})}$$

$$\times \int (1 + |y|^2)^{-n} \, dy$$

$$\le c[\zeta]^{-2l} c_l(a)[\xi]^\mu$$

for

$$c_l(a) = \sup_{\xi\in\mathbb{R}^q} \sup_{y\in\mathbb{R}^q} \|\tilde{\kappa}^{-1}(\xi)(1 + |y|^2)^n(1 - \Delta_y)^l a(y, \xi)\kappa(\xi)\|_{\mathcal{L}(E,\tilde{E})} [\xi]^{-\mu}.$$
$$\tag{1.3.62}$$

Inserting $\zeta = \eta - \xi$ we get from (1.3.61)

$$m(\eta) \le c \int [\eta - \xi]^{\tilde{M}+|s-\mu|-2l}[\xi]^s \|\kappa^{-1}(\xi)\hat{u}(\xi)\|_E \, d\xi.$$

For $2l > \tilde{M} + |s - \mu| + q$ we can apply Young's inequality and obtain

$$\|m(\eta)\|_{L^2(\mathbb{R}^q)} \le cc_l(a) \left\{ \int [\xi]^{2s} \|\kappa^{-1}(\xi)\hat{u}(\xi)\|_E^2 \, d\xi \right\}^{\frac{1}{2}}$$

$$= cc_l(a)\|u\|_{W^s(\mathbb{R}^q,E)}.$$

In view of (1.3.59) this is just the continuity of (1.3.56).

For $a(y', \eta)$ we can proceed in an analogous manner. We have

$$
\begin{aligned}
(F \operatorname{Op}(a)u)(\eta) &= \int e^{-iy'\eta} a(y', \eta) u(y')\, dy' \\
&= \int e^{-iy'\eta} \left\{ \int e^{iy'\zeta} \psi(\zeta, \eta)\, d\zeta \right\} u(y')\, dy' \\
&= \int \psi(\zeta, \eta) \hat{u}(\eta - \zeta)\, d\zeta \\
&= \int \psi(\eta - \zeta, \eta) \hat{u}(\xi)\, d\xi.
\end{aligned}
$$

Now

$$
\begin{aligned}
\| \operatorname{Op}(a)u \|^2_{W^{s-\mu}(\mathbb{R}^q, \tilde{E})} &= \int [\eta]^{2(s-\mu)} \| \tilde{\kappa}^{-1}(\eta)\, (F \operatorname{Op}(a)u)(\eta) \|^2_{\tilde{E}}\, d\eta \\
&= \int [\eta]^{2(s-\mu)} \| \tilde{\kappa}^{-1}(\eta) \int \psi(\eta - \xi, \eta) \hat{u}(\xi)\, d\xi \|^2_{\tilde{E}}\, d\eta \\
&= \| m \|^2_{L^2(\mathbb{R}^q_\eta)}
\end{aligned}
$$

for $m(\eta) = \| [\eta]^{s-\mu} \tilde{\kappa}^{-1}(\eta) \int \psi(\eta - \xi, \eta) \hat{u}(\xi)\, d\xi \|_{\tilde{E}}$. Similarly to the above we have

$$
\psi(\zeta, \eta) = (1 + |\zeta|^2)^{-l} \int e^{-iy'\xi} (1 + |y'|^2)^{-n} (1 + |y'|^2)^n (1 - \Delta_{y'})^l a(y', \eta)\, dy'.
$$

This yields for $n > \frac{q}{2}$

$$
\| \tilde{\kappa}^{-1}(\eta) \psi(\zeta, \eta) \kappa(\eta) \|_{\mathcal{L}(E, \tilde{E})} \le c[\zeta]^{-2l} c_l(a)[\eta]^\mu.
$$

Thus

$$
\begin{aligned}
m(\eta) &\le \int \| [\eta]^{s-\mu} \tilde{\kappa}^{-1}(\eta) \psi(\eta - \xi, \eta) \kappa(\eta) \kappa^{-1}(\eta) \kappa(\xi) \kappa^{-1}(\xi) \hat{u}(\xi) \|_{\tilde{E}}\, d\xi \\
&\le c c_l(a) \int [\eta - \xi]^{-2l} [\eta]^s \| \kappa_{[\xi]/[\eta]} \|_{\mathcal{L}(E)} \| \kappa^{-1}(\xi) \hat{u}(\xi) \|_E\, d\xi.
\end{aligned}
$$

Applying now $[\eta]^s \le c^{|s|} [\eta - \xi]^{|s|} [\xi]^s$ and $\| \kappa_{[\xi]/[\eta]} \|_{\mathcal{L}(E)} \le c[\eta - \xi]^M$ for certain constants $c > 0$ it follows that

$$
m(\eta) \le c c_l(a) \int [\eta - \xi]^{-2l + M + |s|} \| [\xi]^s \kappa^{-1}(\xi) \hat{u}(\xi) \|_E\, d\xi.
$$

Again, by Young's inequality it follows for $-2l + M + |s| < -q$ that (1.3.56) is continuous for the given s, and we also obtain the asserted estimate for the operator norm of $\operatorname{Op}(a)$. $\qquad \square$

Theorem 1.3.60 *Let* $a(\eta) \in S^{\mu}(\mathbb{R}^q; E, \tilde{E})$, *then* $\mathrm{Op}(a)$ *induces a continuous operator*

$$\mathrm{Op}(a) : \mathcal{S}(\mathbb{R}^q, E) \to \mathcal{S}(\mathbb{R}^q, \tilde{E}).$$

Proof. Let $F^k(E)$ denote the closure of $\mathcal{S}(\mathbb{R}^q, E)$ with respect to the norm

$$\|u\|_{F^k(E)} = \left\{ \int [\eta]^{2k} \sum_{|\alpha| \leq k} \|D_\eta^\alpha \hat{u}(\eta)\|_E^2 \, d\eta \right\}^{\frac{1}{2}},$$

where $\hat{u}(\eta) = (F_{y \to \eta} u)(\eta)$. We then have $\mathcal{S}(\mathbb{R}^q, E) = \varprojlim_{k \in \mathbb{N}} F^k(E)$ and analogously $\mathcal{S}(\mathbb{R}^q, \tilde{E}) = \varprojlim_{k \in \mathbb{N}} F^k(\tilde{E})$. To prove the continuity of $\mathrm{Op}(a)$ we have to show that for every $k \in \mathbb{N}$ there is an $l \in \mathbb{N}$ such that $\mathrm{Op}(a) : F^l(E) \to F^k(\tilde{E})$ is continuous. We shall obtain this for $l = k + \mu + M + \tilde{M}$ with the constants $M, \tilde{M}$ from the estimate in Proposition 1.3.1. We have

$$\|\mathrm{Op}(a)u\|_{F^k(\tilde{E})}^2 = \int [\eta]^{2k} \sum_{|\alpha| \leq k} \|D_\eta^\alpha (a(\eta)\hat{u}(\eta))\|_{\tilde{E}}^2 \, d\eta$$

and $D_\eta^\alpha(a(\eta)\hat{u}(\eta)) = \sum_{\beta \leq \alpha} d_\beta D_\eta^\beta a(\eta) D_\eta^{\alpha-\beta} \hat{u}(\eta)$ for suitable constants d_β. Moreover,

$$\|D_\eta^\beta a(\eta) D_\eta^{\alpha-\beta} \hat{u}(\eta)\|_{\tilde{E}}^2 = \|\tilde{\kappa}(\eta)\tilde{\kappa}^{-1}(\eta) \left\{ D_\eta^\beta a(\eta) \right\} \kappa(\eta)\kappa^{-1}(\eta) D_\eta^{\alpha-\beta} \hat{u}(\eta)\|_{\tilde{E}}^2$$
$$\leq c[\eta]^{2(M+\tilde{M})}[\eta]^{2(\mu-|\beta|)}\|D_\eta^{\alpha-\beta}\hat{u}(\eta)\|_E^2.$$

This yields for different constants c

$$\|\mathrm{Op}(a)u\|_{F^k(\tilde{E})}^2 \leq c \int [\eta]^{2(k+\mu+M+\tilde{M})} \sum_{|\alpha| \leq k} \|D_\eta^\alpha \hat{u}(\eta)\|_E^2 \, d\eta$$
$$\leq c\|u\|_{F^{k+\mu+M+\tilde{M}}(E)}^2.$$

$\square$

Theorem 1.3.61 *Let* $a(y, \eta) \in S^{\nu}(\mathbb{R}^q \times \mathbb{R}^q; E, \tilde{E})$, $\nu < 0$, *and let* $a(y, \eta) : E \to \tilde{E}$ *be a compact operator for every* $(y, \eta) \in \mathbb{R}^q \times \mathbb{R}^q$. *Assume that there is a* $\delta > 0$ *and an* $n \in \mathbb{R}$ *with* $n > \frac{q}{2}$ *such that*

$$\|\tilde{\kappa}^{-1}(\eta) \left\{ D_y^\alpha a(y, \eta) \right\} \kappa(\eta)\|_{\mathcal{L}(E, \tilde{E})} \leq c(1 + |y|^2)^{-n+\delta}[\eta]^{\nu} \quad (1.3.63)$$

for all $\alpha \in \mathbb{N}^q$ *with* $|\alpha| \leq 2l$, *where* $2l > \tilde{M} + |s - \nu| + q$ *for a given* $s \in \mathbb{R}$. *Then*

$$\mathrm{Op}(a) : \mathcal{W}^s(\mathbb{R}^q, E) \to \mathcal{W}^s(\mathbb{R}^q, \tilde{E})$$

is a compact operator.

Proof. Let us first assume that $\operatorname{supp} a$ is compact with respect to $(y, \eta) \in \mathbb{R}^q \times \mathbb{R}^q$. Choose a partition of unity on $\mathbb{R}^q \times \mathbb{R}^q$ of the form $\{\varphi_i(y)\psi_j(\eta)\}_{i,j \in \mathbb{N}}$ such that there are only finitely many i, j for every compact set $K \subset\subset \mathbb{R}^q \times \mathbb{R}^q$ with $K \cap \operatorname{supp}\varphi_i \times \operatorname{supp}\psi_j \neq \emptyset$. Let us write

$$p(y, \eta) = \sum p_{ij}(y, \eta)$$

for $p_{ij}(y, \eta) = \varphi_i(y)\psi_j(\eta)a(y_i, \eta_j)$ for certain $y_i \in \operatorname{supp}\varphi_i$, $\eta_j \in \operatorname{supp}\psi_j$. Then, for arbitrary given $\varepsilon > 0$, $l \in \mathbb{N}$, the partition of unity can be chosen so fine on $\operatorname{supp} a$ that

$$\left\| \tilde{\kappa}^{-1}(\eta) \left\{ D_y^\alpha (p(y, \eta) - a(y, \eta)) \right\} \kappa(\eta) \right\|_{\mathcal{L}(E, \tilde{E})} \leq \varepsilon \qquad (1.3.64)$$

holds for all $(y, \eta) \in \mathbb{R}^q \times \mathbb{R}^q$ and $|\alpha| \leq 2l$. In fact, we can achieve first

$$\left\| D_y^\alpha (p(y, \eta) - a(y, \eta)) \right\|_{\mathcal{L}(E, \tilde{E})} \leq \varepsilon_0$$

for an arbitrary $\varepsilon_0 > 0$ and all $|\alpha| \leq 2l$. Then (1.3.64) follows from the fact that the operator norms of $\tilde{\kappa}(\eta)$, $\kappa(\eta)$ as well as of the inverses are uniformly bounded for η varying over a bounded set in $\mathbb{R}^q$. Now, according to Theorem 1.3.59, we obtain

$$\| \operatorname{Op}(p - a) \|_{\mathcal{L}(\mathcal{W}^s(\mathbb{R}^q, E), \mathcal{W}^s(\mathbb{R}^q, \tilde{E}))} \leq c\varepsilon \qquad (1.3.65)$$

for a $c > 0$, provided we have chosen $2l$ large enough for the given s. Let us verify that $\operatorname{Op}(p_{ij}) : \mathcal{W}^s(\mathbb{R}^q, E) \to \mathcal{W}^s(\mathbb{R}^q, \tilde{E})$ is compact for every i, j. This will imply that also $\operatorname{Op}(p) = \sum \operatorname{Op}(p_{ij})$ is a compact operator. For notational convenience we set $i = j = 0$ and $Q = \operatorname{Op}(p_{00})$. Then

$$Qu(y) = \int e^{iy\eta} \varphi_0(y) a(y_0, \eta_0) \psi_0(\eta) \hat{u}(\eta) \, d\eta.$$

We have $\varphi_0(y)e^{iy\eta}\psi_0(\eta) \in C_0^\infty(B, \mathcal{S}(\mathbb{R}_\eta^q))$ for $B := \{y \in \mathbb{R}^q : |y| \leq r\}$ for a sufficiently large $r > 0$. In view of $C_0^\infty(B, \mathcal{S}(\mathbb{R}_\eta^q)) = C_0^\infty(B) \otimes_\pi \mathcal{S}(\mathbb{R}_\eta^q)$ we can write

$$\varphi_0(y)e^{iy\eta}\psi_0(\eta) = \sum_{j=0}^\infty \lambda_j f_j(y) g_j(\eta)$$

for $\sum |\lambda_j| < \infty$, $f_j \in C_0^\infty(B)$, $g_j \in \mathcal{S}(\mathbb{R}^q)$ tending to zero in the corresponding spaces for $j \to \infty$. This yields

$$Qu(y) = \sum_{j=0}^\infty \lambda_j f_j(y) a(y_0, \eta_0) \int g_j(\eta) \hat{u}(\eta) \, d\eta.$$

Every summand can be regarded as a composition

$$Q : \mathcal{W}^s(\mathbb{R}^q, E) \to E \to \tilde{E} \to \mathcal{W}^s(\mathbb{R}^q, \tilde{E}),$$

where the operator in the middle $a(y_0, \eta_0) : E \to \tilde{E}$ is compact. It is obvious that the operators $T_j : u \to \int g_j(\eta) \hat{u}(\eta) \, d\eta$ and $K_j : e \to f_j(y) a(y_0, \eta_0) e$ for $e \in E$ tend to zero in $\mathcal{L}(\mathcal{W}^s(\mathbb{R}^q, E), E)$ and $\mathcal{L}(\tilde{E}, \mathcal{W}^s(\mathbb{R}^q, \tilde{E}))$ respectively, for $j \to \infty$. Thus $Q = \sum \lambda_j K_j T_j$ converges in $\mathcal{L}(\mathcal{W}^s(\mathbb{R}^q, E), \mathcal{W}^s(\mathbb{R}^q, \tilde{E}))$. Since every summand is compact, Q is also a compact operator. Thus $\mathrm{Op}(p)$ is compact and (1.3.65) yields for $\varepsilon \to 0$ that also $\mathrm{Op}(a)$ is a compact operator.

For the original symbol $a(y, \eta)$ satisfying the conditions of the theorem we get from Theorem 1.3.59

$$\| \mathrm{Op}(a) \|_{\mathcal{L}(\mathcal{W}^s(\mathbb{R}^q, E), \mathcal{W}^s(\mathbb{R}^q, \tilde{E}))} \leq c_s c_l(a)$$

for a constant $c_s > 0$, not dependent on a, where

$$c_l(a) = \sup_{y, \eta \in \mathbb{R}^q} \| \tilde{\kappa}^{-1}(\eta)(1 + |y|^2)^n (1 - \Delta_y)^l a(y, \eta) \kappa(\eta) \|_{\mathcal{L}(E, \tilde{E})}.$$

Let $\omega(y, \eta)$ be a function in $C_0^\infty(\mathbb{R}^q \times \mathbb{R}^q)$ with $\omega(y, \eta) = 1$ for $|y, \eta| < c_1$, $\omega(y, \eta) = 0$ for $|y, \eta| > c_0$ for certain $0 < c_1 < c_0 < \infty$. Set $\omega_k(y, \eta) = \omega \left(\frac{y}{k+1}, \frac{\eta}{k+1} \right)$ for $k \in \mathbb{N}$ and $a_k(y, \eta) := \omega_k(y, \eta) a(y, \eta)$. Then $a_k(y, \eta)$ has compact support in (y, η). For every $\varepsilon > 0$ we find a k such that

$$c_l(a_k - a) < \varepsilon. \tag{1.3.66}$$

In fact, we have

$$\begin{aligned}
c_l(a_k - a) &= \sup_{y, \eta \in \mathbb{R}^q} \| \tilde{\kappa}^{-1}(\eta)(1 + |y|^2)^n (1 - \Delta_y)^l (a_k - a)(y, \eta) \kappa(\eta) \|_{\mathcal{L}(E, \tilde{E})} \\
&\leq c \sup_{y, \eta \in \mathbb{R}^q} \| \tilde{\kappa}^{-1}(\eta)(1 + |y|^2)^n \\
&\quad \times (1 - \Delta_y)^l (\omega_k(y, \eta) - 1) a(y, \eta) \kappa(\eta) \|_{\mathcal{L}(E, \tilde{E})} [y]^{-\delta} [\eta]^{-\nu}.
\end{aligned}$$

The expression on the right is finite, by Theorem 1.3.59. Moreover the supremum is taken over $|y, \eta| > c_1(k + 1)$. This implies (1.3.66) for sufficiently large k. In other words, $\mathrm{Op}(a_k)$ tends to $\mathrm{Op}(a)$ in $\mathcal{L}(\mathcal{W}^s(\mathbb{R}^q, E), \mathcal{W}^s(\mathbb{R}^q, \tilde{E}))$ for $k \to \infty$. Thus $\mathrm{Op}(a)$ as a limit of compact operators is again compact. $\square$

Let us now formulate some generalisations of Theorem 1.3.61 which only play an auxiliary role here.

Theorem 1.3.62 *Let* $a(y, y', \eta) \in S^\nu(\mathbb{R}^q \times \mathbb{R}^q \times \mathbb{R}^q; E, \tilde{E})$, $\nu < 0$, *and let* $a(y, y', \eta) : E \to \tilde{E}$ *be a compact operator for every* $(y, y', \eta) \in \mathbb{R}^q \times \mathbb{R}^q \times \mathbb{R}^q$. *Assume that* $a(y, y', \eta)$ *vanishes for* $|y'| > r$ *for a constant* $r > 0$ *and that there is a* $\delta < 0$ *and an* $n \in \mathbb{R}$ *with* $n > \frac{q}{2}$ *such that*

$$\left\| \tilde{\kappa}^{-1}(\eta) \left\{ D_y^\alpha D_{y'}^{\alpha'} a(y, y', \eta) \right\} \kappa(\eta) \right\|_{\mathcal{L}(E, \tilde{E})} \leq c(1 + |y|^2)^{-n+\delta} [\eta]^\nu$$

for all $\alpha \in \mathbb{N}^q$ *with* $|\alpha| \leq 2l$ *with* $2l > \tilde{M} + |s - \nu| + q$ *for a given* $s \in \mathbb{R}$, *and all* $\alpha' \in \mathbb{N}^q$, *with constants* $c = c(\alpha') > 0$. *Then*

$$\mathrm{Op}(a) : \mathcal{W}^s(\mathbb{R}^q, E) \to \mathcal{W}^s(\mathbb{R}^q, \tilde{E})$$

is a compact operator.

Proof. Let us denote by $S^{\nu, \delta - n}$ the subspace of all symbols $a(y, \eta) \in S^\nu(\mathbb{R}^q \times \mathbb{R}^q; E, \tilde{E})$ which satisfy the conditions of Theorem 1.3.61. This is a Fréchet space, and the symbols $a(y, y', \eta)$ that are assumed in the present theorem belong to $C_0^\infty(B, S^{\nu, \delta - n})$ for $B = \{y' \in \mathbb{R}^q : |y'| \leq r\}$. In view of $C_0^\infty(B, S^{\nu, \delta - n}) = C_0^\infty(B) \otimes_\pi S^{\nu, \delta - n}$ we have a convergent sum

$$a(y, y', \eta) = \sum_{j=0}^\infty \lambda_j a_j(y, \eta) \varphi_j(y')$$

with $\sum |\lambda_j| < \infty$, $a_j \in S^{\nu, \delta - n}$, $\varphi_j \in C_0^\infty(B)$ tending to zero in the corresponding spaces. Then also

$$\mathrm{Op}(a) = \sum_{j=0}^\infty \lambda_j \, \mathrm{Op}(a_j) \mathcal{M}_{\varphi_j} \qquad (1.3.67)$$

converges in $\mathcal{L}(\mathcal{W}^s(\mathbb{R}^q, E), \mathcal{W}^s(\mathbb{R}^q, \tilde{E}))$. Since $\mathrm{Op}(a_j)$ is a compact operator, the summands in (1.3.67) are also compact, and hence $\mathrm{Op}(a)$ is a compact operator. $\qquad \square$

Proposition 1.3.63 *Let* $a(y', \eta) \in S^\nu(\mathbb{R}^q \times \mathbb{R}^q; E, \tilde{E})$, $\nu < 0$, *let* $a(y', \eta) : E \to \tilde{E}$ *be a compact operator for every* $(y', \eta) \in \mathbb{R}^q \times \mathbb{R}^q$, *and assume that* $a(y', \eta)$ *vanishes for* $|y'| > r$ *for an* $r > 0$. *Then*

$$\mathrm{Op}(a) : \mathcal{W}^s(\mathbb{R}^q, E) \to \mathcal{W}^s(\mathbb{R}^q, \tilde{E})$$

is a compact operator.

Proof. Let $\omega(y) \in C_0^\infty(\mathbb{R}^q)$ be a function with $\omega(y) = 1$ for $|y| \leq 2r$, $\omega(y) = 0$ for $|y| \geq 3r$. Then the symbol $\omega(y) a(y', \eta)$ satisfies the conditions of Theorem 1.3.62 for all l, and hence $\mathrm{Op}(\omega a)$ is compact.

It remains to consider $\mathrm{Op}(b)$ for $b(y, y', \eta) := (1 - \omega(y))a(y', \eta)$. Since $b(y, y', \eta)$ vanishes in a neighbourhood of diag $\mathbb{R}^q \times \mathbb{R}^q$ we can write

$$\mathrm{Op}(b)u(y) = \iint e^{i(y-y')\eta} |y - y'|^{-2N} (-\Delta_\eta)^N b(y, y', \eta) u(y') \, dy' d\eta.$$

Set $c_N(y, y', \eta) = |y - y'|^{-2N}(-\Delta_\eta)^N b(y, y', \eta)$. Then $c_N(y, y', \eta)$ has compact support with respect to y', and it also satisfies the other conditions of Theorem 1.3.62 when N is chosen sufficiently large. This shows the assertion.   □

Theorem 1.3.64 *Let* $K \subset\subset \mathbb{R}^q$ *be a compact set and let* E', E *be Banach spaces, where* E' *is a compactly embedded subspace of* E, *and* $\kappa'_\lambda = \kappa_\lambda|_{E'}$ *for all* $\lambda \in \mathbb{R}_+$. *Then, for*

$$\mathcal{W}^{s'}(K, E') = \left\{ u \in \mathcal{W}^{s'}(\mathbb{R}^q, E') : \ \mathrm{supp}\, u \subseteq K \right\}$$

we have a canonical compact embedding

$$i : \mathcal{W}^{s'}(K, E') \hookrightarrow \mathcal{W}^s(\mathbb{R}^q, E)$$

for every $s' > s$.

Proof. We can write $i = \mathrm{Op}([\eta]^{s'-s} \, \mathrm{id}_E) \, \mathrm{Op}([\eta]^{s-s'} j_{E,E'} \mathcal{M}_\varphi)$ for any $\varphi(y') \in C_0^\infty(\mathbb{R}^q)$ with $\varphi(y') \equiv 1$ in a neighbourhood of K. Here $j_{E,E'} : E' \to E$ is the compact embedding operator. Then, applying Proposition 1.3.63 for the symbol $[\eta]^{s-s'} j_{E,E'} \mathcal{M}_\varphi$ we get a compact operator

$$\mathrm{Op}([\eta]^{s-s'} j_{E,E'} \mathcal{M}_\varphi) : \mathcal{W}^{s'}(K, E') \to \mathcal{W}^{s'}(\mathbb{R}^q, E).$$

The composition with $\mathrm{Op}([\eta]^{s'-s} \, \mathrm{id}_E) : \mathcal{W}^{s'}(K, E) \to \mathcal{W}^s(\mathbb{R}^q, E)$ gives i which is also compact.   □

We now turn to pseudo–differential operators with operator–valued symbols on a paracompact C^∞ manifold Y under the same assumptions as in the beginning of Section 1.1.5. If $\kappa : U \to \Omega$ is a chart on Y, $\Omega \subseteq \mathbb{R}^q$ open, we set

$$L^\mu(U; E, \tilde{E}) = (\kappa_*)^{-1} L^\mu(\Omega; E, \tilde{E}),$$

where $(\kappa_*)^{-1}$ is the inverse of the operator push-forward under κ, cf. Theorem 1.3.57.

We fix a locally finite open covering $\mathcal{U} = \{U_j\}_{j \in \mathbb{Z}}$ on Y, a subordinate partition of unity $\{\varphi_j\}_{j \in \mathbb{Z}}$ and a system $\{\psi_j\}_{j \in \mathbb{Z}}$ of functions

$\psi_j \in C_0^\infty(U_j)$ with $\varphi_j \psi_j = \varphi_j$ for all j. Let $\kappa_j : U_j \to \Omega_j, j \in \mathbb{Z}$, be a system of charts, with open $\Omega_j \subseteq \mathbb{R}^q$.

We have on Y the space $\mathcal{D}'(Y, E) = \mathcal{L}(C_0^\infty(Y), E)$ of E-valued distributions, the subspace $\mathcal{E}'(Y, E) = \mathcal{L}(C^\infty(Y), E)$ of elements with compact support and we have embeddings

$$C^\infty(Y, E) \hookrightarrow \mathcal{D}'(Y, E) \qquad C_0^\infty(Y, E) \hookrightarrow \mathcal{E}'(Y, E)$$

by a fixed choice of a Riemannian metric on Y with the associated measure dy. This allows us to form

$$\langle f, \varphi \rangle = \int f(y)\varphi(y)\, dy \qquad \text{for } f(y) \in C^\infty(Y, E)$$

for every $\varphi \in C_0^\infty(Y)$. Moreover, we have the space $C^\infty(Y \times Y, \mathcal{L}(E, \tilde{E}))$ and to every element $c(y, y')$ an operator

$$C : u \to \int c(y, y')u(y')\, dy', \qquad u \in C_0^\infty(Y, E)$$

which defines a continuous map $C : C_0^\infty(Y, E) \to C^\infty(Y, \tilde{E})$. The space of all those operators will also be denoted by $L^{-\infty}(Y; E, \tilde{E})$.

Definition 1.3.65 $L^\mu(Y; E, \tilde{E})$ *for* $\mu \in \mathbb{R}$ *is the space of all operators*

$$A = \sum_{j \in \mathbb{Z}} \varphi_j A_j \psi_j + C$$

for arbitrary $A_j \in L^\mu(U_j; E, \tilde{E})$, $j \in \mathbb{Z}$, *and* $C \in L^{-\infty}(Y; E, \tilde{E})$. *In an analogous manner we define* $L_{cl}^\mu(Y; E, \tilde{E})$, *the subspace of classical pseudo-differential operators on* Y.

Note that this is a correct definition, because of the invariance, expressed by the above Theorem 1.3.57.

Let us now denote by

$$S^{(\mu)}(T^*Y \setminus 0; E, \tilde{E})$$

the space of all $p(y, \eta) \in C^\infty(T^*Y \setminus 0, \mathcal{L}(E, \tilde{E}))$ satisfying

$$p(y, \lambda\eta) = \lambda^\mu \tilde{\kappa}_\lambda p(y, \eta) \kappa_\lambda^{-1}$$

for all $\lambda \in \mathbb{R}_+$ and all $(y, \eta) \in T^*Y \setminus 0$. Theorem 1.3.57 allows us to define to every $A \in L_{cl}^\mu(Y; E, \tilde{E})$ a global homogeneous principal symbol of order μ, namely $\sigma_\Lambda^\mu(A)(y, \eta) \in S^{(\mu)}(T^*Y \setminus 0; E, \tilde{E})$. This gives rise to an exact sequence

$$0 \to L_{cl}^{\mu-1}(Y; E, \tilde{E}) \hookrightarrow L_{cl}^\mu(Y; E, \tilde{E}) \xrightarrow{\sigma_\Lambda^\mu} S^{(\mu)}(T^*Y \setminus 0; E, \tilde{E}) \to 0.$$

The essential elements from the local pseudo–differential calculus with operator–valued symbols have straightforward extensions to the calculus over Y. We will not formulate all details here that may easily be added by the reader.

Let us only mention that we have a notion of properly supported pseudo–differential operators, and we can introduce in $L^\mu(Y; E, \tilde{E})$ and $L^\mu_{cl}(Y; E, \tilde{E})$ canonical Fréchet topologies. Moreover, we have a global version of the above Theorem 1.3.56.

We now pass to the invariance of the abstract wedge Sobolev spaces under diffeomorphisms of the underlying open set. We have to impose some additional conditions on the parameter spaces E.

Let $\{E, E^0, E'; \{\kappa_\lambda\}_{\lambda \in \mathbb{R}_+}\}$ be a Hilbert space triple with unitary actions, cf. Definition 1.3.41. To every $a_0 \in \mathcal{L}(E, E^0)$ we have the formal adjoint $a_0^* \in \mathcal{L}(E', E^0)$, defined by $(a_0 u, v) = (u, a_0^* v)$ for all $u \in E$, $v \in E'$. Let $a_0 : E \to E^0$ be an isomorphism satisfying $\kappa_\lambda a_0 \kappa_\lambda^{-1} \in C^\infty(\mathbb{R}_+, \mathcal{L}(E, E^0))$. Then $a_0^* : E' \to E^0$ is also an isomorphism with $\kappa_\lambda a_0^* \kappa_\lambda^{-1} \in C^\infty(\mathbb{R}_+, \mathcal{L}(E', E^0))$. The operator–valued symbols

$$a^s(\eta) := [\eta]^s \kappa_{[\eta]} a_0 \kappa_{[\eta]}^{-1} \in S^s_{cl}(\mathbb{R}^q; E, E^0),$$

$$a^s(\eta)^* := [\eta]^s \kappa_{[\eta]} a_0^* \kappa_{[\eta]}^{-1} \in S^s_{cl}(\mathbb{R}^q; E', E^0)$$

are families of isomorphisms, and we have $a^{-s}(\eta) = a^s(\eta)^{-1}$, $(a^s(\eta)^*)^{-1}$ $= a^{-s}(\eta)^*$, for all $s \in \mathbb{R}$.

Remark 1.3.66 *The spaces E, E^0, E' in our applications, together with the natural choices of $\{\kappa_\lambda\}_{\lambda \in \mathbb{R}_+}$, will satisfy the assumption for the existence of an operator a_0 with the indicated properties. This will be the case, in particular, for the cone Sobolev spaces $E = \mathcal{K}^{s,\gamma}(X^\wedge)$ below, where $E^0 = \mathcal{K}^{0,0}(X^\wedge)$, cf. also Proposition 2.1.60. Recall that we already used a similar assumption above in Remark 1.3.13, cf. the subsequent Proposition 2.1.60 below. Note also that when X is a closed compact C^∞ manifold and $E = H^s(X)$, $E^0 = H^0(X)$, $E' = H^{-s}(X)$ for $s \geq 0$, then $\{\kappa_\lambda\}_{\lambda \in \mathbb{R}_+}$ constructed in Example 1.3.48 to any vector field v on X also has an isomorphism $a_0 : H^s(X) \to H^0(X)$ for which $\kappa_\lambda a_0 \kappa_\lambda^{-1} \in C^\infty(\mathbb{R}_+, \mathcal{L}(H^s(X), H^0(X)))$. In fact, the latter relation is true for every $a_0 \in L^s(X)$, and then it suffices to choose for a_0 an order reducing operator in the sense of Theorem 1.1.83.*

Proposition 1.3.67 *Let $\{E, E^0, E'; \{\kappa_\lambda\}_{\lambda \in \mathbb{R}_+}\}$ be a Hilbert space triple with unitary actions and assume that there is an isomorphism $a_0 : E \to E^0$ with $\kappa_\lambda a_0 \kappa_\lambda^{-1} \in C^\infty(\mathbb{R}_+, \mathcal{L}(E, E^0))$. Then we have*

$$\mathcal{W}^s_{comp}(\Omega, E) = \{u \in \mathcal{E}'(\Omega, E) :$$

$$Au \in \mathcal{W}^0_{loc}(\Omega, E^0) \text{ for every } A \in L^s(\Omega; E, E^0)\}$$

and analogously

$$\mathcal{W}^s_{\mathrm{comp}}(\Omega, E') = \big\{u \in \mathcal{E}'(\Omega, E') :$$
$$Bu \in \mathcal{W}^0_{\mathrm{loc}}(\Omega, E^0)\,for\ every\,B \in L^s(\Omega; E', E^0)\big\},$$

for each $s \in \mathbb{R}$.

Proof. First we have for every $A \in L^s(\Omega; E, E^0)$ a continuous operator $A : \mathcal{W}^s_{\mathrm{comp}}(\Omega, E) \to \mathcal{W}^0_{\mathrm{loc}}(\Omega, E^0)$, cf. Theorem 1.3.58. Conversely, let $u \in \mathcal{E}'(\Omega, E)$ and set $A = \mathrm{Op}(a^s)$ for the above symbol $a^s(\eta)$. Let $P \in L^{-s}(\Omega; E^0, E)$ be a properly supported representative of $\mathrm{Op}(a^{-s})$. Then $Au = f \in \mathcal{W}^s_{\mathrm{loc}}(\Omega, E^0)$ implies $PAu = Pf \in \mathcal{W}^s_{\mathrm{loc}}(\Omega, E)$, and we have $PA = 1 + C$ for a $C \in L^{-\infty}(\Omega; E, E)$. Thus $u = -Cu + Pf$. In view of $C : \mathcal{E}'(\Omega, E) \to \mathcal{W}^\infty_{\mathrm{loc}}(\Omega, E)$ we obtain $u \in \mathcal{W}^s_{\mathrm{loc}}(\Omega, E)$. The assertion for E' can be proved in an analogous manner. □

Theorem 1.3.68 *Let* $\Omega, \tilde{\Omega} \subseteq \mathbb{R}^q$ *be open sets and* $\chi : \Omega \to \tilde{\Omega}$ *be a diffeomorphism. Then, under the assumptions of Proposition* 1.3.67, *the pull-backs* $\chi^* : C^\infty_0(\tilde{\Omega}, E) \to C^\infty_0(\Omega, E)$ *and* $\chi^* : C^\infty(\tilde{\Omega}, E) \to C^\infty(\Omega, E)$ *extend to isomorphisms*

$$\chi^* : \mathcal{W}^s_{\mathrm{comp}}(\tilde{\Omega}, E) \to \mathcal{W}^s_{\mathrm{comp}}(\Omega, E)$$

and

$$\chi^* : \mathcal{W}^s_{\mathrm{loc}}(\tilde{\Omega}, E) \to \mathcal{W}^s_{\mathrm{loc}}(\Omega, E),$$

respectively. An analogous assertion holds for the spaces with E' *instead of* E.

Proof. Let us consider the comp-spaces. For each $A \in L^s(\Omega; E, E^0)$ we have the push-forward $\chi_* A \in L^s(\tilde{\Omega}; E, E^0)$, cf. Theorem 1.3.57. Then the diagram

$$
\begin{array}{ccc}
\mathcal{E}'(\tilde{\Omega}, E) & \xrightarrow{\;\chi^*\;} & \mathcal{E}'(\Omega, E) \\
{\scriptstyle \chi_* A}\big\downarrow & & \big\downarrow{\scriptstyle A} \\
\mathcal{D}'(\tilde{\Omega}, E^0) & \xrightarrow{\;\chi^*\;} & \mathcal{D}'(\Omega, E^0)
\end{array}
$$

commutes. Now $u \in \mathcal{W}^s_{\mathrm{comp}}(\tilde{\Omega}, E)$ is equivalent to $\tilde{A}u \in \mathcal{W}^s_{\mathrm{comp}}(\tilde{\Omega}, E^0)$ for every $\tilde{A} \in L^s(\tilde{\Omega}; E, E^0)$. The operator push-forward defines an isomorphism $\chi_* : L^s(\Omega; E, E^0) \to L^s(\tilde{\Omega}; E, E^0)$. In view of Remark 1.3.40 we get $\chi^* \chi_* Au \in \mathcal{W}^0_{\mathrm{loc}}(\Omega, E^0)$ for every $u \in \mathcal{W}^s_{\mathrm{comp}}(\Omega, E)$, or, equivalently, $A\chi^* u \in \mathcal{W}^0_{\mathrm{loc}}(\Omega, E^0)$. Since this is true for every $A \in$

$L^s(\Omega; E, E^0)$, the relation $\chi^* u \in \mathcal{W}^s_{\text{comp}}(\Omega, E)$ follows from Proposition 1.3.67. By analogous arguments we get the assertion for E' instead of E. The corresponding properties for the loc-spaces are then immediate consequences. □

Next we form the global abstract wedge Sobolev spaces under the condition that the parameter space E is a Hilbert space and satisfies the conditions of Proposition 1.3.67.

Let us fix a compact set $K \subset\subset Y$. Then $\mathcal{W}^s(K, E)$ for $s \in \mathbb{R}$ is defined as the closure of $C_0^\infty(K, E)$ ($= \{u \in C^\infty(Y, E) : \text{supp } u \subseteq K\}$) with respect to the norm

$$\left\{ \sum_{\substack{j \in \mathbb{Z} \\ U_j \cap K \neq \emptyset}} \|(\kappa_j^*)^{-1} \varphi_j u\|^2_{\mathcal{W}^s(\mathbb{R}^q, E)} \right\}^{\frac{1}{2}},$$

where $\kappa_j : U_j \to \Omega_j \subseteq \mathbb{R}^q$ are the given charts and $\kappa_j^* : C_0^\infty(\Omega_j, E) \to C_0^\infty(Y, E)$ the function pull–back. In view of Theorem 1.3.68 this is a correct definition, i.e., independent of the particular choice of the charts and the partition of unity. $\mathcal{W}^s(K, E)$ is a Banach space for each s, and we define

$$\mathcal{W}^s_{\text{comp}}(Y, E) = \varinjlim_{K \subset\subset Y} \mathcal{W}^s(K, E).$$

Moreover, we set

$$\mathcal{W}^s_{\text{comp}}(Y, E)$$
$$= \left\{ u \in \mathcal{D}'(Y, E) : \varphi u \in \mathcal{W}^s_{\text{comp}}(Y, E) \text{ for every } \varphi \in C_0^\infty(Y) \right\}.$$

This is a Fréchet space with the semi–norm system $u \to \|\varphi u\|_{\mathcal{W}^s(K,E)}$ for every $\varphi \in C_0^\infty(Y)$ and $K = \text{supp } \varphi$. If Y is a closed compact C^∞ manifold, we have, in particular, the spaces

$$\mathcal{W}^s(Y, E), \quad s \in \mathbb{R}. \tag{1.3.68}$$

Theorem 1.3.69 *Let E, $\tilde{E}$ be Hilbert spaces satisfying the conditions of Proposition 1.3.67. Then each $A \in L^\mu(Y; E, \tilde{E})$ induces a continuous operator*

$$A : \mathcal{W}^s_{\text{comp}}(Y, E) \to \mathcal{W}^{s-\mu}_{\text{loc}}(Y, \tilde{E}) \tag{1.3.69}$$

for every $s \in \mathbb{R}$. If A is properly supported we get continuous operators

$$A : \mathcal{W}^s_{\text{comp}}(Y, E) \to \mathcal{W}^{s-\mu}_{\text{comp}}(Y, \tilde{E}), \quad A : \mathcal{W}^s_{\text{loc}}(Y, E) \to \mathcal{W}^{s-\mu}_{\text{loc}}(Y, \tilde{E})$$

for every $s \in \mathbb{R}$. In particular, for compact Y we get

$$A : \mathcal{W}^s(Y, E) \to \mathcal{W}^{s-\mu}(Y, \tilde{E}). \tag{1.3.70}$$

Proof. The continuity of (1.3.69) is an immediate consequence of Definition 1.3.65, Theorem 1.3.58, Theorem 1.3.68, together with the fact that smoothing operators C are continuous as $C : \mathcal{W}^s_{\text{comp}}(Y, E) \to \mathcal{W}^\infty_{\text{loc}}(Y, \tilde{E})$ for all $s \in \mathbb{R}$. The second assertion is trivial. □

We stop at this point the discussion of the generalities of the calculus of pseudo–differential operators with operator–valued symbols in the Fourier-edge approach. It is possible to introduce ellipticity and to carry out parametrix constructions, similarly to the results of Luke [70] and Dorschfeldt, Grieme Schulze [21]. Under appropriate conditions to the symbols we could also obtain Fredholm operators and start discussing the index. However we are more interested in specific subclasses of pseudo–differential operators with operator–valued symbols with more symbolic levels that take part in the ellipticity, combined with the "scalar" theory. This will be the subject of Chapter 3 below.

1.4 Pseudo–differential operators with cone–shaped exits to infinity

The following material can be regarded as an aspect of the pseudo–differential calculus on non–compact manifolds, here in the sense of exits (outlets) "to infinity". This theory plays an auxiliary role in the edge symbolic calculus in Chapter 3 below.

The basic ideas of this calculus go back to Shubin [137], Parenti [82], Feygin [34], Grushin [49], Cordes [17], Schrohe [101].

1.4.1 Global pseudo–differential operators in $\mathbb{R}^n$

The local model of an exit to infinity in our sense is an open set $V \subset \mathbb{R}^n$ that is conical for large $|x|$, i.e. there is a conical set $\Gamma \subset \mathbb{R}^n \setminus \{0\}$ (i.e. $x \in \Gamma$ implies $\lambda x \in \Gamma$ for all $\lambda \in \mathbb{R}_+$) such that $V \setminus \Gamma$ is a bounded set. The global theory on manifolds will be reduced to that over V by corresponding charts. We shall start with the case $V = \mathbb{R}^n$.

Definition 1.4.1 $S^{\mu,\varrho}(\mathbb{R}^n \times \mathbb{R}^n)$ *for* $\mu, \varrho \in \mathbb{R}$ *is the space of all symbols* $a(x,\xi) \in C^\infty(\mathbb{R}^n \times \mathbb{R}^n)$ *with*

$$|D_x^\alpha D_\xi^\beta a(x,\xi)| \le c(1 + |\xi|)^{\mu - |\beta|}(1 + |x|)^{\varrho - |\alpha|} \qquad (1.4.1)$$

for all $x, \xi \in \mathbb{R}^n$ *and all* $\alpha, \beta \in \mathbb{N}^n$, *for constants* $c = c(\alpha, \beta) > 0$ *(not dependent on* x, ξ*)*.

This definition is symmetric in x and ξ. Thus we can often reduce assertions with respect to x to corresponding ones with respect to ξ. It is useful from time also to consider the spaces

$$S^{\mu,\mu',\varrho,\varrho'}(\mathbb{R}^{2n} \times \mathbb{R}^{2n}) \tag{1.4.2}$$

consisting of all $a(x, x', \xi, \xi') \in C^\infty(\mathbb{R}^{2n}_{x,x'} \times \mathbb{R}^{2n}_{\xi,\xi'})$ satisfying

$$|D_x^\alpha D_{x'}^{\alpha'} D_\xi^\beta D_{\xi'}^{\beta'} a(x, x', \xi, \xi')|$$
$$\leq c(1 + |\xi|)^{\mu - |\beta|}(1 + |\xi'|)^{\mu' - |\beta'|}(1 + |x|)^{\varrho - |\alpha|}(1 + |x'|)^{\varrho' - |\alpha'|} \tag{1.4.3}$$

for all $x, x', \xi, \xi' \in \mathbb{R}^n$ and all $\alpha, \alpha', \beta, \beta' \in \mathbb{N}^n$, for certain constants $c(\alpha, \alpha', \beta, \beta') > 0$. Analogously we can form

$$S^{\mu,\varrho,\varrho'}(\mathbb{R}^{2n}_{x,x'} \times \mathbb{R}^n_\xi), \qquad S^{\mu,\mu',\varrho}(\mathbb{R}^n_x \times \mathbb{R}^{2n}_{\xi,\xi'}) \tag{1.4.4}$$

with corresponding modified estimates, e.g. , for an element $a(x, x', \xi) \in S^{\mu,\varrho,\varrho'}(\mathbb{R}^{2n}_{x,x'} \times \mathbb{R}^n_\xi)$,

$$|D_x^\alpha D_{x'}^{\alpha'} D_\xi^\beta a(x, x', \xi)| \leq c(1 + |\xi|)^{\mu - |\beta|}(1 + |x|)^{\varrho - |\alpha|}(1 + |x'|)^{\varrho' - |\alpha'|}$$

for all $x, x', \xi \in \mathbb{R}^n, \alpha, \alpha', \beta \in \mathbb{N}^n$, for $c = c(\alpha, \alpha', \beta) > 0$. For simplicity we shall mainly discuss $S^{\mu,\varrho}(\mathbb{R}^n \times \mathbb{R}^n)$. The properties of the more general spaces (1.4.2), (1.4.4) are analogous and shall tacitly be used below.

The space $S^{\mu,\varrho}(\mathbb{R}^n \times \mathbb{R}^n)$ is Fréchet with the best possible constants c in (1.4.1) as semi–norm system, $\alpha, \beta \in \mathbb{N}^n$. It is then obvious that there are continuous embeddings

$$S^{\mu,\varrho}(\mathbb{R}^n \times \mathbb{R}^n) \subseteq S^{\mu',\varrho'}(\mathbb{R}^n \times \mathbb{R}^n) \tag{1.4.5}$$

for $\mu \leq \mu', \varrho \leq \varrho'$. Furthermore, $a(x, \xi) \in S^{\mu,\varrho}(\mathbb{R}^n \times \mathbb{R}^n), b(x, \xi) \in S^{\nu,\sigma}(\mathbb{R}^n \times \mathbb{R}^n)$ implies $(ab)(x, \xi) \in S^{\mu+\nu,\varrho+\sigma}(\mathbb{R}^n \times \mathbb{R}^n)$. Another immediate consequence of Definition 1.4.1 is that

$$D_x^\alpha D_\xi^\beta S^{\mu,\varrho}(\mathbb{R}^n \times \mathbb{R}^n) \subseteq S^{\mu-|\beta|,\varrho-|\alpha|}(\mathbb{R}^n \times \mathbb{R}^n). \tag{1.4.6}$$

Remark 1.4.2 *The elements* $a^{\nu,\sigma}(x, \xi) = (1 + |x|^2)^{\frac{\sigma}{2}}(1 + |\xi|^2)^{\frac{\nu}{2}}$ *belong to* $S^{\nu,\sigma}(\mathbb{R}^n \times \mathbb{R}^n)$ *for every* $\nu, \varrho \in \mathbb{R}$. *The multiplication by* $a^{\nu,\sigma}(x, \xi)$ *induces isomorphisms*

$$a^{\nu,\sigma} : S^{\mu,\varrho}(\mathbb{R}^n \times \mathbb{R}^n) \to S^{\mu+\nu,\varrho+\sigma}(\mathbb{R}^n \times \mathbb{R}^n)$$

for all $\mu, \varrho \in \mathbb{R}$.

We have

$$S^{-\infty,-\infty}(\mathbb{R}^n \times \mathbb{R}^n) := \bigcap_{\mu,\varrho \in \mathbb{R}} S^{\mu,\varrho}(\mathbb{R}^n \times \mathbb{R}^n) = \mathcal{S}(\mathbb{R}^n \times \mathbb{R}^n)$$

and

$$S^{-\infty,\varrho}(\mathbb{R}^n \times \mathbb{R}^n) := \bigcap_{\mu \in \mathbb{R}} S^{\mu,\varrho}(\mathbb{R}^n \times \mathbb{R}^n) = \mathcal{S}(\mathbb{R}^n_\xi, S^\varrho(\mathbb{R}^n_x)),$$

$$S^{\mu,-\infty}(\mathbb{R}^n \times \mathbb{R}^n) := \bigcap_{\varrho \in \mathbb{R}} S^{\mu,\varrho}(\mathbb{R}^n \times \mathbb{R}^n) = \mathcal{S}(\mathbb{R}^n_x, S^\mu(\mathbb{R}^n_\xi)),$$

cf. also the notation in Section 1.2.2.

Theorem 1.4.3 *Let* $a_j(x,\xi) \in S^{\mu_j,\varrho_j}(\mathbb{R}^n \times \mathbb{R}^n)$, $j \in \mathbb{N}$, *be an arbitrary sequence with* $\mu_j \to -\infty$, $\varrho_j \to -\infty$. *Then there is an* $a(x,\xi) \in S^{\mu,\varrho}(\mathbb{R}^n \times \mathbb{R}^n)$ *for* $\mu = \max\{\mu_j\}, \varrho = \max\{\varrho_j\}$ *that is unique modulo* $S^{-\infty,-\infty}(\mathbb{R}^n \times \mathbb{R}^n)$ *such that for every* $M \in \mathbb{R}$ *there is an* $N(M)$ *such that for all* $N \geq N(M)$

$$a(x,\xi) - \sum_{j=0}^{N} a_j(x,\xi) \in S^{\mu-M,\varrho-M}(\mathbb{R}^n \times \mathbb{R}^n).$$

Moreover, if $b_k(x,\xi) \in S^{\mu_k,\varrho}(\mathbb{R}^n \times \mathbb{R}^n), k \in \mathbb{N}$, *is an arbitrary sequence with* $\mu_k \to -\infty$ *as* $k \to \infty$, *then there is a* $b(x,\xi) \in S^{\mu,\varrho}(\mathbb{R}^n \times \mathbb{R}^n)$ *for* $\mu = \max\{\mu_k\}$ *that is unique mod* $S^{-\infty,\varrho}(\mathbb{R}^n \times \mathbb{R}^n)$ *such that for every* $M \in \mathbb{R}$ *there is an* $N(M)$ *such that for all* $N \geq N(M)$

$$b(x,\xi) - \sum_{k=0}^{N} a_k(x,\xi) \in S^{\mu-M,\varrho}(\mathbb{R}^n \times \mathbb{R}^n).$$

An analogous result holds after interchanging the role of x *and* ξ.

According to standard notation we write

$$a(x,\xi) \sim \sum_{j=0}^{\infty} a_j(x,\xi), \qquad b(x,\xi) \sim \sum_{k=0}^{\infty} b_k(x,\xi)$$

and talk about asymptotic sums in the corresponding sense.

The proof of Theorem 1.4.3 follows by adapting the general idea of Proposition 1.1.17 to the present situation. If $\chi(x,\xi)$ is an excision function in $\mathbb{R}^n \times \mathbb{R}^n \ni (x,\xi)$ then there are constants c_j tending to ∞ sufficiently fast for $j \to \infty$ such that

$$a(x,\xi) = \sum_{j=0}^{\infty} \chi(c_j^{-1}x, c_j^{-1}\xi)a_j(x,\xi)$$

converges in $S^{\mu,\varrho}(\mathbb{R}^n \times \mathbb{R}^n)$, as in (1.1.33). Similarly if $\chi(\xi)$ is an excision function in $\mathbb{R}^n \ni \xi$, then

$$b(x,\xi) = \sum_{k=0}^{\infty} \chi(c_k^{-1}\xi) b_k(x,\xi)$$

converges in $S^{\mu,\varrho}(\mathbb{R}^n \times \mathbb{R}^n)$ for a suitable choice of constants $c_k, k \in \mathbb{N}$.
In view of $S^{\mu,\varrho}(\mathbb{R}^n \times \mathbb{R}^n) \subset S^{\mu}(\mathbb{R}^n \times \mathbb{R}^n)$ we can apply the local calculus from the Sections 1.1.2 and 1.1.3 . In particular, we have the pseudo–differential operators $\mathrm{Op}(a) : C_0^{\infty}(\mathbb{R}^n) \to C^{\infty}(\mathbb{R}^n)$ with symbols $a(x,\xi) \in S^{\mu,\varrho}(\mathbb{R}^n \times \mathbb{R}^n)$. The smoothing operators of the present theory in $\mathbb{R}^n$ have the form

$$Cu(x) = \int_{\mathbb{R}^n} c(x,x')u(x')\,dx' \quad \text{with} \quad c(x,x') \in \mathcal{S}(\mathbb{R}^n \times \mathbb{R}^n).$$

Denote by $L^{-\infty,-\infty}(\mathbb{R}^n)$ the space of all such operators. Moreover, we set

$$L^{\mu,\varrho}(\mathbb{R}^n) =$$
$$\{\mathrm{Op}(a) + C : a(x,\xi) \in S^{\mu,\varrho}(\mathbb{R}^n \times \mathbb{R}^n), C \in L^{-\infty,-\infty}(\mathbb{R}^n)\}. \quad (1.4.7)$$

Observe that $a(x,\xi) \in S^{-\infty,-\infty}(\mathbb{R}^n \times \mathbb{R}^n)$ implies $\mathrm{Op}(a) \in L^{-\infty,-\infty}(\mathbb{R}^n)$

Theorem 1.4.4 *Each $A \in L^{\mu,\varrho}(\mathbb{R}^n)$ for $\mu, \varrho \in \mathbb{R}$ induces a continuous operator*

$$A : \mathcal{S}(\mathbb{R}^n) \to \mathcal{S}(\mathbb{R}^n). \quad (1.4.8)$$

Proof. The assertion is obvious for $A \in L^{-\infty,-\infty}(\mathbb{R}^n)$. So we can assume that $Au(x) = \int e^{ix\xi} a(x,\xi)(Fu)(\xi)\,d\xi$, $a(x,\xi) \in S^{\mu,\varrho}(\mathbb{R}^n \times \mathbb{R}^n)$. Let $u \in \mathcal{S}(\mathbb{R}^n)$, $\varphi \in C_0^{\infty}(\mathbb{R}^n)$. Then $b(x,\xi) := \varphi(x)a(x,\xi)Fu(\xi) \in \mathcal{S}(\mathbb{R}^n \times \mathbb{R}^n)$ shows that $Au(x) = F_{\xi \to x}^{-1} b(x,\xi) \in C^{\infty}(\mathbb{R}^n)$. Now let us fix multi–indices $\alpha, \beta \in \mathbb{N}^n$ and show estimates for $\sup_{x \in \mathbb{R}^n} |x^\beta D_x^\alpha Au(x)|$. First we have

$$D_x^\alpha e^{ix\xi} a(x,\xi) = e^{ix\xi} \sum_{\gamma \leq \alpha} c_{\alpha\gamma} \xi^\gamma D_x^{\alpha-\gamma}\, a(x,\xi) \quad (1.4.9)$$

with suitable constants $c_{\alpha\gamma}$. Choose $k, l \in \mathbb{N}$ such that $2k > |\beta| + \varrho + n$, $2l > |\alpha| + \mu + q$, and write

$$(1 + |\xi|^2)^{-l}(1 - \Delta_{x'})^l e^{i(x-x')\xi} = e^{i(x-x')\xi}, \quad (1.4.10)$$
$$(1 + |x - x'|^2)^{-k}(1 - \Delta_\xi)^k e^{i(x-x')\xi} = e^{i(x-x')\xi}. \quad (1.4.11)$$

Then integration by parts gives us

$$x^\beta D_x^\alpha Au(x)$$

$$= \sum_{\gamma \leq \alpha} c_{\alpha\gamma} \iint e^{i(x-x')\xi} \xi^\gamma (1 + |\xi|^2)^{-l} x^\beta D_x^{\alpha-\gamma} a(x,\xi)(1 - \Delta_{x'})^l u(x')\, dx'd\xi$$

$$= \sum_{\gamma \leq \alpha} c_{\alpha\gamma} \iint e^{i(x-x')\xi} (1 + |x - x'|^2)^{-k}$$

$$\{(1 - \Delta_\xi)^k \xi^\gamma (1 + |\xi|^2)^{-l} D_x^{\alpha-\gamma} a(x,\xi)\}(1 - \Delta_{x'})^l u(x')\, dx'd\xi.$$

Using $(1 + |x - x'|^2)^{-k} \leq c_k(1 + |x'|^2)^k(1 + |x|^2)^{-k}$ for suitable c_k we obtain

$$\sup_{x \in \mathbb{R}^n} |x^\beta D_x^\alpha Au(x)|$$

$$\leq \sum_{\gamma \leq \alpha} \sup_{x \in \mathbb{R}^n} c_{\alpha\gamma} c_k \iint |x^\beta|(1 + |x'|^2)^k (1 + |x|^2)^{-k}$$

$$\times |(1 - \Delta_\xi)^k \xi^\gamma (1 + |\xi|^2)^{-l} D_x^{\alpha-\gamma} a(x,\xi)||(1 - \Delta_{x'})^l u(x')|\, dx d\xi$$

$$\leq \sum_{\gamma \leq \alpha} \sup_{x \in \mathbb{R}^n} c_{\alpha\gamma} c_k \iint |x^\beta|(1 + |x|^2)^{-k} \langle x \rangle^\varrho \langle \xi \rangle^{\mu+|\alpha|-2l} p(a)|$$

$$\times (1 + |x'|^2)^l u(x')|\, dx d\xi$$

$$\leq \tilde{c}_{\alpha\beta,k} p(a) \int (1 + |x'|^2)^k (1 - \Delta_{x'})^l u(x')\, dx'.$$

Here p is a semi-norm on $S^{\mu,\varrho}(\mathbb{R}^n \times \mathbb{R}^n)$ and $\tilde{c}_{\alpha\beta,k}$ a constant, only dependent on $\alpha, \beta \in \mathbb{N}^n$, $k \in \mathbb{N}$. For every $k, l \in \mathbb{N}$ there is a semi-norm π_ι on $\mathcal{S}(\mathbb{R}^q)$ and that

$$\sup_{x \in \mathbb{R}^n} |(1 + |x|^2)^k (1 - \Delta_x)^l u(x)| \leq (1 + |x|^2)^{-n} \pi_\iota(u)$$

for all $u \in \mathcal{S}(\mathbb{R}^n)$. Hence $\sup_{y \in \mathbb{R}^n} |x^\beta D_x^\alpha Au(x)| \leq cp(a)\pi_\iota(u)$, for some constant $c > 0$ which proves the continuity of $A : \mathcal{S}(\mathbb{R}^n) \to \mathcal{S}(\mathbb{R}^n)$. $\square$

Remark 1.4.5 *Let us set* $r^\nu(\xi) = (1 + |\xi|^2)^{\frac{\nu}{2}}$, $b^\sigma(x) = (1 + |x|^2)^{\frac{\sigma}{2}}$ *for* $\nu, \sigma \in \mathbb{R}$. *Then* $A \to \mathrm{Op}(b^\sigma) A \mathrm{Op}(r^\nu)$ *induces isomorphisms* $L^{\mu,\varrho}(\mathbb{R}^n) \to L^{\mu+\nu,\varrho+\sigma}(\mathbb{R}^n)$ *for all* $\mu, \varrho \in \mathbb{R}$. *For convenience from time to time we restrict considerations to* $L^{\mu,0}(\mathbb{R}^n)$ *or* $L^{0,0}(\mathbb{R}^n)$, *obtained from* $L^{\mu,\varrho}(\mathbb{R}^n)$ *by a corresponding order reduction .*

Let $C_b^\infty(\mathbb{R}^n)$ be the subspace of all $u \in C^\infty(\mathbb{R}^n)$ which obey $\sup_{x \in \mathbb{R}^n} |D_x^\alpha u(x)| < \infty$ for all $\alpha \in \mathbb{N}^n$. With these semi-norms $C_b^\infty(\mathbb{R}^n)$ is a Fréchet space.

Theorem 1.4.6 *Each $A \in \mathbf{L}^{\mu,0}(\mathbb{R}^n)$, $\mu \in \mathbb{R}$, induces a continuous operator*

$$A : C_b^\infty(\mathbb{R}^n) \to C_b^\infty(\mathbb{R}^n).$$

Proof. The assertion for $\mathbf{L}^{\mu,0}(\mathbb{R}^n)$ is obvious. So let us assume $A = \mathrm{Op}(a)$ for $a(x,\xi) \in \mathbf{S}^{\mu,0}(\mathbb{R}^n \times \mathbb{R}^n)$. Using the above relations (1.4.9), (1.4.10), (1.4.11), we obtain for $u \in C_0^\infty(\mathbb{R}^n)$, $\alpha \in \mathbb{N}^n$

$$D_x^\alpha Au(x) = \sum_{\gamma \le \alpha} c_{\alpha\gamma} \iint e^{i(x-x')\xi}(1 + |x - x'|^2)^{-k}$$

$$\{(1 - \Delta_\xi)^k \xi^\gamma (1 + |\xi|^2)^{-l} D_x^{\alpha-\gamma} a(x,\xi)\}(1 - \Delta_{x'})^l u(x')\,dx'd\xi.$$

Choose $k, l \in \mathbb{N}$ with $k > n$, $2l > |\alpha| + \mu + n$. Then it is easy to verify that the latter integral also converges for $u \in C_b^\infty(\mathbb{R}^n)$. We have

$$|D_x^\alpha Au(x)| \le \sum_{\gamma \le \alpha} c_{\alpha\gamma} \iint (1 + |x - x'|^2)^{-k}|(1 - \Delta_\xi)^k \xi^\gamma (1 + |\xi|^2)^{-l}$$

$$D_x^{\alpha-\gamma} a(x,\xi)||(1 - \Delta_{x'})^l u(x')|\,dx'd\xi.$$

$$\le c_\alpha \iint (1 + |x - x'|^2)^{-k} \langle\xi\rangle^{\mu-2l} p(a)\pi_\iota(u)\,dx'd\xi$$

for all $x \in \mathbb{R}^n$, where p is a suitable semi–norm on $\mathbf{S}^{\mu,0}(\mathbb{R}^n \times \mathbb{R}^n)$, π_ι a semi–norm on $C_b^\infty(\mathbb{R}^n)$, and c_α a constant, only dependent on α. Thus

$$|D_x^\alpha Au(x)| \le c_\alpha p(a)\pi_\iota(u) \int (1 + |x - x'|^2)^{-k}\,dx'$$

$$\le \tilde{c}_\alpha p(a)\pi_\iota(u)$$

for all $x \in \mathbb{R}^n$, for a constant $\tilde{c}_\alpha$, only dependent on $\alpha \in \mathbb{N}^n$. This shows that $A : C_b^\infty(\mathbb{R}^n) \to C_b^\infty(\mathbb{R}^n)$ is continuous. $\square$

Note, in particular, that every $A \in \mathbf{L}^{\mu,0}(\mathbb{R}^n)$ can be applied to $e_\xi(x) = e^{ix\xi} \in C_b^\infty(\mathbb{R}^n)$. For $A = \mathrm{Op}(a)$, $a(x,\xi) \in \mathbf{S}^{\mu,0}(\mathbb{R}^n \times \mathbb{R}^n)$, we have

$$a(x,\xi) = e_{-\xi}(x)(Ae_\xi)(x), \qquad (1.4.12)$$

cf. also (1.1.59). In fact, using $\langle v, F_{x'\to\xi}w\rangle_\xi = \langle F_{\xi\to x'}v, w\rangle_{x'}$, where the subscripts indicate the integration variables, we obtain for $v(\xi) = e^{ix\xi}a(x,\xi)$ for fixed x and $w(x') = e_\xi(x')$

$$(\mathrm{Op}(a)e_\xi) = (2\pi)^{-n}\langle v(\xi), F_{x'\to\xi}e_\xi(x')\rangle_\xi \qquad (1.4.13)$$

$$= (2\pi)^{-n}\langle(F_{\xi\to x'}v)(x'), e_\xi(x')\rangle_{x'} \qquad (1.4.14)$$

$$= v(\xi). \qquad (1.4.15)$$

Here we used the Fourier transform on $\mathcal{S}'(\mathbb{R}^n) \supset C_b^\infty(\mathbb{R}^n)$ and in the last equation the Fourier inversion formula. (1.4.13) yields immediately (1.4.12).

Proposition 1.4.7 *The map*

$$\mathrm{Op} : \boldsymbol{S}^{\mu,\varrho}(\mathbb{R}^n \times \mathbb{R}^n) \to \boldsymbol{L}^{\mu,\varrho}(\mathbb{R}^n)$$

is an isomorphism for all $\mu, \varrho \in \mathbb{R}$.

Proof. The relation (1.4.12) shows that

$$\mathrm{Op} : \boldsymbol{S}^{\mu,0}(\mathbb{R}^n \times \mathbb{R}^n) \to \boldsymbol{L}^{\mu,0}(\mathbb{R}^n) \tag{1.4.16}$$

is an injective map. Moreover, by definition every $A \in \boldsymbol{L}^{\mu,0}(\mathbb{R}^n \times \mathbb{R}^n)$ has the form $\mathrm{Op}(a) + C$ for an $a(x,\xi) \in \boldsymbol{S}^{\mu,0}(\mathbb{R}^n \times \mathbb{R}^n), C \in \boldsymbol{L}^{-\infty,-\infty}(\mathbb{R}^n)$. By $c(x,\xi) = e_{-\xi}(x)(Ce_\xi)(x)$ which is in $\boldsymbol{S}^{-\infty,-\infty}(\mathbb{R}^n \times \mathbb{R}^n)$ we get by analogous arguments an injective map

$$\mathrm{Op} : \boldsymbol{S}^{-\infty,-\infty}(\mathbb{R}^n \times \mathbb{R}^n) \to \boldsymbol{L}^{-\infty,-\infty}(\mathbb{R}^n).$$

However this already shows that (1.4.16) is surjective, i.e., an isomorphism. For arbitrary $\varrho \in \mathbb{R}$ it suffices to apply Remark 1.4.2 and Remark 1.4.5. $\qquad\square$

Remark 1.4.8 *Each $A \in \boldsymbol{L}^{-\infty,-\infty}(\mathbb{R}^n)$ can also be written as $A = \mathrm{Op}(a)$ for some $a(x,x',\xi) \in \boldsymbol{S}^{-\infty,-\infty,-\infty}(\mathbb{R}^{2n} \times \mathbb{R}^n)$. In fact, A has a kernel $c(x,x') \in \mathcal{S}(\mathbb{R}^n \times \mathbb{R}^n)$, and it suffies to set $a(x,x',\xi) = c(x,x')e^{i(x'-x)\xi}\psi(\xi)$ for any $\psi(\xi) \in C_0^\infty(\mathbb{R}^n)$ satisfying $\int \psi(\xi)d\xi = 1$. As we saw we also find a unique $a_0(x,\xi) \in \boldsymbol{S}^{-\infty,-\infty}(\mathbb{R}^n \times \mathbb{R}^n)$ such that $A = \mathrm{Op}(a_0)$.*

Now consider an element $a(x,x',\xi,\xi') \in \boldsymbol{S}^{\mu,\mu',\varrho,\varrho'}(\mathbb{R}^{2n} \times \mathbb{R}^{2n})$ and form

$$b(x,\xi) = \iint e^{-iy\eta}a(x,x+y,\xi+\eta,\xi)\,dy\,d\eta \tag{1.4.17}$$

in the sense of oscillatory integrals. With $a(x,x',\xi,\xi')$ we can associate an operator by

$$\mathrm{Op}(a)u(x) = \iint e^{-i(y\eta+y'\eta')}a(x,x+y,\eta,\eta')u(x+y+y')\,dy\,dy'\,d\eta\,d\eta' \tag{1.4.18}$$

for $u \in \mathcal{S}(\mathbb{R}^n)$, also interpreted in terms of oscillatory integrals.

Theorem 1.4.9 $a(x, x', \xi, \xi') \in S^{\mu,\mu',\varrho,\varrho'}(\mathbb{R}^{2n} \times \mathbb{R}^{2n})$ *implies* $b(x, \xi) \in$ $S^{\mu+\mu',\varrho+\varrho'}(\mathbb{R}^n \times \mathbb{R}^n)$; *we have*

$$\mathrm{Op}(a)u = \mathrm{Op}(b)u, \qquad (1.4.19)$$

and $b(x, \xi)$ *has the asymptotic expansion*

$$b(x, \xi) \sim \sum_\alpha \frac{1}{\alpha!} (\partial_\xi^\alpha D_{x'}^\alpha a)(x, x', \xi, \xi')|_{x'=x, \xi'=\xi}. \qquad (1.4.20)$$

Proof. Let

$$b_\theta(x, \xi) = \iint e^{-iy\eta} a(x, x + y, \xi + \theta\eta, \xi) \, dy d\xi,$$

$\theta \in \mathbb{R}$, and show that $\{b_\theta\}_{|\theta| \le 1}$ is a bounded set in $S^{\mu+\mu',\varrho+\varrho'}(\mathbb{R}^n \times \mathbb{R}^n)$. That means, in particular,

$$|b_\theta(x, \xi)| \le c\langle\xi\rangle^{\mu+\mu'}\langle x\rangle^{\varrho+\varrho'} \qquad (1.4.21)$$

for constants $c > 0$ independent of $|\theta| \le 1$. Choose $p, q \in \mathbb{N}$ such that $2p > n + |\mu|$, $2q > n + |\varrho'|$ and set

$$r_\theta(x, y, \xi, \eta) = \langle\eta\rangle^{-2p}\langle D_y\rangle^{2p}\langle y\rangle^{-2q}\langle D_\eta\rangle^{2q} a(x, x + y, \xi + \theta\eta, \xi).$$

Then

$$b_\theta(x, \xi) = \iint e^{-iy\eta} r_\theta(x, y, \xi, \eta) \, dy d\eta.$$

We have

$$|r_\theta(x, y, \xi, \eta)| \le c\langle\eta\rangle^{-2p}\langle y\rangle^{-2q} \max_{|\gamma| \le 2p} |D_y^\gamma \langle D_\eta\rangle^{2q} a(x, x + y, \xi + \theta\eta, \xi)|$$

$$\le c\langle\eta\rangle^{-2p}\langle y\rangle^{-2q}\langle\xi + \theta\eta\rangle^\mu\langle\xi\rangle^{\mu'}\langle x\rangle^\varrho\langle x + y\rangle^{\varrho'}.$$

Thus

$$|b_\theta(x, \xi)| \le c\left\{\int \langle x + y\rangle^{\varrho'}\langle y\rangle^{-2q}\langle x\rangle^\varrho \, dy\right\}\left\{\int \langle\xi + \theta\eta\rangle^\mu\langle\eta\rangle^{-2p}\langle\xi\rangle^{\mu'} \, d\eta\right\}$$

$$(1.4.22)$$

Let us write $\int \ldots dy = \int_{D_1} \ldots dy + \int_{D_2} \ldots dy$ for $D_1 = \{y : |y| \le \langle x\rangle/2\}$, $D_2 = \{y : |y| > \langle x\rangle/2\}$. First we look at the integral over D_1. Using $\frac{d}{dt}\langle x + ty\rangle = y(x + ty)\langle x + ty\rangle^{-1}$ we get

$$|\langle x + y\rangle - \langle x\rangle| = |\int_0^1 y(x + ty)\langle x + ty\rangle^{-1} \, dt| \le |y|.$$

Then $y \in D_1$ implies $|\langle x + y \rangle| - |\langle x \rangle| \leq \langle x \rangle/2$, and hence $\langle x \rangle/2 \leq |\langle x + y \rangle| \leq 3\langle x \rangle/2$. It follows that

$$\int_{D_1} \langle x + y \rangle^{\varrho'} \langle y \rangle^{-2q} \langle x \rangle^{\varrho} \, dy \leq \langle x \rangle^{\varrho + \varrho'} \int_{D_1} \langle x + y \rangle^{\varrho'} \langle x \rangle^{-\varrho'} \langle y \rangle^{-2q} \, dy$$

$$\leq c \langle x \rangle^{\varrho + \varrho'} \int_{D_1} \langle y \rangle^{-2q} \, dy \leq c \langle x \rangle^{\varrho + \varrho'},$$

for different constants $c > 0$. For the integral over D_2 we write $\langle x + y \rangle^2 = \langle x \rangle^2 + |y|^2$. It follows that $\langle x + y \rangle \leq \langle x \rangle + |y| \leq 3|y|$ for $y \in D_2$. Thus

$$\int_{D_2} \langle x + y \rangle^{\varrho'} \langle y \rangle^{-2q} \langle x \rangle^{\varrho} \, dy \leq \langle x \rangle^{\varrho} \int_{D_2} (3|y|)^{\varrho'} \langle y \rangle^{-2q} \, dy$$

$$= c \langle x \rangle^{\varrho} \int_{\langle x \rangle/2} t^{\varrho' + n - 1} (1 + t^2)^{-q} \, dt$$

$$\leq c \langle x \rangle^{\varrho} \int_{\langle x \rangle/2} t^{\varrho' + n - 1 - 2q} \, dt$$

$$= c \langle x \rangle^{\varrho} \langle x \rangle^{\varrho' + n - 2q} \leq c \langle x \rangle^{\varrho + \varrho'}.$$

In the last integral q was chosen sufficiently large . In an analogous manner we get

$$\int \langle \xi + \theta \eta \rangle^{\mu} \langle \eta \rangle^{-2p} \langle \xi \rangle^{\mu'} \, d\eta \leq c \langle \xi \rangle^{\mu + \mu'}.$$

Then (1.4.22) yields (1.4.21). The next step is to prove

$$|D_x^{\alpha} D_{\xi}^{\beta} b_{\theta}(x, \xi)| \leq c \langle \xi \rangle^{\mu + \mu' - |\beta|} \langle x \rangle^{\varrho + \varrho' - |\alpha|} \qquad (1.4.23)$$

for all $\alpha, \beta \in \mathbb{N}^n$. To this end we observe that $a_{x,\xi,\theta}(y, \eta) := a(x, x + y, \xi + \theta \eta, \xi)$ varies over a bounded set in $S^{\mu, \varrho'}(\mathbb{R}_y^n \times \mathbb{R}_{\eta}^n)$ for (x, ξ) running over any $K \subset\subset \mathbb{R}^{2n}$, $|\theta| \leq 1$. This allows us to write

$$D_x^{\alpha} D_{\xi}^{\beta} b_{\theta}(x, \xi) = \iint e^{-iy\eta} D_x^{\alpha} D_{\xi}^{\beta} a(x, x + y, \xi + \theta \eta, \xi) \, dy d\eta,$$

and we can argue as before, now with the symbol function $D_x^{\alpha} D_{\xi}^{\beta} a(\dots)$ under the oscillatory integral. The orders of growth in (1.4.22) are diminished in the expected way because of the symbol estimates for a, such that (1.4.23) is true. Now we look at

$$\begin{aligned}
&\mathrm{Op}(b)u(x)\\
&= \iint e^{i(x-x')\xi}b(x,\xi)u(x')\,dx'd\xi\\
&= \iint e^{i(x-x')\xi}\left\{\iint e^{-iy\eta}a(x,x+y,\xi+\eta,\xi)dyd\eta\right\}u(x')\,dx'd\xi.
\end{aligned}$$

Set $x' = x+y+y'$, i.e., $x-x' = -(y+y')$ and $\eta' = \xi$, $\tilde{\eta} = \eta+\eta'$. Then

$$\begin{aligned}
&\mathrm{Op}(b)u(x)\\
&= \iint e^{-i(y+y')\eta'}\iint e^{-iy\eta}a(x,x+y,\eta'+\eta,\eta')\,dyd\eta\\
&\quad \times u(x+y+y')\,dy'd\eta'\\
&= \iiiint e^{-i(y\tilde{\eta}+y'\eta')}a(x,x+y,\tilde{\eta},\eta')u(x+y+y')dydy'd\tilde{\eta}\,d\eta'
\end{aligned}$$

that is just the formula (1.4.19). It remains to show (1.4.20). To this end we apply the Taylor expansion of $a(x,x+y,\xi+\eta,\xi)$ in η at $\eta = 0$:

$$\begin{aligned}
&a(x,x+y,\xi+\eta,\xi)\\
&= \sum_{|\alpha|<N}\frac{\eta^\alpha}{\alpha!}\partial_\zeta^\alpha a(x,x+y,\zeta,\xi)|_{\zeta=\xi}\\
&\quad + N\sum_{|\beta|=N}\frac{\eta^\beta}{\beta!}\int_0^1(1-\theta)^{N-1}(\partial_\zeta^\beta a)(x,x+y,\zeta,\xi)|_{\zeta=\xi+\theta\eta}\,d\theta.
\end{aligned}$$

Inserting this in (1.4.17) it follows that

$$\begin{aligned}
b(x,\xi) &= \sum_{|\alpha|<N}\iint e^{-iy\eta}\frac{\eta^\alpha}{\alpha!}\partial_\zeta^\alpha a(x,x+y,\zeta,\xi)|_{\zeta=\xi}\,dyd\eta\\
&\quad + N\sum_{|\beta|=N}\int_0^1(1-\theta)^{N-1}\iint e^{-iy\eta}\frac{\eta^\beta}{\beta!}\\
&\quad \times (\partial_\zeta^\beta a)(x,x+y,\zeta,\xi)|_{\zeta=\xi+\theta\eta}\,dyd\eta d\theta.
\end{aligned}$$

The first sum on the right of the latter equation equals

$$\begin{aligned}
&\sum_{|\alpha|<N}\frac{1}{\alpha!}\iint e^{-iy\eta}(D_{x'}^\alpha\partial_\zeta^\alpha a)(x,x',\zeta,\xi)|_{x'=x+y,\zeta=\xi}\,dyd\eta\\
&= \sum_{|\alpha|<N}\frac{1}{\alpha!}(\partial_\zeta^\alpha D_{x'}^\alpha a)(x,x',\zeta,\xi)|_{x'=x,\zeta=\xi}.
\end{aligned}$$

Here we used $\iint e^{-iy\eta} f(y)\,dy d\eta = f(0)$. Recall that the integrals are interpreted in the oscillatory sense. Finally we get for the remainder

$$N \sum_{|\beta|=N} \int_0^1 (1-\theta)^{N-1} \frac{1}{\beta!}$$

$$\times \iint e^{-iy\eta} \partial_\zeta^\alpha D_{x'}^\alpha a(x,x',\zeta,\xi)|_{x'=x+y,\zeta=\xi+\theta\eta}\,dy d\eta d\theta$$

that belongs to $\boldsymbol{S}^{\mu+\mu'-N,\varrho+\varrho'-N}(\mathbb{R}^n \times \mathbb{R}^n)$ by the first part of the proof. Thus we have obtained (1.4.20). $\qquad\square$

Corollary 1.4.10 *Let* $a(x,x',\xi) \in \boldsymbol{S}^{\mu,\varrho,\varrho'}(\mathbb{R}^{2n} \times \mathbb{R}^n)$ *and* $A = \mathrm{Op}(a)$. *Then there is an* $\boldsymbol{a}(x,\xi) \in \boldsymbol{S}^{\mu,\varrho+\varrho'}(\mathbb{R}^n \times \mathbb{R}^n)$ *such that* $\mathrm{Op}(a) = \mathrm{Op}(\boldsymbol{a})$ *and* $\boldsymbol{a}(x,\xi)$ *has the asymptotic expansion*

$$\boldsymbol{a}(x,\xi) \sim \sum_\alpha \frac{1}{\alpha!} \partial_{x'}^\alpha D_\xi^\alpha a(x,x',\xi)|_{x'=x}. \qquad (1.4.24)$$

In fact, we have $a(x,x',\xi) \in \boldsymbol{S}^{\mu,0,\varrho,\varrho'}(\mathbb{R}^{2n} \times \mathbb{R}^{2n})$ with a fictious additional ξ'–dependence of order zero, and then $\mathrm{Op}(a)$ coincides with (1.4.18). Hence Theorem 1.4.9 can be applied, and we get, in particular, (1.4.24).

Proposition 1.4.11 *Let* $a(x,x',\xi) \in \boldsymbol{S}^{\mu,\varrho,\varrho'}(\mathbb{R}^{2n} \times \mathbb{R}^n)$ *and* $\omega(t) \in C_0^\infty(\mathbb{R})$ *with* $\omega(t) = 1$ *in a neighbourhood of* $t = 0$. *Then the operator*

$$Au(x) = \iint e^{i(x-x')\xi}(1-\omega(|x-x'|))a(x,x',\xi)u(x')dx'd\xi$$

belongs to $\boldsymbol{L}^{-\infty,-\infty}(\mathbb{R}^n)$.

Proof. Using $(-\Delta_\xi)^N e^{i(x-x')\xi} = |x-x'|^{2N} e^{i(x-x')\xi}$ and integration by parts we get for every $N \in \mathbb{N}$

$$Au(x) = \iint e^{i(x-x')\xi}|x-x'|^{-2N}$$

$$\times (1-\omega(|x-x'|))(-\Delta_\xi)^N a(x,x',\xi)u(x')\,dx'd\xi.$$

For sufficiently large N we can form

$$c_N(x,x') = |x-x'|^{-2N}(1-\omega(|x-x'|)) \int e^{i(x-x')\xi}(-\Delta_\xi)^N a(x,x',\xi)\,d\xi$$

as a convergent integral, since $(-\Delta_\xi)^N a(x,x',\xi) \in S^{\mu-2N,\varrho,\varrho'}(\mathbb{R}^{2n} \times \mathbb{R}^n)$. At the same time, for every $\alpha,\beta \in \mathbb{N}^n$ we can choose $N = N(\alpha,\beta)$ so large that

$$\sup_{x\in\mathbb{R}^n} |x^\alpha D_x^\beta Au(x)| = \sup_{x\in\mathbb{R}^n} \left| \int x^\alpha D_x^\beta c_N(x,x')u(x')\right| dx'$$

$$\leq \sup_{x\in\mathbb{R}^n} \int |x^\alpha D_x^\beta c_N(x,x')u(x')| \, dx'$$

$$\leq \sup_{x\in\mathbb{R}^n} \left\{ \int |x^\alpha D_x^\beta N_n(x,x')|^2 \, dx' \right\}^{\frac{1}{2}} \|u\|_{L^2(\mathbb{R}^n)},$$

where $\sup_{x\in\mathbb{R}^n} \{\int |x^\alpha D_x^\beta c_N(x,x')|^2 dx'\}^{\frac{1}{2}} < \infty$. Hence A induces a continuous operator $L^2(\mathbb{R}^n) \to \mathcal{S}(\mathbb{R}^n)$. For the adjoint A^* we can argue in an analogous manner, i.e., we also get the continuity of $A : L^2(\mathbb{R}^n) \to \mathcal{S}(\mathbb{R}^n)$. Applying Theorem 1.1.3 it follows that $A \in \boldsymbol{L}^{-\infty,-\infty}(\mathbb{R}^n)$. $\qquad\square$

Theorem 1.4.12 *Given* $A \in \boldsymbol{L}^{\mu,\varrho}(\mathbb{R}^n)$ *we have* $A^* \in \boldsymbol{L}^{\mu,\varrho}(\mathbb{R}^n)$ *for the formal adjoint* A^*, *defined by* $(Au,v)_{L^2(\mathbb{R}^n)} = (u, A^*v)_{L^2(\mathbb{R}^n)}$ *for all* $u,v \in C_0^\infty(\mathbb{R}^n)$. *If* $A = \mathrm{Op}(a)$ *for a symbol* $a(x,\xi) \in \boldsymbol{S}^{\mu,\varrho}(\mathbb{R}^n \times \mathbb{R}^n)$ *then* $A^* = \mathrm{Op}(a^*)$ *has a symbol* $a^*(x,\xi) \in \boldsymbol{S}^{\mu,\varrho}(\mathbb{R}^n \times \mathbb{R}^n)$ *with the asymptotic expansion*

$$a^*(x,\xi) \sim \sum_\alpha \frac{1}{\alpha!} \partial_\xi^\alpha D_x^\alpha \overline{a(x,\xi)}.$$

Similarly, for the formal transposed ${}^tA$, *defined by* $\langle Au,v\rangle = \langle u, {}^tAv\rangle$ *for all* $u,v \in C_0^\infty(\mathbb{R}^n)$, *we have* ${}^tA \in \boldsymbol{L}^{\mu,\varrho}(\mathbb{R}^n)$, *and* ${}^tA = \mathrm{Op}(c)$ *for a* $c(x,\xi) \in \boldsymbol{S}^{\mu,\varrho}(\mathbb{R}^n \times \mathbb{R}^n)$,

$$c(x,\xi) \sim \sum_\alpha \frac{1}{\alpha!} \partial_\xi^\alpha D_x^\alpha a(x,-\xi).$$

Proof. We can write $A^* = \mathrm{Op}(b)$ for $b(x',\xi) = \overline{a}(x',\xi)$. Then the result is an immediate consequence of Theorem 1.4.9. Analogously we can argue for ${}^tA$. $\qquad\square$

Corollary 1.4.13 *Each* $A \in \boldsymbol{L}^{\mu,\varrho}(\mathbb{R}^n)$ *extends to an operator*

$$A : \mathcal{S}'(\mathbb{R}^n) \to \mathcal{S}'(\mathbb{R}^n).$$

In fact it suffices to apply ${}^tA \in \boldsymbol{L}^{\mu,\varrho}(\mathbb{R}^n)$ and Theorem 1.4.4.

Theorem 1.4.14 $A \in \boldsymbol{L}^{\mu,\varrho}(\mathbb{R}^n)$, $B \in \boldsymbol{L}^{\nu,\sigma}(\mathbb{R}^n)$ *implies* $AB \in \boldsymbol{L}^{\mu+\nu,\varrho+\sigma}(\mathbb{R}^n)$ *for all* $\mu,\nu,\varrho,\sigma \in \mathbb{R}$. *Moreover,* $A = \mathrm{Op}(a)$, $a(x,\xi) \in \boldsymbol{S}^{\mu,\varrho}(\mathbb{R}^n \times \mathbb{R}^n)$, $B = \mathrm{Op}(b)$, $b(x,\xi) \in \boldsymbol{S}^{\nu,\sigma}(\mathbb{R}^n \times \mathbb{R}^n)$ *implies* $AB = \mathrm{Op}(c)$ *for a symbol* $c(x,\xi) \in \boldsymbol{S}^{\mu+\nu,\varrho+\sigma}(\mathbb{R}^n \times \mathbb{R}^n)$ *with the asymptotic expansion*

$$c(x,\xi) \sim \sum_\alpha \frac{1}{\alpha!} \partial_\xi^\alpha a(x,\xi) D_x^\alpha b(x,\xi). \tag{1.4.25}$$

Proof. In view of (1.4.17) we can assume $A = \mathrm{Op}(a)$, $B = \mathrm{Op}(b)$ with corresponding symbols $a(x, \xi)$, $b(x, \xi)$. Then ABu equals

$$\iint e^{i(x-x')\xi} a(x, \xi) \Big\{ \iint e^{i(x'-x'')\xi'} b(x', \xi') u(x'') \, dx'' d\xi' \Big\} \, dx' d\xi$$

$$= \iint e^{i(y\xi + y'\xi')} a(x, \xi) b(x + y, \xi') u(x + y + y') \, dy dy' d\xi d\xi'.$$

Here we have substituted $y = x' - x$, $y' = x'' - x'$, such that $x' = x + y$, $x'' = x + y + y'$. In other words, for $p(x, x', \xi, \xi') = a(x, \xi) b(x', \xi')$ we obtain $\mathrm{Op}(a) \, \mathrm{Op}(b) = \mathrm{Op}(p)$ with $\mathrm{Op}(p)$ being understood in the sense (1.4.18). Now (1.4.25) is an immediate consequence of (1.4.20). Further $\mathrm{Op}(a) \in L^{\mu+\nu, \varrho+\sigma}(\mathbb{R}^n)$ holds by Theorem 1.4.9.   □

From (1.4.17) it follows, in particular, that $\mathrm{Op}(a) \, \mathrm{Op}(b) = \mathrm{Op}(c)$ for every $a(x, \xi) \in S^{\mu, \varrho}(\mathbb{R}^n \times \mathbb{R}^n)$, $b(x, \xi) \in S^{\nu, \sigma}(\mathbb{R}^n \times \mathbb{R}^n)$ with a unique $c(x, \xi) \in S^{\mu+\nu, \varrho+\sigma}(\mathbb{R}^n \times \mathbb{R}^n)$.

We now turn to invariance of pseudo–differential operators of the class $L^{\mu, \varrho}(\mathbb{R}^n)$ under diffeomorphisms

$$\kappa : \mathbb{R}^n \to \mathbb{R}^n$$

satisfying

$$\kappa(\lambda x) = \lambda \kappa(x) \qquad \text{for all } |x| \geq c, \lambda \geq 1 \qquad (1.4.26)$$

with some constants $c > 0$. Analogously to (1.1.71) we can form

$$\Phi_\alpha(x, \eta) = D_z^\alpha e^{i\delta(x, z)\eta}\big|_{z=x}, \qquad \alpha \in \mathbb{N}^n, \qquad (1.4.27)$$

which is a polynomial in η of degree $\leq |\alpha|/2$. Here, as above, $\delta(x, z) = \kappa(z) - \kappa(x) - d\kappa(x)(z - x)$. In the present case we have

$${}^t d\kappa(\lambda x) = {}^t d\kappa(x), \qquad \text{for } |x| \geq c, \lambda \geq 1$$

for some $c > 0$, such that $a(x, \xi) \in S^{\mu, \varrho}(\mathbb{R}^n_x \times \mathbb{R}^n_\xi)$ implies the relation $(\partial_\xi^\alpha a)(x, {}^t d\kappa(x)\eta) \in S^{\mu-|\alpha|, \varrho}(\mathbb{R}^n_x \times \mathbb{R}^n_\eta)$, $\alpha \in \mathbb{N}^n$. In addition

$$\Phi_\alpha(x, \lambda\eta) = D_z^\alpha e^{i\delta(x, z)\lambda\eta}\big|_{z=x} = D_z^\alpha e^{i\delta(\lambda x, \lambda z)\eta}\big|_{z=x}$$

$$= \lambda^{|\alpha|} D_{z'}^\alpha e^{i\delta(\lambda x, z')\eta}\big|_{z'=\lambda x} = \lambda^{|\alpha|} \Phi_\alpha(\lambda x, \eta)$$

for $|x| \geq c$, $\lambda \geq 1$. In other words we have

$$\Phi_\alpha(\lambda x, \eta) = \lambda^{-|\alpha|} \Phi_\alpha(x, \lambda\eta)$$

for $|x| \geq c$, $\lambda \geq 1$. Since $\Phi_\alpha(x, \eta)$ is a polynomial in η of degree $\leq |\alpha|/2$, $\Phi_\alpha(\lambda x, \eta)$ can also be regarded as a symbol in λ of order $\leq |\alpha|/2$, and hence

$$(\partial_\xi^\alpha a)(x, {}^t d\kappa(x)\eta) \Phi_\alpha(x, \eta) \in S^{\mu-|\alpha|/2, \varrho-|\alpha|/2}(\mathbb{R}^n_x \times \mathbb{R}^n_\eta) \qquad (1.4.28)$$

for $a(x, \xi) \in S^{\mu, \varrho}(\mathbb{R}^n_x \times \mathbb{R}^n_\xi)$.

Theorem 1.4.15 *The operator push–forward under a diffeomorphism* $\kappa : \mathbb{R}^n \to \mathbb{R}^n$ *satisfying* (1.4.28) *induces isomorphisms*

$$\kappa_* : \boldsymbol{L}^{\mu,\varrho}(\mathbb{R}^n) \to \boldsymbol{L}^{\mu,\varrho}(\mathbb{R}^n) \qquad (1.4.29)$$

for all $\mu, \varrho \in \mathbb{R}$. *If* $A \in \boldsymbol{L}^{\mu,\varrho}(\mathbb{R}^n)$ *is written* $A = \mathrm{Op}(a)$, *for a symbol* $a(x, \xi) \in \boldsymbol{S}^{\mu,\varrho}(\mathbb{R}^n \times \mathbb{R}^n)$, *then* $B = \kappa_* A$ *has the form* $B = \mathrm{Op}(b)$, *for a* $b(y, \eta) \in \boldsymbol{S}^{\mu,\varrho}(\mathbb{R}^n \times \mathbb{R}^n)$, *where*

$$b(y, \eta)|_{y=\kappa(x)} \sim \sum_\alpha \frac{1}{\alpha!} (\partial_\xi^\alpha a)(x, {}^t d\kappa(x)\eta) \Phi_\alpha(x, \eta). \qquad (1.4.30)$$

Proof. Let $A = \mathrm{Op}(a)$ with $a(x, \xi) \in \boldsymbol{S}^{\mu,\varrho}(\mathbb{R}^n \times \mathbb{R}^n)$. Then $B = \kappa_* A$ is of the form

$$Bv(y) = \iint e^{i\varphi(y,y',\xi)} a(\kappa^{-1}(y), \xi) |\det d\kappa^{-1}(y')| v(y') \, dy' d\xi,$$

where $y = \kappa(x)$, $y' = \kappa(x')$, $v = (\kappa^*)^{-1} u$, $\varphi(y, y', \xi) = (\kappa^{-1}(y) - \kappa^{-1}(y'))\xi$. Like in the proof of Theorem 1.1.37 there exists a neighbourhood V of $\mathrm{diag}\,\mathbb{R}^n \times \mathbb{R}^n$ and a function $g(y, y') \in C^\infty(V, GL(n, \mathbb{R}))$ such that

$$\varphi(y, y', g(y, y')\xi) = (y - y')\xi \qquad \text{for all } (y, y') \in V. \qquad (1.4.31)$$

Since by assumption the diffeomorphism κ is homogeneous in x of order 1 for large $|x|$, we can assume $V \supseteq \{(y, y') : |y - y'| < \epsilon\}$ for an $\epsilon > 0$. There is then an $\omega(t) \in C_0^\infty(\mathbb{R})$ with $\omega(t) = 1$ in a neighbourhood of $t = 0$ such that $\omega(|\kappa^{-1}(y) - \kappa^{-1}(y')|)$ is supported in V. Writing $A = A_0 + A_1$ with $A_i = \mathrm{Op}(a_i)$, $a_0(x, x', \xi) = (1 - \omega(|x - x'|))a(x, \xi)$, $a_1(x, x', \xi) = \omega(|x - x'|)a(x, \xi)$, we have $A_0 \in \boldsymbol{L}^{-\infty,-\infty}(\mathbb{R}^n)$, because of Proposition 1.4.11, and it suffices to show that $B_1 = \kappa_* A_1$ belongs to $\boldsymbol{L}^{\mu,\varrho}(\mathbb{R}_y^n)$. We have

$$B_1 v(y) = \iint e^{i\varphi(y,y',\xi)} a_1(\kappa^{-1}(y), \kappa^{-1}(y'), \xi) |\det d\kappa^{-1}(y')| v(y') \, dy' d\xi.$$

Similarly to the proof of Theorem 1.1.38 we get from (1.4.31)

$$B_1 v(y) = \iint e^{i(y-y')\eta} b_1(y, y', \eta) v(y') \, dy' d\eta$$

with $b_1(y, y', \eta) = a_1(\kappa^{-1}(y), \kappa^{-1}(y'), g(y, y')\eta)h(y, y')$, $h(y, y') = |\det g(y, y')||\det d\kappa^{-1}(y')|$, $\eta = g^{-1}(y, y')\xi$. The symbol $b_1(y, y', \eta)$ is not necessarily in $\boldsymbol{S}^{\mu,\varrho,\varrho'}(\mathbb{R}^{2n} \times \mathbb{R}^n)$ for certain ϱ, ϱ', since the original symbol was cut off near the diagonal. Nevertheless we can write

$$\mathrm{Op}(b_1) = \sum_{|\alpha| \leq M} \frac{1}{\alpha!} \mathrm{Op}(D_\eta^\alpha \partial_{y'}^\alpha b_1|_{y'=y}) + \mathrm{Op}(c_N).$$

for every $N \in \mathbb{N}$ with some $M = M(N)$ such that $c_N(y, y', \eta) = |y - y'|^{-2N} r_N(y, y', \eta)$ for some $r_N(y, y', \eta) \in \boldsymbol{S}^{\mu-2N}(\mathbb{R}^{2n}_{y,y'} \times \mathbb{R}^n_\eta)$, supported near the diagonal.

The terms on the right of

$$b(y, \eta) \sim \sum_\alpha \frac{1}{\alpha!} \partial^\alpha_\eta D^\alpha_{y'} b_1(y, y', \eta)|_{y'=y}$$

correspond to those of (1.4.30) by the same arguments as in Theorem 1.1.38. In view of (1.4.28) it follows that $\mathrm{Op}(b_1) - \mathrm{Op}(b) = \mathrm{Op}(\tilde{c}_N)$ for a symbol $\tilde{c}_N$ of analogous behaviour as the above c_N, for arbitrary N. The mapping properties of $\mathrm{Op}(\tilde{c}_N)$ and $\mathrm{Op}(\tilde{c}_N)^*$ can be checked similarly as in the above proof of Proposition 1.4.11. This yields $\mathrm{Op}(b_1) - \mathrm{Op}(b) \in \boldsymbol{L}^{-\infty,-\infty}(\mathbb{R}^n)$ which completes the proof. $\square$

Note that in view of (1.4.17) the operator push–forward (1.4.29) also induces isomorphisms of the symbol spaces

$$\kappa_* : \boldsymbol{S}^{\mu,\varrho}(\mathbb{R}^n \times \mathbb{R}^n) \to \boldsymbol{S}^{\mu,\varrho}(\mathbb{R}^n \times \mathbb{R}^n), \qquad (1.4.32)$$

where $\kappa_* \mathrm{Op}(a) = \mathrm{Op}(\kappa_* a)$, cf. (1.4.17).

Remark 1.4.16 *The space $\boldsymbol{L}^{\mu,\varrho}(\mathbb{R}^n)$ has a natural Fréchet topology, induced by the Fréchet topology of $\boldsymbol{S}^{\mu,\varrho}(\mathbb{R}^n \times \mathbb{R}^n)$ and the bijection of Proposition 1.4.7.*

1.4.2   Continuity in Sobolev spaces

Set

$$H^{s,\delta}(\mathbb{R}^n) = \{(1 + |x|^2)^{-\frac{\delta}{2}} u(x) : u \in H^s(\mathbb{R}^n)\} \qquad (1.4.33)$$

for $s, \delta \in \mathbb{R}$; $H^{s,0}(\mathbb{R}^n) = H^s(\mathbb{R}^n)$, $\|v\|_{H^{s,\delta}(\mathbb{R}^n)} = \|(1 + |x|^2)^{\frac{\delta}{2}} v\|_{H^s(\mathbb{R}^n)}$. The Schwartz space $\mathcal{S}(\mathbb{R}^n)$ is dense in $H^{s,\delta}(\mathbb{R}^n)$ as a consequence of the density of $\mathcal{S}(\mathbb{R}^n)$ in $H^s(\mathbb{R}^n)$. Note that $\mathcal{S}(\mathbb{R}^n) = H^{\infty,\infty}(\mathbb{R}^n)$.

Theorem 1.4.17 *Each $a \in \boldsymbol{L}^{\mu,\varrho}(\mathbb{R}^n)$, $\mu, \varrho \in \mathbb{R}$, extends to continuous operators $A : H^{s,\delta}(\mathbb{R}^n) \to H^{s-\mu,\delta-\varrho}(\mathbb{R}^n)$ for all $s, \delta \in \mathbb{R}$.*

Proof. First recall that Theorem 1.1.54 yields the continuity of $A_0 : L^2(\mathbb{R}^n) \to L^2(\mathbb{R}^n)$ for every $A_0 \in \boldsymbol{L}^{0,0}(\mathbb{R}^n)$. Further let $a^{s,\delta}(x, \xi) := (1 + |\xi|^2)^{\frac{s}{2}}(1 + |x|^2)^{\frac{\delta}{2}} \in \boldsymbol{S}^{s,\delta}(\mathbb{R}^n \times \mathbb{R}^n)$. Then $R := \mathrm{Op}(a^{s,\delta})$ induces an isomorphism $R : H^{s,\delta}(\mathbb{R}^n) \to L^2(\mathbb{R}^n)$ with $R^{-1} = \mathrm{Op}(a^{-s,-\delta})$. Analogously $\tilde{R} = \mathrm{Op}(a^{s-\mu,\delta-\varrho}) : L^2(\mathbb{R}^n) \to H^{s-\mu,\delta-\varrho}(\mathbb{R}^n)$ is an isomorphism. In view of Theorem 1.4.14 we have $A_0 = \tilde{R} A R^{-1} \in \boldsymbol{L}^{0,0}(\mathbb{R}^n)$, and the continuity then follows from $A = \tilde{R}^{-1} A_0 R$. $\square$

Theorem 1.4.18 *A subset $K \subset L^2(\mathbb{R}^n)$ is relatively compact if and only if it satisfies the following conditions:*

(i) K *is bounded in* $L^2(\mathbb{R}^n)$,

(ii) $\|u_h - u\|_{L^2(\mathbb{R}^n)} \to 0$ *for* $u_h(x) := u(x+h)$ *and* $h \to 0$, *uniformly in* $u \in K$,

(iii) $\int_{|x| \geq c} |u(x)|^2 dx \to 0$ *as* $c \to \infty$, *uniformly in* $u \in K$.

This theorem (of Kolmogorov) is standard and will not be proved here.

Theorem 1.4.19 *The canonical embeddings*

$$H^{s,\delta}(\mathbb{R}^n) \to H^{s',\delta'}(\mathbb{R}^n) \tag{1.4.34}$$

for $s \geq s'$, $\delta \geq \delta'$ are compact for $s > s'$, $\delta > \delta'$.

Proof. Write simply e for the embedding operator (1.4.34), and let e_0 be the embedding $H^{s-s',\delta-\delta'}(\mathbb{R}^n) \to L^2(\mathbb{R}^n)$. Then, using the symbols $a^{s,\delta}(x,\xi)$ from the proof of Theorem 1.4.17, we obtain

$$e = \mathrm{Op}((a^{s',\delta'})^{-1}) e_0 \, \mathrm{Op}(a^{s',\delta'}).$$

Thus e is a compact operator when e_0 is compact. In order to show this for $s > s'$, $\delta > \delta'$ we set $r = s - s'$, $\gamma = \delta - \delta'$ and consider $K = \{u \in L^2(\mathbb{R}^n) : \|u\|_{H^{r,\gamma}(\mathbb{R}^n)} \leq 1\}$. It suffices to verify that K is relatively compact in $L^2(\mathbb{R}^n)$. This will be done by Theorem 1.4.18. The condition (i) is satisfied because of the continuity of e_0. For (ii) we observe $\widehat{(u_h)}(\xi) = e^{ih\xi}\hat{u}(\xi)$ for $u_h(x) = u(x+h), h \in \mathbb{R}^n$ (^indicates the Fourier transform). This yields

$$\int_{\mathbb{R}^n} |u(x+h) - u(x)|^2 \, dx = \|u_h - u\|^2_{L^2(\mathbb{R}^n)} = \|\hat{u}_h - \hat{u}\|^2_{L^2(\mathbb{R}^n)}$$

$$= \|(e^{ih\xi} - 1)\hat{u}\|^2_{L^2(\mathbb{R}^n)}$$

$$= \|(e^{ih\xi} - 1)\langle \xi \rangle^{-r} \langle \xi \rangle^r \hat{u}\|^2_{L^2(\mathbb{R}^n)}$$

$$\leq b(h)\|u\|^2_{H^r(\mathbb{R}^n)} \leq mb(h)$$

for a constant m for all $u \in K$ and $b(h) = \sup_{\xi \in \mathbb{R}^n} |(e^{ih\xi} - 1)\langle \xi \rangle^{-r}| \to 0$ as $h \to 0$. Finally (iii) follows from

$$\int_{|x| \geq c} |u(x)|^2 \, dx = \int_{|x| \geq c} |\langle x \rangle^{-\gamma} \langle x \rangle^\gamma u(x)|^2 \, dx$$

$$\leq \sup_{|x| \geq c} \langle x \rangle^{-2\gamma} \||\langle x \rangle^\gamma u\|^2_{L^2(\mathbb{R}^n)}$$

$$\leq (1 + c^2)^{-\gamma} \|u\|^2_{H^{0,\gamma}(\mathbb{R}^n)} \leq m'(1 + c^2)^{-\gamma}$$

for a constant m' for all $u \in K$. $\square$

Theorem 1.4.20 *Let $A \in L^{\mu,\varrho}(\mathbb{R}^n)$, $\mu, \varrho \in \mathbb{R}$. Then*

$$A : H^{s,\delta}(\mathbb{R}^n) \to H^{s-\mu-\varepsilon, \delta-\varrho-\varepsilon}(\mathbb{R}^n)$$

is a compact operator for every $\varepsilon > 0$ and all $s, \delta \in \mathbb{R}$.

Proof. The assertion is a consequence of the Theorems 1.4.17, 1.4.19. □

Proposition 1.4.21 *Let*

$$A : H^{s,\delta}(\mathbb{R}^n) \to H^{s-\mu, \delta-\varrho}(\mathbb{R}^n)$$

be continuous for all $s, \delta \in \mathbb{R}$, for fixed $\mu, \varrho \in \mathbb{R}$. Then, the formal adjoint A^* of A induces continuous operators

$$A^* : H^{s,\delta}(\mathbb{R}^n) \to H^{s-\mu, \delta-\mu}(\mathbb{R}^n)$$

for all $s, \delta \in \mathbb{R}$. Moreover, for any such operator A the following conditions are equivalent:

(i) $A \in L^{-\infty,-\infty}(\mathbb{R}^n)$.

(ii) *The operators $A, A^* : H^{s,\delta}(\mathbb{R}^n) \to \mathcal{S}(\mathbb{R}^n)$ are continuous for all $s, \delta \in \mathbb{R}$.*

Proof. The first statement is evident. Moreover, by Remark 1.4.14 and Theorem 1.4.17 we can reduce the statement to $\mu = \varrho = 0$. In view of Theorem 1.1.3 we know that $A \in L^{-\infty,-\infty}(\mathbb{R}^n)$ is equivalent to the continuity of $A, A^* : H^{0,0}(\mathbb{R}^n) \to \mathcal{S}(\mathbb{R}^n)$. It remains to observe that $A : H^{s,\delta}(\mathbb{R}^n) \to \mathcal{S}(\mathbb{R}^n)$ is continuous for all $s, \delta \in \mathbb{R}$ when $A \in L^{-\infty,-\infty}(\mathbb{R}^n)$. □

1.4.3   Classical operators with exit symbols

Definition 1.4.1 contains the variable $x \in \mathbb{R}^n$ like a covariable. It is often interesting to know whether a symbol is classical, cf. Definition 1.1.12. The corresponding operators then form a subalgebra. Here we shall consider symbols that are classical both in x and ξ.

Let $S_\xi^{(\mu)}$ be the space of all $a(x, \xi) \in C^\infty(\mathbb{R}^n \times (\mathbb{R}^n \setminus \{0\}))$ with $a(x, \lambda\xi) = \lambda^\mu a(x, \xi)$ for all $\lambda > 0$ and all $(x, \xi) \in \mathbb{R}^n \times (\mathbb{R}^n \setminus \{0\})$, $\mu \in \mathbb{R}$, and define analogously $S_x^{(\mu)}$ by interchanging the role of x and ξ. Then $S_\xi^{(\mu)} \cap S_x^{(\varrho)}$ consists of the space of all $a(x, \xi) \in C^\infty((\mathbb{R}^n \setminus \{0\}) \times (\mathbb{R}^n \setminus \{0\}))$ such that

$$a(\delta x, \lambda\xi) = \delta^\varrho \lambda^\mu a(x, \xi) \text{ for all } \delta, \lambda > 0, (x, \xi) \in (\mathbb{R}^n \setminus \{0\}) \times (\mathbb{R}^n \setminus \{0\}).$$

Let us further define $S_\xi^{[\mu]}$ as the space of all $a(x,\xi) \in C^\infty(\mathbb{R}^n \times \mathbb{R}^n)$ such that

$$a(x, \lambda\xi) = \lambda^\mu a(x,\xi) \text{ for all } \lambda \geq 1, x \in \mathbb{R}^n, |\xi| \geq c$$

for a $c > 0$, dependent on a, and let $S_x^{[\mu]}$ be the analogous space obtained by interchanging the role of x and ξ. Set

$$\boldsymbol{S}^{\mu,[\varrho]} := \boldsymbol{S}^{\mu,\varrho} \cap S_x^{[\varrho]}, \qquad \boldsymbol{S}^{[\mu],\varrho} := \boldsymbol{S}^{\mu,\varrho} \cap S_\xi^{[\mu]},$$

and let

$$\boldsymbol{S}_{\mathrm{cl}(\xi)}^{\mu,[\varrho]} \subset \boldsymbol{S}^{\mu,[\varrho]}$$

be the set of all $a(x,\xi) \in \boldsymbol{S}^{\mu,[\varrho]}$ such that there are elements $a_k \in S_\xi^{[\mu-k]} \cap S_x^{[\varrho]}, k \in \mathbb{N}$, with

$$a(x,\xi) - \sum_{k=0}^{N} a_k(x,\xi) \in \boldsymbol{S}^{\mu-(N+1),\varrho}$$

for every $N \in \mathbb{N}$. Moreover, define

$$\boldsymbol{S}_{\mathrm{cl}(\xi)}^{\mu,\varrho} \subset \boldsymbol{S}^{\mu,\varrho}$$

as the subspace of all $a(x,\xi) \in \boldsymbol{S}^{\mu,\varrho}$ such that there are elements $a_k \in \boldsymbol{S}^{[\mu-k],\varrho}, k \in \mathbb{N}$, with

$$a(x,\xi) - \sum_{k=0}^{N} a_k(x,\xi) \in \boldsymbol{S}^{\mu-(N+1),\varrho}$$

for every $N \in \mathbb{N}$.

By interchanging the role of x and ξ we obtain analogously the spaces $\boldsymbol{S}_{\mathrm{cl}(x)}^{[\mu],\varrho}, \boldsymbol{S}_{\mathrm{cl}(x)}^{\mu,\varrho}$.

Definition 1.4.22 *A symbol* $a(x,\xi) \in \boldsymbol{S}^{\mu,\varrho}(\mathbb{R}^n \times \mathbb{R}^n)$ *is called classical in x and ξ if it has the follwoing properties:*

(i) *There are symbols* $a_k \in \boldsymbol{S}_{\mathrm{cl}(x)}^{[\mu-k],\varrho}, k \in \mathbb{N}$, *such that*

$$a(x,\xi) - \sum_{k=0}^{N} a_k(x,\xi) \in \boldsymbol{S}_{\mathrm{cl}(x)}^{\mu-(N+1),\varrho}$$

for all $N \in \mathbb{N}$.

(ii) *There are symbols* $b_j \in S_{\mathrm{cl}(\xi)}^{\mu,[\varrho-j]}, j \in \mathbb{N}$, *such that*

$$a(x,\xi) - \sum_{j=0}^{N} b_j(x,\xi) \in S_{\mathrm{cl}(\xi)}^{\mu,\varrho-(N+1)}$$

for all $N \in \mathbb{N}$.

The space of all classical symbols in x,ξ *will be denoted by*

$$S_{\mathrm{cl}(\xi,x)}^{\mu,\varrho} := S_{\mathrm{cl}(\xi,x)}^{\mu,\varrho}(\mathbb{R}^n \times \mathbb{R}^n).$$

Remark 1.4.23 *We have* $S_{\mathrm{cl}(x)}^{[\mu],\varrho} \subset S_{\mathrm{cl}(\xi,x)}^{\mu,\varrho}$, $S_{\mathrm{cl}(\xi)}^{\mu,[\varrho]} \subset S_{\mathrm{cl}(\xi,x)}^{\mu,\varrho}$ *for every* $\mu, \varrho \in \mathbb{R}$.

In fact, $a \in S_{\mathrm{cl}(x)}^{[\mu],\varrho}$ means by definition that there are elements $b_j \in S_\xi^{[\mu]} \cap S_x^{[\varrho-j]}$ with $a - \sum_{j=0}^{N} b_j \in S^{\mu,\varrho-(N+1)}$ for all N. Those b_j belong to $S_{\mathrm{cl}(\xi)}^{\mu,[\varrho-j]}$ which shows that (ii) is satisfied. The second relation can be proved in an analogous manner.

Remark 1.4.24 *Let* χ *be an excision function in* $\mathbb{R}^n$, *and let* $c(x,\xi) \in S_\xi^{(\mu)} \cap S_x^{(\varrho)}$. *Then* $\chi(x)\chi(\xi)c(x,\xi) \in S_{\mathrm{cl}(\xi,x)}^{\mu,\varrho}$.

Another simple observation is

$$S_{\mathrm{cl}(x)}^{-\infty,\varrho} \subset S_{\mathrm{cl}(\xi,x)}^{-\infty,\varrho}, \qquad S_{\mathrm{cl}(\xi)}^{\mu,-\infty} \subset S_{\mathrm{cl}(\xi,x)}^{\mu,-\infty}. \tag{1.4.35}$$

To every $a_k \in S^{[\mu-k],\varrho}$ there is a unique $a_{(k)} \in S_\xi^{(\mu-k)}$ such that $a_{(k)} = a_k$ for $|\xi| \geq c$ with some $c > 0$. We have, in particular,

$$a_{(0)}(x,\xi) = \lim_{\lambda\to\infty} \lambda^{-\mu} a(x,\lambda\xi),$$
$$a_{(1)}(x,\xi) = \lim_{\lambda\to\infty} \lambda^{-\lambda+1}(a(x,\lambda\xi) - \chi(\lambda\xi)a_{(0)}(x,\lambda\xi)),$$

and so on, where $\chi(\xi)$ is an excision function in $\mathbb{R}^n$. This gives us unique mappings

$$\sigma_\psi^{\mu-k} : S_{\mathrm{cl}(\xi,x)}^{\mu,\varrho} \to S_\xi^{(\mu-k)}, \qquad k \in \mathbb{N}.$$

In an analogous manner we get a sequence of unique mappings

$$\sigma_e^{\varrho-j} : S_{\mathrm{cl}(\xi,x)}^{\mu,\varrho} \to S_x^{(\varrho-j)}, \qquad j \in \mathbb{N}.$$

We even obtain

$$\sigma_\psi^{\mu-k}(S_{\mathrm{cl}(\xi,x)}^{\mu,\varrho}) \subset C^\infty(\mathbb{R}_\xi^n \setminus \{0\}, S_{\mathrm{cl}}^\varrho(\mathbb{R}_x^n)), \tag{1.4.36}$$
$$\sigma_e^{\varrho-j}(S_{\mathrm{cl}(\xi,x)}^{\mu,\varrho}) \subset C^\infty(\mathbb{R}_x^n \setminus \{0\}, S_{\mathrm{cl}}^\mu(\mathbb{R}_\xi^n)), \tag{1.4.37}$$

for the corresponding homogeneities of order $\mu - k$ in ξ and of order $\varrho - j$ in x, respectively. Thus, $\sigma_\psi^{\mu-k}(a)$ for $a \in \boldsymbol{S}^{\mu,\varrho}_{\mathrm{cl}(\xi,x)}$ has homogeneous components in x of order $\varrho - j, j \in \mathbb{N}$. This gives us unique mappings

$$\sigma_e^{\varrho-j}\sigma_\psi^{\mu-k} : \boldsymbol{S}^{\mu,\varrho}_{\mathrm{cl}(\xi,x)} \to S_\xi^{(\mu-k)} \cap S_x^{(\varrho-j)}, \qquad j,k \in \mathbb{N}. \tag{1.4.38}$$

In an analogous manner we get

$$\sigma_\psi^{\mu-k}\sigma_e^{\varrho-j} : \boldsymbol{S}^{\mu,\varrho}_{\mathrm{cl}(\xi,x)} \to S_\xi^{(\mu-k)} \cap S_x^{(\varrho-j)}, \qquad j,k \in \mathbb{N}.$$

Now it is easy to show that

$$\sigma_e^{\varrho-j}\sigma_\psi^{\mu-k}(a) = \sigma_\psi^{\mu-k}\sigma_e^{\varrho-j}(a)$$

for all $j,k \in \mathbb{N}$, $a \in \boldsymbol{S}^{\mu,\varrho}_{\mathrm{cl}(\xi,x)}$. We simply set $\sigma_{\psi,e}^{\mu-k,\varrho-j} = \sigma_\psi^{\mu-k}\sigma_e^{\varrho-j}$.

Definition 1.4.25 *Let* $a(x,\xi) \in \boldsymbol{S}^{\mu,\varrho}_{\mathrm{cl}(\xi,x)}$; *then* $\sigma_\psi^\mu(a)$ *is called the homogeneous principal interior symbol and the pair* $\{\sigma_e^\varrho(a), \sigma_{\psi,e}^{\mu,\varrho}(a)\}$ *the homogeneous principal exit symbol of* a.

Note that $a \in \boldsymbol{S}^{\mu,\varrho}_{\mathrm{cl}(\xi,x)}$, $b \in \boldsymbol{S}^{\nu,\kappa}_{\mathrm{cl}(\xi,x)}$ implies $ab \in \boldsymbol{S}^{\mu+\nu,\varrho+\kappa}_{\mathrm{cl}(\xi,x)}$ for every μ, ϱ, ν, $\kappa \in \mathbb{R}$, and

$$\sigma_\psi^{\mu+\nu}(ab) = \sigma_\psi^\mu(a)\sigma_\psi^\nu(b),$$
$$\sigma_e^{\varrho+\kappa}(ab) = \sigma_e^\varrho(a)\sigma_e^\kappa(b),$$
$$\sigma_{\psi,e}^{\mu+\nu,\varrho+\kappa}(ab) = \sigma_{\psi,e}^{\mu,\varrho}(a)\sigma_{\psi,e}^{\nu,\kappa}(b).$$

We have $a^{\nu,\kappa}(x,\xi) = (1+|x|^2)^{\kappa/2}(1+|\xi|^2)^{\nu/2} \in \boldsymbol{S}^{\nu,\kappa}_{\mathrm{cl}(\xi,x)}$ and the operator of multiplication by $a^{\nu,\kappa}$ induces isomorphisms $a^{\nu,\kappa} : \boldsymbol{S}^{\mu,\varrho}_{\mathrm{cl}(\xi,x)} \to \boldsymbol{S}^{\mu+\nu,\varrho+\kappa}_{\mathrm{cl}(\xi,x)}$ for all $\nu, \kappa \in \mathbb{R}$.

Proposition 1.4.26 *Let* $a_j(x,\xi) \in \boldsymbol{S}^{\mu-j,\varrho-j}_{\mathrm{cl}(\xi,x)}, j \in \mathbb{N}$, *be an arbitrary sequence and* $a(x,\xi) \sim \sum_{j=0}^\infty a_j(x,\xi)$ *be the asymptotic sum in the sense of Theorem 1.4.3. Then* $a(x,\xi) \in \boldsymbol{S}^{\mu,\varrho}_{\mathrm{cl}(\xi,x)}$.

We omit the proof of this simple assertion.

Proposition 1.4.27 $a(x,\xi) \in \boldsymbol{S}^{\mu,\varrho}_{\mathrm{cl}(\xi,x)}$ *and* $\sigma_\psi^\mu(a) = 0$, $\sigma_e^\varrho(a) = 0$ *imply* $a(x,\xi) \in \boldsymbol{S}^{\mu-1,\varrho-1}_{\mathrm{cl}(\xi,x)}$.

Proof. Applying the condition (i) of Definition 1.4.22 in the case $k = 0$ we may choose $a_0 \in \boldsymbol{S}^{[\mu],\varrho}_{\mathrm{cl}(x)}$ equal to zero. This shows $a \in \boldsymbol{S}^{\mu-1,\varrho}_{\mathrm{cl}(\xi,x)}$. Analogously it follows that $a \in \boldsymbol{S}^{\mu,\varrho-1}_{\mathrm{cl}(\xi,x)}$, i.e., $a \in \boldsymbol{S}^{\mu-1,\varrho}_{\mathrm{cl}(\xi,x)} \cap \boldsymbol{S}^{\mu,\varrho-1}_{\mathrm{cl}(\xi,x)}$. Using again Definition 1.4.22, again, we can write

$$a = \tilde{b}_0 + \tilde{b} \quad \text{with} \quad \tilde{b}_0 \in \boldsymbol{S}^{\mu-1,[\varrho]}_{\mathrm{cl}(\xi)}, \ \tilde{b} \in \boldsymbol{S}^{\mu-1,\varrho-1}_{\mathrm{cl}(\xi,x)},$$
$$a = \tilde{a}_0 + \tilde{a} \quad \text{with} \quad \tilde{a}_0 \in \boldsymbol{S}^{[\mu],\varrho-1}_{\mathrm{cl}(x)}, \ \tilde{a} \in \boldsymbol{S}^{\mu-1,\varrho-1}_{\mathrm{cl}(\xi,x)}.$$

This shows that $\tilde{a}_0 \in S_\xi^{[\mu]}$ is at the same time of order $\mu - 1$ in ξ, since it equals $\tilde{b}_0 + \tilde{b} - \tilde{a}$. This implies $\tilde{a}_0(x,\xi) = 0$ for $|\xi| \geq c$ for some $c > 0$, and hence $\tilde{a}_0 \in S_{\text{cl}(\xi,x)}^{-\infty,\varrho-1}$. In an analogous manner it follows that $\tilde{b}_0 \in S_{\text{cl}(\xi,x)}^{\mu-1,-\infty}$ and thus, using (1.4.35), $a \in S_{\text{cl}(\xi,x)}^{\mu-1,\varrho-1}$. $\square$

We denote by $\Sigma^{\mu,\varrho}$ the space of all pairs (p_ψ, p_e) with

$$p_\psi(x,\xi) \in C^\infty(\mathbb{R}_\xi^n \setminus \{0\}, S_{\text{cl}}^\varrho(\mathbb{R}_x^n)), \qquad p_e(x,\xi) \in C^\infty(\mathbb{R}_x^n \setminus \{0\}, S_{\text{cl}}^\mu(\mathbb{R}_\xi^n))$$

satisfying

$$p_\psi(x,\lambda\xi) = \lambda^\mu p_\psi(x,\xi) \qquad \text{for all } \lambda > 0, (x,\xi) \in \mathbb{R}^n \times (\mathbb{R}^n \setminus \{0\}),$$
$$p_e(\delta x, \xi) = \delta^\varrho p_e(x,\xi) \qquad \text{for all } \delta > 0, (x,\xi) \in (\mathbb{R}^n \setminus \{0\}) \times \mathbb{R}^n$$

and $\sigma_\psi^\mu(p_e) = \sigma_e^\varrho(p_\psi)$.

Proposition 1.4.28 *To every* $(p_\psi, p_e) \in \Sigma^{\mu,\varrho}$ *there exists an* $a(x,\xi) \in S_{\text{cl}(\xi,x)}^{\mu,\varrho}$ *with* $\sigma_\psi^\mu(a) = p_\psi$, $\sigma_e^\varrho(a) = p_e$.

Proof. Let us set $r_0(x,\xi) = \sigma_\psi^\mu(p_e)$ (which equals $\sigma_e^\varrho(p_\psi)$) and choose an excision function χ in $\mathbb{R}^n$. Then $p(x,\xi) := \chi(\xi)p_\psi(x,\xi) \in S_{\text{cl}(\xi,x)}^{\mu,\varrho}$, $q(x,\xi) := \chi(x)p_e(x,\xi) \in S_{\text{cl}(\xi,x)}^{\mu,\varrho}$, cf. Remark 1.4.23, and $r(x,\xi) := \chi(x)\chi(\xi)r_0(x,\xi) \in S_{\text{cl}(\xi,x)}^{\mu,\varrho}$, cf. Remark 1.4.24. We have $r_0 = \sigma_\psi^\mu(q) = \sigma_e^\varrho(p) = \sigma_\psi^\mu(r) = \sigma_e^\varrho(r)$. The symbol $a(x,\xi) := p(x,\xi) + q(x,\xi) - r(x,\xi)$ then satisfies $\sigma_\psi^\mu(a) = \sigma_\psi^\mu(p) + \sigma_\psi^\mu(q) - \sigma_\psi^\mu(r) = p_\psi$, $\sigma_e^\varrho(a) = \sigma_e^\varrho(p) + \sigma_e^\varrho(q) - \sigma_e^\varrho(r) = p_e$. $\square$

Note that we also have $r_0 = \sigma_{\psi,e}^{\mu,\varrho}(a)$.

Definition 1.4.29 *We set*

$$L_{\text{cl}}^{\mu,\varrho}(\mathbb{R}^n) = \{\text{Op}(a) : a(x,\xi) \in S_{\text{cl}(\xi,x)}^{\mu,\varrho}(\mathbb{R}^n \times \mathbb{R}^n)\}.$$

The operators in $L_{\text{cl}}^{\mu,\varrho}(\mathbb{R}^n)$ *are called classical, with exit symbols.*

Analogously to (1.4.16) we have an isomorphism

$$\text{Op} : S_{\text{cl}(\xi,x)}^{\mu,\varrho}(\mathbb{R}^n \times \mathbb{R}^n) \to L_{\text{cl}}^{\mu,\varrho}(\mathbb{R}^n). \tag{1.4.39}$$

Denote by symb the inverse of (1.4.39). Then, as a consequence of Proposition 1.4.28, $(\sigma_\psi^\mu, \sigma_e^\varrho) \circ \text{symb} : L_{\text{cl}}^{\mu,\varrho}(\mathbb{R}^n) \to \Sigma^{\mu,\varrho}$ is surjective. Write for simplicity $(\sigma_\psi^\mu, \sigma_e^\varrho)$ instead of $(\sigma_\psi^\mu, \sigma_e^\varrho) \circ \text{symb}$. Then, from Proposition 1.4.26 it follows that the sequence

$$0 \to L_{\text{cl}}^{\mu-1,\varrho-1}(\mathbb{R}^n) \to L_{\text{cl}}^{\mu,\varrho}(\mathbb{R}^n) \to \Sigma^{\mu,\varrho} \to 0$$

is exact.

Theorem 1.4.30 $A \in L_{\mathrm{cl}}^{\mu,\varrho}(\mathbb{R}^n)$ *implies* $A^* \in L_{\mathrm{cl}}^{\mu,\varrho}(\mathbb{R}^n)$ *and* $^t A \in L_{\mathrm{cl}}^{\mu,\varrho}(\mathbb{R}^n)$ *for every* $\mu, \varrho \in \mathbb{R}$. *We have*

$$\sigma_\psi^\mu(A^*)(x,\xi) = \overline{\sigma_\psi^\mu(A)(x,\xi)}, \qquad \sigma_e^\varrho(A^*)(x,\xi) = \overline{\sigma_e^\varrho(A)(x,\xi)}$$

and

$$\sigma_\psi^\mu(^t A)(x,\xi) = \sigma_\psi^\mu(A)(x,-\xi), \qquad \sigma_e^\varrho(^t A)(x,\xi) = \sigma_e^\varrho(A)(x,-\xi).$$

A consequence of the symbolic rules is

$$\sigma_{\psi,e}^{\mu,\varrho}(A^*)(x,\xi) = \overline{\sigma_{\psi,e}^{\mu,\varrho}(A)(x,\xi)} \tag{1.4.40}$$

and

$$\sigma_{\psi,e}^{\mu,\varrho}(^t A)(x,\xi) = \sigma_{\psi,e}^{\mu,\varrho}(A)(x,-\xi). \tag{1.4.41}$$

Theorem 1.4.30 follows immediately from Theorem 1.4.12. From Theorem 1.4.14 we easily obtain the following result:

Theorem 1.4.31 $A \in L_{\mathrm{cl}}^{\mu,\varrho}(R^n)$, $B \in L_{\mathrm{cl}}^{\nu,\kappa}(\mathbb{R}^n)$ *implies the relation* $AB \in L_{\mathrm{cl}}^{\mu+\nu,\varrho+\kappa}(\mathbb{R}^n)$ *for all* $\mu, \varrho, \nu, \kappa \in \mathbb{R}$. *We have*

$$\sigma_\psi^{\mu+\nu}(AB) = \sigma_\psi^\mu(A)\sigma_\psi^\nu(B),$$
$$\sigma_e^{\varrho+\kappa}(AB) = \sigma_e^\varrho(A)\sigma_e^\kappa(B),$$
$$\sigma_{\psi,e}^{\mu+\nu,\varrho+\kappa}(AB) = \sigma_{\psi,e}^{\mu,\varrho}(A)\sigma_{\psi,e}^{\nu,\kappa}(B).$$

Remark 1.4.32 *The space* $S_{\mathrm{cl}(\xi,x)}^{\mu,\varrho}(\mathbb{R}^n \times \mathbb{R}^n)$ *can be endowed with a natural Fréchet topology. In fact, we have the system of mappings* (1.4.36), (1.4.37), (1.4.38) *on* $S_{\mathrm{cl}(\xi,x)}^{\mu,\varrho}$ *for* $j + k \leq N$ *for arbitrary* N *which determines a symbol* $a(x,\xi)$ *up to an element in* $S^{\mu-(N+1),\varrho-(N+1)}$. *Then the Fréchet topology in* $S_{\mathrm{cl}(\xi,x)}^{\mu,\varrho}$ *can be defined in an analogous manner as projective limit as above in the usual classical symbol spaces, cf. Section 1.1.2. From the bijection of Proposition 1.4.7, restricted to classical elements, we then obtain also a Fréchet topology on the space* $L_{\mathrm{cl}}^{\mu,\varrho}(\mathbb{R}^n)$.

1.4.4 Ellipticity

Definition 1.4.33 *A symbol* $a(x,\xi) \in S^{\mu,\varrho}(\mathbb{R}^n \times \mathbb{R}^n)$ *is called elliptic (of order* (μ,ϱ)*) if there is an* $R > 0$ *such that*

$$a(x,\xi) \neq 0 \text{ for all } (x,\xi) \in \mathbb{R}^n \times \mathbb{R}^n \text{ with } |x| + |\xi| \geq R \tag{1.4.42}$$

and

$$\langle \xi \rangle^{\mu} \langle x \rangle^{\varrho} |a^{-1}(x,\xi)| \leq c \text{ for all } (x,\xi) \in \mathbb{R}^n \times \mathbb{R}^n \text{ with } |x| + |\xi| \geq R$$
(1.4.43)

for a constant $c > 0$. An operator $A \in \boldsymbol{L}^{\mu,\varrho}(\mathbb{R}^n)$, written $A = \mathrm{Op}(a)$ for an $a(x,\xi) \in \boldsymbol{S}^{\mu,\varrho}(\mathbb{R}^n \times \mathbb{R}^n)$, is called elliptic (of order (μ,ϱ)) if $a(x,\xi)$ is elliptic (of order (μ,ϱ)).

Remark 1.4.34 *Let $A \in \boldsymbol{L}^{\mu,\varrho}(\mathbb{R}^n)$ be elliptic and $\chi(x,\xi)$ an excision function in $\mathbb{R}^{2n}$ with $\chi(x,\xi) = 0$ for $|x| + |\xi| \leq R$. Then $\chi(x,\xi)a^{-1}(x,\xi) \in \boldsymbol{S}^{-\mu,-\varrho}(\mathbb{R}^n \times \mathbb{R}^n)$.*

Example 1.4.35 *Let $\xi \to [\xi]$ be an arbitrary strictly positive C^{∞} function in $\mathbb{R}^n$ with $[\xi] = |\xi|$ for $|\xi| > \mathrm{const}$ for some constant > 0. Then the symbol $a^{\mu,\delta}(x,\xi) = [\xi]^{\mu}[x]^{\varrho}$ is elliptic of order (μ,ϱ).*

Theorem 1.4.36 *Let $A \in \boldsymbol{L}^{\mu,\varrho}(\mathbb{R}^n)$ be elliptic. Then*

$$A : H^{s,\delta}(\mathbb{R}^n) \to H^{s-\mu,\delta-\varrho}(\mathbb{R}^n)$$
(1.4.44)

is a Fredholm operator for all $s,\delta \in \mathbb{R}$. Moreover, there exists a $P \in \boldsymbol{L}^{-\mu,-\varrho}(\mathbb{R}^n)$ such that

$$AP - 1, PA - 1 \in \boldsymbol{L}^{-\infty,-\infty}(\mathbb{R}^n),$$
(1.4.45)

and $Au \in H^{s,\delta}(\mathbb{R}^n)$ for arbitrary fixed $s,\delta \in \mathbb{R}$ and $u \in H^{-\infty,-\infty}(\mathbb{R}^n)$ imply $u \in H^{s+\mu,\delta+\varrho}(\mathbb{R}^n)$.

Proof. Set $b(x,\xi) = \chi(x,\xi)a^{-1}(x,\xi)$ for an excision function χ like in Remark 1.4.34. Then $B = \mathrm{Op}(b) \in \boldsymbol{L}^{-\mu,-\varrho}(\mathbb{R}^n)$ satisfies $AB - 1 \in \boldsymbol{L}^{-1,-1}(\mathbb{R}^n)$, cf. Theorem 1.4.14. In view of Theorem 1.4.20 the operator $K_r = AB - 1$ is compact in $H^{s-\mu,\delta-\varrho}(\mathbb{R}^n)$. Analogously we get $BA = 1 + K_l$ with a compact operator K_l in $H^{s,\delta}(\mathbb{R}^n)$. Thus (1.4.44) is a Fredholm operator. Using Proposition 1.4.7 we find a $c(x,\xi) \in \boldsymbol{S}^{-1,-1}(\mathbb{R}^n \times \mathbb{R}^n)$ with $K_r = \mathrm{Op}(c)$. Let $c^{(N)} \in \boldsymbol{S}^{-N,-N}(\mathbb{R}^n \times \mathbb{R}^n)$ denote the N-fold Leibniz product $c\#c\#\ldots\#c$. Then the formal Neumann series $(1 + K_r)^{-1} = \sum_{N=0}^{\infty} (-1)^N K_r^N$ can be carried out in terms of an asymptotic sum of symbols. This yields a $d(x,\xi) \in \boldsymbol{S}^{-1,-1}(\mathbb{R}^n \times \mathbb{R}^n)$ with $1 + d(x,\xi) \sim \sum_{N=0}^{\infty} (-1)^N c^{(N)}(x,\xi)$. Then $(1 + c)\#(1 + d) - 1 \in \boldsymbol{S}^{-\infty,-\infty}(\mathbb{R}^n \times \mathbb{R}^n)$. It follows that $AB(1 + \mathrm{Op}(d)) = (1 + K_r)(1 + \mathrm{Op}(d)) = 1 \bmod \boldsymbol{L}^{-\infty,-\infty}(\mathbb{R}^n)$. Thus $P_r := B(1 + \mathrm{Op}(d)) \in \boldsymbol{L}^{-\mu,-\varrho}(\mathbb{R}^n)$ satisfies the first relation in (1.4.45). Analogously we get an operator $P_l \in \boldsymbol{L}^{-\mu,-\varrho}(\mathbb{R}^n)$ satisfying the second relation. Then, a simple algebraic argument shows $P_l - P_r \in \boldsymbol{L}^{-\infty,-\infty}(\mathbb{R}^n)$. Thus we can set $P = P_r$ or $P = P_l$. Finally, let $Au \in H^{s,\delta}(\mathbb{R}^n)$, $u \in H^{-\infty,-\infty}(\mathbb{R}^n)$. Then $PAu \in H^{s+\mu,\delta+\varrho}(\mathbb{R}^n)$, according to Theorem 1.4.17. Using the second relation of (1.4.45) we get for $C = PA - 1$ that $u + Cu \in H^{s+\mu,\delta+\varrho}(\mathbb{R}^n)$, where $Cu \in H^{\infty,\infty}(\mathbb{R}^n) \subset H^{s+\mu,\delta+\varrho}(\mathbb{R}^n)$. This yields $u \in H^{s+\mu,\delta+\varrho}(\mathbb{R}^n)$. $\qquad\square$

An operator $P \in \boldsymbol{L}^{-\mu,-\varrho}(\mathbb{R}^n)$ satisfying (1.4.45) is called a parametrix of A.

Proposition 1.4.37 *An* $a(x,\xi) \in \boldsymbol{S}^{\mu,\varrho}_{\text{cl}(\xi,x)}(\mathbb{R}^n \times \mathbb{R}^n)$ *is elliptic if and only if*

$$\sigma^\mu_\psi(a)(x,\xi) \neq 0 \text{ for all } x \in \mathbb{R}^n, \xi \in \mathbb{R}^n \setminus \{0\}, \qquad (1.4.46)$$

$$\sigma^\varrho_e(a)(x,\xi) \neq 0 \text{ for all } x \in \mathbb{R}^n \setminus \{0\}, \xi \in \mathbb{R}^n, \qquad (1.4.47)$$

$$\sigma^{\mu,\varrho}_{\psi,e}(a)(x,\xi) \neq 0 \text{ for all } x \in \mathbb{R}^n \setminus \{0\}, \xi \in \mathbb{R}^n \setminus \{0\}. \qquad (1.4.48)$$

Proof. We first show that $a(x,\xi)$ satisfies (1.4.46), (1.4.47), (1.4.48) if and only if

$$|\sigma^\mu_\psi(a)(x,\xi)| \geq \delta |\xi|^\mu [x]^\varrho \text{ for all } \quad x \in \mathbb{R}^n, \xi \in \mathbb{R}^n \setminus \{0\}, \quad (1.4.49)$$

$$|\sigma^\varrho_e(a)(x,\xi)| \geq \delta [\xi]^\mu |x|^\varrho \text{ for all } \quad x \in \mathbb{R}^n \setminus \{0\}, \xi \in \mathbb{R}^n \quad (1.4.50)$$

for some $\delta > 0$. Without loss of generality we can assume $\mu = \varrho = 0$, since the corresponding relations for $a(x,\xi)[\xi]^{-\mu}[x]^{-\varrho}$ and order $(0,0)$ are equivalent to those for $a(x,\xi)$. In view of Definition 1.4.22 we can write

$$a(x,\xi) = a_0(x,\xi) + a_1(x,\xi) \qquad \text{for } a_0 \in \boldsymbol{S}^{[0],0}_{\text{cl}(x)}, a_1 \in \boldsymbol{S}^{-1,0}_{\text{cl}(\xi,x)}$$

and

$$a_0(x,\xi) = a_{00}(x,\xi) + r(x,\xi) \qquad \text{for } a_{00} \in S^{[0]}_x \cap S^{[0]}_\xi, r \in \boldsymbol{S}^{[0],-1}_{\text{cl}(x)}.$$

Because $\sigma^0_\psi(a)(x,\xi) = \sigma^0_\psi(a_0)(x,\xi)$, it suffices to look at a_0. Using the homogeneity $\sigma^0_\psi(a_0)(x,\xi/|\xi|) = \sigma^0_\psi(a_0)(x,\xi)$ we obtain from (1.4.46) that for every $c_1 > 0$ there is a $\delta > 0$ with

$$|\sigma^0_\psi(a_0)(x,\xi)| \geq \delta \qquad \text{for all } \xi \in \mathbb{R}^n \setminus \{0\}, |x| \leq c_1. \quad (1.4.51)$$

Further we have

$$|\sigma^0_\psi(a_0)(x,\xi)| \geq |\sigma^0_\psi(a_{00})(x,\xi)| - |\sigma^0_\psi(r)(x,\xi)|. \qquad (1.4.52)$$

For $h(x,\xi) := \sigma^{0,0}_{\psi,e}(a_{00})(x,\xi)$ the homogeneity $h(x/|x|,\xi/|\xi|) = h(x,\xi)$ and (1.4.48) imply $|h(x,\xi)| \geq 3\varepsilon$ for all $x \in \mathbb{R}^n \setminus \{0\}$, $\xi \in \mathbb{R}^n \setminus \{0\}$, for an $\varepsilon > 0$. Given ε we can choose a sufficiently large $c_1 > 0$ such that

$$|\sigma^0_\psi(a_{00})(x,\xi)| \geq |h(x,\xi)| + \varepsilon \text{ for all } \xi \in \mathbb{R}^n \setminus \{0\}, |x| > c_1. \quad (1.4.53)$$

Analogously, since $\sigma_\psi^0(r)(x, \xi/|\xi|) = \sigma_\psi^0(r)(x, \xi)$, we get

$$|\sigma_\psi^0(r)(x, \xi)| \leq c_0[x]^{-1} \text{ for all } \xi \in \mathbb{R}^n \setminus \{0\} \qquad (1.4.54)$$

for some constant $c_0 > 0$. For $|x| \geq c_1$ for sufficiently large $c_1 > 0$ we get $c_0[x]^{-1} \leq \varepsilon$. Thus (1.4.52), (1.4.53), (1.4.54) yield

$$|\sigma_\psi^0(a_0)(x, \xi)| \geq \varepsilon \text{ for all } \xi \in \mathbb{R}^n \setminus \{0\}, \qquad |x| \geq c_1. \qquad (1.4.55)$$

Then (1.4.51) and (1.4.55) imply (1.4.49) with a suitable $\delta > 0$. The relation (1.4.50) follows in an analogous manner by interchanging the role of x and ξ.

Conversely assume that (1.4.49) and (1.4.50) are satisfied. Then (1.4.46) and (1.4.47) are obvious consequences. Moreover, for $|x|, |\xi| \geq c$ for $c > 0$ so large that $c_0[x]^{-1} \leq \delta/2$ holds for the constant c_0 from (1.4.54), we get

$$|a_{00}(x, \xi)| \geq |a_0(x, \xi)| - |r(x, \xi)| \geq \delta - c_0[x]^{-1} \geq \delta/2.$$

Thus $|\sigma_{\psi,e}^{0,0}(a)(x, \xi)| \geq \delta/2$ for all $x, \xi \in \mathbb{R}^n \setminus \{0\}$.

It remains to show that the ellipticity of $a(x, \xi)$ of order (μ, ϱ) is equivalent to (1.4.49), (1.4.50). This may be done again for $\mu = \varrho = 0$. Assume first that the relations (1.4.49), (1.4.50) hold. Writing $a = a_0 + a_1 = b_0 + b_1$ with $a_0 \in \mathbf{S}_{\mathrm{cl}(x)}^{[0],0}$, $a_1 \in \mathbf{S}_{\mathrm{cl}(x,\xi)}^{-1,0}$, $b_0 \in \mathbf{S}_{\mathrm{cl}(\xi)}^{0,[0]}$, $b_1 \in \mathbf{S}_{\mathrm{cl}(\xi,x)}^{0,-1}$ we obtain for sufficiently large $c > 0$

$$|a_0(x, \xi)| \geq \delta, \qquad |a_1(x, \xi)| \leq \mathrm{const}[\xi]^{-1} \leq \delta/2 \text{ for all } x \in \mathbb{R}^n, |\xi| \geq c,$$
$$|b_0(x, \xi)| \geq \delta, \qquad |b_1(x, \xi)| \leq \mathrm{const}[x]^{-1} \leq \delta/2 \text{ for all } \xi \in \mathbb{R}^n, |x| \geq c.$$

This gives us

$$|a(x, \xi)| \geq |a_0(x, \xi)| - |a_1(x, \xi)| \geq \delta/2 \text{ for all } x \in \mathbb{R}^n, |\xi| \geq c,$$
$$|a(x, \xi)| \geq |b_0(x, \xi)| - |b_1(x, \xi)| \geq \delta/2 \text{ for all } \xi \in \mathbb{R}^n, |x| \geq c.$$

For $R = 2c$ and $|x| + |\xi| \geq R$ we have $|\xi| \geq c$ or $|x| \geq c$. Hence it follows that $|a(x, \xi)| \geq \delta/2$ for all $x, \xi \in \mathbb{R}^n$ with $|x| + |\xi| \geq R$. Conversely let $a(x, \xi)$ be elliptic. Then

$$|a_0(x, \xi)| \geq |a(x, \xi)| - |a_1(x, \xi)| \geq \delta - \mathrm{const}[\xi]^{-1} \geq \delta/2$$
$$\text{for all } x \in \mathbb{R}^n, |\xi| \geq c,$$
$$|b_0(x, \xi)| \geq |a(x, \xi)| - |b_1(x, \xi)| \geq \delta - \mathrm{const}[x]^{-1} \geq \delta/2$$
$$\text{for all } \xi \in \mathbb{R}^n, |x| \geq c$$

when $c > 0$ is large enough. This yields (1.4.49), (1.4.50). $\qquad \square$

In Section 1.1.6 we have employed the operator family

$$R_\lambda(x_0, \xi_0)u(x) = \lambda^{\frac{n}{4}} e^{i\lambda x \xi_0} u(\lambda^{\frac{1}{2}}(x - x_0)),$$

$\lambda \in \mathbb{R}_+$, $(x_0, \xi_0) \in T^*\mathbb{R}^n \setminus \{0\}$, in proving that a pseudo–differential operator with the Fredholm property (when acting on Sobolev spaces) is necessarily elliptic. For analogous reasons we define here

$$S_\lambda(x_0, \xi_0) = F^{-1} R_\lambda(\xi_0, -x_0) F$$

for $\lambda \in \mathbb{R}_+$, $x_0 \in \mathbb{R}^n$, with the Fourier transform $F = F_{x \to \xi}$ in $\mathbb{R}^n$. From Lemma 1.1.84 we obtain the following result:

Lemma 1.4.38   (i) $S_\lambda(x_0, \xi_0)$ *is a unitary operator in* $L^2(\mathbb{R}^n)$.

(ii) *For every* $u \in L^2(\mathbb{R}^n)$ *we have* $S_\lambda(x_0, \xi_0)u \to 0$ *weakly in* $L^2(\mathbb{R}^n)$ *when* $\lambda \to \infty$.

Theorem 1.4.39 *Let* $a \in \boldsymbol{L}_{cl}^{0,0}(\mathbb{R}^n)$ *be given and* C *be a compact operator in* $L^2(\mathbb{R}^n)$. *Then, for every* $u \in L^2(\mathbb{R}^n)$ *we have*

$$\|S_\lambda^{-1}(x_0, \xi_0)(A + C)S_\lambda(x_0, \xi_0)u - \sigma_e^0(A)(x_0, \xi_0)u\|_{L^2(\mathbb{R}^n)} \to 0$$

for $\lambda \to \infty$, *for arbitrary fixed* $x_0 \in \mathbb{R}^n \setminus \{0\}$, $\xi_0 \in \mathbb{R}^n$.

Proof. Since C is compact, we get from Lemma 1.1.84 that

$$S_\lambda^{-1}(x_0, \xi_0)C S_\lambda(x_0, \xi_0)u \to 0$$

for $\lambda \to \infty$. Moreover, we have by definition

$$S_\lambda^{-1}(x_0, \xi_0)A S_\lambda(x_0, \xi_0)u = F^{-1} R_\lambda^{-1}(\xi_0, -x_0) F A F^{-1} R_\lambda(\xi_0, -x_0) F u.$$

Let us form $Bv(z) = (F_{\zeta \to z} A F_{z' \to \zeta}^{-1} v)(z)$ for $v \in \mathcal{S}(\mathbb{R}^n)$. Using

$$Aw(\zeta) = \int e^{i\zeta \varrho} a(\zeta, \varrho) \hat{w}(\varrho) \, d\varrho$$

$$= \int e^{i\zeta \varrho} a(\zeta, \varrho) v(\varrho) \, d\varrho$$

for $w = F^{-1}v$ we get

$$Bv(z) = \int e^{-i\zeta z} \left\{ \int e^{i\zeta \varrho} a(\zeta, \varrho) v(\varrho) \, d\varrho \right\} d\zeta$$

$$= \iint e^{i(\varrho - z)\zeta} a(\zeta, \varrho) v(\varrho) \, d\varrho d\zeta.$$

Inserting $v(\varrho) = \int e^{i\varrho\xi}\hat{v}(\xi)\,d\xi$ we obtain

$$
\begin{aligned}
Bv(z) &= \iiint e^{i(\varrho-z)\zeta} e^{i\varrho\xi} a(\zeta,\varrho)\hat{v}(\xi)\,d\varrho d\zeta d\xi \\
&= \int e^{iz\xi}\left\{ \iint e^{-iz\xi} e^{i(\varrho-z)\zeta} e^{i\varrho\xi} a(\zeta,\varrho)\,d\varrho d\zeta \right\}\hat{v}(\xi)\,d\xi \\
&= \mathrm{Op}(b)v(z),
\end{aligned}
$$

where

$$
\begin{aligned}
b(z,\xi) &= \iint e^{i(\varrho-z)(\xi+\zeta)} a(\zeta,\varrho)\,d\varrho d\zeta \qquad (1.4.56)\\
&= \iint e^{-iy\eta} a(-(\xi+\eta), z+y)\,dy d\eta.
\end{aligned}
$$

Here we have used the substitution $-\eta = \xi + \zeta$, $y = \varrho - z$. Let us set $\tilde{a}(\xi,z) = a(-\xi,z)$. Then, comparing (1.4.56) with Theorem 1.4.9, we see that $b(z,\xi)$ belongs to $\boldsymbol{S}^{0,0}_{\mathrm{cl}(\xi,z)}(\mathbb{R}^n \times \mathbb{R}^n)$ and has an asymptotic expansion $b(z,\xi) \sim \sum \frac{1}{\alpha!} b_\alpha(z,\xi)$ for $b_\alpha(z,\xi) = \partial_\xi^\alpha D_z^\alpha \tilde{a}(\xi,z) \in \boldsymbol{S}^{-|\alpha|,-|\alpha|}(\mathbb{R}^n \times \mathbb{R}^n)$. We have

$$
\sigma^0_\psi(b)(z,\xi) = \sigma^0_\psi(b_0)(z,\xi) = \sigma^0_e(\tilde{a})(\xi,z) = \sigma^0_e(a)(-\xi,z).
$$

From Theorem 1.1.85 we obtain for $v = Fu$, $u \in L^2(\mathbb{R}^n)$, for $\lambda \to \infty$

$$
R^{-1}_\lambda(\xi_0,-x_0) B R_\lambda(\xi_0,-x_0)v \to \sigma^0_\psi(B)(\xi_0,-x_0)v,
$$

which implies

$$
R^{-1}_\lambda(\xi_0,-x_0) B R_\lambda(\xi_0,-x_0)Fu \to \sigma^0_\psi(B)(\xi_0,-x_0)Fu,
$$

and thus

$$
\begin{aligned}
F^{-1}R^{-1}_\lambda(\xi_0,x_0) B R_\lambda(\xi_0,-x_0)Fu &= S^{-1}_\lambda(x_0,\xi_0) A S_\lambda(x_0,\xi_0) \\
&\to \sigma^0_\psi(B)(\xi_0,-x_0)u = \sigma^0_e(A)(x_0,\xi_0)u. \qquad (1.4.57)
\end{aligned}
$$

$\square$

Theorem 1.4.40 *Let $A \in \boldsymbol{L}^{\mu,\varrho}_{\mathrm{cl}}(\mathbb{R}^n)$ be a Fredholm operator*

$$
A : H^{s,\delta}(\mathbb{R}^n) \to H^{s-\mu,\delta-\varrho}(\mathbb{R}^n) \qquad (1.4.58)
$$

for an $s = s_0 \in \mathbb{R}$, $\delta = \delta_0 \in \mathbb{R}$. Then A is elliptic of order (μ, ϱ), and A induces Fredholm operators (1.4.58) for all $s, \delta \in \mathbb{R}$.

Proof. Using Example 1.4.35, Theorem 1.4.36 and Theorem 1.4.31 it suffices to consider the operator

$$A_0 = \mathrm{Op}(a^{s-\mu,\delta-\varrho})A\,\mathrm{Op}(a^{-s,-\delta}) : L^2(\mathbb{R}^n) \to L^2(\mathbb{R}^n)$$

which belongs to $\boldsymbol{L}_{\mathrm{cl}}^{0,0}(\mathbb{R}^n)$. For simplicity denote A_0 again by A, again. The assumption that $A : L^2(\mathbb{R}^n) \to L^2(\mathbb{R}^n)$ is a Fredholm operator implies the existence of an operator $P \in \mathcal{L}(L^2(\mathbb{R}^n))$, $PA = 1 + C$ for some compact operator C in $L^2(\mathbb{R}^n)$. For every fixed $(x_0,\xi_0) \in \mathbb{R}^n \times (\mathbb{R}^n \setminus \{0\})$ we set $R_\lambda = R_\lambda(x_0,\xi_0)$, $\lambda \in \mathbb{R}_+$, and we choose any $u \in L^2(\mathbb{R}^n)$, $u \neq 0$. Then

$$0 < \|u\| = \|R_\lambda u\| = \|(PA - C)R_\lambda u\| \qquad (1.4.59)$$
$$\leq \|PR_\lambda R_\lambda^{-1} AR_\lambda u\| + \|CR_\lambda u\|$$
$$\leq \|P\|\|R_\lambda\|\|R_\lambda^{-1} AR_\lambda u\| + \|CR_\lambda u\|.$$

From Lemma 1.1.84 we get $\|R_\lambda\| = 1$ and $\|CR_\lambda u\| \to 0$ for $\lambda \to \infty$. Moreover, Theorem 1.1.85 gives us $\|R_\lambda^{-1} AR_\lambda u - \sigma_\psi^0(A)(x_0,\xi_0)u\| \to 0$ for $\lambda \to \infty$. This yields for every $\varepsilon > 0$ the estimate

$$0 < \|u\| \leq \|P\|\|\sigma_\psi^0(A)(x_0,\xi_0)\|\|u\| + \varepsilon$$

and hence

$$|\sigma_\psi^0(A)(x_0,\xi_0)| \geq \frac{1}{\|P\|} > 0 \text{ for all } x_0 \in \mathbb{R}^n, \xi_0 \in \mathbb{R}^n \setminus \{0\}. \qquad (1.4.60)$$

Next we fix $(x_0,\xi_0) \in (\mathbb{R}^n \setminus \{0\}) \times \mathbb{R}^n$. Set $v = Fu$, $E = FPF^{-1}$, $B = FAF^{-1}$, $G = FCF^{-1}$. Then $PA = 1 + C$ implies $EB = 1 + G$, and we can argue in an analogous manner as before, namely

$$0 < \|v\| = \|R_\lambda(\xi_0,-x_0)v\| = \|(EB - G)R_\lambda(\xi_0,-x_0)v\|$$
$$\leq \|ER_\lambda(\xi_0,-x_0)R_\lambda^{-1}(\xi_0,-x_0)BR_\lambda(\xi_0,-x_0)v\|$$
$$\quad + \|GR_\lambda(\xi_0,-x_0)v\|$$
$$\leq \|E\|\|R_\lambda^{-1}(\xi_0,-x_0)BR_\lambda(\xi_0,-x_0)Fu\| + \|GR_\lambda(\xi_0,-x_0)v\|$$
$$= \|E\|\|F^{-1}R_\lambda^{-1}(\xi_0,-x_0)BR_\lambda(\xi_0,-x_0)Fu\|$$
$$\quad + \|GR_\lambda(\xi_0,-x_0)v\|.$$

Using the relation (1.4.57) we obtain as above

$$|\sigma_e^0(A)(x_0,\xi_0)| \geq \frac{1}{\|P\|} > 0 \text{ for all } x_0 \in \mathbb{R}^n \setminus \{0\}, \xi_0 \in \mathbb{R}^n. \qquad (1.4.61)$$

The estimates (1.4.60), (1.4.61) correspond to the relations (1.4.49), (1.4.50) for $\mu = \varrho = 0$ which is the ellipticity of $a(x,\xi) \in \boldsymbol{S}_{\mathrm{cl}(\xi,x)}^{0,0}(\mathbb{R}^n \times$

$\mathbb{R}^n$) for $A = \mathrm{Op}(a)$. Thus we finally obtain that the Fredholm property of (1.4.58) for fixed $s, \delta \in \mathbb{R}$ implies the ellipticity of the operator. This yields then the Fredholm property of (1.4.58) for all $s, \delta \in \mathbb{R}$ according to Theorem 1.4.36. □

Remark 1.4.41 *Theorem 1.4.40 has an analogue for non–classical* $A \in L^{\mu,\varrho}_{\mathrm{cl}}(\mathbb{R}^n)$. *A proof was given by Schrohe [102] by other methods, cf. also Grieme [45].*

1.4.5 The calculus on manifolds with exits

The operator calculus of the preceding sections allows us to formulate a global variant on a non–compact C^∞ manifold M with exits to infinity. We will need the case of cylindrical exits of the form $(c, \infty) \times X \ni (t, x)$ with the cone metric, where the base X is a closed compact C^∞ manifold with a fixed Riemannian metric. A Riemannian manifold M has then such an exit if $M \setminus \{(c, \infty) \times X\}$ is compact. Analogously we can consider the case of finitely many exits. A special case is $M = \mathbb{R}^{n+1} \ni \tilde{x}$, where $\{\tilde{x} \in \mathbb{R}^{n+1} : |\tilde{x}| > 1\}$ can be interpreted as $(1, \infty) \times S^n$. For convenience we will consider here the case of an infinite cylinder $\mathbb{R} \times X$ with the product metric.

The statements in this section are simple consequences of the material of Sections 1.4.1–1.4.4; so some obvious details are left to the reader.

A subset $U \subseteq \mathbb{R}^{n+1}$ is called conical for large $|\tilde{x}|$ if there is a conical set $V \subset \mathbb{R}^{n+1} \setminus \{0\}$ (i.e., $\tilde{x} \in V$ implies $\lambda\tilde{x} \in V$ for all $\lambda \in \mathbb{R}_+$) such that $\{\tilde{x} \in U : |\tilde{x}| \geq c\} = \{\tilde{x} \in V : |\tilde{x}| \geq c\}$ for some constant $c > 0$. If U is of this kind, then the symbol space $\boldsymbol{S}^{\mu,\varrho}(U \times \mathbb{R}^{n+1})$ for $\mu, \varrho \in \mathbb{R}$ is defined as the subspace of all $a(\tilde{x}, \tilde{\xi}) \in C^\infty(U \times \mathbb{R}^{n+1})$ such that for every open $U_0 \subset U$, conical for large $|\tilde{x}|$, with $\overline{U}_0 \subset U$, there is an $a_0(\tilde{x}, \tilde{\xi}) \in \boldsymbol{S}^{\mu,\varrho}(\mathbb{R}^{n+1} \times \mathbb{R}^{n+1})$ such that $a_0|_{U_0 \times \mathbb{R}^{n+1}} = a|_{U_0 \times \mathbb{R}^{n+1}}$ holds. In an analogous manner we can introduce the subclass $\boldsymbol{S}^{\mu,\varrho}_{\mathrm{cl}}(U \times \mathbb{R}^{n+1})$ of classical symbols in $(\tilde{x}, \tilde{\xi})$ by requiring that all $a_0(\tilde{x}, \tilde{\xi})$ in the above condition belong to $\boldsymbol{S}^{\mu,\varrho}_{\mathrm{cl}}(\mathbb{R}^{n+1} \times \mathbb{R}^{n+1})$. Moreover, the Schwartz space $\mathcal{S}(U \times V)$ for open sets $U, V \subseteq \mathbb{R}^{n+1}$, which are conical for large $|\tilde{x}|$, is defined as the set of all $c(\tilde{x}, \tilde{x}') \in C^\infty(U \times V)$ such that for every open $U_0 \subset U$, $V_0 \subset V$, conical for large $|\tilde{x}|$ and satisfying $\overline{U}_0 \subset U$, $\overline{V}_0 \subset V$, there is a $c_0(\tilde{x}, \tilde{x}') \in \mathcal{S}(\mathbb{R}^{n+1} \times \mathbb{R}^{n+1})$ with $c_0|_{U_0 \times V_0} = c|_{U_0 \times V_0}$. The spaces

$$\boldsymbol{S}^{\mu,\varrho}(U \times \mathbb{R}^{n+1}), \qquad \boldsymbol{S}^{\mu,\varrho}_{\mathrm{cl}}(U \times \mathbb{R}^n), \qquad \mathcal{S}(U \times V)$$

are Fréchet in a natural way, and we have

$$\mathcal{S}(U \times \mathbb{R}^{n+1}) = \boldsymbol{S}^{-\infty,-\infty}(U \times \mathbb{R}^{n+1}).$$

The general considerations from the beginning of Section 1.4.1 on the symbol spaces have obvious analogues in the present situation. We will use here tacitly the corresponding results, in particular, the analogue of Theorem 1.4.3.

Definition 1.4.42 *Let $U \subseteq \mathbb{R}^{n+1}$ be an open set, conical for large $|\tilde{x}|$. Denote by $L^{-\infty,-\infty}(U)$ the space of all integral operators with kernels in $\mathcal{S}(U \times U)$. Moreover,*

$$L^{\mu,\varrho}(U) = \{\mathrm{Op}(A) + C : a(\tilde{x}, \tilde{\xi}) \in S^{\mu,\varrho}(U \times \mathbb{R}^{n+1}),\ C \in L^{-\infty,-\infty}(U)\}.$$

Here, $\mathrm{Op}(a) = F_{\tilde{\xi} \to \tilde{x}}^{-1} a(\tilde{x}, \tilde{\xi}) F_{\tilde{x}' \to \tilde{\xi}}$, for the Fourier transform F in $\mathbb{R}^{n+1}$. In an analogous manner the subspace $L_{\mathrm{cl}}^{\mu,\varrho}(U)$ of classical pseudo-differential operators on U is defined in terms of $S_{\mathrm{cl}}^{\mu,\varrho}(U \times \mathbb{R}^{n+1})$.

Proposition 1.4.43 *Let $U, V \subseteq \mathbb{R}^{n+1}$ be open sets that are conical for large $|\tilde{x}|$, and let $\kappa : U \to V$ be a diffeomorphism satisfying $\kappa(\lambda \tilde{x}) = \lambda \kappa(\tilde{x})$ for all $\lambda \geq 1$, $|\tilde{x}| \geq c$ for some constant $c > 0$. Then the push-forward κ_* of pseudo-differential operators under κ induces isomorphisms*

$$\kappa_* : L^{\mu,\varrho}(U) \to L^{\mu,\varrho}(V),$$
$$\kappa_* : L_{\mathrm{cl}}^{\mu,\varrho}(U) \to L_{\mathrm{cl}}^{\mu,\varrho}(V)$$

for all $\mu, \varrho \in \mathbb{R}$. For $A \in L_{\mathrm{cl}}^{\mu,\varrho}(U)$ we have

$$\sigma_\psi^\mu(\kappa_* A)(\tilde{y}, \tilde{\eta}) = \sigma_\psi^\mu(A)(\tilde{x}, \tilde{\xi}), \tag{1.4.62}$$
$$\sigma_e^\varrho(\kappa_* A)(\tilde{y}, \tilde{\eta}) = \sigma_e^\varrho(A)(\tilde{x}, \tilde{\xi}),$$
$$\sigma_{\psi,e}^{\mu,\varrho}(\kappa_* A)(\tilde{y}, \tilde{\eta}) = \sigma_{\psi,e}^{\mu,\varrho}(A)(\tilde{x}, \tilde{\xi})$$

for $(\tilde{y}, \tilde{\eta}) = (\kappa(\tilde{x}), {}^t d\kappa(\tilde{x})^{-1} \tilde{\xi})$.

Proof. The assertions follow from the arguments for Theorem 1.4.15 and from the symbolic rule for push-forwards, cf. (1.4.30). □

Remark 1.4.44 *If $U \subseteq \mathbb{R}^{n+1}$ is an open set which is conical for large $|\tilde{x}|$ the ellipticity of order (μ, ϱ) of an $A \in L^{\mu,\varrho}(U)$, written $A = \mathrm{Op}(a) \bmod L^{-\infty,-\infty}(U)$ for $a(\tilde{x}, \tilde{\xi}) \in S^{\mu,\varrho}(U \times \mathbb{R}^{n+1})$ is defined as follows: For every open $U_0 \subset U$ which is conical for large $|\tilde{x}|$ and $\overline{U}_0 \subset U$ there is an $R = R(U_0)$ with $a(\tilde{x}, \tilde{\xi}) \neq 0$ for all $(\tilde{x}, \tilde{\xi}) \in U_0 \times \mathbb{R}^{n+1}$ with $|\tilde{x}| + |\tilde{\xi}| \geq R$, and $\langle \tilde{\xi} \rangle^\mu \langle \tilde{x} \rangle^\varrho |a^{-1}(\tilde{x}, \tilde{\xi})| \leq c$ for all $(\tilde{x}, \tilde{\xi}) \in U_0 \times \mathbb{R}^{n+1}$ with $|\tilde{x}| + |\tilde{\xi}| \geq R$, for a constant $c = c(U_0) > 0$, cf. Definition 1.4.33. This is obviously an invariant condition with respect to the push-forward of Proposition 1.4.43.*

Let X be a compact Riemannian manifold. Choose an open covering by coordinate neighbourhoods $\{U_1, \ldots, U_N\}$ and fix charts $\chi_j : U_j \to \Omega_j$ for open $\Omega_j \subset S^n = \{\tilde{x} \in \mathbb{R}^{n+1} : |\tilde{x}| = 1\}$. Let us define $\boldsymbol{L}^{\mu,\varrho}(\mathbb{R} \times U_j)$ as the subspace of all $A \in L^\mu(\mathbb{R} \times U_j)$ such that for the diffeomorphisms $\kappa_{j,\pm} : \mathbb{R}_\pm \times U_j \to \Omega_j^\wedge := \{\tilde{x} \in \mathbb{R}^{n+1} : \tilde{x}/|\tilde{x}| \in \Omega_j\}$, defined by $\kappa_{j,\pm}(t,x) = |t|\chi_j(x)$, we have $(\kappa_{j,\pm})_*(A|_{\mathbb{R}_\pm \times U_j}) \in \boldsymbol{L}^{\mu,\varrho}(\Omega_j^\wedge)$. In an analogous manner we define $\boldsymbol{L}_{\mathrm{cl}}^{\mu,\varrho}(\mathbb{R} \times U_j)$. To $\{U_1, \ldots, U_N\}$ we fix a subordinate partition of unity $\{\varphi_1, \ldots, \varphi_N\}$ and another function system $\{\psi_1, \ldots, \psi_N\}$, $\psi_j \in C_0^\infty(U_j))$, with $\varphi_j \psi_j = \varphi_j$ for all j.

Definition 1.4.45 *Let X be a compact Riemannian manifold. Then $\boldsymbol{L}^{-\infty,-\infty}(\mathbb{R} \times X)$ denotes the space of all integral operators C*

$$Cu(t,x) = \int_\mathbb{R} \int_X c(t,t',x,x')u(t',x')\,dt'dx'$$

with kernels in $\mathcal{S}(\mathbb{R} \times \mathbb{R}, C^\infty(X \times X))$ (with dt being the Lebesgue measure on $\mathbb{R}$ and dx the measure on X from the Riemannian metric). $\boldsymbol{L}^{\mu,\varrho}(\mathbb{R} \times X)$ for $\mu, \varrho \in \mathbb{R}$ is the subspace of all $A \in L^\mu(\mathbb{R} \times X)$ of the form

$$A = \sum_{j=1}^N \varphi_j A_j \psi_j + C \tag{1.4.63}$$

for arbitrary $A_j \in \boldsymbol{L}^{\mu,\varrho}(\mathbb{R} \times U_j)$, $j = 1, \ldots, N$, and $C \in \boldsymbol{L}^{-\infty,-\infty}(\mathbb{R} \times X)$. In an analogous manner we define $\boldsymbol{L}_{\mathrm{cl}}^{\mu,\varrho}(\mathbb{R} \times X)$ in terms of the local classes $\boldsymbol{L}_{\mathrm{cl}}^{\mu,\varrho}(\mathbb{R} \times U_j)$.

Remark 1.4.46 *Proposition 1.4.43 shows that Definition 1.4.45 is correct in the sense that it does not depend on the particular choice of the covering of X, the charts and the functions φ_j, ψ_j. So we can choose $U_1, \ldots, U_N$ in such a way that the open sets $U_j \cup U_k$ are also coordinate neighbourhoods on X for all j, k.*

Note that for a classical operator $A \in \boldsymbol{L}_{\mathrm{cl}}^{\mu,\varrho}(\mathbb{R} \times X)$ the principal symbols are invariant in the sense

$$\sigma_\psi^\mu(A) \in C^\infty(T^*(\mathbb{R} \times X) \setminus 0), \tag{1.4.64}$$

$$\sigma_e^\varrho(A) \in C^\infty(T^*(\mathbb{R} \times X) \setminus \{t = 0\}), \tag{1.4.65}$$

$$\sigma_{\psi,e}^{\mu,\varrho}(A) \in C^\infty(T^*(\mathbb{R} \times X) \setminus (0 \cup \{t = 0\})), \tag{1.4.66}$$

where 0 indicates the zero section in $T^*(\mathbb{R} \times X)$.

Remark 1.4.47 *Let $\varphi, \psi \in C^\infty(X)$ and $\operatorname{supp}\varphi \cap \operatorname{supp}\psi = \emptyset$. Then $A \in \boldsymbol{L}^{\mu,\varrho}(\mathbb{R} \times X)$ implies $\varphi A\psi \in \boldsymbol{L}^{-\infty,-\infty}(\mathbb{R} \times X)$.*

Remark 1.4.48 *The spaces* $L^{\mu,\varrho}(\mathbb{R} \times X)$ *and* $L^{\mu,\varrho}_{\mathrm{cl}}(\mathbb{R} \times X)$ *are Fréchet in a canonical way. Let us define the Fréchet topology of* $L^{\mu,\varrho}(\mathbb{R} \times X)$; *that of the subspace of classical operators can be introduced in an analogous manner. First,* $L^{-\infty,-\infty}(\mathbb{R} \times X)$ *is a Fréchet space, endowed with the topology from the bijection to* $\mathcal{S}(\mathbb{R} \times \mathbb{R}, C^{\infty}(X \times X))$. *Moreover,* $L^{\mu,\varrho}(\mathbb{R} \times U_j)$ *is Fréchet with the projective limit topology with respect to the mappings*

$$L^{\mu,\varrho}(\mathbb{R} \times U_j) \to L^{\mu}(\mathbb{R} \times U_j),$$

$$(\kappa_{j,\pm})_* : L^{\mu,\varrho}(\mathbb{R} \times U_j)|_{\mathbb{R}_{\pm} \times U_j} \to L^{\mu,\varrho}(\Omega_j^{\wedge}),$$

$j = 1, \ldots, N$, *using Remark 1.4.16 (cf. also Remark 1.4.32 in the classical case). Then we can set*

$$L^{\mu,\varrho}(\mathbb{R} \times X) = \sum_{j=1}^{N} [\varphi_j] L^{\mu,\varrho}(\mathbb{R} \times U_j)[\psi_j] + L^{-\infty,-\infty}(\mathbb{R} \times X)$$

endowed with the Fréchet topology of the non–direct sum (cf. the corresponding notation in Section 1.1.1). The topology is independent of the particular choice of $\{U_j\}$, $\{\varphi_j\}$, $\{\psi_j\}$ *and of the charts.*

We now turn to the definition of the weighted Sobolev spaces on $\mathbb{R} \times X$. Let us take the above system of charts $\chi_j : U_j \to \Omega_j$ for open $\Omega_j \subset S^n$ and consider the associated maps $\kappa_{j,\pm} : \mathbb{R}_{\pm} \times U_j \to \Omega_j^{\wedge} \subset \mathbb{R}^{n+1}$. Let us fix a cut–off function $\omega(t)$. Then $[\omega] H^s_{\mathrm{loc}}(\mathbb{R} \times X)$ is a well–defined (Hilbertisable) Banach space for every $s \in \mathbb{R}$.

Definition 1.4.49 *The space* $H^s_{\mathrm{cone}}(\mathbb{R} \times X)$, $s \in \mathbb{R}$, *is the closure of* $C_0^{\infty}(\mathbb{R} \times X)$ *with respect to the norm*

$$\|u\|_{H^s_{\mathrm{cone}}(\mathbb{R} \times X)}$$

$$= \left\{ \|\omega u\|^2_{[\omega] H^s_{\mathrm{loc}}(\mathbb{R} \times X)} + \sum_{j=1}^{N} \|(\kappa_{j,-}^{-1})^*(1-\omega)\varphi_j u\|^2_{H^s(\mathbb{R}^{n+1})} + \right.$$

$$\left. + \sum_{j=1}^{N} \|(\kappa_{j,+}^{-1})^*(1-\omega)\varphi_j u\|^2_{H^s(\mathbb{R}^{n+1})} \right\}^{\frac{1}{2}}.$$

Moreover, for arbitrary $s, \delta \in \mathbb{R}$, *we set*

$$H^{s,\delta}_{\mathrm{cone}}(\mathbb{R} \times X) = \{(1 + |t|^2)^{-\frac{\delta}{2}} v(t, x) : v \in H^s_{\mathrm{cone}}(\mathbb{R} \times X)\},$$

endowed with the norm

$$\|u\|_{H^{s,\delta}_{\mathrm{cone}}(\mathbb{R} \times X)} = \|(1 + |t|^2)^{\frac{\delta}{2}} u\|_{H^s_{\mathrm{cone}}(\mathbb{R} \times X)}.$$

It can easily be verified that the spaces $H^{s,\delta}_{\mathrm{cone}}(\mathbb{R} \times X)$ are independent of the particular choice of data such as the open covering $\{U_j\}$ of X, the partition of unity $\{\varphi_j\}$ and the charts $\{\chi_j\}$. We have

$$H^{0,0}_{\mathrm{cone}}(\mathbb{R} \times X) = (1 + |t|^2)^{-\frac{n}{4}} L^2(\mathbb{R} \times X),$$

with $L^2(\mathbb{R} \times X)$ being the space of square integrable functions on $\mathbb{R} \times X$ with respect to the measure $dt dx$.

Theorem 1.4.50 *We have canonical continuous embeddings*

$$H^{s',\delta'}_{\mathrm{cone}}(\mathbb{R} \times X) \hookrightarrow H^{s,\delta}_{\mathrm{cone}}(\mathbb{R} \times X)$$

for every $s' \geq s$, $\delta' \geq \delta$ that are compact for $s' > s$, $\delta' > \delta$.

The proof is a consequence of the corresponding results on weighted Sobolev spaces in $\mathbb{R}^{n+1}$, cf. Theorem 1.4.19.

Theorem 1.4.51 *Each $A \in L^{\mu,\varrho}(\mathbb{R} \times X)$ induces a continuous operator*

$$A : \mathcal{S}(\mathbb{R}, C^\infty(X)) \to \mathcal{S}(\mathbb{R}, C^\infty(X))$$

and

$$A : H^{s,\delta}_{\mathrm{cone}}(\mathbb{R} \times X) \to H^{s-\mu,\delta-\varrho}_{\mathrm{cone}}(\mathbb{R} \times X)$$

for every $s, \delta \in \mathbb{R}$.

This result follows in an obvious manner from Theorem 1.4.17 and from Definition 1.4.42.

Remark 1.4.52 *Let $\mu, \varrho \in \mathbb{R}$ be fixed and*

$$A : H^{s,\delta}_{\mathrm{cone}}(\mathbb{R} \times X) \to H^{s-\mu,\delta-\varrho}_{\mathrm{cone}}(\mathbb{R} \times X), \qquad (1.4.67)$$

an operator which is continuous for all $s, \delta \in \mathbb{R}$. Define the formal adjoint A^* and the transposed $\,^t A$ by

$$(Au, v)_{H^{0,0}_{\mathrm{cone}}(\mathbb{R} \times X)} = (u, A^* v)_{H^{0,0}_{\mathrm{cone}}(\mathbb{R} \times X)}$$

and

$$(Au, \overline{v})_{H^{0,0}_{\mathrm{cone}}(\mathbb{R} \times X)} = (u, \overline{\,^t A v})_{H^{0,0}_{\mathrm{cone}}(\mathbb{R} \times X)}$$

for all $u, v \in C_0^\infty(\mathbb{R} \times X)$. Then

$$A^* : H^{s,\delta}_{\mathrm{cone}}(\mathbb{R} \times X) \to H^{s-\mu,\delta-\varrho}_{\mathrm{cone}}(\mathbb{R} \times X)$$

and

$$^t A : H^{s,\delta}_{\mathrm{cone}}(\mathbb{R} \times X) \to H^{s-\mu,\delta-\varrho}_{\mathrm{cone}}(\mathbb{R} \times X)$$

are continuous for all $s, \delta \in \mathbb{R}$.

Theorem 1.4.53 $A \in \boldsymbol{L}^{\mu,\varrho}(\mathbb{R} \times X)$ implies $A^* \in \boldsymbol{L}^{\mu,\varrho}(\mathbb{R} \times X)$ and $^{t}A \in \boldsymbol{L}^{\mu,\varrho}(\mathbb{R} \times X)$ for every $\mu, \varrho \in \mathbb{R}$. In the case $A \in \boldsymbol{L}^{\mu,\varrho}_{\mathrm{cl}}(\mathbb{R} \times X)$ we have $A^*, {}^{t}A \in \boldsymbol{L}^{\mu,\varrho}_{\mathrm{cl}}(\mathbb{R} \times X)$, with the corresponding analogues of the symbolic rules from Theorem 1.4.30 and (1.4.40), (1.4.37).

Proof. Writing A in the form (1.4.63) it suffices to consider $\varphi_j A_j \psi_j$. We can apply the arguments for Theorem 1.4.12 and Theorem 1.4.30, in a slight (obvious) modification which show $(\varphi_j A_j \psi_j)^* \in \boldsymbol{L}^{\mu,\varrho}(\mathbb{R} \times U_j)$ for every j.   □

Proposition 1.4.54 Let (1.4.67) be an operator which is continuous for all $s, \delta \in \mathbb{R}$. Then the following conditions are equivalent:

(i)
$$A \in \boldsymbol{L}^{-\infty,-\infty}(\mathbb{R} \times X),$$

(ii) A and A^* induces continuous operators

$$A, A^* : H^{s,\delta}_{\mathrm{cone}}(\mathbb{R} \times X) \to \mathcal{S}(\mathbb{R}, C^\infty(X))$$

for all $s, \delta \in \mathbb{R}$.

Proof. The result follows easily by appropriate localisations of A in coordinate neighbourhoods from Proposition 1.4.21.   □

Theorem 1.4.55 $A \in \boldsymbol{L}^{\mu,\varrho}(\mathbb{R} \times X)$, $B \in \boldsymbol{L}^{\nu,\kappa}(\mathbb{R} \times X)$ implies $AB \in \boldsymbol{L}^{\mu+\nu,\varrho+\kappa}(\mathbb{R} \times X)$ for every $\mu, \varrho, \nu, \kappa \in \mathbb{R}$. For $A \in \boldsymbol{L}^{\mu,\varrho}_{\mathrm{cl}}(\mathbb{R} \times X)$, $B \in \boldsymbol{L}^{\nu,\kappa}_{\mathrm{cl}}(\mathbb{R} \times X)$ we have $AB \in \boldsymbol{L}^{\mu+\nu,\varrho+\kappa}_{\mathrm{cl}}(\mathbb{R} \times X)$, and $\sigma^{\mu+\nu}_\psi(AB) = \sigma^\mu_\psi(A)\sigma^\nu_\psi(B)$, $\sigma^{\varrho+\kappa}_e(AB) = \sigma^\varrho_e(A)\sigma^\kappa_e(B)$, $\sigma^{\mu,\varrho+\kappa}_{\psi,e}(AB) = \sigma^{\mu,\varrho}_{\psi,e}(A)\sigma^{\nu,\kappa}_{\psi,e}(B)$.

Proof. Writing

$$A = \sum_{j=1}^{N} \varphi_j A_j \psi_j + C, \qquad B = \sum_{k=1}^{N} \varphi_k A_k \psi_k + D$$

for $C, D \in \boldsymbol{L}^{-\infty,-\infty}(\mathbb{R} \times X)$, according to the notation in Definition 1.4.45, it suffices to show

$$(\varphi_j A_j \psi_j)D, C(\varphi_k B_k \psi_k) \in \boldsymbol{L}^{-\infty,-\infty}(\mathbb{R} \times X) \qquad (1.4.68)$$

for all j, k and $(\varphi_j A_j \psi_j)(\varphi_k B_k \psi_k) \in \boldsymbol{L}^{\mu+\nu,\varrho+\kappa}(\mathbb{R} \times X)$. The latter relation is a modification of Theorem 1.4.14 if we choose the covering $\{U_1, \ldots, U_N\}$ as above in Remark 1.4.46 and employ Proposition 1.4.43, cf. also Remark 1.4.47. The relation (1.4.68) follows from Theorems 1.4.51, 1.4.53 and Proposition 1.4.54.   □

Definition 1.4.56 *An operator* $A \in \boldsymbol{L}^{\mu,\varrho}(\mathbb{R} \times X)$ *is called elliptic of order* (μ, ϱ) *if it is elliptic of order* μ *in the class* $L^{\mu}(\mathbb{R} \times X)$, *cf. Definition 1.1.67, and if the operators* A_j *in Definition 1.4.45 are elliptic in the sense that their local representatives* $(\kappa_{j,\pm})_*(A_j|_{\mathbb{R}_\pm \times U_j}) \in \boldsymbol{L}^{\mu,\varrho}(\Omega_j^\wedge)$ *are elliptic in* $\Omega_j^\wedge$ *of order* (μ, ϱ) *for all* j, *cf. Remark 1.4.44.*

Remark 1.4.57 *The ellipticity of an operator* $A \in \boldsymbol{L}_{\mathrm{cl}}^{\mu,\varrho}(\mathbb{R} \times X)$ *is equivalent to the following conditions:*

$$\sigma_\psi^\mu(A) \neq 0 \qquad \text{on } T^*(\mathbb{R} \times X) \setminus 0,$$
$$\sigma_e^\varrho(A) \neq 0 \qquad \text{on } T^*(\mathbb{R} \times X) \setminus \{t = 0\},$$
$$\sigma_{\psi,e}^{\mu,\varrho}(A) \neq 0 \qquad \text{on } T^*(\mathbb{R} \times X) \setminus (0 \cup \{t = 0\}),$$

cf. (1.4.64), (1.4.65), (1.4.66). This is a consequence of Proposition 1.4.37.

Definition 1.4.58 *An operator* $B \in \boldsymbol{L}^{-\mu,-\varrho}(\mathbb{R} \times X)$ *is called a parametrix of* $A \in \boldsymbol{L}^{\mu,\varrho}(\mathbb{R} \times X)$ *if* $AB - 1, BA - 1 \in \boldsymbol{L}^{-\infty,-\infty}(\mathbb{R} \times X)$.

Proposition 1.4.59 *For every* $s, \delta \in \mathbb{R}$ *there exists an operator* $A_{s,\delta} \in \boldsymbol{L}_{\mathrm{cl}}^{s,\delta}(\mathbb{R} \times X)$ *which is elliptic of order* (s, δ).

This is obvious by Example 1.4.35 and Definition 1.4.56.

Theorem 1.4.60 *Let* $A \in \boldsymbol{L}^{\mu,\varrho}(\mathbb{R} \times X)$ *for* $\mu, \varrho \in \mathbb{R}$ *be given. Then the following conditions are equivalent:*

(i) *A is elliptic of order* (μ, ϱ).

(ii) *The operator*

$$A : H_{\mathrm{cone}}^{s,\delta}(\mathbb{R} \times X) \to H_{\mathrm{cone}}^{s-\mu,\delta-\varrho}(\mathbb{R} \times X) \qquad (1.4.69)$$

is Fredholm for certain reals $s = s_0$, $\delta = \delta_0$.

Moreover, the ellipticity of A implies the Fredholm property of (1.4.69) for all $s, \delta \in \mathbb{R}$, *and A has a parametrix* $B \in \boldsymbol{L}^{-\mu,-\varrho}(\mathbb{R} \times X)$. *Finally* $Au \in H_{\mathrm{cone}}^{s,\delta}(\mathbb{R} \times X)$ *for arbitrary fixed* $s, \delta \in \mathbb{R}$ *and* $u \in H_{\mathrm{cone}}^{-\infty,-\infty}(\mathbb{R} \times X)$ *imply* $u \in H_{\mathrm{cone}}^{s+\mu,\delta+\varrho}(\mathbb{R} \times X)$.

We shall show Theorem 1.4.60 in several steps. The assertion (ii) $\Rightarrow$ (i) will be proved only for operators in $\boldsymbol{L}_{\mathrm{cl}}^{\mu,\varrho}(\mathbb{R} \times X)$ which is the situation in the applications below.

Proof of Theorem 1.4.60. Suppose $A \in \boldsymbol{L}^{\mu,\varrho}(\mathbb{R} \times X)$ is elliptic of order (μ, ϱ). We shall first construct a parametrix $B \in \boldsymbol{L}^{-\mu,-\varrho}(\mathbb{R} \times X)$. Writing the operator A in the form (1.4.63), where we may assume $A_j|_{\mathbb{R} \times (U_j \cap U_k)} - A_k|_{\mathbb{R} \times (U_j \cap U_k)} \in \boldsymbol{L}^{-\infty,-\infty}(\mathbb{R} \times (U_j \cap U_k))$, it suffices to construct parametrices $B_j \in \boldsymbol{L}^{-\mu,-\varrho}(\mathbb{R} \times U_j)$ for all j. Then $B = \sum_{j=1}^{N} \varphi_j B_j \psi_j$ is a parametrix of A. The operators B_j can be constructed in local coordinates. On $(-c, c) \times U_j$ for arbitrary $c > 0$ we can apply the standard arguments to construct a parametrix for an operator in $L^\mu((-c, c) \times U_j)$ of order μ. Therefore it remains to construct B_j over $(-\infty, -c) \times U_j$ and $(c, \infty) \times U_j$. However this can be done in a similar manner as in Theorem 1.4.36. The only new aspect here is that we have to consider a set in $\mathbb{R}^{n+1}$ which is conical for large $|\tilde{x}|$ instead of $\mathbb{R}^{n+1}$ itself. The corresponding modifications of the arguments are trivial. Also the last statement of Theorem 1.4.60 follows from the scheme to prove the corresponding elliptic regularity of Theorem 1.1.71. From the existence of a parametrix B to A and from the compactness of the operators $BA - 1$, $AB - 1$ in the spaces $H_{\text{cone}}^{s,\delta}(\mathbb{R} \times X)$ for all $s, \delta \in \mathbb{R}$, cf. Theorems 1.4.50, 1.4.51, we get the Fredholm property of (1.4.69) for all $s, \delta \in \mathbb{R}$.

Now let $A \in \boldsymbol{L}_{\text{cl}}^{\mu,\varrho}(\mathbb{R} \times X)$ be given such that (1.4.69) is a Fredholm operator for certain $s, \delta \in \mathbb{R}$. Using Proposition 1.4.59 and the result of the first part of this proof we obtain that also

$$A_0 = A_{s-\mu,\delta-\varrho} A A_{-s-\delta} : H_{\text{cone}}^{0,0}(\mathbb{R} \times X) \to H_{\text{cone}}^{0,0}(\mathbb{R} \times X)$$

is a Fredholm operator. From Theorem 1.4.55 we see that the ellipticity of A is equivalent to that of A_0. Thus it suffices to consider A_0. For simplicity denote A_0 again by A. In view of the Fredholm property of A there is an operator $P \in \mathcal{L}(H_{\text{cone}}^{0,0}(\mathbb{R} \times X))$, with $PA = 1 + K$ for some compact operator K in $H_{\text{cone}}^{0,0}(\mathbb{R} \times X)$. Now we can write $u = PAu - Ku$ for every $u \in H_{\text{cone}}^{0,0}(\mathbb{R} \times X)$. In particular, take a function u with compact support and supported with respect to x in a coordinate neighbourhood on X, with $u(t_0, x_0) = 1$ for a given $(t_0, x_0) \in \mathbb{R} \times X$. Moreover, let $\varphi, \psi \in C_0^\infty(\mathbb{R} \times U)$ be functions satisfying $\varphi u = u$ and $\varphi \psi = \varphi$. Then

$$u = \varphi P A u - \varphi K u = \tilde{P} \tilde{A} u + C u \qquad (1.4.70)$$

for $\tilde{P} = \varphi P \psi$, $\tilde{A} = \varphi A \varphi$ and $C = \varphi P(1 - \psi)A\varphi - \varphi K\varphi$, where C is compact. The relation (1.4.70) can be transformed in local coordinates in $\mathbb{R}^{n+1}$ by a map $1 \times \chi : \mathbb{R} \times U \to \mathbb{R} \times \mathbb{R}^n$ for a chart $\chi : U \to \mathbb{R}^n$. In order to prove the ellipticity of A with respect to σ_ψ^0 and σ_ϱ^0 we can apply analogous conclusions as in the corresponding parts of the proof of Theorem 1.4.40. □

Chapter 2

Mellin pseudo–differential operators on manifolds with conical singularities

This chapter studies algebras of pseudo–differential operators on manifolds with conical singularities. These so–called cone algebras (with discrete and continuous asymptotics) contain the algebra of differential operators of Fuchs type. Ellipticity with respect to the interior and the conormal symbol implies the existence of a parametrix in the cone algebra. Globally we get Fredholm operators in weighted cone Sobolev spaces. Inverses of cone operators that are bijective between the weighted Sobolev spaces belong again to the cone algebras. The symbolic structure reflects the asymptotics of solutions to elliptic equations, dependent on the asymptotics of the right hand sides and on the meromorphic structure of the inverse of the principal conormal symbol. (The case of differential operators and discrete asymptotics was treated by Kondrat'ev [58]). The cone algebra in this form relies on a suitable Mellin operator convention for symbols of Fuchs type that are smooth up to the vertex, cf. Schulze [118], and on a control of residual operators of smoothing Mellin and Green type with asymptotics, cf. Rempel, Schulze [95], [96] and Schulze [114]. Those residual operators form ideals in the cone algebras. They are compact in the weighted Sobolev spaces when their principal conormal symbol vanishes. For the edge algebras below in Chapter 3 they become essential in non–compact smoothing contributions when they appear as values of operator–valued symbols. Other methods to treat Fuchs pseudo–differential operators are due to Plamenevskij [85], [86] and Melrose, Mendoza [78] (in the framework of the "b–calculus" of Melrose [76]).

2.1 The typical differential operators

2.1.1 Manifolds with conical singularities

Let X be a topological space. Then

$$X^\Delta := (\overline{\mathbb{R}}_+ \times X)/(\{0\} \times X) \qquad (2.1.1)$$

is the infinite (topological) cone with base X . The vertex of (2.1.1) corresponds to $\{0\} \times X$, identified with a point. With X we also associate the open stretched cone

$$X^\wedge := \mathbb{R}_+ \times X \ni (t, x). \qquad (2.1.2)$$

We assume here that X is a closed compact C^∞ manifold.

If X is given as a submanifold of the unit sphere S^{N-1} of $\mathbb{R}^N$, then

$$X^\Delta = \left\{ \tilde{x} \in \mathbb{R}^N : \ \tilde{x} = 0 \text{ or } \tilde{x}/|\tilde{x}| \in X \right\} \qquad (2.1.3)$$

is homeomorphic to (2.1.1). In particular, $(S^{N-1})^\Delta = \mathbb{R}^N$.

Definition 2.1.1 *A manifold B with conical singularities*

$$S = \{b_1, \ldots, b_M\} \subset B$$

is a topological space with the following properties:

(i) *$B \setminus S$ is a C^∞ manifold.*

(ii) *Each $b \in S$ has a neighbourhood V in B such that for some closed compact C^∞ manifold $X = X(b)$ there is given a non–empty system $\Phi(V)$ of homeomorphisms*

$$\varphi : V \to X^\Delta.$$

(iii) *Every $\varphi \in \Phi(V)$ restricts to a diffeomorphism*

$$\varphi_0 : V \setminus \{b\} \to X^\wedge,$$

and every two $\varphi, \psi \in \Phi(V)$ satisfy $\varphi_0 \psi_0^{-1} = \delta|_{X^\wedge}$ for some diffeomorphism $\delta : \mathbb{R} \times X \to \mathbb{R} \times X$.

The elements of S are called the conical singularities (conical points) of B.

Remark 2.1.2 *The notation manifold in Definition 2.1.1 is chosen for convenience. B is not a C manifold in the topological sense unless the bases of the cones near the conical singularities are $\cong$ spheres.*

Note that the system of "charts" in Definition 2.1.1 (ii), (iii) pre-scribes the cone geometry near the singularities in a way that excludes cusp singularities relative to the fixed charts.

Remark 2.1.3 *The systems Φ_j of homeomorphisms to $b_j \in S$ in Def-inition 2.1.1 (ii) may be taken maximal. There are different non-equivalent choices of such systems for the same B, first being given as a topological space, where $B \setminus S$ is a C^∞ manifold for a finite sub-set S. For instance, for $B = \mathbb{R}_+$ with $S = \{0\}$, we may have an infinite number of homeomorphisms from $V := [0, \varepsilon)$ to $\overline{\mathbb{R}}_+$ which are never equivalent. E.g., if $\varphi(t) = t^\gamma$, $\psi(t) = t^\delta$ for $0 \leq t \leq \varepsilon/2$, then $\varphi_0 \psi_0^{-1} = t^{\gamma - \delta}$ is smooth up to $t = 0$ only when $\gamma = \delta$.*

Let B be a manifold with conical singularities $S = \{b_1, \ldots, b_M\}$ and denote by Φ_j the above system of homeomorphisms belonging to b_j, $j = 1, \ldots, M$. Let us fix $\varphi_j \in \Phi_j$ for every j and set $W = S \cup \bigcup_{j=1}^{M}(\varphi_j)_0^{-1}((0, 1+\varepsilon) \times X(b_j))$ with some $\varepsilon > 0$. Then $B \setminus W$ is a C^∞ manifold with boundary $\cong \bigcup_{j=1}^{M} X(b_j)$. By construction the difference between $B \setminus S$ and $B \setminus W$ is the union of cylinders $(\varphi_j)_0^{-1}((0, 1+\varepsilon) \times X(b_j))$. Topologically the cylinders can be identified with $((0, 1+\varepsilon) \times X(b_j))$. This allows us to form

$$\mathbb{B} := (B \setminus W) \cup_\varphi \bigcup_{j=1}^{M} ([0, 1+\varepsilon] \times X(b_j)),$$

where $\cup_\varphi$ indicates the union together with the identification of $\{1 + \varepsilon\} \times X(b_j)$ via $(\varphi_j)_0^{-1}$ with the component $X(b_j)$ of $\partial(B \setminus W)$ for $j = 1, \ldots, M$. We endow $\mathbb{B}$ with the structure of a C^∞ manifold with boundary in an obvious manner. Moreover, we fix a Riemannian metric on $\mathbb{B}$ that equals the product metric of $[0, 1) \times \partial\mathbb{B}$ in a collar neighbourhood of $\partial\mathbb{B} \cong \bigcup_{j=1}^{M} X(b_j)$.

$\mathbb{B}$ is called the stretched manifold to B. It is independent of the particular choice of φ_j, $j = 1, \ldots, M$ and ε, up to diffeomorphisms. Clearly $\mathbb{B}$ is diffeomorphic to $B \setminus W$.

From the construction we can find a continuous map

$$\pi : \mathbb{B} \to B \tag{2.1.4}$$

that restricts to a diffeomorphism

$$\pi_0 : \mathbb{B} \setminus \partial\mathbb{B} \to B \setminus S \tag{2.1.5}$$

and to the map $\pi' : \partial\mathbb{B} \to S$ defined by $\pi'(X_j) = b_j$ for the component $X_j \cong X(b_j)$ of $\partial\mathbb{B}$, $j = 1, \ldots, M$.

Without loss of generality we assume that the diffeomorphism π_0 is compatible with the diffeomorphisms in Definition 2.1.1 (ii) in the following sense. For every neighbourhood V of $b \in S$ in the above definition we have the system Φ of diffeomorphisms $\varphi : V \setminus \{b\} \to X^\wedge$. For every $\delta > 0$ there is a smaller neighbourhood V_δ of b given by $V_\delta = \varphi_0^{-1}((0, \delta) \times X) \cup \{b\}$. The assumption is then that π_0^{-1} induces a map $\psi_\delta : V_\delta \setminus \{b\} \to (0, \varepsilon) \times X$ for some $0 < \varepsilon < 1$ with suitable $\delta = \delta(\varepsilon)$ and that ψ_δ extends to an element $\psi \in \Phi$.

Example 2.1.4 *Let $B = X^\Delta$ for a closed compact C^∞ manifold X. Then $\mathbb{B} = \overline{\mathbb{R}}_+ \times X$. In particular, if $\dim X = 0$, then $B = \overline{\mathbb{R}}_+$ coincides with $\mathbb{B}$. On the other hand if we consider*

$$B := \left\{ x = (x_1, x_2) \in \mathbb{R}^2 : \{0 \leq x_1 \leq a \text{ and } x_2 = 0\} \right.$$
$$\left. \text{or } \{0 \leq x_2 \leq b \text{ and } x_1 = 0\} \right\}$$

then B has three conical singularities $\{(0,0), (a,0), (0,b)\}$ and $\mathbb{B}$ consists of two disjoint finite intervals.

Remark 2.1.5 *It may happen that for two manifolds B and $\tilde{B}$ with conical singularities we have $\mathbb{B} \cong \tilde{\mathbb{B}}$. This is the case, for instance, if we pass from B to $\tilde{B} = B/S$. Then $\tilde{B}$ has only one conical singularity $\tilde{S} = \{b\}$ and the base of the cone near $\{b\}$ for $\tilde{B}$ is $\bigcup_{j=1}^N X(b_j)$. Since the analysis for B will be performed on $\mathbb{B} \setminus S \cong \tilde{B} \setminus \tilde{S}$ or likewise on $\text{int } \mathbb{B}$ it is often sufficient to assume that there is only one conical point.*

Remark 2.1.6 *Every C^∞ manifold $\mathbb{B}$ with compact boundary can be regarded as the stretched manifold of some manifold B with conical singularities. In fact, it suffices to set $B = \mathbb{B}/\partial \mathbb{B}$.*

Let us finally introduce morphisms in the category of manifolds with conical singularities. Recall that C^∞ manifolds in consideration are always supposed to be paracompact.

Definition 2.1.7 *Let B and C be manifolds with conical singularities S and T, respectively, identified with pairs (B, S), (C, T), and let $\mathbb{B}$ and $\mathbb{C}$ be the associated stretched manifolds with the corresponding diffeomorphisms $\pi_0 : \mathbb{B} \setminus \partial \mathbb{B} \to B \setminus S$ and $\rho_0 : \mathbb{C} \setminus \partial \mathbb{C} \to C \setminus T$, respectively, cf. (2.1.5). Then a continuous map $\alpha : B \to C$ is a morphism if it has the following properties:*

(i) *α restricts to a differentiable map $\alpha_0 : B \setminus S \to C \setminus T$.*

(ii) *α induces a map $\alpha' : S \to T$.*

(iii) $\kappa_0 := \rho_0^{-1}\alpha_0\pi_0 : \mathbb{B} \setminus \partial\mathbb{B} \to \mathbb{C} \setminus \partial\mathbb{C}$ *is the restriction of a differentiable map* $\kappa : \mathbb{B} \to \mathbb{C}$ *(i.e.,* κ *is smooth up to the boundary).*

In particular, if the inverse mapping $\alpha^{-1} : C \to B$ exists with analogous properties, then α is called an isomorphism between the corresponding manifolds with conical singularities.

2.1.2 Differential operators of Fuchs type

This section studies the class of those differential operators that play the role of the typical ones in a calculus on manifolds with conical singularities.

If Ω is a C^∞ manifold we denote by

$$\mathrm{Diff}^\mu(\Omega) \tag{2.1.6}$$

the space of all differential operators on Ω of order μ with smooth coefficients in local coordinates. (2.1.6) is a Fréchet space in a natural way.

Definition 2.1.8 *Let B be a manifold with conical singularities. Then*

$$\mathrm{Diff}^\mu(B)_{\mathrm{Fuchs}} \tag{2.1.7}$$

denotes the subspace of all $A \in \mathrm{Diff}^\mu(B \setminus S)$ with the following properties: For every neighbourhood V of $b \in S$, cf. Definition 2.1.1, the operator A has the form

$$A = t^{-\mu} \sum_{j=0}^{\mu} a_j(t) \left(-t\frac{\partial}{\partial t}\right)^j \tag{2.1.8}$$

in the variables $(t,x) \in X^\wedge$, with $a_j(t) \in C^\infty(\overline{\mathbb{R}}_+, \mathrm{Diff}^{\mu-j}(X))$, $j = 0,\ldots,\mu$. The elements of (2.1.7) are called differential operators of Fuchs type.

Since the calculations refer to int $\mathbb{B}$ for the stretched manifold $\mathbb{B}$ to B we also employ the notation

$$\mathrm{Diff}^\mu(\mathbb{B})_{\mathrm{Fuchs}} \tag{2.1.9}$$

for the class of all $A \in \mathrm{Diff}^\mu(\mathrm{int}\,\mathbb{B})$ which are of the form (2.1.8) in a collar neighbourhood $V \cong X^\wedge$ of the component of $\partial\mathbb{B}$ corresponding to X, and this is required for all components of $\partial\mathbb{B}$.

Remark 2.1.9 *Let X be closed compact and $\chi : X^\wedge \to X^\wedge$ be a diffeomorphism which is the restriction of a diffeomorphism $\tilde{\chi} : \mathbb{R} \times X \to \mathbb{R} \times X$ to $\tilde{\chi}$. As usual, χ_* denotes the operator push–forward under χ. Then χ_* induces an isomorphism*

$$\chi_* : \mathrm{Diff}^\mu(\overline{X}^\wedge)_{\mathrm{Fuchs}} \to \mathrm{Diff}^\mu(\overline{X}^\wedge)_{\mathrm{Fuchs}}.$$

Analoguosly if $\chi : \mathbb{B} \to \mathbb{D}$ is a diffeomorphism between compact C^∞ manifolds with boundary, $\mathbb{B}$ and $\mathbb{D}$ being interpreted as the stretched manifolds to manifolds B and D, respectively, with conical singularities, then the operator push-forward induces an isomorphism

$$\chi_* : \mathrm{Diff}^\mu(\mathbb{B})_{\mathrm{Fuchs}} \to \mathrm{Diff}^\mu(\mathbb{D})_{\mathrm{Fuchs}}$$

for every $\mu \in \mathbb{N}$.

Let us fix a strictly positive function $h^\gamma \in C^\infty(\mathrm{int}\,\mathbb{B})$, $\gamma \in \mathbb{R}$, such that

$$h^\gamma = t^\gamma \qquad \mathrm{in} \qquad V \setminus \partial\mathbb{B} \qquad\qquad (2.1.10)$$

for the collar neighbourhood V of $\partial\mathbb{B}$ diffeomorphic to $[0,1) \times \partial\mathbb{B}$.

Next we turn to some examples of differential operators of Fuchs type.

Example 2.1.10 *Let v be an arbitrary C^∞ vector field on $\mathbb{B}$ that is tangential to $\partial\mathbb{B}$, regarded as a first order differential operator. Then $h^{-1}v \in \mathrm{Diff}^1(\mathbb{B})_{\mathrm{Fuchs}}$.*

Example 2.1.11 *Let $\tilde{A} \in \mathrm{Diff}^\mu(\mathbb{R}^{n+1})$ and express $\tilde{A}|_{\mathbb{R}^{n+1}\setminus\{0\}}$ in polar coordinates $(t,x) \in \mathbb{R}_+ \times S^n$, $t = |\tilde{x}|$, $S^n = \{\tilde{x} \in \mathbb{R}^{n+1} : |\tilde{x}| = 1\}$. Then, for the resulting operator, we get an element in $\mathrm{Diff}^\mu(\overline{\mathbb{R}}_+ \times S^\mu)_{\mathrm{Fuchs}}$.*

A Riemannian metric g on X corresponds to a C^∞ section in $TX \otimes TX$. This is locally an x–dependent family of symmetric positive definite $n \times n$–matrices, $n = \dim X$. The C^∞ sections in a vector bundle form a Fréchet space in a natural way. In particular, we can talk about C^∞ functions in $t \in \overline{\mathbb{R}}_+$ with values in Riemannian metrics on X.

Example 2.1.12 *Let $g(t)$ be a t–dependent family of Riemannian metrics on X, C^∞ in $t \in \overline{\mathbb{R}}_+$. Then the Laplace-Beltrami operator to the Riemannian metric $dt^2 + t^2 g(t)$ on $X^\wedge$ belongs to $\mathrm{Diff}^2(\overline{\mathbb{R}}_+ \times X)_{\mathrm{Fuchs}}$.*

Let us now turn to natural scales of weighted Sobolev spaces here, for integer smoothness. The general case will be considered in Section 2.1.4 below.

On X we fixed a Riemannian metric with an associated measure dx. Let $L^2(X^\wedge)$ be the space of square integrable functions on $X^\wedge = \mathbb{R}_+ \times X$ with respect to $dt\, dx$. Then, for $s \in \mathbb{N}$, $\gamma \in \mathbb{R}$, $n = \dim X$, we define

$$\mathcal{H}^{s,\gamma}(X^\wedge) = \left\{ u \in \mathcal{D}'(X^\wedge) : \left(t\frac{\partial}{\partial t}\right)^{\alpha_0} v^\alpha u \in t^{\gamma - \frac{n}{2}} L^2(X^\wedge) \text{ for arbitrary} \right.$$

$$\alpha_0 \in \mathbb{N}, \ \alpha \in \mathbb{N}^n, \ \alpha_0 + |\alpha| \le s, \text{ and arbitrary } C^\infty \text{ vector}$$

$$\left. \text{fields } v_1, \ldots, v_n \text{ on } X, \ v^\alpha = v_1^{\alpha_1} \cdot \ldots \cdot v_n^{\alpha_n} \right\}.$$

It can easily be proved that $\mathcal{H}^{s,\gamma}(X^\wedge)$ is a Banach space with a norm generated by a Hilbert space scalar product. The space $C_0^\infty(X^\wedge)$ is dense in $\mathcal{H}^{s,\gamma}(X^\wedge)$.

Remark 2.1.13 *The space $\mathcal{H}^{s,\gamma}(X^\wedge)$ for $s \in \mathbb{N}$, $\gamma \in \mathbb{R}$ consists of all $u \in \mathcal{D}'(X^\wedge)$ such that for every operator $A_k = \sum_{j=0}^k a_{kj}\left(-t\frac{\partial}{\partial t}\right)^j$ with $a_{kj} \in \mathrm{Diff}^{k-j}(X)$ we have $A_k u \in t^{\gamma-\frac{n}{2}}L^2(X^\wedge)$, $0 \le k \le s$. In particular, if $X^\wedge = \mathbb{R}_+ \times S^n$ is identified with $\mathbb{R}^{n+1} \setminus \{0\}$ we also write $\mathcal{H}^{s,\gamma}(\mathbb{R}^{n+1} \setminus \{0\})$. In view of Example 2.1.11 this is then the space of all $u \in \mathcal{D}'(\mathbb{R}^{n+1} \setminus \{0\})$ such that*

$$|\tilde{x}|^\alpha D_{\tilde{x}}^\alpha u(\tilde{x}) \in |\tilde{x}|^\gamma L^2(\mathbb{R}^{n+1})$$

for all $\alpha \in \mathbb{N}^{n+1}$ with $|\alpha| \le s$. Here it was used that $L^2(\mathbb{R}^{n+1}) = t^{-\frac{n}{2}}L^2(\mathbb{R}_+ \times S^n)$ for $t = |\tilde{x}|$.

The scalar product of $\mathcal{H}^0(X^\wedge) := \mathcal{H}^{0,0}(X^\wedge)$ will be regarded as a reference scalar product for pairings. Let us set for $s \in \mathbb{N}$, $\gamma \in \mathbb{R}$

$$\|u\|_{\mathcal{H}^{-s,-\gamma}(X^\wedge)} = \sup |(u,w)_{\mathcal{H}^0(X^\wedge)}|/\|w\|_{\mathcal{H}^{s,\gamma}(X^\wedge)}$$

with sup being taken over all $w \in C_0^\infty(X^\wedge)$, $w \ne 0$. By definition $\mathcal{H}^{-s,-\gamma}(X^\wedge)$ is then the closure of $C_0^\infty(X^\wedge)$ with respect to this norm. The scalar product

$$(\cdot, \cdot)_{\mathcal{H}^0(X^\wedge)} : C_0^\infty(X^\wedge) \times C_0^\infty(X^\wedge) \to \mathbb{C}$$

extends to a non–degenerate sesquilinear pairing

$$(\cdot, \cdot) : \mathcal{H}^{-s,-\gamma}(X^\wedge) \times \mathcal{H}^{s,\gamma}(X^\wedge) \to \mathbb{C}.$$

Moreover, let $\mathbb{B}$ be a compact C^∞ manifold with boundary, regarded as the stretched manifold of some manifold B with conical singularities.

Then we have first the space $L^2(\mathbb{B})$ of square integrable functions on $\mathbb{B}$, based on a Riemannian metric on $\mathbb{B}$ which is the product metric of $[0,1) \times \partial\mathbb{B}$ in a collar neighbourhood of $\partial\mathbb{B}$. We then set $\mathcal{H}^0(\mathbb{B}) = h^{-\frac{n}{2}}L^2(\mathbb{B})$, $n = \dim \partial\mathbb{B}$, and more generally, for $s \in \mathbb{N}$, $\gamma \in \mathbb{R}$

$$\mathcal{H}^{s,\gamma}(\mathbb{B}) = \{u \in \mathcal{D}'(\mathrm{int}\,\mathbb{B}) : v_1 \cdots\cdots v_s u \in h^\gamma \mathcal{H}^0(\mathbb{B})\,\text{for arbitrary}$$
$$C^\infty\text{ vector fields } v_j \text{ on } \mathbb{B} \text{ that are tangent to } \partial\mathbb{B}\}.$$

Concerning $\mathcal{H}^0(\mathbb{B})$ we can make analogous remarks as before. In particular, $C_0^\infty($
$B)$ is dense in $\mathcal{H}^{s,\gamma}(\mathbb{B})$, and we can introduce these spaces also for $-s \in \mathbb{N}$, $\gamma \in \mathbb{R}$.

Throughout this exposition a cut-off function $\omega(t)$ on $\overline{\mathbb{R}}_+$ is any real-valued $C_0^\infty(\overline{\mathbb{R}}_+)$ function that equals 1 in a neighbourhood of $t = 0$.

Theorem 2.1.14 *Let $\omega(t)$, $\tilde\omega(t)$ be arbitrary cut-off functions. Then, for every $a \in \mathrm{Diff}^\mu(X^\wedge)_{\mathrm{Fuchs}}$*

$$\omega A \tilde\omega : \mathcal{H}^{s,\gamma}(X^\wedge) \to \mathcal{H}^{s-\mu,\gamma-\mu}(X^\wedge)$$

is continuous for every $s \in \mathbb{Z}$, $\gamma \in \mathbb{R}$. Analogously every element $A \in \mathrm{Diff}^\mu(\mathbb{B})_{\mathrm{Fuchs}}$ induces continuous operators

$$A : \mathcal{H}^{s,\gamma}(\mathbb{B}) \to \mathcal{H}^{s-\mu,\gamma-\mu}(\mathbb{B})$$

for all $s \in \mathbb{Z}$, $\gamma \in \mathbb{R}$.

The simple proof will be omitted here. We will return below to results of this kind for arbitrary $s \in \mathbb{R}$ and general pseudo-differential operators of Fuchs type.

We now define the principal symbolic structure of a differential operator A of Fuchs type. First we have the homogeneous principal symbol of A of order μ

$$\sigma_\psi^\mu(A) \in S^{(\mu)}(T^*(\mathrm{int}\,\mathbb{B}) \setminus 0).$$

In local coordinates (t,x) near $\partial\mathbb{B}$ it is of the form

$$\sigma_\psi^\mu(A)(t,x,\tau,\xi) = t^{-\mu}a_{(\mu)}(t,x,\tilde\tau,\xi)|_{\tilde\tau=t\tau},$$

where $a_{(\mu)}$ is homogeneous in $(\tilde\tau,\xi) \neq 0$ of order μ and C^∞ in $t \in \overline{\mathbb{R}}_+$.

Let $T_b^*(\mathrm{int}\,\mathbb{B})$ be the compressed cotangent bundle of $\mathrm{int}\,\mathbb{B}$ which is uniquely determined by the condition that the C^∞-sections in $T_b^*(\mathrm{int}\,\mathbb{B})$ are 1-forms on $\mathrm{int}\,\mathbb{B}$ which are in local coordinates (t,x) near $\partial\mathbb{B}$ of the form

$$t^{-1}p_0(t,x)dt + \sum_{j=1}^n p_j(t,x)dx_j$$

for functions $p_0, \ldots, p_n$ that are C^∞ in (t, x) up to $t = 0$. Note that the sections in the dual bundle $T_b(\text{int } \mathbb{B})$ are just the vector fields on $\mathbb{B}$ that are tangent to $\partial\mathbb{B}$, cf. Example 2.1.10.

Now $h^\mu \sigma_\psi^\mu(A)$ can be interpreted as a function

$$\sigma_{\psi,b}^\mu(A) \in S^{(\mu)}(T_b^*(\text{int } \mathbb{B}) \setminus 0)$$

which is smooth up to $\partial\mathbb{B}$ (recall that the weight function h^μ is kept fixed).

If $A \in \text{Diff}^\mu(\mathbb{B})_{\text{Fuchs}}$ is written as (2.1.8) near $\partial\mathbb{B}$ we set

$$\sigma_M^\mu(A)(z) = \sum_{j=0}^{\mu} a_j(0) z^j \qquad (2.1.11)$$

for $z \in \mathbb{C}$. We call (2.1.11) the conormal symbol of A of conormal order μ. It can be regarded as

$$\sigma_M^\mu(A) \in \mathcal{A}(\mathbb{C}, \text{Diff}^\mu(X)),$$

though below it will often be restricted to suitable parallels of the imaginary axis.

Remark 2.1.15 *The symbols $\sigma_{\psi,b}^\mu(A)$ and $\sigma_M^\mu(A)$ depend to some extent on the local coordinates of $\mathbb{B}$ near $\partial\mathbb{B}$. The invariance can easily be described in precise form. Without loss of generality, for notational convenience, we will assume in the sequel that the charts for $\mathbb{B}$ in a collar neighbourhood of $\partial\mathbb{B}$ map to $\overline{\mathbb{R}}_+ \times U$ for open $U \subseteq \mathbb{R}^n$ and that the transition diffeomorphisms are independent of t for $0 \leq t < \varepsilon$ for some $\varepsilon > 0$. Then we have the invariance of $\sigma_{\psi,b}^\mu(A)$, $\sigma_M^\mu(A)$.*

Theorem 2.1.16 $A \in \text{Diff}^\mu(\mathbb{B})_{\text{Fuchs}}$, $B \in \text{Diff}^\nu(\mathbb{B})_{\text{Fuchs}}$ *implies* $AB \in \text{Diff}^{\mu+\nu}(\mathbb{B})_{\text{Fuchs}}$, *and we have*

$$\sigma_\psi^{\mu+\nu}(AB) = \sigma_\psi^\mu(A)\sigma_\psi^\nu(B),$$
$$\sigma_{\psi,b}^{\mu+\nu}(AB) = \sigma_{\psi,b}^\mu(A)\sigma_{\psi,b}^\nu(B),$$
$$\sigma_M^{\mu+\nu}(AB)(z) = \sigma_M^\mu(A)(z+\nu)\sigma_M^\nu(B)(z).$$

The proof of this theorem is obvious as well as that of the result that $\text{Diff}^\mu(\mathbb{B})_{\text{Fuchs}}$ is preserved under formal adjoints with respect to the scalar product of $\mathcal{H}^0(\mathbb{B})$. We will return below to analogous properties in the pseudo–differential set–up.

The pseudo–differential operators of Fuchs type will be necessary to describe parametrices of elliptic differential operators in $\text{Diff}^\mu(\mathbb{B})_{\text{Fuchs}}$.

The ellipticity here relies on the condition of non–vanishing of $\sigma_\psi^\mu(A)$ and of $\sigma_{\psi,b}^\mu(A)$ up to $\partial\mathbb{B}$. In addition the operators

$$\sigma_M^\mu(A)(z) : H^s(X) \to H^{s-\mu}(X)$$

have to be isomorphisms for all z on a fixed parallel to the imaginary axis and for some $s \in \mathbb{R}$ (which implies then the same for all $s \in \mathbb{R}$).

2.1.3   The Mellin transform

The Mellin transform is defined by the formula

$$Mu(z) = \int_0^\infty t^{z-1} u(t)\, dt. \qquad (2.1.12)$$

Here we assume first $u \in C_0^\infty(\mathbb{R}_+)$ and $z \in \mathbb{C}$. Then (2.1.12) represents a continuous operator

$$M : C_0^\infty(\mathbb{R}_+) \to \mathcal{A}(\mathbb{C}). \qquad (2.1.13)$$

Below we shall extend M to more general distribution spaces on $\mathbb{R}_+$. Then z is restricted to a suitable subset of $\mathbb{C}$, e.g., to

$$\Gamma_\beta = \{z \in \mathbb{C} : \ \mathrm{Re}\, z = \beta\} \qquad (2.1.14)$$

for some $\beta \in \mathbb{R}$, or to $\{z \in \mathbb{C} : \ \mathrm{Re}\, z \leq \beta\}$, $\{z \in \mathbb{C} : \ \mathrm{Re}\, z \geq \beta\}$, and we then consider the Mellin transforms in the corresponding sense.

Remark 2.1.17 *We have*

$$M\left(-t\frac{d}{dt}u\right)(z) = z Mu(z), \qquad (2.1.15)$$

$$M(t^{-p}u)(z) = (Mu)(z-p), \ p \in \mathbb{C}, \qquad (2.1.16)$$

$$M((\log t)u)(z) = \left(\frac{d}{dz}Mu\right)(z), \qquad (2.1.17)$$

$$M(u(t^\beta))(z) = \beta^{-1}(Mu)(\beta^{-1}z), \ \beta \in \mathbb{R} \setminus \{0\}. \qquad (2.1.18)$$

Remark 2.1.18 *For every* $k \in \mathbb{N} \setminus \{0\}$ *there are constants* c_{kj}, d_{kj} *such that*

$$(t\partial_t)^k = \sum_{j=1}^k c_{kj} t^j \partial_t^j,$$

$$t^k \partial_t^k = \sum_{j=1}^k d_{kj} (t\partial_t)^j.$$

The coefficients d_{jk} *are the Stirling numbers of first kind,* c_{jk} *the Stirling numbers of second kind.*

Let us introduce the weighted Mellin transform

$$M_\gamma u = Mu|_{\Gamma_{\frac{1}{2}-\gamma}}, \tag{2.1.19}$$

and write simply $M = M_0$. Then

$$M_\gamma u(z) = M(t^{-\gamma}u)(z+\gamma) \tag{2.1.20}$$

for $z \in \Gamma_{\frac{1}{2}-\gamma}$. Let us set

$$L^{2,\gamma}(\mathbb{R}_+) = t^\gamma L^2(\mathbb{R}_+), \tag{2.1.21}$$

endowed with the scalar product

$$(u,v)_{L^{2,\gamma}(\mathbb{R}_+)} = (t^{-\gamma/2}u, t^{-\gamma/2}v)_{L^2(\mathbb{R}_+)}.$$

The space $C_0^\infty(\mathbb{R}_+)$ is dense in $L^{2,\gamma}(\mathbb{R}_+)$ for every $\gamma \in \mathbb{R}$.

Function and distribution spaces on Γ_β, $\beta \in \mathbb{R}$, will be denoted analogously to the corresponding ones on $\mathbb{R} \ni \tau \to \beta + i\tau \in \Gamma_\beta$, in particular,

$$\mathcal{S}(\Gamma_\beta), \ \mathcal{S}'(\Gamma_\beta), \ H^s(\Gamma_\beta), \ L^2(\Gamma_\beta) \dots .$$

Observe that

$$M_\gamma C_0^\infty(\mathbb{R}_+) \subset \mathcal{S}(\Gamma_{\frac{1}{2}-\gamma}) \tag{2.1.22}$$

for every $\gamma \in \mathbb{R}$.

Theorem 2.1.19 *The operator* $M_\gamma : C_0^\infty(\mathbb{R}_+) \to \mathcal{S}(\Gamma_{\frac{1}{2}-\gamma})$ *extends by continuity to an isomorphism*

$$M_\gamma : L^{2,\gamma}(\mathbb{R}_+) \to L^2(\Gamma_{\frac{1}{2}-\gamma}) \tag{2.1.23}$$

for all $\gamma \in \mathbb{R}$, *where*

$$\|u\|_{L^{2,\gamma}(\mathbb{R}_+)} = (2\pi)^{-\frac{1}{2}}\|M_\gamma u\|_{L^2(\Gamma_{\frac{1}{2}-\gamma})}, \tag{2.1.24}$$

and

$$(M_\gamma^{-1}g)(t) = \frac{1}{2\pi i} \int_{\Gamma_{\frac{1}{2}-\gamma}} t^{-z}g(z)\,dz. \tag{2.1.25}$$

Proof. We use the fact that the Fourier transform on $\mathbb{R} \ni r$

$$F : v(r) \to \int_{-\infty}^{\infty} e^{-ir\tau}v(r)\,dr \tag{2.1.26}$$

induces an isomorphism $F : L^2(\mathbb{R}_r) \to L^2(\mathbb{R}_\tau)$, with the inverse

$$F^{-1} : f(\tau) \to \frac{1}{2\pi} \int\limits_{-\infty}^{\infty} e^{ir\tau} f(\tau)\, d\tau. \qquad (2.1.27)$$

Let $u(t) \in C_0^\infty(\mathbb{R}_+)$ and set

$$(S_\gamma u)(r) = e^{-(1/2-\gamma)r} u(e^{-r}). \qquad (2.1.28)$$

This induces an isomorphism $S_\gamma : C_0^\infty(\mathbb{R}_+) \to C_0^\infty(\mathbb{R})$ and we have

$$\int\limits_0^\infty |t^{-\gamma} u(t)|^2\, dt = \int\limits_{-\infty}^\infty |(S_\gamma u)(r)|^2\, dr.$$

Thus S_γ extends to an isometric isomorphism $S_\gamma : L^{2,\gamma}(\mathbb{R}_{+,t}) \to L^2(\mathbb{R}_r)$. Hence $FS_\gamma : L^{2,\gamma}(\mathbb{R}_+) \to L^2(\mathbb{R}_\tau)$ is also an isomorphism. Now we have obviously for $u \in C_0^\infty(\mathbb{R}_+)$

$$(M_\gamma u)(1/2 - \gamma + i\tau) = (FS_\gamma u)(\tau). \qquad (2.1.29)$$

This yields immediately the extension (2.1.23). The formula (2.1.25) is a consequence of $M_\gamma^{-1} = S_\gamma^{-1} F^{-1}$ and of (2.1.27), (2.1.28). Finally, (2.1.24) follows from

$$\|M_\gamma u\|^2_{L^2(\Gamma_{\frac{1}{2}-\gamma})} = \|FS_\gamma u\|^2_{L^2(\mathbb{R}_\tau)} = 2\pi \|S_\gamma u\|^2_{L^2(\mathbb{R}_r)} = 2\pi \|u\|^2_{L^{2,\gamma}(\mathbb{R}_+)}.$$

Here we used a correspondig property of the Fourier transform, cf. Section 1.1.1. $\qquad\qquad\qquad\qquad\qquad\qquad\qquad\qquad\qquad\qquad\square$

Remark 2.1.20 *Using* $(f,g)_{L^2(\mathbb{R})} = (2\pi)^{-1}(Ff, Fg)_{L^2(\mathbb{R})}$ *with the Fourier transform (2.1.26) we get*

$$(u,v)_{L^{2,\gamma}(\mathbb{R}_+)} = (2\pi)^{-1}(M_\gamma u, M_\gamma v)_{L^2(\Gamma_{1/2-\gamma})},$$

where the scalar product in $L^2(\Gamma_{1/2-\gamma})$ *is defined by*

$$\int \varphi(1/2 - \gamma + i\tau)\overline{\psi(1/2 - \gamma + i\tau)}\, d\tau.$$

Note that the various extensions M_γ and M_δ of M for $\gamma \neq \delta$ cannot be compared, since they map to L^2 spaces on non–intersecting lines. Often we may restrict ourselves to Mellin transforms of functions or distributions with bounded support, more precisely to

$$r^+ \mathcal{E}' := \left\{ u \in \mathcal{D}'(\mathbb{R}_+) : u = v|_{\mathbb{R}_+} \text{ for some } v \in \mathcal{E}'(\mathbb{R}) \right\}, \qquad (2.1.30)$$

which contains $\mathcal{E}'(\mathbb{R}_+)$ as a subspace.

Remark 2.1.21 *The Mellin transform* (2.1.13) *extends to* $\mathcal{E}'(\mathbb{R}_+)$ *by* $Mu(z) = \langle t^{z-1}, u \rangle$ *and induces a linear operator* $M : \mathcal{E}'(\mathbb{R}_+) \to \mathcal{A}(\mathbb{C})$.

In view of (2.1.20) the Mellin transform on $L^{2,\gamma}(\mathbb{R}_+)$ can be expressed in terms of $M = M_0$. Therefore we often restrict ourselves to the weight $\gamma = 0$. Let $a > 0$ and

$$L^2(0, a) = \left\{ u \in L^2(\mathbb{R}_+) : u(t) = 0 \text{ for all } t > a \right\}.$$

We want to formulate a characterisation of the elements in $ML^2(0, a)$. To this end we denote by

$$\mathcal{A}^2(\{\operatorname{Re} z \le 1/2\}) \tag{2.1.31}$$

the subspace of all $f(z) \in \mathcal{A}(\{\operatorname{Re} z < \frac{1}{2}\})$ with the following properties

(i) $f_\beta := f|_{\Gamma_{1/2+\beta}} \in L^2(\Gamma_{1/2+\beta})$ for all $\beta > 0$.

(ii) f_β, $\beta > 0$ converges to some $f_0 \in L^2(\Gamma_{1/2})$ for $\beta \to 0$.

(iii) $f_\beta \in L^2(\Gamma_{1/2+\beta})$ holds uniformly in $0 \le \beta \le c$ for every $c > 0$.

Theorem 2.1.22 *The following conditions are equivalent:*

(i) $f_0(z) \in L^2(\Gamma_{1/2})$ *belongs to* $ML^2(0, a)$ *for* $a > 0$.

(ii) $f_0(z)$ *extends to an* $f(z) \in \mathcal{A}^2(\{\operatorname{Re} z \le \frac{1}{2}\})$ *and there is a constant* $d > 0$ *with*

$$\|f_\beta\|_{L^2(\Gamma_{1/2+\beta})} \le da^\beta \qquad \text{for all } \beta > 0.$$

A proof of Theorem 2.1.22 may be found in the paper of Jeanquartier [56].

In view of Theorem 2.1.22 the Mellin transform can be regarded as an operator
$$M : L^2(0, a) \to \mathcal{A}^2(\{\operatorname{Re} z \le 1/2\}).$$
For every $f = Mu$, $u \in L^2(0, a)$, we have

$$\|f_\beta\|_{L^2(\Gamma_{1/2+\beta})} = (2\pi)^{1/2}\|t^\beta u\|_{L^2(\mathbb{R}_+)},$$

cf. (2.1.24).

Proposition 2.1.23 *Let* $p \in \mathbb{C}$, $k \in \mathbb{N}$, *and fix an arbitrary cut-off function* $\omega(t)$. *Set*

$$g^{p,k}(z) = M(t^{-\gamma}\omega(t)t^{-p}\log^k t)(z + \gamma) \tag{2.1.32}$$

for any $\gamma \in \mathbb{R}$ with $\operatorname{Re} p < 1/2 - \gamma$. Then

$$g^{p,k}(z) - (-1)^k k! (z - p)^{-(k+1)} \in \mathcal{A}(\mathbb{C}), \qquad (2.1.33)$$

i.e., (2.1.32) is independent of the choice of γ. For arbitrary $\chi(z) \in C^\infty(\mathbb{C})$ with $\chi(z) = 0$ for $|z - p| < \varepsilon_0$, $\chi(z) = 1$ for $|z - p| > \varepsilon_1$ with certain $0 < \varepsilon_0 < \varepsilon_1$ we have

$$\chi(z) g^{p,k}(z)|_{\Gamma_\varrho} \in \mathcal{S}(\Gamma_\varrho) \qquad (2.1.34)$$

uniformly in ϱ for $c \leq \varrho \leq c'$ for arbitrary reals $c < c'$.

Proof. Without loss of generality it suffices to consider the case $p = 0$, since the assertion for general p then follows by a translation in the complex plane, cf. (2.1.16). Let first $k = 0$. Denote by $\varepsilon(t)$ the characteristic function of the interval $(0, 1)$. Then

$$M\varepsilon(z) = \int\limits_0^1 t^{z-1}\, dt = z^{-1}.$$

Since $\omega(t) - \varepsilon(t)$ has compact support on $\mathbb{R}_+$, the relation (2.1.33) is a consequence of Remark 2.1.21, in other words

$$M(\omega - \varepsilon)(z) = g^{0,0}(z) - z^{-1} \in \mathcal{A}(\mathbb{C}). \qquad (2.1.35)$$

For $\operatorname{Re} z > 0$ we have for all $j, l \in \mathbb{N}$

$$\frac{d^l}{dz^l} z^j \int\limits_0^\infty t^{z-1} \omega(t)\, dt = \int\limits_0^\infty t^{z-1} \log^l t \left(-t\frac{d}{dt} \right)^j \omega(t)\, dt. \qquad (2.1.36)$$

cf. (2.1.15), (2.1.17). It is clear that

$$\sup \left| \frac{d^l}{dz^l} z^j \int\limits_0^\infty t^{z-1} \omega(t)\, dt \right| < \infty$$

with the supremum over all $z \in \Gamma_\varrho$ and $c \leq \varrho \leq c' < \infty$ when $c > 0$. Furthermore, for $j > 0$, those supremums are finite when z varies over $c \leq \operatorname{Re} z \leq c'$ for arbitrary finite $c < c'$, since then the derivatives of ω have compact support on $\mathbb{R}_+$. Thus, to prove (2.1.34), it remains to consider (2.1.36) for $j = 0$ and arbitrary $\operatorname{Re} z$. Here we use $g^{0,0}(z) = M(\omega - \varepsilon)(z) + z^{-1}$ and consider the items on the right separately. We have

$$\frac{d^l}{dz^l} g^{0,0}(z) = \int\limits_0^\infty t^{z-1} \log^l t (\omega(t) - \varepsilon(t))\, dt + \frac{d^l}{dz^l} z^{-1}.$$

Now

$$\sup_{c \leq \operatorname{Re} z \leq c'} |\frac{d^l}{dz^l} \chi(z) \frac{1}{z}| < \infty$$

as well as

$$\sup_{c \leq \operatorname{Re} z \leq c'} |\chi(z) \int_0^\infty t^{z-1} \log^l t (\omega(t) - \varepsilon(t)) \, dt| < \infty \qquad (2.1.37)$$

are obvious. Similar estimates remain valid when we replace the excision function in (2.1.37) by $(d/dz)^m \chi(z)$ for any m. Here $c < c'$ are arbitrary. In (2.1.37) we used the fact that $\omega(t) - \varepsilon(t)$ is of compact support on $\mathbb{R}_+$. The assertion for general k follows easily by differentiating the relation for $k = 0$ with respect to z. The independence of (2.1.32) of γ is obvious. $\qquad \square$

Remark 2.1.24 *Let $f(z)$ be a meromorphic function in $\mathbb{C}$ with poles p_l of multiplicities $m_l + 1$, $l \in \mathbb{Z}$. Let $\{r_0, \ldots, r_N\}$ be a finite subsystem of the poles and $n_j + 1$ the multiplicity of r_j. Set $C = \bigcup_{j=0}^N \{z : |z - r_j| = \varepsilon\}$ with an $\varepsilon > 0$ so small that C does not surround any other pole of f. Then, there are unique coefficients d_{jk}, $0 \leq k \leq n_j$, $j = 0, \ldots, N$, with*

$$\frac{1}{2\pi i} \int_C t^{-z} f(z) \, dz = \sum_{j=0}^N \sum_{k=0}^{n_j} d_{jk} t^{-r_j} \log^k t.$$

In fact, assume for instance $N = 0$, and set $r = r_0$, $n = n_0$. Consider the Laurent expansion of $f(z)$ at $z = r$

$$f(z) = \sum_{k=0}^n \varphi_k (z - r)^{-(k+1)} + f_0(z),$$

where $f_0(z)$ is holomorphic in a neighbourhood of r. Then, if $h(z)$ is an arbitrary holomorphic function and

$$h(z) = \sum_{j=0}^\infty \frac{1}{j!} (z - r)^j \left(\frac{d^j}{dz^j} h \right) (r)$$

its Taylor expansion at $z = r$, the Residue Theorem gives us

$$\frac{1}{2\pi i} \int_C f(z) h(z) \, dz = \sum_{k=0}^n \frac{1}{k!} \varphi_k \left(\frac{d^k}{dz^k} h \right) (r) \frac{1}{2\pi i} \int_C (z - r)^{-1} \, dz$$

$$= \sum_{k=0}^n \psi_k \left(\frac{d^k}{dz^k} h \right) (r)$$

for certain constants ψ_k. Now the result follows by inserting $h(z) = t^{-z}$. The arguments for an arbitrary finite number of poles are analogous.

Definition 2.1.25 *If $A \subseteq \mathbb{C}$ is a set then a $\chi(z) \in C^\infty(\mathbb{C})$ is called an A–excision function if for certain $0 < \varepsilon_0 < \varepsilon_1$,*

$$\chi(z) = 0 \qquad \text{for } \operatorname{dist}(z, \overline{A}) < \varepsilon_0,$$
$$\chi(z) = 1 \qquad \text{for } \operatorname{dist}(z, \overline{A}) > \varepsilon_1.$$

Theorem 2.1.26 *The restriction of*

$$M : L^2(\mathbb{R}_+) \to L^2(\Gamma_{1/2}) \tag{2.1.38}$$

to $\mathcal{S}(\overline{\mathbb{R}}_+) := \mathcal{S}(\mathbb{R})|_{\overline{\mathbb{R}}_+}$ maps $\mathcal{S}(\overline{\mathbb{R}}_+)$ to a space of meromorphic functions in $\mathbb{C}$ with simple poles at $z = -k$ for $k \in \mathbb{N}$. If $\chi(z)$ is any A-excision function for $A = \{-k\}_{k \in \mathbb{N}}$, then we have for every $u \in \mathcal{S}(\overline{\mathbb{R}}_+)$

$$\chi(z)Mu(z)|_{\Gamma_\varrho} \in \mathcal{S}(\Gamma_\varrho) \tag{2.1.39}$$

uniformly in ϱ for $c \leq \varrho \leq c'$ for arbitrary $c < c'$. Further, for every $\delta < 1$, we have for all $k \in \mathbb{N}$

$$\frac{1}{k!} \frac{d^k u}{dt^k}\bigg|_{t=0} = \frac{1}{2\pi i} \int_{|z+k|=\delta} (Mu)(z)\, dz. \tag{2.1.40}$$

Proof. Let $u(t) \in \mathcal{S}(\overline{\mathbb{R}}_+)$ and $N \in \mathbb{N}$ be arbitrary. Then for every fixed cut–off function $\omega(t)$ we have

$$u(t) = \omega(t) \sum_{k=0}^{N} \frac{1}{k!} \frac{d^k u}{dt^k}(0) t^k + t^N v_N(t)$$

for some $v_N(t) \in \mathcal{S}(\overline{\mathbb{R}}_+)$. From Proposition 2.1.23 it follows that

$$M\left\{ \omega(t) \sum_{k=0}^{N} c_k t^k \right\}(z) \tag{2.1.41}$$

is meromorphic with simple poles at $z = -k$ and residues c_k. In view of the obvious relation $M(t^N v_N)(z) \in \mathcal{A}(\operatorname{Re} z > 1/2 - N)$ for all N we obtain (2.1.40) from the Residue Theorem. (2.1.39) is a consequence of (2.1.34). $\qquad\qquad \square$

Example 2.1.27 *Euler's Γ–function*

$$\Gamma(z) = \int_0^\infty t^{z-1} e^{-t}\, dt$$

is the Mellin transform of $e^{-t}|_{t \geq 0} \in \mathcal{S}(\overline{\mathbb{R}}_+)$.

Remark 2.1.28 *As noted in* (2.1.15) *we have* $(-t\frac{\partial}{\partial t})t^{-z} = zt^{-z}$. *If* $p(z) = \sum_{j=0}^{\mu} a_j z^j$ *is a polynomial with constant coefficients and* C *an arbitrary closed compact* C^{∞} *curve in* $\mathbb{C}$ *with* $p(z) \neq 0$ *for all* $z \in \mathbb{C}$, *then*

$$u(t) = \int_C t^{-z} p^{-1}(z)\, dz \qquad (2.1.42)$$

is a solution of $\sum_{j=0}^{\mu} a_j(-t\frac{\partial}{\partial t})^j u(t) = 0$. *In particular, let* $z_1, \ldots, z_N$ *be the zeros of* $p(z)$ *and* $C_k = \{z : |z - z_k| < \varepsilon\}$ *for sufficiently small* $\varepsilon > 0$. *Then by integrating in* (2.1.42) *over* C_k *we obtain different solutions* $u_k(t)$ *for* $k = 1, \ldots, N$.

Similarly to the Sobolev spaces in $\mathbb{R}^n$, defined in terms of the Fourier transform, cf. Section 1.1.4, we now introduce spaces on $\mathbb{R}_+$ based on the Mellin transform. They will contain the smoothness $s \in \mathbb{R}$ and in addition a weight $\gamma \in \mathbb{R}$.

Definition 2.1.29 $\mathcal{H}^{s,\gamma}(\mathbb{R}_+)$ *for* $s, \gamma \in \mathbb{R}$ *is the closure of* $C_0^{\infty}(\mathbb{R}_+)$ *with respect to the norm*

$$u \to \left\{ \frac{1}{2\pi i} \int_{\Gamma_{1/2-\gamma}} (1 + |z|^2)^s |M_\gamma u(z)|^2\, dz \right\}^{1/2}. \qquad (2.1.43)$$

We set

$$\mathcal{H}^s(\mathbb{R}_+) := \mathcal{H}^{s,0}(\mathbb{R}_+). \qquad (2.1.44)$$

Note that in view of (2.1.22) the norm (2.1.43) is finite on $C_0^{\infty}(\mathbb{R}_+)$ for all $s, \gamma \in \mathbb{R}$. From (2.1.24) it follows that

$$\mathcal{H}^{0,\gamma} = L^{2,\gamma}(\mathbb{R}_+), \quad \mathcal{H}^0 = L^2(\mathbb{R}_+) \qquad (2.1.45)$$

In (2.1.43) we can replace $(1 + |z|^2)^s$ by $(1 + |\tau|^2)^s$, up to equivalence of norms. From (2.1.29) we see that the norm

$$u \to \left\{ \int_{\mathbb{R}} (1 + |\tau|^2)^s |F S_\gamma u(\tau)|^2\, d\tau \right\}^{1/2}$$

is equivalent to (2.1.43). That means

$$u \in \mathcal{H}^{s,\gamma}(\mathbb{R}_+) \Leftrightarrow S_\gamma u \in H^s(\mathbb{R}) \qquad (2.1.46)$$

(extensions or restrictions of operators will often be denoted by the same letter). Thus, S_γ induces an isomorphism

$$S_\gamma : \mathcal{H}^{s,\gamma}(\mathbb{R}_+) \to H^s(\mathbb{R}). \qquad (2.1.47)$$

It follows, in particular, that

$$\mathcal{S}(\Gamma_{1/2-\gamma}) \subset M_\gamma \mathcal{H}^{\infty,\gamma}(\mathbb{R}_+). \qquad (2.1.48)$$

In view of $t = e^{-r}$ we have $S_\gamma = S_0 t^{-\gamma}$ and hence

$$\mathcal{H}^{s,\gamma}(\mathbb{R}_+) = t^\gamma \mathcal{H}^s(\mathbb{R}_+) \qquad (2.1.49)$$

or more generally

$$\mathcal{H}^{s,\gamma+\varrho}(\mathbb{R}_+) = t^\varrho \mathcal{H}^{s,\gamma}(\mathbb{R}_+) \qquad (2.1.50)$$

for all $s, \gamma, \varrho \in \mathbb{R}$. From (2.1.28) it follows that

$$\mathcal{H}^{s,\gamma}(\mathbb{R}_+) \subset H^s_{\mathrm{loc}}(\mathbb{R}_+). \qquad (2.1.51)$$

In fact, if we set $\psi_\gamma(r) = e^{-(1/2-\gamma)r}$ which belongs to $C^\infty(\mathbb{R})$, then $(S_\gamma u)(r) = \psi_\gamma(r)(S_{1/2}u)(r)$. Moreover, $S_{1/2}$ equals the pull–back κ^* with respect to the diffeomorphism $\kappa : \mathbb{R} \to \mathbb{R}_+, \kappa(r) = e^{-r}$. In other words

$$\mathcal{H}^{s,\gamma}(\mathbb{R}_+) = \psi_\gamma^{-1}(-\log t)(\kappa^*)^{-1} H^s(\mathbb{R}). \qquad (2.1.52)$$

Since the H^s_{loc} spaces are invariant under diffeomorphisms and pre-served under multiplication by non–vanishing C^∞ functions, we obtain (2.1.51).

From a fixed Hilbert space scalar product $(\cdot, \cdot)_{H^s(\mathbb{R})}$ we can easily pass to a corresponding scalar product in $\mathcal{H}^{s,\gamma}(\mathbb{R}_+)$ by setting

$$(u, v)_{\mathcal{H}^{s,\gamma}(\mathbb{R}_+)} := (S_\gamma u, S_\gamma v)_{H^s(\mathbb{R})}.$$

In other words $\mathcal{H}^{s,\gamma}(\mathbb{R}_+)$ has the structure of a Hilbert space.

Let us now return to the symbol spaces $S^\mu(\mathbb{R}), \mu \in \mathbb{R}$, with constant coefficients, cf. the notation of Section 1.2.2. We set for any $\varrho \in \mathbb{R}$

$$S^\mu_{(\mathrm{cl})}(\Gamma_\varrho) := \left\{ h(z) \in C^\infty(\Gamma_\varrho) : h(\varrho + i\tau) \in S^\mu_{(\mathrm{cl})}(\mathbb{R}_\tau) \right\}. \qquad (2.1.53)$$

In a similar sense we use the notation

$$S^\mu(Q \times \Gamma_\varrho), \ S^\mu_{\mathrm{cl}}(\overline{Q} \times \Gamma_\varrho), \qquad (2.1.54)$$

for $Q = \mathbb{R}_+$ or $\mathbb{R}_+ \times \mathbb{R}_+$, where the symbols are denoted by $h(t, z)$ and $h(t, t', z)$, respectively, with z varying along Γ_ϱ.

We set, first for $u \in C_0^\infty(\mathbb{R}_+)$,

$$\mathrm{op}_M^\gamma(h)u(t) = M^{-1}_{\gamma, z \to t} \{ M_{\gamma, t' \to z} h(t, t', z)u(t') \} \qquad (2.1.55)$$

for $h(t, t', z) \in S^\mu(\mathbb{R}_+ \times \mathbb{R}_+ \times \Gamma_{1/2-\gamma})$, in particular, write

$$\mathrm{op}_M(h) := \mathrm{op}_M^0(h). \qquad (2.1.56)$$

Remark 2.1.30 *Let* $h(z)$ *be independent of* t, t'. *Set* $(\kappa_\lambda u)(t) = \lambda^m u(\lambda t)$ *for* $\lambda \in \mathbb{R}_+$ *and arbitrary fixed* $m \in \mathbb{R}$. *Then*

$$\mathrm{op}_M(h)\kappa_\lambda = \kappa_\lambda \, \mathrm{op}_M(h)$$

for all $\lambda \in \mathbb{R}_+$.

The operator (2.1.19) can be regarded as a Mellin pseudo–differential operator with symbol h. We shall return in detail below to the specific points of the corresponding calculus. Here we only use a few obvious properties. From (2.1.48) we see that every $h(z) \in S^\mu(\Gamma_{1/2-\gamma})$ gives rise to an operator

$$\mathrm{op}_M^\gamma(h) : C_0^\infty(\mathbb{R}_+) \to \mathcal{H}^{\infty,\gamma}(\mathbb{R}_+). \tag{2.1.57}$$

Theorem 2.1.31 *Let* $h(z) \in S^\mu(\Gamma_{1/2-\gamma})$. *Then* $\mathrm{op}_M^\gamma(h)$ *extends to continuous operators*

$$\mathrm{op}_M^\gamma(h) : \mathcal{H}^{s,\gamma}(\mathbb{R}_+) \to \mathcal{H}^{s,\gamma}(\mathbb{R}_+) \tag{2.1.58}$$

for all $s \in \mathbb{R}$. *In particular, for* $h(z) = p^{-s}(z) := (1 + |z|^2)^{-s/2}$ *we get the isomorphisms*

$$\mathrm{op}_M^\gamma(p^{-s}) : L^{2,\gamma}(\mathbb{R}_+) \to \mathcal{H}^{s,\gamma}(\mathbb{R}_+).$$

The proof is an obvious consequence of the definitions.

Corollary 2.1.32 *The* $L^2(\mathbb{R}_+)$-*scalar product*

$$(\cdot,\cdot) : C_0^\infty(\mathbb{R}_+) \times C_0^\infty(\mathbb{R}_+) \to \mathbb{C} \tag{2.1.59}$$

extends to a non–degenerate sesquilinear form

$$(\cdot,\cdot) : \mathcal{H}^{s,\gamma}(\mathbb{R}_+) \times \mathcal{H}^{-s,-\gamma}(\mathbb{R}_+) \to \mathbb{C}$$

for all $s, \gamma \in \mathbb{R}$.

In fact, for $u, v \in C_0^\infty(\mathbb{R}_+)$ we have

$$
(u, v) = \int\limits_0^\infty u\bar{v}\, dx
$$

$$
= (2\pi)^{-1} \int\limits_{\mathbb{R}} Mu(z)\overline{Mv(z)}\, d\tau
$$

$$
= (2\pi)^{-1} \int\limits_{\mathbb{R}} (1 + |z|^2)^{s/2} Mu(z)(1 + |z|^2)^{-s/2}\overline{Mv(z)}\, d\tau
$$

$$
= \int\limits_0^\infty \operatorname{op}_M(p^s)u\overline{\operatorname{op}_M(p^{-s})v}\, dt
$$

$$
= \int\limits_0^\infty t^{-\gamma} \operatorname{op}_M(p^s)u\overline{t^\gamma \operatorname{op}_M(p^{-s})v}\, dt.
$$

Here $z = 1/2 + i\tau$, $\mathbb{R} = \mathbb{R}_\tau$. Now

$$
|(u, v)| \leq \|t^{-\gamma} \operatorname{op}_M(p^s)u\|_{L^2(\mathbb{R}_+)} \|t^\gamma \operatorname{op}_M(p^{-s})v\|_{L^2(\mathbb{R}_+)}
$$

yields the assertion, since $\|f\|_{\mathcal{H}^{s,\gamma}(\mathbb{R}_+)} \sim \|t^{-\gamma} \operatorname{op}_M(p^s)f\|_{L^2(\mathbb{R}_+)}$. It follows also that

$$
\|w\|_{\mathcal{H}^{s,\gamma}(\mathbb{R}_+)} \sim \sup_{\substack{v \in \mathcal{H}^{-s,-\gamma}(\mathbb{R}_+) \\ v \neq 0}} \left\{ |(w, v)_{L^2(\mathbb{R}_+)}| / \|v\|_{\mathcal{H}^{-s,-\gamma}(\mathbb{R}_+)} \right\}. \qquad (2.1.60)
$$

Theorem 2.1.33 *For $s \in \mathbb{N}$ we have*

$$
\mathcal{H}^{s,\gamma}(\mathbb{R}_+) = \left\{ u \in L^{2,\gamma}(\mathbb{R}_+) : \left(-t\frac{d}{dt}\right)^k u \in L^{2,\gamma}(\mathbb{R}_+) \text{ for } k = 0, \ldots, s \right\}.
$$

Proof. We shall prove the assertion for $\gamma = 0$. The general case is completely analogous. It suffices to show

$$
M\mathcal{H}^s(\mathbb{R}_+) \subseteq \left\{ f \in L^2(\Gamma_{1/2}) : z^k f(z) \in L^2(\Gamma_{1/2}), k = 0, \ldots, s \right\} \tag{2.1.61}
$$

as well as the converse inclusion, cf. (2.1.15). The condition (2.1.61) is equivalent to

$$
|\tau|^k f\left(\frac{1}{2} + i\tau\right) \in L^2(\mathbb{R}) \qquad \text{for } k = 0, \ldots, s, \tag{2.1.62}
$$

$\tau = \operatorname{Im} z$, and the norm

$$\left\{ \int_{\mathbb{R}} (1 + |\tau|^2)^s |(Mu)(\tfrac{1}{2} + i\tau)|^2 \, d\tau \right\}^{\frac{1}{2}} \tag{2.1.63}$$

is equivalent to that in Definition 2.1.29 (for $\gamma = 0$). It is then evident that (2.1.63) is finite just in case (2.1.62). □

Lemma 2.1.34 *Let $\omega(t)$ be an arbitrary cut–off function. Then $u \in \mathcal{H}^{s,\gamma}(\mathbb{R}_+)$ implies $\omega u \in \mathcal{H}^{s,\gamma}(\mathbb{R}_+)$ for all $s, \gamma \in \mathbb{R}$. The operator of multiplication by ω belongs to $\mathcal{L}(\mathcal{H}^{s,\gamma}(\mathbb{R}_+))$.*

Proof. In view of (2.1.50) it suffices to show this assertion for a special weight, e.g., $\gamma = 1/2$. In that case we have

$$u(t) \in \mathcal{H}^{s,1/2}(\mathbb{R}_+) \Leftrightarrow u(e^{-r}) \in H^s(\mathbb{R}). \tag{2.1.64}$$

Thus $\omega(t)u(t)$ corresponds to $\omega(e^{-r})u(e^{-r})$. Now $\varphi(r) := \omega(e^{-r}) \in C^\infty(\mathbb{R})$ is constant for $|r| > \text{const}$ for a suitable constant > 0. Since the multiplication by those φ is a continuous operator in $H^s(\mathbb{R})$ we obtain the assertion immediately from (2.1.64). □

Definition 2.1.35 *For every $s, \gamma \in \mathbb{R}$ we set*

$$\mathcal{K}^{s,\gamma}(\mathbb{R}_+) = [\omega]\mathcal{H}^{s,\gamma}(\mathbb{R}_+) + [1 - \omega]H^s(\mathbb{R}_+),$$

for any fixed cut–off function $\omega(t)$.

From (2.1.51) it follows that this is a correct definition, i.e., independent of the choice of ω.

Proposition 2.1.36 *The operator $\mathcal{M}_\varphi$ of multiplication by a function $\varphi(t) \in C_0^\infty(\overline{\mathbb{R}}_+)$ induces a continuous operator*

$$\mathcal{M}_\varphi : \mathcal{H}^{s,\gamma}(\mathbb{R}_+) \to \mathcal{H}^{s,\gamma}(\mathbb{R}_+),$$

and $\varphi \to \mathcal{M}_\varphi$ is continuous in the sense $C_0^\infty(\overline{\mathbb{R}}_+) \to \mathcal{L}(\mathcal{H}^{s,\gamma}(\mathbb{R}_+))$ for every $s, \gamma \in \mathbb{R}$.

Proof. In view of Lemma 2.1.34 we may assume $\varphi(0) = 0$, since it suffices to consider $\varphi(t) - \varphi(0)\omega(t)$ instead of φ. Moreover, because of $t^\gamma \mathcal{H}^{s,0}(\mathbb{R}_+) = \mathcal{H}^{s,\gamma}(\mathbb{R}_+)$, we may restrict ourselves to the case $\gamma = 0$. Using (2.1.25) we can write $\mathcal{M}_\varphi u(t) = (2\pi)^{-1} \int_{\Gamma_{\frac{1}{2}}} t^{-z} \varphi(t) g(z) \, dz \, dt$ for

$g(z) = (Mu)(z)$. Thus

$$(M_{t \to w}(\mathcal{M}_\varphi u))(w) = \frac{1}{2\pi i} \int_{\Gamma_{\frac{1}{2}}} b(w - z) g(z) \, dz,$$

where $b(\zeta) = (M_{t\to\zeta}\varphi)(\zeta)$, $\zeta = w - z$, $\mathrm{Re}\,\zeta = 0$. Using Theorem 2.1.26 we see that $b(\zeta) \in \mathcal{S}(\Gamma_0)$. In particular, there is a constant $c_N = c_N(\varphi)$ for each $N \in \mathbb{N}$ such that $|b(\zeta)| \leq c_N(1 + |\zeta|)^{-N}$. Set $K(z,w) = b(w - z)(1 + |z|)^{-s}(1 + |w|)^s$. In view of formula (1.1.78) we have $(1 + |z|)^{-s}(1 + |w|)^s \leq (1 + |z - w|)^{|s|}$ and hence $|K(z,w)| \leq (1 + |z - w|)^{-N+|s|}$. Then

$$\int_{\Gamma_{\frac{1}{2}}} |K(z,w)|\,|dz| < C_N, \quad \int_{\Gamma_{\frac{1}{2}}} |K(z,w)|\,|dw| < \tilde{C}_N$$

for every $N > |s| + 1$, with constants $C_N, \tilde{C}_N > 0$, independent of $z, w \in \Sigma_{\frac{1}{2}}$. Let $v \in C_0^\infty(\mathbb{R}_+)$ and $f(w) = (M_{t\to w})(w)$. Then

$$(\mathcal{M}_\varphi u, v)_{L^2(\mathbb{R}_+)} = \int_{\Gamma_{\frac{1}{2}}} M(\mathcal{M}_\varphi u)(w)f(w)\,dw$$

$$= \frac{1}{2\pi i} \iint_{\Gamma_{\frac{1}{2}}\Gamma_{\frac{1}{2}}} b(w - z)g(z)f(w)\,dz\,dw$$

$$= \frac{1}{2\pi i} \iint_{\Gamma_{\frac{1}{2}}\Gamma_{\frac{1}{2}}} K(z,w)f(w)(1 + |w|)^{-s}g(z)(1 + |z|)^s\,dz\,dw$$

which implies

$$|(\mathcal{M}_\varphi u, v)_{L^2(\mathbb{R}_+)}|$$
$$\leq (2\pi)^{-1} \iint |K(z,w)f(w)(1 + |w|)^{-s}g(z)(1 + |z|)^s|\,|dz||dw|$$
$$\leq (2\pi)^{-1}\left\{ \iint |K(z,w)||f(w)(1 + |w|)^{-s}|^2\,|dz||dw| \right\}^{\frac{1}{2}}$$
$$\left\{ \iint |K(z,w)||g(z)(1 + |z|)^s|^2\,|dz||dw| \right\}^{\frac{1}{2}}$$
$$\leq (2\pi)^{-1}d_N\|v\|_{\mathcal{H}^{-s,0}(\mathbb{R}_+)}\|u\|_{\mathcal{H}^{s,0}(\mathbb{R}_+)}.$$

Using the formula (2.1.60) it follows that

$$\|\mathcal{M}_\varphi u\|_{\mathcal{H}^{s,0}(\mathbb{R}_+)} \leq c_N\|u\|_{\mathcal{H}^{s,0}(\mathbb{R}_+)}$$

for a constant $c_N = c_N(\varphi) > 0$. This is the continuity of $\mathcal{M}_\varphi$. It is now an easy exercise to see that $c_N(\varphi)$ tends to zero when φ tends to zero in $C_0^\infty(\overline{\mathbb{R}}_+)$. This proves the second statement. $\square$

2.1.4   Weighted Sobolev spaces

We now introduce the weighted Sobolev spaces $\mathcal{H}^{s,\gamma}(X^\wedge)$ for $s,\gamma \in \mathbb{R}$ as a generalisation of the corresponding spaces of Definition 2.1.29. Here X is a closed compact C^∞ manifold of dimension n. First we shall consider the local version on $\mathbb{R}_+ \times \mathbb{R}^n$.

Definition 2.1.37 *We denote by $\mathcal{H}^{s,\gamma}(\mathbb{R}_+ \times \mathbb{R}^n)$ for $s,\gamma \in \mathbb{R}$ the closure of $C_0^\infty(\mathbb{R}_+ \times \mathbb{R}^n)$ with respect to the norm*

$$\left\{ \frac{1}{2\pi i} \int_{\Gamma_{\frac{n+1}{2}-\gamma}} \int_{\mathbb{R}^n} (1 + |z|^2 + |\xi|^2)^s |(M_{\gamma-\frac{n}{2},t\to z} F_{x\to\xi} u)(z,\xi)|^2 \, dz \, d\xi \right\}^{\frac{1}{2}},$$

with $M_{\gamma-\frac{n}{2}}$ being the weighted Mellin transform and F the Fourier transform in $\mathbb{R}^n$.

Remark 2.1.38 *The transformation $u(t,x) \to (S_{\gamma-\frac{n}{2}}u)(r,x)$, cf. equation (2.1.28), $S_{\gamma-\frac{n}{2}} : C_0^\infty(\mathbb{R}_+ \times \mathbb{R}^n) \to C_0^\infty(\mathbb{R}^{n+1})$, extends to an isomorphism*

$$S_{\gamma-\frac{n}{2}} : \mathcal{H}^{s,\gamma}(\mathbb{R}_+ \times \mathbb{R}^n) \to H^s(\mathbb{R}^{n+1})$$

for every $s,\gamma \in \mathbb{R}$. In other words we have

$$\|u\|_{\mathcal{H}^{s,\gamma}(\mathbb{R}_+\times\mathbb{R}^n)} \sim \|S_{\gamma-\frac{n}{2}}u\|_{H^s(\mathbb{R}^{n+1})}$$

in the sense of equivalence of norms.

This observation allows us to employ standard invariance properties of the Sobolev spaces with respect to diffeomorphisms in x–coordinates. If X is a closed compact C^∞ manifold and $\mathcal{U} = \{U_1,\dots,U_N\}$ an open covering of X by coordinate neighbourhoods and if we fix a subordinate partition of unity $\{\varphi_1,\dots,\varphi_N\}$ and charts $\chi_j : U_j \to \mathbb{R}^n$, $j=1,\dots,N$, then $\mathcal{H}^{s,\gamma}(X^\wedge)$ denotes the closure of $C_0^\infty(X^\wedge)$ with respect to the norm

$$\left\{ \sum_{j=1}^N \|(1 \times \chi_j^*)^{-1}\varphi_j u\|^2_{\mathcal{H}^{s,\gamma}(\mathbb{R}_+\times\mathbb{R}^n)} \right\}^{\frac{1}{2}}. \tag{2.1.65}$$

Here $1 \times \chi_j^* : C_0^\infty(\mathbb{R}_+ \times \mathbb{R}^n) \to C_0^\infty(\mathbb{R}_+ \times U_j)$ is the function pull–back with respect to $1 \times \chi_j : \mathbb{R}_+ \times U_j \to \mathbb{R}_+ \times \mathbb{R}^n$.

We want to give an equivalent definition in terms of the order reducing operator families of Section 1.2.3. Recall that for every $\mu \in \mathbb{R}$ there exists an element $R^\mu(\lambda) \in L_{cl}^\mu(X;\mathbb{R}^l)$ such that

$$R^\mu(\lambda) : H^s(X) \to H^{s-\mu}(X) \tag{2.1.66}$$

is an isomorphism for all $\lambda \in \mathbb{R}^l$ and $s \in \mathbb{R}$. In particular, for $l = 1$ we denote such an operator family by $R^\mu(z)$, $z \in \Gamma_{\frac{n+1}{2}-\gamma}$, where $\operatorname{Im} z \in \mathbb{R}$ is interpreted as parameter.

If $\{U_1, \ldots, U_N\}$ is an open covering of X by coordinate neighbourhoods, $\{\varphi_1, \ldots, \varphi_N\}$ a subordinate partition of unity, $\chi_j : U_j \to \mathbb{R}^n$ a system of charts, $\psi_j \in C_0^\infty(U_j)$ functions with $\varphi_j \psi_j = \varphi_j$ for $j = 1, \ldots, N$, then $R^\mu(\lambda)$ can be chosen in the form

$$R^\mu(\lambda) = \sum_{j=1}^{N} \varphi_j R_j^\mu(\lambda) \psi_j \qquad (2.1.67)$$

with $R_j^\mu = (\chi_j^{-1})_* \operatorname{op}((|\xi|^2 + |\lambda|^2 + c)^{\mu/2})$ for sufficiently large $c > 0$, with $(\chi_j^{-1})_*$ being the operator pull–back.

Functions or distributions on $X^\wedge = \mathbb{R}_+ \times X \ni (t, x)$ will also be denoted by $u(t)$, interpreted as $\mathcal{D}'(X)$–valued objects on $\mathbb{R}_+$.

Theorem 2.1.39 $\mathcal{H}^{s,\gamma}(X^\wedge)$ *for* $s, \gamma \in \mathbb{R}$ *is the closure of* $C_0^\infty(X^\wedge)$ *with respect to the norm*

$$\|u\|_{\mathcal{H}^{s,\gamma}(X^\wedge)} = \left\{ \frac{1}{2\pi i} \int_{\Gamma_{\frac{n+1}{2}-\gamma}} \|R^s(z)(M_{\gamma-\frac{n}{2}}u)(z)\|_{L^2(X)}^2 \, dz \right\}^{\frac{1}{2}}. \qquad (2.1.68)$$

If $\tilde{R}^s(z)$ *is another order reducing family of the above kind then the analogous norm in terms of* $\tilde{R}^s(z)$ *is equivalent to* (2.1.68).

Proof. Let us begin with the second assertion of the theorem. If $\tilde{R}^s(z)$ is another order reducing family we have for $f(z) := (M_{\gamma-\frac{n}{2}}u)(z)$ and $\|\cdot\| := \|\cdot\|_{L^2(X)}$

$$\|R^s(z)f(z)\| = \|R^s(z)(\tilde{R}^s(z))^{-1}\tilde{R}^s(z)f(z)\|$$
$$\leq c\|\tilde{R}^s(z)f(z)\|$$
$$\leq \tilde{c}\|R^s(z)f(z)\|$$

with constants c and $\tilde{c}$ satisfying

$$\|R^s(z)(\tilde{R}^s(z))^{-1}\|_{\mathcal{L}(L^2(X))} \leq c, \quad \|\tilde{R}^s(z)(R^s(z))^{-1}\|_{\mathcal{L}(L^2(X))} \leq \tilde{c}. \qquad (2.1.69)$$

The latter estimates hold for all $z \in \Gamma_{\frac{n+1}{2}-\gamma}$, cf. Section 1.2.2. This implies the asserted equivalence of norms. In order to prove that the norms (2.1.68) and (2.1.65) are equivalent it suffices to restrict the consideration to functions u which are supported in a coordinate neighbourhood U_0 on X. According to the choice of U_0 we can fix an open covering $\{U_1, \ldots, U_N\}$ such that $\overline{U}_0 \subset U_1$ and choose the functions ψ_j from (2.1.67) in such a way that $\operatorname{supp} \psi_j \cap U_0 = \emptyset$ for $j = 2, \ldots, N$. Then for all $u \in C_0^\infty(X^\wedge)$ which are supported by U_0 with respect to x, for all t, and $f(z) = (M_{\gamma - \frac{n}{2}} u)(z)$, using (2.1.67) for $\mu = s$ and $\lambda = \operatorname{Im} z$, we have

$$\|R^s(z)f(z)\| = \|\varphi_1 R_1^s(z)\psi_1 f(z)\|.$$

This equals $\|\operatorname{op}(\tilde{r}^s)(z)g(z)\|_{L^2(\mathbb{R}^n)}$ in local coordinates, where

$$\tilde{r}^s(x, x', z, \xi) = \varphi(x)(1 + |\operatorname{Im} z|^2 + |\xi|^2 + c)^{s/2} \psi(x')$$

for $\varphi = (\chi_1^*)^{-1}\varphi_1$, $\psi = (\chi_1^*)^{-1}\psi_1$, $g(z) = (\chi_1^*)^{-1}f(z)$, and

$$\operatorname{op}(\tilde{r}^s)g(z) = F_{\xi \to x}^{-1} F_{x' \to \xi} \left\{\tilde{r}^s(x, x', z, \xi)g(z)\right\}.$$

Now, comparing this with the norm expression in Definition 2.1.37, we obtain for $r^s(z, \xi) = (1 + |z|^2 + |\xi|^2)^{s/2}$

$$\|\operatorname{op}(\tilde{r}^s)(z)g(z)\|_{L^2(\mathbb{R}^n)} = \|\operatorname{op}(\tilde{r}^s)(z)\operatorname{op}(r^{-s})(z)\operatorname{op}(r^s)(z)g(z)\|_{L^2(\mathbb{R}^n)}$$
$$\leq c\|\operatorname{op}(r^s)(z)g(z)\|_{L^2(\mathbb{R}^n)}$$

for a constant $c > 0$, independent of z. Here we have applied an analogous conclusion as for (2.1.69). We thus obtain

$$\|u\|_{\mathcal{H}^{s,\gamma}(X^\wedge)} \sim \|(1 \times \chi_1^*)^{-1}u\|_{\mathcal{H}^{s,\gamma}(\mathbb{R}_+ \times \mathbb{R}^n)}$$

for the functions u with the indicated support. This gives us the first assertion of the theorem. $\square$

Let us also consider the standard Sobolev space $H^s(\mathbb{R} \times X)$, $s \in \mathbb{R}$, on the cylinder $\mathbb{R} \times X \ni (r, x)$. We can obtain the space $H^s(\mathbb{R} \times X)$, for instance, as the closure of $C_0^\infty(\mathbb{R} \times X)$ with respect to the norm

$$\left\{\int_{\mathbb{R}} \|R^s(\varrho)(F_{r \to \varrho} u)(\varrho)\|_{L^2(X)}^2 \, d\varrho\right\}^{\frac{1}{2}}, \tag{2.1.70}$$

for the one–dimensional Fourier transform $F_{r \to \varrho}$ and some choice of an order reducing family (2.1.66), here for $\lambda = \varrho \in \mathbb{R}$. The specific order reducing family only affects (2.1.70) up to equivalence of norms.

Remark 2.1.40 *If $s \in \mathbb{N}$ the space $H^s(\mathbb{R} \times X)$ coincides with the subspace of all $u \in L^2(\mathbb{R} \times X)$ such that $D(\varrho)(F_{r \to \varrho} u)(\varrho) \in L^2(\mathbb{R}_\varrho \times X)$ for every parameter–dependent differential operator $D(\varrho)$ of order $\leq s$, with the parameter $\varrho \in \mathbb{R}$ (parameter–dependent differential operators of order $\mu \in \mathbb{N}$ are elements of $L_{cl}^\mu(X; \mathbb{R})$ for which the amplitude functions are polynomials in (ξ, ϱ) of order μ).*

Remark 2.1.41 *The transformation* $u(t,x) \to (S_{\gamma-\frac{n}{2}}u)(r,x)$, $S_{\gamma-\frac{n}{2}} :$ $C_0^\infty(\mathbb{R}_+ \times X) \to C_0^\infty(\mathbb{R} \times X)$, *cf.* (2.1.28), *has an extension to an isomorphism*

$$S_{\gamma-\frac{n}{2}} : \mathcal{H}^{s,\gamma}(\mathbb{R}_+ \times X) \to H^s(\mathbb{R} \times X) \qquad (2.1.71)$$

for every $s, \gamma \in \mathbb{R}$, *and we obtain*

$$\|u\|_{\mathcal{H}^{s,\gamma}(X^\wedge)} \sim \|S_{\gamma-\frac{n}{2}}u\|_{H^s(\mathbb{R} \times X)}.$$

It follows that

$$\mathcal{H}^{s,\gamma+\delta}(X^\wedge) = t^\delta \mathcal{H}^{s,\gamma}(X^\wedge)$$

for all $s, \gamma, \delta \in \mathbb{R}$. *This is analogous to Remark 2.1.38 above.*

Note that

$$\mathcal{H}^{0,0}(X^\wedge) = t^{-\frac{n}{2}} L^2(\mathbb{R}_+ \times X),$$

where $L^2(\mathbb{R}_+ \times X)$ is the space of all square integrable functions with respect to $dt\,dx$ for a positive measure dx on X which is associated with a fixed Riemannian metric on X. In $\mathcal{H}^{0,0}(X^\wedge)$ we thus have a fixed scalar product

$$(u, v)_{\mathcal{H}^{0,0}(X^\wedge)} = \int_{\mathbb{R}_+ \times X} t^n u(t,x) \overline{v(t,x)}\, dt\,dx. \qquad (2.1.72)$$

Incidentally we call (2.1.72) the reference scalar product for the scale $\mathcal{H}^{s,\gamma}(X^\wedge)$, $s, \gamma \in \mathbb{R}$. This can also be written

$$(u, v)_{\mathcal{H}^{0,0}(X^\wedge)} = \frac{1}{2\pi i} \int_{\Gamma_{\frac{n+1}{2}}} ((M_{-\frac{n}{2}}u)(z), (M_{-\frac{n}{2}}v)(z))_{L^2(X)}\, dz. \qquad (2.1.73)$$

Proposition 2.1.42 *We have* $\mathcal{H}^{s,\gamma}(X^\wedge) \subset H^s_{\text{loc}}(X^\wedge)$ *for all* $s, \gamma \in \mathbb{R}$.

Proof. The $H^s_{\text{loc}}(\cdot)$–spaces are invariant under diffeomorphisms of the underlying manifold. The isomorphism (2.1.71) is generated by a substitution and a multiplication by a smooth factor. This yields the assertion. ☐

From the results on the norm growth of parameter–dependent families of pseudo–differential operators on X, cf. Theorem 1.2.19 and the formula (2.1.68), we obtain continuous embeddings

$$\mathcal{H}^{s',\gamma}(X^\wedge) \hookrightarrow \mathcal{H}^{s,\gamma}(X^\wedge) \qquad (2.1.74)$$

for all $s, s' \in \mathbb{R}$ with $s' \geq s$. They are not compact for $s' > s$.

Set

$$\hat{H}^s(\mathbb{R}_\varrho \times X) = F_{r \to \varrho} H^s(\mathbb{R} \times X), \qquad (2.1.75)$$

and define $\hat{H}^s(\Gamma_\beta \times X)$ for any $\beta \in \mathbb{R}$ as the corresponding distribution space on $\Gamma_\beta \times X$, obtained from $\hat{H}^s(\mathbb{R} \times X)$ by identifying $\mathbb{R} \ni \varrho$ and $\Gamma_\beta \ni z$ by $\varrho \to \operatorname{Im} z$.

Remark 2.1.43 *As a consequence of Theorem 2.1.39 we obtain equivalence of norms*

$$\|M_{\gamma-\frac{n}{2}} u\|_{\hat{H}^s(\Gamma_{\frac{n+1}{2}-\gamma} \times X)} \sim \|u\|_{\mathcal{H}^{s,\gamma}(X^\wedge)}$$

for all $s \in \mathbb{R}$. In other words, $M_{\gamma-\frac{n}{2}}$ induces an isomorphism

$$M_{\gamma-\frac{n}{2}} : \mathcal{H}^{s,\gamma}(X^\wedge) \to \hat{H}^s(\Gamma_{\frac{n+1}{2}-\gamma} \times X)$$

for all $s \in \mathbb{R}$.

Remark 2.1.44 *From $H^s(\mathbb{R} \times X) = [H^{s_0}(\mathbb{R} \times X), H^{s_1}(\mathbb{R} \times X)]_\theta$ for all $0 \le \theta \le 1$ and $s_0 \le s_1$, $s = \theta s_0 + (1 - \theta) s_1$, with $[\cdot, \cdot]_\theta$ being the interpolation functor with respect to the complex interpolation, we obtain*

$$\mathcal{H}^{s,\gamma}(X^\wedge) = [\mathcal{H}^{s_0,\gamma}(X^\wedge), \mathcal{H}^{s_1,\gamma}(X^\wedge)]_\theta$$

for every $\gamma \in \mathbb{R}$.

Proposition 2.1.45 *For $s \in \mathbb{N}$ and $\gamma \in \mathbb{R}$ the space $\mathcal{H}^{s,\gamma}(X^\wedge)$ coincides with the subspace of all $u \in t^{\gamma-\frac{n}{2}} L^2(X^\wedge)$ for which $(tD_t)^k Bu \in t^{\gamma-\frac{n}{2}} L^2(X^\wedge)$ for all $0 \le k \le s$ and all $B \in \operatorname{Diff}^{s-k}(X)$.*

Proof. In view of Remark 2.1.43 it suffices to characterise $\mathcal{H}^{s,\gamma}(X^\wedge)$ in the image with respect to $M_{\gamma-\frac{n}{2}}$ in terms of the above Remark 2.1.40. In fact, $M_{\gamma-\frac{n}{2}}((-t\partial_t)^k Bu) = z^k Bf(z) \in \hat{H}^0(\Gamma_{\frac{n+1}{2}-\gamma} \times X)$ with $f(z) = (M_{\gamma-\frac{n}{2}} u)(z)$, for all $0 \le k \le s$ and $B \in \operatorname{Diff}^{s-k}(X)$, is equivalent to $D(\varrho)f \in \hat{H}^0(\Gamma_{\frac{n+1}{2}-\gamma} \times X)$ for every ϱ-dependent differential operator $D(\varrho)$ on X of order $\le s$, $\varrho = \operatorname{Im} z$. This is just a characterisation of $\hat{H}^s(\Gamma_{\frac{n+1}{2}-\gamma} \times X)$. But then $\mathcal{H}^{s,\gamma}(X^\wedge)$ equals the subspace of all $u \in t^{\gamma-\frac{n}{2}} L^2(X^\wedge)$ for which $(-t\partial_t)^k Bu = M_{\gamma-\frac{n}{2}}^{-1}(z^k Bf) \in M_{\gamma-\frac{n}{2}}^{-1} \hat{H}^0(\Gamma_{\frac{n+1}{2}-\gamma} \times X) = t^{\gamma-\frac{n}{2}} L^2(X^\wedge)$, $0 \le k \le s$. $\qquad \square$

Theorem 2.1.46 *The scalar product $(u,v)_{\mathcal{H}^{0,0}(X^\wedge)}$ for $u,v \in C_0^\infty(X^\wedge)$ extends to a non–degenerate sesquilinear pairing*

$$(\cdot,\cdot) : \mathcal{H}^{s,\gamma}(X^\wedge) \times \mathcal{H}^{-s,-\gamma}(X^\wedge) \to \mathbb{C}$$

for every $s,\gamma \in \mathbb{R}$, and we have

$$\|u\|_{\mathcal{H}^{s,\gamma}(X^\wedge)} \sim \sup\{|(u,v)|/\|v\|_{\mathcal{H}^{-s,-\gamma}(X^\wedge)} : v \in \mathcal{H}^{-s,-\gamma}(X^\wedge), v \ne 0\}.$$

Proof. In view of $\mathcal{H}^{s,\gamma}(X^\wedge) = t^\gamma \mathcal{H}^{s,0}(X^\wedge)$ we may assume $\gamma = 0$. In fact, for $u = t^\gamma u_0$, $v = t^{-\gamma} v_0$ with the corresponding elements $u_0 \in \mathcal{H}^{s,0}(X^\wedge)$, $v_0 \in \mathcal{H}^{-s,0}(X^\wedge)$ we have $(u,v) = (u_0, v_0)$ and

$$\|u\|_{\mathcal{H}^{s,\gamma}(X^\wedge)} = \|u_0\|_{\mathcal{H}^{s,0}(X^\wedge)}, \quad \|v\|_{\mathcal{H}^{-s,-\gamma}(X^\wedge)} = \|v_0\|_{\mathcal{H}^{-s,0}(X^\wedge)}.$$

Hence the asserted result is a consequence of

$$\|u_0\|_{\mathcal{H}^{s,0}(X^\wedge)} \sim \sup\{|(u_0, v_0)|/\|v_0\|_{\mathcal{H}^{-s,0}(X^\wedge)} : v_0 \in \mathcal{H}^{-s,0}(X^\wedge), \ v_0 \neq 0\}.$$

In view of (2.1.73) and Remark 2.1.43 such an identity is equivalent to

$$\|f\|_{\hat{H}^s(\Gamma_{\frac{n+1}{2}} \times X)} \sim \sup\{|(f,g)_{L^2(\Gamma_{\frac{n+1}{2}} \times X)}|/\|g\|_{\hat{H}^{-s}(\Gamma_{\frac{n+1}{2}} \times X)} : \qquad (2.1.76)$$

$$g \in \hat{H}^{-s}(\Gamma_{\frac{n+1}{2}} \times X), \ g \neq 0\} \qquad (2.1.77)$$

for $f = M_{-\frac{n}{2}} u_0$, $g = M_{-\frac{n}{2}} v_0$. However this is a well–known property of the standard Sobolev spaces, written in the image with respect to the Fourier transform in the variable along $\Gamma_{\frac{n+1}{2}}$. $\qquad \square$

Proposition 2.1.47 *The substitution* $u(t,x) \rightarrow t^{-(n+1)} u(t^{-1}, x)$ *induces isomorphisms*

$$\mathcal{H}^{s,\gamma}(X^\wedge) \rightarrow \mathcal{H}^{s,-\gamma}(X^\wedge)$$

for all $s, \gamma \in \mathbb{R}$.

Proof. Let us first assume $s = 0$. Then $u(t,x) \in \mathcal{H}^{0,\gamma}(X^\wedge)$ is equivalent to

$$\int t^n t^{-2\gamma} |u(t,x)|^2 \, dt \, dx < \infty.$$

It follows for $t^{-(n+1)} u(t^{-1}, x)$ and $\tilde{t} = t^{-1}$, $\tilde{t}^{-2} d\tilde{t} = dt$, that

$$\int t^n t^{2\gamma} |t^{-(n+1)} u(t^{-1}, x)|^2 \, dt \, dx = \int \tilde{t}^{-n} \tilde{t}^{-2\gamma} \tilde{t}^{2(n+1)} \tilde{t}^{-2} |u(\tilde{t}, x)|^2 \, d\tilde{t} \, dx$$

$$= \int \tilde{t}^n \tilde{t}^{-2\gamma} |u(\tilde{t}, x)|^2 \, d\tilde{t} \, dx < \infty.$$

An analogous calculation in the reverse direction yields that $u(t,x) \in \mathcal{H}^{0,\gamma}(X^\wedge)$ is equivalent to $t^{-(n+1)} u(t^{-1}, x) \in \mathcal{H}^{0,-\gamma}(X^\wedge)$. For the general case $u(t,x) \in \mathcal{H}^{s,\gamma}(X^\wedge)$ we use $\mathcal{H}^{s,\gamma}(X^\wedge) = t^\gamma \mathcal{H}^{s,0}(X^\wedge)$, i.e., $u(t,x) = t^\gamma u_0(t,x)$ for a corresponding $u_0 \in \mathcal{H}^{s,0}(X^\wedge)$. For proving $t^{-(n+1)} u(t^{-1}, x) \in \mathcal{H}^{s,-\gamma}(X^\wedge)$ it suffices to show that $t^{-(n+1)} u_0(t^{-1}, x) \in \mathcal{H}^{s,0}(X^\wedge)$. To compare norms it suffices to insert elements in $C_0^\infty(X^\wedge)$

which is a dense subspace. In other words we take $u_0 \in C_0^\infty(X^\wedge)$. First we have

$$\|u_0\|^2_{\mathcal{H}^{s,0}(X^\wedge)} = \frac{1}{2\pi i} \int_{\Gamma_{\frac{n+1}{2}}} \|R^s(z)(M_{-\frac{n}{2}}u_0)(z)\|^2_{L^2(X)}dz.$$

Set $v_0(t,x) = t^{-(n+1)}u_0(t^{-1},x)$ and $\check{u}_0(t,x) = u_0(t^{-1},x)$. Then

$$
\begin{aligned}
(M_{-\frac{n}{2}}v_0)(z) &= M(t^{\frac{n}{2}}v_0)(z - \frac{n}{2}) \\
&= M(t^{\frac{n}{2}-(n+1)}\check{u}_0)(z - \frac{n}{2}) \\
&= M(\check{u}_0)(z - (n+1)) \\
&= -M(u_0)(-z + n + 1) \\
&= -M(t^{\frac{n}{2}}u_0)(-z + n + 1 - \frac{n}{2}) \\
&= -M_{-\frac{n}{2}}(u_0)(-z + n + 1).
\end{aligned}
$$

We then obtain

$$\|v_0\|^2_{\mathcal{H}^{s,0}(X^\wedge)} = \frac{1}{2\pi i} \int_{\Gamma_{\frac{n+1}{2}}} \|R^s(z)M_{-\frac{n}{2}}(u_0)(-z + n + 1)\|^2_{L^2(X)}\,dz.$$

We may assume up to equivalence of norms that $R^s(z)$ only depends on $|\operatorname{Im}z|$ as parameter. Inserting $w = -z + n + 1$ we then obtain

$$\|v_0\|^2_{\mathcal{H}^{s,0}(X^\wedge)} = \frac{1}{2\pi i} \int_{\Gamma_{\frac{n+1}{2}}} \|R^s(|\operatorname{Im}w|)M_{-\frac{n}{2}}(u_0)(w)\|^2_{L^2(X)}\,dw,$$

i.e., $\|v_0\|^2_{\mathcal{H}^{s,0}(X^\wedge)} = \|u_0\|^2_{\mathcal{H}^{s,0}(X^\wedge)}$ (under the above choice of $R^s(z)$). Thus, the operator $u_0(t,x) \to t^{-(n+1)}u_0(t^{-1},x)$ extends to an isomorphism $\mathcal{H}^{s,0}(X^\wedge) \to \mathcal{H}^{s,0}(X^\wedge)$ for every $s \in \mathbb{R}$. $\qquad\square$

Theorem 2.1.48 *Let $\varphi(t,x) \in C_0^\infty(\overline{\mathbb{R}}_+ \times X)$ and $\mathcal{M}_\varphi$ be the operator of multiplication by φ. Then $\mathcal{M}_\varphi$ induces continuous operators*

$$\mathcal{M}_\varphi : \mathcal{H}^{s,\gamma}(X^\wedge) \to \mathcal{H}^{s,\gamma}(X^\wedge) \qquad (2.1.78)$$

for every $s,\gamma \in \mathbb{R}$. Moreover, $\varphi \to \mathcal{M}_\varphi$ induces a continuous operator

$$C_0^\infty(\overline{\mathbb{R}}_+ \times X) \to \mathcal{L}(\mathcal{H}^{s,\gamma}(X^\wedge))$$

for every $s,\gamma \in \mathbb{R}$.

Proof. Let us first assume that φ is independent of t and show that $\mathcal{M}_\varphi : \mathcal{H}^{s,\gamma}(X^\wedge) \to \mathcal{H}^{s,\gamma}(X^\wedge)$ is continuous as well as $C^\infty(X) \to \mathcal{L}(\mathcal{H}^{s,\gamma}(X^\wedge))$, induced by $\varphi \to \mathcal{M}_\varphi$. We have

$$\|R^s(z)(M_{\gamma-\frac{n}{2}}\mathcal{M}_\varphi u)(z)\|^2_{L^2(X)}$$
$$= \|R^s(z)\mathcal{M}_\varphi(M_{\gamma-\frac{n}{2}}u)(z)\|^2_{L^2(X)}$$
$$\leq \|R^s(z)\mathcal{M}_\varphi(R^s(z))^{-1}\|_{\mathcal{L}(L^2(X))}\|R^s(z)(M_{\gamma-\frac{n}{2}}u)(z)\|^2_{L^2(X)}.$$

Now $R^s(z)\mathcal{M}_\varphi(R^s(z))^{-1} \in L^0_{cl}(X; \Gamma_{\frac{n+1}{2}-\gamma})$ implies

$$\|R^s(z)\mathcal{M}_\varphi(R^s(z))^{-1}\|_{\mathcal{L}(L^2(X))} \leq c_\varphi$$

for some $c_\varphi > 0$ independent of z. Using (2.1.68) we get immediately the continuity of (2.1.78) for $\varphi \in C^\infty(X)$. To prove the continuity of $\varphi \to \mathcal{M}_\varphi$ it suffices to show that $\varphi \to 0$ in $C^\infty(X)$ implies $c_\varphi \to 0$. However $R^s(z)\mathcal{M}_\varphi$ tends to zero in $L^s_{cl}(X; \Gamma_{\frac{n+1}{2}-\gamma})$ which implies $R^s(z)\mathcal{M}_\varphi(R^s(z))^{-1} \to 0$ in $L^0_{cl}(X; \Gamma_{\frac{n+1}{2}-\gamma})$ for $\varphi \to 0$. Since then the operator norm in $\mathcal{L}(L^2(X))$ tends to zero uniformly in $z \in \Gamma_{\frac{n+1}{2}-\gamma}$, we obtain $c_\varphi \to 0$ as $\varphi \to 0$.

Next we consider x–independent functions $\varphi(t) \in C_0^\infty(\overline{\mathbb{R}}_+)$. For $\varphi(t) = c\omega(t)$ for a cut–off function $\omega(t)$ and a constant c, the continuity of $\mathcal{M}_\varphi$ follows from Remark 2.1.41, since the corresponding result on the standard Sobolev space $H^s(\mathbb{R} \times X)$ is known. At the same time we see that $\mathcal{M}_\varphi$ tends to zero in the operator norm when $c \to 0$. Thus, for $\varphi(t) \in C_0^\infty(\overline{\mathbb{R}}_+)$ in general, we may assume that $\varphi(0) = 0$. Here we can apply analogous arguments as for Proposition 2.1.36. This also implies the continuity of $\varphi \to \mathcal{M}_\varphi$. $\square$

Let us also introduce certain comp– and loc–versions of the spaces $\mathcal{H}^{s,\gamma}(X^\wedge)$, namely

$$\mathcal{H}^{s,\gamma}_{[comp)}(X^\wedge), \quad \mathcal{H}^{s,\gamma}_{[loc)}(X^\wedge), \qquad\qquad (2.1.79)$$

where $\mathcal{H}^{s,\gamma}_{[comp)}(X^\wedge)$ is defined as the subspace of all $u(t,x) \in \mathcal{H}^{s,\gamma}(X^\wedge)$ with $u(t,x) = 0$ for $t > c$ for a constant $c > 0$, dependent on u, while $\mathcal{H}^{s,\gamma}_{[loc)}(X^\wedge)$ is the subspace of all $u(t,x) \in H^s_{loc}(X^\wedge)$ for which $\omega(t)u(t,x) \in \mathcal{H}^{s,\gamma}(X^\wedge)$ for every cut–off function $\omega(t)$ (it suffices to require the latter property for one cut–off function). $\mathcal{H}^{s,\gamma}_{[loc)}(X^\wedge)$ is a Fréchet space and $\mathcal{H}^{s,\gamma}_{[comp)}(X^\wedge)$ is an inductive limit of Hilbert spaces.

Definition 2.1.49 *Let $\mathbb{B}$ be the stretched manifold to a manifold B with conical singularities, cf. (2.1.4). Then $\mathcal{H}^{s,\gamma}(\mathbb{B})$ for $s, \gamma \in \mathbb{R}$ denotes the subspace of all $u \in H^s_{loc}(\text{int }\mathbb{B})$ such that for any cut–off function ω supported by a collar neighbourhood of $[0,1) \times \partial\mathbb{B}$ we have $\omega u \in \mathcal{H}^{s,\gamma}(X^\wedge)$ (in the coordinates $(t,x) \in X^\wedge$, $X \cong \partial\mathbb{B}$).*

Comparing the notation with $\mathcal{H}^{s,\gamma}(X^\wedge)$ it would be more consistent to write $\mathcal{H}^{s,\gamma}(\mathrm{int}\,\mathbb{B})$ instead of $\mathcal{H}^{s,\gamma}(\mathbb{B})$; however we prefer to use the simpler notation in the case of compact $\mathbb{B}$.

The spaces $\mathcal{H}^{s,\gamma}(\mathbb{B})$ can be endowed with the adequate Banach space norms that are generated by Hilbert space scalar products. They arise directly if we write

$$\mathcal{H}^{s,\gamma}(\mathbb{B}) = [\omega]\mathcal{H}^{s,\gamma}(X^\wedge) + [1 - \omega]H_0^s(\mathbb{B}) \qquad (2.1.80)$$

for some cut–off function ω, supported by a collar neighbourhood of $\partial\mathbb{B}$, where $H_0^s(\mathbb{B})$ denotes the closure of $C_0^\infty(\mathrm{int}\,\mathbb{B})$ in $H^s(\tilde{X})$ when $\tilde{X}$ is a closed compact C^∞ manifold of dimension $n + 1$ that contains $\mathbb{B}$ as a submanifold with boundary.

We will usually assume here that B is compact (except for the case that $B = X^\Delta$ for a closed compact C^∞ manifold X). Recall that $\mathbb{B}$ is then a compact C^∞ manifold with boundary. On $\mathbb{B}$ we fix a Riemannian metric (which is C^∞ and non–degenerate up to $\partial\mathbb{B}$), and define the space $L^2(\mathbb{B})$ with respect to the associated measure. Let $h^\varrho \in C^\infty(\mathrm{int}\,\mathbb{B})$ for $\varrho \in \mathbb{R}$ be any fixed strictly positive function with $h^\varrho = t^\varrho$ in a collar neighbourhood of $\partial\mathbb{B}$. Then

$$\mathcal{H}^{0,0}(\mathbb{B}) = h^{-\frac{n}{2}}L^2(\mathbb{B})$$

for $n = \dim\partial\mathbb{B}$ will be taken as the reference space for the scale $\mathcal{H}^{s,\gamma}(\mathbb{B})$, $s,\gamma \in \mathbb{R}$, with the scalar product

$$(u, v)_{\mathcal{H}^{0,0}(\mathbb{B})} = (h^{\frac{n}{2}}u, h^{\frac{n}{2}}v)_{L^2(\mathbb{B})}.$$

Proposition 2.1.50 *For $s \in \mathbb{N}$ and $\gamma \in \mathbb{R}$ the space $\mathcal{H}^{s,\gamma}(\mathbb{B})$ coincides with the subspace of all $u \in \mathcal{H}^{0,\gamma}(\mathbb{B})$ for which $h^\mu Au \in \mathcal{H}^{0,\gamma}(\mathbb{B})$ for every $A \in \mathrm{Diff}^\mu(\mathbb{B})_{\mathrm{Fuchs}}$ and every integer μ, $0 \le \mu \le s$.*

Proof. The result is an obvious consequence of the above Proposition 2.1.45. □

Theorem 2.1.51 *The scalar product $(u,v)_{\mathcal{H}^{0,0}(X^\wedge)}$ for elements $u, v \in C_0^\infty(\mathrm{int}\,\mathbb{B})$ extends to a non–degenerate sesquilinear pairing*

$$(\cdot,\cdot) : \mathcal{H}^{s,\gamma}(\mathbb{B}) \times \mathcal{H}^{-s,-\gamma}(\mathbb{B}) \to \mathbb{C}$$

for every $s,\gamma \in \mathbb{R}$, and we have

$$\|u\|_{\mathcal{H}^{s,\gamma}(\mathbb{B})} \sim \sup\{|(u,v)|/\|v\|_{\mathcal{H}^{-s,-\gamma}(\mathbb{B})} : v \in \mathcal{H}^{-s,-\gamma}(\mathbb{B}), v \ne 0\}.$$

The proof can easily be reduced to Theorem 2.1.46 and to a corresonding result for the standard Sobolev spaces. Details are omitted.

Remark 2.1.52 *An easy consequence of Remark* 2.1.44 *and of the interpolation properties of the standard Sobolev spaces is the following relation:*

$$\mathcal{H}^{s,\gamma}(\mathbb{B}) = [\mathcal{H}^{s_0,\gamma}(\mathbb{B}), \mathcal{H}^{s_1,\gamma}(\mathbb{B})]_\theta$$

for $0 \le \theta \le 1$, $s_0 \le s_1$, $s = (1 - \theta)s_0 + \theta s_1$ *and arbitrary* $\gamma \in \mathbb{R}$, *where* $[\cdot, \cdot]_\theta$ *is the interpolation functor with respect to the complex interpolation method.*

Theorem 2.1.53 *There are canonical continuous embeddings*

$$\mathcal{H}^{s',\gamma'}(\mathbb{B}) \hookrightarrow \mathcal{H}^{s,\gamma}(\mathbb{B})$$

for all $s' \ge s$, $\gamma' \ge \gamma$ *which are compact for* $s' > s$, $\gamma' > \gamma$.

Proof. Write the spaces $\mathcal{H}^{s,\gamma}(\mathbb{B})$ in the form (2.1.80). Then it suffices to show that the embedding

$$[\omega]\mathcal{H}^{s',\gamma'}(X^\wedge) \hookrightarrow [\omega]\mathcal{H}^{s,\gamma}(X^\wedge) \qquad (2.1.81)$$

is continuous for $s' \ge s$, $\gamma' \ge \gamma$ and compact for $s' > s$, $\gamma' > \gamma$, since $[1 - \omega]H_0^{s'}(\mathbb{B}) \hookrightarrow [1 - \omega]H_0^s(\mathbb{B})$ is known to be continuous for $s' \ge s$ and compact for $s' > s$. From Remark 2.1.41 we get an isomorphism

$$S_{\gamma - \frac{n}{2}} : [\omega]\mathcal{H}^{s,\gamma}(X^\wedge) \to [\tilde{\omega}]H^s(\mathbb{R} \times X) \qquad (2.1.82)$$

with $\tilde{\omega}(r) = \omega(e^{-r})$, $S_{\gamma - \frac{n}{2}}(\omega(t)u(t,x)) = \omega(e^{-r})e^{-(\frac{1}{2} - (\gamma - \frac{n}{2}))r}u(e^{-r}, x)$. Moreover, $S_{\gamma - \frac{n}{2}}$ induces an isomorphism

$$S_{\gamma - \frac{n}{2}} : [\omega]\mathcal{H}^{s',\gamma'}(X^\wedge) \to [\tilde{\omega}]e^{(\gamma - \gamma')r}H^{s'}(\mathbb{R} \times X). \qquad (2.1.83)$$

This shows that $[\omega]\mathcal{H}^{s',\gamma'}(X^\wedge)$ is a subspace of $[\omega]\mathcal{H}^{s,\gamma}(X^\wedge)$ which is continuously embedded for $s' \ge s$, $\gamma' \ge \gamma$. Furthermore the embeddings

$$[\tilde{\omega}]e^{(\gamma - \gamma')r}H^{s'}(\mathbb{R} \times X) \hookrightarrow [\tilde{\omega}]H^s(\mathbb{R} \times X)$$

are compact for $s' > s$, $\gamma' > \gamma$, since $e^{(\gamma - \gamma')r}r^\delta = \varphi_\delta(r)$ and all derivatives in r are uniformly bounded on $\operatorname{supp}\tilde{\omega}$ for every $\delta > 0$. This yields

$$[\tilde{\omega}]e^{(\gamma - \gamma')r}H^{s'}(\mathbb{R} \times X) = [\tilde{\omega}][\varphi_\delta]r^{-\delta}H^{s'}(\mathbb{R} \times X)$$
$$\hookrightarrow [\tilde{\omega}]r^{-\delta}H^{s'}(\mathbb{R} \times X)$$
$$\hookrightarrow [\tilde{\omega}]H^s(\mathbb{R} \times X),$$

where the last map is compact. Thus the preimage with respect to (2.1.83) is compactly embedded into the preimage with respect to (2.1.82) which is the desired compact embedding (2.1.81) for $s' > s$, $\gamma' > \gamma$. $\qquad \square$

The method to prove the compactness of the embedding in Theorem 2.1.53 suggests a generalisation of the spaces $\mathcal{H}^{s,\gamma}(\mathbb{B})$ by an additional weight. Let us set

$$H^s_\varrho(\mathbb{R} \times X) = (1 + r^2)^{-\frac{\varrho}{2}} H^s(\mathbb{R} \times X) \qquad \text{for } \rho \in \mathbb{R}$$

and define

$$\mathcal{H}^{s,\gamma}_\varrho(\mathbb{B}) = S^{-1}_{\gamma - \frac{n}{2}} [\tilde{\omega}] H^{s,\gamma}_\varrho(\mathbb{R} \times X) + [1 - \tilde{\omega}] H^s_{\text{loc}}(\text{int }\mathbb{B}).$$

In other words, if $l^\varrho \in C^\infty(\text{int }\mathbb{B})$ is a strictly positive function with $l^\varrho = (-\log t)^{-\varrho}$ in a neighourhood of $\partial\mathbb{B}$ (in the splitting into (t,x)) then

$$\mathcal{H}^{s,\gamma}_\varrho(\mathbb{B}) = l^\varrho \mathcal{H}^{s,\gamma}(\mathbb{B})$$

for arbitrary $s, \gamma, \varrho \in \mathbb{R}$.

Remark 2.1.54 *There are canonical continuous embeddings*

$$\mathcal{H}^{s',\gamma'}_{\varrho'}(\mathbb{B}) \hookrightarrow \mathcal{H}^{s,\gamma}_\varrho(\mathbb{B})$$

for all $s' \geq s$, $\gamma' \geq \gamma$, $\varrho' \geq \varrho$, which are compact for $s' > s$, $\gamma' > \gamma$ or $s' > s$, $\gamma' \geq \gamma$, $\varrho' > \varrho$. Observe also that

$$\mathcal{H}^{s',\gamma'}_\delta(\mathbb{B}) \hookrightarrow \mathcal{H}^{s,\gamma}_\varrho(\mathbb{B})$$

is compact for $s' > s$, $\gamma' > \gamma$ and for arbitrary $\delta, \varrho \in \mathbb{R}$.

Let us now pass to the invariance of the $\mathcal{H}^{s,\gamma}(\mathbb{B})$–spaces under diffeomorphisms. If $\mathbb{B}$ and $\mathbb{D}$ are the stretched manifolds belonging to compact manifolds B and D, respectively, with conical singularities, then an isomorphism $B \to D$ in the sense of Definition 2.1.1 gives rise to a diffeomorphism

$$\chi : \mathbb{B} \to \mathbb{D} \tag{2.1.84}$$

between the associated C^∞ manifolds with boundary. On the other hand, every such (2.1.84) corresponds to an isomorphism $B \to D$.

Theorem 2.1.55 *If $\chi : \mathbb{B} \to \mathbb{D}$ is a diffeomorphism then the function pull–back $\chi^* : C^\infty_0(\text{int }\mathbb{D}) \to C^\infty_0(\text{int }\mathbb{B})$ extends to isomorphisms*

$$\chi^* : \mathcal{H}^{s,\gamma}(\mathbb{D}) \to \mathcal{H}^{s,\gamma}(\mathbb{B})$$

for all $s, \gamma \in \mathbb{R}$.

Proof. In view of $\mathcal{H}^{0,\gamma}(\mathbb{B}) = h^{\gamma}L^2(\mathbb{B})$ the result is clear for $s = 0$. For arbitrary $s \in \mathbb{N}$ it suffices to apply Proposition 2.1.50 and Remark 2.1.9. For negative integers s we can apply Theorem 2.1.51 and for arbitrary $s \in \mathbb{R}$ the interpolation property from Remark 2.1.52. □

Remark 2.1.56 *If $\varphi \in C^{\infty}(\mathbb{B})$ and $\mathcal{M}_{\varphi}$ denotes the operator of multiplication by φ then $\mathcal{M}_{\varphi}$ induces continuous operators*

$$\mathcal{M}_{\varphi} : \mathcal{H}^{s,\gamma}(\mathbb{B}) \to \mathcal{H}^{s,\gamma}(\mathbb{B})$$

for every $s, \gamma \in \mathbb{R}$. Moreover, $\varphi \to \mathcal{M}_{\varphi}$ induces a continuous operator $C^{\infty}(\mathbb{B}) \to \mathcal{L}(\mathcal{H}^{s,\gamma}(\mathbb{B}))$ for every $s, \gamma \in \mathbb{R}$.

This follows easily from Theorem 2.1.48 for $[\omega]\mathcal{H}^{s,\gamma}(\mathbb{B})$ for a cut–off function ω supported by a collar neighbourhood of $\partial\mathbb{B}$ and from an analogous property in the context of standard Sobolev spaces, applied to $[1 - \omega]H_0^s(\mathbb{B})$. Another argument for Remark 2.1.56 is to first consider $s \in \mathbb{N}$, for which the result is relatively trivial, and then to pass to $s \in \mathbb{Z}$ by duality and to $s \in \mathbb{R}$ by interpolation.

For purposes below it will be useful to introduce an analogue of the spaces $\mathcal{K}^{s,\gamma}(\mathbb{R}_+)$, cf. Definition 2.1.35, for the case $X^{\wedge} = \mathbb{R}_+ \times X$. To this end we choose an open covering $\mathcal{U} = \{U_1, \ldots, U_N\}$ of X by coordinate neighbourhoods and a subordinate partition of unity $\{\varphi_1, \ldots, \varphi_N\}$. Let $\eta_j : U_j \to E_j$ be diffeomorphisms to open subsets $E_j \subset S^n = \{\tilde{x} \in \mathbb{R}^{n+1} : |\tilde{x}| = 1\}$, $j = 1, \ldots, N$. Moreover, set $\kappa_j(t, x) = t\eta_j(x)$ for $x \in U_j$, $t \in \mathbb{R}_+$, which defines a diffeomorphism $\kappa_j : U_j^{\wedge} \to E_j^{\wedge} := \{\tilde{x} \in \mathbb{R}^{n+1} : \tilde{x}/|\tilde{x}| \in E_j\}$ for $j = 1, \ldots, N$. Then, if $\omega(t)$ is any fixed cut–off function, $H_{\text{cone}}^s(X^{\wedge})$ for $s \in \mathbb{R}$ denotes the completion of $C_0^{\infty}(X^{\wedge})$ with respect to the norm

$$\|v\|_{H_{\text{cone}}^s(X^{\wedge})} := \left\{ \|\omega v\|_{H^s(\mathbb{R}_+ \times X)}^2 + \sum_{j=1}^{N} \|(\kappa_j^*)^{-1}(1 - \omega)\varphi_j v\|_{H^s(\mathbb{R}^{n+1})}^2 \right\}^{\frac{1}{2}},$$

κ_j^* being the pull–back of functions with respect to κ_j, and $H^s(\mathbb{R}_+ \times X) = H^s(\mathbb{R} \times X)|_{\mathbb{R}_+ \times X}$. It can easily be proved that this is a correct definition, i.e., independent of the system of charts, of the partition of unity and of ω.

Definition 2.1.57 *Let $s, \gamma \in \mathbb{R}$ be arbitrary and $\omega(t)$ be a fixed cut–off function. Then we set*

$$\mathcal{K}^{s,\gamma}(X^{\wedge}) = [\omega]\mathcal{H}^{s,\gamma}(X^{\wedge}) + [1 - \omega]H_{\text{cone}}^s(X^{\wedge})$$

endowed with the topology of the non–direct sum.

In view of Proposition 2.1.42 the space $\mathcal{K}^{s,\gamma}(X^{\wedge})$ is independent of the particular choice of ω. Each $\mathcal{K}^{s,\gamma}(X^{\wedge})$ can be endowed with a Banach space norm which is generated by a Hilbert space scalar product. If $k^{\delta} \in C^{\infty}(\mathbb{R}_{+})$ is a strictly positive function with

$$k^{\delta}(t) = \begin{cases} t^{\delta} & \text{for } 0 < t < c_0, \\ 1 & \text{for } t > c_1, \end{cases} \tag{2.1.85}$$

with constants $c_0 < c_0 < \infty$, then

$$\mathcal{K}^{s,\gamma}(X^{\wedge}) = k^{\gamma}\mathcal{K}^{s,0}(X^{\wedge}) \tag{2.1.86}$$

for all $s, \gamma \in \mathbb{R}$. Moreover, we have

$$\mathcal{K}^{0,0}(X^{\wedge}) = t^{-\frac{n}{2}}L^2(\mathbb{R}_{+} \times X) \tag{2.1.87}$$

for the L^2–space with respect to $dt\,dx$ on $\mathbb{R}_{+} \times X$.

Remark 2.1.58 *The space $C_0^{\infty}(X^{\wedge})$ is dense in $\mathcal{K}^{s,\gamma}(X^{\wedge})$ for each $s, \gamma \in \mathbb{R}$.*

This property as well as the following observations on the spaces $\mathcal{K}^{s,\gamma}(X^{\wedge})$ are easy consequences of corresponding facts on the $\mathcal{H}^{s,\gamma}(\cdot)$ and $H^s(\cdot)$ spaces.

We will take $\mathcal{K}^{0,0}(X^{\wedge}) = \mathcal{H}^{0,0}(X^{\wedge})$ as the reference space with the scalar product from (2.1.87). Then

$$(\cdot,\cdot)_{\mathcal{K}^{0,0}(X^{\wedge})} : C_0^{\infty}(X^{\wedge}) \times C_0^{\infty}(X^{\wedge}) \to \mathbb{C}$$

extends to a non–degenerate sesquilinear pairing

$$\mathcal{K}^{s,\gamma}(X^{\wedge}) \times \mathcal{K}^{-s,-\gamma}(X^{\wedge}) \to \mathbb{C} \tag{2.1.88}$$

for every $s, \gamma \in \mathbb{R}$.

Remark 2.1.59 *There are canonical continuous embeddings*

$$\mathcal{K}^{s',\gamma'}(X^{\wedge}) \hookrightarrow \mathcal{K}^{s,\gamma}(X^{\wedge})$$

for all $s' \geq s$, $\gamma' \geq \gamma$.

Let us set

$$(\kappa_{\lambda}u)(t,x) = \lambda^{\frac{n+1}{2}}u(\lambda t, x)$$

for $\lambda \in \mathbb{R}_{+}$, $u(t,x) \in \mathcal{K}^{s,\gamma}(X^{\wedge})$, $n = \dim X$. We then obtain a group of isomorphisms

$$\kappa_{\lambda} : \mathcal{K}^{s,\gamma}(X^{\wedge}) \to \mathcal{K}^{s,\gamma}(X^{\wedge})$$

with $\kappa_{\lambda}\kappa_{\delta} = \kappa_{\lambda\delta}$ for all $\lambda, \delta \in \mathbb{R}_{+}$ and every $s, \gamma \in \mathbb{R}$.

Proposition 2.1.60 *We have*

$$\{\kappa_\lambda\}_{\lambda\in\mathbb{R}_+} \in C\left(\mathbb{R}_+, \mathcal{L}_\sigma(\mathcal{K}^{s,\gamma}(X^\wedge))\right) \qquad (2.1.89)$$

for every $s, \gamma \in \mathbb{R}$. *Here* $\mathcal{L}_\sigma(\cdot)$ *is the space of all linear continuous operators in the corresponding space, endowed with the strong operator topology.*

The proof is obvious and will be omitted.

2.2 Cone pseudo–differential operators without asymptotics

2.2.1 Mellin pseudo–differential operators

The parameter–dependent pseudo–differential operators from Section 1.2 will now be interpreted as $L^\mu(X)$–valued symbols $h(z)$ of Mellin pseudo–differential operators, here with the parameter space $\mathbb{R} \ni \tau$, identified with $\Gamma_\beta = \{z \in \mathbb{C} : \operatorname{Re} z = \beta\}$ for any $\beta \in \mathbb{R}$, with $z = \beta + i\tau$ being the Mellin covariable. We denote by $L^\mu(X; \Gamma_\beta)$, $L_{\mathrm{cl}}^\mu(X; \Gamma_\beta)$ the corresponding (Fréchet) spaces of parameter– dependent pseudo–differential operators. X is a closed compact C^∞ manifold, $n = \dim X$.

More generally we also employ (t, t')–dependent Mellin symbols

$$h(t, t', z) \in C^\infty(\mathbb{R}_+ \times \mathbb{R}_+, L^\mu(X; \Gamma_\beta)) \; (\in C^\infty(\mathbb{R}_+ \times \mathbb{R}_+, L_{\mathrm{cl}}^\mu(X; \Gamma_\beta))).$$

It is interesting to control the smoothness up to $t = 0$, $t' = 0$, i.e., to consider

$$h(t, t', z) \in C^\infty(\overline{\mathbb{R}}_+ \times \overline{\mathbb{R}}_+, L^\mu(X; \Gamma_\beta)) \; (\in C^\infty(\overline{\mathbb{R}}_+ \times \overline{\mathbb{R}}_+, L_{\mathrm{cl}}^\mu(X; \Gamma_\beta))).$$

Functions or distributions on $X^\wedge = \mathbb{R}_+ \times X \ni (t, x)$ are also denoted by $u(t)$ when they are regarded as $\mathcal{D}'(X)$–valued ones on $\mathbb{R}_+ \ni t$. Then $h(t, t', z)u(t')$ can be interpreted as the pseudo–differential action on X, applied to the values of $u(t')$ for every fixed t, t', z.

Analogously to (2.1.55) we set

$$\operatorname{op}_M^\gamma(h)u(t) = M_{\gamma, z \to t}^{-1}\{M_{\gamma, t' \to z} h(t, t', z)u(t')\}$$

for $h(t, t', z) \in C^\infty(\mathbb{R}_+ \times \mathbb{R}_+, L^\mu(X; \Gamma_{\frac{1}{2} - \gamma}))$, $\gamma \in \mathbb{R}$, and $\operatorname{op}_M(\cdot) = \operatorname{op}_M^0(\cdot)$. Let us first assume $u \in C_0^\infty(\mathbb{R}_+, C^\infty(X))$. The particular choice of the weight γ will not be important here. Therefore we shall often restrict the calculations to a convenient weight, e.g., $\gamma = \frac{1}{2}$.

Note that

$$\mathrm{op}_M^\gamma(h) : C_0^\infty(\mathbb{R}_+, C^\infty(X)) \to C^\infty(\mathbb{R}_+, C^\infty(X))$$

is continuous for every $h(t,t',z) \in C^\infty(\mathbb{R}_+ \times \mathbb{R}_+, L^\mu(X; \Gamma_{\frac{1}{2}-\gamma}))$.
The expression for $\mathrm{op}_M^\gamma(h)u(t)$ can be written as a double integral

$$\mathrm{op}_M^\gamma(h)u(t) = \int\limits_{\mathbb{R}} \int\limits_0^\infty t^{-(\frac{1}{2}-\gamma+i\tau)}(t')^{\frac{1}{2}-\gamma+i\tau} h(t,t',\tfrac{1}{2}-\gamma+i\tau)u(t')\frac{dt'}{t'}d\tau$$

$$(2.2.1)$$

as far as we assume μ to be negative enough. Setting $h_\gamma(t,t',z) = (t/t')^{-(\frac{1}{2}-\gamma)}h(t,t',z)$ which is again an element of the space $C^\infty(\mathbb{R}_+ \times \mathbb{R}_+, L^\mu(X; \Gamma_{\frac{1}{2}-\gamma}))$, we obtain

$$\mathrm{op}_M^\gamma(h)u(t) = \int\limits_{\mathbb{R}} \int\limits_0^\infty e^{-i(\log t - \log t')\tau} h_\gamma(t,t',\tfrac{1}{2}-\gamma+i\tau)u(t')\frac{dt'}{t'}d\tau.$$

The phase function $\varphi(t,t',\tau) = -(\log t - \log t')\tau$ satisfies the conditions of Theorem 1.1.37 with $\Omega = \mathbb{R}_+$. It is a pseudo–differential phase function in this sense. The same is true of $-(\log t - \log t')\tau + (x - x')\xi$ which arises as phase function after localising the operators along X in a coordinate neighbourhood, with local coordinates $x \in \mathbb{R}^n$ and co-variables $\xi \in \mathbb{R}^n$. This yields immediately $\mathrm{op}_M^\gamma(h) \in L^\mu(\mathbb{R}_+ \times X)$, cf. also Proposition 2.2.1 below. Thus the arguments on oscillatory integrals from Section 1.1.2 are valid also here in the situation when the Fourier transform in the t–direction is replaced by the Mellin transform. In particular, the double integrals in (2.2.1) make sense in the usual distributional sense for each order μ.

From the simple connection between the Mellin and the Fourier transform, cf. (2.1.29), we get

$$\mathrm{op}_M^\gamma(h)u(t) = S_\gamma^{-1}F_{\tau\to r}^{-1}\{F_{r'\to\tau}S_\gamma h(t,t',z)u(t')\}.$$

In particular, if $h(z)$ is independent of t,t', it follows that

$$\mathrm{op}_M^\gamma(h)u(t) = S_\gamma^{-1}\{F_{\tau\to r}^{-1}h(1/2-\gamma+i\tau)F_{r'\to\tau}\}(S_\gamma u)(r'), \quad (2.2.2)$$

i.e., a substitution of new coordinates on $\mathbb{R}_+$ turns the Mellin pseudo–differential operators to pseudo–differential operators based on the Forier transform. Although the calculus could be formulated by using the Fourier transform as well we prefer to preserve the geometry of the underlying space.

An important aspect is the control of symbols and distributions up to $t = 0$ (in particular, smoothness or more general asymptotics near $t = 0$). This is another reason why we avoid substituting coordinates $\mathbb{R}_+ \ni t \to r \in \mathbb{R}$; it is more intuitive not to replace everywhere the Mellin transform by the Fourier transform. In the following Section 2.2.2 we shall return to this discussion under the point of view of operator conventions.

Let us set

$$M_\gamma L^\mu(X^\wedge) = \left\{ \mathrm{op}_M^{\gamma - \frac{n}{2}}(h) : h(t,t',z) \in C^\infty(\mathbb{R}_+ \times \mathbb{R}_+, L^\mu(X; \Gamma_{\frac{n+1}{2} - \gamma})) \right\}$$

and define $M_\gamma L_{\mathrm{cl}}^\mu(X^\wedge)$ analogously by the condition

$$h(t,t',z) \in C^\infty(\mathbb{R}_+ \times \mathbb{R}_+, L_{\mathrm{cl}}^\mu(X; \Gamma_{\frac{n+1}{2} - \gamma})).$$

Proposition 2.2.1 *For every $\gamma \in \mathbb{R}$ we have*

$$M_\gamma L^\mu(X^\wedge) = L^\mu(X^\wedge), \quad M_\gamma L_{\mathrm{cl}}^\mu(X^\wedge) = L_{\mathrm{cl}}^\mu(X^\wedge).$$

Proof. Let us first consider $\gamma = \frac{n+1}{2}$. Then

$$\mathrm{op}_M^{\frac{1}{2}}(h)u(t) = \frac{1}{2\pi i} \int_{\Gamma_0} t^{-z} \left\{ \int_{\mathbb{R}_+} (t')^{z-1} h(t,t',z) u(t') \, dt' \right\} dz$$

$$= \int_{\mathbb{R}} \int_{\mathbb{R}_+} \left(\frac{t}{t'} \right)^{-i\tau} h(t,t',i\tau) u(t') \, \frac{dt'}{t'} \, d\tau.$$

Setting $y = -\log t$, $y' = -\log t'$ it follows that $(t/t')^{-i\tau} = e^{i(y-y')\tau}$ and thus

$$(\mathrm{op}_M^{\frac{1}{2}}(h)u)(e^{-y}) = \iint e^{i(y-y')\tau} h(e^{-y}, e^{-y'}, i\tau) u(e^{-y'}) \, dy' d\tau.$$

For $p(y, y', \tau) = h(e^{-y}, e^{-y'}, i\tau)$ and $y = \chi(t) = -\log t$, $\chi : \mathbb{R}_+ \to \mathbb{R}$, we thus obtain

$$\mathrm{op}(p) = \chi_* \, \mathrm{op}_M^{\frac{1}{2}}(h)$$

with

$$\mathrm{op}(p)v(y) = \iint e^{i(y-y')\tau} p(y, y', \tau) v(y') \, dy' d\tau,$$

where χ_* is the operator push–forward of operators under χ. Since the space of all operators $\mathrm{op}(p)$ arbitrary $p(y, y', \tau) \in C^\infty(\mathbb{R} \times \mathbb{R}, L^\mu(X; \mathbb{R}))$ coincides with $L^\mu(\mathbb{R} \times X)$ and since

$$\chi_* : L^\mu(\mathbb{R}_+ \times X) \to L^\mu(\mathbb{R} \times X)$$

is an isomorphism, we obtain $M_{\frac{n+1}{2}} L^\mu(X^\wedge) = L^\mu(X^\wedge)$, and the same for classical operators.

Concerning arbitrary $\gamma \in \mathbb{R}$ we can use the identity

$$\mathrm{op}_M^{\gamma-\frac{n}{2}}(h) = t^{\gamma-\frac{n+1}{2}} \,\mathrm{op}_M^{\frac{1}{2}}(T^{-\gamma+\frac{n+1}{2}} h) t^{-\gamma+\frac{n+1}{2}}$$

with the multiplication operators by the corresponding powers of t. This implies

$$M_\gamma L^\mu(X^\wedge) = t^{\gamma-\frac{n+1}{2}} M_{\frac{n+1}{2}} L^\mu(X^\wedge) t^{-\gamma+\frac{n+1}{2}}$$
$$= t^{\gamma-\frac{n+1}{2}} L^\mu(X^\wedge) t^{-\gamma+\frac{n+1}{2}} = L^\mu(X^\wedge),$$

and the same for classical operators. $\square$

Theorem 2.2.2 *Let* $h(t,t',z) \in C^\infty(\overline{\mathbb{R}}_+ \times \overline{\mathbb{R}}_+, L^\mu(X; \Gamma_{\frac{n+1}{2}-\gamma}))$ *be independent of* t, t' *for* $t > c$, $t' > c'$ *for certain constants* $c, c' > 0$. *Then* $\mathrm{op}_M^{\gamma-\frac{n}{2}}(h)$ *induces continuous operators*

$$\mathrm{op}_M^{\gamma-\frac{n}{2}}(h) : \mathcal{H}^{s,\gamma}(X^\wedge) \to \mathcal{H}^{s-\mu,\gamma}(X^\wedge)$$

for all $s \in \mathbb{R}$.

Proof. Let us first assume that h is independent of t, t'. Then

$$\mathrm{op}_M^{\gamma-\frac{n}{2}}(h)u(t) = M_{\gamma-\frac{n}{2},z\to t}^{-1} h(z)\{M_{\gamma-\frac{n}{2},t'\to z} u(t')\}.$$

By definition we have

$$\|f\|_{\mathcal{H}^{s,\gamma}(X^\wedge)} = \left\{ \frac{1}{2\pi i} \int_{\Gamma_{\frac{n+1}{2}-\gamma}} \|R^s(z)(M_{\gamma-\frac{n}{2}}f)(z)\|_{L^2(X)}^2 \, dz \right\}^{\frac{1}{2}}$$

with an $R^s(z) \in L_{cl}^\mu(X; \Gamma_{\frac{n+1}{2}-\gamma})$, $s \in \mathbb{R}$, that is parameter–dependent elliptic, with parameter $\tau = \mathrm{Im}\, z$, and which induces isomorphisms $R^s(z) : H^r(X) \to H^{r-s}(X)$ for all $z \in \Gamma_{\frac{n+1}{2}-\gamma}$, $r \in \mathbb{R}$.

Thus

$$\| \mathrm{op}_M^{\gamma-\frac{n}{2}}(h)u\|_{\mathcal{H}^{s-\mu,\gamma}(X^\wedge)}^2$$
$$= \frac{1}{2\pi i} \int_{\Gamma_{\frac{n+1}{2}-\gamma}} \|R^{s-\mu}(z)h(z)(M_{\gamma-\frac{n}{2}}u)(z)\|_{L^2(X)}^2 \, dz$$
$$\leq \sup_{z \in \Gamma_{\frac{n+1}{2}-\gamma}} \|R^{-\mu}(z)h(z)\|_{\mathcal{L}(L^2(X))}^2$$
$$\times \frac{1}{2\pi i} \int_{\Gamma_{\frac{n+1}{2}-\gamma}} \|R^s(z)(M_{\gamma-\frac{n}{2}}u)(z)\|_{L^2(X)}^2 \, dz.$$

This shows the asserted continuity, since

$$\| \operatorname{op}_M^{\gamma-\frac{n}{2}}(h)\|_{s,s-\mu}^2 \leq \sup_{z \in \Gamma_{\frac{n+1}{2}-\gamma}} \|R^{-\mu}(z)h(z)\|_{\mathcal{L}(L^2(X))}^2 < \infty,$$

cf. Corollary 1.2.20. In this proof $\|.\|_{s,r}$ denotes the operator norm in the space $\mathcal{L}(\mathcal{H}^{s,\gamma}(X^\wedge), \mathcal{H}^{r,\gamma}(X^\wedge))$. At the same time we see that the operator norm of $\operatorname{op}_M^{\gamma-\frac{n}{2}}(h)$ tends to zero when $h(z)$ tends to zero in the space $L^\mu(X; \Gamma_{\frac{n+1}{2}-\gamma})$.

We now assume $h(t,t',z) \in C^\infty(\overline{\mathbb{R}}_+ \times \overline{\mathbb{R}}_+, L^\mu(X; \Gamma_{\frac{n+1}{2}-\gamma}))$ and write $h(t,t',z) = f(t,t',z) + f_\infty(z)$ with (t,t')–independent $f_\infty(z)$ and $f(t,t',z)$ vanishing for $t \geq c$, $t' \geq c'$. In view of the first part of the proof it remains to consider $f(t,t',z)$. Let

$$C_0^\infty([0,c]_0) \tag{2.2.3}$$

denote the (Fréchet) space of all $\varphi(t) \in C_0^\infty(\overline{\mathbb{R}}_+)$ with $\varphi(t) = 0$ for $t \geq c$. Then $f(t,t',z)$ can be regarded as an element of $C_0^\infty([0,c]_0) \otimes_\pi L^\mu(X; \Gamma_{\frac{n+1}{2}-\gamma}) \otimes_\pi C_0^\infty([0,c']_0)$. According to Theorem 1.1.6 we then have a representation

$$f(t,t',z) = \sum_{j=0}^\infty \lambda_j \varphi_j(t) f_j(z) \psi_j(t')$$

as a convergent sum with $\lambda_j \in \mathbb{C}$, $\sum |\lambda_j| < \infty$, and elements $\varphi_j \in C_0^\infty([0,c]_0)$, $\psi_j \in C_0^\infty([0,c']_0)$, $f_j \in L^\mu(X; \Gamma_{\frac{n+1}{2}-\gamma})$ tending to zero in the corresponding spaces for $j \to \infty$. From Theorem 2.1.48 we obtain

$$\|\mathcal{M}_{\varphi_j}\|_{s,s} \to 0, \quad \|\mathcal{M}_{\psi_j}\|_{r,r} \to 0 \qquad \text{for } j \to \infty$$

for each $s,r \in \mathbb{R}$. Thus

$$\| \operatorname{op}_M^{\gamma-\frac{n}{2}}(f)\|_{s,s-\mu} = \| \sum_{j=0}^\infty \lambda_j \mathcal{M}_{\varphi_j} \operatorname{op}_M^{\gamma-\frac{n}{2}}(f_j) \mathcal{M}_{\psi_j}\|_{s,s-\mu}$$

$$\leq \sum_{j=0}^\infty |\lambda_j| \|\mathcal{M}_{\varphi_j}\|_{s-\mu,s-\mu} \| \operatorname{op}_M^{\gamma-\frac{n}{2}}(f_j)\|_{s,s-\mu} \|\mathcal{M}_{\psi_j}\|_{s,s}$$

$$< \infty$$

which is the desired continuity in the general case. □

Definition 2.2.3 $M_\gamma L^{-\infty}(\overline{\mathbb{R}}_+ \times X)$ *denotes the space of all operators*

$$A : \mathcal{H}_{[\text{comp}]}^{s,\gamma}(X^\wedge) \to \mathcal{H}_{[\text{loc}]}^{\infty,\gamma}(X^\wedge)$$

that are continuous for all $s \in \mathbb{R}$, and for which the formal adjoint A^* (with respect to the $\mathcal{H}^{0,0}(X^\wedge)$–scalar product) induces continuous operators

$$A^* : \mathcal{H}^{s,-\gamma}_{[\text{comp}]}(X^\wedge) \to \mathcal{H}^{\infty,-\gamma}_{[\text{loc}]}(X^\wedge)$$

for all $s \in \mathbb{R}$. Moreover,

$$M_\gamma L^\mu(\overline{\mathbb{R}}_+ \times X) \qquad \left(M_\gamma L^\mu_{\text{cl}}(\overline{\mathbb{R}}_+ \times X) \right),$$

$\gamma, \mu \in \mathbb{R}$, *denotes the space of all operators* $\text{op}_M^{\gamma-\frac{n}{2}}(h) + C$ *for arbitrary symbols* $h(t, t', z) \in C^\infty(\overline{\mathbb{R}}_+ \times \overline{\mathbb{R}}_+, L^\mu(X; \Gamma_{\frac{n+1}{2}-\gamma}))$ *and* $h(t, t', z) \in C^\infty(\overline{\mathbb{R}}_+ \times \overline{\mathbb{R}}_+, L^\mu_{\text{cl}}(X; \Gamma_{\frac{n+1}{2}-\gamma}))$, *respectively, and* $C \in M_\gamma L^{-\infty}(\overline{\mathbb{R}}_+ \times X)$.

It can be proved that the condition for A^* in Definition 2.2.3 is automatically satisfied if the first condition for A is required.

Remark 2.2.4 *Let* $l(t, t', z) \in C^\infty(\overline{\mathbb{R}}_+ \times \overline{\mathbb{R}}_+, L^{-\infty}(X; \Gamma_{\frac{n+1}{2}-\gamma}))$. *Then* $\text{op}_M^{\gamma-\frac{n}{2}}(l) \in M_\gamma L^{-\infty}(\overline{\mathbb{R}}_+ \times X)$.

The specific point in the consideration of the operator classes $M_\gamma L^\mu(\overline{\mathbb{R}}_+ \times X)$ and $M_\gamma L^\mu_{\text{cl}}(\overline{\mathbb{R}}_+ \times X)$ is a control of the properties in a neighbourhood of $t = 0$. We shall systematically investigate the elements of the corresponding calculus in the following Sections 2.2.2, 2.2.3, in particular, the compositions and formal adjoints.

Formal adjoints of operators will always be defined in terms of a reference scalar product. Let

$$A : \mathcal{H}^{s,\gamma}_{[\text{comp}]}(X^\wedge) \to \mathcal{H}^{s-\mu,\varrho}_{[\text{loc}]}(X^\wedge)$$

be a continuous operator for all $s \in \mathbb{R}$ with a certain order $\mu \in \mathbb{R}$ and fixed weights $\gamma, \varrho \in \mathbb{R}$. Then, if we do not indicate otherwise, the formal adjoint A^* will be understood in the sense

$$(Au, v)_{\mathcal{H}^{0,0}(X^\wedge)} = (u, A^*v)_{\mathcal{H}^{0,0}(X^\wedge)}$$

for all $u, v \in C_0^\infty(X^\wedge)$. In this case the reference scalar product is $(\cdot, \cdot)_{\mathcal{H}^{0,0}(X^\wedge)}$. Then, the operator A^* extends to continuous operators

$$A^* : \mathcal{H}^{s,-\varrho}_{[\text{comp}]}(X^\wedge) \to \mathcal{H}^{s-\mu,-\gamma}_{[\text{loc}]}(X^\wedge)$$

for all $s \in \mathbb{R}$. It can be convenient in some cases also to allow another reference scalar product $(f, g)_{(\delta)} := (f, g)_{\mathcal{H}^{0,\delta}(X^\wedge)}$ for some fixed $\delta \in \mathbb{R}$. Define A^*_δ, the formal adjoint of A with respect to $(\cdot, \cdot)_{(\delta)}$ as

$$(Au, v)_{(\delta)} = (u, A^*_\delta v)_{(\delta)}$$

for all $u, v \in C_0^\infty(X^\wedge)$. In view of $(f, g)_{(\delta)} = (t^{-\delta/2} f, t^{-\delta/2} g)_{(0)}$ we then obtain

$$
\begin{aligned}
(Au, v)_{(\delta)} &= (t^{-\delta/2} Au, t^{-\delta/2} v)_{(0)} \\
&= (u, A^* t^{-\delta} v)_{(0)} \\
&= (t^{\delta/2} u, t^{\delta/2} A^* t^{-\delta} v)_{(0)} \\
&= (u, t^\delta A^* t^{-\delta} v)_{(\delta)},
\end{aligned}
$$

i.e., $A_\delta^* = t^\delta A^* t^{-\delta}$. In this case A_δ^* extends to continuous operators

$$
A^* : \mathcal{H}_{[\mathrm{comp}]}^{s, \delta - \varrho}(X^\wedge) \to \mathcal{H}_{[\mathrm{loc}]}^{s-\mu, \delta-\gamma}(X^\wedge)
$$

for all $s \in \mathbb{R}$.

Let us consider an example. If $f(z) \in L^\mu(X; \Gamma_{\frac{n+1}{2} - \gamma})$ then

$$
\mathrm{op}_M^{\gamma - \frac{n}{2}}(f) : \mathcal{H}^{s, \gamma}(X^\wedge) \to \mathcal{H}^{s-\mu, \gamma}(X^\wedge)
$$

is continuous for all $s \in \mathbb{R}$. We then have $(\mathrm{op}_M^{\gamma - \frac{n}{2}}(f))^* = \mathrm{op}_M^{-\gamma - \frac{n}{2}}(f^*)$ for $f^*(z) = f^{(*)}(n + 1 - \bar{z})$, where $(*)$ indicates the pointwise formal adjoint in the sense of the scalar product of $L^2(X)$, $n = \dim X$.

More generally it can easily be proved that for every $f(t, t', z) \in C^\infty(\overline{\mathbb{R}}_+ \times \overline{\mathbb{R}}_+, L^\mu(X; \Gamma_{\frac{n+1}{2} - \gamma}))$ we have

$$
\left(\mathrm{op}_M^{\gamma - \frac{n}{2}}(f) \right)^* = \mathrm{op}_M^{-\gamma - \frac{n}{2}}(f^*) \tag{2.2.4}
$$

for $f^*(t, t', z) = f^{(*)}(t', t, n + 1 - \bar{z})$. This gives us the following result:

Theorem 2.2.5 $A \in M_\gamma L^\mu(\overline{\mathbb{R}}_+ \times X)$ *implies* $A^* \in M_{-\gamma} L^\mu(\overline{\mathbb{R}}_+ \times X)$ *for every* $\gamma, \mu \in \mathbb{R}$. *An analogous result holds for classical operators.*

Theorem 2.2.6 *Let* $h(t, t', z) \in C^\infty(\mathbb{R}_+ \times \mathbb{R}_+, N_O^\mu(X))$ *and* $\delta, \varrho \in \mathbb{R}$. *Then*

$$
\mathrm{op}_M^\delta(h)u = \mathrm{op}_M^\varrho(h)u
$$

for all $u \in C_0^\infty(\mathbb{R}_+, C^\infty(X))$.

Proof. We have

$$
\mathrm{op}_M^\delta(h)u(t) = \frac{1}{2\pi i} \int\limits_0^\infty \int\limits_{\Gamma_{\frac{1}{2} - \delta}} t^{-z}(t')^{z-1} h(t, t', z) u(t')\, dz\, dt',
$$

interpreted as an oscillatory integral for every fixed t. For every polynomial $p(z)$ we have $(t')^{z-1} = p^{-1}(z)p(t'\partial_{t'})(t')^{z-1}$. Integration by parts yields

$$\mathrm{op}_M^\delta(h)u(t)$$

$$= \frac{1}{2\pi i}\int_0^\infty \int_{\Gamma_{\frac{1}{2}-\delta}} t^{-z}(t')^{z-1}p(-t'\partial_{t'})\{p^{-1}(z)h(t,t',z)u(t')\}\,dz\,dt'.$$

The same can be done for the weight ϱ. Choosing $p(z)$ in such a way that $p^{-1}(z)h(t,t',z)$ is of order < -1 with respect to $|\operatorname{Im} z|$ the integral converges in the usual sense. Set, for instance, $p(z) = (c - z)^N$ for $N \in \mathbb{N}$, $N > \mu+1$ and a real constant c such that $p^{-1}(z)$ is holomorphic in a vertical strip containing $\Gamma_{\frac{1}{2}-\gamma}$ and $\Gamma_{\frac{1}{2}-\varrho}$ (say $c < \min\{\frac{1}{2}-\delta, \frac{1}{2}-\varrho\}$). Then we obtain the result by Cauchy's theorem. □

Proposition 2.2.7 *To every* $f(t,t',z) \in C^\infty(\mathbb{R}_+ \times \mathbb{R}_+, L^\mu(X;\Gamma_0))$ *there exists an* $\boldsymbol{f}(t,z) \in C^\infty(\mathbb{R}_+, L^\mu(X;\Gamma_0))$ *such that*

$$\mathrm{op}_M^{\frac{1}{2}}(f) = \mathrm{op}_M^{\frac{1}{2}}(\boldsymbol{f}) \bmod L^{-\infty}(X^\wedge),$$

and $\boldsymbol{f}(t,z)$ *allows the asymptotic expansion*

$$\boldsymbol{f}(t,z) \sim \sum_{k=0}^\infty \frac{1}{k!}\partial_z^k\left(-t'\frac{\partial}{\partial t'}\right)^k f(t,t',z)|_{t'=t}.$$

Proof. We have

$$\mathrm{op}_M^{\frac{1}{2}}(f)u(t)|_{t=e^{-y}} = \iint (t/t')^{-i\tau} f(t,t',i\tau)u(t')\frac{dt'}{t'}\,d\tau|_{t=e^{-y}}$$

$$= \iint e^{i(y-y')\tau} f(e^{-y}, e^{-y'}, i\tau)u(e^{-y'})\,dy'\,d\tau.$$

The right hand side equals $\mathrm{op}(q)v$ for $v(y) = u(e^{-y})$ and $q(y,y',\tau) = f(e^{-y}, e^{-y'}, i\tau)$. Now, from standard pseudo–differential calculus we know that there is a $\boldsymbol{q}(y,\tau)$ such that $\mathrm{op}(q) = \mathrm{op}(\boldsymbol{q}) \bmod L^{-\infty}(\mathbb{R} \times X)$, where $\boldsymbol{q}(y,\tau) \sim \sum_k \frac{1}{k!}D_\tau^k\partial_{y'}^k q(y,y',\tau)|_{y'=y}$. It is now clear that

$$D_\tau^k\partial_{y'}^k q(y,y',\tau)|_{y'=y} = \{\partial_z^k(-t'\partial_{t'})^k f(t,t',z)|_{t'=t}\}|_{t=e^{-y}}.$$

This gives us the desired result. □

Below in Proposition 2.2.16 we will obtain an analogous relation for arbitrary weights, i.e., for $\mathrm{op}_M^\delta(\cdot)$, $\delta \in \mathbb{R}$, instead of $\mathrm{op}_M^{\frac{1}{2}}(\cdot)$.

2.2.2   Kernel cut–off and Mellin operator convention

The kernel cut–off constructions from Section 1.2 also plays a role for Mellin pseudo–differential operators.

In the following we assume for simplicity $\gamma = \frac{1}{2}$; the case of arbitrary weights is completely analogous. Let $a(t, t', z) \in C^\infty(\mathbb{R}_+ \times \mathbb{R}_+, L^\mu(X, \Gamma_0))$ and set

$$k(a)(t, t', z) = \frac{1}{2\pi i} \int\limits_{\Gamma_0} \varrho^{-z} a(t, t', z) \, dz = (M^{-1}_{\frac{1}{2}, z \to \varrho} a)(t, t', \varrho).$$

Then

$$\mathrm{op}_M^{\frac{1}{2}}(a)u(t) = \int\limits_0^\infty k(a)(t, t', \frac{t}{t'})u(t') \frac{dt'}{t'},$$

$u \in C_0^\infty(\mathbb{R}_+, C^\infty(X))$. As noted above, the expressions can be interpreted in the distributional sense via oscillatory integrals.

Define

$$\mathcal{T}(\mathbb{R}_+, L^{-\infty}(X)) = \{(M^{-1}_{\frac{1}{2}, z \to \varrho} f)(\varrho) : f(z) \in L^{-\infty}(X; \Gamma_0)\},$$

endowed with the (nuclear) Fréchet topology from the bijection

$$\mathcal{T}(\mathbb{R}_+, L^{-\infty}(X)) \simeq L^{-\infty}(X; \Gamma_0)$$

which is induced by $M_{\frac{1}{2}}$. The scalar analogue of this space is

$$\mathcal{T}(\mathbb{R}_+) = \{(M^{-1}_{\frac{1}{2}, z \to \varrho} f)(\varrho) : f(z) \in \mathcal{S}(\Gamma_0)\}.$$

We then have

$$\mathcal{T}(\mathbb{R}_+, L^{-\infty}(X)) = \mathcal{T}(\mathbb{R}_+) \otimes_\pi L^{-\infty}(X). \qquad (2.2.5)$$

Note that

$$\mathcal{T}^\gamma(\mathbb{R}_+) := \{(M^{-1}_{\frac{1}{2} - \gamma, z \to \varrho} f)(\varrho) : f(z) \in \mathcal{S}(\Gamma_{\frac{1}{2} - \gamma})\} = \varrho^\gamma \mathcal{T}(\mathbb{R}_+)$$

for arbitrary $\gamma \in \mathbb{R}$, with $\mathcal{T}^0(\mathbb{R}_+) = \mathcal{T}(\mathbb{R}_+)$. An analogous relation holds in the $L^{-\infty}(X)$–valued case.

Theorem 2.2.8 *For every* $a(z) \in L^\mu(X; \Gamma_0)$ *and every* $\psi(\varrho) \in C_0^\infty(\mathbb{R}_+)$ *with* $\psi(\varrho) = 1$ *in a neighbourhood of* $\varrho = 1$ *we have*

$$(1 - \psi(\varrho))k(a)(\varrho) \in \mathcal{T}(\mathbb{R}_+, L^{-\infty}(X)) \qquad (2.2.6)$$

and

$$a_0(z) := (M_{\frac{1}{2},\varrho\to z}\psi(\varrho)k(a)(\varrho))(z) \in L^\mu(X;\Gamma_0), \qquad (2.2.7)$$

where $a(z)-a_0(z) \in L^{-\infty}(X;\Gamma_0)$. *Moreover, there is an element* $h(z) \in N_O^\mu(X)$ *with* $a_0(z) = h(z)|_{\Gamma_0}$. *An analogous result holds for* $a(z) \in L^\mu_{cl}(X;\Gamma_0)$; *then* $h(z) \in M_O^\mu(X)$.

We shall prove Theorem 2.2.8 in several steps, where we obtain more generally the following result:

Theorem 2.2.9 *Let* $a(z) \in L^\mu(X;\Gamma_0)$, $\varphi(\varrho) \in C_0^\infty(\mathbb{R}_+)$, *and*

$$H(\varphi)a)(z) = M_{\frac{1}{2},\varrho\to z}\{\varphi(\varrho)k(a)(\varrho)\}.$$

Then $(\varphi,a) \to H(\varphi)a$ *defines a separately continuous operator*

$$C_0^\infty(\mathbb{R}_+) \times L^\mu(X;\Gamma_0) \to N_O^\mu(X)$$

which restricts to a separately continuous operator

$$C_0^\infty(\mathbb{R}_+) \times L^\mu_{cl}(X;\Gamma_0) \to M_O^\mu(X).$$

We call

$$H(\varphi) : L^\mu(X;\Gamma_0) \to N_O^\mu(X)$$

a kernel cut–off operator.

Theorem 2.2.10 *Let* $a(t,t',z) \in C^\infty(\overline{\mathbb{R}}_+\times\overline{\mathbb{R}}_+, L^\mu(X,\Gamma_0))$ *and* $\varphi(\varrho) \in C_0^\infty(\mathbb{R}_+)$. *Then* $(\varphi,a) \to H(\varphi)a$ *with*

$$(H(\varphi)a)(t,t',z) := M_{\varrho\to z}(\varphi(\varrho)k(a)(t,t',\varrho))(z)$$

induces a separately continuous operator

$$C_0^\infty(\mathbb{R}_+) \times C^\infty(\overline{\mathbb{R}}_+ \times \overline{\mathbb{R}}_+, L^\mu(X,\Gamma_0)) \to C^\infty(\overline{\mathbb{R}}_+ \times \overline{\mathbb{R}}_+, N_O^\mu(X))$$

which restricts to a separately continuous operator

$$C_0^\infty(\mathbb{R}_+) \times C^\infty(\overline{\mathbb{R}}_+ \times \overline{\mathbb{R}}_+, L^\mu_{cl}(X,\Gamma_0)) \to C^\infty(\overline{\mathbb{R}}_+ \times \overline{\mathbb{R}}_+, M_O^\mu(X)).$$

If $\psi \in C_0^\infty(\mathbb{R}_+)$ *equals* 1 *in a neighbourhood of* $\varrho = 1$ *we have*

$$(1 - \psi(\varrho))k(a)(t,t',\varrho) \in C^\infty(\overline{\mathbb{R}}_+ \times \overline{\mathbb{R}}_+, \mathcal{T}(\mathbb{R}_+, L^{-\infty}(X)))$$

and

$$a(t,t',z) - H(\psi)a(t,t',z)|_{\Gamma_0} \in C^\infty(\overline{\mathbb{R}}_+ \times \overline{\mathbb{R}}_+, L^{-\infty}(X,\Gamma_0)).$$

An analogous result holds for $a(t,t',z) \in C^\infty(\overline{\mathbb{R}}_+ \times \mathbb{R}_+, L^\mu(X,\Gamma_0))$ $(a(t,t',z) \in C^\infty(\overline{\mathbb{R}}_+ \times \mathbb{R}_+, L^\mu_{cl}(X,\Gamma_0)))$.

Theorem 2.2.10 is a corollary of Theorems 2.2.8 and 2.2.9.

In order to interpret the results it is useful to observe that for every $\varphi(\varrho) \in C_0^\infty(\mathbb{R}_+)$ and $a(z) \in L^\mu(X; \Gamma_0)$ we have

$$\varphi(\varrho)k(a)(\varrho) \in \mathcal{E}'(\mathbb{R}_+, L^\mu(X))$$

for $\mathcal{E}'(\mathbb{R}_+, L^\mu(X)) = \mathcal{L}(C^\infty(\mathbb{R}_+), L^\mu(X))$, and the map

$$C_0^\infty(\mathbb{R}_+) \times L^\mu(X; \Gamma_0) \to \mathcal{E}'(\mathbb{R}_+, L^\mu(X))$$

given by $(\varphi, a) \to \varphi k(a)$ is separately continuous. Moreover, we have $H(\varphi)a(z) \in \mathcal{A}(\mathbb{C}, L^\mu(X))$, and the map

$$C_0^\infty(\mathbb{R}_+) \times L^\mu(X; \Gamma_0) \to \mathcal{A}(\mathbb{C}, L^\mu(X)),$$

given by $(\varphi, a) \to H(\varphi)a$ is separately continuous. Analogous relations hold in the classical case. The holomorphic dependence of $H(\varphi)a(z)$ on z is clear, since it is defined as the Mellin transform of a distribution with compact support. Moreover, we see from (2.2.6) that the singular support of $k(a)(\varrho) \in \mathcal{D}'(\mathbb{R}_+, L^\mu(X))$ is contained in $\{1\}$. Analogous relations for the case of pseudo–differential boundary value problems with the transmission property were discussed in Schrohe and Schulze [106].

Proof of Theorem 2.2.8. In order to prove the relation (2.2.6) we employ (2.2.5). A semi–norm system for the Fréchet topology of $\mathcal{T}(\mathbb{R}_+)$ follows by a corresponding semi–norm system on $\mathcal{S}(\Gamma_0)$ and from the isomorphism $M_{\frac{1}{2}}^{-1} : \mathcal{S}(\Gamma_0) \to \mathcal{T}(\mathbb{R}_+)$. For $f(z) \in \mathcal{S}(\Gamma_0)$ we take the semi–norms

$$p_{M,N}(f) = \left\{ \int_{-\infty}^{\infty} |D_\tau^M \tau^N f(i\tau)|^2 \, d\tau \right\}^{\frac{1}{2}},$$

$M, N \in \mathbb{N}$. This induces on $\mathcal{T}(\mathbb{R}_+) \ni u(\varrho)$ the semi–norm

$$q_{M,N}(u) = \left\{ \int_0^\infty |\log^M \varrho \left(\varrho \frac{\partial}{\partial \varrho} \right)^N u(\varrho)|^2 \frac{d\varrho}{\varrho} \right\}^{\frac{1}{2}},$$

cf. Remark 2.1.17. Then, if $\{\pi_\iota\}_{\iota \in \mathbb{N}}$ is a semi–norm system on the space $L^{-\infty}(X)$, the topology on $\mathcal{T}(\mathbb{R}_+) \otimes_\pi L^{-\infty}(X)$ is given by

$$\alpha_{M,N,\iota}(k) = \left\{ \int_0^\infty \pi_\iota^2 \left(\log^M \varrho \left(\varrho \frac{\partial}{\partial \varrho} \right)^N k(\varrho) \right) \frac{d\varrho}{\varrho} \right\}^{\frac{1}{2}}, \qquad (2.2.8)$$

$M, N, \iota \in \mathbb{N}$. Now we have for arbitrary $m \in \mathbb{N}$ the relation

$$
\begin{aligned}
(1 - \psi(\varrho))k(a)(\varrho) &= (1 - \psi(\varrho))\frac{1}{2\pi}\int \varrho^{i\tau} a(i\tau)\, d\tau \\
&= (1 - \psi(\varrho))\log^{-m}\varrho\frac{1}{2\pi}\int \{-D_\tau^m \varrho^{-i\tau}\}a(i\tau)\, d\tau \\
&= (1 - \psi(\varrho))\log^{-m}\varrho\frac{1}{2\pi}\int \varrho^{-i\tau} D_\tau^m a(i\tau)\, d\tau.
\end{aligned}
$$

This shows that $(1 - \psi(\varrho))k(a)(\varrho)$ belongs to $L^{-\infty}(X)$ for every fixed $\varrho \in \mathbb{R}_+$. Now let us show that the semi–norms for the $\mathcal{T}$–space with respect to ϱ for $(1 - \psi)k(a)$ are finite.

For arbitrary M, N, m we have

$$
\log^M \varrho \left(\varrho\frac{\partial}{\partial\varrho}\right)^N (1 - \psi(\varrho))k(a)(\varrho)
$$
$$
= \log^M \varrho\frac{1}{2\pi}\int \left(\varrho\frac{\partial}{\partial\varrho}\right)^N \{\varrho^{-i\tau}(1 - \psi(\varrho))\log^{-m}\varrho\}D_\tau^m a(i\tau)\, d\tau.
$$

Applying the Leibniz rule we see that the integral is a linear combination of terms of the form

$$
\log^{M-m-l}\varrho\psi_j(\varrho)\int_{-\infty}^{\infty} \varrho^{-i\tau}\tau^h D_\tau^m a(i\tau)\, d\tau
$$

with $\psi_j(\varrho) = (\varrho\partial_\varrho)^j(1-\psi(\varrho)) \in C_b^\infty(\mathbb{R}_+)$, $h+j+l = N$. Choose a semi–norm system $\{\pi_\iota\}_{\iota\in\mathbb{N}}$ on $L^{-\infty}(X)$ such that each π_ι is a semi–norm on $L^{\mu-\iota}(X)$. If we fix N, M and ι and choose $m > \mu + N + \iota + 2$ then $(1 + \tau^2)\tau^h D_\tau^m a(i\tau) \in L^{-\iota}(X; \mathbb{R}_\tau)$ and $\tau^h D_\tau^m a(i\tau) \in L^2(\mathbb{R}_\tau, L^{-\iota}(X))$. In view of $M_{\frac12}^{-1}: L^2(\mathbb{R}_\tau, L^{-\iota}(X)) \to \varrho^{\frac12}L^2(\mathbb{R}_\tau, L^{-\iota}(X))$ we obtain

$$
M_{\frac12,\tau\to\varrho}^{-1}(\tau^h D_\tau a(i\tau)) = \frac{1}{2\pi i}\int \varrho^{-i\tau}\tau^h D_\tau^m a(i\tau)\, d\tau
$$
$$
\in \varrho^{\frac12}L^2(\mathbb{R}_{+,\varrho}, L^{-\iota}(X)).
$$

This yields for all h, j, l

$$
\pi_\iota\left(\log^{M-m-l}\varrho\psi_j(\varrho)M_{\frac12,\tau\to\varrho}^{-1}\{\tau^h D_\tau^m a(i\tau)\}\right) \in L^2\left(\mathbb{R}_{+,\varrho}, \frac{d\varrho}{\varrho}\right).
$$

Thus all semi–norms in (2.2.8) are finite and depend continuously on the semi–norms for a in $L^\mu(X; \Gamma_0)$. This completes the proof of the relation (2.2.6).

It follows that $r(z) = M_{\frac{1}{2}, \varrho \to z}((1-\psi)k(a))(z)$ belongs to $L^{-\infty}(X; \Gamma_0)$ which means $a(z) - a_0(z) \in L^{-\infty}(X; \Gamma_0)$ and $a_0(z) \in L^\mu(X; \Gamma_0)$, cf. (2.2.7). It is also obvious that $a_0(z)$ is classical if $a(z)$ is classical.

In order to prove the last statement of Theorem 2.2.8 we can pass at once to the more general Theorem 2.2.9. We have

$$a_0(z) = M_{\varrho \to z}\{\varphi(\varrho) M_{\frac{1}{2}, \tilde{z} \to \varrho}^{-1} a(\tilde{z})\}(z),$$

$$(M_\beta u)(\frac{1}{2} - \beta + i\tau) = (F S_\beta u)(\tau), \qquad \beta \in \mathbb{R},$$

with $(S_\beta u)(r) = e^{-(\frac{1}{2}-\beta)r} u(e^{-r})$ and the Fourier transform $(F_{r \to \tau} v)(\tau)$ $= \int e^{-ir\tau} v(r)\, dr$, cf. (2.1.29). This yields

$$M\varphi(\varrho) M_{\frac{1}{2}}^{-1} a(i\tau) = F_{r \to \tau}\{S_{\frac{1}{2}, \varrho \to r} \varphi(\varrho) S_{\frac{1}{2}, r \to \varrho}^{-1}\} F_{\tilde{\tau} \to r}^{-1} a(i\tilde{\tau})(\tau)$$

$$= F_{r \to \tau}(\tilde{\varphi}(r) F_{\tilde{\tau} \to r}^{-1} a(i\tilde{\tau}))(\tau)$$

for $\tilde{\varphi}(r) = \varphi(e^{-r}) \in C_0^\infty(\mathbb{R})$. The desired property was formulated above in Theorem 1.2.47. This completes the proof of the Theorems 2.2.8, 2.2.9.   □

Remark 2.2.11 *Using the notation of Theorem 2.2.10 and setting* $\psi_R(\varrho) = \psi(\varrho/R)$ *for* $R > 0$ *one can easily prove the following approximation property of the kernel cut-off construction (cf. also the above Remark 1.1.51). Set* $a_R(t, t', z) = H(\psi_R)a(t, t', z)|_{\Gamma_0}$. *Then*

$$a_R(t, t', z) \to a(t, t', z) \qquad \text{for } R \to \infty$$

in the space $C^\infty(\overline{\mathbb{R}}_+ \times \overline{\mathbb{R}}_+, L^\mu(X; \Gamma_0))$. *An analogous result holds in the classical case.*

Remark 2.2.12 *The Theorems* 2.2.8, 2.2.9, 2.2.10 *and Remark* 2.2.11 *have obvious analogues for arbitrary weights, i.e., with* $\Gamma_{\frac{1}{2} - \delta}$, $\delta \in \mathbb{R}$, *instead of* Γ_0.

Corollary 2.2.13 *The operator class* $M_\gamma L^\mu(\overline{\mathbb{R}}_+ \times X)$ *can equivalently be described as the set of all* $\mathrm{op}_M^{\gamma - \frac{n}{2}} + C$ *with* $h(t, t', z) \in C^\infty(\overline{\mathbb{R}}_+ \times \overline{\mathbb{R}}_+, N_O^\mu(X))$ *and* $C \in M_\gamma L^{-\infty}(\overline{\mathbb{R}}_+ \times X)$. *Similarly* $M_\gamma L_{cl}^\mu(\overline{\mathbb{R}}_+ \times X)$ *can be written in the indicated way with* $M_O^\mu(X)$ *instead of* $N_O^\mu(X)$.

In fact, from Theorem 2.2.10 and Remark 2.2.12 we obtain that to every $f(t, t', z) \in C^\infty(\overline{\mathbb{R}}_+ \times \overline{\mathbb{R}}_+, L^\mu(X; \Gamma_{\frac{n+1}{2} - \gamma}))$ there is an $h(t, t', z) \in C^\infty(\overline{\mathbb{R}}_+ \times \overline{\mathbb{R}}_+, N_O^\mu(X))$ such that

$$f(t, t', z) - h(t, t', z)|_{\Gamma_{\frac{n+1}{2} - \gamma}}$$

$$=: l(t, t', z) \in C^\infty(\overline{\mathbb{R}}_+ \times \overline{\mathbb{R}}_+, L^{-\infty}(X; \Gamma_{\frac{n+1}{2} - \gamma})).$$

In view of Remark 2.2.4 we obtain $\operatorname{op}_M^{\gamma-\frac{n}{2}}(f) = \operatorname{op}_M^{\gamma-\frac{n}{2}}(h)$ mod $M_\gamma L^{-\infty}(\overline{\mathbb{R}}_+ \times X)$.

Remark 2.2.14 *It may happen that an operator* $\operatorname{op}_M^{\gamma-\frac{n}{2}}(f)$ *belongs to* $M_\gamma L^\mu(\overline{\mathbb{R}}_+ \times X)$ *although* $f(t,t',z) \in C^\infty(\overline{\mathbb{R}}_+ \times \overline{\mathbb{R}}_+, L^\mu(X; \Gamma_{\frac{n+1}{2}-\gamma}))$ *is not* C^∞ *in* t, t' *up to* $t = 0$, $t' = 0$. *In fact, if we start with an* $l(z) \in L^\mu(X; \Gamma_{\frac{n+1}{2}-\gamma})$ *and choose a function* $\psi(\varrho) \in C_0^\infty(\mathbb{R}_+)$ *with* $\psi(\varrho) = 1$ *in a neighbourhood of* $\varrho = 1$, *we obtain by the constructions in the proof of Theorem 2.2.10*

$$\operatorname{op}_M^{\gamma-\frac{n}{2}}(l) = \operatorname{op}_M^{\gamma-\frac{n}{2}}(h) + C$$

for $C \in M_\gamma L^{-\infty}(\overline{\mathbb{R}}_+ \times X)$ *and* $h(z) = (H(\psi)f)(z)$. *However, if we set* $f(t,t',z) = \psi(t/t')l(z)$ *we obtain* $\operatorname{op}_M^{\gamma-\frac{n}{2}}(h) = \operatorname{op}_M^{\gamma-\frac{n}{2}}(f)$, *and* $f(t,t',z) \notin C^\infty(\overline{\mathbb{R}}_+ \times \overline{\mathbb{R}}_+, L^\mu(X; \Gamma_{\frac{n+1}{2}-\gamma}))$.

Proposition 2.2.15 *For every* $\gamma, \varrho \in \mathbb{R}$ *we have*

$$t^\varrho M_\gamma L^\mu(\overline{\mathbb{R}}_+ \times X)t^{-\varrho} = M_{\gamma+\varrho}L^\mu(\overline{\mathbb{R}}_+ \times X).$$

Proof. The relation $t^\varrho M_\gamma L^{-\infty}(\overline{\mathbb{R}}_+ \times X)t^{-\varrho} = M_{\gamma+\varrho}L^{-\infty}(\overline{\mathbb{R}}_+ \times X)$ is obvious because of $t^\varrho \mathcal{H}^{s,\gamma}(X^\wedge) = \mathcal{H}^{s,\gamma+\varrho}(X^\wedge)$. It remains to charac-terise $t^\varrho \operatorname{op}_M^{\gamma-\frac{n}{2}}(h)t^{-\varrho}$. Because of Corollary 2.2.13 it suffices to consider $h(t,t',z) \in C^\infty(\overline{\mathbb{R}}_+ \times \overline{\mathbb{R}}_+, N_O^\mu(X))$. Then

$$t^\varrho \operatorname{op}_M^{\gamma-\frac{n}{2}}(h)t^{-\varrho} = \operatorname{op}_M^{\gamma+\varrho-\frac{n}{2}}(T^\varrho h)$$

with $(T^\varrho h)(t,t',z) = h(t,t',z + \varrho)$, cf. Theorem 2.2.6. This shows $\operatorname{op}_M^{\gamma+\varrho-\frac{n}{2}}(T^\varrho h) \in M_{\gamma+\varrho}L^\mu(\overline{\mathbb{R}}_+ \times X)$. □

Proposition 2.2.16 *Let* $\delta \in \mathbb{R}$ *be arbitrary and fixed. Then to ev-ery* $f(t,t',z) \in C^\infty(\mathbb{R}_+ \times \mathbb{R}_+, L^\mu(X; \Gamma_{\frac{1}{2}-\delta}))$ *there exists an* $\boldsymbol{f}(t,z) \in C^\infty(\mathbb{R}_+, L^\mu(X; \Gamma_{\frac{1}{2}-\delta}))$ *such that*

$$\operatorname{op}_M^\delta(f) = \operatorname{op}_M^\delta(\boldsymbol{f}) \bmod L^{-\infty}(X^\wedge)$$

with

$$\boldsymbol{f}(t,z) \sim \sum_{k=0}^\infty \frac{1}{k!}\partial_z^k \left(-t'\frac{\partial}{\partial t'}\right)^k f(t,t',z)|_{t'=t}.$$

Proof. Applying the analogue of Theorem 2.2.10 for $\Gamma_{\frac{1}{2}-\delta}$ instead of Γ_0 we find an $h(t,t',z) \in C^\infty(\mathbb{R}_+ \times \mathbb{R}_+, N_O^\mu(X))$ such that

$$\operatorname{op}_M^\delta(f) = \operatorname{op}_M^\delta(h) \bmod L^{-\infty}(X^\wedge).$$

Thus it suffices to show the result for $h(t, t', z)$. In view of Theorem 2.2.6 we have $\mathrm{op}_M^\delta(h) = \mathrm{op}_M^{\frac{1}{2}}(h)$ on $C_0^\infty(\mathbb{R}_+, C^\infty(X))$. Applying Proposition 2.2.7 we obtain an $\boldsymbol{h}(t, z) \in C^\infty(\mathbb{R}_+, L^\mu(X; \Gamma_0))$ such that $\mathrm{op}_M^{\frac{1}{2}}(h) = \mathrm{op}_M^{\frac{1}{2}}(\boldsymbol{h}) \bmod L^{-\infty}(X^\wedge)$, where

$$\boldsymbol{h}(t, t) \sim \sum \frac{1}{k} \partial_z^k \left(-t' \frac{\partial}{\partial t'} \right)^k h(t, t', z)|_{t'=t}.$$

This result can also be understood as follows. To every $M \in \mathbb{N}$ there is an $N \in \mathbb{N}$ such that

$$\mathrm{op}_M^{\frac{1}{2}}(h) - \mathrm{op}_M^{\frac{1}{2}} \left(\sum_{k=0}^{N} \frac{1}{k!} \partial_z^k \left(-t' \frac{\partial}{\partial t'} \right)^k h(t, t', z)|_{t'=t} \right) \in L^{\mu-M}(X^\wedge).$$

Since $h(t, t', z)$ is holomorphic in z, we get by Theorem 2.2.6

$$\mathrm{op}_M^\delta(h) - \mathrm{op}_M^\delta \left(\sum_{k=0}^{N} \frac{1}{k!} \partial_z^k \left(-t' \frac{\partial}{\partial t'} \right)^k h(t, t', z)|_{t'=t} \right) \in L^{\mu-M}(X^\wedge).$$

Since $h(t, t', z)|_{\Gamma_{\frac{1}{2}-\delta}} - f(t, t', z) \in C^\infty(\mathbb{R}_+ \times \mathbb{R}_+, L^{-\infty}(X; \Gamma_0))$, we may replace $h|_{\frac{1}{2}-\delta}$ by f which means

$$\mathrm{op}_M^\delta(f) - \mathrm{op}_M^\delta \left(\sum_{k=0}^{N} \frac{1}{k!} \partial_z^k \left(-t' \frac{\partial}{\partial t'} \right)^k f(t, t', z)|_{t'=t} \right) \in L^{\mu-M}(X^\wedge).$$

This completes the proof.   □

We now turn to the natural symbols for the corresponding pseudo–differential calculus near conical singularities.

Definition 2.2.17 $\tilde{S}^\mu(\overline{\mathbb{R}}_+ \times \Omega \times \mathbb{R}^{n+1})$ *for an open set* $\Omega \subseteq \mathbb{R}^n$ *and* $\mu \in \mathbb{R}$ *denotes the space of all* $p(t, x, \tau, \xi) \in C^\infty(\mathbb{R}_+ \times \Omega \times \mathbb{R}^{n+1})$ *for which there is a* $\tilde{p}(t, x, \tilde{\tau}, \xi) \in S^\mu(\overline{\mathbb{R}}_+ \times \Omega \times \mathbb{R}_{\tilde{\tau}, \xi}^{n+1})$ *with*

$$p(t, x, \tau, \xi) = \tilde{p}(t, x, t\tau, \xi). \tag{2.2.9}$$

In an analogous manner we define the subspaces $\tilde{S}_{\mathrm{cl}}^\mu(\overline{\mathbb{R}}_+ \times \Omega \times \mathbb{R}^{n+1})$ *by requiring* $\tilde{p}(t, x, \tilde{\tau}, \xi) \in S_{\mathrm{cl}}^\mu(\overline{\mathbb{R}}_+ \times \Omega \times \mathbb{R}_{\tilde{\tau}, \xi}^{n+1})$.

In view of the obvious isomorphisms $\tilde{S}^\mu(\ldots) \cong S^\mu(\ldots)$, $\tilde{S}_{\mathrm{cl}}^\mu(\ldots) \cong S_{\mathrm{cl}}^\mu(\ldots)$ the spaces $\tilde{S}^\mu(\ldots)$ and $\tilde{S}_{\mathrm{cl}}^\mu(\ldots)$ are Fréchet in canonical way.

The symbols $p \in \tilde{S}^\mu(\ldots)$ or $\in \tilde{S}_{\mathrm{cl}}^\mu(\ldots)$ are also said to be of Fuchs type. In the calculus this will usually occur in combination with a weight factor $t^{-\mu}$.

If $\chi(\tilde{\tau}, \xi)$ is an excision function in $\mathbb{R}^{n+1}_{\tilde{\tau}, \xi}$ (i.e., $\chi \in C^{\infty}(\mathbb{R}^{n+1})$, $\chi(\tilde{\tau}, \xi) = 0$ for $|\tilde{\tau}, \xi| < c_0$, $\chi(\tilde{\tau}, \xi) = 1$ for $|\tilde{\tau}, \xi| > c_1$ with constants $0 < c_0 < c_1 < \infty$) then $p(t, x, \tau, \xi) \in \tilde{S}^{\mu}(\overline{\mathbb{R}}_+ \times \Omega \times \mathbb{R}^{n+1})$ implies $\chi(t\tau, \xi)p(t, x, \tau, \xi) \in \tilde{S}^{\mu}(\overline{\mathbb{R}}_+ \times \Omega \times \mathbb{R}^{n+1})$ and $(1 - \chi(t\tau, \xi))p(t, x, \tau, \xi) \in \tilde{S}^{-\infty}(\overline{\mathbb{R}}_+ \times \Omega \times \mathbb{R}^{n+1})$.

Proposition 2.2.18 *To every sequence* $p_j \in \tilde{S}^{\mu_j}(\overline{\mathbb{R}}_+ \times \Omega \times \mathbb{R}^{n+1})$, $j \in \mathbb{N}$, *with* $\mu_j \to -\infty$ *as* $j \to \infty$ *there exists a* $p \in \tilde{S}^{\mu}(\overline{\mathbb{R}}_+ \times \Omega \times \mathbb{R}^{n+1})$, $\mu = \max\{\mu_j\}$, *such that for each* M *there exists an* $N(M)$ *with*

$$p - \sum_{j=0}^{N} p_j \in \tilde{S}^{\mu-M}(\overline{\mathbb{R}}_+ \times \Omega \times \mathbb{R}^{n+1}),$$

for all $N \geq N(M)$, *and every such* p *is unique* $\mathrm{mod}\tilde{S}^{-\infty}(\overline{\mathbb{R}}_+ \times \Omega \times \mathbb{R}^{n+1})$.

We write $p \sim \sum_{j=0}^{\infty} p_j$ and call p the asymptotic sum of the p_j.

The proof of Proposition 2.2.18 is straightforward. It suffices to consider $\tilde{p}_j(t, x, \tilde{\tau}, \xi) \in S^{\mu_j}(\overline{\mathbb{R}}_+ \times \Omega \times \mathbb{R}^{n+1}_{\tilde{\tau}, \xi})$ associated with $p_j(t, x, \tau, \xi)$ and to form the usual asymptotic sum $\tilde{p} \sim \sum_{j=0}^{\infty} \tilde{p}_j$, $\tilde{p}(t, x, \tilde{\tau}, \xi) \in S^{\mu}(\overline{\mathbb{R}}_+ \times \Omega \times \mathbb{R}^{n+1}_{\tilde{\tau}, \xi})$. Then p obtained by (2.2.9) satisfies $p \sim \sum_{j=0}^{\infty} p_j$.

Remark 2.2.19 $p \in \tilde{S}^{\mu}(\overline{\mathbb{R}}_+ \times \Omega \times \mathbb{R}^{n+1})$, $q \in \tilde{S}^{\nu}(\overline{\mathbb{R}}_+ \times \Omega \times \mathbb{R}^{n+1})$ *implies the existence of an* $r \in \tilde{S}^{\mu+\nu}(\overline{\mathbb{R}}_+ \times \Omega \times \mathbb{R}^{n+1})$ *with*

$$r(t, x, \tau, \xi) \sim \sum_{\alpha} \frac{1}{\alpha!} \{D^{\alpha}_{\tau, \xi} p(t, x, \tau, \xi)\} \partial^{\alpha}_{t, x} q(t, x, \tau, \xi).$$

Analogously we find to the given p, q *an* r *in the above symbol space such that*

$$t^{-(\mu+\nu)} r(t, x, \tau, \xi) \sim \sum_{\alpha} \frac{1}{\alpha!} \{D^{\alpha}_{\tau, \xi} t^{-\mu} p(t, x, \tau, \xi)\} \partial^{\alpha}_{t, x} (t^{-\nu} q(t, x, \tau, \xi)).$$

It suffices to carry out the differentiations and to apply Proposition 2.2.18.

Definition 2.2.20 *A symbol* $p(t, x, \tau, \xi) \in \tilde{S}^{\mu}(\overline{\mathbb{R}}_+ \times \Omega \times \mathbb{R}^{n+1})$ *is called elliptic of order* μ *if for the associated* $\tilde{p}(t, x, \tilde{\tau}, \xi)$ *via* (2.2.9) *there are constants* $c > 0$, $\tilde{c} > 0$, *with*

$$|\tilde{p}(t, x, \tilde{\tau}, \xi)| \geq \tilde{c}|\tilde{\tau}, \xi|^{\mu} \qquad (2.2.10)$$

for all $(\tilde{\tau}, \xi) \in \mathbb{R}^{n+1}$ *with* $|\tilde{\tau}, \xi| \geq c$ *and all* $(t, x) \in \overline{\mathbb{R}}_+ \times \Omega$. *Analogously* $t^{-\mu} p(t, x, \tau, \xi)$ *for* $p(t, x, \tau, \xi) \in \tilde{S}^{\mu}(\overline{\mathbb{R}}_+ \times \Omega \times \mathbb{R}^{n+1})$ *is called elliptic of order* μ *if* $p(t, x, \tau, \xi)$ *is elliptic in the above sense.*

We can easily weaken the definition of ellipticity by requiring the estimate (2.2.10) only for $(t, x) \in [0, t_1] \times K$ for arbitrary $t_1 > 0$ and $K \subset\subset \Omega$, where the constants c, $\tilde{c}$ may depend on t_1, K. However the simpler definition suffices for our purposes, since the symbols are the local representatives for global operators on $\mathbb{R}_+ \times X$ in a neighbourhood of $t = 0$, and then Ω could be taken as a relatively compact subset, if necessary.

Note that the ellipticity of $p \in \tilde{S}^\mu_{\mathrm{cl}}(\mathbb{R}_+ \times \Omega \times \mathbb{R}^{n+1})$ is equivalent to

$$\tilde{p}_{(\mu)}(t, x, \tilde{\tau}, \xi) \neq 0 \qquad \text{for all } (\tilde{\tau}, \xi) \neq 0, \ (t, x) \in \overline{\mathbb{R}}_+ \times \Omega.$$

Here subscript (μ) indicates the homogeneous principal part of order μ.

Remark 2.2.21 *If $p(t, x, \tau, \xi) \in \tilde{S}^\mu(\overline{\mathbb{R}}_+ \times \Omega \times \mathbb{R}^{n+1})$ is elliptic of order μ there is a $q(t, x, \tau, \xi) \in \tilde{S}^{-\mu}(\overline{\mathbb{R}}_+ \times \Omega \times \mathbb{R}^{n+1})$, elliptic of order $-\mu$, such that*

$$1 \sim \sum_\alpha \frac{1}{\alpha!} \left\{ D^\alpha_{\tau,\xi} p(t, x, \tau, \xi) \right\} \partial^\alpha_{t,x} q(t, x, \tau, \xi);$$

$$1 \sim \sum_\alpha \frac{1}{\alpha!} \left\{ D^\alpha_{\tau,\xi} q(t, x, \tau, \xi) \right\} \partial^\alpha_{t,x} p(t, x, \tau, \xi).$$

Analogous relations hold if we replace p and q by $t^{-\mu} p$ and q by $t^\mu q$ with an appropriate $q(t, x, \tau, \xi) \in \tilde{S}^{-\mu}(\overline{\mathbb{R}}_+ \times \Omega \times \mathbb{R}^{n+1})$.

Let us now express the operator push–forward under diffeomorphisms on the level of symbols by the asymptotic formula (1.1.72). If Ω, $\Sigma \subseteq \mathbb{R}^n$ are open sets, $\chi : \mathbb{R}_+ \times \Omega \to \mathbb{R}_+ \times \Sigma$ a diffeomorphism, then this formula can be understood as a "non–canonical" map

$$\chi_* : S^\mu(\mathbb{R}_+ \times \Omega \times \mathbb{R}^{n+1}) \to S^\mu(\mathbb{R}_+ \times \Sigma \times \mathbb{R}^{n+1})$$

under which $\chi_* \mathrm{op}(a) - \mathrm{op}(\chi_* a) \in L^{-\infty}(\mathbb{R}_+ \times \Sigma)$, where the first χ_* indicates the operator push–forward. Non–canonical means that the choice of $\chi_* a$ is unique only modulo $S^{-\infty}(\mathbb{R}_+ \times \Sigma \times \mathbb{R}^{n+1})$.

Remark 2.2.22 *Let $\chi : \overline{\mathbb{R}}_+ \times \Omega \to \overline{\mathbb{R}}_+ \times \Sigma$ be a diffeomorphism. Then the symbol push–forward χ_* can be chosen in such a way that*

$$\chi_* : \tilde{S}^\mu(\overline{\mathbb{R}}_+ \times \Omega \times \mathbb{R}^{n+1}) \to \tilde{S}^\mu(\overline{\mathbb{R}}_+ \times \Sigma \times \mathbb{R}^{n+1})$$

and also

$$\chi_* : t^{-\mu} \tilde{S}^\mu(\overline{\mathbb{R}}_+ \times \Omega \times \mathbb{R}^{n+1}) \to r^{-\mu} \tilde{S}^\mu(\overline{\mathbb{R}}_+ \times \Sigma \times \mathbb{R}^{n+1}).$$

Here $(r, y) = \chi(t, x)$.

Let X be a closed compact C^∞ manifold, $n = \dim X$. Let us fix an open covering $\mathcal{U} = \{U_1, \ldots, U_N\}$ of X, a subordinate partition of unity $\{\varphi_j\}_{j=1,\ldots,N}$ and a system $\{\psi_j\}_{j=1,\ldots,N}$ of functions $\psi_j \in C_0^\infty(U_j)$ such that $\varphi_j \psi_j = \varphi_j$ for all j. Let $\kappa_j : U_j \to \Omega_j$ be a system of charts with open $\Omega_j \subseteq \mathbb{R}^n$.

Consider a differential operator of Fuchs type on $\mathbb{R}_+ \times X = X^\wedge$

$$A = t^{-\mu} \sum_{k=0}^{\mu} a_k(t) \left(-t \frac{\partial}{\partial t}\right)^k,$$

with $a_k(t) \in C^\infty(\overline{\mathbb{R}}_+, \mathrm{Diff}^{\mu-k}(X))$. Denote by $t^{-\mu} p_j(t, x, \tau, \xi)$ complete symbols of A in the local coordinates $(t, x) \in \mathbb{R}_+ \times \Omega_j$, such that

$$A = t^{-\mu} \sum_{j=0}^{N} \varphi_j (1 \times \kappa_j)_*^{-1} \mathrm{op}(p_j) \psi_j,$$

with $(1 \times \kappa_j)_*^{-1}$ being the operator pull–back with respect to $1 \times \kappa_j :$ $\mathbb{R}_+ \times U_j \to \mathbb{R}_+ \times \Omega_j$, $j = 1, \ldots, N$. It is clear that $p_j(t, x, \tau, \xi) \in \tilde{S}_{\mathrm{cl}}^\mu(\overline{\mathbb{R}}_+ \times \Omega \times \mathbb{R}^{n+1})$ for all j. If $\omega(t), \tilde{\omega}(t)$ are arbitrary cut–off functions then

$$\omega A \tilde{\omega} : \mathcal{H}^{s, \gamma}(X^\wedge) \to \mathcal{H}^{s-\mu, \gamma-\mu}(X^\wedge)$$

is continuous for every $s, \gamma \in \mathbb{R}$. Let A be elliptic, i.e., all $p_j(t, x, \tau, \xi)$ are elliptic in the sense of Definition 2.2.20 (this is an invariant condition with respect to diffeomorphisms $\overline{\mathbb{R}}_+ \times \Omega_{jk} \to \overline{\mathbb{R}}_+ \times \Omega_{kj}$, where $\Omega_{jk} = \Omega_j \cap \kappa_k U_k$, $\Omega_{kj} = \Omega_k \cap \kappa_j U_j$). Then we may ask for a pseudo–differential operator $B \in L_{\mathrm{cl}}^{-\mu}(X^\wedge)$ which is a parametrix of A in the sense $AB - 1, BA - 1 \in L^{-\infty}(X^\wedge)$ and such that

$$\omega B \tilde{\omega} : \mathcal{H}^{s-\mu, \gamma-\mu}(X^\wedge) \to \mathcal{H}^{s, \gamma}(X^\wedge) \qquad (2.2.11)$$

is continuous for arbitrary cut–off functions ω, $\tilde{\omega}$, every $s \in \mathbb{R}$ and a fixed weight $\gamma \in \mathbb{R}$ (or, if possible, for all $\gamma \in \mathbb{R}$).

Using Remark 2.2.21 we can form the inverses q_j of p_j with respect to the Leibniz multiplication and try to set

$$B = \left\{ \sum_{j=1}^{N} \varphi_j (1 \times \kappa_j)_*^{-1} \mathrm{op}(q_j) \psi_j \right\} t^\mu,$$

where $\mathrm{op}(q_j) u(t, x) = F_{(\tau, \xi) \to (t, x)}^{-1} q_j(t, x, \tau, \xi) F_{(t', x') \to (\tau, \xi)} u(t', x')$ for $u \in C_0^\infty(X^\wedge)$.

Since A is elliptic on $X^\wedge$ in the classical sense, though degenerate for $t \to 0$, the operator B is necessarily a parametrix in the standard meaning, cf. Section 1.1.5. However it is by no means obvious how to choose B in a way that the operators (2.2.11) are continuous. The result that such a choice is actually possible is the consequence of the so–called Mellin operator conventions which reformulate B modulo $L^{-\infty}(X^{\wedge})$ in terms of the Mellin transform on $\mathbb{R}_+$ with the desired control up to $t = 0$.

Recall that for $h(t, z) = \sum_{k=0}^{\mu} a_k(t) z^k$ we have

$$h(t, z) \in C^{\infty}(\overline{\mathbb{R}}_+, M_O^{\mu}(X))$$

and that the differential operator A has the form $A = t^{-\mu} \operatorname{op}_M^{\delta}(h)$ for arbitrary $\delta \in \mathbb{R}$. Our Mellin operator conventions will show that B may be found in the form $B = \operatorname{op}_M^{\delta}(l) t^{\mu}$ for a suitable $l(t, z) \in C^{\infty}(\overline{\mathbb{R}}_+, M_O^{-\mu}(X))$.

Mellin operator conventions were constructed in Schulze [118]. The method was generalised later on in Dorschfeldt, Schulze [22], Schrohe, Schulze [106] (for symbols with values in pseudo–differential boundary value problems with the transmission property) and in the joint paper with Gil, Seiler [38].

Note that for an arbitrary system of symbols $p_j(t, x, \tau, \xi) \in \tilde{S}^{\mu}(\overline{\mathbb{R}}_+ \times \Omega_j \times \mathbb{R}^{n+1})$, $j = 1, \dots, N$, and

$$\operatorname{op}_{(x)}(p_j)(t, \tau) v(x) = F_{\xi \to x}^{-1} p_j(t, x, \tau, \xi) F_{x' \to \xi} v(x')$$

we obtain by

$$p(t, \tau) = \sum_{j=1}^{N} \varphi_j (\kappa_j)_*^{-1} \operatorname{op}_{(x)}(p_j)(t, \tau) \psi_j$$

an operator family in $C^{\infty}(\mathbb{R}_+, L^{\mu}(X; \mathbb{R}_{\tau}))$ with

$$p(t, t^{-1}\tau) \in C^{\infty}(\overline{\mathbb{R}}_+, L^{\mu}(X; \mathbb{R}_{\tau})).$$

The idea to construct Mellin operator conventions is relevant already on $\mathbb{R}_+$, and we also want to consider this case. Let us set

$$T(t, t') = (t - t')(\log t - \log t')^{-1}.$$

We then have $T(t, t') \in C^{\infty}(\mathbb{R}_+ \times \mathbb{R}_+)$, $T(t, t') > 0$ for all $t, t' \in \mathbb{R}_+$, and $T(t, t) = t$. Let $\kappa(y) = t = e^{-y}$, $y \in \mathbb{R}$, and consider the operator push–forward $\kappa_* : L^{\mu}(\mathbb{R}) \to L^{\mu}(\mathbb{R}_+)$ under the diffeomorphism $\kappa : \mathbb{R} \to \mathbb{R}_+$.

Lemma 2.2.23 *We have*

$$\kappa_* \operatorname{op}(q) = \operatorname{op}(p)$$

for every $q(y, y', \eta) \in S^{\mu}(\mathbb{R} \times \mathbb{R} \times \mathbb{R})$ *with*

$$p(t, t', \tau) = q(-\log t, -\log t', -T(t, t')\tau)T(t, t')(t')^{-1} \qquad (2.2.12)$$

and for every $p(t, t', \tau) \in S^{\mu}(\mathbb{R}_+ \times \mathbb{R}_+ \times \mathbb{R})$ *with*

$$q(y, y', \eta) = p(e^{-y}, e^{-y'}, -T^{-1}(e^{-y}, e^{-y'})\eta)T^{-1}(e^{-y}, e^{-y'})e^{-y'}. \quad (2.2.13)$$

An analogous assertion holds for $q(y, y', \eta) \in C^{\infty}(\mathbb{R} \times \mathbb{R}, L^{\mu}(X; \mathbb{R}_{\eta}))$ *with a resulting* $p(t, t', \tau) \in C^{\infty}(\mathbb{R}_+ \times \mathbb{R}_+, L^{\mu}(X; \mathbb{R}_{\tau}))$ *and the reverse result.*

Proof. We have

$$\mathrm{op}(q)v(y) = \iint e^{i(y-y')\eta} q(y, y', \eta)v(y') \, dy' d\eta,$$

$v(y) \in C_0^{\infty}(\mathbb{R})$. Inserting $y = -\log t$, $y' = -\log t'$, $v = \kappa^* u$ for a corresponding $u(t) \in C_0^{\infty}(\mathbb{R}_+)$, i.e., $v(y) = u(e^{-y})$, we get

$$\mathrm{op}(q)(\kappa^* u)(-\log t)$$

$$= \iint e^{-i(\log t - \log t')\eta} q(-\log t, -\log t', -\eta)u(t')\frac{dt'}{t'}d\eta$$

$$= \iint e^{i(t-t')\tau} q(-\log t, -\log t', -T(t, t')\tau)T(t, t')(t')^{-1}u(t') \, dt' d\tau.$$

The reverse assertion follows in an analogous manner. The generalisation to $L^{\mu}(X)$–valued symbols is straightforward. $\qquad \Box$

Lemma 2.2.24 *We have*

(i) $\partial_{t'}^k T^{-1}(t, t')|_{t'=t} = c_k t^{-k-1}$ *for certain constants* c_k, $k \in \mathbb{N}$, *and, in particular,*

$$(t'\partial_{t'})^k \{t'T^{-1}(t, t')\}|_{t'=t} \in C^{\infty}(\overline{\mathbb{R}}_+), \qquad k \in \mathbb{N},$$

(ii)

$$t^{k-1}\partial_{t'}^k \{T(t, t')\}|_{t'=t} \in C^{\infty}(\overline{\mathbb{R}}_+), \qquad k \in \mathbb{N}.$$

Proof. Let $u, v \in \mathbb{R}_+$ and $1 + x = uv^{-1}$, $|x| < 1$. Then

$$\log u - \log v = \log(uv^{-1}) = \log(1 + x) = \sum_{j=1}^{\infty} \frac{(-1)^{j+1}}{j}x^j$$

$$= \sum_{j=1}^{\infty} \frac{(-1)^{j+1}}{j}\frac{(u-v)^j}{v^j}.$$

This yields

$$T^{-1}(u,v) = \frac{\log u - \log v}{u - v} = \sum_{k=0}^{\infty} \frac{(-1)^k}{k+1} \frac{(u-v)^k}{v^{k+1}}.$$

It follows that $\partial_u^k T^{-1}(u,v)|_{u=v} = k!\frac{(-1)^k}{k+1}v^{-k-1}$ which implies the first relation of (i). The second one is a consequence of the fact that, for $k \geq 1$, $(t\partial_t)^k$ is a linear combination of operators $t^j\partial_t^j$, $j = 1, \ldots, k$. For (ii) we argue by induction. $\partial_{t'}^k T(t,t')$ is a linear combination of terms

$$T^{r+1}(t,t') \prod_{m=1}^{r} \partial_{t'}^{j_m} T^{-1}(t,t')$$

for $r \leq k$ and $\sum_{m=1}^{r} j_m = k$. This implies that $\partial_{t'}^k T(t,t')|_{t'=t}$ is a linear combination of terms $t^{r+1}t^{-r-k}$, $0 \leq r \leq k$. $\Box$

Let us set

$$\tilde{L}^\mu(\overline{\mathbb{R}}_+ \times X) = \Big\{ \mathrm{op}_{(t)}(p) + c : \; p(t,\tau) \in C^\infty(\mathbb{R}_+, L^\mu(X;\mathbb{R}_\tau)) \text{ such that}$$

$$p(t,t^{-1}\tau) \in C^\infty(\overline{\mathbb{R}}_+, L^\mu(X;\mathbb{R}_\tau)), \, c \in L^{-\infty}(X^\wedge) \Big\}$$

$$(2.2.14)$$

and define analogously $\tilde{L}_{\mathrm{cl}}^\mu(\overline{\mathbb{R}}_+ \times X)$ in terms of $L_{\mathrm{cl}}^\mu(X;\mathbb{R}_\tau)$.

Theorem 2.2.25 *For arbitrary $\mu \in \mathbb{R}$ we have*

$$\tilde{L}^\mu(\overline{\mathbb{R}}_+ \times X)$$
$$= \Big\{ \mathrm{op}_M^{\frac{1}{2}}(f) + r : \; f(t,z) \in C^\infty(\overline{\mathbb{R}}_+, L^\mu(X;\Gamma_0)), r \in L^{-\infty}(X^\wedge) \Big\}.$$

An analogous result holds for $\tilde{L}_{\mathrm{cl}}^\mu(\overline{\mathbb{R}}_+ \times X)$.

Proof. Let $f(t,z) \in C^\infty(\overline{\mathbb{R}}_+, L^\mu(X;\Gamma_0))$, $\mu \in \mathbb{R}$, and set $y = \chi(t) = -\log t$, $\chi : \mathbb{R}_+ \to \mathbb{R}$. Then, similarly to Proposition 2.2.1 we have $\mathrm{op}_{(y)}(q) = \chi_* \mathrm{op}_M^{\frac{1}{2}}(f)$ with the operator push–forward $\chi_* : L^\mu(\mathbb{R}_+ \times X) \to L^\mu(\mathbb{R} \times X)$ (here tensorised by the identity on X), where $q(y,\tau) = f(e^{-y}, i\tau)$. According to Lemma 2.2.23 with $\kappa_* = (\chi^{-1})_*$ we obtain $\kappa_* \mathrm{op}_{(y)}(q) = \mathrm{op}_{(t)}(b) = \mathrm{op}_M^{\frac{1}{2}}(f)$ with

$$b(t,t',\tau) = q(-\log t, -T(t,t')\tau)T(t,t')(t')^{-1}$$
$$= f(t, -iT(t,t')\tau)T(t,t')(t')^{-1}.$$

We now apply the standard rule to convert $b(t, t', \tau)$ to a t'–independent symbol

$$c(t, \tau) \sim \sum_{k=0}^{\infty} \frac{1}{k!} D_\tau^k \partial_{t'}^k b(t, t', \tau)|_{t'=t} \qquad (2.2.15)$$

under which $\mathrm{op}(b) = \mathrm{op}(c) \bmod L^{-\infty}(X^\wedge)$. Then

$$D_\tau^k \partial_{t'}^k b(t, t', \tau) = \partial_{t'}^k \{ (-i)^k (\partial_z^k f)(t, -iT(t,t')\tau) T(t,t')^{k+1} (t')^{-1} \}. \qquad (2.2.16)$$

By induction this is a linear combination of terms of the form

$$(\partial_z^{k+j} f)(t, -iT(t,t')\tau) \tau^j w_{kj}(t, t'), \quad j = 0, \ldots, k, \qquad (2.2.17)$$

where $w_{kj}(t, t')$ is a linear combination of functions like

$$(t')^{-1-l_0} \prod_{m=1}^{r} \partial_{t'}^{l_m} T(t, t')$$

with $r = k + 1 + j$, $l_0 + \sum_{m=1}^{r} l_m = k$. From Lemma 2.2.24 we know that $t^{-j} w_{kj}(t, t') \in C^\infty(\overline{\mathbb{R}}_+)$. Combining (2.2.16), (2.2.17) we see that $D_\tau^k \partial_{t'}^k b(t, t', \tau)|_{t'=t}$ is a linear combination of terms of the form $(\partial_z^{k+j} f)(t, -it\tau)(t\tau)^j s_{kj}(t)$ with certain $s_{kj} \in C^\infty(\overline{\mathbb{R}}_+)$. Thus it follows that

$$\frac{1}{k!} D_\tau^k \partial_{t'}^k b(t, t', \tau)|_{t'=t} =: c_k(t, \tau) \in C^\infty(\mathbb{R}_+, L^{\mu-k}(X; \mathbb{R}))$$

with $c_k(t, t^{-1}\tau) \in C^\infty(\overline{\mathbb{R}}_+, L^{\mu-k}(X; \mathbb{R}))$. Similarly to the above Proposition 2.2.18 we obtain a

$$p(t, \tau) \sim \sum c_k(t, \tau), \qquad p(t, \tau) \in C^\infty(\mathbb{R}_+, L^\mu(X; \mathbb{R})),$$

with $p(t, t^{-1}\tau) \in C^\infty(\overline{\mathbb{R}}_+, L^\mu(X; \mathbb{R}))$. In other words, $\mathrm{op}_M^{\frac{1}{2}}(f) = \mathrm{op}_{(t)}(p) \bmod L^{-\infty}(X^\wedge)$.

In order to prove the reverse inclusion we start with $p(t, \tau)$ such that $p(t, t^{-1}\tau)$ is as assumed. Then, from Lemma 2.2.23 we get a $q(y, y', \eta)$ with $\kappa_* \mathrm{op}(q) = \mathrm{op}(p)$, cf. formula (2.2.13). From the proof of Proposition 2.2.1 we have an isomorphism $M_{\frac{1}{2}} L^\mu(X^\wedge) \to L^\mu(\mathbb{R} \times X)$ induced by $g(t, t', i\tau) \to q(y, y', \eta) = g(e^{-y}, e^{-y'}, i\eta)$, where $\chi_* \mathrm{op}_M^{\frac{1}{2}}(g) = \mathrm{op}(q)$, $\chi_* = (\kappa_*)^{-1}$. In other words, it follows that $\mathrm{op}(p) = \mathrm{op}_M^{\frac{1}{2}}(g)$ for $g(t, t', i\tau) = p(t, -T(t,t')^{-1}\tau)t'T^{-1}(t,t')$. It is now sufficient to pass to an $f(t, z) \in C^\infty(\overline{\mathbb{R}}_+, L^\mu(X; \Gamma_0))$ such that

$\operatorname{op}_M^{\frac{1}{2}}(f) = \operatorname{op}_M^{\frac{1}{2}}(g) \bmod L^{-\infty}(X^\wedge)$. Such an $f(t,z)$ can be obtained by an asymptotic expansion

$$f(t,z) \sim \sum_{k=0}^\infty \frac{1}{k!} \partial_z^k (-t'\partial_{t'})^k g(t,t',z)|_{t'=t},$$

cf. Proposition 2.2.7. By assumption the symbol $p(t,\tau)$ has the form $\tilde{p}(t,t\tau)$ with $\tilde{p}(t,\tilde{\tau}) \in C^\infty(\overline{\mathbb{R}}_+, L^\mu(X;\mathbb{R}))$. Thus it follows that

$$f(t,z) \sim \sum_{k=0}^\infty \frac{1}{k!} \partial_z^k (-t'\partial_{t'})^k \tilde{p}(t, -T(t,t')^{-1}t\tau)t'T(t,t')^{-1}|_{t'=t}. \quad (2.2.18)$$

By Lemma 2.2.24 the functions $(t'\partial_{t'})^j T(t,t')^{-1}t'|_{t'=t}$ are C^∞ up to $t = 0$ for all j. So all terms on the right hand side of (2.2.18) are C^∞ up to $t = 0$. Since the asymptotic sum can be carried out in $C^\infty(\overline{\mathbb{R}}_+, L^\mu(X;\mathbb{R}))$ we get the desired $f(t,z)$, $z = i\tau$. □

Let us sketch an alternative method to obtain to every $p(t,\tau) \in C^\infty(\overline{\mathbb{R}}_+, L^\mu(X;\mathbb{R}))$ with $p(t,t^{-1}\tau) \in C^\infty(\overline{\mathbb{R}}_+, L^\mu(X;\mathbb{R}))$ an $f(t,z) \in C^\infty(\overline{\mathbb{R}}_+, L^\mu(X;\Gamma_0))$ with

$$\operatorname{op}_{(t)}(p) = \operatorname{op}_M^{\frac{1}{2}}(f) \bmod L^{-\infty}(X^\wedge). \quad (2.2.19)$$

Set $p_0(t,\tau) = p(t,t\tau)$, $\tilde{p}_0(t,\tau) = p_0(t,t^{-1}\tau)$ and define $f_0(t,i\tau) = \tilde{p}_0(t,-\tau)$, $f_0(t,z) \in C^\infty(\overline{\mathbb{R}}_+, L^\mu(X;\Gamma_0))$. Then we can show that there exists a $\tilde{p}_1(t,\tau) \in C^\infty(\overline{\mathbb{R}}_+, L^{\mu-1}(X;\mathbb{R}))$ such that $p_1(t,\tau) = \tilde{p}_1(t,t\tau)$ satisfies

$$\operatorname{op}_{(t)}(p) = \operatorname{op}_M^{\frac{1}{2}}(f_0) + \operatorname{op}_{(t)}(p_1) \bmod L^{-\infty}(X^\wedge). \quad (2.2.20)$$

Setting $f_1(t,i\tau) = \tilde{p}_1(t,-\tau)$ which belongs to $C^\infty(\overline{\mathbb{R}}_+, L^{\mu-1}(X;\Gamma_0))$ we then get analogously a $\tilde{p}_2(t,\tau) \in C^\infty(\overline{\mathbb{R}}_+, L^{\mu-2}(X;\mathbb{R}))$ with

$$\operatorname{op}_{(t)}(p_1) = \operatorname{op}_M^{\frac{1}{2}}(f_1) + \operatorname{op}_{(t)}(p_2) \bmod L^{-\infty}(X^\wedge)$$

for $p_2(t,\tau) = \tilde{p}_2(t,t\tau)$. It follows inductively that there is a sequence $f_k(t,z) \in C^\infty(\overline{\mathbb{R}}_+, L^{\mu-k}(X;\Gamma_0))$

$$\operatorname{op}_{(t)}(p) = \sum_{k=0}^M \operatorname{op}_M^{\frac{1}{2}}(f_k) \bmod L^{\mu-(M+1)}(X^\wedge)$$

for every $M \in \mathbb{N}$. The asymptotic sum $f \sim \sum_{k=0}^\infty f_k$ can be carried out in $C^\infty(\overline{\mathbb{R}}_+, L^\mu(X;\Gamma_0))$ and yields then an $f(t,z)$ that satisfies (2.2.19). In other words the main step of the construction is to show (2.2.20).

A method to obtain p_1 is to look at the diffeomorphism $\kappa : \mathbb{R} \rightarrow \mathbb{R}_+$, $\kappa(y) = e^{-y} = t$ and to form the symbol $q(y, \eta) := f_0(e^{-y}, i\eta) = \tilde{p}_0(e^{-y}, -\eta) \in C^\infty(\mathbb{R}, L^\mu(X; \mathbb{R}_\eta))$ such that $\kappa_* \operatorname{op}_{(y)}(q) = \operatorname{op}^{\frac{1}{2}}_M(f_0)$, cf. the proof of Proposition 2.2.1. Applying now the symbolic rule for the operator push–forward $\kappa_* : L^\mu(\mathbb{R} \times X) \rightarrow L^\mu(\mathbb{R}_+ \times X)$ we get a $b(t, \tau) \in C^\infty(\mathbb{R}_+, L^\mu(X; \mathbb{R}_\tau))$ with

$$\kappa_* \operatorname{op}_{(y)}(q) = \operatorname{op}_{(t)}(b) \bmod L^{-\infty}(X^\wedge),$$

$$b(t, \tau)|_{t=\kappa(y)} \sim \sum_{k=0}^{\infty} \frac{1}{k!}(\partial_\eta^k q)(y, {}^t d\kappa(y)\tau)\Phi_k(y, \tau),$$

cf. Theorem 1.1.38. We have ${}^t d\kappa(y)\tau = -e^{-y}\tau$ and hence $q(y, {}^t d\kappa(y)\tau) = \tilde{p}_0(e^{-y}, e^{-y}\tau) = p(e^{-y}, \tau)$. Since $\Phi_0(t, \tau) = 1$, the term of highest order in the asymptotic sum coincides with $p(e^{-y}, \tau)$ which gives us $b(t, \tau) - p(t, \tau) \in C^\infty(\mathbb{R}_+, L^{\mu-1}(X; \mathbb{R}_\tau))$. The remaining part of the asymptotic sum, i.e., over $k \geq 1$, gives just the desired p_1 if we verify that the polynomials $\phi_k(y, \tau)$ are polynomials in $e^{-y}\tau$. We omit here the corresponding elementary calculation. For more details cf. Schulze [118], Dorschfeldt and Schulze [22].

The main point of Theorem 2.2.25 is the result that to every $p(t, \tau) \in C^\infty(\mathbb{R}_+, L^\mu(X; \mathbb{R}_\tau))$ with $p(t, t^{-1}\tau) \in C^\infty(\overline{\mathbb{R}}_+, L^\mu(X; \mathbb{R}_\tau))$ there exists an $f(t, z) \in C^\infty(\overline{\mathbb{R}}_+, L^\mu(X; \Gamma_0))$ with

$$\operatorname{op}(p) = \operatorname{op}^{\frac{1}{2}}_M(f) \bmod L^{-\infty}(X^\wedge).$$

The correspondence $p \rightarrow f$ is called a Mellin operator convention. It is non–canonical in the sense that any other $\tilde{f}(t, z) \in C^\infty(\overline{\mathbb{R}}_+, L^\mu(X; \Gamma_0))$ with

$$f - \tilde{f} \in C^\infty(\overline{\mathbb{R}}_+, L^\mu(X; \Gamma_0))$$

is also a possible choice for the given p.

We will obtain below an analogue of Theorem 2.2.25 with $\operatorname{op}^\gamma_M(f)$ for arbitrary weights $\gamma \in \mathbb{R}$. This can be proved in various ways. The most immediate argument is based on the following:

Corollary 2.2.26 *The spaces* $\tilde{L}^\mu(\mathbb{R}_+ \times X)$ *and*

$$\left\{ \operatorname{op}^{\frac{1}{2}}_M(h) + r : \; h(t, z) \in C^\infty(\overline{\mathbb{R}}_+, N^\mu_O(X)), \, r \in L^{-\infty}(X^\wedge) \right\}$$

coincide for every $\mu \in \mathbb{R}$. *An analogous result holds for* $\tilde{L}^\mu_{cl}(\mathbb{R} \times X)$, *where* $N^\mu_O(X)$ *is to be replaced by* $M^\mu_O(X)$.

In fact, by Theorem 2.2.10 to every $f(t,z) \in C^\infty(\overline{\mathbb{R}}_+, L^\mu(X; \Gamma_0))$ we fixed an $h(t,z) \in C^\infty(\overline{\mathbb{R}}_+, N_O^\mu(X))$ such that $f(t,z) - h(t,z)|_{\Gamma_0} \in C^\infty(\overline{\mathbb{R}}_+, L^{-\infty}(X; \Gamma_0))$ which implies $\mathrm{op}_M^{\frac{1}{2}}(f - h) \in L^{-\infty}(X^\wedge)$.

Remark 2.2.27 *The arguments in the above correspondence $p \to f$ (the Mellin operator convention) show that $p(t,\tau) = \tilde{p}(t\tau)$ for $\tilde{p}(\tilde{\tau}) \in L^\mu(X; \mathbb{R}_{\tilde{\tau}})$ implies the existence of an $f(z) \in L^\mu(X; \Gamma_0)$ with the property (2.2.19). In other words $f(z)$ can be chosen to be independent of t. Note that a simple formula for f was obtained in a joint paper with Gil and Seiler [38] which shows this fact directly.*

Theorem 2.2.28 *The spaces $\tilde{L}^\mu(\overline{\mathbb{R}}_+ \times X)$ coincide with*

$$\left\{ \mathrm{op}_M^\delta(f) + r : \ f(t,z) \in C^\infty(\overline{\mathbb{R}}_+, L^\mu(X; \Gamma_{\frac{1}{2}-\delta})),\ r \in L^{-\infty}(X^\wedge) \right\}$$

as well as with

$$\left\{ \mathrm{op}_M^\delta(h) + r : \ h(t,z) \in C^\infty(\overline{\mathbb{R}}_+, N_O^\mu(X)),\ r \in L^{-\infty}(X^\wedge) \right\}$$

for every $\mu, \delta \in \mathbb{R}$. An analogous result holds for $\tilde{L}_{cl}^\mu(\overline{\mathbb{R}}_+ \times X)$, where $N_O^\mu(X)$ is to be replaced by $M_O^\mu(X)$.

This is a consequence of Theorem 2.2.6.

Note that an alternative method to obtain Theorem 2.2.28 from Theorem 2.2.25 is to construct a $g(t,z) \in C^\infty(\overline{\mathbb{R}}_+, L^\mu(X; \Gamma_{\frac{1}{2}-\delta}))$ to a given $f(t,z) \in C^\infty(\overline{\mathbb{R}}_+, L^\mu(X; \Gamma_0))$ such that $\mathrm{op}_M^{\frac{1}{2}}(f) = \mathrm{op}_M^\delta(g)$ mod $L^{-\infty}(X^\wedge)$. First it is obvious that for

$$f_\delta(t,t',z) := (t/t')^{\frac{1}{2}-\delta} f(t, z - \frac{1}{2} + \delta)$$

we have $\mathrm{op}_M^{\frac{1}{2}}(f) = \mathrm{op}_M^\delta(f_\delta)$. It is now sufficient to apply Proposition 2.2.16 and to observe that the asymptotic sum

$$g(t,z) \sim \sum_{k=0}^\infty \frac{1}{k!} \partial_z^k (-t'\partial_{t'})^k f_\delta(t,t',z)|_{t=t'}$$

can be carried out in $C^\infty(\overline{\mathbb{R}}_+, L^\mu(X; \Gamma_{\frac{1}{2}-\delta}))$. In fact, using

$$(-t'\partial_{t'})^k (t/t')^{1/2-\delta}|_{t'=t} = (t\partial_t)^k t^{\frac{1}{2}-\delta}|_{t=1} = (\frac{1}{2} - \delta)^k,$$

we get

$$g(t,z) \sim \sum_{k=0}^\infty \frac{1}{k!} (-t'\partial_{t'})^k (t/t')^{\frac{1}{2}-\delta}|_{t'=t} \partial_z^k f(t, z - 1/2 + \delta)$$

$$\sim \sum_{k=0}^\infty \frac{1}{k!} (1/2 - \delta)^k \partial_z^k f(t, z - 1/2 + \delta), \tag{2.2.21}$$

which shows that the summands are C^∞ up to $t = 0$, and the smoothness up to $t = 0$ can be ensured for the asymptotic sum.

Remark 2.2.29 $h(z) \in N_O^\mu(X)$ *and* $h(z)|_{\Gamma_\beta} \in L^{-\infty}(X; \Gamma_\beta)$ *for any fixed* $\beta \in \mathbb{R}$ *implies* $h(z) \in M_O^{-\infty}(X)$. *In fact,* (2.2.21) *gives us an asymptotic expansion for* $h(z)|_{\Gamma_\varrho}$ *in terms of* $h(z)|_{\Gamma_\beta}$ *for every* $\varrho \in \mathbb{R}$ *which yields* $h(z)|_{\Gamma_\varrho} \in L^{-\infty}(X; \Gamma_\varrho)$.

2.2.3 Operator calculus

Proposition 2.2.30 *Let* $f(t, t', z) \in C^\infty(\overline{\mathbb{R}}_+ \times \overline{\mathbb{R}}_+, L^\mu(X; \Gamma_0))$ *and assume that for some* $N \in \mathbb{N}$ *we have* $(t - t')^{-N} f(t, t', z) \in C^\infty(\overline{\mathbb{R}}_+ \times \overline{\mathbb{R}}_+, L^\mu(X; \Gamma_0))$. *Then there is a symbol*

$$f_N(t, t', z) \in C^\infty(\overline{\mathbb{R}}_+ \times \overline{\mathbb{R}}_+, L^\mu(X; \Gamma_0))$$

such that $\mathrm{op}_M^{\frac{1}{2}}(f) = \mathrm{op}_M^{\frac{1}{2}}(f_N)$.

Proof. Let us set $g(t, t', z) = (t')^N (t - t')^{-N} f(t, t', z)$. Then $g(t, t', z) \in C^\infty(\overline{\mathbb{R}}_+ \times \overline{\mathbb{R}}_+, L^\mu(X; \Gamma_0))$ and $f(t, t', z) = (\frac{t}{t'} - 1)^N g(t, t', z)$. Choose a $\psi(\varrho) \in C_0^\infty(\mathbb{R}_+)$ with $\psi \equiv 1$ in a neighbourhood of 1. Let $h(t, t', z) = (H(\psi)f)(t, t', z)$ and set $h_0 = f - h$. Then $h_0(t, t', z) \in C^\infty(\overline{\mathbb{R}}_+ \times \overline{\mathbb{R}}_+, L^{-\infty}(X; \Gamma_0))$, cf. Theorem 2.2.10, and $\mathrm{op}_M^{\frac{1}{2}}(f) = \mathrm{op}_M^{\frac{1}{2}}(h) + \mathrm{op}_M^{\frac{1}{2}}(h_0)$. Hence it is sufficient to consider $h(t, t', z)$. We have

$$k(h)(t, t', \varrho) = \psi(\varrho) M_{\frac{1}{2}, z \to \varrho}^{-1} f(t, t', z) = \varphi(\varrho) \log^N \varrho \, k(g)(t, t', \varrho)$$

for $\varphi(\varrho) = (1 - \varrho)^N \log^{-N} \varrho \, \psi(\varrho) \in C_0^\infty(\mathbb{R}_+)$. Furthermore,

$$\log^N \varrho \, k(g)(t, t', \varrho) = k(\partial_z^N g)(t, t', \varrho).$$

Thus we get $h(t, t', z) = (M_{\frac{1}{2}, \varrho \to z} \varphi(\varrho) k(\partial_z^N g))(t, t', z)$ which is in $C^\infty(\overline{\mathbb{R}}_+ \times \overline{\mathbb{R}}_+, L^{\mu - N}(X; \Gamma_0))$, cf. Theorem 2.2.10. $\qquad\square$

Theorem 2.2.31 *For every* $f(t, t', z) \in C^\infty(\overline{\mathbb{R}}_+ \times \overline{\mathbb{R}}_+, L^\mu(X; \Gamma_0))$ *there is an* $\boldsymbol{f}(t, z) \in C^\infty(\overline{\mathbb{R}}_+, L^\mu(X; \Gamma_0))$, *where*

$$\boldsymbol{f}(t, z) \sim \sum_{j=0}^\infty \frac{1}{j!} (-t' \partial_{t'})^j \, \partial_z^j f(t', t, z)|_{t'=t}, \qquad (2.2.22)$$

such that $\mathrm{op}_M^{\frac{1}{2}}(f) - \mathrm{op}_M^{\frac{1}{2}}(\boldsymbol{f}) \in M_{\frac{n+1}{2}} L^{-\infty}(\overline{\mathbb{R}}_+ \times X)$.

Proof. Applying the Taylor expansion in t' near t we obtain for every $N \in \mathbb{N}$, $N \geq 1$

$$f(t,t',z) = \sum_{j=0}^{N-1} \frac{1}{j!}(t'-t)^j \partial_{t'}^j f(t,t',z)|_{t'=t} + f_{(N)}(t,t',z),$$

where

$$(t-t')^{-N} f_{(N)}(t,t',z)$$

$$= \frac{1}{(N-1)!} \int_0^1 (1-\theta)^{N-1} \partial_{t'}^N f(t,t'+\theta(t-t'),z)\,d\theta \quad (2.2.23)$$

belongs to $C^\infty(\overline{\mathbb{R}}_+ \times \overline{\mathbb{R}}_+, L^\mu(X;\Gamma_0))$. By Proposition 2.2.30 we get a $g_{(N)}(t,t',z) \in C^\infty(\overline{\mathbb{R}}_+ \times \overline{\mathbb{R}}_+, L^{\mu-N}(X;\Gamma_0))$ such that $\mathrm{op}_M^{\frac{1}{2}}(f_{(N)}) = \mathrm{op}_M^{\frac{1}{2}}(g_{(N)})$. In order to treat the terms under the summation we set

$$f_j(t,t',z) = \frac{1}{j!}(t'-t)^j (\partial_{t'}^j f(t,t',z)|_{t'=t}),$$

$$h_j(t,z) = \frac{1}{j!}t^j \partial_{t'}^j f(t,t',z)|_{t'=t}.$$

Let us choose a $\psi(\varrho) \in C_0^\infty(\mathbb{R}_+)$ with $\psi \equiv 1$ near $\varrho = 1$ and set $\varphi_j(\varrho) = (\varrho^{-1}-1)^j \log^j \varrho \psi(\varrho)$, $\varphi_j \in C_0^\infty(\mathbb{R}_+)$. Then, according to the construction in the proof of Proposition 2.2.30, we have $\mathrm{op}_M^{\frac{1}{2}}(f_j) = \mathrm{op}_M^{\frac{1}{2}}(m_j) + \mathrm{op}_M^{\frac{1}{2}}(l_j)$ with $l_j(t,t',z) = (M_{\frac{1}{2}}(1-\psi(\varrho))M_{\frac{1}{2}}^{-1}f_j)(t,t',z)$ and $m_j(t,z) = (M_{\frac{1}{2}}\varphi_j(\varrho)\log^j \varrho M_{\frac{1}{2}}^{-1}h_j)(t,z)$. We have $\mathrm{op}_M^{\frac{1}{2}}(l_j) \in M_{\frac{n+1}{2}}L^{-\infty}(\overline{\mathbb{R}}_+ \times X)$, cf. Remark 2.2.4. Thus it suffices to look at $m_j(t,z)$. Since

$$\log^j \varrho M_{\frac{1}{2},z\to\varrho}^{-1} h_j = M_{\frac{1}{2},z\to\varrho}^{-1} \partial_z^j h_j$$

and because of $\partial_z^j h_j \in C^\infty(\overline{\mathbb{R}}_+, L^{\mu-j}(X;\Gamma_0))$, we obtain from Theorem 2.2.10 that

$$m_j(t,z) = (H(\varphi_j)(\partial_z^j h_j))(t,z) \in C^\infty(\overline{\mathbb{R}}_+, L^{\mu-j}(X;\Gamma_0)).$$

This yields altogether

$$\mathrm{op}_M^{\frac{1}{2}}\left(f - \sum_{j=0}^{N-1} m_j\right) = \mathrm{op}_M^{\frac{1}{2}}(g_{(N)}) \bmod M_{\frac{n+1}{2}}L^{-\infty}(\overline{\mathbb{R}}_+ \times X).$$

Choosing the asymptotic sum

$$\boldsymbol{f}(t,z) \sim \sum_{j=0}^{\infty} m_j(t,z) \tag{2.2.24}$$

in $C^{\infty}(\overline{\mathbb{R}}_+, L^{\mu}(X;\Gamma_0))$ we thus obtain

$$\mathrm{op}_M^{\frac{1}{2}}(f) - \mathrm{op}_M^{\frac{1}{2}}(\boldsymbol{f}) \in M_{\frac{n+1}{2}}L^{-\infty}(\overline{\mathbb{R}}_+ \times X).$$

It remains to show that (2.2.24) corresponds to the asserted asymptotic expansions (2.2.22). By construction we have

$$\boldsymbol{f}(t,z) \sim \sum_{j=0}^{\infty} \frac{1}{j!} M_{\frac{1}{2}}\{\varphi_j(\varrho)M_{\frac{1}{2}}^{-1}(t^j \partial_{t'}^j \partial_z^j f(t,t',z)|_{t'=t}\}.$$

This is not very explicit, and so we will derive another representation. In view of

$$\mathrm{op}_M^{\frac{1}{2}}(\log^k(t/t')l(t,t',z)) = \mathrm{op}_M^{\frac{1}{2}}(\partial_z^k l(t,t',z))$$

for every $l(t,t',z) \in C^{\infty}(\mathbb{R}_+ \times \mathbb{R}_+, L^{\nu}(X;\Gamma_0))$ we have

$$\mathrm{op}_M^{\frac{1}{2}}(f) \tag{2.2.25}$$

$$= \sum_{j=0}^{N-1} \mathrm{op}_M^{\frac{1}{2}} \left(\frac{1}{j!}(t-t')^j \partial_{t'}^j f(t,t',z)|_{t'=t} \right) + R_N^{(1)} \tag{2.2.26}$$

$$= \sum_{j=0}^{N-1} \mathrm{op}_M^{\frac{1}{2}} \left(\frac{1}{j!} \left(\frac{t'}{t} - 1 \right)^j \log^{-j} \left(\frac{t'}{t} \right) \{(t')^j \partial_{t'}^j \partial_z^j f(t,t',z)\} |_{t'=t} \right)$$

$$+ R_N^{(1)}$$

with the remainder $R_N^{(1)} = \mathrm{op}_M^{\frac{1}{2}}(f_{(N)})$. Let us now introduce the operators

$$L_j : C^{\infty}(\mathbb{R}_+ \times \mathbb{R}_+, L^{\nu}(X;\Gamma_0)) \to C^{\infty}(\mathbb{R}_+ \times \mathbb{R}_+, L^{\nu-j}(X;\Gamma_0))$$

by

$$(L_j h)(t,t',z) = \frac{1}{j!} \left(\frac{t'}{t} - 1 \right)^j \log^{-j} \left(\frac{t'}{t} \right) \{(t')^j \partial_{t'}^j \partial_z^j h(t,t',z)\}|_{t'=t}.$$

In particular, $L_0 : h(t,t',z) \to h(t,t,z)$. For (2.2.26) we then obtain

$$\mathrm{op}_M^{\frac{1}{2}}(f) = \sum_{j=0}^{N-1} \mathrm{op}_M^{\frac{1}{2}}(L_j f) + R_N^{(1)} = \mathrm{op}_M^{\frac{1}{2}} \left(L_0 f + \sum_{j=1}^{N-1} L_j f \right) + R_N^{(1)}.$$

By iterating the procedure for $L_j f$, $j \geq 1$, we get

$$
\mathrm{op}_M^{\frac{1}{2}}(f) = \mathrm{op}_M^{\frac{1}{2}} \left(L_0 f + \sum_{j_1=1}^{N-1} L_0 L_{j_1} f + \sum_{j_2=1}^{N-j_1-1} \sum_{j_1=1}^{N-1} L_{j_2} L_{j_1} f \right) + R_N^{(2)}
$$

$$
= \mathrm{op}_M^{\frac{1}{2}} \left(L_0 f + \sum_{j_1=1}^{N-1} L_0 L_{j_1} f + \cdots + \right.
$$

$$
\left. + \sum_{j_{N-1}=1}^{N-j_1-\cdots-j_{N-2}-1} \cdots \sum_{j_1=1}^{N-1} L_0 L_{j_{N-1}} \ldots L_{j_1} f \right) + R_N^{(N-1)}.
$$

$$(2.2.27)$$

The remainders are defined by

$$
R_N^{(r)} = R_N^{(r-1)} + \sum \mathrm{op}_M^{\frac{1}{2}} \frac{(t-t')^N}{(N-1)!}
$$

$$
\times \int\limits_0^1 (1-\theta)^{N-1} \partial_{t'}^N L_{j_r} \ldots L_{j_1} f(t, t' + \theta(t'-t), z) \, d\theta.
$$

Let us check that

$$
L_0 L_{j_r} \cdot \cdots \cdot L_{j_1} f \in C^\infty(\overline{\mathbb{R}}_+, L^{\mu-r}(X; \Gamma_0)).
$$

$$(2.2.28)$$

First observe that the operator ∂_z commutes with powers of $(\frac{t'}{t} - 1)/\log(\frac{t'}{t})$. Furthermore, $(t')^k \partial_{t'}^k$ is a linear combination of operators $(t'\partial_{t'})^j$ for $j = 1, \ldots, k$, cf. Remark 2.1.18. For arbitrary $l \in \mathbb{N}$ induction shows

$$
(t'\partial_{t'})^j [(\frac{t'}{t} - 1)^l \log^{-l}(\frac{t'}{t})] = (s\partial_s)^j [(s-1)^l \log^{-l} s]|_{s=\frac{t'}{t}}. \qquad (2.2.29)
$$

Evaluating this function at $t = t'$ gives a constant $c_{jl} \in \mathbb{R}$. Therefore

$$
(L_{j_r} \cdot \cdots \cdot L_{j_1} f)(t, t', z)
$$

$$
= \left(\frac{t'}{t} - 1 \right)^{j_r} \log^{-j_r} \left(\frac{t'}{t} \right) c_{j_1,\ldots,j_r} \{ (t')^{j_1} \partial_{t'}^{j_1} \partial_z^{j_1+\cdots+j_r} f(t, t', z) \}|_{t'=t}
$$

$$(2.2.30)$$

for suitable constants $c_{j_1,\ldots,j_r}$, which shows (2.2.28). We now consider the remainder $R_N^{(N-1)}$. According to (2.2.27) and (2.2.28) we have $R_N^{(N-1)} = \mathrm{op}_M^{\frac{1}{2}}(r_{N-1})$ for some $r_{N-1}(t, t', z) \in C^\infty(\overline{\mathbb{R}}_+ \times \overline{\mathbb{R}}_+, L^\mu(X; \Gamma_0))$.

Choose a function $\psi(\varrho) \in C_0^\infty(\mathbb{R}_+)$ with $\psi \equiv 1$. Then, by Theorem 2.2.10 and Remark 2.2.12 we get

$$R_N^{(N-1)} = \mathrm{op}_M^{\frac{1}{2}}(M_{\frac{1}{2},\varrho\to z}\psi(\varrho)M_{\frac{1}{2},\tilde{z}\to\varrho}^{-1}r_{N-1}(t,t',\tilde{z})) + R$$

for some $R \in M_{\frac{n+1}{2}}L^{-\infty}(\overline{\mathbb{R}}_+ \times X)$. By construction the symbol $r_N(t,t',z)$ is a sum of terms of the form

$$\frac{1}{(N-1)!}(t-t')^N \int_0^1 (1-\theta)^{N-1}\partial_{t'}^N L_{j_r}\dots L_{j_1} f(t,t'+\theta(t-t'),z)\,d\theta$$

$$= \frac{(-1)^N}{(N-1)!}\left(\frac{t'}{t}-1\right)^N \left(\frac{t'}{t}\right)^{-N}$$

$$\times \int_0^1 (1-\theta)^{N-1}\left(\frac{t}{s}\right)^N (s^N\partial_s^N)\left(\frac{s}{t}-1\right)^{j_r}\log^{-j_r}\left(\frac{s}{t}\right)|_{s=t'+\theta(t-t')}\,d\theta$$

$$\times c_{j_1,\dots,j_r}\{(t')^{j_1}\partial_{t'}^{j_1}\partial_z^{j_1+\dots+j_r}f(t,t',z)\}|_{t'=t}, \qquad (2.2.31)$$

where $j_1 + \dots + j_r \geq N$. Using (2.2.29), Remark 2.1.18, and the fact that $s/t = t'/t+\theta(1-t'/t)$, the integral turns out to be a function of t'/t, C^∞ on $\mathbb{R}_+$ though not necessarily on $\overline{\mathbb{R}}_+$. The last factor in (2.2.31) is in $C^\infty(\overline{\mathbb{R}}_+, L^{\mu-N}(X;\Gamma_0))$. Hence $\mathrm{op}_M^{\frac{1}{2}}(M_{\frac{1}{2},\varrho\to z}\psi(\varrho)M_{\frac{1}{2},\tilde{z}\to\varrho}^{-1}r_{N-1}(t,t',\tilde{z}))$ is a sum of operators of the form $\mathrm{op}_M^{\frac{1}{2}}(M_{\frac{1}{2},\varrho\to z}\psi(\varrho)M_{\frac{1}{2},\tilde{z}\to\varrho}^{-1}\alpha(t'/t)h(t,\tilde{z}))$ for certain $\alpha(\varrho) \in C^\infty(\mathbb{R}_+)$ and $h(t,z) \in C^\infty(\overline{\mathbb{R}}_+, L^{\mu-N}(X;\Gamma_0))$. According to the kernel manipulations of Section 2.2.2. these operators can be rewritten in the form $\mathrm{op}_M^{\frac{1}{2}}(M_{\frac{1}{2},\varrho\to z}\psi(\varrho)\alpha(\varrho)M_{\frac{1}{2},\tilde{z}\to\varrho}^{-1}h(t,\tilde{z}))$. In view of $\psi\alpha \in C_0^\infty(\mathbb{R}_+)$ we obtain from Theorem 2.2.10 and Theorem 2.2.5 that $R_N^{(N-1)} \in M_{\frac{n+1}{2}}L^{\mu-N}(\overline{\mathbb{R}}_+ \times X)$. We have obtained altogether an asymptotic expansion for $f(t,z)$. According to (2.2.27), (2.2.30), this is of the form

$$\sum_{k,l=0}^\infty d_{kl}(t'\partial_{t'})^k\partial_z^l f(t,t',z)|_{t'=t}$$

for certain constants d_{kl}. These constants are independent of f. Therefore, we can choose a particularly simple f to determine them. For

$f(t, t', z) = \varphi(t')z^k$ we have

$$
\begin{aligned}
\operatorname{op}_M^{\frac{1}{2}}(f)u(t) &= (-t\partial_t)^k(\varphi u)(t) \\
&= \sum_{l=0}^{k} \binom{k}{l}(-t\partial_t)^l \varphi(t)(-t\partial_t)^{k-l}u(t) \\
&= \operatorname{op}_M^{\frac{1}{2}} \left(\sum_{l=0}^{k} \binom{k}{l}(-t\partial_t)^l \varphi(t)z^{k-l} \right) u(t) \\
&= \operatorname{op}_M^{\frac{1}{2}} \left(\sum_{l=0}^{k} \frac{1}{l!}(-t'\partial_t')^l \partial_z^l f(t, t', z)|_{t'=t} \right) u(t),
\end{aligned}
$$

noting that $\partial_z^l z^k = k \cdot \ldots \cdot (k-l+1)z^{k-l} = \frac{k!}{(k-l)!}z^{k-l}$. Hence, $d_{kl} = \frac{1}{l!}\delta_{kl}$, just as assserted. Clearly, this coincides with the expansion from Proposition 2.2.7. $\square$

Theorem 2.2.32 *For every $f(t, t', z) \in C^\infty(\overline{\mathbb{R}}_+ \times \overline{\mathbb{R}}_+, L^\mu(X; \Gamma_0))$ there is a $\boldsymbol{g}(t', z) \in C^\infty(\overline{\mathbb{R}}_+, L^\mu(X; \Gamma_0))$, obtained by an asymptotic expansion*

$$
\boldsymbol{g}(t', z) \sim \sum_{j=0}^{\infty} \frac{1}{j!}(-t\partial_t)^j(-\partial_z)^j f(t, t', z)|_{t=t'}, \tag{2.2.32}
$$

such that $\operatorname{op}_M^{\frac{1}{2}}(f) - \operatorname{op}_M^{\frac{1}{2}}(\boldsymbol{g}) \in M_{\frac{n+1}{2}}L^{-\infty}(\overline{\mathbb{R}}_+ \times X)$.

Proof. The arguments are analogous to those for Theorem 2.2.31 by interchanging the role of t and t'. $\square$

Remark 2.2.33 *The Theorems 2.2.31, 2.2.32 hold in analogous form for $f(t, t', z) \in C^\infty(\overline{\mathbb{R}}_+ \times \overline{\mathbb{R}}_+, L^\mu(X; \Gamma_{\frac{n+1}{2}-\gamma}))$ with*

$$
\begin{aligned}
\boldsymbol{f}(t, z) &\in C^\infty(\overline{\mathbb{R}}_+, L^\mu(X; \Gamma_{\frac{n+1}{2}-\gamma})), \\
\boldsymbol{g}(t', z) &\in C^\infty(\overline{\mathbb{R}}_+, L^\mu(X; \Gamma_{\frac{n+1}{2}-\gamma})),
\end{aligned}
$$

where

$$
\begin{aligned}
\operatorname{op}_M^{\gamma-\frac{n}{2}}(f) &= \operatorname{op}_M^{\gamma-\frac{n}{2}}(\boldsymbol{f}) \bmod M_\gamma L^{-\infty}(\overline{\mathbb{R}}_+ \times X), \\
\operatorname{op}_M^{\gamma-\frac{n}{2}}(f) &= \operatorname{op}_M^{\gamma-\frac{n}{2}}(\boldsymbol{g}) \bmod M_\gamma L^{-\infty}(\overline{\mathbb{R}}_+ \times X),
\end{aligned}
$$

and $\boldsymbol{f}(t, z)$ and $\boldsymbol{g}(t', z)$ allow the asymptotic expansions (2.2.22) and (2.2.32), respectively.

In fact, using Corollary 2.2.13, we can assume $f(t, t', z) \in C^\infty(\overline{\mathbb{R}}_+ \times \overline{\mathbb{R}}_+, N_O^\mu(X))$, and then, because of Theorem 2.2.6, $\mathrm{op}_M^{\frac{1}{2}}(f) = \mathrm{op}_M^{\gamma-\frac{n}{2}}(f)$. Applying Theorem 2.2.10 the symbols $\boldsymbol{f}(t, z)$ and $\boldsymbol{g}(t', z)$ can be chosen as elements in $C^\infty(\overline{\mathbb{R}}_+, N_O^\mu(X))$. Then, the generalisation of the Theorems 2.2.31, 2.2.32 follow immediately from $\mathrm{op}_M^{\frac{1}{2}}(\boldsymbol{f}) = \mathrm{op}_M^{\gamma-\frac{n}{2}}(\boldsymbol{f})$ and the analogous relation for $\boldsymbol{g}$.

Corollary 2.2.34 *Every $A \in M_\gamma L^\mu(\overline{\mathbb{R}}_+ \times X)$ can be written in the form*

$$A = \mathrm{op}_M^{\gamma-\frac{n}{2}}(f) + C = \mathrm{op}_M^{\gamma-\frac{n}{2}}(g) + D$$

for suitable $f(t, z) \in C^\infty(\overline{\mathbb{R}}_+, N_O^\mu(X))$, $g(t', z) \in C^\infty(\overline{\mathbb{R}}_+, N_O^\mu(X))$ and $C, D \in M_\gamma L^{-\infty}(\overline{\mathbb{R}}_+ \times X)$. An analogous assertion holds for the classical operators with $M_O^\mu(X)$ instead of $N_O^\mu(X)$.

In fact, it suffices to apply the Theorems 2.2.31, 2.2.32 and 2.2.10, and Remark 2.2.12.

Remark 2.2.35 *From Theorem 2.2.5 we know that $A \in M_\gamma L^\mu(\overline{\mathbb{R}}_+ \times X)$ implies $A^* \in M_{-\gamma} L^\mu(\overline{\mathbb{R}}_+ \times X)$ for the formal adjoint with respect to the $\mathcal{H}^{0,0}(X^\wedge)$-scalar product. Writing $A = \mathrm{op}_M^{\gamma-\frac{n}{2}}(f) + C$ for an $f(t, z) \in C^\infty(\overline{\mathbb{R}}_+, L^\mu(X; \Gamma_{\frac{n+1}{2}-\gamma}))$, $C \in M_\gamma L^{-\infty}(\overline{\mathbb{R}}_+ \times X)$ we obtain $A^* = \mathrm{op}_M^{-\gamma-\frac{n}{2}}(f^*) + C^*$ with $f^*(t', z) \in C^\infty(\overline{\mathbb{R}}_+, L^\mu(X; \Gamma_{\frac{n+1}{2}+\gamma}))$,*

$$f^*(t', z) = f^{(*)}(t', n + 1 - \overline{z}).$$

Then Theorem 2.2.32 gives us $A^* = \mathrm{op}_M^{-\gamma-\frac{n}{2}}(h)$ mod $M_{-\gamma} L^{-\infty}(\overline{\mathbb{R}}_+ \times X)$ with an $h(t, z) \in C^\infty(\overline{\mathbb{R}}_+, L^\mu(X; \Gamma_{\frac{n+1}{2}+\gamma}))$,

$$h(t, z) \sim \sum_{j=0}^\infty \frac{1}{j!}(-t\partial_t)^j(-\partial_z)^j f^*(t, z).$$

Theorem 2.2.36 *$A \in M_\gamma L^\mu(\overline{\mathbb{R}}_+ \times X)$, $B \in M_\gamma L^\nu(\overline{\mathbb{R}}_+ \times X)$ implies $A\omega B \in M_\gamma L^{\mu+\nu}(\overline{\mathbb{R}}_+ \times X)$ for every $\mu, \nu \in \mathbb{R}$ and every cut–off function $\omega(t)$. Writing $A = \mathrm{op}_M^{\gamma-\frac{n}{2}}(f) + C$, $B = \mathrm{op}_M^{\gamma-\frac{n}{2}}(g) + D$ with certain $f(t, z) \in C^\infty(\overline{\mathbb{R}}_+, L^\mu(X; \Gamma_{\frac{n+1}{2}-\gamma}))$, $g(t, z) \in C^\infty(\overline{\mathbb{R}}_+, L^\nu(X; \Gamma_{\frac{n+1}{2}-\gamma}))$ and $C, D \in M_\gamma L^{-\infty}(\overline{\mathbb{R}}_+ \times X)$ we obtain modulo $M_\gamma L^{-\infty}(\overline{\mathbb{R}}_+ \times X)$ the relation $A\omega B = \mathrm{op}_M^{\gamma-\frac{n}{2}}(h)$ with an $h(t, z) \in C^\infty(\overline{\mathbb{R}}_+, L^\mu(X; \Gamma_{\frac{n+1}{2}-\gamma}))$,*

$$h(t, z) \sim \sum_{j=0}^\infty \frac{1}{j!}(\partial_z^j f)(t, z)(-t\partial_t)^j(\omega(t)g(t, z)). \tag{2.2.33}$$

Proof. Applying the above Remark 2.2.33 we can assume that $A = \mathrm{op}_M^{\gamma-\frac{n}{2}}(f) + C$, $B = \mathrm{op}_M^{\gamma-\frac{n}{2}}(g) + D$ for certain $C, D \in M_\gamma L^{-\infty}(\overline{\mathbb{R}}_+ \times X)$. Then Theorem 2.2.2 together with Definition 2.2.3 and Remark 2.2.35 shows that the operators $\mathrm{op}_M^{\gamma-\frac{n}{2}}(f)\omega D$, $C\omega\,\mathrm{op}_M^{\gamma-\frac{n}{2}}(g)$ and $C\omega D$ belong to $M_\gamma L^{-\infty}(\overline{\mathbb{R}}_+ \times X)$. Thus it suffices to consider $\mathrm{op}_M^{\gamma-\frac{n}{2}}(f)\{\omega\,\mathrm{op}_M^{\gamma-\frac{n}{2}}(g)\}$. Writing $\tilde{g}(t,z) = \omega(t)g(t,z)$, the second factor has the form $\mathrm{op}_M^{\gamma-\frac{n}{2}}(\boldsymbol{g}) + G$ for certain $\boldsymbol{g}(t',z) \in C^\infty(\overline{\mathbb{R}}_+, L^\mu(X; \Gamma_{\frac{n+1}{2}-\gamma}))$ and $G \in M_\gamma L^{-\infty}(\overline{\mathbb{R}}_+ \times X)$.

It is easy to see that the compositions

$$\mathrm{op}_M^{\gamma-\frac{n}{2}}(f)\,\mathrm{op}_M^{\gamma-\frac{n}{2}}(\boldsymbol{g}) \quad \text{and} \quad \mathrm{op}_M^{\gamma-\frac{n}{2}}(f)G$$

make sense separately, where the latter once again belongs to the space $M_\gamma L^{-\infty}(\overline{\mathbb{R}}_+ \times X)$. Now $\mathrm{op}_M^{\gamma-\frac{n}{2}}(f)\,\mathrm{op}_M^{\gamma-\frac{n}{2}}(\boldsymbol{g}) = \mathrm{op}_M^{\gamma-\frac{n}{2}}(d)$ for $d(t,t',z) = f(t,z)\boldsymbol{g}(t',z) \in C^\infty(\overline{\mathbb{R}}_+ \times \overline{\mathbb{R}}_+, L^{\mu+\nu}(X; \Gamma_{\frac{n+1}{2}-\gamma}))$. This yields $AB \in M_\gamma L^{\mu+\nu}(\overline{\mathbb{R}}_+ \times X)$, cf. Definition 2.2.3. Theorem 2.2.31 applied to $d(t,t',z)$ generates a (t,z)–dependent Mellin symbol $h(t,z)$ that represents the operator modulo $M_\gamma L^{-\infty}(\overline{\mathbb{R}}_+ \times X)$. The arising asymptotic formula simplifies to (2.2.33) which can be obtained directly by elementary calculations or by applying the proof of Theorem 1.1.32 combined with a reduction of pseudo–differential operators based on the Mellin transform to operators based on the Fourier transform, cf. formula (2.1.29). □

2.2.4 Polar coordinates in pseudo–differential operators

This section studies pseudo–differential operators in polar coordinates in $\mathbb{R}^{n+1} \setminus \{0\}$, which lead to symbols of Fuchs type in $(t,x) \in \mathbb{R} \times S^n$ for $t \to 0$. In addition we analyse Fuchs type symbols on $X^\wedge = \mathbb{R}_+ \times X \ni (t,x)$ for $t \to \infty$ under the aspect of the calculus with exit to infinity.

Let $V \subseteq \mathbb{R}_{\tilde{x}}^{n+1} \setminus \{0\}$ be a conical set, $V_1 := \{\tilde{x} \in V : |\tilde{x}| = 1\}$, let $\gamma : V_1 \to \Sigma$ be a diffeomorphism to an open set $\Sigma \subseteq \mathbb{R}^n$, and consider the associated diffeomorphism

$$\kappa : V \to \mathbb{R}_+ \times \Sigma, \quad \kappa(\tilde{x}) = (t, \gamma(\tilde{x}/|\tilde{x}|)) \qquad \text{for } t = |\tilde{x}|. \quad (2.2.34)$$

Then we have the push–forward of pseudo–differential operators

$$\kappa_* : L^\mu(V) \to L^\mu(\mathbb{R}_+ \times \Sigma), \qquad \kappa_* : L^\mu_{\mathrm{cl}}(V) \to L^\mu_{\mathrm{cl}}(\mathbb{R}_+ \times \Sigma).$$

The particular choice of Σ is not essential for the sequel; we assume that $\Sigma = \{x \in \mathbb{R}^n : |x| < 1\}$. Moreover, for simplicity we suppose that there is a larger conical set V and an open $\Xi \subset \mathbb{R}^n$ with $\overline{\Sigma} \subset \Xi$ such that κ is the restriction of a corresponding diffeomorphism $V \to \mathbb{R}_+ \times \Xi$ to V.

Proposition 2.2.37 *To every $a(\tilde{x}, \tilde{\xi}) \in S_{\mathrm{cl}}^\mu(V \times \mathbb{R}^{n+1})$ with $a(\tilde{x}, \tilde{\xi}) = a_1(\tilde{x}, \tilde{\xi})|_{V \times \mathbb{R}^{n+1}}$ for some $a_1(\tilde{x}, \tilde{\xi}) \in S_{\mathrm{cl}}^\mu(\mathbb{R}^{n+1} \times \mathbb{R}^{n+1})$ there exists a $b(t, x, \tau, \xi) \in S_{\mathrm{cl}}^\mu(\mathbb{R}_+ \times \Sigma \times \mathbb{R}^{n+1})$ of the form*

$$b(t, x, \tau, \xi) = t^{-\mu} \tilde{p}(t, x, \tilde{\tau}, \xi)|_{\tilde{\tau} = t\tau},$$

where $\tilde{p}(t, x, \tilde{\tau}, \xi) \in S_{\mathrm{cl}}^\mu(\overline{\mathbb{R}}_+ \times \Sigma \times \mathbb{R}^{n+1})$, such that

$$\kappa_* \mathrm{Op}_{\tilde{x}}(a) = \mathrm{Op}_{(t,x)}(b) \bmod L^{-\infty}(\mathbb{R}_+ \times \Sigma).$$

Proof. The standard formula for complete symbols under operator push–forwards gives us

$$b(t, x, \tau, \xi)|_{(t,x) = \kappa(\tilde{x})} \sim \sum_\alpha \frac{1}{\alpha!} (\partial_\xi^\alpha a)(\tilde{x}, {}^t d\kappa(\tilde{x}) \begin{pmatrix} \tau \\ \xi \end{pmatrix}) \Phi_\alpha(\tilde{x}, \tau, \xi),$$

$$(2.2.35)$$

cf. Theorem 1.1.38. Here $\Phi_\alpha(\tilde{x}, \tau, \xi) = D_{\tilde{z}}^\alpha e^{i\delta(\tilde{x}, \tilde{z})(\tau, \xi)}|_{\tilde{z} = \tilde{x}}$, $\alpha \in \mathbb{N}^{n+1}$, and $\delta(\tilde{x}, \tilde{z}) = \kappa(\tilde{z}) - \kappa(\tilde{x}) - d\kappa(\tilde{x})(\tilde{z} - \tilde{x})$, $\delta(\tilde{x}, \tilde{z})(\tau, \xi) = \delta_0(\tilde{x}, \tilde{z})\tau + \sum_{j=1}^n \delta_j(\tilde{x}, \tilde{z})\xi_j$. For $\kappa(\tilde{x}) = (\kappa_0(\tilde{x}), \kappa_1(\tilde{x}), \ldots, \kappa_n(\tilde{x}))$, where $\kappa_0(\tilde{x}) = t$, we obtain $\kappa_0(\lambda\tilde{x}) = \lambda\kappa_0(\tilde{x})$, $\kappa_j(\lambda\tilde{x}) = \kappa_j(\tilde{x})$ for $j = 1, \ldots, n$, $\lambda \in \mathbb{R}_+$. Then the entries of the Jacobi matrix

$$d\kappa(\tilde{x}) = \begin{pmatrix} j_{0,0}(\tilde{x}) & j_{0,1}(\tilde{x}) & \cdots & j_{0,n}(\tilde{x}) \\ j_{1,0}(\tilde{x}) & j_{1,1}(\tilde{x}) & \cdots & j_{1,n}(\tilde{x}) \\ \vdots & \vdots & \vdots & \\ j_{n,0}(\tilde{x}) & j_{n,1}(\tilde{x}) & \cdots & j_{n,n}(\tilde{x}) \end{pmatrix},$$

$j_{0,k-1}(\tilde{x}) = \partial\kappa_0(\tilde{x})/\partial\tilde{x}_k$, $j_{l,k-1}(\tilde{x}) = \partial\kappa_l(\tilde{x})/\partial\tilde{x}_k$, $k = 1, \ldots, n+1$, $l = 1, \ldots, n$, satisfy the homogeneity relations

$$j_{0,k-1}(\lambda\tilde{x}) = j_{0,k-1}(\tilde{x}), \quad j_{l,k-1}(\lambda\tilde{x}) = \lambda^{-1} j_{l,k-1}(\tilde{x}) \quad \text{for all } \lambda \in \mathbb{R}_+,$$

$$(2.2.36)$$

$k = 1, \ldots, n+1$, $l = 1, \ldots, n$. This gives us

$${}^t d\kappa(\tilde{x}) \begin{pmatrix} \tau \\ \xi \end{pmatrix} = \psi(x) \begin{pmatrix} \tau \\ t^{-1}\xi \end{pmatrix}$$

for a smooth $(n+1) \times (n+1)$–matrix function $\psi(x)$ on Σ.

The function $\Phi_\alpha(\tilde{x},\tau,\xi)$ is a polynomial in (τ,ξ) of degree $d(\alpha) \leq |\alpha|/2$. Let us set $\delta_l^{(\beta)}(\tilde{x}) = D_{\tilde{z}}^\beta \delta_l(\tilde{x},\tilde{z})|_{\tilde{z}=\tilde{x}}$, $l = 0,\dots,n$. From the homogeneity properties of $\kappa(\tilde{x})$ it follows that

$$\delta_0^{(\beta)}(\lambda\tilde{x}) = \lambda^{1-|\beta|}\delta_0(\tilde{x}), \quad \delta_l^{(\beta)}(\lambda\tilde{x}) = \lambda^{-|\beta|}\delta_l(\tilde{x}) \quad \text{for } l = 1,\dots,n \tag{2.2.37}$$

for all $\lambda \in \mathbb{R}_+$. Thus

$$\delta_0^{(\beta)}(\tilde{x})\tau = t^{1-|\beta|}\delta_0^{(\beta)}\left(\frac{\tilde{x}}{|\tilde{x}|}\right)\tau, \quad \delta_l^{(\beta)}(\tilde{x})\xi_l = t^{-|\beta|}\delta_l^{(\beta)}\left(\frac{\tilde{x}}{|\tilde{x}|}\right)\xi_l$$

for $l = 1,\dots,n$. We obtain $\Phi_\alpha(\tilde{x},\tau,\xi) = t^{-d(\alpha)}P_\alpha(\frac{\tilde{x}}{|\tilde{x}|},t\tau,\xi)$, where P_α is a polynomial in $(t\tau,\xi)$ of degree $d(\alpha)$ with coefficients in $C^\infty(V_1)$.

Thus, to evaluate (2.2.35) we have to form the asymptotic sum of

$$\frac{1}{\alpha!}(\partial_\xi^\alpha a)\left(\tilde{x},\psi(x)\binom{\tau}{t^{-1}\xi}\right)t^{-d(\alpha)}P_\alpha(x,t\tau,\xi).$$

By assumption the symbol a is classical. For the homogeneous component $a_{(\mu-j)}(\tilde{x},\tilde{\xi})$ of $a(\tilde{x},\tilde{\xi})$ of order $\mu - j$ we have

$$\left(\partial_\xi^\alpha a_{(\mu-j)}\right)(\tilde{x},\psi(x)\binom{\tau}{t^{-1}\xi}) = t^{-\mu+j+|\alpha|}(\partial_\xi^\alpha a_{(\mu-j)})\left(\tilde{x},\psi(x)\binom{t\tau}{\xi}\right).$$

We have $d(\alpha) \leq |\alpha|/2$. Thus, if we fix an excision function $\chi(\tilde{\tau},\xi)$ we may regard (2.2.35) (after rearranging summands) as an asymptotic sum

$$t^{-\mu}\sum_\alpha \chi(\tilde{\tau},\xi)c_\alpha(t,x,\tilde{\tau},\xi)|_{\tilde{\tau}=t\tau},$$

where $c_\alpha(t,x,\tilde{\tau},\xi)$ is homogeneous in $(\tilde{\tau},\xi)$ of order $\mu - |\alpha|$ and C^∞ in $(t,x) \in \overline{\mathbb{R}}_+ \times \Sigma$. This gives us the assertion. $\square$

Let us set $L_{\mathrm{cl}}^\mu(V)_e = L_{\mathrm{cl}}^\mu(V) \cap \boldsymbol{L}^{\mu,0}(V)$, cf. the notation of Section 1.4.5, and

$$S_{\mathrm{cl}}^\mu(V \times \mathbb{R}^{n+1})_e = S_{\mathrm{cl}}^\mu(V \times \mathbb{R}^{n+1}) \cap \boldsymbol{S}_{\mathrm{cl}}^{\mu,0}(V \times \mathbb{R}^{n+1}). \tag{2.2.38}$$

Analogously we obtain the space

$$L_{\mathrm{cl}}^\mu(X^\wedge)_e := L_{\mathrm{cl}}^\mu(X^\wedge) \cap \boldsymbol{L}^{\mu,0}(X^\wedge) \tag{2.2.39}$$

for any closed compact C^∞ manifold X.

Proposition 2.2.38 Let $\tilde{p}(t,x,\tilde{\tau},\xi) \in S_{\mathrm{cl}}^\mu(\mathbb{R}_+ \times \Sigma \times \mathbb{R}_{\tilde{\tau},\xi}^{n+1})$ be a symbol that is independent of t for $t > $ const for a constant > 0, and set $b(t,x,\tau,\xi) = t^{-\mu}\tilde{p}(t,x,t\tau,\xi)$. Then

$$(\kappa^{-1})_* \mathrm{Op}(b) \in L_{\mathrm{cl}}^\mu(V)_e.$$

Proof. Each symbol $a(\tilde{x}, \tilde{\xi}) \in S^{\mu}_{\mathrm{cl}}(V \times \mathbb{R}^{n+1})$ with the property $\mathrm{Op}_{\tilde{x}}(a) = (\kappa^{-1})_{*} \mathrm{Op}(b)$ modulo smoothing operators has the asymptotic expansion

$$a(\tilde{x}, \tilde{\xi})|_{\tilde{x}=\kappa^{-1}(t,x)} \sim \sum_{\alpha} (\partial^{\alpha}_{r,\xi} b)(t, x, {}^{t}d\kappa^{-1}(t, x)\tilde{\xi}) \Psi_{\alpha}(t, x, \tilde{\xi}), \quad (2.2.40)$$

where $\Psi_{\alpha}(t, x, \tilde{\xi}) = D^{\alpha}_{r,z} e^{i\delta(t,x,r,z)\tilde{\xi}}|_{(r,z)=(t,x)}$, $\alpha \in \mathbb{N}^{n+1}$, $\delta(t, x, r, z) = \kappa^{-1}(r, z) - \kappa^{-1}(t, x) - d\kappa^{-1}(t, x)((r, z) - (t, x))$. Set

$$\delta(t, x, r, z) = \{\delta_{l}(t, x, r, z)\}_{l=1,\ldots,n+1},$$

$$\delta^{(\beta)}_{l}(t, x) = D^{\beta}_{r,z} \delta_{l}(t, x, r, z)|_{(r,z)=(t,x)}$$

for $l = 1, \ldots, n+1$. We then have

$$\delta^{(\beta)}_{l}(\lambda t, x) = \lambda^{1-\beta_0} \delta^{(\beta)}_{l}(t, x) \qquad \text{for all } \lambda \in \mathbb{R}_{+},$$

$l = 1, \ldots, n+1$, $\beta = (\beta_0, \beta_1, \ldots, \beta_n)$. The function $\Psi_{\alpha}(t, x, \tilde{\xi})$ is a polynomial in the variables $\delta^{(\beta)}_{l}(t, x)\tilde{\xi}_{l}$, $1 \leq l \leq n+1$, of degree $d(\alpha) \leq |\alpha|/2$, where $\Psi_0(t, x, \tilde{\xi}) = 1$. From the above homogeneities we see that the coefficients are classical symbols in t for $t \to \infty$ of order ≥ 0. For $\alpha = (\alpha_0, 0)$, where α_0 refers to the t–variable, we have

$$\Psi_{\alpha}(t, x, \tilde{\xi}) = 0 \qquad \text{for } \alpha_0 = 1. \qquad (2.2.41)$$

Moreover, for every $N \in \mathbb{N}$ there exists an $\alpha_0 \in \mathbb{N}$ such that

$$t^{N} |\Psi_{\alpha}(t, x, \tilde{\xi})| < c \qquad \text{for } \alpha = (\alpha_0, 0) \qquad (2.2.42)$$

for all $t \in \mathbb{R}_{+}$, $x \in K$ for arbitrary $K \subset\subset \Sigma$ and all $\tilde{\xi}$ with $|\tilde{\xi}| \leq R$, for a constant $c = c(K, R, N, \alpha) > 0$.

The entries of the Jacobi matrix

$$d\kappa^{-1}(t, x) = \begin{pmatrix} m_{0,0}(t, x) & m_{0,1}(t, x) & \ldots & m_{0,n}(t, x) \\ m_{1,0}(t, x) & m_{1,1}(t, x) & \ldots & m_{1,n}(t, x) \\ \vdots & \vdots & \vdots & \\ m_{n,0}(t, x) & m_{n,1}(t, x) & \ldots & m_{n,n}(t, x) \end{pmatrix},$$

have the homogeneities

$$m_{k-1,0}(\lambda t, x) = m_{k-1,0}(t, x), \quad m_{k-1,l}(\lambda t, x) = \lambda m_{k-1,l}(t, x)$$

for all $\lambda \in \mathbb{R}_{+}$, $k = 1, \ldots, n+1$, $l = 1, \ldots, n$. Let $\psi_0(t, x)$, $\psi_1(t, x), \ldots, \psi_n(t, x)$ be the rows of ${}^{t}d\kappa^{-1}(t, x)$. Then

$${}^{t}d\kappa^{-1}(t, x)\tilde{\xi} = \left\{ \langle \psi_0(t, x), \tilde{\xi} \rangle, \langle \psi_1(t, x), \tilde{\xi} \rangle, \ldots, \langle \psi_n(t, x), \tilde{\xi} \rangle \right\}.$$

Writing $\psi_j(t, x) = \psi_j(\kappa(\tilde{x}))$ we see that

$$\psi_0(t, x) = \psi_0(\kappa(\tilde{x}/|\tilde{x}|)), \quad \psi_j(t, x) = |\tilde{x}|\psi_j(\kappa(\tilde{x}/|\tilde{x}|)), \quad j = 1, \ldots, n.$$

In order to evaluate (2.2.40) we consider

$$(\partial_{\tau,\xi}^\alpha b)(t, x, {}^t d\kappa^{-1}(t, x)\tilde{\xi}) = t^{-\mu}t^{\alpha_0}(\partial_{\tilde{\tau},\xi}^\alpha \tilde{p})(t, x, t\psi(x)\tilde{\xi}),$$

where $\psi(x)$ denotes the matrix with the rows $\psi_i(\kappa(\tilde{x}/|\tilde{x}|))$, $i = 0, \ldots, n$. For the homogeneous component $\tilde{p}_{(\mu-j)}(t, x, \tilde{\tau}, \xi)$ of $\tilde{p}(t, x, \tilde{\tau}, \xi)$ of order $\mu - j$ we obtain

$$t^{\alpha_0}(\partial_{\tilde{\tau},\xi}^\alpha \tilde{p}_{(\mu-j)})(t, x, t\psi(x)\tilde{\xi}) = t^{\alpha_0+\mu-|\alpha|-j}(\partial_{\tilde{\tau},\xi}^\alpha \tilde{p})(t, x, t\psi(x)\tilde{\xi}),$$

$j \in \mathbb{N}$. Now if $\chi(t, \tilde{\xi})$ is an excision function in $(t, \tilde{\xi})$ we obtain for (2.2.40) the asymptotic sum

$$\sum_{k=0}^\infty \sum_{|\alpha|+j=k} \frac{1}{\alpha!}\chi(t, \tilde{\xi})t^{\alpha_0-|\alpha|-j}(\partial_{\tilde{\tau},\xi}^\alpha \tilde{p}_{(\mu-j)})(t, x, \psi(x)\tilde{\xi})\Psi_\alpha(t, x, \tilde{\xi}).$$

$$(2.2.43)$$

Applying (2.2.41), (2.2.42) we see that the kth summand, expressed in the coordinates $\tilde{x} = \kappa^{-1}(t, x) \in V$, belongs to $S_{\mathrm{cl}}^{\mu-k}(V \times \mathbb{R}^{n+1})_e$, where at the same time the order with respect to $t = |\tilde{x}|$ tends to $-\infty$ for $t \to \infty$ as $k \to \infty$. Thus the asymptotic sum can be carried out in $S_{\mathrm{cl}}^\mu(V \times \mathbb{R}^{n+1})_e$, and it is unique modulo smoothing symbols in $\tilde{\xi}$ that are Schwartz functions in $t = |\tilde{x}|$ for $t \to \infty$. Moreover, since the diffeomorphisms are of the same kind as in Section 1.4.5, the remainders in the coordinate substitution in $\mathrm{Op}(b)$ also behave in the required way.   □

Below in Proposition 3.2.19 we will obtain a generalisation of Proposition 2.2.38. The alternative arguments there can be specialised to another proof of Proposition 2.2.38.

Let us denote by $H_{\{0\}}^s(\mathbb{R}^{n+1})$ the subspace of all $u \in H^s(\mathbb{R}_{\tilde{x}}^{n+1})$ such that $u(\tilde{x}) = 0$ for $|\tilde{x}| < 1$. Moreover, if $V \subset \mathbb{R}^{n+1} \setminus \{0\}$ is a conical set, $L_{\{0\}}^2(\mathbb{R}_+ \times \Sigma)$ denotes the subspace of all $v \in L^2(\mathbb{R}_+ \times \Sigma)$ with $v(t, x) = 0$ for $t < 1$.

Lemma 2.2.39 *Let $s \in \mathbb{N}$. Then the following conditions are equivalent:*

(i) $u(\tilde{x}) \in H_{\{0\}}^s(\mathbb{R}^{n+1})$.

(ii) *For every conical set $V \subset \mathbb{R}^{n+1} \setminus \{0\}$ and every diffeomorphism $\kappa : V \to \mathbb{R}_+ \times \Sigma$, cf. (2.2.34), the function $v = (\kappa^{-1})^* u$ satisfies*

$$t^{-s} \sum_{|\beta|+j \leq s} a_{\beta j}(x) \left(-t \frac{\partial}{\partial t} \right)^j D_x^\beta \left\{ (\kappa^{-1})^* u \right\} \in t^{-\frac{n}{2}} L_{\{0\}}^2 (\mathbb{R}_+ \times \Sigma)$$

for all $a_{\alpha j}(x) \in C_b^\infty(\Sigma)$ ($C_b^\infty(\Sigma)$ is the space of all $f \in C^\infty(\Sigma)$ with $\sup_{x \in \Sigma} |D_x^\beta f(x)| \leq c_\beta$ for all $\beta \in \mathbb{N}^n$).

Proof. First we have $u \in H_{\{0\}}^s(\mathbb{R}^{n+1}) \Leftrightarrow D_{\tilde{x}}^\alpha u \in H_{\{0\}}^0(\mathbb{R}^{n+1})$ for all $|\alpha| \leq s$. Substituting polar coordinates gives us $H_{\{0\}}^0(\mathbb{R}^{n+1}) = t^{-\frac{n}{2}} L_{\{0\}}^2(\mathbb{R}_+ \times S^n)$ or $H_{\{0\}}^0(\mathbb{R}^{n+1})|_V = t^{-\frac{n}{2}} L_{\{0\}}^2(\mathbb{R}_+ \times \Sigma)$ for every conical set V. To prove (ii) $\Rightarrow$ (i) it suffices to observe that $D_{\tilde{x}}^\alpha$ in polar coordinates takes the form $\sum_{|\beta|+j \leq s} a_{\beta j}(x)(-t \frac{\partial}{\partial t})^j D_x^\beta$ with functions $a_{\beta j}$ of the required kind. For (i) $\Rightarrow$ (ii) we have to verify that the derivatives $\frac{\partial}{\partial t}$, $t^{-1} \frac{\partial}{\partial x_k}$, $1 \leq k \leq n$, and the operators of multiplication by functions in $C_0^\infty(\Sigma)$, transformed to V, are compatible with requiring $D_{\tilde{x}}^\alpha u \in H_{\{0\}}^s$ for all $|\alpha| \leq s$ in the conical sets V. This is an obvious consequence of the relations

$$\frac{\partial}{\partial t} g(\tilde{x}(t,x)) = \sum_{j=1}^{k+1} \frac{\partial \tilde{x}_j}{\partial t} \frac{\partial}{\partial \tilde{x}_j} g(\tilde{x}(t,x)),$$

$$\frac{\partial}{\partial x_k} g(\tilde{x}(t,x)) = \sum_{j=1}^{k+1} \frac{\partial \tilde{x}_j}{\partial x_k} \frac{\partial}{\partial \tilde{x}_j} g(\tilde{x}(t,x)),$$

$1 \leq k \leq n$, and of the homogeneity of the coefficients $\partial \tilde{x}_j / \partial t$ (of order zero) and $\partial \tilde{x}_j / \partial x_k$ (of order 1). □

For purposes below we introduce the space $\tilde{L}_{\mathrm{cl}}^\mu(\Sigma^\wedge)_e$ as the set of all operator families $t^{-\mu} \mathrm{Op}(\tilde{p}) + C$ for arbitrary $\tilde{p}(t,x,t\tau,\xi)$ with $\tilde{p}(t,x,\tilde{\tau},\xi) \in S_{\mathrm{cl}}^\mu(\overline{\mathbb{R}}_+ \times \Sigma \times \mathbb{R}^{n+1})$ which are independent of t for $t > \mathrm{const}$ for a constant > 0 and C having a kernel in $\mathcal{S}(\Sigma^\wedge \times \Sigma^\wedge)$. Here $\mathcal{S}(\Sigma^\wedge \times \Sigma^\wedge)$ is the subspace of all $c(t,x,t',x') \in C^\infty(\Sigma^\wedge \times \Sigma^\wedge)$ with $(1 - \omega(t))(1 - \omega(t'))c(t,x,t',x') \in \mathcal{S}(\mathbb{R} \times \mathbb{R}, C^\infty(\Sigma \times \Sigma))|_{\mathbb{R}_+ \times \mathbb{R}_+}$ for an arbitrary cut–off function $\omega(t)$.

Let X be a closed compact C^∞ manifold, $n = \dim X$, and choose a chart $\kappa : U \to \Sigma$ on X, $\Sigma \subseteq \mathbb{R}^n$ open. Then we define the space $\tilde{L}_{\mathrm{cl}}^\mu(U^\wedge)_e$ as the operator push–forward of $\tilde{L}_{\mathrm{cl}}^\mu(\Sigma^\wedge)_e$ under $\Sigma^\wedge \to U^\wedge$, $(t,x) \to (t, \kappa^{-1}(x))$. Finally, if $\{U_1, \ldots, U_N\}$ is an open covering of X by coordinate neighbourhoods, $\{\varphi_j\}_{j=1,\ldots,N}$ a subordinate partition of unity and $\{\psi_j\}_{j=1,\ldots,N}$ a system of functions $\psi_j \in C_0^\infty(U_j)$ with

$\varphi_j \psi_j = \varphi_j$ for all j, we set

$$\tilde{L}^{\nu}_{\mathrm{cl}}(X^{\wedge})_{\mathrm{e}} = \Big\{ \sum_{j=1}^{N} \varphi_j B_j \psi_j + C : \ B_j \in \tilde{L}^{\mu}_{\mathrm{cl}}(U_j^{\wedge})_{\mathrm{e}}, C \in L^{-\infty}(X^{\wedge})_{\mathrm{e}} \Big\}.$$

$$(2.2.44)$$

Here $L^{-\infty}(X^{\wedge})_{\mathrm{e}}$ is the space of all integral operators with kernels in $\mathcal{S}(X^{\wedge} \times X^{\wedge})$, i.e., in the space of all $c(t, x, t', x') \in C^{\infty}(X^{\wedge} \times X^{\wedge})$ such that $(1 - \omega(t))(1 - \omega(t'))c(t, x, t', x') \in \mathcal{S}(\mathbb{R} \times \mathbb{R}, C^{\infty}(X \times X))|_{\mathbb{R}_+ \times \mathbb{R}_+}$ for an arbitrary cut–off function $\omega(t)$.

As a corollary of Propostion 2.2.38 we then obtain $\tilde{L}^{\mu}_{\mathrm{cl}}(X^{\wedge})_{\mathrm{e}} \subset L^{\mu}_{\mathrm{cl}}(X^{\wedge})_{\mathrm{e}}$, cf. (2.2.39).

Remark 2.2.40 *Let $\omega(t)$, $\omega_1(t)$ be arbitrary cut–off functions. Then every $A \in \tilde{L}^{\mu}_{\mathrm{cl}}(X^{\wedge})_{\mathrm{e}}$ gives rise to continuous operators*

$$(1 - \omega)A(1 - \omega_1) : \mathcal{K}^{s,\gamma}(X^{\wedge}) \to \mathcal{K}^{s-\mu,\delta}(X^{\wedge})$$

for every $s, \gamma, \delta \in \mathbb{R}$. This is an easy consequence of the results of Section 1.4.5.

2.3   Discrete and continuous asymptotics

2.3.1   Weighted Sobolev spaces with discrete asymptotics

We now return to the weighted Sobolev spaces of Section 2.1.4 under the aspect of subspaces with discrete asymptotics along the cone axis variable t for $t \to 0$. Below in Section 2.3.3 we will study continuous analogues of such asymptotics.

For convenience we shall concentrate on the spaces $\mathcal{K}^{s,\gamma}(X^{\wedge})$ with asymptotics for $t \to 0$. It is sufficient to consider a neighbourhood of $t = 0$. Then we easily obtain corresponding asymptotics in $\mathcal{H}^{s,\gamma}(\mathbb{B})$ near $\partial \mathbb{B}$.

Discrete asymptotics are of the form

$$u(t, x) \sim \sum_{j} \sum_{k=0}^{m_j} c_{jk}(x) t^{-p_j} \log^k t \qquad (2.3.1)$$

for $p_j \in \mathbb{C}$, $j \in \mathbb{N}$, $\operatorname{Re} p_j \to -\infty$ as $j \to \infty$, $m_j \in \mathbb{N}$, and coefficients $c_{jk} \in C^{\infty}(X)$. In particular, the elements of the subspace $C_0^{\infty}(\overline{\mathbb{R}}_+ \times X) \subset \mathcal{K}^{0,\frac{n}{2}}(X^{\wedge})$ have Taylor expansions

$$u(t, x) \sim \sum_{j} c_j(x) t^j$$

which have the form (2.3.1), where $p_j = -j$, $m_j = 0$ for all $j \in \mathbb{N}$. In this case we also talk about Taylor asymptotics.

Asymptotic expansions of the form (2.3.1) will be regarded as an aspect of the elliptic regularity for solutions of elliptic equations on a manifold with conical singularities or edges. In particular, the solutions of the homogeneous equation $Au = 0$ for an elliptic differential operator A on $X^\wedge$ which is of Fuchs type near $t = 0$

$$A = t^{-\mu} \sum_{j=0}^{\mu} a_j(t) \left(-t \frac{\partial}{\partial t} \right)^j ,$$

$a_j(t) \in C^\infty(\overline{\mathbb{R}}_+, \mathrm{Diff}^{\mu-j}(X))$, will have such asymptotics. We shall see that the asymptotic data such as the sequence $\{(p_j, m_j)\}_{j \in \mathbb{N}}$ and the particular coefficients $c_{jk}(x)$ are linked to the poles of the meromorphic operator family $h^{-1}(z)$, $z \in \mathbb{C}$, for $h(z) = \sum_{j=0}^{\mu} a_j(0)z^j : H^s(X) \rightarrow H^{s-\mu}(X)$.

For the precise calculus it is convenient to consider not only infinite expansions like (2.3.1) but also finite ones with sums over all j for which p_j belongs to

$$\left\{ z \in \mathbb{C} : \frac{n+1}{2} - \gamma + \vartheta < \mathrm{Re}\, z < \frac{n+1}{2} - \gamma \right\} \qquad (2.3.2)$$

for some $\vartheta > 0$. Set $\Theta = (\vartheta, 0]$ and $\boldsymbol{g} = (\gamma, \Theta)$, called the weight data for the asymptotics, Θ the weight interval.

For finite Θ we shall classify the asymptotics by sequences

$$P = \{(p_j, m_j, L_j)\}_{j=0,\ldots,N} \qquad (2.3.3)$$

for some $N = N(P)$ and $p_j \in (2.3.2)$, $m_j \in \mathbb{N}$, for $j = 0, \ldots, N$, and for $\Theta = (-\infty, 0]$ by

$$P = \{(p_j, m_j, L_j)\}_{j \in \mathbb{N}} \qquad (2.3.4)$$

with $\mathrm{Re}\, p_j < \frac{n+1}{2} - \gamma$ for all $j \in \mathbb{N}$, $\mathrm{Re}\, p_j \rightarrow -\infty$ as $j \rightarrow \infty$, $m_j \in \mathbb{N}$. In both cases L_j is a finite–dimensional subspace of $C^\infty(X)$. Set

$$\pi_\mathbb{C} P = \bigcup_j \{p_j\}, \qquad (2.3.5)$$

where j runs over $\{0, \ldots, N(P)\}$ for (2.3.3) and over $\mathbb{N}$ for (2.3.4). The spaces L_j consist of the coefficients c_{jk} for $0 \leq k \leq m_j$ which occur in the expression (2.3.1). Let us also set

$$\pi_{\mathbb{C} \times \mathbb{N}} P = \bigcup_j \{(p_j, m_j)\} \qquad (2.3.6)$$

for P being of the form (2.3.3) or (2.3.4).

Definition 2.3.1 *The sequences* (2.3.3) *and* (2.3.4) *are called discrete asymptotic types, associated with the weight data* $\boldsymbol{g} = (\gamma, \Theta)$ *for finite and infinite* Θ*, respectively. We denote by* $\mathrm{As}(X, \boldsymbol{g}^{\bullet})$ *the set of all those* P*. The trivial asymptotic type, i.e., when* $\pi_{\mathbb{C}} P = \emptyset$*, will be indicated by* Θ*.*

There is a natural semi–ordering in $\mathrm{As}(X, \boldsymbol{g}^{\bullet})$. Write $Q \geq P$ if for every $(q, n, M) \in Q$ there is a $(p, m, L) \in P$ with $p = q$, $n \leq m$, $M \subseteq L$.

Note that in (2.3.3) or (2.3.4) we have not assumed $p_j \neq p_l$ for $j \neq l$, though for the asymptotics themselves it will be unnecessary to consider sequences for which different triples have the same first component.

Let $-\infty < \vartheta < 0$, $s, \gamma \in \mathbb{R}$, and set

$$\mathcal{K}_{\Theta}^{s,\gamma}(X^{\wedge}) = \bigcap_{\varepsilon > 0} \mathcal{K}^{s,\gamma-\vartheta-\varepsilon}(X^{\wedge}), \tag{2.3.7}$$

endowed with the Fréchet topology of the intersection (it can equivalently be defined by a countable sequence, with $\varepsilon = \{\varepsilon_k\}_{k \in \mathbb{N}}$ for $\varepsilon_k \to 0$ as $k \to \infty$). The elements of $\mathcal{K}_{\Theta}^{s,\gamma}(X^{\wedge})$ may be interpreted to be of flatness δ for every $0 < \delta < -\vartheta$, with respect to the fixed reference weight γ. They will play the role of remainders in the asymptotic expansions of distributions in $\mathcal{K}_{\Theta}^{s,\gamma}(X^{\wedge})$. Note that $\mathcal{K}_{\Theta}^{s,\gamma}(X^{\wedge}) = \mathcal{K}_{\Delta}^{s,\varrho}(X^{\wedge})$ for $\varrho = \gamma + \delta$ and $\Delta = (\vartheta + \delta, 0]$, for $0 < \delta < -\vartheta$.

Let $P = \mathrm{As}(X, \boldsymbol{g}^{\bullet})$ for $\boldsymbol{g} = (\gamma, \Theta)$ with finite Θ, and set

$$\mathcal{E}_P(X^{\wedge}) = [\![c(x) t^{-p} \log^k t : \ (p, m, L) \in P, 0 \leq k \leq m, c \in L]\!], \tag{2.3.8}$$

$[\![\dots]\!]$ denoting the linear span of the functions in the brackets. If we fix a cut–off function $\omega(t)$ then $\omega \mathcal{E}_P(X^{\wedge})$ is a finite–dimensional subspace of $\mathcal{K}^{\infty,\gamma}(X^{\wedge})$, and the induced topology is equivalent to the Euclidean one under a corresponding linear isomorphism $\omega \mathcal{E}_P(X^{\wedge}) \cong \mathbb{C}^L$ for $L = \dim \omega \mathcal{E}_P(X^{\wedge})$. Moreover, we have

$$\mathcal{K}_{\Theta}^{s,\gamma}(X^{\wedge}) \cap \omega \mathcal{E}_P(X^{\wedge}) = \{0\}$$

for all $s \in \mathbb{R}$.

We now introduce

$$\mathcal{K}_P^{s,\gamma}(X^{\wedge}) := \mathcal{K}_{\Theta}^{s,\gamma}(X^{\wedge}) + \omega \mathcal{E}_P(X^{\wedge}), \tag{2.3.9}$$

endowed with the Fréchet topology of the direct sum. It can easily be proved that the particular choice of the cut–off function ω does not affect the Fréchet topology.

For the case $P \in \mathrm{As}(X, (\gamma, \Theta)^\bullet)$ with $\Theta = (-\infty, 0]$ we proceed as follows. We form a sequence of finite asymptotic types P_k for $\Theta_k = (-(k+1), 0]$, $k \in \mathbb{N}$, by $P_k = \{(p, m, L) \in P : (n+1)/2 - \gamma - (k+1) > \mathrm{Re}\, p\}$ and obtain

$$\mathcal{K}_{P_k}^{s,\gamma}(X^\wedge) = \mathcal{K}_{\Theta_k}^{s,\gamma}(X^\wedge) + \omega \mathcal{E}_{P_k}(X^\wedge) \qquad (2.3.10)$$

with continuous embeddings $\mathcal{K}_{P_{k+1}}^{s,\gamma}(X^\wedge) \hookrightarrow \mathcal{K}_{P_k}^{s,\gamma}(X^\wedge)$ for all $k \in \mathbb{N}$. Then we define

$$\mathcal{K}_P^{s,\gamma}(X^\wedge) = \bigcap_{k \in \mathbb{N}} \mathcal{K}_{P_k}^{s,\gamma}(X^\wedge), \qquad (2.3.11)$$

endowed with the Fréchet topology of the projective limit.

We also write

$$\mathcal{K}_\Theta^{s,\gamma}(X^\wedge) = \mathcal{K}^{s,\infty}(X^\wedge)$$

for $\Theta = (-\infty, 0]$. Moreover, we set $\mathcal{K}_P^{\infty,\gamma}(X^\wedge) = \bigcap_{s \in \mathbb{R}} \mathcal{K}_P^{s,\gamma}(X^\wedge)$ for arbitrary $P \in \mathrm{As}(X, \boldsymbol{g}^\bullet)$, $\boldsymbol{g} = (\gamma, \Theta)$, $\Theta = (\vartheta, 0]$, $-\infty \leq \vartheta < 0$.

Remark 2.3.2 *For every $k \in \mathbb{N}$ we can choose an $\varepsilon_k > 0$ such that $\pi_\mathbb{C} P_k \subset \{z : (n+1)/2 - \gamma - (k+1) + \varepsilon_k > \mathrm{Re}\, z\}$. Then, if we set*

$$E^k = \mathcal{K}^{s, \gamma + (k+1) - \varepsilon_k}(X^\wedge) + \omega \mathcal{E}_{P_k}(X^\wedge),$$

we obtain a sequence of (Hilbertisable) Banach spaces E^k, $k \in \mathbb{N}$, with continuous embeddings $E^{k+1} \hookrightarrow E^k$ for all k, and we have $\mathcal{K}_P^{s,\gamma}(X^\wedge) = \varprojlim_{k \in \mathbb{N}} E^k$. Moreover, $\{\kappa_\lambda\}_{\lambda \in \mathbb{R}_+}$ acting on $\mathcal{K}^{s,\gamma}(X^\wedge)$, cf. Proposition 2.1.60, restricts to a group of isomorphisms $\{\kappa_\lambda\}_{\lambda \in \mathbb{R}_+} \in C(\mathbb{R}_+, \mathcal{L}_\sigma(E^k))$ for every $k \in \mathbb{N}$. Analogous representations hold for $\mathcal{K}_P^{s,\gamma}(X^\wedge)$ for $P \in \mathrm{As}(X, (\gamma, \Theta)^\bullet)$ with finite Θ.

The Fréchet spaces E here can often be written as projective limits of Banach spaces $E = \varprojlim_{k \in \mathbb{N}} E^k$, with continuous embeddings $E^{k+1} \hookrightarrow E^k$ for all k. Set $p_k(e) = \|e\|_{E^k}$. Without loss of generality we can assume that $p_{k+1}(e) \geq p_k(e)$ for all $e \in E$, $k \in \mathbb{N}$. Let F be another countably normed Fréchet space with the norm system $\{q_l\}_{l \in \mathbb{N}}$ satisfying $q_{l+1}(f) \geq q_l(f)$ for all $f \in F$, $l \in \mathbb{N}$. Denote by $\mathcal{L}(E, F)$ the space of all linear continuous operators $A : E \to F$. Then $A \in \mathcal{L}(E, F)$ means that for every $l \in \mathbb{N}$ there is an $m(l) \in \mathbb{N}$ such that

$$q_l(Ae) \leq c_l p_{m(l)}(e) \quad \text{for all } e \in E, \qquad (2.3.12)$$

with a constant $c_l > 0$ The choice of a function $m : \mathbb{N} \to \mathbb{N}$ with (2.3.12) for all $l \in \mathbb{N}$ gives rise to a subspace $\mathcal{L}(E, F)_m$ of $\mathcal{L}(E, F)$ which is Fréchet with the norm system

$$\|A\|_l = \sup \{q_l(Ae)/p_{m(l)}(e) : e \in E, \ e \neq 0\},$$

$l \in \mathbb{N}$, and we have

$$\mathcal{L}(E, F) = \bigcup_m \mathcal{L}(E, F)_m,$$

where the union is taken over all functions $m : \mathbb{N} \to \mathbb{N}$.

Remark 2.3.3 *In order to verify that a linear operator $A : \mathcal{K}_P^{s,\gamma}(X^\wedge)$
$\to \mathcal{K}_Q^{r,\delta}(X^\wedge)$ is continuous we can choose for $E = \mathcal{K}_P^{s,\gamma}(X^\wedge)$ a representation as in Remark 2.3.2 and similarly for $F = \mathcal{K}_Q^{r,\delta}(X^\wedge) = \varprojlim_{k \in \mathbb{N}} F^k$,
where the numbers ε_k can be taken independently of the asymptotic types P, Q. Set*

$$E_0^k = \mathcal{K}^{s,\gamma+(k+1)-\varepsilon_k}(X^\wedge); \quad F_0^k = \mathcal{K}^{r,\delta+(k+1)-\varepsilon_k}(X^\wedge).$$

Since E_0^k can be obtained from E^k as the image under a projection that removes the finite–dimensional space $\omega\mathcal{E}_{P_k}(X^\wedge)$ and similarly F_0^k, it is often possible to reduce the consideration to the case of empty asymptotic types.

Remark 2.3.4 *Let $P, P' \in \mathrm{As}(X, \boldsymbol{g}^\bullet)$ for $\boldsymbol{g} = (\gamma, \Theta)$. Then we have continuous embeddings*

$$\mathcal{K}_{P'}^{s',\gamma}(X^\wedge) \hookrightarrow \mathcal{K}_P^{s,\gamma}(X^\wedge)$$

for $s' \geq s$, $P' \leq P$. There are also natural continuous embeddings under varying γ or Θ; the obvious details are omitted. Note that for each $P \in \mathrm{As}(X, \boldsymbol{g}^\bullet)$ there is an $\varepsilon > 0$ such that

$$\mathcal{K}_P^{s,\gamma}(X^\wedge) \hookrightarrow \mathcal{K}^{s,\gamma+\varepsilon}(X^\wedge)$$

is continuous for all $s \in \mathbb{R}$. It suffices to take $\varepsilon < \mathrm{dist}(\pi_\mathbb{C} P; \Gamma_{\frac{n+1}{2}-\gamma})$.

Remark 2.3.5 *An alternative definition of the space $\mathcal{K}_P^{s,\gamma}(X^\wedge)$ for $P \in \mathrm{As}(X, \boldsymbol{g}^\bullet)$ with $\boldsymbol{g} = (\gamma, (-\infty, 0])$ is the following. $\mathcal{K}_P^{s,\gamma}(X^\wedge)$ is the subspace of all $u \in \mathcal{K}^{s,\gamma}(X^\wedge)$ such that for every $M \in \mathbb{N}$ there is an $N \in \mathbb{N}$ such that for suitable coefficients $c_{jk} \in L_j$, $0 \leq k \leq m_j$,*

$$u(t, x) - \omega(t) \sum_{j=0}^N \sum_{k=0}^{M_j} c_{jk}(x) t^{-p_j} \log^k t \in \mathcal{K}^{s,\gamma+M}(X^\wedge) \qquad (2.3.13)$$

for any cut–off function $\omega(t)$.

Lemma 2.3.6 *For every $s, \gamma \in \mathbb{R}$, $k \in \mathbb{N}$, $p \in \mathbb{C}$ with $\mathrm{Re}\, p < \frac{n+1}{2} - \gamma$, and $\varphi(x) \in C^\infty(X)$, $\omega(t) \in C_0^\infty(\overline{\mathbb{R}}_+)$, we have*

$$\|\varphi(x) t^{-p} \log^k t\, \omega(ct)\|_{\mathcal{K}^{s,\gamma}(X^\wedge)} \to 0 \quad \text{for } c \to \infty. \qquad (2.3.14)$$

Proof. In view of Remark 2.1.44 it suffices to consider $s \in \mathbb{N}$. We have to show that

$$\| t^{-\gamma + \frac{n}{2}} (tD_t)^j B \left\{ \varphi(x) t^{-p} \log^k t \omega(ct) \right\} \|_{L^2(X^\wedge)}$$

tends to zero for $c \to \infty$, for $0 \le j \le s$ and for every $B \in \text{Diff}^{s-j}(X)$, cf. Proposition 2.1.45. This is an obvious consequence of

$$\| t^{-\gamma + \frac{n}{2}} (tD_t)^j \left\{ t^{-p} \log^k t \omega(ct) \right\} \|_{L^2(\mathbb{R}_+)} \to 0 \quad \text{for } c \to \infty$$

for all $0 \le j \le s$. Let us consider for simplicity $k = 0$. The case $k > 0$ only causes trivial modifications of the estimates.

We have

$$t^{-\gamma + \frac{n}{2}} (tD_t)^j \left\{ t^{-p} \omega(ct) \right\} = t^{-\gamma + \frac{n}{2} - p} \sum_{l=0}^{j} \beta_l (ct)^l (\partial_t^l \omega)(ct)$$

for certain constants β_l. Thus, it follows that

$$\left\{ \int |t^{-\gamma + \frac{n}{2}} (tD_t)^j \left\{ t^{-p} \omega(ct) \right\} |^2 \, dt \right\}^{\frac{1}{2}}$$

$$\le \sum_{l=0}^{j} |\beta_l| \left\{ \int |t^{-\gamma + \frac{n}{2} - p} (ct)^l (\partial_t \omega)(ct)|^2 \, dt \right\}^{\frac{1}{2}}$$

$$= \sum_{l=0}^{j} |\beta_l| c^{-1 + 2\gamma - n + 2\,\text{Re}\,p} \left\{ \int |t^{-\gamma + \frac{n}{2} - p + l} (\partial_t \omega)(t)|^2 \, dt \right\}^{\frac{1}{2}}$$

which tends to zero for $c \to \infty$ as soon as $-1 + 2\gamma - n + 2\,\text{Re}\,p < 0$, i.e., $\text{Re}\,p < \frac{n+1}{2} - \gamma$. $\qquad\square$

Below in Section 2.3.3. we will considerably generalise the asymptotics to the so–called continuous case. This will also cover the following variant. Let us denote by $\text{As}(X, \boldsymbol{g})^\bullet$ for $\boldsymbol{g} = (\gamma, \Theta)$ the set of all sequences $P_0 = \{(p_j, m_j)\}$ occurring as $\pi_{\mathbb{C} \times \mathbb{N}} P$ for a certain $P \in \text{As}(X, \boldsymbol{g}^\bullet)$, cf. (2.3.6).

Definition 2.3.7 *Let* $P_0 = \{(p_j, m_j)\}_{j=0,\dots,N} \in \text{As}(X, \boldsymbol{g})^\bullet$ *for* $\boldsymbol{g} = (\gamma, \Theta)$ *and finite* Θ. *Then* $\mathcal{K}_{P_0}^{s,\gamma}(X^\wedge)$ *denotes the subspace of all* $u(t, x) \in \mathcal{K}^{s,\gamma}(X^\wedge)$ *such that there are coefficients* $c_{jk}(x) \in C^\infty(X)$ *for* $0 \le k \le m_j$, $j = 0, \dots, N$, *with*

$$u(t, x) - \omega(t) \sum_{j=0}^{N} \sum_{k=0}^{m_j} c_{jk}(x) t^{-p_j} \log^k t \in \mathcal{K}_\Theta^{s,\gamma}(X^\wedge)$$

for any cut–off function $\omega(t)$. *Analogously we define* $\mathcal{K}_P^{s,\gamma}(X^\wedge)$ *for* $P_0 \in \text{As}(X, \boldsymbol{g})^\bullet$ *for infinite* Θ *by the condition* (2.3.13) *for every* M *with suitable* N, *for coefficients* $c_{jk}(x) \in C^\infty(X)$, $0 \le k \le m_j$, $j \in \mathbb{N}$.

The spaces $\mathcal{K}_{P_0}^{s,\gamma}(X^\wedge)$ are Fréchet in natural way. The properties that we formulate for spaces with asymptotics in $\mathrm{As}(X, \boldsymbol{g}^\bullet)$ hold in a corresponding modified form also for $\mathrm{As}(X, \boldsymbol{g})^\bullet$. In particular, the coefficients $c_{jk} \in C^\infty(X)$ are uniquely determined by u, cf. also Remark 2.3.10 below. The obvious details will not be explicitly formulated here. Let us only note that

$$\mathcal{K}_P^{s,\gamma}(X^\wedge) \subset \mathcal{K}_{P_0}^{s,\gamma}(X^\wedge)$$

is a closed subspace for every $P \in \mathrm{As}(X, \boldsymbol{g}^\bullet)$, $P_0 \in \mathrm{As}(X, \boldsymbol{g}^\bullet)$ when $P_0 = \pi_{\mathbb{C} \times \mathbb{N}} P$.

Let us choose an operator family $R^s(\tau) \in L_{\mathrm{cl}}^s(X; \mathbb{R})$ which induces isomorphisms

$$R^s(\tau) : H^r(X) \to H^{r-s}(X)$$

for all $r \in \mathbb{R}$, $\tau \in \mathbb{R}$, cf. Section 1.2.3. Recall that an A–excision function for a set $A \subseteq \mathbb{C}$ is any $\chi(z) \in C^\infty(\mathbb{C})$ with $\chi(z) = 0$ for $\mathrm{dist}(z, \overline{A}) < \varepsilon_0$, $\chi(z) = 1$ for $\mathrm{dist}(z, \overline{A}) > \varepsilon_1$, for certain $0 < \varepsilon_0 < \varepsilon_1 < \infty$.

Definition 2.3.8 *Let* $\boldsymbol{g} = (\gamma, \Theta)$, $\gamma \in \mathbb{R}$, $\Theta = (\vartheta, 0]$, $-\infty < \vartheta < 0$, *and* $P \in \mathrm{As}(X, \boldsymbol{g}^\bullet)$, *given as a sequence of triples* (p, m, L) *in the sense of Definition 2.3.1. Then* $\mathcal{A}_P^{s,\gamma}(X)$ *denotes the space of all*

$$f(z) \in \mathcal{A} \left(\left\{ \frac{n+1}{2} - \gamma + \vartheta < \mathrm{Re}\, z < \frac{n+1}{2} - \gamma \right\} \setminus \pi_{\mathbb{C}} P, H^s(X) \right)$$

with the following properties:

(i) $f(z)$ *is meromorphic in the strip* $\frac{n+1}{2} - \gamma + \vartheta < \mathrm{Re}\, z < \frac{n+1}{2} - \gamma$ *with poles in the points* $p \in \pi_{\mathbb{C}} P$ *of multiplicities* $m + 1$ *and Laurent coefficients at* $(z - p)^{-(k+1)}$ *belonging to* L *for* $0 \leq k \leq m$.

(ii) *If* $\chi(z)$ *is any* $\pi_{\mathbb{C}} P$–*excision function, then*

$$g(z) := \chi(z) R^s(\mathrm{Im}\, z) f(z)$$

satisfies $g(z)|_{\Gamma_\beta} \in L^2(\Gamma_\beta \times X)$ *for all* β *with* $\frac{n+1}{2} - \gamma + \vartheta < \beta < \frac{n+1}{2} - \gamma$, *uniformly in compact subintervals, and* $g(\frac{n+1}{2} - \gamma - \varepsilon + i\tau) \in L^2(\mathbb{R}_\tau \times X)$, $\varepsilon > 0$, *converges in* $L^2(\mathbb{R}_\tau \times X)$ *for* $\varepsilon \to 0$.

$\mathcal{A}_P^{s,\gamma}(X)$ is a Fréchet space in canonical way. It is independent of the particular choice of the order reducing family $R^s(\tau)$. For $\pi_{\mathbb{C}} P = \emptyset$ we denote the corresponding space by $\mathcal{A}_\Theta^{s,\gamma}(X)$. The Fréchet topology of $\mathcal{A}_\Theta^{s,\gamma}(X)$ can be described by the system of semi–norms of the space

$$\mathcal{A} \left(\left\{ \frac{n+1}{2} - \gamma + \vartheta < \mathrm{Re}\, z < \frac{n+1}{2} - \gamma \right\}, H^s(X) \right)$$

together with the semi–norm system

$$\sup \| R^s (\operatorname{Im} z) f(z)|_{\Gamma_\beta} \|_{L^2(\Gamma_\beta \times X)}$$

where sup is taken over all β with $\delta < \beta \leq \frac{n+1}{2} - \gamma$ for arbitrary $\delta > \frac{n+1}{2} - \gamma + \vartheta$.

For $\mathcal{A}_P^{s,\gamma}(X)$ in general it suffices to observe that for every fixed cut–off function $\omega(t)$ the weighted Mellin transform $M_{\gamma - \frac{n}{2}}$ induces an isomorphism of $\omega \mathcal{E}_P(X^\wedge)$ to a finite–dimensional subspace

$$M_{\gamma - \frac{n}{2}}(\omega \mathcal{E}_P(X^\wedge)) \subset \mathcal{A}_P^{\infty,\gamma}(X)$$

with

$$\mathcal{A}_\Theta^{s,\gamma}(X) \cap M_{\gamma - \frac{n}{2}}(\omega \mathcal{E}_P(X^\wedge)) = \{0\}, \tag{2.3.15}$$

$$\mathcal{A}_P^{s,\gamma}(X) = \mathcal{A}_\Theta^{s,\gamma}(X) \oplus M_{\gamma - \frac{n}{2}}(\omega \mathcal{E}_P(X^\wedge)). \tag{2.3.16}$$

This gives the adequate topology of $\mathcal{A}_P^{s,\gamma}(X)$.

If $P \in \operatorname{As}(X, \boldsymbol{g}^\bullet)$ $(\in \operatorname{As}(X, \boldsymbol{g})^\bullet)$ with $\boldsymbol{g} = (\gamma, (-\infty, 0])$ we can form as above $P_k \in \operatorname{As}(X, (\gamma, \Theta_k)^\bullet)$ $(\in \operatorname{As}(X, (\gamma, \Theta_k))^\bullet)$, $k \in \mathbb{N}$, and obtain continuous embeddings $\mathcal{A}_{P_{k+1}}^{s,\gamma}(X) \hookrightarrow \mathcal{A}_{P_k}^{s,\gamma}(X)$ for all k. We then set

$$\mathcal{A}_P^{s,\gamma}(X) = \bigcap_{k \in \mathbb{N}} \mathcal{A}_{P_k}^{s,\gamma}(X),$$

endowed with the Fréchet topology of the projective limit.

Theorem 2.3.9 *For every cut–off function ω we have continuous operators*

$$M_{\gamma - \frac{n}{2}} \omega : \mathcal{K}_P^{s,\gamma}(X^\wedge) \to \mathcal{A}_P^{s,\gamma}(X)$$

$$\omega M_{\gamma - \frac{n}{2}}^{-1} : \mathcal{A}_P^{s,\gamma}(X) \to \mathcal{K}_P^{s,\gamma}(X^\wedge)$$

for all $s, \gamma \in \mathbb{R}$, $P \in \operatorname{As}(X, \boldsymbol{g}^\bullet)$ $(\in \operatorname{As}(X, \boldsymbol{g})^\bullet)$ for $\boldsymbol{g} = (\gamma, \Theta)$.

Proof. The assertion is a consequence of the definitions of the involved spaces and of Remark 2.3.2 together with the above mapping properties of the Mellin transform on $\omega \mathcal{E}_P(X^\wedge)$–spaces. $\quad\square$

Remark 2.3.10 *The coefficients $c_{jk}(x)$ of the asymptotics of a given $u(t, x) \in \mathcal{K}_P^{s,\gamma}(X^\wedge)$ are uniquely determined by u. The map $u(t, x) \to c_{jk}(x)$ gives rise to a linear continuous operator*

$$\zeta_{jk} : \mathcal{K}_P^{s,\gamma}(X^\wedge) \to L_j \tag{2.3.17}$$

for every $0 \leq k \leq m_j$ and every j.

An example for the map that assigns the coefficients of the asymptotics is the formula for the Taylor coefficients of a function in $\mathcal{S}(\overline{\mathbb{R}}_+ \times X) = \mathcal{S}(\mathbb{R}, C^\infty(X))|_{\overline{\mathbb{R}}_+} \subset \mathcal{K}^{0,\frac{n}{2}}(X^\wedge)$ at $t = 0$, namely

$$\mathcal{S}(\overline{\mathbb{R}}_+ \times X) \ni u(t,x) \to \frac{1}{j!}\left(\frac{\partial^j u}{\partial t^j}\right)(0,x) \qquad (2.3.18)$$

for $j \in \mathbb{N}$, cf. also Theorem 2.1.26. These are the standard trace operators at $t = 0$. Clearly it is not always necessary to specify certain finite–dimensional subspaces $L_j \subset C^\infty(X)$ as the allowed coefficients of the asymptotics, cf. Definition 2.3.7, though we intend to control these subspaces through the calculus.

The mappings (2.3.17) can be interpreted as generalised trace operators. They allow us to replace the usual traces (2.3.18) in cases when for some natural reason the functions $u(t,x)$ are not smooth in t up to $t = 0$ but have more general asymptotics, as it happens for solutions of Fuchs type equations on a manifold with conical singularities.

The following result is a natural generalisation of the classical Borel theorem which asserts that to every sequence $c_\alpha \in \mathbb{C}$, $\alpha \in \mathbb{N}^n$ there is an $f(x) \in C^\infty(\mathbb{R}^n)$ with $(D_x^\alpha f)(0) = c_\alpha$ for all $\alpha \in \mathbb{N}^n$.

Theorem 2.3.11 *Let* $P = \{(p_j, m_j, L_j)\}_{j\in\mathbb{N}} \in \mathrm{As}(X, \boldsymbol{g}^\bullet)$ *for* $\boldsymbol{g} = (\gamma, (-\infty, 0])$ *and* $c_{jk}(x) \in L_j$ *for* $0 \leq k \leq m_j$, $j \in \mathbb{N}$, *be an arbitrary sequence. Then there exists a* $u(t,x) \in \mathcal{K}_P^{\infty,\gamma}(X^\wedge)$ *such that* $\zeta_{jk}(u) = c_{jk}$ *holds for every* j, k.

Proof. Let $\omega(t)$ be a cut–off function. It suffices to show that there are constants $d_j > 0$, $j \in \mathbb{N}$, such that

$$u(t,x) = \sum_{j=0}^{\infty}\sum_{k=0}^{m_j} c_{jk}(x) t^{-p_j} \log^k t\, \omega(d_j t) \qquad (2.3.19)$$

converges in $\mathcal{K}_P^{\infty,\gamma}(X^\wedge)$. In view of $\mathrm{Re}\,p_j \to -\infty$ for $j \to \infty$ we have for every $M \in \mathbb{N}$ a $j(M) \in \mathbb{N}$ such that $\mathrm{Re}\,p_j < \frac{n+1}{2} - \gamma - M$ for all $j \geq j(M)$. Let us show that there are constants $d_j(M,s)$ such that

$$\sum_{j=j(M)}^{\infty}\sum_{k=0}^{m_j} c_{jk}(x) t^{-p_j} \log^k t\, \omega(d_j(M,s)t) \qquad (2.3.20)$$

converges in $\mathcal{K}^{s,\gamma+M}(X^\wedge)$ for every $s \in \mathbb{R}$. Because of Remark 2.1.52 it is sufficient to consider $s \in \mathbb{N}$. From the above Lemma 2.3.6 we obtain constants $d_j(M,s)$ with

$$\|c_{jk}(x) t^{-p_j} \log^k t\, \omega(d_j(M,s)t)\|_{\mathcal{K}^{s,\gamma+M}(X^\wedge)} < 2^{-j}$$

for all $j \geq j(M)$ and $0 \leq k \leq m_j$. This gives us the desired convergence. Setting $d_j(M) = d_j(M, j)$ we obtain the convergence of (2.3.20) for all $s \in \mathbb{R}$. Choosing $d_j(M)$ for $j < j(M)$ arbitrary, this result can also be interpreted in the sense that (2.3.19) converges for $d_j(M)$ instead of d_j in the space E^k for all $s \in \mathbb{R}$; here E^k is the space from Remark 2.3.2, where we take M so large that $M > k + 1 - \varepsilon_k$. Thus, if we finally set $d_j = d_j(j)$, we get the convergence of (2.3.19) in the space $\mathcal{K}_P^{\infty,\gamma}(X^\wedge)$. $\qquad\square$

Theorem 2.3.12 *The space $\mathcal{K}_P^{s,\gamma}(X^\wedge)$ can be written as a non–direct sum of Fréchet spaces*

$$\mathcal{K}_P^{s,\gamma}(X^\wedge) = \mathcal{K}_\Theta^{s,\gamma}(X^\wedge) + \mathcal{K}_P^{\infty,\gamma}(X^\wedge)$$

for every $P \in \mathrm{As}(X, \boldsymbol{g}^\bullet)$, $\boldsymbol{g} = (\gamma, \Theta)$, $s, \gamma \in \mathbb{R}$.

Proof. The assertion is trivial for finite Θ. Thus we have to consider $\Theta = (-\infty, 0]$. In this case we have continuous embeddings

$$\mathcal{K}_\Theta^{s,\gamma}(X^\wedge) \hookrightarrow \mathcal{K}_P^{s,\gamma}(X^\wedge), \qquad \mathcal{K}_P^{\infty,\gamma}(X^\wedge) \hookrightarrow \mathcal{K}_P^{s,\gamma}(X^\wedge)$$

which imply a continuous embedding

$$\mathcal{K}_\Theta^{s,\gamma}(X^\wedge) + \mathcal{K}_P^{\infty,\gamma}(X^\wedge) \hookrightarrow \mathcal{K}_P^{s,\gamma}(X^\wedge)$$

that is obviously injective. On the other hand it is surjective, because of the above Theorem 2.3.11. $\qquad\square$

We will say that an asymptotic type $P \in \mathrm{As}(X, \boldsymbol{g}^\bullet)$ for $\boldsymbol{g} = (\gamma, (\vartheta, 0])$, $-\infty \leq \vartheta < 0$, satisfies the shadow condition if $(p, m, L) \in P$ and $j \in \mathbb{N}$ with $\mathrm{Re}\, p - j > \frac{n+1}{2} - \gamma + \vartheta$ imply $(p - j, m(j), L(j)) \in P$ for all those j, for suitable $m(j) \geq m$, $L(j) \supseteq L$.

Let us write

$$T^\beta P = \{(p_j + \beta, m_j, L_j)\}$$

for $\beta \in \mathbb{R}$, $P \in \mathrm{As}(X, \boldsymbol{g}^\bullet)$, where P is written in the form (2.3.3) or (2.3.4). Then $T^\beta P \in \mathrm{As}(X, \boldsymbol{g}_\beta^\bullet)$ for $\boldsymbol{g}_\beta = (\gamma - \beta, \Theta)$.

Theorem 2.3.13 *The operator $\mathcal{M}_\varphi$ of multiplication by a function $\varphi(t, x) \in C_0^\infty(\overline{\mathbb{R}}_+ \times X)$ induces a continuous operator*

$$\mathcal{M}_\varphi : \mathcal{K}_P^{s,\gamma}(X^\wedge) \to \mathcal{K}_Q^{s,\gamma}(X^\wedge)$$

for every $P \in \mathrm{As}(X, \boldsymbol{g}^\bullet)$, $\boldsymbol{g} = (\gamma, \Theta)$, for some resulting asymptotic type $Q \in \mathrm{As}(X, \boldsymbol{g}^\bullet)$, for every $s \in \mathbb{R}$.

Proof. It suffices to assume $\Theta = (\vartheta, 0]$ for finite $\vartheta < 0$. For $P = 0$ we have $\mathcal{K}_P^{s,\gamma}(X^\wedge) = \mathcal{K}_\Theta^{s,\gamma}(X^\wedge)$. Then the continuity of $\mathcal{M}_\varphi : \mathcal{K}_\Theta^{s,\gamma}(X^\wedge) \to \mathcal{K}_\Theta^{s,\gamma}(X^\wedge)$ is an easy consequence of Theorem 2.1.45, cf. the representation (2.3.7). In view of (2.3.9) and Theorem 2.3.12 it remains to show that for general P the operator $\mathcal{M}_\varphi : \omega \mathcal{E}_P(X^\wedge) \to \mathcal{K}_Q^{\infty,\gamma}(X^\wedge)$ is continuous for suitable Q. Writing $\varphi(t, x)$ as a finite Taylor expansion $\varphi(t,x) = \sum_{j=0}^N t^j \varphi_j(x) + t^N \varphi_{(N)}(t,x)$ with unique $\varphi_j(x) \in C^\infty(X)$, $\varphi_{(N)}(t,x) \in C^\infty(\overline{\mathbb{R}}_+ \times X)$, we obtain for sufficiently large N that the operator of multiplication by $t^N \varphi_{(N)}$ maps the space $\omega \mathcal{E}_P(X^\wedge)$ continuously to $\mathcal{K}_\Theta^{\infty,\gamma}(X^\wedge)$. Finally it is obvious that the multiplication by the finite sum $\sum t^j \varphi_j(x)$ maps $\omega \mathcal{E}(X^\wedge)$ continuously to $\mathcal{K}_Q^{\infty,\gamma}(X^\wedge)$ for a certain Q. Here it suffices again to apply (2.3.9). The resulting asymptotic type Q is determined by the translations from the multiplications by t^j, including the transformation of the spaces of coefficients by $\mathcal{M}_{\varphi_j}$. □

Remark 2.3.14 *Let us write* $\mathcal{K}_P^{s,\gamma}(X^\wedge) = \varprojlim_{k \in \mathbb{N}} E^k$, $\mathcal{K}_Q^{s,\gamma}(X^\wedge) = \varprojlim_{j \in \mathbb{N}} F^j$ *in the sense of Remark 2.3.2, with P, Q given as in Theorem 2.3.13 for a fixed $\varphi(t,x) \in C_0^\infty(\overline{\mathbb{R}}_+ \times X)$. Then, if we define $\varphi_\eta(t,x) = \varphi(t[\eta]^{-1}, x)$ for $\eta \in \mathbb{R}^q$, we get the same Q for all η when we multiply by φ_η, and for every j there is a $k(j)$ such that the operator norm of M_{φ_η} in $\mathcal{L}(E^{k(j)}, F^j)$ is uniformly bounded in $\eta \in \mathbb{R}^q$.*

For purposes below we set

$$\mathcal{S}^\gamma(X^\wedge) = [\omega]\mathcal{K}^{\infty,\gamma}(X^\wedge) + [1 - \omega]\mathcal{S}(\overline{\mathbb{R}}_+ \times X) \qquad (2.3.21)$$

and

$$\mathcal{S}_P^\gamma(X^\wedge) = [\omega]\mathcal{K}_P^{\infty,\gamma}(X^\wedge) + [1 - \omega]\mathcal{S}(\overline{\mathbb{R}}_+ \times X) \qquad (2.3.22)$$

for every $P \in \mathrm{As}(X, \boldsymbol{g}^\bullet)$ ($\in \mathrm{As}(X, \boldsymbol{g})^\bullet$) and any cut–off function $\omega(t)$. Recall that $\mathcal{S}(\overline{\mathbb{R}}_+ \times X) = \mathcal{S}(\mathbb{R}, C^\infty(X))|_{\overline{\mathbb{R}}_+}$. The spaces (2.3.21) and (2.3.22) are endowed with the corresponding Fréchet topologies of the direct sums. They are independent of the particular choice of ω.

It is obvious that $t^\rho \mathcal{S}^\gamma(X^\wedge) = \mathcal{S}^{\gamma+\rho}(X^\wedge)$ for every $\rho, \gamma \in \mathbb{R}$. In other words the operator of multiplication by t^ρ induces an isomorphism $\mathcal{S}(X^\wedge) \to \mathcal{S}^{\gamma+\rho}(X^\wedge)$. Let us set

$$\mathcal{S}^{\gamma-}(X^\wedge) = \bigcap_{\varepsilon > 0} \mathcal{S}^{\gamma-\varepsilon}(X^\wedge), \qquad (2.3.23)$$

endowed with the Fréchet topology of the projective limit. Then we have a continuous embedding

$$\mathcal{S}^\gamma(X^\wedge) \hookrightarrow \mathcal{S}^{\gamma-}(X^\wedge).$$

Note also that when $\rho < \gamma$ and $\vartheta = \rho - \gamma$, $\Theta = (\vartheta, 0]$, we have

$$S_\Theta^\rho(X^\wedge) = S^{\gamma-}(X^\wedge).$$

Here Θ indicates, as usual, the trivial asymptotic type with respect to the reference weight ρ.

Proposition 2.3.15 *There is an isomorphism*

$$S^{\gamma-}(X^\wedge) \cong S^\infty(X^\wedge),$$

where $S^\infty(X^\wedge) = \bigcap_{k \in \mathbb{N} \setminus \{0\}} S^k(X^\wedge) \cong \{u \in S(\mathbb{R}, C^\infty(X)) : u(t, x) = 0$ *for* $t \leq 0\}$.

Proof. First it is obvious that $S^{\gamma-}(X^\wedge) \cong S^{\delta-}(X^\wedge)$ for every $\gamma, \delta \in \mathbb{R}$. Thus we may set $\gamma = 0$. Let us write for the corresponding space $S^-(X^\wedge)$. Then $S^-(X^\wedge) = \varprojlim_{k \in \mathbb{N} \setminus \{0\}} S^{-\frac{1}{k}}(X^\wedge)$. On the other hand, $S^\infty(X^\wedge) = \varprojlim_{k \in \mathbb{N} \setminus \{0\}} S^k(X^\wedge)$. Since there is an isomorphism $S^{-\frac{1}{k}}(X^\wedge) \cong S^k(X^\wedge)$ for every k, it follows the assertion. $\square$

Theorem 2.3.16 *Let* $\boldsymbol{g} = (\gamma, (-\infty, 0])$, $\boldsymbol{h} = (\delta, (-\infty, 0])$ *for* $\gamma, \delta \in \mathbb{R}$ *and* $P \in \mathrm{As}(X, \boldsymbol{g}^\bullet)$, $Q \in \mathrm{As}(X, \boldsymbol{h}^\bullet)$ *be arbitrary such that in the representation*

$$P = \{(p_j, m_j, L_j)\}_{j \in \mathbb{N}}, \quad Q = \{(q_j, n_j, H_j)\}_{j \in \mathbb{N}},$$

cf. (2.3.4), *both* $\pi_\mathbb{C} P$ *and* $\pi_\mathbb{C} Q$ *contain infinitely many points for which the associated spaces* L_j *and* H_j *are not equal to* $\{0\}$. *Then there is an isomorphism*

$$S_P^\gamma(X^\wedge) \cong S_Q^\delta(X^\wedge) \tag{2.3.24}$$

in the sense of Fréchet spaces. Analogously we have isomorphism

$$\mathcal{K}_P^{s,\gamma}(X^\wedge) \cong \mathcal{K}_Q^{s,\delta}(X^\wedge) \tag{2.3.25}$$

for every $s \in \mathbb{R}$.

Proof. First we can reduce the assertion to $\gamma = \delta = 0$ by multiplying the spaces by corresponding powers of t and shifting the asymptotic types in the complex plane. In other words, without loss of generality we may assume that $\gamma = \delta = 0$. Let us first consider the case $P = \{(p_j, 0, L_j)\}_{j \in \mathbb{N}}$, $Q = \{(q_j, 0, H_j)\}_{j \in \mathbb{N}}$ where $\dim L_j = \dim H_j - 1$. Choose a sequence $\{\vartheta_k\}_{k \in \mathbb{N} \setminus \{0\}}$, $-\infty < \vartheta_k < 0$, such that $\pi_\mathbb{C} P \cap \{z : \frac{n+1}{2} + \vartheta_k < \mathrm{Re}\, z < \frac{n+1}{2}\}$ contains exactly k points, and set $P_k = \{(p_j, 0, L_j)\}_{j=0,\ldots,k-1}$. Analogously we choose $\{\lambda_l\}_{l \in \mathbb{N} \setminus \{0\}}$,

$-\infty < \lambda_l < 0$, such that $\pi_{\mathbb{C}}Q \cap \{z : \frac{n+1}{2} + \lambda_l < \operatorname{Re} z < \frac{n+1}{2}\}$ contains exactly l points, and set $Q_l = \{(q_j, 0, H_j)\}_{j=0,\dots,l-1}$. Then $P_k \in$ $\operatorname{As}(X,(0,\Theta_k))$ for $\Theta_k = (\vartheta_k, 0]$, $Q_l \in \operatorname{As}(X,(0,\Lambda_l))$ for $\Lambda_l = (\lambda_l, 0]$. By definition we have

$$\mathcal{S}_P^0(X^\wedge) = \varprojlim_{k\in\mathbb{N}\setminus\{0\}} \mathcal{S}_{P_k}^0(X^\wedge), \quad \mathcal{S}_Q^0(X^\wedge) = \varprojlim_{k\in\mathbb{N}\setminus\{0\}} \mathcal{S}_{Q_k}^0(X^\wedge).$$

Now $\mathcal{S}_P^0(X^\wedge) \cong \mathcal{S}_{\Theta_k}^0(X^\wedge) \oplus \omega\mathcal{E}_{P_k}(X^\wedge)$, $\mathcal{S}_{Q_k}^0(X^\wedge) \cong \mathcal{S}_{\Lambda_k}^0(X^\wedge) \oplus$ $\omega\mathcal{E}_{Q_k}(X^\wedge)$, $\omega(t)$ being any fixed cut–off function. By construction we have isomorphisms $\omega\mathcal{E}(X^\wedge) \cong \omega\mathcal{E}_{Q_k}(X^\wedge)$, since the dimensions of spaces are equal, and moreover $\mathcal{S}_{\Theta_k}^0(X^\wedge) \cong \mathcal{S}_{\Lambda_k}^0(X^\wedge)$ for all k. This yields $\mathcal{S}_{P_k}^0(X^\wedge) \cong \mathcal{S}_{Q_k}^0(X^\wedge)$ for all k and hence (2.3.24). For the general case it suffices to choose to a given $P = \{(p_j, m_j, L_j)\}_{j\in\mathbb{N}}$ an $R = \{(r_j, 0, M_j)\}_{j\in\mathbb{N}}$ such that $\dim M_j = 1$, where $\mathcal{S}_P^0(X^\wedge) \cong$ $\mathcal{S}_R^0(X^\wedge)$. Let $\{\vartheta_k\}_{k\in\mathbb{N}\setminus\{0\}}$ be a sequence, $-\infty < \vartheta_k < 0$, such that $\pi_{\mathbb{C}}P \cap \{z : \frac{n+1}{2} + \vartheta_k < \operatorname{Re} z < \frac{n+1}{2}\}$ contains $N(k) + 1$ points. Let $d_j = \dim L_j$. Choose arbitrary points $\{r_l\}_{l=0,\dots,l(k)}$ in $\{z : \frac{n+1}{2} + \vartheta_k < \operatorname{Re} z < \frac{n+1}{2}\}$, $l(k) = \sum_{j=0}^{N(k)}(m_j+1)d_j$, with $r_l \neq r_j$ for $j \neq l$, and choose an arbitrary one–dimensional subspace M_l to every l. Then, $R_k = \{(r_j, 0, M_j)\}_{j=0,\dots,N(k)}$ is an asymptotic type for which we obviously have

$$\mathcal{S}_{P_k}^0(X^\wedge) \cong \mathcal{S}_{R_k}^0(X^\wedge), \quad P_k = \{(p_j, m_j, L_j)\}_{j=0,\dots,N(k)}$$

for every k. Moreover, $R = \bigcup_{k=1}^\infty R_k$ belongs to $\operatorname{As}(X,(0,(-\infty,0]))$ and satisfies the assumptions of the first part of the proof. Then the desired reduction follows by passing to the projective limits

$$\mathcal{S}_P^0(X^\wedge) = \varprojlim \mathcal{S}_{P_k}^0(X^\wedge) \cong \varprojlim \mathcal{S}_{R_k}^0(X^\wedge) = \mathcal{S}_R^0(X^\wedge).$$

The modification of the arguments for (2.3.25) is obvious. $\qquad\square$

Corollary 2.3.17 *Let* $\dim X = 0$ *and* $P \in \operatorname{As}((\gamma,(-\infty,0]))$. *Then*

$$\mathcal{S}_P^\gamma(\mathbb{R}_+) \cong \mathcal{S}(\overline{\mathbb{R}}_+).$$

More generally, if $n = \dim X$ *is arbitrary, to every asymptotic type* $P \in \operatorname{As}(X,(\gamma,(-\infty,0]))$ *there is a closed subspace* $F \subset \mathcal{S}(\overline{\mathbb{R}}_+, C^\infty(X))$ *such that*

$$\mathcal{S}_P^\gamma(X^\wedge) \cong F.$$

2.3.2 Analytic functionals

A motivation to study distribution spaces with asymptotics on a configuration with singularities is to control the solutions of elliptic equations up to the singular points under the aspect of elliptic regularity.

In classical elliptic boundary value problems on a C^∞ manifold with C^∞ boundary we have smoothness of solutions up to the boundary, when the data (right hand sides, boundary conditions) are smooth. However, if we consider a wedge shaped domain with an edge $Y \ni y$ then we have to expect non–trivial asymptotics of the form

$$u(t,x,y) \sim \sum_j \sum_{k=0}^{m_j(y)} c_{jk}(x,y) t^{-p_j(y)} \log^k t \qquad (2.3.26)$$

as $t \to 0$. Here $(t,x) \in X^\wedge = \mathbb{R}_+ \times X$ are the points in the model cone with base X which is in this case a manifold with boundary. To illustrate the qualitative properties it suffices to assume that X is closed and compact. In addition, since the main contributions to the asymptotics are local in nature, we assume that the edge variable varies in a neighbourhood of Y with local coordinates $y \in \Omega$ for an open set $\Omega \subseteq \mathbb{R}^q$.

In the simplest cases it is known that smooth solutions have asymptotic expansions (2.3.26) for every $y \in \Omega$, with y–dependent asymptotic types $P(y) \in As(X, \boldsymbol{g}^\bullet)$. In particular, the exponents $p_j(y)$ as well as the numbers $m_j(y)$ may depend on y. Consider, for instance, an elliptic edge–degenerate differential operator on $X^\wedge \times \Omega \ni (t,x,y)$

$$A = t^{-\mu} \sum_{j+|\alpha| \leq \mu} a_{j\alpha}(t,y) \left(-t \frac{\partial}{\partial t} \right)^j (t D_y)^\alpha$$

with coefficients $a_{j\alpha}(t,y) \in C^\infty(\overline{\mathbb{R}}_+ \times \Omega, \mathrm{Diff}^{\mu-(j+|\alpha|)}(X))$, cf. Section 3.1.2 below. Assume for simplicity that the coefficients $a_{j\alpha}(t,y)$ are independent of t for $t > $ const. With A we can associate a (y,η)–dependent family of Fuchs type operators

$$t^{-\mu} \sum_{j+|\alpha| \leq \mu} a_{j\alpha}(t,y) \left(-t \frac{\partial}{\partial t} \right)^j (t\eta)^\alpha : \mathcal{K}^{s,\gamma}(X^\wedge) \to \mathcal{K}^{s-\mu,\gamma-\mu}(X^\wedge).$$

$$(2.3.27)$$

This gives rise to a conormal symbol

$$h(y,z) := \sum_{j=0}^{\mu} a_{j0}(0,y) z^j : H^s(X) \to H^{s-\mu}(X), \qquad (2.3.28)$$

obtained from (2.3.27) by omitting $t^{-\mu}$, replacing $-t\frac{\partial}{\partial t}$ by $z \in \mathbb{C}$ and freezing the coefficients at $t = 0$. For every $y \in \Omega$ we thus obtain a holomorphic family of Fredholm opertors $H^s(X) \to H^{s-\mu}(X)$, cf. Section

1.2.4 and Section 2.4.3 below. We shall see that $\{(p_j(y), m_j(y))\}_{j\in\mathbb{N}}$ as well as the coefficients of the asymptotics are determined by the non–bijectivity points $z = p_j(y) \in \mathbb{C}$ of (2.3.28), more precisely, on the poles including the Laurent coefficients of the meromorphic operator function $h^{-1}(y, z)$. In general these data depend on y, in particular, the poles $p_j(y)$ vary with $y \in \Omega$, with changing multiplicities and jumping Laurent coefficients. Also the numeration of poles may depend on y. These phenomena as well as further essential structures of the edge pseudo–differential calculus require a more general and flexible concept of asymptotics than the discrete ones, based on (vector–valued) analytic functionals. The present section will prepare the corresponding material. Proofs, as far as they belong to the functional analytic generalities, will not be given here. More details and background on analytic functionals can be found in Schapira [99], Hörmander, Vol. 1 [54], and, in a form more adapted for our purposes, in Schulze [119], [122] and Hirschmann [52].

Recall that $\mathcal{A}(U)$ for an open set $U \subseteq \mathbb{C}$ is the space of all holomorphic functions in U. We consider $\mathcal{A}(U)$ in the (nuclear) Fréchet topology, defined by the semi–norm system

$$\mathcal{A}(U) \ni h \to \sup_{z\in K} |h(z)|$$

for arbitrary $K\Subset U$. The dual $\mathcal{A}'(U)$ is the space of all continuous linear functionals $\zeta : \mathcal{A}(U) \to \mathbb{C}$, called analytic functionals on U. Write $\langle\zeta, h\rangle$ (or if necessary $\langle\zeta_z, h\rangle$) for the value of ζ on $h(z) \in \mathcal{A}(U)$.

For every two open sets $U, V \subseteq \mathbb{C}$ with $U \subseteq V$ there are canonical mappings $\mathcal{A}'(U) \hookrightarrow \mathcal{A}'(V)$, defined by restricting $\zeta \in \mathcal{A}'(U)$ to $h|_U$ for arbitrary $h \in \mathcal{A}(V)$.

If a given element $\lambda \in \mathcal{A}'(\mathbb{C})$ belongs to the image under $\mathcal{A}'(U) \hookrightarrow \mathcal{A}'(\mathbb{C})$ for an open set $U \subseteq \mathbb{C}$ then U is called a carrier of λ. A compact set $K \subset \mathbb{C}$ is called a carrier of $\lambda \in \mathcal{A}'(\mathbb{C})$ if every open U with $K \subset U$ is a carrier of λ in the above sense. We denote by $\mathcal{A}'(K)$, $K \subset \mathbb{C}$ compact, the subspace of all $\lambda \in \mathcal{A}'(\mathbb{C})$ with carrier K.

The carrier of an analytic functional substitutes the support from the distribution theory, though the carrier is not necessarily unique. For instance, every C^∞ curve L connecting two points $z_0, z_1 \in \mathbb{C}$ defines an element $\zeta \in \mathcal{A}'(\mathbb{C})$ by

$$\langle\zeta, h\rangle = \frac{1}{2\pi i} \int_L h(z)\, dz \qquad \text{for } h(z) \in \mathcal{A}(\mathbb{C}),$$

where L is a carrier of ζ, but the functional only depends on z_0, z_1.

Lemma 2.3.18 *A compact set $K \subset \mathbb{C}$ is a carrier of an element $\lambda \in \mathcal{A}'(\mathbb{C})$ if and only if for every open $U \subseteq \mathbb{C}$ with $K \subset U$ there is a constant $c_U \geq 0$ such that*

$$|\langle \lambda, h \rangle| \leq c_U \sup_{z \in U} |h(z)| \quad \text{for } h(z) \in \mathcal{A}(\mathbb{C}).$$

The best constants in the latter estimates

$$p_U(\lambda) = \sup \left\{ |\langle \lambda, h \rangle| / \sup_{z \in U} |h(z)| : h \in \mathcal{A}(\mathbb{C}), h \neq 0 \right\}$$

for arbitrary open $U \supset K$ form a semi–norm system on the space $\mathcal{A}'(K)$ that defines a (nuclear) Fréchet topology in $\mathcal{A}(K)$. It is stronger than the topology induced by $\mathcal{A}'(\mathbb{C})$. For every two compact sets K, $\tilde{K}$ with $K \subseteq \tilde{K}$ we have a canonical continuous embedding $\mathcal{A}'(K) \hookrightarrow \mathcal{A}'(\tilde{K})$.

Remark 2.3.19 *We have a canonical continuous embedding $\mathcal{A}(\mathbb{C}) \hookrightarrow C(\mathbb{C})$, and the dual map $C'(\mathbb{C}) \hookrightarrow \mathcal{A}'(C)$ is surjective. $C'(\mathbb{C})$ consists of all (complex) measures with compact support. If $K = \operatorname{supp} \mu$ for a $\mu \in C'(\mathbb{C})$ then K is also a carrier of the associated analytic functional*

$$\mathcal{A}(\mathbb{C}) \ni h \to \int h(z) \, d\mu(z).$$

Note also that each distribution in $\mathbb{C}$ with compact support induces an analytic functional.

If $A \subseteq \mathbb{C}$ is a bounded set we denote by A^c the complement of the union of the unbounded connected components of $\mathbb{C} \setminus \overline{A}$. For unbounded A we will always assume that

$$A_{a,a'} := A \cap \{ z \in \mathbb{C} : a \leq \operatorname{Re} z \leq a' \}$$

is bounded for all reals $a \leq a'$. In that case we set $A^c = \bigcup_{a \leq a'} A_{a,a'}$. Moreover, if $\{A_\iota\}_{\iota \in I}$ is a family of subsets in $\mathbb{C}$, we define

$$\sum_{\iota \in I} A_\iota = \left(\bigcup_{\iota \in I} A_\iota \right)^c.$$

Note that $K = K^c$ implies $K = (\partial K)^c$.

Theorem 2.3.20 *For every compact set $K \subset \mathbb{C}$ we have $\mathcal{A}'(K) = \mathcal{A}'(K^c)$.*

We then obtain $\mathcal{A}'(K) = \mathcal{A}'(\partial(K^c))$. For simplicity we often assume in the sequel that a given compact set $K \subset \mathbb{C}$ satisfies $K = K^c$.

Theorem 2.3.21 *Let K_1, $K_2 \subset \mathbb{C}$ be compact sets. Then*

$$\mathcal{A}'(K_1 + K_2) = \mathcal{A}'(K_1) + \mathcal{A}'(K_2),$$

where the right hand side is valid in the sense of a non–direct sum of Fréchet spaces.

Let $K \subset \mathbb{C}$ be compact, $K = K^c$, $w \notin K$, and choose an open set U with $K \subset U$, $w \notin U$. Then the function $E(w, z) := (w - z)^{-1}$ belongs to $\mathcal{A}(U)$ with respect to z. Let $\zeta \in \mathcal{A}'(K)$ and $\Psi(\zeta)(w) = \langle \zeta_z, E(w, z) \rangle$. We then have $\Psi(\zeta)(w) \in \mathcal{A}(\mathbb{C} \setminus K)$ and Ψ defines a linear continuous operator

$$\Psi : \mathcal{A}'(K) \to \mathcal{A}(\mathbb{C} \setminus K). \tag{2.3.29}$$

We can also define a linear continuous operator

$$\Pi : \mathcal{A}(\mathbb{C} \setminus K) \to \mathcal{A}'(K) \tag{2.3.30}$$

by setting

$$\langle \Pi f, h \rangle = \frac{1}{2\pi i} \int\limits_C f(z) h(z) \, dz \quad \text{for } h \in \mathcal{A}(\mathbb{C}),$$

$f \in \mathcal{A}(\mathbb{C} \setminus K)$. Here C is an arbitrary (say C^∞) curve clockwise surrounding K. For C we can take, for instance, a circle with sufficiently large radius.

Note that $\mathcal{A}(\mathbb{C})|_{\mathbb{C}\setminus K}$ (the subspace of all $f \in \mathcal{A}(\mathbb{C}\setminus K)$ which extends to an element of $\mathcal{A}(\mathbb{C})$) is canonically isomorphic to $\mathcal{A}(\mathbb{C})$.

Theorem 2.3.22 *We have $\Pi \circ \Psi = \mathrm{id}_{\mathcal{A}'(K)}$, $\mathrm{im}\, \Pi = \mathcal{A}'(K)$, $\ker \Pi = \mathcal{A}(\mathbb{C})|_{\mathbb{C}\setminus K}$. There is a topologically direct decomposition*

$$\mathrm{im}\, \Psi \oplus \mathcal{A}(\mathbb{C})|_{\mathbb{C}\setminus K} = \mathcal{A}(\mathbb{C} \setminus K),$$

and we have a canonical isomorphism

$$\mathcal{A}'(K) \cong \mathcal{A}(\mathbb{C} \setminus K)/\mathcal{A}(\mathbb{C})|_{\mathbb{C}\setminus K}.$$

Remark 2.3.23 *Let $v(z) \in \mathcal{A}(\mathbb{C} \setminus \{0\})$ be an arbitrary element with $v(z) = z^{-1} \bmod \mathcal{A}(\mathbb{C})$, and set $\tilde{E}(w, z) = v(w - z)$. Similarly to Ψ and Π we can form operators $\tilde{\Psi}$ and $\tilde{\Pi}$ in terms of $\tilde{E}$ instead of E. Then the continuous operators*

$$\tilde{\Psi} : \mathcal{A}'(K) \to \mathcal{A}(\mathbb{C} \setminus K), \quad \tilde{\Pi} : \mathcal{A}'(\mathbb{C} \setminus K) \to \mathcal{A}(K)$$

satisfy a corresponding analogue of Theorem 2.3.22. In particular, if $\omega(t)$ is an arbitrary cut–off function then $v(z) = (M_{t\to z}\omega)(z)$ equals $z^{-1} \bmod \mathcal{A}(\mathbb{C})$, cf. Proposition 2.1.23.

Remark 2.3.24 *Let $K \subset \mathbb{C}$ be a compact set with $K = K^c$. Choose a system of C^∞ curves C_k clockwise surrounding K in the strip $\{z : 2^{-(k+1)} < \mathrm{dist}(z, K) < 2^{-k}\}$, $k \in \mathbb{N}$, and set*

$$p_k(\zeta) = \|\Psi\zeta|_{C_k}\|_{L^2(C_k)} \quad \text{for } \zeta \in \mathcal{A}'(K).$$

Then $\{p_k\}_{k\in\mathbb{N}}$ is a norm system on $\mathcal{A}'(K)$ which defines the Fréchet topology of $\mathcal{A}'(K)$.

Remark 2.3.25 *Theorem 2.3.21 can be regarded as a corollary of the solution of the Cousin problem. The main point of Theorem 2.3.21 is to find a representation $\zeta = \zeta_1 + \zeta_2$ for every $\zeta \in \mathcal{A}'(K_1 + K_2)$ with suitable $\zeta_i \in \mathcal{A}'(K_i)$, $i = 1, 2$. We have $\Psi\zeta =: f \in \mathcal{A}(\mathbb{C} \setminus (K_1 + K_2))$. Then it suffices to write $f = f_1 + f_2$ for certain $f_i \in \mathcal{A}(\mathbb{C} \setminus K_i)$ and to set $\zeta_i = \Pi f_i$, $i = 1, 2$. If U, U_1, U_2 are open sets with $U = U_1 \cap U_2$ the Cousin theorem now allows us to decompose each $f \in \mathcal{A}(U)$ as $f = f_1 + f_2$ with $f_i \in \mathcal{A}(U_i)$, $i = 1, 2$. This corresponds to our situation for $U = \mathbb{C} \setminus (K_1 + K_2)$, $U_i = \mathbb{C} \setminus K_i$, $i = 1, 2$.*

Example 2.3.26 *Let $f(z)$ be a meromorphic function with poles $p_0, \ldots, p_N$ and set $K = \bigcup_{j=0}^N \{p_j\}$. Then, if the multiplicity of p_j is $m_j + 1$ and c_{jk} the Laurent coefficient of f at $(z-p_j)^{-(k+1)}$, $0 \le k \le m_j$, we have*

$$\langle \Pi f, h \rangle = \sum_{j=0}^N \sum_{k=0}^{m_j} \frac{1}{k!} c_{jk} \left(\frac{d}{dz}\right)^k h(z)|_{z=p_j}$$

for $h \in \mathcal{A}(\mathbb{C})$. In other words the element $\zeta \in \mathcal{A}'(K)$ represented by f is a finite linear combination of derivatives of the Dirac measures at the poles. The carrier of ζ is in this case unique and contained in K (and equal to K if $c_{jk} \ne 0$ for at least one k for every j).

Remark 2.3.27 *There are elements $\zeta \in \mathcal{A}'(\{p\})$ for fixed $p \in \mathbb{C}$ which are not a finite linear combination of derivatives of the Dirac measure p. For instance, let $u(z) \in \mathcal{A}(\mathbb{C})$ and set $f(z) = u((z - p)^{-1})$ which belongs to $\mathcal{A}(\mathbb{C} \setminus \{p\})$. Then, if $u(z)$ is not a polynomial, $\zeta = \Pi f \in \mathcal{A}'(\{p\})$ has the form*

$$\langle \zeta, h \rangle = \sum_{k=0}^\infty \frac{1}{k!} c_k \left(\frac{d}{dz}\right)^k h(z)|_{z=p}$$

with an infinite sequence of coefficients $\{c_k\}_{k\in\mathbb{N}}$ such that the series converges for every $h \in \mathcal{A}(U_\varepsilon)$ for $U_\varepsilon = \{z : |z - p| < \varepsilon\}$.

Proposition 2.3.28 *Let* $K \subset \mathbb{C}$ *be compact,* $K = K^c$, *and let* $f(y, z) \in C^\infty(\Omega, \mathcal{A}(\mathbb{C} \setminus K))$ *for an open set* $\Omega \subseteq \mathbb{R}^q_y$. *Then*

$$\langle (\Pi f)(y), h \rangle = \frac{1}{2\pi i} \int\limits_C f(y, z) h(z) \, dz \quad for \ h \in \mathcal{A}(\mathbb{C})$$

represents an element $(\Pi f)(y) \in C^\infty(\Omega, \mathcal{A}'(K))$. *Conversely every* $\zeta(y) \in C^\infty(\Omega, \mathcal{A}'(K))$ *can be written* $\zeta(y) = (\Pi f)(y)$ *for some* $f(y, z) \in C^\infty(\Omega, \mathcal{A}(\mathbb{C} \setminus K))$.

$C^\infty(\Omega, \mathcal{A}'(K))$ is a nuclear Fréchet space in the semi–norm system

$$\zeta(y) \to \sup_{y \in M} \| \Psi(D^\alpha_y \zeta)(y)|_{C_k} \|_{L^2(C_k)},$$

where C_k, $k \in \mathbb{N}$, is the curve system of Remark 2.3.24, $\alpha \in \mathbb{N}^q$ an arbitrary multi–index and M an arbitrary compact subset of Ω.

Proposition 2.3.29 *For every two compact sets* $K_1, K_2 \subset \mathbb{C}$ *and open* $\Omega \subseteq \mathbb{R}^q$ *we have*

$$C^\infty(\Omega, \mathcal{A}'(K_1 + K_2)) = C^\infty(\Omega, \mathcal{A}'(K_1)) + C^\infty(\Omega, \mathcal{A}'(K_2))$$

in the sense of a non–direct sum of Fréchet spaces, cf. also Proposition 1.1.8.

An interesting class of elements $f(y, z) \in C^\infty(\Omega, \mathcal{A}'(\mathbb{C} \setminus K))$ is the following. Let $f(y, z)$ be meromorphic with poles $p_j(y) \in K$, $j = 0, \ldots, N(y)$, of multiplicities $m_j(y) + 1$, such that for every $M \subset\subset \Omega$ there exists a constant $c_M = c_M(f) > 0$ such that

$$\sup_{y \in M} \sum_{j=0}^{N(y)} (m_j(y) + 1) \leq c_M.$$

Denote by $C^\infty(\Omega, \mathcal{A}(\mathbb{C} \setminus K))^\bullet$ the subspace of all those elements $f(y, z) \in C^\infty(\Omega, \mathcal{A}(\mathbb{C} \setminus K))$.

An example is

$$f(y, z) = (a(y) - z)^{-1}(b(y) - z)^{-1}$$

with $a, b \in C^\infty(\Omega)$, $a(y), b(y) \in K$ for all $y \in \Omega$.

Every $f(y, z) \in C^\infty(\Omega, \mathcal{A}(\mathbb{C} \setminus K))^\bullet$ represents an element $(\Pi f)(y) \in C^\infty(\Omega, \mathcal{A}'(K))$. For every fixed $y \in \Omega$ it has a form as in the above Example 2.3.26.

Theorem 2.3.30 *We have*

$$D_y^\alpha : C^\infty(\Omega, \mathcal{A}(\mathbb{C} \setminus K))^\bullet \to C^\infty(\Omega, \mathcal{A}(\mathbb{C} \setminus K))^\bullet$$

for every multi–index $\alpha \in \mathbb{N}^q$.

The applications sketched in the beginning of this section require vector–valued analytic functionals, in particular, with values in a Fréchet space E. In concrete cases we will have $E = C^\infty(X)$ or $E = L^{-\infty}(X)$ for a compact C^∞ manifold X. We set

$$\mathcal{A}'(K, E) = \mathcal{A}'(K) \otimes_\pi E$$

where $\otimes_\pi$ denotes the completed projective tensor product. Analogously we set $\mathcal{A}(U, E) = \mathcal{A}(U) \otimes_\pi E$ for an open set $U \subseteq \mathbb{C}$.

Some results of the above scalar theory immediately imply the corresponding generalisations to the E–valued case. We have

$$\mathcal{A}'(K, E) = \mathcal{A}'(K^c, E)$$

for every compact $K \subset \mathbb{C}$, cf. Theorem 2.3.20, moreover

$$\mathcal{A}'(K_1 + K_2, E) = \mathcal{A}'(K_1, E) + \mathcal{A}'(K_2, E)$$

for every two compact sets $K_1, K_2 \subset \mathbb{C}$, cf. Theorem 2.3.21. The formula

$$\langle \Pi f, h \rangle = \frac{1}{2\pi i} \int_C f(z) h(z) \, dz \quad \text{for } h \in \mathcal{A}(\mathbb{C})$$

yields for every $f \in \mathcal{A}(\mathbb{C} \setminus K, E)$ for any compact $K \subset \mathbb{C}$ and a curve C surrounding K an element $\Pi f \in \mathcal{A}'(K, E)$. This gives rise to a continuous map $\Pi : \mathcal{A}(\mathbb{C} \setminus K, E) \to \mathcal{A}'(K, E)$. Moreover, for every fixed $v(z) \in \mathcal{A}(\mathbb{C} \setminus \{0\})$ with $v(z) = z^{-1} \bmod \mathcal{A}(\mathbb{C})$ we get by $\langle \zeta_z, v(w - z) \rangle$ for $\zeta \in \mathcal{A}'(K, E)$ a continuous map

$$\Psi : \mathcal{A}'(K, E) \to \mathcal{A}(\mathbb{C} \setminus K, E).$$

We then obtain $\Pi \circ \Psi = \mathrm{id}_{\mathcal{A}'(K,E)}$,

$$\mathrm{im}\, \Pi = \mathcal{A}'(K, E), \quad \ker \Pi = \mathcal{A}(\mathbb{C}, E)|_{\mathbb{C} \setminus K}.$$

There is a topologically direct decomposition

$$\mathrm{im}\, \Psi \oplus \mathcal{A}(\mathbb{C}, E)|_{\mathbb{C} \setminus K} = \mathcal{A}(\mathbb{C} \setminus K, E),$$

and a canonical isomorphism

$$\mathcal{A}'(K, E) \cong \mathcal{A}(\mathbb{C} \setminus K, E) / \mathcal{A}(\mathbb{C}, E)|_{\mathbb{C} \setminus K}, \qquad (2.3.31)$$

cf. Theorem 2.3.22.

Remark 2.3.31 *If $K \subset \mathbb{C}$ is a compact set with $K = K^c$, then, analogously to the above Theorem 2.3.24, we get a semi–norm system on the space $\mathcal{A}'(K, C^\infty(X))$ for a compact C^∞ manifold X by*

$$p_{k,s}(\zeta) = \|\Psi\zeta|_{C_k}\|_{L^2(C_k, H^s(X))}$$

for $k, s \in \mathbb{N}$, with the system C_k, $k \in \mathbb{N}$ from the above Remark 2.3.24.

For every open set $\Omega \subseteq \mathbb{R}^q$ and $f(y, z) \in C^\infty(\Omega, \mathcal{A}(\mathbb{C} \setminus K, E))$, $K \subset \mathbb{C}$ compact, we get $(\Pi f)(y) \in C^\infty(\Omega, \mathcal{A}'(K, E))$, cf. Proposition 2.3.28. For every two compact sets K_1, K_2 we have

$$C^\infty(\Omega, \mathcal{A}'(K_1 + K_2, E)) = C^\infty(\Omega, \mathcal{A}'(K_1, E)) + C^\infty(\Omega, \mathcal{A}'(K_2, E))$$

in the sense of a non–direct sum of Fréchet spaces, cf. Proposition 2.3.29.

2.3.3   Weighted Sobolev spaces with continuous asymptotics

This section studies continuous aymptotics, a generalisation of the discrete ones, formulated in terms of analytic functionals. Continuous asymptotics in this form were first introduced in Rempel and Schulze [94], [93] to investigate discrete asymptotics in boundary and edge problems, where the poles in the image under the Mellin transform vary along the edge variables, including their multiplicities, cf. also the papers of the author [114], [115], [123], [125].

Let $f(z)$ be a scalar meromorphic function with poles $p_j \in \mathbb{C}$ of multiplicities $m_j + 1$ and Laurent coefficients $(-1)^k k! c_{jk}$ at $(z - p_j)^{-(k+1)}$, $0 \le k \le m_j$, $j = 0, \ldots, N$, cf. Example 2.3.26 . Then, if we set $K = \bigcup_{j=0}^{N}\{p_j\}$ and choose a curve C surrounding K as in the construction of (2.3.30), we obtain by

$$\langle \zeta, h \rangle := \frac{1}{2\pi i} \int_C f(z) h(z)\, dz \quad \text{for } h \in \mathcal{A}(\mathbb{C})$$

an element $\zeta \in \mathcal{A}'(K)$. In particular, for $h(z) = t^{-z}$, $t \in \mathbb{R}_+$, it follows that

$$\langle \zeta, t^{-z} \rangle = \sum_{j=0}^{N}\sum_{k=0}^{m_j} c_{jk} t^{-p_j} \log^k t,$$

cf. also Remark 2.1.28. Define $\zeta_j \in \mathcal{A}'(K_j)$ for $K_j = \{p_j\}$ by

$$\langle \zeta_j, h \rangle = \sum_{k=0}^{m_j} (-1)^k c_{jk} \left(\frac{d}{dz}\right)^k h(z)|_{z=p_j}. \qquad (2.3.32)$$

Then $\zeta = \sum_{j=0}^{k} \zeta_j$. If we now allow $\zeta_j \in \mathcal{A}'(K_j, C^\infty(X))$ with coefficients $c_{jk}(x) \in C^\infty(X)$ we see that the discrete asymptotics (2.3.1) can be written

$$u(t,x) \sim \sum_j \langle \zeta_j, t^{-z} \rangle.$$

This yields the idea to generalise the asymptotics (2.3.1) in the following way.

Definition 2.3.32 *An element $u(t,x) \in \mathcal{K}^{s,\gamma}(X^\wedge)$ is said to have continuous asymptotics, associated with the weight data $\boldsymbol{g} = (\gamma, \Theta)$, $\Theta = (\vartheta, 0]$, if it has the following properties:*

(i) *For $\vartheta = -\infty$ there are compact sets $K_j \subset \mathbb{C}$, $j \in \mathbb{N}$, with $\sup\{\operatorname{Re} z : z \in K_j\} < \frac{n+1}{2} - \gamma$ for all j, $n = \dim X$, and*

$$\sup\{\operatorname{Re} z : z \in K_j\} \to -\infty \quad as\, j \to \infty \qquad (2.3.33)$$

and elements $\zeta_j \in \mathcal{A}'(K_j, C^\infty(X))$, $j \in \mathbb{N}$, such that for every $\beta \geq 0$ there is an $N = N(\beta)$ such that

$$u(t,x) - \omega(t) \sum_{j=0}^{N} \langle \zeta_j, t^{-z} \rangle \in \mathcal{K}^{s,\beta}(X^\wedge)$$

for any cut–off function $\omega(t)$.

(ii) *For $0 > \vartheta > -\infty$ there is a compact set $K \in \{z : \frac{n+1}{2} - \gamma + \vartheta \leq \operatorname{Re} z < \frac{n+1}{2} - \gamma\}$, $n = \dim X$, and an element $\zeta \in \mathcal{A}'(K, C^\infty(X))$ such that*

$$u(t,x) - \omega(t)\langle \zeta, t^{-z} \rangle \in \mathcal{K}_{\Theta}^{s,\gamma}(X^\wedge)$$

for any cut–off function $\omega(t)$.

Let $\mathcal{V}$ be the system of all closed subsets $V \subset \mathbb{C}$ with $V^c = V$ and $V \cap \{z : c \leq \operatorname{Re} z \leq c'\}$ compact for every $c \leq c'$. Then each $V \in \mathcal{V}$ with $V \subset \{z : \operatorname{Re} z < \frac{n+1}{2} - \gamma\}$ and

$$\sum_{j \in \mathbb{N}} K_j \subseteq V \quad \text{for } \vartheta = -\infty, \quad K^c \subseteq V \quad \text{for } \vartheta > -\infty$$

is called a carrier of the asymptotics of $u \in \mathcal{K}^{s,\gamma}(X^\wedge)$. Clearly for $\vartheta > -\infty$ it suffices to consider such sets V with $V \subseteq \{z : \operatorname{Re} z \geq \frac{n+1}{2} - \gamma + \vartheta\}$.

Proposition 2.3.33 *Let $K \subset \{z : \operatorname{Re} z < \frac{n+1}{2} - \gamma\}$ be a compact set and $\zeta \in \mathcal{A}'(K, C^\infty(X))$. Then we have $\omega(t)\langle \zeta, t^{-z} \rangle \in \mathcal{K}^{\infty,\gamma}(X^\wedge)$.*

Proof. Writing $\mathcal{A}'(K, C^\infty(X)) = \mathcal{A}'(K) \otimes_\pi C^\infty(X)$ every element $\zeta \in \mathcal{A}'(K, C^\infty(X))$ can be written as a convergent sum $\zeta = \sum_{j=0}^\infty \lambda_j \sigma_j \otimes \pi_j$ for $\lambda_j \in \mathbb{C}$, $\sum |\lambda_j| < \infty$, $\sigma_j \in \mathcal{A}'(K)$, $\varphi_j \in C^\infty(X)$ with $\sigma_j \to 0$, $\varphi_j \to 0$ in the corresponding spaces, for $j \to -\infty$, cf. Theorem 1.1.6. For every j we have for any smooth curve $C \subset \{z : \text{Re}\, z < \frac{n+1}{2} - \gamma\}$ surrounding K

$$\omega(t)\langle \sigma_j \otimes \varphi_j, t^{-z}\rangle = \omega(t)\frac{1}{2\pi i} \int_C f_j(z) t^{-z} \varphi_j(x)\, dz =: u_j(t), \quad (2.3.34)$$

where $f_j(z) := \Psi \sigma_j$, cf. (2.3.29). In view of the continuity of the operator Ψ we have $f_j(z) \to 0$ in $C^\infty(C)$ for $j \to \infty$. Now $u_j(t)$ belongs to $\mathcal{K}^{\infty,\gamma}(X^\wedge)$, since it is a superposition of the elements $\omega(t) t^{-z} \varphi_j(x) \in \mathcal{K}^{\infty,\gamma}(X^\wedge)$ with the density $\frac{1}{2\pi i} f_j(z)$ over C. We have $u_j(t, x) = \psi_j(t) \varphi_j(x)$ with

$$\psi_j(t) = \omega(t)\frac{1}{2\pi i} \int_C f_j(z) t^{-z}\, dz$$

and $\|u_j\|_{\mathcal{K}^{r,\gamma}(X^\wedge)} = \|\psi_j\|_{\mathcal{K}^{r,\gamma - \frac{n}{2}}(\mathbb{R}_+)} \|\varphi_j\|_{H^r(X)}$ for every $r \in \mathbb{R}$, up to equivalence of norms. Now the properties of φ_j imply $\varphi_j \to 0$ in $H^r(X)$ for all $r \in \mathbb{R}$, for $j \to \infty$. Moreover, we have $\psi_j \to 0$ in $\mathcal{K}^{r,\gamma - \frac{n}{2}}(\mathbb{R}_+)$ for all $r \in \mathbb{R}$, for $j \to \infty$. This follows from the description of $\mathcal{H}^{s,\gamma}(\mathbb{R}_+)$ norms (that are here sufficient because of the factor ω) in form of the Mellin transform by $\hat{H}^r(\Gamma_{\frac{n+1}{2} - \gamma})$ norms, cf. (2.1.75). In fact, we have

$$(M_{t \to w} \psi_j)(w) = \frac{1}{2\pi i} \int_C f_j(z) \left\{ \int_0^\infty t^{w-z-1} \omega(t)\, dt \right\} dz.$$

This is a superposition of Schwartz functions on $\Gamma_{\frac{n+1}{2} - \gamma} \ni w$ with a density tending to zero. Now it is evident that

$$\omega(t)\langle \zeta, t^{-z}\rangle = \sum_{j=0}^\infty \lambda_j u_j(t, x)$$

converges in $\mathcal{K}^{\infty,\gamma}(X^\wedge)$. $\qquad\qquad\qquad\qquad\qquad\qquad\qquad \square$

Remark 2.3.34 *If* $K_0 \subset \Gamma_{\frac{n+1}{2} - \gamma + \vartheta}$ *for* $-\infty < \vartheta < 0$ *is a compact set and* $\zeta_0 \in \mathcal{A}'(K_0)$ *then we have* $\omega(t)\langle \zeta_0, t^{-z}\rangle \in \mathcal{K}_\Theta^{\infty,\gamma}(X^\wedge)$ *for* $\Theta = (\vartheta, 0]$.

Definition 2.3.35 *Let* $\boldsymbol{g} = (\gamma, \Theta)$ *be weight data,* $\gamma \in \mathbb{R}$, $\Theta = (\vartheta, 0]$, $-\infty \leq \vartheta < 0$, *and denote by* $\text{As}(X, \boldsymbol{g})$

(i) *for $\vartheta > -\infty$ the set of all equivalence classes of pairs*

$$\{V, \mathcal{A}'(V, C^\infty(X))\} \quad \text{for arbitrary } V \in \mathcal{V},$$

$$V \subset \{z : \frac{n+1}{2} - \gamma + \vartheta \le \operatorname{Re} z < \frac{n+1}{2} - \gamma\},$$

where $\{V, \mathcal{A}'(V, C^\infty(X))\} \sim \{W, \mathcal{A}'(W, C^\infty(X))\}$ *means* $V \setminus \Gamma_{\frac{n+1}{2} - \gamma + \vartheta} = W \setminus \Gamma_{\frac{n+1}{2} - \gamma + \vartheta}$,

(ii) *for $\vartheta = -\infty$ the set of all equivalence classes of pairs*

$$\{V, \{\mathcal{A}'(K_j, C^\infty(X))\}_{j \in \mathbb{N}}\}$$

for arbitrary $V \in \mathcal{V}$, $V \subset \{z : \operatorname{Re} z < \frac{n+1}{2} - \gamma\}$, and compact sets $K_j \subseteq V$ with $K_j = K_j^c$, $j \in \mathbb{N}$, with $\{z : c \le \operatorname{Re} z \le c'\} \cap K_j = \emptyset$ only for finitely many j, for arbitrary reals $c \le c'$, where

$$\{V, \{\mathcal{A}('K_j, C^\infty(X))\}_{j \in \mathbb{N}}\} \sim \{W, \{\mathcal{A}'(L_k, C^\infty(X))\}_{j \in \mathbb{N}}\}$$

means that $V = W$.

The elements P of $\operatorname{As}(X, \boldsymbol{g})$ are called continuous asymptotic types.

The points (i), (ii) of the latter definition could be formally unified by replacing the pairs $\{V, \mathcal{A}'(V, C^\infty(X))\}$ in (i) by finite sequences $\{V, \{\mathcal{A}'(K_j, C^\infty(X))\}_{j=0,\dots,N}\}$, where $K_j \subseteq V$, $K_j = K_j^c$ for all j.

The compact sets K_j only play the role of a decomposition of the corresponding set V into pieces on which we can talk about analytic functionals $\zeta_j \in \mathcal{A}'(K_j, C^\infty(X))$ such that for a given $u(t, x) \in \mathcal{K}^{s,\gamma}(X^\wedge)$ the remainder

$$u(t, x) - \omega(t) \sum_{j=0}^{N} \langle \zeta_j, t^{-j} \rangle$$

is of a required flatness order at $t = 0$.

From the definition it follows that there is a bijection

$$\operatorname{As}(X, \boldsymbol{g}) \cong \left\{ V \in \mathcal{V} : V \subset \left\{ z : \frac{n+1}{2} - \gamma + \vartheta \le \right. \right.$$

$$\left. \left. \operatorname{Re} z < \frac{n+1}{2} - \gamma \right\} \right\} \quad \text{for } \vartheta > -\infty, \qquad (2.3.35)$$

$$\operatorname{As}(X, \boldsymbol{g}) \cong \left\{ V \in \mathcal{V} : V \subset \left\{ z : \operatorname{Re} z < \frac{n+1}{2} - \gamma \right\} \right\} \text{ for } \vartheta = -\infty. \qquad (2.3.36)$$

The reason for our notation is to emphasise the relation to the discrete case, indicated by $\boldsymbol{g}^\bullet$. The set $V \in \mathcal{V}$ associated with $P \in \operatorname{As}(X, \boldsymbol{g})$

is called the carrier of P, written $V = \operatorname{carrier} P$. For every $P^\bullet \in$ As$(X, \boldsymbol{g}^\bullet)$ we get a unique $P \in$ As$(X, \boldsymbol{g})$ with $\pi_{\mathbb{C}} P^\bullet = \operatorname{carrier} P$, i.e., there is a canonical map

$$\operatorname{As}(X, \boldsymbol{g}^\bullet) \to \operatorname{As}(X, \boldsymbol{g}).$$

We will indicate by subscript Θ the trivial asymptotic type, i.e., when the carrier is contained in $\Gamma_{\frac{n+1}{2} - \gamma + \vartheta}$ for finite Θ or empty for infinite Θ.

For $P_1, P_2 \in$ As$(X, \boldsymbol{g})$ we form $P_1 + P_2 \in$ As$(X, \boldsymbol{g})$ by carrier$(P_1 + P_2) = \operatorname{carrier} P_1 + \operatorname{carrier} P_2$ via (2.3.35), (2.3.36).

Let us define for every $P \in$ As$(X, \boldsymbol{g})$ with finite Θ the function space

$$\mathcal{E}_P(X^\wedge) = \left\{ \langle \zeta, t^{-z} \rangle : \zeta \in \mathcal{A}'(V, C^\infty(X)) \right\}, \quad V = \operatorname{carrier} P.$$

From the results of the preceding section we have an isomorphism

$$\mathcal{E}_P(X^\wedge) \cong \mathcal{A}'(V, C^\infty(X)). \tag{2.3.37}$$

The inverse of the map $\zeta \to \langle \zeta, t^{-z} \rangle$ comes from

$$\langle \zeta, t^{-z} \rangle \to M_{\gamma - \frac{n}{2}, t \to w} \left(\omega(t) \langle \zeta, t^{-z} \rangle \right) \in \mathcal{A}(\mathbb{C}_w \setminus V, C^\infty(X)) \tag{2.3.38}$$

for every fixed cut–off function $\omega(t)$, with the weighted Mellin transform $M_{\gamma - \frac{n}{2}}$, together with the relation (2.3.31). Thus $\mathcal{E}_P(X^\wedge)$ is a (nuclear) Fréchet space with the topology from the bijection (2.3.37). From (2.3.38) it follows that $\mathcal{E}_P(X^\wedge)$ is isomorphic to $\omega\mathcal{E}_P(X^\wedge)$, so also $\omega\mathcal{E}_P(X^\wedge)$ is Fréchet. In view of Proposition 2.3.33 we have

$$\omega\mathcal{E}_P(X^\wedge) \subset \mathcal{K}^{\infty,\gamma}(X^\wedge).$$

Let us set

$$\mathcal{K}_P^{s,\gamma}(X^\wedge) := \mathcal{K}_\Theta^{s,\gamma}(X^\wedge) + \omega\mathcal{E}_P(X^\wedge), \tag{2.3.39}$$

endowed with the Fréchet topology of the non–direct sum. For carrier $P \subset \Gamma_{\frac{n+1}{2} - \gamma + \vartheta}$ we have $\mathcal{K}_P^{s,\gamma}(X^\wedge) = \mathcal{K}_\Theta^{s,\gamma}(X^\wedge)$, i.e., in this case P represents the trivial asymptotic type with respect to $\boldsymbol{g}$ with the finite weight interval $\Theta = (\vartheta, 0]$.

For $P \in$ As$(X, \boldsymbol{g})$ with $\Theta = (-\infty, 0]$ and carrier $P = V$ we set $V_k = V \cap \{z : \operatorname{Re} z \geq \frac{n+1}{2} - \gamma + \vartheta\}$, $k \in \mathbb{N}$, which represents an element $P_k \in$ As$(X, (\gamma, (-k, 0]))$, cf. (2.3.35). Thus we can form the space $\mathcal{K}_{P_k}^{s,\gamma}(X^\wedge)$, according to (2.3.39). It is obvious that there are continuous embeddings $\mathcal{K}_{P_{k+1}}^{s,\gamma}(X^\wedge) \hookrightarrow \mathcal{K}_{P_k}^{s,\gamma}(X^\wedge) \hookrightarrow \mathcal{K}^{s,\gamma}(X^\wedge)$ for all k. Now we set

$$\mathcal{K}_P^{s,\gamma}(X^\wedge) = \bigcap_{k \in \mathbb{N}} \mathcal{K}_{P_k}^{s,\gamma}(X^\wedge), \tag{2.3.40}$$

endowed with the Fréchet topology of the projective limit.

We will say that an asymptotic type $P \in \text{As}(X, g)$ for $g = (\gamma, \Theta)$ with $\Theta = (-\infty, 0]$ is quasi–discrete, if $V = \text{carrier } P$ can be written $V = \bigcup_{k=0}^{\infty} K_k$ for compact sets $K_k \subset \{z : \text{Re } z < \frac{n+1}{2} - \gamma\}$ with

$$\sup\{\text{Re } z : z \in K_{k+1}\} < \inf\{\text{Re } z : z \in K_k\} \qquad (2.3.41)$$

for all $k \in \mathbb{N}$, and $\sup\{\text{Re } z : z \in K_k\} \to -\infty$ as $k \to \infty$. An analogous notion makes sense for finite Θ when we require $V = (V \cap \Gamma_{\frac{n+1}{2} - \gamma + \vartheta}) \cup \bigcup_{k=0}^{\infty} K_k$ for compact sets $K_k \subset \{z : \frac{n+1}{2} - \gamma + \vartheta < \text{Re } z < \frac{n+1}{2} - \gamma\}$ satisfying the condition (2.3.41). It is clear then that $K_k \subset \{z : c_k \leq \text{Re } z \leq c'_k\}$ for suitable constants $\frac{n+1}{2} - \gamma + \vartheta < c_k \leq c'_k$ with $c'_{k+1} < c_k$ for all k, where $c'_k - c_k \to 0$ for $k \to \infty$.

Proposition 2.3.36 *For every* $P \in \text{As}(X, g)$, $g = (\gamma, \Theta)$, $s \in \mathbb{R}$, *we have*

$$\mathcal{K}_P^{s,\gamma}(X^\wedge) = \mathcal{K}_\Theta^{s,\gamma}(X^\wedge) + \mathcal{K}_P^{\infty,\gamma}(X^\wedge)$$

with the Fréchet topology of the non–direct sum. Moreover,

$$\mathcal{K}_{P_1}^{s,\gamma}(X^\wedge) + \mathcal{K}_{P_2}^{s,\gamma}(X^\wedge) = \mathcal{K}_{P_1 + P_2}^{s,\gamma}(X^\wedge)$$

for every $P_1, P_2 \in \text{As}(X, g)$, *in the sense of non–direct sums of Fréchet spaces.*

Proof. Let first Θ be finite. Then, by (2.3.39), we have

$$\mathcal{K}_P^{\infty,\gamma}(X^\wedge) = \mathcal{K}_\Theta^{\infty,\gamma}(X^\wedge) + \omega \mathcal{E}_P(X^\wedge).$$

Thus $\mathcal{K}_P^{s,\gamma}(X^\wedge) = \mathcal{K}_\Theta^{s,\gamma}(X^\wedge) + \mathcal{K}_P^{\infty,\gamma}(X^\wedge)$ implies the asserted relation. For infinite Θ it suffices to apply (2.3.40) and to use

$$\varprojlim_{k \in \mathbb{N}} \mathcal{K}_{P_k}^{s,\gamma}(X^\wedge) = \varprojlim_{k \in \mathbb{N}} \mathcal{K}_{\Theta_k}^{s,\gamma}(X^\wedge) + \varprojlim_{k \in \mathbb{N}} \mathcal{K}_{P_k}^{\infty,\gamma}(X^\wedge)$$

for $\Theta_k = (-k, 0]$. $\square$

Remark 2.3.37 *Every* $P \in \text{As}(X, g)$ *can be written* $P = P_1 + P_2$ *for quasi–discrete* $P_1, P_2 \in \text{As}(X, g)$. *In fact, it suffices to decompose* $V = P$ *as a corresponding sum* $V = V_1 + V_2$. *For* $\Theta = (-\infty, 0]$ *we may set, for instance,*

$$V_1 = V \cap \bigcup_{j \in \mathbb{Z}} \{z : 2j - 2/3 \leq \text{Re } z \leq 2j + 2/3\},$$

$$V_2 = V \cap \bigcup_{j \in \mathbb{Z}} \{z : 2j + 1 - 2/3 \leq \text{Re } z \leq 2j + 1 + 2/3\}.$$

An alternative choice would be $V_1 = \bigcup \{K_j : j \text{ odd}\}$, $V_2 = \bigcup \{K_j : j \text{ even}\}$ *with the sets* K_j *of Definition 2.3.32 (ii).*

Remark 2.3.38 *Let $s, \gamma \in \mathbb{R}$ and $P \in \mathrm{As}(X, \boldsymbol{g})$ for $\boldsymbol{g} = (\gamma, \Theta)$. Then, there is a sequence of (Hilbertisable) Banach spaces E^k with continuous embeddings $E^{k+1} \hookrightarrow E^k$ for all k such that*

$$\mathcal{K}_P^{s,\gamma}(X^\wedge) = \varprojlim_{k \in \mathbb{N}} E^k.$$

The spaces E^k, $k \in \mathbb{N}$, can be chosen in such a way that $\{\kappa_\lambda\}_{\lambda \in \mathbb{R}_+}$, first acting on $\mathcal{K}^{s,\gamma}(X^\wedge)$, cf. Proposition 2.1.60, restricts to a group of isomorphisms $\{\kappa_\lambda\}_{\lambda \in \mathbb{R}_+} \in C(\mathbb{R}_+, \mathcal{L}_\sigma(E^k))$ for every $k \in \mathbb{N}$.

Let $P \in \mathrm{As}(X, \boldsymbol{g})$ for $\boldsymbol{g} = (\gamma, \Theta)$ for finite Θ, and define the space

$$\mathcal{A}_P^{s,\gamma}(X) = \mathcal{A}_\Theta^{s,\gamma}(X) + M_{\gamma - \frac{n}{2}} \omega \mathcal{E}_P(X^\wedge)|_{\frac{n+1}{2} - \gamma + \vartheta < \mathrm{Re}\, z < \frac{n+1}{2} - \gamma},$$

endowed with the Fréchet topology of the non–direct sum, cf. analogously (2.3.15). In the case $P \in \mathrm{As}(X, \boldsymbol{g})$ for infinite Θ we set (for P_k from (2.3.40))

$$\mathcal{A}_P^{s,\gamma}(X) = \bigcap_{k \in \mathbb{N}} \mathcal{A}_{P_k}^{s,\gamma}(X),$$

endowed with the Fréchet topology of the projective limit.

Remark 2.3.39 *For every $P \in \mathrm{As}(X, \boldsymbol{g})$, $\boldsymbol{g} = (\gamma, \Theta)$, $s \in \mathbb{R}$, we have*

$$\mathcal{A}_P^{s,\gamma}(X) = \mathcal{A}_\Theta^{s,\gamma}(X) + \mathcal{A}_P^{\infty,\gamma}(X)$$

in the Fréchet topology of the non–direct sum. Moreover, we have

$$\mathcal{A}_{P_1}^{s,\gamma}(X) + \mathcal{A}_{P_2}^{s,\gamma}(X) = \mathcal{A}_{P_1 + P_2}^{s,\gamma}(X)$$

for all $P_1, P_2 \in \mathrm{As}(X, \boldsymbol{g})$ in the sense of non–direct sums of Fréchet spaces.

Theorem 2.3.40 *For every cut–off function ω we have continuous operators*

$$M_{\gamma - \frac{n}{2}} \omega : \mathcal{K}_P^{s,\gamma}(X^\wedge) \to \mathcal{A}_P^{s,\gamma}(X),$$
$$\omega M_{\gamma - \frac{n}{2}}^{-1} : \mathcal{A}_P^{s,\gamma}(X) \to \mathcal{K}_P^{s,\gamma}(X^\wedge)$$

for all $s, \gamma \in \mathbb{R}$, $P \in \mathrm{As}(X, \boldsymbol{g})$, $\boldsymbol{g} = (\gamma, \Theta)$.

Proof. The result is clear for trivial P, indicated by subscripts Θ, cf. Theorem 2.3.9. Furthermore, for finite Θ, we have the continuous operators

$$M_{\gamma - \frac{n}{2}} \omega : \mathcal{K}_P^{\infty,\gamma}(X^\wedge) \to \mathcal{A}_P^{\infty,\gamma}(X),$$
$$\omega M_{\gamma - \frac{n}{2}}^{-1} : \mathcal{A}_P^{\infty,\gamma}(X) \to \mathcal{K}_P^{\infty,\gamma}(X^\wedge)$$

as a consequence of the definition and of the properties of the image of $\omega\mathcal{E}_P(X^\wedge)$ under the Mellin transform, cf. (2.3.37), (2.3.38). Then the assertion follows by applying Proposition 2.3.36 and Remark 2.3.39. The case of infinite Θ can be treated by passing to the projective limits. □

Theorem 2.3.41 *The multiplication* $\mathcal{M}_\varphi$ *by a function* $\varphi(t,x) \in C_0^\infty(\overline{\mathbb{R}}_+ \times X)$ *induces a continuous operator*

$$\mathcal{M}_\varphi : \mathcal{K}_P^{s,\gamma}(X^\wedge) \to \mathcal{K}_Q^{s,\gamma}(X^\wedge)$$

for every $P \in \mathrm{As}(X,\boldsymbol{g})$, $\boldsymbol{g} = (\gamma,\Theta)$, *with some resulting asymptotic type* $Q \in \mathrm{As}(X,\boldsymbol{g})$, *for every* $s \in \mathbb{R}$.

Proof. The arguments are completely analogous to those for Theorem 2.3.13. □

Remark 2.3.42 *For the operator of multiplication by*

$$\varphi_\eta(t,x) = \varphi(t[\eta]^{-1}, x),$$

$\eta \in \mathbb{R}^q$, *we have an analogue of Remark 2.3.14 here, for continuous asymptotics.*

Let us now consider a special kind of continuous asymptotic that extends the discrete ones of Section 2.3.1 in a particularly simple way.

Definition 2.3.43 $\mathrm{As}(X,\boldsymbol{g})^\bullet$ *for* $\boldsymbol{g} = (\gamma,\Theta)$, $\gamma \in \mathbb{R}$, $\Theta = (\vartheta, 0]$, *is the set of all sequences*

$$P = \{(p_j, m_j)\}_{j=0,\dots,N} \quad \text{for some } N = N(P) \text{ when } \Theta \text{ is finite,}$$

where $\frac{n+1}{2} - \gamma - \vartheta < \mathrm{Re}\, p_j < \frac{n+1}{2} - \gamma$ *for all* j, *and*

$$P = \{(p_i, m_j)\}_{j\in\mathbb{N}} \quad \text{when } \Theta = (-\infty, 0],$$

where $\mathrm{Re}\, p_j < \frac{n+1}{2} - \gamma$ *for all* j, $\mathrm{Re}\, p_j \to -\infty$ *as* $j \to \infty$; $n = \dim X$.

Let $\mathcal{A}'(\{p\}, E)^{(m)}$ for $(p,m) \in \mathbb{C} \times \mathbb{N}$ and some Fréchet space E be the subspace of all $\zeta \in \mathcal{A}'(\{p\}, E)$ of the form

$$\langle \zeta, h \rangle = \sum_{k=0}^{m} c_k \frac{d^k}{dz^k} h(z)|_{z=p}, \quad h \in \mathcal{A}(\mathbb{C}),$$

for arbitrary $c_k \in E$, $k = 0, \dots, m$. For $P \in \mathrm{As}(X, \boldsymbol{g})^\bullet$ for $\boldsymbol{g} = (\gamma, \Theta)$ with finite Θ we set

$$\mathcal{E}_P(X^\wedge) = \Big\{ \sum_{j=0}^{N} \langle \zeta_j, t^{-z} \rangle : \ \zeta_j \in \mathcal{A}'(\{p_j\}, C^\infty(X))^{(m_j)} \Big\},$$

for $N = N(P)$ as in Definition 2.3.43. $\mathcal{E}_P(X^\wedge)$ is Fréchet as a closed subspace of

$$\left\{ \langle \zeta_j, t^{-z} \rangle : \ \zeta_j \in \mathcal{A}'\Big(\bigcup_{j=0}^{N} \{p_j\}, C^\infty(X) \Big) \right\}.$$

The following results are analogous to earlier ones on discrete and continuous asymptotics. So the obvious proofs will be omitted.

Set $\omega \mathcal{E}_P(X^\wedge) = \{\omega v : \ v \in \mathcal{E}_P(X^\wedge)\}$ for any fixed cut–off function $\omega(t)$. The weighted Mellin transform $M_{\gamma-\frac{n}{2}}$ induces an isomorphism of $\omega \mathcal{E}_P(X^\wedge)$ to a closed subspace of $\mathcal{A}(\mathbb{C}_w \setminus V, C^\infty(X))$ for $V = $ carrier $P = \bigcup_{j=0}^{N}\{p_j\}$, where

$$M_{\gamma-\frac{n}{2}}\omega \mathcal{E}_P(X^\wedge) \cong \underset{j=0}{\overset{N}{\times}} \mathcal{A}'(\{p_j\}, C^\infty(X))^{(m_j)}.$$

This gives us a Fréchet topology in $\mathcal{E}_P(X^\wedge)$. Note that the operator of multiplication by ω yields an isomorphism $\mathcal{E}_P(X^\wedge) \to \omega \mathcal{E}_P(X^\wedge)$ between Fréchet spaces. We have $\omega \mathcal{E}_P(X^\wedge) \subset \mathcal{K}^{\infty,\gamma}(X^\wedge)$. By the formula (2.3.39) we then obtain the subspace $\mathcal{K}_P^{s,\gamma}(X^\wedge)$ of $\mathcal{K}^{s,\gamma}(X^\wedge)$ with asymptotics of type $P \in \mathrm{As}(X,g)^\bullet$. Then, similarly to (2.3.40) it follows a definition also for $\Theta = (-\infty, 0]$.

Remark 2.3.44 *There is a canonical map* $\pi : \mathrm{As}(X,g)^\bullet \to \mathrm{As}(X,g)$ *defined by forgetting the finite orders* m_j *of the analytic functionals over* $\{p_j\} \in \pi_\mathbb{C} P$. *In other words,* $\pi_\mathbb{C} P = $ *carrier* P *induces an element* $\pi P \in \mathrm{As}(X,g)$. *Then*

$$\mathcal{K}_P^{s,\gamma}(X^\wedge) \subset \mathcal{K}_{\pi P}^{s,\gamma}(X^\wedge)$$

is a closed subspace for every $s \in \mathbb{R}$.

Remark 2.3.45 *There is an analogue of Proposition 2.3.36 for the asymptotic types in* $\mathrm{As}(X,g)^\bullet$. *Also Remark 2.3.2 remains in force. Moreover, for* $P \in \mathrm{As}(X,g)^\bullet$ *we can define associated spaces* $\mathcal{A}_P^{s,\gamma}(X)$ *for which the corresponding versions of Remark 2.3.37 and Theorem 2.3.9 hold. Finally we have a corresponding analogue of Theorem 2.3.13 and Remark 2.3.14 for asymptotic types in* $\mathrm{As}(X,g)^\bullet$.

Both for $P \in \mathrm{As}(X,g)$ and $P \in \mathrm{As}(X,g)^\bullet$ we set

$$\mathcal{S}_P^\gamma(X^\wedge) = [\omega]\mathcal{K}_P^{\infty,\gamma}(X^\wedge) + [1-\omega]\mathcal{S}(\overline{\mathbb{R}}_+ \times X)$$

with the corresponding Fréchet topology of the non–direct sum.

2.3.4 Operator–valued Mellin symbols with asymptotics

We will now study suitable subclasses of operator–valued Mellin symbols $f(z) \in L^\mu(X; \Gamma_\beta)$, where the associated pseudo–differential operators transform the distributions with asymptotics near $t = 0$ in a controlled manner. The additional properties of $f(z)$ will consist of meromorphic (or more general) extensions into $\mathbb{C} \ni z$ with some analogy to the Mellin transformed discrete (or continuous) asymptotics.

Definition 2.3.46 (i) $\mathbf{As}^\bullet(X)$ *is defined as the set of all sequences* $R = \{(r_j, m_j, M_j)\}_{j \in \mathbb{Z}}$ *with* $r_j \in \mathbb{C}$, $|\operatorname{Re} r_j| \to \infty$ *as* $|j| \to \infty$, $m_j \in \mathbb{N}$, *and finite-dimensional subspaces* $M_j \subset L^{-\infty}(X)$ *of finite-dimensional operators. The elements of* $\mathbf{As}^\bullet(X)$ *are called discrete asymptotic types of Mellin symbols. If* $\dim X = 0$ *the spaces* M_j *disappear, and we write* $\mathbf{As}^\bullet$ *for the corresponding set of discrete asymptotic types.*

(ii) $\mathbf{As}(X)$ *is defined as the set of all equivalence classes of pairs* $\{V, \{\mathcal{A}'(K_j, L^{-\infty}(X))\}_{j \in \mathbb{Z}}\}$ *for arbitrary* $V \in \mathcal{V}$ *and compact sets* $K_j \subseteq V$ *with* $K_j = K_j^c$, $j \in \mathbb{Z}$, *with* $\{z : c \le \operatorname{Re} z \le c'\} \cap K_j \ne \emptyset$ *only for finitely many* j, *for arbitrary real* $c \le c'$, *where*

$$\{V, \{\mathcal{A}'(K_j, L^{-\infty}(X))\}_{j \in \mathbb{Z}}\} \sim \{W, \{\mathcal{A}'(L_k, L^{-\infty}(X))\}_{k \in \mathbb{Z}}\}$$

means that $V = W$. *The elements of* $\mathbf{As}(X)$ *are called continuous asymptotic types of Mellin symbols. For* $\dim X = 0$ *we write* $\mathbf{As}$.

(iii) $\mathbf{As}(X)^\bullet$ *denotes the set of all sequences*

$$R = \{r_j, \mathcal{A}'(\{r_j\}, L^{-\infty}(X))^{(m_j)}\}_{j \in \mathbb{Z}}$$

for arbitrary $r_j \in \mathbb{C}$, $|\operatorname{Re} r_j| \to \infty$ *as* $|j| \to \infty$, *and* $m_j \in \mathbb{N}$. *The elements of* $\mathbf{As}(X)^\bullet$ *are called weakly discrete asymptotic types of Mellin symbols. If* $\dim X = 0$ *the spaces* $\mathcal{A}'(\{r_j\}, L^{-\infty}(X))^{(m_j)}$ *are to be replaced by* m_j, *and we write* $\mathbf{As}^\bullet$ *as in* (i).

The set V for $R \in \mathbf{As}(X)$ in the notation of Definition 2.3.46 (ii) is called a carrier of R, written carrier R. We call an $R \in \mathbf{As}(X)$ quasi-discrete when there are real β_j, $j \in \mathbb{Z}$, with $|\beta_j| \to \infty$ as $|j| \to \infty$ such that carrier $R \cap \Gamma_{\beta_j} = \emptyset$ for all $j \in \mathbb{Z}$. Incidentally we also talk about a quasi-discrete carrier in this case. For discrete or weakly discrete asymptotic types we set carrier $R = \pi_{\mathbb{C}} R = \{r_j\}_{j \in \mathbb{Z}}$. Note that there are canonical maps

$$\mathbf{As}^\bullet(X) \to \mathbf{As}(X)^\bullet \to \mathbf{As}(X), \tag{2.3.42}$$

where the second one is an inclusion.

If $A \subset \mathbb{C}$ is a closed set we call a $\chi(z) \in C^\infty(\mathbb{C})$ an A-excision function if $\chi(z) = 0$ for $\mathrm{dist}(z, A) \leq \varepsilon_0$, $\chi(z) = 1$ for $\mathrm{dist}(z, A) \geq \varepsilon_1$ for certain constants $0 < \varepsilon_0 < \varepsilon_1 < \infty$.

Let $K \subset \mathbb{C}$ be a compact set with $K = K^c$. For every $\zeta \in \mathcal{A}'(K, L^{-\infty}(X))$ we form a function

$$f_\zeta(z) := M_{\frac{1}{2}-\delta}(\omega(t)\langle \zeta_w, t^{-w} \rangle)$$

for any $\delta \in \mathbb{R}$ with $K \subset \{z : \mathrm{Re}\, z < \frac{1}{2} - \delta\}$ and some fixed cut-off function $\omega(t)$. We identify $f_\zeta(z)$ with its unique analytic extension to an $f_\zeta(z) \in \mathcal{A}(\mathbb{C} \setminus K, L^{-\infty}(X))$ that is independent of the choice of δ. For every K-excision function $\chi(z)$ we have

$$\chi(z) f_\zeta(z)|_{\Gamma_\beta} \in \mathcal{S}(\Gamma_\beta, L^{-\infty}(X))$$

for every $\beta \in \mathbb{R}$, uniformly in $c \leq \beta \leq c'$ for arbitrary $c \leq c'$.

Remark 2.3.47 *Let $(r, m) \in \mathbb{C} \times \mathbb{N}$ and $\zeta \in \mathcal{A}'(\{r\}, M)^{(m)}$ for any Fréchet subspace $M \subseteq L^{-\infty}(X)$. Then the function $f_\zeta(z)$ is a meromorphic M-valued function with a pole in p of multiplicity $m + 1$ and Laurent coefficients belonging to M.*

Definition 2.3.48 *Let $\mu \in \mathbb{R}$ and $R \in \mathbf{As}(X)$, represented by a pair $\{V, \{\mathcal{A}'(K_j, L^{-\infty}(X))\}_{j \in \mathbb{Z}}\}$ as in Definition 2.3.46 (ii). Then $N_R^\mu(X)$ is defined as the space of all $f(z) \in \mathcal{A}(\mathbb{C} \setminus V, L^\mu(X))$ such that for every $c < c'$ there exist finitely many sets $K_{j_0}, \ldots, K_{j_N}$, $N = N(c, c')$, and elements $\zeta_{j_k} \in \mathcal{A}'(K_{j_k}, L^{-\infty}(X))$, $k = 0, \ldots, N$, such that*

$$f_{(c,c')}(z) := \left\{ f(z) - \sum_{k=0}^{N} f_{\zeta_{j_k}}(z) \right\}|_{\{z:\ c < \mathrm{Re}\, z < c'\}}$$

$$\in \mathcal{A}\left(\{z : c < \mathrm{Re}\, z < c'\}, L^\mu(X)\right)$$

and $f_{c,c'}(z)|_{\Gamma_\beta} \in L^\mu(X; \Gamma_\beta)$ uniformly in $c + \varepsilon \leq \beta \leq c' - \varepsilon$ for every $0 < \varepsilon < (c' - c)/2$. Moreover, define $N_R^\mu(X)$ for $R \in \mathbf{As}(X)^\bullet$ ($\in \mathbf{As}^\bullet(X)$) as the space of all $f(z) \in \mathcal{A}(\mathbb{C} \setminus \pi_\mathbb{C} R, L^\mu(X))$ such that for every $c < c'$ there are elements

$$\zeta_{j_k} \in \mathcal{A}'(\{r_{j_k}\}, L^{-\infty}(X))^{(m_j)}) \qquad (\in \mathcal{A}'(\{r_{j_k}\}, M_{j_k})^{(m_{j_k})}),$$

$k = 0, \ldots, N = N(c, c')$, with $\cup_{k=0}^{N}\{r_{j_k}\} = \pi_\mathbb{C} R \cap \{z : c < \mathrm{Re}\, z < c'\}$, such that $f_{(c,c')}(z)$ satisfies the above conditions.

Define $M_R^\mu(X)$ for $R \in \mathbf{As}(X)$ ($\in \mathbf{As}(X)^\bullet, \in \mathbf{As}^\bullet(X)$) in an analogous manner by inserting $L_{cl}^\mu(X)$ instead of $L^\mu(X)$. The corresponding spaces with empty carrier are denoted by $N_O^\mu(X)$ and $M_O^\mu(X)$, respectively.

For $\dim X = 0$ we write N_R^μ, M_R^μ for the corresponding spaces of Mellin symbols of order μ with asymptotics of type R. In this case, e.g., for $R \in \mathbf{As}$, the space N_R^μ is the space of all $f(z) \in \mathcal{A}(\mathbb{C}\setminus V)$ such that, in the notation of Definition 2.3.46 (ii), $L^{-\infty}(X)$ is to be replaced by $\mathbb{C}$, and $f_{(c,c')}(z)|_{\Gamma_\beta} \in S^\mu(\Gamma_\beta)$ holds uniformly in $c + \varepsilon \leq \beta \leq c' - \varepsilon$ for every $0 < \varepsilon < (c' - c)/2$.

Note that $N_R^\mu(X)$ and $M_R^\mu(X)$ only depend on carrier R. In particular, for $R \in \mathbf{As}(X)$ the spaces are independent of the particular choice of $\{K_j\}_{j \in \mathbb{Z}}$. Observe also that $N_R^{-\infty}(X) = M_R^{-\infty}(X)$, and

$$N_R^\mu(X) = N_O^\mu(X) + M_R^{-\infty}(X), \quad M_R^\mu(X) = M_O^\mu(X) + M_R^{-\infty}(X)$$
$$(2.3.43)$$

for $R \in \mathbf{As}(X)$, $\in \mathbf{As}(X)^\bullet$ and $\in \mathbf{As}^\bullet(X)$, in the sense of vector spaces. $N_O^\mu(X)$ and $M_O^\mu(X)$ were introduced above in Section 1.2.4 as well as the spaces (2.3.43) in the particular case $R \in \mathbf{As}^\bullet(X)$, including their Fréchet topologies. We now define natural Fréchet topologies in $M_R^{-\infty}(X)$ for general R and then endow the spaces in (2.3.43) with the Fréchet topologies of the corresponding non-direct sums. Let us first assume that $R \in \mathbf{As}^\bullet(X)$ is quasi-discrete. In this case we can write $V = \text{carrier } R = \sum_{j \in \mathbb{Z}} K_j$ with compact sets K_j , where

$$\sup\{\text{Re } z : z \in K_{j+1}\} < \beta_{j+1} < \inf\{\text{Re } z : z \in K_j\}$$

for certain $\beta_j \in \mathbb{R}$, $j \in \mathbb{Z}$ with $\beta_j \to \mp\infty$ as $j \to \pm\infty$. Choose arbitrary reals $c < c'$ with $\Gamma_c \cap V = \Gamma_{c'} \cap V = \emptyset$ and set

$$M_R^{-\infty}(X)_{(c,c')} = \left\{ f(z)|_{\{z:c<\text{Re } z<c'\}} : f(z) \in M_R^{-\infty}(X) \right\},$$

$V_{(c,c')} = V \cap \{z : c < \text{Re } z < c'\}$, which is a compact set. We then have a canonical isomorphism

$$M_R^{-\infty}(X)_{(c,c')} \cong M_O^{-\infty}(X)_{(c,c')} \oplus \mathcal{A}'(V_{(c,c')}, L^{-\infty}(X)). \qquad (2.3.44)$$

Both summands on the right are Fréchet spaces in a natural way. This gives us also a Fréchet topology in $M_R^{-\infty}(X)$. Next choose a sequence of pairs $c_k < c_k'$, $k \in \mathbb{N}$, with $c_k \to -\infty$, $c_k' \to \infty$ for $k \to \infty$, such that $\Gamma_{c_k} \cap V = \Gamma_{c_k'} \cap V = \emptyset$ for all k. Then we endow

$$M_R^{-\infty}(X) = \varprojlim {}_{k \in \mathbb{N}} M_R^{-\infty}(X)_{(c_k, c_k')} \qquad (2.3.45)$$

with the Fréchet topology of the projective limit that is independent of the particular choice of the sequence.

For general $R \in \mathbf{As}(X)$ which is not necessarily quasi-discrete we can choose quasi-discrete asymptotic types $R_1, R_2 \in \mathbf{As}(X)$ such that $V = V_1 + V_2$ for $V = \text{carrier } R$, $V_i = \text{carrier } R_i$, $i = 1, 2$. Setting

$$M_R^{-\infty}(X) = M_{R_1}^{-\infty}(X) + M_{R_2}^{-\infty}(X),$$

endowed with the Fréchet topology of the non-direct sum, we easily see that this is independent of the specific choice of the decomposition of V into quasi-discrete sets. This relies on a Cousin problem argument as it plays a role also in the proof of Theorem 2.3.21. For the weakly discrete case, i.e., when $R \in \mathbf{As}^{\bullet}(X)$, we have carrier $R = \bigcup_{j \in \mathbb{Z}} K_j$ for $K_j = \{r_j\}$, and instead of (2.3.44) we obtain

$$M_R^{-\infty}(X)_{(c,c')} \cong M_O^{-\infty}(X)_{c,c'} \oplus \bigoplus_{k=0}^{N} \mathcal{A}'(\{r_{j_k}\}, L^{-\infty}(X))^{(m_{j_k})}. \quad (2.3.46)$$

Here $\cup_{k=0}^{N}\{r_{j_k}\} = \operatorname{carrier} R \cap \{z : c < \operatorname{Re} z < c'\}$, $N = N(c, c')$. In view of $\bigoplus_{k=0}^{N} \mathcal{A}'(\{r_{j_k}\}, L^{-\infty}(X))^{(m_{j_k})} \cong \bigoplus_{l=0}^{L} L^{-\infty}(X)$ for $L = \sum_{k=0}^{N}(m_{j_k} + 1)$, we obtain a Fréchet topology in (2.3.46). Applying (2.3.45) we then get a Fréchet topology in $M_R^{-\infty}(X)$ which is independent of the particular choice of the sequence $\{(c_k, c_k')\}_{k \in \mathbb{N}}$.

Remark 2.3.49 *The Fréchet topology of $M_R^{-\infty}(X)$ is nuclear for $R \in \mathbf{As}^{\bullet}(X)$, $\in \mathbf{As}(X)^{\bullet}$ or $\in \mathbf{As}(X)$.*

Recall that $M_R^{-\infty}(X)$ for $R \in \mathbf{As}^{\bullet}(X)$ was already Fréchet topologised in Section 1.2.4. In the formalism here we might also consider

$$M_R^{-\infty}(X)_{(c,c')} \cong M_O^{-\infty}(X)_{(c,c')} \oplus \bigoplus_{k=0}^{N} \mathcal{A}'(\{r_{j_k}\}, M_{j_k})^{(m_{j_k})}$$

and form a projective limit like (2.3.45). Here we employ the fact that R can be identified with the pair $\{V, \{\mathcal{A}'(\{r_j\}, M_j)^{(m_j)}\}_{j \in \mathbb{Z}}\}$ for $V = \cup_{j \in \mathbb{Z}}\{r_j\} = \operatorname{carrier} R$. The image of R under the first map of (2.3.42) is an element $R_1 \in \mathbf{As}(X)^{\bullet}$, represented by a pair

$$\{V, \{\mathcal{A}'(\{r_j\}, L^{-\infty}(X))^{(m_j)}\}_{j \in \mathbb{Z}}\}.$$

Applying the second map of (2.3.42) to R_1 we get an $R_2 \in \mathbf{As}(X)$, represented by $\{V, \{\mathcal{A}'(\{r_j\}, L^{-\infty}(X))\}_{j \in \mathbb{Z}}\}$. This yields canonical continuous embeddings

$$M_R^{\mu}(X) \hookrightarrow M_{R_1}^{\mu}(X) \hookrightarrow M_{R_2}^{\mu}(X),$$

where $M_R^{\mu}(X)$ is closed in $M_{R_1}^{\mu}(X)$ and $M_{R_1}^{\mu}(X)$ closed in $M_{R_2}^{\mu}(X)$. Analogous relations hold for the non-classical spaces $N_R^{\mu}(X)$, $N_{R_1}^{\mu}(X)$, $N_{R_2}^{\mu}(X)$, $\mu \in \mathbb{R}$.

Remark 2.3.50 *For every $h(z) \in N_R^{\mu}(X)$, $R \in \mathbf{As}(X)$, and arbitrary $\beta_1, \beta_2 \in \mathbb{R}$ with $\beta_1 \neq \beta_2$ there exist asymptotic types $R_1, R_2 \in \mathbf{As}(X)$ with $\Gamma_\beta \cap \operatorname{carrier} R_1 = \emptyset$ for $i = 1, 2$, and elements $h_i(z) \in N_{R_i}^{\mu}(X)$, $i =$*

$1, 2$, such that $h(z) = h_1(z) + h_2(z)$. An analogous assertion holds for $R \in \mathbf{As}^{\bullet}(X)$ ($\in \mathbf{As}(X)^{\bullet}$) for certain $R_i \in \mathbf{As}^{\bullet}(X)$ ($\in \mathbf{As}(X)^{\bullet}$) as well as for the classical case $M_R^{\mu}(X)$. Similarly, for arbitrary $h(t, t', z) \in C^{\infty}(\overline{\mathbb{R}}_+ \times \overline{\mathbb{R}}_+, N_R^{\mu}(X))$ and $\beta_1, \beta_2 \in \mathbb{R}$ with $\beta_1 \neq \beta_2$ there are R_1, R_2 as above and $h_i(t, t', z) \in C^{\infty}(\overline{\mathbb{R}}_+ \times \overline{\mathbb{R}}_+, N_{R_i}^{\mu}(X))$, $i = 1, 2$, such that $h(t, t', z) = h_1(t, t', z) + h_2(t, t', z)$. Analogous decompositions are possible in the classical case and for the various discrete asymptotic types.

In the following we will mainly discuss $M_R^{\mu}(X)$. The non-classical case is completely analogous and will not always be separately commented upon.

Example 2.3.51 We have for every polynomial $p(z) = \sum_{j=0}^{\mu} a_j z^j$ with $a_{\mu} \neq 0$

$$p^{-1}(z) \in M_R^{-\mu}$$

for a discrete asymptotic type $R = \{(r_j, n_j)\}_{j=0,\dots,N}$. Here r_j are the zeros of $p(z)$ and $n_j + 1$ the corresponding multiplicities.

Example 2.3.52 The meromorphic functions

$$g^+(z) = (1 - e^{-2\pi i z})^{-1}, \qquad g^-(z) = (1 - e^{2\pi i z})^{-1}$$

belong to M_P^0 for $P = \{(j, 0)\}_{j \in \mathbb{Z}}$. Note that $g^+(z) + g^-(z) = 1$. Moreover, we have $g^+(z)g^-(z) \in M_Q^{-\infty}$ for $Q = \{(j, 1)\}_{j \in \mathbb{Z}}$.

Example 2.3.53 Let X be a closed compact C^{∞} manifold with a fixed Riemannian metric. Then, if Δ is the Laplace-Beltrami operator, $h(z) = (\Delta + z^2)^{-1}$ is a meromorphic operator function which belongs to $M_R^{-2}(X)$ for some discrete asymptotic type $R \in \mathbf{As}(X)^{\bullet}$.

Proposition 2.3.54 We have

$$C^{\infty}(\overline{\mathbb{R}}_+ \times \overline{\mathbb{R}}_+, M_R^{\mu}(X))$$
$$= C^{\infty}(\overline{\mathbb{R}}_+ \times \overline{\mathbb{R}}_+, M_O^{\mu}(X)) + C^{\infty}(\overline{\mathbb{R}}_+ \times \overline{\mathbb{R}}_+, M_R^{-\infty}(X))$$

as a non-direct sum of Fréchet spaces. Analogous decompositions hold for $C^{\infty}(\Omega, M_R^{\mu}(X))$ for any open set $\Omega \subseteq \mathbb{R}^q$.

Proof. This is a consequence of the relation (2.3.43) together with the tensor product representation of C^{∞} functions on a corresponding set with values in Fréchet spaces, and of Proposition 1.1.8. $\qquad \square$

Theorem 2.3.55 *Let $h(z) \in N_R^\mu(X)$ for $\mu \in \mathbb{R}$, $R \in \mathbf{As}(X)$, and assume that $\Gamma_{\frac{n+1}{2}-\gamma} \cap$ carrier $R = \emptyset$ for a $\gamma \in \mathbb{R}$. Then, if $\omega(t)$, $\tilde{\omega}(t)$ are cut-off functions,*

$$\omega \, \mathrm{op}_M^{\gamma-\frac{n}{2}}(h)\tilde{\omega} : \mathcal{K}^{s,\gamma}(X^\wedge) \to \mathcal{K}^{s-\mu,\gamma}(X^\wedge)$$

induces continuous operators

$$\omega \, \mathrm{op}_M^{\gamma-\frac{n}{2}}(h)\tilde{\omega} : \mathcal{K}_P^{s,\gamma}(X^\wedge) \to \mathcal{K}_Q^{s-\mu,\gamma}(X^\wedge)$$

for each asymptotic type $P \in \mathrm{As}(X,\mathbf{g})$ for some resulting $Q \in \mathrm{As}(X,\mathbf{g})$, $\mathbf{g} = (\gamma,\Theta)$, $\Theta = (\vartheta,0]$, $-\infty \leq \vartheta < 0$. For $R \in \mathbf{As}^\bullet(X)$ $(\in \mathbf{As}(X)^\bullet)$ the same is true for every $P \in \mathrm{As}(X,\mathbf{g}^\bullet)$ $(\in \mathrm{As}(X,\mathbf{g})^\bullet)$ for some resulting $Q \in \mathrm{As}(X,\mathbf{g}^\bullet)$ $(\in \mathrm{As}(X,\mathbf{g})^\bullet)$. This holds for all $s \in \mathbb{R}$.

Proof. It is sufficient to look at a fixed finite $\vartheta < 0$. The assertions in the general case then follow by writing the corresponding spaces for $\vartheta = -\infty$ as projective limits of spaces with finite weight intervals, cf. (2.3.11) and (2.3.40). The operator in question has the form

$$\mathcal{M}_\omega M_{\gamma-\frac{n}{2}}^{-1} \mathcal{M}_h M_{\gamma-\frac{n}{2}} \mathcal{M}_{\tilde{\omega}}, \qquad (2.3.47)$$

where $\mathcal{M}_\varphi$ is the operator of multiplication by a function φ. It is obvious that for $P \in \mathrm{As}(X,\mathbf{g})$ there is a $Q \in \mathrm{As}(X,\mathbf{g})$ such that $\mathcal{M}_h : \mathcal{A}_P^{s,\gamma}(X) \to \mathcal{A}_Q^{s-\mu,\gamma}(X)$ is continuous. Then, from Theorem 2.3.40, we get the continuity of (2.3.47) as operator $\mathcal{K}_P^{s,\gamma}(X^\wedge) \to \mathcal{K}_Q^{s-\mu,\gamma}(X^\wedge)$. Analogous arguments apply for $P \in \mathrm{As}(X,\mathbf{g}^\bullet)$ $(\in \mathrm{As}(X,\mathbf{g})^\bullet)$ with resulting $Q \in \mathrm{As}(X,\mathbf{g}^\bullet)$ $(\in \mathrm{As}(X,\mathbf{g})^\bullet)$. $\quad\square$

Remark 2.3.56 *Let $h_j(z) \in N_R^\mu(X)$, $j \in \mathbb{N}$, be a sequence that tends to zero for $j \to \infty$. Then $\omega \, \mathrm{op}_M^{\gamma-\frac{n}{2}}(h_j)\tilde{\omega}$ tends to zero for $j \to \infty$ in the sense that for $\mathcal{K}_P^{s,\gamma}(X^\wedge) = \varprojlim_{k\in\mathbb{N}} E^k$, $\mathcal{K}_Q^{s-\mu,\gamma}(X^\wedge) = \varprojlim_{l\in\mathbb{N}} \tilde{E}^l$, cf. Remark 2.3.38, for every l there is a $k(l)$ such that $\omega \, \mathrm{op}_M^{\gamma-\frac{n}{2}}(h_j)\tilde{\omega}$ tends to zero in $\mathcal{L}(E^{k(l)}, \tilde{E}^l)$ for $j \to \infty$ for every $l \in \mathbb{N}$. This follows easily from a corresponding observation in connection with $\mathcal{M}_{h_j} : \mathcal{A}_P^{s,\gamma}(X) \to \mathcal{A}_Q^{s-\mu,\gamma}(X)$.*

Theorem 2.3.57 *Let $h(z) \in N_R^\mu(X)$ for some $R \in \mathbf{As}(X)$ with carrier $R \cap \Gamma_{\frac{n+1}{2}-\gamma} = \emptyset$ for some $\gamma \in \mathbb{R}$. Then the formal adjoint A^* of $A = \omega \, \mathrm{op}_M^{\gamma-\frac{n}{2}}(h)\tilde{\omega}$ in the sense*

$$(u, A^*v)_{\mathcal{K}^{0,0}(X^\wedge)} = (Au, v)_{\mathcal{K}^{0,0}(X^\wedge)}$$

for all $u, v \in C_0^\infty(\mathbb{R}_+, C^\infty(X))$ has the form

$$A^* = \tilde{\omega} \, \mathrm{op}_M^{-\gamma - \frac{n}{2}} (T^{-n} h^{[*]}) \omega$$

for $h^{[*]}(z) = h(1 - \overline{z})^{(*)}$, where $(*)$ indicates the point-wise formal adjoint with respect to the $L^2(X)$-scalar product. Thus $T^{-n} h^{(*)} \in N_Q^\mu(X)$ for some $Q \in \mathbf{As}(X)$ with carrier $Q \cap \Gamma_{\frac{n+1}{2} + \gamma} = \emptyset$. An analogous result holds for $h(z) \in M_R^\mu(X)$ as well as for $R \in \mathbf{As}^\bullet(X)$ and $\in \mathbf{As}(X)^\bullet$, for the resulting $Q \in \mathbf{As}^\bullet(X)$ and $\in \mathbf{As}(X)^\bullet$, respectively.

Proof. We have $(f, g)_{\mathcal{K}^{0,0}(X^\wedge)} = (t^{\frac{n}{2}} f, t^{\frac{n}{2}} g)_{L^2(X^\wedge)}$ for $f, g \in C_0^\infty(X^\wedge)$, $n = \dim X$. Thus, denoting by $*_0$ the adjoints with respect to the $L^2(X^\wedge)$ scalar products, we obtain

$$
\begin{aligned}
(Au, v)_{\mathcal{K}^{0,0}(X^\wedge)} &= \left(t^{\frac{n}{2}} Au, t^{\frac{n}{2}} v\right)_{L^2(X^\wedge)} \\
&= \left(t^n Au, v\right)_{L^2(X^\wedge)} \\
&= \left(u, (t^n A)^{*_0} v\right)_{L^2(X^\wedge)} \\
&= \left(t^{\frac{n}{2}} t^{-\frac{n}{2}} u, t^{\frac{n}{2}} t^{-\frac{n}{2}} (t^n A)^{*_0} v\right)_{L^2(X^\wedge)} \\
&= \left(t^{-\frac{n}{2}} u, t^{-\frac{n}{2}} (t^n A)^{*_0} v\right)_{\mathcal{K}^{0,0}(X^\wedge)} \\
&= \left(u, t^{-n} (t^n A)^{*_0} v\right)_{\mathcal{K}^{0,0}(X^\wedge)}.
\end{aligned}
$$

Thus it suffices to calculate $(t^n A)^{*_0}$. We have

$$
\begin{aligned}
\left(\omega t^n \, \mathrm{op}_M^{\gamma - \frac{n}{2}} (h) \tilde{\omega}\right)^{*_0} &= \tilde{\omega} \left(t^{\gamma - \frac{n}{2}} \, \mathrm{op}_M(T^{-\gamma + \frac{n}{2}} h) t^{-\gamma + \frac{n}{2}}\right)^{*_0} t^n \omega \\
&= \tilde{\omega} t^{-\gamma + \frac{n}{2}} \, \mathrm{op}_M(T^{\gamma - \frac{n}{2}} h^{(*)}) t^{\gamma - \frac{n}{2}} t^n \omega
\end{aligned}
$$

for $h^{[*]}(z) = h(1 - \overline{z})^{(*)}$. The last expression, interpreted as an operator on $C_0^\infty(\mathbb{R}_+, C^\infty(X))$, can be written

$$\tilde{\omega} t^n t^{-\gamma - \frac{n}{2}} \, \mathrm{op}_M \left(T^{\gamma + \frac{n}{2}} (T^{-n} h^{[*]})\right) t^{\gamma + \frac{n}{2}} \omega = \tilde{\omega} t^n \, \mathrm{op}_M^{-\gamma - \frac{n}{2}} \left(T^{-n} h^{[*]}\right) \omega,$$

which yields $A^* = \tilde{\omega} \, \mathrm{op}_M^{-\gamma - \frac{n}{2}} (T^{-n} h^{[*]}) \omega$. Now $T^{-n} h^{[*]}$ is obviously of the asserted kind with respect to some asymptotic type Q. $\qquad \square$

Theorem 2.3.58 *Let $h_j(t, t', z) \in C^\infty(\overline{\mathbb{R}}_+ \times \overline{\mathbb{R}}_+, N_O^{\mu_j}(X))$, $j \in \mathbb{N}$, be an arbitrary sequence with $\mu_j \to -\infty$ as $j \to \infty$. Then there is an $h(t, t', z) \in C^\infty(\overline{\mathbb{R}}_+ \times \overline{\mathbb{R}}_+, N_O^\mu(X))$ for $\mu = \max\{\mu_j\}$, such that for every $k \in \mathbb{N}$ there is an $l \in \mathbb{N}$ such that*

$$h(t, t', z) - \sum_{j=0}^{l} h_j(t, t', z) \in C^\infty(\overline{\mathbb{R}}_+ \times \overline{\mathbb{R}}_+, N_O^{\mu-k}(X)). \quad (2.3.48)$$

Every $h(t,t',z)$ with this property is unique modulo

$$C^\infty(\overline{\mathbb{R}}_+ \times \overline{\mathbb{R}}_+, M_O^{-\infty}(X)).$$

An analogous result holds for $h_j(t,t',z) \in C^\infty(\overline{\mathbb{R}}_+ \times \overline{\mathbb{R}}_+, M_O^{\mu-j}(X))$, $j \in \mathbb{N}$, where $h(t,t',z) \in C^\infty(\overline{\mathbb{R}}_+ \times \overline{\mathbb{R}}_+, M_O^\mu(X))$.

Proof. Let us fix the weight line Γ_0 and interpret $h_j(t,t',z)$ as an element of $C^\infty(\overline{\mathbb{R}}_+ \times \overline{\mathbb{R}}_+, L^{\mu_j}(X;\Gamma_0))$ by restriction of $\mathbb{C}$ to Γ_0. Then we find an asymptotic sum $f(t,t',z) \in C^\infty(\overline{\mathbb{R}}_+ \times \overline{\mathbb{R}}_+, L^\mu(X;\Gamma_0))$ in the sense that for every $k \in \mathbb{N}$ there is an $l \in \mathbb{N}$ such that $f(t,t',z) - \sum_{j=0}^l h_j(t,t',z) \in C^\infty(\overline{\mathbb{R}}_+ \times \overline{\mathbb{R}}_+, L^{\mu-k}(X;\Gamma_0))$. Applying Theorem 2.2.10 to $f(t,t',z)$ we obtain an $h(t,t',z) = H(\psi)f(t,t',z) \in C^\infty(\overline{\mathbb{R}}_+ \times \overline{\mathbb{R}}_+, N_O^\mu(X))$ with $h(t,t',z) - f(t,t',z) \in C^\infty(\overline{\mathbb{R}}_+ \times \overline{\mathbb{R}}_+, L^{-\infty}(X;\Gamma_0))$. Now $h(t,t',z)$ satisfies the relations

$$\left(h(t,t',z) - \sum_{j=0}^l h_j(t,t',z) \right)\Big|_{\Gamma_0} \in C^\infty(\overline{\mathbb{R}}_+ \times \overline{\mathbb{R}}_+, L^{\mu-k}(X;\Gamma_0))$$

for every k with suitable $l(k)$. This implies (2.3.48) as we can see from the above relation (2.2.21).   □

2.3.5   Green and Mellin cone operators with asymptotics

The distributions with asymptotics on a (stretched) manifold $\mathbb{B}$ with conical singularities give rise to a class of smoothing operators on int $\mathbb{B}$, called Green operators. The notation is motivated by Green's function of an elliptic boundary value problem. The connection to conical singularities is that in the special case $\mathbb{B} = \overline{\mathbb{R}}_+$, interpreted as the inner normal to a boundary, Green's function, modulo a parametrix, is a pseudo–differential operator along the boundary, with operator–valued symbol, acting along $\mathbb{R}_+$ as a family of particular Green operators in the present sense. We restrict the consideration to compact $\mathbb{B}$ or to the infinite stretched cone $\overline{\mathbb{R}}_+ \times X$.

Definition 2.3.59 *Let B be a compact manifold with conical singularities in the sense of Definition 2.1.1 and $\mathbb{B}$ be the associated stretched manifold. An operator*

$$G \in \bigcap_{s\in\mathbb{R}} \mathcal{L}(\mathcal{H}^{s,\gamma}(\mathbb{B}), \mathcal{H}^{\infty,\delta}(\mathbb{B}))$$

is called a Green operator   on $\mathbb{B}$ with the weight data $\boldsymbol{g} = (\gamma,\delta,\Theta)$ for the given fixed weights $\gamma, \delta \in \mathbb{R}$ and $\Theta = (\vartheta,0]$, $-\infty \leq \vartheta < 0$, *if there are (continuous) asymptotic types* $P \in \mathrm{As}(X, (\delta, \Theta))$ *and* $Q \in \mathrm{As}(X, (-\gamma, \Theta))$ *such that* G *induces continuous operators*

$$G : \mathcal{H}^{s,\gamma}(\mathbb{B}) \to \mathcal{H}^{\infty,\delta}_P(\mathbb{B}),$$
$$G^* : \mathcal{H}^{s,-\delta}(\mathbb{B}) \to \mathcal{H}^{\infty,-\gamma}_Q(\mathbb{B})$$

for all $s \in \mathbb{R}$. *Here* G^* *is the formal adjoint to* G *with respect to the* $\mathcal{H}^{0,0}(\mathbb{B})$*-scalar product.*

The space of all Green operators on $\mathbb{B}$ for the weight data $\boldsymbol{g} = (\gamma, \delta, \Theta)$ will be denoted by $C_G(\mathbb{B}, \boldsymbol{g})$. The subspace defined by the conditions

$$P \in \mathrm{As}(X, (\delta, \Theta)^\bullet) \qquad (\in \mathrm{As}(X, (\delta, \Theta))^\bullet),$$
$$Q \in \mathrm{As}(X, (-\gamma, \Theta)^\bullet) \qquad (\in \mathrm{As}(X, (-\gamma, \Theta))^\bullet)$$

will be denoted by $C_G(\mathbb{B}, \boldsymbol{g}^\bullet)$ $(C_G(\mathbb{B}, \boldsymbol{g})^\bullet)$.

Analogous notions make sense on the infinite (open stretched) cone with base X:

Definition 2.3.60 *An operator*

$$G \in \bigcap_{s \in \mathbb{R}} \mathcal{L}\left(\mathcal{K}^{s,\gamma}(X^\wedge), \mathcal{K}^{\infty,\delta}(X^\wedge)\right)$$

is called a Green operator on $X^\wedge$ *with the weight data* $\boldsymbol{g} = (\gamma, \delta, \Theta)$ *if there are (continuous) asymptotic types* $P \in \mathrm{As}(X, (\delta, \Theta))$ *and* $Q \in \mathrm{As}(X, (-\gamma, \Theta))$ *such that* G *induces continuous operators*

$$G : \mathcal{K}^{s,\gamma}(X^\wedge) \to \mathcal{S}^\delta_P(X^\wedge), \qquad (2.3.49)$$
$$G^* : \mathcal{K}^{s,-\delta}(X^\wedge) \to \mathcal{S}^{-\gamma}_Q(X^\wedge) \qquad (2.3.50)$$

for all $s \in \mathbb{R}$. *Here* G^* *is the formal adjoint to* G *with respect to the* $\mathcal{K}^{0,0}(X^\wedge)$*-scalar product.*

The space of all Green operators on $X^\wedge$ for the weight data $\boldsymbol{g} = (\gamma, \delta, \Theta)$ is denoted by $C_G(X^\wedge, \boldsymbol{g})$, and the subspace defined by the conditions

$$P \in \mathrm{As}(X, (\delta, \Theta)^\bullet) \qquad (\in \mathrm{As}(X, (\delta, \Theta))^\bullet), \qquad (2.3.51)$$
$$Q \in \mathrm{As}(X, (-\gamma, \Theta)^\bullet) \qquad (\in \mathrm{As}(X, (-\gamma, \Theta))^\bullet) \qquad (2.3.52)$$

by $C_G(X^\wedge, \boldsymbol{g}^\bullet)$ $(C_G(X^\wedge, \boldsymbol{g})^\bullet)$. Incidentally we write

$$C_G(\mathbb{B}, \boldsymbol{g})_{P,Q} \quad \text{and} \quad C_G(X^\wedge, \boldsymbol{g})_{P,Q}, \qquad (2.3.53)$$

respectively, for the corresponding subclasses with fixed asymptotic
types P, Q.

Note that the mapping properties of the formal adjoints of the Green
operators depend on the choice of the scalar product in $\mathcal{H}^{0,0}(\mathbb{B})$ and
$\mathcal{K}^{0,0}(X^{\wedge})$, respectively. Other allowed scalar products are associated
with different Riemannian metrics on $\mathbb{B}$ and $\overline{\mathbb{R}}_+ \times X$, respectively,
related to the original ones by diffeomorphisms that are smooth up to
$t = 0$.

Green operators in this section will be considered mainly for $X^{\wedge}$.
The properties of Green operators on $\mathbb{B}$ are analogous to those on $X^{\wedge}$
and left to the reader as far as they are not formulated explicitly here.

Remark 2.3.61 *We have*

$$C_G(\mathbb{B}, \boldsymbol{g}) \subset L^{-\infty}(\text{int } \mathbb{B}), \qquad C_G(X^{\wedge}, \boldsymbol{g}) \subset L^{-\infty}(X^{\wedge})$$

for arbitrary $\boldsymbol{g} = (\gamma, \delta, \Theta)$.

Remark 2.3.62 *The properties (2.3.49), (2.3.50) imply that for $G \in C_G(X^{\wedge}, \boldsymbol{g})$ and every cut–off function $\omega(t)$ also the operators ωG, $(1 - \omega)G$ and $G\omega$, $G(1 - \omega)$ belong to $C_G(X^{\wedge}, \boldsymbol{g})$. The same is true for the other asymptotic types.*

Green operators will be generated as the smoothing elements of the
cone pseudo–differential calculus below. Let us only give two simple
examples here.

Example 2.3.63 *Let $g(t, t') \in \mathcal{S}(\overline{\mathbb{R}}_+ \times \overline{\mathbb{R}}_+) \ (= \mathcal{S}(\mathbb{R}^2)|_{\overline{\mathbb{R}}_+ \times \overline{\mathbb{R}}_+})$. Then the integral operator*

$$Gu(t) = \int\limits_0^\infty g(t, t')u(t')(t')^n(t')^n \, dt'$$

is a Green operator on $\mathbb{R}_+$ with the weight data $\boldsymbol{g} = (0, 0, (-\infty, 0])$ and discrete asymptotic types $P = Q = \{(-j, 0)\}_{j \in \mathbb{N}}$, that correspond to the Taylor asymptotics (cf. also Theorem 1.1.3).

Example 2.3.64 *Let $\omega(t), \tilde{\omega}(t)$ be cut–off functions and set*

$$g(t, x, t', x') = \omega(t)t^{-p}\log^k t\varphi(x)\tilde{\omega}(t')(t')^{-q}\log^l t'\psi(x')$$

for certain fixed $\varphi, \psi \in C^\infty(X)$ and $k, l \in \mathbb{N}$, $p, q \in \mathbb{C}$, $\operatorname{Re} p < \frac{n+1}{2} - \delta$, $\operatorname{Re} q < \frac{n+1}{2} + \gamma$ for $n = \dim X$. Then, the integral operator

$$Gu(t, x) = \int\limits_X \int\limits_0^\infty g(t, x, t', x')u(t', x') \, dt'dx'$$

is a Green operator on $X^\wedge$ with the weight data $\boldsymbol{g} = (\gamma, \delta, (-\infty, 0])$ and discrete asymptotic types P and Q, that are determined in an obvious way by p, k, φ and q, l, ψ, respectively.

Remark 2.3.65 *The definition of the spaces (2.3.53) for fixed asymptotic types P, Q (discrete as well as continuous ones) gives rise to natural Fréchet topologies in $C_G(\mathbb{B}, \boldsymbol{g})_{P,Q}$ and $C_G(X^\wedge, \boldsymbol{g})_{P,Q}$. Consider, for instance, the case $X^\wedge$. The continuity of (2.3.49) can be regarded as the continuity of $G : \mathcal{K}^{s,\gamma}(X^\wedge) \to E^k$ when $\mathcal{S}_P^\rho(X^\wedge)$ is written as a projective limit of Hilbert spaces E^k, $k \in \mathbb{N}$. Then, for every fixed s, k we can consider the operator norm of G. Since (2.3.49) for all $s \in \mathbb{N}$ implies the corresponding continuities for all $s \in \mathbb{R}$, we get in this way a countable semi-norm system for G, for $s, k \in \mathbb{N}$. Similarly (2.3.50) leads to a semi-norm system. This yields altogether a countable system. It can easily be verified then that $C_G(X^\wedge, \boldsymbol{g})_{P,Q}$ is a Fréchet space in the corresponding topology.*

Proposition 2.3.66 *Every $G \in C_G(X^\wedge, \boldsymbol{g})$ for $\boldsymbol{g} = (\gamma, \delta, \Theta)$ induces compact operators*

$$G : \mathcal{K}^{s,\gamma}(X^\wedge) \to \mathcal{K}^{s,\delta}(X^\wedge)$$

for all $s \in \mathbb{R}$. Analogously, every $G \in C_G(\mathbb{B}, \boldsymbol{g})$ induces compact operators

$$G : \mathcal{H}^{s,\gamma}(\mathbb{B}) \to \mathcal{H}^{s,\delta}(\mathbb{B})$$

for every $s \in \mathbb{R}$.

Proof. Consider, for instance, the case $X^\wedge$. In view of (2.3.49) there is an $\varepsilon > 0$ such that $G : \mathcal{K}^{s,\gamma}(X^\wedge) \to \langle t \rangle^{-N} \mathcal{K}^{s+N,\delta+\varepsilon}(X^\wedge)$ is continuous for every $N \in \mathbb{R}_+$. However $\langle t \rangle^{-N} \mathcal{K}^{s+N,\delta+\varepsilon}(X^\wedge) \to \mathcal{K}^{s,\delta}(X^\wedge)$ is compact. $\qquad\square$

Let k^β, $\beta \in \mathbb{R}$, denote a strictly positive function in $C^\infty(\mathbb{R}_+)$ with

$$k^\beta(t) = \begin{cases} t^\beta & \text{for } 0 < t < c_0, \\ 1 & \text{for } c_1 < t \leq \infty \end{cases}$$

for certain constants $0 < c_0 < c_1 < \infty$. Analogously denote by h^β a strictly positive function in $C^\infty(\text{int }\mathbb{B})$ with

$$h^\beta(t) = t^\beta \quad \text{in a collar neighbourhood of} \quad \partial\mathbb{B} \qquad (2.3.54)$$

in the given splitting of coordinates (t, x).

Remark 2.3.67 *Let* $g = (\gamma, \delta, \Theta)$. *For every* $\alpha, \beta \in \mathbb{R}$ *we have*

$$t^\beta C_G(X^\wedge, g) t^{-\alpha} = C_G(X^\wedge, g_{\alpha,\beta}),$$
$$k^\beta C_G(X^\wedge, g) k^{-\alpha} = C_G(X^\wedge, g_{\alpha,\beta})$$

for $g_{\alpha,\beta} = (\gamma + \beta, \delta + \alpha, \Theta)$, *and*

$$h^\beta C_G(\mathbb{B}, g) h^{-\alpha} = C_G(\mathbb{B}, g_{\alpha,\beta}).$$

The same is true for the subclasses of Green operators with discrete or weakly discrete asymptotics.

Proposition 2.3.68 *Let* $G \in C_G(X^\wedge, g)$ *for* $g = (\gamma, \gamma, \Theta)$ *and assume that the operator*

$$1 + G : \mathcal{K}^{s,\gamma}(X^\wedge) \to \mathcal{K}^{s,\gamma}(X^\wedge) \tag{2.3.55}$$

is invertible for an $s = s_0 \in \mathbb{R}$. *Then* (2.3.55) *is invertible for all* $s \in \mathbb{R}$, *and there exists a* $D \in C_G(X^\wedge, g)$ *such that* $(1 + G)^{-1} = 1 + D$. *Analogous relations hold for* $G \in C_G(X^\wedge, g^\bullet)$ ($\in C_G(X^\wedge, g)^\bullet$) *with resulting* $D \in C_G(X^\wedge, g^\bullet)$($\in C_G(X^\wedge, g)^\bullet$), *and similarly for* $\mathbb{B}$.

Proof. Consider, for instance, the case $X^\wedge$. The multiplication by k^β induces isomorphisms $\mathcal{K}^{s,\gamma}(X^\wedge) \to \mathcal{K}^{s,\gamma+\beta}(X^\wedge)$ for every $\beta, \gamma \in \mathbb{R}$ and $s \in \mathbb{R}$, as well as isomorphisms between corresponding subspaces with asymptotics, where the asymptotic types are shifted in the complex plane by $-\beta$. In view of Remark 2.3.67 we may assume without loss of generality that $\gamma = 0$. Let us first observe that kernel and cokernel of $1 + G : \mathcal{K}^{s,0}(X^\wedge) \to \mathcal{K}^{s,0}(X^\wedge)$ are independent of s. In fact $(1 + G)u = 0$ for $u \in \mathcal{K}^{s,0}(X^\wedge)$ implies $u = -Gu$ which belongs by (2.3.49) to $\mathcal{S}_P^0(X^\wedge)$ for some asymptotic type P, and we have $\mathcal{S}_P^0(X^\wedge) \subset \mathcal{K}^{\infty,0}(X^\wedge)$. In an analogous manner we can argue for the cokernel as the kernel of the formal adjoint. In other words the invertibility of $1 + G : \mathcal{K}^{s,0}(X^\wedge) \to \mathcal{K}^{s,0}(X^\wedge)$ is independent of s. Let us set $H = \mathcal{K}^{s,0}(X^\wedge)$. By assumption there is a $D \in \mathcal{L}(H)$ such that $(1+G)^{-1} = 1 + D$. This yields $(1+G)(1+D) = 1$, i.e., $D = -G - GD$. Thus, the continuity of $G : H \to \mathcal{S}_P^0(X^\wedge)$ implies the continuity of $D : H \to \mathcal{S}_P^0(X^\wedge)$. For the formal adjoints we can argue in an analogous manner. $\qquad\square$

Proposition 2.3.69 *Let* $h(t, t', z) \in C^\infty(\overline{\mathbb{R}}_+ \times \overline{\mathbb{R}}_+, N_R^\mu(X))$, $\mu \in \mathbb{R}$, $R \in \mathrm{As}(X)$, *and let* $\beta, \gamma \in \mathbb{R}$ *be given such that the carrier* R *does not intersect* $\Gamma_{\frac{n+1}{2}-(\gamma+\beta)}$ *and* $\Gamma_{\frac{n+1}{2}-\gamma}$. *Then, for arbitrary cut-off functions* $\omega(t), \tilde{\omega}(t)$ *the operator*

$$G := \omega \operatorname{op}_M^{\gamma - \frac{n}{2}}(h) t^\beta \tilde{\omega} - \omega t^\beta \operatorname{op}_M^{\gamma - \frac{n}{2}}(T^{-\beta} h) \tilde{\omega}$$

belongs to $C_G(X^\wedge, g^\bullet)$ for $g = (\gamma, \gamma, (-\infty, 0])$ when $\beta \geq 0$, $g = (\gamma - \beta, \gamma + \beta, (-\infty, 0])$ when $\beta \leq 0$. Analogous relations hold for $R \in \mathbf{As}^\bullet(X)$ ($\in \mathbf{As}(X)^\bullet$) where then $G \in C_G(X^\wedge, g^\bullet)$ ($\in C_G(X^\wedge, g)^\bullet$).

Proof. Let first h be independent of t, t', i.e., $h(z) \in N_R^\mu(X)$. We have $N_R^\mu(X) = N_O^\mu(X) + M_R^{-\infty}(X)$. For $h(z) \in N_O^\mu(X)$ we easily obtain $G \equiv 0$, by applying Cauchy's theorem. So we may assume $h(z) \in M_R^{-\infty}(X)$. Let us consider $\beta \geq 0$; the case $\beta < 0$ is completely analogous and left to the reader. Let us set $v = \tilde{\omega} u$ for $u \in \mathcal{K}^{s,\gamma}(X^\wedge)$. Then

$$\omega t^\beta \operatorname{op}_M^{\gamma - \frac{n}{2}} (T^{-\beta} h) v(t) = \frac{1}{2\pi i} \int_{\Gamma_{\frac{n+1}{2} - \gamma}} t^{-z+\beta} h(z - \beta)(M_{\gamma - \frac{n}{2}} v)(z) \, dz$$

$$= \frac{1}{2\pi i} \omega \int_{\Gamma_{\frac{n+1}{2} - (\gamma + \beta)}} t^{-w} h(w)(M_{\gamma - \frac{n}{2}} t^\beta v)(w) \, dw$$

$$= \frac{1}{2\pi i} \omega \int_{\Gamma_{\frac{n+1}{2} - \gamma}} t^{-z} h(z)(M_{\gamma - \frac{n}{2}} t^\beta v)(z) \, dz$$

$$+ \frac{1}{2\pi i} \int_{\Delta_{\beta\gamma}} t^{-z} h(z)(M_{\gamma - \frac{n}{2}} t^\beta v)(z) \, dz, \quad (2.3.56)$$

where $\Delta_{\beta\gamma}$ equals $\Gamma_{\frac{n+1}{2} - \gamma} \cup \Gamma_{\frac{n+1}{2} - (\gamma + \beta)}$ in the orientation that z runs on $\Gamma_{\frac{n+1}{2} - \gamma}$ from $\operatorname{Im} z = +\infty$ to $\operatorname{Im} z = -\infty$ and on $\Gamma_{\frac{n+1}{2} - (\gamma + \beta)}$ from $\operatorname{Im} z = -\infty$ to $\operatorname{Im} z = +\infty$. Applying Cauchy's theorem we may replace $\Delta_{\beta\gamma}$ by a smooth finite curve in the strip $S_{\beta\gamma} = \{z : \frac{n+1}{2} - (\gamma + \beta) < \operatorname{Re} z < \frac{n+1}{2} - \gamma\}$ surrounding $K := S_{\beta\gamma} \cap$ carrier R in the corresponding orientation. The function $f(z) = h(z)(Mt^\beta v)(z)|_{S_{\beta\gamma}}$ belongs to $\mathcal{A} = (\{z : \operatorname{Re} z > \frac{n+1}{2} - (\gamma + \beta)\}, C^\infty(X))$. Thus, according to (2.3.38), the second summand on the right of (2.3.56) has the form $(2\pi i)^{-1} \omega(t) \langle \zeta, t^{-z} \rangle =: g$ for some $\zeta \in \mathcal{A}'(K, C^\infty(X))$. Hence, because of $K \subset \{z : \operatorname{Re} z < \frac{n+1}{2} - \gamma\}$, g is an element in $\mathcal{S}_P^\gamma(X^\wedge)$ for a corresponding asymptotic type $P \in \mathrm{As}(X, g)$, $g = (\gamma, (-\infty, 0])$. The continuity $u \to g$ as a map $G : \mathcal{K}^{s,\gamma}(X^\wedge) \to \mathcal{S}_P^\gamma(X^\wedge)$ is evident. For the formal adjoints we can argue in an analogous manner. Finally it is obvious that $R \in \mathbf{As}^\bullet(X)$ ($\in \mathbf{As}(X)^\bullet$) implies $P \in \mathrm{As}(X, g^\bullet)$ ($\in \mathrm{As}(X, g)^\bullet$). The (t, t')-dependent case can be treated by a slight extension of the above method. First we may again insert $\mu = -\infty$, since $h(t, t', z)$ can be written $h(t, t', z) = h_0(t, t', z) + h_1(t, t', z)$ for $h_0(t, t', z) \in C^\infty(\overline{\mathbb{R}}_+ \times \overline{\mathbb{R}}_+, N_O^\mu(X))$, $h_1(t, t', z) \in C^\infty(\overline{\mathbb{R}}_+ \times \overline{\mathbb{R}}_+, M_R^{-\infty}(X))$. The contribution from $h_0(t, t', z)$ vanishes, and it remains $h_1(t, t', z)$. Applying the Taylor expansion in t' near $t' = 0$ we get a finite sum of Mellin symbols of the form $f_j(t,z)(t')^j$ for $f_j(t,z) \in C^\infty(\overline{\mathbb{R}}_+, M_R^{-\infty}(X))$ and a remainder $f_{(N)}(t,t',z)(t')^N$ for $f_{(N)}(t,t',z) \in C^\infty(\overline{\mathbb{R}}_+ \times \overline{\mathbb{R}}_+, M_R^{-\infty}(X))$. The contributions from the summands $f_j(t,z)(t')^j$ can be treated as above; the additional t-dependence here can also be understood in terms of a Taylor expansion near $t = 0$, remainders with large powers of t also give rise to Green operators. So it remains to deal with $f_{(N)}(t,t',z)(t')^N$. This can be treated by a tensor product argument with respect to $f_{(N)}(t,t',z) \in C^\infty(\overline{\mathbb{R}}_{+,t}) \otimes_\pi C^\infty(\overline{\mathbb{R}}_{+,t'}, M_R^{-\infty}(X))$, combined with Taylor expansions in t. $\qquad\square$

More generally we easily obtain the following:

Remark 2.3.70 *Let $h(t,t',z) \in C^\infty(\overline{\mathbb{R}}_+ \times \overline{\mathbb{R}}_+, N_R^\mu(X))$ and assume that $\{\Gamma_{\frac{n+1}{2}-\gamma} \cup \Gamma_{\frac{n+1}{2}-\tilde\gamma}\} \cap \operatorname{carrier} R = \emptyset$ for certain fixed $\gamma, \tilde\gamma \in \mathbb{R}$. Then for arbitrary fixed cut-off functions $\omega, \tilde\omega$ we have*

$$G = \omega \operatorname{op}_M^{\gamma-\frac{n}{2}}(h)\tilde\omega - \omega \operatorname{op}_M^{\tilde\gamma-\frac{n}{2}}(h)\tilde\omega \in C_G(X^\wedge, \boldsymbol{g})$$

for $\boldsymbol{g} = (\max(\gamma,\tilde\gamma), \min(\gamma,\tilde\gamma), (-\infty, 0])$. In particular, $h(t,t',z) \in C^\infty(\overline{\mathbb{R}}_+ \times \overline{\mathbb{R}}_+, N_O^\mu(X))$ implies $G = 0$. Analogous results hold for $R \in \mathbf{As}^\bullet(X)$ $(\in \mathbf{As}(X)^\bullet)$, where $G \in C_G(X^\wedge, \boldsymbol{g}^\bullet)$ $(\in C_G(X^\wedge, \boldsymbol{g})^\bullet)$. In particular, if $\{\Gamma_{\frac{n+1}{2}-\gamma} \cup \Gamma_{\frac{n+1}{2}-(\gamma+\beta)}\} \cap \operatorname{carrier} R = \emptyset$ for given $\gamma, \beta \in \mathbb{R}$ we obtain that

$$\omega t^{-\beta} \operatorname{op}_M^{\gamma-\frac{n}{2}}(h)\tilde\omega - \omega \operatorname{op}_M^\gamma(T^{-\beta}h)t^{-\beta}\tilde\omega$$

is of Green type with the weight data $\boldsymbol{g} = (\max(\gamma, \gamma+\beta), \min(\gamma, \gamma+\beta) - \beta, (-\infty, 0])$.

Remark 2.3.71 *Let $h(t,t',z) \in C^\infty(\overline{\mathbb{R}}_+ \times \overline{\mathbb{R}}_+, N_R^\mu(X))$, $\mu \in \mathbb{R}$, $R \in \mathbf{As}(X)$, and let $\Gamma_{\frac{n+1}{2}-\gamma} \cap \operatorname{carrier} R = \emptyset$ for some given $\gamma \in \mathbb{R}$. Then, if $\omega(t)$ is a cut-off function and $\varphi(t) \in C_0^\infty(\mathbb{R}_+)$ with $\operatorname{supp} \omega \cap \operatorname{supp} \varphi = \emptyset$ we have*

$$\omega \operatorname{op}_M^{\gamma-\frac{n}{2}}(h)\varphi, \quad \varphi \operatorname{op}_M^{\gamma-\frac{n}{2}}(h)\omega \in C_G(X^\wedge, \boldsymbol{g})$$

for $\boldsymbol{g} = (\gamma, \gamma, (-\infty, 0])$. In particular, for $R \in \mathbf{As}^\bullet(X)$ $(\in \mathbf{As}(X)^\bullet)$ the corresponding operators belong to $C_G(X^\wedge, \boldsymbol{g}^\bullet)$ $(C_G(X^\wedge, \boldsymbol{g})^\bullet)$.

Lemma 2.3.72 *Let $h(z) \in M_R^{-\infty}(X)$, $R \in \mathbf{As}(X)$, such that $\Gamma_{\frac{n+1}{2}-\gamma} \cap \operatorname{carrier} R = \emptyset$ for a $\gamma \in \mathbb{R}$. Then, if $\omega_i(t)$ for $i = 1, \ldots, 4$ are arbitrary cut-off functions, we have*

$$G = \omega_1 \operatorname{op}_M^{\gamma-\frac{n}{2}}(h)\omega_2 - \omega_3 \operatorname{op}_M^{\gamma-\frac{n}{2}}(h)\omega_4 \in C_G(X^\wedge, \boldsymbol{g})$$

for $\boldsymbol{g} = (\gamma, \gamma, (-\infty, 0])$. For $R \in \mathbf{As}^\bullet(X)$ $(\in \mathbf{As}(X)^\bullet)$ we obtain $G \in C_G(X^\wedge, \boldsymbol{g}^\bullet)$ $(\in C_G(X^\wedge, \boldsymbol{g})^\bullet)$.

Proof. Setting $H = \operatorname{op}_M^{\gamma-\frac{n}{2}}(h)$ we obtain $\omega_1 H\omega_2 - \omega_3 H\omega_4 = \omega_3 H(\omega_2 - \omega_4) + (\omega_1 - \omega_3)H\omega_2$. Then it suffices to apply Remark 2.3.71. $\qquad\square$

Lemma 2.3.73 *Let* $f(z) \in M_Q^{-\infty}(X)$, $h(z) \in M_R^{-\infty}(X)$, $Q, R \in$ $\mathbf{As}(X)$, *and assume* $\Gamma_{\frac{n+1}{2}-\gamma} \cap \operatorname{carrier} Q = \Gamma_{\frac{n+1}{2}-\gamma} \cap \operatorname{carrier} R = \emptyset$ *for a* $\gamma \in \mathbb{R}$. *Then, if* $\omega(t), \omega_1(t), \omega_2(t)$ *are arbitrary cut-off functions, we have*

$$G = \omega_1 \operatorname{op}_M^{\gamma-\frac{n}{2}}(f)(1-\omega)\operatorname{op}_M^{\gamma-\frac{n}{2}}(h)\omega_2 \in C_G(X^\wedge, \boldsymbol{g}) \qquad (2.3.57)$$

for $\boldsymbol{g} = (\gamma, \gamma, (-\infty, 0])$. *If more generally* $f(z) \in M_Q^\nu(X)$, $h(z) \in$ $M_R^\mu(X)$ *for arbitrary* $\nu, \mu \in \mathbb{R}$, *then* (2.3.57) *holds under the condition* $\operatorname{supp}\omega_1 \cap \operatorname{supp}(1-\omega) = \emptyset$ *or* $\operatorname{supp}(1-\omega) \cap \operatorname{supp}\omega_2 = \emptyset$. *For* $Q, R \in$ $\mathbf{As}^\bullet(X)$ $(\in \mathbf{As}(X)^\bullet)$ *it follows that* $G \in C_G(X^\wedge, \boldsymbol{g}^\bullet)$ $(\in C_G(X^\wedge, \boldsymbol{g})^\bullet)$.

Proof. Set $H = \operatorname{op}_M^{\gamma-\frac{n}{2}}(h)$, $F = \operatorname{op}_M^{\gamma-\frac{n}{2}}(f)$. The operator

$$(1-\omega)H\omega_2 : \mathcal{K}^{s,\gamma}(X^\wedge) \to \omega\mathcal{K}^{\infty,\gamma+N}(X^\wedge) + (1-\omega)\mathcal{H}^{\infty,\gamma}(X^\wedge)$$

is continuous for every $N \in \mathbb{N}$. Thus, in view of Theorem 2.3.55, to every $\Theta_k = (-k, 0]$, $k \in \mathbb{N} \setminus \{0\}$, there is a $P_k \in \mathbf{As}(X, (\gamma, \Theta_k))$ such that

$$\omega_1 F : \omega\mathcal{K}^{\infty,\gamma+N}(X^\wedge) + (1-\omega)\mathcal{H}^{\infty,\gamma}(X^\wedge) \to \mathcal{S}_{P_k}^\gamma(X^\wedge)$$

is continuous, as soon as $N = N(k)$ is taken sufficiently large. It follows that $G : \mathcal{K}^{s,\gamma}(X^\wedge) \to \mathcal{S}_{P_k}^\gamma(X^\wedge)$ is continuous. This holds for every k. It is clear that there is a $P \in \mathbf{As}(X, (\gamma, (-\infty, 0]))$ that restricts to P_k for every k. Then $G : \mathcal{K}^{s,\gamma}(X^\wedge) \to \mathcal{S}_P^\gamma(X^\wedge)$ is also continuous. For the formal adjoint of G we can do the same, where we use Theorem 2.3.57.

In the case $f(z) \in M_Q^\nu(X)$, $h(z) \in M_R^\mu(X)$ and $(1-\omega)\omega_2 = 0$ we first note that the operator

$$(1-\omega)H\omega_2 : \mathcal{K}^{s,\gamma}(X^\wedge) \to \omega\mathcal{K}^{\infty,\gamma+N}(X^\wedge) + (1-\omega)\mathcal{H}^{\infty,\gamma}(X^\wedge)$$

is continuous for every N. Then, similarly to before, $G : \mathcal{K}^{s,\gamma}(X^\wedge) \to$ $\mathcal{S}_P^\gamma(X^\wedge)$ is continuous. On the other hand, for $\omega_1(1-\omega) = 0$ we choose a cut-off function $\tilde{\omega}(t)$ with $(1-\omega)(1-\tilde{\omega}) = (1-\omega)$. Then the operator

$$(1-\tilde{\omega})H\omega_2 : \mathcal{K}^{s,\gamma}(X^\wedge) \to \omega\mathcal{K}^{s-\mu,\gamma+N}(X^\wedge) + (1-\omega)\mathcal{H}^{s-\mu,\gamma}(X^\wedge)$$

is continuous for every $N \in \mathbb{N}$, and also

$$\omega_1 F(1-\omega) : \omega\mathcal{K}^{s-\mu,\gamma+N}(X^\wedge) + (1-\omega)\mathcal{H}^{s-\mu,\gamma}(X^\wedge) \to \mathcal{S}_{P_k}^\gamma(X^\wedge)$$

is continuous for every $k \in \mathbb{N} \setminus \{0\}$ for some $P_k \in \mathrm{As}(X,(\gamma,(-k,0]))$ sufficiently large for $N = N(k)$. Then, as above,

$$G = \omega_1 F(1-\omega)H\omega_2 : \mathcal{K}^{s,\gamma}(X^\wedge) \to \mathcal{S}_P^\gamma(X^\wedge)$$

is continuous for a corresponding $P \in \mathrm{As}(X,(\gamma,(-\infty,0]))$. For the formal adjoint G^* we can argue in an analogous manner. $\qquad\square$

Lemma 2.3.74 *Let* $\boldsymbol{g} = (\gamma,\delta,\Theta)$, $\gamma,\delta,\nu \in \mathbb{R}$, $\delta + \nu \le \gamma$, $\Theta = (-k,0]$, *and let* $M = \omega(t)t^{-\nu+j}\,\mathrm{op}_M^{\rho-\frac{n}{2}}(h)\omega_0(t)$ *for an* $h(z) \in M_R^{-\infty}(X)$, $R \in$ $\mathrm{As}(X)$, *with* $\Gamma_{\frac{n+1}{2}-\rho} \cap \mathrm{carrier}\,R = \emptyset$ *and* $\delta + \nu - j \le \rho \le \gamma$. *Then* $j \ge \delta-\gamma+\nu+k$ *implies* $M \in C_G(X^\wedge,\boldsymbol{g})$. *In particular, for* $R \in \mathrm{As}^\bullet(X)$ ($\in \mathrm{As}(X)^\bullet$) *it follows that* $M \in C_G(X^\wedge,\boldsymbol{g}^\bullet)$ ($\in C_G(X^\wedge,\boldsymbol{g})^\bullet$).

Proof. First note that for arbitrary $\varepsilon_l > 0$, $l = 1,2$, with $\varepsilon_1 \ne \varepsilon_2$ and $\delta + \nu - j \le \gamma - \varepsilon_l =: \rho_l$ the Mellin symbol $h(z)$ can be written $h(z) = h_1(z) + h_2(z)$ for certain $h_l(z) \in M_{R_l}^{-\infty}(X)$, $R_l \in \mathrm{As}(X)$, satisfying $\{\Gamma_{\frac{n+1}{2}-\rho} \cup \Gamma_{\frac{n+1}{2}-\rho_l}\} \cap \mathrm{carrier}\,R = \emptyset$, $l = 1,2$, cf. Remark 2.3.50. Then applying Proposition 2.3.69, we can write $M = M_1 + M_2 + G$ for

$$M_l = \omega(t)t^{-\nu+j}\,\mathrm{op}_M^{\rho-\frac{n}{2}}(h_l)\omega_0(t),$$

$l = 1,2$, and some $G \in C_G(X^\wedge,\boldsymbol{g})$. In view of the particular choice of ρ_1,ρ_2 the operator

$$M_1 + M_2 : \mathcal{K}^{s,\gamma}(X^\wedge) \to \mathcal{S}^{\delta+k-\max\{\varepsilon_1,\varepsilon_2\}}(X^\wedge)$$

is continuous for every $s \in \mathbb{R}$. Here we use that the resulting weight under M_l is $\delta_l := -\nu + j + \gamma - \varepsilon_l$, and $j \ge \delta - \gamma + \nu + k$ implies $\delta_l \ge -\nu + \delta - \gamma + \nu + k + \gamma - \varepsilon_l = \delta + k - \varepsilon_l$. Since $\varepsilon_1,\varepsilon_2 > 0$ can be taken $< \varepsilon$ for an arbitrary fixed $\varepsilon > 0$, we obtain from the latter mapping property that in fact M satisfies the first condition (2.3.49) in Definition 2.3.60. The formal adjoint can be treated in an analogous manner. This yields $M \in C_G(X^\wedge,\boldsymbol{g})$. Analogous conclusions hold for the other asymptotic types. $\qquad\square$

Definition 2.3.75 *Fix weight data* $\boldsymbol{g} = (\gamma,\delta,\Theta)$ *for* $\gamma,\delta \in \mathbb{R}$, $\Theta = (-k,0]$, $k \in \mathbb{N} \setminus \{0\}$. *Then* $C_{M+G}(X^\wedge,\boldsymbol{g}^\bullet)$ ($C_{M+G}(X^\wedge,\boldsymbol{g})^\bullet$) *will denote the space of all operators of the form* $M + G$ *for arbitrary* $G \in C_G(X^\wedge,\boldsymbol{g}^\bullet)$ ($\in C_G(X^\wedge,\boldsymbol{g})^\bullet$) *and*

$$M = \omega(t)t^{\delta-\gamma}\sum_{j=0}^{k-1} t^j\,\mathrm{op}_M^{\rho_j-\frac{n}{2}}(h_j)\omega_0(t) \qquad (2.3.58)$$

for arbitrary cut-off functions $\omega(t)$, $\omega_0(t)$ *and for symbols* $h_j(z) \in M_{R_j}^{-\infty}(X)$ *for certain* $R_j \in \mathbf{As}^\bullet(X)$ $(\in \mathbf{As}(X)^\bullet)$ *and* $\rho_j \in \mathbb{R}$ *satisfying* $\pi_{\mathbb{C}} R_j \cap \Gamma_{\frac{n+1}{2}-\rho_j} = \emptyset$ *and* $\gamma - j \le \rho_j \le \gamma$ *for all* $j = 0, \dots, k-1$.

Moreover, $C_{M+G}(X^\wedge, \boldsymbol{g})$ *denotes the space of all* $M + N + G$ *for arbitrary* $G \in C_G(X^\wedge, \boldsymbol{g})$ *and operators* M, N, *where* M *is of the form* (2.3.58) *and analogously*

$$N = \tilde{\omega}(t) t^{\delta-\gamma} \sum_{j=0}^{k-1} t^j \operatorname{op}_M^{\delta_j - \frac{n}{2}}(f_j) \tilde{\omega}_0(t)$$

for arbitrary cut-off functions $\omega(t)$, $\omega_0(t)$, $\tilde{\omega}(t)$, $\tilde{\omega}_0(t)$, *and* $h_j(z) \in M_{R_j}^{-\infty}(X)$, $f_j(z) \in M_{Q_j}^{-\infty}(X)$, $R_j, Q_j \in \mathbf{As}(X)$, *and weights* ρ_j, δ_j *satisfying* carrier $R_j \cap \Gamma_{\frac{n+1}{2}-\rho_j} = \emptyset$, carrier $Q_j \cap \Gamma_{\frac{n+1}{2}-\delta_j} = \emptyset$ *and* $\gamma - j \le \rho_j$, $\delta_j \le \gamma$ *for* $j = 0, \dots, k-1$.

If we set for a moment $\boldsymbol{g}_k = (\gamma, \delta, (-k, 0])$, $k \in \mathbb{N} \setminus \{0\}$, then we have obviously

$$C_{M+G}(X^\wedge, \boldsymbol{g}_{k+1}) \subset C_{M+G}(X^\wedge, \boldsymbol{g}_k) \qquad (2.3.59)$$

for all k, and we then define

$$C_{M+G}(X^\wedge, \boldsymbol{g}) = \bigcap_k C_{M+G}(X^\wedge, \boldsymbol{g}_k) \qquad (2.3.60)$$

for the case $\boldsymbol{g} = (\gamma, \delta, (-\infty, 0])$. For the dotted subclasses we define $C_{M+G}(X^\wedge, \boldsymbol{g}^\bullet)$ for $\Theta = (-\infty, 0]$ analogously to (2.3.60).

Definition 2.3.76 *We denote by* $C_{M+G}(\mathbb{B}, \boldsymbol{g})$ *for* $\boldsymbol{g} = (\gamma, \delta, \Theta)$, $\Theta = (-k, 0]$, $k \in (\mathbb{N} \setminus \{0\}) \cup \{\infty\}$, *the space of all operators* $A = \omega A_0 \tilde{\omega} + G$ *for arbitrary* $G \in C_G(\mathbb{B}, \boldsymbol{g})$, $A_0 \in C_{M+G}(X^\wedge, \boldsymbol{g})$, *and cut-off functions* $\omega, \tilde{\omega}$ *supported by a collar neighbourhood of* $\partial \mathbb{B}$ *that is identified with* $X \times [0, c)$ *for some* $c > 0$. *In an analogous manner we define* $C_{M+G}(\mathbb{B}, \boldsymbol{g}^\bullet)$ $(C_{M+G}(\mathbb{B}, \boldsymbol{g})^\bullet)$ *by assuming* $A_0 \in C_{M+G}(X^\wedge, \boldsymbol{g}^\bullet)$ $(\in C_{M+G}(X^\wedge, \boldsymbol{g})^\bullet)$.

Note that

$$C_{M+G}(X^\wedge, \boldsymbol{g}^\bullet) \subset C_{M+G}(X^\wedge, \boldsymbol{g})^\bullet \subset C_{M+G}(X^\wedge, \boldsymbol{g}),$$

and the same for $\mathbb{B}$. For $A \in C_{M+G}(X^\wedge, \boldsymbol{g})$ $(\in C_{M+G}(\mathbb{B}, \boldsymbol{g}))$ we set

$$\sigma_M^{\gamma-\delta-j}(A)(z) = h_j(z) + f_j(z) \qquad (2.3.61)$$

with the notation of Definition 2.3.75 and Definition 2.3.76, respectively. We call $\sigma_M^{\gamma-\delta-j}(A)(z)$ the conormal symbol of A to the conormal order $\gamma - \delta - j$, $j \in \mathbb{N}$, $j < k$.

In the sequel we concentrate on the C_{M+G}-classes on $X^\wedge$. The notions and results for $\mathbb{B}$ are analogous.

Proposition 2.3.77 $A \in C_{M+G}(X^\wedge, \boldsymbol{g})$ for $\boldsymbol{g} = (\gamma, \delta, \Theta)$ with $\Theta = (-k, 0]$, and $\sigma_M^{\gamma-\delta-j}(A) = 0$ for $j \in \mathbb{N}$, $j < k$, implies that $A \in C_G(X^\wedge, \boldsymbol{g})$. Analogous relations hold for $A \in C_{M+G}(X^\wedge, \boldsymbol{g}^\bullet)$ and $\in C_{M+G}(X^\wedge, \boldsymbol{g})^\bullet$, respectively.

Proof. Proposition 2.3.69 shows that the particular choice of the weights ρ_j, δ_j only changes A by a Green operator. Lemma 2.3.72 yields that also the specific cut-off functions are unimportant modulo Green operators. Finally, if $\sigma_M^{\gamma-\delta-j}(M) = 0$ for $j \in \mathbb{N}$, $j < k$, it follows from Lemma 2.3.74 that A is a Green operator. In fact, if $\sigma_M^{\gamma-\delta-j}(M)$ only consists of one summand, this is clear. If h_j and f_j in (2.3.61) are non-trivial, $h_j + f_j = 0$ implies that the weights ρ_j and δ_j can be taken to express the Mellin operators by h_j as well as by f_j, modulo Green operators. This reduces $\sigma_M^{\gamma-\delta-j}(A)$ to one summand and we can argue as before.   □

Proposition 2.3.78 $A \in C_{M+G}(X^\wedge, \boldsymbol{g})$ for $\boldsymbol{g} = (\gamma, \delta, \Theta)$ implies $t^\alpha A t^{-\beta} \in C_{M+G}(X^\wedge, \boldsymbol{c})$ for $\boldsymbol{c} = (\gamma+\beta, \delta+\alpha, \Theta)$, for arbitrary $\alpha, \beta \in \mathbb{R}$, where

$$\sigma_M^{\gamma-\delta-j-(\alpha-\beta)}(t^\alpha A t^{-\beta})(z) = \sigma_M^{\gamma-\delta-j}(A)(z+\beta)$$

for all j. Moreover, if $\Theta = (-k, 0]$ is finite and $\alpha \geq 0$, $\beta \leq 0$, then $t^\alpha A t^{-\beta} \in C_G(X^\wedge, \boldsymbol{g})$ for $\alpha - \beta > k$.

Proof. The first statement is evident. The second one is a consequence of Lemma 2.3.74.   □

Proposition 2.3.79 Let $A \in C_{M+G}(X^\wedge, \boldsymbol{g})$ for $\boldsymbol{g} = (\gamma, \delta, \Theta)$, $\Theta = (-k, 0]$. Then the conormal symbols $\sigma_M^{\gamma-\delta-j}(A) = 0$ for $j \in \mathbb{N}$, $j < k$, are uniquely determined by A. In other words, $A \in C_{M+G}(X^\wedge, \boldsymbol{g}) \cap C_G(X^\wedge, \boldsymbol{g})$ implies $\sigma_M^{\gamma-\delta-j}(A) = 0$ for $j \in \mathbb{N}$, $j < k$.

Proof. Let us fix a cut-off function $\omega_1(t)$ and set $u_w(t) = \omega_1(t) t^{-w}$, $w \in \mathbb{C}$. Then for $\operatorname{Re} w < \frac{n+1}{2} - \gamma$ we have $u_w(t) \in \mathcal{K}^{\infty,\gamma}(X^\wedge)$ (even $u_w \in \mathcal{A}(\{w : \operatorname{Re} w < \frac{n+1}{2} - \gamma\}, \mathcal{K}^{\infty,\gamma}(X^\wedge))$, and we can form $A u_w$. Set $\nu = \gamma - \delta$. By assumption the operator A has the form $A = t^{-\nu} \sum_{j=0}^{l}(M_j + N_j) + G$ for the maximal $l \in \mathbb{N}$ with $l < k$, where G is a Green operator, and

$$M_j = t^j \omega(t) \operatorname{op}_M^{\rho_j - \frac{n}{2}}(h_j)\omega_0(t), \qquad N_j = t^j \tilde\omega(t) \operatorname{op}_M^{\delta_j - \frac{n}{2}}(f_j)\tilde\omega_0(t),$$

cf. the notation in Definition 2.3.75. The function $G u_w$ has some asymptotic type P. Thus $(M_{\delta-\frac{n}{2}} G u_w)(z) \in \mathcal{A}(\mathbb{C}_z \setminus \text{carrier } P, C^\infty(X))$, and $M_{\delta-\frac{n}{2}}(G u_w)$ is also holomorphic in $w \in \mathbb{C}$. Assume without loss of generality that $\omega_1(t)$ satisfies $\omega_0\omega_1 = \omega_1$, $\tilde{\omega}_0\omega_1 = \omega_1$. Applying the weighted Mellin transform to $t^{-\nu}\sum_{j=0}^{l}(M_j + N_j)u_w$ we obtain

$$\sum_{j=0}^{l}\{h_j(z+j-\nu) + f_j(z+j-\nu)\}M_{\gamma-\frac{n}{2}}(u_w)(z+j-\nu) \quad (2.3.62)$$

modulo a holomorphic $C^\infty(X)$-valued function in $\operatorname{Re} z < \frac{n+1}{2} - \delta$. Now $M_{\gamma-\frac{n}{2}}(u_w)(z)$ equals $(z-w)^{-1}$ modulo a holomorphic function in $z, w \in \mathbb{C}$. Thus, for $z+j-\nu$ outside the carriers of asymptotics of $(T^{j-\nu}h_j)(t)$ and $(T^{j-\nu}f_j)(z)$, the expansion (2.3.62) equals $\sum_{j=0}^{l}m_j(z,w)$, where

$$m_j(z,w) = \{h_j(z+j-\nu) + f_j(z+j-\nu)\}(z+j-\nu-w)^{-1},$$

modulo a holomorphic $C^\infty(X)$-valued function in that region. Choosing $w_0 \in \mathbb{C}$, $j_0 \in \mathbb{N}$, with $|\operatorname{Re} w_0 - (j_0 - \nu)| < \varepsilon$ for $\varepsilon > 0$ sufficiently small, $|\operatorname{Im} w_0|$ sufficiently large, and integrating $\sum_{j=0}^{l}m_j(z,w_0)$ over a sufficiently small circle around w_0, Cauchy's integral formula yields $h_{j_0}(w_0+j-\nu) + f_{j_0}(w_0+j-\nu)$, up to a constant, since the summands for $j \neq j_0$ are holomorphic in a neighbourhood of the circle. In this way we have recovered $\sigma_M^{\nu-j_0}(A)$ in a small open set of the complex plane outside the carrier of asymptotics. This determines $\sigma_M^{\nu-j_0}(A)$ in a unique way. □

Remark 2.3.80 *The class $C_{M+G}^\nu(X^\wedge, \boldsymbol{g})$ can equivalently be defined as the set of all operators $\sum_{l=0}^{N}M_l + G$ for arbitrary $N \geq 1$, $G \in C_G(X^\wedge, \boldsymbol{g})$ and M_l being of the form (2.3.58) for every l with l-dependent ρ_j, h_j, satisfying the above conditions. In view of Proposition 2.3.69 and Remark 2.3.50 we can choose the involved weights ρ_j in such a way that $0 < \gamma - \rho_j < \varepsilon$ for some arbitrary $\varepsilon > 0$, independent of j for $j \geq 1$, only dependent on l.*

Proposition 2.3.81 *Let $f(t,t',z) \in C^\infty(\overline{\mathbb{R}}_+ \times \overline{\mathbb{R}}_+, M_R^{-\infty}(X))$ for some $R \in \mathbf{As}(X)$, and let $\gamma \in \mathbb{R}$ with carrier $R \cap \Gamma_{\frac{n+1}{2}-\gamma} = \emptyset$. Then*

$$\omega \operatorname{op}_M^{\gamma-\frac{n}{2}}(f)\tilde{\omega} \in C_{M+G}(X^\wedge, \boldsymbol{g}) \qquad \text{for } \boldsymbol{g} = (\gamma, \gamma, (-\infty, 0]),$$

for arbitrary cut-off functions $\omega(t)$, $\tilde{\omega}(t)$. Analogous relations hold for $R \in \mathbf{As}^\bullet(X)$ $(\in \mathbf{As}(X)^\bullet)$. Then the resulting operator belongs to $C_{M+G}(X^\wedge, \boldsymbol{g}^\bullet)$ $(C_{M+G}(X^\wedge, \boldsymbol{g})^\bullet)$.

Proof. Applying the Taylor expansion in t for $t \to 0$ it follows that

$$f(t,t',z) = \sum_{j=0}^{N}t^j f_j(t',z) + t^N f_{(N)}(t,t',z)$$

for every $N \in \mathbb{N}$, for certain $f_j(t', z) \in C^\infty(\overline{\mathbb{R}}_+, M_R^{-\infty}(X))$ and

$$f_{(N)}(t, t', z) \in C^\infty(\overline{\mathbb{R}}_+ \times \overline{\mathbb{R}}_+, M_R^{-\infty}(X)).$$

Analogously we obtain

$$f_j(t', z) = \sum_{k=0}^{N} (t')^k f_{jk}(z) + (t')^N f_{j,(N)}(t', z),$$

$$f_{(N)}(t, t', z) = \sum_{k=0}^{N} (t')^k f_{(N),k}(t, z) + (t')^N f_{(N),(N)}(t, t', z)$$

for $f_{(N),k}(t, z) \in C^\infty(\overline{\mathbb{R}}_+, M_R^{-\infty}(X))$,

$$f_{(N),(N)}(t, t', z) \in C^\infty(\overline{\mathbb{R}}_+ \times \overline{\mathbb{R}}_+, M_R^{-\infty}(X)).$$

Inserting the Taylor expansions in t' into the expansions with respect to t we obtain a decomposition of $f(t, t', z)$ into a sum with summands of the form $t^j f_{jk}(z)(t')^k$, $t^j f_{j,(N)}(t', z)(t')^N$, $t^N f_{(N),k}(t, z)(t')^k$, $t^N f_{(N),(N)}(t, t', z)(t')^N$, where N is assumed to be sufficiently large. In view of Proposition 2.3.69 the operators $\omega t^j \operatorname{op}_M^{\gamma - \frac{n}{2}}(f_{jk})(t')^k \tilde{\omega}$ belong to $C_{M+G}^0(X^\wedge, \boldsymbol{g})$. Also, to

$$\omega t^j \operatorname{op}_M^{\gamma - \frac{n}{2}}(f_{j,(N)})(t')^N \tilde{\omega},$$

$$\omega t^N \operatorname{op}_M^{\gamma - \frac{n}{2}}(f_{(N),k})(t')^k \tilde{\omega},$$

$$\omega t^N \operatorname{op}_M^{\gamma - \frac{n}{2}}(f_{(N),(N)})(t')^N \tilde{\omega}$$

we can apply Proposition 2.3.69, which allows us to commute powers of t' with sufficiently large exponents to the left. The same can be done for the formal adjoints. This shows that the operators belong to $C_G(X^\wedge, (\gamma, \gamma, (-k, 0]))$, for every k and suitable $N = N(k)$. Applying (2.3.60) we finally obtain the assertion. $\qquad \square$

Theorem 2.3.82 *Let $A \in C_{M+G}(X^\wedge, \boldsymbol{g})$ for $\boldsymbol{g} = (\gamma, \delta, \Theta)$, $\Theta = (-k, 0]$, $k \in \mathbb{N} \cup \{\infty\}$. Then*

$$A : \mathcal{K}^{s,\gamma}(X^\wedge) \to \mathcal{K}^{\infty,\delta}(X^\wedge)$$

is a continuous operator for every $s \in \mathbb{R}$ that induces continuous operators

$$A : \mathcal{K}_P^{s,\gamma}(X^\wedge) \to \mathcal{K}_Q^{\infty,\delta}(X^\wedge) \qquad (2.3.63)$$

for every asymptotic type $P \in \operatorname{As}(X, (\gamma, \Theta))$ for some resulting $Q \in \operatorname{As}(X, (\delta, \Theta))$. In particular, for $A \in C_{M+G}(X^\wedge, \boldsymbol{g}^\bullet)$ ($\in$

$C_{M+G}(X^\wedge, \boldsymbol{g})^\bullet)$ *we have continuous operators for every asymptotic type* $P \in \mathrm{As}(X, (\gamma, \Theta)^\bullet)$ $(\in \mathrm{As}(X, (\gamma, \Theta))^\bullet)$ *for a resulting* $Q \in \mathrm{As}(X, (\delta, \Theta)^\bullet)$ $(\in \mathrm{As}(X, (\delta, \Theta))^\bullet)$. *Analogous assertions hold for* $A \in C_{M+G}(\mathbb{B}, \boldsymbol{g})$ *for any compact (stretched) manifold* $\mathbb{B}$ *with conical singularities with respect to the spaces* $\mathcal{H}^{s,\gamma}(\mathbb{B})$, $\mathcal{H}^{s,\gamma}_P(\mathbb{B})$, ...

Proof. The assertion is a consequence of the Definitions 2.3.59, 2.3.60 and of Theorem 2.3.55, together with the fact that multiplication of operators by functions that are powers of t near $t = 0$ shifts the asymptotic types in the complex plane by the corresponding exponents. $\quad\square$

Theorem 2.3.83 $A \in C_{M+G}(X^\wedge, \boldsymbol{g})$ *for* $\boldsymbol{g} = (\gamma, \delta, \Theta)$ *implies* $A^* \in C_{M+G}(X^\wedge, \boldsymbol{g}^*)$ *for* $\boldsymbol{g}^* = (-\delta, -\gamma, \Theta)$, *where*

$$\sigma_M^{\gamma-\delta-j}(A^*)(t) = \sigma_M^{\gamma-\delta-j}(A)(n+1+\delta-\gamma+j-\overline{z})^{(*)},$$

$(*)$ *indicating the point-wise adjoint with respect to the* $L^2(X)$*-scalar product*, $j \in \mathbb{N}$, $j < k$, $\Theta = (-k, 0]$. *Analogous relations hold for* $A \in C_{M+G}(X^\wedge, \boldsymbol{g}^\bullet)$ $(\in C_{M+G}(X^\wedge, \boldsymbol{g})^\bullet)$ *with* $A^* \in C_{M+G}(X^\wedge, (\boldsymbol{g}^*)^\bullet)$ $(\in C_{M+G}(X^\wedge, (\boldsymbol{g}^*))^\bullet)$ *and similarly over any (stretched) manifold* $\mathbb{B}$ *with conical singularities.*

Proof. The assertion follows from Theorem 2.3.57 and from Proposition 2.3.69. $\quad\square$

Theorem 2.3.84 $A \in C_{M+G}(X^\wedge, \boldsymbol{g})$, $\boldsymbol{g} = (\gamma, \delta, \Theta)$, *and* $B \in C_{M+G}(X^\wedge, \boldsymbol{c})$, $\boldsymbol{c} = (\beta, \gamma, \Theta)$, *implies* $AB \in C_{M+G}(X^\wedge, \boldsymbol{h})$ *for* $\boldsymbol{h} = (\beta, \delta, \Theta)$, *and we have*

$$\sigma_M^{\beta-\delta-l}(AB)(z) = \sum_{p+r=l} (T^{\beta-\gamma-r}\sigma_M^{\gamma-\delta-p}(A))(z)\sigma_M^{\beta-\gamma-r}(B)(z) \quad (2.3.64)$$

for all $l \in \mathbb{N}$ *for which the conormal symbols of the factor on the right are defined. Analogous relations hold for* $A \in C_{M+G}(X^\wedge, \boldsymbol{g}^\bullet)$ $(\in C_{M+G}(X^\wedge, \boldsymbol{g})^\bullet)$, $B \in C_{M+G}(X^\wedge, \boldsymbol{c}^\bullet)$ $(\in C_{M+G}(X^\wedge, \boldsymbol{c})^\bullet)$, *with* $AB \in C_{M+G}(X^\wedge, \boldsymbol{g}^\bullet)$ $(\in C_{M+G}(X^\wedge, \boldsymbol{g})^\bullet)$ *and similarly over any (stretched) manifold* $\mathbb{B}$ *with conical singularities.*

Proof. First observe that $A \in C_G(X^\wedge, \boldsymbol{g})$ or $B \in C_G(X^\wedge, \boldsymbol{c})$ implies $AB \in C_G(X^\wedge, \boldsymbol{h})$. This is an easy consequence of Definition 2.3.60 and of the Theorems 2.3.82, 2.3.83. This reduces the assertion to compositions between operators of the form

$$A = \omega t^{-\nu+p} \mathrm{op}_M^{\rho-\frac{n}{2}}(h)\omega_0, \quad B = \tilde{\omega} t^{-\mu+r} \mathrm{op}_M^{\kappa-\frac{n}{2}}(f)\tilde{\omega}_0,$$

for $\nu := \gamma - \delta$, $\mu := \beta - \gamma$, and $\delta + \nu - p \leq \rho \leq \gamma$, $\gamma + \mu - r \leq \kappa \leq \beta$, and

$$h(z) \in M_R^{-\infty}(X), \qquad \Gamma_{\frac{n+1}{2}-\rho} \cap \text{carrier } R = \emptyset,$$

$$f(z) \in M_Q^{-\infty}(X), \qquad \Gamma_{\frac{n+1}{2}-\kappa} \cap \text{carrier } Q = \emptyset.$$

By Lemma 2.3.72 the particular choice of the cut-off functions is unimportant modulo Green remainders. Therefore we may assume $\tilde{\omega}_0 = \omega$, $\omega_0 \tilde{\omega} = \omega$. For convenience we restrict ourselves to $n = \dim X = 0$. The calculations in general are completely analogous and left to the reader. Set $\tilde{\nu} = \nu - p$, $\tilde{\mu} = \mu - r$. Then it remains to analyse

$$AB = \omega t^{-\tilde{\nu}+\rho} \operatorname{op}_M(T^{-\rho}h) \omega t^{-\rho-\tilde{\mu}+\kappa} \operatorname{op}_M(T^{-\kappa}f) t^{-\kappa} \omega. \qquad (2.3.65)$$

From the conditions on the involved weights we know that $\rho + \tilde{\mu} \leq \kappa$, i.e., $-\rho - \tilde{\mu} + \kappa \geq 0$. From (2.3.65) we will produce expressions of the form

$$\omega t^{-(\tilde{\nu}+\tilde{\mu})} \operatorname{op}_M^\lambda(d_1) \omega \operatorname{op}_M^\lambda(d_2) \omega,$$

applying Proposition 2.3.69, and then omitting the factor ω in the middle by Lemma 2.3.73. The Green remainders will lead to Green operators, again. So they may be ignored. We finally will obtain an operator $\omega t^{-(\tilde{\nu}+\tilde{\mu})} \operatorname{op}_M^\lambda(d) \omega$ for a Mellin symbol $d(z) = d_1(z)d_2(z)$ for an asymptotic type P with $\Gamma_{\frac{1}{2}-\lambda} \cap \text{carrier } P = \emptyset$, and $\delta + \tilde{\mu} + \tilde{\nu} \leq \lambda \leq \beta$. In the case $\kappa = \rho + \tilde{\mu}$ we can set $\lambda = \kappa$, $\rho = \lambda - \tilde{\mu}$, and then (2.3.65) takes the form

$$AB = \omega t^{-(\tilde{\nu}+\tilde{\mu})} \operatorname{op}_M^\lambda(T^{\tilde{\mu}}h) \omega \operatorname{op}_M^\lambda(f) \omega = \omega t^{-(\tilde{\nu}+\tilde{\mu})} \operatorname{op}_M^\lambda((T^{\tilde{\mu}}h)f) \omega$$

modulo a Green operator. This is already the assertion in the case $\kappa = \rho + \tilde{\mu}$. For $\zeta := \kappa - \rho - \tilde{\mu} > 0$ we can choose an $\varepsilon > 0$, $\varepsilon < \zeta$, with $\Gamma_{\frac{1}{2}-(\rho+\varepsilon)} \cap \text{carrier } R = \emptyset$ for the asymptotic type R of h (recall that we consider the case $\dim X = 0$). Moreover, we can decompose f into a sum $f = f_1 + f_2$, $f_i(y,z) \in M_{Q_i}^{-\infty}(X)$, $i = 1,2$, such that $\Gamma_{\frac{1}{2}+(\zeta-\kappa)} \cap \text{carrier } Q_1 = \Gamma_{\frac{1}{2}+(\zeta-\varepsilon-\kappa)} \cap \text{carrier } Q_2 = \emptyset$. Then, in view of Proposition 2.3.69 and Lemma 2.3.72, the following equations are valid modulo Green operators:

$$\begin{aligned} AB &= \omega t^{-\tilde{\nu}+\rho} \operatorname{op}_M(T^{-\rho}h) \omega t^{\zeta} \operatorname{op}_M(T^{-\kappa}f) t^{-\kappa} \omega \\ &= \omega t^{-\tilde{\nu}+\rho} \operatorname{op}_M(T^{-\rho}h) \omega \operatorname{op}_M(T^{\zeta-\kappa}f_1) t^{\zeta-\kappa} \omega \\ &\quad + \omega t^{-\tilde{\nu}+\rho} \operatorname{op}_M(T^{-\rho}h) \omega t^{\varepsilon} \operatorname{op}_M(T^{\zeta-\varepsilon-\kappa}f_2) t^{\zeta-\varepsilon-\kappa} \omega \\ &= \omega t^{-\tilde{\nu}+\rho} \operatorname{op}_M(T^{-\rho}h) \omega \operatorname{op}_M(T^{\zeta-\kappa}f_1) t^{\zeta-\kappa} \omega \\ &\quad + \omega t^{-\tilde{\nu}+\rho+\varepsilon} \operatorname{op}_M(T^{-\rho-\varepsilon}h) \omega \operatorname{op}_M(T^{\zeta-\varepsilon-\kappa}f_2) t^{\zeta-\varepsilon-\kappa} \omega \\ &= \omega t^{-\tilde{\nu}+\rho} \operatorname{op}_M((T^{-\rho}h)(T^{\zeta-\kappa}f_1)) t^{\zeta-\kappa} \omega \\ &\quad + \omega t^{-\tilde{\nu}+\rho+\varepsilon} \operatorname{op}_M((T^{-\rho-\varepsilon}h)(T^{\zeta-\varepsilon-\kappa}f_2)) t^{\zeta-\varepsilon-\kappa} \omega. \end{aligned}$$

Writing $T^{-\rho}h = T^{\zeta-\kappa}T^{-\zeta+\kappa-\rho}h = T^{-\rho-\bar{\mu}}T^{\bar{\mu}}h$ and analogously $T^{-\rho-\varepsilon}h$ $= T^{\zeta-\varepsilon-\kappa}T^{-\rho-\varepsilon-\zeta+\varepsilon+\kappa}h = T^{-\rho-\bar{\mu}-\varepsilon}T^{\bar{\mu}}h$ we obtain (modulo Green operators)

$$(mn)(y,\eta) = \omega t^{-(\bar{\nu}+\bar{\mu})}t^{\rho+\bar{\mu}} \operatorname{op}_M(T^{-\rho-\bar{\mu}}(T^{\bar{\mu}}h)f_1)t^{-\rho-\bar{\mu}}\omega$$
$$+ \omega t^{-(\bar{\nu}+\bar{\mu})}t^{\rho+\varepsilon+\bar{\mu}} \operatorname{op}_M(T^{-\rho-\varepsilon-\bar{\mu}}(T^{\bar{\mu}}h)f_2)t^{-\rho-\varepsilon-\bar{\mu}}\omega$$
$$= \omega t^{-(\bar{\nu}+\bar{\mu})}\left\{\operatorname{op}_M^{\rho+\bar{\mu}}((T^{\bar{\mu}}h)f_1) + \operatorname{op}_M^{\rho+\bar{\mu}+\varepsilon}((T^{\bar{\mu}}h)f_2)\right\}\omega.$$

This shows $AB \in C_{M+G}(X^\wedge, \boldsymbol{h})$. Moreover, we have

$$\sigma_M^{\bar{\nu}+\bar{\mu}}(AB)(z) = (T^{\bar{\mu}}h)(z)\{f_1 + f_2\}(z) = (T^{\bar{\mu}}h)(z)f(z),$$

which yields the symbol relation (2.3.64). The considerations for the other asymptotic types are easier but analogous and left to the reader. The corresponding assertions for $\mathbb{B}$ are then clear. $\qquad\square$

The formula (2.3.64) is called the Mellin translation product for the conormal symbols.

2.4   The cone algebras with asymptotics

2.4.1   Cone pseudo–differential operators with asymptotics

We now turn to spaces of pseudo–differential operators on a stretched manifold $\mathbb{B}$ associated with a compact manifold B with conical singularities in the sense of Definition 2.1.1. As usual we assume that $\mathbb{B}$ has only one conical singularity. This is practically no restriction of generality, since the base X may have several connected components. Compared with Section 2.2 the new aspect here is that the symbolic structures of the operators are enriched with asymptotic information. This causes a specific behaviour on spaces with asymptotics. Concerning the involved asymptotic types we will discuss continuous, weakly discrete and discrete ones. The scheme of considerations will be similar for all these cases; so the assertions will be obtained in unified form, only by varying the inserted asymptotic types. We will also study the case of an infinite (open stretched) cone $X^\wedge = \mathbb{R}_+ \times X \ni (t,x)$ with base X. This will employ the calculus with respect to the conical exit $t \to \infty$, that was prepared in Section 1.4.

Recall that we fixed a collar neighbourhood V of $\partial\mathbb{B} \cong X$ in $\mathbb{B}$ and a diffeomorphism $V \cong [0,1) \times X$. Notation for pull–backs (or push–forwards) under this identification is suppressed for convenience. So, in particular, a cut–off function $\omega(t)$ supported by $[0, 1 - \varepsilon)$ for some $0 < \varepsilon < 1$ is interpreted both as a function on $\mathbb{B}$ and on $X^\wedge$. Moreover, we identify $L_{\mathrm{cl}}^\mu(\operatorname{int} \mathbb{B})|_{\operatorname{int} V}$ with a subspace of $L_{\mathrm{cl}}^\mu(\operatorname{int} V)$. Let us choose cut–off functions $\omega(t)$, $\omega_0(t)$, $\omega_1(t)$ with

$$\omega\omega_0 = \omega, \quad \omega\omega_1 = \omega_1. \tag{2.4.1}$$

Definition 2.4.1 *Fix weight data* $\boldsymbol{g} = (\gamma, \delta, \Theta)$, $\Theta = (-k, 0]$, $k \in (\mathbb{N} \setminus \{0\}) \cup \{\infty\}$, $\gamma, \delta \in \mathbb{R}$, *and let* $\nu \in \mathbb{R}$. *Then* $C^\nu(\mathbb{B}, \boldsymbol{g})$ *denotes the subspace of all* $A \in L_{\mathrm{cl}}^\nu(\operatorname{int} \mathbb{B})$ *of the form*

$$A = A_0 + A_1 + M + G, \tag{2.4.2}$$

where

$$A_0 = \omega t^{-(\gamma-\delta)} \operatorname{op}_M^{\gamma-\frac{n}{2}}(h)\omega_0 \quad \text{for } h(t, z) \in C^\infty(\overline{\mathbb{R}}_+, M_O^\nu(X)), \tag{2.4.3}$$
$$A_1 = (1 - \omega)A_\psi(1 - \omega_1) \quad \text{for } A_\psi \in L_{\mathrm{cl}}^\nu(\operatorname{int} \mathbb{B}), \tag{2.4.4}$$

and $M+G \in C_{M+G}(\mathbb{B}, \boldsymbol{g})$, *cf. Definition 2.3.76. Analogously* $C^\nu(\mathbb{B}, \boldsymbol{g}^\bullet)$ *is the set of all* $A \in L_{\mathrm{cl}}^\nu(\operatorname{int} \mathbb{B})$ *of the form* (2.4.2) *with* (2.4.3), (2.4.4) *and* $M + G \in C_{M+G}(\mathbb{B}, \boldsymbol{g}^\bullet)(\in C_{M+G}(\mathbb{B}, \boldsymbol{g})^\bullet)$. *Moreover,* $C^\nu(X^\wedge, \boldsymbol{g})$ *($C^\nu(X^\wedge, \boldsymbol{g}^\bullet)$, $C^\nu(X^\wedge, \boldsymbol{g})^\bullet$) denotes the subspace of all* $A \in L_{\mathrm{cl}}^\nu(X^\wedge)$ *of the form* (2.4.2) *with* (2.4.3), *further* $A_1 = (1 - \omega)A_\psi(1 - \omega_1)$ *for* $A_\psi \in L_{\mathrm{cl}}^\nu(X^\wedge)$ *such that* $A_1 \in L_{\mathrm{cl}}^\nu(X^\wedge)_e$, *cf.* (2.2.44), *and* $M + G \in C_{M+G}(X^\wedge, \boldsymbol{g})(\in C_{M+G}(X^\wedge, \boldsymbol{g}^\bullet), \in C_{M+G}(X^\wedge, \boldsymbol{g})^\bullet)$.

The elements of the spaces $C^\nu(\mathbb{B}, \boldsymbol{g}), \ldots, C^\nu(X^\wedge, \boldsymbol{g}), \ldots$ are called cone pseudo–differential operators with asymptotics of the corresponding type. Below we will also talk about the cone operator algebras over $\mathbb{B}$ and $X^\wedge$, respectively. As in the weighted Sobolev spaces it would be more consistent to write $C^\nu(\operatorname{int} \mathbb{B}, \boldsymbol{g}), \ldots$, but we prefer the simpler notation $\mathbb{B}$ which should not lead to confusion.

Remark 2.4.2 *From Theorem 2.2.25 and Theorem 2.2.28 it follows that* $A_0 \in \tilde{L}_{\mathrm{cl}}^\nu(\overline{\mathbb{R}}_+ \times X)$, *cf.* (2.2.14). *Thus, for instance, for the case* $X^\wedge$, *we obtain* $A_0 + A_1 \in \tilde{L}_{\mathrm{cl}}^\nu(\overline{\mathbb{R}}_+ \times X)$. *Therefore without loss of generality we can assume in Definition 2.4.1 that*

$$A_M := t^{-(\gamma-\delta)} \operatorname{op}_M^{\gamma-\frac{n}{2}}(h) = A_\psi \bmod L^{-\infty}(X^\wedge)$$

(or on $\mathbb{B}$ *that* $\{A_M - A_\psi\}|_{\operatorname{int} V} \in L^{-\infty}(\operatorname{int} V)$*). Let us say in this case that* A_ψ *and* A_M *are compatible via the Mellin operator convention. If this is satisfied and if we choose other cut–off functions* $\tilde{\omega}, \tilde{\omega}_0, \tilde{\omega}_1$ *with* $\tilde{\omega}\tilde{\omega}_0 = \tilde{\omega}, \tilde{\omega}\tilde{\omega}_1 = \tilde{\omega}_1$, *then*

$$\omega A_M \omega_0 - (1 - \omega)A_\psi(1 - \omega_1)$$
$$= \tilde{\omega}A_M\tilde{\omega}_0 + (1 - \tilde{\omega})A_\psi(1 - \tilde{\omega}_1) \bmod C_G(X^\wedge, \boldsymbol{g}^\bullet).$$

Remark 2.4.3 *The assumption* (2.4.1) *was made for convenience.*
We have

$$\omega A_M \omega_0, \quad (1 - \tilde{\omega}) A_\psi (1 - \tilde{\omega}_1) \in C^\nu(\mathbb{B}, \boldsymbol{g}^\bullet)$$

$(\in \underline{C^\nu}(X^\wedge, \boldsymbol{g}^\bullet))$ *whenever* $A_M = t^{-(\gamma-\delta)} \operatorname{op}_M^{\gamma-\frac{n}{2}}(h)$ *for* $h(t,z) \in C^\infty(\overline{\mathbb{R}}_+, M_O^\nu(X))$ *and* $A_\psi \in L_{\mathrm{cl}}^\nu(\operatorname{int}\mathbb{B})(\in A_\psi \in L_{\mathrm{cl}}^\nu(X^\wedge))$ *with the indicated exit property for* $t \to \infty$) *are given, for arbitrary cut–off functions* $\omega(t)$, $\omega_0(t)$, $\tilde{\omega}(t)$, $\tilde{\omega}_1(t)$.

Theorem 2.4.4 *Each* $A \in C^\nu(\mathbb{B}, \boldsymbol{g})$ *for* $\boldsymbol{g} = (\gamma, \delta, \Theta)$ *induces a continuous operator*

$$A : \mathcal{H}^{s,\gamma}(\mathbb{B}) \to \mathcal{H}^{s-\nu,\delta}(\mathbb{B}) \tag{2.4.5}$$

for every $s \in \mathbb{R}$. *It restricts to a continuous operator*

$$A : \mathcal{H}_P^{s,\gamma}(\mathbb{B}) \to \mathcal{H}_Q^{s-\nu,\delta}(\mathbb{B}) \tag{2.4.6}$$

for every $P \in \operatorname{As}(X, (\gamma, \Theta))$ *for some* $Q \in \operatorname{As}(X, (\delta, \Theta))$. *In particular, for* $A \in C^\nu(\mathbb{B}, \boldsymbol{g}^\bullet)(\in C^\nu(\mathbb{B}, \boldsymbol{g})^\bullet)$ *we obtain a continuous operator* (2.4.6) *for every* $P \in \operatorname{As}(X, (\delta, \Theta)^\bullet)(\in \operatorname{As}(X, (\gamma, \Theta))^\bullet)$ *for some* $Q \in \operatorname{As}(X, (\delta, \Theta)^\bullet)(\in \operatorname{As}(X, (\gamma, \Theta))^\bullet)$.

Proof. Writing $A = A_0 + A_1 + M + G$, cf. Definition 2.4.1, we can treat the summands separately. For $M + G$ we can apply Theorem 2.3.82, and the assertion for A_1 is obvious. The asserted continuity of A_0 follows by a standard tensor product argument, cf. Theorem 1.1.6, using a representation $h(t,z) = \sum_{j=0}^\infty \lambda_j \varphi_j(t) h_j(z)$, convergent in $C^\infty(\overline{\mathbb{R}}_+, M_O^\nu(X))$, where we may assume that $h(t,z)$ and $\varphi_j(t)$ vanish for $t > \operatorname{const}$ for a constant > 0. To $h_j(z) \in M_O^\nu(X)$ we can apply Theorem 2.3.55, where the associated operators tend to zero for $j \to \infty$. Finally Theorem 2.3.13 yields the continuity of the multiplication operator $\mathcal{M}_{\varphi_j}$ that tends to zero for $j \to \infty$. $\square$

Theorem 2.4.5 *Each* $A \in C^\nu(X^\wedge, \boldsymbol{g})$ *for* $\boldsymbol{g} = (\gamma, \delta, \Theta)$ *induces a continuous operator*

$$A : \mathcal{K}^{s,\gamma}(X^\wedge) \to \mathcal{K}^{s-\nu,\delta}(X^\wedge) \tag{2.4.7}$$

for every $s \in \mathbb{R}$. *It restricts to a continuous operator*

$$A : \mathcal{K}_P^{s,\gamma}(X^\wedge) \to \mathcal{K}_Q^{s-\nu,\delta}(X^\wedge) \tag{2.4.8}$$

for every $P \in \operatorname{As}(X, (\gamma, \Theta))$ *for some* $Q \in \operatorname{As}(X, (\delta, \Theta))$. *In particular, for* $A \in C^\nu(X^\wedge, \boldsymbol{g}^\bullet)(\in C^\nu(X^\wedge, \boldsymbol{g})^\bullet)$ *we obtain a continuous operator* (2.4.8) *for every* $P \in \operatorname{As}(X, (\gamma, \Theta)^\bullet)(\in \operatorname{As}(X, (\gamma, \Theta))^\bullet)$ *for some* $Q \in \operatorname{As}(X, (\delta, \Theta)^\bullet)(\in \operatorname{As}(X, (\delta, \Theta))^\bullet)$.

Proof. The arguments are analogous to those for Theorem 2.3.55, except for A_1, where we can apply Remark 2.2.40. □

We now turn to the symbolic structure of the cone pseudo-differential operators. Recall that

$$S^{(\nu)}(T^*(\text{int } \mathbb{B}) \setminus 0)$$

denotes the space of all $p(\tilde{x}, \tilde{\xi}) \in C^\infty(T^*(\text{int } \mathbb{B}) \setminus 0)$ with $p(\tilde{x}, \lambda \tilde{\xi}) = \lambda^\nu p(\tilde{x}, \tilde{\xi})$ for all $\lambda \in \mathbb{R}_+$, $(\tilde{x}, \tilde{\xi}) \in T^*(\text{int } \mathbb{B}) \setminus 0$. Moreover, let

$$S^{(\nu)}(T_b^* \mathbb{B} \setminus 0)$$

be the subspace of all $p(\tilde{x}, \tilde{\xi}) \in S^{(\nu)}(T^*(\text{int } \mathbb{B}) \setminus 0)$ that can be written over $V \cong [0, 1) \times X$ in the splitting of coordinates $\tilde{x} \to (t, x)$ as $p(t, x, t\tau, \xi)$ for some $p(t, x, \tilde{\tau}, \xi) \in S^{(\nu)}(T^* V \setminus 0)$.

Definition 2.4.1 implies $C^\nu(\mathbb{B}, \boldsymbol{g}) \subset L^\nu_{\text{cl}}(\text{int } \mathbb{B})$. Thus every $A \in C^\nu(\mathbb{B}, \boldsymbol{g})$ has a homogeneous principal symbol of order ν

$$\sigma^\nu_\psi(A) \in S^{(\nu)}(T^*(\text{int } \mathbb{B}) \setminus 0).$$

Moreover, from Theorem 2.2.25, Theorem 2.2.28 and (2.2.14) we see that with every $A \in C^\nu(\mathbb{B}, \boldsymbol{g})$ there is associated an element

$$\sigma^\nu_{\psi,b}(A) \in S^{(\nu)}(T_b^* \mathbb{B} \setminus 0)$$

such that $\sigma^\nu_\psi(A)(t, x, \tau, \xi) = t^{-(\gamma-\delta)} \sigma^\nu_{\psi,b}(A)(t, x, t\tau, \xi)$ over V. In an analogous manner we have for every $A \in C^\nu(X^\wedge, \boldsymbol{g})$ elements

$$\sigma^\nu_\psi(A) \in S^{(\nu)}(T^* X^\wedge \setminus 0), \quad \sigma^\nu_{\psi,b}(A) \in S^{(\nu)}(T_b^*(\overline{\mathbb{R}}_+ \times X^\wedge) \setminus 0).$$

Moreover, we set for $A \in C^\nu(\mathbb{B}, \boldsymbol{g})$ or $A \in C^\nu(X^\wedge, \boldsymbol{g})$

$$\sigma_M^{\gamma-\delta-j}(A)(z) = \frac{1}{j!} \left(\frac{\partial^j}{\partial t^j} h \right)(0, z) + \sigma_M^{\gamma-\delta-j}(M + G)(z), \qquad (2.4.9)$$

$j \in \mathbb{N}$, $j < k$, where we use the notation from (2.4.2), (2.4.3) and (2.3.56). We call $\sigma_M^{\gamma-\delta-j}(A)(z)$ the conormal symbol of A to the conormal order $\gamma - \delta - j$. Note that there is a compatibility condition between $\sigma^\nu_{\psi,b}(A)$ and $\sigma^\nu_M(A)$. Interpreting $\sigma_M^{\gamma-\delta}(A)$ as an element of $L^\nu_{\text{cl}}(X; \Gamma_{\frac{n+1}{2}-\gamma})$ we obtain a parameter–dependent homogeneous principal symbol of order ν, with the parameter $\tau = \text{Im } z$. Denoting this by $\sigma^\nu_{\psi,\tau} \sigma_M^{\gamma-\delta}(A)(x, \tau, \xi)$ the compatibility is

$$\sigma^\nu_{\psi,\tau} \sigma_M^{\gamma-\delta}(A)(x, \tau, \xi) = \sigma^\nu_{\psi,b}(A)(0, x, -\tau, \xi). \qquad (2.4.10)$$

This is a consequence of Theorem 2.2.25 (cf. also the alternative arguments after Theorem 2.2.28).

Proposition 2.4.6 *The conormal symbols* $\sigma_M^{\gamma-\delta-j}(A)$, $j \in \mathbb{N}$, $j < k$, *of* $A \in C^\nu(\mathbb{B}, \boldsymbol{g})$ $(\in C^\nu(X^\wedge, \boldsymbol{g}))$ *are uniquely determined by* A.

Proof. The arguments are analogous to those for Propositon 2.3.79, i.e., the conormal symbols can be recovered by the action of A on functions of the form $u_w(t) = \omega_1(t)t^{-w}$, $w \in \mathbb{C}$. □

Let us set

$$\Sigma^{(\nu)}(\mathbb{B}, \boldsymbol{g}) = \{(\sigma_\psi^\nu(A), \sigma_M^{\gamma-\delta}(A)) : A \in C^\nu(\mathbb{B}, \boldsymbol{g})\}.$$

Here $\sigma_M^\nu(A)(z)$ is interpreted as an element of $L_{cl}^\nu(X; \Gamma_{\frac{n+1}{2}-\gamma})$. The space $\Sigma^{(\nu)}(\mathbb{B}, \boldsymbol{g})$ consists of all pairs

$$(p_\psi, p_M) \in S^{(\nu)}(T^*(\text{int}\,\mathbb{B}) \setminus 0) \times L_{cl}^\nu(X; \Gamma_{\frac{n+1}{2}-\gamma})$$

for which there is a $p_{\psi,b} \in S^{(\nu)}(T^*\mathbb{B} \setminus 0)$ such that

$$p_\psi(t, x, \tau, \xi) = t^{-(\gamma-\delta)}p_{\psi,b}(t, x, t\tau, \xi) \quad \text{over} \quad V, \quad (2.4.11)$$

$$\sigma_{\psi,\tau}^\nu p_M(x, \tau, \xi) = p_{\psi,b}(0, x, -\tau, \xi), \quad (2.4.12)$$

and where $p_M(z)$ extends to an element in $M_R^\nu(X)$ for some $R \in \mathbf{As}(X)$ satisfying carrier $R \cap \Gamma_{\frac{n+1}{2}-\gamma} = \emptyset$.

Applying $(\sigma_\psi^\nu, \sigma_M^{\gamma-\delta})$ on $C^\nu(\mathbb{B}, \boldsymbol{g}^\bullet)$ and $C^\nu(\mathbb{B}, \boldsymbol{g})^\bullet$ we get by definition the symbol spaces $\Sigma^{(\nu)}(\mathbb{B}, \boldsymbol{g}^\bullet)$ and $\Sigma^{(\nu)}(\mathbb{B}, \boldsymbol{g})^\bullet$, respectively.

Proposition 2.4.7 *There is a map*

$$\text{op} : \Sigma^{(\nu)}(\mathbb{B}, \boldsymbol{g}) \to C^\nu(\mathbb{B}, \boldsymbol{g}) \quad (2.4.13)$$

(called an operator convention) that is right inverse to $\sigma = (\sigma_\psi^\nu, \sigma_M^{\gamma-\delta})$: $C^\nu(\mathbb{B}, \boldsymbol{g}) \to \Sigma^{(\nu)}(\mathbb{B}, \boldsymbol{g})$, *and similarly*

$$\text{op} : \Sigma^{(\nu)}(\mathbb{B}, \boldsymbol{g}^\bullet) \to C^\nu(\mathbb{B}, \boldsymbol{g}^\bullet), \quad \Sigma^{(\nu)}(\mathbb{B}, \boldsymbol{g})^\bullet \to C^\nu(\mathbb{B}, \boldsymbol{g})^\bullet,$$

obtained by restricting (2.4.13) *to the corresponding subspaces.*

Proof. Let us start with an element $p_\psi \in S^{(\nu)}(T^*(\text{int}\,\mathbb{B}) \setminus 0)$ for which there is a $p_{\psi,b} \in S^{(\nu)}(T_b^*\mathbb{B} \setminus 0)$ satisfying (2.4.11). Then we find a pseudo–differential operator $A_\psi \in L_{cl}^\nu(\text{int}\,\mathbb{B})$ with $\sigma_\psi^\nu(A_\psi) = p_\psi$. If ω, ω_0, ω_1 are fixed cut–off functions, satisfying (2.4.2) we form $A_1 = (1 - \omega)A_\psi(1 - \omega_1)$. Set $V_\varepsilon = [0, \varepsilon) \times X$ for any fixed $0 < \varepsilon < 1$ and denote by $\tilde{S}_{cl}^\nu([0, \varepsilon) \times \Omega \times \mathbb{R}^{n+1})$ for open $\Omega \subseteq \mathbb{R}^n$ the space of all $q(t, x, t\tau, \xi)$ such that $q(t, x, \tilde{\tau}, \xi) \in S_{cl}^\nu([0, \varepsilon) \times \Omega \times \mathbb{R}^{n+1})$. Then, with $p_\psi|_{V_\varepsilon}$ in the coordinates $(t, x) \in V_\varepsilon$ we can associate a system of local symbols $p_j \in \tilde{S}^{\nu}_{\mathrm{cl}}([0,\varepsilon) \times \Omega_j \times \mathbb{R}^{n+1})$, $\Omega_j \subseteq \mathbb{R}^n$ open, such that the homogeneous principal symbol of $t^{-(\gamma-\delta)}p_j$ of order ν equals p_ψ in the local coordinates $(t,x) \in (0,\varepsilon) \times \Omega_j$. This refers to charts $\kappa_j : U_j \to \Omega_j$, $j = 1, \ldots, N$, where $\{U_1, \ldots, U_N\}$ is an open covering of X by coordinate neighbourhoods. If $\{\varphi_j\}_{j=1,\ldots,N}$ is a subordinate partition of unity and $\{\psi_j\}_{j=1,\ldots,N}$ another system of functions $\psi_j \in C_0^\infty(U_j)$ with $\varphi_j \psi_j = \varphi_j$ for all j, then $P := \sum_{j=0}^N \varphi_j(1 \times \kappa_j)_*^{-1} \mathrm{op}(p_j)\psi_j$ is the restricton of an element of $\tilde{L}^{\nu}_{\mathrm{cl}}(\mathbb{R}_+ \times X)$ to $\mathrm{int}\, V_\varepsilon$. Applying Theorem 2.2.28 we find an $f(t,z) \in C^\infty([0,\varepsilon), M_O^\nu(X))$ such that $P - \mathrm{op}_M^{\gamma - \frac{n}{2}}(f)|_{\mathrm{int}\, V_\varepsilon} \in L^{-\infty}(\mathrm{int}\, V_\varepsilon)$. It satisfies $\sigma^\nu_{\psi,\tau}(f(0,z)) = p_{\psi,b}(0,x,-\tau,\xi)$ and $f_1(z) := p_M(z) - f(0,z) \in M_R^{\nu-1}(X)$ for some $R \in \mathbf{As}(X)$ with carrier $R \cap \Gamma_{\frac{n+1}{2}-\gamma} = \emptyset$. Applying (2.3.43) we have a decompsition $f_1(z) = h_1(z) + h_0(z)$ for certain $h_1(z) \in M_O^{\nu-1}(X)$, $h_0(z) \in M_R^{-\infty}(X)$. Set $h(t,z) = f(t,z) + h_1(z)$ and $A_0 = \omega t^{-(\gamma-\delta)} \mathrm{op}_M^{\gamma-\frac{n}{2}}(h)\omega_0$, $M = \omega t^{-\nu} \mathrm{op}_M^{\gamma-\frac{n}{2}}(h_0)\omega_0$. Then we can define $\mathrm{op}(p_\psi, p_M) = A_0 + A_1 + M$, since $\sigma^\nu_\psi(A) = p_\psi$ and $\sigma^{\gamma-\delta}_M = h(0,z) + h_0(z) = f(0,z) + h_1(z) + h_0(z) = p_M$. In this construction we have tacitly used that the cut–off functions ω, ω_0, ω_1 are supported by $[0,\varepsilon)$, cf. Remark 2.4.2. $\qquad\qquad \square$

For $C^\nu(X^\wedge, \boldsymbol{g})$ we can also define a symbolic structure by setting

$$\Sigma^{(\nu)}(X^\wedge, \boldsymbol{g}) = \{(\sigma^\nu_\psi(A), \sigma^{\gamma-\delta}_M(A), \sigma^0_{\mathrm{ce}}(A)) : A \in C^\nu(X^\wedge, \boldsymbol{g})\}.$$

Analogously we form $\Sigma^{(\nu)}(X^\wedge, \boldsymbol{g}^\bullet)$ and $\Sigma^{(\nu)}(X^\wedge, \boldsymbol{g})^\bullet$, respectively. Here $\sigma^\nu_\psi(A) \in S^{(\nu)}(T^*X^\wedge \setminus 0)$ is the standard homogeneous principal symbol of A of order ν; $\sigma^{\gamma-\delta}_M(A)$ is defined in the same manner as for the case $\mathbb{B}$, while $\sigma^0_{\mathrm{ce}}(A)$ is the complete exit symbol of A in the following sense. Choosing an atlas for $X^\wedge$ of the form $\chi_j : U_j^\wedge \to V_j$, $j = 1, \ldots, N$, where $U_j^\wedge = \mathbb{R}_+ \times U_j$ with U_j as above, $V_j \subset \mathbb{R}^{n+1} \setminus \{0\}$ a conical set, $\chi_j(\lambda t, x) = \lambda \chi_j(t,x)$ for $\lambda \in \mathbb{R}_+$, we can associate with every $A \in L^\nu_{\mathrm{cl}}(X^\wedge)_{\mathrm{e}}$ symbols $e_j(\tilde{x}, \tilde{\xi}) \in S^\nu_{\mathrm{cl}}(V_j \times \mathbb{R}^{n+1})_{\mathrm{e}}$, cf. (2.2.38), that are transformed to each other (modulo symbols of order $(-\infty, -\infty)$) under the asymptotic rule for symbol push–forwards with respect to the diffeomorphisms $\chi_k\chi_j^{-1} : V_{jk} \to V_{kj}$, $V_{jk} = V_j \cap \chi_k U_k^\wedge$, $j, k = 1, \ldots, N$. Then $\sigma^0_{\mathrm{ce}}(A)$ is defined as the equivalence class of the tuple of such symbols $\{e_1(\tilde{x}, \tilde{\xi}), \ldots, e_N(\tilde{x}, \tilde{\xi})\}$ for A, where the equivalence to another $\{f_1(\tilde{x}, \tilde{\xi}), \ldots, f_N(\tilde{x}, \tilde{\xi})\}$ means that $(e_j - f_j)(\tilde{x}, \tilde{\xi}) \in S^{-\infty, -\infty}(V \times \mathbb{R}^{n+1})$ for all j. The charts are fixed; otherwise we have to modify the notion of complete exit symbols by saying what is the equivalence relation between two different systems of charts. However this is evident, so for simplicity we keep the atlas fixed.

Proposition 2.4.8 *There is a map*

$$\text{op} : \Sigma^{(\nu)}(X^\wedge, \boldsymbol{g}) \to C^\nu(X^\wedge, \boldsymbol{g}) \qquad (2.4.14)$$

(called an operator convention) that is right inverse to the symbol tuple
$\sigma = (\sigma^\nu_\psi, \sigma^{\gamma-\delta}_M, \sigma^0_{ce}) : C^\nu(X^\wedge, \boldsymbol{g}) \to \Sigma^{(\nu)}(X^\wedge, \boldsymbol{g})$, *and similarly*

$$\text{op} : \Sigma^{(\nu)}(X^\wedge, \boldsymbol{g}^\bullet) \to C^\nu(X^\wedge, \boldsymbol{g}^\bullet), \quad \Sigma^{(\nu)}(X^\wedge, \boldsymbol{g}) \to C^\nu(X^\wedge, \boldsymbol{g})^\bullet,$$

obtained by restricting (2.4.14) to the corresponding subspaces.

Proof. The construction of (2.4.14) with respect to the symbol components σ^ν_ψ and σ^ν_M is the same as in the preceding proof, while for σ^0_{ce} we can proceed in a standard manner by forming the local pseudo–diffential operators $\text{Op}(e_j) \in \tilde{L}^\nu_{cl}(V_j)_e$ and glue them together by $P := \sum_{j=1}^N \varphi_j(\chi_j^{-1})_* \text{Op}(e_j)\psi_j$. This is compatible with the construction for small t. So we can set $A_1 = (1-\omega)P(1-\omega_1)$, cf. the notation of Definition 2.4.1. $\qquad\qquad\qquad\qquad\qquad \Box$

It will be necessary from time to time also to observe the complete interior symbols of the operators in $C^\nu(\mathbb{B}, \boldsymbol{g})$, $\boldsymbol{g} = (\gamma, \delta, \Theta)$.

We fix an open covering of $\mathbb{B}$ by sets $\{V_1, \ldots, V_N, W_{N+1}, \ldots, W_{N+M}\}$ for $V_j = [0, \varepsilon) \times U_j$, with $\{U_1, \ldots, U_N\}$ being an open covering of X by coordinate neighbourhoods, $\varepsilon > 0$, and $\{W_{N+1}, \ldots, W_{N+M}\}$ an open covering of $\mathbb{B}\setminus\{\bigcup_{j=1}^N [0, \frac{\varepsilon}{2}) \times U_j\}$ by coordinate neighbourhoods. Choose a system of charts

$$\kappa_j : U_j \to \Omega_j, \ \Omega_j \subseteq \mathbb{R}^n \text{ open}, \ \beta_k : W_k \to \Delta_k, \ \Delta_k \subseteq \mathbb{R}^{n+1} \text{ open},$$

$j = 1, \ldots, N$, $k = N+1, \ldots, N+M$. Then each $A \in C^\nu(\mathbb{B}, \boldsymbol{g})$ has a system of complete symbols in local coordinates $(t, x) \in (0, \varepsilon) \times \Omega_j$ and $\tilde{x} \in \Delta_k$, respectively, of the form $t^{-(\gamma-\delta)}p_j(t, x, \tau, \xi)$ for $p_j \in \tilde{S}^\nu_{cl}([0, \varepsilon) \times \Omega_j \times \mathbb{R}^{n+1})$, $j = 1, \ldots, N$, and $p_j(\tilde{x}, \tilde{\xi}) \in S^\nu_{cl}(\Delta_j \times \mathbb{R}^{n+1})$, $j = N+1, \ldots, N+M$. Set $\Delta_j = (0, \varepsilon) \times \Omega_j$ for $j = 1, \ldots, N$. Without loss of generality we assume that the symbols are compatible for different j_1, j_2 in the following sense: The symbol push–forwards, given by the standard asymptotic formula for the operator push–forwards under the transition diffeomorphisms between corresponding open subsets of $\Delta_{j_1}, \Delta_{j_2}$ transform the symbols into each other modulo symbols of order $-\infty$. There is an equivalence relation between such symbol tuples $\{p_1, \ldots, p_{N+M}\} \sim \{q_1, \ldots, q_{N+M}\}$ consisting of equality of the components modulo symbols of order $-\infty$.

Denote the equivalence class of any symbol tuple belonging to a given $A \in C^\nu(\mathbb{B}, \boldsymbol{g})$ the complete symbol of A, written $\sigma^\nu_{c\psi}(A)$.

Analogously we can introduce a complete symbol $\sigma^\nu_{c\psi}(A)$ to every $A \in C^\nu(X^\wedge, \boldsymbol{g})$ as the equivalence class of complete symbol tuples in local coordinates in conical subsets of $\mathbb{R}^{n+1}_{\tilde{x}} \setminus \{0\}$, where a symbol of order $-\infty$ for $|\tilde{x}| \to \infty$ means of order $(-\infty, -\infty)$. Note that this coincides with the above $\sigma^0_{ce}(A)$.

Remark 2.4.9 *Let* $A \in C^\nu(\mathbb{B}, \boldsymbol{g})(\in C^\nu(\mathbb{B}, \boldsymbol{g}^\bullet), \in C^\nu(\mathbb{B}, \boldsymbol{g})^\bullet)$. *Then* $\sigma^\nu_{c\psi}(A) = 0$ *implies* $A \in C_{M+G}(\mathbb{B}, \boldsymbol{g})(\in C_{M+G}(\mathbb{B}, \boldsymbol{g}^\bullet), \in C_{M+G}(\mathbb{B}, \boldsymbol{g})^\bullet)$. *Similarly* $A \in C^\nu(X^\wedge, \boldsymbol{g})(\in C^\nu(X^\wedge, \boldsymbol{g}^\bullet), \in C^\nu(X, \boldsymbol{g})^\bullet)$ *and* $\sigma^\nu_{c\psi}(A) = 0$ *implies* $A \in C_{M+G}(X^\wedge, \boldsymbol{g})(\in C_{M+G}(X^\wedge, \boldsymbol{g}^\bullet), \in C_{M+G}(X^\wedge, \boldsymbol{g})^\bullet)$. *If we set* $\sum^\nu_{c\psi}(\mathbb{B}) = \{\sigma^\nu_{c\psi}(A) : A \in C^\nu(\mathbb{B}, \boldsymbol{g})\}$ *and* $\sum^\nu_{c\psi}(X^\wedge) = \{\sigma^\nu_{c\psi}(A) : A \in C^\nu(X^\wedge, \boldsymbol{g})\}$ *then there are maps*

$$\operatorname{op}_{c\psi} : \Sigma^\nu_{c\psi}(\mathbb{B}) \to C^\nu(\mathbb{B}, \boldsymbol{g}^\bullet) \quad and \quad \operatorname{op}_{c\psi} : \Sigma^\nu_{c\psi}(X^\wedge) \to C^\nu(X^\wedge, \boldsymbol{g}^\bullet),$$

such that $\sigma^\nu_{c\psi} \circ \operatorname{op}_{c\psi} = \operatorname{id}$ *on the corresponding symbol spaces.*

Proposition 2.4.10 *Let* $A \in C^\nu(\mathbb{B}, \boldsymbol{g})$ *for* $\boldsymbol{g} = (\gamma, \delta, \Theta)$, *and assume that* $\sigma^\nu_\psi(A) = 0$ *and* $\sigma^{\gamma-\delta}_M(A) = 0$. *Then*

$$A : \mathcal{H}^{s,\gamma}(\mathbb{B}) \to \mathcal{H}^{s-\nu,\delta}(\mathbb{B}) \tag{2.4.15}$$

is a compact operator for each $s \in \mathbb{R}$. *Analogously* $A \in C^\nu(X^\wedge, \boldsymbol{g})$ *for* $\boldsymbol{g} = (\gamma, \delta, \Theta)$ *and* $\sigma^\nu_\psi(A) = 0$, $\sigma^{\gamma-\delta}_M(A) = 0$ *and* $\sigma^0_{ce}(A) = 0$ *implies that*

$$A : \mathcal{K}^{s,\gamma}(X^\wedge) \to \mathcal{K}^{s-\nu,\delta}(X^\wedge) \tag{2.4.16}$$

is a compact operator for each $s \in \mathbb{R}$.

Proof. $A \in C^\nu(\mathbb{B}, \boldsymbol{g})$ and $\sigma^\nu_\psi(A) = 0$, $\sigma^{\gamma-\delta}_M(A) = 0$ imply that A induces a continuous operator $A : \mathcal{H}^{s,\gamma}(\mathbb{B}) \to \mathcal{H}^{s-\nu+1,\delta+\varepsilon}(\mathbb{B})$ for some $\varepsilon > 0$. Since the embedding $\mathcal{H}^{s-\nu+1,\delta+\varepsilon}(\mathbb{B}) \hookrightarrow \mathcal{H}^{s-\nu,\delta}(\mathbb{B})$ is compact, cf. Theorem 2.1.53, we obtain the first assertion. The second one follows in an analogous manner, where we use that $\sigma^\nu_\psi(A) = 0$, $\sigma^{\gamma-\delta}_M(A) = 0$, $\sigma^0_{cl}(A) = 0$ implies the continuity of $A : \mathcal{K}^{s,\gamma}(X^\wedge) \to \langle t \rangle^{-\varepsilon} \mathcal{K}^{s-\nu+1,\delta+\varepsilon}(X^\wedge)$ for an $\varepsilon > 0$, and that $\langle t \rangle^{-\varepsilon} \mathcal{K}^{s-\nu+1,\delta+\varepsilon}(X^\wedge) \hookrightarrow \mathcal{K}^{s-\nu,\delta}(X^\wedge)$ is compact. $\square$

Remark 2.4.11 *For the symbolic structure of* $A \in C^\nu(X^\wedge, \boldsymbol{g})$ *for* $t \to \infty$ *in terms of* $\sigma^0_{ce}(\cdot)$ *we have imposed conditions for the complete symbols in local coordinates of conical subsets. This can be modified by allowing only classical symbols for* $|\tilde{x}| \to \infty$, *cf. Section 1.4.3, and by taking the principal exit symbols* $\sigma_e = (\sigma^0_e, \sigma^\nu_\psi \sigma^0_e)$ *of order zero in* $|\tilde{x}|$. *Then* $\sigma^\nu_\psi(A) = 0$, $\sigma^{\gamma-\delta}_M(A) = 0$, $\sigma_e(A) = 0$ *implies the continuity of*

$$A : \mathcal{K}^{s,\gamma}(X^\wedge) \to \langle t \rangle^{-1} \mathcal{K}^{s-\nu+1,\delta+\varepsilon}(X^\wedge)$$

and hence the compactness of (2.4.16) *for every* $s \in \mathbb{R}$.

2.4.2   The cone algebras

We now study the cone pseudo–differential operators under formal adjoints and compositions. This will be done on $\mathbb{B}$ and on $X^\wedge$ for the various types of asymptotics.

Proposition 2.4.12 *Let* $h(t,z) \in C^\infty(\overline{\mathbb{R}}_+, M_O^\nu(X))$, $\nu \in \mathbb{R}$, *and let* $\omega(t)$, $\omega_0(t)$ *be arbitrary cut–off functions. Then the formal adjoint of*

$$A := \omega \operatorname{op}_M^{\gamma-\frac{n}{2}}(h)\omega_0 : \mathcal{K}^{s,\gamma}(X^\wedge) \to \mathcal{K}^{s-\nu,\gamma}(X^\wedge)$$

in the sense

$$(u, A^*v)_{\mathcal{K}^{0,0}(X^\wedge)} = (Au, v)_{\mathcal{K}^{0,0}(X^\wedge)}$$

for all $u, v \in C_0^\infty(X^\wedge)$ *can be written*

$$A^* = \omega_0 \operatorname{op}_M^{-\gamma-\frac{n}{2}}(f)\omega + M + G \qquad (2.4.17)$$

for some $f(t,z) \in C^\infty(\overline{\mathbb{R}}_+, M_O^\nu(X))$ *and certain*

$$M + G \in C_{M+G}(X^\wedge, (-\gamma, -\gamma, (-\infty, 0])^\bullet).$$

Proof. By definition we have

$$A = \omega(t)t^{\gamma-\frac{n}{2}} \operatorname{op}_M(T^{-\gamma+\frac{n}{2}}h)(t')^{-\gamma+\frac{n}{2}}\omega_0(t').$$

Then the formal adjoint equals

$$A^* = \omega_0(t)t^{-\gamma+\frac{n}{2}} \operatorname{op}_M((T^{-\gamma+\frac{n}{2}}h)^{[*]})(t')^{\gamma-\frac{n}{2}}\omega(t'),$$

$l(t',z)^{[*]} = l(t', 1 - \overline{z})^{(*)}$, where $(*)$ denotes the point–wise formal adjoint with respect to the $L^2(X)$–scalar product. Setting $k_0(t',z) = (T^{-\gamma+\frac{n}{2}}h)^{(*)}(t',z)$ which belongs to $C^\infty(\overline{\mathbb{R}}_+, M_O^\nu(X))$ we obtain by applying Proposition 2.3.68 that

$$A^* = \omega(t) \operatorname{op}_M(T^{-\gamma+\frac{n}{2}}k_0(t',z))\omega(t').$$

Here we used the holomorphy of k_0 with respect to z which shows that the Green remainder of Proposition 2.3.68 vanishes. We have $k(t',z) := T^{-\gamma+\frac{n}{2}}k_0(t',z) \in C^\infty(\overline{\mathbb{R}}_+, M_O^\nu(X))$. Thus we have to characterise $A^* = \omega_0 \operatorname{op}_M(k)\omega$ with t'–dependent k. Applying the Taylor expansion with respect to t' we get

$$k(t',z) = \sum_{j=0}^{N} k_j(z)(t')^j + k_{(N)}(t',z)(t')^N$$

for every N, where $k_j(z) \in M_O^\nu(X)$, $k_{(N)}(t', z) \in C^\infty(\overline{\mathbb{R}}_+, M_O^\nu(X))$. Then

$$\omega_0 \operatorname{op}_M(k)\omega = \sum_{j=0}^{N} \omega_0 \operatorname{op}_M(k_j)(t')^j \omega + \omega_0 \operatorname{op}_M(k_{(N)})(t')^N \omega.$$

Proposition 2.3.68 gives us that the sum on the right of the latter equation equals $\omega_0 \operatorname{op}_M(\sum_{j=0}^{N} t^j T^{-j} k_j)\omega$. In view of $\sum_{j=0}^{N} t^j (T^{-j} k_j)(z) \in C^\infty(\overline{\mathbb{R}}_+, M_O^\nu(X))$ this item is already of the desired form. Let us now assume that $N = 2L$ for an $L \in \mathbb{N}$. Then, applying again Proposition 2.3.68, it follows that

$$\omega_0 \operatorname{op}_M(k_{(N)})(t')^N \omega = \omega_0 t^L \operatorname{op}_M(T^{-L} k_{(N)})(t')^L \omega,$$

where $(T^{-1} k_{(N)})(t', z) \in C^\infty(\overline{\mathbb{R}}_+, M_O^\nu(X))$. Applying now Theorem 2.2.32 together with Remark 2.2.35 we obtain that there exists an $l(t, z) \in C^\infty(\overline{\mathbb{R}}_+, L^\nu(X, \Gamma_{\frac{1}{2}}))$ such that

$$\operatorname{op}_M(T^{-L} k_{(N)}) = \operatorname{op}_M(l_{(N)}) + G_{(N)}$$

for some $G \in M_{\frac{1}{2}} L^{-\infty}(\overline{\mathbb{R}}_+ \times X)$. In view of Corollary 2.2.34 we can choose $l_{(N)}$ as an element in $C^\infty(\overline{\mathbb{R}}_+, M_O^\nu(X))$ by modifying $G_{(N)}$ within $M_{\frac{1}{2}} L^{-\infty}(\overline{\mathbb{R}}_+ \times X)$. This yields the relation

$$\begin{aligned} \omega_0 \operatorname{op}_M(k_{(N)})(t')^N \omega &= \omega_0 t^L \operatorname{op}_M(l_{(N)})(t')^L \omega + \omega_0 t^L G_{(N)}(t')^L \omega \\ &= \omega_0 t^N \operatorname{op}_M(T^{-L} l)\omega + \omega_0 t^L G_{(N)}(t')^L \omega. \end{aligned}$$

In the latter equation we applied Proposition 2.3.68. Summing up we obtain

$$A^* = \omega_0 \operatorname{op}_M(m_N)\omega + \omega_0 t^L G_{(N)}(t')^L \omega$$

for

$$m_N(t, z) = \sum_{j=0}^{N} t^j (T^{-j} k_j)(z) + t^N T^{-L} l_{(N)}(t, z) \in C^\infty(\overline{\mathbb{R}}_+, M_O^\nu(X)).$$

Now we employ the fact that when we enlarge $N = 2L$, e.g., pass to $N + 2K$ for some $K \in \mathbb{N}$, we have

$$d_{2K}(t, z) := m_N(t, z) - m_{N+2K}(t, z) \in C^\infty \overline{\mathbb{R}}_+, M_O^{-\infty}(X)).$$

This allows us to write

$$A^* = \omega_0 \operatorname{op}_M(m_{N+2K})\omega + \omega_0 \operatorname{op}_M(d_{2k})\omega + \omega_0 t^L G_{(N)}(t')^L \omega.$$

For arbitrary $k \in \mathbb{N} \setminus \{0\}$ we can choose $L = L(k)$ so large that $\omega_0 t^L G_{(N)}(t')^L \omega \in C_G(X^\wedge, (0, 0, (-k, 0]))$. Applying Proposition 2.3.81 we see that $\omega_0 \operatorname{op}_M(d_{2k})\omega \in C_{M+G}(X^\wedge, \boldsymbol{g}^\bullet)$ for $\boldsymbol{g} = (0, 0, (-\infty, 0])$. We finally arrive at a representation of A^* in the form $\omega_0 \operatorname{op}_M(m)\omega + M + G$ for $m(t, z) \in C^\infty(\mathbb{R}_+, M_O^\nu(X))$ and $M + G \in C_{M+G}(X^\wedge, (0, 0, (-k, 0]))$, where all involved Mellin symbols are holomorphic with respect to z. By virtue of Proposition 2.3.78 the conormal symbols $\sigma_M^{-j}(A)(z)$ are unique, $j = 0, \ldots, k - 1$. Therefore, by enlarging K once again and also k the only new contributions come from the C_{M+G}–part, i.e., the length of the finite sums of smoothing Mellin operators increases where the conormal symbols for the weight interval $(-k, 0]$ remain untouched and there arise new smoothing Mellin symbols for the conormal orders larger than $k - 1$. In other words we may set $f_0(t, z) = m_N(t, z)$ which yields $A^* = \omega_0 \operatorname{op}_M(f_0)\omega + C_k$ for $C_k \in C_{M+G}(X^\wedge, (0, 0, (-k, 0]))$ for every $k \in \mathbb{N} \setminus \{0\}$. This gives us $A^* \in C^\nu(X^\wedge, \boldsymbol{g}^\bullet)$ for $\boldsymbol{g} = (0, 0, (-\infty, 0])$. Since the involved Mellin symbols are holomorphic we can shift to an arbitrary weight to get A^* in the form (2.4.17). Here it suffices to apply Proposition 2.3.68 that yields $f(t, z) = T^{\gamma + \frac{n}{2}} f_0(t, z)$, and to use Proposition 2.3.77. $\qquad\square$

Theorem 2.4.13 $A \in C^\nu(\mathbb{B}, \boldsymbol{g})$ for $\boldsymbol{g} = (\gamma, \delta, \Theta)$ implies for the formal adjoint $A^* \in C^\nu(\mathbb{B}, \boldsymbol{g}^*)$, $\boldsymbol{g}^* = (-\delta, -\gamma, \Theta)$, and we have

$$\sigma_\psi^\nu(A^*) = \overline{\sigma_\psi^\nu(A)}, \quad \sigma_M^{\gamma - \delta}(A^*) = T^{-n + \gamma - \delta} \sigma_M^{\gamma - \delta}(A)^{[*]}, \qquad (2.4.18)$$

$(f^{[*]})(z) = f(1 - \bar{z})^{(*)}$, $(*)$ indicating the point–wise formal adjoint in $L^2(X)$. In particular, $A \in C^\nu(\mathbb{B}, \boldsymbol{g}^\bullet)(\in C^\nu(\mathbb{B}, \boldsymbol{g})^\bullet)$ implies $A^* \in C^\nu(\mathbb{B}, (\boldsymbol{g}^*)^\bullet)(\in C^\nu(\mathbb{B}, \boldsymbol{g}^*)^\bullet)$.

Proof. Set $A = A_0 + A_1 + M + G$ with the notation of Definition 2.4.1. Then $(M + G)^* \in C^\nu(\mathbb{B}, \boldsymbol{g}^*)$ by Theorem 2.3.83, $A_0^* \in C^\nu(\mathbb{B}, \boldsymbol{g}^*)$ by Proposition 2.4.12, cf. also Remark 2.4.3, and $A_1^* \in (1 - \omega_1)L_{cl}^\nu(\operatorname{int}\mathbb{B})(1 - \omega) \subset C^\nu(\mathbb{B}, \boldsymbol{g}^*)$. It is clear that the various dotted subclasses are preserved under $*$. The formal adjoint is induced by that in the larger operator space $L_{cl}^\nu(\operatorname{int}\mathbb{B})$. This gives us the first symbol relation of (2.4.18). The second one is an easy consequence of the calculations in the preceding proof and of the corresponding relation in Theorem 2.3.83. $\qquad\square$

In an analogous manner we obtain a correponding result over $X^\wedge$:

Theorem 2.4.14 $A \in C^\nu(X^\wedge, \boldsymbol{g})$ for $\boldsymbol{g} = (\gamma, \delta, \Theta)$ implies $A^* \in C^\nu(X^\wedge, \boldsymbol{g}^*)$ for $\boldsymbol{g}^* = (-\delta, -\gamma, \Theta)$, and we have

$$\sigma_\psi^\nu(A^*) = \overline{\sigma_\psi^\nu(A)}, \quad \sigma_M^{\gamma - \delta}(A^*) = T^{-n + \gamma - \delta} \sigma_M^{\gamma - \delta}(A)^{[*]}, \quad \sigma_{ce}^0(A^*) = \sigma_{ce}^0(A)^*,$$

where [∗] *is of the same meaning as in Theorem 2.4.13, while* ∗ *at* $\sigma_{\text{ce}}^0(\cdot)$ *means the equivalence class obtained from the rule for complete symbols under formal adjoints of operators, for every component in the tuple.*

Note that for the lower order conormal symbols we have

$$\sigma_M^{\gamma-\delta-j}(A^*) = T^{-n+\gamma-\delta-j}\sigma_M^{\gamma-\delta-j}(A)^{[*]}$$

for all j.

Theorem 2.4.15 $A \in C^{\nu}(\mathbb{B}, \boldsymbol{g})$ *for* $\nu \in \mathbb{R}$, $\boldsymbol{g} = (\gamma, \delta, \Theta)$ *and* $B \in C^{\mu}(\mathbb{B}, \boldsymbol{c})$ *for* $\mu \in \mathbb{R}$, $\boldsymbol{c} = (\beta, \gamma, \Theta)$, *implies* $AB \in C^{\nu+\mu}(\mathbb{B}, \boldsymbol{h})$ *for* $\boldsymbol{h} = (\beta, \delta, \Theta)$, *and we have*

$$\sigma_{\psi}^{\nu+\mu}(AB) = \sigma_{\psi}^{\nu}(A)\sigma_{\psi}^{\mu}(B), \tag{2.4.19}$$

$$\sigma_M^{\beta-\delta-l}(AB) = \sum_{p+r=l} (T^{\beta-\gamma-r}\sigma_M^{\gamma-\delta-p}(A))\sigma_M^{\beta-\gamma-r}(B) \tag{2.4.20}$$

for all $l \in \mathbb{N}$ *for which the conormal symbols of the factors on the right are defined. Analogous relations hold for* $A \in C^{\nu}(\mathbb{B}, \boldsymbol{g}^{\bullet})(\in C^{\nu}(\mathbb{B}, \boldsymbol{g})^{\bullet})$, $B \in C^{\mu}(\mathbb{B}, \boldsymbol{c}^{\bullet})(\in C^{\mu}(\mathbb{B}, \boldsymbol{c})^{\bullet})$, *where* $AB \in C^{\nu+\mu}(\mathbb{B}, \boldsymbol{h}^{\bullet})(\in C^{\mu}(\mathbb{B}, \boldsymbol{h})^{\bullet})$. *If* A *or* B *belong to the class with subscript* $M + G$ *(G) then they also belong to the composition.*

Proof. It suffices to consider the case of a finite weight interval Θ. Then the result for $\Theta = (-\infty, 0]$ follows by taking the intersection over all $\Theta_k = (-k, 0]$, $k \in \mathbb{N} \setminus \{0\}$. Let us write

$$A = \omega A_M \omega_0 + \chi A_{\psi} \chi_1 + M + G, \quad B = \tilde{\omega} B_M \tilde{\omega}_0 + \tilde{\chi} B_{\psi} \tilde{\chi}_1 + W + C,$$

according to the notation in Definition 2.4.1 and Remark 2.4.2, where $\chi = 1 - \omega$, $\chi_1 = 1 - \omega_1$, $\tilde{\chi} = 1 - \tilde{\omega}$, $\tilde{\chi}_1 = 1 - \tilde{\omega}_1$. In particular, we have

$$A_M = t^{-(\gamma-\delta)} \operatorname{op}_M^{\gamma-\frac{n}{2}}(h), \quad B_M = t^{-(\beta-\gamma)} \operatorname{op}_M^{\beta-\frac{n}{2}}(\tilde{h})$$

for $h(t, z) \in C^{\infty}(\overline{\mathbb{R}}_+, M_O^{\nu}(X))$, $\tilde{h}(t, z) \in C^{\infty}(\overline{\mathbb{R}}_+, M_O^{\mu}(X))$. Then

$$AB = K_{00} + K_{01} + K_{10} + K_{11} + D_0 + D_1 + D_2,$$

where

$$K_{00} = \omega A_M \omega_0 \tilde{\omega} B_M \tilde{\omega}_0, \qquad K_{01} = \omega A_M \omega_0 \tilde{\chi} B_{\psi} \tilde{\chi}_1,$$
$$K_{10} = \chi A_{\psi} \chi_1 \tilde{\omega} B_M \tilde{\omega}_0, \qquad K_{11} = \chi A_{\psi} \chi_1 \tilde{\chi} B_{\psi} \tilde{\chi}_1,$$
$$D_0 = (M + G)(W + C), \qquad D_1 = (M + G)(\tilde{\omega} B_M \tilde{\omega}_0 + \tilde{\chi} B_{\psi} \tilde{\chi}_1),$$
$$D_2 = (\omega A_M \omega_0 + \chi A_{\psi} \chi_1)(W + C).$$

Let us restrict ourselves to the case of operators with continuous asymptotics. The arguments for the various dotted subclasses are analogous and will be omitted. By virtue of Theorem 2.3.83 we obtain $D_0 \in C_{M+G}(\mathbb{B}, \boldsymbol{h})$. Moreover, it is obvious that $(M + G)\tilde{\chi} B_\psi \tilde{\chi}_1$, $\chi A_\psi \chi_1(W + C) \in C_G(\mathbb{B}, \boldsymbol{h})$. Here it suffices to apply the mapping properties of the factors and Theorem 2.4.13. Let us show that $(M + G)\tilde{\omega} B_M \tilde{\omega}_0, \omega A_M \omega_0(W + C) \in C_{M+G}(\mathbb{B}, \boldsymbol{h})$, for instance, for the second operator. Consider the Taylor expansion for $h(t, z)$

$$h(t, z) = \sum_{j=0}^{N} t^j h_j(z) + t^N h_{(N)}(t, z), \qquad (2.4.21)$$

where $h_j(z) \in M_O^\nu(X)$, $h_{(N)}(t, z) \in C^\infty(\overline{\mathbb{R}}_+, M_O^\nu(X))$. Then, by analogous calculations as for Theorem 2.3.84 it follows that

$$\omega t^{-(\gamma-\delta)} \operatorname{op}_M^{\gamma-\frac{n}{2}} \left(\sum_{j=0}^{N} t^j h_j \right) \omega_0(W + C) \in C_{M+G}(\mathbb{B}, \boldsymbol{h}).$$

Let us choose $N = 2L$ for some $L \in \mathbb{N}$. Then, applying Proposition 2.3.69, we can write

$$\omega t^{-(\gamma-\delta)} \operatorname{op}_M^{\gamma-\frac{n}{2}} (t^N h_{(N)}) \omega_0 = \omega t^{-(\gamma-\delta)} t^L \operatorname{op}_M^{\gamma-\frac{n}{2}} (T^L h_{(N)}) t^L \omega_0.$$

Choosing L sufficiently large (dependent on the length of Θ) we see that $t^L(W + G)$ is a Green operator, and hence, because of the mapping properties of the factors and of Proposition 2.4.12 it follows that

$$\omega t^{-(\gamma-\delta)} \operatorname{op}_M^{\gamma-\frac{n}{2}} (t^N h_{(N)}) \omega_0(W + G) \in C_G(\mathbb{B}, \boldsymbol{h}).$$

It remains to treat K_{ij}, $i, j = 1, 2$, where $K_{11} \in C^{\nu+\mu}(\mathbb{B}, \boldsymbol{h})$ is trivial. Let us now look at K_{10}; the considerations for K_{01} are analogous and will be omitted. We have $\chi_1 \tilde{\omega}_0 =: \varphi \in C_0^\infty(\operatorname{int} \mathbb{B})$. Choose a cut–off function ω_3 such that $\varphi \omega_3 = 0$. Then $\varphi B_M \tilde{\omega}_0 = \varphi B_M \omega_3 + \varphi B_M(\tilde{\omega}_0 - \omega_3)$. Here $\varphi B_M \tilde{\omega}_0 \in C_G(\mathbb{B}, \boldsymbol{c})$ and $\varphi B_M(\tilde{\omega}_0 - \omega_3)$ belongs to $L_{\mathrm{cl}}^\mu(\operatorname{int} \mathbb{B})$ and is supported in a compact subset with positive distance to $\partial \mathbb{B}$. Therefore $\chi A_\psi \chi_1 \varphi B_M(\tilde{\omega}_0 - \omega_3) \in L_{\mathrm{cl}}^{\nu+\mu}(\operatorname{int} \mathbb{B})$ is also supported away from $\partial \mathbb{B}$, so it belongs to $C^{\nu+\mu}(\mathbb{B}, \boldsymbol{h})$, and moreover, $\chi A_\psi \chi_1 \varphi B_M \tilde{\omega}_0 \in C_G(\mathbb{B}, \boldsymbol{h})$. To characterise K_{00} we write $\tilde{h}(t, z) = \sum_{j=0}^{N} t^j \tilde{h}_j(z) + t^N \tilde{h}_{(N)}(t, z)$, $\tilde{h}_j(z) \in M_O^\mu(X)$, $\tilde{h}_{(N)}(t, z) \in$

$C^\infty(\overline{\mathbb{R}}_+, M_O^\mu(X))$. Then $K_{00} = M_0 + M_1 + M_2 + M_3$, where

$$M_0 = \omega t^{-(\gamma-\delta)} \operatorname{op}_M^{\gamma-\frac{n}{2}} \Big(\sum_{j=0}^{N} t^j h_j\Big) \omega_0 \tilde{\omega} t^{-(\beta-\gamma)} \operatorname{op}_M^{\beta-\frac{n}{2}} \Big(\sum_{j=0}^{N} t^j \tilde{h}_j\Big) \tilde{\omega}_0,$$

$$(2.4.22)$$

$$M_1 = \omega t^{-(\gamma-\delta)} \operatorname{op}_M^{\gamma-\frac{n}{2}} \Big(\sum_{j=0}^{N} t^j h_j\Big) \omega_0 \tilde{\omega} t^{-(\beta-\gamma)} \operatorname{op}_M^{\beta-\frac{n}{2}} (t^N \tilde{h}_{(N)}) \tilde{\omega}_0,$$

$$(2.4.23)$$

$$M_2 = \omega t^{-(\gamma-\delta)} \operatorname{op}_M^{\gamma-\frac{n}{2}} (t^N h_{(N)}) \omega_0 \tilde{\omega} t^{-(\beta-\gamma)} \operatorname{op}_M^{\beta-\frac{n}{2}} \Big(\sum_{j=0}^{N} t^j \tilde{h}_j\Big) \tilde{\omega}_0,$$

$$(2.4.24)$$

$$M_3 = \omega t^{-(\gamma-\delta)} \operatorname{op}_M^{\gamma-\frac{n}{2}} (t^N h_{(N)}) \omega_0 \tilde{\omega} t^{-(\beta-\gamma)} \operatorname{op}_M^{\beta-\frac{n}{2}} (t^N \tilde{h}_{(N)}) \tilde{\omega}_0. \quad (2.4.25)$$

Using the involved weight assumptions, namely that $\varrho := -\gamma - \mu + \beta \geq 0$, we can write

$$\omega t^{-(\gamma-\delta)} \operatorname{op}_M^{\gamma-\frac{n}{2}} (t^j h_j) \omega_1 t^{-(\beta-\gamma)} \operatorname{op}_M^{\beta-\frac{n}{2}} (t^k \tilde{h}_k) \tilde{\omega}_0$$
$$= \omega t^{\delta-\frac{n}{2}+j+k} \operatorname{op}_M (T^{-\gamma+\frac{n}{2}-k} h_j) \omega_1 \operatorname{op}_M (T^{-\beta+\frac{n}{2}} \tilde{h}_k) t^{-\beta+\frac{n}{2}} \tilde{\omega}_0$$

for $\omega_1 = \omega_0 \tilde{\omega}$. Thus the characterisation of the composition is reduced to

$$N_0 := \omega \operatorname{op}_M(h_0) \omega_1 \operatorname{op}_M(\tilde{h}_0) \omega_2,$$

up to appropriate powers of t on both sides, for certain $h_0(z) \in M_O^\nu(X)$, $\tilde{h}_0(z) \in M_O^\mu(X)$. Analogously M_1, M_2 and M_3 reduce to

$$N_1 = \omega t^L \operatorname{op}_M(h_0) \omega_1 \operatorname{op}_M(\tilde{h}_1) t^L \omega_2, \ N_2 = \omega t^L \operatorname{op}_M(h_1) \omega_1 \operatorname{op}_M(\tilde{h}_0) t^L \omega_2,$$

and

$$N_3 = \omega t^L \operatorname{op}_M(h_1) \omega_1 \operatorname{op}_M(\tilde{h}_1) t^L \omega_2,$$

respectively. Here we have set $N = 2L$ for some $L \in \mathbb{N}$ and commuted corresponding powers of t through the Mellin actions; the involved Mellin symbols are certain $h_0(z) \in M_O^\nu(X)$, $h_1(t,z) \in C^\infty(\overline{\mathbb{R}}_+, M_O^\nu(X))$, $\tilde{h}_0(z) \in M_O^\mu(X)$, $\tilde{h}_1(t,z) \in C^\infty(\overline{\mathbb{R}}_+, M_O^\mu(X))$. For N_0 we choose another cut–off function ω_3 and write

$$N_0 = \omega_3 \operatorname{op}_M(h_0) \omega_1 \operatorname{op}_M(\tilde{h}_0) \omega_2 + N_4$$
$$= \omega_3 \operatorname{op}_M(h_0) \operatorname{op}_M(\tilde{h}_0) \omega_2 + \omega_3 \operatorname{op}_M(h_0)(1 - \omega_1) \operatorname{op}_M(\tilde{h}_0) \omega_2 + N_4,$$

where $N_4 = (\omega - \omega_3) \operatorname{op}_M(h_0) \omega_1 \operatorname{op}_M(\tilde{h}_0) \omega_2$. Assuming $\omega_3(1 - \omega_1) = 0$ the summand in the middle on the right hand side is of Green type, while the first summand equals $\omega_3 \operatorname{op}_M(h_0 \tilde{h}_0)\omega_2$, hence it is also of the required kind. For N_4 we decompose ω_2 into $\omega_2 = (\omega_2 - \omega_4) + \omega_4$ for some cut–off function ω_4 satisfying $(\omega - \omega_3)\omega_4 = 0$. Then

$$N_4 = (\omega - \omega_3)\operatorname{op}_M(h_0)\omega_1 \operatorname{op}_M(\tilde{h}_0)\omega_4$$
$$+ (\omega - \omega_3)\operatorname{op}_M(h_0)\omega_1 \operatorname{op}_M(\tilde{h}_0)(\omega_2 - \omega_4).$$

The first summand on the right is of Green type, while the second one is a pseudo–differential operator of order $\nu + \mu$, supported away from $\partial \mathbb{B}$, such that N_4 also belongs to the cone operator class. The operator N_1, N_2, N_3 can be treated by Theorem 2.2.36. Consider, for instance, N_3. We have

$$N_3 = \omega t^L \operatorname{op}_M(h_1 \#_M(\omega_1 \tilde{h}_1)) t^L \omega_2 + \omega t^L G t^L \omega_2, \qquad (2.4.26)$$

with the Mellin Leibniz product $\#_M$, and an operator $G \in M_0 L^{-\infty}(\overline{\mathbb{R}}_+ \times X)$. For sufficiently large L we then obtain that $\omega t^L G t^L \omega_2$ is of Green type. The Leibniz product can be first carried out in the space $C^\infty(\overline{\mathbb{R}}_+, L_{\mathrm{cl}}^{\nu+\mu}(X; \Gamma_{\frac{1}{2}}))$. Then from the proof of Theorem 2.2.28 it follows that there is an $h(t,z) \in C^\infty(\overline{\mathbb{R}}_+, M_O^{\nu+\mu}(X))$ such that $h(t,z) - (h_1 \#_M(\omega_1 \tilde{h}_1))(t,z) =: h_{-\infty}(t,z) \in C^\infty(\overline{\mathbb{R}}_+, L^{-\infty}(X; \Gamma_{\frac{1}{2}}))$. Thus the first summand on the right of (2.4.26) equals

$$\omega t^L \operatorname{op}_M(h) t^L \omega_2 + \omega t^L \operatorname{op}_M(h_{-\infty}) t^L \omega_2 = \omega t^{2L} \operatorname{op}_M(T^{-L}h)\omega_2 + G_1,$$

where $G_1 = \omega t^L \operatorname{op}_M(h_{-\infty}) t^L \omega_2$ is of Green type and the operator $\omega t^{2L} \operatorname{op}_M(T^{-L}h)\omega_2$ is as required. Summing up we see that $AB \in C^{\nu+\mu}(\mathbb{B}, \boldsymbol{h})$. The relation (2.4.19) is obvious, since the operators A and B are classical pseudo–differential operators in int $\mathbb{B}$. The Mellin translation product (2.4.20) can be obtained analogously to that of Theorem 2.3.84. In the present case it suffices to write

$$\omega A_M \omega_0 = \omega t^{-\nu} \sum_{j=0}^{N} t^j \operatorname{op}_M^{\gamma - \frac{n}{2}}(h_j)\omega_0 + R_N,$$

$$\tilde{\omega} B_M \tilde{\omega}_0 = \tilde{\omega} \sum_{k=0}^{N} t^k \operatorname{op}_M^{\beta - \frac{n}{2}}(\tilde{h}_k)\tilde{\omega}_0 + \tilde{R}_N$$

for sufficiently large N and remainders R_N and $\tilde{R}_N$, respectively. The remainders do not contribute to the finitely many conormal symbols of the composition, so they may be ignored in that calculation and the arguments are then similar to Theorem 2.3.84. $\qquad \square$

Theorem 2.4.16 $A \in C^\nu(X^\wedge, \boldsymbol{g})$ for $\nu \in \mathbb{R}$, $\boldsymbol{g} = (\gamma, \delta, \Theta)$ and $B \in C^\mu(X^\wedge, \boldsymbol{c})$ for $\mu \in \mathbb{R}$, $\boldsymbol{c} = (\beta, \gamma, \Theta)$, implies $AB \in C^{\nu+\mu}(X^\wedge, \boldsymbol{h})$ for $\boldsymbol{h} = (\beta, \delta, \Theta)$, and we have the symbolic rules (2.4.19), (2.4.20) and

$$\sigma_{ce}^0(AB) = \sigma_{ce}^0(A) \# \sigma_{ce}^0(B), \tag{2.4.27}$$

where $\#$ is the Leibniz product between the components in the corresponding local coordinates. Analogous relations hold for $A \in C^\nu(X^\wedge, \boldsymbol{g}^\bullet)$ $(\in C^\nu(X^\wedge, \boldsymbol{g}^\bullet))$, $B \in C^\mu(X^\wedge, \boldsymbol{c}^\bullet)(\in C^\mu(X^\wedge, \boldsymbol{c})^\bullet)$. If A or B belong to the class with subscript $M+G$ (G) then also the composition.

Remark 2.4.17 It is clear that with the notation of Theorem 2.4.15 and Theorem 2.4.16 for the complete symbols we have

$$\sigma_{c\psi}^{\nu+\mu}(AB) = \sigma_{c\psi}^\nu(A) \# \sigma_{c\psi}^\mu(B),$$

where $\#$ is the Leibniz product between the components. This is compatible with (2.4.27) in the case $X^\wedge$. Moreover, we have

$$D := AB - \mathrm{op}_{c\psi}(\sigma_{c\psi}^\nu(A) \# \sigma_{c\psi}^\mu(B)) \in C_{M+G}(\mathbb{B}, \boldsymbol{g})(\in C_{M+G}(X^\wedge, \boldsymbol{g})),$$

cf. Remark 2.4.9. If A and B belong to the classes with discrete asymptotics then so does D.

2.4.3   Ellipticity, parametrices, Fredholm property

The ellipticity of an operator $A \in C^\mu(\mathbb{B}, \boldsymbol{g})$ for $\mu \in \mathbb{R}$, $\boldsymbol{g} = (\gamma, \delta, \Theta)$, $\gamma, \delta \in \mathbb{R}$, will be formulated in terms of the pair of symbols $(\sigma_\psi^\mu(A), \sigma_M^{\gamma-\delta}(A))$. Analogously the ellipticity for $A \in C^\mu(X^\wedge, \boldsymbol{g})$ will be a condition on $(\sigma_\psi^\mu(A), \sigma_M^{\gamma-\delta}(A), \sigma_{ce}^0(A))$.

Recall that a neighbourhood of $\partial\mathbb{B}$ is identified with $V = [0,1) \times X \ni (t,x)$, where $\sigma_\psi^\mu(A)$ has the form

$$\sigma_\psi^\mu(A)(t,x,\tau,\xi) = t^{\delta-\gamma} \sigma_{\psi,b}^\mu(A)(t,x,t\tau,\xi),$$

with $\sigma_{\psi,b}^\mu(A)(t,x,\tilde\tau,\xi)$ being C^∞ in t up to $t = 0$.

Let us study a relation between the conormal symbol $\sigma_M^{\gamma-\delta}(A)(z)$ of an operator $A \in C^\mu(\mathbb{B}, \boldsymbol{g})$ and $\sigma_{\psi,b}^\mu(A)(0,x,t\tau,\xi)$. From Definition 2.3.48 it follows that

$$\sigma_M^{\gamma-\delta}(A) \in M_R^\mu(X) \tag{2.4.28}$$

for some $R \in \mathbf{As}(X)$ with $\Gamma_{\frac{n+1}{2}-\gamma} \cap \operatorname{carrier} R = \emptyset$, cf. Definition 2.4.1. In particular, for $A \in C^\mu(\mathbb{B}, \boldsymbol{g}^\bullet)$ we have (2.4.28) for some $R \in \mathbf{As}^\bullet(X)$ with $\Gamma_{\frac{n+1}{2}-\gamma} \cap \pi_{\mathbb{C}}R = \emptyset$. In this case the conormal symbol of A is a meromorphic $\mathcal{L}(H^s(X), H^{s-\mu}(X))$-valued function, $s \in \mathbb{R}$, where the poles are described by R.

Theorem 2.4.18 *Let* $A \in C^\mu(\mathbb{B}, \boldsymbol{g}^\bullet)$ *with*

$$\sigma_{\psi,b}^\mu(A)(0, x, \tilde{\tau}, \xi) \neq 0 \quad \text{for all} \quad x \in X \quad \text{and} \quad (\tilde{\tau}, \xi) \neq 0. \quad (2.4.29)$$

Then there is a countable subset $D \subset \mathbb{C}$, *where* $D \cap K$ *is finite for every* $K \subset\subset \mathbb{C}$, *such that*

$$\sigma_M^{\gamma - \delta}(A)(z) : H^s(X) \to H^{s-\mu}(X)$$

is an isomorphism for every $z \in \mathbb{C} \setminus D$ *and all* $s \in \mathbb{R}$.

Proof. By definition the conormal symbol $f(z) := \sigma_M^{\gamma-\delta}(A)(z)$ has the form $h(0, z) + l(z)$ for certain $h(t, z) \in C^\infty(\overline{\mathbb{R}}_+, M_O^\mu(X))$, $l(z) \in M_R^{-\infty}(X)$, $R \in \mathbf{As}^\bullet(X)$. The carrier of R consists of a discrete set V in $\mathbb{C}$ with finite intersections $V \cap \{z : c \leq \operatorname{Re} z \leq c'\}$ for every $c \leq c'$. Let us fix a V-excision function $\chi(z)$. Then $\chi(z)f(z)|_{\Gamma_\beta} \in L_{cl}^\mu(X; \Gamma_\beta)$ is parameter–dependent elliptic of order μ, which is a consequence of (2.4.29) and of the relation (2.4.10). This holds for each $\beta \in \mathbb{R}$, uniformly in $c \leq \beta \leq c'$ for every $c \leq c'$. Thus, applying Theorem 1.2.26, we see that $f(z) : H^s(X) \to H^{s-\mu}(X)$ is an isomorphism for every $z \in \mathbb{C}$ with $|\operatorname{Im} z| > N$ for sufficiently large $N = N(c, c') > 0$. Since $f(z)$ is a Fredholm operator for every $z \in \mathbb{C} \setminus V$, it follows from Theorem 1.2.23 that $f(z)$ is an isomorphism for a countable subset of $\mathbb{C} \setminus V$ without accumulation points in $\{z : \operatorname{dist}(z, V) > \varepsilon\}$ for every $\varepsilon > 0$. However, $f(z)$ is also a meromorphic Fredholm family in the sense of Definition 2.4.35. Therefore Theorem 1.2.37 applies and hence the above countable set has no accumulation points in $\{z : \operatorname{dist}(z, V) < \varepsilon\} \setminus V$. This completes the proof. $\qquad\square$

Lemma 2.4.19 *Let* $f(z) \in M_R^{-\infty}(X)$ *for some* $R \in \mathbf{As}(X)$. *Then there is an* $h(z) \in M_Q^{-\infty}(X)$ *for a certain* $Q \in \mathbf{As}(X)$ *such that*

$$(1 + f(z))(1 + h(z)) = (1 + h(z))(1 + f(z)) = 1$$

for all $z \in \mathbb{C}$. *In this relation* $R \in \mathbf{As}^\bullet(X)$ *implies* $Q \in \mathbf{As}^\bullet(X)$.

Proof. First it is easy to see that for every $c, c' \in \mathbb{R}$ with $c \leq c'$ there is an $N = N(c, c') > 0$ such that

$$(1 + f(z))^{-1} = \sum_{j=0}^\infty (-1)^j f^j(z) \quad (2.4.30)$$

converges in $\{z : c \leq \operatorname{Re} z \leq c'\} \cap \{z : |\operatorname{Im} z| > N\}$ to an $h_1(z)$ with $\chi(z)h_1(z)|_{\Gamma_\beta} \in \mathcal{S}(\Gamma_\beta, L^{-\infty}(X))$ for every $c \leq \beta \leq c'$, uniformly in β in this interval, where $\chi(z) \in C^\infty(\mathbb{C})$, $\chi(z) = 0$ for $|\operatorname{Im} z| < N$, $\chi(z) = 1$

for $|\operatorname{Im} z| > 2N$. In addition, $h_1(z)$ belongs to $\mathcal{A}(U(c,c'), L^{-\infty}(X))$
for $U(c,c') = \{z : c < \operatorname{Re} z < c'\} \cap \{z : |\operatorname{Im} z| > N\}$. Moreover, for
$V(c,c') = \{z : c < \operatorname{Re} z < c'\} \cap \{z : |\operatorname{Im} z| < 2N\}$ the constructions to
prove the Theorems 1.2.33, 1.2.37, specialised to $a(z) = 1 + f(z)$, yield a
meromorphic $L^{-\infty}(X)$–valued function $a^{-1}(z)$ with finitely many poles
$p_0, \dots, p_L$ in $V(c,c')$, and Laurent-coefficients at the negative powers of
$z - p_j, j = 0, \dots, L$, that are finite-dimensional operators in $L^{-\infty}(X)$.
This, together with the first part of the construction, shows the existence of an $h(z) \in M_Q^{-\infty}(X)$ of the asserted kind. For the continuous
asymptotics the first part of the proof is the same. Since $1 + f(z)$
is a holomorphic family of Fredholm operators $H^s(X) \to H^s(X)$ on
$\mathbb{C} \setminus \operatorname{carrier} R$ we find a meromorphic Fredholm family $(1 + f(z))^{-1}$
on $\mathbb{C} \setminus \operatorname{carrier} R$, where the poles cannot have accumulation points in
$\{z : \operatorname{dist}(z, \operatorname{carrier} R) > \varepsilon\}$ for every $\varepsilon > 0$. The set R together with
the poles of $(1 + f(z))^{-1}$ in $\mathbb{C} \setminus \operatorname{carrier} R$ form a set $W \in \mathcal{V}$. So
$(1 + f(z))^{-1}$ can be interpreted as an element of the form $1 + h(z)$ for
some $Q \in \mathbf{As}(X)$ with $W = \operatorname{carrier} Q$. $\square$

Theorem 2.4.20 *Let $A \in C^\mu(\mathbb{B}, \boldsymbol{g}^\bullet)$ and assume that the condition
(2.4.29) is satisfied. Then $(\sigma_M^{\gamma-\delta})(A)(z))^{-1}$ extends from $\mathbb{C} \setminus D$ to an
element in $M_Q^{-\mu}(X)$ for a certain $Q \in \mathbf{As}^\bullet(X)$. More generally, for
$A \in C^\mu(\mathbb{B}, \boldsymbol{g})$ we obtain $(\sigma_M^{\gamma-\delta}(A)(z))^{-1} \in M_Q^{-\mu}(X)$ for a certain
$Q \in \mathbf{As}(X)$.*

Proof. We have $f(z) := \sigma_M^{\gamma-\delta}(A)(z) \in M_R^\mu(X)$ for an $R \in \mathbf{As}^\bullet(X)$.
By definition we can write $f(z) = h(z) + l(z)$ for certain $h(z) \in M_O^\mu(X)$,
$l(z) \in M_R^{-\infty}(X)$. Set $h_\beta(z) = h(z)|_{\Gamma_\beta}$ for any $\beta \in \mathbb{R}$. Then
$h_\beta(z) \in L_{\operatorname{cl}}^\mu(X; \Gamma_\beta)$ is parameter–dependent elliptic with the parameter
$\tau = \operatorname{Im} z$. According to Theorem 1.2.25 we can choose a parameter-
dependent parametrix $g_\beta(z) \in L_{\operatorname{cl}}^{-\mu}(X; \Gamma_\beta)$ of $h_\beta(z)$. From Theorem
2.2.10 we get an $m(z) \in M_O^{-\mu}(X)$ with $m(z)|_{\Gamma_\beta} - g_\beta(z) \in L^{-\infty}(X; \Gamma_\beta)$.
Thus, also $m_\beta(z) := m(z)|_{\Gamma_\beta}$ is a parameter–dependent parametrix
of $h_\beta(z)$, i.e., $m_\beta(z)h_\beta(z) - 1 \in L^{-\infty}(X; \Gamma_\beta)$. Here β was fixed.
However the latter relation implies the same for all $\beta \in \mathbb{R}$, i.e.,
$m(z)h(z) = 1 + m_0(z)$ for an $m_0(z) \in M_O^{-\infty}(X)$, cf. Remark
2.2.29. This yields $m(z)f(z) = m(z)h(z) + m(z)l(z) = 1 + f_0(z)$
for $f_0(z) = m_0(z) + m(z)l(z) \in M_{R_0}^{-\infty}(X)$ for some $R_0 \in \mathbf{As}^\bullet(X)$.
Lemma 2.4.19 shows the existence of an $h_0(z) \in M_{Q_0}^{-\infty}(X)$ for a certain $Q_0 \in \mathbf{As}^\bullet(X)$ such that $(1 + h_0(z))(1 + f_0(z)) = 1$. Then it follows
that $(1 + h_0(z))m(z) \in f^{-1}(z) \in M_Q^{-\mu}(X)$ for a $Q \in \mathbf{As}^\bullet(X)$. For
the case with continuous asymptotics we can proceed in an analogous
manner. $\square$

Lemma 2.4.21 *Let $A \in C_{M+G}(\mathbb{B}, \boldsymbol{g})$ for $\boldsymbol{g} = (\gamma, \gamma, \Theta)$ with finite Θ, and assume $\sigma_M^0(A) = 0$. Then there is an operator $B \in C_M(\mathbb{B}, \boldsymbol{g})$ such that*

$$(1 + B)(1 + A), (1 + A)(1 + B) \in C_G(\mathbb{B}, \boldsymbol{g}).$$

An analogous result holds for $A \in C_{M+G}(\mathbb{B}, \boldsymbol{g}^\bullet)$ $(\in C_{M+G}(\mathbb{B}, \boldsymbol{g})^\bullet)$, where $(1 + B)(1 + A), (1 + A)(1 + B) \in C_G(\mathbb{B}, \boldsymbol{g}^\bullet)$ $(\in C_G(\mathbb{B}, \boldsymbol{g})^\bullet)$. The same is true over $X^\wedge$.

Proof. It suffices to set $1 + B = \sum_{j=0}^{N}(-1)^j A^j$ for sufficiently large N, dependent on the length of Θ. In fact, we have

$$(1 + A)(1 + B) = \sum_{j=0}^{N}(-1)^j A^j + \sum_{j=0}^{N}(-1)^j A^{j+1} = 1 + (-1)^N A^{N+1}$$

which is of Green type for large N, cf. Theorem 2.3.84 and Proposition 2.3.77. The multiplication from the left can be discussed in an analogous manner. $\qquad\square$

Remark 2.4.22 *Lemma 2.4.21 can be generalised to the case $\Theta = (-\infty, 0]$, where the operator A in $C_{M+G}(\mathbb{B}, \boldsymbol{g})$ is a sum $M + G$ for $G \in C_G(\mathbb{B}, \boldsymbol{g})$ and M consists of an infinite sum*

$$M = t^{-(\gamma-\delta)} \sum_{j=0}^{\infty} t^j \omega(c_j t) \left\{ \mathrm{op}_M^{\rho_j - \frac{n}{2}}(h_j) + \mathrm{op}_M^{\delta_j - \frac{n}{2}}(f_j) \right\} \omega_0(c_j t) \quad (2.4.31)$$

for constants c_j tending to ∞ sufficiently fast and sequences $\{h_j\}_{j\in\mathbb{N}}$, $\{f_j\}_{j\in\mathbb{N}}$ as above in Definition 2.3.75. Now the weights ρ_j and δ_j are chosen in such a way that $j + \rho_j \to \infty$, $-\rho_j \to \infty$, $j + \delta_j \to \infty$, $-\delta_j \to \infty$ as $j \to \infty$, such that the sums M_k over $j \geq k$ converge in

$$\bigcap_{s\in\mathbb{R}} \mathcal{L}(\mathcal{H}^{s,\gamma}(\mathbb{B}), \mathcal{H}^{\infty,\delta+\beta(k)}(\mathbb{B}))$$

and M_k^* in

$$\bigcap_{s\in\mathbb{R}} \mathcal{L}(\mathcal{H}^{s,-\delta}(\mathbb{B}), \mathcal{H}^{\infty,-\gamma+\beta(k)}(\mathbb{B}))$$

for suitable constants $\beta(k) \geq 0$ with $\beta(k) \to \infty$ as $k \to \infty$. Such a choice of c_j is possible, indeed, cf. Schulze [119]. We do not go into the details here, but the formal Neumann series $\sum_{j=0}^{\infty}(-1)^j A^j$ for $A \in C_{M+G}(\mathbb{B}, (\gamma, \gamma, (-\infty, 0]))$ produces a sequence of conormal symbols that can be decomposed into $h_j + f_j$ for which the associated sum $(2.4.31)$ converges in the above sense. This gives then a construction for B in Lemma 2.4.21 for the case $\Theta = (-\infty, 0]$.

Definition 2.4.23 *Let $A \in C^\mu(\mathbb{B}, \boldsymbol{g})$ for $\mu \in \mathbb{R}$ and $\boldsymbol{g} = (\gamma, \delta, \Theta)$, $\gamma, \delta \in \mathbb{R}$, $\Theta = (-k, 0]$, $k \in (\mathbb{N} \setminus \{0\}) \cup \{\infty\}$. Then A is called elliptic if*

(i) *$\sigma_\psi^\mu(A) \neq 0$ on $T^*(\mathrm{int}\,\mathbb{B}) \setminus 0$ and, near $\partial\mathbb{B}$ in the splitting of coordinates into $(t, x) \in [0, 1) \times X = V$, $t^{\gamma-\delta}\sigma_\psi^\mu(A)(t, x, t^{-1}\tau, \xi) \neq 0$ on $T^*V \setminus 0$,*

(ii) *the operator*

$$\sigma_M^{\gamma-\delta}(A)(z) : H^s(X) \to H^{s-\mu}(X)$$

is an isomorphism for every $z \in \Gamma_{\frac{n+1}{2}-\gamma}$, for some $s = s_0 \in \mathbb{R}$.

Note that the condition (ii) for a given $s = s_0 \in \mathbb{R}$ implies the same for all $s \in \mathbb{R}$. In fact, $\sigma_M^{\gamma-\delta}(A)(z)$ is an elliptic operator in the class $L_{\mathrm{cl}}^\mu(X)$ for each $z \in \Gamma_{\frac{n+1}{2}-\gamma}$, but kernel and cokernel are independent of s, cf. Remark 1.1.79.

Theorem 2.4.24 *To each $\mu \in \mathbb{R}$ and $\boldsymbol{g} = (\gamma, \delta, \Theta)$, for $\gamma, \delta \in \mathbb{R}$ there exists an elliptic operator $A \in C^\mu(\mathbb{B}, \boldsymbol{g}^\bullet)$.*

Proof. First we can choose an element $\Sigma_{c\psi}^\mu(\mathbb{B})$ for which the homogeneous principal symbols in the local representatives are elliptic. Applying $\mathrm{op}_{c\psi}$ from Remark 2.4.9 we can form an associated operator $\tilde{A} \in C^\mu(\mathbb{B}, \boldsymbol{g}^\bullet)$. In particular, we obtain an $h(t, z) \in C^\infty(\overline{\mathbb{R}}_+, M_O^\mu(X))$ for which

$$\tilde{A} = \omega t^{\delta-\gamma} \mathrm{op}_M^{\gamma-\frac{n}{2}}(h)\omega_0 + (1 - \omega)A_\psi(1 - \omega_1) \qquad (2.4.32)$$

is elliptic, $A_\psi = \tilde{A}|_{\mathrm{int}\,\mathbb{B}} \in L_{\mathrm{cl}}^\mu(\mathrm{int}\,\mathbb{B})$. In view of (2.4.10) the operator family $h(0, z) : H^s(X) \to H^{s-\mu}(X)$ is parameter–dependent elliptic with the parameter $z \in \Gamma_\beta$ for every $\beta \in \mathbb{R}$. So it is a holomorphic family of Fredholm operators. Using Theorem 2.4.18 we find a β for which $h(0, z)$ is a family of isomorphisms for all $z \in \Gamma_\beta$. Thus, $h_1(t, z) := h(t, z - \frac{n+1}{2} + \gamma + \beta) \in C^\infty(\overline{\mathbb{R}}_+, M_O^\mu(X))$, where the parameter–dependent homogeneous principal symbol is the same as before (on every parallel to the imaginary axis and for all t), so

$$A := \omega t^{\delta-\gamma} \mathrm{op}_M^{\gamma-\frac{n}{2}}(h_1)\omega_0 + (1 - \omega)A_\psi(1 - \omega_1)$$

is elliptic with respect to σ_ψ^μ, and in addition we have the bijectivity of $h_1(z)$ for $z \in \Gamma_{\frac{n+1}{2}-\gamma}$ which is the second condition of Definition 2.4.23. $\qquad\square$

Remark 2.4.25 Let $A \in C^\mu(\mathbb{B}, \boldsymbol{g})$ for $\mu \in \mathbb{R}$ and $\boldsymbol{g} = (\gamma, \delta, \Theta)$, $\gamma, \delta \in \mathbb{R}$, be elliptic. Moreover, let $h^\beta \in C^\infty(\text{int }\mathbb{B})$ be a strictly positive function with $h^\beta = t^\beta$ near $\partial\mathbb{B}$, $\beta \in \mathbb{R}$. Then $h^\alpha A h^{-\beta} \in C^\mu(\mathbb{B}, \boldsymbol{g}_{\alpha\beta})$ is again elliptic, where $\boldsymbol{g}_{\alpha\beta} = (\gamma + \alpha, \delta + \beta, \Theta)$. Note that $A \in C^\mu(\mathbb{B}, \boldsymbol{g}^\bullet)$ implies $h^\alpha A h^{-\beta} \in C^\mu(\mathbb{B}, \boldsymbol{g}^\bullet_{\alpha\beta})$.

Remark 2.4.26 If $A \in C^\mu(\mathbb{B}, \boldsymbol{g})$ for $\mu \in \mathbb{R}$, $g = (\gamma, \delta, \Theta)$, $\gamma, \delta \in \mathbb{R}$, is elliptic then also the formal adjoint $A^* \in C^\mu(\mathbb{B}, \boldsymbol{g}^*)$, $\boldsymbol{g}^* = (-\delta, -\gamma, \Theta)$. This is obvious with respect to the interior symbol, while $\sigma_M^{\gamma-\delta}(A^*)(z) : H^s(X) \to H^{s-\mu}(X)$ has to be an isomorphism for all $z \in \Gamma_{\frac{n+1}{2}+\delta}$ which follows from the formula (2.4.18). In fact, we have

$$\sigma_M^{\gamma-\delta}(A^*)(z) = \sigma_M^{\gamma-\delta}(A)(n+1-\gamma+\delta-\bar{z})^{(*)}$$

with $(*)$ indicating the point-wise formal $L^2(X)$-adjoint. Then the invertibility of $\sigma_M^{\gamma-\delta}(A)$ on $z \in \Gamma_{\frac{n+1}{2}-\gamma}$ implies the same for $\sigma_M^{\gamma-\delta}(A^*)(z)$ on $z \in \Gamma_{\frac{n+1}{2}-\delta}$. Let us also note that the ellipticity remains preserved under compositions.

Definition 2.4.27 Let $A \in C^\mu(\mathbb{B}, \boldsymbol{g})$ for $\mu \in \mathbb{R}$, and $\boldsymbol{g} = (\gamma, \delta, \Theta)$ for $\gamma, \delta \in \mathbb{R}$. Then an operator $B \in C^{-\mu}(\mathbb{B}, \boldsymbol{g}^{-1})$ for $\boldsymbol{g}^{-1} = (\delta, \gamma, \Theta)$ is called a parametrix of A if

$$G_l := BA - 1 \in C_G(\mathbb{B}, \boldsymbol{g}_l), \quad G_r := AB - 1 \in C_G(\mathbb{B}, \boldsymbol{g}_r) \quad (2.4.33)$$

for $\boldsymbol{g}_l = (\gamma, \gamma, \Theta)$, $\boldsymbol{g}_r(\delta, \delta, \Theta)$.

Note that when $B \in C^{-\mu}(\mathbb{B}, \boldsymbol{g}^{-1})$ is a parametrix of A then so are $B + G$ for every $G \in C_G(\mathbb{B}, \boldsymbol{g}^{-1})$.

Theorem 2.4.28 Let $A \in C^\mu(\mathbb{B}, \boldsymbol{g})$ be elliptic, $\mu \in \mathbb{R}$, $\boldsymbol{g} = (\gamma, \delta, \Theta)$, $\gamma, \delta \in \mathbb{R}$. Then

$$A : \mathcal{H}^{s,\gamma}(\mathbb{B}) \to \mathcal{H}^{s-\mu,\delta}(\mathbb{B}) \quad (2.4.34)$$

is a Fredholm operator for every $s \in \mathbb{R}$, and A has a parametrix $B \in C^{-\mu}(\mathbb{B}, \boldsymbol{g}^{-1})$.

In particular, for $A \in C^\mu(\mathbb{B}, \boldsymbol{g}^\bullet)$ we obtain $B \in C^{-\mu}(\mathbb{B}, (\boldsymbol{g}^{-1})^\bullet)$, and $G_l \in C_G(B, \boldsymbol{g}_l^\bullet)$, $G_r \in C_G(B, \boldsymbol{g}_r^\bullet)$. We have

$$\sigma_\psi^{-\mu}(B) = \sigma_\psi^\mu(A)^{-1}, \sigma_M^{\delta-\gamma}(B) = (T^{\gamma-\delta}\sigma_M^{\gamma-\delta}(A))^{-1}.$$

Proof. In order to construct a parametrix B we first consider the case that the weight interval Θ is finite. Let $\sigma_{c\psi}^\mu(A)$ be the complete symbol of A in the sense of the notation of Remark 2.4.9. Then, by forming component-wise the Leibniz inverses in the corresponding local coordinates we obtain an element $\sigma_{c\psi}^{-\mu}(B) \in \Sigma_{c\psi}^{-\mu}(\mathbb{B})$. Then $B_0 := \mathrm{op}_{c\psi}(\sigma_{c\psi}^{-\mu}(B)) \in C^{-\mu}(\mathbb{B}, \boldsymbol{g}^{-1})$ satisfies $AB_0 - 1 \in C_{M+G}(\mathbb{B}, \boldsymbol{g}_r)$. Set $h(z) = \sigma_M^{\gamma-\delta}(A)(z)$, $l(z) = \sigma_M^{\delta-\gamma}(B_0)(z)$. From Theorem 2.4.18 we obtain an $f(z) \in M_Q^{-\mu}(X)$, $Q \in \mathbf{As}(X)$, such that $h(z)f(z) = 1$. Thus $(T^{\gamma-\delta}h(z))(T^{\gamma-\delta}f(z)) = 1$. Moreover, we have $(T^{\gamma-\delta}h(z))l(z) = 1 + h_0(z)$ for an $h_0(z) \in M_{R_0}^{-\infty}(X)$, $R_0 \in \mathbf{As}(X)$. This gives us $(T^{\gamma-\delta}h(z))f_0(z) = -h_0(z)$, where $f_0(z) := T^{\gamma-\delta}f(z) - l(z) \in M_{R_1}^{-\infty}(X)$, $R_1 \in \mathbf{As}(X)$, i.e., $(T^{\gamma-\delta}h(z))^{-1} = l(z) + f_0(z)$. The operator

$$D_0 = \omega(t)t^{\delta-\gamma}\,\mathrm{op}_M^{\delta-\frac{n}{2}}(f_0)\omega_0(t)$$

for arbitrary fixed cut–off functions $\omega(t)$, $\omega_0(t)$ belongs to the space $C_{M+G}(\mathbb{B}, \boldsymbol{g}^{-1})$, and $B_1 := B_0 + D_0 \in C^{-\mu}(\mathbb{B}, \boldsymbol{g}^{-1})$ satisfies $AB_1 - 1 \in C_{M+G}(\mathbb{B}, \boldsymbol{g}_r)$, $\sigma_M^0(AB_1 - 1) = 0$. From Lemma 2.4.21 we get a $D \in C_{M+G}(\mathbb{B}, \boldsymbol{g}_r)$ such that $B_r := B_1(1 + D) \in C^{-\mu}(\mathbb{B}, \boldsymbol{g}^{-1})$ satisfies $AB_r - 1 =: G_r \in C_G(\mathbb{B}, \boldsymbol{g}_r)$. Analogously we find a $B_l \in C^{-\mu}(\mathbb{B}, \boldsymbol{g}^{-1})$ with $B_l A - 1 =: G_l \in C_G(\mathbb{B}, \boldsymbol{g}_l)$. Now $AB_r = 1 + G_r$ implies $B_l + B_l G_r = B_l A B_r = B_r + G_l B_r$, i.e., $B_l = B_r + G_l B_r - B_l G_r$. Since $G_l B_r - B_l G_r \in C_G(\mathbb{B}, \boldsymbol{g}^{-1})$, we may set $B = B_l$ or $= B_r$. It is obvious that the subclass of cone operators with discrete asymptotics is preserved under the parametrix construction. The Fredholm property of (2.4.34) for every $s \in \mathbb{R}$ follows from the existence of a parametrix, since the Green operators are compact, cf. Proposition 2.3.66. For $\Theta = (-\infty, 0]$ we can argue in the same manner by using Remark 2.4.2. $\qquad\square$

Remark 2.4.29 *It can be proved that the Fredholm property of the operator (2.4.34) for an $s = s_0 \in \mathbb{R}$ implies the ellipticity of A. Then (2.4.36) is Fredholm for all $s \in \mathbb{R}$.*

Theorem 2.4.30 *Let $A \in C^{\mu}(\mathbb{B}, \boldsymbol{g})$ be elliptic, $\mu \in \mathbb{R}$, $\boldsymbol{g} = (\gamma, \delta, \Theta)$, $\gamma, \delta \in \mathbb{R}$. Then*

$$Au \in \mathcal{H}^{s-\mu,\delta}(\mathbb{B}), \qquad u \in \mathcal{H}^{-\infty,\gamma}(\mathbb{B})$$

implies $u \in \mathcal{H}^{s,\gamma}(\mathbb{B})$. Moreover

$$Au \in \mathcal{H}_Q^{s-\mu,\delta}(\mathbb{B}) \qquad \text{for some } Q \in \mathrm{As}(X, (\delta, \Theta))$$

and $u \in \mathcal{H}^{-\infty,\gamma}(\mathbb{B})$ imply $u \in \mathcal{H}_P^{s,\gamma}(\mathbb{B})$ for some asymptotic type $P \in \mathrm{As}(X, (\gamma, \Theta))$. For $A \in C^{\mu}(\mathbb{B}, \boldsymbol{g}^{\bullet})$ we have the same for every $Q \in \mathrm{As}(X, (\delta, \Theta^{\bullet}))$ with some $P \in \mathrm{As}(X, (\gamma, \Theta)^{\bullet})$. This holds for every $s \in \mathbb{R}$.

Proof. The ellipticity of A entails the existence of a parametrix $B \in C^{-\mu}(\mathbb{B}, \boldsymbol{g}^{-1})$ which yields $BA = 1 + G$ for a $G \in C_G(\mathbb{B}, (\gamma, \gamma, \Theta))$. Applying B to $Au = f \in \mathcal{H}_Q^{s-\mu,\delta}(\mathbb{B})$ it follows that $BAu = u + Gu = Bf \in \mathcal{H}_{P_1}^{s,\gamma}(\mathbb{B})$, where $Gu \in \mathcal{H}_{P_2}^{\infty,\gamma}(\mathbb{B})$, for $P_1, P_2 \in \mathrm{As}(X, (\gamma, \Theta))$, cf. Theorem 2.4.4, Definition 2.3.59. Thus $u \in \mathcal{H}_{P_1}^{s,\gamma}(\mathbb{B}) + \mathcal{H}_{P_2}^{\infty,\gamma}(\mathbb{B}) \subset \mathcal{H}_P^{s,\gamma}(\mathbb{B})$ for some $P \in \mathrm{As}(X, (\gamma, \Theta))$. The same can be done for the case with discrete asymptotics. The first part of the theorem can be obtained in an analogous manner by omitting the asymptotic types. $\square$

Proposition 2.4.31 *Let $A \in C^{\mu}(\mathbb{B}, \boldsymbol{g})$ be elliptic, $\mu \in \mathbb{R}$, $\boldsymbol{g} = (\gamma, \delta, \Theta)$, $\gamma, \delta \in \mathbb{R}$. Then there are finite-dimensional subspaces $\mathcal{N}_+ \subset \mathcal{H}_R^{\infty,\gamma}(\mathbb{B})$, $\mathcal{N}_- \subset \mathcal{H}_Q^{\infty,\delta}(\mathbb{B})$ for certain $R \in \mathrm{As}(X, (\gamma, \Theta))$, $Q \in \mathrm{As}(X, (\delta, \Theta))$ such that*

$$\ker A = \mathcal{N}_+, \quad \mathrm{im}\, A \oplus \mathcal{N}_- = \mathcal{H}^{s-\mu,\delta}(\mathbb{B})$$

for all $s \in \mathbb{R}$, cf. (2.4.34). In particular, for $A \in C^{\mu}(\mathbb{B}, \boldsymbol{g}^{\bullet})$ we have $R \in \mathrm{As}(X, (\gamma, \Theta)^{\bullet})$, $Q \in \mathrm{As}(X, (\delta, \Theta)^{\bullet})$. The index

$$\mathrm{ind}\, A = \dim \ker A - \dim \mathrm{coker}\, A$$

is independent of s.

Proof. Consider the isomorphisms $h^{-\gamma} : \mathcal{H}^{s,\gamma}(\mathbb{B}) \to \mathcal{H}^{s,0}(\mathbb{B})$, $h^{-\delta} : \mathcal{H}^{s-\mu,\delta}(\mathbb{B}) \to \mathcal{H}^{s-\mu,0}(\mathbb{B})$ and set $A_0 = h^{-\delta}Ah^{\gamma}$. Then A_0 and A_0^* are elliptic in the class $C^{\mu}(\mathbb{B}, (0, 0, \Theta))$. Thus $\ker A_0$ and $\mathrm{coker}\, A_0 = \ker A_0^*$ are finite-dimensional subspaces of $\mathcal{H}^{\infty,0}(\mathbb{B})$ with asymptotics, independent of s. Thus $\ker A = h^{\gamma} \ker A_0$ is of finite dimension with asymptotics (which is also known from Theorem 2.4.30). Moreover, we may set $\mathcal{N}_- = h^{\delta} \ker A_0^*$. $\square$

Remark 2.4.32 *Let $A, \tilde{A} \in C^{\mu}(\mathbb{B}, \boldsymbol{g})$ be elliptic and assume that $\sigma_\psi^{\mu}(A) = \sigma_\psi^{\mu}(\tilde{A})$, $\sigma_M^{\mu}(A) = \sigma_M^{\mu}(\tilde{A})$ Then $\mathrm{ind}\, A = \mathrm{ind}\, \tilde{A}$. In fact, $A - \tilde{A}$ is then a compact operator. Note that in the paper [113] there was given a characterisation of the set of all stable homotopy classes of elliptic symbol tuples $(\sigma_\psi^{\mu}(A), \sigma_M^{\mu}(A))$.*

Remark 2.4.33 *Let $A \in C^{\mu}(\mathbb{B}, \boldsymbol{g})$ ($\in C^{\mu}(\mathbb{B}, \boldsymbol{g}^{\bullet})$) for $\mu \in \mathbb{R}$, $\boldsymbol{g} = (\gamma, \delta, \Theta)$, $\gamma, \delta \in \mathbb{R}$, be an operator satisfying the condition (i) of Definition 2.4.23. Then there exists an $R \in C^{-\mu}(\mathbb{B}, \boldsymbol{g}^{-1})$ ($\in C^{\mu}(\mathbb{B}, (\boldsymbol{g}^{-1})^{\bullet})$) such that the remainders $RA - 1$ and $AR - 1$ belong to $C_{M+G}(\mathbb{B}, \boldsymbol{g}_l)$ ($\in C_{M+G}(\mathbb{B}, \boldsymbol{g}_l^{\bullet})$) and $C_{M+G}(\mathbb{B}, \boldsymbol{g}_r)$ ($\in C_{M+G}(\mathbb{B}, \boldsymbol{g}_r^{\bullet})$), respectively, cf. the notation of Definition 2.4.27. In fact, it suffices to choose an arbitrary R in the asserted class associated with the system of Leibniz inverses of the complete interior symbol of A.*

Theorem 2.4.34 *To each $\mu \in \mathbb{R}$ and $\boldsymbol{g} = (\gamma, \delta, \Theta)$ for $\gamma, \delta \in \mathbb{R}$ there exists an elliptic operator $A \in C^\mu(\mathbb{B}, \boldsymbol{g}^\bullet)$ which induces isomorphisms*

$$A : \mathcal{H}^{s,\gamma}(\mathbb{B}) \to \mathcal{H}^{s-\mu,\delta}(\mathbb{B})$$

for all $s \in \mathbb{R}$.

Proof. The composition by h^ρ induces isomorphisms $\mathcal{H}^{s,\gamma}(\mathbb{B}) \to \mathcal{H}^{s,\gamma+\rho}(\mathbb{B})$ for all $\gamma, \rho \in \mathbb{R}$ and $s \in \mathbb{R}$. In view of Remark 2.4.25 we may assume $\gamma = \delta = 0$. Applying Theorem 2.4.4 to $\mu/2$ and $\boldsymbol{g} = (0, 0, \Theta)$ we obtain an elliptic operator $R \in C^{\mu/2}(\mathbb{B}, \boldsymbol{g}^\bullet)$. Then Remark 2.4.26 shows that also $R^* \in C^{\mu/2}(\mathbb{B}, \boldsymbol{g}^\bullet)$ is elliptic as well as $RR^* \in C^\mu(\mathbb{B}, \boldsymbol{g}^\bullet)$, and RR^* is formally self-adjoint. This shows that $\operatorname{ind} RR^* = 0$, and $\ker RR^*$ is a finite-dimensional subspace $\mathcal{N}$ of $\mathcal{H}_P^{\infty,0}(\mathbb{B})$ for some $P \in \operatorname{As}(X, (0, \Theta)^\bullet)$. In addition $\mathcal{N} \oplus \operatorname{im} RR^* = \mathcal{H}^{s-\mu,0}(\mathbb{B})$ for every s, where RR^* is regarded as operator $\mathcal{H}^{s,0}(\mathbb{B}) \to \mathcal{H}^{s-\mu,0}(\mathbb{B})$. The orthogonal projection $T : \mathcal{H}^{0,0}(\mathbb{B}) \to \mathcal{N}$ is a Green operator of the class $C_G(\mathbb{B}, (0, 0, \Theta)^\bullet)$. Then $A_0 := RR^*(1 - T) + T : \mathcal{H}^{s,0}(\mathbb{B}) \to \mathcal{H}^{s-\mu,0}(\mathbb{B})$ is an isomorphism, and we have $A_0 \in C^\mu(\mathbb{B}, (0, 0, \Theta)^\bullet)$ which is elliptic. To complete the proof it suffices to set $A = h^\delta A_0 h^{-\gamma}$. $\qquad\square$

Next we study the ellipticity of operators $A \in C^\mu(X^\wedge, \boldsymbol{g})$ under analogous aspects as for $\mathbb{B}$. Here we observe the equivalence classes $\sigma_{\mathrm{ce}}(A)$ of symbol tuples

$$\{p_1(\tilde{x}, \tilde{\xi}), \dots, p_N(\tilde{x}, \tilde{\xi})\} \tag{2.4.35}$$

given in conical subsets $V_j \subset \mathbb{R}^{n+1}_{\tilde{x}} \setminus \{0\}$ to which there are given diffeomorphisms $\chi_j : \mathbb{R}_+ \times U_j \to V_j$ for an open covering $\{U_1, \dots, U_N\}$ of X by coordinate neighbourhoods, satisfying $\chi_j(\lambda t, x) = \lambda \chi_j(t, x)$ for all $\lambda \in \mathbb{R}_+$ and all j, cf. the notation of Section 2.4.1. Recall that $p_j(\tilde{x}, \tilde{\xi}) \in S^\mu_{\mathrm{cl}}(V_j \times \mathbb{R}^{n+1})_{\mathrm{e}}$, cf. (2.2.37).

Definition 2.4.35 *Let $A \in C^\mu(X^\wedge, \boldsymbol{g})$ for $\mu \in \mathbb{R}$ and $\boldsymbol{g} = (\gamma, \delta, \Theta)$, $\gamma, \delta \in \mathbb{R}$, $\Theta = (-k, 0]$, $k \in (\mathbb{N} \setminus \{0\}) \cup \{\infty\}$. Then A is called elliptic if*

(i) $\sigma^\mu_\psi(A) \neq 0$ *on* $T^* X^\wedge \setminus 0$ *and* $t^{\gamma-\delta} \sigma^\mu_\psi(A)(t, x, t^{-1}\tau, \xi) \neq 0$ *on* $T^*(\overline{\mathbb{R}}_+ \times X) \setminus 0$,

(ii) *the operator*

$$\sigma^{\gamma-\delta}_M(A)(z) : H^s(X) \to H^{s-\mu}(X)$$

is an isomorphism for every $z \in \Gamma_{\frac{n+1}{2} - \gamma}$, for some $s = s_0 \in \mathbb{R}$,

(iii) $\sigma_{ce}^0(A)$ *is invertible in the sense that, for any system of local representatives* (2.4.35), *there are symbols*

$$\{p_1^{(-1)}(\tilde{x},\tilde{\xi}),\ldots,p_N^{(-1)}(\tilde{x},\tilde{\xi})\},$$

where $p_j^{(-1)}(\tilde{x},\tilde{\xi}) \in S_{cl}^{-\mu}(V_j \times \mathbb{R}^{n+1})$, *such that* $p_j(\tilde{x},\tilde{\xi}) \# p_j^{(-1)}(\tilde{x},\tilde{\xi})$
$= 1 \bmod S^{-\infty,-\infty}(V_j \times \mathbb{R}^{n+1})$ *for all* $j = 1,\ldots,N$.

Theorem 2.4.36 *For each* $\mu \in \mathbb{R}$ *and* $\boldsymbol{g} = (\gamma,\delta,\Theta)$ *for* $\gamma,\delta \in \mathbb{R}$ *there exists an elliptic operator* $A \in C^\mu(X^\wedge,\boldsymbol{g}^\bullet)$.

Proof. It is easy to see that there exists a tuple (2.4.35) of symbols $p_j(\tilde{x},\tilde{\xi}) \in S_{cl}^\mu(V_j \times \mathbb{R}^{n+1})_e$ for which

$$p_j(\tilde{x},\tilde{\xi}) = (1 + |\tilde{\xi}|^2)^{\mu/2} \bmod \boldsymbol{S}^{\mu-1,-1}(V_j \times \mathbb{R}^{n+1}) \cap S_{cl}^{\mu-1}(V_j \times \mathbb{R}^{n+1}),$$

cf. also Section 1.4.5. The Leibniz inverses exist modulo $S^{-\infty,-\infty}(V_j \times \mathbb{R}^{n+1})$ for all j. Moreover, we can choose the symbols $p_j(\tilde{x},\tilde{\xi})$ in such a way that they form an element in $\Sigma_{c\psi}^\mu(X^\wedge)$, cf. the notations in Remark 2.4.9. This allows us to proceed similarly to the proof of the above Theorem 2.4.4, namely to form an operator (2.4.32), where now $A_\psi \in L_{cl}^\mu(X^\wedge)_e$ is associated with the system (2.4.35) in the standard way, while near $t = 0$ the Mellin symbol $h(t,z)$ is obtained from these symbols with respect to the (t,x) coordinates, cf. Proposition 2.2.37, by applying the Mellin operator convention. It remains to arrange the conormal symbol in the same manner as in the proof of Theorem 2.4.4. $\qquad\square$

Remark 2.4.37 *Let* $A \in C^\mu(X^\wedge,\boldsymbol{g})$ *for* $\mu \in \mathbb{R}$ *and* $\boldsymbol{g} = (\gamma,\delta,\Theta)$, $\gamma,\delta \in \mathbb{R}$, *be elliptic. Moreover, let* $k^\beta \in C^\infty(X^\wedge)$ *be a strictly positive function with* $k^\beta = t^\beta$ *for* $0 < t < c_0$, $k^\beta = 1$ *for* $c_1 < t \leq \infty$, *for certain constants* $c_0 < c_1$, $\beta \in \mathbb{R}$. *Then* $k^\alpha A k^{-\beta} \in C^\mu(X^\wedge,\boldsymbol{g}_{\alpha\beta})$ *is again elliptic, where* $\boldsymbol{g}_{\alpha\beta} = (\gamma+\alpha,\delta+\beta,\Theta)$. *Here* $A \in C^\mu(X^\wedge,\boldsymbol{g}^\bullet)$ *implies* $k^\alpha A k^{-\beta} \in C^\mu(X^\wedge,\boldsymbol{g}_{\alpha\beta}^\bullet)$.

Remark 2.4.38 *If* $A \in C^\mu(X^\wedge,\boldsymbol{g})$ *for* $\mu \in \mathbb{R}$, $\boldsymbol{g} = (\gamma,\delta,\Theta)$, $\gamma,\delta \in \mathbb{R}$, *is elliptic, then also the formal adjoint* $A^* \in C^\mu(X^\wedge,\boldsymbol{g}^*)$ *for* $\boldsymbol{g}^* = (-\delta,-\gamma,\Theta)$. *Moreover, the ellipticity remains preserved under compositions.*

Definition 2.4.39 *Let* $A \in C^\mu(X^\wedge,\boldsymbol{g})$ *for* $\mu \in \mathbb{R}$, $\boldsymbol{g} = (\gamma,\delta,\Theta)$ *for* $\gamma,\delta \in \mathbb{R}$. *Then an operator* $B \in C^{-\mu}(X^\wedge,\boldsymbol{g})$ *for* $\boldsymbol{g}^{-1} = (\delta,\gamma,\Theta)$ *is called a parametrix of* A *if* $G_l = BA - 1 \in C_G(X^\wedge,\boldsymbol{g}_l)$, $G_r = AB - 1 \in C_G(X^\wedge,\boldsymbol{g}_r)$ *for* $\boldsymbol{g}_l = (\gamma,\gamma,\Theta)$, $\boldsymbol{g}_r = (\delta,\delta,\Theta)$.

Theorem 2.4.40 *Let* $A \in C^\mu(X^\wedge, \boldsymbol{g})$ *be elliptic,* $\mu \in \mathbb{R}$, $\boldsymbol{g} = (\gamma, \delta, \Theta)$, $\gamma, \delta \in \mathbb{R}$. *Then*

$$A : \mathcal{K}^{s,\gamma}(X^\wedge) \to \mathcal{K}^{s-\mu,\delta}(X^\wedge) \qquad (2.4.36)$$

is a Fredholm operator for every $s \in \mathbb{R}$, *and* A *has a parametrix* $B \in C^{-\mu}(X^\wedge, \boldsymbol{g}^{-1})$.

In particular, for $A \in C^\mu(X^\wedge, \boldsymbol{g}^\bullet)$ we obtain $B \in C^{-\mu}(X^\wedge, (\boldsymbol{g}^{-1})^\bullet)$, and $G_l \in C_G(X^\wedge, \boldsymbol{g}_l^\bullet)$, $G_r \in C_G(X^\wedge, \boldsymbol{g}_r^\bullet)$. We have

$$\sigma_\psi^\mu(B) = \sigma_\psi^\mu(A)^{-1}, \sigma_M^{\delta-\gamma}(B) = (T^{\gamma-\delta}\sigma_M^{\gamma-\delta}(A))^{-1}, \sigma_{ce}(B) = \sigma_{ce}(A)^{-1},$$

where the inverse of σ_{ce} *is taken in the sense of Definition 2.4.35 (iii).*

Proof. A parametrix P can be constructed locally near $t = 0$ by a simple modification of the construction in the proof of Theorem 1.4.60 which yields an operator P_0, and locally near $t = \infty$ by applying the construction in the proof of Theorem 1.4.60 which yields an operator P_∞. Then, we can set, $P = \omega P_0 \omega_0 + (1 - \omega) P_\infty (1 - \omega_1)$ for cut–off functions ω, ω_0, ω_1 satisfying $\omega \omega_0 = \omega$, $\omega \omega_1 = \omega_1$. This implies then the Fredholm property of (2.4.36), since the Green remainders are compact. $\square$

Remark 2.4.41 *It can be proved that the Fredholm property of the operator* (2.4.36) *for an* $s = s_0 \in \mathbb{R}$ *implies the ellipticity of* A. *Then* (2.4.36) *is a Fredholm operator for all* $s \in \mathbb{R}$.

Theorem 2.4.42 *Let* $A \in C^\mu(X^\wedge, \boldsymbol{g})$ *be elliptic,* $\mu \in \mathbb{R}$, $\boldsymbol{g} = (\gamma, \delta, \Theta)$, $\gamma, \delta \in \mathbb{R}$. *Then*

$$Au \in \mathcal{K}^{s-\mu,\delta}(X^\wedge), \quad u \in \mathcal{K}^{-\infty,\gamma}(X^\wedge)$$

implies $u \in \mathcal{K}^{s,\delta}(X^\wedge)$. *Moreover,*

$$Au \in \mathcal{K}_Q^{s-\mu,\delta}(X^\wedge) \quad \text{for some} \quad Q \in \mathrm{As}(X, (\delta, \Theta))$$

and $u \in \mathcal{K}^{-\infty,\gamma}(X^\wedge)$ *implies* $u \in \mathcal{K}_P^{s,\gamma}(X^\wedge)$ *for some* $P \in \mathrm{As}(X, (\gamma, \Theta))$. *For* $A \in C^\mu(X^\wedge, \boldsymbol{g}^\bullet)$ *we have the same for every* $Q \in \mathrm{As}(X, (\delta, \Theta)^\bullet)$ *with some* $P \in \mathrm{As}(X, (\gamma, \Theta)^\bullet)$. *This holds for every* $s \in \mathbb{R}$.

Proof. The arguments are obvious modifications of those for Theorem 2.4.30. $\square$

Proposition 2.4.43 *Let* $A \in C^\mu(X^\wedge, \boldsymbol{g})$ *be elliptic,* $\mu \in \mathbb{R}$, $\boldsymbol{g} = (\gamma, \delta, \Theta)$, $\gamma, \delta \in \mathbb{R}$. *Then there are finite-dimensional subspaces*

$\mathcal{N}_+ \subset \mathcal{S}_R^\gamma(X^\wedge)$, $\mathcal{N}_- \subset \mathcal{S}_Q^\delta(X^\wedge)$ for certain asymptotic types $R \in$ As$(X,(\gamma,\Theta))$, $Q \in$ As$(X,(\delta,\Theta))$ such that

$$\ker A = \mathcal{N}_+, \qquad \text{im } A \oplus \mathcal{N}_- = \mathcal{K}^{s-\mu,\delta}(X^\wedge)$$

for all $s \in \mathbb{R}$, cf. (2.4.36). In particular, for $A \in C^\mu(X^\wedge, g^\bullet)$ we have $R \in$ As$(X,(\gamma,\Theta)^\bullet)$, $Q \in$ As$(X,(\delta,\Theta)^\bullet)$. The index ind A is independent of s.

Proof. The proof is analogous to that of Proposition 2.4.31. The only new aspect here is that we have Schwartz functions in the kernels and cokernels for $t \to \infty$. This follows from the mapping properties of Green operators on $X^\wedge$. □

Remark 2.4.44 Let $A \in C^\mu(X^\wedge, g)$ $(\in C^\mu(X^\wedge, g^\bullet))$ for $\mu \in \mathbb{R}$, $g = (\gamma,\delta,\Theta)$, $\gamma,\delta \in \mathbb{R}$, be an operator satisfying the conditions (i), (iii) of Definition 2.4.35. Then there exists an $R \in C^{-\mu}(X^\wedge, g^{-1})$ $(\in C^{-\mu}(X^\wedge, (g^{-1})^\bullet))$ such that $RA-1$ and $AR-1$ belong to $C_{M+G}(X^\wedge, g_l)$ $(\in C_{M+G}(X^\wedge, g_l^\bullet))$ and $C_{M+G}(X^\wedge, g_r)$ $(\in C_{M+G}(X^\wedge, g_r^\bullet))$, respectively, cf. the notation of Definition 2.4.39. In fact, it suffices to choose an arbitrary R in the asserted classes associated with the system of Leibniz inverses of the complete interior symbol of A, both in the finite region as well as for $t \to \infty$.

Theorem 2.4.45 To every $\mu \in \mathbb{R}$ and $g = (\gamma,\delta,\Theta)$ for $\gamma,\delta \in \mathbb{R}$ there exists an elliptic operator $A \in C^\mu(X^\wedge, g^\bullet)$ which induces isomorphisms

$$A: \mathcal{K}^{s,\gamma}(X^\wedge) \to \mathcal{K}^{s-\mu,\delta}(X^\wedge)$$

for all $s \in \mathbb{R}$.

Proof. The construction of A is completely analogous to that in the proof of Theorem 2.4.34. □

Theorem 2.4.46 For each $k \in \mathbb{Z}$, $\gamma \in \mathbb{R}$, there exists an operator $A_k \in C_{M+G}(\mathbb{B},(\gamma,\gamma,\Theta)^\bullet)$ such that

$$\text{ind}(1 + A_k) = k \qquad (2.4.37)$$

in the sense $1 + A_k: \mathcal{H}^{s,\gamma}(\mathbb{B}) \to \mathcal{H}^{s,\gamma}(\mathbb{B})$, $s \in \mathbb{R}$. Similarly there exists an $A_k \in C_{M+G}(X^\wedge,(\gamma,\gamma,\Theta)^\bullet)$ such that (2.4.37) holds in the sense of $1 + A_k: \mathcal{K}^{s,\gamma}(X^\wedge) \to \mathcal{K}^{s,\gamma}(X^\wedge)$ for all $s \in \mathbb{R}$.

To prove this theorem we first want to show an auxiliary result:

Lemma 2.4.47 *To each $k \in \mathbb{Z}$ there exists an $f(z) \in \mathcal{A}(\{z : \frac{1}{2} - \delta < \mathrm{Re}\, z < \frac{1}{2} + \delta\})$ for some $\delta > 0$ with $f|_{\Gamma_\beta} \in S^{-\infty}(\Gamma_\beta)$ for every $\frac{1}{2} - \delta < \beta < \frac{1}{2} + \delta$, uniformly in β for $\frac{1}{2} - \delta + \varepsilon < \beta < \frac{1}{2} + \delta - \varepsilon$ for every $0 < \varepsilon < \delta$, such that $1 + f(z) \neq 0$ for all $z \in \Gamma_{\frac{1}{2}}$ and*

$$\mathrm{ind}(1 + \omega\, \mathrm{op}_M(f)\tilde{\omega}) = k$$

in the sense $1 + \omega\, \mathrm{op}_M(f)\tilde{\omega} : L^2(\mathbb{R}_+) \to L^2(\mathbb{R}_+)$, for arbitrary cut–off functions $\omega(t)$, $\tilde{\omega}(t)$.

Proof. It is well–known, cf., for instance, the book [25], that

$$\mathrm{ind}(1 + \omega\, \mathrm{op}_M(f)\tilde{\omega}) = \frac{1}{2\pi}\Delta \arg(1 + f(z))\Big|_{z=\frac{1}{2}-i\infty}^{z=\frac{1}{2}+i\infty},$$

where $\Delta \arg\, (\ldots)|_{z=\frac{1}{2}-i\infty}^{z=\frac{1}{2}+i\infty}$ means the change of the argument when z runs on $\Gamma_{\frac{1}{2}}$ from $\mathrm{Im}\, z = +\infty$ to $\mathrm{Im}\, z = -\infty$. Thus it suffices to construct an $f(z)$ for which this change equals $2\pi k$. Let us first consider the case $k = 1$. Set

$$f(z) = -e^{-1}z^{-1}e^{(z-\frac{1}{2})^2}. \tag{2.4.38}$$

This function is never real on $\Gamma_{\frac{1}{2}} \setminus \{\frac{1}{2}\}$, and we have $f(\frac{1}{2}) = -2$. Since $|f(z)| \to 0$ for $|\mathrm{Im}\, z| \to \infty$, it follows that $1 + f(z) \neq 0$ for all $z \in \Gamma_{\frac{1}{2}}$. Moreover, it is obvious that $f(z) \in \mathcal{A}(\{z : \frac{1}{2} - \delta < \mathrm{Re}\, z < \frac{1}{2} + \delta\})$ for $0 < \delta < \frac{1}{2}$ and that it has the property $f|_{\Gamma_\beta} \in S^{-\infty}(\Gamma_\beta)$ uniformly in compact subintervals. We see even that $f(z) \in M_R^{-\infty}$ for some discrete asymptotic type R, where $z = 0$ is the only (simple) pole. Now $\arg(1 + f(z)) \lessgtr 0$ once $\arg z \lessgtr 0$ for $z \in \Gamma_{\frac{1}{2}}$. Thus $\arg(1 + f(z))$ grows by 2π when $z \in \Gamma_{\frac{1}{2}}$ runs from $\frac{1}{2} + i\infty$ to $\frac{1}{2} - i\infty$. Next observe that $(1 + f(z))^{-1} = 1 + h(z)$, where $h(z)$ belongs to $M_P^{-\infty}$ for a discrete asymptotic type P with $\pi_\mathbb{C} P \cap \Gamma_{\frac{1}{2}} = \emptyset$, and $1 + h(z) \neq 0$ on $\Gamma_{\frac{1}{2}}$. From

$$(1 + \omega\, \mathrm{op}_M(f)\tilde{\omega})(1 + \omega\, \mathrm{op}_M(h)\tilde{\omega}) = 1 + G$$

for an operator $G \in C_G(\mathbb{R}_+, (0, 0, (-\infty, 0]))$ which is compact in $L^2(\mathbb{R}_+)$ we obtain that $\mathrm{ind}(1 + \omega\, \mathrm{op}_M(h)\tilde{\omega}) = -1$. Now

$$\mathrm{ind}(1 + \omega\, \mathrm{op}_M(f)\tilde{\omega})^k = k, \quad \mathrm{ind}(1 + \omega\, \mathrm{op}_M(h)\tilde{\omega})^k = -k$$

for every $k \in \mathbb{N}$. It remains to note that

$$(1 + \omega\, \mathrm{op}_M(f)\tilde{\omega})^k = 1 + \omega\, \mathrm{op}_M(f_{(k)})\tilde{\omega} + G_{(k)}$$

for certain $G_{(k)} \in C_G(\mathbb{R}_+, (0, 0, (-\infty, 0]))$ that are compact in $L^2(\mathbb{R}_+)$. Thus $\mathrm{ind}(1 + \omega\, \mathrm{op}_M(f_{(k)})\tilde{\omega}) = k$. Similarly we can argue for $h(z)$ and obtain Mellin symbols $h_{(k)}(z)$ for which the associated operator has index $-k$. Clearly $f_{(k)}(t)$ and $h_{(k)}(t)$ also have the asserted symbol properties. □

Remark 2.4.48 *The function $f(z)$ given by (2.4.38) satisfies $(1 + f(z)) \neq 0$ for all $z \in \mathbb{C} \setminus \{0\}$. Hence, because of its simple pole at $z = 0$ we have $(1 + f(z))^{-1} = 1 + h(z) \neq 0$ for all $z \in \mathbb{C} \setminus \{0\}$ and $z = 0$ is a simple zero of $1 + h(z)$. We have $h(z) \in M_O^{-\infty}$.*

Proof of Theorem 2.4.46. Let us consider first the case $X^\wedge$ and look at $\mathcal{K}^{0,\frac{n}{2}}(X^\wedge) = L^2(\mathbb{R}_+ \times X)$. Fix a one-dimensional subspace H of $L^2(X)$ spanned by a non-vanishing function in $C^\infty(X)$. Let $P : L^2(X) \to H$ be the orthogonal projection. Then, if $f(z)$ is the function in Lemma 2.4.47 to the given $k \in \mathbb{Z}$, we have $h(z) := f(z)P \in M_R^{-\infty}(X)$ for some discrete asymptotic type R, and

$$1 + \omega\, \mathrm{op}_M(h)\tilde{\omega} : L^2(\mathbb{R}_+ \times X) \to L^2(\mathbb{R}_+ \times X)$$

has the index k. Thus we can set $A_k = \omega\, \mathrm{op}_M(h)\tilde{\omega}$ which belongs to

$$C_{M+G}\left(X^\wedge, \left(\frac{n}{2}, \frac{n}{2}, (-\infty, 0]\right)\right).$$

In view of Proposition 2.4.43 it follows that $1 + A_k : \mathcal{K}^{s,\frac{n}{2}}(X^\wedge) \to \mathcal{K}^{s,\frac{n}{2}}(X^\wedge)$ has the same index for all $s \in \mathbb{R}$. The assertion for a general weight $\gamma \in \mathbb{R}$ follows by conjugating $1 + A_k$ by the corresponding weight shifts that induce isomorphisms $\mathcal{K}^{s,\gamma}(X^\wedge) \to \mathcal{K}^{s,\frac{n}{2}}(X^\wedge)$ for all s and preserve C_{M+G} up to the shift of weight data, cf. Remark 2.4.37. The case $\mathbb{B}$ can be treated in an analogous manner. In other words, by interpreting $\omega, \tilde{\omega}$ as cut-off functions on $\mathbb{B}$ supported by a neighbourhood of $\partial\mathbb{B}$ we can take formally the same operator as for $X^\wedge$, where now 1 means the identical operator on $\mathbb{B}$. □

Theorem 2.4.49 *Let $A \in C^\mu(\mathbb{B}, \boldsymbol{g})$ for $\mu \in \mathbb{R}$ and $\boldsymbol{g} = (\gamma, \delta, \Theta)$, $\gamma, \delta \in \mathbb{R}$, be an elliptic operator. Then there exists a $W \in C_{M+G}(\mathbb{B}, \boldsymbol{g})$ such that*

$$A + W : \mathcal{H}^{s,\gamma}(\mathbb{B}) \to \mathcal{H}^{s-\mu,\delta}(\mathbb{B})$$

is an isomorphism. Analogously if $A \in C^\mu(X^\wedge, \boldsymbol{g})$ is elliptic there exists a $W \in C_{M+G}(X^\wedge, \boldsymbol{g})$ such that

$$A + W : \mathcal{K}^{s,\gamma}(X^\wedge) \to \mathcal{K}^{s-\mu,\delta}(X^\wedge)$$

is an isomorphism. We then have $(A + W)^{-1} \in C^{-\mu}(\mathbb{B}, \boldsymbol{g}^{-1})$ and $(A + W)^{-1} \in C^{-\mu}(X^\wedge, \boldsymbol{g}^{-1})$, respectively, for $\boldsymbol{g}^{-1} = (\delta, \gamma, \Theta)$. The same is true within the subclasses with discrete asymptotics.

Proof. Let us consider the case $X^\wedge$; the considerations for $\mathbb{B}$ are analogous and left to the reader. In view of Theorem 2.4.40 the given operator A defines a Fredholm operator (2.4.36). According to Theorem 2.4.46 there is an operator $W_0 \in C_{M+G}(X^\wedge, (\gamma, \gamma, \Theta)^\bullet)$ such that $\operatorname{ind}(1 + W_0) = -\operatorname{ind} A$. Thus $\operatorname{ind} A(1 + W_0) = 0$. However $A(1 + W_0) = A + W_1$ for some $W_1 \in C_{M+G}(X^\wedge, \boldsymbol{g})$. It remains to choose a finite-dimensional Green operator G such that $A + W_1 + G$ is an isomorphism. Assume without loss of generality $\gamma = \delta = \frac{n}{2}$, cf. Remark 2.4.37, and consider $A + W_1 : L^2(\mathbb{R}_+ \times X) \to L^2(\mathbb{R}_+ \times X)$ which is of index zero. Then, applying Proposition 2.4.43 there are finite-dimensional subspaces $\mathcal{N}_+$, $\mathcal{N}_-$ such that $(\ker A)^\perp \oplus \mathcal{N}_+ = L^2(\mathbb{R}_+ \times X)$, $\operatorname{im} A \oplus \mathcal{N}_- = L^2(\mathbb{R}_+ \times X)$. It is not necessary here to represent $\mathcal{N}_+$, $\mathcal{N}_-$ in terms of Schwartz functions with asymptotics. Since $C_0^\infty(\mathbb{R}_+ \times X)$ is dense in $L^2(\mathbb{R}_+ \times X)$ it suffices to choose $\mathcal{N}_\pm$ as finite–dimensional spaces spanned by $C_0^\infty(\mathbb{R}_+ \times X)$ functions. Now the orthogonal projection $P : L^2(\mathbb{R}_+ \times X) \to \mathcal{N}_+$ is an element in $C_G(X^\wedge, (\frac{n}{2}, \frac{n}{2}, (-\infty, 0])^\bullet)$. Let $D : \mathcal{N}_+ \to \mathcal{N}_-$ be an arbitrary isomorphism. Then $DP =: G$ is also a Green operator of that class and it has obviously the desired property. Clearly the discrete asymptotics remain preserved under this construction. To prove that $(A+W)^{-1}$ is of the same structure as $A+W$ we first construct a parametrix $B_0 \in C^{-\mu}(X^\wedge, \boldsymbol{g}^{-1})$, cf. Theorem 2.4.40. Since $\operatorname{ind}(A+W) + \operatorname{ind} B_0 = 0$ and $\operatorname{ind}(A+W) = 0$, it follows that $\operatorname{ind} B_0 = 0$. Applying the arguments of the first part of the proof we find a finite–dimensional Green operator G_0 such that $B := B_0 + G_0$ is an isomorphism. Then $B(A + W) = 1 + G$ for a certain $G \in C_G(X^\wedge, (\gamma, \gamma, \Theta))$. Then $(A+W)^{-1} = (1+G)^{-1}B \in C^{-\mu}(X^\wedge, \boldsymbol{g}^{-1})$, cf. Proposition 2.3.68 and Theorem 2.4.15. Analogously we can prove the assertion for the subclass with discrete asymptotics. $\qquad\square$

More precisely, the following result can be proved:

Remark 2.4.50 *Let $A \in C^\mu(\mathbb{B}, \boldsymbol{g})$ for $\boldsymbol{g} = (\gamma, \delta, \Theta)$ be an operator for which*

$$A : \mathcal{H}^{s,\gamma}(\mathbb{B}) \to \mathcal{H}^{s-\mu,\delta}(\mathbb{B}) \tag{2.4.39}$$

is an isomorphism for an $s = s_0 \in \mathbb{R}$. Then (2.4.39) is invertible for all $s \in \mathbb{R}$ and we have $A^{-1} \in C^{-\mu}(\mathbb{B}, \boldsymbol{g}^{-1})$. An analogous result holds for $A \in C^\mu(X^\wedge, \boldsymbol{g})$ and for the subclasses with continuous asymptotics. This relies on the above Remarks 2.4.29, 2.4.41.

Chapter 3

Pseudo–differential calculus on manifolds with edges

A manifold with edges is locally a wedge, i.e., a Cartesian product between a model cone and an open set in Euclidean space. This chapter treats edge–degenerate pseudo–differential operators in an algebra with additional trace and potential operators along the edge, locally being pseudo–differential operators on the edge with cone operator–valued symbols. Ellipticity with respect to the interior and the edge symbol implies the existence of a parametrix in the algebra. Globally we obtain Fredholm operators in weighted edge Sobolev spaces. Asymptotics of solutions to elliptic equations are obtained as a refinement of elliptic regularity, using left parametrices and their continuity in subspaces with (weakly discrete or continuous) asymptotics. Typical geometric wedge operators, in particular, Laplace–Beltrami–operators to warped wedge metrics, are covered by the calculus. The operator algebra of this chapter was introduced by the author in [117] and then further developed in [119], [121], [122] and in a joint book with Egorov [24].

Typical elements of the approach are the Mellin operator conventions for edge–degenerate symbols that are smooth up to the edge, and a control of residual operators of Mellin and Green type, smooth in model cone–direction and in the interior. Moreover, the additional edge operators which satisfy an analogue of the Shapiro–Lopatinskij condition in the elliptic case are specific for our edge algebra. Operators in the edge pseudo–differential algebra are determined by the pair of {interior/edge} symbols modulo compact operators when the manifold with edges is compact (otherwise after localisation). For the index theory it is an interesting open problem to characterise the space of stable homotopy classes of elliptic {interior/edge} symbols through elliptic elements. An important analytic aspect is that the components of elliptic principal symbol tuples can interchange essential index information along the homotopies.

Edge–degenerate differential operators occur in many applications, in particular, in boundary and transmission problems, cf. Grisvard [46]. They were intensively investigated by numerous authors, cf. the monograph of Maz'ja, Kozlov, Rossmann [74] and the references there. Also Sobolev type problems, cf. Sternin [142], can be regarded as edge problems.

3.1   Manifolds with edges and wedge Sobolev spaces

3.1.1   Manifolds with edges

A typical example for a manifold with edges is the wedge $W = X^\Delta \times \Omega$ for an open set $\Omega \subseteq \mathbb{R}^q$ as edge and the model cone

$$X^\Delta = (\overline{\mathbb{R}}_+ \times X)/(\{0\} \times X) \qquad (3.1.1)$$

with base X.

Definition 3.1.1 *A manifold W with edges $Y \subset W$ is a topological space with the following properties:*

(i) *$W\backslash Y$ and Y are C^∞ manifolds; $q = \dim Y$, $n+1+q = \dim W\backslash Y$.*

(ii) *Every $y \in Y$ has a neighbourhood V in W with an associated non-empty system $\Phi(V)$ of homeomorphisms*

$$\varphi : V \to X^\Delta \times \Omega \qquad (3.1.2)$$

for a certain closed compact C^∞ manifold $X = X(y)$, $n = \dim X$, and an open set $\Omega \subseteq \mathbb{R}^q$, where $\varphi_0 = \varphi|_{V\backslash Y}$, $\varphi' = \varphi|_{V \cap Y}$ induce diffeomorphisms

$$\varphi_0 : V \backslash Y \to X^\wedge \times \Omega, \qquad \varphi' : V \cap Y \to \Omega. \qquad (3.1.3)$$

(iii) *Let V, $\tilde{V}$ be neighbourhoods of y and $\varphi \in \Phi(V)$, $\tilde{\varphi} \in \Phi(\tilde{V})$ arbitrary homeomorphisms $\varphi : V \to X^\Delta \times \Omega$, $\tilde{\varphi} : \tilde{V} \to X^\Delta \times \tilde{\Omega}$ as in (ii). Then for $U = V \cap \tilde{V}$ and $\Xi = \varphi'|_{U \cap Y}$, $\tilde{\Xi} = \tilde{\varphi}'|_{U \cap Y}$ the correponding homeomorphisms $\psi := \varphi|_U : U \to X^\Delta \times \Xi$, $\tilde{\psi} := \tilde{\varphi}|_U : U \to X^\Delta \times \tilde{\Xi}$ have the following property: The composition $\tilde{\psi}_0 \psi_0^{-1} : X^\wedge \times \Xi \to X^\wedge \times \tilde{\Xi}$ is the restriction of some diffeomorphism $\delta : \mathbb{R} \times X \times \Xi \to \mathbb{R} \times X \times \tilde{\Xi}$ to $X^\wedge \times \Xi$.*

Recall that all manifolds here are assumed to be paracompact.

Remark 3.1.2 *Some conditions in Definition 3.1.1 were imposed for convenience. For instance, it is not necessary to fix the edge dimension q. We might admit edges with components of different dimensions as* *well without affecting the essential aspects of the pseudo–differential analysis here. The assumption that X is closed and compact is more relevant. The case of a compact C^∞ manifold with boundary as base of the model cone corresponds to pseudo–differential boundary value problems on a manifold with edges and boundary.*

Example 3.1.3 *Let B be a manifold with conical singularities S in the sense of Definition 2.1.1. Then, if M is an arbitrary C^∞ manifold, $W = B \times M$ is a manifold with edges $Y = S \times M$.*

A manifold W with edges Y is not necessarily a C manifold, (i.e., with continuous transition maps) though we call W a manifold, since the analytic objects are considered on $W \setminus Y$. Intuitively, the local model of W near every $y \in Y$ is a wedge $X^\Delta \times G$, i.e., a Cartesian product between the model cone X^Δ and a neighbourhood G of y on Y. In local coordinates G corresponds to $\Omega \subseteq \mathbb{R}^q$. We will often pass to the open stretched wedge

$$X^\wedge \times \Omega \ni (t, x, y). \tag{3.1.4}$$

Globally a manifold W with edges Y gives rise to an X^Δ–bundle K over Y: Every $y \in Y$ has a neighbourhood G in Y such that $V = K|_G$ is trivial. The mappings (3.1.2) are just trivialisations with the induced charts $\varphi' : G \to \Omega$. The assumptions show that K is determined by a cylinder–bundle $\mathbb{K}$ over Y with fibres $\overline{\mathbb{R}}_+ \times X$ by factorising the fibre over every $y \in Y$ to the cone X^Δ. The transition mappings for $\mathbb{K}$ follow from Definition 3.1.1 (iii) as

$$\overline{\mathbb{R}}_+ \times X \times \Xi \to \overline{\mathbb{R}}_+ \times X \times \tilde{\Xi}. \tag{3.1.5}$$

The restrictions of (3.1.5) to $\{0\} \times X \times \Xi$ yield diffeomorphisms

$$X \times \Xi \to X \times \tilde{\Xi} \tag{3.1.6}$$

that project to the transition diffeomorphisms $\Xi \to \tilde{\Xi}$ of the corresponding system of charts on the edge. (3.1.6) are the transition maps of an X–bundle over Y, namely $\partial\mathbb{K}$. Similarly to the local wedge neighbourhood V of Definition 3.1.1 (ii), there is a twisted wedge neighbourhood T of Y in W and a homeomorphism $\tau : T \to K$ which restricts to diffeomorphisms

$$\tau_0 : T \setminus Y \to \mathbb{K} \setminus \partial\mathbb{K}, \qquad \tau' : T \cap Y \to Y.$$

Now $T \setminus Y$ can be regarded as the interior of a C^∞ manifold $\mathbb{T}$ with boundary $\partial\mathbb{T} \cong \partial\mathbb{K}$ such that there is a diffeomorphism

$$\tau : \mathbb{T} \to \mathbb{K} \tag{3.1.7}$$

in the sense of C^∞ manifolds with boundary, and $\tau|_{T\backslash Y} = \tau_0$. Since $T \backslash Y$ is a subset of $W \backslash Y$, we can complete $W \backslash Y$ by $\partial \mathbb{T}$ to a C^∞ manifold $\mathbb{W}$ with boundary. $\mathbb{W}$ is called the stretched manifold with edges, associated with W. We have by construction

$$\partial \mathbb{W} = \partial \mathbb{T}$$

which is an X–bundle over Y, isomorphic to $\partial \mathbb{K}$. The canonical projection $\partial \mathbb{W} \to Y$ gives rise to a continuous map $\mathbb{W} \to W$ which induces a diffeomorphism $\mathbb{W} \backslash \partial \mathbb{W} \to W \backslash Y$.

Example 3.1.4 *Let* $W = X^\Delta \times G$ *for an open set* $G \subseteq Y$. *Then* $\mathbb{W} = K$, *and* $\mathbb{W} = \mathbb{K} = \overline{\mathbb{R}}_+ \times X \times G$. *Another example is* $W = B \times M$, *cf. Example 3.1.3, and we have* $\mathbb{W} = \mathbb{B} \times M$ *with the stretched manifold* $\mathbb{B}$ *to* B.

Remark 3.1.5 *The transition mappings* (3.1.5) *for* $\mathbb{K}$ *can be chosen in such a way that they commute with the canonical* $\mathbb{R}_+$ *actions on the fibres* $\overline{\mathbb{R}}_+ \times X$, *given by* $(t,x) \to (\lambda t, x)$, $\lambda \in \mathbb{R}_+$. *They induce an* $\mathbb{R}_+$–*action on* $\mathbb{K}$. *The bundle* $\mathbb{K}$ *is isomorphic to* $\overline{\mathbb{R}}_+ \times \partial \mathbb{K}$. *This enables us to define* $\mathbb{K}_1 = \{(t,x,y) \in \mathbb{K} : 0 \leq t < 1\}$ *and* $\mathbb{T}_1 = \tau^{-1}\mathbb{K}_1$. *Then* $\mathbb{T}_1$ *is a collar neighbourhood of* $\partial \mathbb{W}$ *in* $\mathbb{W}$.

We fix a Riemannian metric on $\mathbb{W}$ that corresponds to the product metric $[0,1) \times \partial \mathbb{W}$ in $\mathbb{T}_1$.

The manifolds with edges form a category with the following morphisms.

Definition 3.1.6 *Let* W *and* $\tilde{W}$ *be manifolds with edges* Y *and* $\tilde{Y}$, *respectively, and* $\mathbb{W}$, $\tilde{\mathbb{W}}$ *be the corresponding stretched manifolds with the associated diffeomorphisms*

$$\pi_0 : \mathbb{W} \backslash \partial \mathbb{W} \to W \backslash Y, \qquad \varrho_0 : \tilde{\mathbb{W}} \backslash \partial \tilde{\mathbb{W}} \to \tilde{W} \backslash \tilde{Y}.$$

Then a continuous map $\alpha : W \to \tilde{W}$ *is a morphism if*

(i) α *restricts to differentiable maps*

$$\alpha_0 : W \backslash Y \to \tilde{W} \backslash \tilde{Y}, \qquad \alpha' : Y \to \tilde{Y},$$

(ii) $\kappa_0 := \varrho_0^{-1}\alpha_0\pi_0 : \mathbb{W} \backslash \partial \mathbb{W} \to \tilde{\mathbb{W}} \backslash \partial \tilde{\mathbb{W}}$ *is the restriction of a differentiable map* $\kappa : \mathbb{W} \to \tilde{\mathbb{W}}$ *(in particular, κ is smooth up to the boundaries).*

If $\alpha : W \to \tilde{W}$ *is a morphism, and if* $\alpha^{-1} : \tilde{W} \to W$ *exists as a morphism, then* α *is called an isomorphism between* W *and* $\tilde{W}$.

3.1.2 Edge–degenerate differential operators

Following the general scheme of developing a calculus of pseudo–differential operators on a manifold with singularities we start with a description of the typical differential operators.

Points on the open stretched wedge $X^\wedge \times \Omega$ with edge $\Omega \subseteq \mathbb{R}^q$ will be denoted by (t, x, y). Recall that the base of the (open stretched) model cone $X^\wedge = \mathbb{R}_+ \times X$ is a closed compact C^∞ manifold X of dimension n.

Definition 3.1.7 *Let W be a manifold with edges Y. Then*

$$\text{Diff}^\mu(W)_{\text{edge}} \tag{3.1.8}$$

denotes the subspace of those differential operators $A \in \text{Diff}^\mu(W \setminus Y)$ such that for every neighbourhood V of $y \in Y$ with $V \setminus Y \cong X^\wedge \times \Omega$, as in Definition 3.1.1, A has the form

$$A = t^{-\mu} \sum_{j+|\alpha|\leq\mu} a_{j\alpha}(t, y) \left(-t\frac{\partial}{\partial t}\right)^j (tD_y)^\alpha \tag{3.1.9}$$

with $a_{j\alpha} \in C^\infty(\overline{\mathbb{R}}_+ \times \Omega, \text{Diff}^{\mu-(j+|\alpha|)}(X))$ for all j, α.

We will also use the notation

$$\text{Diff}^\mu(\mathbb{W})_{\text{edge}} \tag{3.1.10}$$

for the space of all $A \in \text{Diff}^\mu(\text{int}\,\mathbb{W})$ of the form (3.1.9) in a collar neighbourhood of $\partial\mathbb{W}$ in the coordinates (t, x, y). The elements of (3.1.10) also are called edge–degenerate differential operators on $\mathbb{W}$.

Note that (3.1.9) can be written

$$A = \sum_{|\alpha|\leq\mu} B_\alpha D_y^\alpha \tag{3.1.11}$$

with coefficients $B_\alpha \in \text{Diff}^{\mu-|\alpha|}(\mathbb{R}_+ \times X)_{\text{Fuchs}}$, cf. Definition 2.1.8. Conversely each operator of the form (3.1.11) belongs to $\text{Diff}^\mu(\overline{\mathbb{R}}_+ \times X \times \Omega)_{\text{edge}}$. Differential operators $A \in \text{Diff}^\mu(W)_{\text{edge}}$ are first interpreted as continuous maps

$$A : C_0^\infty(W \setminus Y) \to C_0^\infty(W \setminus Y) .$$

Alternatively, each $A \in \text{Diff}^\mu(\mathbb{W})_{\text{edge}}$ represents a continuous operator

$$A : C_0^\infty(\text{int}\,\mathbb{W}) \to C_0^\infty(\text{int}\,\mathbb{W}).$$

Occasionally we formulate definitions or assertions in the variant for W without the obvious analogues for $\mathbb{W}$, or vice–versa.

Proposition 3.1.8 *Let $\chi : W \to \tilde{W}$ be an isomorphism of manifolds with edges, cf. Definition 3.1.1. Then the operator push–forward χ_*, defined by $(\chi_* A)v = (\chi^*)^{-1} A \chi^* v$, $v \in C_0^\infty(W \setminus Y)$ (with χ^* being the function pull–back) induces an isomorphism*

$$\chi_* : \mathrm{Diff}^\mu(W)_{\mathrm{edge}} \to \mathrm{Diff}^\mu(\tilde{W})_{\mathrm{edge}}.$$

In local considerations on $W \setminus Y$ near Y it is convenient to assume that the transition maps on the trivialisations of the $\overline{\mathbb{R}}_+ \times X$–bundle $\mathbb{K}$ over Y are independent of t, cf. Remark 3.1.5. This will be done in the sequel unless we indicate more general coordinate changes. In particular, we can fix a function $h^\gamma \in C^\infty(\mathrm{int}\,W)$, $\gamma \in \mathbb{R}$, that is strictly positive, and

$$h^\gamma(t) = t^\gamma \qquad \text{for } 0 < t < \varepsilon$$

with some $0 < \varepsilon < 1$ (recall that after Remark 3.1.5 we fixed a Riemannian metric on W as the product metric $[0,1) \times \partial W$ near ∂W). Let us now turn to some typical examples of edge–degenerate differential operators on W.

Example 3.1.9 *Let v be an arbitrary C^∞ vector field on W that is on $[0,1) \times \partial W$ in local coordinates $(t,x,y) \in [0,1) \times \Sigma \times \Omega$ (with respect to charts $D \to \Sigma$ on X, $G \to \Omega$ on Y, $\Sigma \subseteq \mathbb{R}^n$, $\Omega \subseteq \mathbb{R}^q$ open) of the form*

$$\alpha(t,x,y)t\frac{\partial}{\partial t} + \sum_{i=1}^n \beta_i(t,x,y)\frac{\partial}{\partial x_i} + \sum_{k=1}^q \delta_k(t,x,y)t\frac{\partial}{\partial y_k},$$

with $\alpha, \beta_i, \delta_k \in C^\infty([0,1) \times \Sigma \times \Omega)$. Then $h^{-1}v \in \mathrm{Diff}^1(W)_{\mathrm{edge}}$.

Example 3.1.10 *Let us interpret the space $\mathbb{R}^{1+n+q} \ni (\tilde{x}, y)$ with $\tilde{x} \in \mathbb{R}^{1+n}$, $y \in \mathbb{R}^q$, as a manifold W with edge $\mathbb{R}^q$, such that $W = \overline{\mathbb{R}}_+ \times S^n \times \mathbb{R}^q$, with $S^n = \{\tilde{x} \in \mathbb{R}^{1+n} : |\tilde{x}| = 1\}$. Then, by substituting polar coordinates $\mathbb{R}^{1+n} \setminus \{0\} \ni \tilde{x} \to (t,x) \in \mathbb{R}_+ \times S^n$ in an arbitrary $\tilde{A} \in \mathrm{Diff}^\mu(\mathbb{R}^{1+n+q})$ the operator $\tilde{A}|_{(\mathbb{R}^{1+n} \setminus \{0\}) \times \mathbb{R}^q}$ takes the form of an element in $\mathrm{Diff}^\mu(\overline{\mathbb{R}}_+ \times S^n \times \mathbb{R}^q)_{\mathrm{edge}}$.*

Example 3.1.11 *Let $g(t,y)$ be a (t,y)–dependent family of Riemannian metrics on X, C^∞ in $(t,y) \in \overline{\mathbb{R}}_+ \times \Omega$, $\Omega \subseteq \mathbb{R}^q$ open. Then, the Laplace–Beltrami operator to the Riemannian metric $dt^2 + t^2 g(t,y) + dy^2$ on $X^\wedge \times \Omega$ belongs to $\mathrm{Diff}^2(\overline{\mathbb{R}}_+ \times X \times \Omega)_{\mathrm{edge}}$.*

3.1.3   Anisotropic descriptions of isotropic objects

An important principle in constructing algebras of operators together
with the adequate analogues of the Sobolev spaces on a manifold with
singularities (here edges) is that the operators and spaces have to coincide with the usual ones after localising outside the singularities.
Conversely, if M is a C^∞ manifold with a fixed C^∞ submanifold Y
of dimension q, regarded as an edge of codimension $\dim M - q$, then
the "isotropic" standard operators and spaces should allow anisotropic
descriptions relative to Y, under which they become specialisations of
the singular objects.

Let us begin this discussion with an anisotropic reformulation of
the standard Sobolev spaces $H^s(\mathbb{R}^{n+1+q})$, $s \in \mathbb{R}$, relative to $Y = \mathbb{R}^q$.
Points in $\mathbb{R}^{n+1}$ will be denoted by $\tilde{x}$, with the covariables $\tilde{\xi}$. Set

$$\|u\|_{H^s(\mathbb{R}^{n+1+q})} = \left\{ \int (|\tilde{\xi}|^2 + [\eta]^2)^s |(Fu)(\tilde{\xi}, \eta)|^2 \, d\tilde{\xi} d\eta \right\}^{\frac{1}{2}}$$

with the Fourier transform $F = F_{(\tilde{x},y)\to(\tilde{\xi},\eta)}$, and

$$\|f\|_{H^s(\mathbb{R}^{n+1})} = \left\{ \int (1 + |\tilde{\xi}|^2)^s |(F_{\tilde{x}\to\tilde{\xi}} f)(\tilde{\xi})|^2 d\tilde{\xi} \right\}^{\frac{1}{2}}.$$

For convenience we often employ the same notations for different equivalent norms.

Lemma 3.1.12 *For every $s \in \mathbb{R}$ we have*

$$\|u\|_{H^s(\mathbb{R}^{n+1+q})} = \left\{ \int [\eta]^{2s} \|\kappa^{-1}(\eta)(F_{y\to\eta} u)(\tilde{x}, \eta)\|^2_{H^s(\mathbb{R}^{n+1})} d\eta \right\}^{\frac{1}{2}}.$$

Here

$$(\kappa_\lambda f)(\tilde{x}) = \lambda^{\frac{n+1}{2}} f(\lambda \tilde{x}), \quad \lambda \in \mathbb{R}_+,$$

$f \in H^s(\mathbb{R}^{n+1})$, and $\kappa(\eta) := \kappa_{[\eta]}$.

Proof. We have

$$\iint (|\tilde{\xi}|^2 + [\eta]^2)^s |(Fu)(\tilde{\xi}, \eta)|^2 d\tilde{\xi} \, d\eta$$

$$= \iint [\eta]^{2s} (1 + (\frac{|\tilde{\xi}|}{[\eta]})^2)^s |(Fu)(\tilde{\xi}, \eta)|^2 \, d\tilde{\xi} d\eta$$

$$= \iint [\eta]^{2s} (1 + |\tilde{\xi}|^2)^s |(Fu)([\eta]\tilde{\xi}, \eta)|^2 [\eta]^{n+1} \, d\tilde{\xi} d\eta$$

$$= \int [\eta]^{2s} \int (1 + |\tilde{\xi}|^2)^s |\kappa(\eta)(Fu)(\tilde{\xi}, \eta)|^2 \, d\tilde{\xi} d\eta$$

$$= \int [\eta]^{2s} \|\kappa^{-1}(\eta)(F_{y\to\eta} u)(\tilde{x}, \eta)\|^2_{H^s(\mathbb{R}^{n+1})} \, d\eta.$$

Here the identity $\kappa_\lambda(F_{\tilde{x} \to \tilde{\xi}} f)(\tilde{\xi}) = F_{\tilde{x} \to \tilde{\xi}}(\kappa_\lambda^{-1} f)(\tilde{\xi})$ was used. $\square$

Let us now reinterpret pseudo–differential operators in $\mathbb{R}^{n+1} \times \Omega$, $\Omega \subseteq \mathbb{R}^q$ open, in anisotropic terms. Given a symbol

$$p(\tilde{x}, y, \tilde{\xi}, \eta) \in S^\mu(\mathbb{R}^{n+1} \times \Omega \times \mathbb{R}_{\tilde{\xi}, \eta}^{n+1+q})$$

of order $\mu \in \mathbb{R}$ we can form a (y, η)–dependent family $a(y, \eta)$ of pseudo–differential operators in $\mathbb{R}^{n+1}$, namely

$$(a(y, \eta) f)(\tilde{x}) = \iint e^{i(\tilde{x} - \tilde{x}')\tilde{\xi}} p(\tilde{x}, y, \tilde{\xi}, \eta) f(\tilde{x}') \, d\tilde{x}' d\eta, \qquad (3.1.12)$$

$f \in C_0^\infty(\mathbb{R}^{n+1})$. Let us set

$$\mathrm{op}_{\psi, \tilde{x}}(p)(y, \eta) = a(y, \eta). \qquad (3.1.13)$$

In Section 1.3.1 we have introduced the spaces of operator–valued symbols $S^\mu(\Omega \times \mathbb{R}^q; E, \tilde{E})$ with Banach spaces E, $\tilde{E}$ and fixed groups of isomorphisms $\{\kappa_\lambda\}_{\lambda \in \mathbb{R}_+} \in C(\mathbb{R}_+, \mathcal{L}_\sigma(E))$, $\{\tilde{\kappa}_\lambda\}_{\lambda \in \mathbb{R}_+} \in C(\mathbb{R}_+, \mathcal{L}_\sigma(\tilde{E}))$. Recall that $S^\mu(\Omega \times \mathbb{R}^q; E, \tilde{E})$ is a Fréchet space. The subspace of y–independent elements $S^\mu(\mathbb{R}^q; E, \tilde{E})$ is closed in the induced topology. Then,

$$\begin{aligned} S^\mu(\Omega \times \mathbb{R}^q; E, \tilde{E}) &= C^\infty(\Omega, S^\mu(\mathbb{R}^q; E, \tilde{E})) \\ &= C^\infty(\Omega) \otimes_\pi S^\mu(\mathbb{R}^q; E, \tilde{E}), \end{aligned} \qquad (3.1.14)$$

where $\otimes_\pi$ is the completed projective tensor product.

In particular, let $E = H^s(\mathbb{R}^{n+1})$, $\tilde{E} = H^{s-\mu}(\mathbb{R}^{n+1})$ for arbitrary fixed $s \in \mathbb{R}$,

$$(\kappa_\lambda f)(\tilde{x}) = \lambda^{\frac{n+1}{2}} f(\lambda \tilde{x}), \qquad \lambda \in \mathbb{R}_+,$$

$f \in H^s(\mathbb{R}^{n+1})$, and define $\tilde{\kappa}_\lambda$ by the same expression for functions $f \in H^{s-\mu}(\mathbb{R}^{n+1})$. Then we have the symbol spaces

$$S^\mu(\Omega \times \mathbb{R}^q; H^s(\mathbb{R}^{n+1}), H^{s-\mu}(\mathbb{R}^{n+1}))$$

as well as the subspaces of classical symbols.

Proposition 3.1.13 *Let* $p(\tilde{x}, y, \tilde{\xi}, \eta) \in S^\mu(\mathbb{R}^{n+1} \times \Omega \times \mathbb{R}_{\tilde{\xi}, \eta}^{n+1+q})$, $\mu \in \mathbb{R}$, *assume that* p *is independent of* $\tilde{x}$ *for* $|\tilde{x}| > R$ *for some* $R > 0$, *and let* $a(y, \eta) = \mathrm{op}_{\psi, \tilde{x}}(p)(y, \eta)$. *Then*

$$a(y, \eta) \in S^\mu(\Omega \times \mathbb{R}^q; H^s(\mathbb{R}^{n+1}), H^{s-\mu}(\mathbb{R}^{n+1}))$$

for every $s \in \mathbb{R}$.

In order to prove Proposition 3.1.13 let us first show the following result.

Lemma 3.1.14 *Let* $p = p(y, \tilde{\xi}, \eta) \in S^\mu(\Omega \times \mathbb{R}^{n+1+q}_{\tilde{\xi}, \eta})$, $\mu \in \mathbb{R}$, *be a symbol that is independent of* $\tilde{x}$. *Then,* $p(y, \tilde{\xi}, \eta) \to a(y, \eta) = \operatorname{op}_{\psi, \tilde{x}}(p)(y, \eta)$ *induces a continuous operator*

$$S^\mu(\Omega \times \mathbb{R}^{n+1+q}) \to S^\mu(\Omega \times \mathbb{R}^q; H^s(\mathbb{R}^{n+1}), H^{s-\mu}(\mathbb{R}^{n+1})) \qquad (3.1.15)$$

for every $s \in \mathbb{R}$.

Proof. Let $r^s(\tilde{\xi}, \eta) = [\tilde{\xi}, \eta]^s$ for any fixed function $(\tilde{\xi}, \eta) \to [\tilde{\xi}, \eta]$ in $C^\infty(\mathbb{R}^{n+1+q})$, strictly positive, with $[\tilde{\xi}, \eta] = |\tilde{\xi}, \eta|$ for $|\tilde{\xi}, \eta| > \text{const}$ for a constant > 0. Set $b^s(\eta) = \operatorname{op}_{\psi, \tilde{x}}(r^s)(\eta)$, $s \in \mathbb{R}$. Then

$$b^s(\eta) : H^s(\mathbb{R}^{n+1}) \to L^2(\mathbb{R}^{n+1})$$

is an isomorphism for every $\eta \in \mathbb{R}^q$, and we have

$$b^s(\lambda \eta) = \lambda^s \kappa_\lambda b^s(\eta) \kappa_\lambda^{-1} \qquad \text{for all } \lambda \geq 1, |\eta| \geq \text{const}$$

for a constant > 0. Moreover, $b^s(\eta) \in C^\infty(\mathbb{R}^q, \mathcal{L}(H^s(\mathbb{R}^{n+1}), L^2(\mathbb{R}^{n+1})))$. This implies $b^s(\eta) \in S^s_{\mathrm{cl}}(\mathbb{R}^q; H^s(\mathbb{R}^{n+1}), L^2(\mathbb{R}^{n+1}))$ and

$$b^s(\eta)^{-1} = b^{-s}(\eta) \in S^{-s}_{\mathrm{cl}}(\mathbb{R}^q; L^2(\mathbb{R}^{n+1}), H^s(\mathbb{R}^{n+1})).$$

Let us set

$$a_0(y, \eta) = b^{s-\mu}(\eta) a(y, \eta) b^{-s}(\eta),$$
$$p_0(y, \tilde{\xi}, \eta) = r^{s-\mu}(\tilde{\xi}, \eta) p(y, \tilde{\xi}, \eta) r^{-s}(\tilde{\xi}, \eta).$$

Then, if we show the property

$$a_0(y, \eta) \in S^0(\Omega \times \mathbb{R}^q; L^2(\mathbb{R}^{n+1}), L^2(\mathbb{R}^{n+1})),$$

we obtain $a(y, \eta) \in S^\mu(\Omega \times \mathbb{R}^q; H^s(\mathbb{R}^{n+1}), H^{s-\mu}(\mathbb{R}^{n+1}))$. Moreover, since

$$a_0(y, \eta) = \operatorname{op}_{\psi, \tilde{x}}(r^{s-\mu} p r^{-s}), \qquad a(y, \eta) = \operatorname{op}_{\psi, \tilde{x}}(r^{-s+\mu} p_0 r^s),$$

we also obtain the asserted continuity if we show it for $s = \mu = 0$. In fact, it suffices to employ the fact that the composition of symbols of different orders is itself a symbol, where the resulting order is the sum of the orders of the factors, and that this yields a separately continuous map between the corresponding symbol spaces. This holds both for scalar and operator–valued symbols. Now

$$\|\kappa^{-1}(\eta) D_y^\alpha D_\eta^\beta a_0(y, \eta) \kappa(\eta)\|_{\mathcal{L}(L^2(\mathbb{R}^{n+1}))} = \|D_y^\alpha D_\eta^\beta a_0(y, \eta)\|_{\mathcal{L}(L^2(\mathbb{R}^{n+1}))},$$

since κ_λ is unitary in $L^2(\mathbb{R}^{n+1})$ for every $\lambda \in \mathbb{R}_+$. We then obtain for $u(\tilde{x}) \in L^2(\mathbb{R}^{n+1})$ and $F = F_{\tilde{x}\to\tilde{\xi}}$

$$\|D_y^\alpha D_\eta^\beta a_0(y,\eta)u\|_{L^2(\mathbb{R}^{n+1})}^2 = \int |D_y^\alpha D_\eta^\beta p_0(y,\tilde{\xi},\eta)(Fu)(\tilde{\xi})|^2 \, d\tilde{\xi}$$

$$\leq \{ \sup_{\tilde{\xi}\in\mathbb{R}^{n+1}} |D_y^\alpha D_\eta^\beta p_0(y,\tilde{\xi},\eta)|\|u\|_{L^2(\mathbb{R}^{n+1})}\}^2.$$

From the symbol estimates for p_0 we get

$$|D_y^\alpha D_\eta^\beta p_0(y,\tilde{\xi},\eta)| \leq c\langle\tilde{\xi},\eta\rangle^{-|\beta|}$$

for all multi–indices α, β and all $y \in K \subset\subset \Omega$ for arbitrary $K \subset\subset \Omega$ and all $(\tilde{\xi},\eta) \in \mathbb{R}^{n+1+q}$, with constants $c = c(\alpha,\beta,K) > 0$. Thus the estimate

$$\sup_{\tilde{\xi}\in\mathbb{R}^{n+1}} |D_y^\alpha D_\eta^\beta p_0(y,\tilde{\xi},\eta)| \leq c(\alpha,\beta,K) \sup_{\tilde{\xi}\in\mathbb{R}^{n+1}} \langle\tilde{\xi},\eta\rangle^{-|\beta|}$$

$$\leq \tilde{c}(\alpha,\beta,K)\langle\eta\rangle^{-|\beta|}$$

for some other constant $\tilde{c}(\alpha,\beta,K)$ gives us the bound

$$\|\kappa^{-1}(\eta)D_y^\alpha D_\eta^\beta a_0(y,\eta)\kappa(\eta)\|_{\mathcal{L}(L^2(\mathbb{R}^{n+1}))} \leq \tilde{c}(\alpha,\beta,K)\langle\eta\rangle^{-|\beta|},$$

where $\tilde{c}(\alpha,\beta,K)$ is proportional to $c(\alpha,\beta,K)$ which shows that $a_0 \in S^0(\Omega\times\mathbb{R}^q; L^2(\mathbb{R}^{n+1}), L^2(\mathbb{R}^{n+1}))$. Since the best constants in the symbol estimates form a system of semi–norms for the Fréchet topology in the corresponding symbol spaces, we see that the map $p_0 \to \mathrm{op}_{\psi,\tilde{x}}(p_0)$ is continuous. □

Remark 3.1.15 *The operator* (3.1.15) *restricts to a continuous operator*

$$S_{\mathrm{cl}}^\mu(\Omega \times \mathbb{R}^{n+1+q}) \to S_{\mathrm{cl}}^\mu(\Omega \times \mathbb{R}^q; H^s(\mathbb{R}^{n+1}), H^{s-\mu}(\mathbb{R}^{n+1}))$$

for each $s \in \mathbb{R}$. *This follows easily by modifying the technique to prove Lemma* 3.1.14 *and from the result that for every* $p_{(\nu)}(y,\tilde{\xi},\eta) \in C^\infty(\Omega \times (\mathbb{R}^{n+1+q} \setminus \{0\}))$ *with* $p_{(\nu)}(y,\lambda\tilde{\xi},\lambda\eta) = \lambda^\nu p_{(\nu)}(y,\tilde{\xi},\eta)$ *for all* $\lambda \in \mathbb{R}_+$, $(y,\tilde{\xi},\eta) \in \Omega \times \mathbb{R}^{n+1+q} \setminus \{0\}$, *we have*

$$\mathrm{op}_{\psi,\tilde{x}}(p_{(\nu)})(y,\lambda\eta) = \lambda^\nu \kappa_\lambda \, \mathrm{op}_{\psi,\tilde{x}}(p_{(\nu)})(y,\eta)\kappa_\lambda^{-1}$$

for all $\lambda \in \mathbb{R}_+$, $(y,\eta) \in \Omega \times (\mathbb{R}^q \setminus \{0\})$.

Lemma 3.1.16 *The operator* $\mathcal{M}_\varphi$ *of multiplication by* $\varphi \in \mathcal{S}(\mathbb{R}^{n+1})$ *represents an element in* $S^0(\mathbb{R}^q; H^s(\mathbb{R}^{n+1}), H^s(\mathbb{R}^{n+1}))$, *and* $\varphi \to \mathcal{M}_\varphi$ *induces a continuous operator*

$$\mathcal{S}(\mathbb{R}^{n+1}) \to S^0(\mathbb{R}^q; H^s(\mathbb{R}^{n+1}), H^s(\mathbb{R}^{n+1}))$$

for each $s \in \mathbb{R}$.

Proof. To verify the result it suffices to look at $\kappa(\eta)^{-1}\mathcal{M}_\varphi\kappa(\eta) = \mathcal{M}_{\varphi_\eta}$ for $\varphi_\eta(\tilde{x}) = \varphi(\tilde{x}[\eta]^{-1})$ and to check the estimate

$$\|\mathcal{M}_{\varphi_\eta}\|_{\mathcal{L}(H^s(\mathbb{R}^{n+1}))} \leq c \qquad \text{for all} \quad \eta \in \mathbb{R}^q$$

for a constant $c = c(s) > 0$. This is an elementary consequence of Theorem 1.1.55 and Corollary 1.1.56, applied in the slightly more general form for functions $\varphi \in \mathcal{S}(\mathbb{R}^q)$. The explicit estimates for the constant C show, in particular, the continuity of $\varphi \to \mathcal{M}_\varphi$. □

Proof of Proposition 3.1.13. By assumption $p(\tilde{x}, y, \tilde{\xi}, \eta)$ has the form $p_1(y, \tilde{\xi}, \eta) + p_2(\tilde{x}, y, \tilde{\xi}, \eta)$, where p_1 is independent of $\tilde{x}$, and p_2 has compact support with respect to $\tilde{x}$. In view of Lemma 3.1.14 we already know that $\mathrm{op}_{\psi,\tilde{x}}(p_1)(y, \eta)$ is an operator–valued symbol as asserted. For p_2 we fix a compact set $K \subset \mathbb{R}^{n+1}_{\tilde{x}}$ such that $p_2(\tilde{x}, y, \tilde{\xi}, \eta) = 0$ for all $\tilde{x} \notin K$. Then p_2 can be regarded as an element of $C_0^\infty(K) \otimes_\pi S^\mu(\Omega \times \mathbb{R}^{n+1+q}_{\tilde{\xi},\eta})$. Hence, applying Theorem 1.1.6, the symbol p_2 can be written

$$p_2(\tilde{x}, y, \tilde{\xi}, \eta) = \sum_{j=0}^\infty \lambda_j \varphi_j(\tilde{x}) r_j(y, \tilde{\xi}, \eta)$$

for $\lambda_j \in \mathbb{C}$, $\sum |\lambda_j| < \infty$, and $\varphi_j \in C_0^\infty(K)$, $r_j \in S^\mu(\Omega \times \mathbb{R}^{n+1+q})$ tending to 0 in the corresponding spaces for $j \to \infty$. Then

$$\mathrm{op}_{\psi,\tilde{x}}(p_2)(y, \eta) = \sum_{j=0}^\infty \lambda_j \mathcal{M}_{\varphi_j}\, \mathrm{op}_{\psi,\tilde{x}}(r_j)(y, \eta)$$

converges in $S^\mu(\mathbb{R}^{n+1} \times \Omega \times \mathbb{R}^{n+1+q})$, since Lemma 3.1.14 and Lemma 3.1.16 imply that $\mathcal{M}_{\varphi_j}\, \mathrm{op}_{\psi,\tilde{x}}(r_j)$ converge in $S^\mu(\mathbb{R}^{n+1} \times \Omega \times \mathbb{R}^{n+1+q})$ to zero as $j \to \infty$. □

Corollary 3.1.17 *Under the conditions of Proposition 3.1.13 we have*

$$D_y^\alpha D_\eta^\beta a(y, \eta) \in \bigcap_{s \in \mathbb{R}} S^{\mu-|\beta|}(\Omega \times \mathbb{R}^q; H^s(\mathbb{R}^{n+1}), H^{s-\mu+|\beta|}(\mathbb{R}^{n+1}))$$

for all multi–indices α, β.

In fact, this follows from $D_y^\alpha D_\eta^\beta\, \mathrm{op}_{\psi,\tilde{x}}(p)(y, \eta) = \mathrm{op}_{\psi,\tilde{x}}(D_y^\alpha D_\eta^\beta p)(y, \eta)$ and from Proposition 3.1.13 together with $D_y^\alpha D_\eta^\beta p \in S^{\mu-|\beta|}(\mathbb{R}^{n+1} \times \Omega \times \mathbb{R}^{n+1+q})$.

3.1.4   Wedge Sobolev spaces

Let E be a Banach space and $\{\kappa_\lambda\}_{\lambda \in \mathbb{R}_+} \in C(\mathbb{R}_+, \mathcal{L}_\sigma(E))$ be a group of isomorphisms, cf. also Section 1.3.2. Recall that $\mathcal{W}^s(\mathbb{R}^q, E)$ for $s \in \mathbb{R}$ is defined as the closure of $\mathcal{S}(\mathbb{R}^q, E)$ with respect to the norm

$$\|u\|_{\mathcal{W}^s(\mathbb{R}^q, E)} = \left\{ \int [\eta]^{2s} \|\kappa^{-1}(\eta)(F_{y \to \eta} u)(\eta)\|_E^2 \, d\eta \right\}^{\frac{1}{2}}. \qquad (3.1.16)$$

We also write $\hat{u}(\eta) = (F_{y \to \eta} u)(\eta)$. In Lemma 3.1.12 we obtained

$$H^s(\mathbb{R}^{n+1+q}) = \mathcal{W}^s(\mathbb{R}^q, H^s(\mathbb{R}^{n+1})) \qquad (3.1.17)$$

with $E = H^s(\mathbb{R}^{n+1})$, $(\kappa_\lambda f)(\tilde{x}) = \lambda^{\frac{n+1}{2}} f(\lambda \tilde{x})$, $\lambda \in \mathbb{R}_+$. The weighted wedge Sobolev spaces will be defined in terms of $E = \mathcal{K}^{r,\gamma}(X^\wedge)$, $r, \gamma \in \mathbb{R}$, with $(\kappa_\lambda f)(t, x) = \lambda^{\frac{n+1}{2}} f(\lambda t, x)$ for $f(t, x) \in \mathcal{K}^{r,\gamma}(X^\wedge)$, $n = \dim X$. Note that

$$u(t, x, y) \in \mathcal{W}^s(\mathbb{R}^q, \mathcal{K}^{r,\gamma}(X^\wedge))$$

is equivalent to

$$(F_{y \to \eta} u)(t, x, \eta) = [\eta]^{\frac{n+1}{2}} (F_{y \to \eta} v)(t[\eta], x, \eta)$$

for some $v(t, x, y) \in H^s(\mathbb{R}^q, \mathcal{K}^{r,\gamma}(X^\wedge))$.

Definition 3.1.18 *The space*

$$\mathcal{W}^{s,\gamma}(X^\wedge \times \mathbb{R}^q) := \mathcal{W}^s(\mathbb{R}^q, \mathcal{K}^{s,\gamma}(X^\wedge))$$

for $s, \gamma \in \mathbb{R}$ is called a weighted wedge Sobolev space of smoothness s and weight γ. We set

$$\mathcal{W}^s_{\mathrm{comp}(y)}(X^\wedge \times \Omega) = \mathcal{W}^s_{\mathrm{comp}}(\Omega, \mathcal{K}^{s,\gamma}(X^\wedge)),$$
$$\mathcal{W}^{s,\gamma}_{\mathrm{loc}(y)}(X^\wedge \times \Omega) = \mathcal{W}^s_{\mathrm{loc}}(\Omega, \mathcal{K}^{s,\gamma}(X^\wedge))$$

for an open set $\Omega \subseteq \mathbb{R}^q$, cf. also Definition 1.3.36.

Recall that the operator $T = F^{-1}_{\eta \to y} \kappa^{-1}(\eta) F_{y' \to \eta}$ induces an isomorphism

$$T : \mathcal{W}^s(\mathbb{R}^q, \mathcal{K}^{s,\gamma}(X^\wedge)) \to H^s(\mathbb{R}^q, \mathcal{K}^{s,\gamma}(X^\wedge)) \qquad (3.1.18)$$

for each $s, \gamma \in \mathbb{R}$.

Remark 3.1.19 *It can easily be proved that the space $C_0^\infty(X^\wedge \times \mathbb{R}^q)$ is dense in $\mathcal{W}^{s,\gamma}(X^\wedge \times \mathbb{R}^q)$ for every $s, \gamma \in \mathbb{R}$. It suffices to verify the elementary fact that $C_0^\infty(X^\wedge \times \mathbb{R}^q)$ is dense in $\mathcal{S}(\mathbb{R}^q, \mathcal{K}^{\infty,\infty}(X^\wedge))$.*

Proposition 3.1.20 *We have*

$$\mathcal{W}^{0,0}(X^\wedge \times \mathbb{R}^q) = t^{-\frac{n}{2}} L^2(\mathbb{R}_+ \times X \times \mathbb{R}^q)$$

for the L^2-space with respect to the measure $dtdxdy$.

Proof. The group $\{\kappa_\lambda\}_{\lambda \in \mathbb{R}_+}$ is unitary on $\mathcal{K}^{0,0}(X^\wedge)$. Thus

$$\|u\|^2_{\mathcal{W}^0(\mathbb{R}^q, \mathcal{K}^{0,0}(X^\wedge))} = \int \|\kappa^{-1}(\eta)\hat{u}(\eta)\|^2_{\mathcal{K}^{0,0}(X^\wedge)}\, d\eta = \int \|\hat{u}(\eta)\|^2_{\mathcal{K}^{0,0}(X^\wedge)}\, d\eta,$$

$\hat{u}(\eta) = (F_{y \to \eta} u)(\eta)$. Writing $\hat{u}(\eta) = f(t, x, \hat{\eta})$ we obtain from $\mathcal{K}^{0,0}(X^\wedge) = t^{-\frac{n}{2}} L^2(X^\wedge)$

$$\|\hat{u}(\eta)\|^2_{\mathcal{K}^{0,0}(X^\wedge)} = \int t^n |f(t, x, \hat{\eta})|^2\, dtdx.$$

Applying Plancherel's theorem we see that $u \in \mathcal{W}^0(\mathbb{R}^q, \mathcal{K}^{0,0}(X^\wedge))$ is equivalent to $t^{\frac{n}{2}} u \in L^2(X^\wedge \times \mathbb{R}^q)$. $\qquad\square$

From Proposition 3.1.20 we have a scalar product in $\mathcal{W}^{0,0}(X^\wedge \times \mathbb{R}^q)$. Also, the spaces $\mathcal{W}^{s,\gamma}(X^\wedge \times \mathbb{R}^q)$ for arbitrary $s, \gamma \in \mathbb{R}$ are Hilbert spaces with the scalar products

$$\int [\eta]^{2s} (\kappa^{-1}(\eta)\hat{u}(\eta), \kappa^{-1}(\eta)\hat{v}(\eta))_{\mathcal{K}^{s,\gamma}(X^\wedge)}\, d\eta$$

based on corresponding scalar products $(\cdot, \cdot)_{\mathcal{K}^{s,\gamma}(X^\wedge)}$.

Note that

$$\left\{ \mathcal{K}^{s,\gamma}(X^\wedge), \mathcal{K}^{0,0}(X^\wedge), \mathcal{K}^{-s,-\gamma}(X^\wedge); \{\kappa_\lambda\}_{\lambda \in \mathbb{R}_+} \right\} \qquad (3.1.19)$$

is a Hilbert space triple with unitary actions in the sense of Definition 1.3.41. Setting

$$(\chi_\lambda u)(y) = \lambda^{\frac{q}{2}} \kappa_\lambda u(\lambda y)$$

for $\lambda \in \mathbb{R}_+$ and $u \in \mathcal{W}^{s,\gamma}(X^\wedge \times \mathbb{R}^q)$ we obtain by

$$\left\{ \mathcal{W}^{s,\gamma}(X^\wedge \times \mathbb{R}^q), \mathcal{W}^{0,0}(X^\wedge \times \mathbb{R}^q), \mathcal{W}^{-s,-\gamma}(X^\wedge \times \mathbb{R}^q); \{\chi_\lambda\}_{\lambda \in \mathbb{R}_+} \right\}$$

another Hilbert space triple with unitary actions.

This gives rise to a non–degenerate sesquilinear pairing

$$(\cdot, \cdot)_{\mathcal{W}^{0,0}(X^\wedge \times \mathbb{R}^q)} : \mathcal{W}^{s,\gamma}(X^\wedge \times \mathbb{R}^q) \times \mathcal{W}^{-s,-\gamma}(X^\wedge \times \mathbb{R}^q) \to \mathbb{C}.$$

It follows that

$$\|u\|_{\mathcal{W}^{s,\gamma}(X^\wedge \times \mathbb{R}^q)} = \sup |(u, v)_{\mathcal{W}^{0,0}(X^\wedge \times \mathbb{R}^q)}| / \|v\|_{\mathcal{W}^{-s,-\gamma}(X^\wedge \times \mathbb{R}^q)},$$

where the supremum is taken over all $v \in \mathcal{W}^{-s,-\gamma}(X^\wedge \times \mathbb{R}^q)$, $v \neq 0$.

Note that Proposition 3.1.20 allows us to define also the space

$$\mathcal{W}^{0,0}(X^\wedge \times \Omega) \tag{3.1.20}$$

for an open set $\Omega \subseteq \mathbb{R}^q$, say with C^∞ boundary, which is isomorphic to

$$\left\{ u \in \mathcal{W}^{0,0}(X^\wedge \times \mathbb{R}^q) : \ \operatorname{supp} u \subseteq \overline{\Omega} \right\}.$$

This also induces a scalar product in (3.1.20).

Proposition 3.1.21 *For every $s, \gamma \in \mathbb{R}$ we have*

$$H^s_{\text{comp}}(\mathbb{R}_+ \times X \times \mathbb{R}^q) \subset \mathcal{W}^{s,\gamma}(X^\wedge \times \mathbb{R}^q) \subset H^s_{\text{loc}}(\mathbb{R}_+ \times X \times \mathbb{R}^q).$$

Proof. We first look at the case $X = S^n = \{\tilde{x} \in \mathbb{R}^{n+1} : |\tilde{x}| = 1\}$, where $X^\wedge \cong \mathbb{R}^{n+1} \setminus \{0\}$. Then

$$\mathcal{K}^{s,\gamma}(X^\wedge) = \omega \mathcal{H}^{s,\gamma}(\mathbb{R}^{n+1} \setminus \{0\}) + (1 - \omega) H^s(\mathbb{R}^{n+1})$$

for any cut–off function $\omega(t)$, $t = |\tilde{x}|$. We shall show the following property. For every $\varepsilon > 0$ there are constants $c_1(\varepsilon)$, $c_2(\varepsilon) > 0$ such that for all $u \in C_0^\infty(\mathbb{R}^{1+n+q})$ with $\operatorname{supp} u \subset \{(\tilde{x}, y) : |\tilde{x}| \geq \varepsilon\}$

$$c_1(\varepsilon)\|u\|^2_{H^s(\mathbb{R}^{1+n+q})} \leq \|u\|^2_{\mathcal{W}^s(\mathbb{R}^q, \mathcal{K}^{s,\gamma}(X^\wedge))} \leq c_2(\varepsilon)\|u\|^2_{H^s(\mathbb{R}^{1+n+q})} \tag{3.1.21}$$

holds. First we know that for every $\delta > 0$ there exist constants $b_1(\delta)$, $b_2(\delta) > 0$ such that

$$b_1(\delta)\|v\|^2_{H^s(\mathbb{R}^{1+n})} \leq \|v\|^2_{\mathcal{K}^{s,\gamma}(X^\wedge)} \leq b_2(\delta)\|v\|^2_{H^s(\mathbb{R}^{1+n})}$$

is satisfied for all $v \in C_0^\infty(\mathbb{R}^{1+n})$ with $\operatorname{supp} v \subset \{\tilde{x} : \ |\tilde{x}| \geq \delta\}$. This can be applied to the η–dependent family of functions $v(\tilde{x}, \eta) = [\eta]^{-\frac{n+1}{2}} f([\eta]^{-1}\tilde{x}, y)$ with $f(\tilde{x}, \eta) = (F_{y \to \eta} u)(\tilde{x}, \eta)$. Then the assumption on $\operatorname{supp} u$ implies the corresponding property on $\operatorname{supp} v$ for all $\eta \in \mathbb{R}^q$, with a suitable $\delta = \delta(\varepsilon) > 0$. It follows that

$$b_1(\delta)[\eta]^{2s}\|v(\cdot, \eta)\|_{H^s(\mathbb{R}^{1+n})} \leq [\eta]^{2s}\|v(\cdot, \eta)\|^2_{\mathcal{K}^{s,\gamma}(X^\wedge)}$$
$$\leq b_2(\delta)[\eta]^{2s}\|v(\cdot, \eta)\|^2_{H^s(\mathbb{R}^{1+n})}$$

for all η. Integrating over $\eta \in \mathbb{R}^q$ yields immediately (3.1.21). This extends to all $u \in \mathcal{W}^s(\mathbb{R}^q, \mathcal{K}^{s,\gamma}(X^\wedge)$ supported by $\{(\tilde{x}, y) : |\tilde{x}| \geq \varepsilon\}$. The general case can be reduced by restricting first the consideration to elements u supported with respect to (t, x) by a set $[\varepsilon, \infty) \times U_0$ with $\overline{U_0} \subset\subset U$ for some coordinate neighbourhood U on X. This corresponds to the above situation, since the set of all u with such a support can be interpreted as a subspace of $H^s(\mathbb{R}^{n+1})$. Finally an arbitrary $u \in \mathcal{W}^{s,\gamma}(X^\wedge \times \mathbb{R}^q)$, supported by $[\varepsilon, \infty) \times X \times \mathbb{R}^q$, can be written as a finite sum of elements supported with respect to x in coordinate neighbourhoods on X. Hence it follows the assertion in general. $\square$

Proposition 3.1.22 *We have continuous embeddings*

$$\mathcal{W}^{s',\gamma'}(X^\wedge \times \mathbb{R}^q) \hookrightarrow \mathcal{W}^{s,\gamma}(X^\wedge \times \mathbb{R}^q)$$

for all $s' \geq s$, $\gamma' \geq \gamma$.

Proof. Let E, E' be Banach spaces, $E' \hookrightarrow E$ a continuous embedding, and assume that $\{\kappa_\lambda\}_{\lambda \in \mathbb{R}_+}$ from E restricts to a strongly continuous group of isomorphisms on E', for simplicity also denoted by $\{\kappa_\lambda\}_{\lambda \in \mathbb{R}_+}$. Then we have

$$\|\kappa^{-1}(\eta)(Fu)(\eta)\|_E \leq c\|\kappa^{-1}(\eta)(Fu)(\eta)\|_{E'},$$

for all $u \in \mathcal{S}(\mathbb{R}^q, E')$ for a constant $c > 0$. Thus $\|u\|_{\mathcal{W}^{s'}(\mathbb{R}^q,E')} \leq c\|u\|_{\mathcal{W}^s(\mathbb{R}^q,E)}$ for all these u. This gives us immediately a continuous embedding $\mathcal{W}^{s'}(\mathbb{R}^q, E') \hookrightarrow \mathcal{W}^s(\mathbb{R}^q, E)$. To complete the proof it suffices to insert $E = \mathcal{K}^{s,\gamma}(X^\wedge)$, $E' = \mathcal{K}^{s',\gamma'}(X^\wedge)$. $\square$

Let $K = K_1 \times K_2$ for $K_1 = \{t \in \overline{\mathbb{R}}_+ : 0 \leq t \leq c_1\}$ and K_2 a compact set in $\mathbb{R}^q$. Set

$$\mathcal{W}^{s,\gamma}(X^\wedge \times \mathbb{R}^q)_K = \{u \in \mathcal{W}^{s,\gamma}(X^\wedge \times \mathbb{R}^q) : \text{supp}\, u \subseteq K_1 \times X \times K_2\}.$$

Theorem 3.1.23 *For $s' > s$, $\gamma' > \gamma$ we have compact embeddings*

$$\mathcal{W}^{s',\gamma'}(X^\wedge \times \mathbb{R}^q)_K \hookrightarrow \mathcal{W}^{s,\gamma}(X^\wedge \times \mathbb{R}^q).$$

Proof. Let $E' = [\omega]\mathcal{K}^{s',\gamma'}(X^\wedge)$, where $\omega(t)$ is a cut–off function with $\omega(t) = 1$ for $0 \leq t \leq t_1$ and set

$$\mathcal{W}^{s'}(K_2, E') = \{u \in \mathcal{W}^{s'}(\mathbb{R}^q, E') : \text{supp}_{(y)}\, u \subseteq K_2\}.$$

Then we have a continuous embedding

$$\mathcal{W}^{s',\gamma'}(X^\wedge \times \mathbb{R}^q)_K \hookrightarrow \mathcal{W}^{s'}(K_2, E').$$

On the other hand from Theorem 1.3.61 we obtain a compact embedding $\mathcal{W}^{s'}(K_2, E') \hookrightarrow \mathcal{W}^s(\mathbb{R}^q, \mathcal{K}^{s,\gamma}(X^\wedge))$. Composing the two embeddings we get the assertion. $\square$

Theorem 3.1.24 *The multiplication $\mathcal{M}_\varphi$ by $\varphi(t,x,y) \in C_0^\infty(\overline{\mathbb{R}}_+ \times X \times \mathbb{R}^q)$ extends to a continuous operator*

$$\mathcal{M}_\varphi : \mathcal{W}^{s,\gamma}(X^\wedge \times \mathbb{R}^q) \to \mathcal{W}^{s,\gamma}(X^\wedge \times \mathbb{R}^q) \qquad (3.1.22)$$

for each $s, \gamma \in \mathbb{R}$. Moreover, $\varphi \to \mathcal{M}_\varphi$ induces a continuous operator

$$C_0^\infty(\overline{\mathbb{R}}_+ \times X \times \mathbb{R}^q) \to \mathcal{L}(\mathcal{W}^{s,\gamma}(X^\wedge \times \mathbb{R}^q)) \qquad (3.1.23)$$

for each $s, \gamma \in \mathbb{R}$.

Proof. Let us fix $K = K_1 \times K_2$ as in Theorem 3.1.23 and set $C_0^\infty(\overline{\mathbb{R}}_+ \times X \times \mathbb{R}^q)_K = \{u \in C_0^\infty(\overline{\mathbb{R}}_+ \times X \times \mathbb{R}^q) : \operatorname{supp} u \subseteq K_1 \times X \times K_2\}$. Then the operator of multiplication by φ can be interpreted as an element m_φ in $S^0(\Omega \times \mathbb{R}^q; \mathcal{K}^{s,\gamma}(X^\wedge), \mathcal{K}^{s,\gamma}(X^\wedge))$ for every open $\Omega \subseteq \mathbb{R}^q$ with $K_2 \subset\subset \Omega$. The corresponding operator $\varphi \to m_\varphi$,

$$C_0^\infty(\overline{\mathbb{R}}_+ \times X \times \mathbb{R}^q)_K \to S^0(\Omega \times \mathbb{R}^q; \mathcal{K}^{s,\gamma}(X^\wedge), \mathcal{K}^{s,\gamma}(X^\wedge))$$

is continuous. Moreover, $\mathcal{M}_\varphi$ can be regarded as $\operatorname{Op}(m_\varphi)$. Applying Theorem 1.3.58 we obtain the continuity and at the same time the continuous dependence of the operator norm of $\mathcal{M}_\varphi$ on m_φ. This yields the assertion. $\qquad\square$

Remark 3.1.25 *Let $k^\gamma(t) = t^\gamma \omega(t) + (1 - \omega(t))$ for a cut–off function $\omega(t)$ and $\gamma \in \mathbb{R}$. Set $a^\gamma(\eta) = k^\gamma(t[\eta])$. Then we have $a^\gamma(\eta) \in S_{cl}^0(\mathbb{R}^q; \mathcal{K}^{s,\varrho}(X^\wedge), \mathcal{K}^{s,\varrho+\gamma}(X^\wedge))$ for each $s, \varrho \in \mathbb{R}$, and $a^\gamma(\eta) : \mathcal{K}^{s,\varrho}(X^\wedge) \to \mathcal{K}^{s,\varrho+\gamma}(X^\wedge)$ is an isomorphism for every η. This yields an isomorphism*

$$\operatorname{Op}(a^\gamma) : \mathcal{W}^{s,\varrho}(X^\wedge \times \mathbb{R}^q) \to \mathcal{W}^{s,\varrho+\gamma}(X^\wedge \times \mathbb{R}^q)$$

for each $s, \varrho \in \mathbb{R}$, and we have $(\operatorname{Op}(a^\gamma))^{-1} = \operatorname{Op}((a^{-\gamma})^{-1})$.

Let us now turn to a concrete version of Proposition 1.3.67. We apply this to the Hilbert space triple (3.1.28) and use the following result.

Lemma 3.1.26 *For every operator $a \in C^\mu(X^\wedge, \boldsymbol{g}^\bullet)$ for $\boldsymbol{g} = (\gamma, \varrho, \Theta)$, regarded as a continuous operator $a : \mathcal{K}^{s,\gamma}(X^\wedge) \to \mathcal{K}^{s-\mu,\delta}(X^\wedge)$ we have*

$$\kappa_\lambda a \kappa_\lambda^{-1} \in C^\infty(\mathbb{R}_+, \mathcal{L}(\mathcal{K}^{s,\gamma}(X^\wedge), \mathcal{K}^{s-\mu,\delta}(X^\wedge)))$$

for each $s \in \mathbb{R}$.

Proof. Let us consider a in the form

$$a = \omega(t) t^{\delta-\gamma} \operatorname{op}_M^{\gamma-\frac{n}{2}}(h)\omega_0(t) + (1 - \omega(t))P(1 - \omega_1(t)) + m + g$$

for cut–off functions $\omega, \omega_0, \omega_1$, a symbol $h(t,z) \in C^\infty(\overline{\mathbb{R}}_+, M_O^\mu(X))$, a pseudo–differential operator P on $X^\wedge$ as it is assumed in the definition of the cone algebra on $X^\wedge$, and $m + g \in C_{M+G}(X^\wedge, \boldsymbol{g}^\bullet)$. Then

$$\kappa_\lambda \left\{ \omega(t) t^{\delta-\gamma} \operatorname{op}_M^{\gamma-\frac{n}{2}}(h)\omega_0(t) \right\} \kappa_\lambda^{-1} = \omega(\lambda t)(\lambda t)^{\delta-\gamma} \operatorname{op}_M^{\gamma-\frac{n}{2}}(h_\lambda)\omega_0(\lambda t)$$

for $h_\lambda(t, z) = h(\lambda t, z)$. Thus for the first summand the assertion is clear. The same is true of the smoothing Mellin operator m. For the second summand we obtain

$$\kappa_\lambda \{(1 - \omega(t))P(1 - \omega_1(t))\} \kappa_\lambda^{-1} = (1 - \omega(\lambda t))P_\lambda(1 - \omega_1(\lambda t)),$$

where the local symbols of P_λ (with respect to the charts on X) have the form $p(\lambda t, x, \lambda^{-1}\tau, \xi)$ when the local symbols of P are $p(t, x, \tau, \xi)$. It is obvious that they are also smooth in λ. Concerning g we use the integral representations

$$(gu)(t, x) = \int\limits_0^\infty k(t, x, t', x')u(t', x')\, dt'dx'$$

for some $k(t, x, t', x') \in \mathcal{S}_R^\delta(X^\wedge) \otimes_\pi \mathcal{S}_Q^{-\gamma}(X^\wedge)$ for certain discrete asymptotic types R, Q. Then $\kappa_\lambda g \kappa_\lambda^{-1}$ has the kernel $k(\lambda t, x, \lambda t', x')$ that is also smooth in λ.   □

Theorem 3.1.27 *For every $s, \gamma \in \mathbb{R}$ there exists an element $a_0 \in C^s(X^\wedge, \boldsymbol{g}^\bullet)$ for $\boldsymbol{g} = (\gamma, 0, (-\infty, 0])$ which induces an isomorphism*

$$a_0 : \mathcal{K}^{s,\gamma}(X^\wedge) \to \mathcal{K}^{0,0}(X^\wedge).$$

The inverse a_0^{-1} belongs to $C^{-s}(X^\wedge, \boldsymbol{h}^\bullet)$ for $\boldsymbol{h} = (0, \gamma, (-\infty, 0])$.

Proof. If suffices to verify the assertion for $\gamma = 0$, since the multiplication by $t^{-\gamma}\omega(t) + (1 - \omega(t))$ allows reducing the weight to 0. Now we form a parameter–dependent symbol of the form $p_0(t, \tau, \xi, \lambda) = (|t\tau|^2 + |\xi|^2 + \lambda^2)^{\frac{s}{4}}$ for $\lambda \in \mathbb{R}$ and pass to an associated $h(z, \xi, \lambda) \in S_{cl}^{\frac{s}{2}}(\mathbb{C} \times \mathbb{R}^n \times \mathbb{R})_{hol}$ such that

$$\mathrm{op}_M^{-\frac{n}{2}}(h)(\xi, \lambda) = \mathrm{op}_t(p_0)(\xi, \lambda) \bmod \mathcal{S}(\mathbb{R}^n \times \mathbb{R}, L^{-\infty}(\mathbb{R}_+)).$$

Set

$$h(z, \lambda) = \sum_{j=1}^N \varphi_j(\chi_j^{-1})_* \mathrm{op}_x(h)(z, \lambda)\psi_j$$

for a system of charts $\chi_j : U_j \to \mathbb{R}^n$ on X, $j = 1, \ldots, N$, a subordinate partition of unity $\{\varphi_1, \ldots, \varphi_N\}$ and functions $\psi_j \in C_0^\infty(U_j)$ with $\varphi_j\psi_j = \varphi_j$ for all j. Let $\omega(t)$ be a cut–off function and define

$$p(t, x, \tau, \xi, \lambda) = \omega(t)p_0(t, \tau, \xi, \lambda) + (1 - \omega(t))p_\infty(t, x, \tau, \xi, \lambda),$$

where p_∞ is a classical symbol of order s which corresponds to the coordinates $\tilde{x}$ in a conical set of $\mathbb{R}^{n+1}$, that is diffeomorphic to $\mathbb{R}_+ \times U$ for open $U \subseteq X$, to a classical parameter–dependent elliptic symbol of order $\frac{s}{2}$ with respect to the covariables $\tilde{\xi}, \lambda$, namely to $(|\tilde{\xi}|^2 + \lambda^2)^{\frac{s}{4}}$. Then

$$P(t, \tau, \lambda) = \sum_{j=1}^N \varphi_j(\chi_j^{-1})_* \mathrm{op}_x(p)(t, \tau, \lambda)$$

is an operator family for which

$$(1 - \omega(t)) \, \mathrm{op}_t(P)(\lambda)(1 - \omega_1(t)) : \mathcal{K}^{\frac{s}{2},0}(X^\wedge) \to \mathcal{K}^{0,0}(X^\wedge)$$

is continuous for every pair of cut–off functions $\omega(t), \omega_1(t)$. Assume $\omega(t)\omega_1(t) = \omega_1(t)$ and choose another cut–off function $\omega_0(t)$ satisfying $\omega(t)\omega_0(t) = \omega(t)$. Then

$$\begin{aligned} a(\lambda) = \, &\omega(t) \, \mathrm{op}_M^{-\frac{n}{2}}(h)(\lambda)\omega_0(t) + \\ &(1 - \omega(t)) \, \mathrm{op}_t(P)(\lambda)(1 - \omega_1(t)) : \mathcal{K}^{\frac{s}{2},0}(X^\wedge) \to \mathcal{K}^{0,0}(X^\wedge) \end{aligned}$$

is parameter–dependent elliptic in $X^\wedge$, including $t \to \infty$ in the sense of the exit symbols. The conormal symbol

$$\sigma_M^0(a)(z,\lambda) : H^s(X) \to H^0(X)$$

is a parameter–dependent elliptic family on X, and it is bijective for $|\lambda| > \mathrm{const}$ for a constant > 0, cf. Theorem 1.2.26. Thus it is elliptic in the cone algebra $C^{\frac{s}{2}}(X^\wedge, \boldsymbol{g}_0^\bullet)$ for $\boldsymbol{g}_0 = (0,0,\Theta)$ for arbitrary Θ. The parameter λ is taken sufficiently large. Now $a^*(\lambda)$ also belongs to $C^{\frac{s}{2}}(X^\wedge, \boldsymbol{g}_0^\bullet)$ and has the same properties. Thus

$$a_1(\lambda_1) = a^*(\lambda_1)a(\lambda_1) \in C^s(X^\wedge, \boldsymbol{g}_0^\bullet)$$

is a Fredholm operator $a_1(\lambda_1) : \mathcal{K}^{s,0}(X^\wedge) \to \mathcal{K}^{0,0}(X^\wedge)$ of index zero. There exists then a Green operator g in $C_G(X^\wedge, \boldsymbol{g}^\bullet)$ such that

$$a_0 := a_1(\lambda_1) + g : \mathcal{K}^{s,0}(X^\wedge) \to \mathcal{K}^{0,0}(X^\wedge)$$

is an isomorphism. $\square$

Note that g in the latter argument can be chosen as an operator with kernel in $C_0^\infty(\mathbb{R}_+ \times X \times \mathbb{R}_+ \times X)$, since arbitrary Green operators can be approximated in the norm topology by such special ones.

Remark 3.1.28 *Let $s, \gamma \in \mathbb{R}$ be fixed and set $a^{s,\gamma}(\eta) = [\eta]^s \kappa_{[\eta]} a_0 \kappa_{[\eta]}^{-1}$ for the operator a_0 from Theorem 3.1.27. Then*

$$\begin{aligned} \mathcal{W}^{s,\gamma}&(X^\wedge \times \mathbb{R}^q) \\ &= \left\{ u \in \mathcal{S}'(\mathbb{R}^q, \mathcal{K}^{-\infty,\gamma}(X^\wedge)) : \mathrm{Op}(a^{s,\gamma})u \in \mathcal{W}^{0,0}(X^\wedge \times \mathbb{R}^q) \right\}. \end{aligned}$$

In fact, we have $a^{s,\gamma}(\eta) \in S_{\mathrm{cl}}^s(\mathbb{R}^q; \mathcal{K}^{r,\gamma}(X^\wedge), \mathcal{K}^{r-s,0}(X^\wedge))$ and

$$(a^{s,\gamma}(\eta))^{-1} \in S_{\mathrm{cl}}^{-s}(\mathbb{R}^q; \mathcal{K}^{r,0}(X^\wedge), \mathcal{K}^{r+s,\gamma}(X^\wedge))$$

for every $r \in \mathbb{R}$. In view of $\mathrm{Op}(a^{s,\gamma}(\eta))^{-1} \, \mathrm{Op}(a^{s,\gamma}) = 1$ and

$$\mathrm{Op}(a^{s,\gamma}(\eta))^{-1} : \mathcal{W}^{0,0}(X^\wedge \times \mathbb{R}^q) \xrightarrow{\cong} \mathcal{W}^{s,\gamma}(X^\wedge \times \mathbb{R}^q)$$

we obtain the assertion.

Theorem 3.1.29 *Let* $\Omega, \tilde{\Omega} \subset \mathbb{R}^q$ *be open sets and* $\chi : \Omega \to \tilde{\Omega}$ *be a diffeomorphism. Then the pull–back* $\chi^* : C_0^\infty(X^\wedge \times \tilde{\Omega}) \to C_0^\infty(X^\wedge \times \Omega)$ *extends to isomorphisms*

$$\chi^* : \mathcal{W}^{s,\gamma}_{\mathrm{comp}(\tilde{y})}(X^\wedge \times \tilde{\Omega}) \to \mathcal{W}^{s,\gamma}_{\mathrm{comp}(y)}(X^\wedge \times \Omega)$$

and

$$\chi^* : \mathcal{W}^{s,\gamma}_{\mathrm{loc}(\tilde{y})}(X^\wedge \times \tilde{\Omega}) \to \mathcal{W}^{s,\gamma}_{\mathrm{loc}(y)}(X^\wedge \times \Omega)$$

for all $s, \gamma \in \mathbb{R}$.

Proof. $C_0^\infty(X^\wedge \times \Omega)$ is dense in $C_0^\infty(\Omega, \mathcal{K}^{s,\gamma}(X^\wedge))$ for every open set $\Omega \subseteq \mathbb{R}^q$. Then $\chi^* : C_0^\infty(X^\wedge \times \tilde{\Omega}) \to C_0^\infty(X^\wedge \times \Omega)$ extends to

$$\chi^* : C_0^\infty(\tilde{\Omega}, \mathcal{K}^{s,\gamma}(X^\wedge)) \to C_0^\infty(\Omega, \mathcal{K}^{s,\gamma}(X^\wedge)). \tag{3.1.24}$$

The Hilbert space triple (3.1.19) satisfies the assumptions of Proposition 1.3.67, here with the isomorphism of Theorem 3.1.27. Thus we can apply Theorem 1.3.68 in the form that (3.1.24) has the corresponding extension to the asserted isomorphism between the weighted Sobolev spaces. $\square$

Now let Y be a paracompact C^∞ manifold, $q = \dim Y$. Then, according to be general contructions of Section 1.3.3, we can form the spaces

$$\mathcal{W}^{s,\gamma}_{\mathrm{comp}(y)}(X^\wedge \times Y) := \mathcal{W}^s_{\mathrm{comp}}(Y, \mathcal{K}^{s,\gamma}(X^\wedge))$$

and analogously for subscripts loc. In particular, for compact Y we get the spaces

$$\mathcal{W}^{s,\gamma}(X^\wedge \times Y) := \mathcal{W}^s(Y, \mathcal{K}^{s,\gamma}(X^\wedge)) \tag{3.1.25}$$

for all $s, \gamma \in \mathbb{R}$.

Let W be a manifold with edges Y in the sense of Definition 3.1.1 and let $\mathbb{W}$ be the associated stretched manifold with the canonical map $\pi : \mathbb{W} \to W$. Assume that Y has a neighbourhood V in W with $V \cong \{([0,1) \times X)/(\{0\} \times X)\} \times Y$, i.e., we assume that W near Y has the structure of a trivial line bundle over Y with the fibre $([0,1) \times X)/(\{0\} \times X)$. This means that $\mathbb{W}$ near $\pi^{-1}Y$ is a Cartesian product $[0,1) \times X \times Y$ or, equivalently $\overline{\mathbb{R}}_+ \times X \times Y$. Now let us choose a cut–off function $\omega(t)$ with support in $[0, 1 - \varepsilon)$ for some $0 < \varepsilon < 1$. Set

$$\mathcal{W}^{s,\gamma}_{\mathrm{loc}}(\mathbb{W}) = \{u \in H^s_{\mathrm{loc}}(\mathrm{int}\, \mathbb{W}) : \omega u \in \mathcal{W}^{s,\gamma}_{\mathrm{loc}(y)}(X^\wedge \times Y)\} \tag{3.1.26}$$

and analogously

$$\mathcal{W}^{s,\gamma}_{\mathrm{comp}}(\mathbb{W}) = \{u \in \mathcal{W}^{s,\gamma}_{\mathrm{loc}}(\mathbb{W}) : \mathrm{supp}\, u \text{ compact in}\, \mathbb{W}\}. \tag{3.1.27}$$

Let us fix a Riemannian metric on $\mathbb{W}$ which corresponds to the product metric of $X^\wedge \times Y$ in a neighbourhood of $\partial\mathbb{W}$ for the metrics $dt\,dx$ on $X^\wedge$ and dy on Y. Then we have the space $L^2(\mathbb{W})$, and we set

$$\mathcal{W}^{0,0}(\mathbb{W}) = h^{-\frac{n}{2}}L^2(\mathbb{W}), \qquad (3.1.28)$$

where h^γ is a function in $C^\infty(\operatorname{int}\mathbb{W})$ which is strictly positive and equals t^γ in the coordinates $(t, x, y) \in [0, 1) \times X \times Y$ for $0 < t < \varepsilon$ for some $0 < \varepsilon < 1$. This gives us a scalar product in (3.1.28).

3.1.5   Wedge Sobolev spaces with asymptotics

The definition of the wedge Sobolov spaces $\mathcal{W}^s(\mathbb{R}^q, E)$ can be extended to Fréchet spaces E, written as projective limits $E = \varprojlim_{j\in\mathbb{N}}E^j$ of Banach spaces E^j, $j \in \mathbb{N}$, with continuous embeddings $E^{j+1} \hookrightarrow E^j$ for all j, where a strongly continuous group $\{\kappa_\lambda\}_{\lambda\in\mathbb{R}_+}$ of isomorphisms on E^0 is given which restricts to E^j as a strongly continuous group of isomorphisms for each j. Then

$$\mathcal{W}^s(\mathbb{R}^q, E) := \varprojlim_{j\in\mathbb{N}}\mathcal{W}^s(\mathbb{R}^q, E^j) \qquad (3.1.29)$$

is also a Fréchet space.

We apply this to the case $E = \mathcal{K}_P^{s,\gamma}(X^\wedge)$ for an asymptotic type $P \in \operatorname{As}(X, \boldsymbol{g})$, $\boldsymbol{g} = (\gamma, \Theta)$ for $\Theta = (\vartheta, 0]$, $-\infty \le \vartheta < 0$. Writing $E^0 = \mathcal{K}^{s,\gamma}(X^\wedge)$ with $(\kappa_\lambda u)(t, x) = \lambda^{\frac{n+1}{2}}u(\lambda t, x)$ for $\lambda \in \mathbb{R}_+$, $u(t, x) \in \mathcal{K}^{s,\gamma}(X^\wedge)$, we find a sequence of Hilbert subspaces E^j of E^0, $j \in \mathbb{N}$, with the above properties. Then we obtain the following wedge Sobolev spaces with edge asymptotics

$$\mathcal{W}_P^{s,\gamma}(X^\wedge \times \mathbb{R}^q) := \mathcal{W}^s(\mathbb{R}^q, \mathcal{K}_P^{s,\gamma}(X^\wedge)). \qquad (3.1.30)$$

Proposition 3.1.30 *Let $E = \varprojlim_{j\in\mathbb{N}}E^j$ be a Fréchet space, E^j Banach spaces with continuous embeddings $E^{j+1} \hookrightarrow E^j$ for all j, where a strongly continuous group $\{\kappa_\lambda\}_{\lambda\in\mathbb{R}_+}$ of isomorphisms on E^0 is given that restricts to a strongly continuous group of isomorphisms on E^j for each j. Moreover, assume that every E^j is a non–direct sum $E_1^j + E_2^j$ for Banach spaces E_i^j with continuous embeddings $E_i^{j+1} \hookrightarrow E_i^j$, where $\{\kappa_\lambda\}_{\lambda\in\mathbb{R}_+}$ restricts to strongly continuous groups of isomorphisms on E_i^j for all j and $i = 1, 2$. Set $E = E_1 + E_2$ which holds as a non–direct sum of the Fréchet spaces $E_i = \varprojlim_{j\in\mathbb{N}}E_i^j$, $i = 1, 2$. Then we have*

$$\mathcal{W}^s(\mathbb{R}^q, E) = \mathcal{W}^s(\mathbb{R}^q, E_1) + \mathcal{W}^s(\mathbb{R}^q, E_2)$$

in the sense of a non–direct sum of Fréchet spaces.

Proposition 3.1.30 is an easy generalisation of Proposition 1.3.39 to Fréchet spaces. The proof is omitted.

Corollary 3.1.31 *We have*

$$\mathcal{W}_P^{s,\gamma}(X^\wedge \times \mathbb{R}^q) = \mathcal{W}^s(\mathbb{R}^q, \mathcal{K}_\Theta^{s,\gamma}(X^\wedge)) + \mathcal{W}^s(\mathbb{R}^q, \mathcal{K}_P^{\infty,\gamma}(X^\wedge))$$

for every $s, \gamma \in \mathbb{R}$, $P \in \mathrm{As}(X, \boldsymbol{g})$, as a non–direct sum of Fréchet spaces. This is a consequence of $\mathcal{K}_P^{s,\gamma}(X^\wedge) = \mathcal{K}_\Theta^{s,\gamma}(X^\wedge) + \mathcal{K}_P^{\infty,\gamma}(X^\wedge)$. In particular, every $f(t, x, y) \in \mathcal{W}_P^{s,\gamma}(X^\wedge \times \mathbb{R}^q)$ has the form

$$f(t, x, y) \in T^{-1} v(t, x, y) \bmod \mathcal{W}^s(\mathbb{R}^q, \mathcal{K}_\Theta^{s,\gamma}(X^\wedge)) \qquad (3.1.31)$$

for a unique $v(t, x, y) \in H^s(\mathbb{R}^q, \mathcal{K}_P^{\infty,\gamma}(X^\wedge))$, cf. (3.1.18).

To interpret this result, we recall that $\mathcal{K}_P^{s,\gamma}(X^\wedge)$ for a continuous asymptotic type P can be defined as the subspace of all $u \in \mathcal{K}^{s,\gamma}(X^\wedge)$ such that for every $\beta \geq 0$ with $\frac{n+1}{2} - \gamma + \vartheta < \frac{n+1}{2} - (\gamma + \beta)$ there is a compact subset $K \subseteq \mathrm{carrier}\, P$ and an element $\zeta = \zeta(u) \in \mathcal{A}'(K, C^\infty(X))$ such that

$$u(t, x) - \omega(t)\langle \zeta, t^{-z} \rangle \in \mathcal{K}^{s,\gamma+\beta}(X^\wedge)$$

for any cut–off function $\omega(t)$. Remember that K can be chosen in such a way that for an arbitrary fixed $\varepsilon > 0$

$$K \subset \left\{ z \in \mathbb{C} : \frac{n+1}{2} - (\gamma + \beta + \varepsilon) \leq \mathrm{Re}\, z < \frac{n+1}{2} - \gamma \right\}.$$

In particular, for discrete asymptotics $P \in \mathrm{As}(X, (\gamma, \Theta)^\bullet)$ with carrier $P = \{p_0, p_1, p_2 \ldots\} \subset \{z : \frac{n+1}{2} - \gamma + \vartheta < \mathrm{Re}\, z < \frac{n+1}{2} - \gamma\}$, $\mathrm{Re}\, p_0 \geq \mathrm{Re}\, p_1 \geq \ldots$, the functional ζ is carried by the finitely many points $\{p_0, \ldots, p_{N(\beta)}\}$ with $\frac{n+1}{2} - (\gamma + \beta) \leq \mathrm{Re}\, p_j$, $j = 0, \ldots, N(\beta)$, where

$$\langle \zeta, t^{-z} \rangle = \sum_{j=0}^{N(\beta)} \sum_{k=0}^{m_j} c_{jk}(x) t^{-p_j} \log^k t$$

for certain $c_{jk} \in M_j$, $0 \leq k \leq m_j$. Here $P = \{(p_j, m_j, M_j)\}$.

Writing $\mathcal{K}_P^{\infty,\gamma}(X^\wedge) = \mathcal{K}^{\infty,\gamma+\beta}(X^\wedge) + \mathcal{V}$ for the Fréchet space $\mathcal{V} = \{\omega(t)\langle \zeta, t^{-z} \rangle : \zeta \in \mathcal{A}'(K, C^\infty(X))\}$, we obtain from the relation (3.1.31)

$$f(t, x, y) \in T^{-1} H^s(\mathbb{R}^q, \mathcal{V}) \bmod \mathcal{W}^s(\mathbb{R}^q, \mathcal{K}^{s,\gamma+\beta}(X^\wedge)).$$

Here we used the decomposition

$$\mathcal{W}^s(\mathbb{R}^q, \mathcal{K}_P^{\infty,\gamma}(X^\wedge)) = T^{-1} H^s(\mathbb{R}^q, \mathcal{V}) + \mathcal{W}^s(\mathbb{R}^q, \mathcal{K}^{\infty,\gamma+\beta}(X^\wedge))$$

which is a consequence of

$$H^s(\mathbb{R}^q, \mathcal{K}_P^{\infty,\gamma}(X^\wedge)) = H^s(\mathbb{R}^q, \mathcal{V}) + H^s(\mathbb{R}^q, \mathcal{K}^{\infty,\gamma+\beta}(X^\wedge)),$$

and we applied $\mathcal{W}^s(\mathbb{R}^q, \mathcal{K}_\Theta^{s,\gamma}(X^\wedge)) \subset \mathcal{W}^s(\mathbb{R}^q, \mathcal{K}^{s,\gamma+\beta}(X^\wedge))$. Note that

$$H^s(\mathbb{R}^q, \mathcal{V}) = \{\omega(t)\langle\zeta, t^{-z}\rangle : \zeta \in \mathcal{A}'(K, C^\infty(X, H^s(\mathbb{R}^q)))\}.$$

Summing up we proved the following result:

Proposition 3.1.32 *Every* $f(t,x,y) \in \mathcal{W}_P^{s,\gamma}(X^\wedge \times \mathbb{R}^q)$ *for a continuous asymptotic type* $P \in \mathrm{As}(X, \boldsymbol{g})$, $\boldsymbol{g} = (\gamma, \Theta)$, *can be written*

$$f(t,x,y) = F_{\eta \to y}^{-1}[\eta]^{\frac{n+1}{2}}\omega(t[\eta])\langle\zeta, (t[\eta])^{-z}\rangle + r_\beta(t,x,y) \qquad (3.1.32)$$

for every $\beta \geq 0$ *with* $\frac{n+1}{2} - \gamma + \vartheta < \frac{n+1}{2} - (\gamma + \beta)$ *for some*

$$\zeta \in \mathcal{A}'(K, C^\infty(X, \hat{H}^s(\mathbb{R}^q))),$$

$\hat{H}^s(\mathbb{R}^q) = F_{y \to \eta}H^s(\mathbb{R}^q)$, *a compact set* $K \subseteq \mathrm{carrier}\, P$, *and a remainder* $r_\beta(t,x,y) \in \mathcal{W}^s(\mathbb{R}^q, \mathcal{K}^{s,\gamma+\beta}(X^\wedge))$. *For discrete asymptotics* $P \in \mathrm{As}(X, \boldsymbol{g}^\bullet)$ *we obtain analogously*

$$f(t,x,y) = \sum_{j=0}^{N(\beta)} \sum_{k=0}^{m_j} F_{\eta \to y}^{-1}[\eta]^{\frac{n+1}{2}}\omega(t[\eta])c_{jk}(x)(t[\eta])^{-p_j}$$
$$\times \log^k(t[\eta])\hat{v}_{jk}(\eta) + r_\beta(t,x,y) \qquad (3.1.33)$$

for a suitable $N(\beta)$, *elements* $v_{jk} \in H^s(\mathbb{R}^q)$, *and* $P = \{(p_j, m_j, M_j)\}$ *and the coefficients* c_{jk} *belong to* M_j *for* $0 \leq k \leq m_j$, *for all* j.

The decomposition of f in Proposition 3.1.32 can be interpreted as an asymptotic expansion of distributions with edge asymptotics, where the first summand is the singular term and the remainder r_β is of edge flatness β relative to the reference weight γ.

Proposition 3.1.33 *To every* $f \in \mathcal{W}_P^{\infty,\gamma}(X^\wedge \times \mathbb{R}^q)$, $P \in \mathrm{As}(X, \boldsymbol{g})$, *and every* $\beta \geq 0$ *with* $\frac{n+1}{2} - \gamma + \vartheta < \frac{n+1}{2} - (\gamma + \beta)$ *there is a compact subset* $K \subseteq \mathrm{carrier}\, P$ *and an element* $\zeta = \zeta(u) \in \mathcal{A}'(K, C^\infty(X, H^\infty(\mathbb{R}^q)))$ *such that*

$$f(t,x,y) - \omega(t)\langle\zeta, t^{-z}\rangle \in \mathcal{W}^{\infty,\gamma+\beta}(X^\wedge \times \mathbb{R}^q) = H^\infty(\mathbb{R}^q, \mathcal{K}^{\infty,\gamma+\beta}(X^\wedge)).$$

In particular, for $P \in \mathrm{As}(X, \boldsymbol{g}^\bullet)$, *written as above, there are unique* $c_{jk}(x) \in M_j$, $v_{jk}(y) \in H^\infty(\mathbb{R}^q)$, $0 \leq k \leq m_j$, $j = 0, \ldots, N(\beta)$, *such that*

$$f(t,x,y) - \sum_{j=0}^{N(\beta)} \sum_{k=0}^{m_j} \omega(t)t^{-p_j}\log^k t c_{jk}(x)v_{jk}(y) \in H^\infty(\mathbb{R}^q, \mathcal{K}^{\infty,\gamma+\beta}(X^\wedge)).$$

Proof. The proof follows easily from $\mathcal{W}^\infty(\mathbb{R}^q, E) = H^\infty(\mathbb{R}^q, E)$ for every E. In particular, in the arguments to prove Proposition 3.1.32 we may omit the isomorphism T^{-1}. In other words in the occurring spaces we can treat the group action $\{\kappa_\lambda\}_{\lambda \in \mathbb{R}_+}$ as the identity. $\square$

It is useful to interpret the singular functions of the edge asymptotics in terms of operator–valued symbols. Let us consider

$$e(\eta) : \mathbb{C} \to \mathcal{K}_P^{\infty,\gamma}(X^\wedge)$$

defined by $e(\eta)c = [\eta]^{\frac{n+1}{2}} \omega(t[\eta]) \langle \zeta, (t[\eta])^{-z} \rangle c$ for $c \in \mathbb{C}$ and an element $\zeta \in \mathcal{A}'(K, C^\infty(X))$, cf. the notation from Propsition 3.1.32. Then

$$e(\eta) \in S^0_{\mathrm{cl}}(\mathbb{R}^q; \mathbb{C}, \mathcal{K}_P^{\infty,\gamma}(X^\wedge))$$

which is a consequence of the homogeneity $e(\lambda\eta) = \kappa_\lambda e(\eta)$ for all $\lambda \geq 1$, $|\eta| > $ const for a constant > 0.

This leads to the following observation:

Remark 3.1.34 *For $P \in \mathrm{As}(X, \boldsymbol{g}^\bullet)$ the singular functions of the edge asymptotics* (3.1.33) *can be written*

$$\sum_{j=0}^{N(\beta)} \sum_{k=0}^{m_j} \mathrm{Op}(e_{jk}) v_{jk}, \qquad v_{jk} \in H^s(\mathbb{R}^q),$$

for suitable $e_{jk}(\eta) \in S^0_{\mathrm{cl}}(\mathbb{R}^q, \mathcal{K}_P^{\infty,\gamma}(X^\wedge))$, *namely*

$$e_{jk}(\eta) = [\eta]^{\frac{n+1}{2}} \omega(t[\eta]) \langle \zeta_{jk}, (t[\eta])^{-z} \rangle$$

for $\langle \zeta_{jk}, h \rangle := (-1)^k c_{jk}(x) (\frac{d^k}{dz^k} h(z))|_{z=p_j}$ *for* $h \in \mathcal{A}(\mathbb{C})$.

The structure of the singular functions in (3.1.32) for the continuous edge asymptotics is more complicated, since the analytic functionals belong to

$$\mathcal{A}'(K, C^\infty(X, \hat{H}^s(\mathbb{R}^q))) = \mathcal{A}'(K, C^\infty(X)) \otimes_\pi \hat{H}^s(\mathbb{R}^q).$$

It follows that every ζ from that space can be written as a convergent sum

$$\zeta = \sum_{j=0}^\infty \lambda_j \zeta_j \hat{v}_j \quad \text{for} \quad \lambda_j \in \mathbb{C}, \quad \sum |\lambda_j| < \infty \qquad (3.1.34)$$

and $\zeta_j \in \mathcal{A}'(K, C^\infty(X))$, $\hat{v}_j \in \hat{H}^s(\mathbb{R}^q)$, both tending to zero in the corresponding spaces for $j \to \infty$.

For every ζ_j we have

$$e_j(\eta) := [\eta]^{\frac{n+1}{2}} \omega(t[\eta]) \langle \zeta_j, (t[\eta])^{-z} \rangle \in S^0_{\mathrm{cl}}(\mathbb{R}^q; \mathbb{C}, \mathcal{K}^{\infty,\gamma}_P(X^\wedge)),$$

where $\zeta_j \to 0$ for $j \to \infty$ implies $e_j(\eta) \to 0$ in this symbol space. It follows that the singular function in (3.1.32) equals

$$F^{-1}_{\eta \to y}[\eta]^{\frac{n+1}{2}} \omega(t[\eta]) \langle \zeta, (t[\eta])^{-z} \rangle = \sum_{j=0}^{\infty} \lambda_j \, \mathrm{Op}(e_j) v_j$$

with convergence in the Fréchet topolopy of $\mathcal{W}^{s,\gamma}_P(X^\wedge \times \mathbb{R}^q)$.

The general spaces $\mathcal{W}^s_{\mathrm{comp}}$ and $\mathcal{W}^s_{\mathrm{loc}}$ can be specified for the present situation, and we obtain

$$\mathcal{W}^s_{\mathrm{comp}}(\Omega, \mathcal{K}^{s,\gamma}_P(X^\wedge)) \quad \text{and} \quad \mathcal{W}^s_{\mathrm{loc}}(\Omega, \mathcal{K}^{s,\gamma}_P(X^\wedge)) \qquad (3.1.35)$$

for every open set $\Omega \subseteq \mathbb{R}^q$. Let us consider the local case. Every $f(t,x,y) \in \mathcal{W}^s_{\mathrm{comp}}(\Omega, \mathcal{K}^{s,\gamma}_P(X^\wedge))$ can be regarded as an element in $\mathcal{W}^{s,\gamma}_P(X^\wedge \times \mathbb{R}^q)$ and the associated singular function from (3.1.32) has the form

$$\varphi(y) F^{-1}_{\eta \to y}[\eta]^{\frac{n+1}{2}} \omega(t[\eta]) \langle \zeta, (t[\eta])^{-z} \rangle \quad \text{for} \quad \zeta \in \mathcal{A}'(K, C^\infty(X, \hat{H}^s(\mathbb{R}^q))),$$

$\varphi \in C^\infty_0(\mathbb{R}^q)$, $\varphi \equiv 1$ on $\mathrm{supp}_{(y)} f$.

It is useful for technical reasons also to allow y–dependent analytic functionals

$$\zeta \in C^\infty_0(\mathbb{R}^q_y, \mathcal{A}'(K, C^\infty(X, \hat{H}^s(\mathbb{R}^q)))) \qquad (3.1.36)$$

in the expressions for the singular functions.

Proposition 3.1.35 *Let $\Omega \subseteq \mathbb{R}^q$ be an open set, $\varphi(y) \in C^\infty_0(\Omega)$, and let $\zeta(y) \in C^\infty_0(\Omega, \mathcal{A}'(K, C^\infty(X, \hat{H}^s(\mathbb{R}^q))))$. Then there exists a $\chi \in \mathcal{A}'(K, C^\infty(X, \hat{H}^s(\mathbb{R}^q)))$ such that*

$$\varphi(y) F^{-1}_{\eta \to y}[\eta]^{\frac{n+1}{2}} \omega(t[\eta]) \langle \zeta(y), (t[\eta])^{-z} \rangle$$
$$= \varphi(y) F^{-1}_{\eta \to y}[\eta]^{\frac{n+1}{2}} \omega(t[\eta]) \langle \chi, (t[\eta])^{-z} \rangle$$

$\mathrm{mod} \mathcal{W}^s_{\mathrm{comp}}(\Omega, \mathcal{K}^{\infty,\gamma}_\Theta(X^\wedge))$. *An analogous result holds if we replace $\zeta(y)$ by $\zeta(y,y') \in C^\infty_0(\Omega \times \Omega, \mathcal{A}'(K, C^\infty(X, \hat{H}^s(\mathbb{R}^q))))$.*

Proof. Let $B \subset\subset \Omega$ be a compact set containing the support of ζ with respect to y and set $C^\infty(B)_0 = \{u \in C^\infty(\mathbb{R}^q) : \mathrm{supp}\, u \subseteq B\}$. Then we have

$$\zeta \in C^\infty(B)_0 \otimes_\pi \mathcal{A}'(K, C^\infty(X)) \otimes_\pi \hat{H}^s(\mathbb{R}^q).$$

Hence ζ can be written as a convergent sum in this space

$$\zeta = \sum_{j=0}^{\infty} \lambda_j \zeta_j \psi_j \hat{v}_j \quad \text{for} \quad \lambda_j \in \mathbb{C}, \quad \sum |\lambda_j| < \infty,$$

$\zeta_j \in \mathcal{A}'(K, C^{\infty}(X))$, $\psi_j \in C^{\infty}(B)_0$, $\hat{v}_j \in \hat{H}^s(\mathbb{R}^q)$, tending to zero in the corresponding spaces for $j \to \infty$.

Setting $w_j = \text{Op}([\eta]^{\frac{n+1}{2}})v_j$, $l_j(y, \eta) = \psi_j(y)\omega(t[\eta])\langle \zeta_j, (t[\zeta])^{-z}\rangle$, we can write

$$\varphi(y)F_{\eta \to y}^{-1}[\eta]^{\frac{n+1}{2}}\omega(t[\eta])\langle \zeta(y), (t[\eta])^{-z}\rangle = \varphi(y) \sum_{j=0}^{\infty} \lambda_j \, \text{Op}(l_j)w_j.$$

$$(3.1.37)$$

According to the general rules of the pseudo–differential calculus we can write

$$\text{Op}(l_j)u = \text{Op}\left(\sum_{|\alpha| \leq N} \frac{1}{\alpha!}(-D_{\eta})^{\alpha}\partial_y^{\alpha}l_j(y, \eta)|_{y=y'}\right)u + \text{Op}(l_{j,(N)}(y, y', \eta))u,$$

$$(3.1.38)$$

where we choose N sufficiently large. We have

$$l_{j,(N)}(y, y', \eta) \in S^{\mu_N}(\mathbb{R}^q \times \mathbb{R}^q \times \mathbb{R}^q; \mathbb{C}, \mathcal{K}_{\Theta}^{\infty,\gamma}(X^{\wedge}))$$

for orders $\mu_N \to -\infty$ as $N \to \infty$. We then obtain a convergent sum

$$\varphi(y) \sum_{j=0}^{\infty} \lambda_j \, \text{Op}(l_{j,(N)})w_j \in \mathcal{W}_{\text{comp}}^s(\Omega, \mathcal{K}_{\Theta}^{\infty,\gamma}(X^{\wedge}))$$

when we fix N so large that $\mu_N \leq 0$. Now the characterisation of (3.1.37) reduces to

$$\varphi(y) \sum_{j=0}^{\infty} \lambda_j \sum_{|\alpha| \leq N} \frac{1}{\alpha!} \text{Op}((-D_{\eta})^{\alpha}\partial_y^{\alpha}l_j(y, \eta)|_{y=y'})w_j$$

$$= \varphi(y) \sum_{j=0}^{\infty} \lambda_j \sum_{|\alpha| \leq N} (-1)^{|\alpha|} \frac{1}{\alpha!} \text{Op}(D_{\eta}^{\alpha}m_j)(\partial_y^{\alpha}\psi_j)w_j \qquad (3.1.39)$$

for $m_j(\eta) = \omega(t[\eta])\langle \zeta_j, (t[\eta])^{-z}\rangle \in S_{\text{cl}}^{-\frac{n+1}{2}}(\mathbb{R}^q; \mathbb{C}, \mathcal{K}_P^{\infty,\gamma}(X^{\wedge}))$. Applying derivatives in η to $\omega(t[\eta])$ contained in $m_j(\eta)$ we produce factors with compact support in $t \in \mathbb{R}_+$. Thus these contributions generate $\mathcal{L}(\mathbb{C}, \mathcal{K}_{\Theta}^{\infty,\gamma}(X^{\wedge}))$–valued symbols. Hence the associated summands in

(3.1.39) yield elements of $\mathcal{W}^s_{\text{comp}}(\Omega, \mathcal{K}^{\infty,\gamma}_\Theta(X^\wedge))$. Therefore there remain $\omega(t[\eta])D^\alpha_\eta\langle\zeta_j, (t[\eta])^{-z}\rangle$ which are of the form $\omega(t[\eta])p_\alpha(\eta)\langle\zeta^\alpha_j, (t[\eta])^{-z}\rangle$ for suitable $p_\alpha(\eta) \in S^{-|\alpha|}_{\text{cl}}(\mathbb{R}^q)$ and $\zeta^\alpha_j \in \mathcal{A}'(K, C^\infty(X))$. Thus (3.1.39) equals

$$\varphi(y)\sum_{j=0}^\infty \lambda_j \sum_{|\alpha|\leq N}(-1)^{|\alpha|}\frac{1}{\alpha!}\operatorname{Op}(m^\alpha_j)\operatorname{Op}(p_\alpha)\psi^\alpha_j w_j \qquad (3.1.40)$$

$\text{mod}\mathcal{W}^s_{\text{comp}}(\Omega, \mathcal{K}^{\infty,\gamma}_\Theta(X^\wedge))$, where $m^\alpha_j(\eta) = \omega(t[\eta])\langle\zeta^\alpha_j, (t[\eta])^{-z}\rangle$, $\psi^\alpha_j = \partial^\alpha_y\psi_j$. We have $\operatorname{Op}(p_\alpha)\psi^\alpha_j w_j \in H^{s-\frac{n+1}{2}}(\mathbb{R}^q)$ for all j, α. Hence by rearranging the sum (3.1.40) and changing the corresponding notation we obtain (3.1.40) in the form

$$\varphi(y)\sum_{j=0}^\infty \tilde{\lambda}_j\operatorname{Op}(\tilde{m}_j)\tilde{w}_j, \quad \tilde{\lambda}_j \in \mathbb{C}, \quad \sum_{j=0}^\infty|\tilde{\lambda}_j| < \infty,$$

for $\tilde{m}_j(\eta) = \omega(t[\eta])\langle\tilde{\zeta}_j, (t[\eta])^{-z}\rangle$, $\tilde{\zeta}_j \in \mathcal{A}'(K, C^\infty(X))$, $\tilde{w}_j \in H^{s-\frac{n+1}{2}}(\mathbb{R}^q)$ tending to zero in the corresponding spaces for $j \to \infty$. Setting $v_j := \operatorname{Op}([\eta]^{-\frac{n+1}{2}})\tilde{w}_j \in H^s(\mathbb{R}^q)$ we finally obtain (3.1.40) in the form

$$\varphi(y)F^{-1}_{\eta\to y}[\eta]^{\frac{n+1}{2}}\omega(t[\eta])\langle\sum_{j=0}^\infty \tilde{\lambda}_j\tilde{\zeta}_j\hat{v}_j, (t[\eta])^{-z}\rangle.$$

It suffices now to set $\chi = \sum_{j=0}^\infty \tilde{\lambda}_j\tilde{\zeta}_j\hat{v}_j$. The arguments for the (y, y')–dependent case are completely analogous and will be omitted. $\square$

Lemma 3.1.36 *Let* $\zeta \in \mathcal{A}'(K, C^\infty(X))$, K *a compact set as in Proposition 3.1.35, associated with a given* $P \in \operatorname{As}(X, \boldsymbol{g})$, $\boldsymbol{g} = (\gamma, \Theta)$. *Then for every* $\varphi, \psi \in C^\infty_0(\Omega)$ *with* $\operatorname{supp}\varphi \cap \operatorname{supp}\psi = \emptyset$ *and* $v \in H^s(\mathbb{R}^q)$ *the element*

$$\varphi(y)F^{-1}_{\eta\to y}[\eta]^{\frac{n+1}{2}}\omega(t[\eta])\langle\zeta, (t[\eta])^{-z}\rangle\{\hat{v}(\eta) - \widehat{\psi v}(\eta)\}$$

belongs to $\mathcal{W}^\infty_{\text{comp}}(\Omega, \mathcal{K}^{\infty,\gamma}_P(X^\wedge))$.
   More generally, $\zeta \in \mathcal{A}'(K, C^\infty(X, ([1 - \psi]H^s(\mathbb{R}^q))^\wedge))$ *implies*

$$\varphi(y)F^{-1}_{\eta\to y}[\eta]^{\frac{n+1}{2}}\omega(t[\eta])\langle\zeta, (t[\eta])^{-z}\rangle \in \mathcal{W}^\infty_{\text{comp}}(\Omega, \mathcal{K}^{\infty,\gamma}_P(X^\wedge)).$$

Proof. We have

$$e(\eta) := [\eta]^{\frac{n+1}{2}}\omega(t[\eta])\langle\zeta, (t[\eta])^{-z}\rangle \in S^0_{\text{cl}}(\mathbb{R}^q; \mathbb{C}, \mathcal{K}^{\infty,\gamma}_P(X^\wedge)).$$

The expression to be characterised is $\varphi \operatorname{Op}(e)(1 - \psi)v = \operatorname{Op}(l)v$ for $l(y, y', \eta) = \varphi(y)(1 - \psi(y'))e(\eta)$. Applying a standard formula from the pseudo–differential calculus, namely

$$\operatorname{Op}(l)v = \sum_{|\alpha| \le N} \frac{1}{\alpha!} \operatorname{Op}(D_\eta^\alpha \partial_{y'}^\alpha l(y, y', \eta)|_{y'=y})v + \operatorname{Op}(l_{(N)})v \quad (3.1.41)$$

for an $l_{(N)}(y, y', \eta) \in S_{\mathrm{cl}}^{-(N+1)}(\mathbb{R}_y^q \times \mathbb{R}_{y'}^q \times \mathbb{R}_\eta^q; \mathbb{C}, \mathcal{K}_P^{\infty,\gamma}(X^\wedge))$ we see that the first sum on the right of the latter equation vanishes, while $\operatorname{Op}(l_{(N)})v \in \mathcal{W}_{\mathrm{comp}}^{s-N}(\Omega, \mathcal{K}_P^{\infty,\gamma}(X^\wedge))$. Since this holds for all N, we obtain the first assertion.

A general element $\zeta \in \mathcal{A}'(K, C^\infty(X, ([1 - \psi]H^s(\mathbb{R}^q))^\wedge))$ can be written as a convergent sum $\sum_{j=0}^\infty \lambda_j \zeta_j \hat{w}_j$ for $\lambda_j \in \mathbb{C}$, $\sum |\lambda_j| < \infty$, and $\zeta_j \in \mathcal{A}'(K, C^\infty(X))$, $w_j \in [1 - \psi]H^s(\mathbb{R}^q)$, again tending to zero in the corresponding spaces for $j \to \infty$. Then

$$\varphi(y)F_{\eta \to y}^{-1}[\eta]^{\frac{n+1}{2}}\omega(t[\eta])\langle \zeta, (t[\eta])^{-z}\rangle = \varphi(y)\sum_{j=0}^\infty \lambda_j \operatorname{Op}(e_j)w_j \quad (3.1.42)$$

for $e_j(\eta) = [\eta]^{\frac{n+1}{2}}\omega(t[\eta])\langle \zeta_j, (t[\eta])^{-z}\rangle$ tending to zero in the space $S_{\mathrm{cl}}^0(\mathbb{R}^q; \mathbb{C}, \mathcal{K}_P^{\infty,\gamma}(X^\wedge))$ for $j \to \infty$. Setting $l_j(y, y', \eta) = \varphi(y)(1 - \tilde{\psi}(y'))e_j(\eta)$ for an arbitrary $\tilde{\psi} \in C_0^\infty(\Omega)$ with $\operatorname{supp} \varphi \cap \operatorname{supp} \tilde{\psi} = \emptyset$ and $(1 - \tilde{\psi})(1 - \psi) = 1 - \psi$ and $w_j = (1 - \tilde{\psi})v_j$ for corresponding elements $v_j \in H^s(\mathbb{R}^q)$, tending to zero for $j \to \infty$, we obtain (3.1.42) in the form $\sum_{j=0}^\infty \lambda_j \operatorname{Op}(l_j)v_j$. Let us now apply the decompositions (3.1.41) to every l_j. Then each sum on the right of (3.1.41) again vanishes, while the symbol $l_{j,(N)}$ of the remainder tends to zero in the corresponding symbol space. Thus $\operatorname{Op}(l_{j,(N)})v_j \to 0$ in $\mathcal{W}_{\mathrm{comp}}^{s-N}(\Omega, \mathcal{K}_P^{\infty,\gamma}(X^\wedge))$ for $j \to \infty$. It follows that the sum

$$\varphi(y)F_{\eta \to y}^{-1}[\eta]^{\frac{n+1}{2}}\omega(t[\eta])\langle \zeta, (t[\eta])^{-z}\rangle = \sum_{j=0}^\infty \lambda_j \operatorname{Op}(l_{j,(N)})v_j$$

converges in $\mathcal{W}_{\mathrm{comp}}^{s-N}(\Omega, \mathcal{K}_P^{\infty,\gamma}(X^\wedge))$. To complete the proof it suffices to note that this is true for all N. $\square$

Theorem 3.1.37 *Let $\Omega, \Omega' \subseteq \mathbb{R}^q$ be open sets and $\chi : \Omega \to \tilde{\Omega}$ be a diffeomorphism. Then for every $P \in \mathrm{As}(X, \boldsymbol{g})$, $\boldsymbol{g} = (\gamma, \Theta)$, $\boldsymbol{g} = (\gamma, \Theta)$, the pull–backs of Theorem 3.1.29 induce isomorphisms*

$$\chi^* : \mathcal{W}_{\mathrm{comp}}^s(\tilde{\Omega}, \mathcal{K}_P^{s,\gamma}(X^\wedge)) \to \mathcal{W}_{\mathrm{comp}}^s(\Omega, \mathcal{K}_P^{s,\gamma}(X^\wedge))$$

and

$$\chi^* : \mathcal{W}_{\mathrm{loc}}^s(\tilde{\Omega}, \mathcal{K}_P^{s,\gamma}(X^\wedge)) \to \mathcal{W}_{\mathrm{loc}}^s(\Omega, \mathcal{K}_P^{s,\gamma}(X^\wedge))$$

for all $s \in \mathbb{R}$. An analogous result holds for the discrete asymptotics.

Proof. The assertion for "comp" immediately implies that for "loc", so it suffices to consider the case with "comp".

First observe that the result is true for $s = \infty$. In fact, in view of Propsition 3.1.33 we have for every $f(t, x, \tilde{y}) \in \mathcal{W}^{\infty}_{\text{comp}}(\tilde{\Omega}, \mathcal{K}^{\infty,\gamma}_P(X^{\wedge}))$ a decomposition

$$f(t, x, \tilde{y}) = \tilde{\varphi}(\tilde{y})\omega(t)\langle \zeta, t^{-z} \rangle + \tilde{\varphi}(\tilde{y})r(t, x, \tilde{y}) \qquad (3.1.43)$$

for any $\tilde{\varphi} \in C^{\infty}(\tilde{\Omega})$, $\tilde{\varphi} \equiv 1$ on $\text{supp}_{(\tilde{y})} f$, and some remainder $r(t, x, \tilde{y}) \in H^{\infty}(\mathbb{R}^q, \mathcal{K}^{\infty,\gamma+\beta}(X^{\wedge}))$ is some remainder. Here we used the notation from Proposition 3.1.33 for an arbitrary fixed $\beta \geq 0$. Now the invariance of the first item on the right of (3.1.43) is obvious, whereas that of the second one follows from Theorem 3.1.29.

Now let $K \subseteq$ carrier P be a compact set contained in the strip $\frac{n+1}{2} - (\gamma + \beta + \varepsilon) \leq \text{Re}\, z < \frac{n+1}{2} - \gamma$ for given $\beta \geq 0$, $\varepsilon > 0$. Then, according to Proposition 3.1.32 every $f(t, x, \tilde{y}) \in \mathcal{W}^s_{\text{comp}}(\tilde{\Omega}, \mathcal{K}^{s,\gamma}_P(X^{\wedge}))$ can be written

$$f(t, x, \tilde{y}) = \tilde{\varphi}(\tilde{y})F^{-1}_{\tilde{\eta} \to \tilde{y}}[\tilde{\eta}]^{\frac{n+1}{2}} \omega(t[\tilde{\eta}])\langle \tilde{\zeta}, (t[\tilde{\eta}])^{-z} \rangle + r_{\beta}(t, x, \tilde{y}) \quad (3.1.44)$$

for a $\tilde{\varphi} \in C^{\infty}_0(\tilde{\Omega})$, $\tilde{\zeta} \in \mathcal{A}'(K, C^{\infty}(X, H^s(\mathbb{R}^q)))$, and

$$r_{\beta}(t, x, \tilde{y}) \in \mathcal{W}^s_{\text{comp}}(\tilde{\Omega}, \mathcal{K}^{\gamma+\beta}(X^{\wedge})).$$

From Theorem 3.1.29 we know that χ^* induces an isomorphism from the space $\mathcal{W}^s_{\text{comp}}(\tilde{\Omega}, \mathcal{K}^{s,\gamma+\beta}(X^{\wedge}))$ to $\mathcal{W}^s_{\text{comp}}(\Omega, \mathcal{K}^{s,\gamma+\beta}(X^{\wedge}))$. Thus it suffices to consider the first item on the right of (3.1.44). Because of Lemma 3.1.36 we may assume $\tilde{\zeta} \in \mathcal{A}'(K, C^{\infty}(X, ([\tilde{\psi}]H^s(\mathbb{R}^q))^{\wedge}))$ for some $\tilde{\psi} \in C^{\infty}_0(\tilde{\Omega})$ with $\text{supp}\,\tilde{\varphi} \cap \text{supp}(1 - \tilde{\psi}) = \emptyset$. Here we apply the first part of the proof which shows that the contribution from $\mathcal{A}'(K, C^{\infty}(X, ([1 - \tilde{\psi}]H^s(\mathbb{R}^q))^{\wedge}))$ in $\mathcal{W}^{\infty}_{\text{comp}}(\tilde{\Omega}, \mathcal{K}^{\infty,\gamma}_P(X^{\wedge}))$ transforms under χ^* to an element in $\mathcal{W}^{\infty}_{\text{comp}}(\Omega, \mathcal{K}^{\infty,\gamma}_P(X^{\wedge}))$. Let first $\tilde{\zeta} = \zeta_1 \hat{v}_1$ for $\zeta_1 \in \mathcal{A}'(K, C^{\infty}(X))$ and $v_1 \in [\tilde{\psi}]H^s(\mathbb{R}^q)$. Then, writing $m_1(\tilde{\eta}) = [\tilde{\eta}]^{\frac{n+1}{2}} \omega(t[\tilde{\eta}])\langle \zeta_1, (t[\tilde{\eta}])^{-z} \rangle$ the first summand on the right of (3.1.44) takes the form

$$f_1(t, x, \tilde{y}) = \tilde{\varphi}(\tilde{y})\,\text{Op}(m_1)v_1, \quad m_1(\tilde{\eta}) \in S^0_{\text{cl}}(\mathbb{R}^q; \mathbb{C}, \mathcal{K}^{\infty,\gamma}_P(X^{\wedge})).$$

It follows that $(\chi^* f)(t, x, y) = \varphi(y)(\chi^* \text{Op}(m_1)(\chi^*)^{-1})w_1$ for $\varphi = \chi^* \tilde{\varphi}$ and $w_1 = \chi^* v_1 \in [\psi]H^s(\mathbb{R}^q)$ for $\psi = \chi^* \tilde{\psi}$. Now $\chi^* \text{Op}(m_1)(\chi^*)^{-1}$ is just the operator push–forward of $\text{Op}(m_1)$ under $\chi^{-1} : \tilde{\Omega} \to \Omega$. Applying the standard arguments to obtain an oscillatory integral representation in the variables y with the covariables η we obtain

$$(\chi^* \text{Op}(m_1)(\chi^*)^{-1}w_1)(y)$$

$$= \iint e^{i(y-y')\eta} m_1(g(y, y')\eta)h(y, y')w_1(y')\, dy'd\eta$$

for some

$$g(y, y') \in C^\infty(\Omega \times \Omega, GL(q, \mathbb{R})), \quad h(y, y') = |\det g(y, y')| |\det d\chi(y')|,$$

cf. the proof of Theorem 1.1.38. It is obvious that $l_1(y, y', \eta) :=$
$m_1(g(y, y')\eta) h(y, y') \in S^0_{\text{cl}}(\Omega \times \Omega \times \mathbb{R}^q; \mathbb{C}, \mathcal{K}^{\infty,\gamma}_P(X^\wedge))$ and hence

$$\varphi\chi^* \operatorname{Op}(m_1)(\chi^*)^{-1} w_1(y) = \varphi \operatorname{Op}(l_1) w_1 \in \mathcal{W}^s_{\text{comp}}(\Omega, \mathcal{K}^{\infty,\gamma}_P(X^\wedge)).$$

This gives us the assertion in the special case $\tilde{\zeta} = \zeta_1 \hat{v}_1$. At the
same time it is evident that when the sequence $\zeta_j \in \mathcal{A}'(K, C^\infty(X))$,
$j \in \mathbb{N}$, tends to zero for $j \to \infty$, then the associated symbols
$l_j(y, y', \eta) = m_j(g(y, y')\eta) h(y, y')$ for $j \to \infty$ also tend to zero in $S^0_{\text{cl}}(\Omega \times$
$\Omega \times \mathbb{R}^q; \mathbb{C}, \mathcal{K}^{\infty,\gamma}_P(X^\wedge))$ for $m_j(\tilde{\eta}) = [\tilde{\eta}]^{\frac{n+1}{2}} \omega(t[\tilde{\eta}]) \langle \zeta_j, (t[\tilde{\eta}])^{-z} \rangle$. Moreover,
$v_j \in [\tilde{\psi}] H^s(\mathbb{R}^q)$, $v_j \to 0$ for $j \to \infty$ implies $w_j = \chi^* v_j \to 0$ in $[\psi] H^s(\mathbb{R}^q)$
for $j \to \infty$, $\psi = \chi^* \tilde{\psi}$. An arbitrary $\zeta \in \mathcal{A}'(K, C^\infty(X, ([\tilde{\psi}] H^s(\mathbb{R}^q))^\wedge))$
can be written as a convergent sum

$$\tilde{\zeta} = \sum_{j=0}^\infty \lambda_j \zeta_j \hat{v}_j, \quad \lambda_j \in \mathbb{C}, \quad \sum_{j=0}^\infty |\lambda_j| < \infty$$

in such a way that both $\zeta_j \in \mathcal{A}'(K, C^\infty(X))$ and $v_j \in [\tilde{\psi}] H^s(\mathbb{R}^q)$ tend
to zero as $j \to \infty$ in the corresponding spaces. Then the sum

$$\chi^* \tilde{\varphi} F^{-1}_{\tilde{\eta} \to \tilde{y}} [\tilde{\eta}]^{\frac{n+1}{2}} \omega(t[\tilde{\eta}]) \langle \tilde{\zeta}, (t[\tilde{\eta}])^{-z} \rangle = \varphi \sum_{j=0}^\infty \lambda_j \operatorname{Op}(l_j) w_j$$

converges in $\mathcal{W}^s_{\text{comp}}(\Omega, \mathcal{K}^{\infty,\gamma}_P(X^\wedge))$. This completes the proof. □

We are now able to define the global weighted Sobolev spaces with
asymptotics on a manifold W with edges Y. First, for each $P \in$
$\text{As}(X, \boldsymbol{g})$, $\boldsymbol{g} = (\gamma, \Theta)$, we denote by $\mathcal{W}^s_{\text{loc}}(Y, \mathcal{K}^{s,\gamma}_P(X^\wedge))$ the subspace of
all $u \in \mathcal{W}^s_{\text{loc}}(Y, \mathcal{K}^{s,\gamma}(X^\wedge))$ such that for every chart $G \to \Omega \subseteq \mathbb{R}^q$ on Y
and every $\varphi \in C^\infty_0(G)$ the function φu, expressed in local coordinates
$y \in \Omega$, belongs to $\mathcal{W}^s_{\text{comp}}(\Omega, \mathcal{K}^{s,\gamma}_P(X^\wedge))$. By $\mathcal{W}^s_{\text{comp}}(Y, \mathcal{K}^{s,\gamma}_P(X^\wedge))$ we
denote the subspace of those elements which have compact support
with respect to $y \in Y$. In view of Theorem 3.1.29 these are correct
definitions. Moreover, if $\mathbb{W}$ is the stretched manifold to W we set

$$\mathcal{W}^{s,\gamma}_{P,\text{loc}}(\mathbb{W}) = \{ u \in \mathcal{W}^{s,\gamma}_{\text{loc}}(\mathbb{W}) : \omega u \in \mathcal{W}^s_{\text{loc}}(Y, \mathcal{K}^{s,\gamma}_P(X^\wedge)) \}$$

for any cut–off function ω as in (3.1.26). Finally we define

$$\mathcal{W}^{s,\gamma}_{P,\text{comp}}(\mathbb{W}) = \mathcal{W}^{s,\gamma}_{\text{comp}}(\mathbb{W}) \cap \mathcal{W}^{s,\gamma}_{P,\text{loc}}(\mathbb{W}).$$

For compact $\mathbb{W}$ we have

$$\mathcal{W}_P^{s,\gamma}(\mathbb{W}) = \mathcal{W}_{P,\mathrm{comp}}^{s,\gamma}(\mathbb{W}) = \mathcal{W}_{P,\mathrm{loc}}^{s,\gamma}(\mathbb{W}).$$

The spaces $\mathcal{W}_{P,\mathrm{loc}}^{s,\gamma}(\mathbb{W})$ are Fréchet; $\mathcal{W}_{P,\mathrm{comp}}^{s,\gamma}(\mathbb{W})$ are inductive limits of Fréchet spaces. For purposes below we now define suitable spaces of smoothing operators, first on $X^\wedge \times \Omega$ for open $\Omega \subseteq \mathbb{R}^q$ and then globally on a manifold $\mathbb{W}$ with edges Y.

Definition 3.1.38 *Let* $\boldsymbol{g} = (\gamma, \delta, \Theta)$ *for* $\gamma, \delta \in \mathbb{R}$ *and* $\Theta = (\vartheta, 0]$, $-\infty \leq \vartheta < 0$. *Then* $Y_G^{-\infty}(X^\wedge \times \Omega, \boldsymbol{g})$ *denotes the space of all operators*

$$G \in \bigcap_{s\in\mathbb{R}} \mathcal{L}(\mathcal{W}_{\mathrm{comp}}^s(\Omega, \mathcal{K}^{s,\gamma}(X^\wedge)), \mathcal{W}_{\mathrm{loc}}^\infty(\Omega, \mathcal{K}^{\infty,\delta}(X^\wedge)))$$

for which there are elements $P \in \mathrm{As}(X, (\delta, \Theta))$, $Q \in \mathrm{As}(X, (-\gamma, \Theta))$ *with*

$$G \in \bigcap_{s\in\mathbb{R}} \mathcal{L}(\mathcal{W}_{\mathrm{comp}}^s(\Omega, \mathcal{K}^{s,\gamma}(X^\wedge)), \mathcal{W}_{\mathrm{loc}}^\infty(\Omega, \mathcal{K}_P^{\infty,\delta}(X^\wedge))),$$

$$G^* \in \bigcap_{s\in\mathbb{R}} \mathcal{L}(\mathcal{W}_{\mathrm{comp}}^s(\Omega, \mathcal{K}^{s,-\delta}(X^\wedge)), \mathcal{W}_{\mathrm{loc}}^\infty(\Omega, \mathcal{K}_Q^{\infty,-\gamma}(X^\wedge))).$$

Here G^* *is the formal adjoint with respect to the scalar product in* $\mathcal{W}^{0,0}(X^\wedge \times \Omega)$, *cf.* (3.1.20). *The subclass of all* $G \in Y_G^{-\infty}(X^\wedge \times \Omega, \boldsymbol{g})$ *for arbitrary* $P \in \mathrm{As}(X, (\delta, \Theta)^\bullet)$, $Q \in \mathrm{As}(X, (-\gamma, \Theta)^\bullet)$ *is denoted by* $Y_G^{-\infty}(X^\wedge \times \Omega, \boldsymbol{g}^\bullet)$.

The elements of $Y_G^{-\infty}(X^\wedge \times \Omega, \boldsymbol{g})(\in Y_G^{-\infty}(X^\wedge \times \Omega, \boldsymbol{g}^\bullet))$ are called smoothing Green operators on the (open stretched) wedge $X^\wedge \times \Omega$, with continuous (discrete) asymptotics. An analogous definition makes sense for a (stretched) manifold $\mathbb{W}$ with edges Y. It will be necessary to allow block matrices

$$\begin{aligned} G = &\begin{pmatrix} G_{11} & G_{12} \\ G_{21} & G_{22} \end{pmatrix} \\ &\in \bigcap_{s\in\mathbb{R}} \mathcal{L}(\mathcal{W}_{\mathrm{comp}}^{s,\gamma}(\mathbb{W}) \oplus H_{\mathrm{comp}}^s(Y, \mathbb{C}^{N_-}), \mathcal{W}_{\mathrm{loc}}^{\infty,\delta}(\mathbb{W}) \oplus H_{\mathrm{loc}}^\infty(Y, \mathbb{C}^{N_+})), \end{aligned}$$

$$(3.1.45)$$

where G_{21} plays the role of a smoothing trace operator, G_{12} of a smoothing potential operator, with respect to the edge Y, while G_{22} is an $N_+ \times N_-$ matrix of smoothing operators on Y.

Definition 3.1.39 $Y^{-\infty}(\mathbb{W}, g; N_-, N_+)$ *for* $g = (\gamma, \delta, \Theta)$ *and* N_-, $N_+ \in \mathbb{N}$ *denotes the subspace of all operators* (3.1.45) *for which there are asymptotic types* $P \in \mathrm{As}(X, (\delta, \Theta))$, $Q \in \mathrm{As}(X, (-\gamma, \Theta))$ *with*

$$G \in \bigcap_{s \in \mathbb{R}} \mathcal{L}(\mathcal{W}^{s,\gamma}_{\mathrm{comp}}(\mathbb{W}) \oplus H^s_{\mathrm{comp}}(Y, \mathbb{C}^{N_-}), \mathcal{W}^{\infty,\delta}_{P,\mathrm{loc}}(\mathbb{W}) \oplus H^\infty_{\mathrm{loc}}(Y, \mathbb{C}^{N_+})),$$

$$G^* \in \bigcap_{s \in \mathbb{R}} \mathcal{L}(\mathcal{W}^{s,-\delta}_{\mathrm{comp}}(\mathbb{W}) \oplus H^s_{\mathrm{comp}}(Y, \mathbb{C}^{N_+}), \mathcal{W}^{\infty,-\gamma}_{Q,\mathrm{loc}}(\mathbb{W}) \oplus H^\infty_{\mathrm{loc}}(Y, \mathbb{C}^{N_-})).$$

Here G^* *is the formal adjoint of* G *in the sense*

$$(Gu, v)_{\mathcal{W}^{0,0}(\mathbb{W}) \oplus L^2(Y, \mathbb{C}^{N_+})} = (u, G^*v)_{\mathcal{W}^{0,0}(\mathbb{W}) \oplus L^2(Y, \mathbb{C}^{N_-})}$$

for all $u \in C_0^\infty(\mathrm{int}\,\mathbb{W}) \oplus C_0^\infty(Y, \mathbb{C}^{N_-})$, $v \in C_0^\infty(\mathrm{int}\,\mathbb{W}) \oplus C_0^\infty(Y, \mathbb{C}^{N_+})$, *cf.* (3.1.28). *The* L^2 *spaces on* Y *refer to a fixed Riemannian metric on* Y *with an associated measure* dy. *The subclass of all operators* $G \in \mathcal{Y}^{-\infty}(\mathbb{W}, g; N_-, N_+)$ *for arbitrary* $P \in \mathrm{As}(X, (\delta, \Theta)^\bullet)$, $Q \in \mathrm{As}(X, (-\gamma, \Theta)^\bullet)$ *will be denoted by* $\mathcal{Y}^{-\infty}(\mathbb{W}, g^\bullet; N_-, N_+)$. *Moreover, we have the operator spaces*

$$Y^{-\infty}(\mathbb{W}, g) \quad and \quad Y^{-\infty}(\mathbb{W}, g^\bullet)$$

respectively, obtained by specialising the above classes to $N_- = N_+ = 0$.

The role of the operator spaces in the latter definitions will become clear below in the calculus of pseudo–differential operators on a manifold with edges; these spaces will consist of the smoothing operators in the corresponding operator algebras. Observe that subscript G in the local classes of Definition 3.1.38 on a wedge $X^\wedge \times \Omega$ indicates more special assumptions. In other words $Y_G^{-\infty}(X^\wedge \times \Omega, g) \subset Y^{-\infty}(\mathbb{W}, g)$ is a proper inclusion for $\mathbb{W} = \overline{\mathbb{R}}_+ \times X \times \Omega$.

Let us now turn to a modification of our notion of asymptotic types. Denote by

$$\mathrm{As}(X, (\gamma, \Theta))^\bullet \tag{3.1.46}$$

the set of all $P \in \mathrm{As}(X, (\gamma, \Theta))$ for which there exists a sequence

$$\{(p_j, m_j)\}_{j=0,\dots,N} \quad \text{for } \Theta \text{ finite,}$$
$$\{(p_j, m_j)\}_{j \in \mathbb{N}} \quad \text{for } \Theta = (-\infty, 0],$$

where $p_j \in \mathbb{C}$, $\mathrm{Re}\, p_j < \frac{n+1}{2} - \gamma$, $m_j \in \mathbb{N}$, and $\mathrm{Re}\, p_j \to -\infty$ for $j \to \infty$ when $\Theta = (-\infty, 0]$, $N = N(P) < \infty$ when Θ is finite, such that the analytic functionals $\zeta_j \in \mathcal{A}'(\{p_j\}, C^\infty(X))$ from the definition of the continuous asymptotics are of the form

$$\langle \zeta_j, h \rangle = \sum_{k=0}^{m_j} c_{jk}(x) \left(\frac{d}{dz} \right)^k h(z)|_{z=p_j}$$

for arbitrary $c_{jk} \in C^\infty(X)$. In other words the only difference from the discrete asymptotics is that here we drop the condition that the coefficients c_{jk} may vary only over a finite–dimensional subspace of $C^\infty(X)$. The above constructions concerning wedge Sobolev spaces with asymptotics can be specialised to $\mathrm{As}(X, (\gamma, \Theta))^\bullet$. In particular, assuming $P \in \mathrm{As}(X, (\delta, \Theta))^\bullet$, $Q \in \mathrm{As}(X, (-\gamma, \Theta))^\bullet$ in the above Definition 3.1.38 we obtain spaces of smoothing operators that we denote by $Y_G^{-\infty}(X^\wedge \times \Omega, g)^\bullet$. In an analogous manner we can introduce $\mathcal{Y}^{-\infty}(\mathbb{W}, g; N_-, N_+)^\bullet$.

The advantage of this slight generalisation of the discrete asymptotics is that it makes the treatment of y–dependent Mellin symbols easier than in the concept of variable discrete asymptotics, where we could allow not only variable spaces of coefficients but also $(p_j, m_j) = (p_j(y), m_j(y))$, including branching and jumping phenomena. For the case of boundary value problems this was treated in Schulze [123], [125] and for edge operators in Schulze [115]. In this book we will content ourselves with the case of pairs (p_j, m_j) which are constant in y.

3.2　Interior symbols and Mellin operator convention

3.2.1　Edge–degenerate symbols

The edge–degenerate differential operators on a (stretched) manifold with edges, cf. Definition 3.1.1, and the program to constructing a "minimal" operator algebra containing these operators including the parametrices of elliptic elements are a motivation to introduce edge–degenerate pseudo–differential symbols. Let us start with the local definition with respect to $(t, x, y) \in \mathbb{R}_+ \times \Sigma \times \Omega$ for open sets $\Sigma \subseteq \mathbb{R}^n$, $\Omega \subseteq R^q$.

Definition 3.2.1 $\tilde{S}^\mu(\overline{\mathbb{R}}_+ \times \Sigma \times \Omega \times \mathbb{R}_{\tau,\xi,\eta}^{1+n+q})$, $\mu \in \mathbb{R}$, *denotes the space of all symbols* $p(t, x, y, \tau, \xi, \eta) \in S^\mu(\overline{\mathbb{R}}_+ \times \Sigma \times \Omega \times \mathbb{R}^{1+n+q})$ *for which there exists an element* $\tilde{p}(t, x, y, \tilde{\tau}, \xi, \tilde{\eta}) \in S^\mu(\overline{\mathbb{R}}_+ \times \Sigma \times \Omega \times \mathbb{R}_{\tilde{\tau},\xi,\tilde{\eta}}^{1+n+q})$ *such that*

$$p(t, x, y, \tau, \xi, \eta) = \tilde{p}(t, x, y, \tilde{\tau}, \xi, \tilde{\eta})|_{\tilde{\tau}=t\tau, \tilde{\eta}=t\eta}. \tag{3.2.1}$$

In an analogous manner we define the subspace of classical symbols $\tilde{S}_{\mathrm{cl}}^\mu(\overline{\mathbb{R}}_+ \times \Sigma \times \Omega \times \mathbb{R}_{\tau,\xi,\eta}^{1+n+q})$ *by requiring* $\tilde{p}(t, x, y, \tilde{\tau}, \xi, \tilde{\eta}) \in S_{\mathrm{cl}}^\mu(\overline{\mathbb{R}}_+ \times \Sigma \times \Omega \times \mathbb{R}_{\tilde{\tau},\xi,\tilde{\eta}}^{1+n+q})$. *The symbols* $p \in \tilde{S}_{(\mathrm{cl})}^\mu(\overline{\mathbb{R}}_+ \times \Sigma \times \Omega \times \mathbb{R}^{1+n+q})$ *are called edge–degenerate.*

The weight factor $t^{-\mu}$ from Definition 3.1.7 is ignored here for a moment, though it will play a role below on the level of operators.

By definition we have isomorphisms $\tilde{S}^\mu \to S^\mu$, $\tilde{S}^\mu_{cl} \to S^\mu_{cl}$. Using the Fréchet topologies in S^μ and S^μ_{cl} we get canonical Fréchet topologies in the spaces $\tilde{S}^\mu(\overline{\mathbb{R}}_+ \times \Sigma \times \Omega \times \mathbb{R}^{1+n+q})$ and $\tilde{S}^\mu_{cl}(\overline{\mathbb{R}}_+ \times \Sigma \times \Omega \times \mathbb{R}^{1+n+q})$, respectively.

For $\beta \in \mathbb{R}$ we define $t^\beta \tilde{S}^\mu = \{t^\beta p : p \in \tilde{S}^\mu\}$ and analogously $t^\beta \tilde{S}^\mu_{cl}$ with canonical Fréchet topologies from the bijections to $\tilde{S}^\mu$ and $\tilde{S}^\mu_{cl}$, respectively.

Proposition 3.2.2 *Let $p_j \in \tilde{S}^{\mu_j}(\overline{\mathbb{R}}_+ \times \Sigma \times \Omega \times \mathbb{R}^{1+n+q})$, $j \in \mathbb{N}$, be an arbitrary sequence with $\mu_j \to -\infty$ as $j \to \infty$. Then there is a $p \in \tilde{S}^\mu(\overline{\mathbb{R}}_+ \times \Sigma \times \Omega \times \mathbb{R}^{1+n+q})$ for $\mu = \max\{\mu_j\}$ such that for every $M \in \mathbb{N}$ there is an $N(M)$ such that for all $N \geq N(M)$*

$$p - \sum_{j=0}^{N} \tilde{S}^{\mu-M}(\overline{\mathbb{R}}_+ \times \Sigma \times \Omega \times \mathbb{R}^{1+n+q}),$$

and p is unique mod $\tilde{S}^{-\infty}(\overline{\mathbb{R}}_+ \times \Sigma \times \Omega \times \mathbb{R}^{1+n+q})$.

We write $p \sim \sum_{j=0}^{\infty} p_j$ and call p the asymptotic sum of the symbols p_j.

If $\tilde{p}_j \in S^{\mu_j}$ is associated with $p_j \in \tilde{S}^{\mu_j}$ in the sense of (3.2.1), then, in order to obtain the asymptotic sum p we can first construct $\tilde{p} \in S^\mu(\overline{\mathbb{R}}_+ \times \Sigma \times \Omega \times \mathbb{R}^{1+n+q})$ in the standard way as a convergent sum

$$\tilde{p}(t,x,y,\tilde{\tau},\xi,\tilde{\eta}) = \sum_{j=0}^{\infty} \chi(\frac{\tilde{\tau},\xi,\tilde{\eta}}{c_j})\tilde{p}_j(t,x,y,\tilde{\tau},\xi,\tilde{\eta})$$

for an excision function $\chi(\tilde{\tau},\xi,\tilde{\eta})$ and a sequence of constants $c_j > 0$ tending to ∞ sufficiently fast for $j \to \infty$. Then it suffices to define p by (3.2.1).

Remark 3.2.3 *Let $\beta,\mu,\varrho,\nu \in \mathbb{R}$ and $p \in \tilde{S}^\mu(\overline{\mathbb{R}}_+ \times \Sigma \times \Omega \times \mathbb{R}^{1+n+q})$, $r \in \tilde{S}^\nu(\overline{\mathbb{R}}_+ \times \Sigma \times \Omega \times \mathbb{R}^{1+n+q})$. Then the Leibniz product*

$$(t^\beta p)\#_{t,x}(t^\varrho r)(t,x,y,\tau,\xi,\eta)$$

$$\sim \sum_\alpha \frac{1}{\alpha!}(D^\alpha_{\tau,\xi}(t^\beta p))(t,x,y,\tau,\xi,\eta) \cdot (\partial^\alpha_{t,x}(t^\varrho r))(t,x,y,\tau,\xi,\eta)$$

can be carried out in $t^{\beta+\varrho}\tilde{S}^{\mu+\nu}(\overline{\mathbb{R}}_+ \times \Sigma \times \Omega \times \mathbb{R}^{1+n+q})$. Similarly the Leibniz product in all variables

$$(t^\beta p)\#_{t,x,y}(t^\varrho r)(t,x,y,\tau,\xi,\eta)$$

$$\sim \sum_\gamma \frac{1}{\gamma!}(D^\gamma_{\tau,\xi,\eta}(t^\beta p))(t,x,y,\tau,\xi,\eta)(\partial^\beta_{t,x,y}(t^\varrho r))(t,x,y,\tau,\xi,\eta)$$

can be carried out in $t^{\beta+\varrho}\tilde{S}^{\mu+\nu}(\overline{\mathbb{R}}_+ \times \Sigma \times \Omega \times \mathbb{R}^{1+n+q})$.

Remark 3.2.4 *Let* $p \in \tilde{S}^{\mu}_{\mathrm{cl}}(\overline{\mathbb{R}}_+ \times \Sigma \times \Omega \times \mathbb{R})$, *and assume that*

$$\tilde{p}_{(\mu)}(t,x,y,\tilde{\tau},\xi,\tilde{\eta}) \neq 0$$

for all $(t,x,y) \in \overline{\mathbb{R}}_+ \times \Sigma \times \Omega$, $(\tilde{\tau},\xi,\tilde{\eta}) \neq 0$. *Then for every* $\beta \in \mathbb{R}$ *there exists an* $\tilde{r}(t,x,y,\tilde{\tau},\xi,\tilde{\eta}) \in \tilde{S}^{-\mu}_{\mathrm{cl}}(\overline{\mathbb{R}}_+ \times \Sigma \times \Omega \times \mathbb{R}^{1+n+q})$ *such that* $t^{-\beta}\tilde{r}(t,x,y,t\tau,\xi,t\eta)$ *is the inverse of* $t^{\beta}\tilde{p}(t,x,y,t\tau,\xi,t\eta)$ *with respect to* $\#_{t,x}$, *modulo a symbol of order* $-\infty$ *in* $(t\tau,\xi,t\eta)$.

Our next objective is to construct (y,η)–dependent operator families on $X^{\wedge}$. To this end we fix the following data:

$$\mathcal{U} = \{U_j\}_{j=1,\dots,N}, \quad \{\varphi_j\}_{j=1,\dots,N}, \quad \{\psi_j\}_{j=1,\dots,N}, \tag{3.2.2}$$

where $\mathcal{U}$ is an open covering of the (closed compact) C^{∞} manifold X by coordinate neighbourhoods, $\{\varphi_j\}$ a subordinate partition of unity and $\{\psi_j\}$ a system of functions $\psi_j \in C^{\infty}_0(U_j)$ with $\varphi_j\psi_j = \varphi_j$ for all j.

Let $\kappa : U \to \Sigma$ be a chart, $U \in \mathcal{U}$, and let $p(t,x,y,\tau,\xi,\eta) \in \tilde{S}^{\mu}(\overline{\mathbb{R}}_+ \times \Sigma \times \Omega \times \mathbb{R}^{1+n+q})$. Set

$$(\mathrm{op}_{\psi,(t,x)}(p)(y,\eta)f)(t,x)$$
$$= \iint e^{i(t-t')\tau+i(x-x')\xi}p(t,x,y,\tau,\xi,\eta)f(t',x')\,dt'dx'd\tau d\xi,$$

$f \in C^{\infty}_0(\Sigma^{\wedge})$, $\Sigma^{\wedge} = \mathbb{R}_+ \times \Sigma$. This is a C^{∞} function of $y \in \Omega$ with values in the space $L^{\mu}(\Sigma^{\wedge}; \mathbb{R}^q)$ of parameter–dependent pseudo–differential operators on $\Sigma^{\wedge}$ with the parameters $q \in \mathbb{R}^q$, cf. Section 1.2.

Moreover, we fix a system of charts

$$\kappa_j : U_j \to \Sigma_j, \quad j = 1,\dots,N,$$

for open $\Sigma_j \subseteq \mathbb{R}^n$. The operator pull–back

$$(1 \times \kappa)^{-1}_* : L^{\mu}(\Sigma^{\wedge}; \mathbb{R}^q) \to L^{\mu}(U^{\wedge}; \mathbb{R}^q)$$

under $1 \times \kappa : \mathbb{R}_+ \times U \to \mathbb{R}_+ \times \Sigma$ gives us the corresponding operator families on $U^{\wedge} = \mathbb{R}_+ \times U$. This extends in a canonical way to the y–dependent case. Note that when $\chi : \Sigma \to \tilde{\Sigma}$ is a diffeomorphism, the (y,η)–wise operator push–forward $\chi_* \mathrm{op}_{\psi,(t,x)}(p)(y,\eta) \in L^{\mu}(\tilde{\Sigma}^{\wedge}; \mathbb{R}^q)$ can be represented modulo $C^{\infty}(\Omega, L^{-\infty}(\tilde{\Sigma}^{\wedge}; \mathbb{R}^q))$ by a symbol in $\tilde{S}^{\mu}(\overline{\mathbb{R}}_+ \times \Sigma \times \Omega \times \mathbb{R}^{1+n+q})$.

To every system of symbols

$$p_j(t,x,y,\tau,\xi,\eta) \in \tilde{S}^{\mu}(\overline{\mathbb{R}}_+ \times \Sigma_j \times \Omega \times \mathbb{R}^{1+n+q}), \quad j = 1,\dots,N \tag{3.2.3}$$

we get an operator family

$$P(y,\eta) = \sum_{j=1}^{N} \varphi_j P_j(y,\eta) \psi_j \qquad (3.2.4)$$

$$P_j(y,\eta) = (1 \times \kappa_j)_*^{-1} \operatorname{op}_{\psi,(t,x)}(p_j)(y,\eta). \qquad (3.2.5)$$

Denote by

$$C^\infty(\Omega, \tilde{L}^\mu(X^\wedge; \mathbb{R}^q)) \qquad (3.2.6)$$

the space of all operator families $P(y,\eta) + C(y,\eta)$ with $P(y,\eta)$ being of the form (3.2.5) for arbitrary p_j of the indicated structure, and $C(y,\eta) \in C^\infty(\Omega, L^{-\infty}(X^\wedge; \mathbb{R}^q))$, cf. Section 1.2.2. Denote by

$$C^\infty(\Omega, \tilde{L}_{cl}^\mu(X^\wedge; \mathbb{R}^q)) \qquad (3.2.7)$$

the subspace of all $P(y,\eta) \in C^\infty(\Omega, \tilde{L}^\mu(X^\wedge; \mathbb{R}^q))$ for which the involved p_j belong to $\tilde{S}_{cl}^\mu(\overline{\mathbb{R}}_+ \times \Sigma \times \Omega \times \mathbb{R}^{1+n+q})$ for all j.

The operator families in (3.2.6), (3.2.7) are of auxiliary character here. In Section 3.2.2 below we shall replace the operators based on the Fourier transform in t near $t = 0$ by operators based on the Mellin transform that will control the mapping properties near $t = 0$ in a more precise way.

Proposition 3.2.5 *Let $P(y,\eta) \in C^\infty(\Omega, \tilde{L}^\mu(X^\wedge; \mathbb{R}^q))$ and let $P^*(y,\eta)$ be the point–wise formal adjoint in the sense $(P(y,\eta)u, v)_{\mathcal{K}^{0,0}(X^\wedge)} = (u, P^*(y,\eta)v)_{\mathcal{K}^{0,0}(X^\wedge)}$ for all $u, v \in C_0^\infty(X^\wedge)$, $(y,\eta) \in \Omega \times \mathbb{R}^q$. Then $P^*(y,\eta) \in C^\infty(\Omega, \tilde{L}^\mu(X^\wedge; \mathbb{R}^q))$. In particular, the relation $P(y,\eta) \in C^\infty(\Omega, \tilde{L}_{cl}^\mu(X^\wedge; \mathbb{R}^q))$ implies $P^*(y,\eta) \in C^\infty(\Omega, \tilde{L}_{cl}^\mu(X^\wedge; \mathbb{R}^q))$.*

Proof. The formal adjoint operators that depend on $(y,\eta) \in \Omega \times \mathbb{R}^q$ obviously belong to $C^\infty(\Omega, L^\mu(X^\wedge; \mathbb{R}^q))$. For the complete symbols in local coordinates $(t, x, y) \in \mathbb{R}_+ \times \Sigma \times \Omega$ we have the standard asymptotic expansion, cf. Theorem 1.1.31. The α th summand then belongs to $\tilde{S}^{\mu - |\alpha|}(\overline{\mathbb{R}}_+ \times \Sigma \times \Omega \times \mathbb{R}^{1+n+q})$, $\alpha \in \mathbb{N}^{1+n}$. For completing the proof it suffices to apply Proposition 3.2.2 which also shows that classical operators remain classical under formal adjoints. $\square$

3.2.2   The Mellin convention in the edge case

The structure of the operator families $P(y,\eta) \in C^\infty(\Omega_y, \tilde{L}^\mu(X^\wedge; \mathbb{R}^q))$ with respect to the edge–degenerate symbols will be of interest, in particular, for $0 < t < c$ for any fixed $c > 0$. With $P(y,\eta)$ we want to associate operator–valued symbols, acting in the cone Sobolev spaces on $X^\wedge$. This will be possible by a Mellin reformulation of $P(y, \eta)$ in t, modulo $C^\infty(\Omega, L^{-\infty}(X^\wedge; \mathbb{R}^q))$, called a Mellin operator convention.

Remember that in Chapter 2 we already solved an analogous problem. Here we need the constructions in the corresponding parameter–dependent form.

Definition 3.2.6 *Let us denote by $N_O^\mu(X; \mathbb{R}^q)$ the subspace of all operator families $h(z, \eta) \in \mathcal{A}(\mathbb{C}_z, L^\mu(X; \mathbb{R}^q_\eta))$ for which*

$$h(\beta + i\tau, \eta) \in L^\mu(X; \mathbb{R}_\tau \times \mathbb{R}^q_\eta)$$

for all $\beta \in \mathbb{R}$, uniformly in $c \le \beta \le c'$ for all reals $c \le c'$. In an analogous manner we define $M_O^\mu(X; \mathbb{R}^q)$ by requiring the property $h(z, \eta) \in \mathcal{A}(\mathbb{C}_z, L^\mu_{cl}(X; \mathbb{R}^q_\eta))$ with $h(\beta + i\tau, \eta) \in L^\mu_{cl}(X; \mathbb{R}_\tau \times \mathbb{R}^q_\eta)$, $\beta \in \mathbb{R}$, uniformly in finite β–intervals.

The definition gives rise to natural Fréchet topologies in the spaces $N_O^\mu(X; \mathbb{R}^q)$ and $M_O^\mu(X; \mathbb{R}^q)$, respectively. For instance, for $N_O^\mu(X; \mathbb{R}^q)$, we endow the space with the weakest locally convex topology for which the natural embedding $\iota : N_O^\mu(X; \mathbb{R}^q) \to \mathcal{A}(\mathbb{C}, L^\mu(X; \mathbb{R}^q))$ is continuous as well as $r_\beta : h(z, \eta) \to h(\beta + i\tau, \eta)$, $r_\beta : N_O^\mu(X; \mathbb{R}^q) \to L^\mu(X; \mathbb{R}_\tau \times \mathbb{R}^q)$, uniformly in arbitrary finite intervals $c \le \beta \le c'$. This relies on the corresponding Fréchet topologies in the spaces of parameter–dependent pseudo–differential operators on X.

The Mellin reformulations will have the form $\mathrm{op}_M^\delta(h)(y, \eta))$ for every fixed $(y, \eta) \in \Omega \times \mathbb{R}^q$, $\delta \in \mathbb{R}$, for a certain operator function $h(t, y, z, \eta) \in C^\infty(\mathbb{R}_+ \times \Omega_y, N_O^\mu(X; \mathbb{R}^q_\eta))$, cf. Section 1.2.4. Similarly to the notation from Chapter 2 we write $(\mathrm{op}_M^\delta(h)(y, \eta))u(t)$ for $\mathcal{D}'(X)$–valued functions $u(t)$, where we assume that the actions along X are applied through the values of h in $L^\mu(X)$.

Theorem 3.2.7 *To each $P(y, \eta) \in C^\infty(\Omega, \tilde{L}^\mu(X^\wedge; \mathbb{R}^q))$ there is an $\tilde{h}(t, y, z, \tilde{\eta}) \in C^\infty(\overline{\mathbb{R}}_+ \times \Omega, N_O^\mu(X; \mathbb{R}^q_{\tilde{\eta}}))$ such that*

$$\mathrm{op}_M^\delta(h)(y, \eta) = P(y, \eta) \bmod C^\infty(\Omega, L^{-\infty}(X^\wedge; \mathbb{R}^q)) \qquad (3.2.8)$$

for $h(t, y, z, \eta) := \tilde{h}(t, y, z, t\eta)$ for every $\delta \in \mathbb{R}$.

Moreover, $P(y, \eta) \in C^\infty(\Omega, \tilde{L}^\mu_{cl}(X^\wedge; \mathbb{R}^q))$ implies $\tilde{h}(t, y, z, \tilde{\eta}) \in C^\infty(\overline{\mathbb{R}}_+ \times \Omega, M_O^\mu(X; \mathbb{R}^q_{\tilde{\eta}}))$.

In order to prove Theorem 3.2.7 we want to formulate a local kernel cut–off result for scalar symbols.

Let $\Sigma \subseteq \mathbb{R}^n$, $\Omega \subseteq \mathbb{R}^q$ be open sets, and denote by

$$S^\mu(\overline{\mathbb{R}}_+ \times \Sigma \times \Omega \times \mathbb{C} \times \mathbb{R}^{n+q})_{\mathrm{hol}}$$

the subspace of all $h(t, x, y, z, \xi, \eta) \in \mathcal{A}(\mathbb{C}, C^\infty(\overline{\mathbb{R}}_+ \times \Sigma \times \Omega \times \mathbb{R}^{n+q}))$
with

$$h(t, x, y, z, \xi, \eta)|_{z=\beta+i\tau} \in S^\mu(\overline{\mathbb{R}}_+ \times \Sigma \times \Omega \times \Gamma_\beta \times \mathbb{R}^{n+q})$$

for every $\beta \in \mathbb{R}$, uniformly in $c \leq \beta \leq c'$ for every $c < c'$. In an analogous manner we introduce

$$S^\mu_{cl}(\overline{\mathbb{R}}_+ \times \Sigma \times \Omega \times \mathbb{C} \times \mathbb{R}^{n+q})_{hol}$$

Proposition 3.2.8 *To each $f(t, x, y, z, \xi, \eta) \in S^\mu(\overline{\mathbb{R}}_+ \times \Sigma \times \Omega \times \Gamma_\beta \times \mathbb{R}^{n+q}_{\xi,\eta})$ there exists an*

$$h(t, x, y, z, \xi, \eta) \in S^\mu(\overline{\mathbb{R}}_+ \times \Sigma \times \Omega \times \mathbb{C} \times \mathbb{R}^{n+q})_{hol}$$

with

$$h(t, x, y, z, \xi, \eta)|_{z=\beta+i\tau} - f(t, x, y, z, \xi, \eta) \in S^{-\infty}(\overline{\mathbb{R}}_+ \times \Sigma \times \Omega \times \Gamma_\beta \times \mathbb{R}^{n+q}).$$

An analogous result holds for classical symbols.

Proof. The assertion is an obvious modification of Theorem 1.1.47, where the new aspect here is that the symbols in question are C^∞ in $(t, x, y) \in \overline{\mathbb{R}} \times \Sigma \times \Omega$. The kernel cut–off is to be applied only with respect to one covariable. The role of $\operatorname{Im} z$ as covariable instead of $\operatorname{Re} z$ in Theorem 1.1.47 only needs a rotation in the complex plane. $\qquad\square$

Proof of Theorem 3.2.7. If $h(t, y, z, \eta)$ is a symbol with the asserted properties, then we have $\operatorname{op}^\delta_M(h)(y, \eta) = \operatorname{op}^\varrho_M(h)(y, \eta)$ on functions with compact support in $t \in \mathbb{R}_+$, for arbitrary $\delta, \varrho \in \mathbb{R}$, cf. Proposition 2.3.69. Thus it suffices to show the assertion for a convenient weight, say $\delta = \frac{1}{2}$. Let us assume that for every open $\Sigma \subseteq \mathbb{R}^n$, $\Omega \subseteq \mathbb{R}^q$ and a given symbol $p(t, x, y, \tau, \xi, \eta) \in \tilde{S}^\mu(\overline{\mathbb{R}}_+ \times \Sigma \times \Omega \times \mathbb{R}^{1+n+q})$ we have constructed an $\tilde{h}(t, x, y, z, \xi, \tilde{\eta}) \in S^\mu(\overline{\mathbb{R}}_+ \times \Sigma \times \Omega \times \mathbb{C} \times \mathbb{R}^{n+q}_{\xi,\tilde{\eta}})_{hol}$ such that $h(t, x, y, z, \xi, \eta) = \tilde{h}(t, x, y, z, \xi, t\eta)$ satisfies the relation

$$\operatorname{op}^{\frac{1}{2}}_{M,t} \operatorname{op}_{\psi,x}(h)(y, \eta) = \operatorname{op}_{\psi,(t,x)}(p)(y, \eta) \qquad (3.2.9)$$

modulo $C^\infty(\Omega, L^{-\infty}(\mathbb{R}_+ \times \Sigma; \mathbb{R}^q))$. Applying this result to a given system of symbols $p_j \in \tilde{S}^\mu(\overline{\mathbb{R}}_+ \times \Sigma_j \times \Omega \times \mathbb{R}^{1+n+q})$, $j = 1, \ldots, N$, for the operator family $P(y, \eta)$ of (3.2.4) we can construct an operator-valued symbol

$$\tilde{h}(t, y, z, \tilde{\eta}) = \sum_{j=0}^N \varphi_j \tilde{h}_j(t, y, z, \tilde{\eta}) \psi_j,$$

where $\tilde{h}_j(t, y, z, \tilde{\eta}) = (\kappa_j)_*^{-1} \operatorname{op}_{\psi,x}(\tilde{h}_j)(t, y, z, \tilde{\eta})$, such that (3.2.8) is fulfilled. In other words it suffices to consider the local case and to start with $p(t, x, y, \tau, \xi, \eta) \in \tilde{S}^\mu(\overline{\mathbb{R}} \times \Sigma \times \Omega \times \mathbb{R}^{1+n+q})$.

For $\tilde{p}(t, x, y, \tilde{\tau}, \xi, \tilde{\eta}) \in S^\mu(\overline{\mathbb{R}}_+ \times \Sigma \times \Omega \times \mathbb{R}^{1+n+q})$ and $b_0(t, x, y, \tau, \xi, \tilde{\eta})$
$:= \tilde{p}(t, x, y, t\tau, \xi, \tilde{\eta})$ we shall construct an $h_0(t, x, y, z, \xi, \tilde{\eta}) \in S^\mu(\overline{\mathbb{R}}_+ \times \Sigma \times \Omega \times \Gamma_0 \times \mathbb{R}^{n+q}_{\xi,\tilde{\eta}})$ with

$$\operatorname{op}_{\psi,(t,x)}(b_0)(y, \tilde{\eta}) \sim \operatorname{op}^{\frac{1}{2}}_{M,t} \operatorname{op}_{\psi,x}(h_0)(y, \tilde{\eta}) + \operatorname{op}_{\psi,(t,x)}(b_1)(y, \tilde{\eta}) \quad (3.2.10)$$

with $b_1(t, x, y, \tau, \xi, \tilde{\eta}) = \tilde{p}_1(t, x, y, t\tau, \xi, \tilde{\eta})$ for a certain symbol

$$\tilde{p}_1(t, x, y, \tilde{\tau}, \xi, \tilde{\eta}) \in S^{\mu-1}(\overline{\mathbb{R}}_+ \times \Sigma \times \Omega \times \mathbb{R}^{1+n+q}).$$

Here and in the sequel $\sim$ indicates equality modulo $C^\infty(\Omega, L^{-\infty}(\overline{\mathbb{R}}_+ \times \Sigma; \mathbb{R}^q_{\tilde{\eta}}))$. Using (3.2.10) we then find inductively sequences

$$h_k(t, x, y, z, \xi, \tilde{\eta}) \in S^{\mu-k}(\overline{\mathbb{R}}_+ \times \Sigma \times \Omega \times \Gamma_0 \times \mathbb{R}^{n+q}_{\xi,\tilde{\eta}})$$

and

$$\tilde{p}_k(t, x, y, \tilde{\tau}, \xi, \tilde{\eta}) \in S^{\mu-k}(\overline{\mathbb{R}}_+ \times \Sigma \times \Omega \times \mathbb{R}^{1+n+q}_{\tilde{\tau},\xi,\tilde{\eta}})$$

with

$$\operatorname{op}_{\psi,(t,x)}(b_k)(y, \tilde{\eta}) \sim \operatorname{op}^{\frac{1}{2}}_{M,t} \operatorname{op}_{\psi,x}(h_k)(y, \tilde{\eta}) + \operatorname{op}_{\psi,(t,x)}(b_{k+1})(y, \tilde{\eta})$$

for $b_k(t, x, y, \tau, \xi, \tilde{\eta}) = \tilde{p}_k(t, x, y, t\tau, \xi, \tilde{\eta})$, $k \in \mathbb{N}$. This gives us

$$\operatorname{op}_{\psi,(t,x)}(b_0)(y, \tilde{\eta}) \sim \sum_{j=0}^{k} \operatorname{op}^{\frac{1}{2}}_{M,t} \operatorname{op}_{\psi,x}(h_j)(y, \tilde{\eta}) + \operatorname{op}_{\psi,(t,x)}(b_{k+1})(y, \tilde{\eta}).$$

Choose an

$$f(t, x, y, z, \xi, \tilde{\eta}) \in S^\mu(\overline{\mathbb{R}}_+ \times \Sigma \times \Omega \times \Gamma_0 \times \mathbb{R}^{n+q}_{\xi,\tilde{\eta}}),$$
$$f(t, x, y, z, \xi, \tilde{\eta}) \sim \sum_{j=0}^{\infty} h_j(t, x, y, z, \xi, \tilde{\eta}).$$

Then it follows that

$$\operatorname{op}_{\psi,(t,x)}(b_0)(y, \tilde{\eta}) \sim \operatorname{op}^{\frac{1}{2}}_{M,t} \operatorname{op}_{\psi,x}(f)(y, \tilde{\eta}).$$

Applying Proposition 3.2.8 we obtain an $\tilde{h}(t, x, y, z, \xi, \tilde{\eta}) \in S^\mu(\overline{\mathbb{R}}_+ \times \Sigma \times \Omega \times \mathbb{C} \times \mathbb{R}^{n+q})_{\text{hol}}$ with

$$\operatorname{op}^{\frac{1}{2}}_{M,t} \operatorname{op}_{\psi,x}(f)(y, \tilde{\eta}) \sim \operatorname{op}^{\frac{1}{2}}_{M,t} \operatorname{op}_{\psi,x}(\tilde{h})(y, \tilde{\eta}).$$

This is exactly what we need, since now it suffices to insert $\tilde{\eta} = t\eta$ to obtain $h(t, x, y, z, \xi, \eta) = \tilde{h}(t, x, y, z, \xi, t\eta)$ with the desired property (3.2.9). It remains the first step, namely to show (3.2.10).

Let us set $h_0(t, x, y, i\tau, \xi, \tilde{\eta}) = b_0(t, x, y, -\tau, \xi, \tilde{\eta})$. Fix for a moment the variables $x, y, \xi, \tilde{\eta}$ and write for abbreviation $h_0(t, i\tau)$ and $b_0(t, -\tau)$, respectively. Then

$$\operatorname{op}_M^{\frac{1}{2}}(h_0)u(t) = \int\limits_{-\infty}^{\infty} \int\limits_{0}^{\infty} \left(\frac{t}{t'}\right)^{-i\tau} h_0(t, i\tau)u(t')\, \frac{dt'}{t'}\, d\tau.$$

Substituting the diffeomorphism $\chi : \mathbb{R} \to \mathbb{R}_+$, $\chi(r) = t = e^{-r}$, yields

$$\operatorname{op}_M^{\frac{1}{2}}(h_0)u(t) = (\chi^*)^{-1} \iint e^{i(r-r')\tau} h_0(e^{-r}, i\tau)v(r')\, dr'\, d\tau$$

for $v = \chi^* u$. Setting $a(r, \varrho) = h_0(e^{-r}, i\varrho)$ it follows that $\operatorname{op}_M^{\frac{1}{2}}(h_0) = \chi_* \operatorname{op}_{\psi, r}(a)$. Hence, by known properties of pseudo–differential operators under push–forwards, there is a $b(t, \tau) \in S^\mu(\mathbb{R}_+ \times \mathbb{R})$ such that $\operatorname{op}_{\psi, t}(b) = \chi_* \operatorname{op}(a) \bmod L^{-\infty}(\mathbb{R}_+)$, where

$$b(t, \tau)|_{t=\chi(r)} \sim \sum_{k=0}^{\infty} \frac{1}{k!}(\partial_\varrho^k a)(r, d\chi(r)\tau)\Phi_k(r, \tau)$$

for $\Phi_k(r, \tau) = D_{\tilde{r}}^k e^{i\delta(r,\tilde{r})\tau}|_{\tilde{r}=r}$, $\delta(r, \tilde{r}) = \chi(\tilde{r}) - \chi(r) - d\chi(r)(\tilde{r} - r)$. We have $d\chi(r) = -e^{-r}$ and $\Phi_0 = 1$. Therefore, the highest order term for $b(t, \tau)$ equals $a(-\log t, -t\tau) = h_0(t, -it\tau) = b_0(t, \tau)$. In other words, it follows that $\operatorname{op}_{\psi, t}(b_0) = \operatorname{op}_M^{\frac{1}{2}}(h_0) + \operatorname{op}_{\psi, t}(b_1)$, where $-b_1(t, \tau)|_{t=\chi(r)} \sim \sum_{k=1}^{\infty} \frac{1}{k!}(\partial_\varrho^k a)(r, e^{-r}\tau)\Phi_k(r, \tau)$. It is clear that $(\partial_\varrho^k a)(-\log t, -t\tau)$ belongs to $\tilde{S}^{\mu-k}(\mathbb{R}_+ \times \mathbb{R})$.

It suffices now to employ that $\Phi_k(-\log t, \tau)$ (which is a polynomial in τ of degree $\leq k/2$) is an element of $\tilde{S}^{k/2}(\mathbb{R}_+ \times \mathbb{R})$, cf. Lemma 3.2.9 below. Then the asymptotic sum for b_1 can be carried out in $\tilde{S}^{\mu-1}$. It is obvious here that the whole procedure also works including the variables $(x, y, \xi, \tilde{\eta})$, where the orders in the items of the asymptotic sum tend to $-\infty$ with respect to the covariables $(\tau, \xi, \tilde{\eta})$. This completes the proof. $\square$

Lemma 3.2.9 We have $\Phi_k(r, \tau)|_{r=-\log t} \in \tilde{S}_{\mathrm{cl}}^{k/2}(\overline{\mathbb{R}}_+ \times \mathbb{R})$.

Proof. We have $\partial_{\tilde{r}}\chi(\tilde{r}) = -\chi(\tilde{r})$ and $d\chi(r) = -\chi(r)$. The function $D_{\tilde{r}}^k e^{i\delta(r,\tilde{r})\tau}$ has the form $\Psi_k(r, \tilde{r}, \tau)e^{i\delta(r,\tilde{r})\tau}$, where $\Psi_k(r, \tilde{r}, \tau)$ is a polynomial in the variables $D_{\tilde{r}}^l(\delta(r, \tilde{r})\tau)$, $l = 0, \ldots, k$. It is now obvious that the only occurring dependence on r and $\tilde{r}$ comes from the factors $\chi(\tilde{r})$

and $\chi(r)$, respectively, where τ^l is multiplied by a product of at least l such factors. By inserting $r = -\log t$, $\tilde{r} = -\log \tilde{t}$ and setting then $t = \tilde{t}$ we get the assertion. $\square$

Remark 3.2.10 *To every $P(y,\eta) \in C^\infty(\Omega, \tilde{L}^\mu(X^\wedge; \mathbb{R}^q))$ we can form an operator family $c_\wedge P(y,\eta) \in C^\infty(\Omega, \tilde{L}^\mu(X^\wedge; \mathbb{R}^q))$, defined by replacing the symbols $p_j(t,x,y,\tau,\xi,\eta) = \tilde{p}_j(t,x,y,\tilde{\tau},\xi,\tilde{\eta})|_{\tilde{\tau}=t\tau, \tilde{\eta}=t\eta}$ of (3.2.3) by $c_\wedge p_j(t,x,y,\tau,\xi,\eta) := \tilde{p}_j(0,x,y,t\eta,\xi,t\eta)$. Then, if we set*

$$(c_\wedge h)(t,y,z,\eta) = \tilde{h}(0,y,z,t\eta)$$

with h and $\tilde{h}$ from Theorem 3.2.7, we have

$$\mathrm{op}_M^\delta(c_\wedge h)(y,\eta) = c_\wedge P(y,\eta) \bmod C^\infty(\Omega, L^{-\infty}(X^\wedge; \mathbb{R}^q)).$$

More generally, if the symbols $\tilde{p}_j(x,y,\tilde{\tau},\xi,\tilde{\eta})$ are independent of t for all j, then $\tilde{h}(y,z,\tilde{\eta})$ can be chosen in t-independent form, cf. analogously Remark 2.2.27.

Proposition 3.2.11 *Let $\varphi(t) \in C^\infty(\overline{\mathbb{R}}_+)$ and $\mathcal{M}_\varphi$ be the operator of multiplication by φ. The $\mathcal{M}_\varphi$ may be regarded as an element of $S^0(\mathbb{R}^q; \mathcal{K}^{s,\gamma}(X^\wedge), \mathcal{K}^{s,\gamma}(X^\wedge))$ for every $s,\gamma \in \mathbb{R}$, and $\varphi \to \mathcal{M}_\varphi$ induces a continuous operator*

$$C_0^\infty(\overline{\mathbb{R}}_+) \to S^0(\mathbb{R}^q; \mathcal{K}^{s,\gamma}(X^\wedge), \mathcal{K}^{s,\gamma}(X^\wedge))$$

for each $s,\gamma \in \mathbb{R}$. Moreover, we have

$$\mathcal{M}_\varphi \in S^0(\mathbb{R}^q; \mathcal{K}_P^{s,\gamma}(X^\wedge), \mathcal{K}_P^{s,\gamma}(X^\wedge))$$

for every $P \in \mathrm{As}(X,\boldsymbol{g})(\in \mathrm{As}(X,\boldsymbol{g}^\bullet)$ or $\in \mathrm{As}(X,\boldsymbol{g})^\bullet)$, $\boldsymbol{g} = (\gamma,\Theta)$, which satisfies the shadow condition, and $\varphi \to \mathcal{M}_\varphi$ induces continuous operators

$$C_0^\infty(\overline{\mathbb{R}}_+) \to S^0(\mathbb{R}^q; \mathcal{K}_P^{s,\gamma}(X^\wedge), \mathcal{K}_P^{s,\gamma}(X^\wedge))$$

for every such P and every $s,\gamma \in \mathbb{R}$.

The simple proof is left to the reader.

Theorem 3.2.12 *Let*

$$\tilde{h}(t,y,z,\tilde{\eta}) \in C^\infty(\overline{\mathbb{R}}_+ \times \Omega, N_O^\mu(X; \mathbb{R}_{\tilde{\eta}}^q))$$

or

$$\tilde{h}(t,y,z,\tilde{\eta}) \in C^\infty(\overline{\mathbb{R}}_+ \times \Omega, M_O^\mu(X; \mathbb{R}^q)),$$

$\mu \in \mathbb{R}$, and set $h(t, y, z, \eta) = \tilde{h}(t, y, z, t\eta)$. Then, if $\omega(t)$, $\tilde{\omega}(t)$ are arbitrary cut–off functions, we have for every $\nu \in \mathbb{R}$

$$a_0(y, \eta) := \omega(t[\eta]) t^{-\nu} \operatorname{op}_M^{\gamma - \frac{n}{2}}(h)(y, \eta) \tilde{\omega}(t[\eta])$$
$$\in S^\nu(\Omega \times \mathbb{R}^q; \mathcal{K}^{s,\gamma}(X^\wedge), \mathcal{K}^{s-\mu,\gamma-\nu}(X^\wedge))$$

for all $s, \gamma \in \mathbb{R}$. If $\tilde{h}$ is independent of t then

$$a_0(y, \eta) \in S_{cl}^\nu(\Omega \times \mathbb{R}^q; \mathcal{K}^{s,\gamma}(X^\wedge), \mathcal{K}^{s-\mu,\gamma-\nu}(X^\wedge))$$

for all $s, \gamma \in \mathbb{R}$.

Proof. If $\tilde{h}(t, y, z, \tilde{\eta})$ is independent of t we have

$$a_0(y, \lambda\eta) = \lambda^\nu \kappa_\lambda a_0(y, \eta) \kappa_\lambda^{-1} \qquad \text{for all } \alpha \geq 1.$$

Moreover, $a_0(y, \eta) \in C^\infty(\Omega \times \mathbb{R}^q, \mathcal{L}(\mathcal{K}^{s,\gamma}(X^\wedge), \mathcal{K}^{s-\mu,\gamma-\nu}(X^\wedge)))$. This gives the second statement of the theorem. It is also easy to see that $a_0(y, \eta)$ depends continuously on $\tilde{h}(y, z, \tilde{\eta})$. This allows to apply a tensor product argument for the general case, writing $\tilde{h}$ as a convergent sum $\sum_{j=0}^\infty \lambda_j \varphi_j(t) \tilde{h}_j(y, z, \tilde{\eta})$ with $\sum |\lambda_j| < \infty$, $\varphi_j \in C^\infty(\overline{\mathbb{R}}_+)$ and $h_j \in C^\infty(\Omega, N_O^\mu(X; \mathbb{R}_\eta^q))$ tending to zero in the corresponding spaces for $j \to \infty$. The operator of multiplication by $\varphi_j \omega_0$ for any cut–off function $\omega(t)$ can be regarded as an operator–valued symbol of order zero, cf. Proposition 3.2.11, that tends to zero for $\varphi_j \to 0$. Thus, choosing ω_0 in such a way that $\omega_0(t) \omega(t[\eta]) = \omega(t[\eta])$, $\omega_0(t) \tilde{\omega}(t[\eta]) = \tilde{\omega}(t[\eta])$ for all t, η, we obtain that $a_0(y, \eta) = \sum_{j=0}^\infty \lambda_j \mathcal{M}_{\varphi_j \omega_0} b_j(y, \eta)$ for $b_j(y, \eta) = \omega(t[\eta]) t^{-\nu} \operatorname{op}_M^{\gamma - \frac{n}{2}}(h_j)(y, \eta) \tilde{\omega}(t[\eta])$ converges in $S^\nu(\Omega \times \mathbb{R}^q; \mathcal{K}^{s,\gamma}(X^\wedge), \mathcal{K}^{s-\mu,\gamma-\nu}(X^\wedge))$ for every $s, \gamma \in \mathbb{R}$. □

Theorem 3.2.13 *Under the conditions of Theorem 3.2.12 for every asymptotic type $P \in \operatorname{As}(X, (\gamma, \Theta))$ there is a $Q \in \operatorname{As}(X; (\gamma - \nu, \Theta))$ for $\Theta = (\vartheta, 0]$, $-\infty \leq \vartheta < 0$, such that*

$$a_0(y, \eta) \in S^\nu(\Omega \times \mathbb{R}^q; \mathcal{K}_P^{s,\gamma}(X^\wedge), \mathcal{K}_Q^{s-\mu,\gamma-\nu}(X^\wedge)),$$

and for t–independent $\tilde{h}$

$$a_0(y, \eta) \in S_{cl}^\nu(\Omega \times \mathbb{R}^q; \mathcal{K}_P^{s,\gamma}(X^\wedge), \mathcal{K}_Q^{s-\mu,\gamma-\nu}(X^\wedge))$$

for all $s \in \mathbb{R}$.

Proof. Assuming first that $\tilde{h}(t, y, z, \tilde{\eta})$ is independent of t, we have the homogeneity as in the proof of the preceding Theorem 3.2.12, for large $|\eta|$, and in addition $a_0(y, \eta) \in C^\infty(\Omega \times \mathbb{R}^q, \mathcal{L}(\mathcal{K}_P^{s,\gamma}(X^\wedge), \mathcal{K}_Q^{s-\mu,\gamma-\nu}(X^\wedge))$ for every P with a resulting Q. This yields the second statement of the theorem. Moreover, $a_0(y, \eta)$ depends continuously on $\tilde{h}(y, z, \tilde{\eta})$. A tensor product argument similarly to the proof of Theorem 3.2.12 then yields the assertion in general. Here we use, in particular, the above Proposition 3.2.11. □

3.2.3   Interior symbols with exit behaviour on the infinite cone

Let X be a closed compact C^∞ manifold, $n = \dim X$. The results of Section 1.4.5 on pseudo–differential operators on a manifold with exits to infinity will be modified now for $X^\wedge = \mathbb{R}_+ \times X \ni (t, x)$ for $t \to \infty$ in a suitable parameter–dependent form, with the parameter space $\mathbb{R}^q \ni \eta$. For the weight $\varrho \in \mathbb{R}$ at infinity we assume here $\varrho = 0$, indicated by subscript e.

$S^{\mu,\delta}(\mathbb{R}^{n+1} \times \mathbb{R}^{n+1+q})$ for $\mu, \delta \in \mathbb{R}$ denotes the subspace of all $a(\tilde{x}, \tilde{\xi}, \eta) \in C^\infty(\mathbb{R}^{n+1}_{\tilde{x}} \times \mathbb{R}^{n+1+q}_{\tilde{\xi},\eta})$ which satisfy the following estimates:

$$|D^\alpha_{\tilde{x}} D^\beta_{\tilde{\xi},\eta} a(\tilde{x}, \tilde{\xi}, \eta)| \le c\langle \tilde{\xi}, \eta \rangle^{\mu-|\beta|} \langle \tilde{x} \rangle^{\delta-|\alpha|}$$

for all $\alpha \in \mathbb{N}^{n+1}$, $\beta \in \mathbb{N}^{n+1+q}$, $(\tilde{x}, \tilde{\xi}, \eta) \in \mathbb{R}^{n+1} \times \mathbb{R}^{n+1+q}$, for constants $c = c(\alpha, \beta) > 0$. Moreover, we set $S^\mu(\mathbb{R}^{n+1} \times \mathbb{R}^{n+1+q})_e := S^{\mu,0}(\mathbb{R}^{n+1} \times \mathbb{R}^{n+1+q})$ and

$$S^\mu_{\mathrm{cl}}(\mathbb{R}^{n+1} \times \mathbb{R}^{n+1+q})_e = S^\mu_{\mathrm{cl}}(\mathbb{R}^{n+1} \times \mathbb{R}^{n+1+q}) \cap S^\mu(\mathbb{R}^{n+1} \times \mathbb{R}^{n+1+q})_e,$$

where the first space on the right is the space of all classical symbols in $(\tilde{\xi}, \eta)$ in the usual sense.

If $U \subseteq \mathbb{R}^{n+1}$ is an open set which is conical, i.e., $\tilde{x} \in U$ implies $\lambda \tilde{x} \in U$ for all $\lambda \in \mathbb{R}_+$, we denote by $S^\mu(U \times \mathbb{R}^{n+1+q})_e$ ($S^\mu_{\mathrm{cl}}(U \times \mathbb{R}^{n+1+q})_e$) the space of all $a(\tilde{x}, \tilde{\xi}, \eta) \in C^\infty(U_{\tilde{x}} \times \mathbb{R}^{n+1+q}_{\tilde{\xi},\eta})$ such that for every open set $U_0 \subset U$ which is conical for large $|\tilde{x}|$, with $\overline{U}_0 \subset U$, there is an $a_0(\tilde{x}, \tilde{\xi}, \eta) \in S^\mu(\mathbb{R}^{n+1} \times \mathbb{R}^{n+1+q})_e$ ($\in S^\mu_{\mathrm{cl}}(\mathbb{R}^{n+1} \times \mathbb{R}^{n+1+q})_e$) with $a(\tilde{x}, \tilde{\xi}, \eta)|_{U_0 \times \mathbb{R}^{n+1+q}} = a_0(\tilde{x}, \tilde{\xi}, \eta)|_{U_0 \times \mathbb{R}^{n+1+q}}$.

Define $L^{-\infty}(U; \mathbb{R}^q)_e$ as the space of all η–dependent operator families on U with kernels in $\mathcal{S}(\mathbb{R}^q_\eta, \mathcal{S}(U \times U))$, cf. the notation in Section 1.4.5. Set

$$L^\mu(U; \mathbb{R}^q)_e = \{\mathrm{Op}(a)(\eta) + C(\eta) : \ a(\tilde{x}, \tilde{\xi}, \eta) \in S^\mu(U \times \mathbb{R}^{n+1+q})_e,$$
$$C(\eta) \in L^{-\infty}(U; \mathbb{R}^q)_e\}$$

and define analogously $L^\mu_{\mathrm{cl}}(U; \mathbb{R}^q)_e$ in terms of $S^\mu_{\mathrm{cl}}(U \times \mathbb{R}^{n+1+q})_e$.

We also can introduce the symbol classes $S^{\mu,\delta,\delta'}(\mathbb{R}^{n+1} \times \mathbb{R}^{n+1} \times \mathbb{R}^{n+1+q}) \ni a(\tilde{x}, \tilde{x}', \xi, \eta)$ for $\mu, \delta, \delta' \in \mathbb{R}$ satisfying the symbol estimates

$$|D^\alpha_{\tilde{x}} D^{\alpha'}_{\tilde{x}'} D^\beta_{\tilde{\xi},\eta} a(\tilde{x}, \tilde{x}', \tilde{\xi}, \eta)| \le c\langle \tilde{\xi}, \eta \rangle^{\mu-|\beta|} \langle \tilde{x} \rangle^{\delta-|\alpha|} \langle \tilde{x}' \rangle^{\delta'-|\alpha'|}$$

for all $\alpha, \alpha' \in \mathbb{N}^{n+1}$, $\beta \in \mathbb{N}^{n+1+q}$, $(\tilde{x}, \tilde{x}', \tilde{\xi}, \eta) \in \mathbb{R}^{2n+n+1+q}$, for constants $c = c(\alpha, \alpha', \beta) > 0$. Similarly, for open conical $U \subseteq \mathbb{R}^{n+1}$ we can introduce the symbol classes $S^{\mu,\delta}(U \times \mathbb{R}^{n+1+q})$ and $S^{\mu,\delta,\delta'}(U \times U \times \mathbb{R}^{n+1+q})$. This enables us to define the parameter–dependent operator classes $L^{\mu,\delta}(U; \mathbb{R}^q) \ni \mathrm{Op}(a)(\eta) + C(\eta)$, $a(\tilde{x}, \tilde{\xi}, \eta) \in S^{\mu,\delta}(U \times \mathbb{R}^{n+1+q})$, $C(\eta) \in L^{-\infty}(U, \mathbb{R}^q)_e$. Clearly we can also take open sets U in $\mathbb{R}^{n+1}$ that are conical only for large $|\tilde{x}|$ and talk about those symbol and operator classes.

Now let us choose an open covering of X by coordinate neighbourhoods $\{U_1, \ldots, U_N\}$, fix a subordinate partition of unity $\{\varphi_1, \ldots, \varphi_N\}$ and a system $\{\psi_1, \ldots, \psi_N\}$ of functions $\psi_j \in C_0^\infty(C_j)$ with $\varphi_j \psi_j = \varphi_j$ for $j = 1, \ldots, N$. Moreover, let $\varepsilon_j : U_j \to V_{1,j}$ be a diffeomorphism to an open subset $V_{1,j} \subset S^n = \{\tilde{x} \in \mathbb{R}^{n+1} : |\tilde{x}| = 1\}$, and define $\varrho_j : \mathbb{R}_+ \times U_j \to V_j := \{\tilde{x} \in \mathbb{R}^{n+1} \setminus \{0\} : \tilde{x}/|\tilde{x}| \in V_{1,j}\}$ by $\varrho_j(t, x) = t\varepsilon_j(x)$, $j = 1, \ldots, N$. Applying the operator push–forward with respect to ϱ_j^{-1} for parameter–dependent pseudo–differential operators we obtain

$$L^\mu(U_j^\wedge; \mathbb{R}^q)_e := (\varrho_j^{-1})_* L^\mu(V_j; \mathbb{R}^q)_e,$$

$U_j^\wedge = \mathbb{R}_+ \times U_j$, and analogously $L_{cl}^\mu(U_j^\wedge; \mathbb{R}^q)_e$. Set

$$\begin{aligned}
\mathcal{S}(X^\wedge \times X^\wedge) = \{c(t, x, t', x') \in C^\infty(X^\wedge \times X^\wedge) : \\
(1 - \omega(t))(1 - \omega(t'))c(t, x, t', x') \\
\in \mathcal{S}(\mathbb{R} \times \mathbb{R}, C^\infty(X \times X))|_{\mathbb{R}_+ \times \mathbb{R}_+} \\
\text{for every cut–off function } \omega(t)\}
\end{aligned}$$

and denote by $L^{-\infty}(X^\wedge; \mathbb{R}^q)_e$ the space of all operator families on $X^\wedge$ with η–dependent kernels in $\mathcal{S}(\mathbb{R}_\eta^q; \mathcal{S}(X^\wedge \times X^\wedge))$.

Define the space of parameter–dependent pseudo–differential operators on $X^\wedge$ with exit behaviour for $t \to \infty$

$$L^\nu(X^\wedge; \mathbb{R}^q)_e = \Big\{ \sum_{j=0}^N \varphi_j A_j(\eta)\psi_j + C(\eta) : A_j(\eta) \in L^\mu(U_j^\wedge; \mathbb{R}^q),$$

$$j = 1, \ldots, N, \ C(\eta) \in L^{-\infty}(X^\wedge; \mathbb{R}^q)_e \Big\}, \qquad (3.2.11)$$

and analogously $L_{cl}^\mu(X^\wedge; \mathbb{R}^q)_e$, the subspace of classical operator families. Clearly these definitions are correct, i.e., independent of the particular choice of data such as the charts or the function systems $\{\varphi_1, \ldots, \varphi_N\}$, $\{\psi_1, \ldots, \psi_N\}$.

By allowing additionally C^∞ dependence on further variables $y \in \Omega$ for an open set $\Omega \subseteq \mathbb{R}^p$, i.e., both for the symbols $a_j(\tilde{x}, y, \tilde{\xi}, \eta)$ involved in A_j as well as for $C(y, \eta) \in C^\infty(\Omega, L^{-\infty}(X^\wedge; \mathbb{R}^q)_e)$, we obtain analogously the spaces of operator families

$$C^\infty(\Omega, L^\mu(X^\wedge; \mathbb{R}^q)_e) \quad \text{and} \quad C^\infty(\Omega, L_{cl}^\mu(X^\wedge; \mathbb{R}^q)_e), \qquad (3.2.12)$$

respectively. The spaces (3.2.12) are Fréchet in a canonical way. The same is true of the corresponding local variants with respect to $x \in X$. The following propositions are straightforward after the material of Section 1.4.5. So the evident proofs will be omitted.

Proposition 3.2.14 $P(y, \eta) \in C^\infty(\Omega, L^\mu(X^\wedge; \mathbb{R}^q)_e)$ *implies* $P^*(y, \eta)$ $\in C^\infty(\Omega, L^\mu(X^\wedge; \mathbb{R}^q)_e)$, *and similarly for the case with subscripts* cl. *Here* $P^*(y, \eta)$ *is the point–wise formal adjoint in the sense*

$$(P(y, \eta)u, v)_{\mathcal{K}^{0,0}(X^\wedge)} = (u, P^*(y, \eta)v)_{\mathcal{K}^{0,0}(X^\wedge)}$$

for all $u, v \in C_0^\infty(X^\wedge)$, $(y, \eta) \in \Omega \times \mathbb{R}^q$.

Proposition 3.2.15 *Let* $P(y, \eta) \in C^\infty(\Omega, L^\mu(X^\wedge; \mathbb{R}^q)_e)$ *and* $Q(y, \eta)$ $\in C^\infty(\Omega, L^\nu(X^\wedge, \mathbb{R}^q)_e)$. *Then*

$$P(y, \eta)(1 - \omega)Q(y, \eta) \in C^\infty(\Omega, L^{\mu+\nu}(X^\wedge; \mathbb{R}^q)_e)$$

for every cut–off function $\omega(t)$. *An analogous result holds for the operator families with subscript* cl.

Remark 3.2.16 *We have*

$$D_y^\alpha D_\eta^\beta C^\infty(\Omega, L^\mu(X^\wedge; \mathbb{R}^q)_e) \subset C^\infty(\Omega, L^{\mu-|\beta|}(X^\wedge; \mathbb{R}^q)_e)$$

for every $\alpha \in \mathbb{N}^p$, $\beta \in \mathbb{N}^q$, *and the same with subscripts* cl.

Let us fix cut–off functions $\omega(t)$, $\omega_1(t)$. Then every $P(y, \eta) \in C^\infty(\Omega, L^\mu(X^\wedge; \mathbb{R}^q)_e)$ gives rise to continuous operators

$$(1 - \omega(t))P(y, \eta)(1 - \omega_1(t)) : H_{\text{cone}}^s(X^\wedge) \to H_{\text{cone}}^{s-\mu}(X^\wedge)$$

or

$$(1 - \omega(t))P(y, \eta)(1 - \omega_1(t)) : \mathcal{K}^{s,\gamma}(X^\wedge) \to \mathcal{K}^{s-\mu,\delta}(X^\wedge)$$

for arbitrary $s, \gamma, \delta \in \mathbb{R}$, C^∞ dependent on $(y, \eta) \in \Omega \times \mathbb{R}^q$. For $(\kappa_\lambda u)(t, x) = \lambda^{\frac{n+1}{2}} u(\lambda t, x)$, $\kappa(\eta) = \kappa_{[\eta]}$, $\eta \in \mathbb{R}^q$, we have

$$\kappa^{-1}(\eta)\{(1 - \omega(t))P(y, \eta)(1 - \omega_1(t))\}\kappa(\eta)$$
$$= (1 - \omega(t[\eta]^{-1}))P_\eta(y, \eta)(1 - \omega_1(t[\eta]^{-1})),$$

where $P_\eta(y, \eta) = \kappa^{-1}(\eta)P(y, \eta)\kappa(\eta)$. It can easily be verified that

$$\|\kappa^{-1}(\eta)\{D_y^\alpha D_\eta^\beta(1 - \omega(t))P(y, \eta)(1 - \omega_1(t))\}$$
$$\times \kappa(\eta)\|_{\mathcal{L}(\mathcal{K}^{s,\gamma}(X^\wedge), \mathcal{K}^{s-\mu+|\beta|,\delta}(X^\wedge))} \leq c[\eta]^{\mu-|\beta|}$$

for all $y \in K$, $K \subset\subset \Omega$, all $\eta \in \mathbb{R}^q$, $\alpha \in \mathbb{N}^p$, $\beta \in \mathbb{N}^q$, for constants $c = c(\alpha, \beta, K) > 0$. This gives us the following result:

Theorem 3.2.17 *Let $P(y,\eta) \in C^\infty(\Omega, L^\mu(X^\wedge; \mathbb{R}^q)_e)$, and fix arbitrary cut–off functions $\omega(t)$, $\omega_1(t)$. Then*

$$(1 - \omega(t))P(y,\eta)(1 - \omega_1(t)) \in S^\mu(\Omega \times \mathbb{R}^q; \mathcal{K}^{s,\gamma}(X^\wedge), \mathcal{K}^{s-\mu,\delta}(X^\wedge))$$

for every $s, \gamma, \delta \in \mathbb{R}$.

For purposes below we shall calculate special push–forwards of parameter–dependent pseudo–differential operators. Let $V \subseteq \mathbb{R}^{n+1}_{\tilde{x}} \setminus \{0\}$ be a conical set, $V_1 := \{\tilde{x} \in V : |\tilde{x}| = 1\}$. Suppose $\gamma : V_1 \to \Sigma$ is a diffeomorphism for an open set $\Sigma \subseteq \mathbb{R}^n_x$, and let

$$\kappa : V \to \mathbb{R}_+ \times \Sigma \qquad (3.2.13)$$

be the diffeomorphism given by $\kappa(\tilde{x}) = (t, \gamma(\tilde{x}/|\tilde{x}|))$ for $t = |\tilde{x}|$. We then have induced isomorphisms

$$\kappa_* : L^\mu(V; \mathbb{R}^q) \to L^\mu(\mathbb{R}_+ \times \Sigma; \mathbb{R}^q), \quad \kappa_* : L^\mu_{cl}(V; \mathbb{R}^q) \to L^\mu_{cl}(\mathbb{R}_+ \times \Sigma; \mathbb{R}^q).$$

Proposition 3.2.18 *Let $a_1(\tilde{x}, \tilde{\xi}, \eta) \in S^\mu_{cl}(\mathbb{R}^{n+1}_{\tilde{x}} \times \mathbb{R}^{n+1+q}_{\tilde{\xi}, \eta})$ and set $a(\tilde{x}, \tilde{\xi}, \eta) := a_1(\tilde{x}, \tilde{\xi}, \eta)|_{V \times \mathbb{R}^{n+1+q}}$ for an open conical set $V \subset \mathbb{R}^{n+1} \setminus \{0\}$. Then we have*

$$\kappa_* \operatorname{Op}_{\tilde{x}}(a)(\eta) = \operatorname{Op}_{(t,x)}(b)(\eta) \bmod L^{-\infty}(\mathbb{R}_+ \times \Sigma; \mathbb{R}^q)$$

for a symbol $b(t, x, \tau, \xi, \eta) \in S^\mu_{cl}(\mathbb{R}_+ \times \Sigma \times \mathbb{R}^{n+1+q})$ of the form

$$b(t, x, \tau, \xi, \tau) = t^{-\mu} \tilde{p}(t, x, \tilde{\tau}, \xi, \tilde{\eta})|_{\tilde{\tau}=t\tau, \tilde{\eta}=t\eta},$$

with $\tilde{p}(t, x, \tilde{\tau}, \xi, \tilde{\eta}) \in S^\mu_{cl}(\overline{\mathbb{R}}_+ \times \Sigma \times \mathbb{R}^{n+1+q})$.

Proof. We have to apply the standard formula for complete symbols under operator push–forwards, namely

$$b(t, x, \tau, \xi, \eta)_{(t,x)=\kappa(\tilde{x})} \sim \sum_\alpha \frac{1}{\alpha!} (\partial^\alpha_\xi a)(\tilde{x}, {}^t d\kappa(\tilde{x})\binom{\tau}{\xi}, \eta)\Phi_\alpha(\tilde{x}, \tau, \xi),$$

$$(3.2.14)$$

cf. Theorem 1.1.38. Here

$$\Phi_\alpha(\tilde{x}, \tau, \xi) = D^\alpha_{\tilde{z}} e^{i\delta(\tilde{x}, \tilde{z})(\tau, \xi)}|_{\tilde{z}=\tilde{x}}, \quad \alpha \in \mathbb{N}^{n+1},$$

where $\delta(\tilde{x}, \tilde{z}) = \kappa(\tilde{z}) - \kappa(\tilde{x}) - d\kappa(\tilde{x})(\tilde{z} - \tilde{x})$, and

$$\delta(\tilde{x}, \tilde{z})(\tau, \xi) =: (\delta_0(\tilde{x}, \tilde{z}), \delta_1(\tilde{x}, \tilde{z}), \dots, \delta_n(\tilde{x}, \tilde{z})),$$

$$\delta(\tilde{x}, \tilde{z})(\tau, \xi) = \delta_0(\tilde{x}, \tilde{z})\tau + \sum_{j=1}^n \delta_j(\tilde{x}, \tilde{z})\xi_j.$$

Setting $\kappa(\tilde{x}) = (\kappa_0(\tilde{x}), \ldots \kappa_n(\tilde{x}))$, where $\kappa_0(\tilde{x}) = t$, we obtain $\kappa_0(\lambda\tilde{x}) = \lambda\kappa_0(\tilde{x})$, $\kappa_j(\lambda\tilde{x}) = \kappa_j(\tilde{x})$ for $j = 1, \ldots, n$, $\lambda \in \mathbb{R}_+$. Then the entries of the Jacobi matrix

$$d\kappa(\tilde{x}) = \begin{pmatrix} j_{0,0}(\tilde{x}) & j_{0,1}(\tilde{x}) & \cdots & j_{0,n}(\tilde{x}) \\ j_{1,0}(\tilde{x}) & j_{1,1}(\tilde{x}) & \cdots & j_{1,n}(\tilde{x}) \\ \vdots & \vdots & \vdots & \vdots \\ j_{n,0}(\tilde{x}) & j_{n,1}(\tilde{x}) & \cdots & j_{n,n}(\tilde{x}) \end{pmatrix},$$

where $j_{0,k-1}(\tilde{x}) = \partial\kappa_0(\tilde{x})/\partial\tilde{x}_k$, $j_{l,k-1}(\tilde{x}) = \partial\kappa_l(\tilde{x})/\partial\tilde{x}_k$, $k = 1, \ldots, n+1$, $l = 1, \ldots, n$, satisfy the homogeneity relations

$$j_{0,k-1}(\lambda\tilde{x}) = j_{0,k-1}(\tilde{x}), \quad j_{l,k-1}(\lambda\tilde{x}) = \lambda^{-1}j_{l,k-1}(\tilde{x}) \qquad \text{for all } \lambda \in \mathbb{R}_+,$$

(3.2.15)

$k = 1, \ldots, n+1$, $l = 1, \ldots, n$. This gives us

$$^t d\kappa(\tilde{x})\begin{pmatrix} \tau \\ \xi \end{pmatrix} = \psi(x)\begin{pmatrix} \tau \\ t^{-1}\xi \end{pmatrix}$$

for a smooth $(n+1) \times (n+1)$–matrix function $\psi(x)$ on Σ.

Concerning $\Phi_\alpha(\tilde{x}, \tau, \xi)$ it is well–known that this is a polynomial in (τ, ξ) of degree $d(\alpha) \leq |\alpha|/2$. Let us set $\delta_l^{(\beta)}(\tilde{x}) = D_z^\beta\delta_l(\tilde{x}, \tilde{z})|_{\tilde{z}=\tilde{x}}$, $l = 0, \ldots, n$. From the homogeneity properties of $\kappa(\tilde{x})$ it follows that

$$\delta_0^{(\beta)}(\lambda\tilde{x}) = \lambda^{1-|\beta|}\delta_0(\tilde{x}), \quad \delta_l^{(\beta)}(\lambda\tilde{x}) = \lambda^{-|\beta|}\delta_l(\tilde{x}) \quad \text{for} \quad l = 1, \ldots, n$$

(3.2.16)

for all $\lambda \in \mathbb{R}_+$. Thus

$$\delta_0^{(\beta)}(\tilde{x})\tau = t^{1-|\beta|}\delta_0^{(\beta)}\left(\frac{\tilde{x}}{|\tilde{x}|}\right)\tau,$$

$$\delta_l^{(\beta)}(\tilde{x})\xi_l = t^{-|\beta|}\delta_l^{(\beta)}(\frac{\tilde{x}}{|\tilde{x}|})\xi_l$$

for $l = 1, \ldots, n$. We obtain $\Phi_\alpha(\tilde{x}, \tau, \xi) = t^{-d(\alpha)}P_\alpha(\frac{\tilde{x}}{|\tilde{x}|}, t\tau, \xi)$, where P_α is a polynomial in $(t\tau, \xi)$ of degree $d(\alpha)$ with coefficients in $C^\infty(V_1)$.

Thus, to evaluate (3.2.14) we have to form the asymptotic sum of

$$\frac{1}{\alpha!}(\partial_\xi^\alpha a)(\tilde{x}, \psi(x)\begin{pmatrix} \tau \\ t^{-1}\xi \end{pmatrix}, \eta)t^{-d(\alpha)}P_\alpha(x, t\tau, \xi).$$

By assumption, the symbol a is classical. For the homogeneous component $a_{(\mu-j)}(\tilde{x}, \tilde{\xi}, \eta)$ of $a(\tilde{x}, \tilde{\xi}, \eta)$ of order $\mu - j$ we can write

$$(\partial_\xi^\alpha a_{(\mu-j)})(\tilde{x}, \chi(x)\begin{pmatrix} \tau \\ t^{-1}\xi \end{pmatrix}, \eta) = t^{-\mu+j+|\alpha|}(\partial_\xi^\alpha a_{(\mu-j)})(\tilde{x}, \psi(x)\begin{pmatrix} t\tau \\ \xi \end{pmatrix}, t\eta).$$

We have $d(\alpha) \leq |\alpha|/2$. Thus, if we fix an excision function $\chi(\tilde{\tau}, \xi, \tilde{\eta})$ we can regard (3.2.14) (after rearranging summands) as an asymptotic sum

$$t^{-\mu} \sum_\alpha \chi(\tilde{\tau}, \xi, \tilde{\eta}) c_\alpha(t, x, \tilde{\tau}, \xi, \tilde{\eta})|_{\tilde{\tau}=t\tau, \tilde{\eta}=t\eta},$$

where $c_\alpha(t, x, \tilde{\tau}, \xi, \tilde{\eta})$ is homogeneous on $(\tilde{\tau}, \xi, \tilde{\eta})$ of order $\mu - |\alpha|$ and C^∞ in $(t, x) \in \overline{\mathbb{R}}_+ \times \Sigma$. This gives us the assertion. □

Proposition 3.2.18 is useful to interpret pseudo–differential operators in polar coordinates with respect to the variables $\tilde{x}$ and to characterise the nature of the resulting symbols near $t = 0$ as edge degenerate ones. We are also interested in the reverse direction, when we start with edge–degenerate symbols and apply κ^{-1} to $(t, x) \in \Sigma^\wedge$. However in this case we want to observe the behaviour near the exit $|\tilde{x}| \to \infty$, $\tilde{x} = \kappa^{-1}(t, x)$.

Proposition 3.2.19 *Let* $\tilde{p}(t, x, \tilde{\tau}, \xi, \tilde{\eta}) \in S^\mu_{cl}(\mathbb{R}_+ \times \Sigma \times \mathbb{R}^{1+n+q}_{\tilde{\tau}, \xi, \tilde{\eta}})$ *be a symbol that is independent of t for $t >$* const *for a constant > 0, and set* $b(t, x, \tau, \xi, \eta) = t^{-\mu}\tilde{p}(t, x, t\tau, \xi, t\eta)$. *Then*

$$(\kappa^{-1})_* \operatorname{Op}(b)(\eta) \in L^\mu_{cl}(V; \mathbb{R}^q)_e.$$

The proof will be given after some auxiliary results.

Lemma 3.2.20 *Let* $a(\tilde{x}, \tilde{x}', \tilde{\xi}, \eta) \in S^{\mu, \delta, \delta'}(V \times V \times \mathbb{R}^{n+1} \times \mathbb{R}^q)$. *Then* $\operatorname{Op}_{\tilde{x}}(a)(\eta) \in L^{\mu, \delta+\delta'}(V; \mathbb{R}^q)$, *and there is a symbol* $\boldsymbol{a}(\tilde{x}, \tilde{\xi}, \eta) \in S^{\mu, \delta+\delta'}(V \times \mathbb{R}^{n+1} \times \mathbb{R}^q)$,

$$\boldsymbol{a}(\tilde{x}, \tilde{\xi}, \eta) \sim \sum_\alpha \frac{1}{\alpha!}(D^\alpha_{\tilde{\xi}} \partial^\alpha_{\tilde{x}'} a)(\tilde{x}, \tilde{x}', \tilde{\xi}, \eta)|_{\tilde{x}'=\tilde{x}},$$

such that $\operatorname{Op}(a)(\eta) - \operatorname{Op}(\boldsymbol{a})(\eta) \in L^{-\infty, -\infty}(V; \mathbb{R}^q)$.

This is a slight modification of the above Theorem 1.4.9, so the proof will be omitted.

Consider a diffeomorphism $\chi : U \to V$ between open sets $U, V \subseteq \mathbb{R}^{n+1}$ that are conical in the large, where there is a constant $c > 0$ such that $\chi(\lambda x) = \lambda\chi(x)$ for all $\lambda \geq 1$, $|x| \geq c$. For $\tilde{x} \in V$ we choose an $\varepsilon > 0$ such that $\{\tilde{x}' : |\tilde{x} - \tilde{x}'| \geq \varepsilon(\tilde{x})\} \subseteq V$. For $\tilde{x}, \tilde{x}' \in V$ with $|\tilde{x} - \tilde{x}'| \leq \varepsilon(x)$ we set

$$\varrho(\tilde{x}, \tilde{x}') = \int_0^1 d\chi^{-1}\left(\tilde{x}' + t(\tilde{x} - \tilde{x}')\right) dt.$$

Then

$$\chi^{-1}(\tilde{x}) - \chi^{-1}(\tilde{x}') = \varrho(\tilde{x}, \tilde{x}')(\tilde{x} - \tilde{x}').$$

Lemma 3.2.21 *There exists a symbol $r(\tilde{x}) \in S^1(V)$ satisfying the following conditions:*

(i) *For each open set $V_0 \subset V$ that is conical for large $|\tilde{x}|$, with $\overline{V}_0 \subset V$, there is a constant $c_0 > 0$ such that $c_0(1 + |\tilde{x}|) \leq r(\tilde{x})$ for all $\tilde{x} \in \overline{V}_0$.*

(ii) *If $\tilde{x} \in V$ then $|\tilde{x} - \tilde{x}'| \leq r(\tilde{x})$ implies that $\varrho(\tilde{x}, \tilde{x}')$ is invertible.*

The simple proof is left to the reader.

Lemma 3.2.22 *There is an element $\omega(\tilde{x}, \tilde{x}') \in S^{0,0}(V \times V)$ such that $\omega(\tilde{x}, \tilde{x}') = 1$ for $|\tilde{x} - \tilde{x}'| \leq \frac{1}{2}r(\tilde{x})$, $\omega(\tilde{x}, \tilde{x}') = 0$ for $|\tilde{x} - \tilde{x}'| \geq \frac{2}{3}r(\tilde{x})$, where $r(\tilde{x})$ is the function from Lemma 3.2.21.*

Proof. Choose a function $\psi \in C^\infty(\mathbb{R})$ with $\psi(t) = 1$ for $t \leq \frac{1}{2}$, $\psi(t) = 0$ for $t > \frac{2}{3}$, and set $\omega(\tilde{x}, \tilde{x}') = \psi(|\tilde{x} - \tilde{x}'|/r(\tilde{x}))$. Then $\omega(\tilde{x}, \tilde{x}')$ has the asserted property. To verify $\omega \in S^{0,0}(V \times V)$ we have to use $r \in S^1(V)$. □

Proof of Proposition 3.2.19. We shall show that

$$(\kappa^{-1})_* \operatorname{Op}(b) \in L^{\mu,0}(V; \mathbb{R}^q). \tag{3.2.17}$$

Let us write $\kappa = (\kappa_0, \kappa')$, $\kappa_0 : V \to \mathbb{R}_+$, $\tilde{x} \to |\tilde{x}|$, $\kappa' : V \to \Sigma$. For $v \in C_0^\infty(V)$ we have

$$(\kappa^{-1})_* \operatorname{Op}(b)(\eta)v(\tilde{x}) = (2\pi)^{-(n+1)/2} \iint e^{i\varphi(\tilde{x}, \tilde{x}', \tau, \xi)} |\det d\kappa(\tilde{x}')| \kappa_0(\tilde{x})^{-\mu}$$
$$\tilde{p}(\kappa_0(\tilde{x}), \kappa'(\tilde{x}), \kappa_0(\tilde{x})\tau, \xi, \kappa_0(\tilde{x})\eta)v(\tilde{x}')\, d\tilde{x}'d\tau d\xi,$$

where $\varphi(\tilde{x}, \tilde{x}', \tau, \xi) = (\kappa_0(\tilde{x}) - \kappa_0(\tilde{x}'))\tau + (\kappa'(\tilde{x}) - \kappa'(\tilde{x}'))\xi$. Let $\omega(\tilde{x}, \tilde{x}')$ be the function from Lemma 3.2.22 and decompose the latter integral into a sum by inserting $1 = \omega(\tilde{x}, \tilde{x}') + (1 - \omega(\tilde{x}, \tilde{x}'))$ under the integral. Then for the contribution from $1 - \omega(\tilde{x}, \tilde{x}')$ we can use the fact that the operator push–forward induces a bijection

$$(\kappa^{-1})_* : L^{-\infty,-\infty}(\mathbb{R}_+ \times \Sigma; \mathbb{R}^q) \to L^{-\infty,-\infty}(V; \mathbb{R}^q)$$

which is easy to verify and left to the reader.

So it remains

$$A(\eta)v(\tilde{x}) = (2\pi)^{-(n+1)/2} \iint e^{i\varphi(\tilde{x}, \tilde{x}', \tau, \xi)} |\det d\kappa(\tilde{x}')| \kappa_0(\tilde{x})^{-\mu}$$
$$\tilde{p}(\kappa_0(\tilde{x}), \kappa'(\tilde{x}), \kappa_0(\tilde{x})\tau, \xi, \kappa_0(\tilde{x})\eta)v(\tilde{x}')\, d\tilde{x}'d\tau d\xi.$$

We have

$$A(\eta)v(\tilde{x}) = \iint e^{i(\tilde{x}-\tilde{x}')\tilde{\xi}} \, d(\tilde{x}, \tilde{x}', \tilde{\xi}, \eta) v(\tilde{x}') \, d\tilde{x}' \, d\tilde{\xi}$$

with $\tilde{\xi} = (\tilde{\xi}_0, \tilde{\xi}') \in \mathbb{R}^{1+n}$ and

$$d(\tilde{x}, \tilde{x}', \tilde{\xi}, \eta) = \kappa_0(\tilde{x})^{-\mu} \omega(\tilde{x}, \tilde{x}') \tilde{p}(\kappa_0(\tilde{x}), \kappa'(\tilde{x}), ({}^t\varrho)^{-1}(\tilde{x}, \tilde{x}')$$
$${}^t(\kappa_0(\tilde{x})\tilde{\xi}_0, \tilde{\xi}', \kappa_0(\tilde{x})\eta)| \det \varrho^{-1}(\tilde{x}, \tilde{x}')|| \det d\kappa(\tilde{x}')|,$$

where $\varrho(\tilde{x}, \tilde{x}') = \int_0^1 d\kappa(\tilde{x}' + s(\tilde{x} - \tilde{x}')) \, ds$. Let us verify that

$$d(\tilde{x}, \tilde{x}', \tilde{\xi}, \eta) = d_0(\tilde{x}, \tilde{x}', \tilde{\xi}, \eta) + \sum_{j=0}^{n} (\tilde{x}_j - \tilde{x}'_j) d_{1,j}(\tilde{x}, \tilde{x}', \tilde{\xi}, \eta) \quad (3.2.18)$$

for symbols

$$d_0 \in S^{\mu,0,0}(V \times V \times \mathbb{R}^{n+1}; \mathbb{R}^q), \qquad d_{1,j} \in S^{\mu,0,0}(V \times V \times \mathbb{R}^{n+1}; \mathbb{R}^q),$$

$j = 0, \ldots, n$. This completes the proof if we use

$$\mathrm{Op}_{\tilde{x}}((\tilde{x}_j - \tilde{x}'_j) d_{1,j}) = \mathrm{Op}_{\tilde{x}}(D_{\xi_j} d_{1,j}), \qquad j = 0, \ldots, n,$$

and the above Lemma 3.2.20.

Consider the Jacobi matrix $d\kappa(\tilde{x})$ of $\kappa(\tilde{x})$ with the above homogeneity relations (3.2.15). For the matrix $\varrho(\tilde{x}, \tilde{x}') = (\varrho_{i,j}(\tilde{x}, \tilde{x}'))_{i,j=0,\ldots,n}$ we then obtain $\varrho_{0,j}(\tilde{x}, \tilde{x}') \in S^{0,0}(V \times V)$, $\varrho_{k,j}(\tilde{x}, \tilde{x}') \in S^{-1,-1}(V \times V)$ for $j = 0, \ldots, n$, $k = 1, \ldots, n$. Then the entries $\sigma_{i,j}(\tilde{x}, \tilde{x}')$ of the matrix $\omega(\tilde{x}, \tilde{x}')({}^t\varrho)^{-1}(\tilde{x}, \tilde{x}')$ satisfy $\sigma_{i,0}(\tilde{x}, \tilde{x}') \in S^{0,0}(V \times V)$, $\sigma_{i,k}(\tilde{x}, \tilde{x}') \in S^{1,1}(V \times V)$ for $i = 0, \ldots, n$, $k = 1, \ldots, n$. From the homogeneity properties of κ_0 and κ' it follows that the components of

$$\omega(\tilde{x}, \tilde{x}')({}^t\varrho)^{-1}(\tilde{x}, \tilde{x}')^t(\kappa_0(\tilde{x}), 1_n)$$

$(1_n = (1, \ldots, 1)$ n times) belong to $S^{1,1}(V \times V)$ and that

$$\kappa_0(\tilde{x})^{-\mu} \omega(\tilde{x}, \tilde{x}') \tilde{p}(\kappa_0(\tilde{x}), \kappa'(\tilde{x}), ({}^t\varrho)^{-1}(\tilde{x}, \tilde{x}')^t(\kappa_0(\tilde{x})\tilde{\xi}_0, \tilde{\xi}', \kappa_0(\tilde{x})\eta)$$
$$\in S^{\mu,\delta,0}(V \times V \times \mathbb{R}^{n+1} \times \mathbb{R}^q). \quad (3.2.19)$$

Next we show that there are symbols $g_j(\tilde{x}, \tilde{x}') \in S^{0,0}(V \times V)$ such that

$$\omega(\tilde{x}, \tilde{x}')| \det({}^t\varrho)^{-1}(\tilde{x}, \tilde{x}')|| \det d\kappa(\tilde{x}')|$$
$$= \alpha \omega(\tilde{x}, \tilde{x}')\Big(1 + \sum_{j=0}^{n} (x_j - x'_j) g_j(\tilde{x}, \tilde{x})\Big) \quad (3.2.20)$$

for some constant α. In fact, the left hand side of (3.2.20) equals

$$\alpha\omega(\tilde{x}, \tilde{x}')(\det d\kappa(\tilde{x}'))(\det \varrho(\tilde{x}, \tilde{x}'))^{-1}$$

for appropriate α. Then, applying the identity

$$f(\tilde{x}' + s(\tilde{x} - \tilde{x}')) = f(\tilde{x}') + \sum_{j=0}^{n} s(\tilde{x}_j - \tilde{x}_j') \int_0^1 (\partial_{\tilde{x}_j} f)(\tilde{x}' + s\lambda(\tilde{x} - \tilde{x}')) \, d\lambda$$

to the entries of the matrix $d\kappa$ we obtain

$$\int_0^1 d\kappa(\tilde{x}' + s(\tilde{x} - \tilde{x}')) \, ds$$

$$= \int_0^1 \left\{ \sum_{j=0}^{n} s(\tilde{x}_j - \tilde{x}_j') \int_0^1 \partial_{\tilde{x}_j} d\kappa(\tilde{x}' + s\lambda(\tilde{x} - \tilde{x}')) \, d\lambda \right\} ds,$$

i.e.,

$$\omega(\tilde{x}, \tilde{x}') \det \varrho(\tilde{x}, \tilde{x}') = \omega(\tilde{x}, \tilde{x}') \left\{ \det d\kappa(\tilde{x}') + \sum_{j=0}^{n} (\tilde{x}_j - \tilde{x}_j') b_j(\tilde{x}, \tilde{x}') \right\}$$

for symbols $b_j(\tilde{x}, \tilde{x}') \in S^{0,0}(V \times V)$, $j = 0, \ldots, n$.
Thus we can form

$$\omega(\tilde{x}, \tilde{x}') = \omega(\tilde{x}, \tilde{x}') \frac{\det d\kappa(\tilde{x}')}{\det \varrho(\tilde{x}, \tilde{x}')} + \omega(\tilde{x}, \tilde{x}') \sum_{j=0}^{n} (\tilde{x}_j - \tilde{x}_j') \frac{b_j(\tilde{x}, \tilde{x}')}{\det \varrho(\tilde{x}, \tilde{x}')}$$

which gives us the relation (3.2.20). Together with (3.2.19) we obtain (3.2.18). Moreover, it is evident that the property of being classical is satisfied for the push–forward of a classical pseudo–differential operator. $\qquad \square$

Remark 3.2.23 *From the proof of Proposition 3.2.19 it follows that the parameter–dependent homogeneous principal symbol $\sigma_{\psi;\eta}^{\mu}(\mathrm{Op}(a)(\eta))$ is elliptic if and only if $\tilde{p}_{(\mu)}(t, x, \tilde{\tau}, \xi, \tilde{\eta}) \neq 0$ for $(\tilde{\tau}, \xi, \tilde{\eta}) \neq 0$. In addition, under the condition of t–independence of $\tilde{p}$ for $t > $ const, there exists the homogeneous principal exit symbol $\sigma_{\mathrm{e}}^0(\mathrm{Op}(a)(\eta))$ for $t \to \infty$ which equals $\sigma_{\psi;\eta}^{\mu}(\mathrm{Op}(a)(\eta))$. Thus, for $\eta \neq 0$ it never vanishes for all $\tilde{\xi} \in \mathbb{R}^{n+1}$. The associated homogeneous principal symbol of $\sigma_{\mathrm{e}}^0(\mathrm{Op}(a)(\eta))$ in $\tilde{\xi}$ of order μ is then $\neq 0$ for $\tilde{\xi} \neq 0$. Hence the ellipticity conditions in the sense of Definition 1.4.56 and Remark 1.4.57 are satisfied.*

Analogously to the construction of $L_{\mathrm{cl}}^{\mu}(X^{\wedge}; \mathbb{R}^q)_{\mathrm{e}}$ at the beginning of this section we can form an operator class

$$\tilde{L}_{\mathrm{cl}}^{\mu}(X^{\wedge}; \mathbb{R}^q)_{\mathrm{e}}. \tag{3.2.21}$$

First let $\Sigma \subseteq \mathbb{R}^n$ be an open set, $\Sigma^{\wedge} = \mathbb{R}_+ \times \Sigma$, and introduce $\tilde{L}_{\mathrm{cl}}^{\mu}(\Sigma^{\wedge}; \mathbb{R}^q)_{\mathrm{e}}$ as the space of all operator families $t^{-\mu} \mathrm{Op}(\tilde{p})(\eta) + C(\eta)$ for arbitary $\tilde{p}(t, x, t\tau, \xi, t\eta)$ with $\tilde{p}(t, x, \tilde{\tau}, \xi, \tilde{\eta}) \in S_{\mathrm{cl}}^{\mu}(\overline{\mathbb{R}}_+ \times \Sigma \times \mathbb{R}^{1+n+q})$ which are independent of t for $t > \mathrm{const}$ for a constant > 0, and $C(\eta)$ having a kernel in $\mathcal{S}(\mathbb{R}_{\eta}^q, \mathcal{S}(\Sigma^{\wedge}))$, where $\mathcal{S}(\Sigma^{\wedge} \times \Sigma^{\wedge})$ is the subspace of all $c(t, x, t', x') \in C^{\infty}(\Sigma^{\wedge} \times \Sigma^{\wedge})$ with $(1 - \omega(t))(1 - \omega(t'))c(t, x, t', x') \in \mathcal{S}(\mathbb{R} \times \mathbb{R}, C^{\infty}(\Sigma^{\wedge} \times \Sigma^{\wedge}))|_{\mathbb{R} \times \mathbb{R}}$ for an arbitrary cut–off function $\omega(t)$. Then, for an arbitrary coordinate neighbourhood U on X and a chart $\kappa : U \to \Sigma$ we can form the space $\tilde{L}_{\mathrm{cl}}^{\mu}(U^{\wedge}; \mathbb{R}^q)_{\mathrm{e}}$ as the operator push-forward of $\tilde{L}_{\mathrm{cl}}^{\mu}(\Sigma^{\wedge}; \mathbb{R}^q)_{\mathrm{e}}$ under $\Sigma^{\wedge} \to U^{\wedge}$, $(t, x) \to (t, \kappa^{-1}(x))$. Finally, using U_j, φ_j, ψ_j as in (3.2.11), we get

$$\tilde{L}_{\mathrm{cl}}^{\mu}(X^{\wedge}; \mathbb{R}^q)_{\mathrm{e}} = \Big\{ \sum_{j=1}^{N} \psi_j B_j(\eta)\psi_j + C(\eta) : B_j(\eta) \in \tilde{L}_{\mathrm{cl}}^{\mu}(U_j^{\wedge}; \mathbb{R}^q)_{\mathrm{e}},$$

$$j = 1, \ldots, N, \quad C(\eta) \in L^{-\infty}(X^{\wedge}; \mathbb{R}^q)_{\mathrm{e}} \Big\}. \tag{3.2.22}$$

As an immediate consequence of Proposition 3.2.19 we now obtain $\tilde{L}_{\mathrm{cl}}^{\mu}(X^{\wedge}; \mathbb{R}^q)_{\mathrm{e}} \subset L_{\mathrm{cl}}^{\mu}(X^{\wedge}; \mathbb{R}^q)_{\mathrm{e}}$. Moreover, from the definition it follows that

$$t^{\mu} \tilde{L}_{\mathrm{cl}}^{\mu}(X^{\wedge}; \mathbb{R}^q)_{\mathrm{e}} \subset \tilde{L}_{\mathrm{cl}}^{\mu}(X^{\wedge}; \mathbb{R}^q), \tag{3.2.23}$$

where the space on the right is the subspace of all y–independent elements of (3.2.7).

Below we will also employ a y–dependent analogue (3.2.21) for y varying in an open set $\Omega \subseteq \mathbb{R}^p$. The only change is that we start the definition with

$$t^{-\mu} \mathrm{Op}(\tilde{p})(y, \eta) + C(y, \eta) \tag{3.2.24}$$

for arbitrary $\tilde{p}(t, x, y, t\tau, \xi, t\eta)$ with $\tilde{p}(t, x, y, \tilde{\tau}, \xi, \tilde{\eta}) \in S_{\mathrm{cl}}^{\mu}(\overline{\mathbb{R}}_+ \times \Sigma \times \Omega \times \mathbb{R}^{1+n+q})$ which are independent of t for $t > \mathrm{const}$, and $C(y, \eta)$ being an operator familiy with kernel in $C^{\infty}(\Omega, \mathcal{S}(\mathbb{R}_{\eta}^q, \mathcal{S}(\Sigma^{\wedge} \times \Sigma^{\wedge})))$. In this way we get

$$C^{\infty}(\Omega, \tilde{L}_{\mathrm{cl}}^{\mu}(X^{\wedge}; \mathbb{R}^q)_{\mathrm{e}}), \tag{3.2.25}$$

and in an analogous manner, of course, $C^{\infty}(\Omega, \tilde{L}^{\mu}(X^{\wedge}; \mathbb{R}^q)_{\mathrm{e}})$.

The following propositions are straightforward. So the proofs will be omitted.

Proposition 3.2.24 *Let $a(y,\eta) \in C^\infty(\Omega, \tilde{L}^\mu(X^\wedge; \mathbb{R}^q)_e)$ and $a^*(y,\eta)$ be the point–wise formal adjoint in the sense $(a(y,\eta)u, v)_{\mathcal{K}^{0,0}(X^\wedge)} = (u, a^*(y,\eta)v)_{\mathcal{K}^{0,0}(X^\wedge)}$ for all $u, v \in C_0^\infty(X^\wedge)$, $(y,\eta) \in \Omega \times \mathbb{R}^q$. Then $a^*(y,\eta) \in C^\infty(\Omega, \tilde{L}^\mu(X^\wedge; \mathbb{R}^q)_e)$. In particular, the relation $a(y,\eta) \in C^\infty(\Omega, \tilde{L}^\mu_{cl}(X^\wedge; \mathbb{R}^q)_e)$ implies $a^*(y,\eta) \in C^\infty(\Omega, \tilde{L}^\mu_{cl}(X^\wedge; \mathbb{R}^q)_e)$.*

Proposition 3.2.25 *Given elements $a(y,\eta) \in C^\infty(\Omega, \tilde{L}^\mu(X^\wedge; \mathbb{R}^q)_e)$, $b(y,\eta) \in C^\infty(\Omega, \tilde{L}^\nu(X^\wedge; \mathbb{R}^q)_e)$ we obtain $a(y,\eta)(1 - \omega)b(y,\eta) \in C^\infty(\Omega, \tilde{L}^{\mu+\nu}(X^\wedge; \mathbb{R}^q)_e)$ for each cut–off function $\omega(t)$. Analogous assertions hold for the corresponding operator families with subscript cl.*

3.3   The algebra of edge symbols

3.3.1   Green symbols

We now turn to the so-called Green symbols as particular (classical) operator–valued symbols belonging to the algebra of edge symbols.

Our notation is motivated by the structure of the Green function of a classical elliptic boundary value problem for an elliptic differential operator. If E is any fundamental solution (or a parametrix) of the operator, then the Green function is modulo E a pseudo–differential operator with a special Green symbol along the boundary. In this interpretation the boundary corresponds to the edge and the inner normal of the domain to the model cone of the "wedge".

Definition 3.3.1 *Let $U \subseteq \mathbb{R}^p$ be an open set, and let $\gamma, \delta \in \mathbb{R}$, $g = (\gamma, \delta, \Theta)$ for $\Theta = (\vartheta, 0]$, $-\infty \leq \vartheta < 0$. Then an operator function*

$$g(y,\eta) \in \bigcap_{s \in \mathbb{R}} C^\infty(U \times \mathbb{R}^q, \mathcal{L}(\mathcal{K}^{s,\gamma}(X^\wedge), \mathcal{K}^{\infty,\delta}(X^\wedge)))$$

is called a Green symbol of order $\nu \in \mathbb{R}$ with continuous (discrete; weakly discrete) asymptotics if there are elements $P \in \mathrm{As}(X, (\delta, \Theta))$, $Q \in \mathrm{As}(X, (-\gamma, \Theta))$ ($P \in \mathrm{As}(X, (\delta, \Theta)^\bullet)$, $Q \in \mathrm{As}(X, (-\gamma, \Theta)^\bullet)$; $P \in \mathrm{As}(X, (\delta, \Theta))^\bullet, Q \in \mathrm{As}(X, (\delta, \Theta))^\bullet)$ such that

$$g(y,\eta) \in \bigcap_{s \in \mathbb{R}} S_{cl}^\nu(U \times \mathbb{R}^q; \mathcal{K}^{s,\gamma}(X^\wedge), \mathcal{S}_P^\delta(X^\wedge)),$$

$$g^*(y,\eta) \in \bigcap_{s \in \mathbb{R}} S_{cl}^\nu(U \times \mathbb{R}^q; \mathcal{K}^{s,-\delta}(X^\wedge), \mathcal{S}_Q^{-\gamma}(X^\wedge)).$$

Here $g^*(y,\eta)$ is the point-wise formal adjoint of $g(y,\eta)$ in the sense

$$(u, g^*v)_{\mathcal{K}^{0,0}(X^\wedge)} = (gu, v)_{\mathcal{K}^{0,0}(X^\wedge)}$$

for all $u, v \in C_0^\infty(X^\wedge)$.

The space of all Green symbols of order ν for fixed asymptotic types P, Q is denoted by

$$R_G^\nu(U \times \mathbb{R}^q, \boldsymbol{g})_{P,Q}. \qquad (3.3.1)$$

Moreover, $R_G^\nu(U \times \mathbb{R}^q, \boldsymbol{g})$ $(R_G^\nu(U \times \mathbb{R}^q, \boldsymbol{g}^\bullet), R_G^\nu(U \times \mathbb{R}^q, \boldsymbol{g})^\bullet)$ denotes the union of all spaces (3.3.1) over P, Q.

In our applications U will be Ω or $\Omega \times \Omega$ for an open set $\Omega \subseteq \mathbb{R}^q$. Points in $\Omega \times \Omega$ will then also be denoted by (y, y').

Remark 3.3.2 $R_G^\nu(U \times \mathbb{R}^q, \boldsymbol{g})_{P,Q}$ *for every fixed P, Q is a Fréchet space in a canonical way. The semi-norm system follows immediately from the conditions of Definition 3.3.1, where it suffices to take the intersections over all $s \in \mathbb{Z}$.*

Note that, in contrast to the cone theory of Chapter 2, here we mainly discuss the more general discrete asymptotic types

$$P \in \operatorname{As}(X, (\delta, \Theta))^\bullet, Q \in \operatorname{As}(X, (-\gamma, \Theta))^\bullet$$

rather than $P \in \operatorname{As}(X, (\delta, \Theta)^\bullet)$, $Q \in \operatorname{As}(X, (-\gamma, \Theta)^\bullet)$. This is more convenient though many results remain true in the more precise form with dots at the weight data. The main reason why we generalise the concept of discrete asymptotics for the edge theory is that the finite-dimensional spaces of coefficients in the asymptotics may vary along y. Thus, without the requirement of y–independent spaces, we have in fact a very mild form of continuous asymptotics.

Example 3.3.3 *Let $\gamma = \delta = 0$, $\Theta = (-\infty, 0]$, $\dim X = 0$, and take for P, Q the Taylor asymptotics T for $t \to 0$. Then the spaces $S_P^0(\mathbb{R}_+) = S_Q^0(\mathbb{R}_+)$ coincide with $\mathcal{S}(\overline{\mathbb{R}}_+) = \mathcal{S}(\mathbb{R})|_{\overline{\mathbb{R}}_+}$, and $R_G^\nu(U \times \mathbb{R}^q, \boldsymbol{g})_{T,T}$ for $\boldsymbol{g} = (0, 0, \Theta)$ equals the space of all*

$$g(y, \eta) \in S_{\mathrm{cl}}^\nu(U \times \mathbb{R}^q; L^2(\mathbb{R}_+), \mathcal{S}(\overline{\mathbb{R}}_+))$$

satisfying

$$g^*(y, \eta) \in S_{\mathrm{cl}}^\nu(U \times \mathbb{R}^q; L^2(\mathbb{R}_+), \mathcal{S}(\overline{\mathbb{R}}_+)),$$

where $*$ refers to the $L^2(\mathbb{R}_+)$-scalar product.

Recall that the space of all operators $a \in \mathcal{L}(L^2(\mathbb{R}_+), \mathcal{S}(\overline{\mathbb{R}}_+))$ with the property $a^* \in \mathcal{L}(L^2(\mathbb{R}_+), \mathcal{S}(\overline{\mathbb{R}}_+))$ coincides with the set of all integral operators

$$(au)(t) = \int_0^\infty a(t, t') u(t') \, dt',$$

$u \in L^2(\mathbb{R}_+)$, where $a(t, t') \in \mathcal{S}(\overline{\mathbb{R}}_+ \times \overline{\mathbb{R}}_+)$ $(= \mathcal{S}(\mathbb{R} \times \mathbb{R})|_{\overline{\mathbb{R}}_+ \times \overline{\mathbb{R}}_+})$.

Note that the symbols from Example 3.3.3 just coincide with the singular Green symbols of type zero in the boundary symbolic algebra for pseudo-differential boundary value problems with the transmisssion property, cf. [10], [91]. We will return in Chapter 4 below to this special case and also give a characterisation for general types.

Remark 3.3.4 *Every operator*

$$a \in \bigcap_{s \in \mathbb{R}} \mathcal{L}(\mathcal{K}^{s,\gamma}(X^\wedge), \mathcal{S}_P^\delta(X^\wedge))$$

with

$$a^* \in \bigcap_{s \in \mathbb{R}} \mathcal{L}(\mathcal{K}^{s,-\delta}(X^\wedge), \mathcal{S}_Q^{-\gamma}(X^\wedge))$$

for $P \in \mathrm{As}(X, (\delta, \Theta)^\bullet)$, $Q \in \mathrm{As}(X, (-\gamma, \Theta)^\bullet)$ *has a kernel*

$$a(t, x, t', x') \in \mathcal{S}_P^\delta(X^\wedge) \otimes_\pi \mathcal{S}_{\overline{Q}}^{-\gamma}(X^\wedge).$$

Here $\overline{Q} = \{(\overline{q}_j, n_j, \overline{L}_j)\}_{j \in \mathbb{N}}$ *when* $Q = \{(q_j, n_j, L_j)\}_{j \in \mathbb{N}}$, *and* $\otimes_\pi$ *is the completed projective tensor product. A proof of this result may be found in* [96], *cf. also* [131].

Example 3.3.5 *Let*

$$a(y, \eta; t, x, t', x') \in C^\infty(U \times S^{q-1}, \mathcal{S}_P^\delta(X^\wedge) \otimes_\pi \mathcal{S}_Q^{-\gamma}(X^\wedge)),$$

where S^{q-1} *is the unit sphere in* $\mathbb{R}^q \ni \eta$ *and let* $\chi(\eta)$ *be an excision function in* $\mathbb{R}^q$. *Then*

$$g(y, \eta)u(t, x) = [\eta]^{\nu+1} \int_0^\infty \int_X \chi(\eta) a(y, \frac{\eta}{|\eta|}; [\eta]t, x, [\eta]t', x') u(t', x') \, dt' dx'$$

defines an element of $R_G^\nu(U \times \mathbb{R}^q, \boldsymbol{g})$ *with* $\boldsymbol{g} = (\gamma, \delta, \Theta)$. *It satisfies*

$$g(y, \lambda\eta) = \lambda^\nu \kappa_\lambda g(y, \eta) \kappa_\lambda^{-1}$$

for all $\lambda \geq 1$, $|\eta| \geq c$ *for some* $c > 0$.

To study elliptic pseudo–differential operators on manifolds with edges it is necessary to formulate additional conditions of trace and potential type with respect to the edges. They have the form of pseudo–differential operators with operator–valued symbols which are similar to the Green symbols from Definition 3.3.1.

Definition 3.3.6 *An operator function*

$$g(y,\eta) \in \bigcap_{s\in\mathbb{R}} C^\infty(U \times \mathbb{R}^q, \mathcal{L}(\mathcal{K}^{s,\gamma}(X^\wedge) \oplus \mathbb{C}^{N_-}, \mathcal{K}^{\infty,\delta}(X^\wedge) \oplus \mathbb{C}^{N_+}))$$

for certain $N_-, N_+ \in \mathbb{N}$ *is called a Green symbol of order* $\nu \in \mathbb{R}$ *if there are asymptotic types* $P \in \mathrm{As}(X,(\delta,\Theta))$, $Q \in \mathrm{As}(X,(-\gamma,\Theta))$ *with*

$$g(y,\eta) \in \bigcap_{s\in\mathbb{R}} S_{\mathrm{cl}}^\nu(U \times \mathbb{R}^q; \mathcal{K}^{s,\gamma}(X^\wedge) \oplus \mathbb{C}^{N_-}, \mathcal{S}_P^\delta(X^\wedge) \oplus \mathbb{C}^{N_+}), \quad (3.3.2)$$

$$g^*(y,\eta) \in \bigcap_{s\in\mathbb{R}} S_{\mathrm{cl}}^\nu(U \times \mathbb{R}^q; \mathcal{K}^{s,-\delta}(X^\wedge) \oplus \mathbb{C}^{N_+}, \mathcal{S}_Q^{-\gamma}(X^\wedge) \oplus \mathbb{C}^{N_-}).$$

$$(3.3.3)$$

The one–parameter groups of isomorphisms on $\mathcal{K}^{s,\rho}(X^\wedge)\oplus\mathbb{C}^N$, $s,\rho \in \mathbb{R}$, $N \in \mathbb{N}$, or on corresponding subspaces are here

$$\kappa_\lambda(u(t,x) \oplus c) = \lambda^{\frac{n+1}{2}} u(\lambda t, x) \oplus c, \quad \lambda \in \mathbb{R}_+,$$

for $n = \dim X$. In other words, on the finite-dimensional components $\mathbb{C}^N$ we always take the identical actions. The $*$ in (3.3.3) means the point-wise formal adjoint in the sense

$$(u, g^*v)_{\mathcal{K}^{0,0}(X^\wedge)\oplus\mathbb{C}^{N_-}} = (gu, v)_{\mathcal{K}^{0,0}(X^\wedge)\oplus\mathbb{C}^{N_+}} \quad (3.3.4)$$

for all $u \in C_0^\infty(X^\wedge) \oplus \mathbb{C}^{N_-}$, $v \in C_0^\infty(X^\wedge) \oplus \mathbb{C}^{N_+}$. Denote the space of all symbols $g(y,\eta)$ in Definition 3.3.6 also by

$$\mathcal{R}_G^\nu(U \times \mathbb{R}^q, \boldsymbol{g}; N_-, N_+)_{P,Q} \quad (3.3.5)$$

for the weight data $\boldsymbol{g} = (\gamma,\delta,\Theta)$. Moreover, $\mathcal{R}_G^\nu(U \times \mathbb{R}^q, \boldsymbol{g}; N_-, N_+)$ denotes the union of all (3.3.5) over P,Q. The case $P \in \mathrm{As}(X,(\delta,\Theta)^\bullet)$, $Q \in \mathrm{As}(X,(-\gamma,\Theta)^\bullet)$ or $P \in \mathrm{As}(X,(\delta,\Theta))^\bullet$, $Q \in \mathrm{As}(X,(-\gamma,\Theta))^\bullet$ will be indicated by a corresponding dot. The occurring cases for U will be Ω or $\Omega \times \Omega$ for an open set $\Omega \subseteq \mathbb{R}^q$; in the latter case the variables will also be denoted by (y,y'). The elements of (3.3.5) are also called Green symbols, though the entries in the block matrices

$$g(y,\eta) = \begin{pmatrix} g_{11} & g_{12} \\ g_{21} & g_{22} \end{pmatrix} (y,\eta)$$

have individual meaning, namely $g_{21}(y,\eta)$ as trace, $g_{12}(y,\eta)$ as potential symbols with respect to the edge, whereas $g_{11}(y,\eta)$ is a Green symbol in the sense of Definition 3.3.1, and $g_{21}(y,\eta)$ is nothing else than an $N_+ \times N_-$ matrix of classical scalar symbols of order ν.

Note that instead of (3.3.3) we could require an analogous condition using the transposed operators with respect to the bilinear pairings

$$\langle u, f\rangle_{\mathcal{K}^{0,0}(X^\wedge)\oplus\mathbb{C}^N} = (u, \overline{f})_{\mathcal{K}^{0,0}(X^\wedge)\oplus\mathbb{C}^N}$$

instead of the sesquilinear pairings. Then the asymptotic type Q would have to be replaced by $\overline{Q}$, the complex conjugate. Now the linear map

$$\mathcal{R}^\nu_G(U \times \mathbb{R}^q, \boldsymbol{g}; N_-, N_+)_{P,Q}$$
$$\to \bigcap_{s\in\mathbb{Z}} S^\nu_{\mathrm{cl}}(U \times \mathbb{R}^q; \mathcal{K}^{s,\gamma}(X^\wedge) \oplus \mathbb{C}^{N_-}, \mathcal{S}^\delta_P(X^\wedge) \oplus \mathbb{C}^{N_+})$$

together with the corresponding one for the transposed gives rise to a Fréchet topology in the space (3.3.5), namely that of the projective limit. We could take equivalently the intersections over all $s \in \mathbb{R}$.

Example 3.3.7 Let $b_{21}(y,\eta;t',x') \in C^\infty(U \times S^{q-1}, \mathcal{S}^{-\gamma}_{\overline{Q}}(X^\wedge))$, $S^{q-1} = \{\eta \in \mathbb{R}^q : |\eta| = 1\}$, choose an excision function $\chi(\eta)$ in $\mathbb{R}^q$, and set

$$g_{21}(y,\eta)u = [\eta]^{\nu+\frac{1-n}{2}} \int_0^\infty \int_X \chi(\eta)b_{21}(y, \frac{\eta}{|\eta|}; [\eta]t', x')u(t', x')\, dt'dx'.$$

Then $g_{21}(y,\eta) \in \mathcal{R}^\nu_G(U \times \mathbb{R}^q, \boldsymbol{g}; 0, 1)$ is a trace symbol for $\boldsymbol{g} = (\gamma, \delta, \Theta)$ for arbitrary δ. It satisfies

$$g_{21}(y, \lambda\eta) = \lambda^\nu g_{21}(y,\eta)\kappa_\lambda^{-1}$$

for all $\lambda \geq 1$, $|\eta| \geq \mathrm{const}$ for a constant > 0. On the other hand, if we choose a function $b_{12}(y,\eta;t,x) \in C^\infty(U \times S^{q-1}, \mathcal{S}^\delta_P(X^\wedge))$ and set

$$(g_{12}(y,\eta)c)(t,x) = [\eta]^{\nu+\frac{1+n}{2}}\chi(\eta)b_{12}(y, \frac{\eta}{|\eta|}; [\eta]t, x)c$$

for $c \in \mathbb{C}$, then $g_{12}(y,\eta) \in \mathcal{R}^\nu_G(U \times \mathbb{R}^q, \boldsymbol{g}; 1, 0)$ is a potential symbol for $\boldsymbol{g} = (\gamma, \delta, \Theta)$ for arbitrary γ. It satisfies

$$g_{12}(y, \lambda\eta) = \lambda^\nu \kappa_\lambda g_{12}(y,\eta)$$

for all $\lambda \geq 1$, $|\eta| \geq \mathrm{const}$ for a constant > 0.

Note that it is typical for the edge pseudo–differential calculus that the trace objects occur in integral form, in contrast to the case of standard boundary value problems, cf. Chapter 4 below. In other words, traces that restrict the argument functions to the edges, possibly after applying differentiations with respect to the t-variable (which is the inner normal in boundary value problems) do not belong to the trace operators here. The Green symbols $g(y, \eta)$ as particular classical operator–valued symbols of order ν have unique homogeneous components $\sigma_\wedge^{\nu-j}(g)(y, \eta)$, $j \in \mathbb{N}$. In particular, the principal homogeneous symbol of order ν can be regarded as an operator family

$$\sigma_\wedge^\nu(g)(y, \eta) : \begin{matrix} \mathcal{K}^{s,\gamma}(X^\wedge) \\ \oplus \\ \mathbb{C}^{N_-} \end{matrix} \to \begin{matrix} \mathcal{K}^{s,\delta}(X^\wedge) \\ \oplus \\ \mathbb{C}^{N_+} \end{matrix}$$

for $(y, \eta) \in U \times (\mathbb{R}^q \setminus \{0\})$ and arbitrary $s \in \mathbb{R}$. Note that $g(y, \eta) \in \mathcal{R}_G^\nu(U \times \mathbb{R}^q, \boldsymbol{g}; N_-, N_+)$ implies $\tilde{g}(y, \eta) = \chi(\eta)\sigma_\wedge^\nu(g)(y, \eta) \in \mathcal{R}_G^\nu(U \times \mathbb{R}^q, \boldsymbol{g}; N_-, N_+)$ for every excision function $\chi(\eta)$ in $\mathbb{R}^q$ and that $\sigma_\wedge^\nu(g) = \sigma_\wedge^\nu(\tilde{g})$.

Remark 3.3.8 *A consequence of Definition 3.3.6 is that*

$$g(y, \eta) \in \mathcal{R}_G^\nu(U \times \mathbb{R}^q, \boldsymbol{g}; N_-, N_+)_{P,Q}$$

implies

$$g^*(y, \eta) \in \mathcal{R}_G^\nu(U \times \mathbb{R}^q, \boldsymbol{g}^*; N_+, N_-)_{Q,P}$$

for $\boldsymbol{g}^* = (-\delta, -\gamma, \Theta)$, and we have $\sigma_\wedge^\nu(g^*)(y, \eta) = \sigma_\wedge^\nu(g)^*(y, \eta)$.

The following assertions are simple consequences of Definition 3.3.6.

Proposition 3.3.9 $g(y, \eta) \in \mathcal{R}_G^\nu(U \times \mathbb{R}^q, \boldsymbol{g}; N_-, N_+)$ for $\boldsymbol{g} = (\gamma, \delta, \Theta)$ *implies*

$$\begin{pmatrix} t^\alpha p(y, \eta) & 0 \\ 0 & r(y, \eta) \end{pmatrix} g(y, \eta) \begin{pmatrix} t^\rho b(y, \eta) & 0 \\ 0 & d(y, \eta) \end{pmatrix}$$
$$\in \mathcal{R}_G^{\nu+\lambda+\tau}(U \times \mathbb{R}^q, \boldsymbol{h}; N_-, N_+)$$

for arbitrary $p(y, \eta) \in S_{\mathrm{cl}}^\beta(U \times \mathbb{R}^q)$, $r(y, \eta) \in S_{\mathrm{cl}}^\lambda(U \times \mathbb{R}^q) \otimes \mathbb{C}^{N_+} \otimes \mathbb{C}^{N_+}$, for $\beta - \alpha = \lambda$, and $b(y, \eta) \in S_{\mathrm{cl}}^\sigma(U \times \mathbb{R}^q)$, $d(y, \eta) \in S_{\mathrm{cl}}^\tau(U \times \mathbb{R}^q) \otimes \mathbb{C}^{N_-} \otimes \mathbb{C}^{N_-}$ for $\sigma - \rho = \tau$, where $\boldsymbol{h} = (\gamma - \rho, \delta + \alpha, \Theta)$.

Proposition 3.3.10 $g(y, \eta) \in \mathcal{R}_G^\nu(U \times \mathbb{R}^q, \boldsymbol{g}; N_-, N_0)$ *for $\boldsymbol{g} = (\gamma, \rho, \Theta)$ and $f(y, \eta) \in \mathcal{R}_G^\mu(U \times \mathbb{R}^q, \boldsymbol{c}; N_0, N_+)$ for $\boldsymbol{c} = (\rho, \delta, \Theta)$ imply for the composition $f(y, \eta)g(y, \eta) \in \mathcal{R}_G^{\mu+\nu}(U \times \mathbb{R}^q, \boldsymbol{h}; N_-, N_+)$, $\boldsymbol{h} = (\gamma, \delta, \Theta)$, and we have*

$$\sigma_\wedge^{\mu+\nu}(fg)(y, \eta) = \sigma_\wedge^\mu(f)(y, \eta)\sigma_\wedge^\nu(g)(y, \eta).$$

Proposition 3.3.11 *Let* $g_j(y,\eta) \in \mathcal{R}_G^{\mu-j}(U \times \mathbb{R}^q, \boldsymbol{g}; N_-, N_+)_{P,Q}$, $j \in$ $\mathbb{N}$, *be an arbitrary sequence, where the asymptotic types* P, Q *are independent of* j. *Then there is a* $g(y,\eta) \in \mathcal{R}_G^\mu(U \times \mathbb{R}^q, \boldsymbol{g}; N_-, N_+)_{P,Q}$ *with the property*

$$g(y,\eta) - \sum_{j=0}^k g_j(y,\eta) \in \mathcal{R}_G^{\mu-(k+1)}(U \times \mathbb{R}^q, \boldsymbol{g}; N_-, N_+)_{P,Q}$$

for every $k \in \mathbb{N}$. *In other words asymptotic sums* $g \sim \sum_{j=0}^\infty g_j$ *can be carried out within the class of Green symbols.*

Proof. Let $\chi(\eta)$ be an excision function in $\mathbb{R}^q$. We shall show that there are constants $c_j > 0$, $j \in \mathbb{N}$, such that

$$g(y,\eta) = \sum_{j=0}^\infty \chi(\frac{\eta}{c_j}) g_j(y,\eta) \tag{3.3.6}$$

converges in $\mathcal{R}_G^\mu(U \times \mathbb{R}^q, \boldsymbol{g}; N_-, N_+)$ with respect to the Fréchet topology mentioned above. We can write the space $\tilde{E} := \mathcal{S}_P^\delta(X^\wedge) \oplus \mathbb{C}^{N_+}$ as a projective limit of Hilbert spaces $\tilde{E}^k$, $k \in \mathbb{N}$, with continuous embeddings $\tilde{E}^{k+1} \hookrightarrow \tilde{E}^k$ for all k, such that $\kappa_\lambda \oplus 1$, $\lambda \in \mathbb{R}_+$, induces a strongly continuous group action on $\tilde{E}^k$ for every k. Setting $E = \mathcal{K}^{s,\gamma}(X^\wedge) \oplus \mathbb{C}^{N_-}$ for fixed $s \in \mathbb{Z}$ we can first choose the constants $c_j = c_j(s,k)$ such that (3.3.6) converges in $S_{cl}^\nu(U \times \mathbb{R}^q; E, \tilde{E}^k)$ for fixed k. By replacing the constants by $c_j(j,j)$ we get the convergence for all $s \in \mathbb{Z}$ and $k \in \mathbb{N}$ at the same time. Doing the same for the (y,η)-wise transposed operators we get other constants ${}^t c_j(j,j)$ for which the corresponding series of transposed operators converges. Then $c_j := \max\{c_j(j,j), {}^t c_j(j,j)\}$, $j \in \mathbb{N}$, are constants for which (3.3.6) converges in $\mathcal{R}_G^\mu(U \times \mathbb{R}^q, \boldsymbol{g}; N_-, N_+)_{P,Q}$. $\square$

Proposition 3.3.12 *Let* $g(y,\eta) \in \mathcal{R}_G^\nu(U \times \mathbb{R}^q, \boldsymbol{g}; N_-, N_+)$ *and* $\varphi(t,y)$, $\psi(t,y) \in C_0^\infty(\overline{\mathbb{R}}_+, C^\infty(U))$, $\varphi'(y) \in C^\infty(U) \otimes \mathbb{C}^{N_+} \otimes \mathbb{C}^{N_+}$, $\psi'(y) \in$ $C^\infty(U) \otimes \mathbb{C}^{N_-} \otimes \mathbb{C}^{N_-}$. *Then*

$$\begin{pmatrix} \varphi & 0 \\ 0 & \varphi' \end{pmatrix} g(y,\eta) \begin{pmatrix} \psi & 0 \\ 0 & \psi' \end{pmatrix} \in \mathcal{R}_G^\nu(U \times \mathbb{R}^q, \boldsymbol{g}; N_-, N_+).$$

Proof. For simplicity let us consider upper left corners, i.e., the case $N_- = N_+ = 0$. The arguments for the general case are completely analogous and left to the reader. We shall consider $\varphi(t,y)g(y,\eta)$; the analogous case $g(y,\eta)\psi(t,y)$ is left to the reader. We use the fact that the operator $\mathcal{M}_\varphi$ of multiplication by φ represents operator–valued symbols

$$\mathcal{M}_\varphi \in S^0(\mathbb{R}^q; \mathcal{K}^{s,\gamma}(X^\wedge), \mathcal{K}^{s,\gamma}(X^\wedge)), \mathcal{M}_\varphi \in S^0(\mathbb{R}^q; \mathcal{S}_P^\gamma(X^\wedge), \mathcal{S}_Q^\gamma(X^\wedge))$$
$$(3.3.7)$$

for all $s, \gamma \in \mathbb{R}$ and arbitrary asymptotic type P with some resulting Q. For brevity we write φ instead of $\mathcal{M}_\varphi$. The symbols (3.3.7) are not classical in general. So the main point of the proof is to show that φg is in fact classical. Applying the Taylor expansion of $\varphi(t,y)$ near $t = 0$

$$\varphi(t,y) = \sum_{j=0}^N t^j \varphi_j(y) + t^{N+1} \varphi_{(N+1)}(t,y),$$

where $\varphi_{(N+1)}(t,y) \in C^\infty(\overline{\mathbb{R}}_+ \times U)$, we have on one hand that the relations (3.3.7) also hold for $\varphi_{(N+1)}(t,y)$ instead of $\varphi(t,y)$ (because of the particularly simple character of the those functions for $t \to \infty$), such that both $\varphi_{(N+1)}$ and $(\varphi_{(N+1)})^*$ are symbols of order 0 between the required spaces, as they occur in Definition 3.3.6, and then $t^{N+1}\varphi_{(N+1)}g$ is again of order $\nu - (N+1)$ between such spaces. However, $t^j \varphi_j(y) g(y,\eta)$ is classical and of order $\nu - j$, i.e., of Green type, cf. Proposition 3.3.9, and hence $\varphi g \sim \sum_{j=0}^\infty t^j \varphi_j g$. This belongs again to the class of Green symbols, cf. Proposition 3.3.11.   □

Remark 3.3.13 *Let* $\varphi(t,y), \psi(t,y) \in C^\infty(\overline{\mathbb{R}}_+ \times U)$ *satisfy* $\varphi(t,y) = \psi(t,y) = 0$ *for* $0 \le t < \varepsilon$, $\varphi(t,y) = \psi(t,y) = 1$ *for* $t >$ const *for a constant* > 0. *Moreover, let* $\varphi'(y), \psi'(y)$ *be as in Proposition 3.3.12. Then* $g(y,\eta) \in \mathcal{R}_G^\nu(U \times \mathbb{R}^q, \boldsymbol{g}; N_-, N_+)$ *implies*

$$\begin{pmatrix} \varphi t^{-k} & 0 \\ 0 & \varphi' \end{pmatrix} g(y,\eta) \begin{pmatrix} \psi t^{-l} & 0 \\ 0 & \psi' \end{pmatrix} \in \mathcal{R}_G^\nu(U \times \mathbb{R}^q, \boldsymbol{g}; N_-, N_+)$$

for every $k, l \in \mathbb{N}$.

The proof is straightforward and will be omitted.

3.3.2 Smoothing Mellin symbols

From the theory of Chapter 2 it follows that the algebra of pseudo–differential operators on the open stretched cone $X^\wedge$ contains another natural class of smoothing operators, namely those of Mellin type. Here in the edge symbolic calculus they will occur as the values of corresponding operator–valued symbols.

Definition 3.3.14 *Let $U \subseteq \mathbb{R}^p$ be an open set, $\nu \in \mathbb{R}$, and $\boldsymbol{g} = (\gamma, \delta, \Theta)$ be weight data, where $\gamma, \delta \in \mathbb{R}$, $\delta + \nu \leq \gamma$, $\Theta = (-k, 0]$, $k \in \mathbb{N} \setminus \{0\}$. Then*

$$R_{M+G}^{\nu}(U \times \mathbb{R}^q, \boldsymbol{g})^{\bullet} \qquad (3.3.8)$$

denotes the space of all operator families $m(y, \eta) + g(y, \eta)$ for arbitrary $g(y, \eta) \in R_G^{\nu}(U \times \mathbb{R}^q, \boldsymbol{g})^{\bullet}$ and

$$m(y, \eta) = \omega(t[\eta])t^{-\nu} \sum_{j=0}^{k-1} t^j \sum_{|\alpha| \leq j} \operatorname{op}_M^{\rho_{j\alpha} - \frac{n}{2}}(h_{j\alpha})(y)\eta^{\alpha}\omega_0(t[\eta]) \qquad (3.3.9)$$

for arbitrary cut–off functions $\omega(t)$, $\omega_0(t)$, and Mellin symbols $h_{j\alpha}(y, z) \in C^{\infty}(U, M_{R_{j\alpha}}^{-\infty}(X))$, for certain $R_{j\alpha} \in \mathbf{As}(X)^{\bullet}$ and $\rho_{j\alpha} \in \mathbb{R}$ satisfying $\pi_{\mathbb{C}}R_{j\alpha} \cap \Gamma_{\frac{n+1}{2} - \rho_{j\alpha}} = \emptyset$ and $\delta + \nu - j \leq \rho_{j\alpha} < \gamma$ for all $j = 0, \ldots, k-1$, and $|\alpha| \leq j$. Moreover,

$$R_{M+G}^{\nu}(U \times \mathbb{R}^q, \boldsymbol{g}) \qquad (3.3.10)$$

denotes the space of all $m(y, \eta) + n(y, \eta) + g(y, \eta)$ for arbitrary $g(y, \eta) \in R_G^{\nu}(U \times \mathbb{R}^q, \boldsymbol{g})$, $m(y, \eta)$ given in the form (3.3.9) and analogously

$$n(y, \eta) = \tilde{\omega}(t[\eta])t^{-\nu} \sum_{j=0}^{k-1} t^j \sum_{|\alpha| \leq j} \operatorname{op}_M^{\delta_{j\alpha} - \frac{n}{2}}(f_{j\alpha})(y)\eta^{\alpha}\tilde{\omega}_0(t[\eta]) \qquad (3.3.11)$$

for arbitrary cut–off functions $\omega(t)$, $\omega_0(t)$, $\tilde{\omega}(t)$, $\tilde{\omega}_0(t)$, and

$$h_{j\alpha}(y, z) \in C^{\infty}(U, M_{R_{j\alpha}}^{-\infty}(X)), \qquad (3.3.12)$$

$$f_{j\alpha}(y, z) \in C^{\infty}(U, M_{Q_{j\alpha}}^{-\infty}(X)), \qquad (3.3.13)$$

for certain $R_{j\alpha}, Q_{j\alpha} \in \mathbf{As}(X)$ and $\rho_{j\alpha}, \delta_{j\alpha} \in \mathbb{R}$ satisfying

$$\text{carrier } R_{j\alpha} \cap \Gamma_{\frac{n+1}{2} - \rho_{j\alpha}} = \text{carrier } Q_{j\alpha} \cap \Gamma_{\frac{n+1}{2} - \delta_{j\alpha}} = \emptyset, \qquad (3.3.14)$$

$$\delta + \nu - j \leq \rho_{j\alpha}, \ \delta_{j\alpha} \leq \gamma \quad \text{for} \quad j = 0, \ldots, k-1, |\alpha| \leq j. \qquad (3.3.15)$$

For $\boldsymbol{g}_k = (\gamma, \delta, \Theta)$ with $\Theta = (-\infty, 0]$ we set

$$R_{M+G}^{\nu}(U \times \mathbb{R}^q, \boldsymbol{g}) = \bigcap_{k \in \mathbb{N} \setminus \{0\}} R_{M+G}^{\nu}(U \times \mathbb{R}^q, \boldsymbol{g}_k)$$

for $\boldsymbol{g}_k = (\gamma, \delta, (-k, 0])$, and similarly for the dotted subclasses. In this section we shall mainly discuss the case of a finite weight interval Θ. However, the results automatically extend to the case $\Theta = (-\infty, 0]$.

In the applications we will have $U = \Omega$ or $U = \Omega \times \Omega$ for an open set $\Omega \subseteq \mathbb{R}^q$. For $U = \Omega \times \Omega$ we will also write (y, y') instead of y. The operator families $a(y, \eta)$ in (3.3.8) and (3.3.10) are called smoothing Mellin symbols with weakly discrete and continuous asymptotics, respectively.

The reason why we require in (3.3.10) different summands of the form (3.3.9), (3.3.11) is that to formulate Mellin pseudo–differential operators we need gaps in the carriers of asymptotics of the Mellin symbols, cf. (3.3.14). In general $V_{j\alpha} = $ carrier $R_{j\alpha} \cup$ carrier $Q_{j\alpha}$ may consist of an infinite band without such gaps. This cannot happen in the weakly discrete case. We will see that the class (3.3.10) is independent of the specific decompositions of the sets $V_{j\alpha}$, based on the relation

$$M_{R_1}^{-\infty}(X) + M_{R_2}^{-\infty}(X) = M_{S_1}^{-\infty}(X) + M_{S_2}^{-\infty}(X) \qquad (3.3.16)$$

for arbitrary $R_j, S_j \in \mathbf{As}(X)$, $j = 1, 2$, satisfying

$$\text{carrier } R_1 + \text{carrier } R_2 = \text{carrier } S_1 + \text{carrier } S_2.$$

This implies

$$\begin{aligned} C^\infty(U, M_{R_1}^{-\infty}(X)) &+ C^\infty(U, M_{R_2}^{-\infty}(X)) \\ &= C^\infty(U, M_{S_1}^{-\infty}(X)) + C^\infty(U, M_{S_2}^{-\infty}(X)), \quad (3.3.17) \end{aligned}$$

also as a non–direct sum of Fréchet spaces, cf. Proposition 1.1.8. Note that the relations (3.3.16), (3.3.17) are valid for arbitrary continuous Mellin asymptotic types R_j, S_j, i.e., it is not necessary that carrier $R_j \cap \Gamma_{\rho_j} = $ carrier $S_j \cap \Gamma_{\sigma_j} = \emptyset$ for certain real ρ_j, σ_j.

Lemma 3.3.15 Let $h(y, z) \in C^\infty(U, M_R^{-\infty}(X))$, $R \in \mathbf{As}(X)$, and assume carrier $R \cap \Gamma_{\frac{n+1}{2} - \rho} = $ carrier $R \cap \Gamma_{\frac{n+1}{2} - \rho + \beta} = \emptyset$ for certain real ρ, β. Then for

$$g(y, \eta) = \omega(t[\eta]) t^{-\nu + j} \left\{ t^\beta \operatorname{op}_M^{\rho - \frac{n}{2}}(h)(y) - \operatorname{op}_M^{\rho - \frac{n}{2}}(T^\beta h) t^\beta \right\} \eta^\alpha \omega_0(t[\eta])$$

for $\nu \in \mathbb{R}$, $j \in \mathbb{N}$, $|\alpha| \leq j$, we have $g(y, \eta) \in R_G^{\nu + j - |\alpha|}(U \times \mathbb{R}^q, \mathbf{r})$ for the weight data

$$\mathbf{r} = (\max(\rho - \beta, \rho), \ \min(\rho + \beta, \rho) - \nu + j, \ (-\infty, 0]).$$

In the case $R \in \mathbf{As}(X)^\bullet$ we get analogously $g(y, \eta) \in R_G^{\nu + j - |\alpha|}(U \times \mathbb{R}^q, \mathbf{r})^\bullet$.

Proof. In view of Proposition 2.3.69 we have $g(y, \eta) \in C_G(X^\wedge, \boldsymbol{r})$ for every fixed $y \in U$, $\eta \in \mathbb{R}^q$. In addition, setting $\gamma = \max(\rho - \beta, \rho)$, $\delta = \min(\rho + \beta, \rho) - \nu + j$, $\Theta = (-\infty, 0]$, there are asymptotic types $P \in \mathrm{As}(X, (\delta, \Theta))$, $Q \in \mathrm{As}(X, (-\gamma, \Theta))$ such that $g(y, \eta) \in C_G(X^\wedge, \boldsymbol{r})_{P,Q}$. Writing

$$\mathcal{S}_P^\delta(X^\wedge) = \varprojlim_{k \in \mathbb{N}} \tilde{E}^k, \qquad \mathcal{S}_Q^{-\gamma}(X^\wedge) = \varprojlim_{l \in \mathbb{N}} \tilde{E}^l$$

for Hilbert spaces $\tilde{E}^k$, E^l, with continuous embeddings

$$\tilde{E}^{k+1} \hookrightarrow \tilde{E}^k \hookrightarrow \tilde{E}^0 = \mathcal{K}^{0,\delta}(X^\wedge), E^{l+1} \hookrightarrow E^l \hookrightarrow E^0 = \mathcal{K}^{0,-\gamma}(X^\wedge),$$

where $\{\kappa_\lambda\}_{\lambda \in \mathbb{R}_+} \in C(\mathbb{R}_+, \mathcal{L}_\sigma(\tilde{E}^k))$, $\{\kappa_\lambda\}_{\lambda \in \mathbb{R}_+} \in C(\mathbb{R}_+, \mathcal{L}_\sigma(E^l))$ for all $k, l \in \mathbb{N}$, we have

$$\begin{aligned} g(y, \eta) &\in C^\infty(U \times \mathbb{R}^q, \mathcal{L}(\mathcal{K}^{s,\gamma}(X^\wedge), \tilde{E}^k), \\ g^*(y, \eta) &\in C^\infty(U \times \mathbb{R}^q, \mathcal{L}(\mathcal{K}^{s,-\delta}(X^\wedge), E^l) \end{aligned} \tag{3.3.18}$$

for all $s \in \mathbb{R}$ and $k, l \in \mathbb{N}$. Moreover, it is obvious that

$$g(y, \lambda\eta) = \lambda^{\nu+j-|\alpha|} \kappa_\lambda g(y, \eta) \kappa_\lambda^{-1} \tag{3.3.19}$$

for all $\lambda \geq 1$, $|\eta| \geq$ const for a constant > 0. This yields the assertion for the continuous asymptotics. Analogous arguments apply for the weakly discrete case. □

Lemma 3.3.16 *Let ω, $\tilde{\omega}$, ω_0, $\tilde{\omega}_0$ be arbitrary cut–off functions, and set*

$$\begin{aligned} m(y, \eta) &= \omega(t[\eta]) t^{-\nu+j} \mathrm{op}_M^{\rho - \frac{n}{2}}(h)(y) \eta^\alpha \omega_0(t[\eta]), \\ \tilde{m}(y, \eta) &= \tilde{\omega}(t[\eta]) t^{-\nu+j} \mathrm{op}_M^{\rho - \frac{n}{2}}(h)(y) \eta^\alpha \tilde{\omega}_0(t[\eta]) \end{aligned}$$

for $h(y, z) \in C^\infty(U, M_R^{-\infty}(X))$, $R \in \mathbf{As}(X)$, carrier $R \cap \Gamma_{\frac{n+1}{2} - \rho} = \emptyset$, where $\delta + \nu - j \leq \rho \leq \gamma$ for the given $\gamma, \delta \in \mathbb{R}$ in the weight data $\boldsymbol{g} = (\gamma, \delta, \Theta)$. Then

$$g(y, \eta) := m(y, \eta) - \tilde{m}(y, \eta) \in R_G^{\nu-j+|\alpha|}(U \times \mathbb{R}^q, \boldsymbol{g}).$$

If $R \in \mathbf{As}(X)^\bullet$ we obtain $g(y, \eta) \in R_G^{\nu-j+|\alpha|}(U \times \mathbb{R}^q, \boldsymbol{g})^\bullet$.

Proof. We know from the cone theory, cf. Lemma 2.3.72, that $g(y, \eta) \in C_G(X^\wedge, \boldsymbol{g})$ for every fixed y, η. In addition we have the relations (3.3.18) for all $s \in \mathbb{R}$, $k, l \in \mathbb{N}$ and (3.3.19) for $\lambda \geq 1$, $|\eta| \geq$ const for our $g(y, \eta)$. This yields the assertion for the continuous asymptotics. Analogous arguments apply for the weakly discrete case. □

Remark 3.3.17 *Let* $b(y), c(y), d(y) \in C^\infty(U) \otimes \mathbb{R}^q \otimes \mathbb{R}^q$, *and set*

$$n(y, \eta) = \omega(t[b(y)\eta]) t^{-\nu+j} \, \mathrm{op}_M^{\rho-\frac{n}{2}}(h)(y)(c(y)\eta)^\alpha \omega_0(t[d(y)\eta]),$$

with the notation from Lemma 3.3.16. Then we have

$$m(y, \eta) - n(y, \eta) \in R_G^{\nu-j+|\alpha|}(U \times \mathbb{R}^q, \boldsymbol{g}),$$

and analogously for the weakly discrete case.

Lemma 3.3.18 *Let* $R \in \mathbf{As}(X)$, $\pi_{\mathbb{C}} R \cap \Gamma_{\frac{n+1}{2}-\rho} = \emptyset$ *for some* $\rho \in \mathbb{R}$, *and let* $h(y, z) \in C^\infty(U, M_R^{-\infty}(X))$. *Then the formal adjoint* $m^*(y, \eta)$ *of the operator*

$$m(y, \eta) := \omega(t[\eta]) t^{-\nu+j} \, \mathrm{op}_M^{\rho-\frac{n}{2}}(h)(y) \eta^\alpha \omega_0(t[\eta]),$$

defined by

$$(u, m^*(y, \eta)v)_{\mathcal{K}^{0,0}(X^\wedge)} = (m(y, \eta)u, v)_{\mathcal{K}^{0,0}(X^\wedge)}$$

for all $u, v \in C_0^\infty(\mathbb{R}_+, C^\infty(X))$, *has the form*

$$m^*(y, \eta) = \omega_0(t[\eta]) t^{-n} \, \mathrm{op}_M^{-\rho+\frac{n}{2}}(h^{[*]})(y) \eta^\alpha t^{-\nu+n+j} \omega(t[\eta]) \qquad (3.3.20)$$

for $h^{[*]}(y, z) = h(y, 1 - \bar{z})^{(*)}$, *where* $(*)$ *means the adjoint with respect to* $L^2(X)$. *In particular, for* carrier $R = \emptyset$, *this can be written*

$$m^*(y, \eta) = \omega_0(t[\eta]) t^{-\nu+j} \, \mathrm{op}_M^{-\rho-\frac{n}{2}}(T^{-n+\nu-j} h^{[*]})(y) \eta^\alpha \omega(t[\eta]). \qquad (3.3.21)$$

Proof. We have $(f, g)_{\mathcal{K}^{0,0}(X^\wedge)} = (t^{\frac{n}{2}} f, t^{\frac{n}{2}} g)_{L^2(X^\wedge)}$ for $f, g \in C_0^\infty(X^\wedge)$, $n = \dim X$. Thus, denoting by $*_0$ the adjoints with respect to the $L^2(X^\wedge)$ scalar product, we get

$$
\begin{aligned}
(mu, v)_{\mathcal{K}^{0,0}(X^\wedge)} &= (t^{\frac{n}{2}} mu, t^{\frac{n}{2}} v)_{L^2(X^\wedge)} \\
&= (t^n mu, v)_{L^2(X^\wedge)} \\
&= (u, (t^n m)^{*_0} v)_{L^2(X^\wedge)} \\
&= (t^{\frac{n}{2}} t^{-\frac{n}{2}} u, t^{\frac{n}{2}} t^{-\frac{n}{2}} (t^n m)^{*_0} v)_{L^2(X^\wedge)} \\
&= (t^{-\frac{n}{2}} u, t^{-\frac{n}{2}} (t^n m)^{*_0} v)_{\mathcal{K}^{0,0}(X^\wedge)} \\
&= (u, t^{-n} (t^n m)^{*_0} v)_{\mathcal{K}^{0,0}(X^\wedge)}.
\end{aligned}
$$

Thus it suffices to calculate $(t^n m)^{*_0}$. We have

$$
\begin{aligned}
(\omega t^{-\nu+n+j} \, \mathrm{op}_M^{\rho-\frac{n}{2}}(h) \omega_0)^{*_0} &= \omega_0 (t^{\rho-\frac{n}{2}} \, \mathrm{op}_M(T^{-\rho+\frac{n}{2}} h) t^{-\rho+\frac{n}{2}})^{*_0} t^{-\nu+n+j} \omega \\
&= \omega_0 t^{-\rho+\frac{n}{2}} \, \mathrm{op}_M(T^{\rho-\frac{n}{2}} h^{[*]}) t^{\rho-\frac{n}{2}} t^{-\nu+n+j} \omega
\end{aligned}
$$

for $h^{[*]}(y, z) = h(y, 1 - \bar{z})^{(*)}$, which yields (3.3.20). Let us now assume carrier $R = \emptyset$ and write

$$t^{-\rho+\frac{n}{2}} \, \mathrm{op}_M(T^{\rho-\frac{n}{2}} h^{[*]}) t^{\rho-\frac{n}{2}} = t^n t^{-\rho-\frac{n}{2}} \, \mathrm{op}_M(T^{\rho+\frac{n}{2}} T^{-n} h^{[*]}) t^{\rho+\frac{n}{2}} t^{-n}$$
$$= t^n \, \mathrm{op}_M^{-\rho-\frac{n}{2}}(T^{-n} h^{[*]}) t^{-n}.$$

Then, applying

$$\omega_0 \, \mathrm{op}_M^\lambda(f) t^\kappa \omega = \omega_0 t^\kappa \, \mathrm{op}_M^\lambda(T^{-\kappa} f) \omega$$

for every holomorphic Mellin symbol f and arbitrary real λ, κ we get

$$(\omega t^{-\nu+n+j} \, \mathrm{op}_M^{\rho-\frac{n}{2}}(h) \omega_0)^{*_0} = \omega_0 t^{n-\nu+j} \, \mathrm{op}_M^{-\rho-\frac{n}{2}}(T^{-n+\nu-j} h^{[*]}) \omega.$$

This yields the relation (3.3.21) in the holomorphic case.   □

Corollary 3.3.19 $a(y, \eta) \in R_{M+G}^\nu(U \times \mathbb{R}^q, \boldsymbol{g})$ for $\boldsymbol{g} = (\gamma, \delta, \Theta)$ implies that the $((y, \eta)$-wise) formal adjoint $a^*(y, \eta)$ with respect to the scalar product $(\cdot, \cdot)_{\mathcal{K}^{0,0}(X^\wedge)}$ belongs to $R_{M+G}^\nu(U \times \mathbb{R}^q, \boldsymbol{g}^*)$ for $\boldsymbol{g}^* = (-\delta, -\gamma, \Theta)$. Moreover, we have $a^*(y, \eta) \in R_{M+G}^\nu(U \times \mathbb{R}^q, \boldsymbol{g}^*)^\bullet$ for $a(y, \eta) \in R_{M+G}^\nu(U \times \mathbb{R}^q, \boldsymbol{g})^\bullet$.

Proposition 3.3.20 Let $\nu \in \mathbb{R}$, $\boldsymbol{g} = (\gamma, \delta, \Theta)$, $\delta + \nu \leq \gamma$. Then we have

$$R_{M+G}^\nu(U \times \mathbb{R}^q, \boldsymbol{g}) \subset S_{\mathrm{cl}}^\nu(U \times \mathbb{R}^q; \mathcal{K}^{s,\gamma}(X^\wedge), \mathcal{K}^{\infty,\delta}(X^\wedge)) \qquad (3.3.22)$$

for every $s \in \mathbb{R}$. Moreover, for every $P \in \mathrm{As}(X, (\gamma, \Theta))$ there is a $Q \in \mathrm{As}(X, (\delta, \Theta))$ such that $a(y, \eta) \in R_{M+G}^\nu(U \times \mathbb{R}^q, \boldsymbol{g})$ implies

$$a(y, \eta) \in S_{\mathrm{cl}}^\nu(U \times \mathbb{R}^q; \mathcal{K}_P^{s,\gamma}(X^\wedge), \mathcal{K}_Q^{\infty,\delta}(X^\wedge)) \qquad (3.3.23)$$

for every $s \in \mathbb{R}$. An analogous relation holds for $a(y, \eta) \in R_{M+G}^\nu(U \times \mathbb{R}^q, \boldsymbol{g})^\bullet$ for every $P \in \mathrm{As}(X, (\gamma, \Theta))^\bullet$ for a resulting $Q \in \mathrm{As}(X, (\delta, \Theta))^\bullet$.

Proof. It suffices to consider a summand in (3.3.9) or (3.3.11), since the Green symbols are as asserted, cf. Definition 3.3.1. In other words we consider

$$a(y, \eta) = \omega(t[\eta]) t^{-\nu+j} \, \mathrm{op}_M^\rho(h)(y) \eta^\alpha \omega_0(t[\eta])$$

for $h(y, z) \in C^\infty(U, M_R^{-\infty}(X))$, $\delta + \nu - j \leq \rho \leq \gamma$, $|\alpha| \leq j$, carrier $R \cap \Gamma_{\frac{n+1}{2} - \rho} = \emptyset$. Then, because of the known mapping properties of operators in the cone algebra, cf. Theorem 2.3.55, we have

$$a(y, \eta) \in C^\infty(U \times \mathbb{R}^q, \mathcal{L}(\mathcal{K}^{s,\gamma}(X^\wedge), \mathcal{K}^{r,\delta}(X^\wedge))$$

for every $s, r \in \mathbb{R}$. In addition

$$a(y, \lambda\eta) = \lambda^{\nu-j+|\alpha|} \kappa_\lambda a(y, \eta) \kappa_\lambda^{-1} \qquad (3.3.24)$$

for all $y \in U$, $|\eta| \geq c$, $\lambda \geq 1$, for some $c > 0$. This yields

$$a(y, \eta) \in S_{\mathrm{cl}}^{\nu-j+|\alpha|}(U \times \mathbb{R}^q; \mathcal{K}^{s,\gamma}(X^\wedge), \mathcal{K}^{\infty,\delta}(X^\wedge)).$$

In view of $-j + |\alpha| \leq 0$ we obtain the relation (3.3.22). For similar reasons we get (3.3.23). In fact, from the mapping properties of operators in the cone algebra between spaces with asymptotics we find to every P a Q such that

$$a(y, \eta) : \mathcal{K}_P^{s,\gamma}(X^\wedge) \to \mathcal{K}_Q^{\infty,\delta}(X^\wedge)$$

is continuous for every $(y, \eta) \in U \times \mathbb{R}^q$ and all $s \in \mathbb{R}$. The arguments show that the spaces $\mathcal{K}_P^{s,\gamma}(X^\wedge)$ and $\mathcal{K}_Q^{\infty,\delta}(X^\wedge)$ can be written

$$\mathcal{K}_P^{s,\gamma}(X^\wedge) = \varprojlim_{k \in \mathbb{N}} E^k, \qquad \mathcal{K}_Q^{\infty,\delta}(X^\wedge) = \varprojlim_{l \in \mathbb{N}} \tilde{E}^l$$

for Hilbert spaces E^k, $\tilde{E}^l$ and continuous embeddings

$$E^{k+1} \hookrightarrow E^k \hookrightarrow E^0 := \mathcal{K}^{s,\gamma}(X^\wedge), \quad \tilde{E}^{l+1} \hookrightarrow \tilde{E}^l \hookrightarrow \tilde{E}^0 := \mathcal{K}^{r,\delta}(X^\wedge)$$

for some r, where

$$\{\kappa_\lambda\}_{\lambda \in \mathbb{R}_+} \in C(\mathbb{R}_+, \mathcal{L}_\sigma(E^k)), \qquad \{\kappa_\lambda\}_{\lambda \in \mathbb{R}_+} \in C(\mathbb{R}_+, \mathcal{L}_\sigma(\tilde{E}^l))$$

for all k, l, and

$$a(y, \eta) \in C^\infty(U \times \mathbb{R}^q, \mathcal{L}(E^{k(l)}, \tilde{E}^l))$$

for every $l \in \mathbb{N}$ with some resulting $k(l) \in \mathbb{N}$. Applying the homogeneity (3.3.24), where $a(y, \eta)$ is regarded as a family of operators $E^{k(l)} \to \tilde{E}^l$ for all $y \in U$, $|\eta| \geq c$, $\lambda \geq 1$, we get $a(y, \eta) \in S_{\mathrm{cl}}^{\nu-j+|\alpha|}(U \times \mathbb{R}^q; E^{k(l)}, \tilde{E}^l)$. This immediately implies (3.3.23). $\qquad\square$

Remark 3.3.21 Let $\boldsymbol{g} = (\gamma, \delta, \Theta)$ for fixed $\gamma, \delta \in \mathbb{R}, \Theta = (-k, 0]$, and let $f(t, y, z) \in C^\infty(\overline{\mathbb{R}}_+ \times U, M_R^{-\infty}(X))$ for an $R \in \mathbf{As}(X)$ with carrier $R \cap \Gamma_{\frac{n+1}{2}-\rho} = \emptyset$, $\rho \leq \gamma$. Then, for arbitrary $\nu \in \mathbb{R}$, $j \in \mathbb{N}$, $|\alpha| \leq j$ we have

$$\omega(t[\eta]) t^{-\nu+j+N} \mathrm{op}_M^{\rho-\frac{n}{2}}(f)(y) \eta^\alpha \omega_0(t[\eta]) \in R_G^{\nu-N-j+|\alpha|}(U \times \mathbb{R}^q, \boldsymbol{g})$$

when $N \in \mathbb{R}$ is sufficiently large. An analogous result is true in the case $R \in \mathbf{As}(X)^\bullet$ for the dotted Green operators.

From Proposition 3.3.20 we get a unique sequence of homogeneous components $\sigma_\wedge^{\nu-l}(a)(y,\eta)$, $l \in \mathbb{N}$, $(y,\eta) \in U \times (\mathbb{R}^q \setminus \{0\})$, for every $a(y,\eta) \in R_{M+G}^\nu(U \times \mathbb{R}^q, g)$, namely

$$\sigma_\wedge^{\nu-l}(a)(y,\eta) = \omega(t|\eta|)t^{-\nu} \sum_{\substack{|\alpha|\leq j \\ j-|\alpha|=l}} t^j \, \mathrm{op}_M^{\rho_{j\alpha}-\frac{n}{2}}(h_{j\alpha})(y)\eta^\alpha \omega_0(t|\eta|)$$

$$+ \tilde{\omega}(t|\eta|) \sum_{\substack{|\alpha|\leq j \\ j-|\alpha|=l}} t^j \, \mathrm{op}_M^{\delta_{j\alpha}-\frac{n}{2}}(f_{j\alpha})(y)\eta^\alpha \tilde{\omega}_0(t|\eta|)$$

$$+ \sigma_\wedge^{\nu-l}(g)(y,\eta) \qquad\qquad (3.3.25)$$

(cf. the notation of Definition 3.3.14; the sums are taken over all j,α such that $|\alpha| \leq j$, $j - |\alpha| = l$). We have

$$\sigma_\wedge^{\nu-l}(a)(y,\lambda\eta) = \lambda^{\nu-l}\kappa_\lambda \sigma_\wedge^{\nu-l}(a)(y,\eta)\kappa_\lambda^{-1}$$

for all $\lambda \in \mathbb{R}_+$ and $(y,\eta) \in U \times (\mathbb{R}^q \setminus \{0\})$.

Definition 3.3.14 shows that every $a(y,\eta) \in R_{M+G}^\nu(U \times \mathbb{R}^q, g)(\in R_{M+G}^\nu(U \times \mathbb{R}^q, g)^\bullet)$ is a (y,η)-dependent family of operators in the space $C_{M+G}(X^\wedge, g)$ $(C_{M+G}(X^\wedge, g)^\bullet)$. Thus we can form the sequence of conormal symbols to the conormal orders $\nu - j$

$$\sigma_M^{\nu-j}(a)(y,z,\eta) = \sum_{|\alpha|\leq j} \{h_{j\alpha}(y,z) + f_{j\alpha}(y,z)\}\eta^\alpha, \qquad (3.3.26)$$

$j = 0,\dots,k-1$. These are polynomials in η of order j; in particular, $\sigma_M^\nu(a)(y,z)$ is independent of η.

Lemma 3.3.22 *Let* $g = (\gamma,\delta,\Theta)$, $\delta + \nu \leq \gamma$, $\Theta = (-k,0]$, *and let*

$$m(y,\eta) = \omega(t[\eta])t^{-\nu+j} \, \mathrm{op}_M^{\rho-\frac{n}{2}}(h)(y)\eta^\alpha \omega_0(t[\eta])$$

for an $h(y,z) \in C^\infty(U, M_R^{-\infty}(X))$, $R \in \mathbf{As}(X)$, *with carrier* $R \cap \Gamma_{\frac{n+1}{2}-\rho} = \emptyset$ *and* $\delta + \nu - j \leq \rho \leq \gamma$, $|\alpha| \leq j$. *Then* $j \geq \delta - \gamma + \nu + k$ *implies* $m(y,\eta) \in R_G^{\nu-j+|\alpha|}(U \times \mathbb{R}^q, g)$. *For* $R \in \mathbf{As}(X)^\bullet$ *it follows that* $m(y,\eta) \in R_G^{\nu-j+|\alpha|}(U \times \mathbb{R}^q, g)^\bullet$.

Proof. First note that for arbitrary $\varepsilon_l > 0$, $l = 1,2$, with $\varepsilon_1 \neq \varepsilon_2$ and $\delta + \nu - j \leq \gamma - \varepsilon_l =: \rho_l$ the Mellin symbol $h(y,z)$ can be written $h(y,z) = h_1(y,z) + h_2(y,z)$ for $h_l(y,z) \in C^\infty(U, M_{R_l}^{-\infty}(X))$ for certain $R_l \in \mathbf{As}(X)$ satisfying carrier $R_l \cap \{\Gamma_{\frac{n+1}{2}-\rho} \cup \Gamma_{\frac{n+1}{2}-\rho_l}\} = \emptyset$, $l = 1,2$, cf. the above relation (3.3.17). Then, applying Lemma 3.3.15, we can write $m(y,\eta) = m_1(y,\eta) + m_2(y,\eta) + g(y,\eta)$ for

$$m_l(y,\eta) = \omega(t[\eta])t^{-\nu+j} \, \mathrm{op}_M^{\rho_l-\frac{n}{2}}(h_l)(y)\eta^\alpha \omega_0(t[\eta]), \quad l = 1,2,$$

and some $g(y,\eta) \in R_G^{\nu-j+|\alpha|}(U \times \mathbb{R}^q, \boldsymbol{g})$. We have for every $(y,\eta) \in U \times \mathbb{R}^q$

$$(m_1 + m_2)(y,\eta) : \mathcal{K}^{s,\gamma}(X^\wedge) \to \mathcal{S}^{\delta+k-\max\{\varepsilon_1,\varepsilon_2\}}(X^\wedge),$$

which follows from the fact that the resulting weight under $m_l(y,\eta)$ is $\delta_l := -\nu + j + \gamma - \varepsilon_l$ and $j \geq \delta - \gamma + \nu + k$ implies $\delta_l \geq -\nu + \delta - \gamma + \nu + k + \gamma - \varepsilon_l = \delta + k - \varepsilon_l$. Since $\varepsilon_1, \varepsilon_2 > 0$ can be taken $< \varepsilon$ for an arbitrary $\varepsilon > 0$, we obtain from the latter mapping property that in fact $m(y,\eta)$ satisfies the first property in Definition 3.3.1. The formal adjoint can be treated in an analogous manner. This yields $m(y,\eta) \in R_G^{\nu-j+|\alpha|}(U \times \mathbb{R}^q, \boldsymbol{g})$. The conclusions for the weakly discrete asymptotics are analogous. $\qquad \square$

Proposition 3.3.23 *Let* $a(y,\eta), \tilde{a}(y,\eta)$ *be elements of* $R_{M+G}^\nu(U \times \mathbb{R}^q, \boldsymbol{g})$ ($\in R_{M+G}^\nu(U \times \mathbb{R}^q, \boldsymbol{g})^\bullet$) *for* $\boldsymbol{g} = (\gamma, \delta, \Theta)$, $\Theta = (-k, 0]$, *and assume*

$$\sigma_M^{\nu-j}(a)(y,z,\eta) = \sigma_M^{\nu-j}(\tilde{a})(y,z,\eta)$$

for all $j \in \mathbb{N}$, $j < \delta - \gamma + \nu + k$. *Then*

$$a(y,\eta) = \tilde{a}(y,\eta) \mod R_G^\nu(U \times \mathbb{R}^q, \boldsymbol{g}) \quad (R_G^\nu(U \times \mathbb{R}^q, \boldsymbol{g})^\bullet).$$

Proof. Lemma 3.3.15 shows that the specific choice of the weights $\rho_{j\alpha}$, $\delta_{j\alpha}$ in Definition 3.3.14 only changes $m(y,\eta)$ by a Green symbol. From Lemma 3.3.16 we see that also the particular cut–off functions do not affect $m(y,\eta)$ up to a Green symbol. Finally from Lemma 3.3.22 we obtain that when $\sigma_M^{\nu-j}(a - \tilde{a})$ vanishes for $j < \delta - \gamma + \nu + k$ the symbol $a - \tilde{a}$ is Green. $\qquad \square$

Remark 3.3.24 *From weight shift constructions like in the proof of Lemma 3.3.22 it follows that* $a_1(y,\eta), a_2(y,\eta) \in R_{M+G}^\nu(U \times \mathbb{R}^q, \boldsymbol{g})$ *implies* $(a_1 + a_2)(y,\eta) \in R_{M+G}^\nu(U \times \mathbb{R}^q, \boldsymbol{g})$. *More precisely, the involved Mellin operators allow representations modulo Green operators such that at most two different weights* $\rho_{j\alpha}$, $\delta_{j\alpha}$ *are necessary for the same* j. *Moreover, they can be chosen to be independent of* α.

Remark 3.3.25 *We have*

$$D_y^\alpha D_\eta^\beta R_{M+G}^\nu(U \times \mathbb{R}^q, \boldsymbol{g}) \subseteq R_{M+G}^{\nu-|\beta|}(U \times \mathbb{R}^q, \boldsymbol{g})$$

for all multi-indices α, β. *In particular, it follows that*

$$D_y^\alpha D_\eta^\beta R_{M+G}^\nu(U \times \mathbb{R}^q, \boldsymbol{g}) \subseteq R_G^{\nu-|\beta|}(U \times \mathbb{R}^q, \boldsymbol{g})$$

for all β *with* $[\delta - \gamma + \nu - |\beta| + k] < 1$, *where* $\boldsymbol{g} = (\gamma, \delta, \Theta)$, $\Theta = (-k, 0]$. *Analogous relations are true for discrete asymptotics.*

Proposition 3.3.26 *Let* $\boldsymbol{g} = (\gamma, \delta, \Theta)$, $\nu \in \mathbb{R}$, $\delta + \nu \leq \gamma$, *and* $h(t, y, z) \in C^\infty(\overline{\mathbb{R}}_+, M_R^{-\infty}(X))$, $R \in \mathbf{As}(X)$, *carrier* $R \cap \Gamma_{\frac{n+1}{2}-\rho} = \emptyset$, *and let* $\delta + \nu - j \leq \rho \leq \gamma$, $|\alpha| \leq j$. *Then*

$$\omega(t[\eta])t^{-\nu+j}\,\mathrm{op}_M^{\rho-\frac{n}{2}}(h)(y)\eta^\alpha\omega_0(t[\eta]) \in R_{M+G}^\nu(U \times \mathbb{R}^q, \boldsymbol{g})$$

and analogously for $R \in \mathbf{As}(X)^\bullet$.

Proof. Applying the Taylor expansion in t we can write

$$h(t, y, z) = \sum_{l=0}^{N} t^l h_l(y, z) + t^{N+1} h_{(N)}(t, y, z)$$

for every $N \in \mathbb{N}$, where $h_j(y, z) \in C^\infty(U, M_R^{-\infty}(X))$ and $h_{(N+1)}(t, y, z) \in C^\infty(\overline{\mathbb{R}}_+ \times U, M_R^{-\infty}(X))$. It is now obvious that the operator family

$$\omega(t[\eta])t^{-\nu+j}\,\mathrm{op}_M^{\rho-\frac{n}{2}}\left(\sum_{l=0}^{N} t^l h_j\right)(y)\eta^\alpha\omega_0(t[\eta])$$

belongs to $R_{M+G}^\nu(U \times \mathbb{R}^q, \boldsymbol{g})$. Moreover, by Remark 3.3.21 we have

$$\omega(t[\eta])t^{-\nu+j+N+1}\,\mathrm{op}_M^{\rho-\frac{n}{2}}(h_{(N)})(y)\eta^\alpha\omega_0(t[\eta]) \in R_G^\nu(U \times \mathbb{R}^q, \boldsymbol{g})$$

for sufficiently large N.   □

Lemma 3.3.27 *Let* $\boldsymbol{g} = (\gamma, \delta, \Theta)$ *and* $h(y, z) \in C^\infty(U, M_R^{-\infty}(X))$, *carrier* $R \cap \Gamma_{\frac{n+1}{2}-\rho} = \emptyset$, $f(y, z) \in C^\infty(U, M_Q^{-\infty}(X))$, *carrier* $Q \cap \Gamma_{\frac{n+1}{2}-\rho} = \emptyset$, *for a* $\rho \in \mathbb{R}$, $\delta \leq \rho \leq \gamma$. *Then, for arbitrary cut–off functions* ω, $\tilde{\omega}$, ω_0 *we have*

$$g(y, \eta) := \omega(t[\eta])\,\mathrm{op}_M^{\rho-\frac{n}{2}}(h)(y)\,(1 - \tilde{\omega}(t[\eta]))\,\mathrm{op}_M^{\rho-\frac{n}{2}}(f)(y)\omega_0(t[\eta])$$
$$\in R_G^0(U \times \mathbb{R}^q, \boldsymbol{g}).$$

An analogous result is true for the subclasses with weakly discrete asymptotics.

Proof. We have for $\lambda \geq 1$, $|\eta| \geq$ const the homogeneity $g(y, \lambda\eta) = \kappa_\lambda g(y, \eta)\kappa_\lambda^{-1}$. Moreover, because of the known behaviour of $g(y, \eta)$ for every fixed (y, η) as a Green operator in the cone algebra on $X^\wedge$, and in view of the smooth dependence of the Mellin symbols on (y, η) we also get the relations (3.3.18) for all $s \in \mathbb{R}$ and $k, l \in \mathbb{N}$. This gives us the assertion.   □

Theorem 3.3.28 *Given* $a(y,\eta) \in R^\nu_{M+G}(U \times \mathbb{R}^q, g)$ *for* $\nu \in \mathbb{R}$, $\quad g = (\gamma, \delta, \Theta)$, $b(y,\eta) \in R^\mu_{M+G}(U \times \mathbb{R}^q, c)$ *for* $\mu \in \mathbb{R}$, $\quad c = (\beta, \gamma, \Theta)$, *we have for the point-wise composition*

$$a(y,\eta)b(y,\eta) \in R^{\mu+\nu}_{M+G}(U \times \mathbb{R}^q, h) \quad for \quad h = (\beta, \delta, \Theta),$$

where

$$\sigma^{\nu+\mu}_\wedge (ab)(y,\eta) = \sigma^\nu_\wedge (a)(y,\eta) \sigma^\mu_\wedge (b)(y,\eta).$$

The conormal symbols satisfy

$$\sigma^{\nu+\mu-l}_M (ab)(y,z,\eta) = \sum_{p+r=l} \left(T^{\mu-r} \sigma^{\nu-p}_M (a) \right)(y,z,\eta) \sigma^{\mu-r}_M (b)(y,z,\eta)$$

(3.3.27)

for all $l \in \mathbb{N}$ *for which the right hand sides are meaningful, cf. Proposition 3.3.23. An analogous result holds for the subspaces with discrete asymptotics. If* $a(y,\eta)$ *or* $b(y,\eta)$ *is of Green type then so is the composition.*

Proof. From Proposition 3.3.20 we obtain that $a(y,\eta)b(y,\eta)$ is of Green type when $a(y,\eta)$ or $b(y,\eta)$ are Green symbols. Thus it suffices to discuss compositions of the form $m(y,\eta)n(y,\eta)$, where

$$m(y,\eta) = \omega(t[\eta])t^{-\nu+p} \operatorname{op}^{\rho-\frac{n}{2}}_M (h)(y)\eta^\alpha \omega_0(t[\eta])$$

for $\delta + \nu - p \le \rho \le \gamma$, $|\alpha| \le p$, $h(y,z) \in C^\infty(U, M^{-\infty}_R(X))$, carrier $R \cap \Gamma_{\frac{n+1}{2}-\rho} = \emptyset$, and

$$n(y,\eta) = \tilde\omega(t[\eta])t^{-\mu+r} \operatorname{op}^{\kappa-\frac{n}{2}}_M (f)(y)\eta^{\tilde\alpha} \tilde\omega_0(t[\eta])$$

for $\gamma + \mu - r \le \kappa \le \beta$, $|\tilde\alpha| \le r$, $f(y,z) \in C^\infty(U, M^{-\infty}_Q(X))$, carrier $Q \cap \Gamma_{\frac{n+1}{2}-\kappa} = \emptyset$. By Lemma 3.3.16 the specific choice of the cut-off functions is unimportant modulo Green remainders. Therefore we may assume $\tilde\omega_0 = \omega$ and $\omega_0\tilde\omega = \omega$. It is obviously sufficient to consider the case $\alpha = \tilde\alpha = 0$. For convenience we shall restrict ourselves to $\dim X = n = 0$. The calculations in general are completely analogous and left to the reader. Set $\tilde\nu = \nu - p$, $\tilde\mu = \mu - r$. Then it remains to analyse

$$(mn)(y,\eta) = \omega t^{-\tilde\nu+p} \operatorname{op}_M(T^{-\rho}h)\omega t^{-\rho-\tilde\mu+\kappa} \operatorname{op}_M(T^{-\kappa}f)t^{-\kappa}\omega. \quad (3.3.28)$$

From the conditions on the involved weights we know that $\rho + \tilde\mu \le \kappa$, i.e., $-\rho - \tilde\mu + \kappa \ge 0$. In order to obtain expressions of the required form from (3.3.28) we have to commute powers of t through the Mellin actions via Lemma 3.3.15 to get summands of the type

$$\omega t^{-(\tilde{\nu}+\tilde{\mu})} \operatorname{op}_M^\lambda(d_1)\omega \operatorname{op}_M^\lambda(d_2)\omega$$

and then to omit the factor ω in the middle, cf. Lemma 3.3.27, that only leads to a Green remainder. This gives then $\omega t^{-(\tilde{\nu}+\tilde{\mu})} \operatorname{op}_M^\lambda(d)\omega$ for a Mellin symbol $d(y,z) = d_1(y,z)d_2(y,z)$ with an asymptotic type P, carrier $P \cap \Gamma_{\frac{1}{2}-\lambda} = \emptyset$, and $\delta + \tilde{\mu} + \tilde{\nu} \le \lambda \le \beta$. In the case $\kappa = \rho + \tilde{\mu}$ we may set $\lambda = \kappa$, $\rho = \lambda - \tilde{\mu}$, and we get for (3.3.28)

$$(mn)(y,\eta) = \omega t^{-(\tilde{\nu}+\tilde{\mu})} \operatorname{op}_M^\lambda(T^{\tilde{\mu}}h)\omega \operatorname{op}_M^\lambda(f)\omega.$$
$$= \omega t^{-(\tilde{\nu}+\tilde{\mu})} \operatorname{op}_M^\lambda((T^{\tilde{\mu}}h)f)\omega$$

modulo Greens symbols. This is already the assertion in the case $\kappa = \rho + \tilde{\mu}$. For $\zeta := \kappa - \rho - \tilde{\mu} > 0$ we can choose an $\varepsilon > 0$, $\varepsilon < \zeta$ with carrier $R \cap \Gamma_{\frac{1}{2}-(\rho+\varepsilon)} = \emptyset$ for the asymptotic type R of h (recall that we consider the case $\dim X = 0$). Moreover, we can decompose f into a sum $f = f_1 + f_2$, $f_i(y,z) \in C^\infty(U, M_{Q_i}^{-\infty}(X))$, $i = 1,2$, with carrier $Q_1 \cap \Gamma_{\frac{1}{2}+(\zeta-\kappa)} = $ carrier $Q_2 \cap \Gamma_{\frac{1}{2}+(\zeta-\varepsilon-\kappa)} = \emptyset$. Then, in view of Lemma 3.3.15 and Lemma 3.3.27, the following equations are valid modulo Green symbols:

$$(mn)(y,\eta) = \omega t^{-\tilde{\nu}+\rho} \operatorname{op}_M(T^{-\rho}h)\omega t^\zeta \operatorname{op}_M(T^{-\kappa}f)t^{-\kappa}\omega$$
$$= \omega t^{-\tilde{\nu}+\rho} \operatorname{op}_M(T^{-\rho}h)\omega \operatorname{op}_M(T^{\zeta-\kappa}f_1)t^{\zeta-\kappa}\omega$$
$$\quad + \omega t^{-\tilde{\nu}+\rho} \operatorname{op}_M(T^{-\rho}h)\omega t^\varepsilon \operatorname{op}_M(T^{\zeta-\varepsilon-\kappa}f_2)t^{\zeta-\varepsilon-\kappa}\omega$$
$$= \omega t^{-\tilde{\nu}+\rho} \operatorname{op}_M(T^{-\rho}h)\omega \operatorname{op}_M(T^{\zeta-\kappa}f_1)t^{\zeta-\kappa}\omega$$
$$\quad + \omega t^{-\tilde{\nu}+\rho+\varepsilon} \operatorname{op}_M(T^{-\rho-\varepsilon}h)\omega \operatorname{op}_M(T^{\zeta-\varepsilon-\kappa}f_2)t^{\zeta-\varepsilon-\kappa}\omega$$
$$= \omega t^{-\tilde{\nu}+\rho} \operatorname{op}_M((T^{-\rho}h)(T^{\zeta-\kappa}f_1))t^{\zeta-\kappa}\omega$$
$$\quad + \omega t^{-\tilde{\nu}+\rho+\varepsilon} \operatorname{op}_M((T^{-\rho-\varepsilon}h)(T^{\zeta-\varepsilon-\kappa}f_2))t^{\zeta-\varepsilon-\kappa}\omega.$$

Writing $T^{-\rho}h = T^{\zeta-\kappa}T^{-\zeta+\kappa-\rho}h = T^{-\rho-\tilde{\mu}}T^{\tilde{\mu}}h$ and analogously

$$T^{-\rho-\varepsilon}h = T^{\zeta-\varepsilon-\kappa}T^{-\rho-\varepsilon-\zeta+\varepsilon+\kappa}h = T^{-\rho-\tilde{\mu}-\varepsilon}T^{\tilde{\mu}}h$$

we obtain modulo Green symbols

$$(mn)(y,\eta) = \omega t^{-(\tilde{\nu}+\tilde{\mu})}t^{\rho+\tilde{\mu}} \operatorname{op}_M(T^{-\rho-\tilde{\mu}}(T^{\tilde{\mu}}h)f_1)t^{-\rho-\tilde{\mu}}\omega$$
$$\quad + \omega t^{-(\tilde{\nu}+\tilde{\mu})}t^{\rho+\varepsilon+\tilde{\mu}} \operatorname{op}_M(T^{-\rho-\varepsilon-\tilde{\mu}}(T^{\tilde{\mu}}h)f_2)t^{-\rho-\varepsilon-\tilde{\mu}}\omega$$
$$= \omega t^{-(\tilde{\nu}+\tilde{\mu})} \left\{ \operatorname{op}_M^{\rho+\tilde{\mu}}((T^{\tilde{\mu}}h)f_1) + \operatorname{op}_M^{\rho+\tilde{\mu}+\varepsilon}(T^{\tilde{\mu}}h)f_2) \right\} \omega.$$

This shows $(mn)(y,\eta) \in R^{\mu+\nu}_{M+G}(U \times \mathbb{R}^q, \boldsymbol{h})$. Moreover, we have

$$\sigma^{\tilde{\nu}+\tilde{\mu}}_M(mn)(y,z,\eta) = (T^{\tilde{\mu}}h)(y,z,\eta)\{f_1 + f_2\}(y,z,\eta)$$
$$= (T^{\tilde{\mu}}h(y,z,\eta))f(y,z,\eta),$$

cf. (3.3.26), which yields the symbol relation (3.3.27). □

The formula (3.3.27) is called the Mellin translation product of the conormal symbols.

Analogously to Definition 3.3.6 we can introduce the space

$$\mathcal{R}^\nu_{M+G}(U \times \mathbb{R}^q, \boldsymbol{g}; N_-, N_+) \qquad (3.3.29)$$

of all block matrices

$$a(y,\eta) = \begin{pmatrix} a_{11} & a_{12} \\ a_{21} & a_{22} \end{pmatrix}(y,\eta),$$

where $a_{11}(y,\eta) \in R^\nu_{M+G}(U \times \mathbb{R}^q, \boldsymbol{g})$, and

$$\begin{pmatrix} 0 & a_{12} \\ a_{21} & a_{22} \end{pmatrix}(y,\eta) \in \mathcal{R}^\nu_G(U \times \mathbb{R}^q, \boldsymbol{g}; N_-, N_+).$$

Here $\boldsymbol{g} = (\gamma, \delta, \Theta)$, $\gamma, \delta, \nu \in \mathbb{R}$, $\Theta = (-k, 0]$, $k \in (\mathbb{N} \setminus \{0\}) \cup \{\infty\}$. Analogously we define the subspaces $\mathcal{R}^\nu_{M+G}(U \times \mathbb{R}^q, \boldsymbol{g}; N_-, N_+)^\bullet$ with weakly discrete asymptotics.

Theorem 3.3.29 *Given $a(y,\eta) \in \mathcal{R}^\nu_{M+G}(U \times \mathbb{R}^q, \boldsymbol{g}; N_0, N_+)$ for $\nu \in \mathbb{R}$, $\boldsymbol{g} = (\gamma, \delta, \Theta)$, $b(y,\eta) \in \mathcal{R}^\mu_{M+G}(U \times \mathbb{R}^q, \boldsymbol{c}; N_-, N_0)$ for $\mu \in \mathbb{R}$, $\boldsymbol{c} = (\beta, \gamma, \Theta)$, then for the point–wise composition we have*

$$a(y,\eta)b(y,\eta) \in \mathcal{R}^{\mu+\nu}_{M+G}(U \times \mathbb{R}^q, \boldsymbol{h}; N_-, N_+) \quad for \quad \boldsymbol{h} = (\beta, \delta, \Theta),$$

and

$$\sigma^{\nu+\mu}_\wedge(ab)(y,\eta) = \sigma^\nu_\wedge(a)(y,\eta)\sigma^\mu_\wedge(b)(y,\eta).$$

Moreover, the conormal symbols of the upper left corners satisfy the relations (3.3.27). An analogous result holds for the subspaces with discrete asymptotics. If $a(y,\eta)$ or $b(y,\eta)$ is of Green type then so is the composition.

Proof. Writing $a(y,\eta)$, $b(y,\eta)$ as block matrices $(a_{ij}(y,\eta))_{i,j=1,2}$ and $(b_{ij}(y,\eta))_{i,j=1,2}$, respectively then the only new terms of the compositions are $(a_{11}b_{12})(y,\eta)$ and $(a_{21}b_{11})(y,\eta)$. The other compositions were characterised in Proposition 3.3.10 and in the above Theorem 3.3.28.

Let us consider, for instance, $(a_{11}b_{12})(y, \eta)$. The remaining compositions can be treated in an analogous manner. Without loss of generality we set

$$a_{11}(y, \eta) = \omega(t[\eta]) t^{-\nu+j} \operatorname{op}_M^{\rho - \frac{n}{2}}(h)(y) \eta^{\alpha} \omega_0(t[\eta])$$

for any $j \in \mathbb{N}$, $|\alpha| \leq j$, $h(y, z) \in C^{\infty}(U, M_R^{-\infty}(X))$, $R \in \mathbf{As}(X)$, $\rho \in \mathbb{R}$, carrier $R \cap \Gamma_{\frac{n+1}{2} - \rho} = \emptyset$, $\delta + \nu - j \leq \rho \leq \gamma$. We have

$$a_{11}(y, \lambda\eta) = \lambda^{\nu - j + |\alpha|} \kappa_{\lambda} a_{11}(y, \eta) \kappa_{\lambda}^{-1}$$

for all $\lambda \geq 1$, $|\eta| \geq$ const for a constant > 0. Furthermore, write $b_{12}(y, \eta)$ as an asymptotic sum

$$b_{12}(y, \eta) \sim \sum_{l=0}^{\infty} b_{12,l}(y, \eta),$$

where $b_{12,l}(y, \eta)$ is a trace symbol of the type of an upper right corner with

$$b_{12,l}(y, \lambda\eta) = \lambda^{\mu - l} \kappa_{\lambda} b_{12,l}(y, \eta)$$

for all $\lambda \geq 1, |\eta| \geq$ const for a constant > 0. Then

$$(a_{11}b_{12,l})(y, \lambda\eta) = \lambda^{\mu + \nu - j + |\alpha| - l} \kappa_{\lambda}(a_{11}b_{12,l})(y, \eta)$$

for all $\lambda \geq 1$, $|\eta| \geq$ const. In addition we have

$$(a_{11}b_{12,l})(y, \eta) \in C^{\infty}(U \times \mathbb{R}^q, \mathcal{L}(\mathbb{C}^{N_-}, \mathcal{S}_P^{\delta}(X^{\wedge}))).$$

This yields that $(a_{11}b_{12,l})(y, \eta)$ belongs to (the space of upper right corners of) $\mathcal{R}_G^{\mu + \nu - j + |\alpha| - l}(U \times \mathbb{R}^q, \mathbf{h}; N_-, N_+)$. Moreover, we obviously have

$$(a_{11}b_{12})(y, \eta) \sim \sum_{l=0}^{\infty}(a_{11}b_{12,l})(y, \eta)$$

which can be carried out in (the space of upper right corners of) $\mathcal{R}_G^{\mu + \nu - j + |\alpha|}(U \times \mathbb{R}^q, \mathbf{h}; N_-, N_+)$, cf. Proposition 3.3.11. Analogous arguments apply for the weakly discrete asymptotics.   □

3.3.3   Edge symbols

This section will study specific subclasses of operator–valued symbols in $S^{\nu}(U \times \mathbb{R}^q; E, \tilde{E})$, the edge symbols. Here

$$E = \mathcal{K}^{s, \gamma}(X^{\wedge}) \oplus \mathbb{C}^{N_-}, \quad \tilde{E} = \mathcal{K}^{s - \nu, \gamma - \mu}(X^{\wedge}) \oplus \mathbb{C}^{N_+},$$

$s \in \mathbb{R}$, and $\kappa_\lambda(u \oplus c) = \lambda^{\frac{n+1}{2}} u(\lambda t, x) \oplus c$, $\lambda \in \mathbb{R}_+$ for $u \oplus c$ in E or $\tilde{E}$. According to the splitting of the spaces E, $\tilde{E}$ the symbols are (y, η)-dependent families of operator block matrices

$$
\begin{pmatrix} a & c_- \\ c_+ & d \end{pmatrix} (y, \eta) : \begin{matrix} \mathcal{K}^{s,\gamma}(X^\wedge) \\ \oplus \\ \mathbb{C}^{N_-} \end{matrix} \to \begin{matrix} \mathcal{K}^{s-\nu, \gamma-\mu}(X^\wedge) \\ \oplus \\ \mathbb{C}^{N_+} \end{matrix}
$$

where we shall assume that

$$
\begin{pmatrix} 0 & c_- \\ c_+ & d \end{pmatrix} (y, \eta) \in \mathcal{R}_G^\nu(U \times \mathbb{R}^q, \boldsymbol{g}; N_-, N_+) \qquad \text{for } \boldsymbol{g} = (\gamma, \gamma - \mu, \Theta)
$$

in the sense of Definition 3.3.1. Therefore we mainly concentrate here on the upper left corners $a(y, \eta)$. For convenience we often assume $U = \Omega \ni y$ for an open set $\Omega \subseteq \mathbb{R}^q$. The case $U = \Omega \times \Omega \ni (y, y')$ is completely analogous, cf. Remark 3.3.46 below. Moreover, in order to control the homogeneous components of symbols we restrict ourselves to classical interior symbols, though many elements of the theory can be generalised to the non-classical case.

For every operator family $a_\psi(y, \eta) \in C^\infty(\Omega, \tilde{L}_{cl}^\nu(X^\wedge; \mathbb{R}^q))$ there exists an $\tilde{h}(t, y, z, \tilde{\eta}) \in C^\infty(\overline{\mathbb{R}}_+ \times \Omega, M_O^\nu(X; \mathbb{R}_{\tilde{\eta}}^q))$ such that $h(t, y, z, \eta) = \tilde{h}(t, y, z, t\eta)$ satisfies

$$
\mathrm{op}_M^\beta(h)(y, \eta) - a_\psi(y, \eta) \in C^\infty(\Omega, L^{-\infty}(X^\wedge; \mathbb{R}^q)) \tag{3.3.30}
$$

for every $\beta \in \mathbb{R}$, cf. Theorem 3.2.7. Let us now choose cut–off functions $\omega(t)$, $\omega_i(t)$ and $\tilde{\omega}(t)$, $\tilde{\omega}_i(t)$, $i = 0, 1$, such that

$$
\omega\omega_0 = \omega, \quad \omega\omega_1 = \omega_1, \quad \tilde{\omega}\tilde{\omega}_0 = \tilde{\omega}, \quad \tilde{\omega}\tilde{\omega}_1 = \tilde{\omega}_1. \tag{3.3.31}
$$

Moreover, let $C^\infty(\Omega, L_{cl}^\nu(X^\wedge; \mathbb{R}^q)_0)$ be the subspace of all elements $p(y, \eta) \in C^\infty(\Omega, L_{cl}^\nu(X^\wedge; \mathbb{R}^q))$ for which there is a constant $c = c(p) > 0$ with $\varphi(t) p(y, \eta) \psi(t) = 0$ for all $\varphi, \psi \in C_0^\infty(\mathbb{R}_+)$ satisfying $\varphi = \psi = 0$ for all $0 < t < c$. Set

$$
C^\infty(\Omega, \tilde{L}_{cl}^\nu(X^\wedge; \mathbb{R}^q)_0) = C^\infty(\Omega, \tilde{L}_{cl}^\nu(X^\wedge; \mathbb{R}^q)) \cap C^\infty(\Omega, L_{cl}^\nu(X^\wedge; \mathbb{R}^q)_0).
$$

Definition 3.3.30 *Let* $\mu, \nu \in \mathbb{R}$, $\mu - \nu \in \mathbb{N}$, $\gamma \in \mathbb{R}$, *and fix the weight data* $\boldsymbol{g} = (\gamma, \gamma - \mu, \Theta)$ *for* $\Theta = (-k, 0]$, $k \in \mathbb{N} \setminus \{0\}$. *Then* $R^\nu(\Omega \times \mathbb{R}^q, \boldsymbol{g})$ *is the space of all operator families*

$$
a(y, \eta) = \tilde{\omega}(t)\{a_0(y, \eta) + a_1(y, \eta)\}\tilde{\omega}_0(t) + m(y, \eta) + g(y, \eta)
$$
$$
+ (1 - \tilde{\omega}(t))a_\infty(y, \eta)(1 - \tilde{\omega}_1(t)), \tag{3.3.32}
$$

where

$$(m + g)(y, \eta) \in R^{\nu}_{M+G}(\Omega \times \mathbb{R}^q, \boldsymbol{g}),$$
$$a_0(y, \eta) = \omega(t[\eta])t^{-\nu}a_M(y, \eta)\omega_0(t[\eta])$$

for $a_M(y, \eta) = \mathrm{op}_M^{\gamma - \frac{n}{2}}(h)(y, \eta)$, $a_1(y, \eta) = (1 - \omega(t[\eta]))t^{-\nu}a_{\psi}(y, \eta)(1 - \omega_1(t[\eta]))$, *for arbitrary* $a_{\psi}(y, \eta) \in C^{\infty}(\Omega_y, \tilde{L}^{\nu}_{\mathrm{cl}}(X^{\wedge}; \mathbb{R}^q_{\eta})_0)$, *cf. (3.2.7),*

$$h(t, y, z, \eta) = \tilde{h}(t, y, z, t\eta) \qquad (3.3.33)$$

for some $\tilde{h}(t, y, z, \tilde{\eta}) \in C^{\infty}(\overline{\mathbb{R}}_+ \times \Omega, M^{\nu}_O(X; \mathbb{R}^q_{\tilde{\eta}}))$, *satisfying*

$$\mathrm{op}_M^{\gamma - \frac{n}{2}}(h)(y, \eta) - a_{\psi}(y, \eta) \in C^{\infty}(\Omega_y, L^{-\infty}(X^{\wedge}; \mathbb{R}^q_{\eta})), \qquad (3.3.34)$$

and $a_{\infty}(y, \eta) \in C^{\infty}(\Omega_y, L^{\nu}_{\mathrm{cl}}(X^{\wedge}; \mathbb{R}^q_{\eta})_0)$. *The subclass of all* $a(y, \eta) \in R^{\nu}(\Omega \times \mathbb{R}^q, \boldsymbol{g})$, *with* $(m + g)(y, \eta) \in R^{\nu}_{M+G}(\Omega \times \mathbb{R}^q, \boldsymbol{g})^{\bullet}$ *will be denoted by* $R^{\nu}(\Omega \times \mathbb{R}^q, \boldsymbol{g})^{\bullet}$. *The elements of* $R^{\nu}(\Omega \times \mathbb{R}^q, \boldsymbol{g})$ $(R^{\nu}(\Omega \times \mathbb{R}^q, \boldsymbol{g})^{\bullet})$ *are called edge symbols with continuous (weakly discrete) asymptotics.*

If we set $\boldsymbol{g}_k = (\gamma, \gamma - \mu, \Theta_k)$ with $\Theta_k = (-k, 0]$, then we have natural embeddings

$$R^{\nu}(\Omega \times \mathbb{R}^q, \boldsymbol{g}_{k+1}) \hookrightarrow R^{\nu}(\Omega \times \mathbb{R}^q, \boldsymbol{g}_k)$$

for all k. For $\Theta = (-\infty, 0]$ and $\boldsymbol{g} = (\gamma, \gamma - \mu, \Theta)$ we then define

$$R^{\nu}(\Omega \times \mathbb{R}^q, \boldsymbol{g}) = \bigcap_{k \in \mathbb{N} \setminus \{0\}} R^{\nu}(\Omega \times \mathbb{R}^q, \boldsymbol{g}_k),$$

and analogously $R^{\nu}(\Omega \times \mathbb{R}^q, \boldsymbol{g})^{\bullet}$ for $\Theta = (-\infty, 0]$.

Remark 3.3.31 *For arbitrary* $a_{\psi}(y, \eta) \in C^{\infty}(\Omega, \tilde{L}^{\nu}_{\mathrm{cl}}(X^{\wedge}; \mathbb{R}^q)_0)$ *and cut–off functions* $\omega(t)$, $\omega_0(t)$ *and* $\omega_1(t)$ *satisfying* $\omega\omega_0 = \omega$, $\omega\omega_1 = \omega_1$, *we have*

$$\omega(t[\eta])a_{\psi}(y, \eta)(1 - \omega_0(t[\eta])), \ (1 - \omega(t[\eta]))a_{\psi}(y, \eta)\omega_1(t[\eta])$$
$$\in C^{\infty}(\Omega, L^{-\infty}(X^{\wedge}; \mathbb{R}^q)).$$

Moreover, the condition (3.3.34) implies

$$\omega(t[\eta])\{a_{\psi}(y, \eta) - a_M(y, \eta)\}\omega_0(t[\eta]) \in C^{\infty}(\Omega, L^{-\infty}(X^{\wedge}; \mathbb{R}^q)).$$

It follows that

$$a_{\psi}(y, \eta) - \{a_0(y, \eta) + a_1(y, \eta)\} \in C^{\infty}(\Omega, L^{-\infty}(X^{\wedge}; \mathbb{R}^q)).$$

Remark 3.3.32 *The operator family* $a_\psi(y, \eta)$ *can always be assumed to be of the form*

$$a_\psi(y, \eta) = \sum_{j=1}^{N} \varphi_j(1 \times \kappa_j^{-1})_* \operatorname{op}_{t,x}(\tilde{p}_j|_{\tilde\tau=t\tau, \tilde\eta=t\eta})(y, \eta)\psi_j \qquad (3.3.35)$$

with respect to a fixed covering of X by coordinate neighbourhoods $\{U_1, \ldots, U_N\}$, charts $\kappa_j : U_j \to \Sigma_j \subseteq \mathbb{R}^n$, a subordinate partition of unity $\{\varphi_1, \ldots, \varphi_N\}$ on X, functions $\psi_j \in C_0^\infty(U_j)$ with $\varphi_j\psi_j = \varphi_j$ for all j, and local symbols

$$\tilde{p}_j(t, x, y, \tilde\tau, \xi, \tilde\eta) \in S_{\mathrm{cl}}^\nu(\overline{\mathbb{R}}_+ \times \Sigma_j \times \Omega \times \mathbb{R}^{1+n+q}). \qquad (3.3.36)$$

Recall that when the symbols $\tilde{p}_j(t, x, y, \tilde\tau, \xi, \tilde\eta)$ are independent of t the functions $\tilde{h}(t, y, z, \tilde\eta)$ involved in (3.3.30) can (and will) also be chosen to be independent of t. For future references we call this the

$$\{\text{case of frozen coefficients at } t = 0\}, \qquad (3.3.37)$$

cf. also Remark 3.2.10. Starting from a general system of symbols $\tilde{p}_j(t, x, y, \tilde\tau, \xi, \tilde\eta)$, $j = 1, \ldots, N$, and an associated operator function $\tilde{h}(t, y, z, \tilde\eta)$, freezing coefficients at $t = 0$ gives us

$$c_\wedge a_\psi(y, \eta) := \sum_{j=1}^{N} \varphi_j(1 \times \kappa_j^{-1})_* \operatorname{op}_{t,x}(\tilde{p}_j|_{t=0, \tilde\tau=t\tau, \tilde\eta=t\eta})(y, \eta)\psi_j \quad (3.3.38)$$

and

$$c_\wedge a_M(y, \eta) := \operatorname{op}_M^{\gamma - \frac{n}{2}}(\tilde{h}|_{t=0, \tilde\eta=t\eta})(y, \eta). \qquad (3.3.39)$$

Let us set

$$c_\wedge a_0(y, \eta) = \omega(t[\eta])t^{-\nu}c_\wedge a_M(y, \eta)\omega_0(t[\eta]), \qquad (3.3.40)$$

$$c_\wedge a_1(y, \eta) = (1 - \omega(t[\eta]))t^{-\nu}c_\wedge a_\psi(y, \eta)(1 - \omega_1(t[\eta])). \qquad (3.3.41)$$

Then we have

$$c_\wedge(a_0 + a_1)(y, \lambda\eta) = \lambda^\nu \kappa_\lambda c_\wedge(a_0 + a_1)(y, \eta)\kappa_\lambda^{-1} \qquad (3.3.42)$$

for all $\lambda \geq 1$ and $|\eta| \geq$ const for a constant > 0.

Proposition 3.3.33 *Let ω', ω_0', ω_1' be another system of cut–off functions on $\mathbb{R}_+$ with $\omega'\omega_0' = \omega'$, $\omega'\omega_1' = \omega_1'$. Then, if $a_0(y,\eta)$, $a_1(y,\eta)$ is given as in Definition 3.3.30, and if we set*

$$a_0'(y,\eta) = \omega'(t[\eta])t^{-\nu}a_M(y,\eta)\omega_0'(t[\eta]),$$
$$a_1'(y,\eta) = (1 - \omega'(t[\eta]))t^{-\nu}a_\psi(y,\eta)(1 - \omega_1'(t[\eta])),$$

we have

$$g(y,\eta) := \{a_0(y,\eta) + a_1(y,\eta)\} - \{a_0'(y,\eta) + a_1'(y,\eta)\} \in R_G^\nu(\Omega \times \mathbb{R}^q, g)^\bullet.$$

Proof. From the cone theory of Chapter 2 it follows that $g(y,\eta)$ is an element of $C_G(X^\wedge, g)^\bullet$ for every fixed $(y,\eta) \in \Omega \times \mathbb{R}^q$. So the only property to be verified is that $g(y,\eta)$ and $g^*(y,\eta)$ are classical operator–valued symbols in the sense of Definition 3.3.1. Let us first assume that the symbols involved in $a_\psi(y,\eta)$ and $a_M(y,\eta)$ are independent of the first variable t such that we have (3.3.37). Then (3.3.42) together with the analogous homogeneity of $a_0'(y,\eta) + a_1'(y,\eta)$ yields $g(y,\lambda\eta) = \lambda^\nu \kappa_\lambda g(y,\eta)\kappa_\lambda^{-1}$ for all $\lambda \geq 1$, $|\eta| \geq$ const, and hence, $g(y,\eta)$ is classical. For the adjoints we can argue in an analogous manner. For the t-dependent case we may assume that the support in t of the involved symbols (3.3.36) as well as of $\tilde{h}(t,y,z,\tilde{\eta})$ is bounded. Then, for every $\varphi(t) \in C_0^\infty(\overline{\mathbb{R}}_+)$ that equals 1 on that support, we obtain by Taylor expansion of the $\tilde{p}_j$ and of $\tilde{h}$ in t near zero

$$g(y,\eta) = \varphi(t)\{\sum_{j=0}^{N} t^j g_j(y,\eta)\} + \varphi(t)t^{N+1}g_{(N)}(y,\eta) \qquad (3.3.43)$$

for every N, where $g_j(y,\eta)$ is as in the first step of the proof, namely a Green symbol that is homogeneous in η of order ν for large $|\eta|$; in the remainder term $\varphi(t)g_{(N)}(y,\eta)$ is operator–valued of order ν, point–wise Green in the cone sense on $X^\wedge$, but not necessarily classical. However $\varphi(t)t^{N+1}g_{(N)}(y,\eta)$ is then of order $\nu - N - 1$. Since $\varphi(t)t^j g_j(y,\eta)$ is a Green symbol of order $\nu - j$, cf. Propositions 3.3.9 and 3.3.12, the formula (3.3.43) is an asymptotic expansion of $g(y,\eta)$ with respect to Green symbols of orders tending to $-\infty$ for $N \to \infty$. Thus, in view of Proposition 3.3.11, we obtain the assertion. $\qquad\square$

Proposition 3.3.34 *Let $a_\psi(y,\eta) \in C^\infty(\Omega, \tilde{L}_{cl}^\nu(X^\wedge; \mathbb{R}^q)_0)$ and suppose both $h(t,y,z,\eta) = \tilde{h}(t,y,z,t\eta)$ and $h'(t,y,z,\eta) = \tilde{h}'(t,y,z,t\eta)$ for $\tilde{h}(t,y,z,\tilde{\eta})$, $\tilde{h}'(t,y,z,\tilde{\eta}) \in C^\infty(\overline{\mathbb{R}}_+ \times \Omega, M_O^\nu(X; \mathbb{R}_{\tilde{\eta}}^q))$ satisfy the relation (3.3.34). Then, for $a_M'(y,\eta) = \mathrm{op}_M^{\gamma-\frac{n}{2}}(h')(y,\eta)$ we have*

$$\omega(t[\eta])t^{-\nu}\{a_M(y,\eta) - a_M'(y,\eta)\}\omega_0(t[\eta]) \in R_{M+G}^\nu(\Omega \times \mathbb{R}^q, g)^\bullet$$

for $g = (\gamma, \gamma - \mu, \Theta)$, $\Theta = (-\infty, 0]$.

Proof. First we have

$$\tilde{h}(t, y, z, \tilde{\eta}) - \tilde{h}'(t, y, z, \tilde{\eta}) \in C^{\infty}(\overline{\mathbb{R}}_+ \times \Omega, M_O^{-\infty}(X; \mathbb{R}^q_{\tilde{\eta}})).$$

Then, analogously to Proposition 3.3.26, we obtain the assertion by applying the Taylor expansion to $(h - h')(t, y, z, t\eta)$ with respect to t, both with respect to the first t-variable as well as for t in the combination $t\eta$. □

We now discuss the symbolic structure of the operator families $a(y, \eta) \in R^{\nu}(\Omega \times \mathbb{R}^q, \boldsymbol{g})$, $\boldsymbol{g} = (\gamma, \gamma - \mu, \Theta)$. First

$$R^{\nu}(\Omega \times \mathbb{R}^q, \boldsymbol{g}) \subset C^{\infty}(\Omega, L^{\nu}_{cl}(X^{\wedge}; \mathbb{R}^q)) \qquad (3.3.44)$$

implies that we have a system of complete symbols in $S^{\nu}_{cl}(\mathbb{R}_+ \times \Sigma_j \times \Omega \times \mathbb{R}^{1+n+q})$ $j = 1, \ldots, N$. By definition they can be chosen in the form $t^{-\nu}p_j(t, x, y, \tau, \xi, \eta)$, where

$$p_j(t, x, y, \tau, \xi, \eta) = \tilde{p}_j(t, x, y, t\tau, \xi, t\eta) \qquad (3.3.45)$$

for certain $\tilde{p}_j(t, x, y, \tilde{\tau}, \xi, \tilde{\eta}) \in S^{\nu}_{cl}(\overline{\mathbb{R}}_+ \times \Sigma_j \times \Omega \times \mathbb{R}^{1+n+q})$, $j = 1, \ldots, N$.

The homogeneous principal symbols of $t^{-\nu}p_j(t, x, y, \tau, \xi, \eta)$ of order ν, $j = 1, \ldots, N$, represent a function $\sigma^{\nu}_{\psi}(a)(t, x, y, \tau, \xi, \eta)$ in the space $S^{(\nu)}(T^*(X^{\wedge} \times \Omega) \backslash 0)$, the subspace of all elements of $C^{\infty}(T^*(X^{\wedge} \times \Omega) \backslash 0)$ that are (positively) homogeneous in $(\tau, \xi, \eta) \neq 0$ of order ν. Without loss of generality we have assumed here that the contributions from the local representatives in the sense of Remark 3.3.32 are invariant under the transition maps for principal symbols with respect to the coordinate diffeomorphisms from the system of charts on X. By definition also the contribution from $a_{\infty}(y, \eta)$ is involved here. In view of (3.3.45) there is an element

$$\sigma^{\nu}_{\psi,b}(a)(t, x, y, \tilde{\tau}, \xi, \tilde{\eta}) \in S^{(\nu)}(T^*(\overline{\mathbb{R}}_+ \times X \times \Omega) \backslash 0) \qquad (3.3.46)$$

$(C^{\infty}$ in t up to $t = 0)$ such that

$$\sigma^{\nu}_{\psi}(a)(t, x, y, \tau, \xi, \eta) = t^{-\nu}\sigma^{\nu}_{\psi,b}(a)(t, x, y, t\tau, \xi, t\eta). \qquad (3.3.47)$$

For purposes below we set

$$\Sigma^{(\nu)}_{\psi} = \{t^{-\nu}\tilde{p}_{(\nu)}(t, x, y, t\tau, \xi, t\eta) : \tilde{p}_{(\nu)}(t, x, y, \tilde{\tau}, \xi, \tilde{\eta})$$
$$\in S^{(\nu)}(T^*(\overline{\mathbb{R}}_+ \times X \times \Omega) \backslash 0)\}.$$

Furthermore, with $a(y, \eta)$ there are associated the operator families $c_{\wedge}a_{\psi}(y, \eta)$ and $c_{\wedge}a_M(y, \eta)$, cf. (3.3.38), (3.3.39), satisfying the relations

$$c_{\wedge}a_{\psi}(y, \eta) - c_{\wedge}a_M(y, \eta) \in C^{\infty}(\Omega, L^{-\infty}(X^{\wedge}; \mathbb{R}^q))$$

and

$$(t^{-\nu}c_\wedge a_\psi)(y, \lambda\eta) = \lambda^\nu \kappa_\lambda (c_\wedge a_\psi)(y, \eta)\kappa_\lambda^{-1},$$
$$(t^{-\nu}c_\wedge a_M)(y, \lambda\eta) = \lambda^\nu \kappa_\lambda (t^{-\nu}c_\wedge a_M)(y, \eta)\kappa_\lambda^{-1}$$

for all $\lambda \in \mathbb{R}_+$, $\eta \in \mathbb{R}^q$.

Definition 3.3.35 *Let $a(y, \eta) \in R^\nu(\Omega \times \mathbb{R}^q, g)$, written in the form (3.3.32) for $g = (\gamma, \gamma - \mu, \Theta)$. Then*

$$\sigma_\psi^\nu(a) \in S^{(\nu)}(T^*(X^\wedge \times \Omega) \setminus 0)$$

is called the homogeneous principal interior symbol of $a(y, \eta)$ of order ν, and

$$\sigma_\wedge^\nu(a)(y, \eta) = \omega(t|\eta|)t^{-\nu}c_\wedge a_M(y, \eta)\omega_0(t|\eta|)$$
$$+ (1 - \omega(t|\eta|))t^{-\nu}c_\wedge a_\psi(y, \eta)(1 - \omega_1(t|\eta|))$$
$$+ \sigma_\wedge^\nu(m + g)(y, \eta)$$

for $\eta \neq 0$ the homogeneous principal edge symbol of $a(y, \eta)$ of order ν.

$\sigma_\wedge^\nu(a)(y, \eta)$ will be regarded as an operator family

$$\sigma_\wedge^\nu(a)(y, \eta) : \mathcal{K}^{s,\gamma}(X^\wedge) \to \mathcal{K}^{s-\nu,\gamma-\mu}(X^\wedge)$$

for every $s \in \mathbb{R}$. It satisfies the homogeneity relation

$$\sigma_\wedge^\nu(a)(y, \lambda\eta) = \lambda^\nu \kappa_\lambda \sigma_\wedge^\nu(a)(y, \eta)\kappa_\lambda^{-1} \qquad (3.3.48)$$

for all $y \in \Omega$, $\eta \in \mathbb{R}^q \setminus \{0\}$, $\lambda \in \mathbb{R}_+$.

The space of all operator families $c_\wedge(a_0 + a_1)(y, \eta)$, cf. (3.3.40), (3.3.41), belonging to certain $a(y, \eta) \in R^\nu(\Omega \times \mathbb{R}^q, g)$, is a subspace of $C^\infty(\Omega, L_{cl}^\nu(X^\wedge; \mathbb{R}^q))$. Therefore, $c_\wedge(a_0 + a_1)(y, \eta)$ has a well-defined parameter-dependent homogeneous principal symbol of order ν of the form $t^{-\nu}\sigma_{\psi,b}^\nu(a)(0, x, y, t\eta, \xi, t\eta)$, cf. (3.3.47). It is evident by construction that every function of this form is generated by a suitable choice of $(a_0 + a_1)(y, \eta)$. Let us set

$$\sigma_\psi^\nu \sigma_\wedge^\nu(a)(t, x, y, \tau, \xi, \eta) := t^{-\nu}\sigma_{\psi,b}^\nu(a)(0, x, y, t\tau, \xi, t\eta)$$
$$=: c_\wedge \sigma_\psi^\nu(a)(t, x, y, \tau, \xi, \eta). \qquad (3.3.49)$$

This defines the subspace $c_\wedge \Sigma_\psi^{(\nu)} \subset \Sigma_\psi^{(\nu)}$. For the space of homogeneous principal edge symbols

$$\Sigma_\wedge^{(\nu)}(g) := \{\sigma_\wedge^\nu(a)(y, \eta) : a(y, \eta) \in R^\nu(\Omega \times \mathbb{R}^q, g)\} \qquad (3.3.50)$$

we now have a surjective operator $\sigma_\psi^\nu : \Sigma_\wedge^{(\nu)}(g) \to c_\wedge \Sigma_\psi^{(\nu)}$. Then, setting

$$\Sigma^{(\nu)}(g) := \{(\sigma_\psi^\nu(a), \sigma_\wedge^\nu(a)) : a(y, \eta) \in R^\nu(\Omega \times \mathbb{R}^q, g)\}, \qquad (3.3.51)$$

it follows that

$$\Sigma^{(\nu)}(g) = \{(p_\psi, p_\wedge) \in \Sigma_\psi^{(\nu)} \times \Sigma_\wedge^{(\nu)}(g) : \sigma_\psi^\nu p_\wedge = c_\wedge p_\psi\}. \qquad (3.3.52)$$

Analogous relations can be formulated for $R^\nu(\Omega \times \mathbb{R}^q, g)^\bullet$, where we write $\Sigma_\wedge^{(\nu)}(g)^\bullet$ and $\Sigma^{(\nu)}(g)^\bullet$ instead of (3.3.50) and (3.3.51), respectively.

Remark 3.3.36 *For every $j \in \mathbb{N}$ we have a canonical embedding*

$$\iota : R^{\nu-j}(\Omega \times \mathbb{R}^q, g) \hookrightarrow R^\nu(\Omega \times \mathbb{R}^q, g),$$

and the same for the weakly discrete asymptotics.

Proposition 3.3.37 *Let us set $\sigma(a) = (\sigma_\psi^\nu(a), \sigma_\wedge^\nu(a))$ for $a(y, \eta) \in R^\nu(\Omega \times \mathbb{R}^q, g)$. Then the following sequence is exact:*

$$0 \to R^{\nu-1}(\Omega \times \mathbb{R}^q, g) \overset{\iota}{\hookrightarrow} R^\nu(\Omega \times \mathbb{R}^q, g) \overset{\sigma}{\to} \Sigma^\nu(g) \to 0.$$

Moreover, there is a linear operator

$$\mathrm{op} : \Sigma^{(\nu)}(g) \to R^\nu(\Omega \times \mathbb{R}^q, g)$$

with $\sigma \circ \mathrm{op} = \mathrm{id}$ on $\Sigma^{(\nu)}(g)$. An analogous result holds for the subclasses with weakly discrete asymptotics.

Proof. The surjectivity of σ is clear by definition. Moreover, $\sigma(a) = (\sigma_\psi^\nu(a), \sigma_\wedge^\nu(a))$ vanishes for $a(y, \eta) \in R^{\nu-1}(\Omega \times \mathbb{R}^q, g)$. In fact, the homogeneous principal interior symbol of order ν of $a(y, \eta)$ vanishes because of (3.3.44), applied for $\nu - 1$, and $\sigma_\wedge^\nu(a)$ vanishes, since the evaluation of the operator families under freezing of coefficients gives zero because of the extra power of t compared with the weight factor $t^{-\nu}$. Thus $R^{\nu-1}(\Omega \times \mathbb{R}^q, g) \subseteq \ker \sigma$. On the other hand $a(y, \eta) \in \ker \sigma$ implies $\sigma_\psi^\nu(a) = 0$ and $\sigma_\wedge^\nu(a) = 0$. This shows, in particular, that the interior order is $\leq \nu - 1$. Moreover, we have flatness of $a_M(y, \eta) + t^\nu m(y, \eta)$ and $a_\psi(y, \eta)$ in t at $t = 0$ of order 1, such that $a(y, \eta) \in R^{\nu-1}(\Omega \times \mathbb{R}^q, g)$. Thus the sequence is exact.

In order to construct a map op, applied to $(p_\psi, p_\wedge) \in \Sigma^{(\nu)}(g)$, that is a right inverse to σ, we start with $p_\psi \in \Sigma_\psi^{(\nu)}$ and form symbols (3.3.36), $j = 1, \dots, N$, for which the homogeneous principal parts of order ν of $t^{-\nu} \tilde{p}_j(t, x, y, t\tau, \xi, t\eta)$ are the local representatives of p_ψ in the corresponding coordinates. Then we can form the operator families

$a_\psi(y,\eta)$, cf. (3.3.35), and $a_\infty(y,\eta)$. Moreover, we associate to $a_\psi(y,\eta)$ an $a_M(y,\eta) = \mathrm{op}_M^{\gamma-\frac{n}{2}}(h)(y,\eta)$ satisfying (3.3.34), like in the general constructions for Definition 3.3.30. Using the notation of Definition 3.3.30 we then obtain an operator family

$$\tilde{a}(y,\eta) = \tilde{\omega}(t)\{a_0(y,\eta) + a_1(y,\eta)\}\tilde{\omega}_0(t) + (1 - \tilde{\omega}(t))a_\infty(y,\eta)(1 - \tilde{\omega}_1(t))$$

that belongs to $R^\nu(\Omega \times \mathbb{R}^q, \boldsymbol{g})$, where $\sigma_\psi^\nu(\tilde{a}) = p_\psi$. Moreover,

$$p_\wedge(y,\eta) - \sigma_\wedge^\nu(\tilde{a})(y,\eta) =: p_{\wedge,\infty}(y,\eta) \in \sigma_\wedge^\nu R_{M+G}^\nu(\Omega \times \mathbb{R}^q, \boldsymbol{g}),$$

cf. also Proposition 3.3.34. It suffices now to choose an $(m+g)(y,\eta) \in R_{M+G}^\nu(\Omega \times \mathbb{R}^q, \boldsymbol{g})$ with $p_{\wedge,\infty}(y,\eta) = \sigma_\wedge^\nu(m+g)(y,\eta)$ which is trivial. Then, since $\sigma_\psi^\nu(m+g) = 0$, we may set $\mathrm{op}(p_\psi, p_\wedge) = \tilde{a}(y,\eta) + (m+g)(y,\eta)$. The construction for the weakly discrete asymptotics is obviously analogous. $\qquad\square$

Let us now recall that both $a(y,\eta) \in R^\nu(\Omega \times \mathbb{R}^q, \boldsymbol{g})$ and $\sigma_\wedge^\nu(a)(y,\eta)$ are (y,η)-dependent families of operators in the cone class $C^\nu(X^\wedge, \boldsymbol{g})$, cf. Section 2.4.1. Thus we can apply the symbolic maps of the cone theory to $a(y,\eta)$ and $\sigma_\wedge^\nu(a)(y,\eta)$, respectively. In particular, we have the conormal symbols which are responsible for the behaviour of the operators with respect to the asymptotics. The conormal symbols of $a(y,\eta)$ are

$$\sigma_M^{\nu-j}(a)(y,z,\eta) = \frac{1}{j!}\frac{\partial^j}{\partial t^j}h(t,y,z,\eta)|_{t=0} + \sigma_M^{\nu-j}(m)(y,z,\eta),$$

$j = 0,\ldots,k-(\mu-\nu)-1$ for $\Theta = (-k,0]$, where $m(y,\eta)$ is the smoothing Mellin part of $a(y,\eta)$, cf. Definition 3.3.30, and $h(t,y,z,\eta)$ the operator–valued Mellin symbol involved in the definition of $a_M(y,\eta)$, cf. the notation (3.3.33). This shows, in particular, that $\sigma_M^{\nu-j}(a)(y,z,\eta)$ is a polynomial in η of degree $\leq j$, and that $\sigma_M^\nu(a)(y,z)$ is independent of η.

Proposition 3.3.38 $a(y,\eta) \in R^\nu(\Omega \times \mathbb{R}^q, \boldsymbol{g})$ for $\boldsymbol{g} = (\gamma, \gamma - \mu, \Theta)$ *implies*
$$a(y,\eta) \in S^\nu(\Omega \times \mathbb{R}^q; \mathcal{K}^{s,\gamma}(X^\wedge), \mathcal{K}^{s-\nu,\gamma-\mu}(X^\wedge))$$

for each $s \in \mathbb{R}$. *Moreover, for every* $P \in \mathrm{As}(X, (\gamma, \Theta))$ *there is a* $Q \in \mathrm{As}(X, (\gamma - \mu, \Theta))$ *such that*

$$a(y,\eta) \in S^\nu(\Omega \times \mathbb{R}^q; \mathcal{K}_P^{s,\gamma}(X^\wedge), \mathcal{K}_Q^{s-\nu,\gamma-\mu}(X^\wedge))$$

for each $s \in \mathbb{R}$. *For* $a(y,\eta) \in R^\nu(\Omega \times \mathbb{R}^q, \boldsymbol{g})^\bullet$ *we obtain the same for every* $P \in \mathrm{As}(X, (\gamma, \Theta))^\bullet$ *with a corresponding resulting* $Q \in \mathrm{As}(X, (\gamma - \mu, \Theta))^\bullet$.

Proof. First we have

$$(1 - \tilde{\omega}(t))a_\infty(y,\eta)(1 - \tilde{\omega}_1(t)) \in S^\nu(\Omega \times \mathbb{R}^q; \mathcal{K}^{s,\gamma}(X^\wedge), \mathcal{K}^{s-\nu,\delta}(X^\wedge))$$

for all $s \in \mathbb{R}$ and arbitrary $\gamma, \delta \in \mathbb{R}$, cf. Theorem 3.2.17. Moreover, the operator of multiplication by a cut–off function $\tilde{\omega}(t)$ belongs to $S^0(\mathbb{R}^q; \mathcal{K}^{s,\gamma}(X^\wedge), \mathcal{K}^{s,\gamma}(X^\wedge))$ as well as to $S^0(\mathbb{R}^q; \mathcal{K}_P^{s,\gamma}(X^\wedge), \mathcal{K}_P^{s,\gamma}(X^\wedge))$ for $s, \gamma \in \mathbb{R}$ and every asymptotic type P. Next, applying Proposition 3.3.20, we may restrict ourselves to $a(y,\eta) = a_0(y,\eta) + a_1(y,\eta)$ in the sense of the notation of Definition 3.3.30. For (3.3.37) we obtain

$$a_0(y,\eta), a_1(y,\eta) \in S_{cl}^\nu(\Omega \times \mathbb{R}^q; \mathcal{K}^{s,\gamma}(X^\wedge), \mathcal{K}^{s-\nu,\gamma-\mu}(X^\wedge))$$

because of (3.3.42). At the same time it is easy to see that when the symbols $\tilde{p}_j(x,y,\tilde{\tau},\xi,\tilde{\eta})$, $j = 1, \ldots, N$, run over sequences that tend to zero in the corresponding symbol spaces and when operator–valued symbols $\tilde{h}$ tend to zero in $C^\infty(\Omega, M_O^\nu(X; \mathbb{R}^q))$, then also $a_1(y,\eta)$ and $a_0(y,\eta)$ tend to zero in $S_{cl}^\nu(\Omega \times \mathbb{R}^q; \mathcal{K}^{s,\gamma}(X^\wedge), \mathcal{K}^{s-\nu,\gamma-\mu}(X^\wedge))$ for all s. This allows us to apply a tensor product argument for the general t-dependent case. Here we also use that the operator of multiplication by an arbitrary $\varphi(t) \in C_0^\infty(\overline{\mathbb{R}}_+)$ is an operator–valued symbol in $S^0(\mathbb{R}^q; \mathcal{K}^{s,\delta}(X^\wedge), \mathcal{K}^{s,\delta}(X^\wedge))$ for every s, δ, tending to zero in this symbol space when φ tends to zero in $C_0^\infty(\overline{\mathbb{R}}_+)$.

The assertion for the spaces with asymptotics is an easy consequence of Proposition 3.3.20 and of a corresponding simpler result for $a_0(y,\eta)$ and $a_1(y,\eta)$. We have $a_1(y,\eta) \in S^\nu(\Omega \times \mathbb{R}^q; \mathcal{K}^{s,\gamma}(X^\wedge), \mathcal{K}^{s-\nu,\infty}(X^\wedge))$ since there appears flatness of infinite order at $t = 0$ under the mappings, and $a_0(y,\eta)$ can be treated along the lines of the proof of Proposition 3.3.20. $\qquad\Box$

Proposition 3.3.39 *We have $D_y^\alpha D_\eta^\beta R^\nu(\Omega \times \mathbb{R}^q, \boldsymbol{g}) \subseteq R^{\nu-|\beta|}(\Omega \times \mathbb{R}^q, \boldsymbol{g})$ for every $\alpha, \beta \in \mathbb{N}^q$. Analogous relations hold for edge symbols with weakly discrete asymptotics.*

Proof. Applying Remark 3.3.25 and in view of the obvious behaviour of $a_\infty(y,\eta)$ under differentiations we may concentrate on $a(y,\eta) = \tilde{\omega}(t)\{a_0(y,\eta) + a_1(y,\eta)\}\tilde{\omega}_0(t)$, cf. the notation of Definition 3.3.30. The discussion of y-derivatives is trivial and will be omitted. Let us consider a first order derivative D_{η_j}, $0 \leq j \leq q$. Then $D_{\eta_j}a(y,\eta) =$

$A(y, \eta) + C(y, \eta) + G(y, \eta)$ for

$$
\begin{aligned}
A(y, \eta) &= \tilde{\omega}(t)\{\omega(t[\eta])[t^{-\nu}D_{\eta_j}a_M(y, \eta)]\omega_0(t[\eta]) \\
&\quad + (1 - \omega(t[\eta]))[D_{\eta_j}t^{-\nu}a_\psi(y, \eta)](1 - \omega_1(t[\eta]))\}\tilde{\omega}_0(t), \\
C(y, \eta) &= \tilde{\omega}(t)\{\omega(t[\eta])t^{-\nu}a_M(y, \eta)D_{\eta_j}\omega_0(t[\eta]) \\
&\quad + (1 - \omega(t[\eta]))t^{-\nu}a_\psi(y, \eta)D_{\eta_j}(1 - \omega_1(t[\eta]))\}\tilde{\omega}_0(t), \\
G(y, \eta) &= \tilde{\omega}(t)\{[D_{\eta_j}\omega(t[\eta])]t^{-\nu}a_M(y, \eta)\omega_0(t[\eta]) \\
&\quad + [D_{\eta_j}(1 - \omega(t[\eta]))]t^{-\nu}a_\psi(y, \eta)(1 - \omega_1(t[\eta]))\}\tilde{\omega}_0(t).
\end{aligned}
$$

First we have $A(y, \eta) \in R^{\nu-1}(\Omega \times \mathbb{R}^q, \boldsymbol{g})$, since the relation (3.3.34) can be differentiated with respect to η_j which produces an additional factor t. It remains to characterise $C(y, \eta)$, $G(y, \eta)$, and we shall see that they belong to $R_G^{\nu-1}(\Omega \times \mathbb{R}^q, \boldsymbol{g})$. By virtue of $\omega_0 \equiv 1$ on $\operatorname{supp}\omega$, $1 - \omega_1 \equiv 1$ on $\operatorname{supp}(1 - \omega)$, we have $\omega(t[\eta])D_{\eta_j}\omega_0(t[\eta]) = 0$, $(1 - \omega(t[\eta]))D_{\eta_j}(1 - \omega_1(t[\eta])) = 0$. Thus $C(y, \eta)$ is a family of operators which are point–wise Green in the cone operator algebra on $X^\wedge$ with the weight data $\boldsymbol{g}$. Let us write $C(y, \eta) = \tilde{\omega}(t)C_0(y, \eta)\tilde{\omega}_0(t)$ for

$$
\begin{aligned}
C_0(y, \eta) &= \omega(t[\eta])t^{-\nu}a_M(y, \eta)D_{\eta_j}\omega_0(t[\eta]) \\
&\quad + (1 - \omega(t[\eta]))t^{-\nu}a_\psi(y, \eta)D_{\eta_j}(1 - \omega_1(t[\eta])).
\end{aligned}
$$

It suffices to show that $C_0(y, \eta) \in R_G^{\nu-1}(\Omega \times \mathbb{R}^q, \boldsymbol{g})$, since then also $C(y, \eta) \in R_G^{\nu-1}(\Omega \times \mathbb{R}^q, \boldsymbol{g})$. For (3.3.37) we have

$$
C_0(y, \lambda\eta) = \lambda^{\nu-1}\kappa_\lambda C_0(y, \eta)\kappa_\lambda^{-1}
$$

for $\lambda \geq 1$, $|\eta| \geq \operatorname{const}$ for a constant > 0. For general dependence on t we can apply Taylor expansions with respect to t near 0 to obtain $C_0(y, \eta)$ as an asymptotic sum of Green symbols of orders $\nu - 1 - j$, $j \in \mathbb{N}$. In other words, it follows that $C_0(y, \eta) \in R_G^{\nu-1}(\Omega \times \mathbb{R}^q, \boldsymbol{g})$ also in this case. For $G(y, \eta)$ we can argue as follows. Setting $\varphi(t, \eta) = D_{\eta_j}\omega(t[\eta])$ we find a function $\psi(t) \in C_0^\infty(\mathbb{R}_+)$ such that $\varphi(t, \eta)\psi(t[\eta]) = \varphi(t, \eta)$ for all $t \in \mathbb{R}_+$, $\eta \in \mathbb{R}^q$. Then, by usual arguments, first for (3.3.37) and then by Taylor expansions in general, we obtain that

$$
\begin{aligned}
G_2(y, \eta) &:= \tilde{\omega}(t)\{\varphi(t, \eta)t^{-\nu}a_M(y, \eta)(\omega_0(t[\eta]) - \psi(t[\eta])) \\
&\quad 1 - \varphi(t, \eta)t^{-\nu}a_\psi(y, \eta)(1 - \omega_1(t[\eta]) - \psi(t[\eta]))\}\tilde{\omega}_0(t)
\end{aligned}
$$

belongs to $R_G^{\nu-1}(\Omega \times \mathbb{R}^q, \boldsymbol{g})$. Here we have employed, in particular, the fact that $\varphi(t, \eta)[\omega_0(t[\eta]) - \psi(t[\eta])] = \varphi(t, \eta)[1 - \omega_1(t[\eta]) - \psi(t[\eta])] = 0$ because of the required properties of the supports of ω, ω_0, ω_1. Thus, writing $G(y, \eta) = G_1(y, \eta) + G_2(y, \eta)$ for

$$
\begin{aligned}
G_1(y, \eta) &= \tilde{\omega}(t)\{[D_{\eta_j}\omega(t[\eta])]t^{-\nu}a_M(y, \eta)\psi(t[\eta]) + [D_{\eta_j}(1 - \omega(t[\eta]))] \\
&\quad (t^{-\nu}a_\psi(y, \eta) - t^{-\nu}a_M(y, \eta) + t^{-\nu}a_M(y, \eta))\psi(t[\eta])\}\tilde{\omega}_0(t) \\
&= \tilde{\omega}(t)\varphi(t, \eta)(t^{-\nu}a_M(y, \eta) - t^{-\nu}a_\psi(y, \eta))\psi(t[\eta])\tilde{\omega}_0(t)
\end{aligned}
$$

THE ALGEBRA OF EDGE SYMBOLS

we have to show that $G_1(y,\eta) \in R_G^{\nu-1}(\Omega \times \mathbb{R}^q, g)$. However

$$G_0(y,\eta) := \varphi(t,\eta)(t^{-\nu}a_M(y,\eta) - t^{-\nu}a_\psi(y,\eta))\psi(t[\eta])$$

can easily be shown to belong to $R_G^{\nu-1}(\Omega \times \mathbb{R}^q, g)$, because of the relation (3.3.34), first for (3.3.37), and then by Taylor expansions in general, and then it also follows that $G_1(y,\eta) = \tilde{\omega}(t)G_0(y,\eta)\tilde{\omega}_0(t) \in R_G^{\nu-1}(\Omega \times \mathbb{R}^q, g)$. The case of general differentiation orders with respect to η can be treated inductively. Finally, the arguments for the weakly discrete asymptotics are the same.   □

Definition 3.3.40 *Let* $\mu, \nu \in \mathbb{R}$, $\mu - \nu \in \mathbb{N}$, $\gamma \in \mathbb{R}$, *and fix the weight data* $g = (\gamma, \gamma - \mu, \Theta)$ *for* $\Theta = (-k, 0]$, $k \in \mathbb{N} \setminus \{0\}$. *Moreover, let* $N_-, N_+ \in \mathbb{N}$. *Then*

$$\mathcal{R}^\nu(\Omega \times \mathbb{R}^q, g; N_-, N_+) \tag{3.3.53}$$

will denote the space of all block matrices

$$a(y,\eta) = \begin{pmatrix} a_{11} & a_{12} \\ a_{21} & a_{22} \end{pmatrix}(y,\eta) \tag{3.3.54}$$

for arbitrary $a_{11}(y,\eta) \in \mathcal{R}^\nu(\Omega \times \mathbb{R}^q, g)$ *and*

$$\begin{pmatrix} 0 & a_{12} \\ a_{21} & a_{22} \end{pmatrix}(y,\eta) \in \mathcal{R}_G^\nu(\Omega \times \mathbb{R}^q, g; N_-, N_+),$$

cf. Definition 3.3.6. In an analogous manner we define the subspace $\mathcal{R}^\nu(\Omega \times \mathbb{R}^q, g; N_-, N_+)^\bullet$ *with weakly discrete asymptotics.*

The definition immediately extends to the case of an infinite weight interval $\Theta = (-\infty, 0]$ as above after Definition 3.3.30.

For $a(y,\eta) \in \mathcal{R}^\nu(\Omega \times \mathbb{R}^q, g; N_-, N_+)$, written in the form (3.3.54), we set

$$\sigma_\psi^\nu(a) = \sigma_\psi^\nu(\text{u.l.c.}\,a),$$

where $a_{11} = \text{u.l.c.}\,a$ (the upper left corner of a), and

$$\sigma_\wedge^\nu(a) = \begin{pmatrix} \sigma_\wedge^\nu(\text{u.l.c.}\,a) & 0 \\ 0 & 0 \end{pmatrix} + \sigma_\wedge^\nu\begin{pmatrix} 0 & a_{12} \\ a_{21} & a_{22} \end{pmatrix}.$$

$\sigma_\wedge^\nu(a)$ is then a (y,η)-dependent operator family

$$\sigma_\wedge^\nu(a)(y,\eta): \begin{array}{c} \mathcal{K}^{s,\gamma}(X^\wedge) \\ \oplus \\ \mathbb{C}^{N_-} \end{array} \to \begin{array}{c} \mathcal{K}^{s-\nu,\gamma-\mu}(X^\wedge) \\ \oplus \\ \mathbb{C}^{N_+} \end{array}$$

satisfying

$$\sigma_\wedge^\nu(a)(y, \lambda\eta) = \lambda^\nu \begin{pmatrix} \kappa_\lambda & 0 \\ 0 & 1 \end{pmatrix} \sigma_\wedge^\nu(a)(y, \eta) \begin{pmatrix} \kappa_\lambda & 0 \\ 0 & 1 \end{pmatrix}^{-1}$$

for all $\lambda \in \mathbb{R}_+$, $y \in \Omega$, $\eta \in \mathbb{R}^q \setminus \{0\}$. Analogously to the above space $\Sigma_\wedge^{(\nu)}(\boldsymbol{g})$, cf. (3.3.50), we can introduce

$$\Sigma_\wedge^{(\nu)}(\boldsymbol{g}; N_-, N_+) := \{\sigma_\wedge^\nu(a)(y, \eta) : a(y, \eta) \in \mathcal{R}^\nu(\Omega \times \mathbb{R}^q, \boldsymbol{g}; N_-, N_+)\} \tag{3.3.55}$$

and define the map $\sigma_\psi^\nu : \Sigma_\wedge^{(\nu)}(\boldsymbol{g}; N_-, N_+) \to c_\wedge \Sigma_\psi^{(\nu)}$ by applying the former σ_ψ^ν, here to the upper left corners of the block matrices. Setting

$$\Sigma^{(\nu)}(\boldsymbol{g}; N_-, N_+)$$
$$:= \{(\sigma_\psi^\nu(a), \sigma_\wedge^\nu(a)) : a(y, \eta) \in \mathcal{R}^\nu(\Omega \times \mathbb{R}^q, \boldsymbol{g}; N_-, N_+)\} \tag{3.3.56}$$

it follows that

$$\Sigma^{(\nu)}(\boldsymbol{g}; N_-, N_+)$$
$$= \left\{ (p_\psi, p_\wedge) \in \Sigma_\psi^{(\nu)} \times \Sigma_\wedge^{(\nu)}(\boldsymbol{g}; N_-, N_+) : \sigma_\psi^{(\nu)} p_\wedge = c_\wedge p_\psi \right\}.$$

Analogous relations may be formulated for $\mathcal{R}^\nu(\Omega \times \mathbb{R}^q, \boldsymbol{g}; N_-, N_+)^\bullet$, where we write $\Sigma_\wedge^{(\nu)}(\boldsymbol{g}; N_-, N_+)^\bullet$ and $\Sigma^{(\nu)}(\boldsymbol{g}; N_-, N_+)^\bullet$, respectively.

The following assertions on formal adjoints and compositions within the symbol spaces are easy modifications of those for cone pseudo–differential operator on $X^\wedge$, cf. Section 2.4.2. So the details will be omitted.

Proposition 3.3.41 $(p_\psi, p_\wedge) \in \Sigma^{(\nu)}(\boldsymbol{g}; N_-, N_+)$ for $\boldsymbol{g} = (\gamma, \gamma - \mu, \Theta)$ implies $(p_\psi^*, p_\wedge^*) \in \Sigma^{(\nu)}(\boldsymbol{g}^*; N_+, N_-)$ for $\boldsymbol{g}^* = (-\gamma + \mu, -\gamma, \Theta)$, where $p_\psi^* = \overline{p}_\psi$, and $p_\wedge^*$ is the point–wise formal adjoint of $p_\wedge$ in the sense

$$(p_\wedge u, v)_{\mathcal{K}^{0,0}(X^\wedge) \oplus \mathbb{C}^{N_+}} = (u, p_\wedge^* v)_{\mathcal{K}^{0,0}(X^\wedge) \oplus \mathbb{C}^{N_-}}$$

for all $u \in C_0^\infty(X^\wedge) \oplus \mathbb{C}^{N_-}$, $v \in C_0^\infty(X^\wedge) \oplus \mathbb{C}^{N_+}$. Analogous relations hold for the subclasses with weakly discrete asymptotics.

Proposition 3.3.42 $(\tilde{p}_\psi, \tilde{p}_\wedge) \in \Sigma^{(\tilde{\nu})}(\tilde{\boldsymbol{g}}; N_-, N_0)$ for $\tilde{\boldsymbol{g}} = (\gamma, \gamma - \tilde{\mu}, \Theta)$ and $(p_\psi, p_\wedge) \in \Sigma^{(\nu)}(\boldsymbol{g}; N_0, N_+)$ for $\boldsymbol{g} = (\gamma - \tilde{\mu}, \gamma - \mu - \tilde{\mu}, \Theta)$ implies $(p_\psi \tilde{p}_\psi, p_\wedge \tilde{p}_\wedge) \in \Sigma^{(\nu+\tilde{\nu})}(\boldsymbol{h}; N_-, N_+)$ for $\boldsymbol{h} = (\gamma, \gamma - \mu - \tilde{\mu}, \Theta)$.

Analogous relations hold for the subclasses with weakly discrete asymptotics.

Proposition 3.3.43 *Let* $\tilde{h}(t,y,z,\tilde{\eta}) \in C^{\infty}(\overline{\mathbb{R}}_{+} \times \Omega, M_{O}^{\nu}(X; \mathbb{R}_{\tilde{\eta}}^{q}))$, $\nu \in$ $\mathbb{R}$, *and let* $\omega(t)$, $\omega_{0}(t)$ *be arbitrary cut–off functions. Then, for*

$$h(t,y,z,\eta) = \tilde{h}(t,y,z,t\eta)$$

the point–wise formal adjoint of the operator

$$a(y,\eta) = \omega(t[\eta])t^{-\nu} \operatorname{op}_{M}^{\gamma-\frac{n}{2}}(h)(y,\eta)\omega_{0}(t[\eta])$$

in the sense

$$(u, a^{*}(y,\eta)v)_{\mathcal{K}^{0,0}(X^{\wedge})} = (a(y,\eta)u, v)_{\mathcal{K}^{0,0}(X^{\wedge})}$$

for all $u,v \in C_{0}^{\infty}(X^{\wedge})$ *can be written*

$$a^{*}(y,\eta) = \omega_{0}(t[\eta])t^{-\nu} \operatorname{op}_{M}^{\gamma-\frac{n}{2}}(f)(y,\eta)\omega(t[\eta]) + (m+g)(y,\eta)$$

for

$$f(t,y,z,\eta) = \tilde{f}(t,y,z,t\eta)$$

for certain $\tilde{f}(t,y,z,\tilde{\eta}) \in C^{\infty}(\overline{\mathbb{R}}_{+} \times \Omega, M_{O}^{\nu}(X; \mathbb{R}_{\tilde{\eta}}^{q}))$ *and* $(m+g)(y,\eta) \in$ $R_{M+G}^{\nu}(\Omega \times \mathbb{R}^{q}, \boldsymbol{g})$, $\boldsymbol{g} = (-\gamma, -\gamma + \mu, (-\infty, 0])$.

Proof. The proof employs the same arguments as those of Proposition 2.4.12. The obvious modifications for the (y,η)–dependent will be omitted. $\qquad\square$

Theorem 3.3.44 $a(y,\eta) \in \mathcal{R}^{\nu}(\Omega \times \mathbb{R}^{q}, \boldsymbol{g}; N_{-}, N_{+})$ *for* $\boldsymbol{g} = (\gamma, \gamma - \mu, \Theta)$ *implies* $a^{*}(y,\eta) \in \mathcal{R}^{\nu}(\Omega \times \mathbb{R}^{q}, \boldsymbol{g}^{*}; N_{+}, N_{-})$ *for* $\boldsymbol{g}^{*} = (-\gamma + \mu, -\gamma, \Theta)$, *where* $a^{*}(y,\eta)$ *is the point–wise formal adjoint of* $a(y,\eta)$, *and we have*

$$\sigma_{\psi}^{\nu}(a^{*}) = \overline{\sigma_{\psi}^{\nu}(a)}, \quad \sigma_{\wedge}^{\nu}(a^{*}) = \sigma_{\wedge}^{\nu}(a)^{*}, \tag{3.3.57}$$

cf. Proposition 3.3.41. Moreover, $a(y,\eta) \in \mathcal{R}^{\nu}(\Omega \times \mathbb{R}^{q}, \boldsymbol{g}; N_{-}, N_{+})^{\bullet}$ *implies* $a^{*}(y,\eta) \in \mathcal{R}^{\nu}(\Omega \times \mathbb{R}^{q}, \boldsymbol{g}^{*}; N_{+}, N_{-})^{\bullet}$.

Proof. In view of Definition 3.3.1 and Corollary 3.3.19 it suffices to consider the case of operator families

$$a(y,\eta) = \tilde{\omega}(t)\left\{a_{0}(y,\eta) + a_{1}(y,\eta)\right\}\tilde{\omega}_{0}(t)$$
$$+ (1 - \tilde{\omega}(t))a_{\infty}(y,\eta)(1 - \tilde{\omega}_{1}(t)), \quad (3.3.58)$$

where we use the notation of Definition 3.3.30. Let us recall

$$a_{0}(y,\eta) = \omega(t[\eta])t^{-\nu}a_{M}(y,\eta)\omega_{0}(t[\eta]),$$
$$a_{1}(y,\eta) = (1 - \omega(t[\eta]))t^{-\nu}a_{\psi}(y,\eta)(1 - \omega_{1}(t[\eta]))$$

with cut–off functions ω, ω_0, ω_1 satisfying $\omega\omega_0 = \omega$, $\omega\omega_1 = \omega_1$. Analogously it was supposed that $\tilde{\omega}\tilde{\omega}_0 = \tilde{\omega}$, $\tilde{\omega}\tilde{\omega}_1 = \tilde{\omega}_1$. Without loss of generality we will assume here that $\varphi(t)\{a_0(y,\eta) + a_1(y,\eta)\}\psi(t) = \varphi(t)a_\infty(y,\eta)\psi(t)$ mod $C^\infty(\Omega, L^{-\infty}(X^\wedge; \mathbb{R}^q))$ for certain functions $\varphi(t), \psi(t) \in C_0^\infty(\mathbb{R}_+)$ with $\varphi \equiv \psi \equiv 1$, for $0 < \varepsilon \leq t \leq \varepsilon^{-1}$ for a sufficiently small $\varepsilon > 0$. Our first simple observation is now that when we form

$$b(y,\eta) := \tilde{\omega}_0(t)\{b_0(y,\eta) + b_1(y,\eta)\}\tilde{\omega}(t)$$
$$+ (1 - \tilde{\omega}_1(t))a_\infty(y,\eta)(1 - \tilde{\omega}(t)), \quad (3.3.59)$$

for

$$b_0(y,\eta) = \omega_0(t[\eta])t^{-\nu}a_M(y,\eta)\omega(t[\eta]),$$
$$b_1(y,\eta) = (1 - \omega_1(t[\eta]))t^{-\nu}a_\psi(y,\eta)(1 - \omega(t[\eta]))$$

it follows that $a(y,\eta) - b(y,\eta) \in R_G^\nu(\Omega \times \mathbb{R}^q, \boldsymbol{g})$. Because of the known properties of Green symbols under formal adjoints it suffices to look at $b(y,\eta)$. Since also $b_0(y,\eta) + b_1(y,\eta) - a_\infty(y,\eta) \in C^\infty(\Omega, L^{-\infty}(X^\wedge; \mathbb{R}^q))$, it follows that

$$\psi(t)[\{b_0(y,\eta) + b_1(y,\eta)\}^* - a_\infty^*(y,\eta)]\varphi(t) \in C^\infty(\Omega, L^{-\infty}(X^\wedge; \mathbb{R}^q)),$$

and hence it suffices to characterise

$$\tilde{\omega}(t)\{b_0(y,\eta) + b_1(y,\eta)\}^*\tilde{\omega}_0(t) + (1 - \tilde{\omega}(t))a_\infty^*(y,\eta)(1 - \tilde{\omega}_1(t))$$

as an element of $R^\nu(\Omega \times \mathbb{R}^q, \boldsymbol{g}^*)$. In view of Proposition 3.2.14 the class $C^\infty(\Omega, L_{\mathrm{cl}}^\nu(X^\wedge; \mathbb{R}^q)_0)$ remains preserved under formal adjoints. This yields the desired form of $a_\infty^*(y,\eta)$. So it remains to consider

$$\{b_0(y,\eta) + b_1(y,\eta)\}^* = \omega(t[\eta])(t^{-\nu}\,\mathrm{op}_M^{\gamma-\frac{n}{2}}(h)(y,\eta))^*\omega_0(t[\eta]) +$$
$$(1 - \omega(t[\eta]))(t^{-\nu}a_\psi)^*(y,\eta)(1 - \omega_1(t[\eta])).$$

As a result of Proposition 3.2.5 and Proposition 3.2.21 we obtain that $(t^{-\nu}a_\psi)^*(y,\eta)$ has the structure that was supposed in Definition 3.3.30. Now the condition (3.3.34) implies $(t^{-\nu}\,\mathrm{op}_M^{\gamma-\frac{n}{2}}(h)(y,\eta))^* - (t^{-\nu}a_\psi)^*(y,\eta) \in C^\infty(\Omega, L^{-\infty}(X^\wedge; \mathbb{R}^q))$. The only point to examine now is that there is an $\tilde{f}(t,y,z,\tilde{\eta}) \in C^\infty(\mathbb{R}_+ \times \Omega, M_O^\nu(X; \mathbb{R}_{\tilde{\eta}}^q))$ such that $f(t,y,z,\eta) := \tilde{f}(t,y,z,t\eta)$ satisfies

$$(\omega(t[\eta])t^{-\nu}\,\mathrm{op}_M^{\gamma-\frac{n}{2}}(h)(y,\eta)\omega_0(t[\eta]))^*$$
$$= \omega_0(t[\eta])t^{-\nu}\,\mathrm{op}_M^{-\gamma+\mu-\frac{n}{2}}(f)(y,\eta)\omega(t[\eta]) + (m+g)(y,\eta).$$

for a certain $(m+g)(y,\eta) \in R^\nu_{M+G}(\Omega \times \mathbb{R}^q, (-\gamma, -\gamma+\mu, (-\infty, 0]))$. However, this was just the result of Proposition 3.3.43. The first relation of (3.3.57) is obvious, while the second one follows from the fact that the procedure of taking formal adjoints commutes with the operator $c_\wedge$ of freezing coefficients. $\qquad\square$

Theorem 3.3.45 *Given* $\tilde{a}(y,\eta) \in \mathcal{R}^{\tilde{\nu}}(\Omega \times \mathbb{R}^q, \tilde{g}; N_-, N_0)$ *for* $\tilde{g} = (\gamma, \gamma - \tilde{\mu}, \Theta)$ *and* $a(y,\eta) \in \mathcal{R}^\nu(\Omega \times \mathbb{R}^q, g; N_0, N_+)$ *for* $g = (\gamma - \tilde{\mu}, \gamma - \tilde{\mu} - \mu, \Theta)$ *we have for the point-wise composition*

$$(a\tilde{a})(y,\eta) \in \mathcal{R}^{\nu+\tilde{\nu}}(\Omega \times \mathbb{R}^q, \boldsymbol{h}; N_-, N_+) \quad \text{for} \quad \boldsymbol{h} = (\gamma, \gamma - \tilde{\mu} - \mu, \Theta),$$

and

$$\sigma_\psi^{\nu+\tilde{\nu}}(a\tilde{a}) = \sigma_\psi^\nu(a)\sigma_\psi^{\tilde{\nu}}(\tilde{a}), \qquad \sigma_\wedge^{\nu+\tilde{\nu}}(a\tilde{a}) = \sigma_\wedge^\nu(a)\sigma_\wedge^{\tilde{\nu}}(\tilde{a}),$$

cf. Proposition 3.3.42.

For $\tilde{a}(y,\eta) \in \mathcal{R}^{\tilde{\nu}}_G(\Omega \times \mathbb{R}^q, \tilde{g}; N_-, N_0)$ $(\in \mathcal{R}^{\tilde{\nu}}_{M+G}(\Omega \times \mathbb{R}^q, \tilde{g}; N_-, N_0))$ *or* $a(y,\eta) \in \mathcal{R}^\nu_G(\Omega \times \mathbb{R}^q, g; N_0, N_+)$ $(\in \mathcal{R}^\nu_{M+G}(\Omega \times \mathbb{R}^q, g; N_0, N_+))$ *it follows that*

$$(a\tilde{a})(y,\eta) \in \mathcal{R}^{\nu+\tilde{\nu}}_G(\Omega \times \mathbb{R}^q, \boldsymbol{h}; N_-, N_+) \quad (\in \mathcal{R}^{\nu+\tilde{\nu}}_{M+G}(\Omega \times \mathbb{R}^q, \boldsymbol{h}; N_-, N_+)).$$

Analogous relations hold for the subclasses with weakly discrete asymptotics.

Proof. Writing the operator families as block matrices

$$a(y,\eta) = (a_{ij}(y,\eta))_{i,j=1,2} \quad \text{and} \quad \tilde{a}(y,\eta) = (\tilde{a}_{jl}(y,\eta))_{j,l=1,2},$$

respectively, we have to characterise the expressions

$$(a_{11}\tilde{a}_{11})(y,\eta), \quad (a_{11}\tilde{a}_{12})(y,\eta), \quad (a_{21}\tilde{a}_{11})(y,\eta) \qquad (3.3.60)$$

as elements within $\mathcal{R}^{\nu+\tilde{\nu}}$. The other occurring compositions were already treated in Theorem 3.3.29. Let us start with $(a_{11}\tilde{a}_{11})(y,\eta)$. Here we will argue in an analogous manner as in the proof of the fact that cone pseudo–differential operators over $X^\wedge$ can be composed within the class, cf. Section 2.4.2. In order to make this precise we fix for a moment $(y,\eta) \in \Omega \times \mathbb{R}^q$, such that $a_{11}(y,\eta) \in C^\nu(X^\wedge, g)$, $\tilde{a}_{11}(y,\eta) \in C^{\tilde{\nu}}(X^\wedge, \tilde{g})$. Then, applying Theorem 2.4.16, it follows that $a_{11}(y,\eta)\tilde{a}_{11}(y,\eta) \in C^{\nu+\tilde{\nu}}(X^\wedge, \boldsymbol{h})$ (the slight modification of the arguments for the weakly discrete asymptotics is obvious). This result was obtained by direct computation. Another interpretation of the compositions was given in Remark 2.4.17. This can be adapted here to the corresponding parameter–dependent variant with $\eta \in \mathbb{R}^q$ as parameter and $y \in \Omega$ as an additional variable on which all operators depend smoothly. We will concentrate on the relations for finite t. The desired behaviour of our operator families under compositions for $t \to \infty$ is clear. In other words, $a_{11}(y, \eta) \tilde{a}_{11}(y, \eta) =: b_\infty(y, \eta) \in C^\infty(\Omega, L^{\nu+\tilde{\nu}}_{cl}(X^\wedge; \mathbb{R}^q_\eta))$. The situation is as follows: $a_{11}(y, \eta)$ has a system of complete symbols

$$t^{-\nu} p_j(t, x, y, \tau, \xi, \eta) = t^{-\nu} \tilde{p}_j(t, x, y, t\tau, \xi, t\eta)$$

for certain $\tilde{p}_j(t, x, y, \tilde{\tau}, \xi, \tilde{\eta}) \in S^\nu_{cl}(\overline{\mathbb{R}}_+ \times \Sigma_j \times \Omega \times \mathbb{R}^{1+n+q})$, $j = 1, \dots, N$, where we may assume that they are invariant with respect to the symbol push-forwards under the coordinate diffeomorphisms with respect to X, up to symbols of order $-\infty$. Analogously $\tilde{a}_{11}(y, \eta)$ has a system of complete symbols

$$t^{-\tilde{\nu}} r_j(t, x, y, \tau, \xi, \eta) = t^{-\tilde{\nu}} \tilde{r}_j(t, x, y, t\tau, \xi, t\eta)$$

for certain $\tilde{r}_j(t, x, y, \tilde{\tau}, \xi, \tilde{\eta}) \in S^{\tilde{\nu}}_{cl}(\overline{\mathbb{R}}_+ \times \Sigma_j \times \Omega \times \mathbb{R}^{1+n+q})$, $j = 1, \dots, N$, also invariant under the symbol push-forwards. Now we can pass to the Leibniz products between these symbols for every j, namely $(t^{-\nu} p_j) \#_{t,x} (t^{-\tilde{\nu}} r_j)$. Applying Remark 3.2.4 the asymptotic sums can be carried out in the class $t^{-(\nu+\tilde{\nu})} \tilde{S}^{\nu+\tilde{\nu}}_{cl}(\overline{\mathbb{R}}_+ \times \Sigma_j \times \Omega \times \mathbb{R}^{1+n+q})$, i.e., we obtain symbols

$$t^{-(\nu+\tilde{\nu})} b_j(t, x, y, \tau, \xi, \eta) = t^{-(\nu+\tilde{\nu})} \tilde{b}_j(t, x, y, t\tau, \xi, t\eta) \qquad (3.3.61)$$

for $\tilde{b}_j(t, x, y, \tilde{\tau}, \xi, \tilde{\eta}) \in S^{\nu+\tilde{\nu}}_{cl}(\overline{\mathbb{R}}_+ \times \Sigma_j \times \Omega \times \mathbb{R}^{1+n+q})$, $j = 1, \dots, N$, also invariant under the symbol push-forwards with respect to X, modulo symbols of order $-\infty$. Now applying the constructions for (3.2.4), using (3.2.1), the symbols (3.3.61) give rise to an operator family $b_\psi(y, \eta) \in C^\infty(\Omega, \tilde{L}^{\nu+\tilde{\nu}}_{cl}(X^\wedge; \mathbb{R}^q))$. According to Theorem 3.2.7, to this we can construct an $\tilde{m}(t, y, z, \tilde{\eta}) \in C^\infty(\overline{\mathbb{R}}_+ \times \Omega, M^{\nu+\tilde{\nu}}_O(X; \mathbb{R}^q_{\tilde{\eta}}))$ such that $m(t, y, z, \eta) = \tilde{m}(t, y, z, t\eta)$ satisfies $\mathrm{op}^\delta_M(m)(y, \eta) = b_\psi(y, \eta) \bmod C^\infty(\Omega, L^{-\infty}(X^\wedge; \mathbb{R}^q))$ for every $\delta \in \mathbb{R}$. All these data enable us to form the operator family

$$b(y, \eta) = \tilde{\omega}(t)\{b_0(y, \eta) + b_1(y, \eta)\}\tilde{\omega}_0(t) + (1 - \tilde{\omega}(t))b_\infty(y, \eta)(1 - \tilde{\omega}_1(t)),$$

using the cut–off functions $\omega, \omega_i, \tilde{\omega}, \tilde{\omega}_i$, $i = 1, 2$, as in Definition 3.3.30, and

$$b_0(y, \eta) = \omega(t[\eta]) t^{-(\nu+\tilde{\nu})} b_M(y, \eta) \omega_0(t[\eta]),$$
$$b_M(y, \eta) = \mathrm{op}^{\gamma-\frac{n}{2}}_M(m)(y, \eta),$$
$$b_1(y, \eta) = (1 - \omega(t[\eta])) t^{-(\nu+\tilde{\nu})} b_\psi(y, \eta)(1 - \omega_1(t[\eta])).$$

Now the above characterisation of compositions through Remark 2.4.17 in the (y,η)-dependent variant is that

$$a_{11}(y,\eta)\tilde{a}_{11}(y,\eta) - b(y,\eta)$$

is an operator–valued symbol in

$$S^{\nu+\tilde{\nu}}(\Omega \times \mathbb{R}^q; \mathcal{K}^{s,\gamma}(X^\wedge), \mathcal{K}^{s-(\nu+\tilde{\nu}),\gamma-(\mu+\tilde{\mu})}(X^\wedge))$$

that is (y,η)-wise in $C_{M+G}(X^\wedge, \mathbf{h})$. On the other hand the explicit computation of the compositions that made this remark possible show in the present situation that all smoothing summands which appear have the form of elements in $R^{\nu+\tilde{\nu}}_{M+G}$ plus $C_G(X^\wedge, \mathbf{h})$–valued operator-valued symbols $g(y,\eta)$ in

$$S^{\nu+\tilde{\nu}}(\Omega \times \mathbb{R}^q; \mathcal{K}^{s,\gamma}(X^\wedge), \mathcal{K}^{s-(\nu+\tilde{\nu}),\gamma-(\mu+\tilde{\mu})}(X^\wedge)).$$

So it remains to see if we even obtain $g(y,\eta) \in R^{\nu+\tilde{\nu}}_G$. But this is easy to verify, since the smoothing summands from the explicit computation that remain after removing the already identified $R^{\nu+\tilde{\nu}}_{M+G}$–parts, certainly have the property

$$g(y,\eta) \in S^{\nu+\tilde{\nu}}(\Omega \times \mathbb{R}^q; \mathcal{K}^{s,\gamma}(X^\wedge), \mathcal{S}^{\gamma-(\mu+\tilde{\mu})}_P(X^\wedge)),$$
$$g^*(y,\eta) \in S^{\nu+\tilde{\nu}}(\Omega \times \mathbb{R}^q; \mathcal{K}^{s,-\gamma+\mu+\tilde{\mu}}(X^\wedge), \mathcal{S}^{-\gamma}_Q(X^\wedge))$$

for all $s \in \mathbb{R}$, for certain asymptotic types P and Q. The only point that remains is to accept that these objects are in fact classical symbols. This can be done by applying Taylor expansions with respect to t near $t = 0$ with respect to the involved t-dependence in the occurring coefficients. They yield asymptotic expansions for the symbols into classical ones of order $\nu + \tilde{\nu} - j$, $j \in \mathbb{N}$, and so classical ones also exist. The second two compositions in (3.3.60) can be treated in an analogous manner, by direct computation, and Taylor expansions in t, to identify Green remainders as classical symbols.   □

Remark 3.3.46 *One can easily generalise Definition 3.3.40 to the case of (y,y')-dependent edge symbols for $(y,y') \in \Omega \times \Omega$. Since the above y only played the role of an additional variable in which the symbols are C^∞ we may switch at once to any open set $U \subseteq \mathbb{R}^p_y$ and repeat all notations and definitions in the obvious modified form. Then we can insert $U = \Omega \times \Omega$. In other words in this way we get the symbol classes*

$$\mathcal{R}^\nu(\Omega \times \Omega \times \mathbb{R}^q, \mathbf{g}; N_-, N_+) \qquad and \qquad \mathcal{R}^\nu(\Omega \times \Omega \times \mathbb{R}^q, \mathbf{g}; N_-, N_+)^*,$$

respectively. The above considerations are valid for them in analogous form.

3.4   Edge pseudo–differential operators

3.4.1   Green, trace and potential operators

Pseudo–differential operators on a manifold with edges will be formulated locally as pseudo–differential operators on Ω with operator–valued symbols of the class (3.3.32), modulo suitable smoothing operators, where $\Omega \subseteq \mathbb{R}^q$ is an open set that corresponds to any chart on the edge Y. Let us start with symbols

$$g(y, y', \eta) = (g_{ij}(y, y', \eta))_{i,j=1,2} \in \mathcal{R}_G^\nu(\Omega \times \Omega \times \mathbb{R}^q, \boldsymbol{g}; N_-, N_+),$$

cf. Section 3.3.1. Recall that $g_{11} = \text{u. l. c.} g$ (the upper left corner of the block matrix g) has the interpretation of a Green, g_{21} of a trace and g_{12} of a potential symbol, while g_{22} is an $N_+ \times N_-$–matrix of classical symbols of order ν on $\Omega \times \Omega$. Also g itself is called a Green symbol, cf. Definition 3.3.1. This should not cause confusion, since the behaviour of the block matrices is very similar to that of the upper left corners.

In Section 3.1.4. we have introduced the space $\mathcal{W}^{0,0}(X^\wedge \times \Omega) = t^{-\frac{n}{2}} L^2(\mathbb{R}_+ \times X \times \Omega)$, with the measure $dt\,dx\,dy$, endowed with the corresponding scalar product. Let us also fix scalar products in the spaces $H^0(\Omega, \mathbb{C}^N) = L^2(\Omega, \mathbb{C}^N)$, $N \in \mathbb{N}$. If

$$G : \mathcal{W}^{s,\gamma}_{\text{comp}(y)}(X^\wedge \times \Omega) \oplus H^s_{\text{comp}}(\Omega, \mathbb{C}^{N_-})$$
$$\rightarrow \mathcal{W}^{r,\varrho}_{\text{loc}(y)}(X^\wedge \times \Omega) \oplus H^r_{\text{loc}}(\Omega, \mathbb{C}^{N_+}) \quad (3.4.1)$$

is a continuous operator for every $s \in \mathbb{R}$ with $r = s - \nu$ for some $\nu \in \mathbb{R}$, then the formal adjoint G^*, defined by

$$(Gu, v)_{\mathcal{W}^{0,0}(X^\wedge \times \Omega) \oplus H^0(\Omega, \mathbb{C}^{N_+})} = (u, G^*v)_{\mathcal{W}^{0,0}(X^\wedge \times \Omega) \oplus H^0(\Omega, \mathbb{C}^{N_-})}$$

for all $u \in C_0^\infty(X^\wedge \times \Omega) \oplus C_0^\infty(\Omega, \mathbb{C}^{N_-})$, $v \in C_0^\infty(X^\wedge \times \Omega) \oplus C_0^\infty(\Omega, \mathbb{C}^{N_+})$ induces continuous operators

$$G^* : \mathcal{W}^{s,-\varrho}_{\text{comp}(y)}(X^\wedge \times \Omega) \oplus H^s_{\text{comp}}(\Omega, \mathbb{C}^{N_+})$$
$$\rightarrow \mathcal{W}^{r,-\gamma}_{\text{loc}(y)}(X^\wedge \times \Omega) \oplus H^r_{\text{loc}}(\Omega, \mathbb{C}^{N_-}).$$

Definition 3.4.1 *Let* $\boldsymbol{g} = (\gamma, \varrho, \Theta)$ *for* $\gamma, \varrho \in \mathbb{R}$, $\Theta = (\vartheta, 0]$, $-\infty \leq \vartheta < 0$, *then*

$$\mathcal{Y}_G^{-\infty}(X^\wedge \times \Omega, \boldsymbol{g}; N_-, N_+)$$

denotes the space of all operators G for which (3.4.1) is continuous for all $s, r \in \mathbb{R}$, such that there exist asymptotic types $P \in \text{As}(X, (\varrho, \Theta))$,

$Q \in \mathrm{As}(X, (-\gamma, \Theta))$ *such that* G *and* G^* *induce continuous operators*

$$G : \mathcal{W}^{s,\gamma}_{\mathrm{comp}(y)}(X^{\wedge} \times \Omega) \oplus H^s_{\mathrm{comp}}(\Omega, \mathbb{C}^N)$$
$$\to \mathcal{W}^{\infty,\varrho}_{P,\mathrm{loc}(y)}(X^{\wedge} \times \Omega) \oplus H^{\infty}_{\mathrm{loc}}(\Omega, \mathbb{C}^{N_+})$$

and

$$G^* : \mathcal{W}^{s,-\varrho}_{\mathrm{comp}(y)}(X^{\wedge} \times \Omega) \oplus H^s_{\mathrm{comp}}(\Omega, \mathbb{C}^{N_+})$$
$$\to \mathcal{W}^{\infty,-\gamma}_{Q,\mathrm{loc}(y)}(X^{\wedge} \times \Omega) \oplus H^{\infty,r}_{\mathrm{loc}}(\Omega, \mathbb{C}^{N_-}),$$

respectively, for all $s \in \mathbb{R}$. *For weakly discrete asymptotic types* $P \in \mathrm{As}(X(\varrho, \Theta))^{\bullet}$, $Q \in \mathrm{As}(X(-\gamma, \Theta))^{\bullet}$ *we denote the corresponding operator class by* $\mathcal{Y}_G^{-\infty}(X^{\wedge} \times \Omega, \mathbf{g}; N_-, N_+)^{\bullet}$. *In the case* $N_- = N_+$ *we write* $Y_G^{-\infty}(X^{\wedge} \times \Omega, \mathbf{g})$ *and* $Y_G^{-\infty}(X^{\wedge} \times \Omega, \mathbf{g})^{\bullet}$, *respectively, cf. Definition 3.1.38.*

We shall see below that $g(y, y', \eta) \in \mathcal{R}_G^{-\infty}(\Omega \times \Omega \times \mathbb{R}^q, \mathbf{g}; N_-, N_+)$ implies $\mathrm{Op}(g) \in \mathcal{Y}_G^{-\infty}(X^{\wedge} \times \Omega, \mathbf{g}; N_-, N_+)$. The operators of Definition 3.4.1 will be the regularising ones in the algebra of pseudo–differential operators on a manifold with edges. Observe that we require much more than the property of being smoothing, namely asymptotics in the image. Since the length of the weight interval Θ is arbitrary, we have in fact a scale of spaces of smoothing operators with canonical embeddings when we shrink the length.

Let $\mathcal{Y}_G^{-\infty}(X^{\wedge} \times \Omega, \mathbf{g}; N_-, N_+)_{0,0}$ be the subspace of all $G \in \mathcal{Y}_G^{-\infty}(X^{\wedge} \times \Omega, \mathbf{g}; N_-, N_+)$, $\mathbf{g} = (\gamma, \varrho, (\vartheta, 0])$, for which the asymptotic types P and Q are empty (recall that empty asymptotics indicate flatness, measured by the length of the weight interval $\Theta = (\vartheta, 0]$). The space $\mathcal{Y}_G^{-\infty}(X^{\wedge} \times \Omega, \mathbf{g}; N_-, N_+)_{0,0}$ is Fréchet in a canonical way. For every $G \in \mathcal{Y}_G^{-\infty}(X^{\wedge} \times \Omega, \mathbf{g}; N_-, N_+)$ there is an $\varepsilon > 0$ such that $G \in \mathcal{Y}_G^{-\infty}(X^{\wedge} \times \Omega, \mathbf{g}_\varepsilon; N_-, N_+)_{0,0}$ for $\mathbf{g}_\varepsilon = (\gamma, \varrho, (-\varepsilon, 0])$. Thus, if we set

$$\mathcal{Y}_G^{-\infty}(X^{\wedge} \times \Omega, \mathbf{g}_0; N_-, N_+) := \bigcup_{\varepsilon > 0} \mathcal{Y}_G^{-\infty}(X^{\wedge} \times \Omega, \mathbf{g}_\varepsilon; N_-, N_+)_{0,0},$$

it follows that

$$\mathcal{Y}_G^{-\infty}(X^{\wedge} \times \Omega, \mathbf{g}; N_-, N_+) \subset \mathcal{Y}_G^{-\infty}(X^{\wedge} \times \Omega, \mathbf{g}_0; N_-, N_+)_{0,0}$$

for $\mathbf{g} = (\gamma, \varrho, (\vartheta, 0])$ for every $-\infty \leq \vartheta < 0$.

Definition 3.4.2 $\mathcal{Y}_G^{\nu}(X^{\wedge} \times \Omega, \mathbf{g}; N_-, N_+)$ *for* $\nu \in \mathbb{R}$, $\mathbf{g} = (\gamma, \varrho, \Theta)$, *denotes the space of all operators* $\mathrm{Op}(g) + C$ *for arbitrary* $g(y, y', \eta) \in$

$\mathcal{R}_G^\nu(\Omega \times \Omega \times \mathbb{R}^q, \boldsymbol{g}; N_-, N_+)$ and $C \in \mathcal{Y}_G^{-\infty}(X^\wedge \times \Omega, \boldsymbol{g}; N_-, N_+)$. For $g(y, y', \eta) \in \mathcal{R}_G^\nu(\Omega \times \Omega \times \mathbb{R}^q, \boldsymbol{g}; N_-, N_+)^\bullet$, $C \in \mathcal{Y}_G^{-\infty}(X^\wedge \times \Omega, \boldsymbol{g}; N_-, N_+)^\bullet$, we denote the corresponding operator class by $\mathcal{Y}_G^\nu(X^\wedge \times \Omega, \boldsymbol{g}; N_-, N_+)^\bullet$. The elements $G \in \mathcal{Y}_G^\nu(X^\wedge \times \Omega, \boldsymbol{g}; N_-, N_+)$ $(\in \mathcal{Y}_G^\nu(X^\wedge \times \Omega, \boldsymbol{g}; N_-, N_+)^\bullet)$ are called Green operators of order ν with continuous (weakly discrete) asymptotics. The corresponding spaces for $N_- = N_+ = 0$ are denoted by $Y_G^\nu(X^\wedge \times \Omega, \boldsymbol{g})$ and $Y_G^\nu(X^\wedge \times \Omega, \boldsymbol{g})^\bullet$, respectively.

Every $G \in \mathcal{Y}_G^\nu(X^\wedge \times \Omega, \boldsymbol{g}; N_-, N_+)$ is of block matrix form

$$G = (G_{ij})_{i,j=1,2}, \tag{3.4.2}$$

where $G_{11} = \text{u.l.c.}\, G$ is also called a Green operator, G_{21} a trace and G_{12} a potential operator. G_{22} is an $N_+ \times N_-$–matrix of classical pseudo–differential operators on Ω of order ν.

As a consequence of the definitions we have

$$\mathcal{Y}_G^\nu(X^\wedge \times \Omega, \boldsymbol{g}; N_-, N_+) \subset L_{cl}^\nu(\Omega; \mathcal{K}^{s,\gamma}(X^\wedge) \oplus \mathbb{C}^{N_-}, \mathcal{K}^{\infty,\varrho}(X^\wedge) \oplus \mathbb{C}^{N_+})$$

for all $s \in \mathbb{R}$. Thus each $G \in \mathcal{Y}_G^\nu(X^\wedge \times \Omega, \boldsymbol{g}; N_-, N_+)$ has a homogeneous principal operator–valued symbol of order ν

$$\sigma_\wedge^\nu(G)(y, \eta) : \mathcal{K}^{s,\gamma}(X^\wedge) \oplus \mathbb{C}^{N_-} \to \mathcal{K}^{\infty,\varrho}(X^\wedge) \oplus \mathbb{C}^{N_+},$$

$(y, \eta) \in \Omega \times (\mathbb{R}^q \setminus \{0\})$. Homogeneity means

$$\sigma_\wedge^\nu(G)(y, \lambda\eta) = \lambda^\nu \begin{pmatrix} \kappa_\lambda & 0 \\ 0 & 1 \end{pmatrix} \sigma_\wedge^\nu(G)(y, \eta) \begin{pmatrix} \kappa_\lambda & 0 \\ 0 & 1 \end{pmatrix}^{-1}$$

for all $\lambda \in \mathbb{R}_+$, with 1 being the identity operators in the corresponding finite–dimensional space. For $G = \text{Op}(g) + C$, $g(y, y', \eta) \in \mathcal{R}_G^\nu(\Omega \times \Omega \times \mathbb{R}^q, \boldsymbol{g}; N_-, N_+)$, $C \in \mathcal{Y}^{-\infty}$, we have

$$\sigma_\wedge^\nu(G)(y, \eta) = \sigma_\wedge^\nu(g)(y, y', \eta)|_{y'=y},$$

cf. the notation of Section 3.3.1.

From the general results on the continuity of pseudo–differential operators with operator–valued symbols and from the specific nature of symbols and smoothing operators here we obtain that each $G \in \mathcal{Y}_G^\nu(X^\wedge \times \Omega, \boldsymbol{g}; N_-, N_+)$ induces a continuous operator

$$C_0^\infty(X^\wedge \times \Omega) \oplus C_0^\infty(\Omega, \mathbb{C}^{N_-}) \to C^\infty(X^\wedge \times \Omega) \oplus C^\infty(\Omega, \mathbb{C}^{N_+}) \tag{3.4.3}$$

that extends to a continuous operator

$$\mathcal{W}_{\text{comp}(y)}^s(\Omega, \mathcal{K}^{s,\gamma}(X^\wedge)) \oplus H_{\text{comp}}^s(\Omega, \mathbb{C}^{N_-})$$
$$\to \mathcal{W}_{\text{loc}(y)}^{s-\nu}(\Omega, \mathcal{K}^{\infty,\varrho}(X^\wedge)) \oplus H_{\text{loc}}^{s-\nu}(\Omega, \mathbb{C}^{N_+})$$

for every $s \in \mathbb{R}$. More precisely we have in the present case the following result:

Theorem 3.4.3 *Every* $G \in \mathcal{Y}_G^\nu(X^\wedge \times \Omega, \boldsymbol{g}; N_-, N_+)$ *for* $\boldsymbol{g} = (\gamma, \varrho, \Theta)$ *induces continuous operators*

$$G : \mathcal{W}^s_{\mathrm{comp}(y)}(\Omega, \mathcal{K}^{r,\gamma}(X^\wedge)) \oplus H^s_{\mathrm{comp}}(\Omega, \mathbb{C}^{N_-})$$
$$\to \mathcal{W}^{s-\nu}_{\mathrm{loc}(y)}(\Omega, \mathcal{K}^{\infty,\varrho}_P(X^\wedge)) \oplus H^{s-\nu}_{\mathrm{loc}}(\Omega, \mathbb{C}^{N_+})$$

for all $r, s \in \mathbb{R}$, *with some asymptotic type* $P \in \mathrm{As}(X, (\varrho, \Theta))$ *(independent of* r, s). *An analogous result is true for* $G \in \mathcal{Y}_G^\nu(X^\wedge \times \Omega, \boldsymbol{g}; N_-, N_+)^*$, *where* $P \in \mathrm{As}(X, (\varrho, \Theta))^*$.

Proof. The proof is an obvious consequence of the above Definition 3.4.1 and of Definition 3.3.1, together with the general mapping properties of pseudo–differential operators with operator–valued symbols, cf. Theorem 1.3.58. □

Corollary 3.4.4 $g(y, y', \eta) \in \mathcal{R}_G^{-\infty}(\Omega \times \Omega \times \mathbb{R}^q, \boldsymbol{g}; N_-, N_+)$ *implies* $\mathrm{Op}(g) \in \mathcal{Y}_G^{-\infty}(X^\wedge \times \Omega, \boldsymbol{g}; N_-, N_+)$ *and analogously for weakly discrete asymptotics.*

Remark 3.4.5 *Let* $G = \mathrm{Op}(g)$ *for* $g(y, \eta) \in \mathcal{R}_G^\nu(\Omega_y \times \mathbb{R}^q, \boldsymbol{g}; N_-, N_+)$. *Then* G *induces continuous operators*

$$G : \mathcal{W}^s(\mathbb{R}^q, \mathcal{K}^{r,\gamma}(X^\wedge)) \oplus H^s(\mathbb{R}^q, \mathbb{C}^{N_-})$$
$$\to \mathcal{W}^{s-\nu}_{\mathrm{loc}(y)}(\Omega, \mathcal{K}^{\infty,\varrho}_P(X^\wedge)) \oplus H^{s-\nu}_{\mathrm{loc}}(\Omega, \mathbb{C}^{N_+})$$

for all $s, r \in \mathbb{R}$ *and some* $P \in \mathrm{As}(X, (\varrho, \Theta))$. *Moreover, if* $\tilde{G} = \mathrm{Op}(\tilde{g})$ *for* $\tilde{g}(y', \eta) \in \mathcal{R}_G^\nu(\Omega_{y'} \times \mathbb{R}^q, \boldsymbol{g}, N_-, N_+)$, *then* $\tilde{G}$ *induces continuous operators*

$$\tilde{G} : \mathcal{W}^s_{\mathrm{comp}(y)}(\Omega, \mathcal{K}^{r,\gamma}(X^\wedge)) \oplus H^s_{\mathrm{comp}}(\Omega, \mathbb{C}^{N_-})$$
$$\to \mathcal{W}^{s-\nu}(\mathbb{R}^q, \mathcal{K}^{\infty,\varrho}_Q(X^\wedge)) \oplus H^{s-\nu}(\mathbb{R}^q, \mathbb{C}^{N_+})$$

for all $r, s \in \mathbb{R}$ *and some* $Q \in \mathrm{As}(X, (\varrho, \Theta))$. *Analogous relations hold for the case of weakly discrete asymptotics.*

Proposition 3.4.6 *Let* $G = (G_{ij})_{i,j=1,2} \in \mathcal{Y}_G^\nu(X^\wedge \times \Omega, \boldsymbol{g}; N_-, N_+)$. *Then* G_{11} *is an integral operator with kernel in* $C^\infty(X^\wedge \times \Omega \times X^\wedge \times \Omega)$, *while* G_{12} *has a kernel in* $C^\infty(X^\wedge \times \Omega \times \Omega) \otimes \mathbb{C}^{N_-}$ *and* G_{21} *a kernel in* $\mathbb{C}^{N_+} \otimes C^\infty(\Omega \times X^\wedge \times \Omega)$.

Proof. Let us consider to G_{11}; the other entries can be discussed in an analogous manner. The elements of $Y^{-\infty}(X^\wedge \times \Omega, \boldsymbol{g})$ have kernels in $C^\infty(X^\wedge \times \Omega \times X^\wedge \times \Omega)$. Thus without loss of generality we may assume $G_{11} = \mathrm{Op}(g)$ for a $g(y, y', \eta) \in \mathcal{R}_G^\nu(\Omega \times \Omega \times \mathbb{R}^q, \boldsymbol{g})$. In order to characterise the kernel of $\mathrm{Op}(g)$ it suffices to look at $\mathrm{Op}(f)$ for $f(y, y', \eta) = (1 - \omega(t))g(y, y', \eta)$ for an arbitrary fixed cut–off function $\omega(t)$, where $(1 - \omega(t))$ is interpreted as the corresponding multiplication operator, cf. Proposition 3.3.12. We then have $f_N(y, y', \eta) := t^{-N}f(y, y', \eta) \in \mathcal{R}_G^\nu(\Omega \times \Omega \times \mathbb{R}^q, \boldsymbol{g})$ for every $N \in \mathbb{N}$, cf. Remark 3.3.13. This implies $f(y, y', \eta) = t^N f_N(y, y', \eta) \in \mathcal{R}_G^{\nu-N}(\Omega \times \Omega \times \mathbb{R}^q, \boldsymbol{g})$, since the multiplication by t^N diminishes the order by N, cf. Proposition 3.3.9 and Remark 3.3.13. Since N is arbitrary, it follows that $\mathrm{Op}(f) \in \mathcal{Y}_G^{-\infty}(X^\wedge \times \Omega, \boldsymbol{g})$ has a kernel in $C^\infty(X^\wedge \times \Omega \times X^\wedge \times \Omega)$. □

Lemma 3.4.7 *Let* $g(y, y', \eta) \in \mathcal{R}_G^\nu(\Omega \times \Omega \times \mathbb{R}^q, \boldsymbol{g}; N_-, N_+)$ *and assume that* $|y - y'|^{-2N}g(y, y', \eta) \in \mathcal{R}_G^\nu(\Omega \times \Omega \times \mathbb{R}^q, \boldsymbol{g}; N_-, N_+)$ *for some* $N \in \mathbb{N}$. *Then there exists a* $g_N(y, y', \eta) \in \mathcal{R}_G^{\nu-2N}(\Omega \times \Omega \times \mathbb{R}^q, \boldsymbol{g}; N_-, N_+)$ *such that* $\mathrm{Op}(g) = \mathrm{Op}(g_N)$. *An analogous result holds for Green symbols with discrete asymptotics.*

The proof is an obvious generalisation of the method for Lemma 1.1.29.

Corollary 3.4.8 *Let* $\varphi(y), \psi(y) \in C^\infty(\Omega)$ *with* $\mathrm{supp}\,\varphi \cap \mathrm{supp}\,\psi = \emptyset$. *Then* $G \in \mathcal{Y}_G^\nu(X^\wedge \times \Omega, \boldsymbol{g}; N_-, N_+)$ *implies* $\varphi G \psi \in \mathcal{Y}_G^{-\infty}(X^\wedge \times \Omega, \boldsymbol{g}; N_-, N_+)$. *Analogous relations are true for the discrete asymptotics.*

Proposition 3.4.9 $g(y, y', \eta) \in \mathcal{R}_G^\nu(\Omega \times \Omega \times \mathbb{R}^q, \boldsymbol{g}; N_-, N_+)$ *implies* $\mathrm{Op}(g) \in \mathcal{Y}_G^\nu(X^\wedge \times \Omega, \boldsymbol{g}; N_-, N_+)$, *and there exists a* $g_1(y, \eta) \in \mathcal{R}_G^\nu(\Omega \times \mathbb{R}^q, \boldsymbol{g}; N_-, N_+)$ *with* $\mathrm{Op}(g) - \mathrm{Op}(g_1) \in \mathcal{Y}_G^{-\infty}(X^\wedge \times \Omega, \boldsymbol{g}; N_-, N_+)$, *where*

$$g_1(y, \eta) \sim \sum_\alpha \frac{1}{\alpha!} D_\eta^\alpha \partial_{y'}^\alpha g(y, y', \eta)|_{y'=y}.$$

An analogous result holds for the weakly discrete asymptotics.

Proof. The scheme to prove Theorem 1.1.30 can be applied in the present situation, and we then obtain for every $N \in \mathbb{N}$ and suitable $M = M(N)$ a decomposition

$$\mathrm{Op}(g) = \sum_{|\alpha| \le M} \mathrm{Op}(g_\alpha) + \mathrm{Op}(g_N)$$

for $g_\alpha(y, \eta) = \frac{1}{\alpha!} D_\eta^\alpha \partial_{y'}^\alpha g(y, y', \eta)|_{y'=y}$. Applying Lemma 3.4.7 it follows that $g_N(y, y', \eta) \in \mathcal{R}_G^{\nu-2N}(\Omega \times \Omega \times \mathbb{R}^q, \boldsymbol{g}; N_-, N_+)$. The asymptotic types in g_α and g_N can be chosen to be independent of α and N. Thus, applying Proposition 3.3.11 we can form $g_1(y, \eta) \sim \sum_\alpha g_\alpha(y, \eta)$ in $\mathcal{R}_G^\nu(\Omega \times \mathbb{R}^q, \boldsymbol{g}; N_-, N_+)$, and then $\mathrm{Op}(g) - \mathrm{Op}(g_1) \in \mathcal{Y}_G^{-\infty}(X^\wedge \times \Omega, \boldsymbol{g}; N_-, N_+)$. The considerations for the weakly discrete asymptotics are the same. □

By interchanging the role of y and y' in the latter proof we obtain the following:

Remark 3.4.10 *For every* $g_1(y, \eta) \in \mathcal{R}_G^\nu(\Omega_y \times \mathbb{R}^q, \boldsymbol{g}; N_-, N_+)$ *there exists a* $g_2(y', \eta) \in \mathcal{R}_G^\nu(\Omega_{y'} \times \mathbb{R}^q, \boldsymbol{g}; N_-, N_+)$ *such that*

$$\mathrm{Op}(g_1) - \mathrm{Op}(g_2) \in \mathcal{Y}_G^{-\infty}(X^\wedge \times \Omega, \boldsymbol{g}; N_-, N_+),$$

where $g_2(y', \eta) \sim \sum_\alpha \frac{1}{\alpha!} (-D_\eta)^\alpha \partial_y^\alpha g_1(y, \eta)|_{y=y'}$. *An analogous result holds for the weakly discrete asymptotics.*

Similarly to the scalar theory we call $g_1(y, \eta)$ a complete symbol and $g_2(y', \eta)$ a dual symbol of the corresponding Green operator.

Proposition 3.4.11 *For every* $G \in \mathcal{Y}_G^\nu(X^\wedge \times \Omega, \boldsymbol{g}; N_-, N_+)$ *there exists a* $G \in \mathcal{Y}_G^\nu(X^\wedge \times \Omega, \boldsymbol{g}; N_-, N_+)$ *which is properly supported with respect to the* y-*variables (i.e., in the sense of a pseudo–differential operator in* $L^\nu(\Omega; \mathcal{K}^{s,\gamma}(X^\wedge) \oplus \mathbb{C}^{N_-}, \mathcal{K}^{\infty,\varrho}(X^\wedge) \oplus \mathbb{C}^{N_+}))$ *with* $G - G_0 \in \mathcal{Y}_G^{-\infty}(X^\wedge \times \Omega, \boldsymbol{g}; N_-, N_+)$. *An analogous result holds for the weakly discrete asymptotics.*

Proof. It suffices to assume $G = \mathrm{Op}(g)$ for a $g(y, y', \eta) \in \mathcal{R}_G^\nu(\Omega \times \Omega \times \mathbb{R}^q, \boldsymbol{g}; N_-, N_+)$. Let us choose a function $\omega(y, y') \in C^\infty(\Omega \times \Omega)$ with proper support which equals 1 in an open neighbourhood of $\mathrm{diag}(\Omega \times \Omega)$. Then we have $g_0(y, y', \eta) := \omega(y, y')g(y, y', \eta)$ and $g_1(y, y', \eta) = (1 - \omega(y, y'))g(y, y', \eta) \in \mathcal{R}_G^\nu(\Omega \times \Omega \times \mathbb{R}^q, \boldsymbol{g}; N_-, N_+)$, and $G_0 = \mathrm{Op}(g_0)$ is properly supported. To $G_1 = G - G_0 = \mathrm{Op}(g_1)$ we can apply Lemma 3.4.7 for arbitrary N. This gives us $\mathrm{Op}(g_1) \in \mathcal{Y}_G^{-\infty}(X^\wedge \times \Omega, \boldsymbol{g}; N_-, N_+)$. The arguments for the weakly discrete asymptotics are the same. $\square$

Remark 3.4.12 *Let* $G \in \mathcal{Y}_G^\nu(X^\wedge \times \Omega, \boldsymbol{g}; N_-, N_+)$ *be properly supported with respect to the* y-*variables. Then (using the notation of Theorem 3.4.3) G induces continuous operators*

$$G : \mathcal{W}_{\mathrm{loc}(y)}^s(\Omega, \mathcal{K}^{r,\gamma}(X^\wedge)) \oplus H_{\mathrm{loc}}^s(\Omega, \mathbb{C}^{N_-})$$
$$\to \mathcal{W}_{\mathrm{loc}(y)}^{s-\nu}(\Omega, \mathcal{K}_P^{\infty,\varrho}(X^\wedge)) \oplus H_{\mathrm{loc}}^s(\Omega, \mathbb{C}^{N_+})$$

as well as between the corresponding spaces with "comp" on both sides, for all $s, r \in \mathbb{R}$. *The same is true for the weakly discrete asymptotics.*

Theorem 3.4.13 $C \in \mathcal{Y}_G^\nu(X^\wedge \times \Omega, \boldsymbol{g}; N_-, N_+)$ *for* $\boldsymbol{g} = (\gamma, \varrho, \Theta)$ *implies for the formal adjoint* $G^* \in \mathcal{Y}_G^\nu(X^\wedge \times \Omega, \boldsymbol{g}^*; N_+, N_-)$ *for* $\boldsymbol{g}^* = (-\varrho, -\gamma, \Theta)$ *and* $\sigma_\wedge^\nu(G^*)(y, \eta) = \sigma_\wedge^\nu(G)^*(y, \eta)$, *cf. also Proposition 3.3.41. Analogous relations hold for the weakly discrete asymptotics.*

Proof. In view of Definition 3.4.1 we may assume $G = \mathrm{Op}(g)$ for a $g(y, y', \eta) \in \mathcal{R}_G^\nu(\Omega \times \Omega \times \mathbb{R}^q, \boldsymbol{g}; N_-, N_+)$. Then $\mathrm{Op}(g)^* = \mathrm{Op}(g^*)$ for a symbol $g^*(y, y', \eta) \in \mathcal{R}_G^\nu(\Omega \times \Omega \times \mathbb{R}^q, \boldsymbol{g}^*; N_+, N_-)$ that follows by applying the point–wise adjoint and interchanging the role of y and y'. Thus $G^* = \mathcal{Y}_G^\nu(X^\wedge \times \Omega, \boldsymbol{g}^*; N_+, N_-)$. Applying Proposition 3.4.9 we obtain the asserted symbolic rule. For the weakly discrete asymptotics we can argue in the same manner. $\square$

Theorem 3.4.14 *Let* $G \in \mathcal{Y}_G^\nu(X^\wedge \times \Omega, \boldsymbol{g}; N_-, N_0)$, $\boldsymbol{g} = (\gamma, \varrho, \Theta)$, *and* $K \in \mathcal{Y}_G^\kappa(X^\wedge \times \Omega, \boldsymbol{h}; N_0, N_+)$, $\boldsymbol{h} = (\varrho, \sigma, \Theta)$, *and let* G *or* K *be properly supported (in the y–variables). Then we have* $KG \in \mathcal{Y}_G^{\kappa+\nu}(X^\wedge \times \Omega, \boldsymbol{c}; N_-, N_+)$ *for* $\boldsymbol{c} = (\gamma, \sigma, \Theta)$, *and*

$$\sigma_\wedge^{\kappa+\nu}(KG)(y, \eta) = \sigma_\wedge^\kappa(K)(y, \eta)\sigma_\wedge^\nu(G)(y, \eta). \qquad (3.4.4)$$

An analogous result holds for the weakly discrete asymptotics.

Proof. If G is properly supported we decompose K as $K_0 + K_1$ with properly supported K_0 and $K_1 \in \mathcal{Y}_G^{-\infty}$, cf. Proposition 3.4.11. Then $K_1 G \in \mathcal{Y}_G^{-\infty}$, according to Definition 3.4.1, Remark 3.4.12 and Theorem 3.4.13, such that it remains to consider $K_0 G$. On the other hand, if K is properly supported we can write $G = G_0 + G_1$ as in Proposition 3.4.11 and conclude $KG_1 \in \mathcal{Y}_G^{-\infty}$, so that it remains KG_0. In other words, without loss of generality we assume that both K and G are properly supported. Applying Proposition 3.4.9 we can write $K = \mathrm{Op}(k) + K_2$ for a $k(y, \eta) \in \mathcal{R}_G^\kappa$ and $K_2 \in \mathcal{Y}_G^{-\infty}$. Moreover, in view of Remark 3.4.10 we can reformulate G as $G = \mathrm{Op}(g) + G_2$ for a $g(y', \eta) \in \mathcal{R}_G^\nu$ and $G_2 \in \mathcal{Y}_G^{-\infty}$. We have $KG = \mathrm{Op}(k)G + K_2 G$. From Definition 3.4.1, Theorem 3.4.13 and Remark 3.4.12 we obtain that $K_2 G \in \mathcal{Y}_G^{-\infty}$. In order to characterise $\mathrm{Op}(k)G$ it suffices to insert functions $u \in C_0^\infty(\Omega, C_0^\infty(X^\wedge) \oplus \mathbb{C}^{N-})$ with $\mathrm{supp}_{(y)} u \subseteq M$ for an arbitrary fixed compact set $M \subset\subset \Omega$. We have $\mathrm{Op}(k)Gu = \mathrm{Op}(k)\mathrm{Op}(g)u + \mathrm{Op}(k)G_2 u$. Since $\mathrm{Op}(k)Gu$ and $\mathrm{Op}(k)\mathrm{Op}(g)u$ are well–defined, cf. Remark 3.4.5, $\mathrm{Op}(k)G_2 u$ also exists. We want to make this more transparent by introducing cut–off factors in the middle. Since G is properly supported, we obtain for a suitable $\psi \in C_0^\infty(\Omega)$ that $Gu = \psi Gu$ for all u with $\mathrm{supp}_{(y)} u \subseteq M$. Thus $\mathrm{Op}(k)Gu = \mathrm{Op}(k)\psi \mathrm{Op}(g)u + \mathrm{Op}(k)\psi G_2 u$. It is clear that $\mathrm{Op}(k)\psi G_2 \in \mathcal{Y}_G^{-\infty}$. Moreover, from Proposition 3.4.9 we get $\mathrm{Op}(k)\psi = \mathrm{Op}(f) + K_3$ for an $f(y, \eta) \in \mathcal{R}_G^\kappa$ and some $K_3 \in \mathcal{Y}_G^{-\infty}$, where here K_3 is defined on $\mathcal{W}^{-\infty}(\mathbb{R}^q, \mathcal{K}^{-\infty, \varrho}(X^\wedge) \oplus \mathbb{C}^{N_0})$. Clearly $\mathrm{Op}(g)u \in \mathcal{W}^\infty(\mathbb{R}^q, \mathcal{K}^{\infty, \varrho}(X^\wedge) \oplus \mathbb{C}^{N_0})$, so $K_3 \mathrm{Op}(g)$ exists and belongs to $\mathcal{Y}_G^{-\infty}$, because of the known mapping properties of the factors, including their formal adjoints. Finally $\mathrm{Op}(f)\mathrm{Op}(g) = \mathrm{Op}(b)$ for

$b(y, y', \eta) = f(y, \eta)g(y', \eta) \in \mathcal{R}_G^{\kappa+\nu}$, i.e., $\mathrm{Op}(f)\,\mathrm{Op}(g) \in \mathcal{Y}_G^{\kappa+\nu}$, as asserted. From this and because of Proposition 3.4.9 we immediately obtain $\sigma_\wedge^{\kappa+\nu}(\mathrm{Op}(f)\,\mathrm{Op}(g))(y, \eta) = \sigma_\wedge^\kappa(f)(y, \eta)\sigma_\wedge^\nu(g)(y, \eta)$. Moreover,

$$\sigma_\wedge^\kappa(f)(y, \eta) = \psi(y)\sigma_\wedge^\kappa(k)(y, \eta) = \psi(y)\sigma_\wedge^\kappa(K)(y, \eta),$$
$$\sigma_\wedge^\nu(g)(y, \eta) = \sigma_\wedge^\nu(G)(y, \eta),$$

so

$$\sigma_\wedge^{\kappa+\nu}(\mathrm{Op}(k)\psi\,\mathrm{Op}(g))(y, \eta) = \psi(y)\sigma_\wedge^\kappa(K)(y, \eta)\sigma_\wedge^\nu(G)(y, \eta).$$

This shows that the asserted symbolic rule is valid on $M_1 := \{y \in \Omega : \psi(y) = 1\}$. However, if the above M runs through an exhaustion of Ω then so does M_1, so that we obtain (3.4.4) in general. $\qquad\square$

Remark 3.4.15 *The operation of taking the formal adjoint $G \to G^*$ and the composition $(K, G) \to KG$, cf. Theorems 3.4.13 and 3.4.14, respectively, imply the known symbolic rules from the general pseudo–differential calculus also for the complete symbols. In other words, for $G = \mathrm{Op}(g) \bmod \mathcal{Y}_G^{-\infty}$, $K = \mathrm{Op}(k) \bmod \mathcal{Y}_G^{-\infty}$, for symbols $g(y, \eta) \in \mathcal{R}_G^\nu$, $k(y, \eta) \in \mathcal{R}_G^\kappa$, it follows that $G^* = \mathrm{Op}(g^*) \bmod \mathcal{Y}_G^{-\infty}$ and $KG = \mathrm{Op}(f) \bmod \mathcal{Y}_G^{-\infty}$ for symbols $g^*(y, \eta) \in \mathcal{R}_G^\nu$ and $f(y, \eta) \in \mathcal{R}_G^{\kappa+\nu}$, respectively. They allow the asymptotic expansions*

$$g^*(y, \eta) \sim \sum_\alpha \frac{1}{\alpha!} D_\eta^\alpha \partial_y^\alpha g^{(*)}(y, \eta),$$

where $(*)$ denotes the point–wise formal adjoint, and

$$f(y, \eta) \sim \sum_\alpha \frac{1}{\alpha!}(D_\eta^\alpha k(y, \eta))\partial_y^\alpha g(y, \eta),$$

which is the Leibniz rule. It is to be noted here that the assumption of Proposition 3.3.11 on the involved asymptotic types of the summands is satisfied.

Remark 3.4.16 *Writing the operators K and G from Theorem 3.4.14 as block matrices*

$$\begin{pmatrix} K_{11} & K_{12} \\ K_{21} & K_{22} \end{pmatrix} \quad and \quad \begin{pmatrix} G_{11} & G_{12} \\ G_{21} & G_{22} \end{pmatrix},$$

respectively, we obtain from Theorem 3.4.14 a number of individual composition results, namely, in particular, that

$$\{potential\ operator\} \cdot \{trace\ operator\} = \{\text{u.l.c. } Green\ operator\}$$

which corresponds to $K_{12}G_{21}$ (u.l.c. Green operator means Green of upper left corner type), moreover

{*trace operator*} · {*potential operator*}
$$= \{pseudo\text{--}differential\ operator\ on\ the\ edge\},$$

corresponding to $K_{21}G_{12}$. For formal adjoints in the sense of Theorem 3.4.13 we have, in particular,

$$\{potential\ operator\}^* = \{trace\ operator\},$$
$$\{trace\ operator\}^* = \{potential\ operator\}.$$

3.4.2   Edge Mellin operators

Definition 3.4.17 *Let $g = (\gamma, \delta, \Theta)$ for $\gamma, \delta \in \mathbb{R}$, $\Theta = (-k, 0]$, $k \in (\mathbb{N} \setminus \{0\}) \cup \{\infty\}$ and $\nu \in \mathbb{R}$, $\delta + \nu \leq \gamma$. Then*

$$\mathcal{Y}^{\nu}_{M+G}(X^{\wedge} \times \Omega, g; N_-, N_+) \qquad (3.4.5)$$

for $N_-, N_+ \in \mathbb{N}$ denotes the space of all operators $\mathrm{Op}(a) + C$ for arbitrary $a(y, \eta) \in \mathcal{R}^{\nu}_{M+G}(\Omega \times \mathbb{R}^q, g; N_-, N_+)$ and $C \in \mathcal{Y}^{-\infty}(X^{\wedge} \times \mathbb{R}^q, g; N_-, N_+)$. The elements of $\mathcal{Y}^{\nu}_{M+G}(X^{\wedge} \times \mathbb{R}^q, g; N_-, N_+)$ are called the smoothing Mellin + Green edge operators of the edge operator algebra with continuous asymptotics. Furthermore we denote by

$$Y^{\nu}_{M+G}(X^{\wedge} \times \Omega, g) \qquad (3.4.6)$$

the space of upper left corners, i.e., the space (3.4.5) for $N_- = N_+ = 0$. Analogously we introduce the subclasses with weakly discrete asymptotics

$$\mathcal{Y}^{\nu}_{M+G}(X^{\wedge} \times \Omega, g; N_-, N_+)^{\bullet} \qquad and \qquad Y^{\nu}_{M+G}(X^{\wedge} \times \Omega, g)^{\bullet},$$

respectively.

In the sequel we mainly concentrate on the case of upper left corners, i.e., when $N_- = N_+ = 0$. The other entries of the block matrices were studied in the preceding section. Morever, the arguments in the proofs will be given for the continuous asymptotics. If we do not say anything else they are valid in an analogous form in the weakly discrete case.

In view of Proposition 3.3.20 we have

$$Y^{\nu}_{M+G}(X^{\wedge} \times \Omega, g) \subset L^{\nu}_{\mathrm{cl}}(\Omega; \mathcal{K}^{s,\gamma}(X^{\wedge}), \mathcal{K}^{\infty,\delta}(X^{\wedge}))$$

for all $s \in \mathbb{R}$. Thus each $W \in Y^\nu_{M+G}(X^\wedge \times \Omega, \boldsymbol{g})$ has a homogeneous principal operator–valued symbol of order ν which is an operator family

$$\sigma^\nu_\wedge(W)(y, \eta) : \mathcal{K}^{s,\gamma}(X^\wedge) \to \mathcal{K}^{\infty,\delta}(X^\wedge)$$

for every $s \in \mathbb{R}$, $(y, \eta) \in \Omega \times (\mathbb{R}^q \setminus \{0\})$, satisfying

$$\sigma^\nu_\wedge(W)(y, \lambda\eta) = \lambda^\nu \kappa_\lambda \sigma^\nu_\wedge(W)(y, \eta) \kappa_\lambda^{-1}$$

for all $\lambda \in \mathbb{R}_+$. If $W = \mathrm{Op}(w) \bmod Y^{-\infty}(X^\wedge \times \Omega, \boldsymbol{g})$ for a $w(y, y', \eta) \in R^\nu_{M+G}(\Omega \times \Omega \times \mathbb{R}^q, \boldsymbol{g})$ then

$$\sigma^\nu_\wedge(W)(y, \eta) = \sigma^\nu_\wedge(w)(y, y', \eta)|_{y'=y},$$

cf. the notation of Section 3.3.2. The explicit form of $\sigma^\nu_\wedge(W)(y, \eta)$ was given above in the formula (3.3.25), up to summands of pure Green type.

Note that we also have the subordinate conormal symbols from the cone theory, namely

$$\sigma^{\nu-j}_M(W)(y, z, \eta) := \sigma^{\nu-j}_M(w)(y, z, \eta),$$

$j = 0, \ldots, k-1$, when $W = \mathrm{Op}(w) \bmod Y^{-\infty}(X^\wedge \times \Omega, \boldsymbol{g})$, $w(y, \eta) \in R^\nu_{M+G}(\Omega \times \mathbb{R}^q, \boldsymbol{g})$.

Theorem 3.4.18 *Each $A \in Y^\nu_{M+G}(X^\wedge \times \Omega, \boldsymbol{g})$ for $\boldsymbol{g} = (\gamma, \delta, \Theta)$ induces continuous operators*

$$A : \mathcal{W}^s_{\mathrm{comp}(y)}(\Omega, \mathcal{K}^{r,\gamma}(X^\wedge)) \to \mathcal{W}^{s-\nu}_{\mathrm{loc}(y)}(\Omega, \mathcal{K}^{\infty,\delta}(X^\wedge)) \qquad (3.4.7)$$

for all $s, r \in \mathbb{R}$. Moreover, for every $P \in \mathrm{As}(X, (\gamma, \Theta))$ there exists a $Q \in \mathrm{As}(X, (\delta, \Theta))$ such that A restricts to continuous operators

$$A : \mathcal{W}^s_{\mathrm{comp}(y)}(\Omega, \mathcal{K}^{r,\gamma}_P(X^\wedge)) \to \mathcal{W}^{s-\nu}_{\mathrm{loc}(y)}(\Omega, \mathcal{K}^{\infty,\delta}_Q(X^\wedge)) \qquad (3.4.8)$$

for all $s, r \in \mathbb{R}$. In particular, $A \in Y^\nu_{M+G}(X^\wedge \times \Omega, \boldsymbol{g})^\bullet$ restricts to continuous operators (3.4.8) for every $P \in \mathrm{As}(X, (\gamma, \Theta))^\bullet$ for some resulting $Q \in \mathrm{As}(X, (\delta, \Theta))^\bullet$, for all $s, r \in \mathbb{R}$.

Proof. The assertion is a consequence of Proposition (3.3.20) and of the continuity of pseudo–differential operators with operator–valued symbols between the corresponding wedge Sobolev spaces, cf. Theorem 3.3.22; this is evidently valid in analogous form also for the corresponding Fréchet spaces E and $\tilde{E}$, respectively. $\square$

Remark 3.4.19 *The restriction of a* $W \in Y^{\nu}_{M+G}(X^{\wedge} \times \Omega, \boldsymbol{g})$ *to* $C_0^{\infty}(X^{\wedge} \times \Omega)$ *induces a continuous operator*

$$W : C_0^{\infty}(X^{\wedge} \times \Omega) \to C^{\infty}(X^{\wedge} \times \Omega). \tag{3.4.9}$$

W *is an integral operator with kernel in* $C^{\infty}(X^{\wedge} \times \Omega \times X^{\wedge} \times \Omega)$.

In fact, in view of Proposition 3.4.6 it suffices to consider the case $W = \text{Op}(m)$ for an operator–valued symbol of the form (3.3.9). Let us choose an $N \in \mathbb{N}$ and write $W = t^{-N} \text{Op}(m_N)$ for $m_N := t^N m$. Then, in view of Remark 3.3.21, we have $m_N(y, \eta) \in \mathcal{R}_G^{\nu-N}$ and hence $\text{Op}(m_N)$ is of Green type. Therefore, because of Proposition 3.4.6 the operator $\text{Op}(m_N)$ has an integral kernel of the asserted kind and then the same is true of $t^{-N} \text{Op}(m_N)$. This gives us the continuity (3.4.9).

Lemma 3.4.20 *Let* $m(y, y', \eta) \in R^{\nu}_{M+G}(\Omega \times \Omega \times \mathbb{R}^q, \boldsymbol{g})$ *and assume that* $|y - y'|^{-2N} m(y, y', \eta) \in R^{\nu}_{M+G}(\Omega \times \Omega \times \mathbb{R}^q, \boldsymbol{g})$ *for some* $N \in \mathbb{N}$. *Then there exists an* $m_N \in R^{\nu-2N}_{M+G}(\Omega \times \Omega \times \mathbb{R}^q, \boldsymbol{g})$ *such that* $\text{Op}(m) = \text{Op}(m_N)$. *An analogous result holds for symbols with weakly discrete asymptotics.*

The proof is an obvious generalisation of the arguments for Lemma 1.1.29, cf. also Remark 3.3.25.

Proposition 3.4.21 $w(y, y', \eta) \in R^{\nu}_{M+G}(\Omega \times \Omega \times \mathbb{R}^q, \boldsymbol{g})$ *implies* $\text{Op}(w) \in Y^{\nu}_{M+G}(X^{\wedge} \times \Omega, \boldsymbol{g})$, *and there exists a* $w_1(y, \eta) \in R^{\nu}_{M+G}(\Omega \times \mathbb{R}^q, \boldsymbol{g})$ *with* $\text{Op}(w) - \text{Op}(w_1) \in Y^{-\infty}(X^{\wedge} \times \Omega, \boldsymbol{g})$, *where*

$$w_1(y, \eta) \sim \sum_{\alpha} \frac{1}{\alpha!} D_{\eta}^{\alpha} \partial_{y'}^{\alpha} w(y, y', \eta)|_{y'=y}.$$

An analogous result holds for the weakly discrete asymptotics.

Proof. We can proceed similarly to Proposition 3.4.9, and write

$$\text{Op}(w) = \sum_{|\alpha| \leq M} \text{Op}(w_{\alpha}) + \text{Op}(w_N)$$

for $w_{\alpha}(y, \eta) = \frac{1}{\alpha!} D_{\eta}^{\alpha} \partial_{y'}^{\alpha} w(y, y', \eta)|_{y'=y}$, $M = M(N)$. From Lemma 3.4.20 it follows that $w_N(y, y', \eta) \in R^{\nu-2N}_{M+G}$, where w_N is even a Green symbol for sufficiently large N, due to Remark 3.3.25. To w_N we can apply Proposition 3.4.9 and obtain the assertion. $\square$

By interchanging the role of y and y' in the latter proof we obtain the following result:

Remark 3.4.22 *For every* $w_1(y,\eta) \in R^\nu_{M+G}(\Omega_y \times \mathbb{R}^q, \boldsymbol{g})$ *there exists a* $w_2(y,\eta) \in R^\nu_{M+G}(\Omega_{y'} \times \mathbb{R}^q, \boldsymbol{g})$ *with* $\mathrm{Op}(w_1) - \mathrm{Op}(w_2) \in Y^{-\infty}_G(X^\wedge \times \Omega, \boldsymbol{g})$, *where* $w_2(y',\eta) \sim \sum_\alpha \frac{1}{\alpha!}(-D_\eta)^\alpha \partial_y^\alpha w_1(y,\eta)|_{y'=y}$. *An analogous result is true for the discrete asymptotics.*

If $W = \mathrm{Op}(w_1) \bmod Y^{-\infty}_G(X^\wedge \times \Omega, \boldsymbol{g})$, for a $w_1(y,\eta) \in R^\nu_{M+G}(\Omega \times \mathbb{R}^q, \boldsymbol{g})$, we call $w_1(y,\eta)$ a complete symbol of W. For $W = \mathrm{Op}(w_2) \bmod Y^{-\infty}_G(X^\wedge \times \Omega, \boldsymbol{g})$ for a symbol $w_2(y',\eta) \in R^\nu_{M+G}(\Omega_{y'} \times \mathbb{R}^q, \boldsymbol{g})$, we call $w_2(y',\eta)$ a dual symbol of W.

Proposition 3.4.23 *To each* $W \in Y^\nu_{M+G}(X^\wedge \times \Omega, \boldsymbol{g})$ *there exists a* $W_0 \in Y^\nu_{M+G}(X^\wedge \times \mathbb{R}^q, \boldsymbol{g})$ *which is properly supported with respect to the y–variables, such that* $W - W_0 \in Y^{-\infty}_G(X^\wedge \times \Omega, \boldsymbol{g})$. *An analogous result holds for the weakly discrete asymptotics.*

Proof. Without loss of generality we assume $W = \mathrm{Op}(w)$ for a $w \in R^\nu_{M+G}$. We can proceed as in the proof of Proposition 3.4.11. With these notations we can set $w_0(y,y',\eta) = \omega(y,y')w(y,y',\eta)$, $w_1(y,y',\eta) = (1 - \omega(y,y'))w(y,y',\eta)$. Then $w_0 \in R^\nu_{M+G}$, and $W_0 = \mathrm{Op}(w_0)$ is properly supported, while $w_1 \in R^\nu_{M+G}$ satisfies the assumptions of Lemma 3.4.20 for each N. For sufficiently large N we obtain a $w_{1,N} \in R^{\nu-2N}_{M+G}$ such that $\mathrm{Op}(w_1) = \mathrm{Op}(w_{1,N})$, where in fact $w_{1,N} \in R^{\nu-2N}_G$, cf. Remark 3.3.25. Thus the proof is complete if we note that $\mathrm{Op}(w_{1,N}) \in Y^{-\infty}_G$. $\qquad\square$

Remark 3.4.24 *Let* $W \in Y^\nu_{M+G}(X^\wedge \times \Omega, \boldsymbol{g})$ *be properly supported with respect to the y–variables. Then (with the notation of Theorem 3.4.18) W induces continuous operators*

$$W : \mathcal{W}^s_{\mathrm{loc}(y)}(\Omega, \mathcal{K}^{r,\gamma}(X^\wedge)) \to \mathcal{W}^{s-\nu}_{\mathrm{loc}(y)}(\Omega, \mathcal{K}^{\infty,\delta}(X^\wedge))$$

and

$$W : \mathcal{W}^s_{\mathrm{loc}(y)}(\Omega, \mathcal{K}^{r,\gamma}_P(X^\wedge)) \to \mathcal{W}^{s-\nu}_{\mathrm{loc}(y)}(\Omega, \mathcal{K}^{\infty,\delta}_Q(X^\wedge))$$

as well as between the corresponding spaces with "comp" on both sides, for all $s, r \in \mathbb{R}$. *Analogous relations hold for the weakly discrete asymptotics.*

Theorem 3.4.25 $W \in Y^\nu_{M+G}(X^\wedge \times \Omega, \boldsymbol{g})$ *for* $\boldsymbol{g} = (\gamma, \delta, \Theta)$ *implies for the formal adjoint* $W^* \in Y^\nu_{M+G}(X^\wedge \times \Omega, \boldsymbol{g}^*)$ *for* $\boldsymbol{g}^* = (-\delta, -\gamma, \Theta)$ *and* $\sigma^\nu_\wedge(W^*)(y,\eta) = \sigma^\nu_\wedge(W)^*(y,\eta)$, *cf. also Proposition 3.3.41. Analogous relations hold for the weakly discrete asymptotics.*

Proof. In view of Theorem 3.4.13 and Proposition 3.4.21 we can assume $W = \mathrm{Op}(w)$ for a $w(y,\eta) \in R^\nu_{M+G}(\Omega \times \mathbb{R}^q, \boldsymbol{g})$. Then $\mathrm{Op}(w)^* =$

$\mathrm{Op}(w^*)$ for a symbol $w^*(y', \eta)$ that follows by taking the point–wise adjoint of $w(y, \eta)$ and then inserting $y = y'$. By Corollary 3.3.19 we have $w^*(y', \eta) \in R_{M+G}^{\nu}(\Omega \times \mathbb{R}^q, \boldsymbol{g}^*)$. Thus $\mathrm{Op}(w^*) \in Y_{M+G}^{\nu}(\Omega \times \mathbb{R}^q, \boldsymbol{g}^*)$. The asserted symbolic rule is now an immediate consequence of the definition. $\qquad\square$

Theorem 3.4.26 *Let* $A \in \mathcal{Y}_{M+G}^{\nu}(X^{\wedge} \times \Omega, \boldsymbol{g}; N_0, N_+)$ *for* $\boldsymbol{g} = (\gamma, \delta, \Theta)$ *and* $B \in \mathcal{Y}_{M+G}^{\mu}(X^{\wedge} \times \Omega, \boldsymbol{c}; N_-, N_0)$ *for* $\boldsymbol{c} = (\beta, \gamma, \Theta)$, *and let* A *or* B *be properly supported (in the* y–*variables). Then* $AB \in \mathcal{Y}_{M+G}^{\nu+\mu}(X^{\wedge} \times \Omega, \boldsymbol{h}; N_-, N_+)$ *for* $\boldsymbol{h} = (\beta, \delta, \Theta)$, *and*

$$\sigma_{\wedge}^{\nu+\mu}(AB)(y, \eta) = \sigma_{\wedge}^{\nu}(A)(y, \eta)\sigma_{\wedge}^{\mu}(B)(y, \eta).$$

$A \in \mathcal{Y}_{G}^{\nu}(\dots)$ *or* $B \in \mathcal{Y}_{G}^{\mu}(\dots)$ *implies* $AB \in \mathcal{Y}_{G}^{\mu+\nu}(\dots)$. *An analogous result holds for discrete asymptotics.*

Proof. The scheme of the proof is the same as for Theorem 3.4.14. We can carry out all formal steps, using Proposition 3.4.23, Remark 3.4.24, Proposition 3.4.21, Remark 3.4.22 in analogous connection as in the proof of Theorem 3.4.14. We finally have to consider a composition of the form $\mathrm{Op}(v)\psi\,\mathrm{Op}(w)$ for $v(y, \eta) \in \mathcal{R}_{M+G}^{\nu}$, $w(y', \eta) \in \mathcal{R}_{M+G}^{\mu}$, and a certain fixed $\psi \in C_0^{\infty}(\Omega)$. This allows us to apply Theorem 3.3.29 and to establish the asserted rule for the principal symbols under compositions. If one factor is of Green type, then so is the composition, cf. Theorem 3.3.29. $\qquad\square$

Remark 3.4.27 *The operations to take the formal adjoints and compositions in the sense of Theorems 3.4.25, 3.4.26 imply symbolic rules also for the complete symbols, cf. analogously Remark 3.4.15. Moreover, since the symbols of the smoothing Mellin+Green operators are families of smoothing Mellin+ Green operators in the sense of the cone algebra over* $X^{\wedge}$, *the subordinate conormal symbols (3.3.19) can also be calculated and their behaviour expressed under formal adjoints and compositions, similarly to the case of the cone algebra in Chapter 2. This is straightforward and will not be done explicitly here. Let us only note that, in particular, we obtain for the behaviour of the sequence of the conormal symbols under compositions, a combination between the Mellin translation product with respect to the cone axis variable* t *and the Leibniz product with respect to the edge variable* y.

Remark 3.4.28 *Writing the operators* A *and* B *as block matrices, similarly to Remark 3.4.16, we obtain from Theorem 3.4.26 a number of individual composition results which show, in particular, that*

$$\{\text{smoothing Mellin } + \text{ Green operator}\}\{\text{potential operator}\}$$
$$= \{\text{potential operator}\},$$

{*trace operator*}{*smoothing Mellin + Green operator*}

$$= \{trace\ operator\}$$

for an arbitrary smoothing Mellin + Green operator of the type of an upper left corner.

3.4.3   The local edge operator algebra

This section will introduce a subalgebra of the edge pseudo-differential operators in local form on $X^{\wedge} \times \Omega$ for an open set $\Omega \subseteq \mathbb{R}^q$. The global operators on a manifold with edges will be discussed in Section 3.4.4 below.

Definition 3.4.29 *Let* $g = (\gamma, \gamma - \mu, \Theta)$ *be fixed weight data for* $\gamma, \mu \in \mathbb{R}$, $\Theta = (-k, 0]$, $k \in (\mathbb{N} \setminus \{0\}) \cup \{\infty\}$, *and let* $\nu \in \mathbb{R}$ *with* $\mu - \nu \in \mathbb{N}$. *Then*

$$\mathcal{Y}_0^{\nu}(X^{\wedge} \times \Omega, g; N_-, N_+) \tag{3.4.10}$$

for $N_-, N_+ \in \mathbb{N}$ *denotes the space of all operators* $\mathrm{Op}(a) + C$ *for arbitrary* $a(y, y', \eta) \in \mathcal{R}^{\nu}(\Omega \times \Omega \times \mathbb{R}^q, g; N_-, N_+)$ *and* $C \in \mathcal{Y}_G^{-\infty}(X^{\wedge} \times \Omega, g; N_-, N_+)$, *cf. Definition 3.3.40, Remark 3.3.46, Definition 3.4.1. The elements of* $\mathcal{Y}_0^{\nu}(X^{\wedge} \times \Omega, g; N_-, N_+)$ *are called edge pseudo-differential operators with continuous asymptotics. Furthermore we denote by*

$$Y_0^{\nu}(X^{\wedge} \times \Omega, g)$$

the space of upper left corner, i.e., the space (3.4.10), *where* $N_- = N_+ = 0$. *Analogously we introduce the subclasses with weakly discrete asymptotics*

$$\mathcal{Y}_0^{\nu}(X^{\wedge} \times \Omega, g; N_-, N_+)^{\bullet} \quad and \quad Y_0^{\nu}(X^{\wedge} \times \Omega, g)^{\bullet},$$

respectively.

Subscript "0" indicates the fact that the complete interior symbols can be chosen in such way that they have bounded support with respect to t, cf. the assumptions in Definition 3.3.30 above.

It will be reasonable in all considerations to first consider the case of a finite weight interval $\Theta = (-k, 0]$. All results of the calculus here then have the corresponding analogues for $\Theta = (-\infty, 0]$.

We will mainly concentrate on the case of upper left corners in the block matrices. The other entries were studied in Section 3.4.1. Moreover, the arguments in the proofs will be given for the continuous asymptotics. If we do not indicate anything to the contrary, they are valid in analogous form also in the weakly discrete case.

In view of Proposition 3.3.38 and because of

$$Y_G^{-\infty}(X^\wedge \times \Omega, \boldsymbol{g}) \subset L^{-\infty}(\Omega; \mathcal{K}^{s,\gamma}(X^\wedge), \mathcal{K}^{\infty,\gamma-\mu}(X^\wedge))$$

we have

$$Y_0^\nu(X^\wedge \times \Omega, \boldsymbol{g}) \subset L^\nu(\Omega; \mathcal{K}^{s,\gamma}(X^\wedge), \mathcal{K}^{s-\nu,\gamma-\mu}(X^\wedge)) \qquad (3.4.11)$$

for all $s \in \mathbb{R}$. For analogous reasons, for every $A \in Y_0^\nu(X^\wedge \times \Omega, \boldsymbol{g})$ and for every $P \in \mathrm{As}(X, (\gamma, \Theta))$ there exists a $Q \in \mathrm{As}(X, (\gamma - \mu, \Theta))$ such that

$$A \in L^\nu(\Omega; \mathcal{K}_P^{s,\gamma}(X^\wedge), \mathcal{K}_Q^{s-\nu,\gamma-\mu}(X^\wedge)).$$

Analogous results hold for the weakly discrete asymptotics. From the general mapping properties of pseudo–differential operators with operator–valued symbols, cf. Theorem 1.3.58, we thus obtain the following result:

Theorem 3.4.30 *Each $A \in Y_0^\nu(X^\wedge \times \Omega, \boldsymbol{g})$ for $\boldsymbol{g} = (\gamma, \gamma - \mu, \Theta)$ induces continuous operators*

$$A : \mathcal{W}_{\mathrm{comp}(y)}^s(\Omega, \mathcal{K}^{r,\gamma}(X^\wedge)) \to \mathcal{W}_{\mathrm{loc}(y)}^{s-\nu}(\Omega, \mathcal{K}^{r-\nu,\gamma-\mu}(X^\wedge)) \qquad (3.4.12)$$

for all $s, r \in \mathbb{R}$. Moreover, for every $P \in \mathrm{As}(X, (\gamma, \Theta))$ there exists a $Q \in \mathrm{As}(X, (\gamma - \mu, \Theta))$ such that (3.4.12) restricts to continuous operators

$$A : \mathcal{W}_{\mathrm{comp}(y)}^s(\Omega, \mathcal{K}_P^{r,\gamma}(X^\wedge)) \to \mathcal{W}_{\mathrm{loc}(y)}^{s-\nu}(\Omega, \mathcal{K}_Q^{r-\nu,\gamma-\mu}(X^\wedge)) \qquad (3.4.13)$$

for all $s, r \in \mathbb{R}$. In particular, $A \in Y_0^\nu(X^\wedge \times \Omega, \boldsymbol{g})^\bullet$ restricts to continuous operators (3.4.12) for every $P \in \mathrm{As}(X, (\gamma, \Theta))^\bullet$ with some resulting $Q \in \mathrm{As}(X, (\gamma - \mu, \Theta))^\bullet$, for all $s, r \in \mathbb{R}$.

Remark 3.4.31 *Let $A \in Y_0^\nu(X^\wedge \times \Omega, \boldsymbol{g})$ be an operator such that for every $N \in \mathbb{N}$ there exists a symbol $a_N(y, y', \eta) \in R^{\nu-N}(\Omega \times \Omega \times \mathbb{R}^q, \boldsymbol{g})$ such that $R_N := A - \mathrm{Op}(a_N) \in Y_0^{\nu-N}(X^\wedge \times \Omega, \boldsymbol{g})$, where the asymptotic types in the Green symbols contained in a_N and those in the Green operators of R_N are independent of N. Then $A \in Y_G^{-\infty}(X^\wedge \times \Omega, \boldsymbol{g})$. An analogous result holds for the weakly discrete asymptotics.*

Remark 3.4.32 *We have $Y_0^\nu(X^\wedge \times \Omega, \boldsymbol{g}) \subset L_{\mathrm{cl}}^\nu(X^\wedge \times \Omega)$.*

Lemma 3.4.33 *Let $a(y, y', \eta) \in R^\nu(\Omega \times \Omega \times \mathbb{R}^q, \boldsymbol{g})$ and assume that*

$$|y - y'|^{-2N} a(y, y', \eta) \in R^\nu(\Omega \times \Omega \times \mathbb{R}^q, \boldsymbol{g})$$

for some $N \in \mathbb{N}$. Then there exists an $a_N(y, y', \eta) \in R^{\nu-2N}(\Omega \times \Omega \times \mathbb{R}^q, \boldsymbol{g})$ such that $\mathrm{Op}(a) = \mathrm{Op}(a_N)$. An analogous result holds for symbols with discrete asymptotics.

Proof. The proof is an obvious generalisation of the arguments for Lemma 1.1.29 together with Proposition 3.3.39. □

Proposition 3.4.34 *For each $A \in Y_0^\nu(X^\wedge \times \Omega, \boldsymbol{g})$ there exists an $a_1(y, \eta) \in R^\nu(\Omega \times \mathbb{R}^q, \boldsymbol{g})$ such that $A - \mathrm{Op}(a_1) \in Y^{-\infty}(X^\wedge \times \Omega, \boldsymbol{g})$. Writing A in the form $A = \mathrm{Op}(a) + C$ for an $a(y, y', \eta) \in R^\nu(\Omega \times \Omega \times \mathbb{R}^q, \boldsymbol{g})$ and $C \in Y_G^{-\infty}(X^\wedge \times \Omega, \boldsymbol{g})$, then the asymptotic expansion*

$$a_1(y, \eta) \sim \sum \frac{1}{\alpha!} D_\eta^\alpha \partial_{y'}^\alpha a(y, y', \eta)|_{y'=y} \qquad (3.4.14)$$

yields such an $a_1(y, \eta)$. An analogous result holds for the weakly discrete asymptotics.

Proof. Applying the scheme used in the proof of Theorem 1.1.30 we obtain for every $N \in \mathbb{N}$ and suitable $M = M(N)$ a decomposition

$$\mathrm{Op}(a) = \sum_{|\alpha| \leq M} \mathrm{Op}(a_\alpha) + \mathrm{Op}(a_N)$$

for $a_\alpha(y, \eta) \in \frac{1}{\alpha!} D_\eta^\alpha \partial_{y'}^\alpha a(y, y', \eta)|_{y'=y}$. Applying Lemma 3.4.33 it follows that $a_N(y, y', \eta) \in R^{\nu-2N}(\Omega \times \Omega \times \mathbb{R}^q, \boldsymbol{g})$. We shall show that the asymptotic sum (3.4.14) can be carried out in $R^\nu(\Omega \times \mathbb{R}^q, \boldsymbol{g})$, and that then $\mathrm{Op}(a) = \mathrm{Op}(a_1) \bmod Y^{-\infty}$.

To this end we first recall that $a(y, y', \eta)$ has local symbols of the form

$$t^{-\nu} p_j(t, x, y, y', \tau, \xi, \eta) = t^{-\nu} \tilde{p}_j(t, x, y, y', t\tau, \xi, t\eta) \qquad (3.4.15)$$

for certain $\tilde{p}_j(t, x, y, y', \tilde{\tau}, \xi, \tilde{\eta}) \in S_{\mathrm{cl}}^\nu(\overline{\mathbb{R}}_+ \times \Sigma_j \times \Omega \times \Omega \times \mathbb{R}^{1+n+q})$, $j = 1, \ldots, N$, cf. Remark 3.3.46. We may assume that these symbols are invariant with respect to the symbol push–forwards under the coordinate diffeomorphisms with respect to x, up to symbols of order $-\infty$. Here Σ_j are open sets in $\mathbb{R}^n$ belonging to charts $U_j \to \Sigma_j$ for an open covering of X by coordinate neighbourhoods $\{U_1, \ldots, U_N\}$.

Without loss of generality we assume that the symbols (3.4.15) vanish for $t > \mathrm{const}$. Recall that we fixed a system of diffeomorphisms $\varepsilon_j : U_j \to V_{1,j}$ for open sets $V_{1,j} \subset S^n$ and associated maps $\varrho_j : \mathbb{R}_+ \times U_j \to V_j = \{\tilde{x} \in \mathbb{R}^{n+1} \setminus \{0\} : \tilde{x}/|\tilde{x}| \in V_{1,j}\}$, with $\varrho_j(t, x) = t\varepsilon_j(x)$, $t = |\tilde{x}|$. Then in the coordinates $\tilde{x} \in V_j$ with the covariables $\tilde{\xi}$ the symbols (3.4.15) turn to symbols

$$p_{j,\infty}(\tilde{x}, y, y', \tilde{\xi}, \eta) \in S_{\mathrm{cl}}^\nu(V_j \times \Omega \times \Omega \times \mathbb{R}_{\tilde{\xi},\eta}^{n+1+q}),$$

$j = 1, \ldots, N$. Now for $a_\alpha(y, \eta)$ we have analogously the symbols

$$\frac{1}{\alpha!} D_\eta^\alpha \partial_{y'}^\alpha \{t^{-\nu} p_j(t, x, y, y', \tau, \xi, \eta)\}|_{y'=y}$$

and

$$\frac{1}{\alpha!}D_\eta^\alpha \partial_{y'}^\alpha \{p_{j,\infty}(\tilde{x},y,y',\tilde{\xi},\eta)\}|_{y'=y}$$

which are invariant with respect to the symbol push–forwards in x and compatible with respect to $(t,x) \to \tilde{x}$, everything modulo symbols of order $-\infty$. Let us form the asymptotic sums

$$t^{-\nu}r_j(t,x,y,\tau,\xi,\eta) \sim \sum_\alpha \frac{1}{\alpha!}D_\eta^\alpha \partial_{y'}^\alpha \{t^{-\nu}p_j(t,x,y,y',\tau,\xi,\eta)\}|_{y'=y}$$

and

$$r_{j,\infty}(\tilde{x},y,\tilde{\xi},\eta) \sim \sum_\alpha \frac{1}{\alpha!}D_\eta^\alpha \partial_{y'}^\alpha p_{j,\infty}(\tilde{x},y,y',\tilde{\xi},\eta)|_{y'=y}.$$

We want to fix a choice of r_j and $r_{j,\infty}$ in such a way that the asymptotic sums converge:

$$t^{-\nu}r_j(t,x,y,\tau,\xi,\eta) = \sum_\alpha t^{-\nu}p_{j,\alpha}(t,x,y,\tau,\xi,\eta) \qquad (3.4.16)$$

for symbols

$$p_{j,\alpha}(t,x,y,\tau,\xi,\eta) := \frac{1}{\alpha!}D_\eta^\alpha \partial_{y'}^\alpha p_j(t,x,y,y',\tau,\xi,\eta)|_{y'=y} \cdot \chi(\frac{t\tau}{c_\alpha},\frac{\xi}{c_\alpha},\frac{t\eta}{c_\alpha}),$$

where χ is an excision function and $c_\alpha > 0$ constants, increasing sufficiently fast for $|\alpha| \to \infty$, and analogously

$$r_{j,\infty}(\tilde{x},y,\tilde{\xi},\eta) = \sum_\alpha p_{j,\infty,\alpha}(\tilde{x},y,\tilde{\xi},\eta) \qquad (3.4.17)$$

for $p_{j,\infty,\alpha}(\tilde{x},y,\tilde{\xi},\eta) := \frac{1}{\alpha!}D_\eta^\alpha \partial_{y'}^\alpha p_{j,\infty}(\tilde{x},y,y',\tilde{\xi},\eta)|_{y'=y}\tilde{\chi}(\frac{\tilde{\xi}}{d_\alpha},\frac{\eta}{d_\alpha})$ for an excision function $\tilde{\chi}$ and constants $d_\alpha > 0$, increasing sufficiently fast as $|\alpha| \to \infty$. We obtain symbols $r_j(t,x,y,\tau,\xi,\eta) = \tilde{r}_j(t,x,y,t\tau,\xi,t\eta)$ for certain $\tilde{r}_j(t,x,y,\tilde{\tau},\xi,\tilde{\eta}) \in S_{cl}^\nu(\mathbb{R}_+ \times \Sigma_j \times \Omega \times \mathbb{R}^{1+n+q})$, and $r_{j,\infty}(\tilde{x},y,\tilde{\xi},\eta) \in S_{cl}^\nu(V_j \times \Omega \times \mathbb{R}^{n+1+q})$ satisfying again the above compatibility condition.

The symbols $p_{j,\infty,\alpha}(\tilde{x},y,\tilde{\xi},\eta)$ and $t^{-\nu}p_{j,\alpha}(t,x,y,\tau,\xi,\eta)$ give rise to operator families $b_{\alpha,\infty}(y,\eta) \in C^\infty(\Omega, L_{cl}^{\nu-|\alpha|}(X^\wedge;\mathbb{R}^q)_0)$, $b_{\alpha,\psi}(y,\eta) \in C^\infty(\Omega, \tilde{L}_{cl}^{\nu-|\alpha|}(X^\wedge;\mathbb{R}^q)_0)$, namely

$$b_{\alpha,\infty}(y,\eta) = \sum_{j=1}^N \varphi_j(\varrho_j^{-1})_* \operatorname{op}_{\psi,\tilde{x}}(p_{j,\infty,\alpha})(y,\eta)\psi_j,$$

$$t^{-\nu}b_{\alpha,\psi}(y,\eta) = \sum_{j=1}^N \varphi_j(1 \times \kappa_j^{-1})_* \operatorname{op}_{\psi,(t,x)}(t^{-\nu}p_{j,\alpha})(y,\eta)\psi_j,$$

cf. the notation in Section 3.2.1, $\kappa_j : U_j \to \Sigma_j$ being the diffeomorphisms from there, and $\{\varphi\}_{j=1,\dots,N}$ a partition of unity on X,
$\{\psi_j\}_{j=1,\dots,N}$ a system of functions $\psi_j \in C_0^\infty(U_j)$ with $\varphi_j\psi_j = \varphi_j$
for all j. Moreover, applying Theorem 3.2.7, we can form elements
$\tilde{f}_\alpha(t,y,z,\tilde{\eta}) \in C^\infty(\overline{\mathbb{R}}_+ \times \Omega, M_O^{\nu-|\alpha|}(X;\mathbb{R}_{\tilde{\eta}}^q))$ such that $f_\alpha(t,y,z,\eta) :=$
$\tilde{f}_\alpha(t,y,z,t\eta)$ satisfies

$$\mathrm{op}_M^\delta(f_\alpha)(y,\eta) = b_{\alpha,\psi}(y,\eta) \bmod C^\infty(\Omega, L^{-\infty}(X^\wedge;\mathbb{R}^q)),$$

for every multi–index α.

Analogously the symbols (3.4.17) for $j = 1,\dots,N$ give rise to an
operator family $b_\infty(y,\eta) \in C^\infty(\Omega, L_{\mathrm{cl}}^\nu(X^\wedge;\mathbb{R}^q)_0)$, and the system of
symbols (3.4.16) for $j = 1,\dots,N$ allows us to construct $b_\psi(y,\eta) \in$
$C^\infty(\Omega, \tilde{L}_{\mathrm{cl}}^\nu(X^\wedge;\mathbb{R}^q)_0)$.

The choice of the $\tilde{f}_\alpha$ is possible in such a way that

$$\tilde{f}(t,y,z,\tilde{\eta}) = \sum_\alpha \tilde{f}_\alpha(t,y,z,\tilde{\eta}) \tag{3.4.18}$$

converges in $C^\infty(\overline{\mathbb{R}}_+ \times \Omega, M_O^\nu(X;\mathbb{R}_{\tilde{\eta}}^q))$. Then, for

$$f(t,y,z,\eta) := \tilde{f}(t,y,z,t\eta),$$

it follows that $\mathrm{op}_M^\delta(f)(y,\eta) = b_\psi(y,\eta) \bmod C^\infty(\Omega, L^{-\infty}(X^\wedge;\mathbb{R}^q))$.

According to the notation of Definition 3.4.2 we fix cut–off functions
$\omega, \omega_i, \tilde{\omega}, \tilde{\omega}_i, i = 0,1$, satisfying $\omega\omega_0 = \omega, \omega\omega_1 = \omega_1, \tilde{\omega}\tilde{\omega}_0 = \tilde{\omega}, \tilde{\omega}\tilde{\omega}_1 = \tilde{\omega}_1$
and form the operator families

$$b_{\alpha,0}(y,\eta) = \omega(t[\eta])t^{-\nu}b_{\alpha,M}(y,\eta)\omega_0(t[\eta])$$

for $b_{\alpha,M}(y,\eta) = \mathrm{op}_M^{\gamma-\frac{n}{2}}(f_\alpha)(y,\eta)$,

$$b_{\alpha,1}(y,\eta) = (1 - \omega(t[\eta]))t^{-\nu}b_{\alpha,\psi}(y,\eta)(1 - \omega_1(t[\eta]))$$

and

$$b_\alpha(y,\eta) = \tilde{\omega}(t)\{b_{\alpha,0}(y,\eta) + b_{\alpha,1}(y,\eta)\}\tilde{\omega}_0(t)$$
$$+ (1 - \tilde{\omega}(t))b_{\alpha,\infty}(y,\eta)(1 - \tilde{\omega}_1(t)).$$

Note that in (3.4.16) we obtain from the differentiations with respect
to η a factor $t^{|\alpha|}$ appearing in front of the symbols, such $b_{\alpha,0}(y,\eta)$
also contains in fact such a factor $t^{-\nu+|\alpha|}$, not only $t^{-\nu}$. Thus
$b_\alpha(y,\eta) \in R^{\nu-|\alpha|}(\Omega \times \mathbb{R}^q, \boldsymbol{g})$ for all α. For the operator families $b_\psi(y,\eta)$,
$f(t,y,z,\eta), b_\infty(y,\eta)$ we can do the same, i.e., form

$$b_0(y,\eta) = \omega(t[\eta])t^{-\nu}b_M(y,\eta)\omega_0(t[\eta])$$

for $b_M(y, \eta) = \mathrm{op}_M^{\gamma - \frac{n}{2}}(f)(y, \eta)$,

$$b_1(y, \eta) = (1 - \omega(t[\eta]))t^{-\nu}b_\psi(y, \eta)(1 - \omega_1(t[\eta]))$$

and

$$b(y, \eta) = \tilde{\omega}(t)\left\{b_0(y, \eta) + b_1(y, \eta)\right\}\tilde{\omega}_0(t)$$
$$+ (1 - \tilde{\omega}(t))b_\infty(y, \eta)(1 - \tilde{\omega}_1(t)).$$

It is now evident by the construction that we have $c_\alpha(y, \eta) := a_\alpha(y, \eta) - b_\alpha(y, \eta) \in C^\infty(\Omega, L^{-\infty}(X^\wedge, \mathbb{R}^q))$ and that

$$\tilde{\omega}c_\alpha(y, \eta)\tilde{\omega}_0 \in R_{M+G}^{\nu - |\alpha|}(\Omega \times \mathbb{R}^q, \boldsymbol{g}),$$
$$\mathrm{Op}((1 - \tilde{\omega})c_\alpha(1 - \tilde{\omega}_1)) \in Y_G^{-\infty}(X^\wedge \times \Omega, \boldsymbol{g}).$$

Thus, for each $N \in \mathbb{N}$ and sufficiently large $M = M(N)$ we get

$$\mathrm{Op}(a) = \sum_{|\alpha| \leq M} \mathrm{Op}(a_\alpha) + \mathrm{Op}(a_N)$$
$$= \sum_{|\alpha| \leq M} \mathrm{Op}(b_\alpha) + \sum_{|\alpha| \leq M} \mathrm{Op}(\tilde{\omega}c_\alpha\tilde{\omega}_0) + \mathrm{Op}(a_N) \bmod Y_G^{-\infty}.$$

According to Remark 3.3.21 and Proposition 3.3.11 there is a symbol $w(y, \eta) \in R_{M+G}^\nu(\Omega \times \mathbb{R}^q, \boldsymbol{g})$ such that $w(y, \eta) \sim \sum_\alpha \tilde{\omega}c_\alpha(y, \eta)\tilde{\omega}_0$. Here we used that the occurring asymptotic types in the Green part of $\tilde{\omega}c_\alpha\tilde{\omega}_0$ are independent of α.

Then, for an arbitrary fixed excision function $\psi(\eta) \in C^\infty(\mathbb{R}^q)$ and a choice of constants $e_\alpha > 0$ increasing sufficiently fast as $|\alpha| \to \infty$, it follows that

$$\mathrm{Op}(a) - \mathrm{Op}(b + w) = \mathrm{Op}\left(\sum_{|\alpha| > M}\left(b_\alpha + \psi\left(\frac{\eta}{e_\alpha}\right)\tilde{\omega}c_\alpha\tilde{\omega}_0\right)\right) + \mathrm{Op}(a_N)$$

$$(3.4.19)$$

modulo $Y_G^{-\infty}$. On the right hand side of the latter identity we used the fact that the infinite sum belongs to $R^{\nu - M}(\Omega \times \mathbb{R}^q, \boldsymbol{g})$, due to the above convergent sums (3.4.16), (3.4.17), (3.4.18). Since (3.4.19) is true for every N, Remark 3.4.31 completes the proof. $\qquad\square$

By interchanging the role of y and y' in the latter proof we obtain the following result:

Remark 3.4.35 *For every* $a_1(y, \eta) \in R^\nu(\Omega_y \times \mathbb{R}^q, \boldsymbol{g})$ *there exists an* $a_2(y', \eta) \in R^\nu(\Omega_{y'} \times \mathbb{R}^q, \boldsymbol{g})$ *such that* $\mathrm{Op}(a_1) - \mathrm{Op}(a_2) \in Y_G^{-\infty}(X^\wedge \times \Omega, \boldsymbol{g})$, *where* $a_2(y', \eta) \sim \sum_\alpha \frac{1}{\alpha!}(-D_\eta)^\alpha \partial_y^\alpha a_1(y, \eta)|_{y'=y}$. *An analogous result holds for the weakly discrete asymptotics.*

As in the general theory we call $a_1(y,\eta)$ a complete symbol and $a_2(y',\eta)$ a dual symbol of the corresponding element of $Y^\nu(X^\wedge \times \Omega, \boldsymbol{g})$.

Proposition 3.4.36 *For every $A \in Y_0^\nu(X^\wedge \times \Omega, \boldsymbol{g})$ there exists an $A_0 \in Y_0^\nu(X^\wedge \times \Omega, \boldsymbol{g})$ which is properly supported with respect to the y–variables (in the sense of a pseudo–differential operator on Ω with operator–valued symbols, cf. (3.4.11)) such that $A - A_0 \in Y_G^{-\infty}(X^\wedge \times \Omega, \boldsymbol{g})$. An analogous result holds for discrete asymptotics.*

Proof. In view of Definition 3.4.2 we may assume $A = \text{Op}(a)$ for an $a(y,y',\eta) \in R^\nu(\Omega \times \Omega \times \mathbb{R}^q, \boldsymbol{g})$. Choose a function $\omega(y,y') \in C^\infty(\Omega \times \Omega)$ with proper support, with $\omega(y,y') \equiv 1$ in an open neighbourhood of $\text{diag}(\Omega \times \Omega)$. Then, if we set $a_0 = \omega a$, $a_1 = (1-\omega)a$, the operator $A_0 = \text{Op}(a_0) \in Y_0^\nu$ is properly supported, while a_1 satisfies the assumptions of Lemma 3.4.7 for every N. Then, using Remark 3.4.31, we obtain $A - A_0 = A_1 = \text{Op}(a_1) \in Y_G^{-\infty}$. $\qquad\square$

Remark 3.4.37 *Let $A \in Y_0^\nu(X^\wedge \times \Omega, \boldsymbol{g})$ be properly supported with respect to the y–variables. Then, with the notation of Theorem 3.4.18, A induces continuous operators*

$$A : \mathcal{W}^s_{\text{loc}(y)}(\Omega, \mathcal{K}^{r,\gamma}(X^\wedge)) \to \mathcal{W}^{s-\nu}_{\text{loc}(y)}(\Omega, \mathcal{K}^{r-\nu,\gamma-\mu}(X^\wedge)),$$

$$A : \mathcal{W}^s_{\text{loc}(y)}(\Omega, \mathcal{K}^{r,\gamma}_P(X^\wedge)) \to \mathcal{W}^{s-\nu}_{\text{loc}(y)}(\Omega, \mathcal{K}^{r-\nu,\gamma-\mu}_Q(X^\wedge))$$

for all $s, r \in \mathbb{R}$ and each asymptotic type P for some resulting asymptotic type Q. Analogous assertions hold with "comp" on both sides.

If an operator $A \in Y^\nu(X^\wedge \times \Omega, \boldsymbol{g})$ is written $A = \text{Op}(a) + C$ in the notation of Definition 3.4.29, we set

$$\sigma_\psi^\nu(A) = \sigma_\psi^\nu(a|_{y=y'}), \qquad \sigma_\wedge^\nu(A) = \sigma_\wedge^\nu(a|_{y=y'}),$$

cf. Definition 3.3.35. Then $\sigma_\psi^\nu(A)$ is called the homogeneous principal interior symbol of A of order ν and $\sigma_\wedge^\nu(a)$ the homogeneous principal edge symbol of A of order ν.

Theorem 3.4.38 *$A \in Y_0^\nu(X^\wedge \times \Omega, \boldsymbol{g})$ for $\boldsymbol{g} = (\gamma, \gamma - \mu, \Theta)$ implies for the formal adjoint $A^* \in Y_0^\nu(X^\wedge \times \Omega, \boldsymbol{g}^*)$ for $\boldsymbol{g}^* = (-\gamma + \mu, -\gamma, \Theta)$, where*

$$\sigma_\psi^\nu(A^*) = \overline{\sigma_\psi^\nu(A)}, \qquad \sigma_\wedge^\nu(A^*) = \sigma_\wedge^\nu(A)^*,$$

cf. also Proposition 3.3.41. Analogous relations hold for the weakly discrete asymptotics.

Proof. In view of Definition 3.4.29 and Definition 3.4.1 we may assume $A = \mathrm{Op}(a)$ for an $a(y, y', \eta) \in R^\nu(\Omega \times \mathbb{R}^q, \boldsymbol{g})$. Then $\mathrm{Op}(a)^* = \mathrm{Op}(a^*)$ for a symbol $a^*(y, y', \eta) \in R^\nu(\Omega \times \mathbb{R}^q, \boldsymbol{g}^*)$ that follows by applying the point–wise adjoint, cf. Theorem 3.3.44 and interchanging the role of y and y'. Thus $A^* \in Y_0^\nu(X^\wedge \times \Omega, \boldsymbol{g}^*)$. The symbolic rules follow by applying Proposition 3.4.34. □

Theorem 3.4.39 *Let* $A \in \mathcal{Y}_0^\nu(X^\wedge \times \Omega, \boldsymbol{g}; N_0, N_+)$ *for* $\boldsymbol{g} = (\gamma - \tilde{\mu}, \gamma - \mu - \tilde{\mu}, \Theta)$ *and* $\tilde{A} \in \mathcal{Y}_0^{\tilde{\nu}}(X^\wedge \times \Omega, \tilde{\boldsymbol{g}}; N_-, N_0)$ *for* $\tilde{\boldsymbol{g}} = (\gamma, \gamma - \tilde{\mu}, \Theta)$, *and let* A *or* $\tilde{A}$ *be properly supported (in the* y*–variables). Then* $A\tilde{A} \in \mathcal{Y}_0^{\nu + \tilde{\nu}}(X^\wedge \times \Omega, \boldsymbol{h}; N_-, N_+)$ *for* $\boldsymbol{h} = (\gamma, \gamma - \mu - \tilde{\mu}, \Theta)$, *and*

$$\sigma_\psi^{\nu + \tilde{\nu}}(A\tilde{A}) = \sigma_\psi^\nu(A)\sigma_\psi^{\tilde{\nu}}(\tilde{A}), \qquad \sigma_\wedge^{\nu + \tilde{\nu}}(A\tilde{A}) = \sigma_\wedge^\nu(A)\sigma_\wedge^{\tilde{\nu}}(\tilde{A}).$$

$A \in \mathcal{Y}_{M+G}^\nu(\dots)$ ($\in \mathcal{Y}_G^\nu(\dots)$) *or* $\tilde{A} \in \mathcal{Y}_{M+G}^{\tilde{\nu}}(\dots)$ ($\in \mathcal{Y}_G^{\tilde{\nu}}(\dots)$) *implies* $A\tilde{A} \in \mathcal{Y}_{M+G}^{\nu + \tilde{\nu}}(\dots)$ ($\in \mathcal{Y}_G^{\nu + \tilde{\nu}}(\dots)$). *Analogous results hold for the weakly discrete asymptotics.*

Proof. The scheme of the proof is the same as for Theorem 3.4.14. The corresponding steps here use Proposition 3.4.36, Remark 3.4.37, Proposition 3.4.34, Remark 3.4.35. We finally have to consider a composition of the form $\mathrm{Op}(a)\psi\,\mathrm{Op}(\tilde{a})$, applied on u which is supported by a fixed compact subset of Ω, where $\psi(y) \in C_0^\infty(\Omega)$, $\psi(y) \equiv 1$ on $\mathrm{supp}\,u$, and $a(y, \eta) \in R^\nu(\Omega \times \mathbb{R}^q, \boldsymbol{g}; N_0, N_+)$, $\tilde{a}(y', \eta) \in R^{\tilde{\nu}}(\Omega \times \mathbb{R}^q, \tilde{\boldsymbol{g}}; N_-, N_0)$. Then $\mathrm{Op}(a)\psi\,\mathrm{Op}(\tilde{a}) = \mathrm{Op}(b)\,\mathrm{Op}(\tilde{a}) = \mathrm{Op}(b\tilde{a})$ mod $\mathcal{Y}_G^{-\infty}$ for some $b(y, \eta) \in R^\nu(\Omega \times \mathbb{R}^q, \boldsymbol{g}; N_0, N_+)$, and hence $\mathrm{Op}(a)\psi\,\mathrm{Op}(\tilde{a}) \in \mathcal{Y}_0^{\nu + \tilde{\nu}}$. In the latter step we applied Theorem 3.3.45 that states that $b(y, \eta)\tilde{a}(y', \eta) \in R^{\nu + \tilde{\nu}}$. At the same time we see that when one factor belongs to the class with subscript $M + G$ (G) then so does the composition. The rule for the principal symbols is also a consequence of Theorem 3.3.45, applied on the set of points in Ω where $\psi(y) \equiv 1$. However ψ is arbitrary and so the result is proved in general. □

Remark 3.4.40 *Assume in Theorem 3.4.39 that* $A = \mathrm{Op}(a)$ *mod* $\mathcal{Y}_G^{-\infty}$ *and* $\tilde{A} = \mathrm{Op}(\tilde{a})$ *for symbols* $a(y, \eta) \in R^\nu$ *and* $\tilde{a}(y, \eta) \in R^{\tilde{\nu}}$, *respectively. Then* $A\tilde{A} = \mathrm{Op}(c)$ *mod* $\mathcal{Y}_G^{-\infty}$ *for a symbol* $c(y, \eta) \in R^{\nu + \tilde{\nu}}$ *that has an asymptotic expansion*

$$c(y, \eta) \sim \sum_\alpha \frac{1}{\alpha!} D_\eta^\alpha a(y, \eta) \partial_y^\alpha \tilde{a}(y, \eta) \tag{3.4.20}$$

which is an analogue of the Leibniz rule. Moreover, the local interior symbols for $\tilde{A}$ *are the Leibniz poducts of those of* A *and* $\tilde{A}$, *respectively, i.e., if the system of complete symbols for* A *and* $\tilde{A}$ *is* $t^{-\nu}p_j(t, x, y, \tau, \xi, \eta)$ *and* $t^{-\tilde{\nu}}\tilde{p}_j(t, x, y, \tau, \xi, \eta)$, *respectively, over* $\mathbb{R}_+ \times$

$\Sigma_j \times \Omega \ni (t,x,y)$ *(cf. the above notations and conventions concerning invariance under the coordinate diffeomorphisms in the x–variables) then $A\tilde{A}$ has the system of complete symbols*

$$\sum_\beta \frac{1}{\beta!}\left\{ D^\beta_{\tau,\xi,\eta}t^{-\nu}p_j(t,x,y,\tau,\xi,\eta)\right\} \partial^\beta_{t,x,y}t^{-\tilde\nu}\tilde{p}_j(t,x,y,\tau,\xi,\eta),$$

cf. Remark 3.2.4. Note that (3.4.20) also implies the symbolic rules for the subordinate symbols under compositions in the cone algebra, pointwise for every (y,η), in particular, the Mellin translation product for the sequences of conormal symbols. Analogous remarks hold for the complete symbols of formal adjoint operators in the sense of Theorem 3.4.38.

3.4.4 Global edge pseudo–differential operators

In this section W is a manifold with edges Y in the sense of Definition 3.1.1 and $\mathbb{W}$ the associated stretched manifold. Let $\pi : \mathbb{W} \to W$ be the canonical projection. We assume as above in the definition of the spaces

$$\mathcal{W}^{s,\gamma}_{\mathrm{loc}}(\mathbb{W}) \quad \text{and} \quad \mathcal{W}^{s,\gamma}_{\mathrm{comp}}(\mathbb{W}) \qquad \text{for } s,\gamma \in \mathbb{R}, \qquad (3.4.21)$$

cf. (3.1.26), (3.1.27), that Y has a neighbourhood V in W with $V \cong \{([0,1) \times X)/(\{0\} \times X)\} \times Y$. This means that $\mathbb{W}$ near $\pi^{-1}Y$ has the structure of a Cartesian product $[0,1) \times X \times Y$, or, equivalently, of $\overline{\mathbb{R}}_+ \times X \times Y$. The cut–off functions ω on $\mathbb{W}$ in this section are assumed to be supported by $[0,1-\varepsilon) \times X \times Y$, for some $0 < \varepsilon < 1$. The edge Y is a paracompact C^∞ manifold of dimension q. We fix a locally finite open covering $\{G_j\}$ by coordinate neighbourhoods and a system of charts $\kappa_j : G_j \to \Omega_j$ for open sets $\Omega_j \subseteq \mathbb{R}^q$. The discussion of invariance of objects on $\mathbb{W}$ near $\pi^{-1}Y$, e.g., of distributions and operators, refers to the coordinate diffeomorphisms with respect to Y. In other words we assume here that the coordinate transformations are independent of $(t,x) \in \overline{\mathbb{R}}_+ \times X$. In this sense the spaces (3.4.21) are invariantly defined as well as the subspaces with asymptotics

$$\mathcal{W}^{s,\gamma}_{P,\mathrm{loc}}(\mathbb{W}) \quad \text{and} \quad \mathcal{W}^{s,\gamma}_{P,\mathrm{comp}}(\mathbb{W}) \qquad \text{for } s,\gamma \in \mathbb{R}$$

for arbitrary (continuous or weakly discrete) asymptotic types P of the classes $\mathrm{As}(X,(\gamma,\Theta))$ or $\mathrm{As}(X,(\gamma,\Theta))^\bullet$, cf. Section 3.1.4.

Recall that we also introduced the spaces of Green smoothing operators on $\mathbb{W}$, cf. Definition 3.1.39. We shall show the invariance of the operators in $\mathcal{Y}^\nu_0(X^\wedge \times \Omega, \boldsymbol{g}; N_-, N_+)$ under a diffeomorphism

$$\chi : \Omega \to \tilde\Omega \quad \text{for open sets } \Omega, \tilde\Omega \subseteq \mathbb{R}^q.$$

The operator push–forward χ_* is first defined in the sense of pseudo–differential operators on Ω with operator–valued symbols. Then the following subspaces are preserved under χ_*:

$$\mathcal{Y}_G^\nu(X^\wedge \times \Omega, \boldsymbol{g}; N_-, N_+) \subset L_{\mathrm{cl}}^\nu(\Omega; \mathcal{K}^{s,\gamma}(X^\wedge) \oplus \mathbb{C}^{N_-}, \mathcal{K}^{\infty,\delta}(X^\wedge) \oplus \mathbb{C}^{N_+}),$$
$$Y_{M+G}^\nu(X^\wedge \times \Omega, \boldsymbol{g}) \subset L_{\mathrm{cl}}^\nu(\Omega; \mathcal{K}^{s,\gamma}(X^\wedge), \mathcal{K}^{\infty,\delta}(X^\wedge))$$

for $\boldsymbol{g} = (\gamma, \delta, \Theta)$, and

$$Y_0^\nu(X^\wedge \times \Omega, \boldsymbol{g}) \subset L^\nu(\Omega; \mathcal{K}^{s,\gamma}(X^\wedge), \mathcal{K}^{s-\mu,\gamma-\mu}(X^\wedge))$$

for $\boldsymbol{g} = (\gamma, \gamma - \mu, \Theta)$ and arbitrary $s \in \mathbb{R}$. We shall employ here the notation of Theorem 1.1.38, in particular,

$$\Phi_\alpha(y, \tilde{\eta}) = D_z^\alpha e^{i\delta(y,z)\tilde{\eta}}|_{z=y}, \quad \alpha \in \mathbb{N}^q,$$

for $\delta(y, z) = \chi(z) - \chi(y) - d\chi(y)(z - y)$. The asymptotic formula for the symbol push–forward with respect to $\Omega \times \mathbb{R}^q \ni (y, \eta) \to (\tilde{y}, \tilde{\eta}) = (\chi(y), {}^t d\chi(y)^{-1}\eta)$ is

$$\tilde{a}(\tilde{y}, \tilde{\eta})|_{\tilde{y}=\chi(y)} \sim \sum_\alpha \frac{1}{\alpha!} (\partial_\eta^\alpha a)(y, {}^t d\chi(y)\tilde{\eta}) \Phi_\alpha(y, \tilde{\eta}). \tag{3.4.22}$$

The proofs of the following theorems will be formulated for the case of continuous asymptotics. Unless we indicate otherwise they are also valid in analogous form for weakly discrete asymptotics of the classes $\mathrm{As}(X, (\gamma, \Theta))^\bullet$. Let us start with the Green operators.

Theorem 3.4.41 *Let* $G \in \mathcal{Y}_G^\nu(X^\wedge \times \Omega, \boldsymbol{g}; N_-, N_+)$ *for* $\nu \in \mathbb{R}$ *and* $\boldsymbol{g} = (\gamma, \varrho, \Theta)$. *Then* $\tilde{G} := \chi_* G \in \mathcal{Y}_G^\nu(X^\wedge \times \tilde{\Omega}, \boldsymbol{g}; N_-, N_+)$. *If* G *is written* $G = \mathrm{Op}(a) + C$ *for an* $a(y, \eta) \in \mathcal{R}_G^\nu(\Omega \times \mathbb{R}^q, \boldsymbol{g}; N_-, N_+)$ *and some* $C \in \mathcal{Y}_G^{-\infty}(X^\wedge \times \Omega, \boldsymbol{g}; N_-, N_+)$ *then it follows that* $\tilde{G} = \mathrm{Op}(\tilde{a}) + \tilde{C}$ *for some* $\tilde{C} \in \mathcal{Y}_G^{-\infty}(X^\wedge \times \tilde{\Omega}, \boldsymbol{g}; N_-, N_+)$ *and a symbol* $\tilde{a}(\tilde{y}, \tilde{\eta}) \in \mathcal{R}_G^\nu(\tilde{\Omega} \times \mathbb{R}^q, \boldsymbol{g}; N_-, N_+)$ *that allows the asymptotic expansion* (3.4.22). *In particular,* $\sigma_\wedge^\nu(\tilde{G})(\tilde{y}, \tilde{\eta}) = \sigma_\wedge^\nu(G)(y, \eta)$ *for* $(\tilde{y}, \tilde{\eta}) = (\chi(y), {}^t d\chi^{-1}(y)\eta)$. *An analogous result holds for Green operators with weakly discrete asymptotics.*

Proof. The invariance of $\mathcal{Y}_G^{-\infty}$ with respect to push–forwards is an obvious consequence of Definition 3.4.2 and of Theorem 3.1.37. Thus we can assume $G = \mathrm{Op}(a)$ for an $a(y, \eta) \in \mathcal{R}_G^\nu(\Omega \times \mathbb{R}^q, \boldsymbol{g}; N_-, N_+)$. Here we can apply the steps of the proof of Theorem 1.1.38 in analogous form. Then $\tilde{G}v(\tilde{y})$ equals

$$\iint e^{i(\tilde{y}-\tilde{y}')\tilde{\eta}} a(\chi^{-1}(\tilde{y}), g(\tilde{y}, \tilde{y}')\tilde{\eta}) h(\tilde{y}, \tilde{y}') v(\tilde{y}') \, d\tilde{y}' d\tilde{\eta} \tag{3.4.23}$$

with $g(\tilde{y}, \tilde{y}')$ being the matrix function occurring in the proof of Theorem 1.1.37 and $h(\tilde{y}, \tilde{y}') = |\det g(\tilde{y}, \tilde{y}')| |\det d\chi^{-1}(\tilde{y}')|$. In view of $a(\chi^{-1}(\tilde{y}), g(\tilde{y}, \tilde{y}')\tilde{\eta})h(\tilde{y}, \tilde{y}') \in \mathcal{R}_G^\nu(\tilde{\Omega} \times \tilde{\Omega} \times \mathbb{R}^q, \boldsymbol{g}; N_-, N_+)$ it follows that $\tilde{G} \in \mathcal{Y}_G^\nu(X^\wedge \times \tilde{\Omega}, \boldsymbol{g}; N_-, N_+)$. Concerning the asymptotic expansion (3.4.22) we can apply the formal constructions of the proof of Theorem 1.1.38, here in the analogous operator-valued case with operators between Hilbert spaces that occur in the representation of $\mathcal{S}_P^\delta(X^\wedge) \oplus \mathbb{C}^{N_+}$ and $\mathcal{S}_Q^{-\gamma}(X^\wedge) \oplus \mathbb{C}^{N_-}$ as countable projective limits, cf. the proof of Proposition 3.3.11. We then consider the case of operator-valued symbols that are families of mappings from $\mathcal{K}^{s,\gamma}(X^\wedge) \oplus \mathbb{C}^{N_-}$ to $\mathcal{S}_P^\delta(X^\wedge) \oplus \mathbb{C}^{N_+}$ for all s and similarly for the point-wise adjoints. Then the assumptions of Proposition 3.3.11 on the summands of (3.4.22) are satisfied, and hence we can form the asymptotic sum $\tilde{a}(\tilde{y}, \tilde{\eta}) \in \mathcal{R}_G^\nu(\tilde{\Omega} \times \mathbb{R}^q, \boldsymbol{g}; N_-, N_+)$. $\square$

Theorem 3.4.42 *For every* $W \in Y_{M+G}^\nu(X^\wedge \times \Omega, \boldsymbol{g})$, $\nu \in \mathbb{R}$, $\boldsymbol{g} = (\gamma, \delta, \Theta)$, *we have* $\tilde{W} := \chi_* W \in Y_{M+G}^\nu(X^\wedge \times \tilde{\Omega}, \boldsymbol{g})$. *If* W *is written* $W = \mathrm{Op}(a) + C$ *for an* $a(y, \eta) \in R_{M+G}^\nu(\Omega \times \mathbb{R}^q, \boldsymbol{g})$ *and some* $C \in Y_G^{-\infty}(X^\wedge \times \Omega, \boldsymbol{g})$ *then* $\tilde{W} = \mathrm{Op}(\tilde{a}) + \tilde{C}$ *for some* $\tilde{C} \in Y_G^{-\infty}(X^\wedge \times \tilde{\Omega}, \boldsymbol{g})$ *and a symbol* $\tilde{a}(\tilde{y}, \tilde{\eta}) \in R_{M+G}^\nu(\tilde{\Omega} \times \mathbb{R}^q, \boldsymbol{g})$ *that allows the asymptotic expansion* (3.4.22). *Then*

$$\sigma_\wedge^\nu(\tilde{W})(\tilde{y}, \tilde{\eta}) = \sigma_\wedge^\nu(W)(y, \eta), \qquad \sigma_M^\nu(\tilde{W})(\tilde{y}, z) = \sigma_M^\nu(W)(y, z)$$

for $(\tilde{y}, \tilde{\eta}) = (\chi(y), {}^t d\chi^{-1}(y)\eta)$ *and all* z. *An analogous result holds for the weakly discrete asymptotics.*

Proof. The scheme of the proof is analogous to that of the preceding theorem. First it suffices to consider $W = \mathrm{Op}(a)$ for $a(y, \eta) \in R_{M+G}^\nu(\Omega \times \mathbb{R}^q, \boldsymbol{g})$. Then we obtain that $\tilde{W}v(\tilde{y})$ equals (3.4.23). Now the way to derive the asymptotic expansion (3.4.22) is formally the same as in the proof of Theorem 1.1.38. In order to obtain $\tilde{W} \in Y_{M+G}^\nu(X^\wedge \times \tilde{\Omega}, \boldsymbol{g})$ it suffices to note that

$$a(\chi^{-1}(\tilde{y}), g(\tilde{y}, \tilde{y}')\tilde{\eta})h(\tilde{y}, \tilde{y}') \in R_{M+G}^\nu(\tilde{\Omega} \times \tilde{\Omega} \times \mathbb{R}^q, \boldsymbol{g}).$$

We know already that this is true for the Green summand involved. Concerning the finite Mellin sums, cf. Definition 3.3.14, it is obvious that the only new effect arises from inserting $\eta = g(\tilde{y}, \tilde{y}')\tilde{\eta}$ in the cut-off factors which is an operation that preserves the symbol class. $\square$

Theorem 3.4.43 *Let* $A \in Y_0^\nu(X^\wedge \times \Omega, \boldsymbol{g})$ *for* $\nu \in \mathbb{R}$, $\boldsymbol{g} = (\gamma, \gamma - \mu, \Theta)$, *cf. Definition 3.4.2. Then* $\tilde{A} = \chi_* A \in Y_0^\nu(X^\wedge \times \tilde{\Omega}, \boldsymbol{g})$. *If* A *is written* $A = \mathrm{Op}(a) + C$ *for an* $a(y, \eta) \in R^\nu(\Omega \times \mathbb{R}^q, \boldsymbol{g})$ *and some* $C \in Y_G^{-\infty}(X^\wedge \times \Omega, \boldsymbol{g})$ *then* $\tilde{A} = \mathrm{Op}(\tilde{a}) + \tilde{C}$ *for some* $\tilde{C} \in Y_G^{-\infty}(X^\wedge \times \tilde{\Omega}, \boldsymbol{g})$ *and a symbol*

$\tilde{a}(\tilde{y}, \tilde{\eta}) \in R^{\nu}(\tilde{\Omega} \times \mathbb{R}^{q}, \boldsymbol{g})$ that allows the asymptotic expansion (3.4.22). Then $\sigma_{\wedge}^{\nu}(\tilde{A})(\tilde{y}, \tilde{\eta}) = \sigma_{\wedge}^{\nu}(A)(y, \eta)$ and $\sigma_{M}^{\nu}(\tilde{A})(\tilde{y}, z) = \sigma_{M}^{\nu}(A)(y, z)$ for $(\tilde{y}, \tilde{\eta}) = (\chi(y), {}^{t}d\chi^{-1}(y)\eta)$ and all z. In addition the complete interior symbols of A and $\tilde{A}$ (in local coordinates) satisfy the standard rule of symbol push-forwards. For the homogeneous principal symbols of order ν we have

$$\sigma_{\psi}^{\nu}(\tilde{A})(t, x, \tilde{y}, \tau, \xi, \tilde{\eta}) = \sigma_{\psi}^{\nu}(A)(t, x, y, \tau, \xi, \eta).$$

An analogous result holds for the operators with weakly discrete asymptotics.

Proof. Without loss of generality we assume $A = \mathrm{Op}(a) + C$ for an $a(y, \eta) \in R^{\nu}(\Omega \times \mathbb{R}^{q}, \boldsymbol{g})$ and some $C \in Y_{G}^{-\infty}(X^{\wedge} \times \Omega, \boldsymbol{g})$. The invariance of C is already known, so we consider $\mathrm{Op}(a)$. The operator-valued symbol $a(y, \eta)$ consists of two parts, namely a contribution containing $(a_0 + a_1 + m + g)(y, \eta)$ and one with $a_{\infty}(y, \eta)$, cf. Definition 3.3.30. We can treat them separately, where $a_{\infty}(y, \eta)$ is not the specific point here. We leave $a_{\infty}(y, \eta)$ to the reader; the details are straightforward. In view of Theorem 3.4.42 there remains $a(y, \eta) = \tilde{\omega}(t)\{a_0(y, \eta) + a_1(y, \eta)\}\tilde{\omega}_0(t)$. As above it follows for $A = \mathrm{Op}(a)$ and $\tilde{A} = \chi_* A$ that $\tilde{A}v(\tilde{y})$ equals (3.4.23). In this case we have $b(\tilde{y}, \tilde{y}', \tilde{\eta}) := a(\chi^{-1}(\tilde{y}), g(\tilde{y}, \tilde{y}')\tilde{\eta})h(\tilde{y}, \tilde{y}') \in R^{\nu}(\Omega \times \Omega \times \mathbb{R}^{q}, \boldsymbol{g})$. Then Proposition 3.4.34 implies $\tilde{A} \in Y_0^{\nu}(X^{\wedge} \times \tilde{\Omega}, \boldsymbol{g})$. We can apply the formal steps to obtain the asymptotic expansion (3.4.22). In the present case it suffices to observe that the summands on the right of (3.4.22) belong to $R^{\nu - |\alpha|/2}(\Omega \times \mathbb{R}^{q}, \boldsymbol{g})$. Here we use the compatibility of the rule of Theorem 3.2.7 (that is a correspondence $P(y, \eta) \to h(t, y, z, \eta)$) under the differentiations and substitutions involved in the items of (3.4.22). Moreover, similarly to the arguments in the proof of Theorem 3.4.42, we can substitute ${}^{t}d\chi(y)\tilde{\eta}$ into the cut-off factors $\omega(t[\eta]), \ldots$, without leaving the symbol class, cf. analogously Proposition 3.3.33. Now the asymptotic sum (3.4.22) can be carried out in a similar manner as for Proposition 3.4.34, where the idea is to first pass to convergent sums of the interior and the Mellin symbols and then to take the asymptotic sum of the remaining smoothing Mellin+Green symbols. □

Let us now define the global pseudo–differential operators on a (stretched) manifold $\mathbb{W}$ with edges. We shall consider, in particular, the case $X^{\wedge} \times Y$ with the edge manifold Y. If G is a coordinate neighbourhood on Y and $\kappa : G \to \Omega$ a chart, $\Omega \subseteq \mathbb{R}^{q}$, we set $Y_0^{\nu}(X^{\wedge} \times G, \boldsymbol{g}) = \{(\kappa_*)^{-1}P : P \in Y_0^{\nu}(X^{\wedge} \times \Omega, \boldsymbol{g})\}$. Then, if $\{G_j\}_{j \in \mathbb{N}}$ is a locally finite covering of Y by coordinate neighbourhoods, $\{\varphi_j\}_{j \in \mathbb{N}}$ a subordinate partition of unity and $\{\psi_j\}_{j\in\mathbb{N}}$ another system of functions $\psi_j \in C_0^\infty(G_j)$ with $\varphi_j\psi_j = \varphi_j$ for all j, we define

$$Y_0^\nu(X^\wedge \times Y, \boldsymbol{g}) = \Big\{\sum_{j\in\mathbb{N}} \varphi_j A_j \psi_j + C : A_j \in Y_0^\nu(X^\wedge \times G_j, \boldsymbol{g}),$$

$$C \in Y_G^{-\infty}(X^\wedge \times Y, \boldsymbol{g})\Big\}.$$

Here $Y_G^{-\infty}(X^\wedge \times Y, \boldsymbol{g})$ is the space of all operators

$$G \in \bigcap_{s\in\mathbb{R}} \mathcal{L}(\mathcal{W}_{\text{comp}}^s(Y, \mathcal{K}^{s,\gamma}(X^\wedge)), \mathcal{W}_{\text{loc}}^\infty(Y, \mathcal{K}^{\infty,\gamma-\mu}(X^\wedge)))$$

such that there are asymptotic types $P \in \text{As}(X, \gamma - \mu, \Theta)$, $Q \in \text{As}(X, (-\gamma, \Theta))$ with

$$G \in \bigcap_{s\in\mathbb{R}} \mathcal{L}(\mathcal{W}_{\text{comp}}^s(Y, \mathcal{K}^{s,\gamma}(X^\wedge)), \mathcal{W}_{\text{loc}}^\infty(Y, \mathcal{K}_P^{\infty,\gamma-\mu}(X^\wedge))),$$

$$G^* \in \bigcap_{s\in\mathbb{R}} \mathcal{L}(\mathcal{W}_{\text{comp}}^s(Y, \mathcal{K}^{s,-\gamma+\mu}(X^\wedge)), \mathcal{W}_{\text{loc}}^\infty(Y, \mathcal{K}_Q^{\infty,-\gamma}(X^\wedge))).$$

In an analogous manner we define $Y_0^\nu(X^\wedge \times Y, \boldsymbol{g})^\bullet$ by assuming that the asymptotic types in the Green operators belong to $\text{As}(X, \dots)^\bullet$ and those in the smoothing Mellin operators to $\mathbf{As}(X)^\bullet$.

Let us denote by $Y_{M+G}^\nu(\dots)$ and $Y_{M+G}^\nu(\dots)^\bullet$ the subspaces only consisting of smoothing Mellin + Green operators. They are characterised by

$$Y_{M+G}^\nu(X^\wedge \times Y, \boldsymbol{g}) = Y_0^\nu(X^\wedge \times Y, \boldsymbol{g}) \cap L^{-\infty}(X^\wedge \times Y),$$

and analogously for the dotted classes.

Next let us consider vector bundles $J^\pm$ over Y with a fixed system of trivialisations $\iota^\pm : J^\pm|_{G_j} \to \Omega_j \times \mathbb{C}^{N\pm}$ and transition maps

$$\lambda_{jk}^\pm : \Omega_{jk} \times \mathbb{C}^{N\pm} \to \Omega_{kj} \times \mathbb{C}^{N\pm} \quad \text{for} \quad \Omega_{jk} = \Omega_k \cap \kappa_k(G_j \cap G_k),$$

compatible with the diffeomorphisms $\kappa_j\kappa_k^{-1} : \Omega_{jk} \to \Omega_{kj}$ from the charts $\kappa_j : G_j \to \Omega_j$. (Without loss of generality we suppose the sets G_j to be contractible). We have associated isomorphisms

$$L_{jk}^\pm : \mathcal{W}_{\text{comp}(y)}^{s,\gamma}(X^\wedge \times \Omega_{jk}) \oplus H_{\text{comp}}^s(\Omega_{jk}, \mathbb{C}^{N\pm}) \to$$

$$\mathcal{W}_{\text{comp}(y)}^{s,\gamma}(X^\wedge \times \Omega_{kj}) \oplus H_{\text{comp}}^s(\Omega_{kj}, \mathbb{C}^{N\pm})$$

for every $s, \gamma \in \mathbb{R}$, where $L_{jk}^\pm$ is defined as a block matrix with the identity in the upper left corner and the second diagonal entry being induced by $\lambda_{jk}^{\pm}$. Analogous isomorphisms hold for the spaces with subscripts "loc". For every $G \in \mathcal{Y}_G^{\nu}(X^{\wedge} \times \Omega_{jk}, \boldsymbol{g}; N_-, N_+)$,

$$G : \mathcal{W}_{\mathrm{comp}(y)}^{s,\gamma}(X^{\wedge} \times \Omega_{jk}) \oplus H_{\mathrm{comp}}^s(\Omega_{jk}, \mathbb{C}^{N_-}) \to$$
$$\mathcal{W}_{\mathrm{loc}(y)}^{s-\nu,\varrho}(X^{\wedge} \times \Omega_{jk}) \oplus H_{\mathrm{loc}}^{s-\nu}(\Omega_{jk}, \mathbb{C}^{N_+})$$

we then obtain $L_{jk}^+ G(L_{jk}^-)^{-1} \in \mathcal{Y}_G^{\nu}(X^{\wedge} \times \Omega_{kj}, \boldsymbol{g}; N_-, N_+)$, as a simple generalisation of Theorem 3.4.41.

This allows us to form the operator classes

$$\mathcal{Y}_G^{\nu}(X^{\wedge} \times G_j, \boldsymbol{g}; J^-|_{G_j}, J^+|_{G_j})$$

as the pull–backs of $\mathcal{Y}_G^{\nu}(X^{\wedge} \times \Omega_j, \boldsymbol{g}; N_-, N_+)$ with respect to the trivialisations $J^{\pm}|_{G_j} \to \Omega_j \times \mathbb{C}^{N\pm}$.

As usual we fix a Riemannian metric on Y and Hermitian metrics in the bundles $J^{\pm}$. Similarly to Definition 3.1.39 we obtain the global smoothing operators $C \in \mathcal{Y}^{-\infty}(\mathbb{W}, \boldsymbol{g}; J^-, J^+)$ for $\boldsymbol{g} = (\gamma, \varrho, \Theta)$ by the conditions

$$C \in \bigcap_{s \in \mathbb{R}} \mathcal{L}(\mathcal{W}_{\mathrm{comp}}^{s,\gamma}(\mathbb{W}) \oplus H_{\mathrm{comp}}^s(Y, J^-), \mathcal{W}_{P,\mathrm{loc}}^{\infty,\varrho}(\mathbb{W}) \oplus H_{\mathrm{loc}}^{\infty}(Y, J^+))$$

and

$$C^* \in \bigcap_{s \in \mathbb{R}} \mathcal{L}(\mathcal{W}_{\mathrm{comp}}^{s,-\varrho}(\mathbb{W}) \oplus H_{\mathrm{comp}}^s(Y, J^+), \mathcal{W}_{Q,\mathrm{loc}}^{\infty,-\gamma}(\mathbb{W}) \oplus H_{\mathrm{loc}}^{\infty}(Y, J^-))$$

for certain $P \in \mathrm{As}(X, (\varrho, \Theta))$, $Q \in \mathrm{As}(X, (-\gamma, \Theta))$. Recall that in this notation "comp" and "loc" only refer to the region of $\mathbb{W}$ far from $\pi^{-1}Y$ or for y outside a compact subset of Y, while $\pi^{-1}Y$ itself is contained in $\mathbb{W}$. Thus, for instance, in the compact case, distributions on $\mathrm{int}\,\mathbb{W}$ may be non–zero up to $\pi^{-1}Y$, though they are of compact support. In the local expressions we used $X^{\wedge} \times \Omega$ instead of the notation $\mathbb{W} = \overline{\mathbb{R}}_+ \times X \times \Omega$, but this should not cause confusion.

Formal adjoints are understood in the sense

$$(Cu, v)_{\mathcal{W}_{\mathrm{loc}}^{0,0}(\mathbb{W}) \oplus H_{\mathrm{loc}}^0(Y, J^+)} = (u, C^* v)_{\mathcal{W}_{\mathrm{loc}}^{0,0}(\mathbb{W}) \oplus H_{\mathrm{loc}}^0(Y, J^-)} \qquad (3.4.24)$$

for all $u \in C_0^{\infty}(\mathrm{int}\,\mathbb{W}) \oplus C_0^{\infty}(Y, J^-)$, $v \in C_0^{\infty}(\mathrm{int}\,\mathbb{W}) \oplus C_0^{\infty}(Y, J^+)$, where the pairings refer to the Riemannian metrics on $\mathbb{W}$ and Y and the Hermitean metrics in the bundles. Moreover, let $\mathcal{Y}_G^{\nu}(\mathbb{W}, \boldsymbol{g}; J^-, J^+)$ be the space of all operators $A = \omega A_0 \tilde{\omega} + C$ for arbitrary $A_0 \in \mathcal{Y}_G^{\nu}(X^{\wedge} \times Y, \boldsymbol{g}; J^-, J^+)$, $C \in \mathcal{Y}^{-\infty}(\mathbb{W}, \boldsymbol{g}; J^-, J^+)$, and cut–off functions ω, $\tilde{\omega}$ (cf. the notation in the beginning of this section). In an analogous manner we introduce $\mathcal{Y}_G^{\nu}(\mathbb{W}, \boldsymbol{g}; J^-, J^+)^{\bullet}$. The elements of $\mathcal{Y}_G^{\nu}(\ldots)$ $(\mathcal{Y}_G^{\nu}(\ldots)^{\bullet})$

are called Green operators in the edge operator algebra. If they are regarded as block matrices $G = (G_{ij})_{i,j=1,2}$ then G_{21} is also called a trace, G_{12} a potential operator with respect to the edge.

Note that the notation here is not completely consistent in the sense that for $\mathbb{W} = \overline{\mathbb{R}}_+ \times X \times Y$ the analogues of the operator spaces $\mathcal{Y}_G^\nu$ in the above local theory contained a smaller class of smoothing elements, namely those of the class $\mathcal{Y}_G^{-\infty}$ with a specific behaviour for $t \to \infty$, while now the smoothing elements are of type $\mathcal{Y}^{-\infty}$ only. The context becomes clear when we say whether we have the $\mathcal{Y}_0^\nu$ or the $\mathcal{Y}^\nu$–spaces in mind. We hope this will not cause confusion.

Definition 3.4.44 *Let $\nu \in \mathbb{R}$ and $\boldsymbol{g} = (\gamma, \gamma - \mu, \Theta)$ be weight data with $\gamma, \mu \in \mathbb{R}$, $\mu - \nu \in \mathbb{N}$. Then $\mathcal{Y}^\nu(\mathbb{W}, \boldsymbol{g}; J^-, J^+)$ for $J^-, J^+ \in \mathrm{Vect}(Y)$ is defined as the space of all operators*

$$
A = \begin{pmatrix} P + G_{11} & G_{12} \\ G_{21} & G_{22} \end{pmatrix} : \begin{array}{c} C_0^\infty(\mathrm{int}\,\mathbb{W}) \\ \oplus \\ C_0^\infty(Y, J^-) \end{array} \to \begin{array}{c} C^\infty(\mathrm{int}\,\mathbb{W}) \\ \oplus \\ C^\infty(Y, J^+) \end{array}
$$

for $(G_{ij})_{i,j=1,2} \in \mathcal{Y}_G^\nu(\mathbb{W}, \boldsymbol{g}; J^-, J^+)$ and $P \in L_{\mathrm{cl}}^\nu(\mathrm{int}\,\mathbb{W})$, where $\omega P \tilde{\omega} \in Y_0^\nu(X^\wedge \times Y, \boldsymbol{g})$ for cut–off functions ω, $\tilde{\omega}$. In an analogous manner we define $\mathcal{Y}^\nu(\mathbb{W}, \boldsymbol{g}; J^-, J^+)^\bullet$ by requiring the asymptotic types of the Green and smoothing Mellin operators to be in $\mathrm{As}(X, \ldots)^\bullet$ and $\mathrm{As}(X)^\bullet$, respectively. In the particular case $\mathbb{W} = \overline{\mathbb{R}}_+ \times X \times Y$ we denote the corresponding operator classes by

$$
\mathcal{Y}^\nu(X^\wedge \times Y, \boldsymbol{g}; J^-, J^+) \quad and \quad \mathcal{Y}^\nu(X^\wedge \times Y, \boldsymbol{g}; J^-, J^+)^\bullet,
$$

respectively. The subclasses, defined by the condition $\omega P \tilde{\omega} \in Y_{M+G}^\nu(X^\wedge \times Y, \boldsymbol{g})$, denoted by $\mathcal{Y}_{M+G}^\nu(\ldots)$, are called Mellin + Green operators. Similar notations are used with upper dots.

Remark 3.4.45 *The operators $A = (A_{ij})_{i,j=1,2} \in \mathcal{Y}_{M+G}^\nu(\mathbb{W}, \boldsymbol{g}; J^-, J^+)$ have C^∞ kernels in the following sense: For convenience, let $J^+ = Y \times \mathbb{C}^{N_+}$, $J^- = Y \times \mathbb{C}^{N_-}$ be trivial bundles. Then, $A_{11} \in L^{-\infty}(\mathrm{int}\,\mathbb{W})$, $A_{22} \in L^{-\infty}(Y) \otimes \mathbb{C}^{N_+} \otimes \mathbb{C}^{N_-}$, A_{12} is an integral operator with kernel in $C^\infty(\mathrm{int}\,\mathbb{W} \times Y) \otimes \mathbb{C}^{N_-}$ and A_{21} an integral operator with kernel in $\mathbb{C}^{N_+} \otimes C^\infty(Y \times \mathrm{int}\,\mathbb{W})$. For the general invariant formulations of kernels, cf. analogously Section 1.1.5.*

Remark 3.4.46 *The concept of pseudo–differential operators on a manifold with edges also makes sense when the upper left corners of the block matrices are operators between distributional sections in vector bundles E, F over $\mathbb{W}$. In other words, for arbitrary $E, F \in \mathrm{Vect}(\mathbb{W})$, $J^-, J^+ \in \mathrm{Vect}(Y)$ we can introduce corresponding spaces of operators*

$$
\mathcal{Y}^\nu(\mathbb{W}, \boldsymbol{g}; E, F; J^-, J^+).
$$

The definition employs the invariance of systems of operators in the local representations both with respect to the transition maps for E, F and J^-, J^+. This is straightforward and we shall not discuss this case in detail. The essential results of the calculus, in particular, concerning ellipticity , cf. Section 3.5 below, are valid in analogous form also for the more general case of non–trivial bundles E, F.

The following results, as far as they are immediate consequences of those from the local theory, will be formulated without further explanations.

Remark 3.4.47 *Every $A \in \mathcal{Y}^\nu(\mathbb{W}, \boldsymbol{g}; J^-, J^+)$ can be written*

$$A = \omega A_0 \omega_0 + \begin{pmatrix} (1-\omega)P_0(1-\omega_1) & 0 \\ 0 & 0 \end{pmatrix} + C$$

for arbitrary fixed cut–off functions ω, ω_0, ω_1 satisfying $\omega\omega_0 = \omega$, $\omega\omega_1 = \omega_1$, where $A_0 \in \mathcal{Y}^\nu(X^\wedge \times Y, \boldsymbol{g}; J^-, J^+)$, $P_0 \in L^\nu_{cl}(\mathrm{int}\,\mathbb{W})$, and $C \in \mathcal{Y}^{-\infty}(\mathbb{W}, \boldsymbol{g}; J^-, J^+)$. Here A_0 can be assumed to be properly supported with respect to the y–variables and P_0 to be properly supported in the standard sense. Analogous assertions hold for the subspaces with weakly discrete asymptotics.

If $A \in \mathcal{Y}^\nu(\mathbb{W}, \boldsymbol{g}; J^-, J^+)$ has the form in Remark 3.4.47, where also C is properly supported, we say that A is properly supported in the $\mathbb{W}$–sense.

Theorem 3.4.48 *Let $A \in \mathcal{Y}^\nu(\mathbb{W}, \boldsymbol{g}; J^-, J^+)$, $\boldsymbol{g} = (\gamma, \gamma - \mu, \Theta)$. Then A induces a continuous operator*

$$A : \mathcal{W}^{s,\gamma}_{\mathrm{comp}}(\mathbb{W}) \oplus H^s_{\mathrm{comp}}(Y, J^-) \to \mathcal{W}^{s-\nu,\gamma-\mu}_{\mathrm{loc}}(\mathbb{W}) \oplus H^{s-\nu}_{\mathrm{loc}}(Y, J^+) \tag{3.4.25}$$

that restricts to a continuous operator

$$A : \mathcal{W}^{s,\gamma}_{P,\mathrm{comp}}(\mathbb{W}) \oplus H^s_{\mathrm{comp}}(Y, J^-) \to \mathcal{W}^{s-\nu,\gamma-\mu}_{Q,\mathrm{loc}}(\mathbb{W}) \oplus H^{s-\nu}_{\mathrm{loc}}(Y, J^+) \tag{3.4.26}$$

for every $s \in \mathbb{R}$ and for every $P \in \mathrm{As}(X, (\gamma, \Theta))$ with some resulting $Q \in \mathrm{As}(X, (\gamma - \mu, \Theta))$. In particular, for $A \in \mathcal{Y}^\nu(\mathbb{W}, \boldsymbol{g}; J^-, J^+)^\bullet$ we obtain (3.4.26) for every $P \in \mathrm{As}(X, (\gamma, \Theta))^\bullet$ with some $Q \in \mathrm{As}(X, (\gamma - \mu, \Theta))^\bullet$.

Corollary 3.4.49 *Let $\mathbb{W}$ be compact. Then Theorem 3.4.48 specialises to the continuity of*

$$A : \mathcal{W}^{s,\gamma}(\mathbb{W}) \oplus H^s(Y, J^-) \to \mathcal{W}^{s-\nu,\gamma-\mu}(\mathbb{W}) \oplus H^{s-\nu}(Y, J^+) \tag{3.4.27}$$

and

$$A : \mathcal{W}_P^{s,\gamma}(\mathbb{W}) \oplus H^s(Y, J^-) \to \mathcal{W}_Q^{s-\nu,\gamma-\mu}(\mathbb{W}) \oplus H^{s-\nu}(Y, J^+)$$

for every $s \in \mathbb{R}$ *and for every* P *with a resulting* Q. *The operator*

$$A : \mathcal{W}^{s,\gamma}(\mathbb{W}) \oplus H^s(Y, J^-) \to \mathcal{W}^{s-\mu,\gamma-\mu}(\mathbb{W}) \oplus H^{s-\mu}(Y, J^+)$$

is compact when $A \in \mathcal{Y}^\nu(\mathbb{W}, \boldsymbol{g}; J^-, J^+)$ *and* $\nu \le \mu - 1$ *for* $\boldsymbol{g} = (\gamma, \gamma - \mu, \Theta)$.

Remark 3.4.50 *If* $A \in \mathcal{Y}^\nu(\mathbb{W}, \boldsymbol{g}; J^-, J^+)$ *has the particular form of Remark 3.4.47 with properly supported operators* C *and* A_0 *in the* y-*variables and properly supported* P_0, *then we get continuous operators*

$$A : \mathcal{W}_{\mathrm{loc}}^{s,\gamma}(\mathbb{W}) \oplus H_{\mathrm{loc}}^s(Y, J^-) \to \mathcal{W}_{\mathrm{loc}}^{s-\nu,\gamma-\mu}(\mathbb{W}) \oplus H_{\mathrm{loc}}^{s-\nu}(Y, J^+),$$
$$A : \mathcal{W}_{P,\mathrm{loc}}^{s,\gamma}(\mathbb{W}) \oplus H_{\mathrm{loc}}^s(Y, J^-) \to \mathcal{W}_{Q,\mathrm{loc}}^{s-\nu,\gamma-\mu}(\mathbb{W}) \oplus H_{\mathrm{loc}}^{s-\nu}(Y, J^+)$$

for all $s \in \mathbb{R}$ *and every* P *with a resulting* Q, *and analogously between the spaces with "comp" on both sides.*

Remark 3.4.51 $A \in \mathcal{Y}^\nu(X^\wedge \times Y, \boldsymbol{g}; J^-, J^+)$ *and* $\varphi, \psi \in C^\infty(Y)$ *with* $\mathrm{supp}\,\varphi \cap \mathrm{supp}\,\psi = \emptyset$ *implies* $\varphi A \psi \in \mathcal{Y}^{-\infty}(X^\wedge \times Y, \boldsymbol{g}; J^-, J^+)$. *Moreover,* $A \in Y^\nu(\mathbb{W}, \boldsymbol{g})$ *and* $a, b \in C^\infty(\mathbb{W})$ *with* $\mathrm{supp}\,a \cap \mathrm{supp}\,b = \emptyset$ *implies* $aAb \in Y^{-\infty}(\mathbb{W}, \boldsymbol{g})$. *Analogous assertions hold for the subspaces with weakly discrete asymptotics.*

The local principal symbols of the edge pseudo–differential operators from Section 3.4.4 are invariant in the following sense. The homogeneous principal interior symbol of $A \in \mathcal{Y}^\nu(\mathbb{W}, \boldsymbol{g}; J^-, J^+)$ of order ν is an element

$$\sigma_\psi^\nu(A) \in S^{(\nu)}(T^*(\mathrm{int}\,\mathbb{W}) \setminus 0),$$

where $S^{(\nu)}(T^*(\mathrm{int}\,\mathbb{W}) \setminus 0)$ is the space of all $p \in C^\infty(T^*(\mathrm{int}\,\mathbb{W}) \setminus 0)$ that are homogeneous of order ν with respect to dilations in the covariables. Moreover, the homogeneous principal edge symbol of A of order ν has the interpretation of a bundle morphism

$$\sigma_\wedge^\nu(A) : \pi_Y^* \begin{pmatrix} \mathcal{K}^{s,\gamma}(X^\wedge) \\ \oplus \\ J^- \end{pmatrix} \to \pi_Y^* \begin{pmatrix} \mathcal{K}^{s-\nu,\gamma-\mu}(X^\wedge) \\ \oplus \\ J^+ \end{pmatrix},$$

for every $s \in \mathbb{R}$, where $\pi_Y : T^*Y \setminus 0 \to Y$ is the canonical projection, and the homogeneity is

$$\sigma_\wedge^\nu(A)(y, \lambda\eta) = \lambda^\nu \begin{pmatrix} \kappa_\lambda & 0 \\ 0 & 1 \end{pmatrix} \sigma_\wedge^\nu(A)(y, \eta) \begin{pmatrix} \kappa_\lambda & 0 \\ 0 & 1 \end{pmatrix}^{-1}$$

for $\lambda \in \mathbb{R}_+$, $(y,\eta) \in T^*Y \setminus 0$.

Finally we have the conormal symbol of order ν, derived from the cone theory,

$$\sigma_M^\nu(A)(y,z) := \sigma_M^\nu(\mathrm{u.\,l.\,c.}\,\sigma_\wedge^\nu(A))(y,z)$$
$$\in C^\infty(Y \times \Gamma_{\frac{n+1}{2}-\gamma}, \mathcal{L}(H^s(X), H^{s-\nu}(X)))$$

for all $s \in \mathbb{R}$. This can also be interpreted as an element in the space $C^\infty(Y, L_{\mathrm{cl}}^\nu(X; \Gamma_{\frac{n+1}{2}-\gamma}))$; $n = \dim X$.

Let us set

$$\Sigma_\psi^{(\nu)} = \left\{ \sigma_\psi^\nu(A) : \ A \in \mathcal{Y}^\nu(\mathbb{W}, \boldsymbol{g}; J^-, J^+) \right\},$$
$$\Sigma_\wedge^{(\nu)}(\boldsymbol{g}; J^-, J^+) = \left\{ \sigma_\wedge^\nu(A) : \ A \in \mathcal{Y}^\nu(\mathbb{W}, \boldsymbol{g}; J^-, J^+) \right\}$$

and

$$\Sigma^{(\nu)}(\boldsymbol{g}; J^-, J^+) = \left\{ (\sigma_\psi^\nu(A), \sigma_\wedge^\nu(A)) : \ A \in \mathcal{Y}^\nu(\mathbb{W}, \boldsymbol{g}; J^-, J^+) \right\}. \tag{3.4.28}$$

According to the local constructions of Section 3.3.3 each element $p_\wedge \in \Sigma_\wedge^{(\nu)}(\boldsymbol{g}; J^-, J^+)$ has a homogeneous principal interior symbol of order ν, denoted by $\sigma_\psi^\nu p_\wedge$, cf. (3.3.49). Moreover, the elements $p_\psi \in \Sigma_\psi^{(\nu)}$ allow the operation of freezing the coefficients on the edge that is locally described by $t = 0$, cf. (3.3.38). The corresponding operation on p_ψ was denoted by $c_\wedge$. Then we have

$$\Sigma^{(\nu)}(\boldsymbol{g}; J^-, J^+) = \left\{ (p_\psi, p_\wedge) \in \Sigma_\psi^{(\nu)} \times \Sigma_\wedge^{(\nu)}(\boldsymbol{g}; J^-, J^+) : \ \sigma_\psi^\nu p_\wedge = c_\wedge p_\psi \right\}, \tag{3.4.29}$$

cf. formula (3.3.52).

Proposition 3.4.52 *There is a canonical embedding*

$$\iota : \mathcal{Y}^{\nu-1}(\mathbb{W}, \boldsymbol{g}; J^-, J^+) \to \mathcal{Y}^\nu(\mathbb{W}, \boldsymbol{g}; J^-, J^+),$$

and the following sequence is exact:

$$0 \to \mathcal{Y}^{\nu-1}(\mathbb{W}, \boldsymbol{g}; J^-, J^+) \xrightarrow{\iota} \mathcal{Y}^\nu(\mathbb{W}, \boldsymbol{g}; J^-, J^+) \xrightarrow{\sigma} \Sigma^{(\nu)}(\boldsymbol{g}; J^-, J^+) \to 0,$$

where $\sigma(A) = (\sigma_\psi^\nu(A), \sigma_\wedge^\nu(A))$. *Moreover, there is an operator*

$$\mathrm{op} : \Sigma^{(\nu)}(\boldsymbol{g}; J^-, J^+) \to \mathcal{Y}^\nu(\mathbb{W}, \boldsymbol{g}; J^-, J^+)$$

such that $\sigma \circ \mathrm{op} = \mathrm{id}$ *on* $\Sigma^{(\nu)}(\boldsymbol{g})$. *An analogous result holds for the discrete asymptotics.*

Proof. The exactness of the sequence is obvious. To construct op it suffices to look at $\mathcal{Y}_G^\nu(\mathbb{W}, \boldsymbol{g}; J^-, J^+)$ and $Y^\nu(\mathbb{W}, \boldsymbol{g})$ separately. We may replace in both cases $\mathbb{W}$ by $X^\wedge \times Y$, since Green operators are smoothing in the interior, while the operators in $Y^\nu(\mathbb{W}, \boldsymbol{g})$ belong to $L_{cl}^\nu(\operatorname{int}\mathbb{W})$, where the op–construction from the scalar theory in the interior can be applied and glued together with the one on $X^\wedge \times Y$ to be constructed below. Now op for $\mathcal{Y}_G^\nu$ is similar to that of the scalar calculus: First if $\kappa_j : G_j \to \Omega_j$ is a chart on Y and $p_{j,\lambda}$ the local representative of the given homogeneous principal operator–valued symbol, then $(\kappa_j^{-1})_* \operatorname{Op}(\chi p_{j,\wedge}) =: A_j$ for any excision function $\chi(\eta)$ belongs to $\mathcal{Y}_G^\nu$ over $X^\wedge \times G_j$ and satisfies $\sigma_\wedge^\nu(A_j) = p_{j,\wedge}$. It suffices then to set $\operatorname{op} p_\wedge = \sum_{j\in\mathbb{N}} \varphi_j A_j \psi_j$. Next consider the case $Y^\nu(\mathbb{W}, \boldsymbol{g})$. Let $(p_{j,\psi}, p_{j,\wedge})$ be the local representative of $(p_\psi, p_\wedge) \in \Sigma^{(\nu)}(\boldsymbol{g}; J^-, J^+)$ over Ω_j. Then, applying Proposition 3.3.37 we find an $a_j(y, \eta) \in R^\nu(\Omega_j \times \mathbb{R}^q, \boldsymbol{g})$ with $(\sigma_\psi^\nu, \sigma_\wedge^\nu)(a_j) = (p_{j,\psi}, p_{j,\wedge})$. Setting $A_j = (\kappa_j^{-1})_* \operatorname{Op}(a_j)$ we may set $\operatorname{op}(p_\psi, p_\wedge) = \sum_{j\in\mathbb{N}} \varphi_j A_j \psi_j$. $\square$

Proposition 3.4.53 *The elements of* (3.4.28) *allow a natural operation of formal adjoints, indicated by* *, *and* $(p_\psi, p_\wedge) \in \Sigma^{(\nu)}(\boldsymbol{g}; J^-, J^+)$ *for* $\boldsymbol{g} = (\gamma, \gamma - \mu, \Theta)$ *implies* $(p_\psi^*, p_\wedge^*) \in \Sigma^{(\nu)}(\boldsymbol{g}^*; J^+, J^-)$ *for* $\boldsymbol{g}^* = (-\gamma + \mu, -\gamma, \Theta)$.

Proof. The assertion is a corollary of Proposition 3.3.41 and of the invariance of the symbols. $\square$

Proposition 3.4.54 $(\tilde{p}_\psi, \tilde{p}_\wedge) \in \Sigma^{(\tilde{\nu})}(\tilde{\boldsymbol{g}}; J^-, J^0)$ *for* $\tilde{\boldsymbol{g}} = (\gamma, \gamma - \tilde{\mu}, \Theta)$ *and* $(p_\psi, p_\wedge) \in \Sigma^{(\nu)}(\boldsymbol{g}; J^0, J^+)$ *for* $\boldsymbol{g} = (\gamma - \tilde{\mu}, \gamma - \mu - \tilde{\mu}, \Theta)$ *implies* $(p_\psi \tilde{p}_\psi, p_\wedge \tilde{p}_\wedge) \in \Sigma^{(\nu+\tilde{\nu})}(\boldsymbol{h}; J^-, J^+)$ *for* $\boldsymbol{h} = (\gamma, \gamma - \mu - \tilde{\mu}, \Theta)$.

Proof. The assertion is a corollary of Proposition 3.3.42 and of the invariance of the symbols. $\square$

Theorem 3.4.55 $A \in \mathcal{Y}^\nu(\mathbb{W}, \boldsymbol{g}; J^-, J^+)$ *for* $\boldsymbol{g} = (\gamma, \gamma - \mu, \Theta)$ *implies* $A^* \in \mathcal{Y}^\nu(\mathbb{W}, \boldsymbol{g}^*; J^+, J^-)$ *for* $\boldsymbol{g}^* = (-\gamma + \mu, -\gamma, \Theta)$, *cf.* (3.4.24), *where*

$$\sigma_\psi^\nu(A^*) = \overline{\sigma_\psi^\nu(A)}, \qquad \sigma_\wedge^\nu(A^*) = \sigma_\wedge^\nu(A)^*.$$

Analogous relations hold for the weakly discrete asymptotics.

Proof. Let $\omega, \omega_0, \omega_1$ be cut–off functions on $\mathbb{W}$ with $\omega\omega_0 = \omega$, $\omega\omega_1 = \omega_1$. Then there is a $C \in \mathcal{Y}^{-\infty}(\mathbb{W}, \boldsymbol{g}; J^-, J^+)$ such that $A = \omega A \omega_0 + (1-\omega)A(1-\omega_1) + C$. Since $C^* \in \mathcal{Y}^{-\infty}(\mathbb{W}, \boldsymbol{g}^*; J^+, J^-)$ by the definition of the smoothing edge pseudo–differential operators, it suffices to look at $\omega A \omega_0 \in \mathcal{Y}^\nu(X^\wedge \times Y, \boldsymbol{g}; J^-, J^+)$ and $(1-\omega)A(1-\omega_1)$. The latter operator equals a diagonal block matrix $\operatorname{mod} \mathcal{Y}^{-\infty}$ with the upper left corner $(1-\omega)P(1-\omega_1)$ for some $P \in L_{cl}^\nu(\text{int}\,\mathbb{W})$, while the other entries vanish. The behaviour under adjoints of $(1-\omega)A(1-\omega_1)$ is therefore known, and it remains $\omega A \omega_0$. In view of Theorem 3.4.13 it suffices to consider the upper left corners. Each $A \in Y^\nu(X^\wedge \times Y, g)$ has the form $A = \sum_{j \in \mathbb{N}} \varphi_j A_j \psi_j + C$ for certain $A_j \in Y^\nu(X^\wedge \times G_j, g)$ and some $C \in Y^{-\infty}(X^\wedge \times Y, g)$. Then the assertion is a consequence of Theorem 3.4.38. $\qquad\square$

Theorem 3.4.56 *Let $A \in \mathcal{Y}^\nu(\mathbb{W}, g; J^0, J^+)$ for $g = (\gamma - \tilde{\mu}, \gamma - \mu - \tilde{\mu}, \Theta)$ and $\tilde{A} \in \mathcal{Y}^{\tilde{\nu}}(\mathbb{W}, \tilde{g}; J^-, J^0)$ for $\tilde{g} = (\gamma, \gamma - \tilde{\mu}, \Theta)$, and assume that A or $\tilde{A}$ is as in Remark 3.4.47, where the part near Y is properly supported in y, and the operator in $\text{int}\,\mathbb{W}$ is properly supported in the standard sense. Then $A\tilde{A} \in \mathcal{Y}^{\nu+\tilde{\nu}}(\mathbb{W}, h; J^-, J^+)$ for $h = (\gamma, \gamma - \mu - \tilde{\mu}, \Theta)$, and we have*

$$\sigma_\psi^{\nu+\tilde{\nu}}(A\tilde{A}) = \sigma_\psi^\nu(A)\sigma_\psi^{\tilde{\nu}}(\tilde{A}), \qquad \sigma_\wedge^{\nu+\tilde{\nu}}(A\tilde{A}) = \sigma_\wedge^\nu(A)\sigma_\wedge^{\tilde{\nu}}(\tilde{A}). \qquad (3.4.30)$$

Moreover, $A \in \mathcal{Y}_{M+G}^\nu(\dots)$ $(\in \mathcal{Y}_G^\nu(\dots))$ or $\tilde{A} \in \mathcal{Y}_{M+G}^{\tilde{\nu}}(\dots)$ $(\in \mathcal{Y}_G^{\tilde{\nu}}(\dots))$ implies $A\tilde{A} \in \mathcal{Y}_{M+G}^{\nu+\tilde{\nu}}(\dots)$ $(\in \mathcal{Y}_G^{\nu+\tilde{\nu}}(\dots))$. An analogous result holds for the weakly discrete asymptotics. If $\mathbb{W}$ is compact the condition concerning proper supports can be omitted.

Proof. Let $\omega, \omega_0, \omega_1$ be cut–off functions on $\mathbb{W}$ and write $A_0 = \omega A \omega_0$, $A_1 = (1-\omega)A(1-\omega_1)$, $\tilde{A}_0 = \omega \tilde{A} \omega_0$, $\tilde{A}_1 = (1-\omega)\tilde{A}(1-\omega_1)$. Then $A = A_0 + A_1 + C$, $\tilde{A} = \tilde{A}_0 + \tilde{A}_1 + \tilde{C}$ for certain $C \in \mathcal{Y}^{-\infty}(\mathbb{W}, g; J^0, J^+)$, $\tilde{C} \in \mathcal{Y}^{-\infty}(\mathbb{W}, \tilde{g}; J^-, J^0)$. From the definition of the $\mathcal{Y}^{-\infty}$–classes and from the Theorems 3.4.48, 3.4.55 it follows that the compositions containing C or $\tilde{C}$ are again in $\mathcal{Y}^{-\infty}$. In other words, it remains to characterise $\{A_0 + A_1\}\{\tilde{A}_0 + \tilde{A}_1\} = A\tilde{A}_0 + B$ for $B = A_0\tilde{A}_1 + A_1\tilde{A}_0 + A_1\tilde{A}_1$.

The following equations are valid modulo $\mathcal{Y}^{-\infty}$, and we shall drop such remainders in the discussion, since smoothing factors produce smoothing operators in the compositions. Let us choose an atlas on Y with the charts $\lambda_k : D_k \to \Delta_k$, $k \in \mathbb{N}$, on Y which is a refinement of the above $\kappa_j : G_j \to \Omega_j$, $j \in \mathbb{N}$, in the sense that when two open sets $D_k, D_l \subset Y$ have a non–empty intersection there is a j such that $D_k \cup D_l \subset G_j$. Let $\{\delta_k\}_{k \in \mathbb{N}}$ be a subordinate partition of unity and choose a second system $\{\gamma_k\}_{k \in \mathbb{N}}$ of functions $\gamma_k \in C_0^\infty(D_k)$ with $\delta_k \gamma_k = \delta_k$ for all k. Then

$$A_0\tilde{A}_0 = \left\{\sum_{k \in \mathbb{N}} \delta_k A_0 \gamma_k\right\}\left\{\sum_{l \in \mathbb{N}} \delta_l \tilde{A}_0 \gamma_l\right\} = \sum_{\substack{k,l \in \mathbb{N} \\ D_k \cap D_l \neq \emptyset}} \delta_k A_0 \gamma_k \delta_l \tilde{A}_0 \gamma_l.$$

From Theorem 3.4.39 and from Theorems 3.4.41, 3.4.42, 3.4.43 it follows that the summands on the right hand side belong to the space

$\mathcal{Y}^{\nu+\tilde{\nu}}(\mathbb{W}, \boldsymbol{h}; J^-, J^+)$, and hence $A_0 \tilde{A}_0 \in \mathcal{Y}^{\nu+\tilde{\nu}}(\mathbb{W}, \boldsymbol{h}; J^-, J^+)$, since this is a locally finite sum.

For $B = (B_{ij})_{i,j=1,2}$ we obtain that B_{ij} for $i = 1, 2$ and $j = 1, 2$, $(i, j) \neq (1, 1)$, have C^∞–kernels, cf. Remark 3.4.45. The same is true of those summands in B_{11} that contain a smoothing Mellin+Green factor. There remain the compositions in which the factors are pseudo–differential on $\operatorname{int} \mathbb{W}$ or Mellin operators (non–smoothing) near $\pi^{-1}Y$. By virtue of Theorem 3.3.14 the latter are also standard pseudo–differential operators on $\operatorname{int} \mathbb{W}$, and every item contains a properly supported factor. This gives us $B_{11} \in L_{\mathrm{cl}}^{\nu+\tilde{\nu}}(\operatorname{int} \mathbb{W})$. The symbolic rules (3.4.30) as well as the remaining statements of the theorem follow from Theorem 3.4.39. $\qquad\square$

3.5 Ellipticity, parametrices, Fredholm property

3.5.1 Elliptic edge symbols

Let us recall that $\mathcal{R}^\nu(\Omega \times \mathbb{R}^q, \boldsymbol{g}; N_-, N_+)$ for $\boldsymbol{g} = (\gamma, \gamma - \mu, \Theta)$, $\Omega \subseteq \mathbb{R}^q$, is the space of all operator families

$$a(y, \eta) = \begin{pmatrix} a_{11} & a_{12} \\ a_{21} & a_{22} \end{pmatrix} (y, \eta) : \begin{matrix} \mathcal{K}^{s,\gamma}(X^\wedge) \\ \oplus \\ \mathbb{C}^{N_-} \end{matrix} \to \begin{matrix} \mathcal{K}^{s-\nu,\gamma-\mu}(X^\wedge) \\ \oplus \\ \mathbb{C}^{N_+} \end{matrix} ,$$

$s \in \mathbb{R}$, where $a_{11}(y, \eta) \in R^\nu(\Omega \times \mathbb{R}^q, \boldsymbol{g})$ and

$$\begin{pmatrix} 0 & a_{12} \\ a_{21} & a_{22} \end{pmatrix} (y, \eta) \in \mathcal{R}_G^\nu(\Omega \times \mathbb{R}^q, \boldsymbol{g}; N_-, N_+),$$

cf. Definition 3.3.30, Definition 3.3.40. We mainly consider here the case of continuous asymptotics.

The upper left corners of the block matrices have the form

$$a_{11}(y, \eta) = \tilde{\omega}(t)\{a_0(y, \eta) + a_1(y, \eta)\}\tilde{\omega}_0(t) + m(y, \eta) + g(y, \eta)$$
$$+ (1 - \tilde{\omega}(t))a_\infty(y, \eta)(1 - \tilde{\omega}_1(t))$$

for cut–off functions $\tilde{\omega}$, $\tilde{\omega}_0$, $\tilde{\omega}_1$ satisfying $\tilde{\omega}\tilde{\omega}_0 = \tilde{\omega}$, $\tilde{\omega}\tilde{\omega}_1 = \tilde{\omega}_1$ and operator families $a_0(y, \eta) = \omega(t[\eta])t^{-\nu}a_M(y, \eta)\omega_0(t[\eta])$ for $a_M(y, \eta) = \operatorname{op}_M^{\gamma - \frac{n}{2}}(h)(y, \eta)$, $a_1(y, \eta) = (1 - \omega(t[\eta]))t^{-\nu}a_\psi(y, \eta)(1 - \omega_1(t[\eta]))$, where $\omega, \omega_0, \omega_1$ are cut–off functions, $\omega\omega_0 = \omega$, $\omega\omega_1 = \omega_1$,

$$a_\psi(y, \eta) \in C^\infty(\Omega, \tilde{L}_{\mathrm{cl}}^\nu(X^\wedge, \mathbb{R}_\eta^q)_0), \tag{3.5.1}$$

$$h(t, y, z, \eta) = \tilde{h}(t, y, z, t\eta) \tag{3.5.2}$$

for $\tilde{h}(t, y, z, \tilde{\eta}) \in C^\infty(\overline{\mathbb{R}}_+ \times \Omega, M_O^\nu(X; \mathbb{R}_{\tilde{\eta}}^q))$ satisfying

$$a_\psi(y, \eta) = \mathrm{op}_M^{\gamma - \frac{n}{2}}(h)(y, \eta) \in C^\infty(\Omega, L^{-\infty}(X^\wedge; \mathbb{R}_\eta^q)),$$

and finally $a_\infty(y, \eta) \in C^\infty(\Omega, L_{\mathrm{cl}}^\nu(X^\wedge; \mathbb{R}_\eta^q)_0)$.

To construct parametrices of elliptic operators on a manifold W with edge singularities Y it is unimportant to observe the contribution from $a_\infty(y, \eta)$ far from the edge, since outside a neighbourhood of the edge the pseudo–differential operators can be treated in the standard manner as elements of $L_{\mathrm{cl}}^\nu(W \setminus Y)$. Of course, it would be possible to apply the parameter–dependent operator calculus with exit behaviour for $t \to \infty$ and with parameters (y, η) in the sense of (3.2.25) and to apply ellipticity and parametrix constructions in terms of operator–valued symbols. However this would enlarge the exposition without contributing new information.

In Section 3.3.3 we introduced the principal symbolic structure of the edge symbols $a(y, \eta) \in \mathcal{R}^\nu(\Omega \times \mathbb{R}^q, \boldsymbol{g}; N_-, N_+)$ as a map

$$(\sigma_\psi^\nu, \sigma_\wedge^\nu) : \mathcal{R}^\nu(\Omega \times \mathbb{R}^q, \boldsymbol{g}, N_-, N_+) \to \Sigma^{(\nu)}(\boldsymbol{g}; N_-, N_+).$$

$\sigma_\psi^\nu(a) = \sigma_\psi^\nu(\mathrm{u.\,l.\,c.}\,a)$ is the homogeneous principal interior symbol of a of order ν,

$$\sigma_\psi^\nu(a)(t, x, y, \tau, \xi, \eta) = t^{-\nu}\sigma_{\psi,b}^\nu(a)(t, x, y, t\tau, \xi, t\eta), \qquad (3.5.3)$$

cf. (3.3.46), where

$$\sigma_{\psi,b}^\nu(a)(t, x, y, \tilde{\tau}, \xi, \tilde{\eta}) \in S^{(\nu)}(T^*(\overline{\mathbb{R}} \times X \times \Omega) \setminus 0) \qquad (3.5.4)$$

is C^∞ in t up to $t = 0$. Moreover,

$$\sigma_\wedge^\nu(a)(y, \eta) : \begin{array}{c} \mathcal{K}^{s,\gamma}(X^\wedge) \\ \oplus \\ \mathbb{C}^{N_-} \end{array} \to \begin{array}{c} \mathcal{K}^{s-\nu,\gamma-\mu}(X^\wedge) \\ \oplus \\ \mathbb{C}^{N_+} \end{array} \qquad (3.5.5)$$

is the homogeneous principal edge symbol of a of order ν, given on $T^*\Omega \setminus 0 \ni (y, \eta)$ which is an operator family (3.5.5) for every $s \in \mathbb{R}$, satisfying

$$\sigma_\wedge^\nu(a)(y, \lambda\eta) = \lambda^\nu \begin{pmatrix} \kappa_\lambda & 0 \\ 0 & 1 \end{pmatrix} \sigma_\wedge^\nu(a)(y, \eta) \begin{pmatrix} \kappa_\lambda & 0 \\ 0 & 1 \end{pmatrix}^{-1}$$

for all $\lambda \in \mathbb{R}_+$. Let us set

$$\sigma_{\wedge,0}^\nu(a_{11})(y, \eta) = \omega(t|\eta|)t^{-\mu}c_\wedge a_M(y, \eta)\omega_0(t|\eta|)$$
$$+ (1 - \omega(t|\eta|))c_\wedge t^{-\mu}a_\psi(y, \eta)(1 - \omega_1(t|\eta|)). \quad (3.5.6)$$

Then, by definition,

$$\sigma_\wedge^\mu(a_{11})(y, \eta) = \sigma_{\wedge,0}^\mu(a_{11})(y, \eta) + \sigma_\wedge^\mu(m + g)(y, \eta).$$

Theorem 3.5.1 *Let* $a(y,\eta) \in R^\mu(\Omega \times \mathbb{R}^q, \boldsymbol{g})$ *for* $\mu \in \mathbb{R}$, $\boldsymbol{g} = (\gamma, \gamma - \mu, \Theta)$, $\gamma \in \mathbb{R}$, *with*

$$\sigma^\mu_{\psi,b}(a)(0,x,y,\tilde{\tau},\xi,\tilde{\eta}) \neq 0 \quad \text{for all} \quad (x,y) \in X \times \Omega \qquad (3.5.7)$$

and $(\tilde{\tau},\xi,\tilde{\eta}) \neq 0$. *Then for every* $y \in \Omega$ *there is a countable subset* $D(y) \subset \mathbb{R}$, *where* $D(y) \cap \{z : c \leq \operatorname{Re} z \leq c'\}$ *is finite for every* $c \leq c'$, *such that*

$$\sigma^\mu_{\wedge,0}(a)(y,\eta) : \mathcal{K}^{s,\varrho}(X^\wedge) \to \mathcal{K}^{s-\mu,\varrho-\mu}(X^\wedge) \qquad (3.5.8)$$

is a Fredholm operator for each $\varrho \in \mathbb{R} \setminus D(y)$ *and all* $\eta \neq 0$, $s \in \mathbb{R}$.

Proof. We have

$$\sigma^\mu_{\wedge,0}(a)(y,\eta) \in C^\mu(X^\wedge, (\varrho, \varrho - \mu, (-\infty, 0])^\bullet)$$

for each $\varrho \in \mathbb{R}$, cf. Definition 2.4.1. In this proof $y \in \Omega$ and $\eta \neq 0$ is kept fixed. In order to prove the Fredholm property of (3.5.8) it suffices to apply Theorem 2.4.40. In other words we have to check Definition 2.4.35. The conditions $\sigma^\mu_\psi(\sigma^\mu_{\wedge,0}(a)) \neq 0$ on $T^*X^\wedge \setminus 0$ and $t^\mu \sigma^\mu_\psi(\sigma^\mu_{\wedge,0}(a))(t,x,t^{-1}\tau,\xi) \neq 0$ on $T^*(\overline{\mathbb{R}}_+ \times X) \setminus 0$ are obviously a consequence of (3.5.7).

Moreover, we have $\sigma^\mu_M(\sigma^\mu_{\wedge,0}(a))(y,z) = h(0,y,z,0)$ in the notation of (3.5.2), and $h(0,y,z,0) \in C^\infty(\Omega, M^\mu_O(X))$. Then, similarly to the proof of Theorem 2.4.18 we see that $h(0,y,z,0)|_{\Gamma_\beta} \in C^\infty(\Omega, L^\mu_{cl}(X; \Gamma_\beta))$ is parameter–dependent elliptic of order μ, for each fixed $y \in \Omega$ and for every $\beta \in \mathbb{R}$. This yields by the same arguments as for Theorem 2.4.18 that $h(0,y,z,0) : H^s(X) \to H^{s-\mu}(X)$ is an isomorphism for all $z \in \mathbb{C} \setminus A(y)$ for some countable subset $A(y) \subset \mathbb{C}$, where $A(y) \cap K$ is finite for every compact $K \subset\subset \mathbb{C}$. This shows that the condition (ii) of Definition 2.4.35 is satisfied for all weights $\gamma \in \mathbb{R} \setminus D(y)$ for some countable set $D(y)$ that has a finite intersection with $\{z : c \leq \operatorname{Re} z \leq c'\}$ for every $c \leq c'$. Finally (iii) of Definition 2.4.35 is fulfilled because of the condition (3.5.7) that allows us to apply Remark 3.2.23. $\square$

Theorem 3.5.2 *Let* $a(y,\eta) \in R^\mu(\Omega \times \mathbb{R}^q, \boldsymbol{g})$ *for* $\mu \in \mathbb{R}$, $\boldsymbol{g} = (\gamma, \gamma - \mu, \Theta)$, $\gamma \in \mathbb{R}$, *and assume* (3.5.7). *Then there exists a* $b(y,\eta) \in R^{-\mu}(\Omega \times \mathbb{R}^q, \boldsymbol{g}^{-1})$ *for* $\boldsymbol{g}^{-1} = (\gamma - \mu, \gamma, \Theta)$ *such that*

$$\sigma^{-\mu}_\wedge(b)(y,\eta)\sigma^\mu_\wedge(a)(y,\eta) = 1 \quad \bmod \sigma^0_\wedge(R^0_{M+G}(\Omega \times \mathbb{R}^q, \boldsymbol{g}_l)), \qquad (3.5.9)$$
$$\sigma^\mu_\wedge(a)(y,\eta)\sigma^{-\mu}_\wedge(b)(y,\eta) = 1 \quad \bmod \sigma^0_\wedge(R^0_{M+G}(\Omega \times \mathbb{R}^q, \boldsymbol{g}_r)), \qquad (3.5.10)$$

where $\boldsymbol{g}_l = (\gamma, \gamma, \Theta)$, $\boldsymbol{g}_r = (\gamma - \mu, \gamma - \mu, \Theta)$.

Proof. Let us fix an open covering $\{U_1, \ldots, U_N\}$ of X by coordinate neighbourhoods and charts $\kappa_j : U_j \to \Sigma_j$ for open $\Sigma_j \subseteq \mathbb{R}^n$. Then $\sigma_\psi^\mu(a)$ consists locally of the homogeneous components of order μ of symbols

$$\{t^{-\mu} p_j(t, x, y, t\tau, \xi, t\eta)\}_{j=1,\ldots,N} \tag{3.5.11}$$

in $S_{\mathrm{cl}}^\mu(\overline{\mathbb{R}}_+ \times \Sigma_j \times \Omega \times \mathbb{R}^{1+n+q})$ that are compatible with respect to the symbol push–forwards of complete symbols associated with the operator push–forwards under the transition diffeomorphisms in the x–coordinates, modulo symbols of order $-\infty$, and $p_j(t, x, y, \tilde{\tau}, \xi, \tilde{\eta}) \in S_{\mathrm{cl}}^\mu(\overline{\mathbb{R}}_+ \times \Sigma_j \times \mathbb{R}_{\tilde{\tau}, \xi, \tilde{\eta}}^{1+n+q})$. Now the condition (3.5.7) entails the Leibniz inverses of $t^{-\mu} p_j(0, x, y, t\tau, \xi, t\eta)$ with respect to (t, x) (i.e., with respect to the Leibniz multiplication $\#_{t,x}$). Let us first show that they have the form $t^\mu r_j(x, y, t\tau, \xi, t\eta)$ modulo symbols of order $-\infty$, for suitable $r_j(x, y, \tilde{\tau}, \xi, \tilde{\eta}) \in S_{\mathrm{cl}}^{-\mu}(\Sigma_j \times \Omega \times \mathbb{R}_{\tilde{\tau}, \xi, \tilde{\eta}}^{1+n+q})$.

Choosing an excision function $\chi(\tilde{\tau}, \xi, \tilde{\eta})$ that vanishes for $|\tilde{\tau}, \xi, \tilde{\eta}| \le c$ and equals 1 for $|\tilde{\tau}, \xi, \tilde{\eta}| \ge 2c$ we find c so large that $d_j(x, y, t\tau, \xi, t\eta) := \chi(t\tau, \xi, t\eta) p_j^{-1}(0, x, y, t\tau, \xi, t\eta)$ exists. Then we obtain

$$t^\mu d_j(x, y, t\tau, \xi, t\eta) t^{-\mu} p_j(0, x, y, t\tau, \xi, t\eta) = 1 + g_j(x, y, t\tau, \xi, t\eta)$$

for a symbol $g_j(x, y, t\tau, \xi, t\eta)$ with $g_j(x, y, \tilde{\tau}, \xi, \tilde{\eta}) \in S^{-\infty}(\Sigma_j \times \Omega \times \mathbb{R}_{\tilde{\tau}, \xi, \tilde{\eta}}^{1+n+q})$. This implies

$$t^\mu d_j(x, y, t\tau, \xi, t\eta) \#_{t,x} t^{-\mu} p_j(0, x, y, t\tau, \xi, t\eta) = 1 + e_j(x, y, t\tau, \xi, t\eta),$$

where $e_j(x, y, \tilde{\tau}, \xi, \tilde{\eta}) \in S_{\mathrm{cl}}^{-1}(\Sigma_j \times \Omega \times \mathbb{R}_{\tilde{\tau}, \xi, \tilde{\eta}}^{1+n+q})$, cf. Proposition 3.2.2 and Remark 3.2.3. Next we carry out the asymptotic sum $\sum_{j=0}^\infty (-1)^j (e_j)^{\#j}$; $\#j$ indicates applying j times the Leibniz product with respect to t, x. The asymptotic sum can be written (uniquely modulo smoothing symbols) as $1 + f_j(x, y, t\tau, \xi, t\eta)$ with $f_j(x, y, \tilde{\tau}, \xi, \tilde{\eta}) \in S_{\mathrm{cl}}^{-1}(\Sigma_j \times \Omega \times \mathbb{R}^{1+n+q})$. Then

$$(1 + f_j(x, y, t\tau, \xi, t\eta)) \#_{t,x} (t^\mu d_j(x, y, t\tau, \xi, t\eta)) = t^\mu c_j(x, y, t\tau, \xi, t\eta)$$

modulo a symbol $t^\mu l_j(x, y, t\tau, \xi, t\eta)$ with $l_j(x, y, \tilde{\tau}, \xi, \tilde{\eta}) \in S^{-\infty}(\Sigma_j \times \Omega \times \mathbb{R}^{1+n+q})$. This gives us

$$t^\mu c_j(x, y, t\tau, \xi, t\eta) \#_{t,x} t^{-\mu} p_j(0, x, y, t\tau, \xi, t\eta) = 1 + n_j(x, y, t\tau, \xi, t\eta) \tag{3.5.12}$$

for an $n_j(x, y, \tilde{\tau}, \xi, \tilde{\eta}) \in S^{-\infty}(\Sigma_j \times \Omega \times \mathbb{R}^{1+n+q})$ (the latter relation means that the asymptotic sum for the $\#_{t,x}$– composition preserves our class of $(x, y, t\tau, \xi, t\eta)$–dependent symbols and is unique modulo a corresponding remainder).

Let $\{\varphi_1,\ldots,\varphi_N\}$ be a partition of unity to $\{U_1,\ldots,U_N\}$, the open covering of X, and choose functions $\psi_j \in C_0^\infty(U_j)$ with $\varphi_j\psi_j = \varphi_j$ for all j. Set

$$b_\psi(y,\eta) = \sum_{j=0}^N \varphi_j(1 \times \kappa_j^{-1})_* \operatorname{op}_{t,x}(c_j)(y,\eta)\psi_j.$$

Then, according to the results of Section 3.2.2, there exists an element $m(t,y,z,\eta) = \tilde{m}(y,z,t\eta)$ for some $\tilde{m}(y,z,\tilde\eta) \in C^\infty(\Omega, M_O^{-\mu}(X;\mathbb{R}_{\tilde\eta}^q))$ such that

$$b_\psi(y,\eta) = \operatorname{op}_M^{\gamma-\mu}(m)(y,\eta) \bmod C^\infty(\Omega, L^{-\infty}(X^\wedge;\mathbb{R}^q)).$$

Let $\omega,\omega_i,\tilde\omega,\tilde\omega_i$, $i = 0,1$, be arbitrary cut–off functions satisfying $\omega\omega_0 = \omega$, $\omega\omega_1 = \omega_1$, $\tilde\omega\tilde\omega_0 = \tilde\omega$, $\tilde\omega\tilde\omega_1 = \tilde\omega_1$, and set

$$b(y,\eta) = \tilde\omega(t)\{b_0(y,\eta) + b_1(y,\eta)\}\tilde\omega_0(t)$$
$$+ (1-\tilde\omega(t))t^\mu b_\psi(y,\eta)(1-\tilde\omega(t)), \quad (3.5.13)$$

where

$$b_0(y,\eta) = \omega(t[\eta])t^\mu b_M(y,\eta)\omega_0(t[\eta])$$

for $b_M(y,\eta) = \operatorname{op}_M^{\gamma-\mu}(m)(y,\eta)$,

$$b_1(y,\eta) = (1-\omega(t[\eta]))t^\mu b_\psi(y,\eta)(1-\omega_1(t[\eta])).$$

We have $b(y,\eta) \in R^{-\mu}(\Omega \times \mathbb{R}^q, \boldsymbol{g}^{-1})$ and

$$\sigma_\wedge^{-\mu}(b)(y,\eta) = \omega(t|\eta|)t^\mu b_M(y,\eta)\omega_0(t|\eta|)$$
$$+ (1-\omega(t|\eta|))t^\mu b_\psi(y,\eta)(1-\omega_1(t|\eta|)).$$

It is now obvious that (3.5.9) holds. In fact, because of the homogeneity in η it suffices to insert $|\eta| = 1$ and to carry out the composition in the sense of cone pseudo–differential operators over $X^\wedge$, where the composition is smoothing in the interior, i.e., of the class $C_{M+G}(X^\wedge, \boldsymbol{g}_l)$. To verify (3.5.9) we can argue in an analogous manner, noting that the symbols $t^\mu c_j(x,y,t\tau,\xi,t\eta)$ can also be Leibniz composed from the right to $t^{-\mu}p_j(0,x,y,t\tau,\xi,t\eta)$ to yield 1 up to a remainder of the above kind, cf. (3.5.12). This completes the proof. □

Lemma 3.5.3 Let $w(y,\eta) \in R_{M+G}^0(\Omega \times \mathbb{R}^q, \boldsymbol{g})$, $\boldsymbol{g} = (\gamma,\gamma,\Theta)$, and assume that $\sigma_M^0(w)(y,z) = 0$. Then there exists an $m(y,\eta) \in R_{M+G}^0(\Omega \times \mathbb{R}^q, \boldsymbol{g})$ such that

$$(1+w(y,\eta))(1+m(y,\eta)) = 1 + g_1(y,\eta),$$
$$(1+m(y,\eta))(1+w(y,\eta)) = 1 + g_2(y,\eta)$$

for certain $g_1,g_2 \in R_G^0(\Omega \times \mathbb{R}^q, \boldsymbol{g})$.

Proof. We can proceed similarly to the proof of Lemma 2.4.47. It suffices to form $1 + m(y, \eta) = \sum_{j=0}^{N}(-1)^j w^j(y, \eta)$ for sufficiently large N. In fact, it follows then that $(1 + w(y, \eta))(1 + m(y, \eta)) = 1 + (-1)^N w^{N+1}(y, \eta)$ which is of Green type for large N, cf. Theorem 3.3.28 and Proposition 3.3.23. The multiplication from the left can be treated in an analogous manner. $\qquad\square$

Theorem 3.5.4 *Let* $a(y, \eta) \in R^\mu(\Omega \times \mathbb{R}^q, \boldsymbol{g})$ *for* $\mu \in \mathbb{R}$, $\boldsymbol{g} = (\gamma, \gamma - \mu, \Theta)$, $\gamma \in \mathbb{R}$. *Assume that* $a(y, \eta)$ *satisfies the condition* (3.5.5), *and let*

$$\sigma_M^\mu(\sigma_\wedge^\mu(a))(y, z) : H^s(X) \to H^{s-\mu}(X) \tag{3.5.14}$$

be an isomorphism for every $y \in \Omega$, $z \in \Gamma_{\frac{n+1}{2} - \gamma}$ *for some* $s \in \mathbb{R}$. *Then there exists a* $b(y, \eta) \in R^{-\mu}(\Omega \times \mathbb{R}^q, \boldsymbol{g}^{-1})$ *such that for suitable cut–off functions* $\vartheta(t)$, $\vartheta_0(t)$ *with* $\vartheta\vartheta_0 = \vartheta$ *we have*

$$\vartheta b(y, \eta) \vartheta_0 a(y, \eta) = \vartheta \cdot 1 \mod R_G^0(\Omega \times \mathbb{R}^q, \boldsymbol{g}_l), \tag{3.5.15}$$
$$\vartheta a(y, \eta) \vartheta_0 b(y, \eta) = \vartheta \cdot 1 \mod R_G^0(\Omega \times \mathbb{R}^q, \boldsymbol{g}_r) \tag{3.5.16}$$

and

$$\sigma_\wedge^{-\mu}(b)(y, \eta) \sigma_\wedge^\mu(a)(y, \eta) = 1 \mod \sigma_\wedge^0 R_G^0(\Omega \times \mathbb{R}^q, \boldsymbol{g}_l), \tag{3.5.17}$$
$$\sigma_\wedge^\mu(a)(y, \eta) \sigma_\wedge^{-\mu}(b)(y, \eta) = 1 \mod \sigma_\wedge^0 R_G^0(\Omega \times \mathbb{R}^q, \boldsymbol{g}_r), \tag{3.5.18}$$

for $\boldsymbol{g}_l = (\gamma, \gamma, \Theta)$, $\boldsymbol{g} = (\gamma - \mu, \gamma - \mu, \Theta)$.

Proof. The condition (3.5.7) implies $\sigma_\psi^\mu(a)(t, x, y, \tau, \xi, \eta) \neq 0$ for all $(t, x, y, \tau, \xi, \eta) \in T^*((0, \varepsilon) \times X \times \Omega) \setminus 0$ for some $\varepsilon > 0$. Thus, with the notation of the proof of Theorem 3.5.2 we can form the Leibniz inverses of the components of (3.5.11) for $0 < t < \varepsilon$, cf. Remark 3.2.4. In this way we obtain a tuple of symbols $\{t^\mu \tilde{r}_j(t, x, y, t\tau, \xi, t\eta)\}_{j=1,\dots,N}$ in $0 < t < \varepsilon$, where the components are compatible with respect to the symbol push–forwards connected with the transition diffeomorphisms in the x–variables. Let $\vartheta^1(t)$ be a cut–off function supported by $[0, \varepsilon)$ and set

$$b_\psi^1(y, \eta) = \sum_{j=0}^{N} \varphi_j (1 \times \kappa_j^{-1})_* \vartheta^1(t) \operatorname{op}_{t,x}(\tilde{r}_j(t, x, y, t\tau, \xi, t\eta)) \psi_j \tag{3.5.19}$$

with the functions φ_j, ψ_j and the charts $\kappa_j : U_j \to \Sigma_j$ from the proof of Theorem 3.5.2. Then, applying the results of Section 3.2.2 we find an $\tilde{m}^1(t, y, z, \tilde{\eta}) \in C^\infty(\overline{\mathbb{R}}_+ \times \Omega, M_O^{-\mu}(X; \mathbb{R}_{\tilde{\eta}}^q))$ such that $m^1(t, y, z, \eta) := \tilde{m}^1(t, y, z, t\eta)$ satisfies the relation

$$b_\psi^1(y, \eta) = \operatorname{op}_M^{\gamma - \mu}(m^1)(y, \eta) \mod C^\infty(\Omega, L^{-\infty}(X^\wedge; \mathbb{R}^q)).$$

This is compatible with the constructions in the proof of Theorem 3.5.2, i.e., the Leibniz inversion of the interior symbols, frozen in the first t–variable at $t = 0$, gives the above construction, modulo symbols of order $-\infty$. Concerning the Mellin operator convention we have, using notation of the proof of Theorem 3.5.2, $\tilde{m}(y, z, \tilde{\eta}) = \tilde{m}^1(0, y, z, \tilde{\eta})$ mod $C^\infty(\Omega, M_O^{-\infty}(X; \mathbb{R}_{\tilde{\eta}}^q))$. We now form an element $b^1(y, \eta) \in R^{-\mu}(\Omega \times \mathbb{R}^q, \boldsymbol{g}^{-1})$ analogously to (3.5.13), where $b_\psi(y, \eta)$ is to be replaced by (3.5.19) and $b_M(y, \eta)$ by $b_M^1(y, \eta) = \operatorname{op}_M^{\gamma-\mu}(m^1)(y, \eta)$.

Choosing the cut–off functions $\vartheta(t)$, $\vartheta_0(t)$ in such a way that $\vartheta \vartheta^1 = \vartheta$, $\vartheta_0 \vartheta^1 = \vartheta_0$, we obtain $\vartheta b^1(y, \eta) \vartheta_0 a(y, \eta) = \vartheta \cdot 1$ mod $R_{M+G}^0(\Omega \times \mathbb{R}^q, \boldsymbol{g}_l)$. An analogous relation holds for $\vartheta a(y, \eta) \vartheta_0 b(y, \eta)$.

Let us set $l(y, z + \mu) = (\sigma_M^\mu(\sigma_\wedge^\mu(a)))^{-1}(y, z)$. Then $\sigma_M^{-\mu}(b^1)(y, z) = l(y, z) + l_0(y, z)$ for some $l_0(y, z) \in C^\infty(\Omega, M_R^{-\infty}(X))$ with a certain asymptotic type R. Define

$$b^2(y, \eta) = b^1(y, \eta) - \omega(t[\eta]) t^\mu \operatorname{op}_M^{\gamma-\mu}(l_0)(y) \omega_0(t[\eta]) \qquad (3.5.20)$$

for cut–off functions ω, ω_0. Then it follows that

$$\vartheta b^2(y, \eta) \vartheta_0 a(y, \eta) = 1 + w^2(y, \eta)$$

for a $w^2(y, \eta) \in R_{M+G}^0(\Omega \times \mathbb{R}^q, \boldsymbol{g}_l)$ with $\sigma_M^0(w^2)(y, z) = 0$. Applying Lemma 3.5.3 we find a $w^3(y, \eta) \in R_{M+G}^0(\Omega \times \mathbb{R}^q, \boldsymbol{g}_l)$ such that $(1 + w^3(y, \eta))(1 + w^2(y, \eta)) = 1 + g(y, \eta)$ for a $g(y, \eta) \in R_G^0(\Omega \times \mathbb{R}^q, \boldsymbol{g}_l)$. Set $b(y, \eta) = (1 + w^3(y, \eta)) b^2(y, \eta)$. Then $b(y, \eta)$ satisfies the relation (3.5.15). The same construction is possible for the composition with $b(y, \eta)$ from the right. Then, using the fact that the Green symbols form an ideal, standard algebraic manipulations show that (3.5.16) also holds. This implies (3.5.17) and (3.5.18). $\qquad \square$

Definition 3.5.5 *Let* $a(y, \eta) \in \mathcal{R}^\mu(\Omega \times \mathbb{R}^q, \boldsymbol{g}; N_-, N_+)$ *for* $\mu \in \mathbb{R}$, $\boldsymbol{g} = (\gamma, \gamma - \mu, \Theta)$, $\gamma \in \mathbb{R}$, $\Theta = (-k, 0]$, $k \in (\mathbb{N} \setminus \{0\}) \cup \{\infty\}$. *Then* $a(y, \eta)$ *is called elliptic if there is a constant* $c > 0$ *such that*

(i) $\sigma_\psi^\mu(a) \neq 0$ *on* $T^*((0, c) \times X \times \Omega) \setminus 0$ *and*

$$t^\mu \sigma_\psi^\mu(a)(t, x, y, t^{-1}\tau, \xi, t^{-1}\eta) \neq 0$$

on $T^*([0, c) \times X \times \Omega) \setminus 0$,

(ii)

$$\sigma_\wedge^\mu(a)(y, \eta) : \quad \begin{matrix} \mathcal{K}^{s,\gamma}(X^\wedge) & & \mathcal{K}^{s-\mu,\gamma-\mu}(X^\wedge) \\ \oplus & \to & \oplus \\ \mathbb{C}^{N_-} & & \mathbb{C}^{N_+} \end{matrix} \qquad (3.5.21)$$

is an isomorphism for every $(y, \eta) \in T^*\Omega \setminus 0$ *and some* $s \in \mathbb{R}$.

It is clear that when (3.5.21) is an isomorphism for an $s = s_0 \in \mathbb{R}$ then for all $s \in \mathbb{R}$.

Remark 3.5.6 *The condition* (ii) *of Definition 3.5.5 implies that for* $a_{11} = \mathrm{u.\,l.\,c.}\,a$

$$\sigma_\wedge^\mu(a_{11})(y,\eta) : \mathcal{K}^{s,\gamma}(X^\wedge) \to \mathcal{K}^{s-\mu,\gamma-\mu}(X^\wedge)$$

is a family of Fredholm operators, where

$$\mathrm{ind}\,\sigma_\wedge^\mu(a_{11})(y,\eta) = N_+ - N_-$$

for every $s \in \mathbb{R}$ *and fixed* $(y,\eta) \in T^*\Omega \setminus 0$. *Moreover, because of*

$$\sigma_\wedge^\mu(a_{11})(y,\lambda\eta) = \lambda^\mu \kappa_\lambda \sigma_\wedge^\mu(a_{11})(y,\eta)\kappa_\lambda^{-1}$$

for all $\lambda \in \mathbb{R}_+$, *we have*

$$\mathrm{ind}\,\sigma_\wedge^\mu(a_{11})(y,\eta) = \mathrm{ind}\,\sigma_\wedge^\mu(a_{11})(y,y/|\eta|),$$

or, more precisely, $\dim \ker \sigma_\wedge^\mu(a_{11})(y,\eta)$ *and* $\dim \mathrm{coker}\,\sigma_\wedge^\mu(a_{11})(y,\eta)$ *only depend on* $\eta/|\eta|$.

If $a(y,\eta) \in R^\mu(\Omega \times \mathbb{R}^q, \boldsymbol{g})$ is an element satisfying the conditions of Theorem 3.5.4, then

$$\sigma_\wedge^\mu(a_{11})(y,\eta) : \mathcal{K}^{s,\gamma}(X^\wedge) \to \mathcal{K}^{s-\mu,\gamma-\mu}(X^\wedge) \tag{3.5.22}$$

is a family of Fredholm operators, parametrised by $(y,\eta) \in T^*\Omega \setminus 0$. The variable y plays the role of local coordinates from a chart on a C^∞ manifold Y (the edge). For the final considerations on Y we only need the properties for open $\Omega_0 \subset \Omega$ with compact $\overline{\Omega}_0 \subset\subset \Omega$ (since the functions from a partition of unity localise the objects on compact sets).

So we fix a compact subset B in Ω where y varies. Since $\mathrm{ind}\,\sigma_\wedge^\mu(a)$ only depends on $\eta/|\eta|$, (3.5.22) may be regarded as a Fredholm family parametrised by the compact set $S^*B = \{(y,\eta) \in T^*\Omega\setminus 0 : y \in B, |\eta| = 1\}$ (recall that there are chosen Riemannian metrics on the manifolds under consideration). This is now a standard situation of the classical K–theory, and we have an index element

$$\mathrm{ind}_{S^*B}\,\sigma_\wedge^\mu(a) \in K(S^*B),$$

cf. also the notation of Section 1.2.4. For S^*B we have a canonical projection

$$\pi_Y : S^*B \to B \quad \text{induced by} \quad \pi_Y : T^*\Omega \setminus 0 \to \Omega,$$

and we can talk about the induced homomorphism of the K groups

$$\pi_Y^* : K(B) \to K(S^*B).$$

Note that when $a(y, \eta) \in \mathcal{R}^\mu(\Omega \times \mathbb{R}^q, \boldsymbol{g}; N_-, N_+)$ is elliptic in the sense of Definition 3.5.5, we have for $B \subset\subset \Omega$ the relation

$$\text{ind}_{S^*B}\, \sigma_\wedge^\mu(a_{11}) \in \pi_Y^* K(B), \qquad (3.5.23)$$

$a_{11} = \text{u.l.c.}\,a$. In fact, from (3.5.21) it follows simply that

$$\text{ind}_{S^*B}\, \sigma_\wedge^\mu(a_{11}) = \pi_Y^* \{[B \times \mathbb{C}^{N_+}, B \times \mathbb{C}^{N_-}]\}.$$

Conversely for a given $a_{11}(y, \eta) \in R^\mu(\Omega \times \mathbb{R}^q, \boldsymbol{g})$ satisfying the assumptions of Theorem 3.5.4 the condition (3.5.23) is necessary for an interpretation as $a_{11} = \text{u.l.c.}\,a$ for a block matrix $a(y, \eta) \in \mathcal{R}^\mu(\Omega \times \mathbb{R}^q, \boldsymbol{g}; N_-, N_+)$ which is elliptic in the sense of Definition 3.5.5.

Here for simplicity we talk about the case of a contractible Ω and take for B the closed subset of all $y \in \Omega$ with $\text{dist}(y, \partial\Omega) > \varepsilon$ for some $\varepsilon > 0$. If the relation (3.5.23) holds we also say that the index obstruction for the existence of additional edge trace and potential conditions vanishes.

Remark 3.5.7 *Let* $a(y, \eta), b(y, \eta) \in R^\mu(\Omega \times \mathbb{R}^q, \boldsymbol{g})$, $\boldsymbol{g} = (\gamma, \gamma - \mu, \Theta)$, *with* $\sigma_\psi^\mu(a) = \sigma_\psi^\mu(b)$ *satisfying the condition* (i) *of Definition 3.5.5, and assume that both* $\sigma_M^\mu(a)(y, z)$ *and* $\sigma_M^\mu(b)(y, z)$ *induce isomorphisms* $H^s(X) \to H^{s-\mu}(X)$ *for all* $y \in \Omega$, $z \in \Gamma_{\frac{n+1}{2}-\gamma}$. *Then* $\text{ind}_{S^*B}\, \sigma_\wedge^\mu(a) \in \pi_Y^* K(B)$ *is equivalent to* $\text{ind}_{S^*B}\, \sigma_\wedge^\mu(b) \in \pi_Y^* K(B)$. *Note, in particular, that for* $a(y, \eta) = b(y, \eta) \bmod R_{M+G}^\mu(\Omega \times \mathbb{R}^q, \boldsymbol{g})$ *the above equivalence holds. More generally, vanishing (or non-vanishing) of the index obstruction only depends on* $\sigma_\psi^\mu(\cdot)$.

Further observations concerning the condition that the index element of an edge symbol belongs to the pull-back of $K(B)$ with respect to $\pi_Y : S^*B \to B$ may be found in Schulze [122] under the key word "index obstruction" for the case $\dim X = 0$. They can easily be generalised to arbitrary X.

Lemma 3.5.8 *Let* $g(y, \eta) = (g_{ij}(y, \eta))_{i,j=1,2} \in \sigma_\wedge^0 \mathcal{R}_G^0(\Omega \times \mathbb{R}^q, \boldsymbol{g}; N, N)$ *for* $\boldsymbol{g} = (\gamma, \gamma, \Theta)$ *and assume that*

$$\begin{pmatrix} 1 + g_{11} & g_{12} \\ g_{21} & g_{22} \end{pmatrix} (y, \eta) : \begin{matrix} \mathcal{K}^{s_0,\gamma}(X^\wedge) \\ \oplus \\ \mathbb{C}^N \end{matrix} \to \begin{matrix} \mathcal{K}^{s_0,\gamma}(X^\wedge) \\ \oplus \\ \mathbb{C}^N \end{matrix} \qquad (3.5.24)$$

is invertible for all $(y, \eta) \in T^*\Omega \setminus 0$. *Then* (3.5.24) *is invertible for all* $s \in \mathbb{R}$ *and* $(y, \eta) \in T^*\Omega \setminus 0$, *and there is an* $h(y, \eta) = (h_{ij}(y, \eta))_{i,j=1,2} \in \sigma_\wedge^0 \mathcal{R}_G^0(\Omega \times \mathbb{R}^q, \boldsymbol{g}, N, N)$ *such that*

$$\begin{pmatrix} 1 + h_{11} & h_{12} \\ h_{21} & h_{22} \end{pmatrix} (y, \eta) = \begin{pmatrix} 1 + g_{11} & g_{12} \\ g_{21} & g_{22}^{-1} \end{pmatrix}^{-1} (y, \eta).$$

Proof. If (3.5.24) is invertible for $s = s_0$ the invertibility for all $s \in \mathbb{R}$ can be obtained in a similar manner as Proposition 2.4.43. Next let us reduce the given operator to the case of weight $\frac{n}{2}$, by multiplication of the block matrix from the left by $t^{-\gamma+\frac{n}{2}} \oplus |\eta|^{-\gamma+\frac{n}{2}} \mathrm{id}_{\mathbb{C}^N}$ and from the right by $t^{\gamma-\frac{n}{2}} \oplus |\eta|^{\gamma-\frac{n}{2}} \mathrm{id}_{\mathbb{C}^N}$, cf. Proposition 3.3.9. In other words, without loss of generality we may assume $\gamma = \frac{n}{2}$, where $\mathcal{K}^{0,\frac{n}{2}}(X^\wedge) = L^2(X^\wedge)$. Since we consider homogeneous operator–valued functions of order zero, it is sufficent to invert (3.5.24) on $S^*\Omega = \{(y, \eta) \in T^*\Omega \setminus 0 : |\eta| = 1\}$ and then to extend the obtained block matrix by homogeneity 0 to $T^*\Omega \setminus 0$. Let us write $\mathcal{H} = L^2(X^\wedge) \oplus \mathbb{C}^N$ and

$$G(y, \eta) = \begin{pmatrix} g_{11} & g_{12} \\ g_{21} & -1 + g_{22} \end{pmatrix} (y, \eta).$$

Here 1 denotes the identity operator in various spaces. Then the question is to characterise the inverse of $1 + G(y, \eta) : \mathcal{H} \to \mathcal{H}$ in the form $1 + H(y, \eta)$, such that $H(y, \eta) \in C^\infty(S^*\Omega, \mathcal{L}(\mathcal{H}, \mathcal{S}_P^{\frac{n}{2}}(X^\wedge) \oplus \mathbb{C}^N))$, $H^*(y, \eta) \in C^\infty(S^*\Omega, \mathcal{L}(\mathcal{H}, \mathcal{S}_Q^{\frac{n}{2}}(X^\wedge) \oplus \mathbb{C}^N))$ for suitable asymptotic types P, Q. For fixed $(y, \eta) \in S^*\Omega$ we can argue in an analogous manner as in the proof of Proposition 2.3.68, using that $G(y, \eta)$ itself is of this kind. We see, in particular, that $H(y, \eta) = -G(y, \eta) - G(y, \eta)H(y, \eta)$, and analogously for the adjoint. This gives us easily that $H(y, \eta)$ and $H^*(y, \eta)$ are C^∞ on $S^*\Omega$ with values in the continuous operators of the required kind. $\square$

Let us assume from now on that $\overline{\Omega}$ is compact and $\overline{\Omega} \subset \tilde{\Omega}$ for some larger domain $\tilde{\Omega}$.

Theorem 3.5.9 *Let* $a(y, \eta) \in \mathcal{R}^\mu(\tilde{\Omega} \times \mathbb{R}^q, \boldsymbol{g}; N_-, N_+)$ *for* $\boldsymbol{g} = (\gamma, \gamma - \mu, \Theta)$ *be elliptic in the sense of Definition 3.5.5. Then, for an appropriate cut–off function* $\omega(t)$, *there exists a*

$$b(y, \eta) \in \mathcal{R}^{-\mu}(\Omega \times \mathbb{R}^q, \boldsymbol{g}^{-1}; N_+, N_-)$$

for $\boldsymbol{g}^{-1} = (\gamma - \mu, \gamma, \Theta)$ *such that*

$$\omega \sigma_\psi^{-\mu}(b) = \omega \sigma_\psi^\mu(a)^{-1}, \quad \sigma_\wedge^{-\mu}(b) = \sigma_\wedge^\mu(a)^{-1}. \tag{3.5.25}$$

Proof. The ellipticity of $a(y, \eta)$ implies that $a_{11}(y, \eta) = \mathrm{u.\,l.\,c.}\, a(y, \eta)$ satisfies the conditions of Theorem 3.5.4 over $\tilde{\Omega}$. In other words we find a $\tilde{b}_{11}(y, \eta) \in R^{-\mu}(\tilde{\Omega} \times \mathbb{R}^q, \boldsymbol{g}^{-1})$ such that

$$\sigma_\wedge^{-\mu}(\tilde{b}_{11})(y, \eta)\sigma_\wedge^\mu(a_{11})(y, \eta) = 1 + g_{(0)}(y, \eta) \tag{3.5.26}$$

for some $g_{(0)}(y, \eta) \in \sigma_\wedge^0 R_G^0(\tilde{\Omega} \times \mathbb{R}^q, \boldsymbol{g}_l)$, and analogously for the composition in the reverse order.

This means that $\operatorname{ind} \sigma_\wedge^{-\mu}(\tilde{b}_{11})(y, \eta) = -\operatorname{ind} \sigma_\wedge^\mu(a_{11}) = N_- - N_+$. The following constructions may be carried out over $S^*\tilde{\Omega}$ which is a compact space. They will be formally analogous to those at the beginning of Section 1.2.4. Let us fix $s = s_0 \in \mathbb{R}$. We find a finite–dimensional subspace $\mathcal{V}_+ \subset \mathcal{K}^{s_0, \gamma}(X^\wedge)$ such that $\operatorname{im} \sigma_\wedge^{-\mu}\tilde{b}_{11}(y, \eta) + \mathcal{V}_+ = \mathcal{K}^{s_0, \gamma}(X^\wedge)$ for all $(y, \eta) \in S^*\tilde{\Omega}$. Since $C_0^\infty(X^\wedge)$ is dense in $\mathcal{K}^{s_0, \gamma}(X^\wedge)$, we may replace $\mathcal{V}_+$ by some finite–dimensional subspace $\mathcal{M}_+ \subset C_0^\infty(X^\wedge)$ such that $\operatorname{im} \sigma_\wedge^{-\mu}(\tilde{b}_{11})(y, \eta) + \mathcal{M}_+ = \mathcal{K}^{s_0, \gamma}(X^\wedge)$. Let $M_+ = \dim \mathcal{M}_+$ and choose an isomorphism $k : \mathbb{C}^{M_+} \to \mathcal{M}_+$.

Then

$$(\sigma_\wedge^{-\mu}(\tilde{b}_{11})(y, \eta) \quad k) : \begin{matrix} \mathcal{K}^{s_0 - \mu, \gamma - \mu}(X^\wedge) \\ \oplus \\ \mathbb{C}^{M_+} \end{matrix} \to \mathcal{K}^{s_0, \gamma}(X^\wedge) \tag{3.5.27}$$

is surjective for all $(y, \eta) \in S^*\overline{\Omega}$. In view of

$$\operatorname{ind}_{S^*\overline{\Omega}} \sigma_\wedge^{-\mu}(\tilde{b}_{11}) = -\operatorname{ind}_{S^*\overline{\Omega}} \sigma_\wedge^{-\mu}(a_{11}) \in \pi_Y^* K(\overline{\Omega}),$$

by choosing M_+ sufficiently large (according to a larger possible subspace $\mathcal{M}_+ \subset C_0^\infty(X^\wedge)$) we see that the kernel of (3.5.27) is a finite–dimensional trivial subbundle $\mathcal{M}_- \subset S^*\overline{\Omega} \times \{\mathcal{K}^{s_0 - \mu, \gamma - \mu}(X^\wedge) \oplus \mathbb{C}^{M_+}\}$. Denoting the fibre dimension of $\mathcal{M}_-$ by M_- we obtain that $M_+ - M_- = N_+ - N_-$. Because of Proposition 2.4.43 the fibres of $\mathcal{M}_-$ are subspaces of $\mathcal{S}^{\gamma - \mu}(X^\wedge) \oplus \mathbb{C}^{M_+}$ (recall that $\mathcal{S}^{\gamma - \mu}(X^\wedge) = \omega \mathcal{K}^{\infty, \gamma - \mu}(X^\wedge) + (1 - \omega)\mathcal{S}(\mathbb{R}_+, C^\infty(X))$ for a cut–off function $\omega(t)$). Applying the scalar product of $\mathcal{K}^{0,0}(X^\wedge)$ on $C_0^\infty(X^\wedge) \times C_0^\infty(X^\wedge)$, extended to a non–degenerate sesquilinear pairing $\mathcal{K}^{s_0, \gamma - \mu}(X^\wedge) \times \mathcal{K}^{-s_0, -\gamma + \mu}(X^\wedge) \to \mathbb{C}$, we can also form the non–degenerate sesquilinear pairing

$$(\mathcal{K}^{s_0, \gamma - \mu}(X^\wedge) \oplus \mathbb{C}^{M_+}) \times (\mathcal{K}^{-s_0, -\gamma + \mu}(X^\wedge) \oplus \mathbb{C}^{M_+}) \to \mathbb{C} \tag{3.5.28}$$

in an obvious manner by adding directly the scalar product of $\mathbb{C}^{M_+}$. This allows us to form a subbundle $\mathcal{M}_*^- \subset S^*\overline{\Omega} \times \{\mathcal{K}^{-s_0, -\gamma + \mu}(X^\wedge) \oplus \mathbb{C}^{M_+}\}$ such that the fibre $(\mathcal{M}_-)_{(y, \eta)}$ of $\mathcal{M}_-$ over $(y, \eta) \in S^*\overline{\Omega}$ is isomorphic to the correponding fibre $(\mathcal{M}_-^*)_{(y, \eta)}$ of $\mathcal{M}_*^-$ via (3.5.28). We can choose a base $(l_1(y, \eta), \ldots, l_{M_-}(y, \eta))$ in $(\mathcal{M}_-^*)_{(y,\eta)}$ smoothly dependent on $(y, \eta) \in S^*\overline{\Omega}$. Then the pairing

$$\mathcal{K}^{s_0, \gamma - \mu}(X^\wedge) \oplus \mathbb{C}^{M_+} \ni u \oplus c \to (u \oplus c, l_j(y, \eta)), \quad j = 1, \ldots, M_- \tag{3.5.29}$$

may be interpreted as a (y, η)–dependent family of maps

$$\begin{matrix} \mathcal{K}^{s_0, \gamma - \mu}(X^\wedge) \\ \oplus \\ \mathbb{C}^{M_+} \end{matrix} \to \mathbb{C}^{M_-}$$

that induce an isomorphism $\mathcal{M}_- \to S^*\overline{\Omega} \times \mathbb{C}^{M_-}$.

Using the density of $C_0^\infty(X^\wedge)$ in $\mathcal{K}^{-s_0, -\gamma + \mu}(X^\wedge)$ we can choose the vectors $l_j(y, \eta)$ by a suitable approximation as elements of the space $C^\infty(S^*\overline{\Omega}, C_0^\infty(X^\wedge) \oplus \mathbb{C}^{M_+})$, such that the pairings (3.5.29) still induce an isomorphism $\mathcal{M}_- \to S^*\overline{\Omega} \times \mathbb{C}^{M_-}$. According to the splittings in (3.5.28) we can write $l_j(y, \eta)$ a rows $l_j(y, \eta) = (t_j(y, \eta), r_j(y, \eta))$, $j = 1, \ldots, M_-$, and form the block matrices

$$t(y, \eta) = \begin{pmatrix} t_1 \\ \vdots \\ t_{M_-} \end{pmatrix}(y, \eta), \qquad r(y, \eta) = \begin{pmatrix} r_1 \\ \vdots \\ r_{M_-} \end{pmatrix}(y, \eta).$$

They yield together with (3.5.27) a family of isomorphisms

$$\begin{pmatrix} \sigma_\wedge^{-\mu}(\tilde{b}_{11}) & k \\ t & r \end{pmatrix}(y, \eta) : \begin{matrix} \mathcal{K}^{s_0 - \mu, \gamma - \mu}(X^\wedge) \\ \oplus \\ \mathbb{C}^{M_+} \end{matrix} \to \begin{matrix} \mathcal{K}^{s_0, \gamma}(X^\wedge) \\ \oplus \\ \mathbb{C}^{M_-} \end{matrix}, \tag{3.5.30}$$

parametrised by $(y, \eta) \in S^*\overline{\Omega}$. Next we form the extension of (3.5.30) by homogeneity $-\mu$ from $S^*\overline{\Omega}$ to $T^*\Omega \setminus 0$ to

$$\sigma_\wedge^{-\mu}(\tilde{b})(y, \eta) = (\sigma_\wedge^{-\mu}(\tilde{b}_{ij})(y, \eta))_{i,j=1,2},$$

setting

$$\sigma_\wedge^{-\mu}(\tilde{b}_{21})(y, \eta) = |\eta|^{-\mu} t(y, \frac{\eta}{|\eta|}) \kappa_{|\eta|}^{-1}, \quad \sigma_\wedge^{-\mu}(\tilde{b}_{12})(y, \eta) = |\eta|^{-\mu} \kappa_{|\eta|}^{-\mu} k,$$

and $\sigma_\wedge^{-\mu}(\tilde{b}_{22})(y, \eta) = |\eta|^{-\mu} r(y, \frac{\eta}{|\eta|})$.

We want to construct the inverse of

$$\sigma_\wedge^{-\mu}(a)(y, \eta) : \begin{matrix} \mathcal{K}^{s, \gamma}(X^\wedge) \\ \oplus \\ \mathbb{C}^{N_-} \end{matrix} \to \begin{matrix} \mathcal{K}^{s - \mu, \gamma - \mu}(X^\wedge) \\ \oplus \\ \mathbb{C}^{N_+} \end{matrix}, \tag{3.5.31}$$

for $s = s_0$. This is then the inverse of (3.5.31) for every $s \in \mathbb{R}$. In fact we first recall that there is an integer L such that $N_+ = M_+ + L$, $N_- = M_- + L$. In the case $L \geq 0$ we choose in the above construction for (3.5.27) the number M_+ at once equal to N_-. Then we have automatically $N_- = M_-$. For $L < 0$ we enlarge $N_\pm$ by $-L$ by replacing (3.5.31) by a direct sum $\sigma_\wedge^\mu(\tilde{a})(y,\eta) := \sigma_\wedge^\mu(a)(y,\eta) \oplus \mathrm{diag}(|\eta|^\mu, \ldots, |\eta|^\mu)$, where $\mathrm{diag}(\ldots)$ denotes a $(-L) \times (-L)$–diagonal matrix with entries $|\eta|^\mu$. Then, if we first construct $\sigma_\wedge^\mu(\tilde{a})^{-1}(y,\eta)$, we also obtain $\sigma_\wedge^\mu(a)^{-1}(y,\eta)$ by omitting the superfluous entries. In other words, to simplify notation we will assume from now on that $N_\pm = M_\pm$.

Using (3.5.26) we obtain

$$\sigma_\wedge^{-\mu}(\tilde{b})(y,\eta)\sigma_\wedge^\mu(a)(y,\eta) = \begin{pmatrix} 1 + g_{11} & g_{12} \\ g_{21} & g_{22} \end{pmatrix}(y,\eta),$$

where $g(y,\eta) = (g_{ij}(y,\eta))_{i,j=1,2}$ belongs to $\sigma_\wedge^0 \mathcal{R}_G^0(\Omega \times \mathbb{R}^q, \boldsymbol{g}_l; N_-, N_+)$. Since the operators are invertible, then

$$\begin{pmatrix} 1 + g_{11} & g_{12} \\ g_{21} & g_{22} \end{pmatrix}(y,\eta) : \begin{array}{c} \mathcal{K}^{s_0,\gamma}(X^\wedge) \\ \oplus \\ \mathbb{C}^{N_-} \end{array} \rightarrow \begin{array}{c} \mathcal{K}^{s_0,\gamma}(X^\wedge) \\ \oplus \\ \mathbb{C}^{N_-} \end{array}$$

is invertible for all $(y,\eta) \in T^*\overline{\Omega} \setminus 0$. From Lemma 3.5.8 we get an element $h(y,\eta) = (h_{ij}(y,\eta))_{i,j=1,2} \in \sigma_\wedge^0 \mathcal{R}_G^0(\Omega \times \mathbb{R}^q, \boldsymbol{g}_l; N_-, N_+)$ such that

$$\sigma_\wedge^\mu(a)^{-1}(y,\eta) = \begin{pmatrix} 1 + h_{11} & h_{12} \\ h_{21} & h_{22} \end{pmatrix}(y,\eta)\sigma_\wedge^{-\mu}(\tilde{b})(y,\eta),$$

$(y,\eta) \in T^*\overline{\Omega} \setminus 0$. Let $\chi(\eta)$ be an excision function in $\mathbb{R}_\eta^q$. Set

$$b_1(y,\eta) = \left\{ \begin{pmatrix} 1 & 0 \\ 0 & 0 \end{pmatrix} + \chi(\eta)h(y,\eta) \right\} b_0(y,\eta)$$

for

$$b_0(y,\eta) = \begin{pmatrix} \tilde{b}_{11}(y,\eta) & \chi(\eta)\tilde{b}_{12}(y,\eta) \\ \chi(\eta)\tilde{b}_{21}(y,\eta) & \chi(\eta)\tilde{b}_{22}(y,\eta) \end{pmatrix}.$$

Then, if we fix cut–off functions ω, ω_0 with $\omega\omega_0 = \omega$, vanishing for $t > \frac{c}{2}$ for the constant c of Definition 3.5.5 the symbol

$$b(y,\eta) = \begin{pmatrix} \omega(t) & 0 \\ 0 & 1 \end{pmatrix} b_1(y,\eta) \begin{pmatrix} \omega_0(t) & 0 \\ 0 & 1 \end{pmatrix}$$

belongs to $\mathcal{R}^{-\mu}(\Omega \times \mathbb{R}^q, \boldsymbol{g}^{-1}; N_+, N_-)$ and satisfies $\sigma_\wedge^{-\mu}(b) = \sigma_\wedge^\mu(a)^{-1}$, $\omega\sigma_\psi^{-\mu}(b) = \omega\sigma_\psi^\mu(a)^{-1}$. $\qquad\square$

In the following we set

$$\boldsymbol{\omega}(t) = \begin{pmatrix} \omega(t) & 0 \\ 0 & 1 \end{pmatrix}, \tag{3.5.32}$$

where 1 is the identity in $\mathbb{C}^N$ for various $N \in \mathbb{N}$. The particular N will be clear from the context; so we do not indicate it explicitly.

Theorem 3.5.10 *Let* $a(y, \eta) \in \mathcal{R}^\mu(\tilde{\Omega} \times \mathbb{R}^q, \boldsymbol{g}; N_-, N_+)$ *for* $\boldsymbol{g} = (\gamma, \gamma - \mu, \Theta)$ *be elliptic in the sense of Definition 3.5.5. Then, for appropriate cut-off functions* $\omega(t)$, $\omega_0(t)$ *with* $\omega\omega_0 = \omega$ *there exists an* $a^{(-1)}(y, y', \eta) \in \mathcal{R}^{-\mu}(\Omega \times \Omega \times \mathbb{R}^q, \boldsymbol{g}^{-1}; N_+, N_-)$ *for* $\boldsymbol{g}^{-1} = (\gamma - \mu, \gamma, \Theta)$ *such that*

$$\boldsymbol{\omega} \operatorname{Op}(a)\boldsymbol{\omega}_0 \operatorname{Op}(a^{(-1)}) = \boldsymbol{\omega} \cdot \operatorname{id} \operatorname{mod} \mathcal{Y}^{-\infty}(X^\wedge \times \Omega, \boldsymbol{g}_r; N_+, N_+),$$

$$\boldsymbol{\omega}_0 \operatorname{Op}(a^{(-1)})\boldsymbol{\omega} \operatorname{Op}(a) = \boldsymbol{\omega} \cdot \operatorname{id} \operatorname{mod} \mathcal{Y}^{-\infty}(X^\wedge \times \Omega, \boldsymbol{g}_l; N_-, N_-),$$

with id *being the identity operators in the corresponding spaces,* $\operatorname{Op}(\dots) = F_{\eta \to y}^{-1}(\dots)F_{y' \to \eta}$.

Proof. Let us modify the arguments at the beginning of the proof of Theorem 3.5.9 in the sense that the corresponding Leibniz inversion is applied to the local tuples of interior symbols near $t = 0$ as they are given by the operator family $a(y, \eta)$ in Theorem 3.5.10. Then the resulting element $b(y, \eta) \in \mathcal{R}^{-\mu}(\Omega \times \mathbb{R}^q, \boldsymbol{g}^{-1}; N_+, N_-)$ satisfies the relations (3.5.25) for some cut-off function $\omega(t)$. Let us choose an element $b_1(y, y', \eta) \in \mathcal{R}^{-\mu}(\Omega \times \Omega \times \mathbb{R}^q; \boldsymbol{g}^{-1}; N_+, N_-)$ which is properly supported with respect to the y-variables such that $\operatorname{Op}(b) = \operatorname{Op}(b_1) \bmod \mathcal{Y}^{-\infty}(X^\wedge \times \Omega^\wedge \times \boldsymbol{g}_l; N_-, N_-)$. Then, if ω_0, ω are cut-off functions vanishing for $t > c/2$ for the constant c of from Definition 3.5.5 with $\omega_0 \omega = \omega$, we get

$$\boldsymbol{\omega}_0 \operatorname{Op}(b_1)\boldsymbol{\omega} \operatorname{Op}(a) = \boldsymbol{\omega} \operatorname{id} + C$$

for some $C \in \mathcal{Y}_{M+G}^{-1}(X^\wedge \times \Omega, \boldsymbol{g}_l; N_-, N_-)$.

It is clear that $\boldsymbol{\omega} \operatorname{id} + C = \boldsymbol{\omega}(\operatorname{id} + C) \bmod \mathcal{Y}^{-\infty}(X^\wedge \times \Omega, \boldsymbol{g}; N_-, N_-)$. Applying a formal Neumann series argument we find an operator $D \in \mathcal{Y}_{M+G}^{-1}(X^\wedge \times \Omega, \boldsymbol{g}_l; N_-, N_-)$ with $(1 + D)(1 + C) = 1 \bmod \mathcal{Y}^{-\infty}(X^\wedge \times \Omega, \boldsymbol{g}_l; N_-, N_-)$. Then we have also

$$(\boldsymbol{\omega}_1 \operatorname{id} + D)(1 + C) = \boldsymbol{\omega} \operatorname{id} \operatorname{mod} \mathcal{Y}^{-\infty}(X^\wedge \times \Omega, \boldsymbol{g}_l; N_-, N_-)$$

for any cut-off function $\omega_1(t)$ satisfying $\omega_1 \omega = \omega$. Choose ω_1 in such a way that $\omega_1 \omega_0 = \omega_0$. Then we may set $a^{(-1)}(y, y', \eta) = (\boldsymbol{\omega}_1, \operatorname{id} + D)b_1(y, y', \eta)$ which yields the second relation of Theorem 3.5.10. The first one can be proved in an analogous manner. $\square$

3.5.2 Ellipticity and parametrices in the edge algebra

Let us recall that $\mathcal{Y}^\nu(X^\wedge \times \Omega, \boldsymbol{g}; N_-, N_+)$ for $\nu \in \mathbb{R}$, $\boldsymbol{g} = (\gamma, \gamma - \mu, \Theta)$, $\gamma, \mu \in \mathbb{R}$, $\mu - \nu \in \mathbb{N}$, is the space of all operators

$$A = \mathrm{Op}(a) + C \quad \text{for} \quad a(y, \eta) \in \mathcal{R}^\nu(\Omega \times \mathbb{R}^q, \boldsymbol{g}; N_-, N_+), \quad (3.5.33)$$

$C \in \mathcal{Y}^{-\infty}(X^\wedge \times \Omega, \boldsymbol{g}; N_-, N_+)$, cf. Definition 3.4.29, Proposition 3.4.34, Proposition 3.4.9. As usual $\Omega \subseteq \mathbb{R}^q$ is an open set and $\mathrm{Op}(a)u(y) = F_{\eta \to y}^{-1} a(y, \eta)(F_{y' \to \eta} u)(\eta)$, $F = F_{y \to \eta}$ the Fourier transform in $\mathbb{R}^q$.

For $A \in \mathcal{Y}^\nu(X^\wedge \times \Omega, \boldsymbol{g}; N_-, N_+)$ we have u. l. c. $A \in L_{\mathrm{cl}}^\nu(X^\wedge \times \Omega)$. This yields a unique homogeneous principal interior symbol of order ν

$$\sigma_\psi^\nu(A)(t, x, y, \tau, \xi, \eta) \in S^{(\nu)}(T^*(X^\wedge \times \Omega) \setminus 0) \quad (3.5.34)$$

with

$$\sigma_{\psi,b}^\nu(A)(t, x, y, \tau, \xi, \eta) = t^\nu \sigma_\psi^\nu(A)(t, x, y, t^{-1}\tau, \xi, t^{-1}\eta)$$
$$\in S^{(\nu)}(T^*(\overline{\mathbb{R}} \times X^\wedge \times \Omega) \setminus 0) \quad (3.5.35)$$

cf. also (3.5.3), (3.5.4). Moreover, $A \in \mathcal{Y}^\nu(X^\wedge \times \Omega, \boldsymbol{g}; N_-, N_+)$ has a unique homogeneous principal edge symbol of order ν that is an operator family

$$\sigma_\wedge^\nu(A)(y, \eta): \quad \begin{array}{c} \mathcal{K}^{s,\gamma}(X^\wedge) \\ \oplus \\ \mathbb{C}^{N_-} \end{array} \quad \to \quad \begin{array}{c} \mathcal{K}^{s-\nu, \gamma-\mu}(X^\wedge) \\ \oplus \\ \mathbb{C}^{N_+} \end{array}$$

for $(y, \eta) \in T^*\Omega \setminus 0$, $s \in \mathbb{R}$, satisfying

$$\sigma_\wedge^\nu(A)(y, \lambda\eta) = \lambda^\nu \begin{pmatrix} \kappa_\lambda & 0 \\ 0 & 1 \end{pmatrix} \sigma_\wedge^\nu(A)(y, \eta) \begin{pmatrix} \kappa_\lambda & 0 \\ 0 & 1 \end{pmatrix}^{-1}$$

for all $\lambda \in \mathbb{R}_+$, $(y, \eta) \in T^*\Omega \setminus 0$.

It is clear that when A is given in the form (3.5.33) we have

$$\sigma_\psi^\nu(A) = \sigma_\psi^\nu(a), \quad \sigma_\wedge^\nu(A) = \sigma_\wedge^\nu(a),$$

cf. (3.3.56).

If $\mathbb{W}$ is a (stretched) manifold with edges Y and $J^\pm \in \mathrm{Vect}(Y)$ we may also consider the space $\mathcal{Y}^\nu(\mathbb{W}, \boldsymbol{g}; J^-, J^+)$ of pseudo–differential operators on $\mathbb{W}$ with edge trace and potential conditions along Y, where $\mathcal{Y}^\nu(X^\wedge \times \Omega, \boldsymbol{g}; N_-, N_+)$ is the local model near $\partial\mathbb{W}$ with respect to local coordinates $y \in \Omega$ on Y, and $N_\pm$ being the fibre demension of $J^\pm$. More generally we have

$$\mathcal{Y}^\nu(\mathbb{W}, \boldsymbol{g}; E, F; J^- J^+) \quad \text{for} \quad E, F \in \text{Vect}(\mathbb{W}), \quad J^\pm \in \text{Vect}(Y), \tag{3.5.36}$$

cf. Remark 3.4.46.

We shall formulate the results on ellipticity in the algebra of pseudo–differential operators on $\mathbb{W}$ for the case (3.5.36) that is necessary in the index theory. The reader who is more interested in the basic analytical ideas may look at trivial bundles $E = F = \mathbb{W} \times \mathbb{C}$ and $J^\pm = Y \times \mathbb{C}^{N_\pm}$.

The constructions and notations from this case allow the corresponding extensions to the more general situation for arbitrary $E, F \in$ Vect($\mathbb{W}$), $J^\pm \in$ Vect(Y). The global invariant interpretations are straightforward, similar to the material in Section 1.1.5. We will tacitly use this here without further comment, in particular, the weighted wedge Sobolev spaces of distributional sections in $E \in$ Vect($\mathbb{W}$):

$$\mathcal{W}^{s,\gamma}_{\text{comp}}(\mathbb{W}, E) \quad \text{and} \quad \mathcal{W}^{s,\gamma}_{\text{loc}}(\mathbb{W}, E),$$

as well as the subspaces with asymptotics

$$\mathcal{W}^{s,\gamma}_{P,\text{comp}}(\mathbb{W}, E) \quad \text{and} \quad \mathcal{W}^{s,\gamma}_{P,\text{loc}}(\mathbb{W}, E),$$

respectively. For compact $\mathbb{W}$ the subscripts comp, loc disappear, and we have the spaces

$$\mathcal{W}^{s,\gamma}(\mathbb{W}, E), \quad \mathcal{W}^{s,\gamma}_{P}(\mathbb{W}, E),$$

for $s, \gamma \in \mathbb{R}$ and (continuous) asymptotic types $P \in \text{As}(X, (\gamma, \Theta))$.

Note that the coefficient spaces in the analytic functionals for the asymptotic types are the C^∞ sections in $E|_{X \times X}$.

The elements of Vect($\mathbb{W}$) $\ni E$ restrict to bundles on $\partial \mathbb{W} \cong X \times Y$ and can be pulled back with repect to $X^\wedge \times Y \to X \times Y$, $(t, x, y) \to (x, y)$, to bundles on $X^\wedge \times Y$. Often we briefly denote all these bundles again by E. Then $\mathcal{K}^{s,\gamma}(X^\wedge, E)$ means the family of spaces $\mathcal{K}^{s,\gamma}(X^\wedge, E_y)$, $y \in Y$, where $E_y \in$ Vect($X^\wedge$) is obtained by restricting E from $X^\wedge \times Y$ to $X^\wedge \times \{y\}$.

Remark 3.5.11 *The elements* $C \in \mathcal{Y}^{-\infty}(\mathbb{W}, \boldsymbol{g}; E, F; J^-, J^+)$ *for* $\boldsymbol{g} = (\gamma, \gamma - \mu, \Theta)$ *induce continuous operators*

$$C : \begin{array}{c} \mathcal{W}^{s,\gamma}_{\text{comp}}(\mathbb{W}, E) \\ \oplus \\ H^s_{\text{comp}}(Y, J^-) \end{array} \rightarrow \begin{array}{c} \mathcal{W}^{\infty,\gamma-\mu}_{P,\text{loc}}(\mathbb{W}, F) \\ \oplus \\ H^\infty_{\text{loc}}(Y, J^+) \end{array}$$

and

$$C^* : \begin{array}{c} \mathcal{W}_{\mathrm{comp}}^{s,-\gamma+\mu}(\mathbb{W}, F) \\ \oplus \\ H_{\mathrm{comp}}^{s}(Y, J^+) \end{array} \to \begin{array}{c} \mathcal{W}_{Q,\mathrm{loc}}^{\infty,-\gamma}(\mathbb{W}, E) \\ \oplus \\ H_{\mathrm{loc}}^{\infty}(Y, J^-) \end{array}$$

for all $s \in \mathbb{R}$, *with* C*-dependent*

$$P \in \mathrm{As}(X, (\gamma - \mu, \Theta)), \qquad Q \in \mathrm{As}(X, (-\gamma, \Theta)).$$

The formal adjoint C^* *relies on fixed Riemannian metrics on* $\mathbb{W}$, Y *and Hermitian metrics in the involved bundles which determine the global scalar products in* $\mathcal{W}_{\mathrm{comp}}^{0,0}(\mathbb{W}, E) \oplus H_{\mathrm{comp}}^{0}(Y, J^-)$ *and* $\mathcal{W}_{\mathrm{comp}}^{0,0}(\mathbb{W}, F) \oplus H_{\mathrm{comp}}^{0}(Y, J^+)$, *respectively,*

$$(u, C^*v)_{\mathcal{W}_{\mathrm{comp}}^{0,0}(\mathbb{W},E) \oplus H_{\mathrm{comp}}^{0}(Y,J^-)} = (Cu, v)_{\mathcal{W}_{\mathrm{comp}}^{0,0}(\mathbb{W},F) \oplus H_{\mathrm{comp}}^{0}(Y,J^+)}$$

for all $u \in C_0^{\infty}(\mathrm{int}\,\mathbb{W}, E) \oplus C_0^{\infty}(Y, J^-)$, $v \in C_0^{\infty}(\mathrm{int}\,\mathbb{W}, F) \oplus C_0^{\infty}(Y, J^+)$.

Every $A \in \mathcal{Y}^{\nu}(\mathbb{W}, \boldsymbol{g}; E, F; J^-, J^+)$ has a global interior principal symbol

$$\sigma_{\psi}^{\nu}(A) : \pi_{\mathrm{int}\,\mathbb{W}}^* E \to \pi_{\mathrm{int}\,\mathbb{W}}^* F,$$

where $\pi_{\mathrm{int}\,\mathbb{W}}^* : T^*(\mathrm{int}\,\mathbb{W}) \setminus 0 \to \mathrm{int}\,\mathbb{W}$ is the canonical projection and $\pi_{\mathrm{int}\,\mathbb{W}}^*$ the associated bundle pull–back. Moreover, A has a global edge principal symbol

$$\sigma_{\wedge}^{\nu}(A) : \pi_Y^* \begin{pmatrix} \mathcal{K}^{s,\gamma}(X^{\wedge}, E) \\ \oplus \\ J^- \end{pmatrix} \to \pi_Y^* \begin{pmatrix} \mathcal{K}^{s-\nu,\gamma-\mu}(X^{\wedge}, F) \\ \oplus \\ J^+ \end{pmatrix}$$

with the canonical projection $\pi_Y : T^*Y \setminus 0 \to Y$ and the bundle pull–back π_Y^*.

Let us fix on $\mathbb{W}$ a neighbourhood $\mathbb{V}$ of the stretched edge $\partial\mathbb{W}$ of the form $[0,1) \times X \times Y$ (in the corresponding splitting of variables (t, x, y)). Then, if we set

$$\sigma_{\psi,b}^{\nu}(A) = t^{\nu}\sigma_{\psi}^{\nu}(A)(t, x, y, t^{-1}\tau, \xi, t^{-1}\eta) \quad \text{for} \quad (t, x, y) \in \mathbb{V}$$

we get a morphism $\sigma_{\psi,b}^{\nu}(A) : \pi_{\mathbb{V}}^* E \to \pi_{\mathbb{V}}^* F$ for $\pi_{\mathbb{V}} : T^*\mathbb{V} \setminus 0 \to \mathbb{V}$ that is smooth up to $t = 0$.

The elements $A \in \mathcal{Y}^{\nu}(\mathbb{W}, \boldsymbol{g}; E, F; J^-, J^+)$, written in block matrix form $A = (A_{ij})_{i,j=1,2}$, induce continuous operators

$$A = \begin{pmatrix} A_{11} & A_{12} \\ A_{21} & A_{22} \end{pmatrix} : \begin{array}{c} \mathcal{W}_{\mathrm{comp}}^{s,\gamma}(\mathbb{W}, E) \\ \oplus \\ H_{\mathrm{comp}}^{s}(Y, J^-) \end{array} \to \begin{array}{c} \mathcal{W}_{\mathrm{loc}}^{s-\nu,\gamma-\mu}(\mathbb{W}, F) \\ \oplus \\ H_{\mathrm{loc}}^{s-\nu}(Y, J^+) \end{array}, \quad (3.5.37)$$

and for every $P \in \mathrm{As}(X, (\gamma, \Theta))$ there is a $Q \in \mathrm{As}(X, (\gamma - \mu, \Theta))$ with

$$
A: \quad
\begin{array}{c}
\mathcal{W}^{s,\gamma}_{P,\mathrm{comp}}(\mathbb{W}, E) \\
\oplus \\
H^{s}_{\mathrm{comp}}(Y, J^{-})
\end{array}
\quad \to \quad
\begin{array}{c}
\mathcal{W}^{s-\nu,\gamma-\mu}_{Q,\mathrm{loc}}(\mathbb{W}, F) \\
\oplus \\
H^{s-\nu}_{\mathrm{loc}}(Y, J^{+})
\end{array}
$$

for all $s \in \mathbb{R}$.

Remark 3.5.12 *Let* $A \in \mathcal{Y}^{\nu}(\mathbb{W}, \boldsymbol{g}; E, F; J^{-}, J^{+})$, $\boldsymbol{g} = (\gamma, \gamma - \mu, \Theta)$, $\gamma, \mu, \nu \in \mathbb{R}$, $\mu - \nu \in \mathbb{N}$, *and let* $\mathbb{W}$ *be compact. Then*

$$
\sigma^{\nu}_{\psi}(A) = 0, \quad \sigma^{\nu}_{\wedge}(A) = 0
$$

implies that the operator

$$
A: \quad
\begin{array}{c}
\mathcal{W}^{s,\gamma}(\mathbb{W}, E) \\
\oplus \\
H^{s}(Y, J^{-})
\end{array}
\quad \to \quad
\begin{array}{c}
\mathcal{W}^{s-\nu,\gamma-\mu}(\mathbb{W}, F) \\
\oplus \\
H^{s-\nu}(Y, J^{+})
\end{array}
\tag{3.5.38}
$$

is compact for each $s \in \mathbb{R}$.

In particular, the elements $A \in \mathcal{Y}^{-\infty}(\mathbb{W}, \boldsymbol{g}; E, F; J^{-}, J^{+})$ *induce compact operators* (3.5.38), *cf. also Corollary 3.4.49 and Proposition 3.4.52.*

Definition 3.5.13 *An operator* $A \in \mathcal{Y}^{\mu}(\mathbb{W}, \boldsymbol{g}; E, F; J^{-}, J^{+})$ *for* $\boldsymbol{g} = (\gamma, \gamma - \mu, \Theta)$, $\gamma, \mu \in \mathbb{R}$, *is called elliptic if*

(i) $\sigma^{\mu}_{\psi}(A) : \pi^{*}_{\mathrm{int}\,\mathbb{W}} E \to \pi^{*}_{\mathrm{int}\,\mathbb{W}} F$ *and* $\sigma^{\mu}_{\psi,b}(A) : \pi^{*}_{\mathbb{V}} E \to \pi^{*}_{\mathbb{V}} F$ *are isomorphisms (the latter one up to* $t = 0$*),*

(ii)

$$
\sigma^{\mu}_{\wedge}(A) : \pi^{*}_{Y}
\begin{pmatrix}
\mathcal{K}^{s,\gamma}(X^{\wedge}, E) \\
\oplus \\
J^{-}
\end{pmatrix}
\to \pi^{*}_{Y}
\begin{pmatrix}
\mathcal{K}^{s-\mu,\gamma-\mu}(X^{\wedge}, F) \\
\oplus \\
J^{+}
\end{pmatrix}
$$

is an isomorphism.

The condition (i) in Definition 3.5.13 is the standard ellipticity of $A_{11} = \mathrm{u.\,l.\,c.}\, A$, with an additional control of $\sigma^{\mu}_{\psi}(A)$ up to $t = 0$. The operators A_{ij} for $(ij) \neq (11)$ have the interpretation of elliptic trace and potential conditions along the edge. Then Definition 3.5.13 (ii) is an analogue of the classical Shapiro–Lopatinskij condition in elliptic boundary value problems, cf. also Chapter 4 below. $N_{-} = 0$ or $N_{+} = 0$ are allowed cases here, in particular, $N_{-} = N_{+} = 0$, cf. the examples below.

In the following, when we have to compose operators, one factor has to satisfy a condition of being properly supported in an adequate sense. In Remark 3.4.47 we have noted that each A can be written in the form

$$A = \omega A_0 \omega_0 + \begin{pmatrix} (1-\omega)P_0(1-\omega_1) & 0 \\ 0 & 0 \end{pmatrix} + C \qquad (3.5.39)$$

for cut–off functions $\omega, \omega_1, \omega_2$, satisfying $\omega\omega_0 = \omega$, $\omega\omega_1 = \omega_1$, $A_0 \in \mathcal{Y}^\nu$ over $X^\wedge \times Y$, $P_0 \in L_{cl}^\nu(\text{int}\,\mathbb{W})$, $C \in \mathcal{Y}^{-\infty}$. When A_0 is properly supported in the y–variables, P_0 properly supported in the standard sense, and C properly supported on $\mathbb{W}$, then we say that A is properly supported (in the $\mathbb{W}$–sense).

Definition 3.5.14 Let $A \in \mathcal{Y}^\mu(\mathbb{W}, \boldsymbol{g}; E, F; J^-, J^+)$, $\boldsymbol{g} = (\gamma, \gamma - \mu, \Theta)$, $B \in \mathcal{Y}^{-\mu}(\mathbb{W}, \boldsymbol{g}^{-1}; F, E; J^+, J^-)$, $\boldsymbol{g}^{-1} = (\gamma - \mu, \gamma, \Theta)$, and let A or B be properly supported (in the $\mathbb{W}$–sense). Then B is called a paramtrix of A if

$$G_l := BA - 1 \in \mathcal{Y}^{-\infty}(\mathbb{W}, \boldsymbol{g}_l; E, E; J^-, J^-), \qquad (3.5.40)$$
$$G_r := AB - 1 \in \mathcal{Y}^{-\infty}(\mathbb{W}, \boldsymbol{g}_r; F, F; J^+, J^+), \qquad (3.5.41)$$

where $\boldsymbol{g}_l = (\gamma, \gamma, \Theta)$, $\boldsymbol{g}_r = (\gamma - \mu, \gamma - \mu, \Theta)$.

Note that when $B \in \mathcal{Y}^{-\mu}(\mathbb{W}, \boldsymbol{g}^{-1}; F, E; J^+, J^-)$ is a parametrix of A then also $B + C$ for each $C \in \mathcal{Y}^{-\infty}(\mathbb{W}, \boldsymbol{g}^{-1}; F, E; J^+, J^-)$ (provided A or $B + C$ is properly supported in the $\mathbb{W}$–sense). Moreover, if $B_l, B_r \in \mathcal{Y}^{-\mu}(\mathbb{W}, \boldsymbol{g}^{-1}; F, E; J^+, J^-)$ satisfy $B_l A - 1 \in \mathcal{Y}^{-\infty}(\mathbb{W}, \boldsymbol{g}_l; E, E; J^-, J^-)$, $AB_r - 1 \in \mathcal{Y}^{-\infty}(\mathbb{W}, \boldsymbol{g}_r; F, F; J^+, J^+)$, then $B_l - B_r$ is an element of $\mathcal{Y}^{-\infty}(\mathbb{W}, \boldsymbol{g}^{-1}; F, E; J^+, J^-)$.

If B is a parametrix of A we have

$$\sigma_\psi^{-\mu}(B) = \sigma_\psi^\mu(A)^{-1}, \quad \sigma_\wedge^{-\mu}(B) = \sigma_\wedge^\mu(A)^{-1}. \qquad (3.5.42)$$

This is a consequence of Theorem 3.4.56.

Theorem 3.5.15 Let $A \in \mathcal{Y}^\mu(\mathbb{W}, \boldsymbol{g}; E, F; J^-, J^+)$ for $\boldsymbol{g} = (\gamma, \gamma - \mu, \Theta)$, $\gamma, \mu \in \mathbb{R}$, be elliptic. Then A has a parametrix $B \in \mathcal{Y}^{-\mu}(\mathbb{W}, \boldsymbol{g}^{-1}; F, E; J^+, J^-)$ (which is properly supported in the $\mathbb{W}$–sense).

Proof. Write A in the form (3.5.39), where

$$A_0 \in \mathcal{Y}^\mu(X^\wedge \times Y, \boldsymbol{g}; E, F; J^-, J^+),$$
$$P_0 \in L_{cl}^\mu(\text{int}\,\mathbb{W}; E, F), \qquad C \in \mathcal{Y}^{-\infty}(\mathbb{W}, \boldsymbol{g}; E, F; J^-, J^+).$$

For the parametrix construction we can ignore C. The operator P_0 is elliptic in the standard sense. Thus there is a parametrix $P_0^{(-1)} \in$

$L_{\mathrm{cl}}^{-\mu}(\mathrm{int}\,\mathbb{W}; F, E)$, cf. Section 1.1.6. Moreover, we will construct below an operator $A_0^{(-1)} \in \mathcal{Y}^{-\mu}(X^{\wedge} \times Y, \boldsymbol{g}^{-1}; F, E; J^{+}, J^{-})$ such that

$$\omega A_0 \omega_0 A_0^{(-1)} = \omega \cdot \mathrm{id} \bmod \mathcal{Y}^{-\infty}(X^{\wedge} \times Y, \boldsymbol{g}_r; F, F; J^{+}, J^{+}),$$
$$A_0^{(-1)} \omega A_0 \omega_0 = \omega \cdot \mathrm{id} \bmod \mathcal{Y}^{-\infty}(X^{\wedge} \times Y, \boldsymbol{g}_l; E, E; J^{-}, J^{-}),$$

with id being various identity operators. Then, choosing $A_0^{(-1)}$ properly supported with respect to y–variables and $P_0^{(-1)}$ properly supported on $\mathrm{int}\,\mathbb{W}$ in the standard sense, we can set

$$A^{(-1)} = \omega A_0^{(-1)} \omega_0 + \begin{pmatrix} (1-\omega) P_0^{(-1)} (1-\omega_1) & 0 \\ 0 & 0 \end{pmatrix}$$

which is a parametrix of A, properly supported (in the $\mathbb{W}$–sense). Thus is remains to construct $A_0^{(-1)}$.

Let G be a coordinate neighbourhood on Y and $\chi : G \to \Omega$ be a chart, $\Omega \subseteq \mathbb{R}^q$ open. Then $A_0|_{X^{\wedge} \times \Omega}$ can be written $\mathrm{Op}(a) \bmod \mathcal{Y}^{-\infty}$, where $a(y, \eta) \in \mathcal{R}^{\mu}(\Omega \times \mathbb{R}^q, \boldsymbol{g}; E, F; N_{-}, N_{+})$ (in evident generalisation of the notation from Definition 3.3.40). We may shrink G to a relatively compact open set the closure of which is contained in G. Thus, denoting the latter set again by G, we can assume that $\overline{\Omega}$ is compact and that our objects are in fact given in a larger open set containing $\overline{\Omega}$. Thus we may apply Theorem 3.5.10 in the corresponding evident generalisation to the case of general bundles E, F. This gives us a symbol $a^{(-1)}(y, \eta) \in \mathcal{R}^{-\mu}(\Omega \times \mathbb{R}^q, \boldsymbol{g}^{-1}; F, E; N_{+}, N_{-})$, where $\mathrm{Op}(a^{(-1)})$ is the local representative of an operator in $\mathcal{Y}^{-\mu}(X^{\wedge} \times Y, \boldsymbol{g}^{-1}; F, E; J^{+}, J^{-})$ over $X^{\wedge} \times G$ in the corresponding local coordinates with respect to y. In the pull–back of $\mathrm{Op}(a^{(-1)})$ under $1 \times \chi : X^{\wedge} \times G \to X^{\wedge} \times \Omega$ there are also involved the chosen trivialisations of the corresponding bundles. Now let $\{G_j\}_{j \in \mathbb{N}}$ be a locally finite open covering of Y by coordinate neighbourhoods, $\{\varphi_j\}_{j \in \mathbb{N}}$ a subordinate partition of unity and $\{\psi_j\}_{j \in \mathbb{N}}$ another system of functions satisfying $\varphi_j \psi_j = \varphi_j$ for all j. Then, denoting the above operator over $X^{\wedge} \times G_j$ by $A_{0,j}^{(-1)}$, we can set $A_0^{(-1)} = \sum_{j \in \mathbb{N}} \varphi_j A_{0,j}^{(-1)} \psi_j$. This completes the proof. $\square$

Theorem 3.5.16 *Let* $A \in \mathcal{Y}^{\mu}(\mathbb{W}, \boldsymbol{g}; E, F; J^{-}, J^{+})$ *for* $\boldsymbol{g} = (\gamma, \gamma - \mu, \Theta)$, $\gamma, \mu \in \mathbb{R}$, *be elliptic. Then*

$$Au = f \in \begin{matrix} \mathcal{W}_{\mathrm{loc}}^{s-\mu, \gamma - \mu}(\mathbb{W}, F) \\ \oplus \\ H_{\mathrm{loc}}^{s-\mu}(Y, J^{+}) \end{matrix} \quad , \quad u \in \begin{matrix} \mathcal{W}_{\mathrm{comp}}^{-\infty, \gamma}(\mathbb{W}, E) \\ \oplus \\ H_{\mathrm{comp}}^{-\infty}(Y, J^{-}) \end{matrix} \quad ,$$

$s \in \mathbb{R}$, implies $u \in \begin{array}{c} \mathcal{W}^{s,\gamma}_{\mathrm{comp}}(\mathbb{W},E) \\ \oplus \\ H^{s}_{\mathrm{comp}}(Y,J^-) \end{array}$. Moreover, for every $Q \in \mathrm{As}(X,(\gamma - \mu, \Theta))$ there is a $P \in \mathrm{As}(X,(\gamma, \Theta))$ such that

$$Au = f \in \begin{array}{c} \mathcal{W}^{s-\mu,\gamma-\mu}_{Q,\mathrm{loc}}(\mathbb{W},F) \\ \oplus \\ H^{s-\mu}_{\mathrm{loc}}(Y,J^+) \end{array}, \quad u \in \begin{array}{c} \mathcal{W}^{-\infty,\gamma}_{\mathrm{comp}}(\mathbb{W},E) \\ \oplus \\ H^{-\infty}_{\mathrm{comp}}(Y,J^-) \end{array}$$

implies $u \in \begin{array}{c} \mathcal{W}^{s,\gamma}_{P,\mathrm{comp}}(\mathbb{W},E) \\ \oplus \\ H^{s}_{\mathrm{comp}}(Y,J^-) \end{array}$ for each $s \in \mathbb{R}$.

Proof. From Theorem 3.5.15 we have a properly supported parametrix B. In view of the continuity of

$$B : \begin{array}{c} \mathcal{W}^{s-\mu,\gamma-\mu}_{Q,\mathrm{loc}}(\mathbb{W},F) \\ \oplus \\ H^{s-\mu}_{\mathrm{loc}}(Y,J^+) \end{array} \to \begin{array}{c} \mathcal{W}^{s,\gamma}_{\tilde{Q},\mathrm{loc}}(\mathbb{W},E) \\ \oplus \\ H^{s}_{\mathrm{loc}}(Y,J^-) \end{array}$$

for every $Q \in \mathrm{As}(X,(\gamma - \mu, \Theta))$ with some resulting $\tilde{Q} \in \mathrm{As}(X,(\gamma, \Theta))$ it follows from (3.5.40) that $BAu = (1 + G_l)u \in \mathcal{W}^{s,\gamma}_{\tilde{Q},\mathrm{loc}}(\mathbb{W},E) \oplus H^{s}_{\mathrm{loc}}(Y,J^-)$. From

$$G_l : \begin{array}{c} \mathcal{W}^{s,\gamma}_{\mathrm{comp}}(\mathbb{W},E) \\ \oplus \\ H^{s}_{\mathrm{comp}}(Y,J^-) \end{array} \to \begin{array}{c} \mathcal{W}^{\infty,\gamma}_{R,\mathrm{loc}}(\mathbb{W},E) \\ \oplus \\ H^{\infty}_{\mathrm{loc}}(Y,J^-) \end{array}$$

for every $s \in \mathbb{R}$, with some asymptotic type $R \in \mathrm{As}(X,(\gamma, \Theta))$, dependent on G_l, cf. Remark 3.5.11, we obtain $u \in \{\mathcal{W}^{s,\gamma}_{\tilde{Q},\mathrm{loc}}(\mathbb{W},E) + \mathcal{W}^{\infty,\gamma}_{R,\mathrm{loc}}(\mathbb{W},E)\} \oplus H^{s}_{\mathrm{loc}}(Y,J^-)$. However the space in $\{\ldots\}$ is contained in $\mathcal{W}^{s,\gamma}_{P,\mathrm{loc}}(\mathbb{W},E)$ for some (minimal) $P \in \mathrm{As}(X,(\gamma, \Theta))$. Moreover, since u was supposed to belong to the space with subscript comp we get the second statement of the theorem. The first one without asymptotic types is easier and can be proved in an analogous manner. $\square$

Theorem 3.5.17 Let $\mathbb{W}$ be compact and $A \in \mathcal{Y}^{\mu}(\mathbb{W}, \boldsymbol{g}; E, F; J^-, J^+)$ be an elliptic operator, $\boldsymbol{g} = (\gamma, \gamma - \mu, \Theta)$, $\gamma, \mu \in \mathbb{R}$. Then

$$A : \begin{array}{c} \mathcal{W}^{s,\gamma}(\mathbb{W},E) \\ \oplus \\ H^{s}(Y,J^-) \end{array} \to \begin{array}{c} \mathcal{W}^{s-\mu,\gamma-\mu}(\mathbb{W},F) \\ \oplus \\ H^{s-\mu}(Y,J^+) \end{array}$$

is a Fredholm operator for every $s \in \mathbb{R}$.

Proof. In view of Theorem 3.5.15 the operator A has a parametrix B, and we have the relations (3.5.40), (3.5.41). The operators

$$G_l : \begin{array}{c} \mathcal{W}^{s,\gamma}(\mathbb{W},E) \\ \oplus \\ H^{s}(Y,J^-) \end{array} \to \begin{array}{c} \mathcal{W}^{s,\gamma}(\mathbb{W},E) \\ \oplus \\ H^{s}(Y,J^-) \end{array}$$

and

$$
G_r : \quad
\begin{array}{ccc}
\mathcal{W}^{s,\gamma-\mu}(\mathbb{W}, F) & & \mathcal{W}^{s,\gamma-\mu}(\mathbb{W}, F) \\
\oplus & \rightarrow & \oplus \\
H^s(Y, J^+) & & H^s(Y, J^+)
\end{array}
$$

are compact for each $s \in \mathbb{R}$, cf. Remark 3.5.12. This implies the Fredholm property of A. $\square$

Corollary 3.5.18 *Let $\mathbb{W}$ be compact and $A \in \mathcal{Y}^\mu(\mathbb{W}, g; E, F; J^-, J^+)$ be elliptic. Then $A_{11} = \mathrm{u.l.c.}\ A : \mathcal{W}^{s,\gamma}(\mathbb{W}, E) \to \mathcal{W}^{s-\mu,\gamma-\mu}(\mathbb{W}, F)$ has a closed image for each $s \in \mathbb{R}$.*

This is a consequence of Theorem 3.5.17 and of the following general result:

Theorem 3.5.19 *Let H_i, L_i be Hilbert spaces, $i = 1, 2$ and let*

$$
\mathcal{A} = \begin{pmatrix} A & K \\ T & Q \end{pmatrix} : \quad
\begin{array}{ccc}
H_1 & & H_2 \\
\oplus & \rightarrow & \oplus \\
L_1 & & L_2
\end{array}
\qquad (3.5.43)
$$

be a Fredholm operator. Then $\mathrm{im}\,\mathcal{A}$ is a closed subspace in H_2.

Proof. Let us first consider Hilbert spaces H, $\tilde{H}$ and continuous operators $A : H \to \tilde{H}$, $B : \tilde{H} \to H$ satisfying $BA = 1$. Then $(AB)^2 = AB$ is a continuous projection to $\mathrm{im}\,A$ and hence $\mathrm{im}\,A$ is closed as the kernel of the complementary projection $1 - AB$. Now consider the operator (3.5.43), first under the assumption that it is an isomorphism. Let the operator

$$
B = \begin{pmatrix} P & C \\ S & R \end{pmatrix} : \quad
\begin{array}{ccc}
H_2 & & H_1 \\
\oplus & \rightarrow & \oplus \\
L_2 & & L_1
\end{array}
$$

be the inverse of $\mathcal{A}$. Then, if we set $\mathbf{B} = \begin{pmatrix} P & C \end{pmatrix}$, $\mathbf{A} = \begin{pmatrix} A \\ T \end{pmatrix}$, and $H = \begin{smallmatrix} H_1 \\ \oplus \\ L_1 \end{smallmatrix}$, $\tilde{H} = \begin{smallmatrix} H_2 \\ \oplus \\ L_2 \end{smallmatrix}$, it follows that $\mathbf{BA} = 1$. Hence $\mathrm{im}\,\mathbf{A}$ is a closed subspace of $\tilde{H}$ which also implies that $\mathrm{im}\,A \cong \mathrm{im}\,\mathbf{A} \cap \ker \pi$ is closed, where $\pi : H_2 \oplus L_2 \to L_2$ is the canonical projection. Set $\tilde{L}_1 := \begin{smallmatrix} L_1 \\ \oplus \\ \mathbb{C}^{N_-} \end{smallmatrix}$, $\tilde{L}_2 := \begin{smallmatrix} L_2 \\ \oplus \\ \mathbb{C}^{N_+} \end{smallmatrix}$ for $N_-, N_+ \in \mathbb{N}$ to be fixed below.

If $\mathcal{A}$ is not an isomorphism there is a surjective operator

$$
\begin{pmatrix} A & K & K_1 \\ T & Q & 0 \end{pmatrix} : \quad
\begin{array}{ccc}
H_1 & & H_2 \\
\oplus & \rightarrow & \oplus \\
\tilde{L}_1 & & L_2
\end{array}
$$

with $N_- = \operatorname{codim} \operatorname{im} \mathcal{A}$ and for $N_+ = \dim \ker \mathcal{A}$ a block matrix

$$\tilde{\mathcal{A}} = \begin{pmatrix} A & K & K_1 \\ T & Q & 0 \\ T_1 & Q_1 & Q_2 \end{pmatrix} : \begin{matrix} H_1 \\ \oplus \\ \tilde{L}_1 \end{matrix} \rightarrow \begin{matrix} H_2 \\ \oplus \\ \tilde{L}_2 \end{matrix}$$

that is an isomorphism. Now set

$$\tilde{T} = \begin{pmatrix} T \\ T_1 \end{pmatrix}, \quad \tilde{K} = \begin{pmatrix} K & K_1 \end{pmatrix}, \quad \tilde{Q} = \begin{pmatrix} Q & 0 \\ Q_1 & Q_2 \end{pmatrix}.$$

Then

$$\tilde{\mathcal{A}} = \begin{pmatrix} A & \tilde{K} \\ \tilde{T} & \tilde{Q} \end{pmatrix} : \begin{matrix} H_1 \\ \oplus \\ \tilde{L}_1 \end{matrix} \rightarrow \begin{matrix} H_2 \\ \oplus \\ \tilde{L}_2 \end{matrix}$$

is an isomorphism, and we can argue as before.   □

Remark 3.5.20 *Let* $A, \tilde{A} \in \mathcal{Y}^\mu(\mathbb{W}, \boldsymbol{g}; E, F; J^-, J^+)$ *be two elliptic operators on compact* $\mathbb{W}$, *cf. the notation of the above Theorem 3.5.17. Then*

$$\sigma_\psi^\mu(A) = \sigma_\psi^\mu(\tilde{A}), \qquad \sigma_\wedge^\mu(A) = \sigma_\wedge^\mu(\tilde{A})$$

implies $\operatorname{ind} A = \operatorname{ind} \tilde{A}$. *In fact, the operator* $A - \tilde{A}$ *is of order* $\mu - 1$, *cf. Proposition 3.4.52, and it is compact by Corollary 3.4.49.*

Chapter 4

Boundary value problems

Boundary value problems play a two–fold role for the pseudo–differential calculus on manifolds with singularities. On one hand they may be regarded as particular edge problems, with the boundary as edge and the inner normal (with respect to some Riemannian metric) as model cone. On the other hand, the above cone and wedge calculi contain the standard pseudo–differential operators on a (closed, compact) cone base X in terms of various kinds of operator–valued symbols. An analogous program on manifolds with conical or edge singularities with boundary requires a base which is a compact manifold X with C^∞ boundary, together with the calculus of boundary value problems on X. In other words this basic calculus on X is to be prepared first.

The present chapter develops a self–contained approach to boundary value problems with the transmission property that can be used then in parameter–dependent form for corresponding cone and edge algebras.

The boundary value problems of that class were first introduced as an algebra by Boutet de Monvel [10] and later on further developed by other authors, in particular, Rempel and Schulze [91], Grubb [48], Schulze [122], Schrohe and Schulze [106]. Joint papers with Schrohe [106], [107], [108], [109], [111], [110] established the algebras of boundary value problems for conical and edge singularities. Another joint paper with Rabinovich and Tarkhanov [90] treated the case of cusp and whirl singularities.

4.1 Pseudo–differential operators on the half–axis

4.1.1 The Mellin operator conventions

The half axis $\mathbb{R}_+ = \{t \in \mathbb{R} : t > 0\}$ is regarded in this chapter as an infinite cone with $t \to 0$ as conical singularity and $t \to \infty$ as exit to infinity. The main point here is to study the operators in a neighbourhood of zero, though it is sometimes convenient for some purposes to observe an exit behaviour for $t \to \infty$.

Definition 4.1.1 $S^{\mu}(\mathbb{R} \times \mathbb{R})_{e}$ *for* $\mu \in \mathbb{R}$ *denotes the subspace of all* $a(t, \tau) \in S^{\mu}(\mathbb{R} \times \mathbb{R})$ *satisfying*

$$|D_t^j D_\tau^k a(t, \tau)| \leq c(j, k)(1 + |\tau|)^{\mu - k}(1 + |t|)^{-j} \qquad (4.1.1)$$

for all $j, k \in \mathbb{N}$, $t, \tau \in \mathbb{R}$ *and constants* $c(j, k) > 0$. *Moreover, we set*

$$S_{\mathrm{cl}}^{\mu}(\mathbb{R} \times \mathbb{R})_{e} = S_{\mathrm{cl}}^{\mu}(\mathbb{R} \times \mathbb{R}) \cap S^{\mu}(\mathbb{R} \times \mathbb{R})_{e}. \qquad (4.1.2)$$

Note that in (4.1.2) the property of being classical here refers to the covariable τ (in contrast to a corresponding notion of Section 1.4.3 which is superfluous here). The subscript e indicates the exit condition with respect to $|t| \to \infty$.

In an analogous manner we can introduce the symbol classes

$$S^{\mu}(\mathbb{R}_+ \times \mathbb{R})_{e}, \quad S_{\mathrm{cl}}^{\mu}(\mathbb{R}_+ \times \mathbb{R})_{e},$$

by replacing $\mathbb{R} \ni t$ in Definition 4.1.1 by $\mathbb{R}_+$ and requiring the estimates (4.1.1) for all $t \geq \varepsilon$ for arbitrary $\varepsilon > 0$, with constants $c(j, k)$ that also depend on ε. We also set

$$S^{\mu}(\overline{\mathbb{R}}_+ \times \mathbb{R})_{e} = S^{\mu}(\overline{\mathbb{R}}_+ \times \mathbb{R}) \cap S^{\mu}(\mathbb{R}_+ \times \mathbb{R})_{e},$$
$$S_{\mathrm{cl}}^{\mu}(\overline{\mathbb{R}}_+ \times \mathbb{R})_{e} = S_{\mathrm{cl}}^{\mu}(\overline{\mathbb{R}}_+ \times \mathbb{R}) \cap S^{\mu}(\mathbb{R}_+ \times \mathbb{R})_{e}.$$

Analogous notations make sense for symbols $a(t, t', \tau) \in S^{\mu}(\mathbb{R}_+ \times \mathbb{R}_+ \times \mathbb{R}_\tau)$. We denote by op($a$) the pseudo–differential operator associated with $a(t, t', \tau) \in S^{\mu}(\mathbb{R}_+ \times \mathbb{R}_+ \times \mathbb{R}_\tau)$ in the sense

$$\mathrm{op}(a)u(t) = \iint e^{i(t - t')\tau} a(t, t', \tau) u(t') \, dt' d\tau,$$

$d\tau = (2\pi)^{-1} d\tau$, $u \in C_0^{\infty}(\mathbb{R}_+)$. Recall that op($a$) extends to op($a$) : $\mathcal{E}'(\mathbb{R}_+) \to \mathcal{D}'(\mathbb{R}_+)$. Incidentally, to indicate the t–variable in the pseudo–differential action, we also write

$$\mathrm{op}_t(a) = \mathrm{op}(a).$$

For $a(t, \tau) \in S^{\mu}(\mathbb{R}_+ \times \mathbb{R})_{e}$ we can form $(1 - \omega) \mathrm{op}(a)(1 - \omega_1)$ which induces continuous operators

$$(1 - \omega) \mathrm{op}(a)(1 - \omega_1) : H^s(\mathbb{R}_+) \to H^{s-\mu}(\mathbb{R}_+) \qquad (4.1.3)$$

for arbitrary cut–off functions $\omega(t)$, $\omega_1(t)$, for all $s \in \mathbb{R}$, cf. the results of Section 1.4.2. Choosing cut–off functions $\omega(t)$, $\omega_0(t)$, $\omega_1(t)$ satisfying

$$\omega\omega_0 = \omega, \qquad \omega\omega_1 = \omega_1$$

we can write

$$\begin{aligned}
\operatorname{op}(a) &= \omega \operatorname{op}(a) + (1 - \omega)\operatorname{op}(a) \\
&= \omega \operatorname{op}(a)\omega_0 + (1 - \omega)\operatorname{op}(a)(1 - \omega_1) + g, \qquad (4.1.4)
\end{aligned}$$

where $g = \omega \operatorname{op}(a)(1 - \omega_0) + (1 - \omega)\operatorname{op}(a)\omega_1 \in L^{-\infty}(\mathbb{R}_+)$. The latter relation follows from the pseudo–locality of pseudo–differential operators. Thus

$$a \to \omega \operatorname{op}(a)\omega_0 + (1 - \omega)\operatorname{op}(a)(1 - \omega_1)$$

can be regarded as an operator convention. The particular choice of the cut–off functions only affects the operator mod $L^{-\infty}(\mathbb{R}_+)$.

In operator conventions for boundary value problems in a domain $\Omega \subseteq \mathbb{R}^n$ with C^∞ boundary we split the variables near the boundary into the normal and tangential components.

Let $U \subseteq \mathbb{R}^n$ be an open set with $\overline{\Omega} \subset U$. Boundary value problems in Ω will be considered for operators induced by $\operatorname{Op}(a) = F_{\xi \to x}^{-1} a(x, \xi) F_{x' \to \xi}$ for symbols $a(x, \xi) \in S^\mu(U \times \mathbb{R}^n)$. Near $\partial\Omega$ we introduce local coordinates such that Ω corresponds to the half space $\mathbb{R}_+^n = \{x = (t, y) \in \mathbb{R}^n : t > 0\}$ and $\mathbb{R}^{n-1} \ni y$ to a neighbourhood on the boundary $\partial\Omega$. Writing the symbol $a(x, \xi)$ in local coordinates we get $a(t, y, \tau, \eta)$ with the covariables (τ, η) to (t, y). The general idea to interpret an operator in the half space is now the following. Set $a_{y,\eta}(t, \tau) = a(t, y, \tau, \eta)$, regarded as a (y, η)–dependent family of symbols in $S^\mu(\overline{\mathbb{R}}_+ \times \mathbb{R})$. Then

$$\operatorname{op}_t(a_{y,\eta}) =: a(y, \eta)$$

is as an operator–valued symbol. Applying the pseudo–differential action to the y–variables $\operatorname{Op}_y(a)v = F_{\eta \to y}^{-1} F_{y' \to \eta} a(y, \eta)v(y')$ we obtain the original operator:

$$\operatorname{op}(a) = \operatorname{op}_y(\operatorname{op}_t(a_{y,\eta})).$$

Hence this is the first necessary step to study the pseudo–differential operators along the inner normal with symbols

$$a(t, \tau) \in S^\mu(\overline{\mathbb{R}}_+ \times \mathbb{R}). \qquad (4.1.5)$$

Recall that the situation over the cone $\overline{\mathbb{R}}_+$ is slightly different: the symbols were assumed to be of the form

$$t^{-\mu}p(t, t\tau) \qquad \text{for some } p(t, \tilde{\tau}) \in S^\mu(\overline{\mathbb{R}}_+ \times \mathbb{R}) \qquad (4.1.6)$$

near $t = 0$. We shall establish a relation between (4.1.5) and (4.1.6) that allows us to apply the operator conventions from the cone (or in the half space from the wedge) theory.

Another aspect in boundary value problems is the so–called transmission property that requires separate discussion if we want to obtain continuous operators $H^s(\mathbb{R}_+) \to H^{s-\mu}(\mathbb{R}_+)$. We shall deal with the transmission property below in the Sections 4.1.3 and 4.1.4.

In the following assertions $\mathrm{op}(a)$ has the meaning of an operator $C_0^\infty(\mathbb{R}_+) \to C^\infty(\mathbb{R}_+)$.

Proposition 4.1.2 *To every* $a(t,\tau) \in S_{\mathrm{cl}}^\mu(\overline{\mathbb{R}}_+ \times \mathbb{R}_\tau)$ *there exists a* $p(t,\tau) \in S_{\mathrm{cl}}^\mu(\mathbb{R}_+ \times \mathbb{R})$ *of the form* $p(t,\tau) = \tilde{p}(t,t\tau)$ *for some* $\tilde{p}(t,\tilde{\tau}) \in S_{\mathrm{cl}}^\mu(\overline{\mathbb{R}}_+ \times \mathbb{R}_{\tilde{\tau}})$ *such that*

$$\mathrm{op}(a) - t^{-\mu}\,\mathrm{op}(p) \in L^{-\infty}(\mathbb{R}_+).$$

Proof. Let $\chi(\tau)$ be an excision function, and let $a_{(\mu-j)}(t,\tau)$ be the homogeneous component of $a(t,\tau)$ of order $\mu - j$ in τ, $j \in \mathbb{N}$. Then $a(t,\tau) \sim \sum_{j=0}^\infty \chi(\tau)a_{(\mu-j)}(t,\tau)$. From the homogeneity we get

$$t^{-\mu+j}a_{(\mu-j)}(t,t\tau) = a_{(\mu-j)}(t,\tau) \qquad \text{for all } j \in \mathbb{N}.$$

Set $\tilde{p}_{(\mu-j)}(t,\tilde{\tau}) = t^j a_{(\mu-j)}(t,\tilde{\tau})$. Then the asymptotic sum $\tilde{p}(t,\tilde{\tau}) \sim \sum_{j=0}^\infty \chi(\tilde{\tau})\tilde{p}_{(\mu-j)}(t,\tilde{\tau})$ can be carried out in $S^\mu(\overline{\mathbb{R}}_+ \times \mathbb{R}_{\tilde{\tau}})$, i.e., we get a $\tilde{p}(t,\tilde{\tau}) \in S^\mu(\overline{\mathbb{R}}_+ \times \mathbb{R}_{\tilde{\tau}})$. For arbitrary $N \in \mathbb{N}$ we now obtain

$$a(t,\tau) - t^{-\mu}\tilde{p}(t,t\tau) = \sum_{j=0}^N \{\chi(\tau)a_{(\mu-j)}(t,\tau) - \chi(t\tau)t^{-\mu}\tilde{p}_{(\mu-j)}(t,t\tau)\}$$

$$\mathrm{mod}\, S^{\mu-(N+1)}(\mathbb{R}_+ \times \mathbb{R}).$$

$$(4.1.7)$$

Since $\chi(\tau)a_{(\mu-j)}(t,\tau) = \chi(t\tau)a_{(\mu-j)}(t,\tau) \bmod S^{-\infty}(\mathbb{R}_+ \times \mathbb{R})$, it follows that the sum on the right hand side of (4.1.7) equals 0 mod $S^{-\infty}(\mathbb{R}_+ \times \mathbb{R})$. Since N is arbitrary, we get $a(t,\tau) - t^{-\mu}\tilde{p}(t,t\tau) \in S^{-\infty}(\mathbb{R}_+ \times \mathbb{R})$. This yields immediately the assertion. $\qquad\square$

Recall from Definition 4.1.1 the notation

$$\tilde{S}^\mu(\overline{\mathbb{R}}_+ \times \mathbb{R}) = \{p(t,\tau) \in S^\mu(\mathbb{R}_+ \times \mathbb{R}) : \text{there exists a}$$
$$\tilde{p}(t,\tilde{\tau}) \in S^\mu(\overline{\mathbb{R}}_+ \times \mathbb{R}) \qquad \text{with } p(t,\tau) = \tilde{p}(t,t\tau)\}$$

and analogously $\tilde{S}_{\mathrm{cl}}^\mu(\overline{\mathbb{R}}_+ \times \mathbb{R})$. In Section 2.2.2 we have obtained the following result:

Theorem 4.1.3 *For every* $p(t, \tau) \in \tilde{S}^\mu(\overline{\mathbb{R}}_+ \times \mathbb{R})$ *there exists an* $h(t, z) \in C^\infty(\overline{\mathbb{R}}_+, N_O^\mu)$ *such that*

$$\mathrm{op}(p) - \mathrm{op}_M^\gamma(h) \in L^{-\infty}(\mathbb{R}_+) \qquad (4.1.8)$$

for every $\gamma \in \mathbb{R}$. *For* $p(t, \tau) \in \tilde{S}_{cl}^\mu(\overline{\mathbb{R}}_+ \times \mathbb{R})$ *we obtain* $h(t, z) \in C^\infty(\overline{\mathbb{R}}_+, M_O^\mu)$ *such that* (4.1.8) *holds.*

Theorem 4.1.3 is a corollary of Theorem 2.2.25 and Theorem 2.2.28 for the case $\dim X = 0$.

Theorem 4.1.4 *For every* $a(t, \tau) \in S_{cl}^\mu(\overline{\mathbb{R}}_+ \times \mathbb{R})$, $\mu \in \mathbb{R}$, *there exists an* $h(t, z) \in C^\infty(\overline{\mathbb{R}}_+, M_O^\mu)$ *with*

$$\mathrm{op}(a) - t^{-\mu} \mathrm{op}_M^\gamma(h) \in L^{-\infty}(\mathbb{R}_+)$$

for every $\gamma \in \mathbb{R}$.

Proof. It suffices to apply Proposition 4.1.2 and Theorem 4.1.3. $\square$

Theorem 4.1.5 *To every* $a(t, \tau) \in S_{cl}^\mu(\overline{\mathbb{R}}_+ \times \mathbb{R})$, $\mu \in \mathbb{R}$, *and arbitrary weight* $\gamma \in \mathbb{R}$ *there exists an operator* $g_\gamma \in L^{-\infty}(\mathbb{R}_+)$ *such that*

$$\mathrm{op}(a) + g_\gamma : C_0^\infty(\mathbb{R}_+) \to C^\infty(\mathbb{R}_+)$$

extends to a continuous operator

$$\mathrm{op}(a) + g_\gamma : \mathcal{K}^{s,\gamma}(\mathbb{R}_+) \to \mathcal{K}^{s-\mu,\gamma-\mu}(\mathbb{R}_+)$$

for each $s \in \mathbb{R}$.

Proof. The relations (4.1.4) and (4.1.3) show that it suffices to consider $\omega \, \mathrm{op}(a) \omega_0$ for cut–off functions $\omega(t)$, $\omega_0(t)$. Using Theorem 4.1.4 we obtain

$$\omega \, \mathrm{op}(a) \omega_0 = t^{-\mu} \omega \, \mathrm{op}_M^\gamma(h) \omega_0 \bmod L^{-\infty}(\mathbb{R}_+)$$

for some $h(t, z) \in C^\infty(\overline{\mathbb{R}}_+, M_O^\mu)$. This yields together with the continuity of

$$t^{-\mu} \omega \, \mathrm{op}_M^\gamma(h) \omega_0 : \mathcal{K}^{s,\gamma}(\mathbb{R}_+) \to \mathcal{K}^{s-\mu,\gamma-\mu}(\mathbb{R}_+)$$

for every $s \in \mathbb{R}$ the assertion. $\square$

Remark 4.1.6 *The proof of* Theorem 4.1.5 *shows that*

$$g_\gamma = \mathrm{op}(a) - \{t^{-\mu} \omega \, \mathrm{op}_M^\gamma(h) \omega_0 + (1 - \omega) \, \mathrm{op}(a)(1 - \omega_1)\}$$

is an adequate choice for the operator g_γ for arbitrary cut–off functions ω, ω_0, ω_1 satisfying $\omega\omega_0 = \omega$, $\omega\omega_1 = \omega_1$. By

$$a(t,\tau) \to t^{-\mu}\omega\,\mathrm{op}_M^\gamma(h)\omega_0 + (1-\omega)\,\mathrm{op}(a)(1-\omega_1)$$

we have thus established an operator convention, i.e., a (non–canonical) map

$$S_{\mathrm{cl}}^\mu(\overline{\mathbb{R}}_+ \times \mathbb{R}) \to \bigcap_{s\in\mathbb{R}} \mathcal{L}(\mathcal{K}^{s,\gamma}(\mathbb{R}_+), \mathcal{K}^{s-\mu,\gamma-\mu}(\mathbb{R}_+)).$$

This is analogous to the corresponding operator convention from the cone theory, though here we start with another symbol class.

4.1.2 Alternative Mellin operator conventions

One may ask whether the remainders in Theorem 4.1.5 can be controlled in a more precise manner. We shall see below in this section that the reformulation of the pseudo–differential action on $\mathbb{R}_+$ in Mellin terms gives rise to meromorphic Mellin symbols and that the remainders are then Green operators on the half axis with discrete asymptotics. Set

$$g^+(z) = (1 - e^{-2\pi i z})^{-1}, \qquad g^-(z) = (1 - e^{2\pi i z})^{-1} \qquad (4.1.9)$$

for $z \in \mathbb{C}$. The functions $g^\pm(z)$ are meromorphic with simple poles in the real integers, and we have

$$g^+(z) + g^-(z) = 1, \quad g^\pm(z+l) = g^\pm(z) \quad \text{for all} \quad l \in \mathbb{Z}.$$

The functions $g^\pm(z)$ have been used in Eskin's book [25] for a Mellin pseudo–differential operator algebra in $\mathcal{L}(L^2(\mathbb{R}_+))$ for symbols $a(\tau) \in S_{\mathrm{cl}}^0(\mathbb{R})$. Corresponding Mellin reformulations of the pseudo–differential actions associated with such symbols are based on the Propositions 4.1.12, 4.1.13, 4.1.14 below, cf. [25]. The material of the present section may also be regarded as an extension of the results of [122], Chapter 2.

Let us choose functions $\chi^\pm(\tau) \in C^\infty(\mathbb{R})$ with

$$\chi^+(\tau) = \begin{cases} 1 & \text{for } \tau > 2\varepsilon, \\ 0 & \text{for } \tau < \varepsilon, \end{cases} \qquad \chi^-(\tau) = \begin{cases} 0 & \text{for } \tau > -\varepsilon, \\ 1 & \text{for } \tau < -2\varepsilon, \end{cases}$$

for some $\varepsilon > 0$. Then we have

$$g^+(\beta + i\tau)\chi^+(\tau), g^-(\beta + i\tau)\chi^-(\tau) \in \mathcal{S}(\mathbb{R}_\tau)$$

for every $\beta \in \mathbb{R}$, uniformly in $c \le \beta \le c'$ for every $c < c'$.

Remark 4.1.7 *We have* $g^{\pm}(z) \in M_P^0$ *for* $P = \{(j,0)\}_{j \in \mathbb{Z}} \in \mathbf{As}^{\bullet}$,
moreover,

$$(g^{\pm}(z))^2 - g^{\pm}(z), \qquad g^+(z)g^-(z) \in M_Q^{-\infty} \qquad for \ Q = \{(k,1)\}_{k \in \mathbb{Z}},$$

and $g(z) := -e^{-i\pi z}(1 - e^{2\pi i z})^{-1} \in M_P^{-\infty}$, $g^2(z) = -g^+(z)g^-(z)$.

For $\mathbb{R}_+ = \{t \in \mathbb{R} : t > 0\}$, $\mathbb{R}_- = \{t \in \mathbb{R} : t < 0\}$ and $u \in L^2(\mathbb{R}_+)$, $v \in L^2(\mathbb{R}_-)$ we set

$$e^+ u = \begin{cases} u & \text{for } t > 0, \\ 0 & \text{for } t > 0, \end{cases} \qquad e^- v = \begin{cases} 0 & \text{for } t > 0, \\ v & \text{for } t > 0. \end{cases}$$

e^+, e^- induce continuous operators

$$e^+ : L^2(\mathbb{R}_+) \to L^2(\mathbb{R}), \quad e^- : L^2(\mathbb{R}_-) \to L^2(\mathbb{R}).$$

Moreover, let

$$r^+ : L^2(\mathbb{R}) \to L^2(\mathbb{R}_+), \quad r^- : L^2(\mathbb{R}) \to L^2(\mathbb{R}_-)$$

be the restriction operators to $\mathbb{R}_+$ and $\mathbb{R}_-$, respectively. For every $a(t, \tau) \in S^0(\mathbb{R} \times \mathbb{R})_e$, cf. Definition 4.1.1 we get continuous operators

$$\begin{array}{rcl} \text{op}(a) & : \ L^2(\mathbb{R}) & \to \ L^2(\mathbb{R}), \\ r^+ \text{op}(a)e^+ & : \ L^2(\mathbb{R}_+) & \to \ L^2(\mathbb{R}_+), \\ r^- \text{op}(a)e^- & : \ L^2(\mathbb{R}_-) & \to \ L^2(\mathbb{R}_-) \end{array}$$

as well as

$$r^{\pm} \text{op}(a)e^{\mp} : L^2(\mathbb{R}_{\mp}) \to L^2(\mathbb{R}_{\pm}).$$

Let us set, in particular,

$$\text{op}^+(a) = r^+ \text{op}(a)e^+. \tag{4.1.10}$$

For $a(\tau)$ we may also insert symbols which are multipliers in $L^2(\mathbb{R}_\tau)$, e.g., $\theta^{\pm}(\tau)$, the characteristic functions of $\mathbb{R}_{\pm,\tau}$. We will employ the notation (4.1.10) also for symbols $a(t, t', \tau) \in S^{\mu}(\mathbb{R}_+ \times \mathbb{R}_+ \times \mathbb{R}), \mu \in \mathbb{R}$, in the sense

$$\text{op}^+(a) : C_0^{\infty}(\mathbb{R}_+) \to C^{\infty}(\mathbb{R}_+) \tag{4.1.11}$$

or

$$\text{op}^+(a) : \mathcal{E}'(\mathbb{R}_+) \to \mathcal{D}'(\mathbb{R}_+).$$

For $a(t, \tau) \in \tilde{S}^{\mu}(\overline{\mathbb{R}}_+ \times \mathbb{R})_e$ we can also form

$$\text{op}^+(a) : \mathcal{S}_0(\overline{\mathbb{R}}_+) \to C^{\infty}(\mathbb{R}_+),$$

where $\mathcal{S}_0(\overline{\mathbb{R}}_+)$ is the subspace of all $u \in \mathcal{S}(\mathbb{R})$ supported by $\overline{\mathbb{R}}_+$ (with the natural identification $\mathcal{S}_0(\overline{\mathbb{R}}_+) = \mathcal{S}_0(\overline{\mathbb{R}}_+)|_{\mathbb{R}_+}$). In other words, various extensions of (4.1.11) are denoted by $\text{op}^+(a)$ if no confusion is possible.

We now recall some well–known facts on Sobolev spaces on the half axis and on continuous operators with symbols

$$l_\pm^\alpha(\delta, \tau) = (\delta \pm i\tau)^\alpha, \qquad \alpha \in \mathbb{C}, \quad \delta \in \mathbb{R}_+,$$

in those spaces. Proofs may be found, for instance, in Eskin's book [25].

Set for $s \in \mathbb{R}$

$$H^s(\mathbb{R}_+) = H^s(\mathbb{R})|_{\mathbb{R}_+},$$

$$H_0^s(\overline{\mathbb{R}}_+) = \{\text{closure of } C_0^\infty(\mathbb{R}_+) \text{ in } H^s(\mathbb{R})\}. \tag{4.1.12}$$

Then

$$H^s(\mathbb{R}_+) = H_0^s(\overline{\mathbb{R}}_+) \quad \text{for } |s| < \frac{1}{2} \tag{4.1.13}$$

(more precisely: $r^+ : H_0^s(\overline{\mathbb{R}}_+) \to H^s(\mathbb{R}_+)$ is an isomorphism for those s) and $e^+ : L^2(\mathbb{R}_+) \to L^2(\mathbb{R})$ restricts (or extends) to continuous operators

$$e^+ : H^s(\mathbb{R}_+) \to H^s(\mathbb{R}) \quad \text{for } |s| < \frac{1}{2}. \tag{4.1.14}$$

Moreover, we have

$$H_0^s(\overline{\mathbb{R}}_+) \subseteq H^s(\mathbb{R}_+) \quad \text{for } s \geq 0, \quad H_0^s(\overline{\mathbb{R}}_+) \supseteq H^s(\mathbb{R}_+) \quad \text{for } s \leq 0.$$

The space $\mathcal{S}(\overline{\mathbb{R}}_+) = \mathcal{S}(\mathbb{R})|_{\overline{\mathbb{R}}_+}$ is dense in $H^s(\mathbb{R}_+)$ for every $s \in \mathbb{R}$. Recall that $F = F_{t \to \tau}$ denotes the Fourier transform on $\mathbb{R} \ni t$.

Proposition 4.1.8 *For* $u(t) \in H^s(\mathbb{R})$, $s \in \mathbb{R}$, *the following conditions are equivalent:*

(i) $u \in H_0^s(\overline{\mathbb{R}}_+)$,

(ii) $f(\tau) = (F_{t \to \tau} u)(\tau)$ *has an extension to a holomorphic and locally integrable function* $h(\zeta)$ *in* $\text{Im}\,\zeta < 0$, $\tau = \text{Re}\,\zeta$, *satisfying*

$$\int (1 + |\zeta|^2)^s |h(\zeta)|^2 \, d\tau \leq c \tag{4.1.15}$$

for a constant c *independent of* $\text{Im}\,\zeta$.

Moreover, every holomorphic and locally integrable function $h(\zeta)$ in $\operatorname{Re}\zeta < 0$ satisfying (ii) *is the extension of some $f(\tau) = (F_{t\to\tau}u)(\tau)$, $u(t) \in H_0^s(\overline{\mathbb{R}}_+)$.*

Remark 4.1.9 $\operatorname{op}(l_+^\alpha) : H^s(\mathbb{R}) \to H^{s-\alpha}(\mathbb{R})$ *for $\alpha \in \mathbb{R}$ induces continuous operators*

$$\operatorname{op}(l_+^\alpha) : H_0^s(\overline{\mathbb{R}}_+) \to H_0^{s-\alpha}(\overline{\mathbb{R}}_+)$$

for all $s \in \mathbb{R}$. In other words, we have $\operatorname{op}(l_+^\alpha)u|_{\mathbb{R}_-} = 0$ for every $u \in H^s(\mathbb{R})$ with $\operatorname{supp} u \subseteq \overline{\mathbb{R}}_+$. This implies, in particular,

$$\operatorname{op}^+(al_+^\alpha)u = \operatorname{op}^+(a)\operatorname{op}^+(l_+^\alpha)u$$

for every $a(t,\tau) \in S^\mu(\mathbb{R} \times \mathbb{R})_e$, $u \in C_0^\infty(\overline{\mathbb{R}}_+)$.

Proposition 4.1.10 *Let $e_s^+ : H^s(\mathbb{R}_+) \to H^s(\mathbb{R})$ be an arbitrary continuous extension operator, i.e., $r^+e_s^+ = \operatorname{id}$ on $H^s(\mathbb{R}_+)$, $s \in \mathbb{R}$. Then*

$$r^+ \operatorname{op}(l_-^\alpha)e_s^+ : H^s(\mathbb{R}_+) \to H^{s-\alpha}(\mathbb{R}_+)$$

is a continuous operator for every $\alpha, s \in \mathbb{R}$ which is independent of the particular choice of e_s^+.

Let us also recall here without proof a relation between the $\mathcal{K}^{s,\gamma}(\mathbb{R}_+)$ spaces and $H^s(\mathbb{R}_+)$, $H_0^s(\overline{\mathbb{R}}_+)$. Let $\tau(s) = \max\{k \in \mathbb{N} : k < |s| - \frac{1}{2}\}$ for $s \in \mathbb{R}$, $|s| > \frac{1}{2}$, and set

$$\mathcal{T}^s = \{\text{linear span of } t^j\omega(t) \text{ for } j = 0,\ldots,\tau(s)\}$$

for $s > \frac{1}{2}$, with a cut–off function $\omega(t)$, $\mathcal{T}^s = \{0\}$ for $s \le \frac{1}{2}$,

$$\mathcal{D}^s = \{\text{ linear span of } \partial_t^j\delta_0 \text{ for } j = 0,\ldots,\tau(s)\}$$

for $s < -\frac{1}{2}$, where δ_0 is the Dirac measure at $t = 0$, $\mathcal{D}^s = \{0\}$ for $s \ge -\frac{1}{2}$.

Theorem 4.1.11 *There are canonical (topological) isomorphisms*

$$H^s(\mathbb{R}_+) = \mathcal{K}^{s,s}(\mathbb{R}_+) + \mathcal{T}^s \quad \text{for } s \in \mathbb{R}, \ s \notin \frac{1}{2} + \mathbb{N},$$

$$H_0^s(\overline{\mathbb{R}}_+) = \mathcal{K}^{s,s}(\mathbb{R}_+) + \mathcal{D}^s \quad \text{for } s \in \mathbb{R}, \ -s \notin \frac{1}{2} + \mathbb{N}.$$

A proof may be found in Schulze [122].

Proposition 4.1.12 *Let $\theta^\pm(\tau) = 1$ for $\tau \gtrless 0$, $= 0$ for $\tau \lessgtr 0$. Then we have*

$$\operatorname{op}^+(\theta^\pm) = \operatorname{op}_M(g^\pm).$$

Proof. In view of $\theta^+ + \theta^- = 1$ it suffices to consider, for instance, $\theta^-(\tau)$. We have for $f \in L^2(\mathbb{R})$

$$\text{op}(\theta^-)f(t) = \lim_{\varepsilon \to 0} \frac{1}{2\pi i} \int\limits_{-\infty}^{\infty} \frac{f(s)}{t - i\varepsilon - s} \, ds. \qquad (4.1.16)$$

For $u \in L^2(\mathbb{R}_+)$ we obtain

$$M(\text{op}^+(\theta^-)u)(z) = M \left\{ \lim_{\varepsilon \to 0} \frac{1}{2\pi i} \int\limits_{0}^{\infty} \frac{u(s)}{t - i\varepsilon - s} \, ds \right\}(z)$$

$$= \lim_{\varepsilon \to 0} \frac{1}{2\pi i} \int\limits_{0}^{\infty} t^{z-1} \left\{ \int \frac{u(s)}{t - i\varepsilon - s} \, ds \right\} dt$$

$$= \lim_{\varepsilon \to 0} \frac{1}{2\pi i} \int\limits_{0}^{\infty} u(s) \left\{ \int \frac{t^{z-1}}{t - i\varepsilon - s} \, dt \right\} ds.$$

A well–known formula says that

$$\frac{1}{2\pi i} \int\limits_{0}^{\infty} \frac{t^{z-1}}{t - a} \, dt = (1 - e^{2\pi i z})^{-1} e^{(z-1)\log a} \qquad (4.1.17)$$

for $a \in \mathbb{C} \setminus \mathbb{R}_+$, $0 < \text{Re } z < 1$, $\log a = \log |a| + i \arg a$, $0 < \arg a < 2\pi$. Thus

$$M(\text{op}^+(\theta^-)u)(z) = \lim_{\varepsilon \to 0} \int\limits_{0}^{\infty} u(s)(s + i\varepsilon)^{z-1} \, ds \, g^-(z)$$

$$= g^-(z)(Mu)(z).$$

This is just the assertion for $\text{op}^+(\theta^-)$.   □

Proposition 4.1.13 *Let*

$$mu(t) = -\frac{1}{2\pi i} \int\limits_{0}^{\infty} \frac{u(s)}{t + s} \, ds$$

for $u \in C_0^\infty(\mathbb{R}_+)$. Then we have

$$mu(t) = -\text{op}_M(g)u(t)$$

for the function $g(z)$ of Remark 4.1.7. Thus m extends to a continuous operator $m : L^2(\mathbb{R}_+) \to L^2(\mathbb{R}_+)$.

Proof. We have

$$
M(mu)(z) = -\frac{1}{2\pi i} \int\limits_0^\infty t^{z-1} \left\{ \int\limits_0^\infty \frac{u(s)}{t-s}\, ds \right\} dt
$$

$$
= -\int\limits_0^\infty \left\{ \frac{1}{2\pi i} \int\limits_0^\infty \frac{t^{z-1}}{t+s}\, dt \right\} u(s)\, ds.
$$

From (4.1.17) it follows that

$$
\frac{1}{2\pi i} \int\limits_0^\infty \frac{t^{z-1}}{t+s}\, dt = g^-(z) e^{(z-1)\log(-s)} = \frac{-e^{\pi i z} s^{z-1}}{1 - e^{2\pi i z}}
$$

which yields the assertion. $\hfill\square$

Proposition 4.1.14 *Set* $(\varepsilon^* u)(t) = u(-t)$ *for* $u \in L^2(\mathbb{R}_+)$. *Then*

$$
\varepsilon^* r^- \operatorname{op}(\theta^\pm) e^+ = \pm \operatorname{op}_M(g).
$$

Proof. The relation (4.1.16) implies

$$
\varepsilon^* r^- \operatorname{op}(\theta^-) e^+ u = \varepsilon^* r^- \lim_{\delta \to +0} \frac{1}{2\pi i} \int\limits_0^\infty \frac{u(s)}{t - i\delta - s}\, ds
$$

$$
= r^+ \lim_{\delta \to +0} \frac{-1}{2\pi i} \int\limits_0^\infty \frac{u(s)}{t + s - i\delta}\, ds = -\operatorname{op}_M(g) u.
$$

Moreover, $\operatorname{op}(\theta^+) + \operatorname{op}(\theta^-) = 1$; then $0 = \varepsilon^* r^- \{\operatorname{op}(\theta^+) + \operatorname{op}(\theta^-)\} e^+$ yields the statement for θ^+. $\hfill\square$

The Propositions 4.1.12, 4.1.13, 4.1.14 are from Eskin's book [25].

Many concrete meromorphic Mellin symbols can be defined in terms of Euler's Γ–function $\Gamma(z) = \int_0^\infty t^{z-1} e^{-t}\, dt$. Recall that $\Gamma(z)$ is holomorphic in $\operatorname{Re} z > 0$ and has simple poles in all $z_k = -k$, $k \in \mathbb{N}$, cf. also Section 2.1.3. We have

$$
\Gamma(z+1) = z\Gamma(z) \qquad \text{for all } z \in \mathbb{C} \tag{4.1.18}
$$

and $\Gamma(n+1) = n!$ for all $n \in \mathbb{N}$. Note that

$$
(1+z)^\alpha = \sum_{k=0}^\infty \binom{\alpha}{k} z^k \qquad \text{for all } -\pi < \arg z < \pi,
$$

with the binomial coefficients

$$\binom{\alpha}{k} = \frac{(-1)^k}{k!} \frac{\Gamma(k - \alpha)}{\Gamma(-\alpha)}, \qquad k \in \mathbb{N}, \ \alpha \in \mathbb{C}.$$

Let us set

$$f_\varrho(z) = \frac{\Gamma(1 - z)}{\Gamma(1 - z + \varrho)} \qquad \text{for } \varrho \in \mathbb{R}, \tag{4.1.19}$$

in particular,

$$f_k(z) = \begin{cases} 1 & \text{for } k = 0, \\ \prod\limits_{p=1}^{k} (p - z)^{-1} & \text{for } k \in \mathbb{N} \setminus \{0\}, \end{cases} \tag{4.1.20}$$

cf. (4.1.18). Since the Γ–function has no zeros, we see that the function

$$f_\varrho(z) \text{ is holomorphic in } \operatorname{Re} z < 1. \tag{4.1.21}$$

More precisely, we have $f_\varrho(z) \in M_R^{-\varrho}$ for every $\varrho \in \mathbb{R}$, where $R = \{(1 + k, 0)\}_{k \in \mathbb{N}}$. This follows from the well–known properties of the Γ–function, cf. also [122].

Proposition 4.1.15 *Let $\alpha \in \mathbb{R}$, $\gamma > -\frac{1}{2}$, and let $\omega, \tilde{\omega}$ be arbitrary cut–off functions. Then for all $m, n \in \mathbb{N}$ there is an $N(m, n) \in \mathbb{N}$ such that for every $N \geq N(m, n)$ the following decomposition holds*

$$\omega \operatorname{op}^+(l_+^\alpha) \tilde{\omega} u = \omega t^{-\alpha} \sum_{k=0}^{N} (t\delta)^k \binom{\alpha}{k} \operatorname{op}_M^\gamma(f_{k-\alpha}) \tilde{\omega} u + G_N u, \tag{4.1.22}$$

$u \in C_0^\infty(\mathbb{R}_+)$, *where G_N is an integral operator with kernel in $C_0^m(\overline{\mathbb{R}}_+ \times \overline{\mathbb{R}}_+)$ which extends to a continuous operator*

$$G_N : L^2(\mathbb{R}_+) \to [\omega] t^n C^m(\overline{\mathbb{R}}_+). \tag{4.1.23}$$

Proof. For every $u \in C_0^\infty(\mathbb{R}_+)$ we have

$$\operatorname{op}^+(l_+^\alpha) u(t) = (2\pi)^{-1} r^+ \int\limits_{-\infty}^{\infty} \left\{ \int\limits_{0}^{\infty} e^{i(t-s)\tau} (\delta + i\tau)^\alpha u(s) \, ds \right\} d\tau$$

$$= (2\pi)^{-1} r^+ \int\limits_{0}^{\infty} \left\{ \int\limits_{-\infty}^{\infty} e^{i(t-s)\tau} (\delta + i\tau)^\alpha \, d\tau \right\} u(s) \, ds.$$

Using the identity

$$\int_{-\infty}^{\infty} e^{it\tau}(\delta + i\tau)^{\alpha}\,d\tau = \begin{cases} \frac{2\pi}{\Gamma(-\alpha)}t^{-\alpha-1}e^{-\delta t} & \text{for } t > 0, \\ 0 & \text{for } t \leq 0, \end{cases}$$

we obtain

$$\text{op}^{+}(l_{+}^{\alpha})u(t)$$

$$= \frac{1}{\Gamma(-\alpha)}\int_{0}^{t}(t-s)^{-\alpha-1}e^{-\delta(t-s)}u(s)\,ds$$

$$= \frac{1}{\Gamma(-\alpha)}\int_{0}^{t}(t-s)^{-\alpha-1}\left\{\sum_{k=0}^{N}\frac{(-\delta)^{k}}{k!}(t-s)^{k} + R_{N}(t,s)\right\}u(s)\,ds$$

$$= \frac{1}{\Gamma(-\alpha)}\left\{\sum_{k=0}^{N}\frac{(-\delta)^{k}}{k!}\int_{0}^{t}(t-s)^{-\alpha-1+k}u(s)\,ds\right.$$

$$+ \left.\int_{0}^{t}(t-s)^{-\alpha-1}R_{N}(t,s)u(s)\,ds\right\}.$$

The remainder $R_{N}(t,s)$ belongs to $C^{\infty}(\mathbb{R}^2)$ and vanishes of order $N+1$ at $s = t$. We have

$$\int_{0}^{t}(t-s)^{k-\alpha-1}u(s)\,ds = \int_{1}^{\infty}\left(t - \frac{t}{r}\right)^{k-\alpha-1}tr^{-2}u\left(\frac{t}{r}\right)\,dr$$

$$= t^{k-\alpha}\int_{1}^{\infty}(r-1)^{k-\alpha-1}r^{-(k-\alpha)}u\left(\frac{t}{r}\right)\frac{dr}{r}.$$

Let us set $b_{k-\alpha}(r) = (r-1)^{k-\alpha-1}r^{-(k-\alpha)}$ for $r > 0$, $b_{k-\alpha}(r) = 0$ for $r \leq 1$, and

$$h_{k-\alpha}(z) = (Mb_{k-\alpha})(z) = \int_{1}^{\infty}r^{z-1-k+\alpha}(r-1)^{k-\alpha-1}\,dr.$$

From

$$\int_{1}^{\infty}r^{-\nu}(r-1)^{\mu-1}\,dr = B(\nu - \mu, \mu) \qquad \text{for } B(\zeta, \mu) = \frac{\Gamma(\zeta)\Gamma(\mu)}{\Gamma(\zeta + \mu)},$$

cf. [42], pp. 298, 964, we obtain

$$h_{k-\alpha}(z) = \frac{\Gamma(1-z)\Gamma(k-\alpha)}{\Gamma(1-z-\alpha+k)} = \Gamma(k-\alpha)f_{k-\alpha}(z).$$

It follows that

$$\int\limits_0^t (t-s)^{k-\alpha-1}u(s)\,ds = \Gamma(k-\alpha)t^{k-\alpha}\operatorname{op}_M(f_{k-\alpha})u$$

and hence

$$\operatorname{op}^+(l_+^\alpha)(\delta)u(t) = t^{-\alpha}\sum_{k=0}^N \frac{(-1)^k}{k!}\frac{\Gamma(k-\alpha)}{\Gamma(-\alpha)}(t\delta)^k\operatorname{op}_M(f_{k-\alpha})u$$

$$+ \frac{1}{\Gamma(-\alpha)}\int\limits_0^t (t-s)^{-\alpha-1}R_N(t,\delta)u(s)\,ds$$

$$= t^{-\alpha}\sum_{k=0}^N \binom{\alpha}{k}(t\delta)^k\operatorname{op}_M(f_{k-\alpha})u$$

$$+ \frac{1}{\Gamma(-\alpha)}\int\limits_0^t (t-s)^{-\alpha-1}R_N(t,s)u(s)\,ds.$$

This gives us the asserted decomposition for $\gamma = 0$, where

$$G_Nu(t) = \omega(t)\frac{1}{\Gamma(-\alpha)}\int\limits_0^t (t-s)^{-\alpha-1}R_N(t,s)\tilde{\omega}(s)u(s)\,ds.$$

The construction of $R_N(t,s)$ shows that we could write $R_N(t,s) = (t-s)^N\tilde{R}_N(t,s)$ with $\tilde{R}_N(t,s) \in C^\infty(\mathbb{R}\times\mathbb{R})$. In view of $(t-s)^{-\alpha-1+N} \in C^m(\mathbb{R}\times\mathbb{R})$ for each m with a sufficiently large $N = N(m)$ we get the asserted smoothness of the kernel of $G_N(t,s)$ in $\overline{\mathbb{R}}_+\times\overline{\mathbb{R}}_+$. At the same time, choosing N sufficiently large, we obtain a zero at $t = s$ of any given order. In other words, we can write $G_N(t,s) = (t-s)^k\tilde{G}_N(t,s)$, where $\tilde{G}_N(t,s)$ has still a kernel in $C_0^m(\overline{\mathbb{R}}_+\times\overline{\mathbb{R}}_+)$. We have

$$G_Nu(t) = \int\limits_0^t (t-s)^k\tilde{G}_N(t,s)u(s)\,ds$$

$$= t^{k+1}\int\limits_0^1 (1-r)^k\tilde{G}_N(t,tr)u(tr)\,dr.$$

This yields the mapping property (4.1.23). Finally, in view of

$$\omega \, \mathrm{op}_M^\gamma(f_{k-\alpha})\tilde{\omega} u = \omega \, \mathrm{op}_M(f_{k-\alpha})\tilde{\omega} u$$

for all $\gamma > -\frac{1}{2}$, cf. Proposition 2.3.69, we obtain the assertion in general. $\square$

Let us now consider an arbitrary $a(t,\tau) \in S_{\mathrm{cl}}^\mu(\overline{\mathbb{R}}_+ \times \mathbb{R})$, $\mu \in \mathbb{R}$. We have an asymptotic expansion

$$a(t,\tau) \sim \sum_{j=0}^\infty \chi(\tau)\{a_j^+(t,\tau)\theta^+(\tau) + a_j^-(t,\tau)\theta^-(\tau)\}$$

for any excision function $\chi(\tau)$, with functions

$$a_j^\pm(t,\tau) = a_j^\pm(t)(i\tau)^{\mu-j}, \qquad j \in \mathbb{N}, \tag{4.1.24}$$

where $a_j^\pm(t) \in C^\infty(\overline{\mathbb{R}}_+)$, $i = \sqrt{-1}$.

Lemma 4.1.16 *For every $a(t,\tau) \in S_{\mathrm{cl}}^\mu(\overline{\mathbb{R}}_+ \times \mathbb{R})$, $\mu \in \mathbb{R}$, there exists a sequence of functions $A_j^\pm(t,\delta) \in C^\infty(\overline{\mathbb{R}}_+)$ (that are polynomials in δ of order j) such that for each $m \in \mathbb{N}$ there is an $N(m) \in \mathbb{N}$ such that the operator $\mathrm{op}^+(a)$ on $u \in C_0^\infty(\mathbb{R}_+)$ has the following decomposition:*

$$\mathrm{op}^+(a)u(t)$$
$$= \sum_{j=0}^N \{A_j^+(t,\delta)\,\mathrm{op}_M(g^+) + A_j^-(t,\delta)\,\mathrm{op}_M(g^-)\}\,\mathrm{op}^+(l_+^{\mu-j})u(t) + C_N u(t)$$

for every $N \geq N(m)$, where C_N is an integral operator with kernel in $C^m(\overline{\mathbb{R}}_+ \times \overline{\mathbb{R}}_+)$.

Proof. For every $N \in \mathbb{N}$ we can write

$$a(t,\tau) = \sum_{j=0}^N \chi(\tau)\{a_j^+(t,\tau)\theta^+(\tau) + a_j^-(t,\tau)\theta^-(\tau)\} + r_{(N)}(t,\tau)$$

$$\tag{4.1.25}$$

for a symbol $r_{(N)}(t,\tau) \in S^{\mu-(N+1)}(\overline{\mathbb{R}}_+ \times \mathbb{R})$. Then, for every $m \in \mathbb{N}$ there is an $N \in \mathbb{N}$ such that $\mathrm{op}^+(r_{(N)})$ has a kernel in $C^m(\overline{\mathbb{R}}_+ \times \overline{\mathbb{R}}_+)$. Thus it suffices to consider the finite sum on the right of (4.1.25). Let us look, for instance, at $\chi(\tau)a_j^+(t,\tau)\theta^+(\tau)$; the case with minus signs is completely analogous. By definition we have $a_j^+(t,\tau) = a_j^+(t)(i\tau)^{\mu-j}$ with $a_j^+(t) \in C^\infty(\overline{\mathbb{R}}_+)$. Let us write

$$(i\tau)^{\mu-j} = (l_+(\delta,\tau) - \delta)^{\mu-j} = (1 - \delta l_+^{-1}(\delta,\tau))^{\mu-j} l_+^{\mu-j}(\delta,\tau)$$

$$= \Big\{ \sum_{k=0}^{\infty} \binom{\mu-j}{k} (-\delta)^k l_+^{-k}(\delta,\tau) \Big\} l_+^{\mu-j}(\delta,\tau).$$

Then

$$\mathrm{op}^+(\chi(\tau)a_j^+(t,\tau)\theta^+(\tau))u(t) = a_j^+(t)\,\mathrm{op}^+(\chi(\tau)\theta^+(\tau)$$

$$\times \sum_{k=0}^{N} \binom{\mu-j}{k} \cdot (-\delta)^k l_+^{\mu-(j+k)}(\delta,\tau))u(t) + R_N(\delta)u(t) \quad (4.1.26)$$

For each m we can choose N such that $R_N(\delta)$ has a kernel in $C^m(\overline{\mathbb{R}}_+ \times \overline{\mathbb{R}}_+)$. Moreover, we have

$$\mathrm{op}^+(\chi(\tau)\theta^+(\tau)l_+^{\mu-l}(\delta,\tau))u = \mathrm{op}^+(\theta^+(\tau)l_+^{\mu-l}(\delta,\tau))u(t) + G_l(\delta)u(t) \quad (4.1.27)$$

for an operator $G_l(\delta)$ with kernel in $C^\infty(\overline{\mathbb{R}}_+ \times \overline{\mathbb{R}}_+)$. From

$$\theta^-(t)\,\mathrm{op}(l_+^{\mu-l})(\delta)u = 0,$$

cf. Remark 4.1.9, if follows that

$$\mathrm{op}^+(\theta^+(\tau)l_+^{\mu-l}(\delta,\tau))u(t) = \mathrm{op}^+(\theta^+)\,\mathrm{op}^+(l_+^{\mu-l})u(t). \quad (4.1.28)$$

Then (4.1.28), (4.1.27) and (4.1.26) yield the assertion. $\qquad \square$

Let $a(t,\tau) \in S_{cl}^\mu(\overline{\mathbb{R}}_+ \times \mathbb{R})$, $\mu \in \mathbb{R}$, with $a_j^\pm(t,\tau) = a_j^\pm(t)(i\tau)^{\mu-j}$ as the homogeneous component of order $\mu - j$ for $\tau \to \pm\infty$, $j \in \mathbb{N}$, cf. (4.1.24). Set

$$\sigma_M^{\mu-j}(a)(t,z) = \{a_j^+(t)g^+(z+\mu) + a_j^-(t)g^-(z+\mu)\}f_{j-\mu}(z), \quad (4.1.29)$$

cf. (4.1.19). Then

$$\sigma_M^{\mu-j}(a)(t,z) \in C^\infty(\overline{\mathbb{R}}_+, M_{R_j}^{\mu-j}), \qquad j \in \mathbb{N},$$

for certain $R_j \in \mathbf{As}^\bullet$, cf. Definition 2.3.48. The explicit form of R_j follows easily from the poles of $f_{j-\mu}(z)$ and of $g^\pm(z+\mu)$, cf. the remarks in the beginning of this section. From Lemma 4.1.16 we get for $u \in C_0^\infty(\mathbb{R}_+)$

$$\omega\,\mathrm{op}^+(a)\tilde{\omega}u(t) = \omega \sum_{j=0}^{N} \{A_j^+(t,\delta)\,\mathrm{op}_M(g^+) + A_j^-(t,\delta)\,\mathrm{op}_M(g^-)\}$$

$$\times \mathrm{op}^+(l_+^{\mu-j})\tilde{\omega}u(t) + \omega C_N\tilde{\omega}u(t) \quad (4.1.30)$$

for cut–off functions $\omega(t)$, $\tilde{\omega}(t)$. Moreover, Proposition 4.1.15 gives us

$$
\omega_0 \operatorname{op}^+(l_+^{\mu-j})\tilde{\omega}u(t) = \omega_0 t^{-\mu+j} \sum_{k=0}^{N} (t\delta)^k \binom{\mu-j}{k}
$$
$$
\times \operatorname{op}_M(f_{k+j-\mu})\tilde{\omega}u(t) + \omega_0 G_N \tilde{\omega}u(t), \quad (4.1.31)
$$

for cut–off functions $\omega_0(t)$, $\tilde{\omega}(t)$. The operators C_N and G_N were characterised above. Next we want to insert (4.1.31) into (4.1.30). Here we have to be careful because of the occurring factor $t^{-\mu}$ in the middle. From (4.1.21) we get

$$
\omega t^{-\mu} \operatorname{op}_M(f_\varrho)\tilde{\omega}u(t) = \omega \operatorname{op}_M(T^{-\mu}f_\varrho)t^{-\mu}\tilde{\omega}u(t) \qquad \text{for } \mu > -\frac{1}{2}
$$
$$(4.1.32)$$

for each ϱ and $u \in C_0^\infty(\mathbb{R}_+)$. This yields for $\mu > -\frac{1}{2}$

$$
\omega \operatorname{op}^+(a)\tilde{\omega}u(t)
$$
$$
= \omega \sum_{j=0}^{N} \sum_{k=0}^{N} \{A_j^+ \operatorname{op}_M(g^+) + A_j^- \operatorname{op}_M(g^-)\}\omega_0 t^j (t\delta)^k \binom{\mu-j}{k}
$$
$$
\times \operatorname{op}_M(T^{-\mu}f_{k+j-\mu})t^{-\mu}\tilde{\omega}u(t) + (R_N^1 + R_N^2 + R_N^3)u(t). \quad (4.1.33)
$$

Here

$$
R_N^1 u(t) = \omega \sum_{j=0}^{N} \{A_j^+ \operatorname{op}_M(g^+) + A_j^- \operatorname{op}_M(g^-)\}\omega_0 G_N \tilde{\omega}u(t),
$$
$$
R_N^2 u(t) = \omega C_N \tilde{\omega}u(t),
$$
$$
R_N^3 u(t) = \omega \sum_{j=0}^{N} \{A_j^+ \operatorname{op}_M(g^+) + A_j^- \operatorname{op}_M(g^-)\}(1 - \omega_0)\operatorname{op}^+(l_+^{\mu-j})\tilde{\omega}u(t).
$$

Choosing ω_0 in such a way that $\omega\omega_0 = \omega$ holds, we obtain that R_N^3 is an operator with kernel in $C^\infty(\mathbb{R}_+ \times \mathbb{R}_+)$ (it can easily be shown that this is a Green operator with appropriate weight data and discrete asymptotics). R_N^1 and R_N^2 have kernels in $C^m(\mathbb{R}_+ \times \mathbb{R}_+)$ for each fixed $m \in \mathbb{N}$, provided $N = N(m)$ is sufficiently large.

In the case $\mu \le 0$ we can commute the factor $t^{-\mu}$ through $\operatorname{op}_M(g^\pm)$ by

$$
\omega t^{-\mu} \operatorname{op}_M(T^\mu g^\pm)\tilde{\omega}u(t) = \omega \operatorname{op}_M(g^\pm)t^{-\mu}\tilde{\omega}u(t) + G_{(\mu)}u(t), \quad (4.1.34)
$$

where $G_{(\mu)}$ is a finite–dimensional Green operator belonging to the space $\mathcal{L}(L^2(\mathbb{R}_+))$. Here we need the condition

$$
-\mu \ne l + \frac{1}{2} \qquad \text{for all } l \in \mathbb{N}. \quad (4.1.35)
$$

In the case $-\mu = l + \frac{1}{2}$ for some $l \in \mathbb{N}$ we first employ

$$\omega \operatorname{op}_M(g^\pm) t^{-\mu} \tilde{\omega} u(t) = \omega \operatorname{op}_M^\beta(g^\pm) t^{-\mu} \tilde{\omega} u(t)$$

for every real β with $|\beta| < \frac{1}{2}$, and then we can commute $t^{-\mu}$ through the Mellin action

$$\omega t^{-\mu} \operatorname{op}_M^\beta(T^\mu g^\pm) \tilde{\omega} u(t) = \omega \operatorname{op}_M^\beta(g^\pm) t^{-\mu} \tilde{\omega} u(t) + G_{(\mu)} u(t)$$

for some other finite–dimensional operator

$$G_{(\mu)} \in \mathcal{L}(t^\beta L^2(\mathbb{R}_+), t^{\beta-\mu} L^2(\mathbb{R}_+)).$$

This gives us

$$\omega t^{-\mu} \operatorname{op}_M^\beta(T^\mu g^\pm) \tilde{\omega} u(t) = \omega \operatorname{op}_M(g^\pm) t^{-\mu} \tilde{\omega} u(t) + G_{(\mu)} u(t)$$

for $-\mu = l + \frac{1}{2}$ for some $l \in \mathbb{N}$, $|\beta| < \frac{1}{2}$, for all $u \in C_0^\infty(\mathbb{R}_+)$. Let us now characterise the compositions between Mellin operators when we insert (4.1.31) into (4.1.30) in the case

$$-\mu = l + \frac{1}{2} \text{ for some } l \in \mathbb{N} \qquad (4.1.36)$$

and $\beta \in \mathbb{R}$, $|\beta| < \frac{1}{2}$. Instead of (4.1.31) we can first write

$$\omega_0 \operatorname{op}^+(l_+^{\mu-j}) \tilde{\omega} u(t) = \omega_0 t^{-\mu+j} \sum_{k=0}^N (t\delta)^k \binom{\mu-j}{k} \operatorname{op}_M^\beta(f_{k+j-\mu}) \tilde{\omega} u(t)$$
$$+ \omega_0 G_N \tilde{\omega} u(t), \qquad (4.1.37)$$

where we employed the identity

$$\omega_0 \operatorname{op}_M(f_{k+j-\mu}) \tilde{\omega} u(t) = \omega_0 \operatorname{op}_M^\beta(f_{k+j-\mu}) \tilde{\omega} u(t)$$

for $|\beta| < \frac{1}{2}$. Now, instead of (4.1.33) we obtain for arbitrary $\mu \leq 0$ and $\beta = 0$ for (4.1.35), $-\frac{1}{2} < \beta < 0$ for (4.1.36)

$$\omega \operatorname{op}^+(a) \tilde{\omega} u(t) = \omega \sum_{j=0}^N \{ A_j^+ \operatorname{op}_M(g^+) + A_j^- \operatorname{op}_M(g^-) \} \omega_0 \operatorname{op}^+(l_+^{\mu-j}) \tilde{\omega} u(t)$$
$$+ (R_N^2 + R_N^3) u(t),$$

where R_N^2, R_N^3 are the operators from (4.1.33). Moreover,

$$\omega \sum_{j=0}^{N} \{A_j^+ \operatorname{op}_M(g^+) + A_j^- \operatorname{op}_M(g^-)\} \omega_0 \operatorname{op}^+(l_+^{\mu-j}) \tilde{\omega} u(t)$$

$$= \omega \sum_{j=0}^{N} \{A_j^+ \operatorname{op}_M^\beta(g^+) + A_j^- \operatorname{op}_M^\beta(g^-)\} t^{-\mu} t^\mu \omega_0 \operatorname{op}^+(l_+^{\mu-j}) \tilde{\omega} u(t)$$

$$= \omega t^{-\mu} \sum_{j=0}^{N} \sum_{k=0}^{N} [A_j^+ \operatorname{op}_M^\beta(T^\mu g^+) + A_j^- \operatorname{op}_M^\beta(T^\mu g^-) + G^{(j)}]$$

$$\cdot \left[\omega_0 t^j (t\delta)^k \binom{\mu-j}{k} \operatorname{op}_M^\beta(f_{k+j-\mu}) \tilde{\omega} u(t) + \omega_0 t^\mu G_N \tilde{\omega} u(t) \right].$$

Here $G^{(j)}$ are various Green operators from the commutator relations (4.1.34) multiplied by the factors A_j^+ and A_j^-, respectively, and N is so large that $\omega_0 t^\mu G_N \tilde{\omega}$ maps $C_0^\infty(\mathbb{R}_+)$ to $L^2(\mathbb{R}_+)$, cf. the characterisation of G_N in Proposition 4.1.15. For sufficiently small β all compositions in the following expression make sense:

$$S_N u(t)$$

$$= \omega t^{-\mu} \sum_{j=0}^{N} \sum_{k=0}^{N} [A_j^+ \operatorname{op}_M^\beta(T^\mu g^+) + A_j^- \operatorname{op}_M^\beta(T^\mu g^-) + G^{(j)}] \omega_0 t^\mu G_N \tilde{\omega} u(t)$$

$$+ \omega t^{-\mu} \sum_{j=0}^{N} \sum_{k=0}^{N} G^{(j)} \omega_0 t^j (t\delta)^k \binom{\mu-j}{k} \operatorname{op}_M^\beta(f_{k+j-\mu}) \tilde{\omega} u(t).$$

$G^{(j)}$ is a Green operator, and we used the fact that $G^{(j)}$ remains well–defined after small perturbations of weights in the argument functions. We then obtain

$$\omega \operatorname{op}^+(a) \tilde{\omega} u(t) = \omega t^{-\mu} \sum_{j=0}^{N} \sum_{k=0}^{N} \{A_j^+ \operatorname{op}_M^\beta(T^\mu g^+) + A_j^- \operatorname{op}_M^\beta(T^\mu g^-)\}$$

$$\cdot \omega_0 t^j (t\delta)^k \binom{\mu-j}{k} \operatorname{op}_M^\beta(f_{k+j-\mu}) \tilde{\omega} u(t)$$

$$+ (S_N + R_N^2 + R_N^3) u(t) \qquad (4.1.38)$$

for $\mu \leq 0$, $-\frac{1}{2} < \beta \leq 0$, $|\beta|$ sufficiently small, and $\beta = 0$ for (4.1.35), $\beta \neq 0$ for (4.1.36). The remainders in (4.1.33) and (4.1.38) have kernels in $C^m(\mathbb{R}_+ \times \mathbb{R}_+)$ for each fixed $m \in \mathbb{N}$ provided $N = N(m)$ is sufficiently large. Now, both in (4.1.33) and (4.1.38) we can commute the factors t^{j+k} between the Mellin actions in the first sums through $\mathrm{op}_M(g^\pm)$ and $\mathrm{op}_M^\beta(T^\mu g^\pm)$, respectively, with Green remainders. In other words, using

$$\omega \, \mathrm{op}_M(g^\pm) t^{j+k} \omega_0 = \omega t^{j+k} \, \mathrm{op}_M(g^\pm) \omega_0 + D$$

with corresponding Green operators D and analogous relations for the commutations with $\mathrm{op}_M^\beta(T^\mu g^\pm)$ we get the factors t^{j+k} to the left. Between the Mellin operators we still have the cut–off function ω_0 that was chosen in such a way that $\omega(1 - \omega_0) = 0$ holds. This implies that when we omit ω_0, the expressions remain the same modulo some Green operator. Summing up we have proved the following result:

Proposition 4.1.17 *Let* $a(t, \tau) \in S_{\mathrm{cl}}^\mu(\overline{\mathbb{R}}_+ \times \mathbb{R})$, $\mu \in \mathbb{R}$, *be arbitrary, and choose cut–off functions* $\omega, \tilde{\omega}$. *Then, on* $C_0^\infty(\mathbb{R}_+)$ *the operator* $\omega \, \mathrm{op}^+(a) \tilde{\omega}$ *has the following decompositions: For every* $m \in \mathbb{N}$ *there is an* $N(m) \in \mathbb{N}$ *with*

$$\omega \, \mathrm{op}^+(a) \tilde{\omega} = \omega \sum_{j=0}^{N} \sum_{k=0}^{N} t^{j+k} \{ B_{kj}^+(t, \delta) \, \mathrm{op}_M(g^+ T^{-\mu} f_{k+j-\mu})$$
$$+ \, B_{kj}^-(t, \delta) \, \mathrm{op}_M(g^- T^{-\mu} f_{k+j-\mu}) \} t^{-\mu} \tilde{\omega} + K_N \quad (4.1.39)$$

for $\mu > -\frac{1}{2}$, *and*

$$\omega \, \mathrm{op}^+(a) \tilde{\omega} = \omega t^{-\mu} \sum_{j=0}^{N} \sum_{k=0}^{N} t^{j+k} \{ B_{kj}^+(t, \delta) \, \mathrm{op}_M^\beta((T^\mu g^+) f_{k+j-\mu})$$
$$+ \, B_{kj}^-(t, \delta) \, \mathrm{op}_M^\beta((T^{-\mu} g^-) f_{k+j-\mu}) \} \tilde{\omega} + K_N \quad (4.1.40)$$

for $\mu \le 0$, $|\beta| < \frac{1}{2}$, *and* $\beta = 0$ *for* (4.1.35), $\beta \neq 0$ *for* (4.1.36), *for every* $N \ge N(m)$, *where* $B_{kj}^\pm(t, \delta) = A_j^\pm(t, \delta) \delta^k \binom{\mu-j}{k}$, *and* K_N *is an operator with kernel in* $C^m(\mathbb{R}_+ \times \mathbb{R}_+)$.

Note that in the above arguments for (4.1.40) in the case (4.1.36) we imposed first the condition $-\frac{1}{2} < \beta < 0$. But from the formula (4.1.40) under this assumption we can easily get an analogous formula for arbitrary $\beta \in \mathbb{R} \setminus \{0\}$, $|\beta| < \frac{1}{2}$. This applies the usual commutator arguments in the Mellin operators of the sum which cause Green remainders of the allowed quality.

The formula (4.1.39) can be transformed to the same form as (4.1.40) by commuting $t^{-\mu}$ through the Mellin operators, again, modulo Green operators which have kernels in $C^\infty(\mathbb{R}_+ \times \mathbb{R}_+)$. In other words we can use

$$\omega \, \mathrm{op}_M(g^\pm T^{-\mu} f_{k+j-\mu}) t^{-\mu} \tilde{\omega} = \omega t^{-\mu} \, \mathrm{op}_M((T^\mu g^\pm) f_{k+j-\mu}) \tilde{\omega} + D$$

for

$$\mu \neq l + \frac{1}{2} \qquad \text{for all } l \in \mathbb{N}, \qquad (4.1.41)$$

and

$$\omega \operatorname{op}_M(g^{\pm} T^{-\mu} f_{k+j-\mu}) t^{-\mu} \tilde{\omega} = \omega \operatorname{op}_M^{\beta}(g^{\pm} T^{-\mu} f_{k+j-\mu}) t^{-\mu} \tilde{\omega}$$
$$= \omega t^{-\mu} \operatorname{op}_M^{\beta}((T^{\mu} g^{\pm}) f_{k+j-\mu}) \tilde{\omega} + D$$

for $|\beta| < \frac{1}{2}$, $\beta \neq 0$, when

$$\mu = l + \frac{1}{2} \qquad \text{for some } l \in \mathbb{N} . \qquad (4.1.42)$$

Here D denotes a Green operator (possibly different in the two occurrences). In other words, the formula (4.1.40) holds for all $\mu \in \mathbb{R}$, where in the case $\mu > -\frac{1}{2}$ we set $\beta = 0$ for (4.1.41), and $|\beta| < \frac{1}{2}$, $\beta \neq 0$ for (4.1.42). Set

$$h_n(t, z; \delta) = \sum_{j+k=n} \{B_{kj}^+(t, \delta) g^+(z + \mu) + B_{kj}^-(t, \delta) g^-(z + \mu)\} f_{n-\mu}(z),$$

$n \in \mathbb{N}$, and denote by $\operatorname{op}_M^{\beta}(h_n)(\delta)$ the corresponding δ-dependent weighted Mellin pseudo-differential operator, defined for arbitrary $\beta \in \mathbb{R}$, for which $\frac{1}{2} - \beta + \mu \notin \mathbb{Z}$ holds. Note that

$$h_n(t, z; \delta) \in C^{\infty}(\overline{\mathbb{R}}_+, M_R^{\mu-n}) \qquad \text{for every } n \in \mathbb{N}$$

for $R = \{(j, 0)\}_{j \in \mathbb{Z}} \cup \{(1 + k, 1)\}_{k \in \mathbb{N}}$.

Theorem 4.1.18 *Let* $a(t, \tau) \in S_{\mathrm{cl}}^{\mu}(\overline{\mathbb{R}}_+ \times \mathbb{R})$, $\mu \in \mathbb{R}$, *and* ω, $\tilde{\omega}$ *be arbitrary cut-off functions. Then, on* $C_0^{\infty}(\mathbb{R}_+)$, *the operator* $\omega \operatorname{op}^+(a)\tilde{\omega}$ *allows the following decompositions: For each* $m \in \mathbb{N}$ *there is an* $N(m) \in \mathbb{N}$ *with*

$$\omega \operatorname{op}^+(a)\tilde{\omega} = \omega t^{-\mu} \sum_{j=0}^{N} t^j \operatorname{op}_M^{\beta}(h_j)(\delta)\tilde{\omega} + K_N \qquad (4.1.43)$$

for every $N \geq N(m)$, *for arbitrary* $\beta \in \mathbb{R}$ *satisfying* $\frac{1}{2} - \beta + \mu \notin \mathbb{Z}$, *where* K_N *is an operator with kernel in* $C^m(\mathbb{R}_+ \times \mathbb{R}_+)$.

Proof. The above arguments showed the result for $\beta = 0$ when $\mu \notin \mathbb{Z} + \frac{1}{2}$ and $\beta \neq 0$, $|\beta| < \frac{1}{2}$, when $\mu \in \mathbb{Z} + \frac{1}{2}$. However, if $\eta \in \mathbb{R}$ is arbitrary and satisfies $\frac{1}{2} - \eta + \mu \notin \mathbb{Z}$, then $\omega \operatorname{op}_M^{\beta}(h_k)(\delta)\tilde{\omega} - \omega \operatorname{op}_M^{\eta}(h_k)(\delta)\tilde{\omega}$ is a (finite-dimensional) Green operator with suitable weights such that we may insert into the formula (4.1.43) arbitrary β under the above condition, where the remainder $K_N = K_N(\delta)$ depends on the particular choice of β. $\square$

Remark 4.1.19 *According to the general structure of Mellin symbols we can write*

$$h_j(t,z;\delta) = m_j(t,z;\delta) + n_j(t,z;\delta)$$

with $m_j(t,z;\delta) \in C^\infty(\overline{\mathbb{R}}_+, M_O^{\mu-n})$ *and* $n_j(t,z;\delta) \in C^\infty(\overline{\mathbb{R}}_+, M_R^{-\infty})$ *for all* j. *There is an* $f(t,z;\delta) \in C^\infty(\overline{\mathbb{R}}_+, M_O^\mu)$ *with*

$$f(t,z;\delta) - \sum_{j=0}^{N} t^j m_j(t,z;\delta) \in C^\infty(\overline{\mathbb{R}}_+, M_O^{\mu-(N+1)})$$

for all $N \in \mathbb{N}$. *Then*

$$\omega \, \mathrm{op}^+(a)\tilde\omega = \omega t^{-\mu} \, \mathrm{op}_M^\beta(f)\tilde\omega \bmod L^{-\infty}(\mathbb{R}_+) \ ,$$

here for arbitrary $\beta \in \mathbb{R}$, *since* f *is holomorphic in* $z \in \mathbb{C}$. *In this way Theorem 4.1.18 gives an alternative Mellin reformulation of the operator* $\omega \, \mathrm{op}^+(a)\tilde\omega$, *compared with Section 4.1.1.*

Note that the remainders K_N are known by the construction. They can be evaluated in more explicit form. Interesting special cases and more details may be found in [122].

Note that the representation (4.1.43) becomes particularly simple if the symbol $a(\tau) \in S_{\mathrm{cl}}^\mu(\mathbb{R})$ is independent of t. In this case we have instead of (4.1.29)

$$\sigma_M^{\mu-j}(a)(z) = \{a_j^+ g^+(z+\mu) + a_j^- g^-(z+\mu)\} f_{j-\mu}(z) \qquad (4.1.44)$$

with the constants $a_j^\pm$ of the asymptotic expansion

$$a(\tau) \sim \sum_{j=0}^{\infty} a_j^\pm (i\tau)^{\mu-j} \qquad \text{for } \tau \to \pm\infty.$$

Theorem 4.1.20 *Let* $a(\tau) \in S_{\mathrm{cl}}^\mu(\mathbb{R})$, $\mu \in \mathbb{R}$, *and fix cut–off functions* $\omega(t)$, $\tilde\omega(t)$. *Then, for every* $\beta \in \mathbb{R}$ *with* $\frac{1}{2} - \beta + \mu \notin \mathbb{Z}$

$$\omega \, \mathrm{op}^+(a)\tilde\omega u(t) = \omega t^{-\mu} \sum_{j=0}^{N} t^j \, \mathrm{op}_M^\beta(\sigma_M^{\mu-j}(a))\tilde\omega u(t) + K_N u(t)$$

for $u \in C_0^\infty(\mathbb{R}_+)$, *for every* $N \geq N(m)$, *with an operator* K_N *with kernel in* $C^m(\mathbb{R}_+ \times \mathbb{R}_+)$.

Theorem 4.1.20 was given for completeness. It may be found in [122] as well as the following result:

Theorem 4.1.21 *For arbitrary $a(\tau) \in S^{\mu}_{cl}(\mathbb{R})$, $\mu \in \mathbb{R}$, and $\beta \in \mathbb{R}$ with $\frac{1}{2} - \beta + \mu \notin \mathbb{Z}$ there is a finite-dimensional Green operator $G_{\beta} \in C_G(\mathbb{R}_+, \boldsymbol{a}^{\bullet})$ for $\boldsymbol{a} = (\alpha, \tilde{\alpha}, (-\infty, 0])$ with suitable $\alpha, \tilde{\alpha} \in \mathbb{R}$, dependent of μ, β, such that*

$$\omega \operatorname{op}^+(a)\tilde{\omega} - G_{\beta} \in C^{\mu}(\mathbb{R}_+, \boldsymbol{g}^{\bullet})$$

for $\boldsymbol{g} = (\beta, \beta - \mu, (-\infty, 0])$, where

$$\sigma^{\mu-j}_M((\omega \operatorname{op}^+(a)\tilde{\omega} - G_{\beta}))(z) = \sigma^{\mu-j}_M(a)(z), \qquad j \in \mathbb{N}.$$

In particular, for $\mu = 0$ this is true for $\beta = 0$, where $G_{\beta} = 0$. Then,

$$\omega \operatorname{op}^+(a)\tilde{\omega} \in C^0(\mathbb{R}_+, \boldsymbol{g}^{\bullet}) \qquad \text{for } a(\tau) \in S^0_{cl}(\mathbb{R}),$$

for $\boldsymbol{g} = (0, 0, (-\infty, 0])$.

It is not hard to prove analogous results for arbitrary $a(t, \tau) \in S^{\mu}_{cl}(\overline{\mathbb{R}}_+ \times \mathbb{R})$. We will not go into the details here. For purposes below we want to formulate an analogue of the above Lemma 4.1.16 in terms of the symbols $l^{\mu}_-(\delta, \tau) = (\delta - i\tau)^{\mu}$.

Lemma 4.1.22 *For every $a(t, \tau) \in S^{\mu}_{cl}(\overline{\mathbb{R}}_+ \times \mathbb{R})$, $\mu \in \mathbb{R}$, there exists a sequence of functions $E^{\pm}_j(t, \delta) \in C^{\infty}(\overline{\mathbb{R}}_+)$ (that are polynomials in δ of order j) such that for each $m \in \mathbb{N}$ there is an $N(m) \in \mathbb{N}$ such that the operator $\operatorname{op}^+(a)$ on $u \in C^{\infty}_0(\mathbb{R}_+)$ allows the following decomposition:*

$$\operatorname{op}^+(a)u(t) = \sum_{j=0}^{N}\{E^+_j(t, \delta) \operatorname{op}^+(l^{\mu-j}_-) \operatorname{op}_M(g^+)$$

$$+ E^-_j(t, \delta) \operatorname{op}^+(l^{\mu-j}_-) \operatorname{op}^+(g^-)\}u(t) + C_N u(t)$$

for every $N \geq N(m)$, where C_N is an integral operator with kernel in $C^m(\overline{\mathbb{R}}_+ \times \overline{\mathbb{R}}_+)$.

Proof. We can proceed analogously to the proof of Lemma 4.1.16. First we write $a(t, \tau)$ in the form (4.1.25) and then insert

$$(i\tau)^{\mu-j} = (-l_-(\delta, \tau) + \delta)^{\mu-j} = (-1)^{\mu-j}(1 - \delta l_-(\delta, \tau))^{\mu-j}l^{\mu-j}_-(\delta, \tau)$$

$$= (-1)^{\mu-j}\sum_{k=0}^{\infty}\binom{\mu-j}{k}(-\delta)^k l^{\mu-(j+k)}_-(\delta, \tau).$$

This gives us

$$\operatorname{op}^+(a)u(t)$$

$$= \operatorname{op}^+(\sum_{j=1}^{N}\sum_{k=1}^{N}\binom{\mu-j}{k}(-\delta)^k(-1)^{\mu-j}[a^+_j(t)l^{\mu-(j+k)}_-(\delta, \tau)\theta^+(\tau)$$

$$+ a^-_j(t)l^{\mu-(j+k)}_-(\delta, \tau)\theta^-(\tau)])u(t) + R_N u(t)$$

for every N, where the remainder R_N has a kernel in $C^m(\overline{\mathbb{R}}_+ \times \overline{\mathbb{R}}_+)$ with $m = m(N) \to \infty$ a $N \to \infty$. Now it suffices to set

$$E_k^\pm(t,\delta) = \sum_{\substack{j+l=k \\ j \leq N,\ k \leq N}} \binom{\mu - j}{l}(-\delta)^l(-1)^{\mu-j}a_j^\pm(t)$$

to obtain the assertion. $\qquad\qquad\qquad\qquad\qquad\qquad\qquad\qquad\square$

4.1.3   Transmission operators

Operators of the form

$$r^- \operatorname{op}(a)e^+ : C_0^\infty(\mathbb{R}_+) \to C^\infty(\mathbb{R}_-), \qquad (4.1.45)$$
$$r^+ \operatorname{op}(a)e^- : C_0^\infty(\mathbb{R}_-) \to C^\infty(\mathbb{R}_+) \qquad (4.1.46)$$

for $a(t,\tau) \in S_{\mathrm{cl}}^\mu(\mathbb{R} \times \mathbb{R})$ will be called transmission operators. They occur in transmission problems as a part of the "transmission symbolic structure" normal to a jump of a symbol in $\mathbb{R}^n \ni (t,y)$, $t \in \mathbb{R}$, $y \in \mathbb{R}^{n-1}$, at $t = 0$, cf. Rempel, Schulze [91]. In view of the pseudo-locality of pseudo-differential operators the operators (4.1.45), (4.1.45) are smoothing, however they are singular in some typical way near $t = 0$. Continuous extensions of these operators to Sobolev spaces can be discussed in a similar way as for the pseudo-differential operators on the half axis. The situation is particularly simple for $\mu = 0$, where we can argue in terms of continuous operators in L^2-spaces. Therefore we shall begin with this case. Let us set

$$(\varepsilon^* u)(t) = u(-t).$$

To characterise the operators (4.1.45), (4.1.45) it suffices to consider $\varepsilon^* r^- \operatorname{op}(a)e^+$. As above in Section 4.1.2 we set

$$g(z) = -e^{i\pi z}(1 - e^{2\pi i z})^{-1},$$

$$f_j(z) = \begin{cases} 1 & \text{for } j = 0, \\ \prod_{p=1}^{j}(p - z)^{-1} & \text{for } j \in \mathbb{N} \setminus \{0\}. \end{cases}$$

Theorem 4.1.23 Let $a(t,\tau) \in S_{\mathrm{cl}}^0(\mathbb{R} \times \mathbb{R})$ be independent of t for $|t| >$ const. Then we have

$$A = \varepsilon^* r^- \operatorname{op}(a)e^+ \in C_{M+G}^0(\mathbb{R}_+, \boldsymbol{g}^\bullet)$$

for $\boldsymbol{g} = (0, 0, (-\infty, 0])$. Here

$$\sigma_M^{-j}(A)(z) = (-1)^j(a_j^+ - a_j^-)g(z)f_j(z), \qquad j \in \mathbb{N},$$

when $a = a(\tau)$ is independent of t, and

$$\sigma_M^{-j}(A)(z) = \left\{ \sum_{l+k=j} (-1)^l \frac{1}{k!} \left(\frac{d}{dt} \right)^k (a_l^+(t) - a_l^-(t))|_{t=0} f_l(z) \right\} g(z),$$

$j \in \mathbb{N}$, in the general case, where $a(t, \tau) \sim \sum\limits_{j=0}^{\infty} a_j^{\pm}(t)(i\tau)^{-j}$ for $\tau \to \pm\infty$.

Proof. Let us first consider the case of t–independent $a(\tau)$. Fix arbitrary cut–off functions $\omega(t)$, $\tilde{\omega}(t)$. Then the operators $\omega A(1 - \tilde{\omega})$, $(1 - \omega)A\tilde{\omega}$ and $(1 - \omega)A(1 - \tilde{\omega})$ have kernels in $\mathcal{S}(\overline{\mathbb{R}}_+ \times \overline{\mathbb{R}}_+)$. In fact, the kernel of op(a) has the form $k(a)(t - t')$ for $k(a)(\zeta) = \int e^{i\zeta\tau} a(\tau) d\tau$. Theorem 1.1.45 gives us $\chi(\zeta)k(a)(\zeta) \in \mathcal{S}(\mathbb{R})$ for every excision function $\chi(\zeta)$. Now the kernel of $\varepsilon^* r^-$ op$(a)e^+$ is $k(a)(-t - t')$, where $(t, t') \in \mathbb{R}_+ \times \mathbb{R}_+$, and hence $\omega(t)k(a)(-t - t')(1 - \tilde{\omega}(t'))$, $(1 - \omega(t))k(a)(-t - t')\tilde{\omega}(t)$, $(1 - \omega(t))k(a)(-t - t')(1 - \tilde{\omega}(t'))$ belong to $\mathcal{S}(\overline{\mathbb{R}}_+ \times \overline{\mathbb{R}}_+)$. Thus, to characterise A, it suffices to consider $\omega A\tilde{\omega}$. Analogously to the calculations in the proof of Lemma 4.1.16 we have

$$a(\tau) = \sum_{j=0}^{N} \{A_j^+(\delta)\theta^+(\tau) + A_j^-(\delta)\theta^-(\tau)\} l_+^{-j}(\delta, \tau) + a_{(N)}(\tau)$$

for every N, where $|a_{(N)}(\tau)(1 + |\tau|)^{-N}| <$ const for all $\tau \in \mathbb{R}$. It follows that

$$\text{op}(a)e^+ = \sum_{j=0}^{N} \{A_j^+ \text{op}(\theta^+) + A_j^- \text{op}(\theta^-)\}e^+ \text{op}^+(l_+^{-j}) + \text{op}(a_{(N)})e^+.$$

$$(4.1.47)$$

Here we used Remark 4.1.9. The operator op$(a_{(N)})$ has a kernel in $C^m(\mathbb{R} \times \mathbb{R})$ with $m(N) \to \infty$ for $N \to \infty$. Using Proposition 4.1.14 we obtain

$$\omega\varepsilon^* r^- \text{op}(a)e^+\tilde{\omega} = \omega \sum_{j=0}^{N} (A_j^+ - A_j^-) \text{op}_M(g) \text{op}^+(l_+^{-j})\tilde{\omega} + R_N^1$$

for $R_N^1 = \omega\varepsilon^* r^- \text{op}(a_{(N)})e^+\tilde{\omega}$, which has a kernel in $C^m(\overline{\mathbb{R}}_+ \times \overline{\mathbb{R}}_+) \cap C^\infty(\mathbb{R}_+ \times \mathbb{R}_+)$ for sufficiently large $N \geq N(m)$. If ω_0 is an arbitrary cut–off function we obtain that

$$\omega \, \text{op}_M(g)(1 - \omega_0) \, \text{op}^+(l_+^{-j})\tilde{\omega}$$

is a Green operator of the class $C_G(\mathbb{R}_+, \boldsymbol{g}^\bullet)$. Thus it remains to characterise

$$\omega \sum_{j=0}^{N} (A_j^+ - A_j^-) \operatorname{op}_M(g) \omega_0 \operatorname{op}^+(l_+^{-j}) \tilde{\omega} \qquad (4.1.48)$$

as an element in $C_{M+G}^0(\mathbb{R}_+, \boldsymbol{g}^\bullet)$. However, in view of the information from the preceding section, in particular, of Theorem 4.1.21, applied to $a(\tau) = l_+^{-j}(\delta, \tau)$, this is already clear, since $\omega_0 \operatorname{op}^+(l_+^{-j}) \tilde{\omega} \in C^0(\mathbb{R}_+, \boldsymbol{g}^\bullet)$ and because of $\omega \operatorname{op}_M(g) \omega_1 \in C_{M+G}^0(\mathbb{R}_+, \boldsymbol{g}^\bullet)$ for every cut–off function ω_1, where we can assume $\omega_1 \omega_0 = \omega_0$. The operator R_N^1 can be characterised as an element in $C_G(\mathbb{R}_+, \boldsymbol{g}_{\tilde{m}}^\bullet)$ for $\boldsymbol{g}_{\tilde{m}} = (0, 0, (-\tilde{m} + 1, 0])$, $\tilde{m}(N) \to \infty$ as $N \to \infty$. Hence $\omega \varepsilon^* r^- \operatorname{op}(a) \tilde{\omega} e^+$ belongs altogether to $C_{M+G}^0(\mathbb{R}_+, \boldsymbol{g}_m^\bullet)$ for every m, i.e., to $C_{M+G}^0(\mathbb{R}_+, \boldsymbol{g}^\bullet)$ for $\boldsymbol{g} = (0, 0, (-\infty, 0])$. The asserted formula for the conormal symbols is a consequence of the fact that they are necessarily independent of δ. So we can pass to $\delta \to 0$ in the formula (4.1.48) and to observe that $A_j^\pm(\delta)|_{\delta=0} = a_j^\pm$. The proof for general $a(t, \tau)$ is completely analogous to the case with constant coefficients. Here the only new aspect is that the coefficients $A_j^\pm(t, \delta)$ are t–dependent, C^∞ up to $t = 0$. The expression for the conormal symbols in this case then follows by Taylor expansion of $A_j^\pm(t, \delta)$ at $t = 0$.   □

Theorem 4.1.24 *Let $a(t, \tau) \in S_{\mathrm{cl}}^\mu(\mathbb{R} \times \mathbb{R})$ be independent of t for $|t| > $ const, $\mu \in \mathbb{R}$. Then for every $\beta \in \mathbb{R}$ with $\frac{1}{2} - \beta + \mu \notin \mathbb{Z}$ there exists a finite–dimensional Green operator $G_\beta \in C_G(\mathbb{R}_+, \boldsymbol{a}^\bullet)$ for $\boldsymbol{a} = (\alpha, \tilde{\alpha}, (-\infty, 0])$ with suitable $\alpha, \tilde{\alpha} \in \mathbb{R}$, dependent on μ, β, such that*

$$A = \varepsilon^* r^- \operatorname{op}(a) e^+$$

satisfies $A - G_\beta \in C_{M+G}^\mu(\mathbb{R}_+, \boldsymbol{g}^\bullet)$ for $\boldsymbol{g} = (\beta, \beta - \mu, (-\infty, 0])$. We have

$$\sigma_M^{\mu-j}(A - G_\beta)(z) = (-1)^j (a_j^+ - a_j^-) g(z + \mu) f_{j-\mu}(z), \qquad j \in \mathbb{N},$$

when $a = a(\tau)$ is independent of t and

$$\sigma_M^{\mu-j}(A - G_\beta)(z)$$

$$= \left\{ \sum_{l+k=j} (-1)^l \frac{1}{k!} \left(\frac{d}{dt}\right)^k (a_l^+(t) - a_l^-(t))|_{t=0} f_{l-\mu}(z) \right\} g(z + \mu),$$

$j \in \mathbb{N}$, in the general case, where $a(t, \tau) \sim \sum_{j=0}^\infty a_j^\pm(t)(i\tau)^{\mu-j}$ for $\tau \to -\infty$.

Proof. After the proof of the above Theorem 4.1.23 and the technique to prove Theorem 4.1.20 it is fairly obvious how to proceed. Therefore we want to sketch only the main steps. First it suffices to consider $\omega A \tilde{\omega}$ for cut–off functions ω, $\tilde{\omega}$, since $\omega A(1-\tilde{\omega})$, $(1-\omega)A\tilde{\omega}$ and $(1-\omega)A(1-\tilde{\omega})$ have kernels in $\mathcal{S}(\overline{\mathbb{R}}_+ \times \overline{\mathbb{R}}_+)$. To each operator K with such a kernel $k(t,t')$ and every β, $\mu \in \mathbb{R}$ there is a finite–dimensional Green operator G such that $K - G$ belongs to $C_G(\mathbb{R}_+, \boldsymbol{g}^\bullet)$ for $\boldsymbol{g} = (\beta, \beta - \mu, (-\infty, 0])$. This follows easily by removing operators with kernels that correspond to finite Taylor expansions in t' and t, respectively, near zero. More precisely, we form

$$\tilde{k}(t,t') = k(t,t') - \sum_{j=0}^{N} \frac{1}{j!}(t')^j (\frac{\partial}{\partial t'})^j k(t,t')|_{t'=0}\tilde{\omega}(t'),$$

$$k_0(t,t') = \tilde{k}(t,t') - \omega(t) \sum_{j=0}^{N} \frac{1}{l!}t^l (\frac{\partial}{\partial t})^l \tilde{k}(t,t')|_{l=0}.$$

Then, the operator K_0 with kernel k_0 belongs to $C_G(\mathbb{R}_+, \boldsymbol{g}^\bullet)$ when N is sufficiently large, and $K - K_0 = G$ is finite–dimensional. Thus it suffices to consider $\omega A \tilde{\omega}$. For simplicity we study the case of t–independent symbols $a(\tau)$. The general case is completely analogous. For every N we can write

$$a(\tau) = \sum_{j=0}^{N} \{A_j^+ \theta^+(\tau) + A_j^- \theta^-(\tau)\} l_+^{\mu-j}(\delta, \tau) + a_{(N)}(\tau),$$

cf. Lemma 4.1.16. The operator $\mathrm{op}(a_{(N)})$ has a kernel in $C^m(\mathbb{R} \times \mathbb{R})$ for each fixed $m \in \mathbb{N}$ when $N \geq N(m)$ is sufficiently large. Thus, analogously to the zero order case, the contribution from this remainder causes no trouble. We get

$$\omega \varepsilon^* r^- \mathrm{op}(a)\tilde{\omega}e^+ u(t) = \omega \sum_{j=0}^{N} (A_j^+ - A_j^-) \mathrm{op}_M(g) \mathrm{op}^+(l_+^{\mu-j})\tilde{\omega}u(t) + R_N^1 u(t)$$

for $u \in C_0^\infty(\mathbb{R}_+)$. Here $R_N^1 = \omega \varepsilon^* r^- \mathrm{op}(a_{(N)})e^+ \tilde{\omega}$. From Remark 4.1.9 we see that $\mathrm{op}_M(g)\mathrm{op}^+(l_+^{\mu-j})$ makes sense on $C_0^\infty(\mathbb{R}_+)$, since $\mathrm{op}^+(l_+^{\mu-j})u \in L^2(\mathbb{R}_+)$ for $u \in C_0^\infty(\mathbb{R}_+)$. The operator $\omega \mathrm{op}_M(g)(1-\omega_0)\mathrm{op}^+(l_+^{\mu-j})\tilde{\omega}$ is of the desired structure; so it remains to characterise $\omega \mathrm{op}_M(g)\omega_0 \mathrm{op}^+(l_+^{\mu-j})\tilde{\omega}$, $j \in \mathbb{N}$, for any cut–off function ω_0. However, using Proposition 4.1.15, we see that we can concentrate on

$$\omega \mathrm{op}_M(g)\omega_0 t^{-\mu+j} \sum_{k=0}^{N} (t\delta)^k \binom{\mu-j}{k} \mathrm{op}_M^\gamma(f_{k+j-\mu})\tilde{\omega} \qquad (4.1.49)$$

for sufficiently large N and suitable $\gamma \geq \max\{\mu, 0\}$, under which the Mellin actions and compositions make sense. The influence of G_N in the formula (4.1.22) disappears for $N \to \infty$, modulo a finite–dimensional Green operator, dependent on β. Now (4.1.49) can be understood completely in terms of the cone algebra on $\mathbb{R}_+$ with discrete asymptotics. We see, in particular, that γ may be replaced by β, provided that $\frac{1}{2} - \beta + \mu \notin \mathbb{Z}$ holds, modulo a finite–dimensional Green operator. This concerns every j. The occurring asymptotic types show that the remainders are in a class of finite–dimensional Green operators for all j. The asserted formulas for the conormal symbols are elementary.   □

Remark 4.1.25 *From the general composition rules in the cone algebras we know how the transmission operators of this section behave under compositions. For the involved conormal symbols we then obtain the Mellin translation product, cf. the formula (2.3.62). Let us consider, for instance, the zero order case and $a(\tau)$, $b(\tau) \in S_{\mathrm{cl}}^0(\mathbb{R})$. Set $\check{a}(\tau) = a(-\tau)$. Then, if $\theta^{\pm}(t)$ is the characteristic function of $\mathbb{R}_{\pm} \ni t$, we have*

$$
\begin{aligned}
r^+ \operatorname{op}(a)\theta^- \operatorname{op}(b)e^+ &= r^+ \operatorname{op}(a)\varepsilon^* \varepsilon^* \theta^- \operatorname{op}(b)e^+ \\
&= r^+ \varepsilon^* \operatorname{op}(\check{a})\theta^+ \varepsilon^* \theta^- \operatorname{op}(b)e^+ \\
&= \{\varepsilon^* r^- \operatorname{op}(\check{a})e^+\}\{\varepsilon^* r^- \operatorname{op}(b)e^+\}, \qquad (4.1.50)
\end{aligned}
$$

where the operators are considered on $C_0^\infty(\mathbb{R}_+)$ (or on $L^2(\mathbb{R}_+)$). This has a relation to the composition $\operatorname{op}^+(a)\operatorname{op}^+(b)$. In fact, we have

$$
\operatorname{op}^+(a)\operatorname{op}^+(b) - \operatorname{op}^+(ab) = \{\varepsilon^* r^- \operatorname{op}(\check{a})e^+\}\{\varepsilon^* r^- \operatorname{op}(b)e^+\}. \quad (4.1.51)
$$

Then, for every $r \in \mathbb{N}$, it follows that

$$
\begin{aligned}
&\sigma_M^{-r}(\operatorname{op}^+(a)\operatorname{op}^+(b) - \operatorname{op}^+(ab))(z) \\
&= \Big\{ \sum_{j+k=r} (-1)^j (a_j^+ - a_j^-)(b_k^+ - b_k^-) \Big\} g^+(z)g^-(z)f_r(z).
\end{aligned}
$$

Here we used $g^2(z) = -g^+(z)g^-(z)$.

Definition 4.1.26 *A symbol $a(t, \tau) \in S_{\mathrm{cl}}^\mu(\overline{\mathbb{R}}_+ \times \mathbb{R})$ for $\mu \in \mathbb{Z}$ is said to have the transmission property (with respect to $t = 0$) if the coefficients $a_j^{\pm}(t) \in C^\infty(\overline{\mathbb{R}}_+)$ in the asymptotic expansion*

$$
a(t, \tau) \sim \sum_{j=0}^\infty a_j^{\pm}(t)(i\tau)^{\mu-j} \qquad \text{for } \tau \to \pm\infty
$$

$i = \sqrt{-1}$, satisfy the conditions

$$\left(\frac{d}{dt}\right)^k (a_j^+(t) - a_j^-(t))|_{t=0} = 0 \qquad for \ all \ j, k \in \mathbb{N}.$$

If $a(t, \tau)$ is given as a symbol in $S_{\mathrm{cl}}^\mu(\mathbb{R} \times \mathbb{R})$, then we talk about the transmission property if $a(t, \tau)|_{\overline{\mathbb{R}}_+ \times \mathbb{R}}$ satisfies the above conditions.

Remark 4.1.27 If $a(t, \tau) \in S_{\mathrm{cl}}^\mu(\overline{\mathbb{R}}_+ \times \mathbb{R})$ has the transmission property, then $\sigma_M^{\mu-j}(a)(t, z) = 0$ for all $j \in \mathbb{N}$, cf. (4.1.29). In particular, if we form to $a(t, \tau)$ and $\beta \in \mathbb{R}$ an appropriate finite–dimensional Green operator G_β such that $\mathrm{op}^+(a) - G_\beta \in C^\mu(\mathbb{R}_+, \boldsymbol{g}^\bullet)$ holds for $\boldsymbol{g} = (\beta, \beta - \mu, (-\infty, 0])$, cf. Theorem 4.1.21, then $\sigma_M^{\mu-j}(\mathrm{op}^+(a) - G_\beta) = 0$ for all $j \in \mathbb{N}$.

Proposition 4.1.28 Let $a_j(t, \tau) \in S_{\mathrm{cl}}^{\mu-j}(\overline{\mathbb{R}}_+ \times \mathbb{R})$, $j \in \mathbb{N}$, be an arbitrary sequence of symbols with the transmission property. Then the asymptotic sum $a(t, \tau) \sim \sum_{j=0}^\infty a_j(t, \tau)$ in $S_{\mathrm{cl}}^\mu(\overline{\mathbb{R}}_+ \times \mathbb{R})$ has also the transmission property.

The proof is trivial and will be omitted.

Note that every element in $S^{-\infty}(\overline{\mathbb{R}}_+ \times \mathbb{R})$ has the transmission property.

Proposition 4.1.29 Let $a(t, \tau) \in S_{\mathrm{cl}}^\mu(\overline{\mathbb{R}}_+ \times \mathbb{R})$ and $b(t, \tau) \in S_{\mathrm{cl}}^\nu(\overline{\mathbb{R}}_+ \times \mathbb{R})$ be symbols with the transmission property. Then the following symbols have also the transmission property:

(i) the pointwise product $a(t, \tau)b(t, \tau) \in S_{\mathrm{cl}}^{\mu+\nu}(\overline{\mathbb{R}}_+ \times \mathbb{R})$,

(ii) $D_t^k D_\tau^l a(t, \tau) \in S_{\mathrm{cl}}^{\mu-l}(\overline{\mathbb{R}}_+ \times \mathbb{R})$ for every $k, l \in \mathbb{N}$,

(iii) the Leibniz product $a(t, \tau) \# b(t, \tau) \sim \sum_{k=0}^\infty \frac{1}{k!} D_\tau^k a(t, \tau) \partial_t^k b(t, \tau) \in S_{\mathrm{cl}}^{\mu+\nu}(\overline{\mathbb{R}}_+ \times \mathbb{R})$,

(iv) $a^*(t, \tau) \sim \sum_{k=0}^\infty \frac{1}{k!} D_\tau^k \partial_t^k \overline{a(t, \tau)}$ (which is associated to the formal adjoint operator).

Proposition 4.1.30 Let $a(t, \tau) \in S_{\mathrm{cl}}^\mu(\overline{\mathbb{R}}_+ \times \mathbb{R})$ be a symbol with the transmission property, and assume that the homogeneous principal part $a_{(\mu)}(t, \tau)$ of $a(t, \tau)$ of order μ satisfies $a_{(\mu)}(t, \tau) \neq 0$ for all $t \in \overline{\mathbb{R}}_+$, $\tau \neq 0$. Then $\chi(\tau) a_{(\mu)}^{-1}(t, \tau) \in S_{\mathrm{cl}}^{-\mu}(\overline{\mathbb{R}}_+ \times \mathbb{R})$ has the transmission property for each excision function $\chi(\tau)$; moreover, every $b(t, \tau) \in S_{\mathrm{cl}}^{-\mu}(\overline{\mathbb{R}}_+ \times \mathbb{R})$ satisfying $a(t, \tau) \# b(t, \tau) - 1 \in S^{-\infty}(\overline{\mathbb{R}}_+ \times \mathbb{R})$ has the transmission property.

The obvious proofs of the Propositions 4.1.29, 4.1.30 will be omitted. Note that every polynomial in τ

$$r^\mu(t, \tau) = \sum_{j=0}^{\mu} c_j(t) \tau^j$$

with $c_j(t) \in C^\infty(\overline{\mathbb{R}}_+)$ has the transmission property. An example is

$$r^\mu(\tau) = (1 + i\tau)^\mu, \qquad \mu \in \mathbb{Z}.$$

This can often be employed as an order reducing symbol, i.e., if $a(t, \tau)$ is an arbitrary symbol of order μ with the transmission property, then $a_0(t, \tau) = a(t, \tau) r^{-\mu}(\tau)$ is of order zero, with the transmission property.

Proposition 4.1.31 *Let* $a(t, \tau) \in S_{cl}^\mu(\mathbb{R} \times \mathbb{R})_e$, $\mu \in \mathbb{Z}$, *have the transmission property. Then the operator*

$$\varepsilon^* r^- \operatorname{op}(a) e^+ : L^2(\mathbb{R}_+) \to L^2(\mathbb{R}_+)$$

has a kernel $g(a)(t, t')$ *in* $\mathcal{S}(\overline{\mathbb{R}}_+ \times \overline{\mathbb{R}}_+)$. *Moreover, if* $\{a_n(t, \tau)\}_{n \in \mathbb{N}}$ *is a sequence of such symbols tending to zero in* $S_{cl}^\mu(\mathbb{R} \times \mathbb{R})_e$ *for* $n \to \infty$ *then* $g(a_n)(t, t')$ *tends to zero in* $\mathcal{S}(\overline{\mathbb{R}}_+ \times \overline{\mathbb{R}}_+)$ *for* $n \to \infty$.

Proof. Let us first show the assertion for $\mu = 0$. Choose a function $\varphi(t) \in C^\infty(\mathbb{R})$ with $\varphi(t) = 1$ for $t < c$, $\varphi(t) = 0$ for $t > 2c$ for a constant $c > 0$. Set $a_0(t, \tau) = \varphi(t) a(t, \tau)$, $a_\infty(t, \tau) = a(t, \tau) - a_0(t, \tau)$. Then $r^- \operatorname{op}(a_\infty) e^+ = 0$. Thus it suffices to consider the symbol $a_0(t, \tau)$. Let $\varphi_0(t) \in C_0^\infty(\mathbb{R})$ be another function with $\varphi_0(t) = 1$ in a neighbourhood of $t = 0$, satisfying $(1 - \varphi_0(t))\varphi(t) = 0$ for $t > \tilde{c}$ for some $\tilde{c} < 0$. Write $a_0(t, \tau) = b(t, \tau) + \tilde{b}(t, \tau)$ for $b(t, \tau) = \varphi_0(t) a_0(t, \tau), \tilde{b}(t, \tau) = (1 - \varphi_0(t)) a_0(t, \tau)$. Moreover, fix a $\psi(\zeta) \in C_0^\infty(\mathbb{R})$ with $\psi(\zeta) = 1$ in a neighbourhood of $\zeta = 0$ and $\psi(\zeta) = 0$ for $|\zeta| > |\tilde{c}|/4$. Set

$$k(t, \zeta) = \int e^{i\zeta\tau} \tilde{b}(t, \tau) \, d\tau$$

and $k_0(t, \zeta) = (1 - \psi(\zeta)) k(t, \zeta)$. Then $k_0(t, \zeta) \in \mathcal{S}(\mathbb{R}_\zeta, S^0(\mathbb{R}_t))$. The kernel of $r^- \operatorname{op}(\tilde{b}) e^+$ in its dependence on (t, t') is non–vanishing only for $t \leq \tilde{c}$, $t' \geq 0$, which is a region outside $\operatorname{diag}(\mathbb{R}_t \times \mathbb{R}_{t'})$, and it equals $k_0(t, t - t')$. The kernel of $\varepsilon^* r^- \operatorname{op}(\tilde{b}) e^+$ then equals $k_0(-t, -t - t')$, and this belongs to $\mathcal{S}(\overline{\mathbb{R}}_+ \times \overline{\mathbb{R}}_+)$. Thus, it remains to consider $b(t, \tau)$.

Applying the decomposition (4.1.47) to $b(t, \tau)$ we get

$$\operatorname{op}(b) e^+ = \sum_{j=0}^{N} A_j^+ \operatorname{op}(l_+^{-j}) e^+ + \operatorname{op}(b_{(N)}) e^+$$

for a symbol $b_{(N)}(t,\tau)$ of order $-N$. Using $r^-\operatorname{op}(l_+^{-j})e^+ = 0$, cf. Remark 4.1.9, we obtain $\varepsilon^* r^-\operatorname{op}(b)e^+ = \varepsilon^* r^-\operatorname{op}(b_{(N)})e^+$ for every N. Using the semi–norm system

$$r_{kl} : u(t) \to \sup_{t\in\overline{\mathbb{R}}_+} |(1+t)^k \left(\frac{d}{dt}\right)^l u(t)|, \qquad k,l \in \mathbb{N},$$

for the space $\mathcal{S}(\overline{\mathbb{R}}_+) = \mathcal{S}(\mathbb{R})|_{\overline{\mathbb{R}}_+}$, we can show

$$r_{kl}(\varepsilon^* r^-\operatorname{op}(b)e^+ u) \leq c_{kl}\|u\|_{L^2(\mathbb{R}_+)} \tag{4.1.52}$$

for all $u \in L^2(\mathbb{R}_+)$ and all $k,l \in \mathbb{N}$, for suitable constants $c_{kl} > 0$. In fact, first we have a continuous operator $\operatorname{op}(b_{(N)})e^+ : L^2(\mathbb{R}_+) \to [\check{\varphi}]H^N(\mathbb{R})$, where $\check{\varphi} \in C_0^\infty(\mathbb{R})$ is any function with $\check{\varphi}\varphi_0 = \varphi_0$. This implies that

$$\varepsilon^* r^-\operatorname{op}(b)e^+ : L^2(\mathbb{R}_+) \to [\tilde{\omega}]H^N(\mathbb{R}_+)$$

is a continuous operator for a suitable cut–off function $\tilde{\omega}$. For given l we can choose N so large that $(1+t)^k \left(\frac{d}{dt}\right)^l : [\tilde{\omega}]H^N(\mathbb{R}_+) \to [\tilde{\omega}]C(\overline{\mathbb{R}}_+)$ is continuous. Thus

$$(1+t)^k \left(\frac{d}{dt}\right)^l \varepsilon^* r^-\operatorname{op}(b)e^+ : L^2(\mathbb{R}_+) \to [\tilde{\omega}]C(\overline{\mathbb{R}}_+)$$

is a continuous operator which yields just the estimate (4.1.52). Analogous arguments can be applied to the adjoint $(\varepsilon^* r^-\operatorname{op}(b)e^+)^*$ which is of analogous structure. This shows that

$$\varepsilon^* r^-\operatorname{op}(b)e^+, (\varepsilon^* r^-\operatorname{op}(b)e^+)^* : L^2(\mathbb{R}_+) \to \mathcal{S}(\overline{\mathbb{R}}_+)$$

are continuous operators. Such operators have kernels in $\mathcal{S}(\overline{\mathbb{R}}_+ \times \overline{\mathbb{R}}_+)$, cf. analogously Theorem 1.1.3. The second statement employs the natural Fréchet topology in $S_{cl}^0(\overline{\mathbb{R}}_+ \times \mathbb{R})_e$. Convergence in this space means convergence of the homogeneous components of all orders and of remainders of each negative order when we substract from the given symbols a finite sum associated with the homogeneous components. It is then elementary to check that $a_n \to 0$ for $n \to \infty$ implies that $\varepsilon^* r^-\operatorname{op}(a_n)e^+$ tends to zero in $\mathcal{L}(L^2(\mathbb{R}_+), \mathcal{S}(\overline{\mathbb{R}}_+))$ for $n \to \infty$ and that the same is true of the adjoints. For the case of arbitrary $\mu \geq 0$ which is the remaining point we write

$$\varepsilon^* r^-\operatorname{op}(a)e^+ = \varepsilon^* r^-\operatorname{op}(l_-^\mu)\operatorname{op}(l_-^{-\mu})\operatorname{op}(a)e^+$$
$$= \varepsilon^* r^-\operatorname{op}(l_-^\mu)\theta^-\operatorname{op}(l_-^{-\mu})\operatorname{op}(a)e^+$$
$$= (\varepsilon^* r^-\operatorname{op}(l_-^\mu)e^-\varepsilon^*)\varepsilon^* r^-\operatorname{op}(l_-^{-\mu})\operatorname{op}(a)e^+.$$

We have
$$\varepsilon^* r^- \operatorname{op}(l_-^{-\mu}) \operatorname{op}(a) e^+ \sim \varepsilon^* r^- \operatorname{op}(l_-^{-\mu} \# a) e^+,$$
where $\sim$ means equality modulo an operator with kernel in $\mathcal{S}(\overline{\mathbb{R}}_+ \times \overline{\mathbb{R}}_+)$. In view of $l_-^{-\mu} \# a \in S_{\mathrm{cl}}^0(\mathbb{R} \times \mathbb{R})_{\mathrm{e}}$ we get from the first part of the proof that $\varepsilon^* r^- \operatorname{op}(l_-^{-\mu}) \operatorname{op}(a) e^+$ has a kernel in $\mathcal{S}(\overline{\mathbb{R}}_+ \times \overline{\mathbb{R}}_+)$. This means, in particular, that it induces a continuous operator $L^2(\mathbb{R}_+) \to \mathcal{S}(\overline{\mathbb{R}}_+)$. The operator $\varepsilon^* r^- \operatorname{op}(l_-^{\mu}) e^- \varepsilon^* : \mathcal{S}(\overline{\mathbb{R}}_+) \to \mathcal{S}(\overline{\mathbb{R}}_+)$ is also continuous, and hence $\varepsilon^* r^- \operatorname{op}(a) e^+ : L^2(\mathbb{R}_+) \to \mathcal{S}(\overline{\mathbb{R}}_+)$ is continuous. For the formal adjoint with respect to the $L^2(\mathbb{R}_+)$–scalar product we can argue in an analogous manner. Thus $\varepsilon^* r^- \operatorname{op}(a) e^+$ has a kernel in $\mathcal{S}(\overline{\mathbb{R}}_+ \times \overline{\mathbb{R}}_+)$. Also the assertion on convergence can be reduced to the case of order zero. Here it suffices to observe that when $\{a_n\}_{n \in \mathbb{N}}$ tends to zero in $S_{\mathrm{cl}}^\mu(\mathbb{R} \times \mathbb{R})_{\mathrm{e}}$ then there is a sequence of representatives for $l_-^{-\mu} \# a_n \in S_{\mathrm{cl}}^0(\mathbb{R} \times \mathbb{R})_{\mathrm{e}} \bmod S^{-\infty}(\mathbb{R} \times \mathbb{R})_{\mathrm{e}}$ that tend to zero in $S_{\mathrm{cl}}^0(\mathbb{R} \times \mathbb{R})_{\mathrm{e}}$ and that in addition the remainders $\operatorname{op}(l_-^{-\mu}) \operatorname{op}(a_n) - \operatorname{op}(l_-^{-\mu} \# a_n)$ tend to zero in the sense that their kernels in $\mathcal{S}(\mathbb{R} \times \mathbb{R})$ tend to zero for $n \to \infty$. $\qquad\square$

4.1.4 Operators of order zero with the transmission property

We now study the transmission property in more detail, in particular, under the aspect of the boundary symbolic calculus for corresponding boundary value problems. Let us first consider symbols of order zero and operators in $L^2(\mathbb{R}_+)$.

Definition 4.1.32 *Denote by $\boldsymbol{D}^0(\mathbb{R}_+)$ the space of all operators*

$$A = \operatorname{op}^+(a) + G \qquad (4.1.53)$$

for arbitrary $a(t, \tau) \in S_{\mathrm{cl}}^0(\overline{\mathbb{R}}_+ \times \mathbb{R})_{\mathrm{e}}$ with the transmission property, cf. Definition 4.1.1 and Definition 4.1.26, and integral operators G with kernels in $\mathcal{S}(\overline{\mathbb{R}}_+ \times \overline{\mathbb{R}}_+)$. Moreover, $\boldsymbol{D}^0(\mathbb{R}_+)$ denotes the subclass of all operators (4.1.53) with t–independent $a(\tau)$.

The space of all operators G with kernels in $\mathcal{S}(\overline{\mathbb{R}}_+ \times \overline{\mathbb{R}}_+)$ is denoted by $\Gamma^0(\mathbb{R}_+)$.

Recall that $\operatorname{op}^+(a) = r^+ \operatorname{op}(a) e^+$ with e^+ being the extension operator by zero to $\mathbb{R}_-$ and r^+ the restriction operator to $\mathbb{R}_+$. Note that we could also take symbols $\tilde{a}(t, \tau) \in S_{\mathrm{cl}}^0(\mathbb{R} \times \mathbb{R})_{\mathrm{e}}$. Of course, the values of $\tilde{a}(t, \tau)$ for $t < 0$ play no role in $\operatorname{op}^+(\tilde{a})$; therefore it suffices to insert $a(t, \tau) = \tilde{a}(t, \tau)|_{t>0}$.

$\boldsymbol{D}^0(\mathbb{R}_+)$ will be regarded first as a subspace of $\mathcal{L}(L^2(\mathbb{R}_+))$. The operators $G \in \Gamma^0(\mathbb{R}_+)$ are called Green operators (of type zero) in the boundary symbolic calculus. Remember that $\Gamma^0(\mathbb{R}_+)$ is characterised as the set of all $G \in \mathcal{L}(L^2(\mathbb{R}_+))$ such that

$$G : L^2(\mathbb{R}_+) \to \mathcal{S}(\overline{\mathbb{R}}_+), \qquad G^* : L^2(\mathbb{R}_+) \to \mathcal{S}(\overline{\mathbb{R}}_+) \qquad (4.1.54)$$

are continuous. Here G^* is the adjoint of G with respect to the $L^2(\mathbb{R}_+)$-scalar product.

Remark 4.1.33 *Let ω, ω_0, ω_1 be cut-off functions satisfying $\omega\omega_0 = \omega$, $\omega\omega_1 = \omega_1$. Then we have $\omega A(1 - \omega_0)$, $(1 - \omega)A\omega_1 \in \Gamma^0(\mathbb{R}_+)$ for every $A \in \boldsymbol{D}^0(\mathbb{R}_+)$.*

Let us set

$$H^+ = \left\{ f(\tau) : f(\tau) = \int\limits_0^\infty e^{-i\tau t} u(t)\, dt \text{ for some } u \in \mathcal{S}(\overline{\mathbb{R}}_+) \right\}.$$

Note that $H^+ \subset C^\infty(\mathbb{R})$ and that the definition induces a canonical isomorphism $H^+ \cong \mathcal{S}(\overline{\mathbb{R}}_+)$. Moreover, let

$$H_0^- = \left\{ g(\tau) : g(\tau) = \int\limits_{-\infty}^0 e^{-i\tau t} v(t)\, dt \text{ for some } v \in \mathcal{S}(\overline{\mathbb{R}}_-) \right\},$$

and set

$$H^- = H_0^- + H', \quad H_0 = H_0^- + H^+, \quad H = H_0 + H',$$

where H' is the space of all polynomials in τ.

The elements of H^- will be called minus functions, those of $H^+ + H'$ plus functions.

Lemma 4.1.34 *The following conditions are equivalent:*

(i) $f(\tau) \in H^+$,

(ii) *for $f(\tau)$ there exists an $h(\zeta) \in \mathcal{A}(\{\operatorname{Im}\zeta < 0\}) \cap C^\infty(\{\operatorname{Im}\zeta \leq 0\})$ with $f(\tau) = h(\tau)$ for $\tau = \operatorname{Re}\zeta$, and there is an asymptotic expansion*

$$h(\zeta) \sim \sum_{k=1}^\infty a_k \zeta^{-k} \qquad \text{for } |\zeta| \to \infty,\, \operatorname{Im}\zeta \leq 0; \qquad (4.1.55)$$

all derivatives $D_\zeta^j h(\zeta)$, $j \in \mathbb{N}$, have analogous asymptotic expansions obtained by formal differentiations of (4.1.55).

Proof. If $f(\tau) \in H^+$ is associated with $u \in \mathcal{S}(\overline{\mathbb{R}}_+)$, the function $h(\zeta) = \int_0^\infty e^{-i\zeta t} u(t)\, dt$, belongs to $\mathcal{A}(\{\operatorname{Im}\zeta < 0\}) \cap C^\infty(\{\operatorname{Im}\zeta \le 0\})$ and extends $f(\tau)$ to the negative complex half plane. Integration by parts gives us for for $\zeta \ne 0$ the relation

$$h(\zeta) = i\zeta^{-1} e^{-i\zeta t} u(t)|_0^\infty - i\zeta^{-1} \int_0^\infty e^{-i\zeta t}(\partial_t u)(t)\, dt$$

$$= -i\zeta^{-1}\{u(0) + \int_0^\infty e^{-i\zeta t}(\partial_t u)(t)\, dt\}$$

$$= \dots$$

$$= \sum_{k=1}^N (-i\zeta^{-1})^k (\partial_t^{k-1} u)(0) + (-i\zeta^{-1})^N \int_0^\infty e^{-i\zeta t}(\partial_t^N u)(t)\, dt$$

for each N. Now we can form for $h'(\zeta) = \int_0^\infty e^{-i\zeta t}(-it)u(t)\, dt$ the analogous asymptotic expansion which is in fact the same as that obtained by the formal differentiation of

$$h(\zeta) \sim \sum_{k=1}^\infty (-i\zeta^{-1})^k (\partial_t^{k-1} u)(0)$$

with respect to ζ.

For the reverse direction we assume that $f(\tau)$ satisfies (ii). Then $u(t) = (F_{\tau \to t}^{-1} f)(t) \in L^2(\mathbb{R})$. By Proposition 4.1.8 the function $u(t)$ vanishes for almost all $t < 0$. In view of the assumptions, the function $\tau^k D_\tau^l f(\tau) \in C^\infty(\mathbb{R})$ is the sum of a polynomial and a function $f_{kl}(\tau)$ satisfying the conditions of (ii) for all $k, l \in \mathbb{N}$. Thus $F^{-1}(\tau^k D_\tau^l f(\tau))$ is the sum of derivatives of the Dirac distribution at the origin and an element in $L^2(\mathbb{R})$ supported by $\overline{\mathbb{R}}_+$. Thus $D_t^k t^l u(t)|_{t>0} \in L^2(\mathbb{R}_+)$. Since k, l are arbitrary, we obtain $u \in \mathcal{S}(\overline{\mathbb{R}}_+)$. □

Remark 4.1.35 $f(\tau) \in H^+$ *is equivalent to* $\overline{f(\tau)} \in H_0^-$. *There is an analogue of* Lemma 4.1.34 *for the space* H_0^-, *where now the extensions concern the upper complex half plane.*

Remark 4.1.36 *Let* $v_- \in H^r(\mathbb{R})$, $r \in \mathbb{R}$, *be an arbitrary element with* $\operatorname{supp} v_- \subseteq \overline{\mathbb{R}}_-$. *Then for every* $p_-(\tau) \in H^-$ *we have*

$$r^+ \operatorname{op}(p_-)v_- = 0.$$

Similarly, for $p_+(\tau) \in H^+ + H'$ *we obtain that* $v_+ \in H^r(\mathbb{R})$ *and* $\operatorname{supp} v_+ \subseteq \overline{\mathbb{R}}_+$ *implies*

$$r^- \operatorname{op}(p_+)v_+ = 0.$$

This is a consequence of Proposition 4.1.8 (also in the analogous version for $H_0^s(\overline{\mathbb{R}}_-)$) and of Lemma 4.1.34, Remark 4.1.35.

Proposition 4.1.37 *Every $G \in \Gamma^0(\mathbb{R}_+)$ induces continuous operators*

$$G : H^s(\mathbb{R}_+) \to \mathcal{S}(\overline{\mathbb{R}}_+) \qquad \text{for all } s \in \mathbb{R}, \ s > -\frac{1}{2}.$$

Proof. For $s \geq 0$ the asserted continuity is a consequence of (4.1.54). For $s \leq 0$, $s > -\frac{1}{2}$, we shall show that G may be regarded as a continuous operator $G : H_0^s(\overline{\mathbb{R}}_+) \to \mathcal{S}(\overline{\mathbb{R}}_+)$. In view of $H_0^s(\overline{\mathbb{R}}_+) = H^s(\mathbb{R}_+)$ for $|s| < \frac{1}{2}$ this completes then the proof. For every function $g_0(t') \in \mathcal{S}(\overline{\mathbb{R}}_+)$ the scalar product $(g_0, \overline{u})_{L^2(\mathbb{R}_+)}$ for $u \in C_0^\infty(\mathbb{R}_+)$ extends to $u \in H_0^s(\overline{\mathbb{R}}_+)$ for $0 \geq s > -\frac{1}{2}$ via the pairing $H^{-s}(\mathbb{R}_+) \times H_0^s(\overline{\mathbb{R}}_+) \to \mathbb{C}$, since $g_0(t') \in H^{-s}(\mathbb{R}_+)$, and

$$|(g_0, \overline{u})| \leq \|g_0\|_{H^{-s}(\mathbb{R}_+)} \|u\|_{H_0^s(\overline{\mathbb{R}}_+)}. \tag{4.1.56}$$

In view of $\mathcal{S}(\overline{\mathbb{R}}_+ \times \overline{\mathbb{R}}_+) = \mathcal{S}(\overline{\mathbb{R}}_+) \otimes_\pi \mathcal{S}(\overline{\mathbb{R}}_+)$ the kernel $g(t, t')$ of G can be written as a convergent sum

$$g(t, t') = \sum_{j=0}^\infty \lambda_j f_j(t) g_j(t')$$

for $\lambda_j \in \mathbb{C}$, $\sum |\lambda_j| < \infty$, $f_j, g_j \in \mathcal{S}(\overline{\mathbb{R}}_+)$, tending to zero for $j \to \infty$, cf. Theorem 1.1.6. Then we get

$$Gu(t) = \sum_{j=0}^\infty \lambda_j f_j(t) (g_j, \overline{u}),$$

which converges first in the sense of $L^2(\mathbb{R}_+)$. Let $\{r_k\}_{k \in \mathbb{N}}$ be a seminorm system for the Fréchet topology of $\mathcal{S}(\overline{\mathbb{R}}_+)$. Then

$$r_k(Gu) \leq \left\{ \sum_{j=0}^\infty |\lambda_j| r_k(f_j) \|g_j\|_{H^{-s}(\mathbb{R}_+)} \right\} \|u\|_{H_0^s(\overline{\mathbb{R}}_+)} \leq c_k \|u\|_{H_0^s(\overline{\mathbb{R}}_+)}$$

with $c_k = \sum_{j=0}^\infty |\lambda_j| r_k(f_j) \|g_j\|_{H^{-s}(\mathbb{R}_+)} < \infty$. Here we have used (4.1.56) and $r_k(f_j) \to 0$, $\|g_j\|_{H^{-s}(\mathbb{R}_+)} \to 0$ for $j \to \infty$, where the latter relation follows from the continuous embedding $\mathcal{S}(\overline{\mathbb{R}}_+) \hookrightarrow H^{-s}(\mathbb{R}_+)$ for $s \leq 0$. $\qquad \square$

Theorem 4.1.38 *Each $A \in \mathbf{D}^0(\mathbb{R}_+)$ induces a continuous operator*

$$A : H^s(\mathbb{R}_+) \to H^s(\mathbb{R}_+)$$

for every $s \in \mathbb{R}$, $s > -\frac{1}{2}$, and a continuous operator

$$A : \mathcal{S}(\overline{\mathbb{R}}_+) \to \mathcal{S}(\overline{\mathbb{R}}_+). \tag{4.1.57}$$

Proof. Let $\omega, \omega_0, \omega_1$ be cut–off functions with $\omega\omega_0 = \omega$, $\omega\omega_1 = \omega_1$. Then

$$A = \omega A \omega_0 + (1 - \omega)A(1 - \omega_1) + G \qquad (4.1.58)$$

for some $G \in \Gamma^0(\mathbb{R}_+)$, cf. Remark 4.1.33. From $a(t, \tau) \in S^0_{\text{cl}}(\overline{\mathbb{R}}_+ \times \mathbb{R})_e$ and 1.4.2 Theorem 1.4.17 it follows that $(1 - \omega)A(1 - \omega_1) : H^s(\mathbb{R}_+) \to H^s(\mathbb{R}_+)$ is continuous for all $s \in \mathbb{R}$. The operator G was treated in Proposition 4.1.37. It remains to consider $\omega A \omega_0$, where we may concentrate on $\omega \operatorname{op}^+(a)\omega_0$. From Lemma 4.1.22 and Definition 4.1.26 we get for $u \in C_0^\infty(\mathbb{R}_+)$

$$\omega \operatorname{op}^+(a)\omega_0 u(t) = \omega \sum_{j=0}^N E_j^+ \operatorname{op}^+(l_-^{-j})\omega_0 u(t) + \omega C_N \omega_0 u(t) \qquad (4.1.59)$$

for every N. Here we used $g^+ + g^- = 1$. The operator $\operatorname{op}^+(l_-^{-j})\omega_0$ can equivalently be written $r^+ \operatorname{op}(l_-^{-j})e_s^+\omega_0$ for an arbitrary extension operator $e_s^+ : H^s(\mathbb{R}_+) \to H^s(\mathbb{R}_+)$, cf. Proposition 4.1.10. Since the multiplication by $\omega E_j^+ \in C_0^\infty(\overline{\mathbb{R}}_+)$ induces continuous operators in $H^s(\mathbb{R}_+)$, we get that $\omega E_j^+ \operatorname{op}^+(l_-^{-j})\omega_0$ extend to continuous operators in $H^s(\mathbb{R}_+)$ for all $s \in \mathbb{R}$ and all j. To treat C_N we choose N so large that C_N has a kernel in $C^m(\overline{\mathbb{R}}_+, C^m(\overline{\mathbb{R}}_+))$ which is a subspace of $C^m(\overline{\mathbb{R}}_+, L^2(\mathbb{R}_+))$ or of $C^m(\overline{\mathbb{R}}_+, H^{-s}(\mathbb{R}_+))$ for $0 \geq s > -\frac{1}{2}$. To prove the continuity of $\omega C_N \omega_0 : H^s(\mathbb{R}_+) \to H^s(\mathbb{R}_+)$ for $s \geq 0$ we can use the first interpretation, when we choose m so large that $C^m(\overline{\mathbb{R}}_+) \hookrightarrow H^s(\mathbb{R}_+)$ is continuous. For $0 \geq s > -\frac{1}{2}$ we employ the second interpretation. Let $c(t, t')$ be the kernel of $\omega C_N \omega_0$. Then, using the duality between $H^{-s}(\mathbb{R}_+)$ and $H_0^s(\overline{\mathbb{R}}_+) = H^s(\mathbb{R}_+)$ for those s we get

$$\sup_t |(D_t^l c(t, t'), \overline{u}(t'))| \leq \sup_t \|D_t^l c(t, t')\|_{H^{-s}(\mathbb{R}_{+,t'})} \|u\|_{H^s(\mathbb{R}_+)}$$

for every $0 \leq l \leq m$, because of the continuity of $D_t^l c(t, t')$ in the first argument as $H^{-s}(\mathbb{R}_{+,t'})$–valued function. It follows altogether that $\omega C_N \omega_0 : H^s(\mathbb{R}_+) \to C^m(\overline{\mathbb{R}}_+)$ is continuous for the above s which completes the proof. □

Theorem 4.1.39 *Every $A \in \boldsymbol{D}^0(\mathbb{R}_+)$ induces continuous operators*

$$A : \mathcal{K}^{s,\gamma}(\mathbb{R}_+) \to \mathcal{K}^{s,\gamma}(\mathbb{R}_+) + \mathcal{S}(\overline{\mathbb{R}}_+),$$

$$A : \mathcal{K}_P^{s,\gamma}(\mathbb{R}_+) \to \mathcal{K}_P^{s,\gamma}(\mathbb{R}_+) + \mathcal{S}(\overline{\mathbb{R}}_+)$$

for all $s \in \mathbb{R}$, $\gamma \in \mathbb{R}$, $\gamma > -\frac{1}{2}$, for every $P \in As(\gamma, \Theta)$ for arbitrary $\Theta = (\vartheta, 0]$, $\vartheta < 0$ (satisfying the shadow condition).

Proof. Write A in the form (4.1.58); then the summand in the middle induces continuous operators

$$(1-\omega)A(1-\omega_1) : \mathcal{K}^{s,\gamma}(\mathbb{R}_+) \to \mathcal{K}^{s,\gamma}_\Theta(\mathbb{R}_+) \qquad \text{for all } s,\gamma \in \mathbb{R}.$$

Thus it remains to consider $\omega A\omega_0$ and G. The property

$$G : \mathcal{K}^{s,\gamma}(\mathbb{R}_+) \to \mathcal{S}(\overline{\mathbb{R}}_+)$$

for $s \in \mathbb{R}$, $\gamma > -\frac{1}{2}$ follows from the fact that $\mathcal{S}(\overline{\mathbb{R}}_+) \subset \mathcal{K}^{\infty,\beta}(\mathbb{R}_+)$ holds for $\beta > \frac{1}{2}$ and from the identification $\mathcal{S}(\overline{\mathbb{R}}_+ \times \overline{\mathbb{R}}_+) = \mathcal{S}(\overline{\mathbb{R}}_+) \otimes_\pi \mathcal{S}(\overline{\mathbb{R}}_+)$ which allows the action of G on $\mathcal{K}^{s,\gamma}(\mathbb{R}_+)$ by the pairing $\mathcal{K}^{\infty,-\gamma}(\mathbb{R}_+) \times \mathcal{K}^{s,\gamma}(\mathbb{R}_+) \to \mathbb{C}$. Concerning $\omega A\omega_0$ we may then assume $A = \mathrm{op}^+(a)$ for an $a(t,\tau) \in S^0_{\mathrm{cl}}(\overline{\mathbb{R}}_+ \times \mathbb{R})_e$ with the transmission property. Lemma 4.1.16 gives us

$$\mathrm{op}^+(a)u(t) = \sum_{j=0}^{N} A_k^+(t,\delta)\,\mathrm{op}^+(l_+^{-j})u(t) + C_N u(t)$$

for an operator C_N with kernel in $C^m(\overline{\mathbb{R}}_+ \times \overline{\mathbb{R}}_+)$ and $m(N) \to \infty$ for $N \to \infty$. Assume for convenience $m = 4k$ for some $k \in \mathbb{N}$. Write the kernel $c_N(t,t')$ of C_N in the form

$$c_N(t,t') = \sum_{j=0}^{k} d_j(t)(t')^j + d_{(k)}(t,t')(t')^k$$

$$= \sum_{j=0}^{k}\sum_{l=0}^{k} t^l d_{jl}(t')^j + \sum_{j=0}^{k} t^k d_{j,(k)}(t)(t')^j$$

$$+ \sum_{l=0}^{k} t^l d_{(k),l}(t')(t')^k + t^k d_{(k),(k)}(t,t')(t')^k.$$

The functions $d_{j,(k)}(t)$, $d_{(k),l}(t')$, $d_{(k),(k)}(t,t')$ are elements of $C^{2k}(\overline{\mathbb{R}}_+)$ and $C^{2k}(\overline{\mathbb{R}}_+ \times \overline{\mathbb{R}}_+)$, respectively. The integral operators with kernels $\omega(t)t^l d_{jl}(t')^j \omega_0(t')$ belong to $\Gamma^0(\mathbb{R}_+)$ and therefore have the desired mapping properties. Concerning the operators with kernels $\omega(t)t^k d_{j,(k)}(t)(t')^j \omega(t')$ we can first carry out the integrations $\int_0^\infty (t')^j \omega(t')u(t')\,dt'$ for the argument functions in question, for the same reason as for the operators in $\Gamma^0(\mathbb{R}_+)$ (where for $s < 0$ this is to be understood as the above pairing). Then the composition by $\omega(t)t^k$ gives rise to a shift of the resulting weight, such that

$$u \to \omega(t)t^k \int_0^\infty (t')^j \omega(t')u(t')\,dt'$$

corresponds to a continuous operator $\mathcal{K}^{s,\gamma}(\mathbb{R}_+) \to \mathcal{K}^{s,\gamma+k}(\mathbb{R}_+)$ for every $s \in \mathbb{R}$ and $\gamma > -\frac{1}{2}$. Finally the multiplication by $d_{j,(k)}(t)\omega_0(t)$ for a cut–off function $\omega_0(t)$ with $\omega_0\omega = \omega$ is continuous in $\mathcal{K}^{s,\gamma+k}(\mathbb{R}_+)$ provided $k = k(s)$ was chosen sufficiently large. For the operators with kernels $\omega(t)t^l d_{(k),l}(t')(t')^k \omega_0(t)$ we can argue in an analogous manner. The operator with kernel in $\omega(t)t^k d_{(k),(k)}(t,t')(t')^k \omega_0(t')$ is simple as well, since for sufficiently large k it suffices to know that

$$u \to \omega(t) \int\limits_0^\infty d_{(k),(k)}(t,t')\omega_0(t')u(t')\,dt'$$

is continuous in $\mathcal{K}^{s,0}(\mathbb{R}_+)$ for every s. For $s \in \mathbb{N}$ fixed this is a trivial calculation, where $k = k(s)$ is chosen in a suitable way; then by duality and interpolation we get the continuity for all $s \in \mathbb{R}$, $|s| < s_0$, for every fixed s_0, for $k \geq k(s_0)$. It remains to consider $\omega \operatorname{op}^+(l_+^{-j})\omega_0$, since the multiplication operator by $\omega A_k^+(t,\delta)\tilde{\omega} \in C_0^\infty(\overline{\mathbb{R}}_+)$ has the desired property. Now we can use Proposition 4.1.15. The remainder G_N can be treated as before such that we have to show

$$\omega \operatorname{op}^\gamma(f_k)\omega_0 : \mathcal{K}^{s,\gamma}(\mathbb{R}_+) \to \mathcal{K}^{s,\gamma}(\mathbb{R}_+), \qquad k \in \mathbb{N},$$

and the same with subscript P, for all $s \in \mathbb{R}$, $\gamma > -\frac{1}{2}$. This is already known from the cone theory, since $f_k(z)$ is a Mellin symbol of order zero which is holomorphic for $\operatorname{Re} z < 1$. $\square$

Theorem 4.1.40 $A \in \boldsymbol{D}^0(\mathbb{R}_+)$ *implies* $A^* \in \boldsymbol{D}^0(\mathbb{R}_+)$ *for the adjoint with respect to* $(\cdot,\cdot)_{L^2(\mathbb{R}_+)}$.

Proof. Let $A = \operatorname{op}^+(a) + G$ with $a(t,\tau) \in S_{\mathrm{cl}}^0(\overline{\mathbb{R}}_+ \times \mathbb{R})_e$ having the transmission property, and $G \in \Gamma^0(\mathbb{R}_+)$. Then $G^* \in \Gamma^0(\mathbb{R}_+)$ is obvious. Moreover, if $u,v \in C_0^\infty(\mathbb{R}_+)$, we have

$$(\operatorname{op}^+(a)u,v)_{L^2(\mathbb{R}_+)} = \int \{ r^+ \iint e^{i(t-t')\tau}a(t,\tau)u(t')\,dt'd\tau\}\overline{v(t)}\,dt$$

$$= \int u(t')\left\{ \overline{r^+ \iint e^{i(t'-t)\tau}\overline{a(t,\tau)}v(t)\,dtd\tau} \right\} dt'.$$

Here we have identified e^+u with u for $u \in C_0^\infty(\mathbb{R}_+)$. Using the asymptotic formula to express adjoints of pseudo–differential operators

$$a^{(*)}(t,\tau) \sim \sum_k \frac{1}{k!}D_\tau^k \partial_t^k \overline{a(t,\tau)}$$

we then obtain

$$r^+ \iint e^{i(t'-t)\tau}\overline{a(t,\tau)}v(t)\,dtd\tau = \operatorname{op}^+(a^{(*)})v + Gv$$

for some $G \in \Gamma^0(\mathbb{R}_+)$ which shows the assertion. $\square$

Theorem 4.1.41 $A, B \in D^0(\mathbb{R}_+)$ *implies* $AB \in D^0(\mathbb{R}_+)$. *Here* $A \in \Gamma^0(\mathbb{R}_+)$ *or* $B \in \Gamma^0(\mathbb{R}_+)$ *implies* $AB \in \Gamma^0(\mathbb{R}_+)$.

Proof. First note that $A \in \Gamma^0(\mathbb{R}_+)$ or $B \in \Gamma^0(\mathbb{R}_+)$ implies $AB \in \Gamma^0(\mathbb{R}_+)$. In fact, this is a consequence of Theorem 4.1.38 for $s = 0$ together with (4.1.57), and of the characterisation of $\Gamma^0(\mathbb{R}_+)$ in terms of the mapping properties (4.1.54). Thus we may assume $A = \mathrm{op}^+(a)$, $B = \mathrm{op}^+(b)$ for certain $a(t,\tau)$, $b(t,\tau) \in S^0_{\mathrm{cl}}(\overline{\mathbb{R}}_+ \times \mathbb{R})_e$ with the transmission property. The Leibniz product $(a \# b)(t, \tau) \in S^0_{\mathrm{cl}}(\overline{\mathbb{R}}_+ \times \mathbb{R})_e$ then also has the transmission property, cf. Proposition 4.1.29, and we have

$$\mathrm{op}^+(a) \, \mathrm{op}^+(b) = r^+ \, \mathrm{op}(a) \, \mathrm{op}(b) e^+ - r^+ \, \mathrm{op}(a) \theta^- \, \mathrm{op}(b) e^+$$
$$= r^+ \, \mathrm{op}(a \# b) e^+ + D - r^+ \, \mathrm{op}(a) \theta^- \, \mathrm{op}(b) e^+$$

for some $D \in \Gamma^0(\mathbb{R}_+)$. Similarly to (4.1.50) we get

$$r^+ \, \mathrm{op}(a) \theta^- \, \mathrm{op}(b) e^+ = \{\varepsilon^* r^- \, \mathrm{op}(\breve{a}) e^+\} \{\varepsilon^* r^- \, \mathrm{op}(b) e^+\} \qquad (4.1.60)$$

for $\breve{a}(t,\tau) = a(-t, -\tau)$. It is now a consequence of Proposition 4.1.31 that (4.1.60) belongs to $\Gamma^0(\mathbb{R}_+)$. $\qquad\square$

Remark 4.1.42 *It can easily be proved that (with the notation of the proof of Theorem 4.1.41)*

$$g := \mathrm{op}^+(a) \, \mathrm{op}^+(b) - \mathrm{op}^+(a \# b)$$

tends to zero in the Fréchet topology of $\Gamma^0(\mathbb{R}_+)$ *when* $a(t,\tau)$ *or* $b(t,\tau)$ *tends to zero in* $S^0_{\mathrm{cl}}(\overline{\mathbb{R}}_+ \times \mathbb{R})_e$, *using a suitable choice of* $a \# b$, *continuously dependent on* a *and* b.

Proposition 4.1.43 *Let* $G \in \Gamma^0(\mathbb{R}_+)$ *be given such that*

$$1 + G : L^2(\mathbb{R}_+) \to L^2(\mathbb{R}_+)$$

is invertible. Then $(1 + G)^{-1} = 1 + C$ *for some* $C \in \Gamma^0(\mathbb{R}_+)$.

Proof. Writing $(1 + G)^{-1} = 1 + C$ we get $C = -G(1 + C)$ and $C^* = -G^*(1 + C^*)$. Thus (4.1.54) implies the analogous mapping properties for C and C^*, i.e., $C \in \Gamma^0(\mathbb{R}_+)$. $\qquad\square$

Note that every $G \in \Gamma^0(\mathbb{R}_+)$ is compact as operator $G : L^2(\mathbb{R}_+) \to L^2(\mathbb{R}_+)$. Thus $1 + G : L^2(\mathbb{R}_+) \to L^2(\mathbb{R}_+)$ is a Fredholm operator, and $\mathrm{ind}(1 + G) = 0$.

Lemma 4.1.44 *Let* $A : L^2(\mathbb{R}_+) \to L^2(\mathbb{R}_+)$ *be a Fredholm operator with* $\mathrm{ind}\, A = 0$. *Then there exists a finite–dimensional operator* $G \in \Gamma^0(\mathbb{R}_+)$ *such that* $A + G : L^2(\mathbb{R}_+) \to L^2(\mathbb{R}_+)$ *is an isomorphism.*

Proof. Let $k = \dim \ker A$ and choose an orthonormal base $\{f_1, \ldots, f_k\}$ in $\ker A$ and an orthonormal base $\{g_1, \ldots, g_k\}$ in the orthogonal complement of $\operatorname{im} A$ in $L^2(\mathbb{R}_+)$. Then the operator

$$G_0 u(t) = \sum_{j=0}^{k} g_j(t)(u, f_j)_{L^2(\mathbb{R}_+)} \qquad \text{for } u \in L^2(\mathbb{R}_+) \qquad (4.1.61)$$

has the property that $A + G_0$ is an isomorphism in $L^2(\mathbb{R}_+)$. The space $C_0^\infty(\mathbb{R}_+)$ is dense in $L^2(\mathbb{R}_+)$. By approximating f_j, g_j by $C_0^\infty(\mathbb{R}_+)$ functions we get an operator $G_\varepsilon \in \Gamma^0(\mathbb{R}_+)$ that can be chosen in such a way that $\|G_\varepsilon - G_0\|_{\mathcal{L}(L^2(\mathbb{R}_+))} < \varepsilon$ for every choice of $\varepsilon > 0$. Since the isomorphisms in a Hilbert space form an open set, we get the assertion for sufficiently small $\varepsilon > 0$. $\qquad\square$

4.1.5 Green operators of type d

Definition 4.1.45 *An operator of the form*

$$Gu(t) = \sum_{k=0}^{d} \int_0^\infty g_k(t, t') \left(\frac{d}{dt'}\right)^k u(t') \, dt', \qquad (4.1.62)$$

$u \in \mathcal{S}((\overline{\mathbb{R}}_+))$ *with* $g_k(t, t') \in \mathcal{S}(\overline{\mathbb{R}}_+ \times \overline{\mathbb{R}}_+)$, $k = 0, \ldots, d$, *is called a Green operator of type d (in the boundary symbolic calculus). The space of all these operators is denoted by* $\Gamma^d(\mathbb{R}_+)$.

In other words, by definition every $G \in \Gamma^d(\mathbb{R}_+)$ is of the form

$$G = \sum_{k=0}^{d} G_k \left(\frac{d}{dt}\right)^k$$

for certain $G_k \in \Gamma^0(\mathbb{R}_+)$.

Proposition 4.1.46 *Every* $G \in \Gamma^d(\mathbb{R}_+)$ *for $d \geq 1$ has the form*

$$Gu(t) = \int_0^\infty f(t, t') u(t') \, dt' + \sum_{j=0}^{d-1} \left(\frac{d}{dt}\right)^j u(t)\big|_{t=0} w_j(t) \qquad (4.1.63)$$

for certain $f(t, t') \in \mathcal{S}(\overline{\mathbb{R}}_+ \times \overline{\mathbb{R}}_+)$, $w_j(t) \in \mathcal{S}(\overline{\mathbb{R}}_+)$, $j = 0, \ldots, d-1$. *The functions* $f(t, t') \in \mathcal{S}(\overline{\mathbb{R}}_+ \times \overline{\mathbb{R}}_+)$ *and* $w_j(t) \in \mathcal{S}(\overline{\mathbb{R}}_+)$, $j = 0, \ldots, d-1$, *are uniquely determined by G. Conversely, every operator of the form (4.1.63) belongs to* $\Gamma^d(\mathbb{R}_+)$.

Proof. If $G \in \Gamma^d(\mathbb{R}_+)$ is given in the form (4.1.62), we can write by integration by parts

$$Gu(t) = \int_0^\infty g_0(t,t')u(t')\,dt' + \sum_{k=1}^d \int_0^\infty \left\{ -\frac{d}{dt'}g_k(t,t') \right\} \left(\frac{d}{dt'} \right)^{k-1} u(t')\,dt'$$

$$+ \sum_{k=1}^d g_k(t,t')|_{t'=0} \left(\frac{d}{dt'} \right)^{k-1} u(t')|_{t'=0}. \qquad (4.1.64)$$

The integrals on the right of (4.1.64) for $k > 1$ can be reformulated again by integration by parts. Then iterating the procedure for the remaining integrals we get the form (4.1.63) after finitely many steps. To prove the inverse relation we first show that for every $w(t) \in \mathcal{S}(\overline{\mathbb{R}}_+)$ there is a $g(t,t') \in \mathcal{S}(\overline{\mathbb{R}}_+ \times \overline{\mathbb{R}}_+)$ which satisfies

$$u(0)w(t) = \int_0^\infty g(t,t') \left(\frac{d}{dt'}u(t') \right) dt' + \int_0^\infty \left(\frac{d}{dt'}g(t,t') \right) u(t')\,dt'.$$

In fact, for every such g we have necessarily

$$(gu)|_0^\infty = \int_0^\infty \left(\frac{d}{dt'}g(t,t') \right) u(t')\,dt' + \int_0^\infty g(t,t')\frac{d}{dt'}u(t')\,dt'.$$

Thus it suffices to choose an arbitrary $g(t,t') \in \mathcal{S}(\overline{\mathbb{R}}_+ \times \overline{\mathbb{R}}_+)$ with $g(t,0) = w(t)$. Analogous arguments apply for $((\frac{d}{dt})^j u)(0)w_j(t)$ for all j. This yields the first assertion. In order to show the uniqueness of $f(t,t')$ and $w_0(t), \ldots, w_{d-1}(t)$ in the representation (4.1.63) we assume that

$$Gu(t) = \int_0^\infty \tilde{f}(t,t')u(t')\,dt' + \sum_{j=0}^{d-1} \left(\frac{d}{dt} \right)^j u(t)|_{t=0}\tilde{w}_j(t)$$

is another such representation of G. Then

$$\int_0^\infty (f - \tilde{f})(t,t')u(t')\,dt' + \sum_{j=0}^{d-1} \left(\frac{d}{dt} \right)^j u(t)|_{t=0}(w_j - \tilde{w}_j)(t) = 0 \quad (4.1.65)$$

for all $u \in \mathcal{S}(\overline{\mathbb{R}}_+)$. Applying the latter relation to all $u \in C_0^\infty(\mathbb{R}_+)$ we obtain

$$\int_0^\infty (f - \tilde{f})(t,t')u(t')\,dt' = 0$$

which implies then the same for all $u \in L^2(\mathbb{R}_+)$, since $C_0^\infty(\mathbb{R}_+)$ is dense in $L^2(\mathbb{R}_+)$. Thus $f = \tilde{f}$. Next we can apply (4.1.65) to $u(t) = w(t)$ for a cut–off function $\omega(t)$. This yields $w_0(t) = \tilde{w}_0(t)$. Inserting $u(t) = t^j\omega(t)$ we get analogously $w_j(t) = \tilde{w}_j(t)$ for all $j = 0, \ldots, d - 1$. $\quad\square$

Note that the formula (4.1.63) can be written

$$G = G_0 + \sum_{j=0}^{d-1} T_j \circ K_j \qquad (4.1.66)$$

for a $G_0 \in \Gamma^0(\mathbb{R}_+)$, $G_0 u(t) = \int_0^\infty f(t,t')u(t')\,dt'$, and operators

$$K_j : \mathbb{C} \to \mathcal{S}(\overline{\mathbb{R}}_+), \quad K_j c = cw_j(t) \qquad \text{for } c \in \mathbb{C}$$

with the given $w_j \in \mathcal{S}(\overline{\mathbb{R}}_+)$,

$$T_j : \mathcal{S}(\overline{\mathbb{R}}_+) \to \mathbb{C}, \quad T_j u = \frac{d^j}{dt^j}u(t)|_{t=0} \qquad \text{for } u \in \mathcal{S}(\overline{\mathbb{R}}_+).$$

K_j and T_j are special potential and trace operators, respectively, in the boundary symbolic calculus .

Proposition 4.1.47 *Each $G \in \Gamma^d(\mathbb{R}_+)$ extends to a continuous operator*

$$G : H^s(\mathbb{R}_+) \to \mathcal{S}(\overline{\mathbb{R}}_+) \qquad \text{for every } s \in \mathbb{R}, \ s > d - \frac{1}{2}.$$

Proof. The assertion is a consequence of Proposition 4.1.37, Proposition 4.1.46 and of the fact that $u \to \frac{d}{dt}u|_{t=0}$ is continuous as a map $H^s(\mathbb{R}_+) \to \mathbb{C}$ for $s > \frac{1}{2}$. $\quad\square$

Lemma 4.1.48 *Let $G \in \Gamma^{\tilde{d}}(\mathbb{R}_+)$, and assume that for a function $u \in H^s(\mathbb{R}_+)$, $s > d - \frac{1}{2}$, for some $d \leq \tilde{d}$ we have $Gu \in H^s(\mathbb{R}_+)$ and that $(1 + G)u \in \mathcal{S}(\overline{\mathbb{R}}_+)$ holds. Then $u \in \mathcal{S}(\overline{\mathbb{R}}_+)$.*

Proof. Applying Proposition 4.1.46 we write G in the form (4.1.63) for $\tilde{d}$ instead of d. Set $G_0 u(t) = \int_0^\infty f(t,t')u(t')\,dt'$. Then, according to Proposition 4.1.47 we have $G_0 u \in \mathcal{S}(\overline{\mathbb{R}}_+)$. Thus, if we set $G_1 = G - G_0$, we get $(1 + G_1)u \in \mathcal{S}(\overline{\mathbb{R}}_+)$. In other words, without loss of generality we assume in the proof $G_0 = 0$. Choose an orthonormal base $\{h_0, \ldots, h_{\tilde{d}-1}\}$ with respect to $(\cdot, \cdot)_{H^s(\mathbb{R}_+)}$ in the linear span of $\{w_0, \ldots, w_{\tilde{d}-1}\}$, $h_j \in \mathcal{S}(\overline{\mathbb{R}}_+)$, $j = 0, \ldots, \tilde{d} - 1$. Then G_1 takes the form

$$G_1 u(t) = \sum_{j=0}^{\tilde{d}-1} h_j \left\{ \sum_{l=0}^{\tilde{d}-1} c_{jl} \left(\frac{d}{dt} \right)^l u|_{t=0} \right\}$$

with constants c_{jl}. We obtain

$$(h_k, G_1 u)_{H^s(\mathbb{R}_+)} = \sum_{l=0}^{\tilde{d}-1} c_{kl} \left(\frac{d}{dt}\right)^l u|_{t=0}.$$

The right hand side exists as a complex number. This implies necessarily $c_{kl} = 0$ for all $l > d-1$ and all k, and $G_1 u \in \mathcal{S}(\overline{\mathbb{R}}_+)$. Thus, since we have by assumption $(1 + G_1)u \in \mathcal{S}(\overline{\mathbb{R}}_+)$, it follows that $u \in \mathcal{S}(\overline{\mathbb{R}}_+)$. □

4.1.6 Operators of higher order with the transmission property

An operator convention in this section is a rule

$$a(t, \tau) \to \mathrm{op}^+(a)$$

to associate an operator with a symbol $a(t, \tau) \in S_{\mathrm{cl}}^\mu(\overline{\mathbb{R}}_+ \times \mathbb{R}_+)$ with the transmission property. Operator conventions are usually not canonical, except for polynomials $a(t, \tau) = \sum_{j=0}^\mu a_j(t)\tau^j$, where we set

$$\mathrm{op}^+(a)u(t) = \sum_{j=0}^\mu a_j(t)D_t^j u(t).$$

For a symbol $a(t, \tau) \in S_{\mathrm{cl}}^\mu(\overline{\mathbb{R}}_+ \times \mathbb{R})$ with the transmission property we define

$$\mathrm{op}^+(a) = r^+ \mathrm{op}(a)e^+, \tag{4.1.67}$$

acting on $H^s(\mathbb{R}_+)$ for $s \in \mathbb{R}$, $s > \frac{1}{2}$, where

$$e^+ : H^s(\mathbb{R}_+) \to \mathcal{S}'(\mathbb{R})$$

is the extension operator by zero. The rule (4.1.67) looks rather artificial at first glance, since, for instance on $\mathcal{S}(\overline{\mathbb{R}}_+)$, the operator e^+ produces a jump at the origin, which affects the nature of the operation when it is not a differentiation composed with r^+. For that reason we shall also permit other choices of operator conventions, in particular, for $\mu \in \mathbb{N}$, where we set

$$\mathrm{op}^+(a) = \mathrm{op}^+(p)\,\mathrm{op}^+(p^{-1}\#a) \tag{4.1.68}$$

for an arbitrary polynomial $p(\tau)$ of order μ with $p^{-1}(\tau) \in S^{-\mu}(\mathbb{R})$ and the Leibniz product $\#$, such that $(p^{-1}\#a)(t, \tau) \in S_{\mathrm{cl}}^0(\overline{\mathbb{R}}_+ \times \mathbb{R})$. If b is an arbitrary symbol of order ≤ 0 we usually set $\mathrm{op}^+(b) = r^+ \mathrm{op}(b)e^+$. The various choices of p will affect $\mathrm{op}^+(a)$ only modulo Green operators of some type d, cf. Remark 4.1.55 below. In particular, we can set

$$p(\tau) = (1 - i\tau)^\mu \qquad \text{for } \mu \in \mathbb{N}. \tag{4.1.69}$$

In this case we have

$$\mathrm{op}^+(p)\,\mathrm{op}^+(p^{-1}\#a) = r^+\,\mathrm{op}(p)\,\mathrm{op}(p^{-1}\#a)e^+$$
$$= r^+\,\mathrm{op}(a)e^+ \bmod \Gamma^0(\mathbb{R}_+)$$

for each symbol $a(t,\tau) \in S^0_{\mathrm{cl}}(\overline{\mathbb{R}}_+ \times \mathbb{R})_e$ with the transmission property. To understand how the Green operators of positive types appear in the operations of the calculus, it is convenient to discuss representations of the operators in the form (4.1.68) for arbitrary $p(\tau)$.

Lemma 4.1.49 *Let $p(\tau)$ and $q(\tau)$ be polynomials of order m and n, respectively, $m \geq n$, and let $p(\tau) \neq 0$ for all $\tau \in \mathbb{R}$, $p^{-1}(\tau) \in S^{-m}_{\mathrm{cl}}(\mathbb{R})$. Then we have*

$$\mathrm{op}^+(p^{-1})\,\mathrm{op}^+(q) - \mathrm{op}^+(p^{-1}q) \in \Gamma^n(\mathbb{R}_+).$$

Proof. For $r(\tau) := (1 - i\tau)^n$ we have

$$\mathrm{op}^+(q) = \mathrm{op}^+(q)\,\mathrm{op}^+(r^{-1})\,\mathrm{op}^+(r) = \mathrm{op}^+(qr^{-1})\,\mathrm{op}^+(r).$$

Theorem 4.1.41 implies

$$\mathrm{op}^+(p^{-1}q) = \mathrm{op}^+(p^{-1}q)\,\mathrm{op}^+(r^{-1})\,\mathrm{op}^+(r)$$
$$= (\mathrm{op}^+(p^{-1}qr^{-1}) + G_1)\,\mathrm{op}^+(r)$$

for a certain $G_1 \in \Gamma^0(\mathbb{R}_+)$ and

$$\mathrm{op}^+(p^{-1})\,\mathrm{op}^+(qr^{-1}) = \mathrm{op}^+(p^{-1}qr^{-1}) + G_2$$

for some $G_2 \in \Gamma^0(\mathbb{R}_+)$. Thus we get

$$\mathrm{op}^+(p^{-1})\,\mathrm{op}^+(q) - \mathrm{op}^+(p^{-1}q)$$
$$= \mathrm{op}^+(p^{-1})\,\mathrm{op}^+(qr^{-1})\,\mathrm{op}^+(r) - (\mathrm{op}^+(p^{-1}qr^{-1}) + G_1)\,\mathrm{op}^+(r)$$
$$= (-G_1 + G_2)\,\mathrm{op}^+(r) \in \Gamma^n(\mathbb{R}_+).$$

$\square$

Lemma 4.1.50 *Let $G \in \Gamma^d(\mathbb{R}_+)$ be given and $a_0(t,\tau) \in S^0_{\mathrm{cl}}(\overline{\mathbb{R}}_+ \times \mathbb{R})_e$ be a symbol with the transmission property. Then*

$$G\,\mathrm{op}^+(a_0) \in \Gamma^d(\mathbb{R}_+). \tag{4.1.70}$$

Moreover, for every polynomial $p(\tau)$ of order m we have

$$\mathrm{op}^+(p)G \in \Gamma^d(\mathbb{R}_+).$$

Proof. The second relation is obvious. To prove (4.1.70) we set $r(\tau) = (1 - i\tau)^d$. By assuption G is a linear combination of operators of the form $G_0 \operatorname{op}^+(p)$ for certain $G_0 \in \Gamma^0(\mathbb{R}_+)$ and polynomials $p(\tau)$ of order $\leq d$. Using $\operatorname{op}^+(r^{-1}) \operatorname{op}^+(r) = 1$ we obtain

$$G_0 \operatorname{op}^+(p) \operatorname{op}^+(a_0) = G_0 \operatorname{op}^+(p) \operatorname{op}^+(a_0) \operatorname{op}^+(r^{-1}) \operatorname{op}^+(r)$$
$$= G_0 \operatorname{op}^+(p)\{\operatorname{op}^+(a_0 r^{-1}) + G_1\} \operatorname{op}^+(r)$$

for some $G_1 \in \Gamma^0(\mathbb{R}_+)$, cf. Theorem 4.1.41, and $\operatorname{op}^+(p) \operatorname{op}^+(a_0 r^{-1}) = \operatorname{op}^+(p\#(a_0 r^{-1})) + G_2$ for some $G_2 \in \Gamma^0(\mathbb{R}_+)$. Since $p(\tau)$ is a polynomial, we can set $G_2 = 0$ because the asymptotic sum for the Leibniz product is finite. This yields $G_0 \operatorname{op}^+(p) \operatorname{op}^+(a_0) = G_3 \operatorname{op}^+(r)$ for $G_3 = G_0\{\operatorname{op}^+(p\#(a_0 r^{-1}) + G_1\} \in \Gamma^0(\mathbb{R}_+)$. $\qquad\square$

Theorem 4.1.51 *Let* $a(t, \tau) \in S_{cl}^\mu(\overline{\mathbb{R}}_+ \times \mathbb{R})_e$ *be a symbol with the transmission property, let* $p(\tau)$ *and* $q(\tau)$ *be polynomials of order* m *and* n, *respectively, satisfying* $p^{-1}(\tau) \in S_{cl}^{-m}(\mathbb{R})$, $q^{-1}(\tau) \in S_{cl}^{-n}(\mathbb{R})$, *and let* $\mu \leq m$, $\mu \leq n$. *Then*

$$\operatorname{op}^+(p) \operatorname{op}^+(p^{-1}\#a) - \operatorname{op}^+(q) \operatorname{op}^+(q^{-1}\#a) \in \Gamma^0(\mathbb{R}_+).$$

Proof. Let us set $r(\tau) = (1 - i\tau)^m$ with $m \geq n$. Then, since $r(\tau)$, $r^{-1}(\tau)$ are minus functions, we have

$$\operatorname{op}^+(r) \operatorname{op}^+(r^{-1}) = \operatorname{op}^+(r^{-1}) \operatorname{op}^+(r) = 1.$$

Thus

$$\operatorname{op}^+(q) \operatorname{op}^+(q^{-1}\#a) = \operatorname{op}^+(r) \operatorname{op}^+(r^{-1}) \operatorname{op}^+(q) \operatorname{op}^+(q^{-1}\#a)$$
$$= \operatorname{op}^+(r) \operatorname{op}^+(r^{-1}q) \operatorname{op}^+(q^{-1}\#a),$$

where we have used $\operatorname{op}^+(r^{-1}) \operatorname{op}^+(q) = \operatorname{op}^+(r^{-1}q)$, that holds since $r^{-1}(\tau)$ is a minus function, cf. Remark 4.1.36. This yields

$$\operatorname{op}^+(q) \operatorname{op}^+(q^{-1}\#a) = \operatorname{op}^+(r)\{\operatorname{op}^+(r^{-1}\#a) + G\}$$

for some $G \in \Gamma^0(\mathbb{R}_+)$, cf. Theorem 4.1.41. From Lemma 4.1.50 we finally obtain

$$\operatorname{op}^+(q) \operatorname{op}^+(q^{-1}\#a) = \operatorname{op}^+(r) \operatorname{op}^+(r^{-1}\#a) + G_0$$

for a $G_0 \in \Gamma^0(\mathbb{R}_+)$. Replacing q by p we get an anlogous relation, namely

$$\operatorname{op}^+(p) \operatorname{op}^+(p^{-1}\#a) = \operatorname{op}^+(r) \operatorname{op}^+(r^{-1}\#a) + G_1$$

for a $G_1 \in \Gamma^0(\mathbb{R}_+)$. This yields the assertion. $\qquad\square$

Corollary 4.1.52 *Let $a(t,\tau) \in S_{cl}^{\mu}(\overline{\mathbb{R}}_+ \times \mathbb{R})_e$ be a symbol with the transmission property. Then* $\mathrm{op}^+(a)$ *induces continuous operators*

$$\mathrm{op}^+(a) : H^s(\mathbb{R}_+) \to H^{s-\mu}(\mathbb{R}_+) \qquad \text{for every } s \in \mathbb{R}, \ s > -\frac{1}{2},$$

and

$$\mathrm{op}^+(a) : \mathcal{S}(\overline{\mathbb{R}}_+) \to \mathcal{S}(\overline{\mathbb{R}}_+).$$

In fact, for $\mu \in \mathbb{N}$ this is a consequence of the definition of the operator convention $\mathrm{op}^+(\cdot)$, cf. 4.1.68, and Theorem 4.1.38. For $-\mu \in \mathbb{N}$ we can write

$$\mathrm{op}^+(a) = \mathrm{op}^+(l_-^{\mu}) \, \mathrm{op}^+(l_-^{-\mu}) \, \mathrm{op}^+(a) = \mathrm{op}^+(l_-^{\mu})\{\mathrm{op}^+(l_-^{-\mu} \# a) + G_0\}$$

for a $G_0 \in \Gamma^0(\mathbb{R}_+)$ and apply $\mathrm{op}^+(l_-^{\mu}) : H^s(\mathbb{R}_+) \to H^{s-\mu}(\mathbb{R}_+)$ together with Theorem 4.1.38.

Definition 4.1.53 *We denote by $\boldsymbol{D}^{\mu,d}(\mathbb{R}_+)$ the space of all operators*

$$A = \mathrm{op}^+(a) + G \qquad\qquad (4.1.71)$$

for arbitrary $a(t,\tau) \in S_{cl}^{\mu}(\overline{\mathbb{R}}_+ \times \mathbb{R})_e$ with the transmission property and $G \in \Gamma^d(\mathbb{R}_+)$. Moreover, $\boldsymbol{D}^{\mu,d}(\mathbb{R}_+)$ denotes the subclass of all operators (4.1.71) with t–independent $a(\tau)$.

Remark 4.1.54 *Each $A \in \boldsymbol{D}^{\mu,d}(\mathbb{R}_+)$ induces continuous operators*

$$A : H^s(\mathbb{R}_+) \to H^{s-\mu}(\mathbb{R}_+) \qquad \text{for every } s \in \mathbb{R}, \ s > d - \frac{1}{2},$$

and

$$A : \mathcal{S}(\overline{\mathbb{R}}_+) \to \mathcal{S}(\overline{\mathbb{R}}_+).$$

This follows from Corollary 4.1.52 and Proposition 4.1.47.

Remark 4.1.55 *For $a(t,\tau) \in S_{cl}^{\mu}(\overline{\mathbb{R}}_+ \times \mathbb{R})_e$ and $p(\tau) = (1 - i\tau)^{\mu^+}$ for $\mu^+ = \max\{\mu, 0\}$ we have*

$$r^+ \mathrm{op}(a) e^+ u = \mathrm{op}^+(p) \, \mathrm{op}^+(p^{-1} \# a) u + G u$$

for all $u \in \mathcal{S}(\overline{\mathbb{R}}_+)$, with some $G \in \Gamma^0(\mathbb{R}_+)$. Thus the extension of

$$\mathrm{op}^+(p) \, \mathrm{op}^+(p^{-1} \# a) + G : \mathcal{S}(\overline{\mathbb{R}}_+) \to \mathcal{S}(\overline{\mathbb{R}}_+)$$

to a continuous operator $H^s(\mathbb{R}_+) \to H^{s-\mu}(\mathbb{R}_+)$ for every $s \in \mathbb{R}$, $s > -\frac{1}{2}$ (which is a consequence of Theorem 4.1.38) may also be regarded as the extension of

$$r^+ \mathrm{op}(a) e^+ : \mathcal{S}(\overline{\mathbb{R}}_+) \to \mathcal{S}(\overline{\mathbb{R}}_+)$$

by continuity to a continuous operator

$$r^+ \operatorname{op}(a) e^+ : H^s(\mathbb{R}_+) \to H^{s-\mu}(\mathbb{R}_+) \tag{4.1.72}$$

for all $s > -\frac{1}{2}$. In other words the interpretation of the operator $\operatorname{op}^+(a)$ in Definition 4.1.53 as (4.1.72) or (4.1.68) does not affect the operator classes $\boldsymbol{D}^{\mu,d}(\mathbb{R}_+)$ or $D^{\mu,d}(\mathbb{R}_+)$.

Lemma 4.1.56 *Let $a(\tau) \in S_{\mathrm{cl}}^0(\mathbb{R})$ be a symbol with the transmission property and $p(\tau)$ be a polynomial in τ of order m. Then*

$$\operatorname{op}^+(p) \operatorname{op}^+(a) - \operatorname{op}^+(a) \operatorname{op}^+(p) \in \Gamma^m(\mathbb{R}_+).$$

In particular, if $a(\tau) = p^{-1}(\tau) \in S_{\mathrm{cl}}^{-m}(\mathbb{R}_+)$ then

$$\operatorname{op}^+(p^{-1}) \operatorname{op}^+(p) = 1 + G \qquad \text{for some } G \in \Gamma^m(\mathbb{R}_+).$$

Proof. Let us set $r(\tau) = (1 - i\tau)^m$. Then, using $\operatorname{op}^+(r^{-1}) \operatorname{op}^+(r) = 1$ we get

$$\begin{aligned}
\operatorname{op}^+(a) \operatorname{op}^+(p) &= \operatorname{op}^+(a) \operatorname{op}^+(p) \operatorname{op}^+(r^{-1}) \operatorname{op}^+(r) \\
&= \operatorname{op}^+(a) \operatorname{op}^+(pr^{-1}) \operatorname{op}^+(r).
\end{aligned}$$

Theorem 4.1.41 gives us $\operatorname{op}^+(a) \operatorname{op}^+(pr^{-1}) = \operatorname{op}^+(apr^{-1}) + G_0$ for a $G_0 \in \Gamma^0(\mathbb{R}_+)$. Thus

$$\begin{aligned}
\operatorname{op}^+(a) \operatorname{op}^+(p) &= \{\operatorname{op}^+(apr^{-1}) + G_0\} \operatorname{op}^+(r) \\
&= \{\operatorname{op}^+(p) \operatorname{op}^+(ar^{-1}) + G_0\} \operatorname{op}^+(r). \tag{4.1.73}
\end{aligned}$$

Again by Theorem 4.1.41 we can write $\operatorname{op}^+(ar^{-1}) = \operatorname{op}^+(a) \operatorname{op}^+(r^{-1}) + G_1$ for a $G_1 \in \Gamma^0(\mathbb{R}_+)$. Thus

$$\begin{aligned}
\operatorname{op}^+(a) \operatorname{op}^+(p) &= \{\operatorname{op}^+(p)[\operatorname{op}^+(a) \operatorname{op}^+(r^{-1}) + G_1] + G_0\} \operatorname{op}^+(r) \\
&= \operatorname{op}^+(p) \operatorname{op}^+(a) + G_2,
\end{aligned}$$

where $G_2 = \{\operatorname{op}^+(p) G_1 + G_0\} \operatorname{op}^+(r) \in \Gamma^m(\mathbb{R}_+)$. $\qquad \square$

Lemma 4.1.57 *Let $a(t, \tau) \in S_{\mathrm{cl}}^0(\overline{\mathbb{R}}_+ \times \mathbb{R})_e$ be a symbol with the transmission property and let $p(\tau)$ be a polynomial of order m with $p^{-1}(\tau) \in S_{\mathrm{cl}}^{-m}(\mathbb{R})$. Then there is an $\tilde{a}(t, \tau) \in S_{\mathrm{cl}}^0(\overline{\mathbb{R}}_+ \times \mathbb{R})_e$ with the transmission property satisfying*

$$\operatorname{op}^+(p) \operatorname{op}^+(\tilde{a}) - \operatorname{op}^+(a) \operatorname{op}^+(p) \in \Gamma^m(\mathbb{R}_+). \tag{4.1.74}$$

Conversely for every $\tilde{a}(t, \tau) \in S_{\mathrm{cl}}^0(\overline{\mathbb{R}}_+ \times \mathbb{R})_e$ there is an $a(t, \tau) \in S_{\mathrm{cl}}^0(\overline{\mathbb{R}}_+ \times \mathbb{R})_e$ such that the relation (4.1.74) holds.

Proof. Setting $\tilde{a}(t, \tau) = p^{-1}(\tau) \# (ap)(t, \tau)$ we obtain $d := p \# \tilde{a} - ap = p \# \tilde{a} - a \# p \in S^{-\infty}(\overline{\mathbb{R}}_+ \times \mathbb{R})_e$. For $r(\tau) = (1 - i\tau)^m$ we have

$$
\operatorname{op}^+(p) \operatorname{op}^+(\tilde{a}) - \operatorname{op}^+(a) \operatorname{op}^+(p)
$$
$$
= \{\operatorname{op}^+(p) \operatorname{op}^+(\tilde{a}) \operatorname{op}^+(r^{-1}) - \operatorname{op}^+(a) \operatorname{op}^+(p) \operatorname{op}^+(r^{-1})\} \operatorname{op}^+(r)
$$
$$
= \{\operatorname{op}^+(p)[\operatorname{op}^+(\tilde{a} r^{-1}) + G_1] - \operatorname{op}^+(a) \operatorname{op}^+(pr^{-1})\} \operatorname{op}^+(r)
$$
$$
= \{\operatorname{op}^+(p) \operatorname{op}^+(\tilde{a} r^{-1}) - \operatorname{op}^+(a) \operatorname{op}^+(pr^{-1})\} \operatorname{op}^+(r) + G_2
$$

for certain $G_1 \in \Gamma^0(\mathbb{R}_+)$, $G_2 \in \Gamma^m(\mathbb{R}_+)$, cf. Theorem 4.1.41 and Lemma 4.1.50. Now

$$
\operatorname{op}^+(p) \operatorname{op}^+(\tilde{a} r^{-1}) = \operatorname{op}^+(p \# (\tilde{a} r^{-1})) = \operatorname{op}^+((p \# \tilde{a}) r^{-1})
$$

(the Leibniz products are finite sums here). Moreover, Theorem 4.1.41 gives us

$$
\operatorname{op}^+(a) \operatorname{op}^+(pr^{-1}) = \operatorname{op}^+(apr^{-1}) + G_0
$$

for a $G_0 \in \Gamma^0(\mathbb{R}_+)$. Thus

$$
\{\operatorname{op}^+(p) \operatorname{op}^+(\tilde{a} r^{-1}) - \operatorname{op}^+(a) \operatorname{op}^+(pr^{-1})\} \operatorname{op}^+(r)
$$
$$
= \{\operatorname{op}^+((p \# \tilde{a}) r^{-1}) - \operatorname{op}^+(apr^{-1}) - G_0\} \operatorname{op}^+(r)
$$
$$
= \{\operatorname{op}^+([p \# \tilde{a} - ap] r^{-1}) - G_0\} \operatorname{op}^+(r) = \{\operatorname{op}^+(dr^{-1}) - G_0\} \operatorname{op}^+(r).
$$

Now the symbol $d(t, \tau)$ as well as $d(t, \tau) r^{-1}(\tau)$ belong to $S^{-\infty}(\overline{\mathbb{R}}_+ \times \mathbb{R})_e$, hence $\operatorname{op}^+(dr^{-1}) \in \Gamma^0(\mathbb{R}_+)$. This shows that the operator on the right of the latter equation belongs to $\Gamma^m(\mathbb{R}_+)$. Thus we obtain the first part of the assertion. The reverse relation can be proved in an analogous manner. $\square$

Lemma 4.1.58 *Let $G \in \Gamma^d(\mathbb{R}_+)$ and $a(t, \tau) \in S^\mu_{\text{cl}}(\overline{\mathbb{R}}_+ \times \mathbb{R})_e$ be a symbol with the transmission property. Then*

$$
G \operatorname{op}^+(a) \in \Gamma^{(\mu+d)^+}(\mathbb{R}_+).
$$

Proof. For $\mu \geq 0$ we choose a polynomial $p(\tau)$ of order μ with $p^{-1}(\tau) \in S_{\text{cl}}^{-\mu}(\mathbb{R})$ and write

$$
\operatorname{op}^+(a) = \operatorname{op}^+(p) \operatorname{op}^+(a_0) \qquad \text{for } a_0 = p^{-1} \# a.
$$

In the case $\mu \leq 0$ we set $q(\tau) = (1 - i\tau)^{-\mu}$ and

$$
\operatorname{op}^+(a) = \operatorname{op}^+(q^{-1}) \operatorname{op}^+(b_0) \qquad \text{for } b_0 = q \# a.
$$

This is allowed, since $q(\tau)$ and $q^{-1}(\tau)$ are minus functions. By definition G is a finite linear combination of compositions of the form

$G_0 \operatorname{op}^+(s)$ for $G_0 \in \Gamma^0(\mathbb{R}_+)$ and polynomials $s(\tau)$ of order $\leq d$. So we may assume G to be of this form. For $\mu \geq 0$ we have

$$G \operatorname{op}^+(a) = G_0 \operatorname{op}^+(s) \operatorname{op}^+(p) \operatorname{op}^+(a_0) = G_0 \operatorname{op}^+(sp) \operatorname{op}^+(a_0)$$
$$= G_0 \{\operatorname{op}^+(\tilde{a}_0) \operatorname{op}^+(sp) + D\} =: D_1$$

for a $D \in \Gamma^{d+\mu}(\mathbb{R}_+)$ and a certain $\tilde{a}_0(t,\tau) \in S^0_{\mathrm{cl}}(\overline{\mathbb{R}}_+ \times \mathbb{R})_e$ with the transmission property, cf. Lemma 4.1.57. In view of $G_0 \operatorname{op}^+(\tilde{a}_0) \in \Gamma^0(\mathbb{R}_+)$ we obtain $D_1 \in \Gamma^{\mu+d}(\mathbb{R}_+)$. For $\mu \leq 0$ and $d + \mu \leq 0$ we have

$$G \operatorname{op}^+(a) = G_0 \operatorname{op}^+(s) \operatorname{op}^+(q^{-1}) \operatorname{op}^+(b_0) = G_0 \operatorname{op}^+(sq^{-1}) \operatorname{op}^+(b_0). \tag{4.1.75}$$

Since both sq^{-1} and b_0 are of order ≤ 0 we get $\operatorname{op}^+(aq^{-1}) \operatorname{op}^+(b_0) = \operatorname{op}^+(c_0) + G_1$ for $c_0 = (sq^{-1}) \# b_0$ and some $G_1 \in \Gamma^0(\mathbb{R}_+)$. This yields obviously $G \operatorname{op}^+(a) \in \Gamma^0(\mathbb{R}_+)$. If $e := d + \mu \geq 0$ we set $r(\tau) = (1 - i\tau)^e$ and write $s(\tau) = r(\tau)\tilde{s}(\tau)$ for $\tilde{s}(\tau) = r^{-1}(\tau)s(\tau)$. Then

$$G \operatorname{op}^+(a) = G_0 \operatorname{op}^+(s) \operatorname{op}^+(q^{-1}) \operatorname{op}^+(b_0)$$
$$= G_0 \operatorname{op}^+(r) \operatorname{op}^+(\tilde{s}q^{-1}) \operatorname{op}^+(b_0)$$
$$= G_0 \operatorname{op}^+(r) \{\operatorname{op}^+(c_0) + G_1\}$$

for $c_0 = (\tilde{s}q^{-1}) \# b_0$ and some $G_1 \in \Gamma^0(\mathbb{R}_+)$. Using again Lemma 4.1.57 we get for a $\tilde{c}_0$ that $\operatorname{op}^+(r) \operatorname{op}^+(c_0) = \operatorname{op}^+(\tilde{c}_0) \operatorname{op}^+(r) + D_1$ for some $D_1 \in \Gamma^e(\mathbb{R}_+)$. This yields

$$G \operatorname{op}^+(a) = G_0 \{\operatorname{op}^+(\tilde{c}_0) \operatorname{op}^+(r) + D_1\} + G_0 \operatorname{op}^+(r)G_1$$
$$= G_2 \operatorname{op}^+(r) + G_0 D_1 + G_0 \operatorname{op}^+(r)G_1 \in \Gamma^e(\mathbb{R}_+).$$

$\square$

Theorem 4.1.59 *Given* $A \in \boldsymbol{D}^{\mu,d}(\mathbb{R}_+)$, $B \in \boldsymbol{D}^{\nu,c}(\mathbb{R}_+)$ *we have* $AB \in \boldsymbol{D}^{\mu+\nu,h}(\mathbb{R}_+)$ *for* $h = \max\{(d + \nu)^+, c\}$, $\varrho^+ = \max\{\varrho, 0\}$. *Moreover,* $A \in \Gamma^d(\mathbb{R}_+)$ *or* $B \in \Gamma^c(\mathbb{R}_+)$ *implies* $AB \in \Gamma^h(\mathbb{R}_+)$.

Proof. By assumption we can write $A = \operatorname{op}^+(a) + G$, $B = \operatorname{op}^+(b) + D$ for $\operatorname{op}^+(a) = \operatorname{op}^+(p) \operatorname{op}^+(p^{-1}\#a)$, $\operatorname{op}^+(b) = \operatorname{op}^+(q) \operatorname{op}^+(q^{-1}\#b)$, with polynomials $p(\tau)$ and $q(\tau)$ of orders μ^+ and ν^+, respectively. In the proof the operators will be regarded as actions on $\mathcal{S}(\overline{\mathbb{R}}_+)$. We have

$$AB = P + D_0 + D_1 + D_2$$

for $D_0 = \operatorname{op}^+(p) \operatorname{op}^+(p^{-1}\#a)D$, $D_1 = G \operatorname{op}^+(q) \operatorname{op}^+(q^{-1}\#b)$, $D_2 = GD$, and $P = \operatorname{op}^+(p) \operatorname{op}^+(p^{-1}\#a) \operatorname{op}^+(q) \operatorname{op}^+(q^{-1}\#b)$. Here we have $D_0, D_2 \in \Gamma^c(\mathbb{R}_+)$ and $D_1 \in \Gamma^{(d+\nu)^+}(\mathbb{R}_+)$, cf. Lemma 4.1.58. It remains to characterise P. It is obviously sufficient to consider $P_0 = \mathrm{op}^+(a_0)\,\mathrm{op}^+(q)\,\mathrm{op}^+(b_0)$ for $a_0 = p^{-1}\#a$, $b_0 = q^{-1}\#b$. Applying Lemma 4.1.57 we get $\mathrm{op}^+(a_0)\,\mathrm{op}^+(q) = \mathrm{op}^+(q)\,\mathrm{op}^+(\tilde{a}_0) + G_1$ for a certain $G_1 \in \Gamma^{\nu^+}(\mathbb{R}_+)$ and a symbol $\tilde{a}_0$ of order zero. Then

$$P_0 = \{\mathrm{op}^+(q)\,\mathrm{op}^+(\tilde{a}_0) + G_1\}\,\mathrm{op}^+(b_0)$$
$$= \mathrm{op}^+(q)\{\mathrm{op}^+(\tilde{a}_0\#b_0) + G_2\} + G_1\,\mathrm{op}^+(b_0)$$

for $G_2 \in \Gamma^0(\mathbb{R}_+)$. Now $\mathrm{op}^+(q)G_2 \in \Gamma^0(\mathbb{R}_+)$ and $G_1\,\mathrm{op}^+(b_0) \in \Gamma^{\nu^+}(\mathbb{R}_+)$, $\nu^+ \le (d+\nu)^+$, gives us the assertion. $\square$

Next we will discuss reduction of orders by means of the isomorphisms

$$r^+\,\mathrm{op}(l_-^\nu)e_s^+ : H^s(\mathbb{R}_+) \to H^{s-\nu}(\mathbb{R}_+), \qquad (4.1.76)$$

$l_-^\nu(\tau) = (1 - i\tau)^\nu$, for a continuous extension operator $e_s^+ : H^s(\mathbb{R}_+) \to H^s(\mathbb{R})$. Recall that (4.1.76) is independent of the particular choice of e_s^+, cf. Proposition 4.1.10. Let us write for a symbol $q(\tau)$

$$\mathrm{op}_s^+(q) = r^+\,\mathrm{op}(q)e_s^+.$$

In the cases occurring here the symbols q are minus functions; this is independent of the choice of e_s^+.

Lemma 4.1.60 *Let $p(\tau)$ be a polynomial of order m. Then we have*

$$\mathrm{op}_{r-m}^+(l_-^{-m})\,\mathrm{op}^+(p) = \mathrm{op}_r^+(l_-^{-m}p)$$

as an operator on on $H^r(\mathbb{R}_+)$, $r \in \mathbb{R}$.

Proof. We have obviously $\mathrm{op}^+(p) = \mathrm{op}_r^+(p)$ on $H^r(\mathbb{R}_+)$ and the relation $\mathrm{op}_{r-m}^+(l_-^{-m})\,\mathrm{op}_r^+(p) = r^+\,\mathrm{op}(l_-^{-m}p)e_r^+$. $\square$

Each $A \in \mathbf{D}^{\mu,d}(\mathbb{R}_+)$ induces continuous operators

$$A : H^s(\mathbb{R}_+) \to H^{s-\mu}(\mathbb{R}_+)$$

for $s > d - \frac{1}{2}$, cf. Remark 4.1.54. Let us fix such an $s \in \mathbb{N}$ and set

$$A_0 = \mathrm{op}_{s-\mu}^+(l_-^{s-\mu})A\,\mathrm{op}_0^+(l_-^{-s}) : H^0(\mathbb{R}_+) \to H^0(\mathbb{R}_+)$$

with the isomorphisms

$$\mathrm{op}_0^+(l_-^{-s}) : H^0(\mathbb{R}_+) \to H^s(\mathbb{R}_+),$$
$$\mathrm{op}_{s-\mu}^+(l_-^{s-\mu}) : H^{s-\mu}(\mathbb{R}_+) \to H^0(\mathbb{R}_+).$$

Theorem 4.1.61 $A \in \boldsymbol{D}^{\mu,d}(\mathbb{R}_+)$ *implies* $A_0 \in \boldsymbol{D}^{0,0}(\mathbb{R}_+)$ *for every fixed* $s \in \mathbb{N}$, $s > d - \frac{1}{2}$, *and we have*

$$A = \mathrm{op}^+(l_-^{-s+\mu})A_0 \, \mathrm{op}_s^+(l_-^s). \tag{4.1.77}$$

In order to prove this theorem we first show some lemmata:

Lemma 4.1.62 $G \in \Gamma^d(\mathbb{R}_+)$ *implies*

$$G_0 := \mathrm{op}_{s-\mu}^+(l_-^{s-\mu})G \, \mathrm{op}_0^+(l_-^{-s}) \in \Gamma^0(\mathbb{R}_+)$$

for every $s \in \mathbb{N}$, $s > d - \frac{1}{2}$.

Proof. Let us first assume $G \in \Gamma^0(\mathbb{R}_+)$, i.e.,

$$Gu(t) = \int\limits_0^\infty g(t,t')u(t') \, dt'$$

for a kernel $g(t,t') \in \mathcal{S}(\overline{\mathbb{R}}_+ \times \overline{\mathbb{R}}_+)$. Then the continuity of $\mathrm{op}_0^+(l_-^{-s})$: $L^2(\mathbb{R}_+) \to H^s(\mathbb{R}_+)$, $G : H^s(\mathbb{R}_+) \to \mathcal{S}(\overline{\mathbb{R}}_+)$, $\mathrm{op}_{s-\mu}^+(l_-^{s-\mu}) : \mathcal{S}(\overline{\mathbb{R}}_+) \to \mathcal{S}(\overline{\mathbb{R}}_+)$ shows the continuity of $G_0 : L^2(\mathbb{R}_+) \to \mathcal{S}(\overline{\mathbb{R}}_+)$. In order to characterise the adjoint G_0^* we first observe that $G_1 := \mathrm{op}_{s-\mu}^+(l_-^{s-\mu})G$ belongs to $\Gamma^0(\mathbb{R}_+)$. Thus it suffices to look at the $L^2(\mathbb{R}_+)$-adjoint

$$(G_1 \, \mathrm{op}_0^+(l_-^{-s}))^* = \mathrm{op}_0^+(l_-^{-s})^* G_1^* = G_0^*. \tag{4.1.78}$$

We have $G_1^* \in \Gamma^0(\mathbb{R}_+)$ and $\mathrm{op}_0^+(l_-^{-s})^* = r^+ \, \mathrm{op}(l_-^{-s})e^+ = \mathrm{op}^+(l_+^{-s})$ for $l_+^{-s}(\tau) = (1 + i\tau)^{-s}$ which has the transmission property. Since the operators $G_1^* : L^2(\mathbb{R}_+) \to \mathcal{S}(\overline{\mathbb{R}}_+)$ and $\mathrm{op}^+(l_+^{-s}) : \mathcal{S}(\overline{\mathbb{R}}_+) \to \mathcal{S}(\overline{\mathbb{R}}_+)$ are continuous, we get also the continuity of $G_0^* : L^2(\mathbb{R}_+) \to \mathcal{S}(\overline{\mathbb{R}}_+)$. Thus $G_0 \in \Gamma^0(\mathbb{R}_+)$. For arbitrary G the operator G_0 is a sum of compositions $\mathrm{op}_{s-\mu}^+(l_-^{s-\mu})G_k(\frac{d}{dt})^k \mathrm{op}_0^+(l_-^{-s})$, $k = 0, \ldots, d$, for certain $G_k \in \Gamma^0(\mathbb{R}_+)$. Here we may again omit $\mathrm{op}_{s-\mu}^+(l_-^{s-\mu})$ for similar reasons as before. But

$$\left(\frac{d}{dt}\right)^k \mathrm{op}^+(l_-^{-s}) = \mathrm{op}^+(a_k)$$

for $a_k(\tau) = (i\tau)^k l_-^{-s}(\tau)$ which belongs to $S_{\mathrm{cl}}^0(\mathbb{R})$ and has the transmission property. In view of Theorem 4.1.41 we see that also in the case of arbitrary d we have $G_0 \in \Gamma^0(\mathbb{R}_+)$. $\qquad\square$

Lemma 4.1.63 $a_0(t,\tau) \in S_{\mathrm{cl}}^0(\overline{\mathbb{R}}_+ \times \mathbb{R})_e$ *implies*

$$\mathrm{op}_s^+(l_-^s) \, \mathrm{op}^+(a_0) \, \mathrm{op}_0^+(l_-^{-s}) \in \boldsymbol{D}^{0,0}(\mathbb{R}_+)$$

for every $s \in \mathbb{N}$.

Proof. For $u \in \mathcal{S}(\overline{\mathbb{R}}_+)$ we have $\operatorname{op}^+(l_-^{-s})u = \operatorname{op}_0^+(l_-^{-s})u \in \mathcal{S}(\overline{\mathbb{R}}_+)$, and we get $v := \operatorname{op}^+(a_0)\operatorname{op}^+(l_-^{-s})u = \operatorname{op}^+(a_0 l_-^{-s})u + G_0 u \in \mathcal{S}(\overline{\mathbb{R}}_+)$ for some $G_0 \in \Gamma^0(\mathbb{R}_+)$, cf. Theorem 4.1.41. Similarly to the proof of Lemma 4.1.62 we obtain $G_1 = \operatorname{op}_s^+(l_-^s)G_0 \in \Gamma^0(\mathbb{R}_+)$. Moreover, we have $\operatorname{op}_s^+(l_-^s)v = \operatorname{op}^+(l_-^s)v = \operatorname{op}^+(l_-^s \#(a_0 l_-^{-s}))u + G_1 u = \operatorname{op}^+(b_0)u + G_1 u$ with $b_0(t,\tau) \in S_{\mathrm{cl}}^0(\overline{\mathbb{R}}_+ \times \mathbb{R})_{\mathrm{e}}$. Thus, on $\mathcal{S}(\overline{\mathbb{R}}_+)$ the operator $\operatorname{op}_s^+(l_-^s)\operatorname{op}^+(a_0)\operatorname{op}_0^+(l_-^{-s})$ has the representation $\operatorname{op}^+(b_0) + G_1$ which is the desired result. $\qquad\square$

Corollary 4.1.64 $A_0 \in \boldsymbol{D}^{0,0}(\mathbb{R}_+)$ *implies*

$$\tilde{A}_0 = \operatorname{op}^+(l_-^s)A_0 \operatorname{op}_s^+(l_-^{-s}) \in \boldsymbol{D}^{0,0}(\mathbb{R}_+)$$

for every $s \in \mathbb{N}$.

Proof of Theorem 4.1.61. Let us first assume $\mu \in \mathbb{N}$. The operator $A \in \boldsymbol{D}^{\mu,d}(\mathbb{R}_+)$ can be written $A = \operatorname{op}^+(p)\operatorname{op}^+(a_0) + G$ for a certain $a_0(t,\tau) \in S_{\mathrm{cl}}^0(\overline{\mathbb{R}}_+ \times \mathbb{R})_{\mathrm{e}}$, a polynomial $p(\tau)$ of order μ and a $G \in \Gamma^d(\mathbb{R}_+)$. By Lemma 4.1.62 we have $G_0 = \operatorname{op}_{s-\mu}^+(l_-^{s-\mu})G\operatorname{op}_0^+(l_-^{-s}) \in \Gamma^0(\mathbb{R}_+)$. Thus we can consider

$$\operatorname{op}_{s-\mu}^+(l_-^{s-\mu})\operatorname{op}^+(p)\operatorname{op}^+(a_0)\operatorname{op}_0^+(l_-^{-s})$$
$$= \operatorname{op}_{s-\mu}^+(l_-^{s-\mu})\operatorname{op}_s^+(p)\operatorname{op}^+(a_0)\operatorname{op}_0^+(l_-^{-s}). \quad (4.1.79)$$

Using $\operatorname{op}_{s-\mu}^+(l_-^{s-\mu})\operatorname{op}_s^+(p) = \operatorname{op}_0^+(l_-^{-\mu}p)\operatorname{op}_s^+(l_-^s)$ we obtain from Lemma 4.1.63 that (4.1.79) equals $\operatorname{op}_0^+(l_-^{-\mu}p)\tilde{A}_0$ for an $\tilde{A}_0 \in \boldsymbol{D}^{0,0}(\mathbb{R}_+)$. Since $l_-^{-\mu}(\tau)p(\tau)$ is of order ≤ 0, by Theorem 4.1.41 we finally get that $A_0 \in \boldsymbol{D}^{0,0}(\mathbb{R}_+)$. For $-\mu \in \mathbb{N}$ we write $A = \operatorname{op}^+(a) + G$, $G \in \Gamma^d(\mathbb{R}_+)$. Since $\operatorname{op}_{s-\mu}^+(l_-^{s-\mu})G\operatorname{op}_0^+(l_-^{-s}) \in \Gamma^0(\mathbb{R}_+)$, by Lemma 4.1.62 it suffices to consider $A = \operatorname{op}^+(a)$ for an $a(t,\tau) \in S_{\mathrm{cl}}^\mu(\overline{\mathbb{R}}_+ \times \mathbb{R})_{\mathrm{e}}$. Then

$$\tilde{A}_0 = \operatorname{op}_s^+(l_-^s)\operatorname{op}^+(l_-^{-\mu})\operatorname{op}^+(a)\operatorname{op}_0^+(l_-^{-s})$$
$$= \operatorname{op}_s^+(l_-^s)\{\operatorname{op}^+(l_-^{-\mu}\#a) + G_0\}\operatorname{op}_0^+(l_-^{-s})$$

with $(l_-^{-\mu}\#a)(t,\tau) \in S_{\mathrm{cl}}^0(\overline{\mathbb{R}}_+ \times \mathbb{R})_{\mathrm{e}}$ and a $G_0 \in \Gamma^0(\mathbb{R}_+)$. Hence we can apply Lemma 4.1.63 and Lemma 4.1.62. In other words in this case we also obtain $A_0 \in \boldsymbol{D}^{0,0}(\mathbb{R}_+)$. $\qquad\square$

Corollary 4.1.65 *Let* $r_-^\nu(\tau)$ *be a family of minus symbols with the transmission property, given for every* $\nu \in \mathbb{Z}$, *with* $r_-^\nu(\tau)^{-1} = r_-^{-\nu}(\tau)$. *Then for every fixed* $s \in \mathbb{N}$ *with* $s > d - \frac{1}{2}$ *we have*

$$\tilde{A}_0 = \operatorname{op}_{s-\mu}^+(r_-^{s-\mu})A\operatorname{op}^+(r_-^{-s}) \in \boldsymbol{D}^{0,0}(\mathbb{R}_+) \quad (4.1.80)$$

and

$$A = \operatorname{op}^+(r_-^{-s+\mu})\tilde{A}_0 \operatorname{op}_s^+(r_-^s). \quad (4.1.81)$$

Proof. Using Theorem 4.1.61 we can write

$$A = \mathrm{op}^+(r_-^{-s+\mu})\,\mathrm{op}_{s-\mu}^+(r_-^{s-\mu})\,\mathrm{op}^+(l_-^{-s+\mu})A_0\,\mathrm{op}_s^+(l_-^s)\,\mathrm{op}^+(r_-^{-s})\,\mathrm{op}_s^+(r_-^s),$$

and we have

$$\mathrm{op}_{s-\mu}^+(r_-^{s-\mu})\,\mathrm{op}^+(l_-^{-s+\mu}) = \mathrm{op}^+(r_-^{s-\mu}l_-^{-s+\mu}),$$
$$\mathrm{op}_s^+(l_-^s)\,\mathrm{op}^+(r_-^{-s}) = \mathrm{op}^+(l_-^s r_-^{-s}),$$

where $r_1 = r_-^{s-\mu}l_-^{-s+\mu}$ and $r_2 = l_-^s r_-^{-s}$ are symbols of order zero with the transmission property. Using Theorem 4.1.41 we thus obtain $\tilde{A}_0 = \mathrm{op}^+(r_1)A_0\,\mathrm{op}^+(r_2) \in \boldsymbol{D}^{0,0}(\mathbb{R}_+)$ and the formula (4.1.81). Moreover, in view of $\mathrm{op}^+(r_-^{-s})\,\mathrm{op}_s^+(r_-^s) = 1$, $\mathrm{op}^+(r_-^{-s+\mu})\,\mathrm{op}_{s-\mu}^+(r_-^{s-\mu}) = 1$ we get the relation (4.1.80). $\qquad\Box$

4.1.7 Ellipticity

Proposition 4.1.66 *Let* $G \in \Gamma^d(\mathbb{R}_+)$ *and assume that*

$$1 + G : \mathcal{S}(\overline{\mathbb{R}}_+) \to \mathcal{S}(\overline{\mathbb{R}}_+)$$

is an isomorphism. Then $(1+G)^{-1} = 1 + C$ *for some* $C \in \Gamma^d(\mathbb{R}_+)$.

Proof. Write G in the form (4.1.66) for a $G_0 \in \Gamma^0(\mathbb{R}_+)$. Then $1 + G_0 : L^2(\mathbb{R}_+) \to L^2(\mathbb{R}_+)$ is a Fredholm operator with index zero. According to Lemma 4.1.44 there is a finite–dimensional operator $G_1 \in \Gamma^0(\mathbb{R}_+)$ such that $1 + G_0 + G_1 : L^2(\mathbb{R}_+) \to L^2(\mathbb{R}_+)$ is an isomorphism. The latter operator also induces an isomorphism $\mathcal{S}(\overline{\mathbb{R}}_+) \to \mathcal{S}(\overline{\mathbb{R}}_+)$. By construction G_1 has the form of a finite linear combination of compositions $G_1 = \sum_{k=1}^m C_k B_k$ for $m = \dim\ker(1 + G_0)$, cf. (4.1.61). This gives us together with (4.1.66) the representation

$$G = G_0 + G_1 + \Big\{ \sum_{j=0}^{d-1} T_j K_j - \sum_{k=1}^m C_k B_k \Big\}.$$

Setting $T_{d-1+k} := -C_k$, $K_{d-1+k} = B_k$ for $k = 1,\ldots,m$ and $N := d+m$, $D = G_0 + G_1$, we get $G = D + TK$, where

$$T = (T_0,\ldots,T_{N-1}) : \mathcal{S}(\overline{\mathbb{R}}_+) \to \mathbb{C}^N,$$
$$K = (K_0,\ldots,K_{N-1}) : \mathbb{C}^N \to \mathcal{S}(\overline{\mathbb{R}}_+).$$

In view of

$$\begin{pmatrix} 1 & K \\ 0 & 1 \end{pmatrix} \begin{pmatrix} 1+D & -K \\ T & 1 \end{pmatrix} \begin{pmatrix} 1 & 0 \\ -T & 1 \end{pmatrix} = \begin{pmatrix} 1+G & 0 \\ 0 & 1 \end{pmatrix} \qquad (4.1.82)$$

the operator $1 + G : \mathcal{S}(\overline{\mathbb{R}}_+) \to \mathcal{S}(\overline{\mathbb{R}}_+)$ is an isomorphism if and only if

$$
\begin{pmatrix} 1+D & -K \\ T & 1 \end{pmatrix} : \begin{array}{c} \mathcal{S}(\overline{\mathbb{R}}_+) \\ \oplus \\ \mathbb{C}^N \end{array} \to \begin{array}{c} \mathcal{S}(\overline{\mathbb{R}}_+) \\ \oplus \\ \mathbb{C}^N \end{array}
$$

is an isomorphism. From Proposition 4.1.43 we get $(1+D)^{-1} = 1 + C_0$ for some $C_0 \in \Gamma^0(\mathbb{R}_+)$. Then

$$
\begin{pmatrix} 1 & 0 \\ -T(1+C_0) & 1 \end{pmatrix} \begin{pmatrix} 1+D & -K \\ T & 1 \end{pmatrix} \begin{pmatrix} 1 & (1+C_0)K \\ 0 & 1 \end{pmatrix}
$$

$$
= \begin{pmatrix} 1+D & 0 \\ 0 & 1+T(1+C_0)K \end{pmatrix} \quad (4.1.83)
$$

shows that $Q := 1 + T(1+C_0)K$ is necessarily an invertible $N \times N$-matrix. The equations (4.1.82), (4.1.83) yield

$$
\begin{pmatrix} 1+G & 0 \\ 0 & 1 \end{pmatrix}^{-1} = \begin{pmatrix} 1 & 0 \\ T & 1 \end{pmatrix} \begin{pmatrix} 1 & (1+C_0)K \\ 0 & 1 \end{pmatrix} \begin{pmatrix} 1+C_0 & 0 \\ 0 & Q^{-1} \end{pmatrix}
$$

$$
\times \begin{pmatrix} 1 & 0 \\ -T(1+C_0) & 1 \end{pmatrix} \begin{pmatrix} 1 & -K \\ 0 & 1 \end{pmatrix},
$$

i.e., $(1+G)^{-1} = 1 + C_0 + (1+C_0)KQ^{-1}T(1+C_0) = 1 + C$ for a $C \in \Gamma^d(\mathbb{R}_+)$. $\qquad \square$

Remark 4.1.67 *The space $\Gamma^d(\mathbb{R}_+)$ is Fréchet in a natural way which follows easily from the identifications $\Gamma^d(\mathbb{R}_+) \cong \Gamma^0(\mathbb{R}_+) \oplus (\mathcal{S}(\overline{\mathbb{R}}_+ \otimes \mathbb{C}^d)$, cf. Proposition 4.1.46, and $\Gamma^0(\mathbb{R}_+) \cong \mathcal{S}(\overline{\mathbb{R}}_+ \times \overline{\mathbb{R}}_+)$. Now the construction of C in the proof of Proposition 4.1.66 shows that when $G(y) \in C^\infty(\Omega, \Gamma^d(\mathbb{R}_+))$ such that*

$$
1 + G(y) : \mathcal{S}(\overline{\mathbb{R}}_+) \to \mathcal{S}(\overline{\mathbb{R}}_+)
$$

is invertible for every $y \in \Omega$, then $(1+G(y))^{-1} = 1 + C(y)$ for $C(y) \in C^\infty(\Omega, \Gamma^d(\mathbb{R}_+))$.

Definition 4.1.68 *An operator $A \in \boldsymbol{D}^{\mu,d}(\mathbb{R}_+)$, written $A = \mathrm{op}^+(a) + G$ for a symbol $a(t,\tau) \in S_{cl}^\mu(\overline{\mathbb{R}}_+ \times \mathbb{R})_e$ with the transmission property and a $G \in \Gamma^d(\mathbb{R}_+)$ is called elliptic if*

(i) *the homogeneous principal part $a_{(\mu)}(t,\tau)$ of $a(t,\tau)$ in τ of order μ is $\neq 0$ for all $\tau \neq 0$, $t \in \overline{\mathbb{R}}_+$,*

(ii) *there are constants c and R such that $|a(t,\tau)| \geq c(1+|\tau|^2)^{\frac{\mu}{2}}$ holds for all $(t,\tau) \in \overline{\mathbb{R}}_+ \times \mathbb{R}$ with $t^2 + |\tau|^2 \geq R$.*

Remark 4.1.69 *The definitions and results of the present section have generalisations to $k \times k$ systems of operators. In particular, in the analogue of the above Definition 4.1.68 we have to require* $\det a_{(\mu)}(t, \tau) \neq 0$ *for all* $\tau \neq 0$, $t \in \overline{\mathbb{R}}_+$ *and* $|\det a(t, \tau)| \geq c(1 + |\tau|^2)^{\mu/2}$ *for all* $(t, \tau) \in \overline{\mathbb{R}}_+ \times \mathbb{R}$ *with* $t^2 + |\tau|^2 \geq R$.

Remark 4.1.70 *Let* $A \in \boldsymbol{D}^{\mu,d}(\mathbb{R}_+)$ *be elliptic and* $s \in \mathbb{N}$, $s > d - \frac{1}{2}$, *be fixed. Set*

$$A_0 = \operatorname{op}^+_{s-\mu}(l^{s-\mu}_-) A \operatorname{op}^+_0(l^{-s}_-) \in \boldsymbol{D}^{0,0}(\mathbb{R}_+),$$

cf. Theorem 4.1.61. Then A_0 *is also elliptic.*

Definition 4.1.71 *Let* $A \in \boldsymbol{D}^{\mu,d}(\mathbb{R}_+)$, $B \in \boldsymbol{D}^{-\mu,c}(\mathbb{R}_+)$ *for certain* $\mu \in \mathbb{Z}$, $d, c \in \mathbb{N}$. *Then* B *is called a parametrix of* A *if* $AB - 1 \in \Gamma^e(\mathbb{R}_+)$, $BA - \in \Gamma^h(\mathbb{R}_+)$ *for certain* $e, h \in \mathbb{N}$.

Proposition 4.1.72 *Each elliptic operator* $A \in \boldsymbol{D}^{\mu,d}(\mathbb{R}_+)$ *has a parametrix* $B \in \boldsymbol{D}^{-\mu,c}(\mathbb{R}_+)$,

$$c = \begin{cases} d - \mu & \text{for } d - \mu \geq 0, \\ (-\mu)^+ & \text{for } d - \mu \leq 0, \end{cases} \tag{4.1.84}$$

with $AB - 1 \in \Gamma^{(d-\mu)^+}(\mathbb{R}_+)$, $BA - 1 \in \Gamma^d(\mathbb{R}_+)$; $\nu^+ = \max\{\nu, 0\}$.
In particular, for $d = \mu^+$ *we have* $c = (-\mu)^+$, $(d - \mu)^+ = (-\mu)^+$.

Proof. Let us fix an $s \in \mathbb{N}$, $s > d - \frac{1}{2}$, and form the operator A_0, according to Remark 4.1.70. Writing $A_0 = \operatorname{op}^+(a_0)$ mod $\Gamma^0(\mathbb{R}_+)$ for an $a_0(t, \tau) \in S^0_{\text{cl}}(\overline{\mathbb{R}}_+ \times \mathbb{R})_e$ with the transmission property, we can form the Leibniz inverse $b_0(t, \tau) \in \mathcal{S}^0_{\text{cl}}(\overline{\mathbb{R}}_+ \times \mathbb{R})_e$ of $a_0(t, \tau)$. Then $B_0 = \operatorname{op}^+(b_0) \in \boldsymbol{D}^{0,0}(\mathbb{R}_+)$ is a parametrix of A_0. Now we set

$$B = \operatorname{op}^+(l^{-s}_-) B_0 \operatorname{op}^+_{s-\mu}(l^{s-\mu}_-). \tag{4.1.85}$$

To characterise the structure of B it suffices to consider the operator on $H^s(\mathbb{R}_+)$ for any sufficiently large s. Let us assume $s = d$ and $-r := d - \mu \leq 0$. Then $-d = r - \mu$, and we have first

$$B = \operatorname{op}^+_{-r}(l^{-\mu}_-) \operatorname{op}^+(l^r_-) B_0 \operatorname{op}^+_{-r}(l^{-r}_-). \tag{4.1.86}$$

Restricting this to $L^2(\mathbb{R}_+)$ we get

$$B = \operatorname{op}^+(l^{-\mu}_-) \operatorname{op}^+_r(l^r_-) B_0 \operatorname{op}^+_0(l^{-r}_-).$$

From Corollary 4.1.65 we obtain

$$\tilde{B}_0 = \operatorname{op}^+_r(l^r_-) B_0 \operatorname{op}^+(l^{-r}_-) \in \boldsymbol{D}^{0,0}(\mathbb{R}_+).$$

It follows that $B \in \boldsymbol{D}^{-\mu,0}(\mathbb{R}_+)$ for $\mu \geq 0$ and $B \in \boldsymbol{D}^{-\mu,-\mu}(\mathbb{R}_+)$ for $\mu \leq 0$. The latter relation is a consequence of Lemma 4.1.49 and Lemma 4.1.57. For $-r = d - \mu \leq 0$ we take (4.1.86) as an operator on $H^{-r}(\mathbb{R}_+)$. We then write $B = \mathrm{op}^+(l_-^{-d})B_0 \,\mathrm{op}_{-r}^+(l_-^{-r})$ and obtain $B \in \boldsymbol{D}^{-\mu,d-\mu}(\mathbb{R}_+)$, using Definition 4.1.45 and Theorem 4.1.59. We obtain altogether $B \in \boldsymbol{D}^{-\mu,c}(\mathbb{R}_+)$ for (4.1.100).

Applying (4.1.77) we now obtain

$$AB = \mathrm{op}^+(l_-^{-s+\mu})A_0 \,\mathrm{op}_s^+(l_-^s)\,\mathrm{op}^+(l_-^{-s})B_0 \,\mathrm{op}_{s-\mu}^+(l_-^{s-\mu})$$
$$= \mathrm{op}^+(l_-^{-s+\mu})A_0 B_0 \,\mathrm{op}_{s-\mu}^+(l_-^{s-\mu}) = 1 + G_r$$

for $G_r = \mathrm{op}^+(l_-^{-s+\mu})G_{r,0}\,\mathrm{op}_{s-\mu}^+(l_-^{s-\mu})$, where $A_0 B_0 = 1 + G_{r,0}$, $G_{r,0} \in \Gamma^0(\mathbb{R}_+)$. Inserting $s = d$ we obtain $G_r \in \Gamma^0(\mathbb{R}_+)$ for $d - \mu \leq 0$ using Lemma 4.1.62. For $d - \mu \geq 0$ we get $G_r \in \Gamma^{d-\mu}(\mathbb{R}_+)$. This shows $G_r \in \Gamma^{(d-\mu)^+}(\mathbb{R}_+)$. On the other hand we have for $s > d - \frac{1}{2}$

$$BA = \mathrm{op}^+(l_-^{-s})B_0 \,\mathrm{op}_{s-\mu}^+(l_-^{s-\mu})\,\mathrm{op}^+(l_-^{-s+\mu})A_0 \,\mathrm{op}_s^+(l_-^s)$$
$$= \mathrm{op}^+(l_-^{-s})B_0 A_0 \,\mathrm{op}_s^+(l_-^s) = 1 + G_l$$

for $G_l = \mathrm{op}^+(l_-^{-s})G_{l,0}\,\mathrm{op}_s^+(l_-^s)$, where $B_0 A_0 = 1 + G_{l,0}$ for a corresponding $G_{l,0} \in \Gamma^0(\mathbb{R}_+)$. Taking $s = d$ we obtain $G_l \in \Gamma^d(\mathbb{R}_+)$. $\qquad\square$

Remark 4.1.73 *Let $A \in \boldsymbol{D}^{\mu,d}(\mathbb{R}_+)$ be elliptic, regarded as a continuous operator*

$$A : H^s(\mathbb{R}_+) \to H^{s-\mu}(\mathbb{R}_+)$$

for fixed $s \in \mathbb{R}$, $s > d - \frac{1}{2}$. Then the parametrix $B \in \boldsymbol{D}^{-\mu,c}(\mathbb{R}_+)$, constructed in Proposition 4.1.72, *induces a continuous operator*

$$B : H^{s-\mu}(\mathbb{R}_+) \to H^s(\mathbb{R}_+)$$

by the representation (4.1.85).

Proposition 4.1.74 *Let $A \in \boldsymbol{D}^{\mu,d}(\mathbb{R}_+)$ be elliptic. Then A extends to a Fredholm operator*

$$A : H^s(\mathbb{R}_+) \to H^{s-\mu}(\mathbb{R}_+)$$

for all $s \in \mathbb{R}$, $s > d - \frac{1}{2}$, and there are finite–dimensional subspaces $L_+, L_- \subset \mathcal{S}(\overline{\mathbb{R}}_+)$ with

$$L_+ = \ker\{A : H^s(\mathbb{R}_+) \to H^{s-\mu}(\mathbb{R}_+)\},$$

$$H^{s-\mu}(\mathbb{R}_+) = AH^s(\mathbb{R}_+) + L_- \qquad\qquad (4.1.87)$$

for all $s \in \mathbb{R}$, $s > d - \frac{1}{2}$, and analogously

$$L_+ = \ker\{A : \mathcal{S}(\overline{\mathbb{R}}_+) \to \mathcal{S}(\overline{\mathbb{R}}_+)\},$$
$$\mathcal{S}(\overline{\mathbb{R}}_+) = A\mathcal{S}(\overline{\mathbb{R}}_+) + L_-.$$

Proof. Let us first assume $A \in \boldsymbol{D}^{0,0}(\mathbb{R}_+)$. From Proposition 4.1.72 we get a parametrix $B \in \boldsymbol{D}^{0,0}(\mathbb{R}_+)$, where $AB = 1 + C_r$, $BA = 1 + C_l$ for certain $C_r, C_l \in \Gamma^0(\mathbb{R}_+)$. Since $A : H^s(\mathbb{R}_+) \to H^s(\mathbb{R}_+)$, $B : H^s(\mathbb{R}_+) \to H^s(\mathbb{R}_+)$ are continuous, and $C_r, C_l : H^s(\mathbb{R}_+) \to H^s(\mathbb{R}_+)$ are compact operators for all $s > -\frac{1}{2}$, we obtain that A is a Fredholm operator. Moreover, we have $L_+ = \ker A \subseteq \ker(1 + C_l) \subset \mathcal{S}(\overline{\mathbb{R}}_+)$, cf. Proposition 4.1.47. This holds for all $s > -\frac{1}{2}$. For the formal adjoint A^*, also being elliptic, defined as the restriction or extension to $H^s(\mathbb{R}_+)$ of the $L^2(\mathbb{R}_+)$-adjoint $A^* : L^2(\mathbb{R}_+) \to L^2(\mathbb{R}_+)$, we can argue in an analogous manner. For $L_- = \ker A^*$, first in the $L^2(\mathbb{R}_+)$-sense, it remains to verify that $AH^s(\mathbb{R}_+) + L_- = H^s(\mathbb{R}_+)$ for every $s > -\frac{1}{2}$. Let first $s \geq 0$, $l = \dim L_-$, and choose an isomorphism $K : \mathbb{C}^l \to L_-$. Then by construction the operator

$$(A \quad K) : \begin{matrix} L^2(\mathbb{R}_+) \\ \oplus \\ \mathbb{C}^l \end{matrix} \to L^2(\mathbb{R}_+) \tag{4.1.88}$$

is surjective. Now $Au + Kv = f \in H^s(\mathbb{R}_+)$ implies $u \in H^s(\mathbb{R}_+)$. In fact, we get $BAu + BKv = Bf \in H^s(\mathbb{R}_+)$, cf. Theorem 4.1.38, $BKv \in \mathcal{S}(\overline{\mathbb{R}}_+)$, and $BAu = (1 + C_l)u$. This gives us immediately $u \in H^s(\mathbb{R}_+)$. In other words the surjectivity of (4.1.88) implies the surjectivity of

$$(A \quad K) : \begin{matrix} H^s(\mathbb{R}_+) \\ \oplus \\ \mathbb{C}^l \end{matrix} \to H^s(\mathbb{R}_+) \tag{4.1.89}$$

for every $s \geq 0$. Analogously we get the surjectivity of

$$(A \quad K) : \begin{matrix} \mathcal{S}(\overline{\mathbb{R}}_+) \\ \oplus \\ \mathbb{C}^l \end{matrix} \to \mathcal{S}(\overline{\mathbb{R}}_+). \tag{4.1.90}$$

Now the closure of (4.1.90) in $H^s(\mathbb{R}_+)$ yields a continuous operator (4.1.89) also for $0 \geq s > -\frac{1}{2}$, and the domain is dense, since $\mathcal{S}(\overline{\mathbb{R}}_+)$ is dense in $H^s(\mathbb{R}_+)$. The operator (4.1.89) is injective. Since it is also a Fredholm operator, the range is closed; thus (4.1.89) is an isomorphism also for the remaining s which shows (4.1.87). For arbitrary $A \in \boldsymbol{D}^{\mu,d}(\mathbb{R}_+)$ it suffices to apply Theorem 4.1.61 and to note that the order reducing operators $\mathrm{op}_s^+(l_-^s)$ and $\mathrm{op}^+(l_-^{-s+\mu})$ induce isomorphisms $\mathcal{S}(\overline{\mathbb{R}}_+) \to \mathcal{S}(\overline{\mathbb{R}}_+)$. $\square$

Proposition 4.1.75 *Let* $a(\tau) = \sum_{j=0}^{\mu} a_j \tau^j$, $a_j \in \mathbb{C}$, $\mu \in \mathbb{N}$, *and assume that* $a(\tau)$ *is elliptic in the sense of* (i), (ii) *of Definition 4.1.68* *(which is equivalent to $a_\mu \neq 0$ and $a(\tau) \neq 0$ for all $\tau \in \mathbb{R}$). Then the operators*

$$\mathrm{op}^+(a) : H^s(\mathbb{R}_+) \to H^{s-\mu}(\mathbb{R}_+). \tag{4.1.91}$$

for all $s \in \mathbb{R}$, $s > -\frac{1}{2}$, and

$$\mathrm{op}^+(a) : \mathcal{S}(\overline{\mathbb{R}}_+) \to \mathcal{S}(\overline{\mathbb{R}}_+) \tag{4.1.92}$$

are surjective.

Proof. In view of Proposition 4.1.74 the operator (4.1.92) is Fredholm, and $\ker \mathrm{op}^+(a)$, $\mathrm{coker}\, \mathrm{op}^+(a)$ are independent of s. They also coincide with $\ker \mathrm{op}^+(a)$ and $\mathrm{coker}\, \mathrm{op}^+(a)$, respectively, as operators in $\mathcal{S}(\overline{\mathbb{R}}_+)$. Therefore it suffices to consider (4.1.92) for some particular s, e.g., $s \geq \mu$. The symbol $a^{-1}(\tau) \in S_{\mathrm{cl}}^{-\mu}(\mathbb{R})$ has the transmission property and we can form $\mathrm{op}^+(a^{-1}) : H^{s-\mu}(\mathbb{R}_+) \to H^s(\mathbb{R}_+)$, where

$$\mathrm{op}^+(a)\,\mathrm{op}^+(a^{-1}) = 1,$$

since $a(\tau)$ is a minus function, cf. Remark 4.1.36. This shows the surjectivity of $\mathrm{op}^+(a)$. □

Definition 4.1.76 *Let $\mu \in \mathbb{Z}$, $d \in \mathbb{N}$ and $N_-, N_+ \in \mathbb{N}$ and denote by $D^{\mu,d}(\mathbb{R}_+; N_-, N_+)$ the space of all operator block matrices*

$$\begin{pmatrix} A & K \\ T & Q \end{pmatrix} : \begin{matrix} \mathcal{S}(\overline{\mathbb{R}}_+) \\ \oplus \\ \mathbb{C}^{N_-} \end{matrix} \to \begin{matrix} \mathcal{S}(\overline{\mathbb{R}}_+) \\ \oplus \\ \mathbb{C}^{N_+} \end{matrix}$$

for arbitrary $A \in D^{\mu,d}(\mathbb{R}_+)$, $Tu = \{T_1 u, \dots, T_{N_+} u\}$ for

$$T_j u = \sum_{k=0}^{d} \int_0^\infty b_{jk}(t') \left(\frac{d}{dt'}\right)^k u(t')\, dt'$$

for arbitrary $b_{jk} \in \mathcal{S}(\overline{\mathbb{R}}_+)$, $j = 1, \dots, N_+$, $k = 0, \dots, d$, and

$$Kc = \sum_{l=1}^{N_-} f_l(t) c_l, \qquad c = (c_1, \dots, c_{N_-}) \in \mathbb{C}^{N_-},$$

for arbitrary $f_l \in \mathcal{S}(\overline{\mathbb{R}}_+)$, $l = 1, \dots, N_-$, and $N_+ \times N_-$–matrices Q. Moreover, $\Gamma^d(\mathbb{R}_+; N_-, N_+)$ denotes the subspace of all operators in $D^{\mu,d}(\mathbb{R}_+; N_-, N_+)$ with $A \in \Gamma^d(\mathbb{R}_+)$.

K is called a potential operator, T a trace operator of type d in the boundary symbolic calculus.

Remark 4.1.77 *Every trace operator* $T : \mathcal{S}(\overline{\mathbb{R}}_+) \to \mathbb{C}$ *of type* d *can be written*

$$Tu = \int\limits_0^\infty b(t')u(t')\,dt' + \sum_{j=0}^{d-1} \beta_j \left(\frac{d}{dt}\right)^j u(t)|_{t=0}$$

for certain $b \in \mathcal{S}(\overline{\mathbb{R}}_+)$ *and* $\beta_j \in \mathbb{C}$, $j = 0,\ldots,d-1$, *where* $b(t')$ *and* $\beta_0,\ldots,\beta_{d-1}$ *are uniquely determined by* T.

This follows similarly to Proposition 4.1.46.

Proposition 4.1.78 *Let* $E \in \Gamma^0(\mathbb{R}_+; N, N)$ *have the property that*

$$1 + E : \begin{array}{c} H^0(\mathbb{R}_+) \\ \oplus \\ \mathbb{C}^N \end{array} \to \begin{array}{c} H^0(\mathbb{R}_+) \\ \oplus \\ \mathbb{C}^N \end{array}$$

is an isomorphism. Then $(1 + E)^{-1} = 1 + C$ *for some* $C \in \Gamma^0(\mathbb{R}_+; N, N)$.

The arguments are similar to those for Proposition 4.1.43. The obvious modifications are left to the reader. The following remark is analogous to the above Proposition 4.1.66:

Remark 4.1.79 *Let* $G \in \Gamma^d(\mathbb{R}_+; N, N)$ *and assume that*

$$1 + G : \begin{array}{c} \mathcal{S}(\overline{\mathbb{R}}_+) \\ \oplus \\ \mathbb{C}^N \end{array} \to \begin{array}{c} \mathcal{S}(\overline{\mathbb{R}}_+) \\ \oplus \\ \mathbb{C}^N \end{array}$$

is an isomorphism. Then $(1 + G)^{-1} = 1 + C$ *for some* $C \in \Gamma^d(\mathbb{R}_+; N, N)$.

Remark 4.1.80 *Similarly to Remark* 4.1.67 *the space* $\Gamma^d(\mathbb{R}_+; N, N)$ *is Fréchet in a canonical way and the invertibility of* $1 + G(y)$ *for a family* $G(y) \in C^\infty(\Omega, \Gamma^d(\mathbb{R}_+; N, N))$ *for open* $\Omega \subseteq \mathbb{R}^p$ *implies* $C(y) \in C^\infty(\Omega, \Gamma^d(\mathbb{R}_+; N, N))$.

Proposition 4.1.81 $A \in \boldsymbol{D}^{\mu,d}(\mathbb{R}_+; N_0, N_+)$, $B \in \boldsymbol{D}^{\nu,c}(\mathbb{R}_+; N_-, N_0)$ *implies* $AB \in \boldsymbol{D}^{\mu+\nu,h}(\mathbb{R}_+; N_-, N_+)$ *for* $h = \max\{(d+\nu)^+, c\}$.

Proof. The composition of upper left corners in the block matrices was treated above in Theorem 4.1.59. In addition the upper left corners contain the compositions {potential} ∘ {trace} which obviously belong to $\Gamma^h(\mathbb{R}_+)$. The remaining compositions can be characterised by using Remark 4.1.54. □

Remark 4.1.82 *Each* $A \in \boldsymbol{D}^{\mu,d}(\mathbb{R}_+; N_-, N_+)$ *induces continuous operators*

$$
A: \quad
\begin{array}{ccc}
H^s(\mathbb{R}_+) & & H^{s-\mu}(\mathbb{R}_+) \\
\oplus & \to & \oplus \\
\mathbb{C}^{N_-} & & \mathbb{C}^{N_+}
\end{array}
\tag{4.1.93}
$$

for all $s \in \mathbb{R}$, $s > d - \frac{1}{2}$. *If the upper left corner of* A *is elliptic in the sense of* Definition 4.1.68, *the operator* (4.1.93) *is Fredholm, and there are finite–dimensional subspaces*

$$
L_+ \subset \mathcal{S}(\overline{\mathbb{R}}_+) \oplus \mathbb{C}^{N_-}, \qquad L_- \subset \mathcal{S}(\overline{\mathbb{R}}_+) \oplus \mathbb{C}^{N_+}
$$

with

$$
L_+ = \{\text{kernel of } (4.1.93)\}, \tag{4.1.94}
$$
$$
\{\text{image of } (4.1.93)\} + L_- = H^{s-\mu}(\mathbb{R}_+) \oplus \mathbb{C}^{N_+} \tag{4.1.95}
$$

for every $s \in \mathbb{R}$, $s > d - \frac{1}{2}$.

Theorem 4.1.83 *Let* $A \in \boldsymbol{D}^{\mu,d}(\mathbb{R}_+; N_-, N_+)$ *be an operator, where the upper left corner is elliptic in* $\boldsymbol{D}^{\mu,d}(\mathbb{R}_+)$. *Then* (4.1.93) *is invertible for any* $s \in \mathbb{R}$, $s > -\frac{1}{2}$, *if and only if*

$$
A: \quad
\begin{array}{ccc}
\mathcal{S}(\overline{\mathbb{R}}_+) & & \mathcal{S}(\overline{\mathbb{R}}_+) \\
\oplus & \to & \oplus \\
\mathbb{C}^{N_-} & & \mathbb{C}^{N_+}
\end{array}
\tag{4.1.96}
$$

is invertible. In that case we have $A^{-1} \in \boldsymbol{D}^{-\mu,(d-\mu)^+}(\mathbb{R}_+; N_+, N_-)$. *In particular, for* $d = \mu^+$ *we have* $(d - \mu)^+ = (-\mu)^+$.

Proof. Similarly to Theorem 4.1.61 we can set

$$
A = \text{diag}(\text{op}^+(l_-^{-s+\mu}), \text{id}_{\mathbb{C}^{N_+}}) A_0 \, \text{diag}(\text{op}_s^+(l_-^s), \text{id}_{\mathbb{C}^{N_-}}),
$$

for $A_0 = \text{diag}(\text{op}_{s-\mu}^+(l_-^{s-\mu}), \text{id}_{\mathbb{C}^{N_+}}) A \, \text{diag}(\text{op}^+(l_-^{-s}), \text{id}_{\mathbb{C}^{N_-}})$, $s \in \mathbb{R}$, $s > d - \frac{1}{2}$. This reduces the question on invertibility of the operator A to $A_0 \in \boldsymbol{D}^{0,0}(\mathbb{R}_+; N_-, N_+)$. Suppose we already have constructed the inverse $A_0^{-1} \in \boldsymbol{D}^{0,0}(\mathbb{R}_+; N_+, N_-)$. Then we get immediately $A^{-1} \in \boldsymbol{D}^{-\mu,(d-\mu)^+}(\mathbb{R}_+; N_+, N_-)$ by composing A_0^{-1} by the corresponding order reducing isomorphisms which are isomorphisms on the Sobolev spaces as well as on the Schwartz spaces in the first components. In other words, we may assume from now on $A \in \boldsymbol{D}^{0,0}(\mathbb{R}_+; N_-, N_+)$. Remark 4.1.82 shows that (4.1.96) is an isomorphism if (4.1.93) is an isomorphism (for $\mu = 0$) for any $s > -\frac{1}{2}$. It remains to construct A^{-1} in

$D^{0,0}(\mathbb{R}_+; N_+, N_-)$. Set $A_{11} = $ u.l.c. A (recall that u.l.c. is an abbreviation of upper left corner), and let $B_{11} \in D^{0,0}(\mathbb{R}_+)$ be a parametrix of A_{11}, cf. Proposition 4.1.72. Then $\mathrm{ind}\, B_{11} = N_- - N_+$. Applying Proposition 4.1.74 to the operator B_{11} we find a subspace $M_- \subset \mathcal{S}(\overline{\mathbb{R}}_+)$ of dimension $\tilde{N}_+ \geq N_+$ and an isomorphism $C : \mathbb{C}^{\tilde{N}_+} \to M_-$ such that

$$
\begin{pmatrix} B_{11} & C \end{pmatrix} : \begin{matrix} H^0(\mathbb{R}_+) \\ \oplus \\ \mathbb{C}^{\tilde{N}_+} \end{matrix} \to H^0(\mathbb{R}_+) \tag{4.1.97}
$$

is surjective. $\ker \begin{pmatrix} B_{11} & C \end{pmatrix}$ is a finite–dimensional subspace $\tilde{M}_+ \subset H^0(\mathbb{R}_+) \oplus \mathbb{C}^{\tilde{N}_+}$, $\tilde{N}_- := \dim \tilde{M}_+ \geq N_-$, consisting of pairs $(v(t), w) \in \mathcal{S}(\overline{\mathbb{R}}_+) \oplus \mathbb{C}^{\tilde{N}_+}$. Let $P : H^0(\mathbb{R}_+) \oplus \mathbb{C}^{\tilde{N}_+} \to \tilde{M}_+$ be the orthogonal projection, and let $\begin{pmatrix} \tilde{D} & \tilde{R} \end{pmatrix} : \tilde{M}_+ \to \mathbb{C}^{\tilde{N}_-}$ be an isomorphism. Then $\begin{pmatrix} D & R \end{pmatrix} := \begin{pmatrix} \tilde{D} & \tilde{R} \end{pmatrix} \circ P : H^0(\mathbb{R}_+) \oplus \mathbb{C}^{\tilde{N}_+} \to \mathbb{C}^{\tilde{N}_-}$ fills (4.1.97) to a block matrix

$$
\begin{pmatrix} B_{11} & C \\ D & R \end{pmatrix} : \begin{matrix} H^0(\mathbb{R}_+) \\ \oplus \\ \mathbb{C}^{\tilde{N}_+} \end{matrix} \to \begin{matrix} H^0(\mathbb{R}_+) \\ \oplus \\ \mathbb{C}^{\tilde{N}_-} \end{matrix} \tag{4.1.98}
$$

which is an isomorphism. In order to obtain (4.1.98) in the space $D^{0,0}(\mathbb{R}_+; \tilde{N}_+, \tilde{N}_-)$ we choose the operators $\begin{pmatrix} \tilde{D} & \tilde{R} \end{pmatrix}$ in such a way that for an orthogonal base $\{(v_j, w_j)\}_{j=1,\dots,\tilde{N}_-}$ in $\tilde{M}_+$ we have

$$
\begin{pmatrix} D & R \end{pmatrix} \begin{pmatrix} f \\ g \end{pmatrix} = \left(\left(\begin{pmatrix} f \\ g \end{pmatrix}, \begin{pmatrix} \overline{v}_j \\ \overline{w}_j \end{pmatrix} \right) \right)_{j=1,\dots,\tilde{N}_-}
$$

for all $(f, g) \in H^0(\mathbb{R}_+) \oplus \mathbb{C}^{\tilde{N}_+}$, where $(\cdot, \cdot)$ is the scalar product in that space. We have $\tilde{N}_+ - N_+ = \tilde{N}_- - N_- =: L$. Let us form the isomorphism

$$
\tilde{A} := \begin{pmatrix} A & 0 \\ 0 & 1_L \end{pmatrix} : \begin{matrix} H^0(\mathbb{R}_+) \\ \oplus \\ \mathbb{C}^{\tilde{N}_-} \end{matrix} \to \begin{matrix} H^0(\mathbb{R}_+) \\ \oplus \\ \mathbb{C}^{\tilde{N}_+} \end{matrix} .
$$

Then $\tilde{A}\tilde{B} = 1 + E : H^0(\mathbb{R}_+) \oplus \mathbb{C}^{\tilde{N}_+} \to H^0(\mathbb{R}_+) \oplus \mathbb{C}^{\tilde{N}_+}$ is also an isomorphism. Here $E \in \Gamma^0(\mathbb{R}_+; \tilde{N}_+, \tilde{N}_+)$, cf. Definition 4.1.76. Applying Proposition 4.1.78 we find a $C \in \Gamma^0(\mathbb{R}_+; \tilde{N}_+, \tilde{N}_+)$ with $(1 + E)(1 + C) = 1$. Thus $\tilde{A}\tilde{B}(1 + C) = 1$ and hence $\tilde{A}^{-1} = \tilde{B}(1 + C) \in D^{0,0}(\mathbb{R}_+; \tilde{N}_+, \tilde{N}_-)$, cf. Proposition 4.1.81. In view of the special form of $\tilde{A}$ the operator $\tilde{B}$ can be written $\mathrm{diag}(B, 1_L)$ for some $B \in D^{0,0}(\mathbb{R}_+; \tilde{N}_+, \tilde{N}_-)$, and then $A^{-1} = B$. $\quad\square$

Let us now turn to some examples of invertible elements in the space $\boldsymbol{D}^{\mu,d}(\mathbb{R}_+; N_-, N_+)$. Consider the operator

$$\begin{pmatrix} \mathrm{op}^+(a) \\ r' \end{pmatrix} : H^s(\mathbb{R}_+) \to \begin{matrix} H^{s-2}(\mathbb{R}_+) \\ \oplus \\ \mathbb{C} \end{matrix} \qquad (4.1.99)$$

for $r'u := u(0)$ and $a(\tau) := -\delta^2 - \tau^2$ for some fixed $\delta > 0$. Then

$$\mathrm{op}^+(a)u(t) = \left(-\delta^2 + \frac{\partial^2}{\partial t^2}\right)u(t)$$

defines a surjective operator $\mathrm{op}^+(a) : H^s(\mathbb{R}_+) \to H^{s-2}(\mathbb{R}_+)$ by Proposition 4.1.75. We have

$$\ker \mathrm{op}^+(a) = \{ce^{-\delta t} : c \in \mathbb{C}\}$$

which is a one–dimensional subspace of $\mathcal{S}(\overline{\mathbb{R}}_+)$. Since $r' : \ker \mathrm{op}^+(a) \to \mathbb{C}$ is an isomorphism, $(4.1.99)$ is also an isomorphism. We have

$$\mathrm{op}^+(a)\,\mathrm{op}^+(a^{-1}) = 1, \qquad (4.1.100)$$

and hence $\mathrm{op}^+(a^{-1}) : H^{s-2}(\mathbb{R}_+) \to H^s(\mathbb{R}_+)$ is injective. Moreover, the relation $(4.1.100)$ shows $\mathrm{im}\,\mathrm{op}^+(a^{-1}) \cap \ker \mathrm{op}^+(a) = \{0\}$ and $\mathrm{im}\,\mathrm{op}^+(a^{-1}) \oplus \ker \mathrm{op}^+(a) = H^s(\mathbb{R}_+)$. Thus, if we define a map $k_0 : \mathbb{C} \to \mathcal{S}(\overline{\mathbb{R}}_+)$ by $k_0 c = ce^{-\delta t}$, $c \in \mathbb{C}$, we get an isomorphism

$$(\mathrm{op}^+(a^{-1}), k_0) : \begin{matrix} H^{s-2}(\mathbb{R}_+) \\ \oplus \\ \mathbb{C} \end{matrix} \to H^s(\mathbb{R}_+).$$

We have $\mathrm{op}^+(a) \circ k_0 = 0$, $r'k_0 = 1$. Moreover, $b := r'\,\mathrm{op}^+(a^{-1})$ is a trace operator of type zero. Now

$$\begin{pmatrix} \mathrm{op}^+(a) \\ r' \end{pmatrix}(\mathrm{op}^+(a^{-1}), k_0) = \begin{pmatrix} 1 & 0 \\ b & 1 \end{pmatrix} : \begin{matrix} H^{s-2}(\mathbb{R}_+) \\ \oplus \\ \mathbb{C} \end{matrix} \to \begin{matrix} H^{s-2}(\mathbb{R}_+) \\ \oplus \\ \mathbb{C} \end{matrix}$$

is an isomorphism. Thus

$$\begin{pmatrix} \mathrm{op}^+(a) \\ r' \end{pmatrix}^{-1} = (\mathrm{op}^+(a^{-1}), k_0)\begin{pmatrix} 1 & 0 \\ -b & 1 \end{pmatrix} = (\mathrm{op}^+(a^{-1}) - k_0 b, k_0).$$

If we assume in this calculation first $s \geq 2$, we obtain the resulting inverse also for all $s > -\frac{1}{2}$. Moreover, we also get the inverse of

$$\begin{pmatrix} \mathrm{op}^+(a) \\ r' \end{pmatrix} : \mathcal{S}(\overline{\mathbb{R}}_+) \to \begin{matrix} \mathcal{S}(\overline{\mathbb{R}}_+) \\ \oplus \\ \mathbb{C} \end{matrix} \; .$$

We shall see below that the inversion of (4.1.99) belongs to the solution of the Dirichlet problem for the Laplace operator and that, in particular, $\mathrm{op}^+(a^{-1}) - k_0 r' \mathrm{op}^+(a^{-1})$ can be interpreted as the principal boundary symbol of Green's function. Analogously we can consider Neumann boundary conditions. In this case we have to invert

$$\begin{pmatrix} \mathrm{op}^+(a) \\ b_1 \end{pmatrix} : H^s(\mathbb{R}_+) \to \begin{matrix} H^{s-2}(\mathbb{R}_+) \\ \oplus \\ \mathbb{C} \end{matrix}$$

for $b_1 := r'\frac{\partial}{\partial t}$, $s > \frac{1}{2}$. We can form the isomorphism

$$\begin{pmatrix} \mathrm{op}^+(a) \\ b_1 \end{pmatrix} \quad (\mathrm{op}^+(a^{-1}), k_0)$$

$$= \begin{pmatrix} 1 & 0 \\ b_2 & b_1 k_0 \end{pmatrix} : \begin{matrix} H^{s-2}(\mathbb{R}_+) \\ \oplus \\ \mathbb{C} \end{matrix} \to \begin{matrix} H^{s-2}(\mathbb{R}_+) \\ \oplus \\ \mathbb{C} \end{matrix}$$

for $b_2 := b_1 \mathrm{op}^+(a^{-1})$, where $c := -b_1 k_0 \neq 0$. Thus

$$\begin{pmatrix} \mathrm{op}^+(a) \\ b_1 \end{pmatrix}^{-1} = (\mathrm{op}^+(a^{-1}), k_0) \begin{pmatrix} 1 & 0 \\ c^{-1}b_2 & -c^{-1} \end{pmatrix}$$

$$= (\mathrm{op}^+(a^{-1}) + c^{-1}k_0 b_2, -c^{-1}k_0).$$

4.2 The symbolic calculus

4.2.1 Interior symbols

The symbolic calculus for boundary value problems consists of two components, namely the symbols in the interior and those along the boundary. The interior symbols are the standard ones, associated with pseudo–differential operators in the open interior of the given manifold with boundary. The boundary symbols are operator–valued, with values in spaces of operators, defined on the inner normal to the boundary, and the variables and covariables are those of the boundary. The following sections will investigate the interactions between the interior and the boundary symbols.

Definition 4.2.1 *A symbol* $a(t, y, \tau, \eta) \in S_{\mathrm{cl}}^\mu(\overline{\mathbb{R}}_+ \times \Omega \times \mathbb{R}^{1+q})$ *for* $\mu \in \mathbb{Z}$ *is said to have the transmission property (with respect to* $t = 0$*) if*

$$a_{\gamma;y,\eta}(t, \tau) := D_\eta^\gamma a(t, y, \tau, \eta)$$

has the transmission property in the sense of Definition 4.1.26 for every $\gamma \in \mathbb{N}^q$, $(y, \eta) \in \Omega \times \mathbb{R}^q$.

Remark 4.2.2 *The definition of the transmission property can be generalised considerably. It is not our aim here to give the most general version. In our formalism with operator–valued symbols it is convenient to require classical symbols and integer orders. All symbols in $S_{cl}^\mu(\overline{\mathbb{R}} \times \Omega \times \mathbb{R}^{1+q})$, $\mu \in \mathbb{R}$, with or without the transmission property, are contained in the edge operator framework, cf. Schulze [122].*

Remark 4.2.3 $a(t, y, \tau, \eta) \in S_{cl}^\mu(\overline{\mathbb{R}}_+ \times \Omega \times \mathbb{R}^{1+q})$ *has the transmission property if and only if for the homogeneous components* $a_{(\mu-j)}(t, y, \tau, \eta)$, $j \in \mathbb{N}$,

$$a_{(\mu-j)}(t, y, -\tau, -\eta) - e^{i\pi(\mu-j)} a_{(\mu-j)}(t, y, \tau, \eta)$$

vanish to infinite order on the set $\{(0, y, \tau, 0) : y \in \Omega, \tau \neq 0\}$, *cf. Boutet de Monvel [10].*

Remark 4.2.4 *The subspace of all symbols with the transmission property is closed in the Fréchet topology induced by $S_{cl}^\mu(\overline{\mathbb{R}}_+ \times \Omega \times \mathbb{R}^{1+q})$.*

The following propositions are very elementary; so the proofs will be omitted.

Proposition 4.2.5 *Let $a(t, y, \tau, \eta) \in S_{cl}^\mu(\overline{\mathbb{R}}_+ \times \Omega \times \mathbb{R}^{1+q})$, $b(t, y, \tau, \eta) \in S_{cl}^\nu(\overline{\mathbb{R}}_+ \times \Omega \times \mathbb{R}^{1+q})$ have the transmission property. Then the following symbols also have the transmission property:*

(i) *the point-wise product $a(t, y, \tau, \eta)b(t, y, \tau, \eta) \in S_{cl}^{\mu+\nu}(\overline{\mathbb{R}}_+ \times \Omega \times \mathbb{R}^{1+q})$.*

(ii) $D_{t,y}^\alpha D_{\tau,\eta}^\beta a(t, y, \tau, \eta) \in S_{cl}^{\mu-|\beta|}(\overline{\mathbb{R}}_+ \times \Omega \times \mathbb{R}^{1+q})$ *for all $\alpha, \beta \in \mathbb{N}^{1+q}$.*

(iii) *If $a_j(t, y, \tau, \eta) \in S_{cl}^{\mu-j}(\overline{\mathbb{R}}_+ \times \Omega \times \mathbb{R}^{1+q})$, $j \in \mathbb{N}$, is an arbitrary sequence with the transmission property, then*

$$a(t, y, \tau, \eta) \sim \sum_{j=0}^\infty a_j(t, y, \tau, \eta)$$

(the asymptotic sum in $S_{cl}^\mu(\overline{\mathbb{R}}_+ \times \Omega \times \mathbb{R}^{1+q})$) also has the transmission property.

(iv) *We have*

$$a(t, y, \tau, \eta)\#b(t, y, \tau, \eta) \sim \sum \frac{1}{\alpha!} D_{\tau,\eta}^\alpha a(t, y, \tau, \eta)\partial_{t,y}^\alpha b(t, y, \tau, \eta)$$
$$\in S_{cl}^{\mu+\nu}(\overline{\mathbb{R}}_+ \times \Omega \times \mathbb{R}^{1+q})$$

(associated with the composition of operators).

(v) $a^*(t, y, \tau, \eta) \sim \sum \frac{1}{\alpha!} D^\alpha_{\tau,\eta} \partial^\alpha_{t,y} \overline{a(t, y, \tau, \eta)}$ *(associated with the formal adjoint operator)*.

Proposition 4.2.6 *Let* $a(t, y, \tau, \eta) \in S^\mu_{\mathrm{cl}}(\overline{\mathbb{R}}_+ \times \Omega \times \mathbb{R}^{1+q})$, *be a symbol with the transmission property, and assume that the homogeneous principal part* $a_{(\mu)}(t, y, \tau, \eta)$ *of* $a(t, y, \tau, \eta)$ *of order* μ *satisfies* $a_{(\mu)}(t, y, \tau, \eta) \neq 0$ *for all* $(y, \eta) \in \overline{\mathbb{R}}_+ \times \Omega$, $(\tau, \eta) \neq 0$. *Then the symbol*

$$\chi(\tau, \eta) a^{-1}_{(\mu)}(t, y, \tau, \eta) \in S^{-\mu}_{\mathrm{cl}}(\overline{\mathbb{R}}_+ \times \Omega \times \mathbb{R}^{1+q})$$

has the transmission property for every excision function $\chi(\tau, \eta)$.

Moreover, every $b(t, y, \tau, \eta) \in S^{-\mu}_{\mathrm{cl}}(\overline{\mathbb{R}}_+ \times \Omega \times \mathbb{R}^{1+q})$, *satisfying* $a(t, y, \tau, \eta) \# b(t, y, \tau, \eta) - 1 \in S^{-\infty}(\overline{\mathbb{R}}_+ \times \Omega \times \mathbb{R}^{1+q})$, *has the transmission property.*

Remark 4.2.7 *Let* $S^{(\mu)}_{\mathrm{trans}}(\overline{\mathbb{R}}_+ \times \Omega \times (\mathbb{R}^{1+q} \setminus \{0\}))$ *denote the subspace of all* $a_{(\mu)}(t, y, \tau, \eta) \in C^\infty(\overline{\mathbb{R}}_+ \times \Omega \times (\mathbb{R}^{1+q} \setminus \{0\})$ *with* $a_{(\mu)}(t, y, \lambda\tau, \lambda\eta) = \lambda^\mu a_{(\mu)}(t, y, \tau, \eta)$ *for all* $\lambda \in \mathbb{R}_+$ *and all* $(t, y, \tau, \eta) \in \overline{\mathbb{R}}_+ \times \Omega \times (\mathbb{R}^{1+q} \setminus \{0\})$, *which are homogeneous principal symbols of certain* $a(t, y, \tau, \eta) \in S^\mu_{\mathrm{cl}}(\overline{\mathbb{R}}_+ \times \Omega \times \mathbb{R}^{1+q})$ *with the transmission property. Then* $S^{(\mu)}_{\mathrm{trans}}(\overline{\mathbb{R}}_+ \times \Omega \times (\mathbb{R}^{1+q} \setminus \{0\}))$ *is a Fréchet space in the topology induced by* $C^\infty(\overline{\mathbb{R}}_+ \times \Omega \times (\mathbb{R}^{1+q} \setminus \{0\}))$.

Let us now construct a useful example for a symbol with the transmission property. Choose a function $f(\zeta) \in \mathcal{S}(\mathbb{R})$ with $\mathrm{supp}\, f \subset \mathbb{R}_-$ and $\int f(\zeta)\, d\zeta = 1$. Then

$$\chi(\tau) := \int e^{-i\tau\zeta} f(\zeta)\, d\zeta \in \mathcal{S}(\mathbb{R}_\tau),$$

and $\chi(0) = 1$. Let $\delta = \sup |\partial_\tau \chi(\tau)|$ and $\rho > \delta$ a constant so large that $\chi(\tau/\rho\langle\eta\rangle)\langle\eta\rangle - i\tau \neq 0$ for all $(\tau, \eta) \in \mathbb{R}^{1+q}$. Define

$$r^\mu_-(\tau, \eta) = \left(\chi\left(\frac{\tau}{\rho\langle\eta\rangle}\right) \langle\eta\rangle - i\tau \right)^\mu \tag{4.2.1}$$

and

$$r^\mu_+(\tau, \eta) = \overline{r^\mu_-(\tau, \eta)} \tag{4.2.2}$$

for all $\mu \in \mathbb{Z}$.

Remark 4.2.8 *We have*

$$r^\mu_-(\tau, \eta) \in H^-, \qquad r^\mu_+(\tau, \eta) \in H^+ + H'$$

for each $\eta \in \mathbb{R}^q$, *cf. Lemma 4.1.34, Remark 4.1.35.*

Proposition 4.2.9 *The symbols* $r_\pm^\mu(\tau,\eta)$ *belong to* $S_{\mathrm{cl}}^\mu(\mathbb{R}^{1+q})$.

Proof. We shall show the assertion for $r_-^\mu(\tau,\eta)$ for $\mu = 1$. Then the general case is then a trivial consequence. It suffices to consider $\chi(\tau/\rho\langle\eta\rangle)\langle\eta\rangle$, since $i\tau$ is classical. Moreover, we may set $\rho = 1$.

Let $\varphi(t)$ be an excision function on $\mathbb{R}$, i.e., $\varphi(t) = 0$ near $t = 0$, $\varphi(t) = 1$ for $|t| > $ const. First we want to show

$$a_k(\xi) := \varphi(|\xi|)\chi\left(\frac{\tau}{|\eta|}\right)|\eta|^k \in S_{\mathrm{cl}}^k(\mathbb{R}^{1+q}) \qquad (4.2.3)$$

for each $k \in \mathbb{Z}$; here $\xi = (\tau,\eta)$. The function $a_k(\xi)$ is homogeneous of order k for large $|\xi|$. In order to prove that $a_k(\xi)$ is a symbol it suffices to show $a_k(\xi) \in C^\infty(\mathbb{R}^{1+q})$. Every derivative is a linear combination of functions of the form $\psi(|\xi|)\chi(\tau/|\eta|)|\eta|^k\xi^\alpha$ for a multi–index α, where $\psi(t)$ equals $\varphi(t)$, or $\psi(t)$ is an element in $C_0^\infty(\mathbb{R})$ vanishing near $t = 0$. So we only have to show continuity of such functions. Let us take, for instance, $a_k(\xi)$ itself. Since φ vanishes near zero, the only points of interest are those of the form $\eta = 0$, $\tau \neq 0$. Now if (η_j,τ_j) is a sequence with $\eta_j \neq 0$, $\eta_j \to 0$, $\tau_j \to c \neq 0$, then $\chi(\tau/|\eta|)|\eta|^k \to 0$, since $\chi(t)t^k \to 0$ for $t \to \infty$ for arbitrary k.

Next we want to show under the same assumptions as above

$$|\chi(\tau/\langle\eta\rangle)\langle\eta\rangle^k - \varphi(|\xi|)\chi(\tau/|\eta|)|\eta|^k| \leq c\langle\xi\rangle^{k-2} \qquad (4.2.4)$$

for all $\xi \in \mathbb{R}^{1+q}$, for a constant $c > 0$. Here we may assume $|\xi| \geq 1$ so large that $\varphi(|\xi|) = 1$. For $|\eta| \leq \frac{1}{2}$ we then have $|\tau| \geq \frac{1}{2}$ and $|\tau| \geq \frac{1}{2}|\xi| \geq \frac{1}{4}\langle\xi\rangle$. For arbitrary $m \in \mathbb{N}$ and suitable constants c, c_1, c_2 we get

$$\chi(\tau/\langle\eta\rangle)\langle\eta\rangle^k \leq c \left\langle \frac{\tau}{\langle\eta\rangle} \right\rangle^{-m} \langle\eta\rangle^k$$
$$\leq c_1\langle\tau\rangle^{-m}\langle\eta\rangle^k$$
$$\leq c_2\langle\xi\rangle^{k-m}.$$

and similarly $\varphi(|\xi|)\chi(\tau/|\eta|)|\eta|^k \leq c\langle\xi\rangle^{k-m}$. So it remains to consider $|\xi| \geq 1$, $|\eta| \geq \frac{1}{2}$, where in the estimates $|\eta|$ may be replaced by $\langle\eta\rangle$. We have for $k \geq 0$

$$a^k - b^k = (a - b)\sum_{j=0}^{k-1} a^j b^{k-1-j}.$$

Setting $a = \langle\eta\rangle$, $b = |\eta|$ this shows that

$$\langle\eta\rangle^k - |\eta|^k = ((\langle\eta\rangle + |\eta|)(\langle\eta\rangle - |\eta|)(\langle\eta\rangle + |\eta|)^{-1}\sum_{j=0}^{k-1}\langle\eta\rangle^j|\eta|^{k-1-j}$$
$$\leq c\langle\eta\rangle^{k-2}, \qquad (4.2.5)$$

where we have used $(\langle\eta\rangle + |\eta|)(\langle\eta\rangle - |\eta|) = \langle\eta\rangle^2 - |\eta|^2 = 1$. For $k < 0$ we take $a = \langle\eta\rangle^{-1}$, $b = |\eta|^{-1}$ and obtain the same result, using that

$$\frac{1}{|\eta|} - \frac{1}{\langle\eta\rangle} = \frac{1}{\langle\eta\rangle|\eta|(\langle\eta\rangle + |\eta|)}.$$

Now the difference in (4.2.4) is $\leq f_1 + f_2$ for

$$f_1 = \left|\chi\left(\frac{\tau}{\langle\eta\rangle}\right) - \chi\left(\frac{\tau}{|\eta|}\right)\right|\langle\eta\rangle^k,$$

$$f_2 = \left|\chi\left(\frac{\tau}{|\eta|}\right)\right||\langle\eta\rangle^k - |\eta|^k|.$$

Let $I(\xi)$ denote the interval between $\tau/\langle\eta\rangle$ and $\tau/|\eta|$. Then

$$\sup_{\tau\in I(\xi)} |\partial_\tau\chi(\tau)| \leq c\left\langle\frac{\tau}{\langle\eta\rangle}\right\rangle^{-N}$$

for each N, with a constant $c = c_N$. Thus

$$|f_1| \leq \sup_{\tau\in I(\xi)} |\partial_\tau\chi(\tau)|\left|\frac{\tau}{\langle\eta\rangle} - \frac{\tau}{|\eta|}\right|\langle\eta\rangle^k$$

$$\leq c\left\langle\frac{\tau}{\langle\eta\rangle}\right\rangle^{-N}\left|\frac{\tau}{\langle\eta\rangle} - \frac{\tau}{|\eta|}\right|\langle\eta\rangle^k. \tag{4.2.6}$$

Moreover, we have $|\tau/\langle\eta\rangle - \tau/|\eta|| = |\tau||\eta|^{-1}\langle\eta\rangle^{-1}(\langle\eta\rangle + |\eta|)^{-1}$. Then $\langle\tau/\langle\eta\rangle\rangle\langle\eta\rangle = \langle\xi\rangle$ implies for N large enough

$$|f_1| \leq c\left\langle\frac{\tau}{\langle\eta\rangle}\right\rangle^{-N} |\tau||\eta|^{-1}\langle\eta\rangle^{-1}(\langle\eta\rangle + |\eta|)^{-1}\langle\eta\rangle^k$$

$$\leq c\langle\xi\rangle^{k-2}.$$

For f_2 we use $\chi(\tau/|\eta|) \leq c\langle\tau/\langle\eta\rangle\rangle^{-N}$ for every N with a suitable $c = c_N$. Then, together with (4.2.5), we obtain $|f_2| \leq c\langle\xi\rangle^{k-2}$. Thus we have proved (4.2.4). Next we show $p_1(\xi) := \chi(\frac{\tau}{\langle\eta\rangle})\langle\eta\rangle \in S_{\mathrm{cl}}^1(\mathbb{R}^n)$. By the above calculations we have $\varphi(|\xi|)\chi(\tau/|\eta|)|\eta| \in S_{\mathrm{cl}}^1(\mathbb{R}^n)$. Moreover, (4.2.4) yields

$$r_0(\xi) := \chi(\tau/\langle\eta\rangle)\langle\eta\rangle - \varphi(|\xi|)\chi(\tau/|\eta|)|\eta| \in S^0(\mathbb{R}^n).$$

This gives us the first term of the asymptotic expansion for $p_1(\xi)$. Let us now write $r_0(\xi) = g_1(\xi) + g_2(\xi)$ for $|\xi| \geq 1$, $|\eta| \neq 0$, where

$$g_1 = \left(\chi\left(\frac{\tau}{\langle\eta\rangle}\right) - \chi\left(\frac{\tau}{|\eta|}\right)\right)\langle\eta\rangle,$$

$$g_2 = \chi\left(\frac{\tau}{|\eta|}\right)(\langle\eta\rangle - |\eta|).$$

By Taylor expansion we get

$$\chi(t) = \sum_{j=0}^{N} \frac{1}{j!} \chi^{(j)}(t_0)(t - t_0)^j + b_N \chi^{(N+1)}(\theta)(t - t_0)^{N+1}$$

for some constant b_N. For $t = \tau/\langle\eta\rangle$, $t_0 = \tau/|\eta|$ we get $t - t_0 = -(\tau/|\eta|)\langle\eta\rangle^{-1}(\langle\eta\rangle + |\eta|)^{-1}$, which follows from $(\langle\eta\rangle + |\eta|)(\langle\eta\rangle - |\eta|) = 1$. Then

$$\chi(\tau/\langle\eta\rangle) = \sum_{j=0}^{N} \frac{(-1)^j}{j!} \chi^{(j)}(\tau/|\eta|) \left(\frac{\tau}{|\eta|}\right)^j \langle\eta\rangle^{-j}(\langle\eta\rangle + |\eta|)^{-j}$$

$$+ \tilde{b}_N \chi^{(N+1)}(\theta) \left(\frac{\tau}{|\eta|}\right)^{N+1} \langle\eta\rangle^{-N-1}(\langle\eta\rangle + |\eta|)^{-N-1}$$

with another constant $\tilde{b}_N$, and θ between $\tau/\langle\eta\rangle$ and $\tau/|\eta|$. Moreover, we have a Taylor expansion

$$(1 + t^2)^{\frac{1}{2}} = 1 + \sum_{j=1}^{N} c_j t^j + c_N(t)t^{N+1}$$

that we use for $t = |\eta|^{-1}$. Hence $\langle\eta\rangle = |\eta| \left\langle \frac{1}{|\eta|} \right\rangle$ gives us

$$\langle\eta\rangle - |\eta| = |\eta|(\left\langle \frac{1}{|\eta|} \right\rangle - 1)$$

$$= \sum_{j=1}^{N} c_j |\eta|^{1-j} + c_N(|\eta|^{-1})|\eta|^{-N}. \qquad (4.2.7)$$

Here $c_N(|\eta|^{-1})$ is, up to constants, a derivative of $(1 + t^2)^{1/2}$, $t = |\eta|^{-1}$, which is bounded for $N \geq 1$.

Let us finally employ the expansion

$$\frac{1}{1 + \langle t \rangle} = \sum_{j=0}^{N} d_j t^j + d_N(t)t^{N+1},$$

$d_N(t)$ being bounded on $\mathbb{R}$. This yields

$$(\langle\eta\rangle + |\eta|)^{-1} = |\eta|^{-1}(1 + \langle|\eta|^{-1}\rangle)^{-1}$$

$$= \sum_{j=0}^{N} d_j |\eta|^{-1-j} + d_N(|\eta|^{-1})|\eta|^{-N-2}.$$

Using now (4.2.3) we obviously get asymptotic expansions both for $g_1(\xi)$ and $g_2(\xi)$ in homogeneous components for large $|\xi|$. Concerning the remainders in the various expansions we have to show that they are symbols of orders tending to $-\infty$ for $N \to \infty$. Let us first look at the estimates without derivatives in the covariables. We then have to consider two types of terms, namely

$$\chi^{(j)}\left(\frac{\tau}{|\eta|}\right)\left(\frac{\tau}{|\eta|}\right)^{j}|\eta|^{-k}R_N(\xi)$$

and

$$\chi^{(j)}(\Theta)\left(\frac{\tau}{|\eta|}\right)^{j}|\eta|^{-k}\tilde{R}_N(\xi)$$

with uniformly bounded functions $R_N, \tilde{R}_N, k \geq 0$, and $\tau/\langle\eta\rangle \leq \theta \leq \tau/|\eta|$. Like in the proof of the relation (4.2.4) these expressions are $\leq c\langle\xi\rangle^{-k}$ for constants $c > 0$. Finally for the derivatives we get essentially the same expressions, cf. the arguments for (4.2.3). Hence we can argue in an analogous manner. $\square$

Proposition 4.2.10 *The symbols $r_-^{\mu}(\xi)$, $\mu \in \mathbb{Z}$, have the transmission property.*

Proof. In view of Proposition 4.2.5, (i), and Proposition 4.2.6 it suffices to set $\mu = 1$. Then, since $-i\tau$ has the transmission property, we only have to consider $\chi(\frac{\tau}{\rho(\eta)})\langle\eta\rangle$. Then the conditions of Definition 4.2.1 are satisfied, since $D_\eta^\gamma\left\{\chi(\frac{\tau}{\rho(\eta)})\langle\eta\rangle\right\}$ vanishes to infinite order at $\tau = \pm\infty$ for every $\eta \in \mathbb{R}^q$. $\square$

4.2.2 Some classes of operator–valued symbols

Proposition 4.2.11 *For each $s \in \mathbb{N}$ there exists a continuous extension operator $e_s^+ : H^s(\mathbb{R}_+) \to H^s(\mathbb{R})$ satisfying*

$$e_s^+\kappa_\lambda = \kappa_\lambda e_s^+ \quad \text{for every } \lambda \in \mathbb{R}_+, \qquad (4.2.8)$$

$$(\kappa_\lambda u)(t) = \lambda^{\frac{1}{2}}u(\lambda t).$$

Proof. For $u(t) \in H^s(\mathbb{R}_+)$ we set

$$(e_s^+u)(t) = \begin{cases} u(t) & \text{for } t > 0, \\ \sum_{j=1}^{s}\alpha_j u(-jt) & \text{for } t < 0, \end{cases}$$

with constants α_j satisfying the conditions

$$\left(\frac{d}{dt}\right)^k (e_s^+ u)(t)|_{t=0} = \left(\frac{d}{dt}\right)^k u(t)|_{t=0}$$

for $k = 0, \dots, s - 1$. This gives rise to the system of equations

$$\sum_{j=1}^{s} (-1)^k j^k \alpha_j = 1, \qquad k = 0, \dots, s - 1,$$

which determines the constants $\alpha_1, \dots, \alpha_s$ uniquely. The relation (4.2.8) is then obviously satisfied. $\qquad\square$

It can be shown that Proposition 4.2.11 has a generalisation to arbitrary $s \in \mathbb{R}$. Corresponding extension operators may be found in Triebel [148], Section 2.9.3. We will content ourselves here with $s \in \mathbb{N}$, since the applications can be reduced to this case. Other more general cases were treated in the article of Burenkov, Schulze and Tarkhanov [14].

Remark 4.2.12 *If* $e_s^+ : H^s(\mathbb{R}_+) \to H^s(\mathbb{R})$ *is a continuous extension operator satisfying* $e_s^+ \kappa_\lambda = \kappa_\lambda e_s^+$ *for all* $\lambda \in \mathbb{R}_+$, *then*

$$e_s^+ \in S_{\mathrm{cl}}^0(\mathbb{R}^q; H^s(\mathbb{R}_+), H^s(\mathbb{R})).$$

In fact, e_s^+ *is independent of* $\eta \in \mathbb{R}^q$; *so we only have to check the first symbol estimate*

$$\|\kappa^{-1}(\eta) e_s^+ \kappa(\eta)\|_{\mathcal{L}(H^s(\mathbb{R}_+), H^s(\mathbb{R}))} < \text{const}$$

for all $\eta \in \mathbb{R}^q$.

As usual we set for every $a(t, y, \tau, \eta) \in S^\mu(\mathbb{R} \times \Omega \times \mathbb{R}^{1+q})$ for open $\Omega \subseteq \mathbb{R}^q$

$$\mathrm{op}(a)(y, \eta) u(t) = \iint e^{i(t-t')\tau} a(t, y, \tau, \eta) u(t') \, dt' d\tau.$$

Recall that $\eta \to [\eta]$ denotes any fixed strictly positive C^∞ function in $\mathbb{R}^q$ satisfying $[\eta] = |\eta|$ for $|\eta| > c$ for some constant $c > 0$. If $\{\kappa_\lambda\}_{\lambda \in \mathbb{R}_+}$ is a group of isomorphisms acting on a space E, then we set

$$\kappa(\eta) := \kappa_{[\eta]}.$$

Proposition 4.2.13 *Let* $a(t,y,\tau,\eta) \in S^\mu(\mathbb{R} \times \Omega \times \mathbb{R}^{1+q})$, $\mu \in \mathbb{R}$, *be a symbol which is independent of t for* $|t| > c_1$ *for a constant* $c_1 > 0$. *Then* $\mathrm{op}(a)(y,\eta)$, *regarded as a family of continuous operators* $H^s(\mathbb{R}) \to H^{s-\mu}(\mathbb{R})$, $s \in \mathbb{R}$, *satisfies*

$$\mathrm{op}(a)(y,\eta) \in S^\mu(\Omega \times \mathbb{R}^q; H^s(\mathbb{R}), H^{s-\mu}(\mathbb{R})),$$

where the operator–valued symbols rely on $(\kappa_\lambda u)(t) = \lambda^{1/2} u(\lambda t)$, $\lambda > 0$, *for* $u \in H^r(\mathbb{R})$, $r \in \mathbb{R}$. *Moreover,* $a(t,y,\tau,\eta) \to \mathrm{op}(a)(y,\eta)$ *induces a continuous operator*

$$S^\mu(\mathbb{R} \times \Omega \times \mathbb{R}^{1+q})_1 \to S^\mu(\Omega \times \mathbb{R}^q; H^s(\mathbb{R}), H^{s-\mu}(\mathbb{R})) \qquad (4.2.9)$$

for every $s, \mu \in \mathbb{R}$. *Here subscript* 1 *indicates the symbols that are independent of t for* $|t| \geq c_1$.

Proof. Let us first assume that the symbol a is independent of t. Then

$$\kappa_\lambda^{-1} \mathrm{op}(a)(y,\eta)\kappa_\lambda = \mathrm{op}(a_\lambda)(y,\eta)$$

for $a_\lambda(y,\tau,\eta) = a(y,\lambda\tau,\eta)$. In fact, we have

$$\kappa_\lambda^{-1} \mathrm{op}(a)(y,\eta)\kappa_\lambda u(t) = \iint e^{i(\lambda^{-1}t - t')\tau} a(y,\tau,\eta) u(\lambda t')\, dt' d\tau$$

$$= \iint e^{i(t-\tilde{t})\tilde{\tau}} a(y,\lambda\tilde{\tau},\eta) u(\tilde{t})\, d\tilde{t}d\tilde{\tau}.$$

Here we have used the substitution $\tilde{t} = \lambda t'$, $\tilde{\tau} = \lambda^{-1}\tau$. Let us now show that $\mathrm{op}(a)(y,\eta)$ for t-independent a is an operator–valued symbol, i.e., satisfies

$$\|\kappa^{-1}(\eta)\{D_y^\alpha D_\eta^\beta \mathrm{op}(a)(y,\eta)\}\kappa(\eta)\|_{\mathcal{L}(H^s(\mathbb{R}), H^{s-\mu}(\mathbb{R}))} \leq c[\eta]^{\mu-|\beta|} \qquad (4.2.10)$$

for all $\alpha, \beta \in \mathbb{N}^q$, $y \in K$, $\eta \in \mathbb{R}^q$, for arbitrary $K \subset\subset \Omega$, with constants $c = c(\alpha, \beta, K) > 0$. Let us take first $\alpha = \beta = 0$. Then we have

$$\|\kappa^{-1}(\eta) \mathrm{op}(a)(y,\eta)\kappa(\eta)\|_{\mathcal{L}(H^s(\mathbb{R}), H^{s-\mu}(\mathbb{R}))}$$
$$= \|\mathrm{op}(a_{[\eta]})(y,\eta)\|_{\mathcal{L}(H^s(\mathbb{R}), H^{s-\mu}(\mathbb{R}))}$$
$$\leq c \sup_{\tau \in \mathbb{R}}[\tau]^{-\mu} |a(y,[\eta]\tau,\eta)|$$
$$\leq c \sup_{\tau \in \mathbb{R}}[\tau]^{-\mu}([\eta]|\tau| + [\eta])^\mu$$
$$\leq c[\eta]^\mu. \qquad (4.2.11)$$

Here c denotes different constants; y varies over an arbitrary $K \subset\subset \Omega$. In an analogous manner we obtain the estimates (4.2.10) in general. The constants c depend on the symbol a and tend to zero when $a(y, \tau, \eta)$ tends to zero in $S^\mu(\Omega \times \mathbb{R}^{1+q})$. This is a consequence of the estimates (4.2.11) and of the analogues for the derivatives and of the semi-norms for the Fréchet topology of the space $S^\mu(\Omega \times \mathbb{R}^{1+q})$. In an analogous manner we can proceed for the derivatives $D_y^\alpha D_\eta^\beta \operatorname{op}(a)(y, \eta) = \operatorname{op}(D_y^\alpha D_\eta^\beta a)(y, \eta)$ for arbitrary α, β. This gives us a continuous operator

$$S^\mu(\Omega \times \mathbb{R}^{1+q}) \to S^\mu(\Omega \times \mathbb{R}^q; H^s(\mathbb{R}), H^{s-\mu}(\mathbb{R}))$$

for every $s \in \mathbb{R}$. Let $C^\infty([-c_1, c_1])_0$ denote the space of all $\varphi(t) \in C^\infty(\mathbb{R})$ that are supported by the interval $[-c_1, c_1]$. Then

$$S^\mu(\mathbb{R} \times \Omega \times \mathbb{R}^{1+q})_1$$
$$= S^\mu(\Omega \times \mathbb{R}^{1+q}) + C^\infty([-c_1, c_1])_0 \otimes_\pi S^\mu(\Omega \times \mathbb{R}^{1+q}),$$

cf. similarly Remark 1.1.10. In view of the above result it is now sufficient to show that we also get a continuous operator

$$C^\infty([-c_1, c_1])_0 \otimes_\pi S^\mu(\Omega \times \mathbb{R}^{1+q}) \to S^\mu(\Omega \times \mathbb{R}^q; H^s(\mathbb{R}), H^{s-\mu}(\mathbb{R})).$$
$$(4.2.12)$$

Here we employ a tensor product argument, cf. Theorem 1.1.6. Write a given $a(t, y, \tau, \eta) \in C^\infty([-c_1, c_1])_0 \otimes_\pi S^\mu(\Omega \times \mathbb{R}^{1+q})$ as a convergent sum

$$a(t, y, \tau, \eta) = \sum_{j=0}^\infty \lambda_j \varphi_j(t) a_j(y, \tau, \eta)$$

with $\lambda_j \in \mathbb{C}$, $\sum |\lambda_j| < \infty$, $\varphi_j \in C^\infty([-c_1, c_1])_0$, $a_j \in S^\mu(\Omega \times \mathbb{R}^{1+q})$, tending to 0 in the corresponding spaces for $j \to \infty$. The operator $\mathcal{M}_\varphi$ of multiplication by $\varphi(t) \in C_0^\infty(\mathbb{R})$ is continuous in the sense $\mathcal{M}_\varphi : H^s(\mathbb{R}) \to H^s(\mathbb{R})$ for each $s \in \mathbb{R}$ and $\mathcal{M}_\varphi \to 0$ in $\mathcal{L}(H^s(\mathbb{R}))$ for $\varphi \to 0$ in $C_0^\infty(\mathbb{R})$, cf. Corollary 1.1.56. In addition, $\mathcal{M}_\varphi$ can be interpreted as an operator–valued symbol in $S^0(\mathbb{R}^q; H^s(\mathbb{R}), H^s(\mathbb{R}))$ for each $s \in \mathbb{R}$, also tending to zero in that space when φ tends to zero in $C_0^\infty(\mathbb{R})$, cf. Example 1.3.10. Now

$$\operatorname{op}(a)(y, \eta) = \sum_{j=0}^\infty \lambda_j \mathcal{M}_{\varphi_j} \operatorname{op}(a_j)(y, \eta)$$

obviously converges in $S^\mu(\Omega \times \mathbb{R}^q; H^s(\mathbb{R}), H^{s-\mu}(\mathbb{R}))$. At the same time we easily see that (4.2.12) is also continuous. $\qquad\square$

Remark 4.2.14 *For* $a(y, \tau, \eta) \in S^{\mu}_{\mathrm{cl}}(\Omega \times \mathbb{R}^{1+q})$, $\mu \in \mathbb{R}$, *we obtain*

$$\mathrm{op}(a)(y, \eta) \in S^{\mu}_{\mathrm{cl}}(\Omega \times \mathbb{R}^{q}; H^{s}(\mathbb{R}), H^{s-\mu}(\mathbb{R})),$$

and $a(y, \tau, \eta) \to \mathrm{op}(a)(y, \eta)$ *induces a continuous operator*

$$S^{\mu}_{\mathrm{cl}}(\Omega \times \mathbb{R}^{1+q}) \to S^{\mu}_{\mathrm{cl}}(\Omega \times \mathbb{R}^{q}; H^{s}(\mathbb{R}), H^{s-\mu}(\mathbb{R}))$$

for each $s \in \mathbb{R}$, $s > -\frac{1}{2}$.

This follows easily from the fact that for the homogeneous components $a_{(\mu-j)}(y, \tau, \eta)$ in $\xi = (\tau, \eta)$ of order $\mu - j$ and every excision function $\chi(\tau, \eta)$ we have

$$a_{\mu-N-1}(y, \tau, \eta)$$
$$= a(y, \tau, \eta) - \chi(\tau, \eta) \sum_{j=0}^{N} a_{(\mu-j)}(y, \tau, \eta) \in S^{\mu-N-1}(\Omega \times \mathbb{R}^{1+q})$$

and

$$\mathrm{op}(\chi a_{(\mu-j)})(y, \lambda \eta) = \lambda^{\mu-j} \kappa_{\lambda} \, \mathrm{op}(\chi a_{(\mu-j)})(y, \eta) \kappa_{\lambda}^{-1}$$

for all $\lambda \geq 1$ and $|\eta| \geq c$ for a constant $c > 0$, where $\chi(\tau, \eta) = 1$ for all $|\tau, \eta| \geq c$.

In Section 1.3.1 we have defined operator–valued symbols between Fréchet spaces $E, \tilde{E}$, written as projective limits of Banach spaces with strongly continuous groups of isomorphisms on each of these Banach spaces, under a natural compatibility condition. Let us write, in particular,

$$\mathcal{S}(\mathbb{R}) = \varprojlim_{k \in \mathbb{N}} \langle t \rangle^{-k} H^{k}(\mathbb{R})$$

with the action $(\kappa_{\lambda} u)(t) = \lambda^{\frac{1}{2}} u(\lambda t)$, $\lambda > 0$ on $\langle t \rangle^{-k} H^{k}(\mathbb{R})$.

Proposition 4.2.15 *Let* $a(t, y, \tau, \eta) \in S^{\mu}_{\mathrm{cl}}(\mathbb{R} \times \Omega \times \mathbb{R}^{1+q})$, $\mu \in \mathbb{R}$, *be a symbol which is independent of* t *for* $|t| > c_1$ *for a constant* $c_1 > 0$. *Then*

$$\mathrm{op}(a)(y, \eta) \in S^{\mu}(\Omega \times \mathbb{R}^{q}; \mathcal{S}(\mathbb{R}), \mathcal{S}(\mathbb{R})).$$

For t–*independent* a *we have*

$$\mathrm{op}(a)(y, \eta) \in S^{\mu}_{\mathrm{cl}}(\Omega \times \mathbb{R}^{q}; \mathcal{S}(\mathbb{R}), \mathcal{S}(\mathbb{R})).$$

Proof. For simplicity we restrict ourselves to y-independent symbols. The general case is completely analogous and will be omitted. Setting $E^k = \langle t \rangle^{-k} H^k(\mathbb{R})$, $k \in \mathbb{Z}$, with the norm $\|u\|_{E^k} = \|\langle t \rangle^k u(t)\|_{H^k(\mathbb{R})}$, we have $\mathcal{S}(\mathbb{R}) = \varprojlim_{k \in \mathbb{N}} E^k$. We then have to verify that for every $k \in \mathbb{N}$ there is an $l \in \mathbb{N}$ such that

$$\|\kappa^{-1}(\eta)\{D_\eta^\beta \operatorname{op}(a)(\eta)\}\kappa(\eta)\|_{\mathcal{L}(E^l, E^k)} \leq c[\eta]^{\mu - |\beta|}$$

holds for all $\beta \in \mathbb{N}^q$ and all $\eta \in \mathbb{R}^q$, with constants $c = c(k, l, \beta) > 0$. Let us first assume that a is independent of t, i.e., $a = a(\tau, \eta)$. Choose an excision function $\chi(\tau, \eta)$ and write

$$a(\tau, \eta) = \sum_{j=0}^N \chi(\tau, \eta) a_{(\mu-j)}(\tau, \eta) + a_N(\tau, \eta)$$

for any N, where $a_{(\mu-j)}$, $j \in \mathbb{N}$, are the homogeneous components of a, i.e.,

$$a_{(\mu-j)}(\lambda\tau, \lambda\eta) = \lambda^{\mu-j} a_{(\mu-j)}(\tau, \eta)$$

for all $(\tau, \eta) \neq 0$, $\lambda > 0$, and $a_N(\tau, \eta) \in S_{\text{cl}}^{\mu-(N+1)}(\mathbb{R}^{1+q})$. Set $h_{\mu-j}(\eta) = \operatorname{op}(\chi a_{(\mu-j)})(\eta)$. For every fixed $\eta \in \mathbb{R}^q$ we have $(\chi a_{(\mu-j)})(\tau, \eta) \in S_{\text{cl}}^{\mu-j}(\mathbb{R})$, and $h_{\mu-j}(\eta) : \mathcal{S}(\mathbb{R}) \to \mathcal{S}(\mathbb{R})$ is continuous. The latter property is elementary and will not be commented on here. In other words we have for every $k \in \mathbb{N}$ an $l \in \mathbb{N}$ such that

$$\|h_{\mu-j}(\eta)\|_{\mathcal{L}(E^l, E^k)} \leq c_{kl}.$$

It can easily be verified that l can be chosen to be independent of j, and in addition that $h_{\mu-j}(\eta) \in C^\infty(\mathbb{R}^q, \mathcal{L}(E^l, E^k))$. In particular, the constants c_{kl} are uniformly bounded for $|\eta| \leq$ const for every fixed constant > 0. In view of $h_{\mu-j}(\lambda\eta) = \lambda^{\mu-j} \kappa_\lambda h_{\mu-j}(\eta) \kappa_\lambda^{-1}$ for $|\eta| \geq$ const for a constant > 0 and all $\lambda \geq 1$ we obtain that

$$h_{\mu-j}(\eta) \in S_{\text{cl}}^{\mu-j}(\mathbb{R}^q; E^l, E^k)$$

holds for all j. Note also that $a \to h_{\mu-j}$ defines a continuous operator

$$S_{\text{cl}}^\mu(\mathbb{R}^{1+q}) \to S_{\text{cl}}^{\mu-j}(\mathbb{R}^q; E^l, E^k)$$

for all j and every k with the resulting l. For the remainder we have to check

$$\|\kappa^{-1}(\eta)\{D_\eta^\beta \operatorname{op}(a_N)(\eta)\}\kappa(\eta)\|_{\mathcal{L}(E^l, E^k)} \leq c[\eta]^{\mu-(N+1)-|\beta|}$$

with the corresponding constants c. Let us consider the case $\beta = 0$. The arguments for arbitrary β are completely analogous, because of $D_\eta^\beta \operatorname{op}(a_N)(\eta) = \operatorname{op}(D_\eta^\beta a_N)(\eta)$ and $D_\eta^\beta(a_N) \in S_{\mathrm{cl}}^{\mu-(N+1)-|\beta|}(\mathbb{R}^{1+q})$. Setting $b(\tau,\eta) = a_N([\eta]\tau,\eta)$ we then have to show

$$\| \operatorname{op}(b)(\eta) \|_{\mathcal{L}(E^l, E^k)} \le c[\eta]^{\mu-(N+1)}$$

for all $\eta \in \mathbb{R}^q$ with a constant $c = c(k,l) > 0$. We will choose N so large that $\mu - (N+1) \le 0$ holds. In the following c denotes different suitable constants. For convenience we now pass to another equivalent scale of spaces F^k, $k \in \mathbb{N}$, with $\mathcal{S}(\mathbb{R}) = \varprojlim_{k \in \mathbb{N}} F^k$, where $\{\kappa_\lambda\}_{\lambda \in \mathbb{R}_+}$ also acts as a strongly continuous group of isomorphisms for each k. We define F^k as the closure of $\mathcal{S}(\mathbb{R})$ with respect to the norm

$$\|u\|_{F^k} = \left\{ \int \langle \tau \rangle^{2k} \sum_{m=0}^{k} |D_\tau^m \hat{u}(\tau)|^2 \, d\tau \right\}^{\frac{1}{2}}.$$

Then

$$\| \operatorname{op}(b)(\eta)u \|_{F^k}^2 = \int \langle \tau \rangle^{2k} \sum_{m=0}^{k} |D_\tau^m (b([\eta]\tau,\eta)\hat{u}(\tau))|^2 \, d\tau.$$

We have $D_\tau^m(b([\eta]\tau,\eta)\hat{u}(\tau)) = \sum_{n=0}^{m} d_{nm} D_\tau^n b([\eta]\tau,\eta) D_\tau^{m-n} \hat{u}(\tau)$ for certain constants d_{nm}. Using

$$|D_\tau^n b([\eta]\tau,\eta)| = |[\eta]^n (D_\tau^n b)([\eta]\tau,\eta)|$$
$$\le c[\eta]^n ([\eta]|\tau| + [\eta])^{\mu-(N+1)-n}$$
$$\le c[\eta]^{\mu-(N+1)} \langle \tau \rangle^{\mu-(N+1)-n}$$

it follows that

$$\| \operatorname{op}(b)(\eta)u \|_{F^k}^2 \le c[\eta]^{2(\mu-N-1)} \int \langle \tau \rangle^{2k} \sum_{m=0}^{k} |\langle \tau \rangle^{\mu-N-1} D_\tau^m \hat{u}(\tau)|^2 d\tau$$
$$\le c[\eta]^{2(\mu-N-1)} \|u\|_{F^k}^2.$$

Here we used $\mu - N - 1 \le 0$. Thus we obtain the required estimate for $l = k$. At the same time we see that $a(\tau,\eta) \to \operatorname{op}(a_N)(\eta)$ induces a continuous operator

$$S_{\mathrm{cl}}^\mu(\mathbb{R}^{1+q}) \to S^{\mu-(N+1)}(\mathbb{R}^q; F^k, F^k).$$

Returning now to the original scale we get by $a \to \operatorname{op}(a)$ continuous operators

$$S_{\mathrm{cl}}^\mu(\mathbb{R}^{1+q}) \to S_{\mathrm{cl}}^\mu(\mathbb{R}^q; E^l, E^k) \tag{4.2.13}$$

for every $k \in \mathbb{N}$ with some $l = l(k)$, where $l(k)$ only depends on μ. The assertion for t-dependent symbols $a(t, \tau, \eta)$ can be proved by a tensor product argument. We can write

$$a(t, \tau, \eta) = a_0(t, \tau, \eta) + \chi^+(t)c^+(\tau, \eta) + \chi^-(t)c^-(\tau, \eta),$$

while $a_0(t, \tau, \eta) \in S^\mu_{cl}(\mathbb{R} \times \mathbb{R}^{1+q})$ vanishes for $|t| \geq c_1$, moreover $c^+(\tau, \eta), c^-(\tau, \eta) \in S^\mu_{cl}(\mathbb{R}^{1+q})$ are t-independent, and $\chi^\pm(t)$ are suitable functions in $C^\infty(\mathbb{R})$ with $\chi^+(t) = 0$ for $t < t_0^+$, $\chi^+(t) = 1$ for $t > t_1^+$, $\chi^-(t) = 0$ for $t > t_0^-$, $\chi^-(t) = 1$ for $t < t_1^-$, for certain $t_0^+ < t_1^+$, $t_0^- > t_1^-$. The operators $\mathrm{op}(c^\pm)(\eta)$ were considered above, and the multiplication operators by $\chi^\pm(t)$ behave like operator–valued symbols in $S(\mathbb{R})$. In order to treat $a_0(t, \tau, \eta)$ we assume for notational convenience $c_1 = 1$ (the generalisation to arbitrary c_1 is trivial). Let $C^\infty([-1,1]_0)$ denote the subspace of all $\varphi \in C^\infty(\mathbb{R})$ with $\varphi(t) = 0$ for $|t| \geq 1$. This is a Fréchet space, and we have

$$a_0(t, \tau, \eta) \in C^\infty([-1,1]_0) \otimes_\pi S^\mu_{cl}(\mathbb{R}^{1+q}).$$

Write $a_0(t, \tau, \eta) = \sum \lambda_j \varphi_j(t) c_j(\tau, \eta)$ with $\sum |\lambda_j| < \infty$, where the sequences $\varphi_j(t) \in C^\infty([-1,1]_0)$, $c_j(\tau, \eta) \in S^\mu_{cl}(\mathbb{R}^{1+q})$ tending to zero in the corresponding spaces for $j \to \infty$, cf. Theorem 1.1.6. It is now easy to see that the operator of multiplication $\mathcal{M}_\varphi$ by $\varphi \in C^\infty([-1,1]_0)$ represents a symbol in $S^0(\mathbb{R}^q; E^k, E^k)$ and that $\varphi \to \mathcal{M}_\varphi$ induces a continuous map $C^\infty([-1,1]_0) \to S^0(\mathbb{R}^q; E^k, E^k)$ for each k. Then we obtain, using the continuity of (4.2.13), that $\mathrm{op}(a_0)(\eta) = \sum \lambda_j \mathcal{M}_{\varphi_j} \mathrm{op}(c_j)(\eta)$ converges in $S^\mu(\mathbb{R}^q; E^l, E^k)$ for every k with $l = l(k)$. This proves the assertion. $\square$

Proposition 4.2.16 *We have*

$$\mathrm{op}^+(r^\mu_-)(\eta) \in S^\mu_{cl}(\mathbb{R}^q; H^s(\mathbb{R}_+), H^{s-\mu}(\mathbb{R}_+)) \qquad (4.2.14)$$

for every $\mu \in \mathbb{Z}$ and arbitrary $s \in \mathbb{R}$, $s > -\frac{1}{2}$.

Proof. First observe that the restriction operator $r^+ : H^s(\mathbb{R}) \to H^s(\mathbb{R}_+)$ may be regarded as an element

$$r^+ \in S^0_{cl}(\mathbb{R}^q; H^s(\mathbb{R}), H^s(\mathbb{R}_+)). \qquad (4.2.15)$$

In fact, since it is independent of η, it suffices to check the symbol estimate

$$\|\kappa^{-1}(\eta) r^+ \kappa(\eta)\|_{\mathcal{L}(H^s(\mathbb{R}), H^s(\mathbb{R}_+))} < c$$

for all $\eta \in \mathbb{R}^q$. However this is trivial because of $r^+ \kappa_\lambda = \kappa_\lambda r^+$ for all $\lambda \in \mathbb{R}$. Moreover, let $e_s^+ : H^s(\mathbb{R}_+) \to H^s(\mathbb{R})$ be an arbitrary continuous extension operator. Then

$$r^+ \operatorname{op}(r_-^\mu)(\eta) e_s^+ : H^s(\mathbb{R}_+) \to H^{s-\mu}(\mathbb{R}_+)$$

is continuous for every $\eta \in \mathbb{R}^q$. For the extension operator $e^+ :$ $H^s(\mathbb{R}_+) \to \mathcal{S}'(\mathbb{R})$ by zero we obtain $v := (e^+ - e_s^+)u \in H^r(\mathbb{R})$ for every $u \in H^s(\mathbb{R}_+)$ for $r = \min\{s, \frac{1}{2} - \varepsilon\}$ for every $\varepsilon > 0$, where $\operatorname{supp} v \subseteq \overline{\mathbb{R}}_-$. In view of Remark 4.1.36 and Remark 4.2.8 we obtain $r^+ \operatorname{op}(r_-^\mu)(\eta)(e^+ - e_s^+)u = 0$, in other words

$$\operatorname{op}^+(r_-^\mu)(\eta) = r^+ \operatorname{op}(r_-^\mu)(\eta)(e_s^+) \tag{4.2.16}$$

for $s > -\frac{1}{2}$. For $|s| < \frac{1}{2}$ we could take $e_s^+ = e^+$, cf. (4.1.13). Using $e^+ \kappa_\lambda = \kappa_\lambda e^+$ on $H^s(\mathbb{R}_+)$ for those s and the above Remark 4.2.14 we obtain (4.2.14) for $|s| < \frac{1}{2}$. Moreover, for every $s \in \mathbb{N}$ we can take the extension operator e_s^+ from Proposition 4.2.11. Then (4.2.16), (4.2.15), Remark 4.2.14 and Remark 4.2.12 yield (4.2.14) for $s \in \mathbb{N}$. For the remaining s we first show

$$\operatorname{op}^+(r_-^\mu)(\eta) \in S^\mu(\mathbb{R}^q; H^s(\mathbb{R}_+), H^{s-\mu}(\mathbb{R}_+)). \tag{4.2.17}$$

For $s \in \mathbb{N}$ we already know

$$\|\kappa^{-1}(\eta)\{(1 + |\eta|^2)^{-\frac{\mu}{2}+|\beta|} D_\eta^\beta r^+ \operatorname{op}(r_-^\mu)(\eta)e^+\}\kappa(\eta)\|_{\mathcal{L}(H^s(\mathbb{R}_+), H^{s-\mu}(\mathbb{R}_+))} < c$$

for arbitrary $\beta \in \mathbb{N}^q$. Analogous estimates then follow for all $s \geq 0$ by interpolation. This is just (4.2.17). Now if $(r_-^\mu)_{(\mu-j)}(\tau, \eta)$, $j \in \mathbb{N}$, are the homogeneous components of $r_-^\mu(\tau, \eta)$ of order $\mu - j$, we get for an arbitrary excision function $\chi(\tau, \eta)$

$$\operatorname{op}^+(r_-^\mu)(\eta) = \sum_{j=0}^{N} \operatorname{op}^+(\chi(r_-^\mu)_{(\mu-j)})(\eta) + \operatorname{op}^+((r_-^\mu)_{N+1})(\eta)$$

for a symbol $(r_-^\mu)_{N+1}(\tau, \eta) \in S_{\mathrm{cl}}^{\mu-N-1}(\mathbb{R}^{1+q})$ with the transmission property. From

$$\operatorname{op}^+(\chi(r_-^\mu)_{(\mu-j)})(\lambda\eta) = \lambda^{\mu-j} \kappa_\lambda \operatorname{op}^+(\chi(r_-^\mu)_{(\mu-j)})(\eta) \kappa_\lambda^{-1}$$

for all $\lambda \geq 1$, $|\eta| \geq \mathrm{const}$, we obtain

$$\operatorname{op}^+(\chi(r_-^\mu)_{(\mu-j)})(\eta) \in S_{\mathrm{cl}}^{\mu-j}(\mathbb{R}^q; H^s(\mathbb{R}_+), H^{s-\mu+j}(\mathbb{R}_+))$$
$$\subseteq S_{\mathrm{cl}}^{\mu-j}(\mathbb{R}^q; H^s(\mathbb{R}_+), H^{s-\mu}(\mathbb{R}_+))$$

for all $s > -\frac{1}{2}$. Here we used, in particular, Corollary 4.1.52. It remains to show

$$\mathrm{op}^+((r_-^\mu)_{N+1})(\eta) \in S^{\mu-N-1}(\mathbb{R}^q; H^s(\mathbb{R}_+), H^{s-\mu}(\mathbb{R}_+)). \qquad (4.2.18)$$

This follows from Theorem 4.2.37 below, to be applied for $a(\tau, \eta) = (r_-^\mu)_{N+1}(\tau, \eta) \in S_{\mathrm{cl}}^{\mu-N-1}(\mathbb{R}^{1+q})$, where the arguments for (4.2.18) only employ the relation (4.2.17). □

Proposition 4.2.17 *We have* $\mathrm{op}^+(r_-^\mu)(\eta) \in S_{\mathrm{cl}}^\mu(\mathbb{R}^q; \mathcal{S}(\overline{\mathbb{R}}_+), \mathcal{S}(\overline{\mathbb{R}}_+))$ *for every* $\mu \in \mathbb{Z}$.

Proof. Write $E^k = \langle t \rangle^{-k} H^k(\mathbb{R}), E_+^k = \langle t \rangle^{-k} H^k(\mathbb{R}_+)$, $k \in \mathbb{N}$, such that $\mathcal{S}(\mathbb{R}) = \varprojlim_{k \in \mathbb{N}} E^k$, $\mathcal{S}(\overline{\mathbb{R}}_+) = \varprojlim_{k \in \mathbb{N}} E_+^k$. For each fixed $l \in \mathbb{N}$ the extension operator $e_l^+ : H^l(\mathbb{R}_+) \to H^l(\mathbb{R})$ has a restriction to $e_l^+ : E_+^l \to E^l$, satisfying $e_l^+ \kappa_\lambda = \kappa_\lambda e_l^+$ for all $\lambda \in \mathbb{R}_+$, with κ_λ acting on E_+^l and E^l, respectively. Thus we have $e_l^+ \in S_{\mathrm{cl}}^0(\mathbb{R}^q; E_+^l, E^l)$ for every $l \in \mathbb{N}$. Moreover, it is obvious $r^+ \in S_{\mathrm{cl}}^0(\mathbb{R}^q; E^k, E_+^k)$, $k \in \mathbb{N}$. From Proposition 4.2.15 we know that $\mathrm{op}(r_-^\mu)(\eta) \in S_{\mathrm{cl}}^\mu(\mathbb{R}^q; E^l, E^k)$ for every $k \in \mathbb{N}$ for a suitable $l = l(k) \in \mathbb{N}$. This gives us $r^+ \mathrm{op}(r_-^\mu)(\eta) e_l^+ \in S_{\mathrm{cl}}^\mu(\mathbb{R}^q; E_+^l, E_+^k)$, $k \in \mathbb{N}$ for $l = l(k)$. In view of $r^+ \mathrm{op}(r_-^\mu)(\eta) e_l^+ = r^+ \mathrm{op}(r_-^\mu)(\eta) e^+$, we get the assertion. □

4.2.3 Green, trace, and potential boundary symbols

Definition 4.2.18 $R_G^{\mu,0}(\Omega \times \Omega \times \mathbb{R}^q)$ *for open* $\Omega \subseteq \mathbb{R}^q$ *and* $\mu \in \mathbb{R}$ *denotes the space of all operator–valued symbols*

$$g(y, y', \eta) \in S_{\mathrm{cl}}^\mu(\Omega \times \Omega \times \mathbb{R}^q; L^2(\mathbb{R}_+), \mathcal{S}(\overline{\mathbb{R}}_+))$$

satisfying

$$g^*(y, y', \eta) \in S_{\mathrm{cl}}^\mu(\Omega \times \Omega \times \mathbb{R}^q; L^2(\mathbb{R}_+), \mathcal{S}(\overline{\mathbb{R}}_+)),$$

where $*$ *denotes the point-wise* $L^2(\mathbb{R}_+)$-*adjoint. Moreover,* $R_G^{\mu,d}(\Omega \times \Omega \times \mathbb{R}^q)$ *for* $d \in \mathbb{N}$ *denotes the space of all operator–valued symbols*

$$g(y, y', \eta) \in \bigcap_{\substack{s \in \mathbb{R} \\ s > d - \frac{1}{2}}} S_{\mathrm{cl}}^\mu(\Omega \times \Omega \times \mathbb{R}^q; H^s(\mathbb{R}_+), \mathcal{S}(\overline{\mathbb{R}}_+)) \qquad (4.2.19)$$

which have the form

$$g(y, y', \eta) = \sum_{k=0}^d g_k(y, y', \eta) \left(\frac{d}{dt} \right)^k \qquad (4.2.20)$$

for arbitrary $g_k(y, y', \eta) \in R_G^{\mu-k,0}(\Omega \times \Omega \times \mathbb{R}^q)$, $k = 0, \ldots, d$.

The corresponding subspaces of y'-independent elements are denoted by $R_G^{\mu,d}(\Omega \times \mathbb{R}^q)$. The elements of $R_G^{\mu,d}(\Omega \times \Omega \times \mathbb{R}^q)$ are called Green boundary symbols of order μ and type d, cf. Definition 4.1.45.

Note that each operator family

$$g(y,y',\eta) : \mathcal{S}(\overline{\mathbb{R}}_+) \to \mathcal{S}(\overline{\mathbb{R}}_+)$$

of the form (4.2.20) has the property (4.2.19) in the sense that it extends to $H^s(\mathbb{R}_+)$ for every $s > d - \frac{1}{2}$ as a classical operator–valued symbol.

Similarly to the above Definition 4.1.76 we will also need trace and potential symbols.

Definition 4.2.19 $\mathcal{R}_G^{\mu,0}(\Omega \times \Omega \times \mathbb{R}^q; N_-, N_+)$ *for open* $\Omega \subseteq \mathbb{R}^q$ *and* $\mu \in \mathbb{R}$ *denotes the space of all operator–valued symbols*

$$g(y,y',\eta) \in S_{\mathrm{cl}}^\mu(\Omega \times \Omega \times \mathbb{R}^q; L^2(\mathbb{R}_+) \oplus \mathbb{C}^{N_-}, \mathcal{S}(\overline{\mathbb{R}}_+) \oplus \mathbb{C}^{N_+}) \quad (4.2.21)$$

satisfying

$$g^*(y,y',\eta) \in S_{\mathrm{cl}}^\mu(\Omega \times \Omega \times \mathbb{R}^q; L^2(\mathbb{R}_+) \oplus \mathbb{C}^{N_+}, \mathcal{S}(\overline{\mathbb{R}}_+) \oplus \mathbb{C}^{N_-}), \quad (4.2.22)$$

where $*$ *denotes the point-wise adjoint in the sense*

$$(gu, v)_{L^2(\mathbb{R}_+) \oplus \mathbb{C}^{N_+}} = (u, g^* v)_{L^2(\mathbb{R}_+) \oplus \mathbb{C}^{N_-}}$$

for all $u \in L^2(\mathbb{R}_+) \oplus \mathbb{C}^{N_-}$, $v \in L^2(\mathbb{R}_+) \oplus \mathbb{C}^{N_+}$. *The strongly continuous groups on* $L^2(\mathbb{R}_+) \oplus \mathbb{C}^N$, $N \in \mathbb{N}$, *that are involved in the definition of the symbol spaces are here* $\kappa_\lambda \oplus \mathrm{id}_{\mathbb{C}^N}$ *for* $(\kappa_\lambda f)(t) = \lambda^{\frac{1}{2}} f(\lambda t)$, $\lambda \in \mathbb{R}_+$. *Moreover,* $\mathcal{R}_G^{\mu,d}(\Omega \times \Omega \times \mathbb{R}^q; N_-, N_+)$ *for* $d \in \mathbb{N}$ *denotes the space of all operator–valued symbols*

$$g(y,y',\eta) \in \bigcap_{\substack{s \in \mathbb{R} \\ s > d - \frac{1}{2}}} S_{\mathrm{cl}}^\mu(\Omega \times \Omega \times \mathbb{R}^q; H^s(\mathbb{R}_+) \oplus \mathbb{C}^{N_-}, \mathcal{S}(\overline{\mathbb{R}}_+) \oplus \mathbb{C}^{N_+})$$

which have the form

$$g(y,y',\eta) = g_0(y,y',\eta) + \sum_{k=1}^d g_k(y,y',\eta) \begin{pmatrix} \left(\frac{d}{dt}\right)^k & 0 \\ 0 & 0 \end{pmatrix} \quad (4.2.23)$$

for arbitrary $g_k(y,y',\eta) \in \mathcal{R}_G^{\mu-k,0}(\Omega \times \Omega \times \mathbb{R}^q; N_-, N_+)$, $k = 0, \ldots, d$. *The corresponding subspaces of* y'-*independent elements will be denoted by* $\mathcal{R}_G^{\mu,d}(\Omega \times \mathbb{R}^q; N_-, N_+)$.

In the sequel, for simplicity we will mainly consider $g(y, \eta) \in$ $\mathcal{R}_G^{\mu,d}(\Omega \times \mathbb{R}^q; N_-, N_+)$. In the pseudo–differential calculus below with respect to the Fourier transform along Ω we can reduce the (y, y', η)-dependent symbols to y'-independent ones modulo elements of order $-\infty$, cf. Proposition 4.3.8 below.

Moreover, we shall often consider the case $N_- = N_+ = 0$. The assertions in general are then easy modifications of the arguments for $N_- = N_+ = 0$. The elements $g(y, \eta) \in \mathcal{R}_G^{\mu,d}(\Omega \times \mathbb{R}^q; N_-, N_+)$ can be written as block matrices

$$g(y, \eta) = \begin{pmatrix} g_{11} & g_{12} \\ g_{21} & g_{22} \end{pmatrix} (y, \eta). \qquad (4.2.24)$$

Here $g_{11}(y, \eta) = \text{u. l. c. } g(y, \eta)$(the upper left corner of $g(y, \eta)$) belongs to $R_G^{\mu,d}(\Omega \times \mathbb{R}^q)$, $g_{21}(y, \eta)$ is called a trace symbol of order μ and type d, $g_{12}(y, \eta)$ a potential symbol of order μ, and $g_{22}(y, \eta)$ is an $N_+ \times N_-$-matrix of elements in $S_{\text{cl}}^\mu(\Omega \times \mathbb{R}^q)$.

We shall endow $\mathcal{R}_G^{\mu,d}(\Omega \times \mathbb{R}^q; N_-, N_+)$ with a natural Fréchet topology. To this end we first observe that $\mathcal{R}_G^{\mu,0}(\Omega \times \mathbb{R}^q; N_-, N_+)$ is a Fréchet space. This follows immediately from Definition 4.2.19 with the Fréchet topologies in

$$S_{\text{cl}}^\mu(\Omega \times \mathbb{R}^q; L^2(\mathbb{R}_+) \oplus \mathbb{C}^{N_\mp}, \mathcal{S}(\overline{\mathbb{R}}_+) \oplus \mathbb{C}^{N_\pm}).$$

By definition every $g(y, \eta) = \mathcal{R}_G^{\mu,d}(\Omega \times \mathbb{R}^q; N_-, N_+)$ is a classical operator–valued symbol and we can talk about the homogeneous principal symbol $\sigma_\wedge^\mu(g)(y, \eta)$ of $g(y, \eta)$ of order μ. This is an operator family

$$\sigma_\wedge^\mu(g)(y, \eta) = \begin{array}{ccc} H^s(\mathbb{R}_+) & & \mathcal{S}(\overline{\mathbb{R}}_+) \\ \oplus & \to & \oplus \\ \mathbb{C}^{N_-} & & \mathbb{C}^{N_+} \end{array} \qquad (4.2.25)$$

for every $s > d - \frac{1}{2}$, satisfying

$$\sigma_\wedge^\mu(g)(y, \lambda\eta) = \lambda^\mu \begin{pmatrix} \kappa_\lambda & 0 \\ 0 & 1 \end{pmatrix} \sigma_\wedge^\mu(g)(y, \eta) \begin{pmatrix} \kappa_\lambda & 0 \\ 0 & 1 \end{pmatrix}^{-1} \qquad (4.2.26)$$

for all $\lambda > 0$. Here 1 denotes the identity operators in the corresponding finite-dimensional spaces. Let us denote by

$$\mathcal{R}_G^{(\mu),d}(\Omega \times (\mathbb{R}^q \setminus \{0\}); N_-, N_+) \qquad (4.2.27)$$

the space of all $\sigma_\wedge^\mu(g)(y,\eta)$ for arbitrary $g(y,\eta) \in \mathcal{R}_G^{\mu,d}(\Omega \times \mathbb{R}^q; N_-, N_+)$. Then we obviously have an exact sequence

$$0 \to \mathcal{R}_G^{\mu-1,d}(\Omega \times \mathbb{R}^q; N_-, N_+) \xrightarrow{\iota} \mathcal{R}_G^{\mu,d}(\Omega \times \mathbb{R}^q; N_-, N_+)$$

$$\xrightarrow{\sigma_\wedge^\mu} \mathcal{R}_G^{(\mu),d}(\Omega \times (\mathbb{R}^q \setminus \{0\}); N_-, N_+)$$

$$\to 0 \qquad\qquad (4.2.28)$$

with the canonical embedding ι. A right inverse of $\sigma_\wedge^\mu$ may be obtained by $g_{(\mu)}(y,\eta) \to \chi(\eta)g_{(\mu)}(y,\eta)$ for any excision function $\chi(\eta)$ in $\mathbb{R}^q$.

Proposition 4.2.20 *Let* $g_j(y,\eta) \in \mathcal{R}_G^{\mu-j,0}(\Omega \times \mathbb{R}^q; N_-, N_+)$, $j \in \mathbb{N}$, *be an arbitrary sequence. Then there is a* $g(y,\eta) \in \mathcal{R}_G^{\mu,0}(\Omega \times \mathbb{R}^q; N_-, N_+)$ *such that*

$$g(y,\eta) - \sum_{j=0}^N g_j(y,\eta) \in \mathcal{R}_G^{\mu-N-1,0}(\Omega \times \mathbb{R}^q; N_-, N_+)$$

holds for every $N \in \mathbb{N}$. *Every such* $g(y,\eta)$ *is unique* $\mathrm{mod}\,\mathcal{R}_G^{-\infty,0}(\Omega \times \mathbb{R}^q; N_-, N_+)$.

Proposition 4.2.20 is a special case of Proposition 3.3.11. An asymptotic sum $g(y,\eta) \sim \sum_{j=0}^\infty g_j(y,\eta)$ in this sense can be obtained as a convergent sum

$$g(y,\eta) = \sum_{j=0}^\infty \chi\left(\frac{\eta}{c_j}\right) g_j(y,\eta), \qquad\qquad (4.2.29)$$

where $\chi(\eta)$ is a fixed excision function and c_j a sequence of positive reals tending to ∞ sufficiently fast as $j \to \infty$. Note that the elements $g(y,\eta) \in \mathcal{R}_G^{\mu,0}(\Omega \times \mathbb{R}^q; N_-, N_+)$ are represented by kernel functions

$$g_{11}(y,\eta;t,t') \in C^\infty(\Omega \times \mathbb{R}^q, \mathcal{S}(\overline{\mathbb{R}}_+ \times \overline{\mathbb{R}}_+)),$$

$$g_{12}(y,\eta;t)_l \in C^\infty(\Omega \times \mathbb{R}^q, \mathcal{S}(\overline{\mathbb{R}}_+)), \qquad l = 1, \dots, N_-,$$

$$g_{21}(y,\eta;t')_m \in C^\infty(\Omega \times \mathbb{R}^q, \mathcal{S}(\overline{\mathbb{R}}_+)), \qquad m = 1, \dots, N_+,$$

and scalar symbols $g_{22}(y,\eta)_{ml} \in S_{cl}^\mu(\Omega \times \mathbb{R}^q)$, such that

$$g_{11}(y,\eta)u(t) = \int_0^\infty g_{11}(y,\eta;t,t')u(t')\,dt',$$

$$(g_{12}(y,\eta)c)(t) = \sum_{l=1}^{N_-} g_{12}(y,\eta;t)_l c_j, \qquad c = (c_1, \dots, c_{N_-}) \in \mathbb{C}^{N_-},$$

$$g_{21}(y,\eta)u = \left\{ \int_0^\infty g_{21}(y,\eta;t')_m u(t')dt' \right\}_{m=1,\dots,N_+}.$$

Let us have a look at the sequence of homogeneous components of order $\mu - j$ of a given $g(y, \eta) \in \mathcal{R}_G^{\mu,0}(\Omega \times \mathbb{R}^q; N_-, N_+)$. Note that

$$\mathcal{R}_G^{(\mu),0}(\Omega \times (\mathbb{R}^q \setminus \{0\}); N_-, N_+),$$

$\mu \in \mathbb{R}$, is the space of all functions

$$g_{(\mu)}(y, \eta) \in C^\infty(\Omega \times (\mathbb{R}^q \setminus \{0\}), \mathcal{L}(L^2(\mathbb{R}_+), \mathcal{S}(\overline{\mathbb{R}}_+)))$$

with

$$g_{(\mu)}^*(y, \eta) \in C^\infty(\Omega \times (\mathbb{R}^q \setminus \{0\}), \mathcal{L}(L^2(\mathbb{R}_+), \mathcal{S}(\overline{\mathbb{R}}_+)))$$

satisfying

$$g_{(\mu)}(y, \lambda\eta) = \lambda^\mu \begin{pmatrix} \kappa_\lambda & 0 \\ 0 & 1 \end{pmatrix} g_{(\mu)}(y, \eta) \begin{pmatrix} \kappa_\lambda & 0 \\ 0 & 1 \end{pmatrix}^{-1}$$

for all $(y, \eta) \in \Omega \times (\mathbb{R}^q \setminus \{0\})$ and all $\lambda \in \mathbb{R}_+$. Every such $g_{(\mu)}(y, \eta)$ is uniquely determined by an element

$$e(y, \eta) \in C^\infty(\Omega \times S^{q-1}; \Gamma^0(\mathbb{R}_+; N_-, N_+)),$$

cf. Definition 4.1.76, satisfying

$$g_{(\mu)}(y, \eta) = |\eta|^\mu \begin{pmatrix} \kappa_{|\eta|} & 0 \\ 0 & 1 \end{pmatrix} e\left(y, \frac{\eta}{|\eta|}\right) \begin{pmatrix} \kappa_{|\eta|} & 0 \\ 0 & 1 \end{pmatrix}^{-1}$$

for all $(y, \eta) \in \Omega \times (\mathbb{R}^q \setminus \{0\})$. Now $e_{11}(y, \eta) = \text{u.l.c.}\, e(y, \eta)$, $|\eta| = 1$, has a representation of the form

$$e_{11}(y, \eta)u(t) = \int_0^\infty e_{11}(y, \eta; t, t')u(t')\, dt'$$

for a function

$$e_{11}(y, \eta; t, t') \in C^\infty(\Omega \times S^{q-1}, \mathcal{S}(\overline{\mathbb{R}}_+ \times \overline{\mathbb{R}}_+)).$$

Then $g_{(\mu),11}(y, \eta) = \text{u.l.c.}\, g_{(\mu)}(y, \eta)$ acts on $u(t) \in L^2(\mathbb{R}_+)$ as

$$g_{(\mu),11}(y, \eta)u(t) = \int_0^\infty |\eta|^{\mu+1} e_{11}\left(y, \frac{\eta}{|\eta|}; t|\eta|, t'|\eta|\right) u(t')\, dt'.$$

In an analogous manner $g_{(\mu),12}(y, \eta)$, $g_{(\mu),21}(y, \eta)$ are determined by vectors

$$\{e_{12}(y, \eta; t)_l\}_{l=1,\ldots,N_-}, \qquad \{e_{21}(y, \eta; t')_m\}_{m=1,\ldots,N_+},$$

of elements in $C^\infty(\Omega \times S^{q-1}, \mathcal{S}(\overline{\mathbb{R}}_+))$, such that

$$(g_{(\mu),12}(y,\eta)c)(t) = \sum_{l=1}^{N_-} |\eta|^{\mu+\frac{1}{2}} e_{12}\left(y, \frac{\eta}{|\eta|}; t|\eta|\right)_l c_l,$$

$c = (c_1, \ldots, c_{N_-}) \in \mathbb{C}^{N_-}$,and

$$g_{(\mu),21}(y,\eta)u = \left\{ \int_0^\infty |\eta|^{\mu+\frac{1}{2}} e_{21}\left(y, \frac{\eta}{|\eta|}; t'|\eta|\right)_m u(t')\, dt' \right\}_{m=1,\ldots,N_+},$$

$u \in L^2(\mathbb{R}_+)$.
Let us write

$$E^k = \langle t \rangle^{-k} H^k(\mathbb{R}_+), \quad k \in \mathbb{N}, \tag{4.2.30}$$

endowed with the norm $\|u\|_{\langle t \rangle^{-k} H^k(\mathbb{R}_+)} := \|\langle t \rangle^k u\|_{H^k(\mathbb{R}_+)}$. If we set $(\kappa_\lambda u)(t) = \lambda^{\frac{1}{2}} u(\lambda t)$, $\lambda \in \mathbb{R}_+$, then $\{\kappa_\lambda\}_{\lambda \in \mathbb{R}_+}$ acts on $H^k(\mathbb{R}_+)$ as a strongly continuous group of isomorphisms for every k. Moreover, we have

$$\mathcal{S}(\overline{\mathbb{R}}_+) = \varprojlim_{k \in \mathbb{N}} E^k.$$

Recall that for $g(y,\eta) \in \mathcal{R}_G^{\mu,0}(\Omega \times \mathbb{R}^q)$ from Definition 4.2.19 it follows that

$$\|\kappa^{-1}(\eta)\{D_y^\alpha D_\eta^\beta g(y,\eta)\}\kappa(\eta)\|_{\mathcal{L}(L^2(\mathbb{R}_+), E^k)} \leq c\langle \eta \rangle^{\mu-|\beta|},$$

$$\|\kappa^{-1}(\eta)\{D_y^\alpha D_\eta^\beta g^*(y,\eta)\}\kappa(\eta)\|_{\mathcal{L}(L^2(\mathbb{R}_+), E^k)} \leq c\langle \eta \rangle^{\mu-|\beta|}$$

for all $y \in K \subset\subset \Omega$, $\eta \in \mathbb{R}^q$, $\alpha, \beta \in \mathbb{N}^q$ and all $k \in \mathbb{N}$, with constants $c = c(\alpha, \beta, K; k) > 0$. Analogous symbol estimates hold for $g(y,\eta) \in \mathcal{R}_G^{\mu,0}(\Omega \times \mathbb{R}^q; N_-, N_+)$. Together with the property of being classical they characterise the space $\mathcal{R}_G^{\mu,0}(\Omega \times \mathbb{R}^q; N_-, N_+)$.

It will be convenient from time to time to talk about E-valued symbols for some Fréchet space E. Let $\{r_k\}_{k \in \mathbb{N}}$ be a countable semi-norm system for E. Denote by $S^\mu(\Omega \times \mathbb{R}^q, E)$ the subspace of all $p(y,\eta) \in C^\infty(\Omega \times \mathbb{R}^q, E)$ satisfying the estimates

$$r_k(D_y^\alpha D_\eta^\beta p(y,\eta)) \leq c\langle \eta \rangle^{\mu-|\beta|}$$

for all $y \in K \subset\subset \Omega$, $\eta \in \mathbb{R}^q$, $\alpha, \beta \in \mathbb{N}^q$ and all $k \in \mathbb{N}$, with constants $c = c(\alpha, \beta, K; k) > 0$.

An element $p(y,\eta) \in S^\mu(\Omega \times \mathbb{R}^q, E)$ is said to be classical if there is a sequence of functions $p_{(\mu-j)}(y,\eta) \in C^\infty(\Omega \times (\mathbb{R}^q \setminus \{0\}), E)$, $j \in \mathbb{N}$, satisfying

$$p_{(\mu-j)}(y, \lambda\eta) = \lambda^{\mu-j} p_{(\mu-j)}(y,\eta) \quad \text{for all} \quad \lambda \in \mathbb{R}_+$$

such that for any excision function $\chi(\eta)$

$$p(y,\eta) - \chi(\eta) \sum_{j=0}^{N} p_{(\mu-j)}(y,\eta) \in S^{\mu-(N+1)}(\Omega \times \mathbb{R}^q, E)$$

for every $N \in \mathbb{N}$. We denote by $S^{\mu}_{\mathrm{cl}}(\Omega \times \mathbb{R}^q, E)$ the subspace of classical elements in $S^{\mu}(\Omega \times \mathbb{R}^q, E)$. Analogously to the scalar case, i.e., $E = \mathbb{C}$, the spaces $S^{\mu}(\Omega \times \mathbb{R}^q, E)$ and $S^{\mu}_{\mathrm{cl}}(\Omega \times \mathbb{R}^q, E)$ are Fréchet spaces in a natural way.

Note, in particular, that we have a canonical isomorphism

$$S^{\mu}(\Omega \times \mathbb{R}^q, \mathcal{S}(\overline{\mathbb{R}}_+ \times \overline{\mathbb{R}}_+)) \cong S^{\mu}(\Omega \times \mathbb{R}^q) \otimes_{\pi} \mathcal{S}(\overline{\mathbb{R}}_+ \times \overline{\mathbb{R}}_+),$$

and the same with subscripts "cl". Analogous relations hold with $\mathcal{S}(\overline{\mathbb{R}}_+)$ instead of $\mathcal{S}(\overline{\mathbb{R}}_+ \times \overline{\mathbb{R}}_+)$.

Proposition 4.2.21 *For every $f(y,\eta;t,t') \in S^{\mu+1}_{\mathrm{cl}}(\Omega \times \mathbb{R}^q) \otimes_{\pi} \mathcal{S}(\overline{\mathbb{R}}_+ \times \overline{\mathbb{R}}_+)$ the operator family*

$$g(y,\eta): \; u(t) \to \int_0^{\infty} f(y,\eta;t[\eta],t'[\eta])u(t')\,dt' \qquad (4.2.31)$$

belongs to $R^{\mu,0}_G(\Omega \times \mathbb{R}^q)$, and every $g(y,\eta) \in R^{\mu,0}_G(\Omega \times \mathbb{R}^q)$ has such a representation.

Proof. Let us first assume $g(y,\eta)$ to be of the form (4.2.31) for an $f(y,\eta;t,t') \in S^{\mu+1}_{\mathrm{cl}}(\Omega \times \mathbb{R}^q) \otimes_{\pi} \mathcal{S}(\overline{\mathbb{R}}_+ \times \overline{\mathbb{R}}_+)$. Define an operator $G(y,\eta) := \kappa^{-1}(\eta)g(y,\eta)\kappa(\eta)$ by

$$G(y,\eta): u(t) \to \int_0^{\infty} [\eta]^{-1} f(y,\eta;t,t')u(t')\,dt'.$$

Writing $f(y,\eta;t,t')$ as a convergent sum

$$f(y,\eta;t,t') = \sum_{j=0}^{\infty} \lambda_j f_j(y,\eta)g_j(t,t')$$

with $\lambda_j \in \mathbb{C}$, $\sum |\lambda_j| < \infty$, $f_j(y,\eta) \in S^{\mu+1}_{\mathrm{cl}}(\Omega \times \mathbb{R}^q)$, $g_j(t,t') \in \mathcal{S}(\overline{\mathbb{R}}_+ \times \overline{\mathbb{R}}_+)$, $f_j \to 0$, $g_j \to 0$ in the corresponding spaces for $j \to \infty$, we obtain for

$$G_j u(t) = \int_0^{\infty} g_j(t,t')u(t')\,dt', \qquad u \in L^2(\mathbb{R}_+)$$

that $\|G_j\|_{\mathcal{L}(L^2(\mathbb{R}_+),E^k)} \to 0$ as $j \to \infty$ for every fixed $k \in \mathbb{N}$. This implies

$$\|f_j(y,\eta)G_j\|_{\mathcal{L}(L^2(\mathbb{R}_+),E^k)} \le c_j \langle \eta \rangle^\mu \|G_j\|_{\mathcal{L}(L^2(\mathbb{R}_+),E^k)}$$

for $y \in K \subset\subset \Omega$, $\eta \in \mathbb{R}^q$, with constants $c_j \to 0$ for $j \to \infty$, because of the symbol estimates for $f_j(y,\eta)$. Thus we obtain

$$\|G(y,\eta)\|_{\mathcal{L}(L^2(\mathbb{R}_+),E^k)} \le c \langle \eta \rangle^\mu$$

for $y \in K \subset\subset \Omega$, $\eta \in \mathbb{R}^q$, with a constant $c = c(K)$, for every k. In an analogous manner we can argue for $D_y^\alpha D_\eta^\beta g(y,\eta)$, $\alpha,\beta \in \mathbb{N}^q$, which yields the corresponding estimates with $\langle \eta \rangle^{\mu-|\beta|}$ on the right. For the adjoints we can do the same. This finally shows $g(y,\eta) \in R_G^{\mu,0}(\Omega \times \mathbb{R}^q)$. Conversely let $g(y,\eta) \in R_G^{\mu,0}(\Omega \times \mathbb{R}^q)$, and write

$$g(y,\eta) \sim \sum_{j=0}^\infty \chi(\eta) g_{(\mu-j)}(y,\eta)$$

for the homogeneous components $g_{(\mu-j)}(y,\eta)$ of $g(y,\eta)$ of order $\mu - j$, $j \in \mathbb{N}$, $\chi(\eta)$ being an excision function. The asymptotic sum can be carried out as a convergent sum

$$\tilde{g}(y,\eta) = \sum_{j=0}^\infty \chi\left(\frac{\eta}{d_j}\right) g_{(\mu-j)}(y,\eta)$$

with constant $d_j \to \infty$ sufficiently fast for $j \to \infty$. Then

$$g(y,\eta) - \tilde{g}(y,\eta) \in R_G^{-\infty,0}(\Omega \times \mathbb{R}^q).$$

As we have seen above $g_{(\mu-j)}(y,\eta)$ is defined by a function of the form

$$|\eta|^{\mu-j} e_{(\mu-j)}\left(y, \frac{\eta}{|\eta|}; t|\eta|, t'|\eta|\right) \tag{4.2.32}$$

for $e_{(\mu-j)}(y,\eta;t,t') \in C^\infty(\Omega \times S^{q-1}, \mathcal{S}(\overline{\mathbb{R}}_+ \times \overline{\mathbb{R}}_+))$. Assume without loss of generality d_j to be so large that $[\eta] = |\eta|$ for $\chi(\eta/d_j) \neq 0$. Then

$$h_j(y,\eta;t[\eta],t'[\eta]) := \chi\left(\frac{\eta}{d_j}\right) |\eta|^{\mu+1-j} e_{(\mu-j)}\left(y, \frac{\eta}{|\eta|}; t|\eta|, t'|\eta|\right)$$

is defined by a function $h_j(y,\eta;t,t') \in S_{cl}^{\mu+1-j}(\Omega \times \mathbb{R}^q) \otimes_\pi \mathcal{S}(\overline{\mathbb{R}}_+ \times \overline{\mathbb{R}}_+)$. Now we have to verify that

$$\sum_{j=0}^\infty h_j(y,\eta;t,t') \tag{4.2.33}$$

converges in $S_{\mathrm{cl}}^{\mu+1}(\Omega \times \mathbb{R}^q) \otimes_\pi \mathcal{S}(\overline{\mathbb{R}}_+ \times \overline{\mathbb{R}}_+)$ for a suitable choice of the constants d_j. Here we may forget about subscript "cl", since the summands are homogeneous of orders $\mu + 1 - j$ in η for $|\eta| \geq$ const, and hence the convergence of the associated series of homogeneous components is trivial. Let us set

$$H_j(y, \eta)u(t) = \int\limits_0^\infty h_j(y, \eta; t, t')u(t')\, dt'.$$

The convergence of (4.2.33) in $S^{\mu+1}(\Omega \times \mathbb{R}^q) \otimes_\pi \mathcal{S}(\overline{\mathbb{R}}_+ \times \overline{\mathbb{R}}_+)$ to a limit $h(y, \eta; t, t')$ is equivalent to the convergence in $S^{\mu+1}(\Omega \times \mathbb{R}^q, \mathcal{S}(\overline{\mathbb{R}}_+ \times \overline{\mathbb{R}}_+))$, where the latter space can be identified with a corresponding space of $\Gamma^0(\mathbb{R}_+)$-valued symbols $H(y, \eta)$ with the semi–norm system

$$\pi_{(\alpha, \beta, K; k)}(h) := \sup_{y \in K} \sup_\eta \langle \eta \rangle^{-\mu - 1 + |\beta|} \| D_y^\alpha D_\eta^\beta H(y, \eta) \|_{\mathcal{L}(L^2(\mathbb{R}_+), E^k)},$$

$$(4.2.34)$$

$$\pi'_{(\alpha, \beta, K; k)}(h) := \sup_{y \in K} \sup_\eta \langle \eta \rangle^{-\mu - 1 + |\beta|} \| D_y^\alpha D_\eta^\beta\, {}^t H(y, \eta) \|_{\mathcal{L}(L^2(\mathbb{R}_+), E^k)}$$

$$(4.2.35)$$

for all k, all $K \subset\subset \Omega$, $\alpha, \beta \in \mathbb{N}^q$, with constants $c = c(\alpha, \beta, K; k) > 0$. Here $H(y, \eta)u(t) := \int_0^\infty h(y, \eta; t, t')u(t')\, dt'$, and superscript t indicates the transposed operator. From the properties of the functions $h_j(y, \eta; t, t')$ we immediately see that the operator families $H_j(y, \eta)$ and ${}^t H_j(y, \eta)$ are elements of

$$S_{\mathrm{cl}}^{\mu+1-j}(\Omega \times \mathbb{R}^q; L^2(\mathbb{R}_+), \mathcal{S}(\overline{\mathbb{R}}_+))_{(1)}$$
$$= \varprojlim{}_{k \in \mathbb{N}} S^{\mu+1-j}(\Omega \times \mathbb{R}^q; L^2(\mathbb{R}_+), E^k)_{(1)}.$$

Subscripts $_{(1)}$ indicate the operator–valued symbol spaces with the identical group actions in the involved spaces. Now the sequence of the $H_j(y, \eta)$ has an asymptotic sum

$$H(y, \eta) \in S^{\mu+1}(\Omega \times \mathbb{R}^q; L^2(\mathbb{R}_+), \mathcal{S}(\overline{\mathbb{R}}_+))_{(1)}$$

such that at the same time ${}^t H(y, \eta)$ is the asymptotic sum of the ${}^t H_j(y, \eta)$ in $S^{\mu+1}(\Omega \times \mathbb{R}^q; L^2(\mathbb{R}_+), \mathcal{S}(\overline{\mathbb{R}}_+))_{(1)}$. We obtain $H(y, \eta)$ as a convergent sum

$$H(y, \eta) = \sum_{j=0}^\infty \chi\left(\frac{\eta}{\tilde{d}_j}\right) H_j(y, \eta) \qquad (4.2.36)$$

for a suitable sequence of constants $\tilde{d}_j$ tending to ∞ for $j \to \infty$ sufficiently fast, with uniqueness mod $S^{-\infty}(\Omega \times \mathbb{R}^q; L^2(\mathbb{R}_+), \mathcal{S}(\overline{\mathbb{R}}_+))$ (recall that the space of symbols of order $-\infty$ is independent of the group actions in the corresponding spaces, such that (1) is unnecessary in this case). Taking $\tilde{d}_j \geq d_j$ for all j which is an allowed choice, we then may modify the above d_j, again, by taking them larger, if necessary, such that it is possible to set $\tilde{d}_j = d_j$. The convergence of (4.2.36) which relies on the semi-norm systems (4.2.34), (4.2.35) now corresponds exactly to the desired convergence of (4.2.33) to a function $h(y, \eta; t, t')$ in the space $S^{\mu+1}(\Omega \times \mathbb{R}^q) \otimes_\pi \mathcal{S}(\overline{\mathbb{R}}_+ \times \overline{\mathbb{R}}_+)$. Then the original operator family $g(y, \eta)$ satisfies

$$g_{-\infty}(y, \eta) := g(y, \eta) - H(y, \eta) \in R_G^{-\infty, 0}(\Omega \times \mathbb{R}^q).$$

It remains to observe that also the latter difference has a representation by some $h_{-\infty}(y, \eta; t, t') \in S^{-\infty}(\Omega \times \mathbb{R}^q) \otimes_\pi \mathcal{S}(\overline{\mathbb{R}}_+ \times \overline{\mathbb{R}}_+)$ in the form

$$g_{-\infty}(y, \eta)u(t) = \int_0^\infty h_{-\infty}(y, \eta; t[\eta], t'[\eta])u(t')\, dt'.$$

$\square$

Remark 4.2.22 *A consequence of Proposition 4.2.21 is that for the operator* $\kappa_\rho^0 : v(t) \to v(\rho t)$, $\rho \in \mathbb{R}_+$, *we have* $\kappa_\rho^0 \circ g(y, \eta), g(y, \eta) \circ \kappa_\rho^0 \in R_G^{\mu,0}(\Omega \times \mathbb{R}^q)$ *for every* $g(y, \eta) \in R_G^{\mu,0}(\Omega \times \mathbb{R}^q)$ *and fixed* $\rho \in \mathbb{R}_+$.

Proposition 4.2.23 *The space* $\mathcal{R}_G^{\mu,0}(\Omega \times \mathbb{R}^q; 1, 1)$ *of block matrices*

$$g(y, \eta) = (g_{ij}(y, \eta))_{i,j=1,2}$$

is characterised as a space of matrices $f = (f_{ij})_{i,j=1,2}$. *Here* $g_{11}(y, \eta) \in R_G^{\mu,0}(\Omega \times \mathbb{R}^q)$ *is given by an* $f_{11}(y, \eta; t, t') \in S_{\mathrm{cl}}^{\mu+1}(\Omega \times \mathbb{R}^q) \otimes_\pi \mathcal{S}(\overline{\mathbb{R}}_+ \times \overline{\mathbb{R}}_+)$ *as in Proposition 4.2.21. Furthermore* $g_{22}(y, \eta) = f_{22}(y, \eta)$, *while the trace symbol* $g_{21}(y, \eta) : L^2(\mathbb{R}_+) \to \mathbb{C}$ *is given by*

$$g_{21}(y, \eta) : u(t) \to \int_0^\infty f_{21}(y, \eta; t'[\eta])u(t')\, dt'$$

for an $f_{21}(y, \eta; t') \in S_{\mathrm{cl}}^{\mu+\frac{1}{2}}(\Omega \times \mathbb{R}^q) \otimes_\pi \mathcal{S}(\overline{\mathbb{R}}_+)$. *The potential symbol* $g_{12}(y, \eta) : \mathbb{C} \to L^2(\mathbb{R}_+)$ *is given by*

$$g_{12}(y, \eta) : c \to f_{12}(y, \eta; t[\eta])c, \quad c \in \mathbb{C},$$

for an $f_{12}(y, \eta; t) \in S_{\mathrm{cl}}^{\mu+\frac{1}{2}}(\Omega \times \mathbb{R}^q) \otimes_\pi \mathcal{S}(\overline{\mathbb{R}}_+)$. *An analogous characterisation holds for* $\mathcal{R}_G^{\mu,0}(\Omega \times \mathbb{R}^q; N_-, N_+)$.

We omit the proof of Proposition 4.2.23 that is completely analogous to that of Proposition 4.2.21.

Remark 4.2.24 *Let* $g(y,\eta) \in S^\mu(\Omega \times \mathbb{R}^q; L^2(\mathbb{R}_+) \oplus \mathbb{C}^{N-}, \mathcal{S}(\overline{\mathbb{R}}_+) \oplus \mathbb{C}^{N+})$ *be a symbol for which there is a sequence* $g_j(y,\eta) \in \mathcal{R}_G^{\mu-j,0}(\Omega \times \mathbb{R}^q; N_-, N_+)$, $j \in \mathbb{N}$, *such that for every* $k \in \mathbb{N}$ *there is an* $N = N(k)$ *such that*

$$r_k(y,\eta) := g(y,\eta) - \sum_{j=0}^{N} g_j(y,\eta)$$

satisfies $r_k(y,\eta) \in S^{-k}(\Omega \times \mathbb{R}^q; L^2(\mathbb{R}_+) \oplus \mathbb{C}^{N-}, E^k \oplus \mathbb{C}^{N+})$ *and* $r_k^*(y,\eta) \in S^{-k}(\Omega \times \mathbb{R}^q; L^2(\mathbb{R}_+) \oplus \mathbb{C}^{N+}, E^k \oplus \mathbb{C}^{N-})$ *cf.* (4.2.30), *then we have* $g(y,\eta) \in \mathcal{R}_G^{\mu,0}(\Omega \times \mathbb{R}^q; N_-, N_+)$ *and* $g(y,\eta) \sim \sum_{j=0}^\infty g_j(y,\eta)$ *in the sense of Proposition 4.2.20.*

In the sequel we occasionally set

$$\gamma^j u = \frac{d^j}{dt^j} u(t)|_{t=0}, \qquad j \in \mathbb{N}.$$

Note that

$$\gamma^j = \lambda^{\frac{1}{2}+j} \gamma^j \kappa_\lambda^{-1} \qquad \text{for all } \lambda \in \mathbb{R}_+.$$

Lemma 4.2.25 *Let* $g(y,\eta) \in R_G^{\mu,0}(\Omega \times \mathbb{R}^q)$ *and form the operator family* $g(y,\eta) \circ \frac{d}{dt'}$ *on* $H^s(\mathbb{R}_+)$ *for* $s > \frac{1}{2}$. *Then there is a decomposition*

$$g(y,\eta) \circ \frac{d}{dt'} = g_0(y,\eta) + k(y,\eta) \circ \gamma^0$$

with elements $g_0(y,\eta) \in R_G^{\mu+1,0}(\Omega \times \mathbb{R}^q)$, $k(y,\eta) \in \mathcal{R}_G^{\mu+\frac{1}{2},0}(\Omega \times \mathbb{R}^q; 1,1)$ *(the latter one being a symbol of potential type).* $g_0(y,\eta)$ *and* $k(y,\eta)$ *are uniquely determined by* $g(y,\eta)$.

Proof. Using the representation (4.2.31) of $g(y,\eta)$ and the calculations from the proof of Proposition 4.1.46 we obtain

$$g(y,\eta)(\frac{d}{dt'}u(t')) = \int_0^\infty f(y,\eta; t[\eta], t'[\eta]) \left(\frac{d}{dt'}u(t')\right) dt'$$
$$= g_0(y,\eta)u(t) + (k(y,\eta)\gamma^0 u)(t)$$

with

$$g_0(y,\eta)u(t) = \int_0^\infty \left\{ -\frac{d}{dt'} f(y,\eta; t[\eta], t'[\eta]) \right\} u(t')\, dt',$$
$$(k(y,\eta)\gamma^0 u)(t) = f(y,\eta; t[\eta], 0)u(0).$$

From Proposition 4.2.23 we now get the asserted properties of $g_0(y, \eta)$ and $k(y, \eta)$. The uniqueness of the decomposition was already obtained in Proposition 4.1.46. $\square$

Lemma 4.2.26 *Let* $b(y, \eta) \in \mathcal{R}_G^{\mu,0}(\Omega \times \mathbb{R}^q; 1, 1)$ *be a trace symbol (recall that it has the type of a lower left corner of a block matrix of the form* (4.2.24)*), and consider the operator family* $b(y, \eta) \circ \frac{d}{dt'} : H^s(\mathbb{R}_+) \to \mathbb{C}$ *for* $s > \frac{1}{2}$. *Then there is a decomposition*

$$b(y, \eta) \circ \frac{d}{dt'} = b_0(y, \eta) + c(y, \eta) \circ \gamma^0$$

with a $b_0(y, \eta) \in \mathcal{R}_G^{\mu+1,0}(\Omega \times \mathbb{R}^q; 1, 1)$ *of trace type and a symbol* $c(y, \eta) \in S_{cl}^{\mu+\frac{1}{2}}(\Omega \times \mathbb{R}^q)$, *where* $b_0(y, \eta)$ *and* $c(y, \eta)$ *are uniquely determined by* $b(y, \eta)$.

Proof. By Proposition 4.2.23 we have

$$b(y, \eta) \circ \frac{d}{dt'} u(t') = \int_0^\infty f(y, \eta; t'[\eta]) \left(\frac{d}{dt'} u(t') \right) dt'$$

for an $f(y, \eta; t') \in S_{cl}^{\mu+\frac{1}{2}}(\Omega \times \mathbb{R}^q) \otimes_\pi \mathcal{S}(\overline{\mathbb{R}}_+)$. Integration by parts gives us

$$b(y, \eta) \circ \frac{d}{dt'} u(t') = b_0(y, \eta)u + c(y, \eta)\gamma^0 u$$

for $b_0(y, \eta)u = \int_0^\infty \left\{ -\frac{d}{dt'} f(y, \eta; t'[\eta]) \right\} u(t')dt'$ and $c(y, \eta) = f(y, \eta; 0)$. The asserted properties of $b_0(y, \eta)$ and $c(y, \eta)$ are then obvious. The uniqueness follows by analogous arguments as for Lemma 4.2.25. $\square$

Theorem 4.2.27 *Every* $g(y, \eta) \in \mathcal{R}_G^{\mu,d}(\Omega \times \mathbb{R}^q; N_-, N_+)$ *has a unique representation of the form*

$$g(y, \eta) = g_0(y, \eta) + g_1(y, \eta)$$

for an element $g_0(y, \eta) \in \mathcal{R}_G^{\mu,0}(\Omega \times \mathbb{R}^q; N_-, N_+)$ *and*

$$g_1(y, \eta) = \sum_{j=0}^{d-1} \begin{pmatrix} 0 & k_j(y, \eta) \\ 0 & c_j(y, \eta) \end{pmatrix} \begin{pmatrix} 0 & 0 \\ \gamma_j & 0 \end{pmatrix}$$

for potential symbols $k_j(y, \eta) \in \mathcal{R}_G^{\mu-j-\frac{1}{2},0}(\Omega \times \mathbb{R}^q; 1, 0)$ *and*

$$c_j(y, \eta) = (c_{j,1}(y, \eta), \dots, c_{j,N_+}(y, \eta))$$

with $c_{j,l}(y, \eta) \in S_{cl}^{\mu-j-\frac{1}{2}}(\Omega \times \mathbb{R}^q)$, $l = 1, \dots, N_+$, $j = 0, \dots, d-1$.

Proof. Writing $g(y,\eta)$ as a block matrix $(g_{ij}(y,\eta))_{i,j=1,2}$, it suffices to consider $g_{11}(y,\eta)$ and $g_{21}(y,\eta)$ separately. By definition we have

$$g_{11}(y,\eta)u(t) = \sum_{k=0}^{d} g_{11,k}(y,\eta) \left(\frac{d}{dt'}\right)^k u,$$

$$g_{21}(y,\eta)u(t) = \sum_{k=0}^{d} g_{21,k}(y,\eta) \left(\frac{d}{dt'}\right)^k u,$$

where $g_{11,k}(y,\eta) \in R_G^{\mu-k,0}(\Omega \times \mathbb{R}^q)$, $g_{21,k}(y,\eta) \in \mathcal{R}_G^{\mu-k,0}(\Omega \times \mathbb{R}^q; 1, 1)$ (the latter ones being of trace type). Then the assertion follows by applying Lemma 4.2.25 and Lemma 4.2.26 and by iterating the arguments. $\square$

Remark 4.2.28 *There is a canonical isomorphism*

$$\mathcal{R}_G^{\mu,d}(\Omega \times \mathbb{R}^q; N_-, N_+) \cong \mathcal{R}_G^{\mu,0}(\Omega \times \mathbb{R}^q; N_-, N_+) \times \mathcal{K}_G^{\mu,d}(\Omega \times \mathbb{R}^q),$$

where $\mathcal{K}_G^{\mu,d}(\Omega \times \mathbb{R}^q)$ is the Cartesian product

$$\underset{j=0}{\overset{d-1}{\times}} \left\{ \mathcal{R}_G^{\mu-j-\frac{1}{2},0}(\Omega \times \mathbb{R}^q; 1, 0)_{\mathrm{pot}} \times S_{\mathrm{cl}}^{\mu-j-\frac{1}{2}}(\Omega \times \mathbb{R}^q) \otimes \mathbb{C}^{N_+} \right\},$$

subscript "pot" indicating the corresponding spaces of potential objects (consisting of the upper right corners of the block matrices). Since all factors in this representation have natural Fréchet topologies, also $\mathcal{R}_G^{\mu,d}(\Omega \times \mathbb{R}^q; N_-, N_+)$ is a Fréchet space in a canonical way.

Moreover, $\mathcal{R}_G^{(\mu),d}(\Omega \times (\mathbb{R}^q \setminus \{0\}); N_-, N_+)$ is isomorphic to the Cartesian product between $\mathcal{R}_G^{(\mu),0}(\Omega \times (\mathbb{R}^q \setminus \{0\}); N_-, N_+)$ and

$$\underset{j=0}{\overset{d-1}{\times}} \left\{ \mathcal{R}_G^{(\mu-j-\frac{1}{2}),0}(\Omega \times (\mathbb{R}^q \setminus \{0\}); 1, 0)_{\mathrm{pot}} \right.$$

$$\left. \times S^{(\mu-j-\frac{1}{2})}(\Omega \times (\mathbb{R}^q \setminus \{0\})) \otimes \mathbb{C}^{N_+} \right\}.$$

In this way also $\mathcal{R}_G^{(\mu),d}(\Omega \times (\mathbb{R}^q \setminus \{0\}); N_-, N_+)$ is a Fréchet space.

Proposition 4.2.29 *Let $g_j(y,\eta) \in \mathcal{R}_G^{\mu-j,d}(\Omega \times \mathbb{R}^q; N_-, N_+)$, $j \in \mathbb{N}$, be an arbitrary sequence. Then there is a $g(y,\eta) \in \mathcal{R}_G^{\mu,d}(\Omega \times \mathbb{R}^q; N_-, N_+)$ such that for every $N \in \mathbb{N}$ with*

$$g(y,\eta) - \sum_{j=0}^{N} g_j(y,\eta) \in \mathcal{R}_G^{\mu-(N+1),d}(\Omega \times \mathbb{R}^q; N_-, N_+),$$

and modulo $\mathcal{R}_G^{-\infty,d}(\Omega \times \mathbb{R}^q; N_-, N_+)$ every such $g(y,\eta)$ is uniquely determined.

Proof. Using Remark 4.2.28 it suffices to carry out the asymptotic sums for the involved factors separately and to apply Proposition 4.2.20. □

Proposition 4.2.30 $f(y, \eta) \in \mathcal{R}_G^{\mu, d}(\Omega \times \mathbb{R}^q; N_0, N_+)$, $g(y, \eta) \in \mathcal{R}_G^{\nu, e}(\Omega \times \mathbb{R}^q; N_-, N_0)$ implies $f(y, \eta)g(y, \eta) \in \mathcal{R}_G^{\mu+\nu, e}(\Omega \times \mathbb{R}^q; N_-, N_+)$, and we have

$$\sigma_\wedge^{\mu+\nu}(fg) = \sigma_\wedge^\mu(f)\sigma_\wedge^\nu(g).$$

Proof. The assertion follows from the composition behaviour of Green operators on the half axis for every fixed (y, η), cf. Proposition 4.1.78 (where in the present case the type h equals e), and from the fact that the involved classical symbols are preserved under compositions; this is compatible with the compositions of the principal symbols. □

Proposition 4.2.31 $g(y, \eta) \in \mathcal{R}_G^{\mu, d}(\Omega \times \mathbb{R}^q; N_-, N_+)$, implies

$$\begin{pmatrix} t^j & 0 \\ 0 & p(y, \eta) \end{pmatrix} g(y, \eta) \begin{pmatrix} t^k & 0 \\ 0 & r(y, \eta) \end{pmatrix} \in \mathcal{R}_G^{\mu-(j+k), d}(\Omega \times \mathbb{R}^q; N_-, N_+)$$

for every $j, k \in \mathbb{N}$ and

$$p(y, \eta) = \begin{pmatrix} p_1(y, \eta) & & 0 \\ & \ddots & \\ 0 & & p_{N_+}(y, \eta) \end{pmatrix},$$

$$r(y, \eta) = \begin{pmatrix} r_1(y, \eta) & & 0 \\ & \ddots & \\ 0 & & r_{N_-}(y, \eta) \end{pmatrix}$$

for arbitrary $p_n(y, \eta) \in S_{cl}^{-j}(\Omega \times \mathbb{R}^q)$, $n = 1, \ldots, N_+$, $r_m(y, \eta) \in S_{cl}^{-k}(\Omega \times \mathbb{R}^q)$, $m = 1, \ldots, N_-$.

The proof is obvious. The only point to note is that the compositions with powers of t contribute the negative exponents to the operator-valued orders. Let us also observe that for arbitrary $a(y, \eta) \in S_{cl}^\alpha(\Omega \times \mathbb{R}^q)$, $b(y, \eta) \in S_{cl}^\beta(\Omega \times \mathbb{R}^q)$, $\alpha, \beta \in \mathbb{R}$, we have

$$g(y, \eta) \in \mathcal{R}_G^{\mu, d}(\Omega \times \mathbb{R}^q) \Rightarrow a(y, \eta)g(y, \eta)b(y, \eta) \in R_G^{\mu+\alpha+\beta, d}(\Omega \times \mathbb{R}^q).$$

Let us denote by Φ the operator of multiplication by the block matrix

$$\begin{pmatrix} \varphi & 0 \\ 0 & 1 \end{pmatrix} \quad \text{for } \varphi(t, y) \in C_0^\infty(\overline{\mathbb{R}}_+ \times \Omega)$$

with 1 being the unit $N \times N$-matrix for a suitable N. Analogously we use the notation Ψ for a $\psi(y, \eta) \in C_0^\infty(\overline{\mathbb{R}}_+ \times \Omega)$.

Theorem 4.2.32 *For every* $\varphi(t,y), \psi(t,y) \in C_0^\infty(\overline{\mathbb{R}}_+ \times \Omega)$ *we have*

$$\Phi \mathcal{R}_G^{\mu,d}(\Omega \times \mathbb{R}^q; N_-, N_+) \Psi \subseteq \mathcal{R}_G^{\mu,d}(\Omega \times \mathbb{R}^q; N_-, N_+).$$

Moreover, $g(y,\eta) \to \Phi g(y,\eta)\Psi$ *is continuous in* φ, g *and* ψ.

Proof. For simplicity we restrict ourselves to the space of upper left corners $R_G^{\mu,0}(\Omega \times \mathbb{R}^q)$. The generalisation to arbitrary d and N_-, N_+ is evident. Consider, for instance, $\varphi(t,y)g(y,\eta)$. Let us apply the Taylor expansion in t

$$\varphi(t,y) = \sum_{j=0}^N t^j \varphi_j(y) + t^{N+1}\varphi_{(N+1)}(t,y),$$

where $\varphi_j \in C^\infty(\Omega)$, $\varphi_{(N+1)} \in C^\infty(\overline{\mathbb{R}}_+ \times \Omega)$. Then, by the previous theorem we get $t^j \varphi_j(y)g(y,\eta) \in R_G^{\mu-j,d}(\Omega \times \mathbb{R}^q)$, $g_0(y,\eta) := t^{N+1}g(y,\eta) \in R_G^{\mu-(N+1),d}(\Omega \times \mathbb{R}^q)$. Now the operator of multiplication by $\varphi_{(N+1)}(t,y)$ is an operator–valued symbol in $S^0(\Omega \times \mathbb{R}^q; L^2(\mathbb{R}_+), L^2(\mathbb{R}_+))$ and in $S^0(\Omega \times \mathbb{R}^q; \mathcal{S}(\overline{\mathbb{R}}_+), \mathcal{S}(\overline{\mathbb{R}}_+))$. This yields $\varphi_{(N+1)}(t,y)g_0(y,\eta) \in S^{\mu-(N+1)}(\Omega \times \mathbb{R}^q, L^2(\mathbb{R}_+), \mathcal{S}(\overline{\mathbb{R}}_+))$ and the same for the adjoint. It follows that

$$\varphi(t,y)g(y,\eta) \sim \sum_{j=0}^\infty t^j \varphi_j(y)g(y,\eta)$$

in the sense of asymptotic sums in $R_G^{\mu,0}(\Omega \times \mathbb{R}^q)$, cf. Proposition 4.2.20. The proof of the separate continuity is straightforward and left to the reader. $\qquad\square$

Theorem 4.2.33 *Let* $a(t,y,\tau,\eta) \in S_{cl}^\mu(\mathbb{R} \times \Omega \times \mathbb{R}^{1+q})$, $\mu \in \mathbb{Z}$, *be a symbol with the transmission property which is independent of* t *for* $|t| > c_1$ *for a constant* $c_1 > 0$. *Then we have*

$$g(y,\eta) := e^* r^- \operatorname{op}(a)(y,\eta)e^+ \in R_G^{\mu,0}(\Omega \times \mathbb{R}^q)$$

and

$$h(y,\eta) := r^+ \operatorname{op}(a)(y,\eta)e^- e^* \in R_G^{\mu,0}(\Omega \times \mathbb{R}^q).$$

Proof. First observe that the proof of Proposition 4.1.31 easily yields that the kernel of $g(y,\eta)$ is a function in $C^\infty(\Omega \times \mathbb{R}^q, \mathcal{S}(\overline{\mathbb{R}}_+ \times \overline{\mathbb{R}}_+))$. For simplicity we assume $c_1 = 1$. Moreover, let us omit y, i.e., assume $a = a(t,\tau,\eta)$. The trivial generalisation to the y-dependent case will be omitted. Then the kernel of $g(\eta)$ belongs to $C^\infty(\mathbb{R}^q, \mathcal{S}(\overline{\mathbb{R}}_+ \times \overline{\mathbb{R}}_+))$, and the point is then to characterise $g(\eta)$ as a classical operator–valued symbol, according to the definition of $R_G^{\mu,0}(\mathbb{R}^q)$. In particular, we may look at $|\eta| > \text{const}$ for a constant > 0. Choosing an excision function $\chi(\tau, \eta)$ in $\mathbb{R}^{1+q}$ we can write

$$a(t, \tau, \eta) = \sum_{j=0}^{N} \chi(\tau, \eta) a_{(\mu-j)}(t, \tau, \eta) + b_N(t, \tau, \eta) \qquad (4.2.37)$$

for every N, where $a_{(\mu-j)}(t, \lambda\tau, \lambda\eta) = \lambda^{\mu-j} a_{(\mu-j)}(t, \tau, \eta)$ for all $\lambda \in \mathbb{R}_+$, $(\tau, \eta) \in \mathbb{R}^{1+q} \setminus \{0\}$, and $b_N(t, \tau, \eta) \in S_{\text{cl}}^{\mu-(N+1)}(\mathbb{R} \times \mathbb{R}^{1+q})$ is a symbol with the transmission property. Write

$$\mathcal{S}(\overline{\mathbb{R}}_+) = \varprojlim_{k \in \mathbb{N}} E^k$$

for $E^k = \langle t \rangle^{-k} H^k(\mathbb{R}_+)$, $k \in \mathbb{N}$. Then, according to the definition of $R_G^{\mu,0}(\mathbb{R}^q)$ we have to show the estimates

$$\left\{ \int \left| \langle t \rangle^m \frac{\partial^j}{\partial t^j} \{\kappa^{-1}(\eta) D_\eta^\beta g(\eta) \kappa(\eta) u\}(t) \right|^2 dt \right\}^{\frac{1}{2}} \leq c[\eta]^{\mu-|\beta|} \|u\|_{L^2(\mathbb{R}_+)}$$

for every $u \in L^2(\mathbb{R}_+)$, all multi-indices $\alpha, \beta \in \mathbb{N}^q$, $\eta \in \mathbb{R}^q$, and for all $m, j \in \mathbb{N}$, with constants $c = c(\alpha, \beta, K, m, j) > 0$. The same is to be done for $g^*(\eta)$. In addition we have to verify that $g(\eta)$ and $g^*(\eta)$ are classical symbols in the sense $S_{\text{cl}}^\mu(\mathbb{R}^q; L^2(\mathbb{R}_+), H^k(\mathbb{R}_+))$ for all k. In the decomposition (4.2.37) we may suppose that for given m, j the number N is chosen sufficiently large.

Let us start the discussion with the summands $\chi a_{(\mu-j)}$, $j = 0, \dots, N$. Denote by $C^\infty([-1, 1]_0)$ the subspace of all functions $\varphi(t) \in C^\infty(\mathbb{R})$ which vanish for $|t| \geq 1$. The space $C^\infty([-1, 1]_0)$ is Fréchet in a canonical way, and we can write

$$a_{(\mu-j)}(t, \tau, \eta) = \tilde{a}_{(\mu-j)}(t, \tau, \eta) + c_{(\mu-j)}(\tau, \eta).$$

Here $c_{(\mu-j)}(\tau, \eta)$ belongs to $S^{(\mu-j)}(\mathbb{R}^{1+q} \setminus \{0\})$ (i.e. $c_{(\mu-j)}$ is independent of t and homogeneous in $(\tau, \eta) \neq 0$ of order $\mu - j$) and $\tilde{a}_{(\mu-j)}(t, \tau, \eta) \in C^\infty([-1, 1]_0) \otimes_\pi S^{(\mu-j)}(\mathbb{R}^{1+q} \setminus \{0\})$. Note that in general $a_{(\mu-j)}(-\infty, \tau, \eta) \neq a_{(\mu-j)}(+\infty, \tau, \eta)$. However, in the present situation this plays no role.

Now $h_{\mu-j}(\eta) := \varepsilon^* r^- \text{op}(\chi c_{(\mu-j)})(\eta) e^+$ has a kernel in $C^\infty(\mathbb{R}^q, \mathcal{S}(\overline{\mathbb{R}}_+ \times \overline{\mathbb{R}}_+))$. We thus obtain

$$h_{\mu-j}(\eta), h_{\mu-j}^*(\eta) \in \bigcap_k C^\infty(\mathbb{R}^q, \mathcal{L}(L^2(\mathbb{R}_+), E^k)),$$

where $*$ indicates the adjoint in $L^2(\mathbb{R}_+)$ for every η. Moreover, we have

$$h_{\mu-j}(\lambda\eta) = \lambda^{\mu-j}\kappa_\lambda h_{\mu-j}(\eta)\kappa_\lambda^{-1}$$

for all $\lambda \geq 1$ and $|\eta| \geq$ const for a constant > 0. This gives us $h_{\mu-j}(\eta) \in R_G^{\mu-j,0}(\mathbb{R}^q)$ for every j. At the same time we easily see that when $c_{(\mu-j)}$ tends to zero in $S^{(\mu-j)}(\mathbb{R}^{1+q} \setminus \{0\})$ (for any fixed j) then $h_{\mu-j}(\eta)$ tends to zero in $R_G^{\mu-j,0}(\mathbb{R}^q)$. Next we set

$$f_{\mu-j}(\eta) := \varepsilon^* r^-\, \mathrm{op}(\chi\tilde{a}_{(\mu-j)})(\eta)e^+. \tag{4.2.38}$$

This can be treated by a tensor product argument. We can write

$$\tilde{a}_{(\mu-j)}(t,\tau,\eta) = \sum_{i=0}^{\infty} \lambda_i \varphi_i(t) d_i(\tau,\eta)$$

converging in $C^\infty([-1,1]_0) \otimes_\pi S^{(\mu-j)}(\mathbb{R}^{1+q} \setminus \{0\})$, where $\lambda_i \in \mathbb{C}$, $\sum |\lambda_i| < \infty$, $\varphi_i \in C^\infty([-1,1]_0)$, $d_i \in S^{(\mu-j)}(\mathbb{R}^{1+q} \setminus \{0\})$ tending to zero in the corresponding spaces. This allows us to write

$$f_{\mu-j}(\eta) = \sum_{i=0}^{\infty} \lambda_i \varphi_i^- l_i(\eta) \quad \text{for} \quad \varphi_i^-(t) = \begin{cases} \varphi_i(-t) & \text{for } t \geq 0, \\ 0 & \text{for } t \leq 0, \end{cases} \tag{4.2.39}$$

with $l_i(\eta) = \varepsilon^* r^-\, \mathrm{op}(\chi d_i)(\eta)e^+$ tending to zero in $R_G^{\mu-j,0}(\mathbb{R}^q)$ (cf. the above argument in connection with $h_{\mu-j}(\eta)$). Since the operator of multiplication by φ_i^- in the sense $g \to \varphi_i^- g$ for a fixed $g \in R_G^{\mu-j,0}(\mathbb{R}^q)$ is continuous as a map $C_0^\infty(\overline{\mathbb{R}}_+) \to R_G^{\mu-j,0}(\mathbb{R}^q)$, cf. Theorem 4.2.32, we obtain the convergence of (4.2.39) in $R_G^{\mu-j,0}(\mathbb{R}^q)$. It follows altogether that

$$\varepsilon^* r^-\, \mathrm{op}(\chi a_{(\mu-j)})(\eta)e^+ \in R_G^{\mu-j,0}(\mathbb{R}^q), \quad j = 0, \dots, N.$$

Let us now return to (4.2.37) and set $g(\eta) = \varepsilon^* r^-\, \mathrm{op}(b_N)(\eta)e^+$. For every k we choose k so large that

$$g(\eta), g^*(\eta) \in S^{\mu-(N+1)}(\mathbb{R}^q; L^2(\mathbb{R}_+), E^k) \tag{4.2.40}$$

holds. Let us set from now on $b(t,\tau,\eta) = b_N(t,\tau,\eta)$. Also here we have independence of t for $|t| \geq 1$. We first look at the t-independent case $b = b(\tau,\eta)$. Similarly to the proof of Proposition 4.2.13 we have

$$B(\eta) := \varepsilon^* r^-\, \mathrm{op}(b_\eta)(\eta)e^+ = \kappa^{-1}(\eta)\varepsilon^* r^-\, \mathrm{op}(b)(\eta)\kappa(\eta)$$

for $b_\eta(\tau,\eta) = b([\eta]\tau,\eta)$. Now let $u(t') \in L^2(\mathbb{R}_+)$ and consider

$$\mathrm{op}(b_\eta)(\eta)e^+u(t) = \int\limits_{-\infty}^{\infty} \int\limits_{0}^{\infty} e^{i(t-t')\tau} b([\eta]\tau,\eta)u(t')\, dt'd\tau.$$

Inserting $e^{i(t-t')\tau} = (1 + (t - t')^2)^{-m}(1 - \Delta_\tau)^m e^{i(t-t')\tau}$ we obtain by integration by parts

$$\mathrm{op}(b_\eta)(\eta)e^+u(t)$$

$$= \int \int_0^\infty (1 + (t - t')^2)^{-m} e^{i(t-t')\tau}(1 - \Delta_\tau)^m b([\eta]\tau, \eta)u(t')\, dt'd\tau.$$

This yields

$$B(\eta)u(t)$$

$$= \int \int_0^\infty (1 + (t - t')^2)^{-m} e^{-i(t+t')\tau}(1 - \Delta_\tau)^m b([\eta]\tau, \eta)u(t')\, dt'd\tau$$

for $t > 0$. Recall that we have $B(\eta)u(t) \in \mathcal{S}(\overline{\mathbb{R}}_+)$ for every η, since $b([\eta]\tau, \eta)$ has the transmission property. Let us now consider

$$\int_0^\infty |\langle t\rangle^k \partial_t^j B(\eta)u(t)|^2 dt = \int_0^\infty |\iint \langle t\rangle^k \partial_t^j \left\{ (1 + (t + t')^2)^{-m} e^{-i(t+t')\tau} \right\}$$

$$(1 - \Delta_\tau)^m b([\eta]\tau, \eta)u(t')\, dt'd\tau|^2\, dt \qquad (4.2.41)$$

for any fixed $k, j \in \mathbb{N}$. We have

$$\partial_t^j \left\{ (1 + (t + t')^2)^{-m} e^{-i(t+t')\tau} \right\} = \sum_{l=0}^j c_l \partial_t^{j-l}(1 + (t + t')^2)^{-m} \partial_t^l e^{-i(t+t')\tau}$$

$$= \sum_{l=0}^j \varphi_{jl}(t, t')\tau^l e^{-i(t+t')\tau}$$

for certain constants c_l, and $\varphi_{jl}(t, t') := c_l \partial_t^{j-l}(1 + (t+t')^2)^{-m}(-i)^l$. For each $n \in \mathbb{N}$ we can choose m so large that $\langle t\rangle^k |\varphi_{jl}(t, t')| \leq c\langle t\rangle^{-n}\langle t'\rangle^{-n}$ for all $t, t' \in \mathbb{R}_+$, $l = 0, \ldots, j$, for a constant $c > 0$. In fact, we have, for instance, for $j = l$ (with different constants c),

$$\langle t\rangle^k |\varphi_{ll}(t, t')|$$

$$\leq c\langle t\rangle^k (1 + (t + t')^2)^{-m}$$

$$\leq c\langle t\rangle^{-n}\{\langle t\rangle^{k+n}(1 + (t + t')^2)^{-\frac{k+n}{2}}\}(1 + (t + t')^2)^{-m+\frac{k+n}{2}}$$

$$\leq c\langle t\rangle^{-n}\langle t'\rangle^{-n} \sup_{t,t'}\{\langle t\rangle^{k+n}(1 + (t + t')^2)^{-\frac{k+n}{2}}\}$$

$$\times \sup_{t,t'}\{\langle t'\rangle^n(1 + (t + t')^2)^{-m+\frac{k+n}{2}}\}$$

$$\leq c\langle t\rangle^{-n}\langle t'\rangle^{-n}$$

for $n - m + \frac{k+n}{2} \leq 0$. In an analogous manner we can argue for φ_{jl} in general. (4.2.41) can be estimated by a sum of expressions of the form

$$\int_0^\infty \left| \iint \langle t \rangle^k \varphi_{jl}(t, t') \tau^l e^{-i(t+t')\tau} (1 - \Delta_\tau)^m b([\eta]\tau, \eta) u(t') \, dt' d\tau \right|^2 dt.$$

We have $(1 - \Delta_\tau)^m b([\eta]\tau, \eta) = (1 - [\eta]^2)^m ((1 - \Delta_\tau)^m b)([\eta]\tau, \eta)$. Using $(1 - \Delta_\tau)^m b \in S^{\mu-N-1-2m}(\mathbb{R}^{1+q})$ we get

$$|(1 - [\eta]^2)^m ((1 - \Delta_\tau)^m b)([\eta]\tau, \eta)|$$
$$\leq c(1 - [\eta]^2)^m (|[\eta]\tau|^2 + [\eta]^2)^{\frac{1}{2}(\mu-N-1-2m)}$$
$$\leq c[\eta]^{\mu-N-1}(1 + |\tau|^2)^{\frac{1}{2}(\mu-N-1-2m)}.$$

Thus, choosing m so large that $l + \frac{1}{2}(\mu - N - 1 - 2m) < -1$ holds, we get for different constants c

$$\left| \iint \langle t \rangle^k \varphi_{jl}(t, t') \tau^l e^{-i(t+t')\tau} (1 - \Delta_\tau)^m b([\eta]\tau, \eta) u(t') \, dt' d\tau \right|$$
$$\leq c[\eta]^{\mu-N-1} \int |\tau|^l (1 + |\tau|^2)^{-\frac{1}{2}(N+1+2m)} d\tau \int \langle t \rangle^k |\varphi_{jl}(t, t')| |u(t')| \, dt'$$
$$\leq c[\eta]^{\mu-N-1} \int \langle t \rangle^{-n} \langle t' \rangle^{-n} |u(t')| \, dt'$$
$$\leq c[\eta]^{\mu-N-1} \langle t \rangle^{-n} \|\langle t' \rangle^{-n}\|_{L^2(\mathbb{R}_+)} \|u\|_{L^2(\mathbb{R}_+)}$$
$$\leq c[\eta]^{\mu-N-1} \|u\|_{L^2(\mathbb{R}_+)}$$

for $n \geq 1$. It follows then that (4.2.41) can be estimated by

$$c[\eta]^{\mu-N-1} \int |\langle t \rangle^{-n}|^2 \, dt \|u\|^2_{L^2(\mathbb{R}_+)} \leq c(b)[\eta]^{\mu-N-1} \|u\|^2_{L^2(\mathbb{R}_+)}$$

for a constant $c(b)$ which tends to zero when b tends to zero in space $S^{\mu-(N+1)}(\mathbb{R}^{q+1})$.

In an analogous manner we can proceed for $D_\eta^\beta b(\tau, \eta)$ for every $\beta \in \mathbb{N}^q$. This yields the first relation of (4.2.40). In a similar way we can treat $g^*(\eta)$. The estimates also show that the maps $b \to g$, $b \to g^*$ are continuous in the sense $S^{\mu-(N+1)}(\mathbb{R}^{q+1}) \to S^{\mu-(N+1)}(\mathbb{R}^q; L^2(\mathbb{R}_+), E^k)$ for all k. This enables us to apply a tensor product argument to see that (4.2.40) also holds for $b_N(t, \tau, \eta) \in S^{\mu-(N+1)}(\mathbb{R} \times \mathbb{R}^{1+q})$ when $b_N(t, \tau, \eta)$ is independent of t for $|t| \geq 1$. In fact, the arguments reduce to the case $b_N(t, \tau, \eta) \in C^\infty([0,1]_0) \otimes_\pi S^{\mu-(N+1)}(\mathbb{R}^{1+q})$, and similarly as above for (4.2.39) we obtain the relation (4.2.40) also in the t-dependent case. The second relation can be obtained in the analogous manner.   □

Remark 4.2.34 *The map $a(t, y, \tau, \eta) \rightarrow g(y, \eta) = \varepsilon^* r^- \operatorname{op}(a)(y, \eta)e^+$ is a continuous operator from the subspace of $S_{\mathrm{cl}}^\mu(\mathbb{R} \times \Omega \times \mathbb{R}^{1+q})$, consisting of those symbols with the transmission property which are t–independent for $|t| \geq c_1$, into $R_G^{\mu,0}(\Omega \times \mathbb{R}^q)$.*

Theorem 4.2.35 *Let $a(t, y, \tau, \eta) \in S_{\mathrm{cl}}^\mu(\mathbb{R} \times \Omega \times \mathbb{R}^{1+q})$, $\mu \in \mathbb{Z}$, be a symbol with the transmission property which is independent of t for $|t| > c_1$, for a constant $c_1 > 0$. Let $e_s^+ : H^s(\mathbb{R}_+) \rightarrow H^s(\mathbb{R})$ be the extension operator of Proposition 4.2.11. Then the operator family*

$$g(y, \eta) := r^+ \operatorname{op}(a)(y, \eta)e^+ u - r^+ \operatorname{op}(a)(y, \eta)e_s^+ u \quad \text{for} \quad u \in H^s(\mathbb{R}_+)$$

belongs to $R_G^{\mu,0}(\Omega \times \mathbb{R}^q)$.

Proof. For $u \in H^s(\mathbb{R}_+)$ we can write $e^- v = (e^+ - e_s^+)u$ for a $v \in H^s(\mathbb{R}_-)$, where e^- is the extension operator by zero to $\mathbb{R}_+$. We then have

$$g(y, \eta)u = r^+ \operatorname{op}(a)(y, \eta)e^- v = \sum_{j=1}^s \alpha_j r^+ \operatorname{op}(a)(y, \eta)\kappa_j^0 e^- \varepsilon^* u$$

for $(\kappa_j^0 f)(t) = f(jt)$ and constants α_j, cf. the proof of Proposition 4.2.11. We have $\operatorname{op}(a)(y, \eta)\kappa_j^0 = \kappa_j^0 \operatorname{op}(a_j)(y, \eta)$ for $a_j(t, y, \tau, \eta) = a(j^{-1}t, y, j\tau, \eta)$ which is also a symbol with the above properties. From the second relation of Theorem 4.2.33 we then obtain

$$g(y, \eta)u = \sum_{j=1}^s \alpha_j \kappa_j^0 h_j(y, \eta)u$$

for $h_j(y, \eta) = r^+ \operatorname{op}(a_j)(y, \eta)e^- \varepsilon^* \in R_G^{\mu,0}(\Omega \times \mathbb{R}^q)$. Now it suffices to use Remark 4.2.22 to obtain the result. □

4.2.4 Boundary symbols

Boundary value problems will be formulated locally near the boundary as pseudo–differential operators with operator–valued symbols, cf. Sections 1.3.1 – 1.3.3. They are called boundary symbols here, with values in the operator algebra on the half axis in the sense of Sections 4.1.1 – 4.1.7.

Let $a(t, y, \tau, \eta) \in S_{\mathrm{cl}}^\mu(\overline{\mathbb{R}}_+ \times \Omega \times \mathbb{R}^{1+q})$, $\mu \in \mathbb{Z}$, be a symbol with the transmission property. For convenience we shall always assume that a is independent of t for $t > c_1$ for a constant $c_1 > 0$. According to (4.1.64) we set

$$\operatorname{op}^+(a)(y, \eta) = r^+ \operatorname{op}(a)(y, \eta)e^+. \tag{4.2.42}$$

This is well–defined as an operator

$$\mathrm{op}^+(a)(y,\eta): H^s(\mathbb{R}_+) \to H^{s-\mu}(\mathbb{R}_+) \qquad (4.2.43)$$

for every $(y,\eta) \in \Omega \times \mathbb{R}^q$ and $s \in \mathbb{R}$, $s > -\frac{1}{2}$. We shall see that $(4.2.43)$ is an operator–valued symbol with respect to (y,η).

Proposition 4.2.36 Let $a(t,y,\tau,\eta) \in S_{\mathrm{cl}}^\mu(\overline{\mathbb{R}}_+ \times \Omega \times \mathbb{R}^{1+q})$, $b(t,y,\tau,\eta)$ $\in S_{\mathrm{cl}}^\nu(\overline{\mathbb{R}}_+ \times \Omega \times \mathbb{R}^{1+q})$ be symbols with the transmission property, and let a and b be independent of t for $t > c_1$ for some constant $c_1 > 0$. Moreover, let $\mu \leq 0$ and $\nu \leq 0$. Then we have

$$g(y,\eta) := \mathrm{op}^+(a)(y,\eta)\,\mathrm{op}^+(b)(y,\eta) - \mathrm{op}^+(a\#b)(y,\eta)$$
$$\in R_G^{\mu+\nu,0}(\Omega \times \mathbb{R}^q),$$

where $\#$ is the Leibniz product with respect to (t,τ).

Proof. Let $\theta^-(t)$ be the characteristic function of $\mathbb{R}_-$. Then $g(y,\eta)$ can be written $g_0(y,\eta) - g_1(y,\eta)$ for

$$g_0(y,\eta) = r^+ \{\mathrm{op}(a)(y,\eta)\,\mathrm{op}(b)(y,\eta) - \mathrm{op}(a\#b)(y,\eta)\}e^+$$
$$\in R_G^{-\infty,0}(\Omega \times \mathbb{R}^q),$$

using the fact that when a symbol $p(t,y,\tau,\eta) \in S^{-\infty}(\mathbb{R}\times\Omega\times\mathbb{R}^{1+q})$ is independent of t for $|t| > c_1$ it follows that $r^+ \mathrm{op}(p)(y,\eta)e^+ \in R_G^{-\infty,0}(\Omega \times \mathbb{R}^q)$. Moreover, we have $g_1(y,\eta) = r^+ \mathrm{op}(a)(y,\eta)\theta^- \mathrm{op}(b)(y,\eta)e^+$. Writing $g_1(y,\eta) = f(y,\eta)h(y,\eta)$ for $f(y,\eta) = \varepsilon^* r^- \mathrm{op}(\check{a})(y,\eta)e^+$, $\check{a}(t,y,\tau,\eta) = a(-t,y,-\tau,\eta)$ and $h(y,\eta) = \varepsilon^* r^- \mathrm{op}(b)(y,\eta)e^+$, we can apply Theorem 4.2.33 to obtain $f(y,\eta) \in R_G^{\mu,0}(\Omega \times \mathbb{R}^q)$, $h(y,\eta) \in R_G^{\nu,0}(\Omega \times \mathbb{R}^q)$, and then $g_1(y,\eta) \in R_G^{\mu+\nu,0}(\Omega \times \mathbb{R}^q)$ from Proposition 4.2.30. $\square$

Theorem 4.2.37 Let $a(t,y,\tau,\eta) \in S_{\mathrm{cl}}^\mu(\overline{\mathbb{R}}_+ \times \Omega \times \mathbb{R}^{1+q})$ be a symbol with the transmission property that is independent of t for $t > c_1$ for some $c_1 > 0$. Then we have

$$\mathrm{op}^+(a)(y,\eta) \in S^\mu(\Omega \times \mathbb{R}^q; H^s(\mathbb{R}_+), H^{s-\mu}(\mathbb{R}_+)) \qquad (4.2.44)$$

for every $s \in \mathbb{R}$, $s > -\frac{1}{2}$. In particular, if $a = a(y,\tau,\eta)$ is independent of t, then

$$\mathrm{op}^+(a)(y,\eta) \in S_{\mathrm{cl}}^\mu(\Omega \times \mathbb{R}^q; H^s(\mathbb{R}_+), H^{s-\mu}(\mathbb{R}_+)) \qquad (4.2.45)$$

for $s \in \mathbb{R}$, $s > -\frac{1}{2}$.

Proof. Let us first assume $|s| < \frac{1}{2}$. Then $\mathrm{op}^+(a)(y,\eta)$ can be interpreted as a composition of the following operator–valued symbols, namely

$$e^+ \in S^0_{\mathrm{cl}}(\mathbb{R}^q; H^s(\mathbb{R}_+), H^s(\mathbb{R})),$$

cf. Remark 4.2.12 for $e^+ = e^+_s$,

$$\mathrm{op}(a)(y,\eta) \in S^\mu(\Omega \times \mathbb{R}^q; H^s(\mathbb{R}), H^{s-\mu}(\mathbb{R})),$$

cf. Proposition 4.2.13, and

$$r^+ \in S^0_{\mathrm{cl}}(\mathbb{R}^q; H^{s-\mu}(\mathbb{R}), H^{s-\mu}(\mathbb{R}_+)),$$

cf . (4.2.15). This yields (4.2.44) for $|s| < \frac{1}{2}$. Next consider the case $s \in \mathbb{N}$ which is sufficient, since the assertion for arbitrary s then follows by interpolation. Similarly to the order reduction procedure of Theorem 4.1.61 we set

$$p_0(y,\eta) = \mathrm{op}^+_{s-\mu}(r_-^{s-\mu})(\eta)\,\mathrm{op}^+(a)(y,\eta)\,\mathrm{op}^+(r_-^{-s})(\eta)$$

with $\mathrm{op}^+_s(r_-^\nu)(\eta) = r^+ \,\mathrm{op}(r_-^\nu)(\eta)e^+_s$ in the sense of Remark 4.2.12. Then

$$\mathrm{op}^+(a)(y,\eta) = \mathrm{op}^+(r_-^{-s+\mu})(\eta)p_0(y,\eta)\,\mathrm{op}^+_s(r_-^s)(\eta),$$

cf. Corollary 4.1.65. It is then clear that

$$\mathrm{op}^+(a)(y,\eta) \in S^\mu(\Omega \times \mathbb{R}^q; H^s(\mathbb{R}_+), H^{s-\mu}(\mathbb{R}_+))$$

is equivalent to $p_0(y,\eta) \in S^0(\Omega \times \mathbb{R}^q; H^0(\mathbb{R}_+), H^0(\mathbb{R}_+))$. In order to show the latter property we first write

$$p_0(y,\eta) = \mathrm{op}^+_s(r_-^s)(\eta)\left\{\mathrm{op}^+(r_-^{-\mu}\#a)(y,\eta) + \mathrm{op}^+(c)(y,\eta)\right\}\mathrm{op}^+(r_-^{-s})(\eta) \tag{4.2.46}$$

for some $c(y,\eta) \in S^{-\infty}(\overline{\mathbb{R}}_+ \times \Omega \times \mathbb{R}^{1+q})$. Here we used that $r_-^{s-\mu}(\tau,\eta)$ is a minus symbol, so the operator

$$\mathrm{op}^+_{s-\mu}(r_-^{s-\mu})(\eta) = \mathrm{op}^+_s(r_-^s)(\eta)\,\mathrm{op}^+_{s-\mu}(r^{-\mu})(\eta)$$

could be composed from the left to $\mathrm{op}^+(a)(y,\eta)\,\mathrm{op}^+(r_-^{-s})(\eta) : H^0(\mathbb{R}_+) \to H^{s-\mu}(\mathbb{R}_+)$. Now, from Proposition 4.2.36 we obtain

$$\mathrm{op}^+(r_-^{-\mu}\#a)(y,\eta)\,\mathrm{op}^+(r_-^{-s})(\eta) = \mathrm{op}^+((r_-^{-\mu}\#a)r_-^{-s})(y,\eta) + g(y,\eta)$$

for a $g(y,\eta) \in R_G^{-s,0}(\Omega \times \mathbb{R}^q)$. Moreover, it is obvious that

$$g_0(y,\eta) := \mathrm{op}^+(c)(y,\eta)\,\mathrm{op}^+(r_-^{-s})(\eta) \in R_G^{-\infty,0}(\Omega \times \mathbb{R}^q)$$

holds. It follows that

$$\{\mathrm{op}^+(r_-^{-\mu}\#a)(y,\eta) + \mathrm{op}^+(c)(y,\eta)\}\,\mathrm{op}^+(r_-^{-s})(\eta)\}$$
$$= r^+\,\mathrm{op}((r_-^{-\mu}\#a)r_-^{-s})(y,\eta)e^+ + g(y,\eta) + g_0(y,\eta). \quad (4.2.47)$$

Here $b := (r_-^{-\mu}\#a)r_-^{-s}$ is a symbol in $S_{\mathrm{cl}}^{-s}(\mathbb{R}\times\Omega\times\mathbb{R}^{1+q})$ that can be chosen to be independent of t for large t (we tacitly used an extension operator to $\mathbb{R}_- \ni t$). Hence, in view of Proposition 4.2.13 we get $\mathrm{op}(b)(y,\eta) \in S^{-s}(\Omega\times\mathbb{R}^q; H^0(\mathbb{R}), H^{-s}(\mathbb{R}))$ which implies

$$(4.2.47) = r^+\,\mathrm{op}(b)e^+ + g + g_0 \in S^{-s}(\Omega\times\mathbb{R}^q; H^0(\mathbb{R}_+), H^{-s}(\mathbb{R}_+)).$$

Using Proposition 4.2.16 we finally obtain the relation

$$(4.2.46) = \mathrm{op}_s^+(r_-^s)\{\mathrm{op}^+(b) + g + g_0\} \in S^0(\Omega\times\mathbb{R}^q; H^0(\mathbb{R}_+), H^0(\mathbb{R}_+)).$$

The property (4.2.45) for t-independent symbols is a consequence of Remark 4.2.14 and Proposition 4.2.16. $\square$

Theorem 4.2.38 *Let $a(t,y,\tau,\eta) \in S_{\mathrm{cl}}^{\mu}(\overline{\mathbb{R}}_+\times\Omega\times\mathbb{R}^{1+q})$ be a symbol with the transmission property that is independent of t for $t > c_1$ for some $c_1 > 0$. Then*

$$\mathrm{op}^+(a)(y,\eta) \in S^{\mu}(\Omega\times\mathbb{R}^q; \mathcal{S}(\overline{\mathbb{R}}_+), \mathcal{S}(\overline{\mathbb{R}}_+)). \quad (4.2.48)$$

In particular, if $a = a(y,\tau,\eta)$ is independent of t, then

$$\mathrm{op}^+(a)(y,\eta) \in S_{\mathrm{cl}}^{\mu}(\Omega\times\mathbb{R}^q; \mathcal{S}(\overline{\mathbb{R}}_+), \mathcal{S}(\overline{\mathbb{R}}_+)). \quad (4.2.49)$$

Proof. We have to show that for each $k \in \mathbb{N}$ there is an $l \in \mathbb{N}$ with

$$\mathrm{op}^+(a)(y,\eta) \in S^{\mu}(\Omega\times\mathbb{R}^q; E_+^l, E_+^k),$$

where $E_+^k = \langle t\rangle^{-k}H^k(\mathbb{R}_+)$, and that the symbols are classical for t-independent a. Let us concentrate on the general case. The property of being classical for t-independent a is then obvious and will be omitted, cf. analogously Theorem 4.2.37. For fixed $k \in \mathbb{N}$ the extension operator e_k^+ from Proposition 4.2.11 has the property $e_k^+ : E_+^k \to E^k$ for $E^k = \langle t\rangle^{-k}H^k(\mathbb{R})$ and $e_k^+ \in S_{\mathrm{cl}}^0(\mathbb{R}^q; E_+^k, E^k)$, because of the homogeneity. Write

$$\mathrm{op}^+(a)(y,\eta) = r^+\,\mathrm{op}(a)(y,\eta)e_k^+ + r^+\,\mathrm{op}(a)(y,\eta)(e^+ - e_k^+).$$

Then Theorem 4.2.35 gives us $r^+\,\mathrm{op}(a)(y,\eta)(e^+-e_k^+) \in R_G^{\mu,0}(\Omega\times\mathbb{R}^q) \subset S_{\mathrm{cl}}^{\mu}(\Omega\times\mathbb{R}^q; \mathcal{S}(\overline{\mathbb{R}}_+), \mathcal{S}(\overline{\mathbb{R}}_+))$. Thus it remains to show that

$$r^+\,\mathrm{op}(a)(y,\eta)e_+^k \in S^{\mu}(\Omega\times\mathbb{R}^q; E_+^l, E_+^k).$$

However, in view of the above symbol property of e_k^+ and of $r^+ \in S_{\mathrm{cl}}^0(\mathbb{R}^q; E^l, E_+^l)$ this follows immediately from Proposition 4.2.15. $\square$

Remark 4.2.39 *Both Theorem 4.2.37 and Theorem 4.2.38 (i.e., the relations (4.2.44) and (4.2.48)) can easily be generalised to symbols $a(t, y, \tau, \eta) \in S^\mu_{cl}(\overline{\mathbb{R}}_+ \times \Omega \times \mathbb{R}^{1+q})$ with the transmission property which are of the form $\omega(t) a_0(t, y, \tau, \eta) + (1 - \omega(t)) t^{-j} a_1(y, \tau, \eta)$ for a cut-off function $\omega(t)$ and arbitrary $j \in \mathbb{N}$. The obvious details are left to the reader.*

Definition 4.2.40 *Let $\Omega \subseteq \mathbb{R}^q$ be an open set. Then $R^{\mu,d}(\Omega \times \mathbb{R}^q)$ for $\mu \in \mathbb{Z}$, $d \in \mathbb{N}$, is defined as the space of all operator families*

$$a(y, \eta) = \mathrm{op}^+(p)(y, \eta) + g(y, \eta)$$

for arbitrary $p(t, y, \tau, \eta) \in S^\mu_{cl}(\overline{\mathbb{R}}_+ \times \Omega \times \mathbb{R}^{1+q})$ with the transmission property, independent of t for $t > c_1$ for a constant $c_1 > 0$, dependent on p, and $g(y, \eta) \in R^{\mu,d}_G(\Omega \times \mathbb{R}^q)$.

More generally $\mathcal{R}^{\mu,d}(\Omega \times \mathbb{R}^q; N_-, N_+)$ denotes the space of all

$$a(y, \eta) = \begin{pmatrix} a_{11} & a_{12} \\ a_{21} & a_{22} \end{pmatrix} (y, \eta)$$

for arbitrary $a_{11}(y, \eta) \in R^{\mu,d}(\Omega \times \mathbb{R}^q)$ and

$$\begin{pmatrix} 0 & a_{12} \\ a_{21} & a_{22} \end{pmatrix} (y, \eta) \in \mathcal{R}^{\mu,d}_G(\Omega \times \mathbb{R}^q; N_-, N_+),$$

cf. Definition 4.2.19. The elements of $\mathcal{R}^{\mu,d}(\Omega \times \mathbb{R}^q; N_-, N_+)$ are be called (complete) boundary symbols. In particular, for $a_{11}(y, \eta) = \mathrm{op}^+(p)(y, \eta) + g(y, \eta) = \mathrm{u.\,l.\,c.}\, a(y, \eta)$ (i.e., upper left corner), $g(y, \eta)$ is called the (complete) Green symbol of $a(y, \eta)$, $a_{21}(y, \eta)$ the (complete) trace symbol and $a_{12}(y, \eta)$ the (complete) potential symbol of $a(y, \eta)$.

$\mathcal{R}^{\mu,d}_G(\Omega \times \mathbb{R}^q; N_-, N_+)$ was treated in detail in Section 4.2.3. Therefore we shall concentrate here on the space $R^{\mu,d}(\Omega \times \mathbb{R}^q)$ of upper left corners.

We can also introduce the class $R^{\mu,d}(\Omega \times \Omega \times \mathbb{R}^q) \ni a(y, y', \eta)$, defined as the space of all operator families of the form

$$a(y, y', \eta) = \mathrm{op}^+(p)(y, y', \eta) + g(y, y', \eta)$$

for arbitrary $p(t, y, y', \eta) \in S^\mu_{cl}(\overline{\mathbb{R}}_+ \times \Omega \times \Omega \times \mathbb{R}^{1+q})$ with the transmission property, independent of t for $t > c_1$, and $g(y, y', \eta) \in R^{\mu,d}_G(\Omega \times \Omega \times \mathbb{R}^q)$. These operator–valued amplitude functions will occur below in the algebra of boundary value problems. Their properties are completely analogous to those of $R^{\mu,d}(\Omega \times \mathbb{R}^q)$. So for notational convenience we mainly consider $R^{\mu,d}(\Omega \times \mathbb{R}^q)$ and $\mathcal{R}^{\mu,d}(\Omega \times \mathbb{R}^q; N_-, N_+)$, respectively.

In view of Theorem 4.2.37 and of Definition 4.2.19 we have

$$R^{\mu,d}(\Omega \times \mathbb{R}^q) \subseteq S^\mu(\Omega \times \mathbb{R}^q; H^s(\mathbb{R}_+), H^{s-\mu}(\mathbb{R}_+)) \qquad (4.2.50)$$

for every $s \in \mathbb{R}$, $s > d - \frac{1}{2}$. Moreover,

$$R^{\mu,d}(\Omega \times \mathbb{R}^q) \subseteq S^\mu(\Omega \times \mathbb{R}^q; \mathcal{S}(\overline{\mathbb{R}}_+), \mathcal{S}(\overline{\mathbb{R}}_+)), \qquad (4.2.51)$$

cf. Theorem 4.2.38. Observe $R_G^{-\infty,d}(\Omega \times \mathbb{R}^q) \subset R^{-\infty,d}(\Omega \times \mathbb{R}^q)$ with proper inclusion, since for a symbol $p(t,y,\tau,\eta) \in S^{-\infty}(\overline{\mathbb{R}}_+ \times \Omega \times \mathbb{R}^{1+q})$ that is t–independent for large t, we do not generally have the relation $\operatorname{op}^+(p)(y,\eta) \in R_G^{-\infty,0}(\Omega \times \mathbb{R}^q)$.

Proposition 4.2.41 *There are canonical embeddings*

$$\mathcal{R}^{\mu-j,d}(\Omega \times \mathbb{R}^q; N_-, N_+) \hookrightarrow \mathcal{R}^{\mu,d}(\Omega \times \mathbb{R}^q; N_-, N_+)$$

for all $j \in \mathbb{N}$, and

$$D_y^\alpha D_\eta^\beta \mathcal{R}^{\mu,d}(\Omega \times \mathbb{R}^q; N_-, N_+) \hookrightarrow \mathcal{R}^{\mu-|\beta|,d}(\Omega \times \mathbb{R}^q; N_-, N_+)$$

for all multi-indices $\alpha, \beta \in \mathbb{N}^q$.

The proof is obvious.

Proposition 4.2.42 *Let $a_j(y,\eta) \in R^{\mu-j,d}(\Omega \times \mathbb{R}^q)$, $j \in \mathbb{N}$, be an arbitrary sequence and assume that the constants c_1 for which the involved interior symbols $p_j(t,y,\tau,\eta)$ are independent of t for $t > c_1$ are independent of j. Then there exists an $a(y,\eta) \in R^{\mu,d}(\Omega \times \mathbb{R}^q)$ with*

$$a(y,\eta) - \sum_{j=0}^N a_j(y,\eta) \in R^{\mu-(N+1),d}(\Omega \times \mathbb{R}^q)$$

for every $N \in \mathbb{N}$, and a is unique mod $R^{-\infty,d}(\Omega \times \mathbb{R}^q)$.

We write $a(y,\eta) \sim \sum_{j=0}^\infty a_j(y,\eta)$, called the asymptotic sum of the $a_j(y,\eta)$.

Proof. By definition we have $a_j(y,\eta) = \operatorname{op}^+(p_j)(y,\eta) + g_j(y,\eta)$ for certain $p_j(t,y,\tau,\eta)$ with the transmission property, independent of t for $t > c_1$ for a constant $c_1 > 0$, and $g_j(y,\eta) \in R_G^{\mu-j,d}(\Omega \times \mathbb{R}^q)$. Let us apply Proposition 4.2.5 and form the asymptotic sum $p \sim \sum_{j=0}^\infty p_j$ in $S_{\mathrm{cl}}^\mu(\overline{\mathbb{R}}_+ \times \Omega \times \mathbb{R}^{1+q})$ which can be chosen as a symbol which is independent of t for $t > c_1$. Write p as a convergent sum

$$p(t,y,\tau,\eta) = \sum_{j=0}^\infty \chi\left(\frac{\tau,\eta}{c_j}\right) p_j(t,y,\tau,\eta)$$

for an excision function $\chi(\tau, \eta)$ and constants c_j tending to infinity for $j \to \infty$ sufficiently fast. Then, because of

$$\mathrm{op}^+(p_j)(y, \eta) - \mathrm{op}^+(\tilde{p}_j)(y, \eta) \in R_G^{-\infty,0}(\Omega \times \mathbb{R}^q)$$

for $\tilde{p}_j(t, y, \tau, \eta) = \chi((\tau, \eta)/c_j)p_j(t, y, \tau, \eta)$, we can set $a(y, \eta) = \mathrm{op}^+(p)(y, \eta) + g(y, \eta)$, where $g(y, \eta) \sim \sum_{j=0}^{\infty} g_j(y, \eta)$, according to Proposition 4.2.20. □

In view of $R^{\mu,d}(\Omega \times \mathbb{R}^q) \subset C^\infty(\Omega, L^\mu(\mathbb{R}_+; \mathbb{R}^q))$ every element $a(y, \eta) \in R^{\mu,d}(\Omega \times \mathbb{R}^q)$ has a (classical) parameter–dependent complete symbol $p(t, y, \tau, \eta)$ which is unique modulo $S^{-\infty}(\mathbb{R}_+ \times \Omega \times \mathbb{R}^{1+q})$. Set

$$\sigma_\psi^\mu(a)(t, y, \tau, \eta) = p_{(\mu)}(t, y, \tau, \eta),$$

where $p_{(\mu)}(t, y, \tau, \eta)$ denotes the unique homogeneous principal symbol of p of order μ which belongs to $S^{(\mu)}(T^*(\overline{\mathbb{R}}_+ \times \Omega) \backslash 0)$. Let us call $\sigma_\psi^\mu(a)$ the homogeneous principal interior symbol of order μ of $a(y, \eta)$, and set

$$\Sigma_\psi^\mu = \left\{ \sigma_\psi^\mu(a) : a(y, \eta) \in R^{\mu,d}(\Omega \times \mathbb{R}^q) \right\}.$$

Note that the operator family $\mathrm{op}^+(p_{(\mu)}|_{t=0})(y, \eta)$, defined for $(y, \eta) \in \Omega \times (\mathbb{R}^q \backslash \{0\})$, satisfies

$$\mathrm{op}^+(p_{(\mu)}|_{t=0})(y, \lambda\eta) = \lambda^\mu \kappa_\lambda \, \mathrm{op}^+(p_{(\mu)}|_{t=0})(y, \eta)\kappa_\lambda^{-1}$$

for all $\lambda \in \mathbb{R}_+$ and all y, η. Moreover, by definition, every $a(y, \eta) \in R^{\mu,d}(\Omega \times \mathbb{R}^q)$ has the form $a(y, \eta) = \mathrm{op}^+(p)(y, \eta) + g(y, \eta)$ for a

$$p(t, y, \tau, \eta) \in S_{\mathrm{cl}}^\mu(\overline{\mathbb{R}}_+ \times \Omega \times \mathbb{R}^{1+q})$$

with the transmission property, and $g(y, \eta) \in R_G^{\mu,d}(\Omega \times \mathbb{R}^q)$. We set

$$\sigma_\wedge^\mu(a)(y, \eta) = \mathrm{op}^+(p_{(\mu)}|_{t=0})(y, \eta) + \sigma_\wedge^\mu(g)(y, \eta),$$

cf. the notation in Section 4.2.3, and call $\sigma_\wedge^\mu(a)(y, \eta)$ the homogeneous principal boundary symbol of $a(y, \eta)$ of order μ. If $a(y, \eta) = \mathrm{op}^+(\tilde{p})(y, \eta) + \tilde{g}(y, \eta)$ is another representation of $a(y, \eta)$ for certain $\tilde{p}(t, y, \tau, \eta) \in S_{\mathrm{cl}}^\mu(\overline{\mathbb{R}}_+ \times \Omega \times \mathbb{R}^{1+q})$ with the transmission property and $\tilde{g}(y, \eta) \in R_G^{\mu,d}(\Omega \times \mathbb{R}^q)$ then $p - \tilde{p} \in S^{-\infty}(\overline{\mathbb{R}}_+ \times \Omega \times \mathbb{R}^{1+q})$ and hence $\tilde{g}(y, \eta) = \mathrm{op}^+(p - \tilde{p})(y, \eta) + g(y, \eta)$, where $\mathrm{op}^+(p - \tilde{p})(y, \eta) \in R^{-\infty,0}(\Omega \times \mathbb{R}^q)$. This implies that $a \to \sigma_\wedge^\mu(a)$ is well-defined. Let us set

$$\Sigma_\wedge^{\mu,d} = \left\{ \sigma_\wedge^\mu(a) : a(y, \eta) \in R^{\mu,d}(\Omega \times \mathbb{R}^q) \right\}.$$

Denote by $\Sigma^{\mu,d}$ the subspace of $\Sigma_\psi^\mu \times \Sigma_\wedge^{\mu,d} \ni (p_\psi, p_\wedge)$ satisfying the relation $p_\wedge - \mathrm{op}^+(p_\psi|_{t=0}) \in R_G^{(\mu),d}(\Omega \times (\mathbb{R}^q \backslash \{0\}))$.

Proposition 4.2.43 *We have an exact sequence*

$$0 \to R^{\mu-1,d}(\Omega \times \mathbb{R}^q) \xrightarrow{\iota} R^{\mu,d}(\Omega \times \mathbb{R}^q) \xrightarrow{\sigma} \Sigma^{\mu,d} \to 0,$$

where ι is the canonical embedding, and $\sigma = (\sigma^\mu_\psi, \sigma^\mu_\wedge)$. There is a map

$$\mathrm{op} : \Sigma^{\mu,d} \to R^{\mu,d}(\Omega \times \mathbb{R}^q)$$

with $\sigma \circ \mathrm{op} = \mathrm{id}$ on $\Sigma^{\mu,d}$.

Proof. Let us first show the relation $\mathrm{im}\,\iota = \ker\sigma$. In fact, $\sigma^\mu_\psi(a) = 0$ implies $a(y,\eta) = \mathrm{op}^+(p)(y,\eta) + g(y,\eta)$ for a $p(t,y,\tau,\eta) \in S^{\mu-1}_{\mathrm{cl}}(\overline{\mathbb{R}}_+ \times \Omega \times \mathbb{R}^{1+q})$ with the transmission property, and $g(y,\eta) \in R^{\mu,d}_G(\Omega \times \mathbb{R}^q)$. Since $\sigma^\mu_\wedge(\mathrm{op}^+(p))(y,\eta) = 0$, it follows from $\sigma^\mu_\wedge(a)(y,\eta) = 0$ that $g(y,\eta) \in R^{\mu-1,d}_G(\Omega \times \mathbb{R}^q)$, cf. (4.2.28), in other words, $\ker\sigma \subseteq \mathrm{im}\,\iota$. On the other hand $\mathrm{im}\,\iota \subseteq \ker\sigma$ is trivial. To prove that σ is surjective it suffices to construct a right inverse

$$\mathrm{op} : \Sigma^{\mu,d} \to R^{\mu,d}(\Omega \times \mathbb{R}^q) \tag{4.2.52}$$

of σ. If $p_{(\mu)}(t,y,\tau,\eta)$ is the first component of a pair $(p_{(\mu)}, a_{(\mu)})$ in $\Sigma^{\mu,d}$, then, for any excision function $\chi(\tau,\eta)$ in $\mathbb{R}^{1+q}$ we can form $\mathrm{op}^+(\chi p_{(\mu)})(y,\eta)$. Then

$$g_{(\mu)}(y,\eta) = a_{(\mu)}(y,\eta) - \sigma^\mu_\wedge(\mathrm{op}^+(\chi p_{(\mu)}))(y,\eta) \in R^{(\mu),d}_G(\Omega \times (\mathbb{R}^q \setminus \{0\})),$$

cf. (4.2.27). Thus we can set

$$\mathrm{op}(p_{(\mu)}, a_{(\mu)}) = \mathrm{op}^+(\chi p_{(\mu)})(y,\eta) + \chi'(\eta) g_{(\mu)}(y,\eta)$$

for any excision function $\chi'(\eta)$ in $\mathbb{R}^q$. $\square$

Let us generalise the above notation to $\mathcal{R}^{\mu,d}(\Omega \times \mathbb{R}^q; N_-, N_+) \ni a(y,\eta)$. Writing $a(y,\eta)$ as a block matrix

$$a(y,\eta) = \begin{pmatrix} a_{11} & a_{12} \\ a_{21} & a_{22} \end{pmatrix} (y,\eta)$$

with $a_{11} = \mathrm{u.l.c.}(a) \in R^{\mu,d}(\Omega \times \mathbb{R}^q)$ we set $\sigma^\mu_\psi(a) = \sigma^\mu_\psi(a_{11})$ and

$$\sigma^\mu_\wedge(a)(y,\eta) = \begin{pmatrix} \sigma^\mu_\wedge(a_{11}) & \sigma^\mu_\wedge(a_{12}) \\ \sigma^\mu_\wedge(a_{21}) & \sigma^\mu_\wedge(a_{22}) \end{pmatrix} (y,\eta)$$

with the above $\sigma^\mu_\wedge(a_{11})$ and $\sigma^\mu_\wedge(a_{ij})$ from Section 4.2.3 for the remaining i,j.

If we denote by $\Sigma^{\mu,d}(N_-, N_+)$ the space of all pairs $(\sigma^\mu_\psi(a), \sigma^\mu_\wedge(a))$ for arbitrary elements $a \in \mathcal{R}^{\mu,d}(\Omega \times \mathbb{R}^q; N_-, N_+)$, then $\Sigma^{\mu,d}$ equals $\mathrm{u.l.c.}\,\Sigma^{\mu,d}(N_-, N_+)$ (the space of upper left corners). Using (4.2.28) we obtain an analogue of Proposition 4.2.43:

Proposition 4.2.44 *We have an exact sequence*

$$0 \to \mathcal{R}^{\mu-1,d}(\Omega \times \mathbb{R}^q; N_-, N_+) \xrightarrow{\iota} \mathcal{R}^{\mu,d}(\Omega \times \mathbb{R}^q; N_-, N_+)$$
$$\xrightarrow{\sigma} \Sigma^{\mu,d}(N_-, N_+) \to 0,$$

where ι is the canonical embedding, and $\sigma = (\sigma^\mu_\psi, \sigma^\mu_\wedge)$. There is a map

$$\mathrm{op} : \Sigma^{\mu,d}(N_-, N_+) \to \mathcal{R}^{\mu,d}(\Omega \times \mathbb{R}^q; N_-, N_+)$$

with $\sigma \circ \mathrm{op} = \mathrm{id}$ on $\Sigma^{\mu,d}(N_-, N_+)$.

Theorem 4.2.45 $a(y,\eta) \in \mathcal{R}^{0,0}(\Omega \times \mathbb{R}^q; N_-, N_+)$ *implies* $a^*(y,\eta) \in \mathcal{R}^{0,0}(\Omega \times \mathbb{R}^q; N_+, N_-)$, *where* $a^*(y,\eta)$ *is defined by*

$$(u, a^*(y,\eta)v)_{L^2(\mathbb{R}_+) \oplus \mathbb{C}^{N_-}} = (a(y,\eta)u, v)_{L^2(\mathbb{R}_+) \oplus \mathbb{C}^{N_+}}$$

for all $u \in C_0^\infty(\mathbb{R}_+) \oplus \mathbb{C}^{N_-}$, $v \in C_0^\infty(\mathbb{R}_+) \oplus \mathbb{C}^{N_+}$ *for every fixed* $(y,\eta) \in \Omega \times \mathbb{R}^q$.

Proof. In view of Definition 4.2.19 it suffices to consider the case $a(y,\eta) = \mathrm{op}^+(p)(y,\eta)$ for a symbol $p(t,y,\tau,\eta) \in S^0_{\mathrm{cl}}(\overline{\mathbb{R}}_+ \times \Omega \times \mathbb{R}^{1+q})$ with the transmission property that is independent of t for $t > \mathrm{const}$. Then

$$(\mathrm{op}^+(p)(y,\eta)u, v)_{L^2(\mathbb{R}_+)} = (u, [\mathrm{op}^+(p^*)(y,\eta) + g(y,\eta)]v)_{L^2(\mathbb{R}_+)}$$

for all $u, v \in C_0^\infty(\mathbb{R}_+)$, for some $g(y,\eta) \in R_G^{-\infty,0}(\Omega \times \mathbb{R}^q)$. Here p^* means a symbol obtained by applying the symbol rule for formal adjoint operators with respect to $L^2(\mathbb{R}_+)$, which also has the transmission property. $\square$

Theorem 4.2.46 $a(y,\eta) \in \mathcal{R}^{\mu,d}(\Omega \times \mathbb{R}^q; N_0, N_+)$, $b(y,\eta) \in \mathcal{R}^{\nu,c}(\Omega \times \mathbb{R}^q; N_-, N_0)$ *implies* $a(y,\eta)b(y,\eta) \in \mathcal{R}^{\mu+\nu,h}(\Omega \times \mathbb{R}^q; N_-, N_+)$ *for* $h = \max\{(d+\nu)^+, c\}$, *where* $\varrho^+ = \max\{\varrho, 0\}$, *and we have*

$$\sigma_\psi^{\mu+\nu}(ab) = \sigma_\psi^\mu(a)\sigma_\psi^\nu(b),$$
$$\sigma_\wedge^{\mu+\nu}(ab) = \sigma_\wedge^\mu(a)\sigma_\wedge^\nu(b).$$

In particular, $a(y,\eta) \in \mathcal{R}_G^{\mu,d}(\Omega \times \mathbb{R}^q; N_0, N_+)$ *or* $b(y,\eta) \in \mathcal{R}_G^{\nu,c}(\Omega \times \mathbb{R}^q; N_-, N_0)$ *imply* $a(y,\eta)b(y,\eta) \in \mathcal{R}_G^{\mu+\nu,h}(\Omega \times \mathbb{R}^q; N_-, N_+)$.

Proof. Let us show the result first for $N_0 = N_- = N_+ = 0$. Suppose $a(y,\eta) = \mathrm{op}^+(p)(y,\eta) + g(y,\eta)$, $b(y,\eta) = \mathrm{op}^+(r)(y,\eta) + l(y,\eta)$ for symbols $p(t,y,\tau,\eta) \in S^\mu_{\mathrm{cl}}(\overline{\mathbb{R}}_+ \times \Omega \times \mathbb{R}^{1+q})$, $r(t,y,\tau,\eta) \in S^\nu_{\mathrm{cl}}(\overline{\mathbb{R}}_+ \times \Omega \times \mathbb{R}^{1+q})$ with the transmission property, independent of t for $t > c_1$ for a constant $c_1 > 0$, and $g(y, \eta) \in R_G^{\mu, d}(\Omega \times \mathbb{R}^q)$, $l(y, \eta) \in R_G^{\nu, h}(\Omega \times \mathbb{R}^q)$. Then

$$a(y, \eta)b(y, \eta) = \mathrm{op}^+(p)(y, \eta)\,\mathrm{op}^+(r)(y, \eta) + \mathrm{op}^+(p)(y, \eta)l(y, \eta)$$
$$+ g(y, \eta)\,\mathrm{op}^+(r)(y, \eta) + g(y, \eta)l(y, \eta).$$

From Proposition 4.2.36 we get

$$\mathrm{op}^+(p)(y, \eta)\,\mathrm{op}^+(r)(y, \eta) = \mathrm{op}^+(p\#r)(y, \eta) \bmod R_G^{\mu+\nu, 0}(\Omega \times \mathbb{R}^q),$$

where the Leibniz product $p\#r$ with respect to (t, τ) can be carried out in the class of symbols with the transmission property, independent of t for $t > c_1$. Moreover, from Proposition 4.2.30 we obtain $g(y, \eta)l(y, \eta) \in R_G^{\mu+\nu, c}(\Omega \times \mathbb{R}^q)$. Let us now consider $\mathrm{op}^+(p)(y, \eta)l(y, \eta)$. By definition, cf. (4.2.23), we have $l(y, \eta) = \sum_{k=0}^d l_k(y, \eta)(\frac{d}{dt})^k$ for certain $l_k(y, \eta) \in R_G^{\nu-k, 0}(\Omega \times \mathbb{R}^q)$. Thus it suffices to show the relation $\mathrm{op}^+(p)(y, \eta)l_k(y, \eta) \in R_G^{\mu+\nu-k, 0}(\Omega \times \mathbb{R}^q)$. If p is independent of t we know from Theorem 4.2.38 that

$$\mathrm{op}^+(p)(y, \eta) \in S_{\mathrm{cl}}^\mu(\Omega \times \mathbb{R}^q; \mathcal{S}(\overline{\mathbb{R}}_+), \mathcal{S}(\overline{\mathbb{R}}_+)).$$

Composition from the right by $l_k(y, \eta) \in S_{\mathrm{cl}}^{\nu-k}(\Omega \times \mathbb{R}^q; L^2(\mathbb{R}_+), \mathcal{S}(\overline{\mathbb{R}}_+))$ then yields $(\mathrm{op}^+(p)l_k)(y, \eta) \in S_{\mathrm{cl}}^{\mu+\nu-k}(\Omega \times \mathbb{R}^q; L^2(\mathbb{R}_+), \mathcal{S}(\overline{\mathbb{R}}_+))$, which is the first property in Definition 4.2.19. Using Theorem 4.2.45 and Definition 4.2.19 we get analogously

$$(\mathrm{op}^+(p)l_k)^*(y, \eta) = l_k^*(y, \eta)\,\mathrm{op}^+(p)^*(y, \eta)$$
$$\in S_{\mathrm{cl}}^{\mu+\nu-k}(\Omega \times \mathbb{R}^q; L^2(\mathbb{R}_+), \mathcal{S}(\overline{\mathbb{R}}_+)),$$

i.e., the desired property is proved. For t-dependent p we apply the Taylor expansion in t near 0

$$p(t, y, \tau, \eta) = \sum_{j=0}^N t^j p_j(y, \tau, \eta) + t^{N+1} p_{(N)}(t, y, \tau, \eta),$$

where p_j are symbols that are t-independent, and $p_{(N)}(t, y, \tau, \eta) \in S_{\mathrm{cl}}^\mu(\overline{\mathbb{R}}_+ \times \Omega \times \mathbb{R}^{1+q})$ is a symbol with the transmission property that has the form

$$p_{(N)}(t, y, \tau, \eta) = p(\infty, y, \tau, \eta) - \sum_{j=0}^N t^{-(N+1)+j} p_j(y, \tau, \eta)$$

for $t > c_1$, $t \to \infty$. Now $(\mathrm{op}^+(p)l_k)(y, \eta) = \sum_{j=0}^N t^j (\mathrm{op}^+(p_j)l_k)(y, \eta) + t^{N+1}(\mathrm{op}^+(p_{(N)})l_k)(y, \eta)$. For the first N summands we can argue as before and apply in addition Proposition 4.2.31. Moreover, $l_k(y, \eta) \in S_{\mathrm{cl}}^{\nu-k}(\Omega \times \mathbb{R}^q; L^2(\mathbb{R}_+), \mathcal{S}(\overline{\mathbb{R}}_+))$ and Remark 4.2.39 yield

$$(\mathrm{op}^+(p_{(N)})l_k)(y, \eta) \in S^{\mu+\nu-k}(\Omega \times \mathbb{R}^q; L^2(\mathbb{R}_+), \mathcal{S}(\overline{\mathbb{R}}_+)).$$

This gives us

$$t^{N+1}(\mathrm{op}^+(p_{(N)})l_k)(y, \eta) \in S^{\nu-k-(N+1)}(\Omega \times \mathbb{R}^q; L^2(\mathbb{R}_+), \mathcal{S}(\overline{\mathbb{R}}_+)),$$

since the operator of multiplication by t^{N+1} diminishes the order as operator–valued symbol by $N+1$. In other words, since N is arbitrary, we also obtain $(\mathrm{op}^+(p)l_k)(y, \eta) \in S_{\mathrm{cl}}^{\mu+\nu-k}(\Omega \times \mathbb{R}^q; L^2(\mathbb{R}_+), \mathcal{S}(\overline{\mathbb{R}}_+))$. For the formal adjoint we proceed in an analogous manner, i.e., it follows that $\mathrm{op}^+(p)(y, \eta)l_k(y, \eta) \in R_G^{\mu+\nu-k,0}(\Omega \times \mathbb{R}^q)$. □

4.3 The algebra of boundary value problems

4.3.1 The local calculus

The pseudo–differential boundary value problems in $\mathbb{R}_+ \times \Omega$ for open $\Omega \subseteq \mathbb{R}^q$ will be defined as pseudo–differential operators on Ω with operator–valued symbols of the class $\mathcal{R}^{\mu,d}(\Omega \times \mathbb{R}^q; N_-, N_+)$, modulo appropriate smoothing operators. As usual we set $H^s(\mathbb{R}_+ \times \mathbb{R}^q) = \{u|_{\mathbb{R}_+ \times \mathbb{R}^q} : u \in H^s(\mathbb{R} \times \mathbb{R}^q)\}$, $s \in \mathbb{R}$.

Let us also introduce the Hilbert spaces

$$H^s(\mathbb{R}_+ \times K) := \{u \in H^s(\mathbb{R}_+ \times \mathbb{R}^q) : u = 0 \text{ on } \mathbb{R}_+ \times (\mathbb{R}^q \setminus K)\}$$

for arbitrary $K \subset\subset \mathbb{R}^q$ and $s \in \mathbb{R}$. Set $H^s_{\mathrm{comp}(y)}(\mathbb{R}_+ \times \Omega) = \varinjlim H^s(\mathbb{R}_+ \times K)$, where the inductive limit is taken over all $K \subset\subset \Omega$. Moreover, $H^s_{\mathrm{loc}(y)}(\mathbb{R}_+ \times \Omega)$ denotes the subspace of all $u \in \mathcal{D}'(\mathbb{R}_+ \times \Omega)$ with $\varphi u \in H^s_{\mathrm{comp}(y)}(\mathbb{R}_+ \times \Omega)$ for every $\varphi \in C_0^\infty(\Omega)$. The space $H^s_{\mathrm{loc}(y)}(\mathbb{R}_+ \times \Omega)$ in Fréchet with the semi–norm system $u \to \|\varphi u\|_{H^s(\mathbb{R}_+ \times K)}$, $K := \mathrm{supp}\,\varphi$, $\varphi \in C_0^\infty(\Omega)$.

Given an operator

$$\mathcal{C} : \begin{array}{c} C_0^\infty(\mathbb{R}_+ \times \Omega) \\ \oplus \\ C_0^\infty(\Omega, \mathbb{C}^{N_-}) \end{array} \to \begin{array}{c} C^\infty(\mathbb{R}_+ \times \Omega) \\ \oplus \\ C^\infty(\Omega, \mathbb{C}^{N_+}) \end{array} \qquad (4.3.1)$$

we define the formal adjoint $\mathcal{C}^*$ by

$$(u, \mathcal{C}^* v)_{L^2(\mathbb{R}_+ \times \Omega) \oplus L^2(\Omega, \mathbb{C}^{N_-})} = (\mathcal{C}u, v)_{L^2(\mathbb{R}_+ \times \Omega) \oplus L^2(\Omega, \mathbb{C}^{N_+})}$$

for all $u \in C_0^\infty(\mathbb{R}_+ \times \Omega) \oplus C_0^\infty(\Omega, \mathbb{C}^{N-})$, $v \in C_0^\infty(\mathbb{R}_+ \times \Omega) \oplus C_0^\infty(\Omega, \mathbb{C}^{N+})$.

The space $\mathcal{B}^{-\infty,0}(\mathbb{R}_+ \times \Omega; N_-, N_+)$ of smoothing operators of type zero is defined as the set of all operators (4.3.1) for which $\mathcal{C}$ and $\mathcal{C}^*$ induce continuous operators

$$
\mathcal{C}, \mathcal{C}^* : \quad
\begin{matrix}
C_0^\infty(\Omega, \mathcal{S}(\overline{\mathbb{R}}_+)) \\
\oplus \\
C_0^\infty(\Omega, \mathbb{C}^{N\mp})
\end{matrix}
\quad \to \quad
\begin{matrix}
C^\infty(\Omega, \mathcal{S}(\overline{\mathbb{R}}_+)) \\
\oplus \\
C^\infty(\Omega, \mathbb{C}^{N\pm})
\end{matrix}
$$

which extend to continuous operators

$$
\mathcal{C}, \mathcal{C}^* : \quad
\begin{matrix}
H^s_{\mathrm{comp}(y)}(\mathbb{R}_+ \times \Omega) \\
\oplus \\
H^s_{\mathrm{comp}}(\Omega, \mathbb{C}^{N\mp})
\end{matrix}
\quad \to \quad
\begin{matrix}
H^\infty_{\mathrm{loc}(y)}(\mathbb{R}_+ \times \Omega) \\
\oplus \\
H^\infty_{\mathrm{loc}}(\Omega, \mathbb{C}^{N\pm})
\end{matrix}
$$

for all $s \in \mathbb{R}$, $s > -\frac{1}{2}$. Clearly the signs in the second components indicate N_-, N_+ for $\mathcal{C}$ and N_+, N_- for $\mathcal{C}^*$. Moreover, the space $\mathcal{B}^{-\infty,d}(\mathbb{R}_+ \times \Omega; N_-, N_+)$ of smoothing operators of type $d \in \mathbb{N}$ is the set of all operators

$$
\mathcal{C} = \sum_{k=0}^d \mathcal{C}_k \begin{pmatrix} (\frac{\partial}{\partial t})^k & 0 \\ 0 & 1 \end{pmatrix}
$$

for arbitrary $\mathcal{C}_k \in \mathcal{B}^{-\infty,0}(\mathbb{R}_+ \times \Omega; N_-, N_+)$, $k = 0, \dots, d$. Here 1 indicates the identity operator in $C_0^\infty(\Omega, \mathbb{C}^{N-})$.

Remark 4.3.1 $g(y, y', \eta) \in \mathcal{R}^{-\infty,d}(\Omega \times \Omega \times \mathbb{R}^q; N_-, N_+)$ *implies* $\mathrm{Op}(g) \in \mathcal{B}^{-\infty,d}(\mathbb{R}_+ \times \Omega; N_-, N_+)$.

Here, as usual,

$$
\mathrm{Op}(a)u(y) = \iint e^{i(y-y')\eta} a(y, y', \eta) u(y') \, dy' d\eta
$$

for every operator–valued symbol $a(y, y', \eta)$.

Definition 4.3.2 $\mathcal{B}^{\mu,d}(\mathbb{R}_+ \times \Omega; N_-, N_+)$ *for* $\mu \in \mathbb{Z}$, $d \in \mathbb{N}$, *is the space of all operators of the form* $\mathcal{A} = \mathrm{Op}(a) + \mathcal{C}$ *for arbitrary* $a(y, \eta) \in \mathcal{R}^{\mu,d}(\Omega \times \mathbb{R}^q; N_-, N_+)$ *and* $\mathcal{C} \in \mathcal{B}^{-\infty,d}(\mathbb{R}_+ \times \Omega; N_-, N_+)$. *The elements* $\mathcal{A} \in \mathcal{B}^{\mu,d}(\mathbb{R}_+ \times \Omega; N_-, N_+)$ *are called pseudo–differential boundary value problems of order* μ *and type* d *in Boutet de Monvel's algebra over* $\mathbb{R}_+ \times \Omega$.

The operators $\mathcal{A} \in \mathcal{B}^{\mu,d}(\mathbb{R}_+ \times \Omega; N_-, N_+)$ are block matrices

$$
\mathcal{A} = \begin{pmatrix} A & K \\ T & Q \end{pmatrix},
$$

and we shall write from time to time $A = \text{u. l. c.} \mathcal{A}$. An element

$$\mathcal{G} = \begin{pmatrix} G & K \\ T & Q \end{pmatrix} = \text{Op}(g) + \mathcal{C}$$

for any $g(y, \eta) \in \mathcal{R}_G^{\mu,d}(\Omega \times \mathbb{R}^q; N_-, N_+)$ and $\mathcal{C} \in \mathcal{B}^{-\infty,d}(\mathbb{R}_+ \times \Omega; N_-, N_+)$ is called a Green operator of order μ and type d. The space of all those $\mathcal{G}$ shall be denoted by $\mathcal{B}_G^{\mu,d}(\mathbb{R}_+ \times \Omega; N_-, N_+)$. In particular, we call T a trace operator of order μ and type d, K a potential operator of order μ. The operator Q is an $N_+ \times N_-$ matrix of elements in $L_{\text{cl}}^\mu(\Omega)$.

The elements $\mathcal{A} \in \mathcal{B}^{\mu,d}(\mathbb{R}_+ \times \Omega; N_-, N_+)$ are regarded as boundary value problems for $A = \text{u. l. c.} \mathcal{A}$.

Let us set

$$B^{\mu,d}(\mathbb{R}_+ \times \Omega) = \mathcal{B}^{\mu,d}(\mathbb{R}_+ \times \Omega; 0, 0),$$
$$B_G^{\mu,d}(\mathbb{R}_+ \times \Omega) = \mathcal{B}_G^{\mu,d}(\mathbb{R}_+ \times \Omega; 0, 0),$$

where matrices in which only the upper left corners are non-zero are canonically identified with the upper left corners themselves.

Remark 4.3.3 *We have*

$$B^{\mu,d}(\mathbb{R}_+ \times \Omega) \subset L_{\text{cl}}^\mu(\mathbb{R}_+ \times \Omega),$$
$$B_G^{\mu,d}(\mathbb{R}_+ \times \Omega) \subset L^{-\infty}(\mathbb{R}_+ \times \Omega).$$

The non-vanishing order μ of the Green operators $G \in B_G^{\mu,d}(\mathbb{R}_+ \times \Omega)$ of the form

$$Gu(t, y) = \iint g(t, y, t', y') u(t', y') \, dt' dy'$$

on $C_0^\infty(\mathbb{R}_+ \times \Omega) \ni u$ for certain $g(t, y, t', y') \in C^\infty(\mathbb{R}_+ \times \Omega \times \mathbb{R}_+ \times \Omega)$ comes from the singular behaviour of the kernels $g(t, y, t', y')$ near $(t, y) = (t', y')$ for $t, t' \to 0$, cf., in particular, Proposition 4.2.21.

Example 4.3.4 *Let Δ be the Laplace operator in $\mathbb{R}^{1+q}$. Then, if we set $r'u = u|_{t=0}$, say for $u \in C_0^\infty(\mathbb{R}^q, \mathcal{S}(\overline{\mathbb{R}}_+))$, the column matrix*

$$\mathcal{A} = \begin{pmatrix} \Delta \\ r' \end{pmatrix}$$

corresponds to the Dirichlet problem for Δ in the half space $\mathbb{R}_+^{1+q} = \{(t, y) : t > 0, y \in \mathbb{R}^q\}$. Here $\Delta \in \mathcal{B}^{2,0}(\mathbb{R}_+ \times \mathbb{R}^q)$, $r' \in \mathcal{B}^{\frac{1}{2},1}(\mathbb{R}_+ \times \mathbb{R}^q; 0, 1)$; so the situation with respect to the orders is not exactly as in Definition 4.3.2. However by a reduction of orders on the boundary $\mathbb{R}^q$ through $\text{op}(\langle \eta \rangle^{\frac{1}{2}})$ which induces isomorphisms between the corresponding Sobolev spaces on the boundary we can replace r' by $T = \mathrm{op}(\langle\eta\rangle^{\frac{3}{2}})r' \in \mathcal{B}^{2,1}(\mathbb{R}_+ \times \mathbb{R}^q; 0, 1)$. This yields an operator $\mathcal{A}_1 \in \mathcal{B}^{2,1}(\mathbb{R}_+ \times \mathbb{R}^q; 0, 1)$ which is equivalent to $\mathcal{A}$ modulo the order reduction, cf. also Remark 4.3.27 below. More general boundary value problems for Δ can be obtained if we replace r' by $r'B$ for some differential operator on $\mathbb{R}_+ \times \mathbb{R}^q$ with coefficients in $C^\infty(\overline{\mathbb{R}}_+ \times \mathbb{R}^q)$. In particular, for $B = \frac{\partial}{\partial t}$ we get in this way the Neumann problem for Δ.

Theorem 4.3.5 Every $G \in \mathcal{B}_G^{\mu,d}(\mathbb{R}_+ \times \Omega; N_-, N_+)$ has a unique representation of the form $G = G_0 + G_1$ for a $G_0 \in \mathcal{B}_G^{\mu,0}(\mathbb{R}_+ \times \Omega; N_-, N_+)$ and

$$G_1 = \sum_{j=0}^{d-1} \begin{pmatrix} 0 & W_j \\ 0 & R_j \end{pmatrix} \begin{pmatrix} 0 & 0 \\ T_j & 0 \end{pmatrix}$$

for $T_j u = \frac{\partial^j}{\partial t^j} u|_{t=0}$, $W_j \in \mathcal{B}^{\mu-j-\frac{1}{2},0}(\mathbb{R}_+ \times \Omega; 1, 0)$ and $R_j \in L_{\mathrm{cl}}^{\mu-j-\frac{1}{2}}(\Omega) \otimes \mathbb{C}^{N_+}$, $j = 0, \ldots, d-1$.

Proof. For simplicity we restrict ourselves to $N_- = N_+ = 0$; the arguments for the general case are completely analogous. By definition we have $G = \mathrm{Op}(g) + C$ for a $g(y, \eta) \in R_G^{\mu,d}(\Omega \times \mathbb{R}^q)$, $C \in B^{-\infty,d}(\mathbb{R}_+ \times \Omega)$. Applying Theorem 4.2.27 we obtain

$$g(y, \eta) = g_0(y, \eta) + \sum_{j=0}^{d-1} k_j(y, \eta) \gamma^j$$

for $g_0(y, \eta) \in \mathcal{R}_G^{\mu,0}(\Omega \times \mathbb{R}^q)$ and $\gamma^j u = (\frac{\partial}{\partial t})^j u|_{t=0}$, where $\gamma^j \in \mathcal{R}_G^{j+\frac{1}{2},j}(\Omega \times \mathbb{R}^q; 0, 1)$ is a trace symbol and $k_j(y, \eta) \in \mathcal{R}_G^{\mu-j-\frac{1}{2},0}(\Omega \times \mathbb{R}^q; 0, 1)$ a potential symbol. Setting $T_j = \mathrm{Op}(\gamma^j)$, $W_j^1 = \mathrm{Op}(k_j)$, $G_0^1 = \mathrm{Op}(g_0)$ we get a decomposition of $\mathrm{Op}(g)$ of the asserted form, namely $\mathrm{Op}(g) = G_0^1 + \sum_{j=0}^{d-1} W_j^1 \circ T_j$. The operator C has the form $C = \sum_{k=0}^{d} C_k(\frac{\partial}{\partial t})^k$, where C_k has a kernel in $C^\infty(\Omega \times \Omega, \mathcal{S}(\overline{\mathbb{R}}_+ \times \overline{\mathbb{R}}_+))$. Here we can argue analogously to Proposition 4.1.46. Integrations by parts with respect to t can be carried out under the integrals; the only modifications are the additional y–variables, but they remain untouched in the procedure. This gives us $C = G_0^2 + \sum_{j=0}^{d-1} W_j^2 \circ T_j$ for certain $G_0^2 \in B_G^{-\infty,0}(\mathbb{R}_+ \times \Omega)$ and potential operators $W_j^2 \in \mathcal{B}_G^{-\infty,0}(\mathbb{R}_+ \times \Omega; 1, 0)$. We thus obtain the asserted decomposition for $G_0 = G_0^1 + G_0^2$, $W_j = W_j^1 + W_j^2$.

In order to prove the uniqueness we assume that we have a second decomposition of G of the above kind. Then considering the difference of both representations, it suffices to show that

$$\left(G_0 + \sum_{j=0}^{d-1} W_j \circ T_j\right) u = 0 \quad \text{for all} \quad u \in C_0^\infty(\Omega, \mathcal{S}(\overline{\mathbb{R}}_+))$$

implies that $G_0 \equiv 0$, $W_j \equiv 0$ for all j. Inserting, in particular, $u \in C_0^\infty(\mathbb{R}_+ \times \Omega)$, we get $T_j u = 0$ for all j, i.e., $G_0 u = 0$ for all $u \in C_0^\infty(\mathbb{R}_+ \times \Omega)$ and thus $G_0 \equiv 0$. On the other hand, inserting $u(t,y) = t^j \omega(t) v(y)$ for a cut-off function $\omega(t)$ and arbitrary $v \in C_0^\infty(\Omega)$ gives us $W_j v = 0$ for all those v which yields $W_j \equiv 0$ for all j. $\qquad\square$

Theorem 4.3.6 *Each $\mathcal{A} \in \mathcal{B}^{\mu,d}(\mathbb{R}_+ \times \Omega; N_-, N_+)$ induces continuous operators*

$$
\mathcal{A} : \begin{array}{ccc}
H_{\text{comp}(y)}^s(\mathbb{R}_+ \times \Omega) & & H_{\text{loc}(y)}^{s-\mu}(\mathbb{R}_+ \times \Omega) \\
\oplus & \to & \oplus \\
H_{\text{comp}}^s(\Omega, \mathbb{C}^{N_-}) & & H_{\text{loc}}^{s-\mu}(\Omega, \mathbb{C}^{N_+})
\end{array}
\tag{4.3.2}
$$

for all $s \in \mathbb{R}$, $s > d - \frac{1}{2}$, and

$$
\mathcal{A} : \begin{array}{ccc}
C_0^\infty(\Omega, \mathcal{S}(\overline{\mathbb{R}}_+)) & & C^\infty(\Omega, \mathcal{S}(\overline{\mathbb{R}}_+)) \\
\oplus & \to & \oplus \\
C_0^\infty(\Omega, \mathbb{C}^{N_-}) & & C^\infty(\Omega, \mathbb{C}^{N_+}).
\end{array}
\tag{4.3.3}
$$

Proof. The asserted mapping properties are obvious for $\mathcal{A} \in \mathcal{B}^{-\infty,d}(\mathbb{R}_+ \times \Omega; N_-, N_+)$. Thus we may assume $\mathcal{A} = \text{Op}(a)$ for $a(y,\eta) \in \mathcal{R}^{\mu,d}(\Omega \times \mathbb{R}^q; N_-, N_+)$. In view of

$$
\mathcal{R}^{\mu,d}(\Omega \times \mathbb{R}^q; N_-, N_+) \\
\subset S^\mu(\Omega \times \mathbb{R}^q; H^s(\mathbb{R}_+) \oplus \mathbb{C}^{N_-}, H^{s-\mu}(\mathbb{R}_+) \oplus \mathbb{C}^{N_+})
$$

for all $s \in \mathbb{R}$, $s > d - \frac{1}{2}$, cf. (4.2.50) and Definition 4.2.19, we may apply the general continuity property of Theorem 1.3.58 for pseudo–differential operators with operator–valued symbols. This yields continuous operators

$$
\text{Op}(a) : \mathcal{W}_{\text{comp}}^s(\Omega, H^s(\mathbb{R}_+) \oplus \mathbb{C}^{N_-}) \to \mathcal{W}_{\text{loc}}^{s-\mu}(\Omega, H^{s-\mu}(\mathbb{R}_+) \oplus \mathbb{C}^{N_+}).
$$

Using

$$
\mathcal{W}_{\text{comp}}^s(\Omega, H^s(\mathbb{R}_+) \oplus \mathbb{C}^N) = H_{\text{comp}(y)}^s(\mathbb{R}_+ \times \Omega) \oplus H_{\text{comp}}^s(\Omega, \mathbb{C}^N)
$$

and an analogous relation for the loc-spaces we obtain the first assertion. Since $\mathcal{R}^{\mu,d}(\Omega \times \mathbb{R}^q; N_-, N_+) \subset S^\mu(\Omega \times \mathbb{R}^q; \mathcal{S}(\overline{\mathbb{R}}_+) \oplus \mathbb{C}^{N_-}, \mathcal{S}(\overline{\mathbb{R}}_+) \oplus \mathbb{C}^{N_+})$, cf. (4.2.51) and Definition 4.2.19, we get the second assertion from Theorem 1.3.58, where

$$
\bigcap_{s \in \mathbb{R}} \bigcap_{k \in \mathbb{N}} \mathcal{W}_{\text{loc}}^s(\Omega, \langle t \rangle^{-k} H^k(\mathbb{R}_+) \oplus \mathbb{C}^N) = C^\infty(\Omega, \mathcal{S}(\overline{\mathbb{R}}_+)) \oplus C^\infty(\Omega, \mathbb{C}^N)
$$

584 BOUNDARY VALUE PROBLEMS

and an analogous relation for the comp-spaces shows that

$$\mathrm{Op}(a): \begin{array}{c} C_0^\infty(\Omega, \mathcal{S}(\overline{\mathbb{R}}_+)) \\ \oplus \\ C_0^\infty(\Omega, \mathbb{C}^{N-}) \end{array} \to \begin{array}{c} C^\infty(\Omega, \mathcal{S}(\overline{\mathbb{R}}_+)) \\ \oplus \\ C^\infty(\Omega, \mathbb{C}^{N+}) \end{array}$$

is continuous. □

Similarly to the pseudo–differential calculus in the scalar case it is convenient to talk about properly supported elements of $\mathcal{B}^{\mu,d}(\mathbb{R}_+ \times \Omega; N_-, N_+)$ with respect to the variables $y \in \Omega$. To this end we use a slight generalisation of Definition 4.2.40, namely the space

$$\mathcal{R}^{\mu,d}(\Omega \times \Omega \times \mathbb{R}^q; N_-, N_+)$$

of operator–valued symbols of the form

$$a(y, y', \eta) = \begin{pmatrix} a_{11}(y, y', \eta) & 0 \\ 0 & 0 \end{pmatrix} + g(y, y', \eta)$$

for arbitrary $g(y, y', \eta) \in \mathcal{R}_G^{\mu,d}(\Omega \times \Omega \times \mathbb{R}^q; N_-, N_+)$, cf. Definition 4.2.19, and $a_{11}(y, y', \eta) = \mathrm{op}^+(p)(y, y', \eta)$ for arbitrary $p(t, y, y', \tau, \eta) \in S_{\mathrm{cl}}^\mu(\overline{\mathbb{R}}_+ \times \Omega \times \Omega \times \mathbb{R}^{1+q})$ with the transmission property, $p(t, y, y', \tau, \eta)$ being independent of t for $t > c_1$ for some $c_1 > 0$. Then

$$R^{\mu,d}(\Omega \times \Omega \times \mathbb{R}^q) = \mathcal{R}^{\mu,d}(\Omega \times \Omega \times \mathbb{R}^q; 0, 0),$$

cf. the notation after Definition 4.2.40.

Remark 4.3.7 *The continuity properties (4.3.2) for $s > d - \frac{1}{2}$ and (4.3.3) also hold for $\mathcal{A} = \mathrm{Op}(a)$ for every $a(y, y', \eta) \in \mathcal{R}^{\mu,d}(\Omega \times \Omega \times \mathbb{R}^q; N_-, N_+)$. The proof is similar to that of Theorem 4.3.6. In particular, for $a(y, y', \eta) \in \mathcal{R}_G^{\mu,d}(\Omega \times \Omega \times \mathbb{R}^q; N_-, N_+)$ we get continuous operators*

$$\mathrm{Op}(a): \begin{array}{c} H_{\mathrm{comp}(y)}^s(\mathbb{R}_+ \times \Omega) \\ \oplus \\ H_{\mathrm{comp}}^s(\Omega, \mathbb{C}^{N-}) \end{array} \to \begin{array}{c} H_{\mathrm{loc}}^{s-\mu}(\Omega, E^k) \\ \oplus \\ H_{\mathrm{loc}}^{s-\mu}(\Omega, \mathbb{C}^{N+}) \end{array}$$

for all $s \in \mathbb{R}$, $s > d - \frac{1}{2}$ and every $k \in \mathbb{N}$, where $E^k = \langle t \rangle^{-k} H^k(\mathbb{R}_+)$.

Proposition 4.3.8 $a(y, y', \eta) \in \mathcal{R}^{\mu,d}(\Omega \times \Omega \times \mathbb{R}^q; N_-, N_+)$ *implies* $\mathcal{A} = \mathrm{Op}(a) \in \mathcal{B}^{\mu,d}(\mathbb{R}_+ \times \Omega; N_-, N_+)$, *and there is an* $a(y, \eta) \in \mathcal{R}^{\mu,d}(\Omega \times \mathbb{R}^q; N_-, N_+)$ *with* $\mathcal{A} = \mathrm{Op}(a) \mod \mathcal{B}^{-\infty,d}(\mathbb{R}_+ \times \Omega; N_-, N_+)$, *where* $a(y, \eta)$ *has the asymptotic expansion*

$$a(y, \eta) \sim \sum_\alpha \frac{1}{\alpha!} D_{y'}^\alpha \partial_\eta^\alpha a(y, y', \eta)|_{y'=y}. \tag{4.3.4}$$

Proof. First observe that in the formula (4.3.4) the corresponding variant of Proposition 4.2.42 for (y, y')-dependent symbols is used. From the general pseudo–differential calculus with operator–valued symbols it is known that the symbol $a(y, \eta)$ satisfies

$$\mathrm{Op}(a) - \mathrm{Op}(\boldsymbol{a}) \in L^{-\infty}(\Omega; H^s(\mathbb{R}_+) \oplus \mathbb{C}^{N_-}, H^{s-\mu}(\mathbb{R}_+) \oplus \mathbb{C}^{N_+})$$

for every $s \in \mathbb{R}$, $s > d - \frac{1}{2}$. The arguments are straightforward generalisations of those to Theorem 1.1.30. In particular, we have obtained a representation

$$\mathrm{Op}(a) = \sum_{|\alpha| \leq M} \frac{1}{\alpha!} \mathrm{Op}(D_{y'}^\alpha \partial_\eta^\alpha a(y, y', \eta)|_{y'=y}) + \mathrm{Op}(a_N) \qquad (4.3.5)$$

for an $a_N(y, y', \eta)$ of order $\mu - 2N$, for every N with suitable M. In the present case, because of Proposition 4.2.41 (in the obvious variant for (y, y')-dependent symbols), we obtain in this way $a_N(y, y', \eta) \in \mathcal{R}^{\mu-2N,d}(\Omega \times \Omega \times \mathbb{R}^q; N_-, N_+)$. Since N is arbitrary, it follows that

$$\mathcal{A} - \mathrm{Op}(\boldsymbol{a}) \in \bigcap_{k \in \mathbb{N}} \mathrm{Op}(\mathcal{R}^{-k,d}) \qquad (4.3.6)$$

for $\mathcal{R}^{-k,d} = \mathcal{R}^{-k,d}(\Omega \times \Omega \times \mathbb{R}^q; N_-, N_+)$. Thus it suffices to characterise the right hand side of (4.3.6). We may consider the entries of the corresponding block matrices separately. Let us first assume $a(y, y', \eta) = \mathrm{op}^+(p)(y, y', \eta)$ for $p(t, y, y', \tau, \eta) \in S_{\mathrm{cl}}^\mu(\overline{\mathbb{R}}_+ \times \Omega \times \Omega \times \mathbb{R}^{1+q})$ with the transmission property. In this case the formula (4.3.5) gives us for every N a symbol $p_N(t, y, y', \tau, \eta) \in S^{\mu-2N}(\overline{\mathbb{R}}_+ \times \Omega \times \Omega \times \mathbb{R}^{1+q})$ with the transmission property for which

$$\mathrm{Op}(p_N) : H_{\mathrm{comp}(y)}^s(\mathbb{R}_+ \times \Omega) \to H_{\mathrm{loc}}^{s-\mu+2N}(\mathbb{R}_+ \times \Omega)$$

is continuous for every $s > -\frac{1}{2}$. In an analogous manner we can deal with the formal adjoint with respect to the $L^2(\mathbb{R}_+ \times \Omega)$–scalar product. This gives us $\mathrm{Op}(a) - \mathrm{Op}(\boldsymbol{a}) \in B^{-\infty,0}(\mathbb{R}_+ \times \Omega)$. It remains to consider the case $a(y, y', \eta) \in \mathcal{R}_G^{\mu,d}(\Omega \times \Omega \times \mathbb{R}^q; N_-, N_+)$. Because of the definition, cf. (4.2.23), we may look at the k th summand separately for every $k = 0, \ldots, d$. This reduces the problem to the case $d = 0$. Using Proposition 4.2.20 in the corresponding (y, y')-dependent variant we get for the asymptotic sum $a(y, \eta) \in \mathcal{R}_G^{\mu,0}(\Omega \times \mathbb{R}^q; N_-, N_+)$. The symbol $a_N(y, y', \eta)$ from (4.3.5) belongs to $\mathcal{R}_G^{\mu-2N,0}(\Omega \times \Omega \times \mathbb{R}^q; N_-, N_+)$. Thus, in particular,

$$\mathrm{Op}(a_N) : \begin{array}{c} H_{\mathrm{comp}(y)}^s(\mathbb{R}_+ \times \Omega) \\ \oplus \\ H_{\mathrm{comp}}^s(\Omega, \mathbb{C}^{N_-}) \end{array} \to \begin{array}{c} H_{\mathrm{loc}(y)}^{s-\mu+2N}(\mathbb{R}_+, E^k) \\ \oplus \\ H_{\mathrm{loc}}^{s-\mu+2N}(\Omega, \mathbb{C}^{N_+}) \end{array}$$

is continuous for every $k \in \mathbb{N}$ and all $s > -\frac{1}{2}$; $E^k = \langle t \rangle^{-k} H^k(\mathbb{R}_+)$. An analogous continuity holds for the formal adjoint $\mathrm{Op}(a_N)^*$ with respect to the scalar product in the corresponding L^2-spaces. Then $\mathrm{Op}(a) - \mathrm{Op}(\boldsymbol{a})$ has these continuity properties for all N and k, which yields $\mathrm{Op}(a) - \mathrm{Op}(\boldsymbol{a}) \in \mathcal{B}_G^{-\infty,0}(\mathbb{R}_+ \times \Omega; N_-, N_+)$.

$\square$

Remark 4.3.9 *To each $a(y, y', \eta) \in \mathcal{R}^{\mu,d}(\Omega \times \Omega \times \mathbb{R}^q; N_-, N_+)$ there is a $d(y', \eta) \in \mathcal{R}^{\mu,d}(\Omega \times \mathbb{R}^q; N_-, N_+)$ such that*

$$\mathrm{Op}(a) = \mathrm{Op}(d) \bmod \mathcal{B}^{-\infty,d}(\mathbb{R}_+ \times \Omega; N_-, N_+),$$

where $d(y', \eta)$ has the asymptotic expansion

$$d(y', \eta) \sim \sum_{\alpha} \frac{1}{\alpha!} (-D_y)^\alpha \partial_\eta^\alpha a(y, y', \eta)|_{y=y'}.$$

This can be obtained by analogous arguments to Proposition 4.3.8.

Remark 4.3.10 *Let $\varphi(y), \psi(y) \in C^\infty(\Omega)$ and set*

$$\Phi = \begin{pmatrix} \varphi & 0 \\ 0 & \varphi \cdot 1_{N_+} \end{pmatrix}, \qquad \Psi = \begin{pmatrix} \psi & 0 \\ 0 & \psi \cdot 1_{N_-} \end{pmatrix}$$

with 1_N being the $N \times N$-unit matrix. Then $\mathcal{A} \in \mathcal{B}^{\mu,d}(\mathbb{R}_+ \times \Omega; N_-, N_+)$ implies $\Phi \mathcal{A} \Psi \in \mathcal{B}^{\mu,d}(\mathbb{R}_+ \times \Omega; N_-, N_+)$. Moreover, for $\mathrm{supp}\, \varphi \cap \mathrm{supp}\, \psi = \emptyset$ we have $\Phi \mathcal{A} \Psi \in \mathcal{B}^{-\infty,d}(\mathbb{R}_+ \times \Omega; N_-, N_+)$.

For every $\mathcal{A} \in \mathcal{B}^{\mu,d}(\mathbb{R}_+ \times \Omega; N_-, N_+)$ we can define a continuous operator

$$K_{\mathcal{A}} : C_0^\infty(\Omega \times \Omega) \to \mathcal{L}(C_0^\infty(\overline{\mathbb{R}}_+) \oplus \mathbb{C}^{N_-}, C^\infty(\overline{\mathbb{R}}_+) \oplus \mathbb{C}^{N_+}) \qquad (4.3.7)$$

in the following way. First, if $\mathcal{C} \in \mathcal{B}^{-\infty,d}(\mathbb{R}_+ \times \Omega; N_-, N_+)$ is of the form (4.3.1), $\mathcal{C} = \begin{pmatrix} G & C \\ B & R \end{pmatrix}$, and $w(y, y') \in C_0^\infty(\Omega \times \Omega)$ we set

$$\langle K_{\mathcal{C}}, w \rangle = \begin{pmatrix} \langle K_G, w \rangle & \langle K_C, w \rangle \\ \langle K_B, w \rangle & \langle K_R, w \rangle \end{pmatrix}.$$

Here

$$\langle K_G, w \rangle(t, t') = \iint g(t, y, t', y') w(y, y') \, dy \, dy'$$

is to be identified with an operator $\langle K_G, w \rangle : C_0^\infty(\overline{\mathbb{R}}_+) \to C^\infty(\overline{\mathbb{R}}_+)$ by

$$\langle K_G, w \rangle u = \int \langle K_G, w \rangle(t, t') u(t') \, dt',$$

moreover,

$$\{\langle K_{B_k}, w\rangle(t')\}_{k=1,\ldots,N_+} = \left\{ \int b_k(y,t',y')w(y,y')\,dydy' \right\}_{k=1,\ldots,N_+}$$

with an operator $\langle K_B, w\rangle : C_0^\infty(\overline{\mathbb{R}}_+) \to \mathbb{C}^{N_+}$ by

$$(\langle K_B, w\rangle u)_k = \int \langle K_{B_k}, w\rangle(t')u(t')\,dt', k = 1,\ldots,N_+,$$

further

$$\langle K_{C_j}, w\rangle(t) = \iint c_j(t,y,y')w(y,y')\,dydy', \quad j = 1,\ldots,N_-$$

with an operator $\langle K_C, w\rangle(t)z = \sum_{j=1}^{N_-}\langle K_{C_j}, w\rangle(t)z_j$, $z = (z_1,\ldots,z_{N_-})$ $\in \mathbb{C}^{N_-}$, i.e. $\langle K_C, w\rangle : \mathbb{C}^{N_-} \to C^\infty(\overline{\mathbb{R}}_+)$, and finally

$$\langle K_R, w\rangle_k z = \sum_{j=1}^{N_-}\left\{ \iint r_{kj}(y,y')w(y,y')\,dydy' \right\} z_j, \quad k = 1,\ldots,N_+$$

with an operator $\langle K_R, w\rangle : \mathbb{C}^{N_-} \to \mathbb{C}^{N_+}$.

For $\mathcal{A} = \mathrm{Op}(a)$ with $a(y,y',\eta) \in \mathcal{R}^{\mu,d}(\Omega \times \Omega \times \mathbb{R}^q; N_-, N_+)$ we define

$$\langle K_\mathcal{A}, w\rangle = \iint e^{i(y-y')\eta}a(y,y',\eta)w(y,y')\,dydy'd\eta$$

in the sense of an oscillatory integral with an $\mathcal{L}(\mathcal{S}(\overline{\mathbb{R}}_+)\oplus\mathbb{C}^{N_-}, \mathcal{S}(\overline{\mathbb{R}}_+)\oplus \mathbb{C}^{N_+})$-valued amplitude function. Using the construction for $\mathcal{B}^{-\infty,d}(\mathbb{R}_+ \times\Omega; N_-, N_+)$ this gives us (4.3.7) in general. We have thus obtained the (operator–valued) distributional kernel of elements in $\mathcal{B}^{\mu,d}(\mathbb{R}_+ \times \Omega; N_-, N_+)$.

We call $\mathcal{A} \in \mathcal{B}^{\mu,d}(\mathbb{R}_+ \times \Omega; N_-, N_+)$ properly supported with respect to the y-variables if

$$K_\mathcal{A} \in \mathcal{D}'(\Omega \times \Omega, \mathcal{L}(C_0^\infty(\overline{\mathbb{R}}_+) \oplus \mathbb{C}^{N_-}, C^\infty(\overline{\mathbb{R}}_+) \oplus \mathbb{C}^{N_+}))$$

is properly supported.

Proposition 4.3.11 *Every $\mathcal{A} \in \mathcal{B}^{\mu,d}(\mathbb{R}_+ \times \Omega; N_-, N_+)$ can be written $\mathcal{A} = \mathcal{A}_0 + \mathcal{C}$ where $\mathcal{A}_0 \in \mathcal{B}^{\mu,d}(\mathbb{R}_+ \times \Omega; N_-, N_+)$ is properly supported with respect to the y-variables and $\mathcal{C} \in \mathcal{B}^{-\infty,d}(\mathbb{R}_+ \times \Omega; N_-, N_+)$.*

Proof. It suffices to consider $\mathcal{A} = \mathrm{Op}(a)$ with $a(y,\eta)$ as in Definition 4.3.2. Let $\omega(y,y') \in C^\infty(\Omega \times \Omega)$ be a function with proper support that equals 1 in an open neighbourhood of $\mathrm{diag}(\Omega \times \Omega)$. Then we may set $\mathcal{A}_0 = \mathrm{Op}(\omega a)$ if we verify that $\mathcal{C} = \mathrm{Op}((1 - \omega)a)$ belongs to $\mathcal{B}^{-\infty,d}(\mathbb{R}_+ \times \Omega; N_-, N_+)$. However the arguments in the proof of Proposition 4.3.8, in particular, the formula (4.3.5) show that $\mathcal{C} \in \mathcal{B}^{-k,d}(\mathbb{R}_+ \times \Omega; N_-, N_+)$ for every $k \in \mathbb{N}$ which yields the assertion. $\square$

Analogously to the scalar pseudo–differential calculus we have the following property:

Proposition 4.3.12 *Let* $\mathcal{A} \in \mathcal{B}^{\mu,d}(\mathbb{R}_+ \times \Omega; N_-, N_+)$ *be properly supported in the y-variables. Then* $\mathcal{A}$ *induces continuous operators*

$$
\mathcal{A}: \quad
\begin{array}{c}
H^s_{\mathrm{loc}(y)}(\mathbb{R}_+ \times \Omega) \\
\oplus \\
H^s_{\mathrm{loc}}(\Omega, \mathbb{C}^{N_-})
\end{array}
\quad \rightarrow \quad
\begin{array}{c}
H^{s-\mu}_{\mathrm{loc}(y)}(\mathbb{R}_+ \times \Omega) \\
\oplus \\
H^{s-\mu}_{\mathrm{loc}}(\Omega, \mathbb{C}^{N_+})
\end{array}
$$

for all $s \in \mathbb{R}$, $s > d - \frac{1}{2}$, *and also between the corresponding spaces with subscript "comp", and*

$$
\mathcal{A}: \quad
\begin{array}{c}
C^\infty(\overline{\mathbb{R}}_+ \times \Omega) \\
\oplus \\
C^\infty(\Omega, \mathbb{C}^{N_-})
\end{array}
\quad \rightarrow \quad
\begin{array}{c}
C^\infty(\overline{\mathbb{R}}_+ \times \Omega) \\
\oplus \\
C^\infty(\Omega, \mathbb{C}^{N_+})
\end{array}
$$

as well as between the corresponding C_0^∞*-spaces.*

Let us now turn to the principal symbolic structure of the operators $\mathcal{A} \in \mathcal{B}^{\mu,d}(\mathbb{R}_+ \times \Omega; N_-, N_+)$. In view of Remark 4.3.3 we have $A = \mathrm{u.\,l.\,c.}\,\mathcal{A} \in L^\mu_{\mathrm{cl}}(\mathbb{R}_+ \times \Omega)$; the (unique) homogeneous components $\sigma_\psi^{\mu-j}(A)(t, y, \tau, \eta)$ of order $\mu - j$, $j \in \mathbb{N}$, of any complete symbol of A are smooth in t up to $t = 0$. In particular, we set

$$
\sigma_\psi^\mu(\mathcal{A}) := \sigma_\psi^\mu(A) \tag{4.3.8}
$$

which belongs to $S^{(\mu)}(T^*(\overline{\mathbb{R}}_+ \times \Omega) \setminus 0)$.

Let us choose an $a(t, y, \tau, \eta) \in S^\mu_{\mathrm{cl}}(\overline{\mathbb{R}}_+ \times \Omega \times \mathbb{R}^{1+q})$ with the transmission property with $\sigma_\psi^{\mu-j}(A)$, $j \in \mathbb{N}$, as the sequence of homogeneous components of order $\mu - j$. Then $G := A - \mathrm{Op}(a)$ belongs to $\mathcal{B}_G^{\mu,d}(\mathbb{R}_+ \times \Omega)$. Thus $G = \mathrm{Op}(g) + C$ for certain $g(y, \eta) \in R_G^{\mu,d}(\Omega \times \mathbb{R}^q)$ and $C \in \mathcal{B}^{-\infty,d}(\mathbb{R}_+ \times \Omega)$. We now obtain unique homogeneous components $\sigma_\wedge^{\mu-j}(g)(y, \eta)$ of order $\mu - j$, $j \in \mathbb{N}$, of g that are independent of the particular choice of g. We then set

$$
\sigma_\wedge^\mu(G) = \sigma_\wedge^\mu(g),
$$

i.e., in particular, $\sigma_\wedge^\mu(C) = 0$, and define

$$
\sigma_\wedge^\mu(A)(y, \eta) = \mathrm{op}^+(\sigma_\psi^\mu(A)|_{t=0})(y, \eta) + \sigma_\wedge^\mu(G)(y, \eta) \tag{4.3.9}
$$

for $(y, \eta) \in \Omega \times (\mathbb{R}^q \setminus \{0\})$. By definition we have

$$
\mathcal{A} = \begin{pmatrix} A & 0 \\ 0 & 0 \end{pmatrix} + \mathrm{Op}(h) + \mathcal{C}
$$

for a symbol $h(y,\eta) \in \mathcal{R}_G^{\mu,d}(\Omega \times \mathbb{R}^q; N_-, N_+)$ and $\mathcal{C} \in \mathcal{B}^{-\infty,d}(\mathbb{R}_+ \times \Omega; N_-, N_+)$. Using the notation of (4.2.25) we can form the homogeneous principal symbol $\sigma_\wedge^\mu(h)(y,\eta)$ of order μ. Setting $\sigma_\wedge^\mu(\mathcal{C}) = 0$ we finally define

$$\sigma_\wedge^\mu(\mathcal{A})(y,\eta) = \begin{pmatrix} \sigma_\wedge^\mu(A)(y,\eta) & 0 \\ 0 & 0 \end{pmatrix} + \sigma_\wedge^\mu(h)(y,\eta).$$

This is an operator family, parametrised by $(y,\eta) \in \Omega \times (\mathbb{R}^q \setminus \{0\})$,

$$\sigma_\wedge^\mu(\mathcal{A})(y,\eta) : \begin{matrix} H^s(\mathbb{R}_+) \\ \oplus \\ \mathbb{C}^{N_-} \end{matrix} \to \begin{matrix} H^{s-\mu}(\mathbb{R}_+) \\ \oplus \\ \mathbb{C}^{N_+} \end{matrix} \quad \text{for} \quad s > d - \frac{1}{2}$$

or

$$\sigma_\wedge^\mu(\mathcal{A})(y,\eta) : \begin{matrix} \mathcal{S}(\overline{\mathbb{R}}_+) \\ \oplus \\ \mathbb{C}^{N_-} \end{matrix} \to \begin{matrix} \mathcal{S}(\overline{\mathbb{R}}_+) \\ \oplus \\ \mathbb{C}^{N_+} \end{matrix}.$$

Recall that the homogeneity μ in the operator–valued sense means

$$\sigma_\wedge^\mu(\mathcal{A})(y,\lambda\eta) = \lambda^\mu \begin{pmatrix} \kappa_\lambda & 0 \\ 0 & 1 \end{pmatrix} \sigma_\wedge^\mu(\mathcal{A})(y,\eta) \begin{pmatrix} \kappa_\lambda & 0 \\ 0 & 1 \end{pmatrix}^{-1}$$

for all $\lambda \in \mathbb{R}_+$ and $(y,\eta) \in \Omega \times (\mathbb{R}^q \setminus \{0\})$.

Definition 4.3.13 *Let $\mathcal{A} \in \mathcal{B}^{\mu,d}(\mathbb{R}_+ \times \Omega; N_-, N_+)$, $\mu \in \mathbb{Z}$, $d \in \mathbb{N}$. Then $\sigma_\psi^\mu(\mathcal{A})(t,y,\tau,\eta)$ is called the homogeneous principal interior symbol of $\mathcal{A}$ of order μ and $\sigma_\wedge^\mu(\mathcal{A})(y,\eta)$ the homogeneous boundary symbol of $\mathcal{A}$ of order μ.*

Let us set

$$\Sigma^{\mu,d}(N_-, N_+) = \left\{ (\sigma_\psi^\mu(\mathcal{A}), \sigma_\wedge^\mu(\mathcal{A})) : \mathcal{A} \in \mathcal{B}^{\mu,d}(\mathbb{R}_+ \times \Omega; N_-, N_+) \right\}.$$

Here and in the sequel we shall assume for simplicity that the operators in $\mathcal{B}^{\mu,d}(\mathbb{R}_+ \times \Omega; N_-, N_+)$ only contain symbols $a(t,y,\tau,\eta)$ with the transmission property for which all homogeneous components of order $\mu - j$, $j \in \mathbb{N}$, are independent of t for $t > 1$. This is convenient for several Fréchet topologies, though the assumption is not very essential, since in the global calculus below the operators will be cut off outside a neighbourhood of the boundary. In order to avoid additional notation here we shall simply denote by $\mathcal{B}^{\mu,d}(\mathbb{R}_+ \times \Omega; N_-, N_+)$ the space in the sense of Definition 4.3.2 with this condition on the symbols. By construction we have the compatibility condition (4.3.9) between the components of the elements in $\Sigma^{\mu,d}(N_-, N_+)$. The subspace

$$\Sigma_G^{\mu,d}(N_-, N_+) := \left\{ (0, \sigma_\wedge^\mu(\mathcal{G}),) : \mathcal{G} \in \mathcal{B}_G^{\mu,d}(\mathbb{R}_+ \times \Omega; N_-, N_+) \right\},$$

has a natural Fréchet topology, cf. Remark 4.2.28. We get a surjective map $\pi : \Sigma^{\mu,d}(N_-, N_+) \to \Sigma_G^{\mu,d}(N_-, N_+)$ by setting

$$\pi(\sigma_\psi^\mu(\mathcal{A}), \sigma_\wedge^\mu(\mathcal{A})) = \left(0, \begin{pmatrix} \sigma_\wedge^\mu(G_{11}) & \sigma_\wedge^\mu(G_{12}) \\ \sigma_\wedge^\mu(G_{21}) & \sigma_\wedge^\mu(G_{22}) \end{pmatrix} \right),$$

where

$$\sigma_\wedge^\mu(\mathcal{A}) = \begin{pmatrix} \mathrm{op}^+(\sigma_\psi^\mu(A)|_{t=0}) + \sigma_\wedge^\mu(G_{11}) & \sigma_\wedge^\mu(G_{12}) \\ \sigma_\wedge^\mu(G_{21}) & \sigma_\wedge^\mu(G_{22}) \end{pmatrix},$$

$A = \mathrm{u.\,l.\,c.}\, A$. Denoting by $S_{\mathrm{trans}}^{(\mu)}(T^*(\overline{\mathbb{R}}_+ \times \Omega) \setminus 0)$ the subspace of all elements in $S^{(\mu)}(T^*(\overline{\mathbb{R}}_+ \times \Omega) \setminus 0)$ which occur as the homogeneous components of order μ of symbols in $S_{\mathrm{cl}}^\mu(\overline{\mathbb{R}}_+ \times \mathbb{R}^q \times \mathbb{R}^{1+q})$ with the transmission property (that are independent of t for $t > 1$) we get $\ker \pi \cong S_{\mathrm{trans}}^{(\mu)}(T^*(\overline{\mathbb{R}}_+ \times \Omega) \setminus 0)$. From the algebraic isomorphism

$$\Sigma^{\mu,d}(N_-, N_+) \cong S_{\mathrm{trans}}^{(\mu)}(T^*(\overline{\mathbb{R}}_+ \times \Omega) \setminus 0) \oplus \Sigma_G^{\mu,d}(N_-, N_+) \quad (4.3.10)$$

and the canonical Fréchet topology in $S_{\mathrm{trans}}^{(\mu)}(T^*(\overline{\mathbb{R}}_+ \times \Omega) \setminus 0)$ (induced by $S^{(\mu)}(T^*(\overline{\mathbb{R}}_+ \times \Omega) \setminus 0)$) we get a Fréchet topology also in $\Sigma^{\mu,d}(N_-, N_+)$.

Proposition 4.3.14 *We have an exact sequence*

$$0 \to \mathcal{B}^{\mu-1,d}(\mathbb{R}_+ \times \Omega; N_-, N_+) \overset{\iota}{\to} \mathcal{B}^{\mu,d}(\mathbb{R}_+ \times \Omega; N_-, N_+)$$
$$\overset{\sigma}{\to} \Sigma^{\mu,d}(N_-, N_+) \to 0, \quad (4.3.11)$$

where ι is the canonical embedding, and $\sigma = (\sigma_\psi^\mu, \sigma_\wedge^\mu)$. There is a map

$$\mathrm{op} : \Sigma^{\mu,d}(N_-, N_+) \to \mathcal{B}^{\mu,d}(\mathbb{R}_+ \times \Omega; N_-, N_+) \quad (4.3.12)$$

with $\sigma \circ \mathrm{op} = \mathrm{id}$ on $\Sigma^{\mu,d}(N_-, N_+)$.

Proof. The exactness of the sequence (4.3.11) is obvious by construction. It remains to construct the mapping op. First, for $(p_\psi, p_\wedge) \in \Sigma^{\mu,d}(N_-, N_+)$ we have $p_G := p_\wedge - \mathrm{op}^+(p_\psi|_{t=0}) \in \Sigma_G^{\mu,d}(N_-, N_+)$; here, for convenience,

$$\mathrm{op}^+(p_\psi|_{t=0}) \quad (4.3.13)$$

is identified with a corresponding block matrix that has (4.3.13) as upper left corner and zeros as the remaining entries. Analogously, if $\chi(\tau, \eta)$ is an excision function in $\mathbb{R}^{1+q}$, we also interpret $\mathrm{Op}(\mathrm{op}^+(\chi p_\psi))$ as a corresponding block matrix. For any excision function $\psi(\eta)$ in $\mathbb{R}^q$ we now set

$$\mathrm{op}(p_\psi, p_\wedge) = \mathrm{Op}(\mathrm{op}^+(\chi p_\psi)) + \mathrm{Op}(\psi p_G).$$

Then $\mathrm{op}(p_\psi, p_\wedge) \in \mathcal{B}^{\mu, d}(\mathbb{R}_+ \times \Omega; N_-, N_+)$, and $\sigma \, \mathrm{op}(p_\psi, p_\wedge) = (p_\psi, p_\wedge)$. □

The map (4.3.12) is also called an operator convention. Note that the choice of the excision functions χ, ψ is arbitrary. If we form (4.3.12) in terms of another pair χ', ψ' of this kind and denote the resulting map by op', then, as a consequence of (4.3.11), we obtain

$$\mathrm{op}(p_\psi, p_\wedge) - \mathrm{op}'(p_\psi, p_\wedge) \in \mathcal{B}^{\mu-1, d}(\mathbb{R}_+ \times \Omega; N_-, N_+).$$

For each $(p_\psi, p_\wedge) \in \Sigma^{\mu, d}(N_-, N_+)$ we can define the formal adjoint $(p_\psi, p_\wedge)^* := (p_\psi^*, p_\wedge^*)$ by $p_\psi^* = \overline{p}_\psi$ (the complex conjugate) and

$$(u, p_\wedge^* v)_{L^2(\mathbb{R}_+) \oplus \mathbb{C}^{N_-}} = (p_\wedge u, v)_{L^2(\mathbb{R}_+) \oplus \mathbb{C}^{N_+}}$$

for all $u \in C_0^\infty(\mathbb{R}_+) \oplus \mathbb{C}^{N_-}$, $v \in C_0^\infty(\mathbb{R}_+) \oplus \mathbb{C}^{N_+}$. Moreover, for each $(p_\psi, p_\wedge) \in \Sigma^{\mu, d}(N_0, N_+)$, $(r_\psi, r_\wedge) \in \Sigma^{\nu, c}(N_-, N_0)$ we can form $(p_\psi, p_\wedge)(r_\psi, r_\wedge) = (p_\psi r_\psi, p_\wedge r_\wedge)$, where the second component is the composition of operator families.

Proposition 4.3.15 $(p_\psi, p_\wedge) \in \Sigma^{0,0}(N_-, N_+)$ *implies the relation* $(p_\psi, p_\wedge)^* \in \Sigma^{0,0}(N_+, N_-)$.

Moreover, $(p_\psi, p_\wedge) \in \Sigma^{\mu, d}(N_0, N_+)$, $(r_\psi, r_\wedge) \in \Sigma^{\nu, c}(N_-, N_0)$ *implies* $(p_\psi r_\psi, p_\wedge r_\wedge) \in \Sigma^{\mu+\nu, h}(N_-, N_+)$ *for* $h = \max\{(d+\nu)^+, c\}$, *where* $\rho^+ = \max\{\rho, 0\}$.

This is a consequence of Proposition 4.1.78.

For each operator $\mathcal{A} \in \mathcal{B}^{\mu, d}(\mathbb{R}_+ \times \Omega; N_-, N_+)$ we can define the formal adjoint $\mathcal{A}^*$ by

$$(u, \mathcal{A}^* v)_{L^2(\mathbb{R}_+ \times \Omega) \oplus L^2(\Omega, \mathbb{C}^{N_-})} = (\mathcal{A}u, v)_{L^2(\mathbb{R}_+ \times \Omega) \oplus L^2(\Omega, \mathbb{C}^{N_+})}$$

for all $u \in C_0^\infty(\mathbb{R}_+ \times \Omega) \oplus C_0^\infty(\Omega, \mathbb{C}^{N_-})$, $v \in C_0^\infty(\mathbb{R}_+ \times \Omega) \oplus C_0^\infty(\Omega, \mathbb{C}^{N_+})$.

Theorem 4.3.16 $\mathcal{A} \in \mathcal{B}^{0,0}(\mathbb{R}_+ \times \Omega; N_-, N_+)$ *implies* $\mathcal{A}^* \in \mathcal{B}^{0,0}(\mathbb{R}_+ \times \Omega; N_+, N_-)$, *and we have*

$$\sigma_\psi^0(\mathcal{A}^*) = \sigma_\psi^0(\mathcal{A})^*, \qquad \sigma_\wedge^0(\mathcal{A}^*) = \sigma_\wedge^0(\mathcal{A})^*. \tag{4.3.14}$$

Proof. For $\mathcal{A} \in \mathcal{B}^{-\infty,0}(\mathbb{R}_+ \times \Omega; N_-, N_+)$ the relation $\mathcal{A}^* \in \mathcal{B}^{-\infty,0}(\mathbb{R}_+ \times \Omega; N_+, N_-)$ is obvious. Thus we may assume $\mathcal{A} = \mathrm{Op}(a)$ for an $a(y,\eta) \in \mathcal{R}^{0,0}(\Omega \times \mathbb{R}^q; N_-, N_+)$. It is then clear that $\mathcal{A}^* = \mathrm{Op}(a^*)$ for $a^*(y',\eta) \in \mathcal{R}^{0,0}(\Omega \times \mathbb{R}^q; N_+, N_-)$, cf. also Theorem 4.2.45. In view of Proposition 4.3.8 we get $\mathcal{A}^* \in \mathcal{B}^{0,0}(\mathbb{R}_+ \times \Omega; N_+, N_-)$. The relation (4.3.4), here applied to $a^*(y',\eta)$, gives us a symbol $\boldsymbol{a}^*(y,\eta)$ with $\mathcal{A}^* = \mathrm{Op}(\boldsymbol{a}^*)$ mod $\mathcal{B}^{-\infty,0}(\mathbb{R}_+ \times \Omega; N_+, N_-)$. For (4.3.14) it suffices now to observe $\sigma_\psi^0(\boldsymbol{a}^*) = \overline{\sigma_\psi^0(a)}$, $\sigma_\wedge^0(\boldsymbol{a}^*) = \sigma_\wedge^0(a)^*$.   □

Theorem 4.3.17 *Let* $\mathcal{A} \in \mathcal{B}^{\mu,d}(\mathbb{R}_+ \times \Omega; N_0, N_+)$, $\mathcal{B} \in \mathcal{B}^{\nu,c}(\mathbb{R}_+ \times \Omega; N_-, N_0)$, *and*

$$\Phi = \begin{pmatrix} \varphi & 0 \\ 0 & \varphi \cdot 1_{N_0} \end{pmatrix}$$

for arbitrary $\varphi \in C_0^\infty(\Omega)$ *with* 1_{N_0} *being the* $N_0 \times N_0$ *unit matrix. Then* $\mathcal{A}\Phi\mathcal{B} \in \mathcal{B}^{\mu+\nu,h}(\mathbb{R}_+ \times \Omega; N_-, N_+)$ *for* $h = \max\{(d+\nu)^+, c\}$, *and we have*

$$\sigma_\psi^{\mu+\nu}(\mathcal{A}\Phi\mathcal{B}) = \sigma_\psi^\mu(\mathcal{A})\varphi\sigma_\psi^\nu(\mathcal{B}), \quad \sigma_\wedge^{\mu+\nu}(\mathcal{A}\Phi\mathcal{B}) = \sigma_\wedge^\mu(\mathcal{A})\Phi\sigma_\wedge^\nu(\mathcal{B}).$$

Moreover, $\mathcal{A} \in \mathcal{B}_G^{\mu,d}(\mathbb{R}_+ \times \Omega; N_0, N_-)$ *or* $\mathcal{B} \in \mathcal{B}_G^{\nu,c}(\mathbb{R}_+ \times \Omega; N_-, N_0)$ *implies* $\mathcal{A}\Phi\mathcal{B} \in \mathcal{B}_G^{\mu+\nu,h}(\mathbb{R}_+ \times \Omega; N_-, N_+)$.

Proof. Let us write $\mathcal{A} = \mathrm{Op}(a) + \mathcal{C}_1$, $\mathcal{B} = \mathrm{Op}(b) + \mathcal{C}_2$, for symbols $a(y,\eta)$ and $b(y,\eta)$ of the classes $\mathcal{R}^{\mu,d}$ and $\mathcal{R}^{\nu,c}$, respectively, and smoothing operators $\mathcal{C}_1, \mathcal{C}_2$. Set $\Phi(y)b(y,\eta) = \tilde{b}(y,\eta)$ which is again a symbol in $\mathcal{R}^{\nu,c}$. Applying Remark 4.3.9 we find a symbol $d(y',\eta) \in \mathcal{R}^{\nu,c}$ with $\mathrm{Op}(b) = \mathrm{Op}(d) + \mathcal{D}$ for some smoothing operator $\mathcal{D}$. Thus $\mathcal{A}\Phi\mathcal{B} = \{\mathrm{Op}(a) + \mathcal{C}_1\}\{\mathrm{Op}(d) + \mathcal{D} + \Phi\mathcal{C}_2\}$.

The summands containing smoothing factors are again smoothing. So it remains to look at $\mathrm{Op}(a)\mathrm{Op}(d) = \mathrm{Op}(c)$ for $c(y,y',\eta) = a(y,\eta)d(y',\eta)$, where $c(y,y',\eta) \in \mathcal{R}^{\mu+\nu,h}(\Omega \times \Omega \times \mathbb{R}^q; N_-, N_+)$. From Proposition 4.3.8 and Theorem 4.2.46 we obtain $\mathrm{Op}(c) \in \mathcal{B}^{\mu+\nu,h}(\mathbb{R}_+ \times \Omega; N_-, N_+)$. Moreover, the asymptotic formula (4.3.4) gives us the asserted symbol rules. If one factor belongs to the class with subscript G then $c(y,y',\eta) \in \mathcal{R}_G^{\mu+\nu,h}$, cf. Theorem 4.2.46, and hence $\mathcal{A}\Phi\mathcal{B} \in \mathcal{B}_G^{\mu+\nu,h}(\mathbb{R}_+ \times \Omega; N_-, N_+)$.   □

Note that, by the same arguments as above for the proof of Theorem 1.1.32 , we obtain $\mathcal{A}\Phi\mathcal{B} = \mathrm{Op}(\boldsymbol{c}) + \mathcal{C}$ for some smoothing $\mathcal{C}$ and a symbol $\boldsymbol{c}(y,\eta) \in \mathcal{R}^{\mu+\nu,h}$ that has the asymptotic expansion

$$\boldsymbol{c}(y,\eta) \sim \sum_\alpha \frac{1}{\alpha!}(D_\eta^\alpha a(y,\eta))\partial_y^\alpha(\Phi b)(y,\eta)$$

which is the Leibniz product.

Remark 4.3.18 $\mathcal{A} \in \mathcal{B}^{\mu,d}(\mathbb{R}_+ \times \Omega; N_-, N_+)$ *implies*

$$\begin{pmatrix} \varphi & 0 \\ 0 & \psi \cdot 1_{N_+} \end{pmatrix} \mathcal{A}, \, \mathcal{A} \begin{pmatrix} \varphi & 0 \\ 0 & \psi \cdot 1_{N_-} \end{pmatrix} \in \mathcal{B}^{\mu,d}(\mathbb{R}_+ \times \Omega; N_-, N_+)$$

for arbitrary $\varphi \in C_0^\infty(\overline{\mathbb{R}}_+ \times \Omega)$, $\psi \in C_0^\infty(\Omega)$.

Let $\Omega, \tilde{\Omega} \subseteq \mathbb{R}^q$ be open sets and $\chi : \Omega \to \tilde{\Omega}$ be a diffeomorphism. Then to every $\mathcal{A} \in \mathcal{B}^{\mu,d}(\mathbb{R}_+ \times \Omega; N_-, N_+)$ we can form the operator push-forward $\chi_* \mathcal{A}$, with $\mathcal{A}$ being interpreted in the sense (4.3.3), and set

$$\chi_* \mathcal{A} = (\chi^*)^{-1} \mathcal{A} \chi^* : \quad \begin{matrix} C_0^\infty(\tilde{\Omega}, \mathcal{S}(\overline{\mathbb{R}}_+)) \\ \oplus \\ C_0^\infty(\tilde{\Omega}, \mathbb{C}^{N_-}) \end{matrix} \quad \to \quad \begin{matrix} C^\infty(\tilde{\Omega}, \mathcal{S}(\overline{\mathbb{R}}_+)) \\ \oplus \\ C^\infty(\tilde{\Omega}, \mathbb{C}^{N_+}) \end{matrix}$$

with the function pull-backs

$$\chi^* : \quad \begin{matrix} C_0^\infty(\tilde{\Omega}, \mathcal{S}(\overline{\mathbb{R}}_+)) \\ \oplus \\ C_0^\infty(\tilde{\Omega}, \mathbb{C}^{N_-}) \end{matrix} \quad \xrightarrow{\cong} \quad \begin{matrix} C_0^\infty(\Omega, \mathcal{S}(\overline{\mathbb{R}}_+)) \\ \oplus \\ C_0^\infty(\Omega, \mathbb{C}^{N_-}) \end{matrix}$$

applied component-wise with respect to the y-variables, and similarly between the corresponding C^∞ spaces.

Theorem 4.3.19 *Let* $\chi : \Omega \to \tilde{\Omega}$ *be a diffeomorphism between open sets* $\Omega, \tilde{\Omega} \subseteq \mathbb{R}^q$. *Then* $\mathcal{A} \in \mathcal{B}^{\mu,d}(\mathbb{R}_+ \times \Omega; N_-, N_+)$ *implies* $\chi_* \mathcal{A} \in \mathcal{B}^{\mu,d}(\mathbb{R}_+ \times \tilde{\Omega}; N_-, N_+)$, *and we have*

$$\sigma_\psi^\mu(\chi_* \mathcal{A})(t, \tilde{y}, \tau, \tilde{\eta}) = \sigma_\psi^\mu(\mathcal{A})(t, y, \tau, \eta), \qquad (4.3.15)$$

$$\sigma_\wedge^\mu(\chi_* \mathcal{A})(\tilde{y}, \tilde{\eta}) = \sigma_\wedge^\mu(\mathcal{A})(y, \eta) \qquad (4.3.16)$$

for $\tilde{y} = \chi(y)$, $\tilde{\eta} = ({}^t d\chi(y))^{-1} \eta$, *with* $d\chi(y)$ *being the Jacobian of* χ, *and upper* t *indicating the transposed matrix.*

Proof. The smoothing operators are obviously invariant under χ_*. Thus we may assume $\mathcal{A} = \mathrm{Op}(a)$ for an $a(y, \eta) \in \mathcal{R}^{\mu,d}(\Omega \times \mathbb{R}^q; N_-, N_+)$. The arguments of the proof of Theorem 1.1.38 can also be applied in the present situation and yield, in particular, for $\chi_* \mathcal{A}$ an amplitude function $b(\tilde{y}, \tilde{y}', \eta) \in \mathcal{R}^{\mu,d}(\tilde{\Omega} \times \tilde{\Omega} \times \mathbb{R}^q; N_-, N_+)$ such that $\chi_* \mathcal{A} = \mathrm{Op}(b)$. This shows $\chi_* \mathcal{A} \in \mathcal{B}^{\mu,d}(\mathbb{R}_+ \times \tilde{\Omega}; N_-, N_+)$. Moreover, we can apply the asymptotic formula for a complete symbol

$$b(\tilde{y}, \eta)|_{\tilde{y}=\chi(y)} \sim \sum \frac{1}{\alpha!} (\partial_\eta^\alpha a)(y, {}^t d\chi(y)\eta) \Phi_\alpha(y, \eta),$$

cf. (1.1.72), which yields, in particular, the asserted symbol rules (4.3.15), (4.3.16).

□

Remark 4.3.20 *It can be proved that* $\mathcal{B}^{\mu,d}(\mathbb{R}_+ \times \Omega; N_-, N_+)$ *is invariant also under every diffeomorphism* $\mathbb{R}_+ \times \Omega \to \mathbb{R}_+ \times \tilde{\Omega}$, $(t,y) \to (\tilde{t}, \tilde{y})$, *that extends to a diffeomorphism* $(-\varepsilon, \infty) \times \Omega \to (-\tilde{\varepsilon}, \infty) \times \tilde{\Omega}$ *for certain* $\varepsilon, \tilde{\varepsilon} > 0$ *and for which* $\tilde{t}(t,y) = t$ *for* $t > c$ *for a constant* $t > 0$. *For* $\mathrm{Op}(\mathrm{op}^+(p))$ *for a symbol* $p(t,y,\tau,\eta)$ *with the transmission property (that is t-independent for large t) this is very easy to show, while for the operators in* $\mathcal{B}_G^{\mu,d}(\mathbb{R}_+ \times \Omega; N_-, N_+)$ *the invariance follows without much effort by the explicit kernel characterisations of* $\mathcal{R}_G^{\mu,d}(\Omega \times \mathbb{R}^q; N_-, N_+)$ *of Section 4.2.3 and by the obvious invariance of* $\mathcal{B}^{-\infty,d}(\mathbb{R}_+ \times \Omega; N_-, N_+)$ *under such diffeomorphisms.*

Remark 4.3.21 *Definition 4.3.2 can easily be extended to the case where the upper left corners are* $l \times k$-*matrices of elements in* $\mathcal{B}^{\mu,d}(\mathbb{R}_+ \times \Omega)$ *and the trace and potential operators are* $N_+ \times k$ *and* $l \times N_-$-*matrices respectively. We obtain in this way a class* $\mathcal{B}^{\mu,d}(\mathbb{R}_+ \times \Omega; k, l; N_-, N_+)$ *of operators*

$$
\mathcal{A}: \quad
\begin{matrix}
H^s_{\mathrm{comp}(y)}(\mathbb{R}_+ \times \Omega, \mathbb{C}^k) \\
\oplus \\
H^s_{\mathrm{comp}}(\Omega, \mathbb{C}^{N_-})
\end{matrix}
\quad \to \quad
\begin{matrix}
H^{s-\mu}_{\mathrm{loc}(y)}(\mathbb{R}_+ \times \Omega, \mathbb{C}^l) \\
\oplus \\
H^{s-\mu}_{\mathrm{loc}}(\Omega, \mathbb{C}^{N_+})
\end{matrix}
\quad ,
$$

$s > d - \frac{1}{2}$. *The above operator calculus has a straightforward extension to this more general case. It will be tacitly used below.*

4.3.2 Boundary value problems on a C^∞ manifold with boundary

We now turn to boundary value problems on a (paracompact) C^∞ manifold X with C^∞ boundary Y. Throughout the exposition we assume that there is fixed a Riemannian metric on X with the associated measure dx and the induced Riemannian metric on Y with the associated measure dy. For convenience we assume that the Riemannian metric on X induces the product metric on a collar neighbourhood $V \cong [0,1) \times Y$ of Y, with the Lebesgue measure dt on the interval. We adopt here the general notation from Section 1.1.5 on complex C^∞ vector bundles on the manifolds in question and the associated spaces of distributional sections. Concerning the definition of the Sobolev spaces $H^s_{\mathrm{comp}}(X,E)$, $H^s_{\mathrm{loc}}(X,E)$ for $E \in \mathrm{Vect}(X)$, $s \in \mathbb{R}$, we can first pass to the double $2X$ of X by glueing together two copies of X along Y which is then a C^∞ manifold without boundary and then set

$$
H^s_{\mathrm{comp}}(X,E) = H^s_{\mathrm{comp}}(2X, 2E)|_{\mathrm{int}\, X},
$$
$$
H^s_{\mathrm{loc}}(X,E) = H^s_{\mathrm{loc}}(2X, 2E)|_{\mathrm{int}\, X},
$$

where $2E$ denotes a bundle in $\mathrm{Vect}(2X)$ that restricts to E over X. In addition we have the spaces

$$H^s_{0,\mathrm{comp}}(X, E) = \left\{ u \in H^s_{\mathrm{comp}}(2X, 2E) : \operatorname{supp} u \subseteq X \right\}$$

and analogously the corresponding $H^s_{0,\mathrm{loc}}(X, E)$ spaces. For compact X with boundary we may omit the subscripts comp, loc and obtain the spaces

$$H^s(X, E) \quad \text{and} \quad H^s_0(X, E),$$

respectively. There is an isomorphism

$$H^s_{\mathrm{loc}}(X, E) \cong H^s_{\mathrm{loc}}(2X, 2E) / H^s_{0,\mathrm{loc}}(X_-, E_-)$$

for $X_- := 2X \setminus (\operatorname{int} X)$, $E_- := 2E|_{X_-}$, which turns $H^s_{\mathrm{loc}}(X, E)$ into a Fréchet space in a natural way. $H^s_{\mathrm{comp}}(X, E)$ is an inductive limit of Hilbert spaces.

We assume that each bundle $E \in \mathrm{Vect}(X)$ is endowed with a Hermitean metric. Then, for compact X we have the space

$$L^2(X, E) = H^0(X, E)$$

with a chosen scalar product. Moreover, we set

$$C^\infty(X, E) = C^\infty(2X, 2E)|_X,$$
$$C^\infty_0(X, E) = C^\infty_0(2X, 2E)|_X.$$

Also for non–compact X we can employ associated local L^2-scalar products when the functions in question have compact support. If we have a continuous operator

$$\mathcal{C} : \begin{array}{ccc} C^\infty_0(\operatorname{int} X, E) & & C^\infty(\operatorname{int} X, F) \\ \oplus & \to & \oplus \\ C^\infty_0(Y, J^-) & & C^\infty(Y, J^+) \end{array} \qquad (4.3.17)$$

for $E, F \in \mathrm{Vect}(X)$, $J^-, J^+ \in \mathrm{Vect}(Y)$, this allows us to define the formal adjoint $\mathcal{C}^*$ in the sense

$$(u, \mathcal{C}^* v)_{L^2_{\mathrm{loc}}(X,E) \oplus L^2_{\mathrm{loc}}(Y,J^-)} = (\mathcal{C}u, v)_{L^2_{\mathrm{loc}}(X,F) \oplus L^2_{\mathrm{loc}}(Y,J^+)} \qquad (4.3.18)$$

for all $u \in C^\infty_0(\operatorname{int} X, E) \oplus C^\infty_0(Y, J^-)$, $v \in C^\infty_0(\operatorname{int} X, F) \oplus C^\infty_0(Y, J^+)$.

For $E, F \in \mathrm{Vect}(X)$, $J^-, J^+ \in \mathrm{Vect}(Y)$, $Y = \partial X$, we denote by $\mathcal{B}^{-\infty,0}(X; E, F; J^-, J^+)$ the set of all operators (4.3.17) for which $\mathcal{C}$

and $\mathcal{C}^*$ extend to continuous operators

$$
\mathcal{C} : \quad
\begin{array}{ccc}
C_0^\infty(X, E) & & C^\infty(X, F) \\
\oplus & \to & \oplus \\
C_0^\infty(Y, J^-) & & C^\infty(Y, J^+)
\end{array}
$$

$$
\mathcal{C}^* : \quad
\begin{array}{ccc}
C_0^\infty(X, F) & & C^\infty(X, E) \\
\oplus & \to & \oplus \\
C_0^\infty(Y, J^+) & & C^\infty(Y, J^-)
\end{array}
$$

and

$$
\mathcal{C} : \quad
\begin{array}{ccc}
H_{\mathrm{comp}}^s(X, E) & & H_{\mathrm{loc}}^\infty(X, F) \\
\oplus & \to & \oplus \quad (Y, J^+) \\
H_{\mathrm{comp}}^s(Y, J^-) & & H_{\mathrm{loc}}^\infty
\end{array}
$$

$$
\mathcal{C}^* : \quad
\begin{array}{ccc}
H_{\mathrm{comp}}^s(X, F) & & H_{\mathrm{loc}}^\infty(X, E) \\
\oplus & \to & \oplus \\
H_{\mathrm{comp}}^s(Y, J^+) & & H_{\mathrm{loc}}^\infty(Y, J^-)
\end{array}
$$

for all $s \in \mathbb{R}$, $s > -\frac{1}{2}$. Moreover, we define $\mathcal{B}^{-\infty,d}(X; E, F; J^-, J^+)$ for $d \in \mathbb{N}$ as the set of all operators

$$
\mathcal{C} = \sum_{k=0}^d \mathcal{C}_k \begin{pmatrix} T_k & 0 \\ 0 & \mathrm{id}_{J^-} \end{pmatrix}
$$

for arbitrary $\mathcal{C}_k \in \mathcal{B}^{-\infty,0}(X; E, F; J^-, J^+)$, $k = 0, \ldots, d$, where T_k is an arbitrary differential operator on X of order k with smooth coefficients which equals $\partial^k / \partial t^k \cdot \mathrm{id}_E$ in a collar neighbourhood V of Y.

The elements in $\mathcal{B}^{-\infty,d}(X; E, F; J^-, J^+)$ are called the smoothing operators of type d in Boutet de Monvel's algebra on X.

Next we choose a locally finite open covering $\mathcal{U} = \{U_j\}_{j \in \mathbb{Z}}$ of X by coordinate neighbourhoods. Then $\mathcal{U}' = \{U_j'\}_{j \in \mathbb{Z}}$ for $U_j' = U_j \cap Y$ is a locally finite open covering of Y by coordinate neighbourhoods. To $\mathcal{U}$ we fix a subordinate partition of unity $\{\varphi_j\}_{j \in \mathbb{Z}}$. Moreover, we choose a system $\{\psi_j\}_{j \in \mathbb{Z}}$ of functions $\psi_j \in C_0^\infty(U_j)$ with $\varphi_j \psi_j = \varphi_j$ for all j.

Let $\chi_j : U_j \to \overline{\mathbb{R}}_+ \times \Omega_j$ be a system of charts for those j where $U_j \cap Y \neq \emptyset$, $\Omega_j \subseteq \mathbb{R}^q$ open, $\dim X = 1 + q$. Then $\chi_j' : U_j' \to \Omega_j$ is a system of charts for Y. Given $E, F \in \mathrm{Vect}(X)$, $J^\pm \in \mathrm{Vect}(Y)$ we choose trivialisations

$$
\tau_U : E|_U \to \overline{\mathbb{R}}_+ \times \Omega \times \mathbb{C}^k,
$$

$$
\rho_U : F|_U \to \overline{\mathbb{R}}_+ \times \Omega \times \mathbb{C}^l
$$

for every $U \in \mathcal{U}$ with $U \cap Y \neq \emptyset$, where k and l are the fibre dimensions of E and F, respectively, and

$$
\delta_{U'}^\pm : J^\pm|_{U'} \to \Omega \times \mathbb{C}^{N\pm}
$$

for $U' = U \cap Y$, with $N_\pm$ being the fibre dimension of $J^\pm$. Moreover, we assume that the trivialisations are compatible with the chosen charts $\chi : U \to \overline{\mathbb{R}}_+ \times \Omega$ and $\chi' : U' \to \Omega$, respectively (i.e., the charts are induced by restricting the trivialisations to the corresponding zero sections). Then, to every $\mathcal{A} \in \mathcal{B}^{\mu,d}(\overline{\mathbb{R}}_+ \times \Omega; k, l; N_-, N_+)$, cf. Remark 4.3.21, first regarded as operator

$$
\mathcal{A} : \begin{array}{ccc} C_0^\infty(\overline{\mathbb{R}}_+ \times \Omega, \mathbb{C}^k) & & C^\infty(\overline{\mathbb{R}}_+ \times \Omega, \mathbb{C}^l) \\ \oplus & \to & \oplus \\ C_0^\infty(\Omega, \mathbb{C}^{N_-}) & & C^\infty(\Omega, \mathbb{C}^{N_+}) \end{array}
$$

we can form

$$
\mathcal{A}_U := \begin{pmatrix} (\rho_U)_*^{-1} & 0 \\ 0 & (\delta_{U'}^+)_*^{-1} \end{pmatrix} \mathcal{A} \begin{pmatrix} (\tau_U)_* & 0 \\ 0 & (\delta_{U'}^-)_* \end{pmatrix}, \tag{4.3.19}
$$

$$
\mathcal{A}_U : \begin{array}{ccc} C_0^\infty(U, E|_U) & & C^\infty(U, F|_U) \\ \oplus & \to & \oplus \\ C_0^\infty(U', J^-|_{U'}) & & C^\infty(U', J^+|_{U'}), \end{array}
$$

where $(\tau_U)_*$, $(\delta_{U'}^-)_*$, ... denote the corresponding push-forwards of sections, e.g.,

$$
(\tau_U)_* : C^\infty(U, E|_U) \xrightarrow{\cong} C^\infty(\overline{\mathbb{R}}_+ \times \Omega, \mathbb{C}^k),
$$
$$
(\rho_U)_* : C^\infty(U, F|_U) \xrightarrow{\cong} C^\infty(\overline{\mathbb{R}}_+ \times \Omega, \mathbb{C}^l).
$$

Definition 4.3.22 $\mathcal{B}^{\mu,d}(X; E, F; J^-, J^+)$ *for* $\mu \in \mathbb{Z}$, $d \in \mathbb{N}$, *is defined as the space of all operators*

$$
\mathcal{A} = \sum \begin{pmatrix} \varphi_j \cdot \mathrm{id}_F & 0 \\ 0 & \varphi_j' \cdot \mathrm{id}_{J^+} \end{pmatrix} \mathcal{A}_{U_j} \begin{pmatrix} \psi_j \cdot \mathrm{id}_E & 0 \\ 0 & \psi_j' \cdot \mathrm{id}_{J^-} \end{pmatrix}
$$
$$
+ \begin{pmatrix} \tilde{\varphi} A_{\mathrm{int}} \tilde{\psi} & 0 \\ 0 & 0 \end{pmatrix} + \mathcal{C}, \tag{4.3.20}
$$

for $\tilde{\varphi} = 1 - \sum \varphi_j$, $\tilde{\psi} \in C_0^\infty(\mathrm{int}\, X)$, $\tilde{\varphi}\tilde{\psi} = \tilde{\varphi}$, *where the sums are taken over all j with $U_j \cap Y \neq \emptyset$, $\varphi_j' = \varphi_j|_{U_j'}$, $\psi_j' = \psi_j|_{U_j'}$; $\mathrm{id}_{J^\pm}$ are the identity operators in the spaces of sections of $J^\pm|_{U_j'}$, and $\mathcal{A}_{U_j}$ are assumed to be of the form (4.3.19) with respect to U_j, moreover, $A_{\mathrm{int}} \in L_{\mathrm{cl}}^\mu(\mathrm{int}\, X; E, F)$, cf. Section 1.1.5, and $\mathcal{C} \in \mathcal{B}^{-\infty,d}(X; E, F; J^-, J^+)$. The elements in $\mathcal{B}^{\mu,d}(X; E, F; J^-, J^+)$ are called pseudo-differential boundary value problems of order μ and type d in Boutet de Monvel's algebra over X.*

Remark 4.3.23 *Definition* 4.3.22 *is correct in the sense that it is independent of the particular choice of the open covering* $\mathcal{U}$ *of* X, *the charts and the system of trivialisations of the involved bundles and of* $\{\varphi_j\}_{j\in\mathbb{Z}}$ *and* $\{\psi_j\}_{j\in\mathbb{Z}}$ *with the indicated properties. This follows easily from the local calculus of Section* 4.3.1, *in particular, from Theorem* 4.3.19.

Let us denote by $\mathcal{B}_G^{\mu,d}(X; E, F; J^-, J^+)$ the subclass of all operators (4.3.20) for which the $\mathcal{A}_{U_j}$ are obtained from corresponding elements in $\mathcal{B}_G^{\mu,d}(\mathbb{R}_+ \times \Omega_j; k, l; N_-, N_+)$ and for which A_{int} vanishes. . It will also be convenient to set

$$\mathcal{B}^{\mu,d}(X) = \bigcup \mathcal{B}^{\mu,d}(X; E, F; J^-, J^+), \qquad (4.3.21)$$

$$\mathcal{B}_G^{\mu,d}(X) = \bigcup \mathcal{B}_G^{\mu,d}(X; E, F; J^-, J^+), \qquad (4.3.22)$$

where the union is taken over all $E, F \in \text{Vect}(X), J^-, J^+ \in \text{Vect}(Y)$. Here and in the sequel we assume for convenience that the fibre dimensions of the involved bundles are independent of the base points $x \in X$ and $y \in Y$, respectively.

Fibre dimensions zero are also allowed; corresponding bundles will then be denoted by 0. Incidentally we set

$$B^{\mu,d}(X; E, F) = \mathcal{B}^{\mu,d}(X; E, F; 0, 0),$$
$$B_G^{\mu,d}(X; E, F) = \mathcal{B}_G^{\mu,d}(X; E, F; 0, 0).$$

The elements $\mathcal{A} \in \mathcal{B}^{\mu,d}(X; E, F; J^-, J^+)$ are block matrices of continuous operators

$$\mathcal{A} = \begin{pmatrix} A & K \\ T & Q \end{pmatrix} : \begin{matrix} C_0^\infty(X, E) \\ \oplus \\ C_0^\infty(Y, J^-) \end{matrix} \to \begin{matrix} C^\infty(X, F) \\ \oplus \\ C^\infty(Y, J^+). \end{matrix}$$

Analogously to the notation of the preceding section we set $A = $ u. l. c. $\mathcal{A}$ (upper left corner of $\mathcal{A}$) and call T a trace operator of order μ and type d, K a potential operator of order μ. For Q we obtain an element in $L^\mu_{\text{cl}}(Y; J^-, J^+)$. The elements $G \in B_G^{\mu,d}(X; E, F)$ are called Green operators of order μ and type d in Boutet de Monvel's algebra. Incidentally we also call the elements of $\mathcal{B}_G^{\mu,d}(X; E, F; J^-, J^+)$ Green operators, since their properties are very similar to those of the corresponding upper left corners.

Note that

$$\text{Diff}^\mu(X; E, F) \subset B^{\mu,0}(X; E, F),$$

where $\mathrm{Diff}^{\mu}(X; E, F)$ is the space of all differential operators

$$A : C_0^{\infty}(X, E) \to C^{\infty}(X, F)$$

(with smooth coefficients in local coordinates).

Theorem 4.3.24 *Each $\mathcal{A} \in \mathcal{B}^{\mu,d}(X; E, F; J^-, J^+)$ extends to a continuous operator*

$$\mathcal{A} : \begin{array}{c} H^s_{\mathrm{comp}}(X, E) \\ \oplus \\ H^s_{\mathrm{comp}}(Y, J^-) \end{array} \to \begin{array}{c} H^{s-\mu}_{\mathrm{loc}}(X, F) \\ \oplus \\ H^{s-\mu}_{\mathrm{loc}}(Y, J^+) \end{array}$$

for every $s \in \mathbb{R}$, $s > d - \frac{1}{2}$.

Proof. The proof is a consequence of Theorem 4.3.6 and of the invariance of the Sobolev spaces under the push-forwards that are involved in Definition 4.3.22. $\qquad\square$

In the beginning we have fixed a collar neighbourhood V of Y, identified with $[0, 1) \times Y$, where $t \in [0, 1)$ plays the role of a global normal variable to the boundary. Let us define the trace operators

$$T_j \cdot \mathrm{id}_E : C^{\infty}(X, E) \to C^{\infty}(Y, E|_Y)$$

for $E \in \mathrm{Vect}(X)$ by $u \to \partial^j / \partial t^j u|_{t=0}$, $j \in \mathbb{N}$. With the above notation we then have $T_j \cdot \mathrm{id}_E \in \mathcal{B}^{j,j+1}_G(X; E, 0; 0, E|_Y)$.

Theorem 4.3.25 *Each $\mathcal{G} \in \mathcal{B}^{\mu,d}_G(X; E, F; J^-, J^+)$ has a unique representation of the form $\mathcal{G} = \mathcal{G}_0 + \mathcal{G}_1$ for a $\mathcal{G}_0 \in \mathcal{B}^{\mu,0}_G(X; E, F; J^-, J^+)$ and*

$$\mathcal{G}_1 = \sum_{j=0}^{d-1} \begin{pmatrix} 0 & W_j \\ 0 & R_j \end{pmatrix} \begin{pmatrix} 0 & 0 \\ T_j \cdot \mathrm{id}_E & 0 \end{pmatrix}$$

for potential operators $W_j \in \mathcal{B}^{\mu-j-\frac{1}{2},0}_G(X; 0, F; E|_Y, 0)$ and pseudo-differential operators $R_j \in L^{\mu-j-\frac{1}{2}}_{\mathrm{cl}}(Y; E|_Y, J^+)$, $j = 0, \dots, d - 1$.

Proof. The proof of Theorem 4.3.25 is an obvious generalisation of the corresponding local variant in Theorem 4.3.5. $\qquad\square$

Remark 4.3.26 *In boundary value problems for differential operators $A \in \mathrm{Diff}^{\mu}(X; E, F)$, say for compact X, it is often natural to consider trace operators of different orders. In such a case, i.e., when we have*

$$\mathcal{A} = \begin{pmatrix} A \\ T \end{pmatrix} : H^s(X, E) \to \begin{array}{c} H^{s-\mu}(X, F) \\ \oplus \\ \bigoplus_{j=1}^{N} H^{s-\mu_j}(Y, G_j) \end{array} \qquad (4.3.23)$$

for a column matrix T of trace operators $T_j : H^s(X, E) \to H^{s-\mu_j}(Y, G_j)$ in Boutet de Monvel's algebra, it is possible to generalise the concept and to allow different orders of the entries of operator block matrices. However if we choose elements $R_j \in L_{cl}^{\mu-\mu_j}(Y; G_j, G_j)$ which are elliptic of order $\mu - \mu_j$ and induce isomorphisms $R_j : H^s(Y, G_j) \to H^{s-\mu+\mu_j}(Y, G_j)$ for all $s \in \mathbb{R}$, cf. Theorem 1.1.83, then for

$$
\mathcal{R} := \begin{pmatrix} 1 & 0 \\ 0 & \mathrm{diag}(R_j) \end{pmatrix} : \quad
\begin{matrix} H^{s-\mu}(X, F) \\ \oplus \\ \bigoplus_{j=1}^{N} H^{s-\mu_j}(Y, G_j) \end{matrix}
\xrightarrow{\cong}
\begin{matrix} H^{s-\mu}(X, F) \\ \oplus \\ \bigoplus_{j=1}^{N} H^{s-\mu}(Y, G_j) \end{matrix}
$$

the operator $\mathcal{A}_1 = \mathcal{R}\mathcal{A}$ belongs to $\mathcal{B}^{\mu,d}(X)$ with $\mathcal{A} = \mathcal{R}^{-1}\mathcal{A}_1$. In other words, a trivial reduction of orders which is bijective allows us to transform $\mathcal{A}$ into an operator in the sense of Definition 4.3.22. Thus the considerations on $\mathcal{B}^{\mu,d}(X)$ have immediate generalisations to the case of trace (or potential) operators of different orders.

Proposition 4.3.27 *Let $\mathcal{A}_k \in \mathcal{B}^{\mu-k,d}(X; E, F; J^-, J^+)$, $k \in \mathbb{N}$, be an arbitrary sequence. Then there is an $\mathcal{A} \in \mathcal{B}^{\mu,d}(X; E, F; J^-, J^+)$ such that*

$$
\mathcal{A} - \sum_{k=0}^{N} \mathcal{A}_k \in \mathcal{B}^{\mu-(N+1),d}(X; E, F; J^-, J^+)
$$

for all $N \in \mathbb{N}$, and $\mathcal{A}$ is unique $\mathrm{mod}\,\mathcal{B}^{-\infty,d}(X; E, F; J^-, J^+)$.

Proof. It suffices to apply the corresponding asymptotic summations for the local representatives of $\mathcal{A}_k$ over U_j, cf. the formula (4.3.20), and to the operators in int X. In other words if we form

$$
\mathcal{A}_{U_j} \sim \sum_{k=0}^{\infty} \mathcal{A}_{k,U_j}, \qquad \mathcal{A}_{\mathrm{int}} \sim \sum_{k=0}^{\infty} \mathcal{A}_{k,\mathrm{int}},
$$

where the first asymptotic sum can be obtained by carrying out the corresponding asymptotic sum for the operator–valued symbols, cf. Proposition 4.2.42, and the second one by the standard procedure, we can form the desired $\mathcal{A}$ by (4.3.20) where we may set $\mathcal{C} = 0$. The uniqueness $\mathrm{mod}\,\mathcal{B}^{-\infty,d}(X; E, F; J^-, J^+)$ is evident. $\square$

Let us now study the symbolic structure of the operators in $\mathcal{B}^{\mu,d}(X)$. Denote by $\pi_X : T^*X \setminus 0 \to X$ the canonical projection. For each $\mathcal{A} \in \mathcal{B}^{\mu,d}(X; E, F; J^-, J^+)$ we have $A := \mathrm{u.l.c.}\,\mathcal{A} \in L_{cl}^{\mu}(\mathrm{int}\ X; E, F)$, and the homogeneous principal symbol $\sigma_\psi^\mu(A)$ of A of order μ is smooth up to the boundary. Thus it has the interpretation of a bundle morphism

$$\sigma_\psi^\mu(A) : \pi_X^* E \to \pi_X^* F, \qquad (4.3.24)$$

where π_X^* indicates the bundle pull–back with respect to $\pi_X : T^*X \setminus 0 \to X$. Let us set

$$\sigma_\psi^\mu(\mathcal{A}) = \sigma_\psi^\mu(A) \quad \text{for } A = \text{u.l.c.}\,\mathcal{A}.$$

Moreover, the local representatives of $\mathcal{A} \in \mathcal{B}^{\mu,d}(X; E, F; J^-, J^+)$ in a collar neighbourhood of Y and the associated homogeneous boundary symbols of order μ, cf. Definition 4.3.13, give rise to a bundle morphism

$$\sigma_\wedge^\mu(\mathcal{A}) : \pi_Y^* \begin{pmatrix} \mathcal{S}(\overline{\mathbb{R}}_+) \otimes E|_Y \\ \oplus \\ J^- \end{pmatrix} \to \pi_Y^* \begin{pmatrix} \mathcal{S}(\overline{\mathbb{R}}_+) \otimes F|_Y \\ \oplus \\ J^+ \end{pmatrix} \qquad (4.3.25)$$

between the corresponding pull-backs with respect to the canonical projection $\pi_Y : T^*Y \setminus 0 \to Y$. This interpretation also employs the invariance properties of Theorem 4.3.19. Alternatively we could also consider the Sobolev space extensions

$$\sigma_\wedge^\mu(\mathcal{A}) : \pi_Y^* \begin{pmatrix} H^s(\mathbb{R}_+) \otimes E|_Y \\ \oplus \\ J^- \end{pmatrix} \to \pi_Y^* \begin{pmatrix} H^{s-\mu}(\mathbb{R}_+) \otimes F|_Y \\ \oplus \\ J^+ \end{pmatrix} \qquad (4.3.26)$$

for any $s \in \mathbb{R}$, $s > d - \frac{1}{2}$, though it will be more convenient to employ (4.3.25), since this is independent of d, and (4.3.26) is uniquely determined by (4.3.25) anyway. From the corresponding local homogeneities we get

$$\sigma_\wedge^\mu(\mathcal{A})(y, \lambda\eta) = \lambda^\mu \begin{pmatrix} \kappa_\lambda & 0 \\ 0 & 1 \end{pmatrix} \sigma_\wedge^\mu(\mathcal{A})(y, \eta) \begin{pmatrix} \kappa_\lambda & 0 \\ 0 & 1 \end{pmatrix}^{-1}$$

for all $\lambda \in \mathbb{R}_+$, $(y, \eta) \in T^*Y \setminus 0$.

Definition 4.3.28 *Let* $\mathcal{A} \in \mathcal{B}^{\mu,d}(X; E, F; J^-, J^+)$, $\mu \in \mathbb{Z}$, $d \in \mathbb{N}$. *Then* $\sigma_\psi^\mu(\mathcal{A})$ *is called the homogeneous principal interior symbol of* $\mathcal{A}$ *of order* μ *and,* $\sigma_\wedge^\mu(\mathcal{A})$ *the homogeneous boundary symbol of* $\mathcal{A}$ *of order* μ.

Let us set

$$\Sigma^{\mu,d}(E, F; J^-, J^+) = \{(\sigma_\psi^\mu(\mathcal{A}), \sigma_\wedge^\mu(\mathcal{A})) : \mathcal{A} \in \mathcal{B}^{\mu,d}(X; E, F; J^-, J^+)\},$$

$$\qquad (4.3.27)$$

$$\Sigma_G^{\mu,d}(E, F; J^-, J^+) = \{(0, \sigma_\wedge^\mu(\mathcal{G})) : \mathcal{G} \in \mathcal{B}_G^{\mu,d}(X; E, F; J^-, J^+)\}.$$

$$\qquad (4.3.28)$$

The components of $(\sigma_\psi^\mu(\mathcal{A}), \sigma_\wedge^\mu(\mathcal{A}))$ satisfy the compatibility condition

$$\sigma_\wedge^\mu(\mathcal{A}) - \begin{pmatrix} \mathrm{op}^+(\sigma_\psi^\mu(\mathcal{A})|_{T^*Y\setminus 0}) & 0 \\ 0 & 0 \end{pmatrix} \in \Sigma_G^{\mu,d}(E, F; J^-, J^+).$$

Remark 4.3.29 *From the local constructions of the preceding section we see that* $\Sigma^{\mu,d}(E, F; J^-, J^+)$ *is a Fréchet space in a canonical way.* $\Sigma_G^{\mu,d}(E, F; J^-, J^+)$ *is a closed subspace.*

Proposition 4.3.30 *We have an exact sequence*

$$0 \to \mathcal{B}^{\mu-1,d}(X; E, F; J^-, J^+) \xrightarrow{\iota} \mathcal{B}^{\mu,d}(X; E, F; J^-, J^+)$$
$$\xrightarrow{\sigma} \Sigma^{\mu,d}(E, F; J^-, J^+) \to 0, \quad (4.3.29)$$

where ι *is the canonical embedding, and* $\sigma = (\sigma_\psi^\mu, \sigma_\wedge^\mu)$. *There is a map*

$$\mathrm{op} : \Sigma^{\mu,d}(E, F; J^-, J^+) \to \mathcal{B}^{\mu,d}(X; E, F; J^-, J^+) \quad (4.3.30)$$

with $\sigma \circ \mathrm{op} = \mathrm{id}$ *on* $\Sigma^{\mu,d}(E, F; J^-, J^+)$.

Proof. The exactness of the sequence (4.3.29) is obvious by construction. In order to construct op we start with a pair $(p_\psi, p_\wedge) \in \Sigma^{\mu,d}(E, F; J^-, J^+)$ and form

$$p_G := p_\wedge - \begin{pmatrix} \mathrm{op}^+(p_\psi|_{T^*Y\setminus 0}) & 0 \\ 0 & 0 \end{pmatrix} \in \Sigma_G^{\mu,d}(E, F; J^-, J^+).$$

Let $\omega(t)$, $\omega_0(t)$, $\omega_1(t)$ be cut-off functions which equal 1 near $t = 0$ and zero for $t > \frac{1}{2}$ and satisfy $\omega\omega_0 = \omega$, $\omega\omega_1 = \omega_1$. Then, in the notation of Definition 4.3.22, we form

$$P_0 = \omega(t) \sum \varphi_j (\varrho_{U_j})_*^{-1} \mathrm{Op}(\mathrm{op}^+(\chi p_{\psi,j}))(\tau_{U_j})_* \psi_j \omega_0(t)$$

for any fixed excision function $\chi(\tau, \eta)$ in $\mathbb{R}^n$ and $p_{\psi,j}$ being the representative of $p_\psi : \pi_X^* E \to \pi_X^* F$ over U_j with respect to the chosen trivialisations of E, F. The sum is taken over all j with $U_j \cap Y \neq \emptyset$. Moreover, according to the constructions of Section 1.1.5, there is an operator $\tilde{P} \in L_{\mathrm{cl}}^\mu(\mathrm{int}\, X; E, F)$ with $\sigma_\psi^\mu(\tilde{P}) = p_\psi$. Then, if we set $P_1 = (1 - \omega)\tilde{P}(1 - \omega_1)$ and $P = P_0 + P_1$, we get an operator $P \in L_{\mathrm{cl}}^\mu(\mathrm{int}\, X; E, F)$ satisfying $\sigma_\psi^\mu(P) = p_\psi$ and $P \in B^{\mu,0}(X; E, F)$.

Now let $p_{G,j}$ be the local representative of p_G with respect to the trivialisations of the bundles E, F over U_j and J^-, J^+ over $U_j' = U_j \cap Y$. Then, setting

$$\mathcal{G}_{U_j} = \begin{pmatrix} (\varrho_{U_i})_*^{-1} & 0 \\ 0 & (\delta_{U_j'}^+)_*^{-1} \end{pmatrix} \mathrm{Op}(\chi' p_{G,j}) \begin{pmatrix} (\tau_{U_i})_* & 0 \\ 0 & (\delta_{U_j'}^-)_* \end{pmatrix}$$

for a fixed excision function $\chi'(\eta)$ in $\mathbb{R}^q$ and

$$\mathcal{G} = \sum \begin{pmatrix} \varphi_j & 0 \\ 0 & \varphi'_j \cdot \mathrm{id}_{J+} \end{pmatrix} \mathcal{G}_{U_j} \begin{pmatrix} \psi_j & 0 \\ 0 & \psi'_j \cdot \mathrm{id}_{J-} \end{pmatrix}$$

with the sum over all, j with $U_j \cap Y \neq \emptyset$, we get an element $\mathcal{G} \in \mathcal{B}_G^{\mu,d}(X; E, F; J^-, J^+)$ with $\sigma_\wedge^\mu(\mathcal{G}) = p_G$. Now the mapping op can be chosen as

$$\mathrm{op}(p_\psi, p_\wedge) = \mathcal{G} + \begin{pmatrix} P & 0 \\ 0 & 0 \end{pmatrix}.$$

$\square$

The map (4.3.30) is called an operator convention for the algebra of boundary value problems over X. If we change the involved arbitrary data such as the charts, the partition of unity, excision functions, ..., and denote the resulting map by op', then

$$\mathrm{op}(p_\psi, p_\wedge) - \mathrm{op}'(p_\psi, p_\wedge) \in \mathcal{B}^{\mu-1,d}(X; E, F; J^-, J^+).$$

To each $(p_\psi, p_\wedge) \in \Sigma^{\mu,d}(E, F; J^-, J^+)$ we can define the formal adjoint $(p_\psi, p_\wedge)^* := (p_\psi^*, p_\wedge^*)$, where p_ψ^* is defined as in Theorem 1.1.64 and $p_\wedge^*$ by

$$(u, p_\wedge^* v)_{L^2(\mathbb{R}_+, E_y) \oplus J_y^-} = (p_\wedge u, v)_{L^2(\mathbb{R}_+, F_y) \oplus J_y^+}$$

for all $u \in C_0^\infty(\mathbb{R}_+, E_y) \oplus J_y^-$, $v \in C_0^\infty(\mathbb{R}_+, F_y) \oplus J_y^+$, where subscript y indicates the fibres over $y \in Y$; the scalar products refer to the chosen Hermitean metrics in the bundles. Moreover, to each $(p_\psi, p_\wedge) \in \Sigma^{\mu,d}(E_0, F, J_0, J^+)$, $(r_\psi, r_\wedge) \in \Sigma^{\nu,c}(E, E_0, J^-, J_0)$ we can form $(p_\psi, p_\wedge)(r_\psi, r_\wedge) = (p_\psi r_\psi, p_\wedge r_\wedge)$, where the second component is the composition of operator families.

Proposition 4.3.31 $(p_\psi, p_\wedge) \in \Sigma^{0,0}(E, F; J^-, J^+)$ *implies* $(p_\psi, p_\wedge)^* \in \Sigma^{0,0}(F, E; J^+, J^-)$. *Moreover,* $(p_\psi, p_\wedge) \in \Sigma^{\mu,d}(E_0, F; J_0, J^+)$, $(r_\psi, r_\wedge) \in \Sigma^{\nu,c}(E, E_0; J^-, J_0)$ *implies* $(p_\psi r_\psi, p_\wedge r_\wedge) \in \Sigma^{\mu+\nu,h}(E, F; J^-, J^+)$ *for* $h = \max\{(d + \nu)^+, c\}$.

This follows from the corresponding local result Proposition 4.3.15.

Remark 4.3.32 *It can easily be proved that the $*$ operation*

$$\Sigma^{0,0}(E, F; J^-, J^+) \to \Sigma^{0,0}(F, E; J^+, J^-)$$

is continuous, and that the composition

$$\Sigma^{\mu,d}(E_0, F; J_0, J^+) \times \Sigma^{\nu,c}(E, E_0; J^-, J_0) \to \Sigma^{\mu+\nu,h}(E, F; J^-, J^+)$$

is continuous.

Theorem 4.3.33 *Given* $\mathcal{A} \in \mathcal{B}^{0,0}(X; E, F; J^-, J^+)$ *we obtain for the formal adjoint* $\mathcal{A}^* \in \mathcal{B}^{0,0}(X; F, E; J^+, J^-)$ *(cf. the formula (4.3.18)), and we have*

$$\sigma_\psi^0(\mathcal{A}^*) = \sigma_\psi^0(\mathcal{A})^*, \qquad \sigma_\wedge^0(\mathcal{A}^*) = \sigma_\wedge^0(\mathcal{A})^*. \tag{4.3.31}$$

The proof follows from Theorem 4.3.16 and Theorem 1.1.64.

Theorem 4.3.34 *For*

$$\mathcal{A} \in \mathcal{B}^{\mu,d}(X; E_0, F; J_0, J^+), \quad \mathcal{B} \in \mathcal{B}^{\nu,c}(X; E, E_0; J^-, J_0)$$

we have $\mathcal{A}\Phi\mathcal{B} \in \mathcal{B}^{\mu+\nu,h}(X; E, F; J^-, J^+)$ *for* $h = \max\{(d+\nu)^+, c\}$, *where*

$$\Phi = \begin{pmatrix} \varphi \cdot \mathrm{id}_{E_0} & 0 \\ 0 & \psi \cdot \mathrm{id}_{J_0} \end{pmatrix}$$

for any $\varphi \in C_0^\infty(X)$, $\psi \in C_0^\infty(Y)$, *and we have*

$$\sigma_\psi^{\mu+\mu}(\mathcal{A}\Phi\mathcal{B}) = \sigma_\psi^\mu(\mathcal{A})\varphi \cdot \mathrm{id}_{E_0} \sigma_\psi^\nu(\mathcal{B}), \quad \sigma_\wedge^{\mu+\mu}(\mathcal{A}\Phi\mathcal{B}) = \sigma_\wedge^\mu(\mathcal{A})\Phi|_Y \sigma_\wedge^\nu(\mathcal{B})$$

for $\Phi|_Y = \mathrm{diag}(\varphi|_Y \cdot \mathrm{id}_{E_0|_Y}, \psi \cdot \mathrm{id}_{J_0})$. *Moreover,* $\mathcal{A} \in \mathcal{B}_G^{\mu,d}$ *or* $\mathcal{B} \in \mathcal{B}_G^{\nu,c}$ *implies* $\mathcal{A}\Phi\mathcal{B} \in \mathcal{B}_G^{\mu+\nu,h}$.

Proof. The proof follows from Theorem 4.3.17, Remark 4.3.18 and Theorem 1.1.65   □

Corollary 4.3.35 *Let* X *be a compact* C^∞ *manifold with boundary. Then* $\mathcal{A} \in \mathcal{B}^{\mu,d}(X; E_0, F; J_0, J^+)$, $\mathcal{B} \in \mathcal{B}^{\nu,c}(X; E, E_0; J^-, J_0)$ *implies* $\mathcal{A}\mathcal{B} \in \mathcal{B}^{\mu+\nu,h}(X; E, F; J^-, J^+)$ *for* $h = \max\{(d+\nu)^+, c\}$, *and we have*

$$\sigma_\psi^{\mu+\nu}(\mathcal{A}\mathcal{B}) = \sigma_\psi^\mu(\mathcal{A})\sigma_\psi^\nu(\mathcal{B}),$$
$$\sigma_\wedge^{\mu+\nu}(\mathcal{A}\mathcal{B}) = \sigma_\wedge^\mu(\mathcal{A})\sigma_\wedge^\nu(\mathcal{B}).$$

Moreover, $\mathcal{A} \in \mathcal{B}_G^{\mu,d}$ *or* $\mathcal{B} \in \mathcal{B}_G^{\nu,c}$ *implies* $\mathcal{A}\mathcal{B} \in \mathcal{B}_G^{\mu+\nu,h}$.

It is useful to introduce natural Fréchet topologies in the spaces $\mathcal{B}^{\mu,d}(X; E, F; J^-, J^+)$. For notational convenience we consider the case of trivial bundles $E = F = X \times \mathbb{C}$. Let us denote the corresponding operator space by

$$\mathcal{B}^{\mu,d}(X; J^-, J^+).$$

The consideration for general $E, F \in \mathrm{Vect}(X)$ is completely analogous and left to the reader.

Denote by $L^\mu_{\mathrm{cl,trans}}(X)$ the space of all operators

$$A = \sum \varphi_j A_{U_j} \psi_j + \tilde\varphi A_{\mathrm{int}} \tilde\psi + C$$

in the sense of the notation in Definition 4.3.22 for which the local representatives of A_{U_j} in $\mathbb{R}_+ \times \Omega_j$ for the corresponding open $\Omega_j \subseteq \mathbb{R}^q$ have the form

$$\mathrm{Op}(\mathrm{op}^+(a)) \quad \text{for any} \quad a(t,y,\tau,\eta) \in S^\mu_{\mathrm{cl,trans}}(\overline{\mathbb{R}}_+ \times \Omega_j \times \mathbb{R}^{1+q}),$$

(subscript "trans" indicates the subspace of all elements of $S^\mu_{\mathrm{cl}}(\overline{\mathbb{R}}_+ \times \Omega_j \times \mathbb{R}^{1+q})$ with the transmission property), moreover, we have $A_{\mathrm{int}} \in L^\mu_{\mathrm{cl}}(\mathrm{int}\,X)$, and $C \in B^{-\infty,0}_G(X)$. Note that $B^{-\infty,0}_G(X)$ is isomorphic $C^\infty(X \times X)$, the space of associated kernels. Let for a moment $\mathcal{L}^\mu_{\mathrm{cl,trans}}(X)$ denote the subspace of all operator block matrices in $\mathcal{B}^{\mu,d}(X; J^-, J^+)$ for which $L^\mu_{\mathrm{cl,trans}}(X) = \mathrm{u.\,l.\,c.}\,\mathcal{L}^\mu_{\mathrm{cl,trans}}(X)$, while the other entries of the elements in $\mathcal{L}^\mu_{\mathrm{cl,trans}}(X)$ vanish. Then we have $L^\mu_{\mathrm{cl,trans}}(X) \cong \mathcal{L}^\mu_{\mathrm{cl,trans}}(X)$ in a canonical way, and

$$\mathcal{B}^{\mu,d}(X; J^-, J^+) = \mathcal{L}^\mu_{\mathrm{cl,trans}}(X) + \mathcal{B}^{\mu,d}_G(X; J^-, J^+) \qquad (4.3.32)$$

in the sense of a non-direct sum of vector spaces. According to Definition 1.1.4 we shall topologise the summands separately and then take in (4.3.32) the Fréchet topology of the non-direct sum. Let $S^{(\mu)}_{\mathrm{trans}}(T^*X \setminus 0)$ be the subspace of all elements of $S^{(\mu)}(T^*X \setminus 0)$ ($=$ the set of all functions in $C^\infty(T^*X \setminus 0)$ which are homogeneous of order μ in the fibre variables) which have the transmission property with respect to Y (the latter condition means that they are locally the homogeneous principal parts of order μ of classical symbols with the transmission property). Then $S^{(\mu)}_{\mathrm{trans}}(T^*X \setminus 0)$ is a Fréchet space in the topology induced by $S^{(\mu)}(T^*X \setminus 0)$, cf. also Remark 4.2.7. The map

$$\sigma^\mu_\psi : L^\mu_{\mathrm{cl,trans}}(X) \to S^{(\mu)}_{\mathrm{trans}}(T^*X \setminus 0)$$

that assigns to an $A \in L^\mu_{\mathrm{cl,trans}}(X)$ its homogeneous principal symbol of order μ is surjective. According to Proposition 4.3.30 we can fix a right inverse of σ^μ_ψ

$$\mathrm{op} : S^{(\mu)}_{\mathrm{trans}}(T^*X \setminus 0) \to L^\mu_{\mathrm{cl,trans}}(X). \qquad (4.3.33)$$

Remark 4.3.36 *The space $L^\mu_{\mathrm{cl,trans}}(X)$ is Fréchet in a natural way, similarly to the constructions above in Remark 1.1.60 that are analogous for paracompact X and operators with the transmission property. Then (4.3.33) is continuous.*

The proof is straightforward and will be omitted.

Let us now turn to $\mathcal{B}_G^{\mu,0}(X; J^-, J^+)$. According to (4.3.28) we have the space $\Sigma_G^{\mu,0}(\mathbb{C}, \mathbb{C}; J^-, J^+) =: \Sigma_G^{\mu,0}(J^-, J^+)$ which is Fréchet in a canonical way, cf. Remark 4.2.28. We shall construct analogous sequences of mappings as above for the interior part, now in terms of principal boundary symbols. Let

$$\sigma_\wedge^\mu : \mathcal{B}_G^{\mu,0}(X; J^-, J^+) \to \Sigma_G^{\mu,0}(J^-, J^+)$$

be the principal boundary symbolic map, which is the second component of σ in the sequence (4.3.29), here on the space $\mathcal{B}_G^{\mu,0}(X; J^-, J^+)$. Denote by

$$\mathrm{op} : \Sigma_G^{\mu,0}(J^-, J^+) \to \mathcal{B}_G^{\mu,0}(X; J^-, J^+)$$

a right a inverse of $\sigma_\wedge^\mu$. Then $\delta^{\mu-1} : \mathcal{A} \to \mathcal{A} - \mathrm{op}(\sigma_\wedge^\mu(\mathcal{A}))$ defines a linear map $\delta^{\mu-1} : \mathcal{B}_G^{\mu,0}(X; J^-, J^+) \to \mathcal{B}_G^{\mu-1,0}(X; J^-, J^+)$.

Set $\sigma^{\mu-1}(\mathcal{A}) = \sigma_\wedge^{\mu-1}(\delta^{\mu-1}(\mathcal{A}))$, $\delta^{\mu-2}(\mathcal{A}) = \delta^{\mu-1}(\mathcal{A}) - \mathrm{op}(\sigma^{\mu-1}(\mathcal{A}))$, and successively

$$\sigma^{\mu-j}(\mathcal{A}) = \sigma_\wedge^{\mu-j}(\delta^{\mu-j}(\mathcal{A})),$$
$$\delta^{\mu-(j+1)}(\mathcal{A}) = \delta^{\mu-j}(\mathcal{A}) - \mathrm{op}(\sigma^{\mu-j}(\mathcal{A}))$$

for all $j \in \mathbb{N}$. This gives us a sequence of linear mappings

$$\sigma^{\mu-j} : \mathcal{B}_G^{\mu,0}(X; J^-, J^+) \quad \to \quad \Sigma_G^{\mu-j}(J^-, J^+),$$
$$\delta^{\mu-j} : \mathcal{B}_G^{\mu,0}(X; J^-, J^+) \quad \to \quad \mathcal{B}_G^{\mu-j,0}(X; J^-, J^+).$$

Define for every compact subset $M \subseteq Y$ and $J \in \mathrm{Vect}(Y)$ the space

$$H^s(M, J) = \{u \in H_{\mathrm{loc}}^s(Y, J) : \ \mathrm{supp}\, u \subseteq M\}.$$

Then Theorem 4.3.24 yields an operator

$$\mathcal{B}_G^{\nu,0}(X; J^-, J^+)$$
$$\to \bigcap_{\substack{s \in \mathbb{R} \\ s > -\frac{1}{2}}} \mathcal{L}(H^s(K) \oplus H^s(M, J^-), H_{\mathrm{loc}}^{s-\nu}(X) \oplus H_{\mathrm{loc}}^{s-\nu}(Y, J^+))$$

for every pair of compact subsets $K \subset\subset X$, $M \subset\subset Y$. It is sufficient to consider

$$\mathcal{L}_{K,M}^\nu = \bigcap_{s \in \mathbb{N}} \mathcal{L}(H^s(K) \oplus H^s(M, J^-), H_{\mathrm{loc}}^{s-\nu}(X) H_{\mathrm{loc}}^{s-\nu}(Y, J^+))$$

which is a Fréchet space in a canonical way. Then we have a map

$$\eta^{\nu}_{K,M} : \mathcal{B}^{\nu,0}_G(X; J^-, J^+) \to \mathcal{L}^{\nu}_{K,M}.$$

Let us form the composition

$$\lambda^{\mu-j}_{K,M} := \eta^{\mu-j}_{K,M} \circ \delta^{\mu-j} : \mathcal{B}^{\mu,0}_G(X; J^-, J^+) \to \mathcal{L}^{\mu-j}_{K,M} \qquad (4.3.34)$$

for every $j \in \mathbb{N}$ and $K \subset\subset X$, $M \subset\subset Y$. Consider now the transposed $^t\mathcal{A} \in \mathcal{B}^{\mu,0}_G(X; \overline{J^-}, \overline{J^+})$ for $\mathcal{A} \in \mathcal{B}^{\mu,0}_G(X; J^+, J^-)$, defined by

$$\left(u, \overline{{}^t\mathcal{A}v}\right)_{L^2_{\mathrm{loc}}(X)\oplus L^2_{\mathrm{loc}}(Y,J^-)} = \left(\mathcal{A}u, \overline{v}\right)_{L^2(X)\oplus L^2_{\mathrm{loc}}(Y,\overline{J^+})}$$

for all $u \in C^\infty_0(\mathrm{int}\, X) \oplus C^\infty_0(Y, J^-)$, $v \in C^\infty_0(\mathrm{int}\, X) \oplus C^\infty_0(Y, \overline{J^+})$. The bar at the bundles indicates those bundles which appear when we replace the cocycles of transition maps between the local trivialisations by their complex conjugates. Then, setting $^t\sigma^{\mu-j}(\mathcal{A}) = \sigma^{\mu-j}({}^t\mathcal{A})$, and $^t\lambda^{\mu-j}_{K,M}(\mathcal{A}) = \lambda^{\mu-j}_{K,M}({}^t\mathcal{A})$, we get a second system of linear mappings on the space $\mathcal{B}^{\mu,0}_G(X; J^-, J^+)$, namely

$$^t\lambda^{\mu-j}_{K,M} : \mathcal{B}^{\mu,0}_G(X; J^-, J^+) \to {}^t\mathcal{L}^{\mu-j}_{K,M},$$

$j \in \mathbb{N}$, $K \subset\subset X$, $M \subset\subset Y$, with the obvious meaning of $^t\mathcal{L}^{\mu-j}_{K,M}$.

We endow the space $\mathcal{B}^{\mu,0}_G(X; J^-, J^+)$ with the topology of the projective limit with respect to the system of maps

$$\{\sigma^{\mu-j}\}_{j\in\mathbb{N}}, \quad \{\lambda^{\mu-j}_{K,M}\}_{j\in\mathbb{N}, K\subset\subset X, M\subset\subset Y}, \quad \{^t\lambda^{\mu-j}_{K,M}\}_{j\in\mathbb{N}, K\subset\subset X, M\subset\subset Y}.$$

In order to generate the topology it suffices to consider countable systems of compact sets $K \subset\subset X$, $M \subset\subset Y$.

Lemma 4.3.37 $\mathcal{B}^{\mu,0}_G(X; J^-, J^+)$ *is a Fréchet space with respect to the above topology.*

The proof is straightforward and will be omitted.

For $\mathcal{B}^{\mu,d}_G(X; J^-, J^+)$, $d \in \mathbb{N}$, we may apply the above Theorem 4.3.25 which yields an isomorphism

$$\mathcal{B}^{\mu,d}_G(X; J^-, J^+) \cong \mathcal{B}^{\mu,0}_G(X; J^-, J^+)$$

$$\times \underset{j=0}{\overset{d-1}{\times}} \left\{ \mathcal{B}^{\mu-j-\frac{1}{2},0}_G(X; 0, \mathbb{C}; \mathbb{C}, 0) \times L^{\mu-j-\frac{1}{2}}_{\mathrm{cl}}(Y; \mathbb{C}, J^+) \right\};$$

$\mathbb{C}$ indicates the trivial line bundle with fibre $\mathbb{C}$ over the corresponding base spaces X and Y, respectively. The factors on the right are already endowed with Fréchet topologies. The algebraic isomorphism then yields a corresponding Fréchet topology in $\mathcal{B}^{\mu,d}_G(X; J^-, J^+)$. Returning now to the case of arbitrary bundles $E, F \in \mathrm{Vect}(X)$, $J^-, J^+ \in \mathrm{Vect}(Y)$ we get by analogous constructions the following result:

Theorem 4.3.38 $\mathcal{B}^{\mu,d}(X; E, F; J^+, J^-)$ *is a Fréchet space in a canonical way. The operators from Proposition* 4.3.30

$$\sigma : \mathcal{B}^{\mu,d}(X; E, F; J^+, J^-) \to \Sigma^{\mu,d}(E, F; J^-, J^+)$$

and

$$\mathrm{op} : \Sigma^{\mu,d}(E, F; J^-, J^+) \to \mathcal{B}^{\mu,d}(X; E, F; J^-, J^+)$$

are continuous.

The following theorems are immediate consequences of the above constructions.

Theorem 4.3.39 *The* $*$ *operation of Theorem* 4.3.33 *defines a continuous map* $\mathcal{B}^{0,0}(X; E, F; J^-, J^+) \to \mathcal{B}^{0,0}(X; F, E; J^+, J^-)$.

Theorem 4.3.40 *For every* Φ *the compositions of Theorem* 4.3.34 *define by* $(\mathcal{A}, \mathcal{B}) \to \mathcal{A}\Phi\mathcal{B}$ *a continuous map*

$$\mathcal{B}^{\mu,d}(X; E_0, F; J_0, J^+) \times \mathcal{B}^{\nu,c}(X; E, E_0; J^-, J_0)$$
$$\to \mathcal{B}^{\mu+\nu,h}(X; E, F; J^-, J^+).$$

In particular, if X *is compact, this holds for* $(\mathcal{A}, \mathcal{B}) \to \mathcal{A}\mathcal{B}$.

4.3.3   Ellipticity, parametrices, Fredholm property

Definition 4.3.41 *An operator* $\mathcal{A} \in \mathcal{B}^{\mu,d}(X; E, F; J^-, J^+)$ *is called elliptic (of order* μ*) if both*

$$\sigma_\psi^\mu(\mathcal{A}) : \pi_X^* E \to \pi_X^* F \tag{4.3.35}$$

and

$$\sigma_\wedge^\mu(\mathcal{A}) : \pi_Y^* \begin{pmatrix} \mathcal{S}(\overline{\mathbb{R}}_+) \otimes E|_Y \\ \oplus \\ J^- \end{pmatrix} \to \pi_Y^* \begin{pmatrix} \mathcal{S}(\overline{\mathbb{R}}_+) \otimes F|_Y \\ \oplus \\ J^+ \end{pmatrix} \tag{4.3.36}$$

are isomorphisms.

The condition that (4.3.35) is an isomorphism is also called the interior ellipticity of $\mathcal{A}$. The isomorphism (4.3.36) is an analogue of the classical Shapiro–Lopatinskij condition. An elliptic operator $\mathcal{A} \in \mathcal{B}^{\mu,d}(X; E, F; J^-, J^+)$ is called an elliptic boundary value problem for $A = \mathrm{u.\,l.\,c.\,} \mathcal{A}$.

Remark 4.3.42 *The condition that* (4.3.36) *is an isomorphism is equivalent to*

$$
\sigma_\wedge^\mu(\mathcal{A}): \; \pi_Y^* \begin{pmatrix} H^s(\mathbb{R}_+) \otimes E|_Y \\ \oplus \\ J^- \end{pmatrix} \xrightarrow{\cong} \pi_Y^* \begin{pmatrix} H^{s-\mu}(\mathbb{R}_+) \otimes F|_Y \\ \oplus \\ J^+ \end{pmatrix}
$$

$$(4.3.37)$$

for a fixed $s = s_0 \in \mathbb{R}$, $s_0 > d - \frac{1}{2}$, *or, which is the same, for all* $s \in \mathbb{R}$, $s > d - \frac{1}{2}$. *This follows from Theorem 4.1.83, applied* (y, η)-*wise to the corresponding* $k \times k$-*system* u.l.c. $\sigma_\wedge^\mu(\mathcal{A})(y, \eta)$ *of operators in* $\mathbf{D}^{\mu,d}(\mathbb{R}_+)$, $k = $ *fibre dimension of* E.

Example 4.3.43 *Let* Δ *be the Laplace-Beltrami operator with respect to a Riemannian metric on* X. *Then, if* r' *is the restriction operator to the boundary,*

$$
\mathcal{A} = \begin{pmatrix} \Delta \\ r' \end{pmatrix}: \; H^s(X) \to \begin{matrix} H^{s-2}(X) \\ \oplus \\ H^{s-\frac{1}{2}}(Y) \end{matrix},
$$

$s > \frac{1}{2}$, *is elliptic. This is just the Dirichlet problem for* Δ. *To be formally precise instead of* $\mathcal{A}$ *we may consider the operator*

$$
\mathcal{A}_1 = \begin{pmatrix} \Delta \\ Rr' \end{pmatrix} \in \mathcal{B}^{2,1}(X)
$$

for any elliptic operator $R \in L^{\frac{3}{2}}_{\mathrm{cl}}(Y)$ *which induces isomorphisms* $R: H^s(Y) \to H^{s-\frac{3}{2}}(Y)$ *for all* $s \in \mathbb{R}$. *Then both components of* $\mathcal{A}$ *are of order* 2, *cf. also Remark* 4.3.26. *Recall that the ellipticity on the level of boundary symbols was discussed at the end of Section 4.1.7.*

Let us assume in this section that X is compact. Then $Y = \partial X$ is also compact, and the unit cosphere bundle S^*Y, induced by the given Riemannian metric in T^*Y, is a compact space. Set

$$
\sigma_\wedge^\mu \sigma_\psi^\mu(\mathcal{A})(y, \eta) = \mathrm{op}^+(\sigma_\psi^\mu(\mathcal{A})|_{t=0})(y, \eta).
$$

Then the condition that (4.3.35) is an isomorphism implies that

$$
\sigma_\wedge^\mu \sigma_\psi^\mu(\mathcal{A})(y, \eta): \; H^s(\mathbb{R}_+) \otimes E_y \to H^{s-\mu}(\mathbb{R}_+) \otimes F_y \qquad (4.3.38)
$$

is a family of Fredholm operators parametrised by $(y, \eta) \in T^*Y \setminus 0$, with

$$
\mathrm{ind}\, \sigma_\wedge^\mu \sigma_\psi^\mu(\mathcal{A})(y, \lambda\eta) = \mathrm{ind}\, \sigma_\wedge^\mu \sigma_\psi^\mu(\mathcal{A})(y, \eta)
$$

for all $\lambda \in \mathbb{R}_+$. The latter property is a consequence of the homogeneity

$$\sigma_\wedge^\mu \sigma_\psi^\mu(\mathcal{A})(y, \lambda\eta) = \lambda^\mu \kappa_\lambda \sigma_\wedge^\mu \sigma_\psi^\mu(\mathcal{A})(y, \eta) \kappa_\lambda^{-1}$$

for all $\lambda \in \mathbb{R}_+$. Thus, under the aspect of the index, it is adequate to restrict $\sigma_\wedge^\mu \sigma_\psi^\mu(\mathcal{A})$ to S^*Y.

Recall that when X is a compact topological space and $a : X \to \mathcal{L}(H_1, H_2)$ a continuous function with values in the space of Fredholm operators between (separable) Hilbert spaces H_1, H_2, then there is defined an element

$$\mathrm{ind}_X a \in K(X), \tag{4.3.39}$$

where $K(X)$ is the K-group of X, cf. Section 1.1.4. For simplicity we will assume here once and for all that X is arc-wise connected, such that $\mathrm{ind}\, a(x) = \dim\ker a(x) - \dim\mathrm{coker}\, a(x)$ is independent of $x \in X$. More generally we may consider bundle morphisms $a : H_1 \to H_2$ when H_i are Hilbert space bundles over X with H_i as fibres, $i = 1, 2$, such that a induces the identical map on the base X, and where a is a Fredholm operator between the corresponding fibres. Since the bundles are trivial by Kuiper's theorem, in this case we also get an element (4.3.39), called the index element of a. This can be applied to (4.3.38) for $(y, \eta) \in S^*Y$, and hence we obtain

$$\mathrm{ind}_{S^*Y}\, \sigma_\wedge^\mu \sigma_\psi^\mu(\mathcal{A}) \in K(S^*Y).$$

Note that this is independent of $s \in \mathbb{R}$ for $s > -\frac{1}{2}$, cf. Remark 4.1.82.

Remark 4.3.44 *Let* $A \in B^{\mu,d}(X; E, F)$ *be an operator for which* $\sigma_\psi^\mu(A) : \pi_X^* E \to \pi_X^* F$ *is an isomorphism,* $\pi_X : T^*X \setminus 0 \to X$. *Then for arbitrary* $G \in B_G^{\mu,d}(X; E, F)$ *we have*

$$\mathrm{ind}_{S^*Y}\, \sigma_\wedge^\mu \sigma_\psi^\mu(A) = \mathrm{ind}_{S^*Y}(\sigma_\wedge^\mu \sigma_\psi^\mu(A) + \sigma_\wedge^\mu(G)).$$

In fact,

$$\sigma_\wedge^\mu(G)(y, \eta) : \quad H^s(\mathbb{R}_+) \otimes E_y \to H^{s-\mu}(\mathbb{R}_+) \otimes F_y$$

is a compact operator for every $(y, \eta) \in S^*Y$; *morphisms consisting fibre–wise of compact operators do not change the index element.*

Proposition 4.3.45 *Let* $A \in B^{\mu,d}(X; E, F)$ *be an operator for which* $\sigma_\psi^\mu(A) : \pi_X^* E \to \pi_X^* F$ *is an isomorphism. Then there is an elliptic operator* $\mathcal{A} \in B^{\mu,d}(X; E, F; J^-, J^+)$ *for suitable* $J^-, J^+ \in \mathrm{Vect}(Y)$ *if and only if* $\mathrm{ind}_{S^*Y}\, \sigma_\wedge^\mu \sigma_\psi^\mu(A) \in \pi_Y^* K(Y)$, *where* $\pi_Y : S^*Y \to Y$ *is the canonical projection. If* $\mathcal{A} \in B^{\mu,d}(X; E, F; J^-, J^+)$ *is elliptic then*

$$\mathrm{ind}_{S^*Y}\, \sigma_\wedge^\mu \sigma_\psi^\mu(\mathrm{u.\,l.\,c.}\, \mathcal{A}) = [\pi_Y^* J^+] - [\pi_Y^* J^-]. \tag{4.3.40}$$

Proof. Let $a : \boldsymbol{H}_1 \to \boldsymbol{H}_2$ be a continuous family of Fredholm operators, cf. the above notation, and $J_1, J_2 \in \mathrm{Vect}(\boldsymbol{X})$ be vector bundles such that this family can be completed to a bundle isomorphism

$$
\begin{pmatrix} a & k \\ t & q \end{pmatrix} : \begin{array}{c} \boldsymbol{H}_1 \\ \oplus \\ J_1 \end{array} \to \begin{array}{c} \boldsymbol{H}_2 \\ \oplus \\ J_2 \end{array} .
$$

Then it is a purely algebraic fact that $\mathrm{ind}_{\boldsymbol{X}}\, a = [J_2] - [J_1]$. This yields, in particular, the relation (4.3.40), i.e., $\mathrm{ind}_{S^*Y}\, \sigma^\mu_\wedge \sigma^\mu_\psi(A) \in \pi_Y^* K(Y)$ for $A = \mathrm{u.l.c.}\, \mathcal{A}$. It remains to show the reverse direction. Here it is clear that when we set $a = \sigma^\mu_\wedge \sigma^\mu_\psi(A)$ there is an $N_- \in \mathbb{N}$ and a linear map $k : \mathbb{C}^N_- \to \mathcal{S}(\overline{\mathbb{R}}_+) \otimes E_y$ such that

$$
\begin{pmatrix} a & k \end{pmatrix} := \begin{array}{c} H^s(\mathbb{R}_+) \otimes E_y \\ \oplus \\ \mathbb{C}^{N_-} \end{array} \to H^{s-\mu}(\mathbb{R}_+) \otimes F_y
$$

is surjective for all $(y, \eta) \in S^*Y$. Now $\ker \begin{pmatrix} a & k \end{pmatrix}$ can be interpreted as a vector subbundle of $\pi_Y^* \{(H^s(\mathbb{R}_+) \otimes E|_Y) \oplus \mathbb{C}^{N_-}\}$ of some fibre dimension N_+, and $\mathrm{ind}_{S^*Y}\, a \in \pi_Y^* K(Y)$ just implies that for a sufficiently large N_0 there is a $J^+ \in \mathrm{Vect}(Y)$ such that $\ker \begin{pmatrix} a & k \end{pmatrix} \oplus \mathbb{C}^{N_0} \cong \pi_Y^* J^+$. By replacing k acting on $\mathbb{C}^{N_-}$ by $k \oplus 0$ on $\mathbb{C}^{N_-} \oplus \mathbb{C}^{N_0}$, in other words by assuming N_- sufficiently large, we obtain that $\ker \begin{pmatrix} a & k \end{pmatrix}$ itself is isomorphic to such a pull-back. This allows us to fill $\begin{pmatrix} a & k \end{pmatrix}$ to a block matrix

$$
\begin{pmatrix} a & k \\ t & q \end{pmatrix} : \begin{array}{c} H^s(\mathbb{R}_+) \otimes E|_Y \\ \oplus \\ \mathbb{C}^{N_-} \end{array} \to \begin{array}{c} H^{s-\mu}(\mathbb{R}_+) \otimes F|_Y \\ \oplus \\ \pi_Y^* J^+ \end{array}
$$

that is an isomorphism, by extending a chosen isomorphism $\begin{pmatrix} t & q \end{pmatrix} : \ker \begin{pmatrix} a & k \end{pmatrix} \to \pi_Y^* J^+$ to a morphism $H^s(\mathbb{R}_+) \otimes E|_Y \to \pi_Y^+ J^+$. It is easy to construct $\begin{pmatrix} t & q \end{pmatrix}$ explicitly in terms of the families of kernels of $\begin{pmatrix} a & k \end{pmatrix}$ which are vectors

$$
b(y, \eta) \oplus r(y, \eta) \in (\mathcal{S}(\overline{\mathbb{R}}_+) \otimes E_y) \oplus \mathbb{C}^{N_-},
$$

smoothly dependent on $(y, \eta) \in S^*Y$. In fact, using the sesquilinear pairing in $(L^2(\mathbb{R}_+) \otimes E_y) \oplus \mathbb{C}^{N_-}$ we can form the map

$$
\begin{array}{c} H^s(\mathbb{R}_+) \otimes E_y \\ \oplus \\ \mathbb{C}^{N_-} \end{array} \ni \begin{pmatrix} u \\ \oplus \\ v \end{pmatrix} \to (u, \overline{b}) + (v, \overline{r}) \qquad (4.3.41)
$$

with the above pairings in $(L^2(\mathbb{R}_+) \otimes E_y)$ and $\mathbb{C}^{N-}$, respectively. The range of (4.3.41) is the fibre of the bundle $\pi_Y^* J^+$ over (y, η), and the representation of $\bar{b}(y, \eta) \oplus \bar{r}(y, \eta)$ by a matrix with entries in $\mathcal{S}(\overline{\mathbb{R}}_+) \oplus \mathbb{C}$ follows from Proposition 4.1.75. Thus, in view of Remark 4.1.77 we can interpret the extension of $(t \quad q)$ by homogeneity μ in the sense

$$(t \quad q)(y, \lambda\eta) = \lambda^\mu (t \quad q)(y, \eta) \begin{pmatrix} \kappa_\lambda & 0 \\ 0 & 1 \end{pmatrix}^{-1}$$

for $\lambda = |\eta|$ from S^*Y to $T^*Y \setminus 0$ as the second row of a homogeneous boundary symbol with the given upper left corner $a = \sigma_\wedge^\mu \sigma_\psi^\mu(A)$. The already constructed potential part k on S^*Y is also to be extended by homogeneity μ to $T^*Y \setminus 0$ by $k(y, \lambda\eta) = \lambda^\mu \kappa_\lambda k(y, \eta)$ for $\lambda = |\eta|$, where here by construction $k = k(y, \eta)$ was independent of η, i.e., we obtain in this case $k(y, \eta) = |\eta|^\mu \kappa_{|\eta|} k$. □

Proposition 4.3.45 is a topological obstruction for the existence of elliptic boundary conditions to an elliptic operator on X, cf. analogously the above discussion in connection with formula (3.5.23) and Remark 3.5.7. It is well-known that for elliptic differential operators such as the Dirac operators or the signature operator on a manifold with boundary which are interesting in geometry and topology the above obstruction does not vanish. So there are no elliptic boundary conditions for these operators. A general analsis of this phenomenon can be found in the paper Schulze, Sternin, Shatalov [129].

Proposition 4.3.46 Let $\mathcal{A} \in \mathcal{B}^{\mu,d}(X; E, F; J^-, J^+)$ be elliptic. Then to

$$(\sigma_\psi^\mu(\mathcal{A}), \sigma_\wedge^\mu(\mathcal{A})) \in \Sigma^{\mu,d}(E, F; J^-, J^+)$$

there exists the (component–wise) inverse

$$(\sigma_\psi^\mu(\mathcal{A}), \sigma_\wedge^\mu(\mathcal{A}))^{-1} \in \Sigma^{-\mu,(d-\mu)^+}(F, E; J^+, J^-).$$

In particular, for $d = \mu^+$ we have $(d-\mu)^+ = (-\mu)^+$.

Proof. The assertion is a consequence of Theorem 4.1.83 for the variant of $k \times k$–matrix–valued upper left corners, k = fibre dimension of E and F, cf. Remark 4.1.69, together with the observation that locally on $T^*Y \setminus 0$ all components of $\sigma_\wedge^\mu(\mathcal{A})^{-1}(y, \eta)$ smoothly depend on (y, η). This follows from Remark 4.1.67 which suffices, since the (smooth) inverse of $\sigma_\psi^\mu(\mathcal{A})$ reduces the representation of $\sigma_\wedge^\mu(\mathcal{A})^{-1}(y, \eta)$ to the case of Remark 4.1.67, according to the strategy in the proof of Theorem 4.1.83. □

Definition 4.3.47 *Let* $\mathcal{A} \in \mathcal{B}^{\mu,d}(X; E, F; J^-, J^+)$, $\mu \in \mathbb{Z}$, $d \in \mathbb{N}$. *Then an operator* $\mathcal{B} \in \mathcal{B}^{-\mu,c}(X; F, E; J^+, J^-)$ *for any* $c \in \mathbb{N}$ *is called a parametrix of* $\mathcal{A}$ *if* $\mathcal{AB} - 1 \in \mathcal{B}^{-\infty,e}(X; F, F; J^+, J^+)$, $\mathcal{BA} - 1 \in \mathcal{B}^{-\infty,h}(X; E, E; J^-, J^-)$ *for certain* $e, h \in \mathbb{N}$.

Theorem 4.3.48 *Every elliptic operator* $\mathcal{A} \in \mathcal{B}^{\mu,\mu^+}(X; E, F; J^-, J^+)$ *has a parametrix* $\mathcal{B} \in \mathcal{B}^{-\mu,(-\mu)^+}(X; F, E; J^+, J^-)$.

Proof. From Proposition 4.3.46 we obtain that the pair of inverse symbols belongs to $\Sigma^{-\mu,(-\mu)^+}(F, E; J^+, J^-)$. Then Proposition 4.3.14 yields an operator

$$\mathcal{B}_0 = \mathrm{op}(\sigma_\psi^\mu(\mathcal{A})^{-1}, \sigma_\wedge^\mu(\mathcal{A})^{-1}) \in \mathcal{B}^{-\mu,(-\mu)^+}(X; F, E; J^+, J^-)$$

which satisfies

$$\mathcal{B}_0\mathcal{A} = 1 + \mathcal{C} \quad \text{for some} \quad \mathcal{C} \in \mathcal{B}^{-1,\mu^+}(X; E, E; J^-, J^-), \qquad (4.3.42)$$

cf. also Theorem 4.3.34 with respect to the rule for evaluating the resulting types of compositions. Thus we obtain

$$\mathcal{C}^j \in \mathcal{B}^{-j,\mu^+}(X; E, E; J^-, J^-)$$

for every $j \in \mathbb{N}$. In view of Proposition 4.3.27 we can form the asymptotic sum $\mathcal{D} = \sum_{j=1}^\infty (-1)^j \mathcal{C}^j \in \mathcal{B}^{-1,\mu^+}(X; E, E; J^-, J^-)$. It satisfies $(1 + \mathcal{D})(1 + \mathcal{C}) = 1 \bmod \mathcal{B}^{-\infty,\mu^+}(X; E, E; J^-, J^-)$. From (4.3.42) we then obtain that $\mathcal{B}_l := (1 + \mathcal{D})\mathcal{B}_0 \in \mathcal{B}^{-\mu,(-\mu)^+}(X; F, E; J^+, J^-)$ is a left parametrix of $\mathcal{A}$ and $\mathcal{R}_l = \mathcal{B}_l\mathcal{A} - 1 \in \mathcal{B}^{-\infty,\mu^+}(X; E, E; J^-, J^-)$. In an analogous manner we can show that $\mathcal{A}$ has a right parametrix $\mathcal{B}_r \in \mathcal{B}^{-\mu,(-\mu)^+}(X; F, E; J^+, J^-)$, with the remainder

$$\mathcal{R}_r = \mathcal{AB}_r - 1 \in \mathcal{B}^{-\infty,(-\mu)^+}(X; F, F; J^+, J^+).$$

Now $\mathcal{B}_l - \mathcal{B}_r = \mathcal{R}_l\mathcal{B}_r - \mathcal{B}_l\mathcal{R}_r \in \mathcal{B}^{-\infty,(-\mu)^+}(X; F, E; J^+, J^-)$, in other words we may set $\mathcal{B} = \mathcal{B}_l$ or $\mathcal{B} = \mathcal{B}_r$. □

Theorem 4.3.49 *Every elliptic* $\mathcal{A} \in \mathcal{B}^{\mu,\mu^+}(X; E, F; J^-, J^+)$ *induces a Fredholm operator*

$$\mathcal{A}: \begin{array}{ccc} H^s(X, E) & & H^{s-\mu}(X, F) \\ \oplus & \to & \oplus \\ H^s(Y, J^-) & & H^{s-\mu}(Y, J^+) \end{array} \qquad (4.3.43)$$

for every $s \in \mathbb{R}$, $s > \mu^+ - \frac{1}{2}$. *Moreover,* $\mathcal{A}u = f \in H^r(X, F) \oplus H^r(Y, J^+)$ *for some* $r > (-\mu)^+ - \frac{1}{2}$ *and* $u \in H^t(X, E) \oplus H^t(Y, J^-)$ *for some* $t \in \mathbb{R}$, $t > \mu^+ - \frac{1}{2}$, *implies* $u \in H^{s+\mu}(X, E) \oplus H^{s+\mu}(Y, J^-)$.

Proof. In view of Theorem 4.3.48 the operator $\mathcal{A}$ has a parametrix

$$\mathcal{B} \in \mathcal{B}^{-\mu,(-\mu)^+}(X; F, E; J^+, J^-).$$

Applying Theorem 4.3.24 we get continuous operators

$$\mathcal{B}: \begin{array}{c} H^r(X,F) \\ \oplus \\ H^r(Y,J^+) \end{array} \rightarrow \begin{array}{c} H^{r+\mu}(X,E) \\ \oplus \\ H^{r+\mu}(Y,J^-) \end{array}$$

for all $r \in \mathbb{R}, r > (-\mu)^+ - \frac{1}{2}$. Since we have $\mathcal{A}u \in H^r(X,F) \oplus H^r(Y,J^+)$ for such an r as soon as $u \in H^t(X,E) \oplus H^t(Y,J^-)$ for some $t \in \mathbb{R}$, $t > \mu^+ - \frac{1}{2}$, we get $\mathcal{B}f = \mathcal{B}\mathcal{A}u \in H^{r+\mu}(X,E) \oplus H^{r+\mu}(Y,J^-)$. From $\mathcal{B}\mathcal{A} = 1 + \mathcal{R}_l$ it follows that $u + \mathcal{R}_l u \in H^{r+\mu}(X,E) \oplus H^{r+\mu}(Y,J^-)$. Then $\mathcal{R}_l u \in H^\infty(X,E) \oplus H^\infty(Y,J^-)$, implies $u \in H^{r+\mu}(X,E) \oplus H^{r+\mu}(Y,J^-)$, which is the asserted regularity of u. Moreover,

$$\mathcal{R}_r = \mathcal{A}\mathcal{B} - 1: \begin{array}{c} H^{s-\mu}(X,F) \\ \oplus \\ H^{s-\mu}(Y,J^+) \end{array} \rightarrow \begin{array}{c} H^{s-\mu}(X,F) \\ \oplus \\ H^{s-\mu}(Y,J^+) \end{array}$$

is compact for every $s > (-\mu)^+ - \frac{1}{2}$ and

$$\mathcal{R}_l = \mathcal{B}\mathcal{A} - 1: \begin{array}{c} H^s(X,E) \\ \oplus \\ H^s(Y,J^-) \end{array} \rightarrow \begin{array}{c} H^s(X,E) \\ \oplus \\ H^s(Y,J^-) \end{array}$$

is compact for every $s > \mu^+ - \frac{1}{2}$. Thus (4.3.43) is a Fredholm operator for every $s > \mu^+ - \frac{1}{2}$. $\square$

Remark 4.3.50 *Definition 4.3.47 can easily be generalised to non-compact X. In this case we say that when $\mathcal{A} \in \mathcal{B}^{\mu,d}(X; E, F; J^-, J^+)$ is given, an operator $\mathcal{B} \in \mathcal{B}^{-\mu,c}(X; F, E; J^+, J^-)$ is a parametrix of $\mathcal{A}$ if*

$$\mathcal{A}\Phi_1\mathcal{B}\Psi_1 - \Psi_1 \in \mathcal{B}^{-\infty,h}(X; F, F; J^+, J^+) \quad and$$
$$\mathcal{B}\Phi_2\mathcal{A}\Psi_2 - \Psi_2 \in \mathcal{B}^{-\infty,e}(X; E, E; J^-, J^-)$$

hold for certain $e, h \in \mathbb{N}$, where

$$\Phi_1 = \begin{pmatrix} \varphi \cdot \mathrm{id}_E & 0 \\ 0 & \varphi' \cdot \mathrm{id}_{J^-} \end{pmatrix}, \quad \Psi_1 = \begin{pmatrix} \psi \cdot \mathrm{id}_F & 0 \\ 0 & \psi' \cdot \mathrm{id}_{J^+} \end{pmatrix},$$

$$\Phi_2 = \begin{pmatrix} \varphi \cdot \mathrm{id}_F & 0 \\ 0 & \varphi' \cdot \mathrm{id}_{J^+} \end{pmatrix}, \quad \Psi_2 = \begin{pmatrix} \psi \cdot \mathrm{id}_E & 0 \\ 0 & \psi' \cdot \mathrm{id}_{J^-} \end{pmatrix}$$

for arbitrary $\varphi, \psi \in C_0^\infty(X)$ *with* $\varphi\psi = \varphi$ *and* $\varphi', \psi' \in C_0^\infty(Y)$
with $\varphi'\psi' = \varphi'$. *Analogously to Theorem 4.3.48 it is then easy to*
show that every elliptic $\mathcal{A} \in \mathcal{B}^{\mu,\mu^+}(X; E, F; J^-, J^+)$ *has a parametrix*
$\mathcal{B} \in \mathcal{B}^{-\mu,(-\mu)^+}(X; F, E; J^+, J^-)$. *Moreover, we get an analogue of the*
elliptic regularity from Theorem 4.3.49 for non-compact X, *namely*
that $\mathcal{A}u = f \in H_{\text{loc}}^r(X, F) \oplus H_{\text{loc}}^r(Y, J^+)$ *for* $r > (-\mu)^+ - \frac{1}{2}$ *and*
$u \in H_{\text{comp}}^t(X, E) \oplus H_{\text{comp}}^t(Y, J^-)$ *for some* $t > \mu^+ - \frac{1}{2}$ *implies for*
the solution $u \in H_{\text{comp}}^{r+\mu}(X, E) \oplus H_{\text{comp}}^{r+\mu}(Y, J^-)$.

Bibliography

[1] M.S. Agranovič and M.I. Višik. Elliptic problems with parameter and parabolic problems of general type. *Uspekhi Mat. Nauk*, 19, 3:53–161, 1964.

[2] R. Airapetyan and I. Witt. Isometric properties of the Hankel transform in weighted Sobolev spaces. Preprint 97/14, Institute for Mathematics, Potsdam, 1997.

[3] H. Amann. *Linear and quasilinear problems*, volume 1: Abstract linear theory. Birkhäuser, Basel, 1995.

[4] H. Amann. Operator–valued Fourier multipliers, vector–valued Besov spaces, and applications. *Math. Nachr.*, 186:5–56, 1997.

[5] M.F. Atiyah, V. Patodi, and I.M. Singer. Spectral asymmetry and Riemannian geometry I, II, III. *Math. Proc. Cambridge Philos. Soc.*, 77,78,79:43–69, 405–432, 315–330, 1975, 1976, 1976.

[6] M.F. Atiyah and I.M. Singer. The index of elliptic operators I. *Ann. of Math.*, 87:484–503, 1968.

[7] J. Bennish. Variable discrete asymptotics of solutions to elliptic boundary–value problems. In *Differential Equations, Asymptotic Analysis, and Mathematical Physics*, volume 100 of *Math. Research*. Akademie Verlag Berlin, 1997.

[8] P. Boggiato, E. Buzano, and L. Rodino. *Global hypoellipticity and spectral theory*, volume 92 of *Math. Research*. Akademie Verlag Berlin, 1996.

[9] L. Boutet de Monvel. Comportement d'un opérateur pseudo-différentiel sur une variété à bord. *J. Anal. Math.*, 17:241–304, 1966.

[10] L. Boutet de Monvel. Boundary problems for pseudo–differential operators. *Acta Math.*, 126:11–51, 1971.

617

[11] J. Brüning and R. Seeley. An index theorem for first order singular operators. *Amer. J. Math.*, 110:659–714, 1988.

[12] T. Buchholz and B.-W. Schulze. Anisotropic edge pseudo-differential operators with discrete asymptotics. *Math. Nachr.*, 184:73–125, 1997.

[13] T. Buchholz and B.-W. Schulze. Volterra operators and parabolicity. Anisotropic pseudo–differential operators. Preprint, Institute for Mathematics, Potsdam, 1997.

[14] V. Burenkov, B.-W. Schulze, and N. Tarkhanov. Extension operators for Sobolev spaces commuting with a given transform. Preprint MPI 96–142, Max–Planck–Institut für Mathematik, Bonn, 1996.

[15] J. Cheeger. On the spectral geometry of spaces with cone–like singularities. *Proc. Nat. Acad. Sci. U.S.A.*, 76:2103–2106, 1979.

[16] J. Cheeger. Spectral geometry of singular Riemannian spaces. *J. Differential Geom.*, 18:575–657, 1983.

[17] H.O. Cordes. A global parametrix for pseudo–differential operators over $\mathbb{R}^n$, with applications. Reprint, SFB 72, Universität Bonn, 1976.

[18] H.O. Cordes. *The technique of pseudodifferential operators*. Cambridge University Press, 1995.

[19] M. Dauge. *Elliptic boundary value problems on corner domains*, volume 1341 of *Lecture Notes in Mathematics*. Springer–Verlag, 1988.

[20] Ch. Dorschfeldt. *Algebras of pseudo–differential operators near edge and corner singularities*, volume 102 of *Math. Research*. Akademie Verlag, Berlin, 1998.

[21] Ch. Dorschfeldt, U. Grieme, and B.-W. Schulze. Pseudo–differential calculus in the Fourier–edge approach on non-compact manifolds. Preprint MPI 96–79, Max–Planck–Institut, Bonn, 1996.

[22] Ch. Dorschfeldt and B.-W. Schulze. Pseudo–differential operators with operator–valued symbols in the Mellin–edge–approach. *Ann. Global Anal. Geom.*, 12, 2:135–171, 1994.

[23] Ch. Dorschfeldt and B.-W. Schulze. Parameter–dependent cone calculus and Mellin corner operators. Preprint, Institute for Mathematics, Potsdam, 1997.

[24] Ju. V. Egorov and B.-W. Schulze. *Pseudo–differential operators, singularities, applications.* Birkhäuser Verlag, Basel, 1997.

[25] G.I. Eskin. *Boundary value problems for elliptic pseudodifferential equations*, volume 52 of *Math. Monographs.* Amer. Math. Soc., Providence, Rhode Island, 1980. Transl. of Nauka, Moskva, 1973.

[26] G.I. Eskin. Boundary–value problems for second–order elliptic equations in domains with corners. *Proc. Sympos. Pure Math.*, 43:105–131, 1985.

[27] G.I. Eskin. Index formulas for elliptic boundary value problems in plane domains with corners. *Trans. Amer. Math. Soc.*, 314, 1:283–348, 1989.

[28] G.I. Eskin. The wave equation in a wedge with general boundary conditions. *Comm. Partial Differential Equations*, 17:99–160, 1992.

[29] B.V. Fedosov. Analytical index formulas for elliptic operators. *Trudy Moskov. Mat. Obshch.*, 30:159–241, 1974.

[30] B.V. Fedosov and B.-W. Schulze. On the index of elliptic operators on a cone. In *Advances in Partial Differential Equations (Schrödinger Operators, Markov Semigroups, Wavelet Analysis, Operator Algebras)*, pages 348–372. Akademie Verlag, Berlin, 1996. Preprint MPI 96-31, Max–Planck Institut, Bonn, 1996.

[31] B.V. Fedosov, B.-W. Schulze, and N.N. Tarkhanov. On the index of elliptic operators on a wedge. Preprint MPI 96-143, Max–Planck-Institut, Bonn, 1996.

[32] B.V. Fedosov, B.-W. Schulze, and N.N. Tarkhanov. The index of elliptic operators on manifolds with conical points. Preprint 97/24, Institute for Mathematics, Potsdam, 1997.

[33] B.V. Fedosov, B.-W. Schulze, and N.N. Tarkhanov. On the index formula for singular surfaces. Preprint 97/31, Institute for Mathematics, Potsdam, 1997.

[34] V.I. Feygin. Two algebras of pseudodifferential operators in $\mathbb{R}^n$ and some applications. *Trudy Moskov. Mat. Obshch.*, 36:155–194, 1977.

[35] I.M. Gelfand and G.I. Shilow. *Verallgemeinerte Funktionen (Distributionen) I*. Deutscher Verlag der Wissenschaften, Berlin, 1967.

[36] P Gérard and G. Lebeau. Diffusion d'une onde par un coin. *J. Amer. Math. Soc.*, 6:341–424, 1993.

[37] J.B. Gil. *Heat equation asymptotics and index for cone differential operators*. PhD thesis, University of Potsdam, 1998.

[38] J.B. Gil, B.-W Schulze, and J. Seiler. Holomorphic operator-valued symbols for edge–degenerate pseudo–differential operators. In *Differential Equations, Asymptotic Analysis, and Mathematical Physics*, volume 100 of *Math. Research*, pages 113–137. Akademie Verlag Berlin, 1997.

[39] J.B. Gil, B.-W Schulze, and J. Seiler. Parameter–dependent cone operators in the edge symbolic calculus. Preprint, Institute for Mathematics, Potsdam, 1997.

[40] P.B. Gilkey. *Invariance theory, the heat equation, and the Atiyah–Singer index theorem*. Studies in Advanced Mathematics. CRC Press, Boca Raton, Ann Arbor, London, Tokyo, 1995.

[41] I. Gohberg and N. Krupnik. *Einführung in die Theorie der eindimensionalen singulären Integraloperatoren*. Birkhäuser Verlag, Basel, Boston, Stuttgart, 1979.

[42] I.C. Gradštein and I.M. Ryžik. *Tablicy integralov, summ, rjadov i proizvedenii*. Gos. isd. fiziko–matem. literatury, Moskva, 1963.

[43] B. Gramsch. Inversion von Fredholmfunktionen bei stetiger und holomorpher Abhängigkeit von Parametern. *Math. Ann.*, 214:95–147, 1975.

[44] B. Gramsch and W. Kaballo. Multiplicative decompositions of holomorphic Fredholm functions an Ψ^*–algebras. Preprint FB Mathematik 9, J. Gutenberg–Universität Mainz, 1996.

[45] U. Grieme. *Pseudo–differential operators with operator–valued symbols on non–compact manifolds*. PhD thesis, University of Potsdam, 1998.

[46] P. Grisvard. *Elliptic problems in nonsmooth domains.*, volume 24 of *Monographs and Studies in Math.* Pitman, Boston, London, Melbourne, 1985.

[47] P. Grisvard. Singularités en élasticité. *Archive for Rational Mech. Anal. (2)*, 107:157–180, 1989.

[48] G. Grubb. *Functional calculus of pseudo–differential boundary problems.* Birkhäuser Verlag, Basel, Boston, Stuttgart, 1986.

[49] V.V. Grushin. Pseudodifferential operators in $\mathbb{R}^n$ with bounded symbols. *Funkc. anal.*, 4, 3:37–50, 1970.

[50] V.P. Havin. Separation of singularities of analytic functions. *Dokl. Akad. Nauk SSSR*, 121:239–242, 1958. (Russian).

[51] T. Hirschmann. Functional analysis in cone and edge Sobolev spaces. *Ann. of Global Anal. Geom.*, 8, 2:167–192, 1990.

[52] T. Hirschmann. Pseudo–differential operators and asymptotics on manifolds with corners. Report R–Math 07/90, Karl–Weierstraß–Institut für Mathematik, Berlin, 1990.

[53] T. Hirschmann. Pseudo–differential operators and asymptotics on manifolds with corners. supplement: Pseudo–differential operator calculus with exit conditions. Report R–Math 04/91, Karl–Weierstraß–Institut für Mathematik, Berlin, 1991.

[54] L. Hörmander. *The analysis of linear partial differential operators*, volume 1 and 2. Springer–Verlag, New York, 1983.

[55] I.L. Hwang. The L^2-boundness of pseudodifferential operators. *Trans. Amer. Math. Soc.*, 302:55–76, 1987.

[56] P. Jeanquartier. Transformation de Mellin et développements asymptotiques. *Enseign. Math.*, 25:285–308, 1979.

[57] J. Kohn and L. Nirenberg. An algebra of pseudo–differential operators. *Comm. Pure Appl. Math.*, 18:269–305, 1965.

[58] V.A. Kondrat'ev. Boundary value problems for elliptic equations in domains with conical points. *Trudy Moskov Mat. Obshch*, 16:209–292, 1967.

[59] V.A. Kondrat'ev and O.A. Oleynik. Boundary problems for partial differential equations on non–smooth domains. *Uspekhi Mat. Nauk*, 38, 2:3–76, 1983.

[60] G. Köthe. Dualität in der Funktionentheorie. *J. Reine Angew. Math.*, 191:30–39, 1953.

[61] H. Kumano-go. *Pseudo–differential operators*. The MIT Press, Cambridge, Massachusetts and London, England, 1981.

[62] R. Lauter. *Holomorphic functional calculus in several variables and Ψ^*-algebras of totally characteristic operators on manifolds with boundary*. PhD thesis, University of Mainz, 1996.

[63] M. Lesch. *Operators of Fuchs type, conical singularities, and asymptotic methods*, volume 136 of *Teubner–Texte zur Mathematik*. B.G. Teubner Verlagsgesellschaft, Stuttgart–Leipzig, 1997.

[64] S.Z. Levendorskii. *Asymptotic distribution of eigenvalues of differential operators*. Kluwer Academic Publishers, Dordrecht, NL, 1990.

[65] S.Z. Levendorskii. *Degenerate elliptic equations*. Kluwer Academic Publishers, Dordrecht, NL, 1993.

[66] O. Liess. *Conical refraction and higher microlocalization*, volume 1555 of *Springer Lecture Notes in Math.* Springer, 1993.

[67] O. Liess. *Pseudo–differential operators; an introduction*, volume 349 of *Pitman Research Notes in Math.*, pages 82–123. Pitman, 1995.

[68] J.-L. Lions and E. Magenes. *Problèmes aux limites non homogènes et applications*, volume 1. Dunod, Paris, 1968.

[69] R.B. Lockhart and R.C. McOwen. Elliptic differential operators on noncompact manifolds. *Ann. Scuola Norm. Sup. Pisa*, 7, 3:409–447, 1985.

[70] G. Luke. Pseudo–differential operators on Hilbert bundles. *J. Differential Equations*, 12:566–589, 1972.

[71] F. Mantlik. Norm closures of operator algebras with symbolic structures. *Math. Nachr.* to appear.

[72] F. Mantlik. Norm closure and extension of the symbolic calculus for the cone algebra. *Ann. Global Anal. Geom.*, 13, 4:339–376, 1995.

[73] F. Mantlik. Tensor products, Fréchet–Hilbert complexes and the Künneth theorem. *Results Math.*, 28:287–302, 1995.

[74] V.G. Maz'ja, V. Kozlov, and J. Rossmannn. Point boundary singularities in elliptic theory. *Amer. Math. Soc. Surveys and Monographs*, 1997. to appear.

[75] R. Mazzeo. Elliptic theory of differential edge operators I. *Comm. Partial Differential Equations*, 16:1615–1664, 1991.

[76] R.B. Melrose. Transformation of boundary problems. *Acta Math.*, 147:149–236, 1981.

[77] R.B. Melrose. *The Atiyah–Patodi–Singer index theorem.* Research Notes in Mathematics. A.K. Peters, Wellesley, 1993.

[78] R.B. Melrose and G.A. Mendoza. Elliptic operators of totally characteristic type. Preprint, MSRI, 1983.

[79] R.B. Melrose and V. Nistor. C^*–algebras of b–pseudodifferential operators and an $\mathbb{R}^k$–equivariant index theorem. PENNSTATE Report No. PM 199, 1996.

[80] W. Müller. *Manifolds with cusps of rank one*, volume 1244 of *Springer Lecture Notes in Math.* Springer, 1987.

[81] B. Paneah and B.-W. Schulze. Edge–degenerate operators with Mellin and Green transmission conditions. Preprint, Institute for Mathematics, Potsdam, 1997.

[82] C. Parenti. Operatori pseudo–differenziali in $\mathbb{R}^n$ e applicazioni. *Annali Mat. Pura Appl.*, 93:359–389, 1972.

[83] P. Piazza. On the index of elliptic operators on manifolds with boundary. *J. Funct. Anal.*, 117:308–359, 1993.

[84] U. Pillat and B.-W. Schulze. Elliptische Randwert–Probleme für Komplexe von Pseudodifferentialoperatoren. *Math. Nachr.*, 94:173–210, 1980.

[85] B.A. Plamenevskij. On the boundedness of singular integrals in spaces with weight. *Mat. Sb.*, 76, 4:573–592, 1968.

[86] B.A. Plamenevskij. Algebras of pseudo–differential operators. *Nauka i Tekhn. Progress*, Moscow, 1986.

[87] B.A. Plamenevskij and V.G. Rozenblum. On the index of pseudodifferential operators with isolated singularities in the symbol. *Leningrad Math. J.* 2, pages 1085–1110, 1991.

[88] P.R.. Popivanov and D.K. Palagachev. *The degenerate oblique derivative problem for elliptic and parabolic equations*, volume 93 of *Math. Research.* Akademie–Verlag, Berlin, 1997.

[89] V.S. Rabinovich. Pseudodifferential operators in non–bounded domains with conical structure at infinity. *Mat. Sb.*, 80:77–97, 1969.

[90] V.S. Rabinovich, B.-W. Schulze, and N. Tarkhanov. A calculus of boundary value problems in domains with non–Lipschitz singular points. Preprint 97/9, Institute for Mathematics, Potsdam, 1997.

[91] S. Rempel and B.-W. Schulze. *Index theory of elliptic boundary problems.* Akademie–Verlag, Berlin, 1982.

[92] S. Rempel and B.-W. Schulze. Parametrices and boundary symbolic calculus for elliptic boundary problems without transmission property. *Math. Nachr.*, 105:45–149, 1982.

[93] S. Rempel and B.-W. Schulze. Branching of asymptotics for elliptic operators on manifolds with edges. In *Proc. "Partial Differential Equations"*, volume 19 of *Banach Center Publ.*, Warsaw, 1984. PWN Polish Scientific Publisher.

[94] S. Rempel and B.-W. Schulze. Complete Mellin symbols and the conormal asymptotics in boundary value problems. In *Proc. Journées "Equ. aux Dériv. Part."*, *Conf. No. V*, St.–Jean de Monts, 1984.

[95] S. Rempel and B.-W. Schulze. Complete Mellin and Green symbolic calculus in spaces with conormal asymptotics. *Ann. Global Anal. Geom.*, 4, 2:137–224, 1986.

[96] S. Rempel and B.-W. Schulze. *Asymptotics for elliptic mixed boundary problems (pseudo–differential and Mellin operators in spaces with conical singularity)*, volume 50 of *Math. Research.* Akademie–Verlag, Berlin, 1989.

[97] G. Rozenblum. The index of cone Mellin operators. Preprint 1996-15, Chalmers Univ. of Techn. & Göteborg Univ., Göteborg, 1996.

[98] H.H. Schäfer. *Topological vector spaces*. MacMillan, New York, 1966.

[99] P. Schapira. *Théorie des hyperfonctions*, volume 126 of *Springer Lecture Notes in Math*. Springer–Verlag, 1970.

[100] E. Schrohe. *Fréchet algebras of pseudodifferential operators and boundary value problems*. Birkhäuser, Boston, Basel. (in preparation).

[101] E. Schrohe. *Spaces of weighted symbols and weighted Sobolev spaces on manifolds*, volume 1256 of *Lecture Notes in Math.*, pages 360–377. Springer-Verlag, Berlin–Heidelberg, 1987.

[102] E. Schrohe. A pseudo–differential calculus for weighted symbols and a Fredholm criterion for boundary value problems on noncompact manifolds. Habilitationsschrift, FB Mathematik, Universität Mainz, 1991.

[103] E. Schrohe. Spectral invariance, ellipticity, and the Fredholm property for pseudo–differential operators on weighted Sobolev spaces. *Ann. Global Anal. Geom.*, 10:237–254, 1992.

[104] E. Schrohe. Invariance of the cone algebra without asymptotics. *Ann. Global Anal. Geom.*, 14:403–425, 1996.

[105] E. Schrohe. Wodzicki's noncommutative residue and traces for operator algebras on manifolds with conical singulaties. In L. Rodino, editor, *Microlocal Analysis and Spectral Theory*, volume 490 of *NATO ASI Series, Series C: Mathematical and Physical Sciences*, pages 227–250. Kluwer Academic Publisher, Dordrecht, Boston, London, 1997.

[106] E. Schrohe and B.-W. Schulze. Boundary value problems in Boutet de Monvel's calculus for manifolds with conical singularities I. In *Advances in Partial Differential Equations (Pseudo–Differential Calculus and Mathematical Physics)*, pages 97–209. Akademie Verlag, Berlin, 1994.

[107] E. Schrohe and B.-W. Schulze. Boundary value problems in Boutet de Monvel's calculus for manifolds with conical singularities II. In *Advances in Partial Differential Equations (Boundary Value Problems, Schrödinger Operators, Deformation Quantization)*, pages 70–205. Akademie Verlag, Berlin, 1995.

[108] E. Schrohe and B.-W. Schulze. Mellin and Green symbols for boundary value problems on manifolds with edges. Preprint MPI 96–173, Max–Planck–Institut, Bonn, 1996.

[109] E. Schrohe and B.-W. Schulze. Mellin operators in a pseudodifferential calculus for boundary value problems on manifolds with edges. Preprint MPI 96–74, Max–Planck–Institut, Bonn, 1996.

[110] E. Schrohe and B.-W. Schulze. Pseudodifferential boundary value problems on manifolds with edges. Preprint, Institute for Mathematics, Potsdam, 1997.

[111] E. Schrohe and B.-W. Schulze. A symbol algebra for pseudodifferential boundary value problems on manifolds with edges. In *Differential Equations, Asymptotic Analysis, and Mathematical Physics*, volume 100 of *Math. Research*, pages 292–324. Akademie Verlag, Berlin, 1997.

[112] E. Schrohe and J. Seiler. An analytical index formula for pseudodifferential operators on wedges. Preprint MPI 96–172, Max–Planck–Institut, Bonn, 1996.

[113] B.-W. Schulze. Mellin expansions of pseudo–differential operators and conormal asymptotics of solutions. In *Pseudo–Differential Operators*, volume 1256 of *Lecture Notes in Math.*, pages 378–401. Springer–Verlag, Berlin–Heidelberg, 1987.

[114] B.-W. Schulze. Ellipticity and continuous conormal asymptotics on manifolds with conical singularities. *Math. Nachr.*, 136:7–57, 1988.

[115] B.-W. Schulze. Regularity with continuous and branching asymptotics for elliptic operators on manifolds with edges. *Integral Equations Operator Theory*, 11:557–602, 1988.

[116] B.-W. Schulze. Corner Mellin operators and reduction of orders with parameters. *Ann. Scuola Norm. Sup. Pisa Cl. Sci.*, 16, 1:1–81, 1989.

[117] B.-W. Schulze. Pseudo–differential operators on manifolds with edges. In *Symposium "Partial Differential Equations", Holzhau 1988*, volume 112 of *Teubner–Texte zur Mathematik*, pages 259–287. Teubner, Leipzig, 1989.

[118] B.-W. Schulze. Mellin representations of pseudo–differential operators on manifolds with corners. *Ann. Glob. Anal. Geom.*, 8, 3:261–297, 1990.

[119] B.-W. Schulze. *Pseudo-differential operators on manifolds with singularities.* North–Holland, Amsterdam, 1991.

[120] B.-W. Schulze. Crack problems in the edge pseudo-differential calculus. *Appl. Anal.*, 45:333–360, 1992.

[121] B.-W. Schulze. The Mellin pseudo-differential calculus on manifolds with corners. In *Symposium: "Analysis in Domains and on Manifolds with Singularities", Breitenbrunn 1990*, volume 131 of *Teubner–Texte zur Mathematik*, pages 208–289, Leipzig, 1992.

[122] B.-W. Schulze. *Pseudo-differential boundary value problems, conical singularities, and asymptotics.* Akademie Verlag, Berlin, 1994.

[123] B.-W. Schulze. The variable discrete asymptotics in pseudo-differential boundary value problems I. In *Advances in Partial Differential Equations (Pseudo–Differential Calculus and Mathematical Physics)*, pages 9–96. Akademie Verlag, Berlin, 1994.

[124] B.-W. Schulze. *Transmission algebras on singular spaces with components of different dimensions*, volume 78, pages 321–342. Birkhäuser Verlag, Basel, 1995.

[125] B.-W. Schulze. The variable discrete asymptotics in pseudo-differential boundary value problems II. In *Advances in Partial Differential Equations (Boundary Value Problems, Schrödinger Operators, Deformation Quantization)*, pages 9–69. Akademie Verlag, Berlin, 1995.

[126] B.-W. Schulze. Boundary value problems and edge pseudo-differential operators. In L. Rodino, editor, *Microlocal Analysis and Spectral Theory*, volume 490 of *NATO ASI Series, Series C: Mathematical and Physical Sciences*, pages 165–226, Dordrecht, Boston, London, 1997. Kluwer Academic Publisher.

[127] B.-W. Schulze, B.Ju. Sternin, and V.E. Shatalov. On some global aspects of the theory of partial differential equations on manifolds with singularities. Preprint MPI 96-28, Max–Planck–Institut, Bonn, 1996.

[128] B.-W. Schulze, B.Ju. Sternin, and V.E. Shatalov. An operator algebra on manifolds with cusp-type singularities. Preprint MPI 96-111, Max–Planck–Institut, Bonn, 1996.

[129] B.-W. Schulze, B.Ju. Sternin, and V.E. Shatalov. On general boundary value problems for elliptic equations. Preprint, Institute for Mathematics, Potsdam, 1997.

[130] B.-W. Schulze, B.Ju. Sternin, and V.E. Shatalov. On the index of differential operators on manifolds with conical singularities. Preprint 97/10, Institute for Mathematics, Potsdam, 1997.

[131] B.-W. Schulze and N.N. Tarkhanov. Green pseudodifferential operators on a manifold with edges. Preprint MPI 96–162, Max–Planck–Institut, Bonn, 1996.

[132] B.-W. Schulze and N.N. Tarkhanov. The Riemann–Roch theorem for manifolds with conical singularities. Preprint 97/18, Institute for Mathematics, Potsdam, 1997.

[133] R. Seeley. Complex powers of an elliptic operator. In *Proc. Sympos. Pure Math.*, volume 10, pages 288–307, 1967.

[134] R. Seeley. Topics in pseudo–differential operators. In *C.I.M.E. Conference on pseudo–differential operators, Stresa 1968*, pages 167–305, Cremonese, Rome, 1969.

[135] J. Seiler. Continuity of edge and corner pseudo–differential operators. Preprint MPI 96–160, Max–Planck–Institut, Bonn, 1996.

[136] J. Seiler. *Pseudodifferential calculus on manifolds with non-compact edges; ellipticity and index problems.* PhD thesis, University of Potsdam, 1997.

[137] M.A. Shubin. Pseudodifferential operators in $\mathbb{R}^n$. *Dokl. Akad. Nauk SSSR*, 196:316–319, 1971.

[138] M.A. Shubin. *Pseudodifferential operators and spectral theory.* Springer Verlag, Berlin, Heidelberg, New York, 1987. (Engl. transl. from Nauka i Tekhn. Progress, Moskva 1978).

[139] J. Sjöstrand. Operators of principal type with interior boundary conditions. *Acta Math.*, 130:1–51, 1973.

[140] J. Sjöstrand. Parametrices for pseudodifferential operators with multiple characteristics. *Ark. Mat.*, 12:85–130, 1974.

[141] B.Ju. Sternin. Elliptic and parabolic equations on manifolds with boundary consisting of components of different dimensions. *Trudy. Moskov. Mat. Obshch.*, 15:346–382, 1966.

[142] B.Ju. Sternin. *Topological aspects of a problem of S. L. Sobolev.* Mosk. inst. elektr. mashinostroenija, Moskva, 1971.

[143] B.Ju. Sternin and V.E. Shatalov. Asymptotic solutions to Fuchsian equations in several variables. Preprint MPI 94–124, Max-Planck–Institut, Bonn, 1996.

[144] B.Ju. Sternin and V.E. Shatalov. *Borel–Laplace transform and asymptotic theory.* CRC Press, Boca Raton, New York, London, Tokyo, 1996.

[145] J.D. Tamarkin. On Fredholm's integral equations whose kernel are analytic in a parameter. *Ann. of Math.*, 28:127–152, 1927.

[146] F. Treves. *Topological vector spaces, distributions and kernels.* Academic–Press, New York, London, 1967.

[147] F. Treves. *Introduction to pseudo–differential and Fourier integral operators*, volume 1 and 2. Plenum, New York, 1985.

[148] H. Triebel. *Interpolation theory, functional spaces, and differential operators.* North–Holland, Amsterdam, 1978. Deutscher Verlag der Wissenschaften Berlin 1978.

[149] A. Unterberger and H. Upmeier. *Pseudodifferential analysis on symmetric cones.* Studies in Advanced Mathematics. CRC Press Boca Raton, New York, London, Tokyo, 1996.

[150] M.I. Višik and G.I. Eskin. Convolution equations in a bounded region. *Uspekhi Mat. Nauk*, 20, 3:89–152, 1965.

[151] M.I. Višik and G.I. Eskin. Convolution equations in bounded domains in spaces with weighted norms. *Mat. Sb.*, 69, 1:65–110, 1966.

[152] H. Widom. *Asymptotic expansions for pseudo–differential operators on bounded domains*, volume 1152 of *Lecture Notes in Mathematics.* Springer Verlag, Berlin–Heidelberg, 1985.

[153] I. Witt. *Non–linear hyperbolic equations in domains with conical points. Existence and regularity of solutions*, volume 84 of *Math. Research.* Akademie Verlag, Berlin, 1995.

Index